20th Innovative Manufacturing Engineering and Energy Conference (IManEE 2016)

IOP Conference Series: Materials Science and Engineering
Volume 161

Kozani, Greece
23 – 25 September 2016

Editors:

Stergios Maropoulos
Panagiotis Kyratsis
Konstantinos Kakoulis
Nikolaos Taousanidis
Gheorghe Oancea
Oana Dodun

ISBN: 978-1-5108-3383-8
ISSN: 1757-8981

Printed from e-media with permission by:

Curran Associates, Inc.
57 Morehouse Lane
Red Hook, NY 12571

Printed by Curran Associates, Inc. (2017)

Institute of Physics
Dirac House, Temple Back
Bristol BS1 6BE UK

Phone: 44 1 17 929 7481
Fax: 44 1 17 920 0979

techtracking@iop.org

Additional copies of this publication are available from:

Curran Associates, Inc.
57 Morehouse Lane
Red Hook, NY 12571 USA
Phone: 845-758-0400
Fax: 845-758-2633
Email: curran@proceedings.com
Web: www.proceedings.com

Table of contents

Volume 161

20th Innovative Manufacturing Engineering and Energy Conference (IManEE 2016)

23–25 September 2016, Kozani, Greece

Accepted papers received: 15 November 2016
Published online: 1 December 2016

Preface

Papers

IOP Conf. Series: Materials Science and Engineering **161** (2016) 011001 doi:10.1088/1757-899X/161/1/011001

INNOVATIVE MANUFACTURING ENGINEERING & ENERGY INTERNATIONAL CONFERENCE

IManEE 2016

September 23-25, 2016 Kallithea Chalkidiki, Greece

Selected, peer revied papers from the Innovative Manufacturing Engineering & Energy International Conference (IManEE 2016)

Edited by

Stergios **MAROPOULOS**
Dept. of Mechanical Engineering and Industrial Design
Technological Education Institute of Western Macedonia, Greece

Panagiotis **KYRATSIS**
Dept. of Mechanical Engineering and Industrial Design
Technological Education Institute of Western Macedonia, Greece

Konstantinos **KAKOULIS**
Dept. of Mechanical Engineering and Industrial Design
Technological Education Institute of Western Macedonia, Greece

Nikolaos **TAOUSANIDIS**
Dept. of Mechanical Engineering and Industrial Design
Technological Education Institute of Western Macedonia, Greece

Gheorghe **OANCEA**
Department of Manufacturing Engineering, Transilvania University of Brasov, Romania.

Oana **DODUN**
Department of Manufacturing Engineering, “Gheorghe Asachi” Technical University of Iaşi, România

Foreword

IManEE 2016 Conference is the result of a collaboration between the Department of Mechanical Engineering & Industrial Design of the Technological Education Institute of West Macedonia, Greece and the Department of Machine Manufacturing Technology of the "Gheorghe Asachi" Technical University of Iasi, Romania. It is the 20th version of a series of conferences that first started in 1996 in Iasi as TMCM 96 and since 2013 adopted the name of ImanE.It is the first time it is held in Greece after a proposal of Professor Oana Dodun.This year the field of Energy was added to the program.

In the ImanEE 2016 Coference 120 papers will be presented in five sessions, Materials & Mechanical Behaviour, Quality of Manufacturing Systems, Manufacturing Processes & Manufacturing, CAD/CAM & Design and Energy & Environment, concerning field such as Advanced Machining and Surface Engineering, Assembling Technologies, Forming Technologies, Non-Conventional Technologies in Manufacture and Industry, Welding, Technologies and Non Destructive Testing (NDT, Advanced Materials and Construction Technologies, Flexible Manufacturing, Automation and Robotics, CAD/CAM/CAE/CAx Technologies, Mechanical and Manufacturing Equipment, Devices and Instrumentation, Design and Analysis Industrial and Product Management, Quality and Evaluation Design, Product Design and Industrial Design Engineering, Bioengineering and Biomaterials Sustainable Design and Sustainability, Biomass Applications, Smart Grids, Energy Efficiency in Buildings, Energy systems and energy management, Energy policies, Environmental technologies and studies and Renewable energy technologies and energy saving materials.

The authors of the accepted papers come from Universities, Research Institutes from countries such as Romania, Poland, Germany, Italy, Turkey, China, Morocco, Republic of Moldova, Ukraine, Hungary, Greece and United Kingdom.

The organisation of the Conference was a joint venture from a team of people from Greece and Romania who worked hard to make sure the event is met with success.

We hope the IManEE 2016 Conference will be a good opportunity for new collaborations among participants to develop that will lead to joint research projects for the benefit of all.

Professor Stergios Maropoulos,
Head of Department of Mechanical Engineering & Industrial Design,
Technological Education Institute of West Macedonia, Greece

IOP Conf. Series: Materials Science and Engineering **161** (2016) 011001 doi:10.1088/1757-899X/161/1/011001

Conference Chair:
Prof. Stergios **MAROPOULOS**, PhD, Head of Department of Mechanical Engineering and Industrial Design, Technological Education Institute of West Macedonia, Greece.

Conference Co-Chairs:
Assoc.Prof. Konstantinos **KAKOULIS**, PhD, Department of Mechanical Engineering and Industrial Design, Technological Education Institute of West Macedonia, Greece.
Assis.Prof. Savvas **DOUVARTZIDIS**, PhD, Department of Mechanical Engineering and Industrial Design, Technological Education Institute of West Macedonia, Greece.
Assis.Prof. Sofoklis **MAKRIDIS**, PhD, Department of Mechanical Engineering, University of West Macedonia.
Assis.Prof. Panagiotis **KYRATSIS**, PhD, Department of Mechanical Engineering and Industrial Design, Technological Education Institute of West Macedonia, Greece.
Assis.Prof. Nikolaos **TAOUSANIDIS**, PhD, Department of Mechanical Engineering and Industrial Design, Technological Education Institute of West Macedonia, Greece.

Honorary Committee
Prof. Paul **MAROPOULOS**, MSc, PhD, CEng, FIMechE, Fellow CIRP Pro-Vice-Chancellor for Research and Enterprise, Aston University, UK.
Prof. Laurenţiu **SLĂTINEANU**, PhD, "Gheorghe Asachi" Technical University of Iaşi, România.
Dr. Norman **RIDLEY**, Honorary Research Fellow, School of Materials, University of Manchester, UK

Scientific Committee
Marioara **ABRUDEANU**, U of Piteşti, România
Harisios **ACHILAS**, International Hellenic University, Greece
Elisavet **AMANATIDOU**, TEI West Macedonia, Greece
Aristomenis **ANTONIADIS**, Technical U of Crete, Greece
Nikolaos **ASIMOPOULOS**, TEI West Macedonia, Greece
Argiro **ASVESTA**, TEI West Macedonia, Greece
George **AVGOUROPOULOS**, University of Patra, Greece
Eugen **AXINTE**, Gheorghe Asachi TU of Iaşi, România
Nicolae **BÂLC**, TU of Cluj-Napoca, România
Mustafa **BAGRIYANIK**, Istanbul Technical U, Tyrkey
Mihai **BĂNICĂ**, TU of Cluj-Napoca, NU Center of Baia Mare
Athena **BARONOU**, TEI West Macedonia, Greece
Katarzyna **BIAŁAS**, Silesian TU, Poland
Nikolaos **BILALIS**, Technical University of Crete, Greece
Antonios **BISBAS**, TEI West Macedonia, Greece
Octavian **BOLOGA**,U Lucian Blaga of Sibiu,România
Gheorghe **BRABIE**, U Vasile Alecsandri of Bacau, România
Paul Dan **BRÎNDASU**, U Lucian Blaga of Sibiu, România
Andrzej **BUCHACZ**, Silesian TU of Gliwice, Poland
Carmen **BUJOREANU**, Gheorghe Asachi TU of Iaşi, Romania
Ian **CAMPBELL**,Loughborough University, UK
George **CHARALAMPIDIS**, TEI West Macedonia, Greece
Francisco **CHINESTA**, Ecole Centrale de Nantes, France
George **CHRISTOFORIDIS**, TEI West Macedonia, Greece
Konstantinos **DAVID**, TEI Central Macedonia, Greece
Evangelos **DIALYNAS**, National Technical U, Greece
Florin **DIMOFTE**, U of Toledo, USA
Christos **DIMOPOULOS**, European U Cyprus, Cyprus
Vaya **DINOPOULOU**, TEI West Macedonia, Greece
Oana **DODUN**, Gheorghe Asachi TU of Iaşi, România
Savvas **DOUVARTZIDIS**, TEI West Macedonia, Greece
George **DRAGHICI**, U Politehnica of Timişoara, România
Petru **DUŞA**, Gheorghe Asachi TU of Iaşi, România

Nicoleta **ENE**, U of Toledo, USA
Catalin **FETECAU**, U Dunarea de Jos U of Galaţi, România
Andronikos **FILIOS**, TEI Pirea, Greece
George **FRAGULIS**, TEI West Macedonia, Greece
Stergios **GANATSIOS**, TEI West Macedonia, Greece
James **GAO**, Greenwich University, UK
Cesar **GARCIA-HERNANDEZ**, University of Zaragoza, Spain
Konstantinos **GAVROS**, TEI West Macedonia, Greece
Daniel **GHICULESCU**, U Politehnica of Bucureşti, România
Dimitrios **GIAGKOPOULOS**, U of Western Macedonia, Greece
Ioannis **GIAPINTZAKIS**, U of Cyprus, Cyprus
António **GONÇALVES-COELHO**, U Nova de Lisboa, Portugal
Maria **GOULA**, TEI West Macedonia, Greece
Ioannis **GOUDAS**, TEI West Macedonia, Greece
Mihaiţă **HORODINCA**, TU Gheorghe Asachi of Iaşi, România
Jose Luis **HUERTAS-TALON**, University of Zaragoza, Spain
Radu **IBĂNESCU**, Gheorghe Asachi TU of Iaşi, România
Andreas **IORDANIDIS**, TEI West Macedonia, Greece
Konstantinos **KAKOULIS**, TEI West Macedonia, Greece
Ioannis **KAPAGERIDIS**, TEI West Macedonia, Greece
Loukas **KARAGIANNAKIS**, TEI West Macedonia, Greece
Vaios **KARAGIANNIS**, TEI West Macedonia, Greece
Nikolaos **KARTALIS**, TEI West Macedonia, Greece
John **KECHAGIAS**, TEI of Thessaly, Greece
Vasileios **KIKIS**, TEI West Macedonia, Greece
Apostolos **KORLOS**, TEI of Thessaloniki, Greece
Theodora **KYRATSI**, University of Cyprus, Cyprus
Panagiotis **KYRATSIS**, TEI West Macedonia, Greece
Nikolaos **KYRATZIS**, TEI West Macedonia, Greece
George **KYRIAZOPOULOS**, TEI West Macedonia, Greece
Stanislaw **LEGUTKO**, Poznan U of Technology, Poland
Yingguang **LI**, Nanjing U of Aeronautics and Astronautics,China
Mircea **LOBONTIU**, NUCenter of Baia Mare, România
Antonis **LONDOS**, Frederick University, Cyprus
Loucas **LOUCA**, University of Cyprus, Cyprus
Octavian **LUPESCU**, Gheorghe Asachi TU of Iaşi, România
Catalina **MAIER**, U Dunarea de Jos of Galaţi, România
Sofoklis **MAKRIDIS**, University of West Macedonia
Athanasios **MANAVIS**, TEI of West Macedonia, Greece
Gabriel **MANSOUR**, Aristotle U of Thessaloniki, Greece
Emmanouil **MARAVELAKIS**, TEI of Crete, Greece
Paul **MAROPOULOS**, Aston University, UK
Stergios **MAROPOULOS**, TEI of West Macedonia, Greece
Ioannis **MASTORAS**, TEI of West Macedonia, Greece
Eugen **MERTICARU**, Gheorghe Asachi TU of Iaşi, România
Vasile **MERTICARU**, Gheorghe Asachi TU of Iaşi, România
Alina **MINEA**, Gheorghe Asachi TU of Iaşi, România
Sevasti **MITSI**, Aristotle U of Thessaloniki, Greece
Athanasios **MICHAILIDIS**, Aristotle University of Thessaloniki, Greece
António **MOURÃO**, U Nova de Lisboa, Portugal
Gavril **MUSCA**, Gheorghe Asachi TU of Iaşi, România
Gheorghe **NAGIT**, Gheorghe Asachi TU of Iaşi, România
Irina **NEAGA**, U of Plymouth, UK
Florin **NEGOESCU**, Gheorghe Asachi TU of Iaşi, România
Konstantinos – Stefanos **NIKAS**, TEI Pirea, Greece
Laurentiu Eduard **NITU**, U of Piteşti, România
Anna **NOCIVIN**, U Ovidius of Constanţa, România

Ioannis **NTINTAKIS**, TEI of Thessaly, Greece.
Asimina **NTOMOPOULOU**, TEI of West Macedonia, Greece
Gheorghe **OANCEA**, Transilvania U of Braşov, România
Nicolae **OANCEA**, U Dunarea de Jos of Galaţi, România
Dumitru **OLARU**, Gheorghe Asachi TU of Iasi, România
Aydogan **OZDEMIR**, Istanbul Technical University, Turkey
Răzvan **PĂCURAR**, TU of Cluj-Napoca, România
Ion **PANA**, Petroleum-Gas U of Ploieşti, România
Carmen Ema **PANAITE**, Gheorghe Asachi TU of Iasi, România
Dimokritos **PANAGIOTOPOULOS**, TEI of West Macedonia, Greece
Sofia **PAPANIKOLAOU**, TEI of West Macedonia, Greece
Konstantinos **PARISIS**, TEI of West Macedonia, Greece
Viorel **PAUNOIU**, U Dunarea de Jos of Galaţi, România
Andreas **PETRAKIS**, TEI of West Macedonia, Greece
Spyridon **PLIATIAS**, TEI of West Macedonia, Greece
Argirios **PLIATSIOS**, TEI of West Macedonia, Greece
Marcel Sabin **POPA**, TU of Cluj-Napoca, România
Nikolaos **POULAKIS**, TEI of West Macedonia, Greece
Fabrizio **QUADRINI**, U Tor Vergata of Rome, Italy
Miroslav **RADOVANOVIC**, U of Nis, Serbia
Minodora **RIPA**, U Dunarea de Jos U of Galaţi, România
Bogdan **RUSU**, Gheorghe Asachi TU of Iasi, România
Konstantinos **SACHPAZIS**, TEI of West Macedonia, Greece
Loredana **SANTO**, U Tor Vergata of Rome, Italy
Francisco **SANTOS**, U of Valladolid, Spain
Nikolaos **SAPIDIS**, University of Western Macedonia, Greece
Ioanis **SARRIS**, TEI ATH, Greece
Serafim **SAVVIDIS**, TEI of West Macedonia, Greece
Carol **SCHNAKOVSZKY**, U Vasile Alecsandri of Bacau, România
Nicolae **SEGHEDIN**, Gheorghe Asachi TU of Iaşi, România
George **SIDERIDIS**, TEI of West Macedonia, Greece
Laurenţiu **SLATINEANU**, Gheorghe Asachi TU of Iaşi, România
Lina **STATHOPOULOU,**TEI of West Macedonia, Greece
Adamos **STIMONIARIS**, TEI of West Macedonia, Greece
Dimitrios **STIMONIARIS**, TEI of West Macedonia, Greece
Birgit Kjærside **STORM**, U of Aalborg, Esbjerg, Denmark
Gabriela **STRNAD**, U Petru Maior, Tg. Mureş, România
Mirela **SUCHEA**, U Alexndru I Cuza of Iaşi, România
Lucian **TABACARU**, Gheorghe Asachi TU of Iaşi, România
Nikolaos **TAOUSANIDIS**, TEI of West Macedonia, Greece
Stavroula **TAVOULTZIDOU**, TEI of West Macedonia, Greece
Irina **TIŢA**, Gheorghe Asachi TU of Iaşi, România
Mihail Aurel **ŢÎŢU**, U Lucian Blaga of Sibiu, România
Alexei **TOCA**, TU of Moldova, Chisinau, Republic of Moldova
Pavlina **TOTEVA**, Technical University of Varna, Bulgaria
Athanasios **TRIANTAFYLLOU**, TEI of West Macedonia, Greece
Apostolos **TSAGARIS**, TEI of Thessaloniki, Greece
Konstantinos **TSANAKTSIDIS**, TEI of West Macedonia, Greece
Dimitrios **TSIAMITROS**, TEI of West Macedonia, Greece
Ionnis **TSIAFIS**, Aristotle University of Thessaloniki, Greece
Lazaros **TSIKRITZIS**, TEI of West Macedonia, Greece
Alexandros **TSOUKNIDAS**, TEI of West Macedonia, Greece
Dimitrios **TZETZIS**, International Hellenic University, Greece
Emmanouil **VARITIS**, TEI of West Macedonia, Greece
Gyula **VARGA**, University of Miskolc, Hungary
Argirios **VATALIS**, TEI of West Macedonia, Greece
Michail **VLACHOGIANNIS**, TEI of Thessaly, Greece

Charitini **VOULGARAKI**, TEI of West Macedonia, Greece
Vaya **ZACHARAKI**, TEI of West Macedonia, Greece
Danut **ZAHARIEA**, Gheorghe Asachi TU of Iaşi, România
Valentin **ZICHIL**, U Vasile Alecsandri of Bacau, România

Reviewers

Dimitris Tzetzis, Miroslav Radovanovic, Konstantinos Kitsakis, Apostolos Korlos, Romeo Cioara, Oana Dodun, Bogdan Rusu, Charitini Voulgaraki, Mihaita Horodinca, Apostolos Tsagkaris, Ion Pana, Gabriela Strnad, Loredana Santo, Charisios Achillas, Anamaria Ilie, Angelos Markopoulos, Miroslav Radovanovic, Apostolos Korlos, Vayos Karayannis, Sofia Papanikolou, Csaba Felho, Serafim Savvidis, Cristian Pisarciuc, Konstantinos Kitsakis, Loukas Karagiannakis, Cristina Oprea, Fabrizio Quadrin, Anna Nocivin, Bogdan Pralea, Virgil Gabriel Teodor, Ioannis Sarris, Sofia Papanikolaou, Catalina Maier, Stylianos Koukoumialos, George Sideridis, John Kechagias, Dimitrios Tsiamitros, Radu Ibanescu, Katarzyna Białas, Alina Minea, Ioannis Antoniadis, Dimitrios Petropoulos, Nikolaos Efkolidis, Alexander Tsouknidas, Eduard Laurentiu Nitu, Adriana Munteanu, Lucian Tabacaru, Vaya Dinopoulou, Elisavet Amanatidou, Laurentiu Slatineanu, Adrian Moise, Ionut Lambrescu, Nikolaos Bilalis, George Fragulis, Mihail Aurel Titu, Dimitrios Stimoniaris, Constantinos Tsanaktsidis, Mihaita Horodinca, Dorin Luca, Alina Minea, Cesar Garcia-Hernandez, Apostolos Tsagkaris, Octavian Bologa,

Organising Committee from TEI of Western Macedonia:

Prof. Stergios **MAROPOULOS**, PhD, Head of Department of Mechanical Engineering and Industrial Design, Technological Education Institute of West Macedonia, Greece.
Assis. Prof. Panagiotis **KYRATSIS**, PhD, Department of Mechanical Engineering and Industrial Design, Technological Education Institute of West Macedonia, Greece.
Applic.Prof. Dimitrios **STIMONIARIS**, MSc, Department of Electrical Engineering, Technological Education Institute of West Macedonia, Greece.
Sofia **PAPANIKOLAOU**, Technological Education Institute of West Macedonia, Greece.
Lina **STATHOPOULOU**, Technological Education Institute of West Macedonia, Greece.
Athanasios **MANAVIS**, Technological Education Institute of West Macedonia, Greece.

Organizing Committee from TU of Iaşi:

Eugen **AXINTE**; Petru **DUŞA**; Octavian **LUPESCU**; Gavril **MUSCĂ**; Gheorghe **NAGIT**;
Lucian **TĂBĂCARU**; Vasile **MERTICARU**;
Florin **NEGOESCU**; Viorel **COHAL**; Laurenţiu **GHENGHEA**; Margareta **COTEAŢĂ**;
Andrei **MIHALACHE**; Mihai **BOCA**; Marius**RÎPANU**

Keynote speakers

1 Paul **MAROPOULOS**, Aston University, UK.
"Digital Manufacturing and Engineering and aspects of research collaboration in Europe – reflection on Brexit"

2 Mihnea **COSTOIU**, Politehnica University of Bucharest, Romania
"The role of structural funds in the development of universities and their correlation with the labor market"

3 Laurenţiu **SLĂTINEANU**, PhD, "Gheorghe Asachi" Technical University of Iaşi, România.
"Technology, history, friendship, evolution"

4 George **SAVVAIDIS**, Aristotle University of Thessaloniki, Greece
"An Energy-Critical Plane based fatigue damage approach for the Life Prediction of Metal Alloys"

5 Gheorghe **OANCEA**, Transilvania University of Brasov, Romania
"***Remanufacturing of parts using innovative technologies"***

6 Daniel **GHICULESCU**,Politehnica University of Bucharest, Romania
"Development and innovation of ultrasonically aided electrical discharge maching and its specific equipment"

7 Aydogan **OZDEMIR,** Istabul Technical University, Tyrkey
"EV Parking Lots: New Storage Assets of Smart Cities"

8 Loredana **SANTO,** University of Rome "Tor Vergata", Italy
"The recent advances and future perspectives of Shape Memory Polymer Composites"

9 Gabriela **STRNAD,** University of Targu Mures, Romania
"Titanium – based micro and nano structured surfaces for biomedical applications"

10 **Cesar GARCIA – HERNANDEZ & Eduardo SANCHEZ – ALVAREZ,** University of Zaragoza, Spain
"Manufacturing of insoles optimized for safety footwear considering extreme temperature working conditions"

IManEE ORGANIZERS:

Department of Mechanical Engineering and Industrial Design, *Tehnological Education Institute of West Macedonia, Kozani, Greece*
Department of Industrial Engineering, *"Georghe Asachi" Technical University of Iaşi, România*

IOP Conf. Series: Materials Science and Engineering **161** (2016) 011002 doi:10.1088/1757-899X/161/1/011002

Peer review statement

All papers published in this volume of *IOP Conference Series: Materials Science and Engineering* have been peer reviewed through processes administered by the proceedings Editors. Reviews were conducted by expert referees to the professional and scientific standards expected of a proceedings journal published by IOP Publishing.

IOP Conf. Series: Materials Science and Engineering **161** (2016) 012001 doi:10.1088/1757-899X/161/1/012001

Silent Flange Coupling Design Used for the Schenck Eddy Current Dynamometer

D Schinteie and V Croitorescu

Scientific Research and Continuous Training Center for Sustainable Automotive Technologies, University POLITEHNICA of Bucharest, Faculty of Transport, room JC004, 313 Splaiul Independentei St., 6th sector, Bucharest, Romania

E-mail: valerian.croitorescu@upb.ro

Abstract. The silent flange used for coupling different machines/systems to an eddy current dynamometer represents one of the modular components each test-bench should use. By introducing a silent flange into a dynamometer, the coupling steps are easier and faster. For an appropriate design, the silent flange was analyzed using dedicated software during different operation procedures and scenarios, for materials that allow easy manufacturing. This study shows that the design for this silent flange model has no danger of failure due to the small deformation and the values for the equivalent stresses. The silent flange coupling is suitable for the dynamometer for his high positioning accuracy, the zero backlash and the fact that there is no motion between the shafts.

1. Introduction

The eddy current dynamometer is one of the best known dynamometers solutions, being used to measure the force, torque or power, produced by an engine, an electric machine or a transmission, by measuring simultaneously the torque and the rotational speed. Its usage is often met to determine the power characteristics of a machine under test, the fuel consumption and the harmful emissions for internal combustion engines on different loads and scenarios, and even if for complete powertrain solution, including hybrid powertrains and electric powertrains. Beside the apparently simple measurements of power, fuel and emissions, the eddy current dynamometer is used for detailed investigation and calibration of the engines. For being able to achieve best results, the eddy current dynamometer has to include different components to be coupled to the component/system under test.

The flanged coupling is used for better connection between the eddy current dynamometer and the component/system under test. It consists of a unit used for joining two different shafts within a component or a system, being able to meet stringent requirements for different testing procedures. Different coupling solutions are possible to be used, depending on the measurement precision, the tolerance to misalignment and the torsional rigidity, having different shapes and design. The disc couplings are able to transmit torque between the bolts through a series of a thin, stainless steel disc assembled in a pack. These bolts connect shaft hubs into one piece construction. Misalignments are absorbed through flexing a disc.

The objective of this paper is to highlight the calculus of a silent flange as major and mandatory component for the Schenk WT190 eddy current dynamometer. Its design was started and achieved during the preliminary development stage, while the structural analysis was performed before the flange final design. The final designed silent flange solution was needed in order to facilitate the mounting and unmounting operations needed for the machines under test to be coupled to the dynamometer. The silent flange represents the mandatory component needed on the test bench, without whom no tests were able to be performed while fast coupling any torque/power machine/system. This study case of the flanged coupling design used for the Schenck Eddy Current Dynamometer in order to carry the torque/the power from the torque/power machine/systems includes in deep research stages and structural analysis for the behavioral functioning under different loads. All the research stages were done in order to manufacture the flange and to make the test bench working for different types of torque/power machineries/systems, from internal combustion engines, to electric machines and to different transmission components that are able to deliver torque/power.

2. The Silent Flange Design

The flange is linking two different components in order to deliver torque/power at different rotational speeds. The disc coupling flange uses different connecting screw-bolts and fixing clamp, being a rigid coupling, mandatory to achieve no axial displacement between the coupling components. The main geometrical dimension for a coupling flange are presented in figure 1, being determined using the equations (1) to (7).

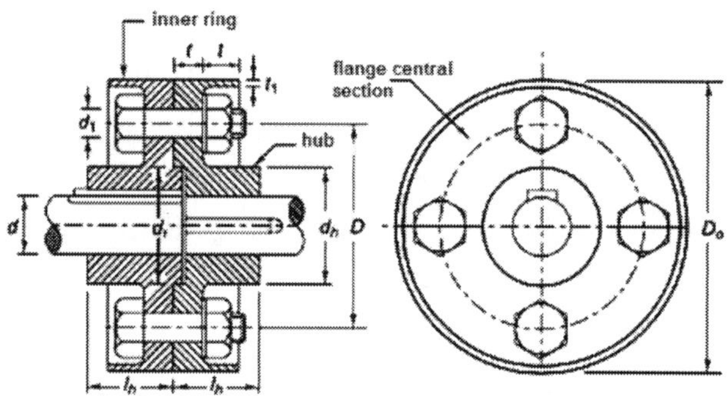

Figure 1. The main dimensions for the coupling flange [3]

$$d_h = 2 \cdot d \tag{1}$$
$$l_h = 1.5 \cdot d \tag{2}$$
$$D = 3 \cdot d \tag{3}$$
$$t = 0.5 \cdot d \tag{4}$$
$$t_1 = 0.25 \cdot d \tag{5}$$
$$d_r = 1.5 \cdot d \tag{6}$$
$$D_0 = (4 \cdot d + 2 \cdot t_1) \tag{7}$$

where d_h is the external diameter of the hub flange, l_h is the effective length of the hub flange, D is the screw-bolt diameter evenly distributed, t is the flange thickness, t_1 is the safety thickness of the outer ring, d_r is the diameter radial bearing, D_0 is the flange external diameter,

The design of the rigid flange has to accomplish the requirements that include the maximum transmitted torque, of 600 Nm, and the maximum transmitted power, of 190 kW. For achieving a safer operation after the final design, the safety values will be 1.5 higher.

The used materials will be selected in order to respect the tensile strength, apparent yield strength, elongation at rupture and get breaking constriction. The maximum allowable stress is determined based on the maximum rupture strength.
In order to determine the running shaft external diameter, the flange do not have a dedicated shaft coupling diameter.
Analysis coupling stiffness and torque transmission can be carried out in two distinct situations. Forces acting circular individually on the screw-bolts are described in figure 2.

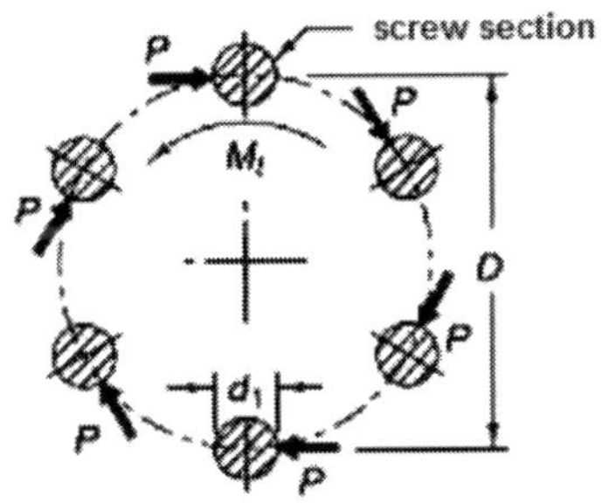

Figure 2. Screw-bolts shear strength [3]

The screw-bolts can be fastened tight and the torque transmission is made only by surface mounting flange coupling bolts. The screw-bolts are subjected to direct shear stress due to the resultant force P and are not subject to torsional stress. The resultant force is the result of shear stress. The transmitted torque is being determined using equation (8).

$$M_{tot} = P \cdot \frac{D}{2} \cdot N_b \quad (8)$$

where: M_{tot} is the transmited torque using the flange, P is the maximum power to be transmited, D is the screw-bolt circle diameter settlement, N_b is the screw-bolts total number.
If the The screws are not mounted fixed and tighted, the previous analysis can not be applied. In this case, the screws are tight enough with pre-tightening and the torque transmission is accomplished by friction. For a uniform distribution of friction, the radius of friction is given by equation (9):

$$R_f = \frac{2}{3} \cdot \frac{(R_0^3 - R_i^3)}{(R_0^2 - R_i^2)} \quad (9)$$

where R_0 is the flange external radius ($D_0/2$), R_i is the shaft diameter (d/2), R_f is the friciton radius.
The friction force and friction torque are determined using equations (10) and (11):

$$F_f = \mu \cdot P_i \cdot N_b \quad (10)$$
$$M_f = F_f \cdot R_f = \mu \cdot P_i \cdot N_b \cdot R_f \quad (11)$$

where: P_i is the initial tensile strnght for each screw-bolt, μ is the friction coefficient between flanges.
The studied eddy current dynamometer has its own primary flange from which the macine/system under test can be coupled. The design of this flange is made by the WT190 Schenck dynamometer manufacturer, as presented in figure 3.a. The new designed silent flange is custom made, allowing the

easier coupling for different systems to be under tests, depending on the application, as presented in figure 3.b.

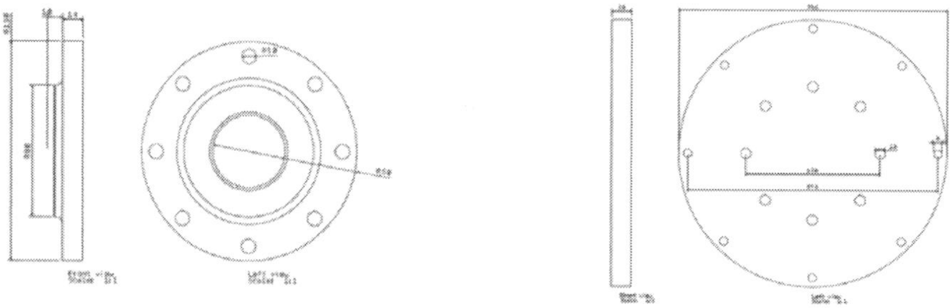

Figure 3. The *a*-conventional flange and the *b*-silent flange

The silent flange was designed using dedicated software for numerical analysis. Due to the complexity of the model, to the geometrical constrains and to the limitation of the post-processing levels, the designed model was simplified.

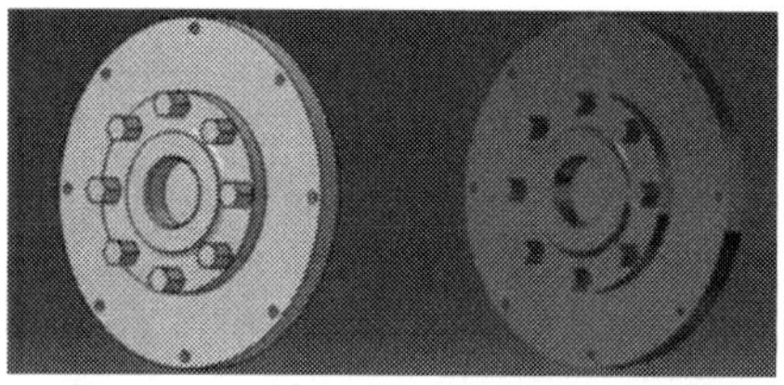

Figure 4. The silent flange sketch

3. The Silent Flange Structural Analysis

The silent flange is the component designed to provide the connection between the torque/power machine/system and the electric machine, part of the test bench. The main purpose of the flange coupling is to transfer the torque from a shaft to another shaft, having the same rotary velocity.

Critical stresses are involved while designing and testing the flange, in adition to its operation near the limit. The analysis can be performed either by finite element method (FEM) either by using empirical formulas. Some assumptions can be made if they are necessary. The chosen method for designing the flange was based on finite element modeling and analysis, using both ANSYS and CATIA. Following the needs of an elaborate study, the method is using more than one level of design. Also, different types of loading were taken into account.

The stress state of the flange, at different levels of loadings require FEM analysis and design (based on modeling and simulation), including setting the dimension and the shape, using different materials for manufacturing, while different selection criteria are taken into account, in addition to the stress evaluation stages of the silent flange virtually manufactured using the chosen materials.

The silent flange destination is defining the analysis type, including an endurance test using a buckling test rig. Using an accurate method during the silent flange stress evaluation, the areas with higher stresses are hightligted. The design requirements are helping to conclude to the silent flange stress behavior on differet stresses situations, including the distributed effort to the screw-bolt holes areas. The deformations that are present in the areas of interest are obtained.

The chosen material for manufacturing the silent flange helps achieving the needed accuracy. The used material properties are presented in Table 1.

Table 1. The material properties

Density	CTE*	Specific Heat	Thermal Conductivity	Resistivity
7850 kg m^-3	1.2e-005 C^-1	434 J kg^-1 C^-1	60.5 W m^-1 C^-1	1.7e-007 Ohm m

*CTE= Coefficient of Thermal Expansion

Table 2. The modeling properties

Model Geometry Parts	
Object Name	*PartBody*
State	Meshed
Graphics Properties	
Visible	Yes
Transparency	1
Definition	
Suppressed	No
Stiffness Behavior	Flexible
Coordinate System	Default Coordinate System
Reference Temperature	By Environment
Material	
Assignment	Structural Steel
Nonlinear Effects	Yes
Thermal Strain Effects	Yes
Bounding Box	
Length X	1.e-002 m
Length Y	0.261 m
Length Z	0.261 m
Properties	
Volume	5.2471e-004 m³
Mass	4.119 kg
Centroid X	5.e-003 m
Centroid Y	-6.0449e-011 m
Centroid Z	-1.9521e-007 m
Moment of Inertia Ip1	3.494e-002 kg·m²
Moment of Inertia Ip2	1.7502e-002 kg·m²
Moment of Inertia Ip3	1.7507e-002 kg·m²
Statistics	
Nodes	30786
Elements	5422
Mesh Metric	None

Table 3. The silent flange analysis: loads distribution

Model (A4) > Static Structural (A5) > Loads Distribution									
Object Name	*Fixed Support*	*Force*	*Force 2*	*Force 3*	*Force 4*	*Force 5*	*Force 6*	*Force 7*	*Force 8*
State	Fully Defined								
Scope									
Scoping Method	Geometry Selection								
Geometry	16 Faces	2 Faces							
Definition									
Type	Fixed Support	Force							
Suppressed	No								
Define By		Components							

The flange analysis includes the mechanical deformation analysis using static structural, being able to provide the stresses and the deformations results.

4. Results

The geometrical model of the flange was developed. Several analysis were performed on the model. Important steps have been completed in designing the flange: setting the dimensions and the material, model development, simulation and validation, structural analysis.

Due to the small deformations and the determined equivalent stress, the designed coupling flange is suitable for being used on the Schenk Dynamometer WT 190. More of that, the zero backlash and the fact that there is no motion between the shafts make the designed coupling flange suitable for high positioning accuracy applications.
The maximum stress reaches the value of around 2.9 MPa which enables to conclude that the silent flange can be fairly used in the application of the engine, electric machine and transmissions testing. New solutions are taken into account for future research and manufacturing possibilities, including composite materials (hard special purpose plastics) in addition to iron.

,

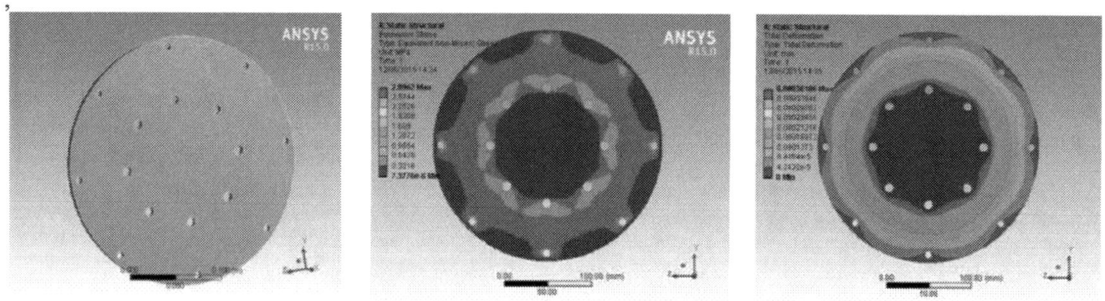

a-sketch b-equivalent stress c-maximum deformation

Figure 5. The silent flange analysis

Table 4. Results

Results		
Object Name	*Equivalent Stress*	*Total Deformation*
Minimum	7.3776 Pa	0. m
Maximum	2.8962e+006 Pa	3.8189e-007 m
Minimum Value Over Time		
Minimum	7.3776 Pa	0. m
Maximum	7.3776 Pa	0. m
Maximum Value Over Time		
Minimum	2.8962e+006 Pa	3.8189e-007 m
Maximum	2.8962e+006 Pa	3.8189e-007 m

5. Conculsions

The above presented approaches contained the stresses evaluation for the silent flange to be used on a Schenk WT 190 dynamometer. Based on its design, the silent flange dimensions were established. In addition, the material to be used for manufacturing was established.
The silent flange study is needed in order to facilitate mounting for different machines/systems under testing procedures. The allignment between the shafts is mandatory to perform safe and secure testing. The stresses behavior and deformations evolution during operation confirm the possibility of using the silent flange following its objectives.

6. References

[1] Aman A., Ashraful A.S., Tausif R.A., Ullah Khan H., Asiful I., *Design & Construction of a Flanged Coupling,* Report Department of Mechanical Engineering, Khulna University of Engineering & Technology, September 2014

[2] Baraa A-B. M.H.., H. Wissam K., Hussam A-M., *An Integrated Approach to Design and Analyze Unprotected Type Flange Coupling*, Journal: AL-TAQANI, ISSN: 1818653X Year: 2005 Volume: 18 Issue: 2 Pages: 133-145 Publisher: Foundation of Technical Education

[3] Bhandari V.B., *Design of Machine Elements*, Tata McGraw - Hill Education, 2010

[4] Dahl G., *Introduction to couplings*, Misumi's Standard Mechanical Components, 2013

[5] Sufiyan D., *Rigid Couplings and It's Design*, August 2013 – Butler Polytechnic Vadodara internal presentation, available at http://www.slideshare.net/sufiyanderaiya/rigid-coupling-and-its-design

IOP Conf. Series: Materials Science and Engineering **161** (2016) 012002 doi:10.1088/1757-899X/161/1/012002

Formation of technical requirements for flexible rotary machine nodes

Y BULENKOV[1] and A MIKHAYLOV[2]

[1] Ass. Prof. Department of Manufacturing Engineering, DonNTU, Donetsk, Ukraine
[2] Prof. Department of Manufacturing Engineering, DonNTU, Donetsk, Ukraine

E-mail: bulenkov@ukr.net

Abstract. The method of parameters determining for the flexible rotary machines and lines and its individual components is described in this article. The method is based on the analysis of the fail safe performance probability. It allows determining the fail safe performance probability for tools, transportation and tool changing device nodes, elements of flexible rotary machine and is based on the analysis of flexible rotor line structure. The relationships between rational flexible rotary line structure and parameters of the individual nodes are shown on the flexible rotor line for the screws processing.

1. Problem

Processing of screws during continuous transportation on flexible rotary lines is the most efficient way to produce them [1-5]. Known design methods for flexible rotor lines allow developing rational structures and schemas for these lines [1, 4], but transitioning from theoretical studies to identify the specific parameters of machines during parametric synthesis is insufficiently studied. It is therefore necessary to improve the process of parametric synthesis for flexible rotary lines.

The purpose of performing this work is to study the relationship between the fail safe performance probability of flexible rotary line and machines for determining their individual parameters that are used as initial data in parametric synthesis.

To achieve this goal it is necessary to investigate the fail safe performance probability of the flexible rotary line and the fail safe performance probability of rotary machines to choose the most rational structure of the flexible rotary machine.

2. Method

Structural analysis is used to figure out the logic scheme of the rotary line. Probability theory is used to determine the fail safe performance probability for the various elements of the line.

3. The method of parameters determining for the flexible rotary machines and lines

Figure 1 shows the constructive-technological scheme of flexible rotary line for the manufacture of rod fasteners - screws. The scheme shows the flexible rotary machine for cutting and alignment end faces TP1, technological rotary machines TP2 and TP3 for preliminary and final shaping of end faces respectively. TP4 is a technological rotary machine for the threaded rod extraction and TP5 is a technological rotary machine for thread rolling. Transportation rotary machines TrR1-TrR5 carry out interrotary transfer of details. Wire on input stream V is supplied to the technological rotary machine TP1. Here the wire cut into individual blanks. There is an alignment of end faces and disembarkation of a facet on one of the ends of a core in the same rotary machine. Next, the workpiece moves to the next technological rotary machines, where a preliminary and final shaping of a head, reducing threaded rod and thread roll forming are performed. A replacement of tool occurs during idling in multinomenclature working positions for processing of other nomenclature products by turning the tool drum. Thereafter parts are discharged from the line by output stream W.

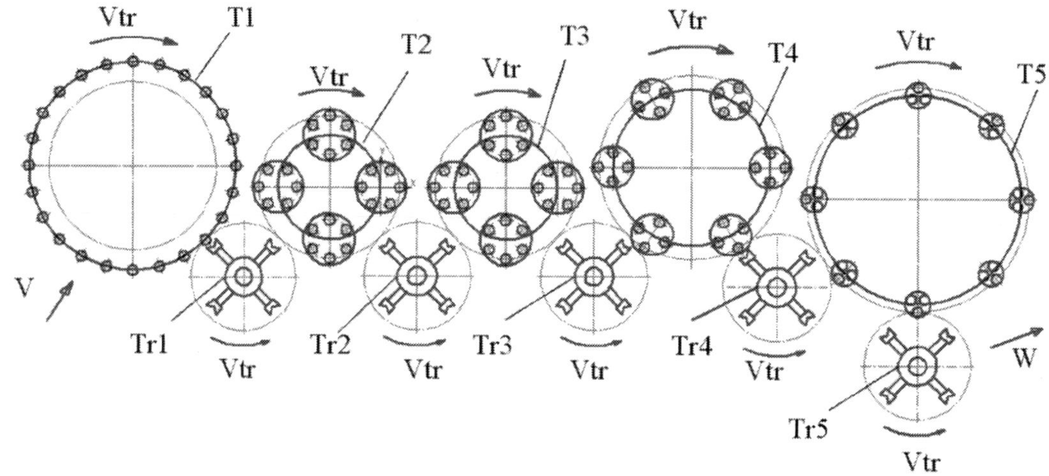

Figure 1. The constructive-technological scheme of flexible rotary line.

Let's look at the fail safe performance probability of flexible rotary line. This line corresponds to the logic scheme (figure 2), which is a mathematical model of the line.

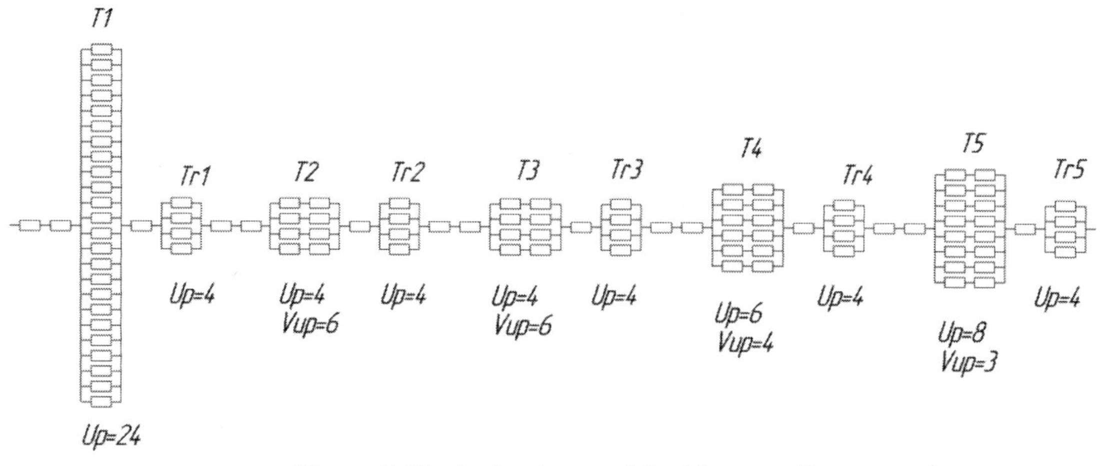

Figure 2. The logic scheme of flexible rotary line.

We must consider that refusal in branched system will lead to loss of parts of the same size. The probability of the release of all types and sizes of products should be assessed during the line design, rather than the probability of sustaining performance above a predetermined value. The fail safe performance probability of flexible rotary line is evaluated under the following conditions.

1. The line stops either at refusal in unbranched system, or in case of failure more than one group in branched system.

2. Features of tools and tool selection devices are the same, i.e. they each have the same performance and have the same fail safe performance probability. Gripping bodies of transport rotary machines also have the adequate characteristics.

Thus, the line can be in one of states:

1. Preservation of full productivity P1
2. Failure of one tool P2
3. Failure of one of the gripping body P3
4. Failure of the control unit in one multinomenclature tool module P4
5. Refusal in unbranched system or refusal more than two elements in the branched system P5.

Since all considered states form full group, the sum of probabilities of these states is equal to unit

$$P1 + P2 + P3 + P4 + P5 = 1. \tag{1}$$

Then fail safe performance probability of flexible rotary line can be determined by the following formula:

$$P = 1 - P5 = P1 + P2 + P3 + P4 \tag{2}$$

$$\begin{aligned} P = {} & Pn^{15} \times Ptr^{20} \times Pu^{46} \times Pi^{120} + Pn^{15} \times Ptr^{20} \times Pu^{46} \times (\sum_{5} C_{24}^{23}) \times (1-Pi) \times Pi^{119} + \\ & + Pn^{15} \times Pu^{46} \times Pi^{120} \times (\sum_{6} C_{4}^{3}) \times (1-Ptr) \times Ptr^{19} + \\ & + Pn^{15} \times Ptr^{20} \times Pi^{120} \times (1-Pu) \times Pu^{45} \times (C_{24}^{23} + 2 \times C_{4}^{3} + C_{6}^{5} + C_{8}^{7}). \end{aligned} \tag{3}$$

where Pn - fail safe performance probability of elements in unbranched system,
Ptr - fail safe performance probability of transport devices,
Pu - fail safe performance probability of control units of the group tool module,
Pi - fail safe performance probability of the tool.

The dependence of fail safe performance probability of the flexible rotary line on fail safe performance probability of the tool and probability of transfer of the product was determined for Pn = 1 and Pu = 0,995 (see figure 3).

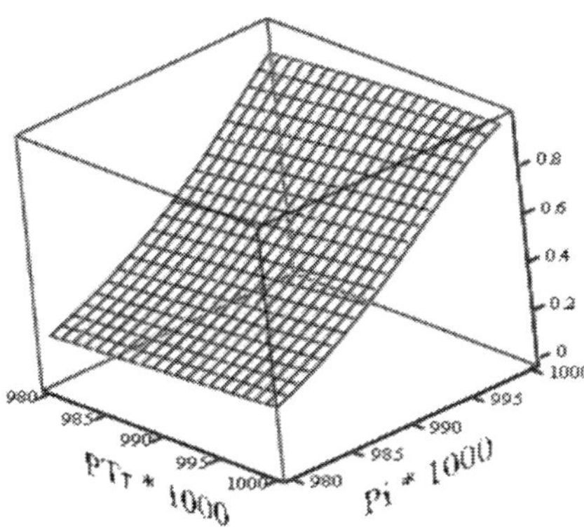

Figure 3. The dependence of fail safe performance probability of the flexible rotary line on fail safe performance probability of the tool and probability of transfer of the product.

Figure 3 shows that fail safe performance probability of the flexible rotary line most affected by fail safe performance probability of the tool. The probability of transfer of products between the rotarys significantly does not influence, since the number of gripping bodies in line considerably smaller than the number of tools.

Let's consider two variants of flexible rotary machine structures (figures 4 and 5). Flexible rotary machine shown in figure 4 is performed on the "classic" scheme. Within this scheme on elements of the rotary housing (RM) are installed a fully self-contained power heads (PH), which containes all the

IOP Conf. Series: Materials Science and Engineering **161** (2016) 012002 doi:10.1088/1757-899X/161/1/012002

necessary elements to carry out their functions. The logical scheme representing mathematical model of flexible rotary machine will include elements of not branching part, - RM, and also elements of the branching subsystem, - PH.

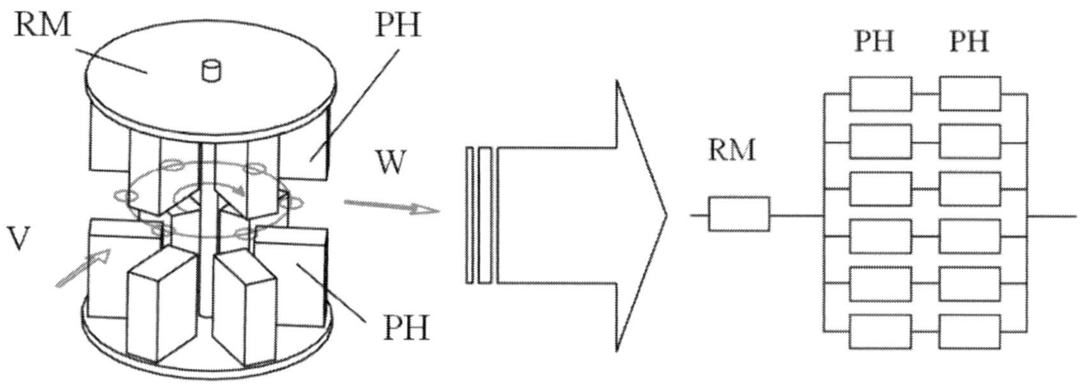

Figure 4. Flexible rotary machine is performed on the "classic" scheme.

Figure 5 shows the structure of flexible rotary machine, formed on the block diagram of new flexible rotary machine [1]. Such rotor machine has the configuration, classical for technological rotors, and includes power system (PS), general for all group tool modules. Group tool module have units for switching characteristics of the power system and stores of tools and devices [1]. Not branching part in the logical scheme of such machine will include the RM and PS elements. Branching part will include group tool modules.

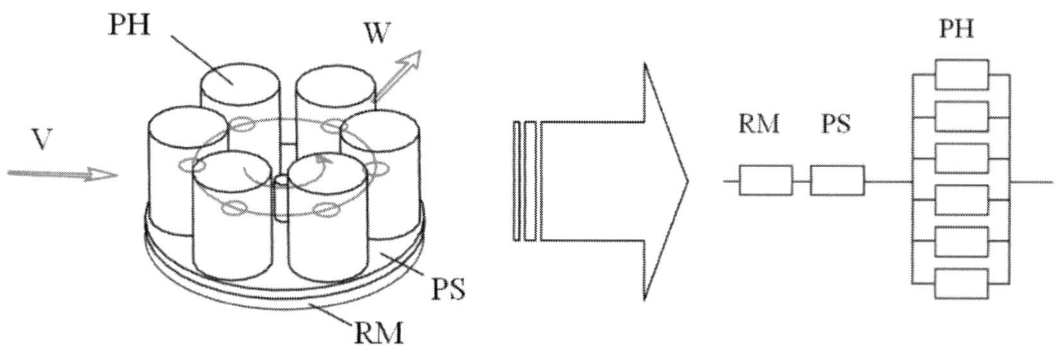

Figure 5. Flexible rotary machine, formed on the block diagram of new flexible rotary machine.

Let's consider fail safe performance probability of flexible rotary machine which scheme is shown in figure 4. The fail safe performance probability of the flexible rotary machine can be defined as the sum of fail safe performance probabilities in all possible states.

We estimate the fail safe performance probability of the flexible rotary machine under following conditions.

1. Flexible rotary machine stops either at refusal in unbranched system, or in case of failure more than one power head or group tool module.

2. Characteristics of power heads and group tool modules are identical, i.e. they have respectively identical productivity and have identical probability of no-failure operation.

According to the first condition flexible rotary machine 1 can be in one of states.

1. Preservation of full productivity P1:

$$P1 = Pa^{12} \cdot Ppm, \tag{4}$$

where Pa - fail safe performance probability of a power head,
Rrt - fail safe performance probability of flexible rotary machine elements.
2. Refusal of one power head P2:

$$P2 = Ppm \cdot (\sum_{6} C_2^1) \cdot (1 - Pa) \cdot Pa^{11} \tag{5}$$

where Cij - quantity of combinations from I on j.
3. Refusal of elements of the flexible rotary machine or refusal more than one group tool module P3.
Since all considered states form full group, the sum of probabilities of these states is equal to unit:

$$P1 + P2 + P3 = 1 \tag{6}$$

Then fail safe performance probability of TP1 can be determined by the following formula:

$$P = 1 - P3 = P1 + P2 \tag{7}$$

$$P = Pa^{12} \cdot Ppm + Ppm \cdot (\sum_{6} C_2^1) \cdot (1 - Pa) \cdot Pa^{11} \tag{8}$$

Dependence of fail safe performance probability of flexible rotary machine 1 on fail safe performance probability of power heads was determined for Rrt = 1 and Pa =0,980 … 1 (figure 6).

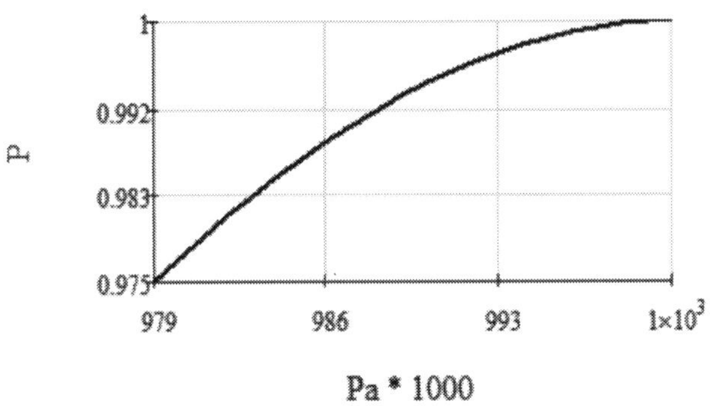

Figure 6. Dependence of fail safe performance probability of flexible rotary machine 1 on fail safe performance probability of power heads.

We estimate the appropriateness of the proposed flexible rotary machine. To do this, compare it fail safe performance probability with the fail safe performance probability of flexible rotary machine 2, formed accordingly to scheme in figure 5 (where the flexible rotary machine has an overall power system and products processed in group tool modules).

According to the given above conditions flexible rotary machine 2 can be in one of 3 states.

1. Preservation of full productivity:

$$P1 = Ppm \cdot Ppr \cdot Pa^{6} \tag{9}$$

where Ppr - fail safe performance probability of elements of power system.

2. Failure of one group tool module:

$$P2 = Ppm \cdot (C_6^5) \cdot (1 - Pa) \cdot Pa^{5} \tag{10}$$

3. Refusal of elements of the rotor machine, elements of power system or refusal more than one group tool module.

Since all considered states form full group, the sum of probabilities of these states is equal to unit:

$$P1 + P2 + P3 = 1 \tag{11}$$

Then the fail safe performance probability of flexible rotary machine 2 can be determined by such formula:

$$P = 1 - P3 = P1 + P2 \quad (12)$$

$$P = Ppm \cdot Ppr \cdot Pa^6 + Ppm \cdot (C_6^5) \cdot (1 - Pa) \cdot Pa^5 \quad (13)$$

In figure 7 dependence of fail safe performance probability of flexible rotary machines 1 and 2 on fail safe performance probabilities of elements of power system and group tool module is presented. It is obvious that both of these indicators influence on fail safe performance probability of flexible rotary machine 2, however reliability of power system will render in this case crucial importance.

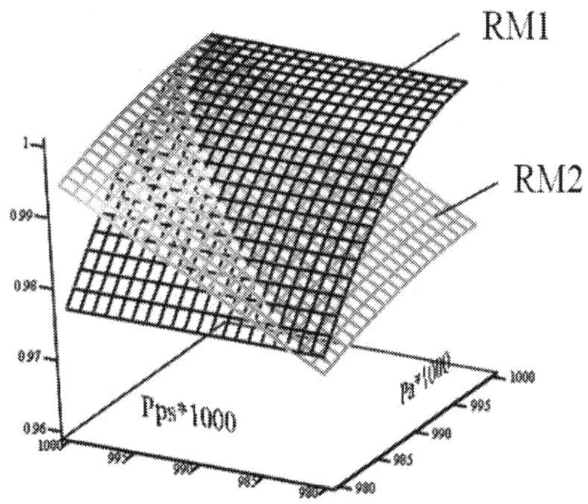

Figure 7. Dependence of fail safe performance probability of flexible rotary machines 1 and 2 on fail safe performance probabilities of elements of power system and group tool module.

4. Results

This work is dedicated to the development of design methods for flexible rotor lines that will combine several large-scale productions of products of different sizes in the mass diversified flow of goods by processing several sizes of goods in each group tool module. The developed method allows to conduct the structural analysis of flexible rotor lines and to determine the requirements on the fail safe performance probability for the line individually. These requirements can be used as input parameters at the parametric synthesis stage during design of flexible rotor machines for screws processing. The developed method can be used to establish connection between the machine structure and the parameters of their individual components and mechanisms.

References

[1] Bulenkov Y and Mikhaylov A 2009 Structural synthesis of flexible rotary machines *Progressive Technologies and Systems for Machine-Building* **37** 32-40.
[2] Bystrov V, Frolovich E, Klusov I and Preuss V 2003Rotor technology, machines and lines at the present stage of industrial development *Journal of Engineering* **10** 43-47
[3] Koshkin L 1986 *Rotary and rotary conveyor lines* (Moscow: Machine building).
[4] Mikhaylov A 2002 *Basics of synthesis for flow-spatial technology systems of continuous action* (Donetsk: Donetsk National Technical University,).
[5] Bulenkov Y 2013 Choice of rational structure for aggregate rotary machine *Progressive Technologies and Systems for Machine-Building* **46** 30-37

IOP Conf. Series: Materials Science and Engineering **161** (2016) 012003 doi:10.1088/1757-899X/161/1/012003

Research and constructive solutions on the reduction of slosh noise

O M Manta (Balas)[1], R Balas[2] and C V Doicin[3]

[1] PhD Student, POLITEHNICA University of Bucharest, Bucharest, Romania
[2] Engineering team leader, Honeywell, Bucharest, Romania
[3] Professor, POLITEHNICA University of Bucharest, Bucharest, Romania

E-mail: oanamaria2004@yahoo.com

Abstract. The paper presents a product design making of, over a "delicate issue" in automotive industry as slosh noise phenomena. Even though the current market tendency shows great achievements over this occurrence, in this study, the main idea is to design concepts of slosh noise baffles adapted for serial life existing fuel tanks in the automotive industry. Moreover, starting with internal and external research, going further through reversed engineering and applying own baffle technical solutions from conceptual sketches to 3D design, the paper shows the technical solutions identified as an alternative to a new development of fuel tank. Based on personal and academic experience there were identified several problematics and the possible answers based on functional analysis, in order to avoid blocking points. The idea of developing baffles adapted to already existent fuel tanks leaded to equivalent solutions analyzed from functional point of view. Once this stage is finished, a methodology will be used so as to choose the optimum solution so as to get the functional design.

1. Introduction

Automotive industry may be considered a highway for new technology various applications. Each salon amazes in terms of design, horsepower, materials and technical solutions. Together with all these characteristics, there are created new needs for drivers and passengers, also their expectations rise, and the critical spirit, too. Such a specific situation is Start and Stop technology [1]. In this case, the client becomes more sensitive to acoustic performance. The noise level rests the same as in the past (per component), but the client perceives it different. When the engine is turned off at the traffic light, he hears the sound of the waves inside the fuel tank. Translated into noise performance characteristic for the fuel tank, it is called "Slosh noise". A well-known phenomena generated by fluid movement inside a fuel tank while various scenarios like acceleration, deceleration, breaking etc. are taking place. In fact, during the process of moving the liquid inside the tank storages kinetic energy and the slosh noise is produced when this type of energy is over limit [2]. This kind of inconvenience can be solved by using filter elements in fixation zones, paint protection with acoustic performance, add of insulators, or to redesign a present fuel tank so as to introduce baffles. To decrease the effect of slosh noise, there were patented different solutions of internal baffles or even baffles integrated in the envelope of the fuel tanks. The main inconvenience of the existing solutions is that the fuel tank needs to be redesigned and the production process rebuild, so a completely new concept is developed.

In a world where the strategy is "design to cost" and in the same time "design to quality", too, the ideal situation is to find a solution to decrease the noise inconvenience generated by the waves inside the fuel tank without modification of its geometry. Therefore, the aim of the study above is to analyze technical solutions of baffles suitable for already existent fuel tanks that have no impact on modification of the production process [3].

2. From concept to design – extensible "wings" baffle and "chain" baffle

Based on the idea to create a baffle that can be used on already existing fuel tanks on the market, took birth the "Extensible Wings Baffle" project. From the conceptual sketch (figure1) and a general model of a fuel tank & fuel pump module took shape a preliminary model (figure 2).

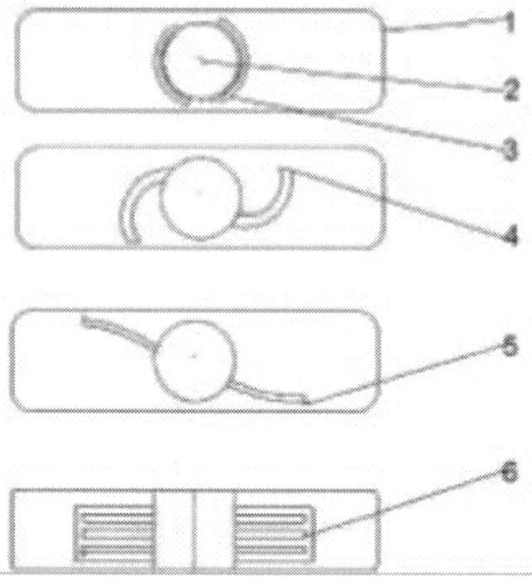

Figure 1. Conceptual sketch for Extensible Wings Baffle (baffle integrated on the fuel pump module, where 1/ Upper view of FT, 2/ Module, 3/ Closed baffle position – circular to the module, 4/ Upper view – Baffle during assembly – opening stage, 5/ Baffle in position, 6/ Front view of baffle – section)

Figure 2. Fuel tank assembly equipped with fuel pump module and Extensible Wings Baffle

The main characteristic for this concept is that it can be inserted directly into the fuel tank at the same time with the fuel pump module. For this, the baffle must be wrapped on the pump module (figure 3, 4, 5 – closed position). Once it is inserted in position, the two wings (figure 4) will open to their final position.

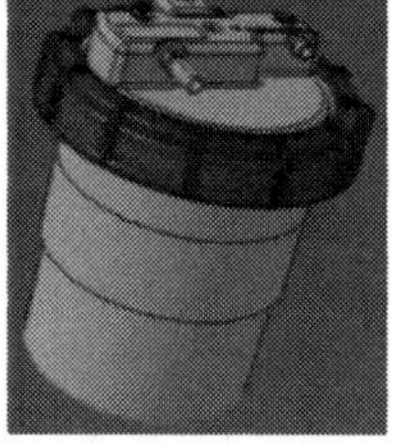

Figure 3. Fuel Pump model

Figure 4. Baffle – open Wings

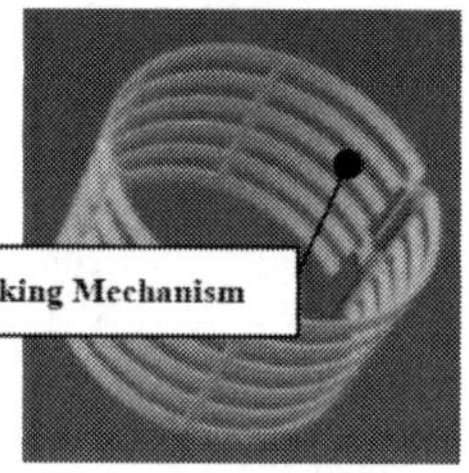

Figure 5. Baffle – Closed Wins

The preliminary design helped understanding the functionality of the concept and highlighted some constructive constrains:

- Presence and movement of the pump module floating device influences the length and shape of the wings;
- Floating device arm movement influences the insertion and locking area of the baffle wings;
- Wings simple structure can present a risk regarding its high flexibility and cannot maintain the open position in function;

- Wings are limited to only two.

To solve this constrains a remodeling was done to include some solutions. In figure 6, some of them are displayed:

- The closing wings system changed so that the wings are closed one on each other by superposition (figure 7). If the floating device was interfering with the wings, applying this solution not only that resolve this problem, but also allows the wings to keep their length (figure 8);
- The baffle with closed wings and pump module fit directly into the fuel tank hole. The insertion is presented in figure 9 and the final assembly in figure 10.

Figure 6. Pump module – Closed wings

Figure 7. Pump module – Closed wings

Figure 8. Pump module – Opened wings

In order to eliminate the risk generated by the flexibility of the wings it was implemented a material structure improvement by adding some multiple ribs at the base of the wings and on their surface (figure 11).

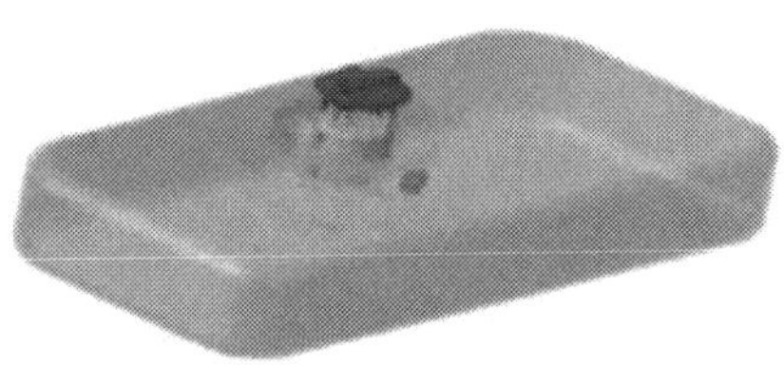

Figure 9. Fuel tank assembly– superposed wings – preassembly

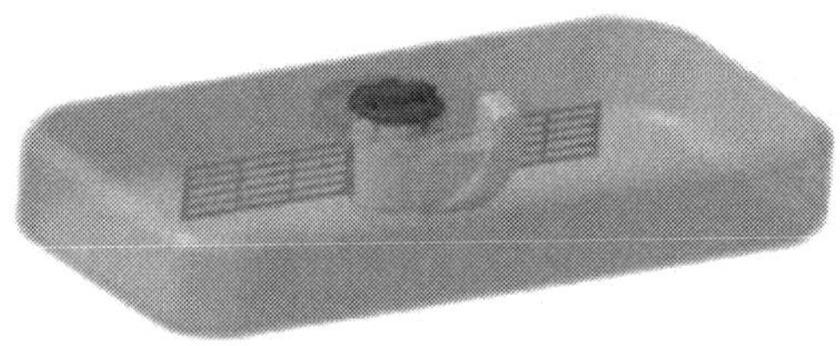

Figure 10. equipped fuel tank assembly – open wings

This solution reduces the flexibility of the wings and has a negative impact for keeping the wings closed on the pump module before the insertion. In this case, the problem will be solved in the manufacturing stage. The baffle will come with opened wings and during the assembly operations, the operator will approach the wings each other to obtain the closed position at the same time with the insertion.

Figure 11. Presence of ribs on the wings of the baffle so as to avoid over opening

The main characteristics for the extensible wings baffle concept are:

- Easy insertion for the operator;
- Simple construction/design;
- Baffle insertion in the same operation with the fuel pump module;
- Can be designed and dimensioned for a high variety of fuel tanks that need to improve the slosh noise (already on the market);

- Represents a study base for other technical solutions;
- Does not add extra cost in the manufacturing area;

In addition, the design of this concept highlighted some constructive limitations:

- The movement of the floating device in the fuel pump can limit the construction of the locking area of the wings for some different design of fuel tanks and fuel pumps;
- The wings structure can present some risk in time for maintaining the opened wings position;
- The wings number limited;
- The length of the wings is limited by the fuel pump diameter and can represent a problem for baffle effects in some cases.

Based on the "Extensible wings baffle" concept and on the main research idea – baffle to be used in already existing fuel tanks – a second concept is generated to resolve some of the limitation of the previous concept.

The new concept is based on a construction similar to a chain (figure 12). Depending on the desired length and other characteristics, the number of links (figure 15) of the chain and even the design (figure 14) can be modified. The links (figure 16) are connected to a central support (figure 13) that will be installed on the fuel pump module.

Figure 12. Fuel tank equipped with chain baffle

Figure 13. Central support

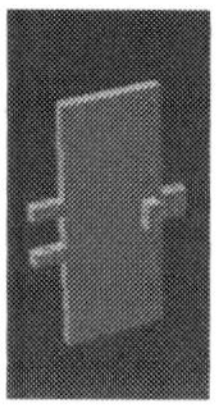

Figure 14. Chain design

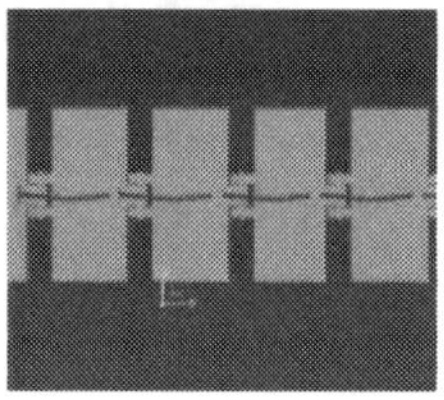

Figure 15. Several links of the chain

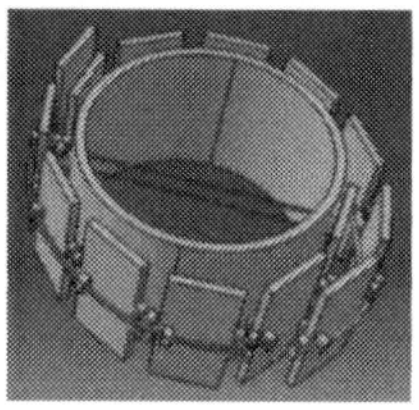

Figure 16. Closed position of the chain baffle

This concept changes one of the main characteristics of the "Extensible Wings Baffle", regarding the insertion into the fuel tank at the same time with the pump module. The baffle will be introduced before the pump module due to the connection system between the links of the chain (figure 17, 18). The links of the chain are connected with some pins and a rigid cable that assures their final position (figure 15). In addition, the chain will be wrapped to the central support or will be insert freely into the fuel tank. The cable will be loosening before the insertion in the fuel tank. After this assembly is in position in the fuel tank, the pump module will be pushed in (figure 19).

Figure 17. Preassembly of the pump module

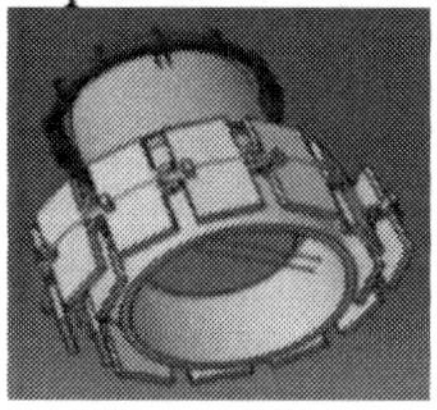

Figure 18. Detail of preassembling the pump module

Figure 19. Detail position of the pump module

Pushing in the fuel pump module will tension the cable. The cable connects the links of the baffle and connects the two wings. By tensioning this cable, the links will connect into a rigid and fixed position assuring a final position for the Chain Baffle.

General characteristic for Chain Baffle concept are:

- Free and easy insertion into the fuel tank;
- Can be used for a high variety of fuel tanks that need slosh noise reduction;
- Possibility to create a high variation of forms depending of the fuel tank design;
- Multiple wings can be created;

Based on sever analysis for this model, the concept was developed further to improve some raised issues:

The connection pins only assures connection and have no influence over position and stability;

- The length of the cable needed to permit the chain to be wrapped up to the central support, is too big and after insertion of the fuel pump module will not assure the sufficient tension for the assembly stability;
- Execution for chain links will require precision for the pin locking system and will add additional cost to the product;
- The cable holes in the chain links is too small and creates a risk of breaking during assembly process.

The elements of the baffle (figure 20, 21) were redesigned to permit assembly for another locking pin (figure 22) on a bigger surface. This way can be obtained an improvement for material resistance. The bigger dimensions reduce the manufacturing precision needed and can reduce cost for execution of each of the links. In addition, the spaces between the links were reduced so that the cable length to reduce too. This improves the stability of the chain baffle after assembly due to the connection on a bigger surface. Regarding the cable, to be more effective, the cable support on each link was modified to obtain a more robust contact.

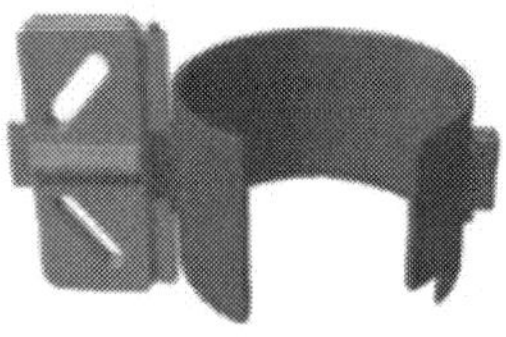

Figure 20. Connection between central support and chain element

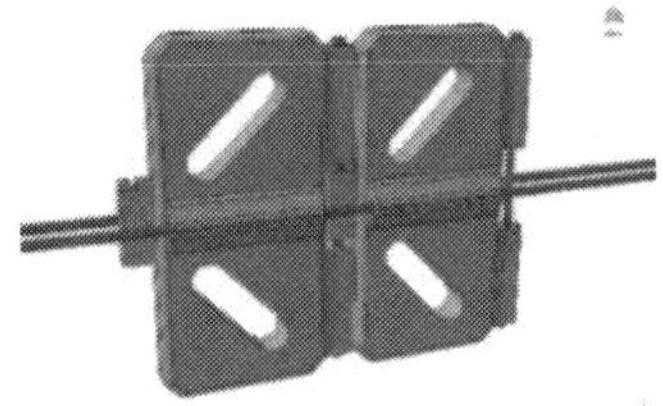

Figure 21. Detail of the cable connection

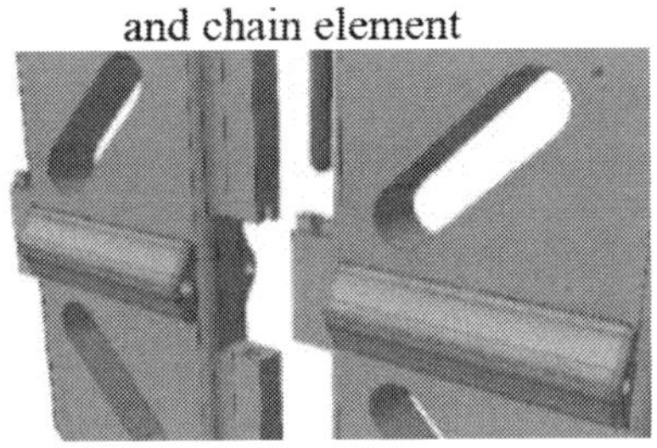

Figure 22. Detail of the chain connection

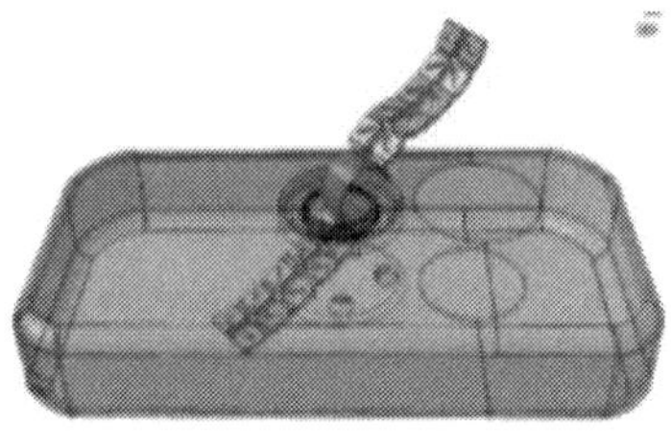

Figure 23. Insertion of the chain baffle in the fuel tank

To obtain all of these changes the wrapping up of the baffle chain on its central support had to be partially eliminated. Before the insertion in the fuel tank, there is still a degree of freedom between the links, which has a positive impact in the assembly process for the operator (figure 23, 24, 25).

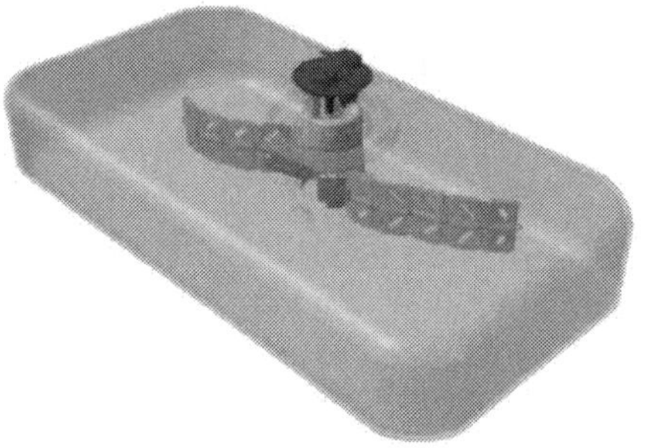

Figure 24. Redesigned chain baffle in pre assembly stage

Figure 25. Equipped fuel tank with redesigned chain baffle in complete assembly position

Conclusion and future work

The 3D models used to highlight the baffle solutions correspond to real dimensions of the fuel pump modules, existent on the current market. Furthermore, the simplified generic 3D model used for fuel tank takes into account the basic dimensions of a serial life fuel tank with a capacity of 50 liters.

The article takes into account the presumption that the technical solutions described above decrease the kinetic energy inside a fuel tank by breaking the waves as much as possible and in the same time, the noise level reduces, too [4]. This hypotheses is going to be confirmed only after static and dynamic numeric calculations and if possible later more by physical tests.

Both extensible baffle and chain baffle are suitable for usage in an already existent fuel tank or serial life fuel tank. While the first efficiency is limited due to the diameter of the module, the second one – the chain baffle can be an alternative that can be adapted as much as possible to the shape of the envelope of the fuel tank. The length of the chain and the possibility to have several chain connections can lead to an increase of efficiency. All this gain in performance of the component has as compromise additional operations on the finishing line of the fuel tank where all the elements are inserted inside. Each additional operation has extra costs, but before choosing a solution it should be prepared a deep analysis that can highlight if the cost generated is worth for a performance issue perceived by the final client – the buyer of the car.

References

[1] De Man, P. and Van Schaftingen, J., *Prediction of Vehicle Fuel Tank Slosh Noise from Component-Level Test Data*, SAE Technical Paper 2012-01-0215, 2012

[2] IU, H, Cleghorn, W., and Mills, J., *Design and Analysis of Fuel Tank Baffles to Reduce the Noise Generated From Fuel Sloshing*, SAE Technical Paper 2004-01-0403, 2004

[3] Manta (Balas) O.M., Balas R., Doicin C.V., *Study regarding the influence of the fuel tank constructive characteristics on slosh-noise effect,* Applied Mechanics and Materials, Vol. 834, pp 22-27, Trans Tech Publications, Switzerland, 2016

[4] Frank, E., Moon, C., Rae, J., and Popovich, M., *Optimization of Test Parameters and Analysis Methods for Fuel Tank Slosh Noise*, SAE Int. J. Passeng. Cars - Mech. Syst. 6(2):1306-1312, 2013

Precission plunge milling for angled vertical walls, on three axis machining center

B Pralea[1], GH Nagit [2]

[1,2]Gheorghe Asachi Technical University of Iasi, Romania

Email:bpralea@gmail.com

Abstract. In this study the authors will construct a type of milling based on the plunge milling and raster milling toolpath strategy, only this time it is not going to be used for roughing but only for finishing vertical and angled surfaces of particular components of injection molds, with the help of a three axis vertical machining center. This particular toolpath strategy designed, with a commercial CAM program, for finishing the vertical and angled walls, offers a better surface roughness, and a better dimensional control (accuracy), both for the milling tool and also for the work piece.

1. Introduction

The advance methods of milling as plunge milling or vertical raster milling have been improved in order to provide a better material removal rate. For some molds an efficient method of material removal is vertical raster milling. A lot of methods have been taken into consideration [1,2,3] but only few of them have been proposed for study and implemented into CAM software and almost none of them have taken into account the quality of angled wall surfaces inside of an injection mold, milled by a three axis CNC [1].

The proposed type of milling is based on the raster milling; only it is not used for roughing but only for finishing of the determined surface. The vertical raster milling it is different from the conventional types [4], of tool-path strategies such as offset area clear model or contour milling. The vertical raster milling or plunge milling is based on three phases: the first phase has its foundation on tool compensation, the second phase is when the tool goes down executing in the same time the milling operation, the third phase is when the tool goes up to a predefined z (inputted by the CAM software user), after this phase it will restart the cycle again with a predefined stepover or the tool will retract to the safe Z point.

The feed rate is considered maximum during the compensation phase; therefore the tool working time is depended on the number of ups and downs and also the depth of of the selected wall. The main force [3] that is involved in this particular strategy is the axial force, reducing in the same time the tool vibrations (compared to the conventional milling strategies). The experiments showed that the angled walls within the deep cavities of the injection molds are more precise than those which are milled with conventional strategies.

2. Methodology and Experimental procedure

The above presented research is based on the plunge milling toolpath strategy, and represents a machining process used to remove material rapidly during roughing operations [1]. It can be used to achieve major increase in productivity especially in the case of deep milled work pieces. Some experiments [4] showed the necessary preparations for building a plunge milling toolpath strategy in a CAM program and they are based on selecting the right tool with the admissible length that offers a safe working environment, the identification of the valid entry points, the depth of the procedure.

Therefore before the milling strategy is generated, the first step is selecting the areas that need to be finished and generating a border that will represent the limit for the working tool. The second step is defining the stock material and the direction of toolpath (along X axis or Y axis). The third step refers to the calculation of the cusp height. This takes into consideration the value of the side step, the diameter of the tool and the type of mill insert. A greater number of side steps improve the milled surface giving it a better roughness coefficient.

2.1. Study Case I

The first study case was conducted on a spacer of an Injection Mold (1.1730 material-C45), which has an angled surface (figure1.). That particular area is going to be processed using CAM software (PowerMill). This area will be subjected to several milling strategies such as roughing, pre-finishing and finishing. The roughing toolpath will be executed by a mill with 4 round inserts, that have a radius of 6mm plate. The speed was set at 1200rpm and a feed rate of 3600mm/min. The side step over is set at 24mm with an axial step of 0.5mm. The wall offset was set at 0.4mm. Time needed to perform the operation was 83 minutes. In the below picture is represented the simulation of toolpath strategy. The areas that have green lines represents the tool trajectories and the red dots, the points were the tool changes the trajectory. The grey area suggests the material that has to be machined.

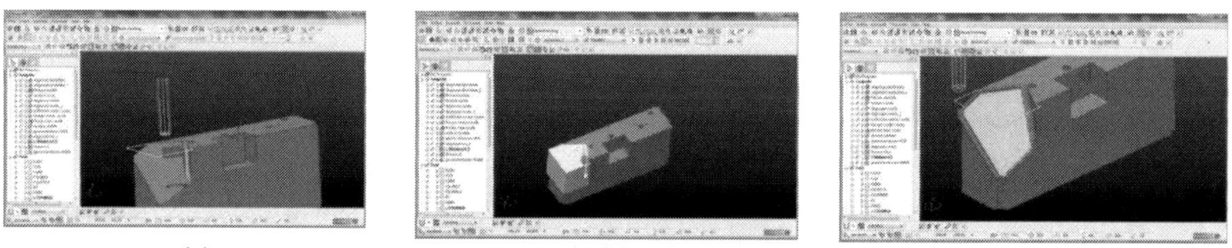

(a) (b) (c)

Figure 1. The simulation of the finishing toolpath strategies of the Mold Spacer, starting with the vertical wall (a), loading up for simulation the stock-model (b) and the final stage, showing the implemented toolpath (c), the feed movement being marked with green lines

Prefinishing toolpath strategy was carried out by a mill with 6 triangular inserts that have a radius of 0.8mm, designed specifically for deep milling areas. The speed and feed were set as follows: 1200 rpm and a feed of 1400 mm / min with a side step of 0.7 mm step, the maximum depth of the milling is 182 mm. The side wall offset was set at 0.15 mm. The entire operation took 55 min. The finishing strategy was carried out with the same tool and the same cutting data. Both strategies were designed like a plunge milling strategy, but in this case were used for finishing the discussed piece.

2.2 Study Case II

So on a plate, (figure 2.) within an injection mold (1,2311-40CrMnMo7 material), which has a cavity, that holds the active parts of the injection mold (figure 3.) will be milled using a CAM software. This particular area is important because it centers and fixes the active part of the mould. It is designed two vertical walls and two incline walls (with an angle of two degrees). This area will be subjected to a roughing, and pre-finishing and finishing strategy.

The assembly of the two components is shown in figure 4. Roughing will be conducted by a mill cutter with four round inserts, and a diameter of 52mm. The radius of the insert is 6mm. The machining parameters are set at speed of 860 rpm and a feed rate of 3600mm / min. The side step is set at 34mm and the step down is set at 1,1mm. The radial and axial offset is set at a thickness of 0,8mm. The time needed to perform the entire operations was 668 minutes. Because of a calculated data that indicates the life span of the mill inserts at 90 min., we divided the toolpath into 8 programs.

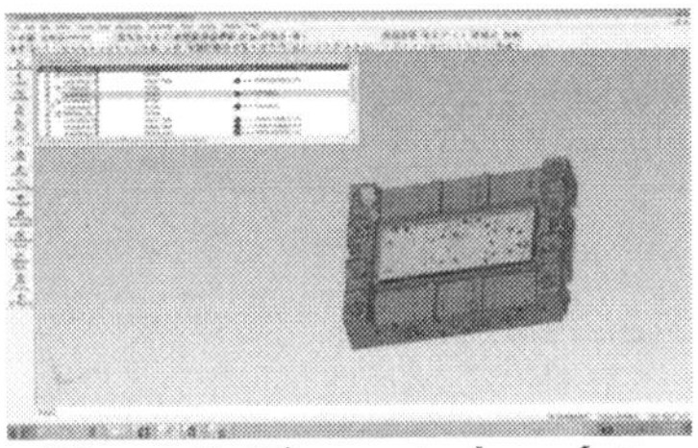

Figure 2. The centering plate that holds the active parts of a mold

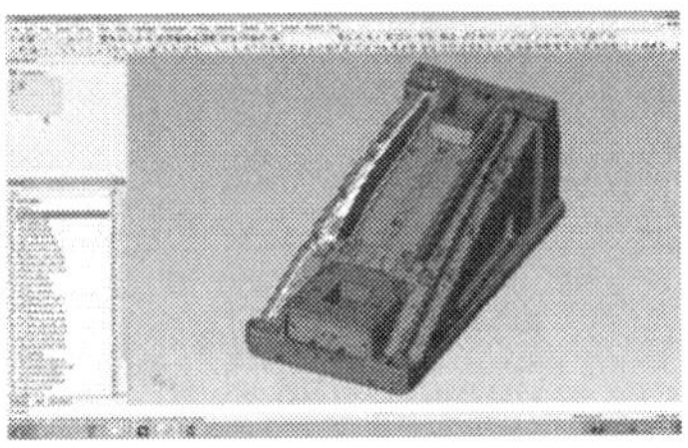

Figure 3. The active part of the mold

Figure 4. The assembly of the two components

The pre finishing toolpath was carried out by a tool with 5 rectangular inserts that have a radius of 0.8mm (figure 5.). The tool has a diameter of 35mm and it designed especially for milling deep cavities.

The actual data for the mill diameter was taken from a tool presetter as shown in figure 6(a). The tool presetter offers the possibility to measure accurate the diameter and the length of the tool. The values appear on the device screen (figure 7(b).) and they can be easily implemented on the CNC machine. The machining parameters: the speed is set at 1500 rpm and the feed rate at 1000 mm / min with a side step of 0.7 mm, the maximum depth of milling being 65 mm. The wall offset was set at 0.15 mm. The time needed to perform the entire operation was 120 minutes. The finishing toolpath was carried out with the same mill and same machining parameters. Both finishing and pre finishing toolpath strategies were based on the plunge milling method.

Figure 5. The Tool with a diameter of 35 mm and a Radius of 0.8

(a) (b)

Figure 6. The Tool presetter (a) with the milling tool prepared for measure with the dimensional parameters being shown on the presetter monitor (b)

As we stated in the introduction these particular finishing strategy based on plunge milling is better in terms of surface quality (figure 7.) and also the dimensional control, both for the cavity walls and the mill diameter.

We took some data, from various molds, for the same centering plate of the active part, and analyzed it, in comparison with the convention milling strategies. The data is shown in table1. The picture below illustrates the differences between the conventional strategy (the surface outside the red square) and the finishing strategy based on plunge milling (the surface inside the red square). The red arrow suggest the vertical toolpath of the cutter.

Figure 7. The differences between conventional milling and the finishing method based on the plunge milling strategy

Table 1. The table shows the differences between measured surfaces, with different types of milling, at a precision level

Nominal Dimensions of the measured part	Surface precision for Conventional milling	Surface precision for Plunge Milling Finishing Strategy
900x800x210	0,07	0,02
1050x900x156	0,08	0,02
800x760x226	0,08	0,02
986x736x166	0,04	0,02
620x520x150	0,07	0,01
918x710x220	0,08	0,02
1396x640x249	0,04	0,01
1100x900x210	0,07	0,01

Figure 8. The RMP60 measuring the angled wall of the plate

The measurements were collected using and inspection probe from Renishaw, model RMP60 (figure 8.). The RMP60 is a compact spindle probe, with radio signal transmission and together with the Inspection Plus software [5] which is installed on DOOSAN 750L, provided accurate and easy to use data.The DOOSAN 750 L is a vertical machining center which is designed for high speed and precision of medium to large work pieces such as the plates presented in the study.[6]

3.Conclusion.

The present study shows that the quality and the dimensional control of the angled walls, which are machined on a 3 axis CNC, are superior in the case of the finishing method based on the plunge milling toolpath strategies in comparison with the conventional milling strategies. Also the thermal stability of the tool and the machinig center is more stable during this type of milling and this offers a better life span for both of them. All these factors give a better advantage in terms of mold manufacturing.

References

[1] Danis I, Wojtowicz N, Monies F, Lamesle P and Lagarrigue P, 2013, Cutting conditions and surface integrity during dry plunge milling of a wroought magnesium alloy, *The Manufacturing Engineering Society International Conference, MESIC 2013*, vol **63**, pp 36-44

[2] Ko, J.H. and Altintas, 2007, Dynamics and stability of plunge milling operations, *Journal of Manufacturing Science and Engineering, Transactions of the ASME*, vol **129**, no 1, pp 32-40

[3] Rafanelli F, Campatelli G and Scippa A, 2015, Effects of cutting conditions on forces and force coefficients in plunge milling operations, *Advances in Mechanical Engineering*, vol **7**, no 6, pp 1-9

[4] Rauch M and Hascoet J, 2012, Selecting a milling strategy with regard to the machine tool capabilities: application to plunge milling, *The International Journal of Advanced Manufacturing Technology*, vol **59**, no 1, pp 47-54

[5] http://www.renishaw.com/en/machine-tool-probes-and-software--6073

[6] http://www.doosanmachinetools.com/en/product/detail.do?CATEGORY_ID=mten020300&PRODUCT_ID=mtenp0071

Acknowledgement

The authors would like to thank TREVIS GRUP ROMANIA and UNITEAM ITALY, for their assistence and for having them enbled the factory equipment.

IOP Conf. Series: Materials Science and Engineering **161** (2016) 012005 doi:10.1088/1757-899X/161/1/012005

Analysis of lead twist in modern high-performance grinding methods

J Kundrák[1], K Gyáni[1], C Felhő[1], AP Markopoulos[2] and I Deszpoth[1]

[1]Institute of Manufacturing Science, University of Miskolc, Miskolc-Egyetemváros H-3515, Hungary
[2]Section of Manufacturing Technology, School of Mechanical Engineering, National Technical University of Athens, Heroon Polytechniou 9, 15780, Athens, Greece

E-mail: kundrak@uni-miskolc.hu

Abstract. According to quality requirements of road vehicles shafts, which bear dynamic seals, twisted-pattern micro-geometrical topography is not allowed. It is a question whether newer modern grinding methods – such as quick-point grinding and peel grinding – could provide twist-free topography. According to industrial experience, twist-free surfaces can be made, however with certain settings, same twist occurs. In this paper it is proved by detailed chip-geometrical analysis that the topography generated by the new procedures is theoretically twist-patterned because of the feeding motion of the CBN tool. The presented investigation was carried out by a single-grain wheel model and computer simulation.

1. Introduction

Vehicle shaft surface texture is critical for evaluating the wear of rotary dynamic seals [1]; given that an automobile has over eighty dynamic seals, possible failures cost the transportation and power industry millions of dollars per year [2]. Prescriptions for optimum performance of radial lip seals require a precise shaft surface texture and zero twist. It is worth noting that if the shaft surface texture is rough, the seal wears rapidly. On the other hand, if the texture is too smooth, the seal will not contact the shaft appropriately, leading again to problematic application. Ideally, the shaft surface should possess a texture that allows a fluid film of about 0.25 μm thick to form in the shaft-seal interface [3].

Of equal importance is the absence of a twist on the shaft surface [4]. This characteristic can initiate a fluid pumping effect, which, depending on the orientation of the twist may result to leakage or dry run [5]. The creation of a twist is inherent in all manufacturing processes, attributed to feed rate of the cutting tool in turning and the part orientation of either the workpiece or grinding disc. Twist pattern can be identified either as macro-geometrical, i.e. an axial periodic structure that is singularly or multiply circumferential, or micro-geometrical that describes high-frequency features on the shaft, either periodic or non-periodic, which are tilted with respect to the workpiece [6].

In traditional procedures, sealing surfaces can be made by plunge grinding and by very long spark-out time. In longitudinal feed grinding, attention should be paid on several conditions such as: wheel dressing, the parallelism of rotation axis of the wheel and the workpiece and the regulatory spiral created on the wheel by single point diamond regulation, see figure 1.

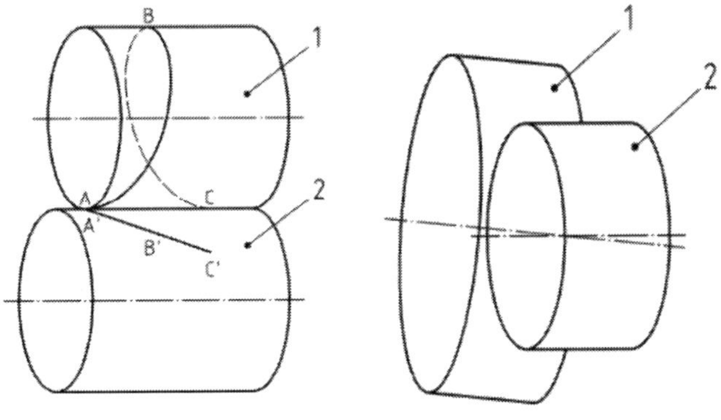

Figure 1 Basic cases of twist formation in grinding: a) Copy of the regulatory spiral b) Non-parallel axis between the grinding wheel and the workpiece

In case of the existence of a twist, the direction of oil flow is determined by the combination of three factors, namely rotation, twist and conveying direction [7-9], according to Table 1.

In advanced grinding procedures, i.e. in case of CBN wheels, the metal tool body made it possible to raise the wheel speed up to 100 to 200 m/s and the width of the wheel's contact surface to be decreased to some mm. However, with narrow wheels, the cylindrical surfaces can be ground only by applying longitudinal feed ($v_{f.L}$).

Table 1. Conveying direction at twist

	Rotation direction		Twist direction		Conveying direction	
	Right	Left	Right	Left	Right	Left
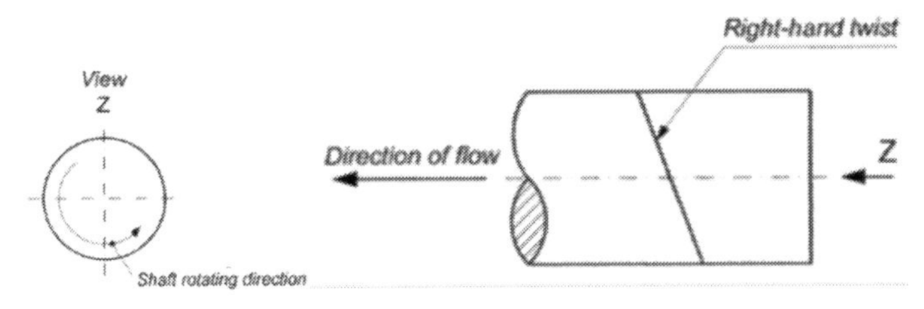	✚			✚		✚
	✚		✚		✚	
		✚	✚			✚
		✚		✚	✚	

Advanced grinding processes must be examined from the point of twist generation, because their productivity is multiplied in comparison to the traditional plunge procedure, even when having moderate grinding conditions, see Table 2.

Table 2. Characteristic data of advanced grinding processes

Parameter \ Procedure	Plunge grinding	Peel grinding	Quick-point grinding
Wheel generatrix	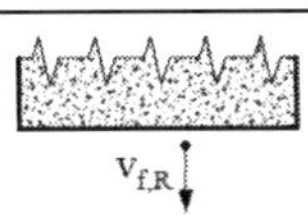 $v_{f.R}$ straight	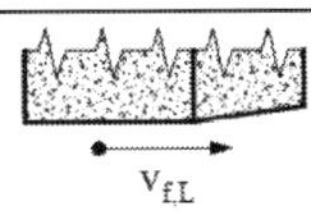 $v_{f.L}$ conical	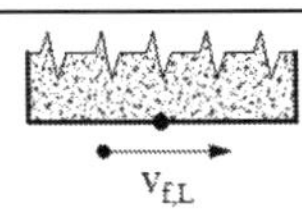 $v_{f.L}$ straight
Material removal rate Q (mm^3/s)	Q	5×Q	5×Q
Main machine time T_m (min)	T_m	$1/5 \times T_m$	$1/5 \times T_m$

2. Parameters of twist topography

The significance of twist has intensified because of the tightening accuracy and surface quality prescriptions, as well the severe requirements of environmental protection. That is why automobile

IOP Conf. Series: Materials Science and Engineering **161** (2016) 012005 doi:10.1088/1757-899X/161/1/012005

factories ascertained and defined the twist parameters (figure 2) first in factory standards, then in ISO and DIN international standards, so that the degree of twist can be measured and thus controlled [10].

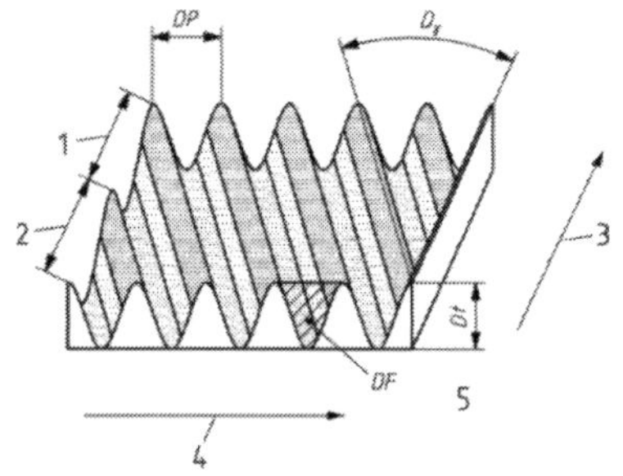

1: twist spiral 1
2: twist spiral 2
3: circumferential direction
4: shaft direction
5:twist orientation

DP period length
Dγ twist angle
Dt twist depth
DG number of threads
DF theoretical supply cross section
DFu theoretical supply cross section per revolution

Figure 2 Twist parameters according to DIN EN ISO 25178-3:2012 [11]

Over the few past years, several twist measuring methods have been proposed. Some of them are the string method, stylus metrology and 3D surface topography based on interferometry or scatterometry. More on these methods can be found in the relevant literature [3, 12-14].

3. The examined grinding processes

Two procedures are wide-spread in advanced grinding processes: peel grinding and quick-point grinding [15].

Peel grinding can be performed by straight or an inclined wheel, as in figure 3. The point is that the CBN wheel is regulated in $v_{f,L}$ longitudinal feed direction in an 6° angle; therefore it can grind off the allowance, which is some tenths of mm, with a single feed rate. The back, cylindrical part of the wheel performs the smoothing grinding task without a separate spark-out phase. Because of the inclined position of the wheel, the lubricant can penetrate in the contact zone much more intensively, in that the wheel regulation is exact and easier. Furthermore, the wheel production is simpler. The angle of the wheel inclination is 15° and the inclination of the wheel shaft is considered in base plane (P_r).

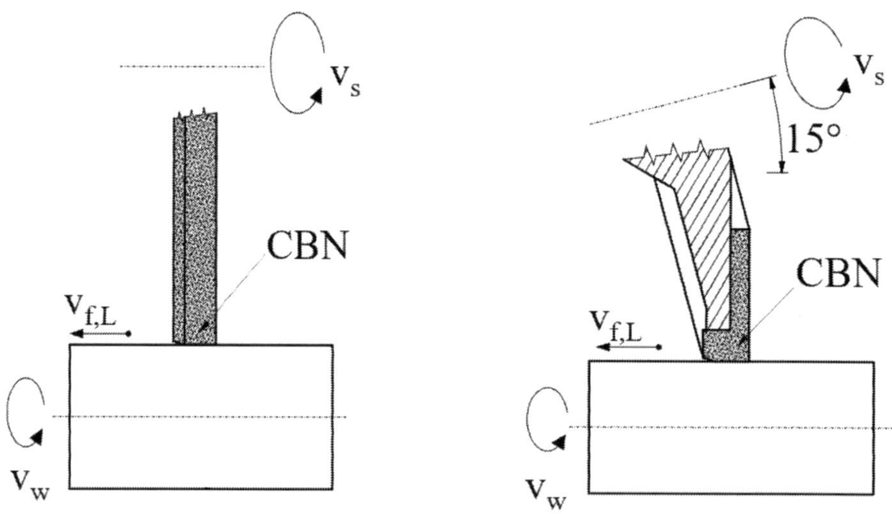

(a) (b)

Figure 3 Peel grinding by a wheel in a) straight and b) inclined position

Quick-point grinding is seemingly similar, but the mechanism of material removal is different. The wheel is inclined, however not only in base plane, but also in perpendicular PS edge plane [16]. The point of the procedure is given by the inclination characterized by angle β. In that way the wheel generatrix is not parallel to the workpiece generatrix, shrinking the contact to a point, as can be seen in figure 4, point D. The wheel generatrix is not conical but straight, parallel with the shaft.

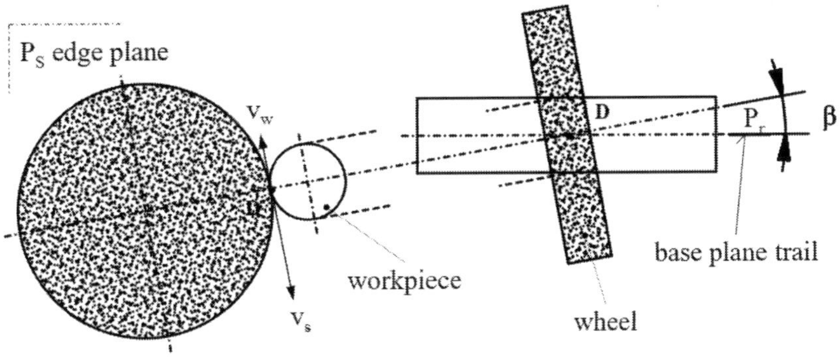

Figure 4 In quick-point grinding wheel is also inclined in the edge plane by angle β

The two procedures can be carried out on the same machine. If only cylindrical surfaces and shoulders must be processed, peel grinding, with its simpler geometry, in enough.

4. Cutting relations of peel grinding

The most important advantage of peel grinding is that the shoulders facing right and left can be ground in one clamp. There are two wheels in grinding-head-stock of the machine, which are positioned to programmable suitable positions as in figure 5. To rotate the shaft there is no need for a driving appliance, because the grinding forces are so small that the frictional torque from the peak is enough.

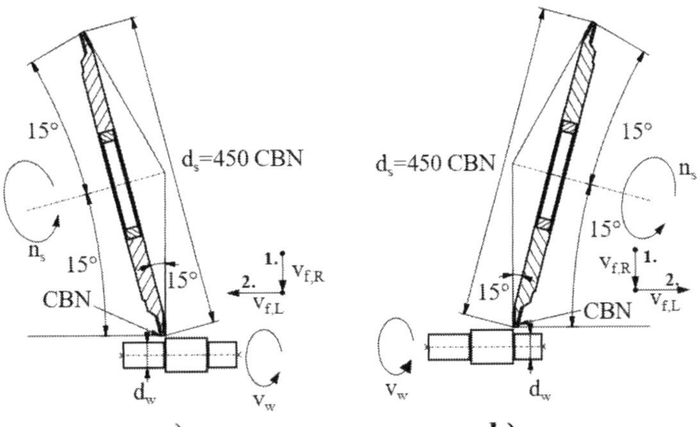

Figure 5 Wheel position in peel grinding

The material removal circumstances of peel grinding and the theoretically undeformed chip-cross-section can be seen in figure 6. Since the emerging grinding force is very small, it may be considered not significant. On the other hand, chip-cross section A_c is removed along a spiral, because of $v_{f,L}$ feed, the value of which is equal to f.

5. Analysis of the formation of twist at a single-grain wheel model

The chip formation and the evolution of the topography and thus the evolution of the twist in grinding is also analyzed by a single-grain wheel model, see figure 7. The grinding of such a cylindrical surface can be seen on the same figure, which starts at a left facing shoulder.

The three positions in figure 7b are distinguished from each other in the characteristics of the rotation of the workpiece compared to the grinding wheel (no rotation, inverse or identical rotational direction).

The two types of twists characterized by angles α and β are theoretically always present in case of peel grinding. However, their emergence is not always clear; in some cases it is not even noticeable. In case of inverse rotations – which is the regular case in grinding – the micro-twist is left-handed and the macro-twist is right-handed [17-19].

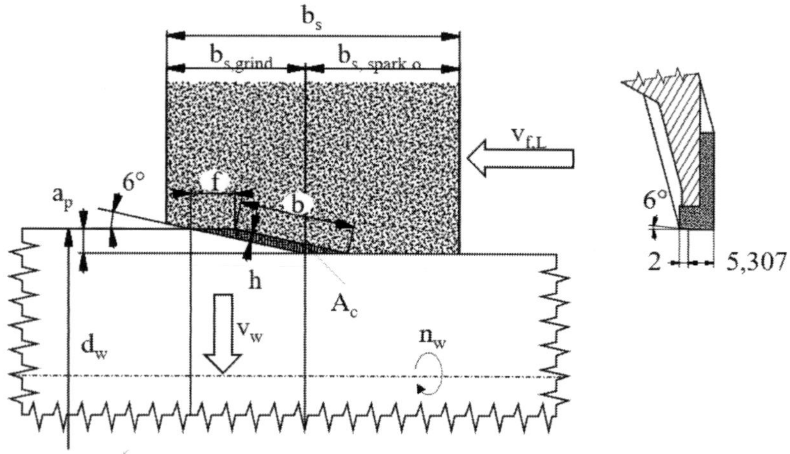

a) b)

Figure 6 The working part of the wheel and the theoretically undeformed chip-cross-section in: a) contact zone; b) the profile of the wheel

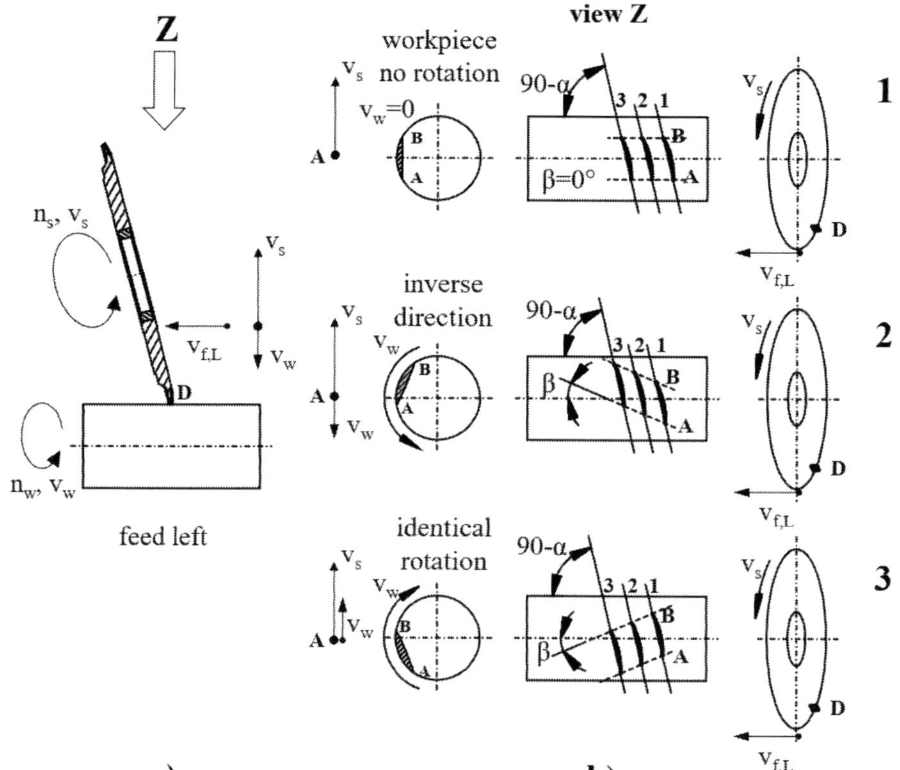

a) b)

Figure 7 Twist generation work of the single grain wheel model a) general layout b) three typical cases of the twist generation

It is advisable to examine the shape and dimensions of the chip which was removed by a single grain (figure 8). The chip is crescent-shaped and curved in two mutually perpendicular shapes.

One plane is perpendicular to the rotation axis, and can be characterized with the trace AB. The other plane is the tangent plane which is perpendicular to the first one. In the drawing, the curvature of chip is shown in this plane. In order to make the investigation simpler, points A and B are projected into the tangent plane (figure 8b). Thus the maximum chip width b_{max} can be defined as well as the chip length ℓ_g, which equals to the contact arc length.

In the second step, the chip is rectified in the tangent plane (figure 8c). The chip thickness is reported in the section of the chip which was transformed into the tangent plane (figure 8d).

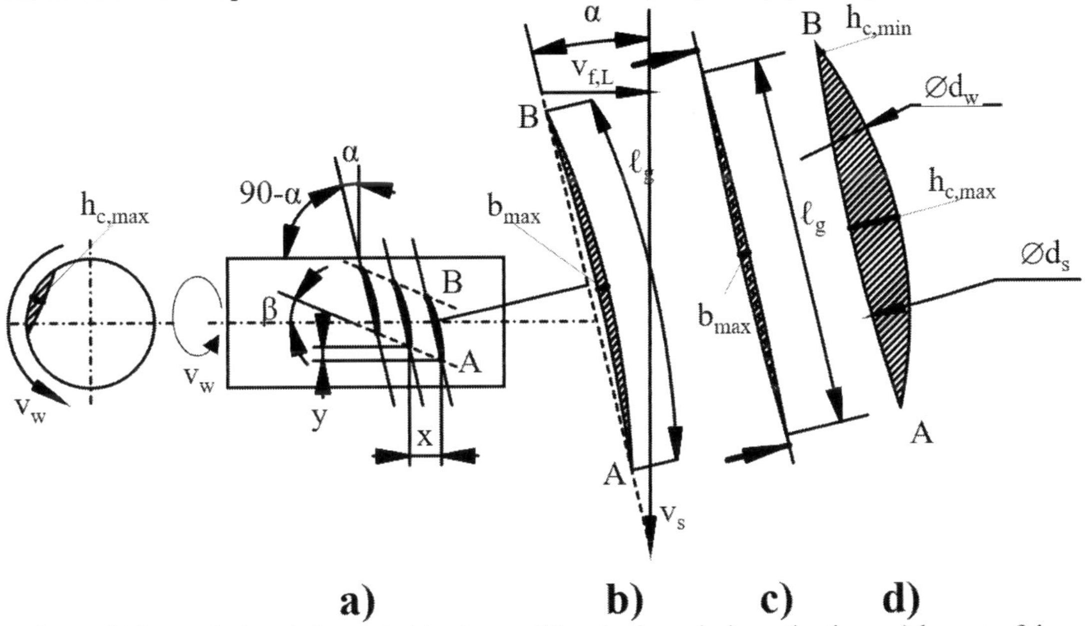

Figure 8 Theoretical undeformed chip shape of the single grain investigations: a) lay out of the scratches on the surface of the workpiece; b) the crescent-shaped chip magnified and transformed into the tangent plane; c) neglecting of the curvature, rectification; d) theoretical undeformed chip cross section

6. CAD based modelling of theoretical twist in peel grinding

The model of the external cylindrical surfaces of the workpiece form simple cylinders, and the base circle is modelled in the X-Y plane. More than one cylindrical surfaces can be on a workpiece, and their diameters and lengths can be defined in parametric form. The next step is the determination of the tangent line between the wheel and the part. Intersection points of the workpiece and the circles which indicate the periphery of the tool are determined for that. This gives the angular movement of the grinding wheel between the entry and exit points of the grain. The actual trajectory of the grinding grain can be calculated later with this knowledge. Although the shape of the grains is not known in grinding, in order to be able to model them, a simple shape was assumed, namely an equilateral triangle.

In view of the arc between the entry and exit points (point A and point B) and on the basis of the wheel diameter the calculation of the trajectory of the wheel can be performed. This is realized in two steps.

In the first step, it is be assumed, that the workpiece does not rotate and there is no linear feed; in this case, the trajectory of the grain is a peripheral circular section of the grinding wheel. A point of the curve is recorded in this first step with the given accuracy (Original Trajectory).

In the second step it is taken into account that the trajectory of the grain changes, because of the peripheral speed of the workpiece (v_w) and the longitudinal feed velocity ($v_{f,L}$). A point of this modified curve is calculated by numerical methods, and the modified trajectory is superposed into them. The next step is the creation of the theoretical undeformed chip shape. The grinding grain shape is guided along the modified trajectory and the excised material (chip) is removed; thus the imprint of the single grain is obtained on the cylindrical surface, which represents the workpiece (figure 9).

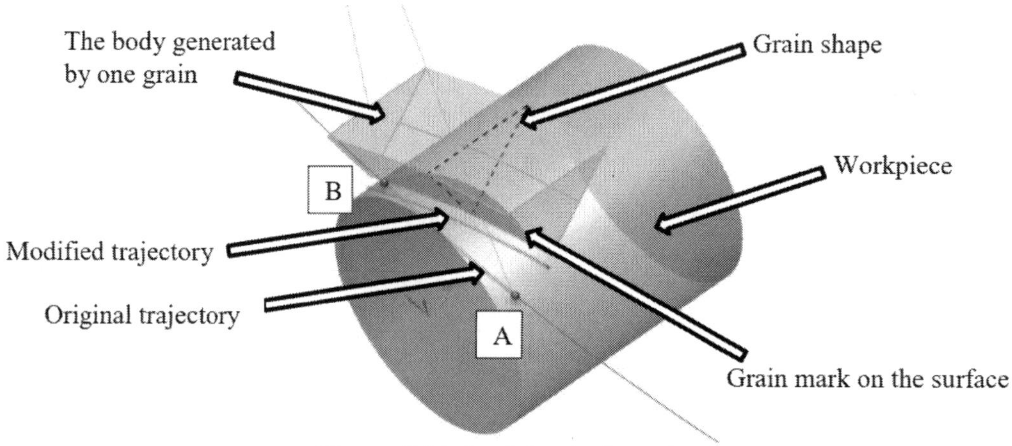

Figure 9 Mark of the single grain

The oblique position is clearly visible which is typical for micro-twist, and which is characterized by angle α at the previous geometrical analysis. The single grain removes a chip at every wheel rotation, and the location of chips positioned to each other translated along a helix shape. This positioning is denoted as macro-twist at the geometrical analysis. The positioning of the helix at the simulation is shown in figure 10. In reality, when there are many grains at the working part of the wheel, the marks are located more densely, partly overlapping each other both in longitudinal and lateral directions, but the orientation which was demonstrated by the simulation still persists.

Figure 10 The successive oblique micro-twists (α) and the macro-twist formed by micro-twists (angles α and β in figure 8)

7. Conclusion

The studies made so far have shown that it is clear, that the peel grinding procedures performed by a conical wheel theoretically grind twisted surfaces. However, it is possible to implement such settings, by which such surfaces can be generated; it can be qualified as twist-free according to the actual factory standards and measurement methods, corresponding to the specifications. It could be a solution that in case of reverse-direction rotations the two types of twists can compensate or even theoretically cancel each other. For the conscious application of this, the additional knowledge of regularities and rules and the deeper exploration of their relations with technological data are needed. Therefore the application of a polishing single-purpose machine is recommended for all cases, when the differences are greater that the allowed deviation.

Acknowledgement
This research was partially carried out in the framework of the Center of Excellence of Innovative Engineering Design and Technologies at the University of Miskolc. Furthermore the authors greatly appreciate the financial support of the National Research, Development and Innovation Office – NKFIH (No. of Agreement: OTKA K 116876).

References
[1] Modeling SR 1997 Rotary Lip Seals *Wear* **207** 92
[2] Seewig J and Hercke T 2009 2nd Generation Lead Measurement *Proceedings XIX IMEKO World Congress*
[3] Cohen DK, Smith S, Novak EL and Masters AT 2011 Quantitatively Measuring Surface Texture and Shaft Lead of Dynamic Sealing Systems Application Note #549 Bruker Corporation
[4] ISO 6194: 2009 (E). Rotary Shaft Lip-Type Seals Incorporating Elastomeric Sealing Elements
[5] Baumann M, Bauer F and Haas W 2013 Comprehensive Lead Analysis of Shaft Counterfaces for the Tribological System Radial Lip Seal *Proceedings 22nd International Conference on Fluid Sealing*
[6] Seewig J and Hercke T 2009 Lead characterisation by an objective evaluation method *Wear* **266** 530–533
[7] Kunstfeld T 2005 Einfluss der Wellenoberfläche auf das Dichtverhalten von Radial-Wellendichtungen *PhD Dissertation* Universität Stuttgart
[8] Buhl S 2006 Wechselbeziehungen im Dichtsystem von Radial-Wellendichtring, Gegenlauffläche und Fluid *PhD Dissertation* Universität Stuttgart
[9] Prem E and Vogt R The Simmerring® Reliability right from the beginning, Basics for preventing damage *Freudenberg Simrit GmbH & Co. KG* 55
[10] Schubert A, Zhang R and Steinert P 2013 Manufacturing of Twist-Free Surfaces by Hard Turning *Procedia CIRP* **7** 294-298
[11] DIN EN ISO 25178-3:2012
[12] Baumann M, Novak E, Bauer F and Haas W 2013 Fast Lead Analysis Method for Shaft Counter Surfaces of Radial Lip Seals *Proceedings STLE Annual Meeting & Exhibition*
[13] Hertzsch A, Kröger K and Großmann M 2013 Optical twist measurement by scatterometry Proceedings AMA Conferences - SENSOR 2013, OPTO 2013, IRS^2 2013
[14] Nagy N and Kundrák J 2016 Measuring Twist on Machined Surfaces *Proceedings MultiScience - XXX. microCAD International, Multidisciplinary Scientific Conference*
[15] Lütjens P and Muschardt H 2000 Hard turning or grinding *European Production Engineering* **July** 10-13
[16] Ma L, Gong Y and Chen X 2014 Study of surface roughness model and surface forming mechanism of ceramics in quick point grinding *International Journal of Machine Tool & Manufacture* **77** 82-92
[17] Felhő C 2014 Investigation of surface roughness in machining by single and multi-point tools *Dr.-Ing. Dissertation* Magdeburg 171
[18] Bauman M, Bauer F and Haas W 2013 How to measure lead in sealing technology *Sealing Technology* **July** 8-12.
[19] Hercke T 2013 Drallmessung von Hommel-Etamic in der Praxis *Daimler AG Stuttgart-Untertürkheim* 14

"Hydraulic Cushion" Type Overload Protection Devices Usable in Mechanical Presses. A Patent Study

R Cioară

Department of Engineering and Industrial Management, Transilvania University, Brasov 500036, Romania

E-mail: cioarar@unitbv.ro

Abstract. The possible consequences of machine-tool overload are well-known. In order to prevent such, machine-tools are equipped with various overload protection devices. Mechanical presses, intensively strained machine-tools, are typically equipped with three protection systems: against accidental access to the working area during machine deployment, against torque overload and force overload. Force overload protection systems include either destructible parts and are used in small to medium nominal force mechanical presses, or non-destructible ones used mostly in medium to large nominal force (H-frame) presses. A particular class of force overload protection systems without destructible parts are "hydraulic cushion" type devices. While such systems do not necessarily cause the machine to stop, the slide's stroke does not reach the initial dead centre and consequently cannot exert the designed technological force on the workpiece. By a patent study referencing 19 relevant patents the paper captures both the diversity of the constrictive solutions of "hydraulic cushion" type protection devices and their positioning modalities within the structure of a mechanical press. An important aim of the study is to highlight the reserve of creativity existing in this field, at least from the viewpoint of the hydraulic cushion positioning, as well as to emphasize the essential requirement of a relative motion between the mobile and the fixed parts of the tool, a motion of opposite sense to that of the slide-crank mechanism.

1. Introduction

Like any technical system, a machine-tool is designed and built to sustain a certain maximum strain level, most often depending on a maximum admissible force, which is a quantity indicated in the technical specifications of the machine. The possible consequences of machine-tool overload are well-known [1], concerning mainly the risk for the human operator and the possible deterioration or even destruction of the machine. In order to prevent such, machine-tools are equipped with various overload protection devices.

This paper deals strictly with mechanical presses, intensively strained machine-tools which are typically equipped with three protection devices:

- a protection device that prevents access to the working area during machine operation (the space between the table and the slide, which is actually the space between the fixed and mobile parts of the tool);
- a torque overload protection device;
- a force overload protection device.

The protection device against access to the working area during machine operation is mainly aimed at protecting the human operator against accidents. A press can be equipped with two or more such systems, usually of different kind, also having a warning function.

The torque overload protection device is aimed at protecting the main shaft and the transmission of the machine against any strain caused by a torque greater than the maximum admitted valued taken into consideration for designing the machine [2]. Torque overload appears when the opposing force at the slide generates in the main shaft an opposing torque greater than the maximum admitted torque to be transmitted by the main shaft. If the press is endowed with a friction clutch, this has to be dimensioned [3] such as to ensure the necessary protection against this type of overload.

Thus the clutch will slip every time the opposing torque at the main shaft of the machine, and implicitly the torque at the clutch shaft exceed the admitted limit. If the friction clutch of the machine does not ensure this function, then the main linkage of the press will include a distinctive dedicated subassembly. Constructions of this type are similar to clutches, an example being the constructive solution proposed by patent US 5542884 [4].

Force overload occurs when the opposing force at the slide exceeds the nominal force taken into account in the design and construction of the machine. Essentially, most force overload safety systems allow the machine to continue its operation, thus the rotation of the main shaft (or of the main shafts, as the case may be), but compensates the relative motion between the mobile and the stationary parts of the tool. Thus the opposing force of the workpiece to the action of the slide is limited or even cancelled. The protection device stroke length needs to be at least equal to the maximum capable load stroke length of the press.

Force overload protection devices include either destructible parts [2] and are typically used in small to medium nominal force mechanical presses, or non-destructible ones used mostly in medium to large nominal force (H-frame) presses.

Protection systems with non-recoverable parts are achieved by placing a calibrated destructible part subject to shearing in the slide assembly, between the slide and seat of the joint that links it to the rod. The press will not cease operation, but will not be able to exert a technological force upon the workpiece. An example is offered by the constructive solution proposed by patent US 3014420 [5]. In case the safety part is destroyed, easy access to it is evidently required. In some cases destructible parts are used designed to shear by two concentric circular contours [2].

An important group of force overload protection systems not deploying destructible parts are those based on measuring the elastic deformation of a reference component of the press structure. The most frequently selected reference component is the frame, part of it or a tie bar used for assembling the frame. These protection systems interrupt the operation of the press, which remains strained pending the intervention of the human operator. The elastic deformation is measured electric contacts, tensometric gauge systems, laser measurement systems, special displacement sensor systems, etc. A fairly recent example of tensometric gauge protection system is proposed in patent US 5125332 [6].

A special class of destructible part free force overload safety systems is that of "hydraulic cushions". While more complex as to their construction, these systems are reliable and highly efficient. In addition the value of the force that triggers the protection system is easily adjusted, and integration of the hydraulic protection system into the automatic control and operation system of the machine is fairly simple. While the system does not necessarily cause the machine to stop, the slide's stroke does not reach the initial dead centre and consequently cannot exert the designed technological force on the workpiece. By a patent study the paper captures both the diversity of the constrictive solutions of "hydraulic cushion" type protection devices and their positioning modalities within the structure of a mechanical press.

2. "Hydraulic cushion" type protection systems placed between the slide and the rod seat

The typical position of many overload protection systems, including the ones with destructible parts is between the slide and the seat of the join linking it to the rod [7], directly or indirectly. This approach is found in a pioneer patent in this field, US 1808606 (figure 1), but also in many subsequent ones, like US 2936055 (figure 2), US 3426873 (figure 3), US 4030336 (figure 4), US 4085669 (figure 5), US 4166415 (figure 6), and US 4827839 (figure 7). To be noticed in the solution out forward by patent US 2936055 is the cylinder membrane used for packing the hydraulic cushion, as well as the mechanism responsible for adjusting the slide position, consisting of an electric motor –cylinder gear – bevel gear – rotating screw nut. Interesting and somewhat atypical is the constructive solution for the adjustment of the slide position presented in patent US 4166415 [8], figure 6.

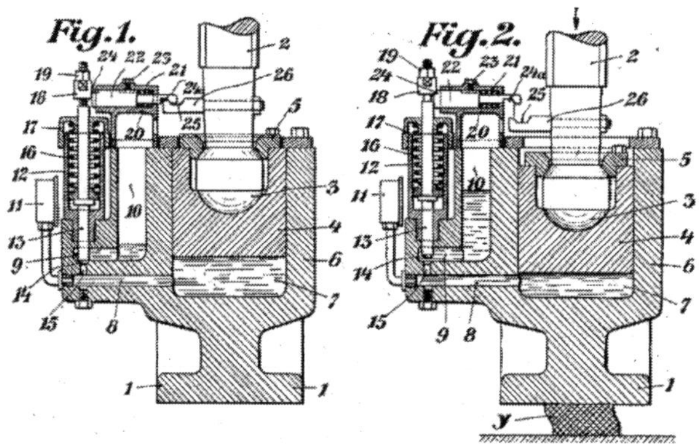

Figure 1. Force overload hydraulic protection system with a hydraulic cushion placed between the seat of the rod nut and the slide, according to patent US 1808606 [9].

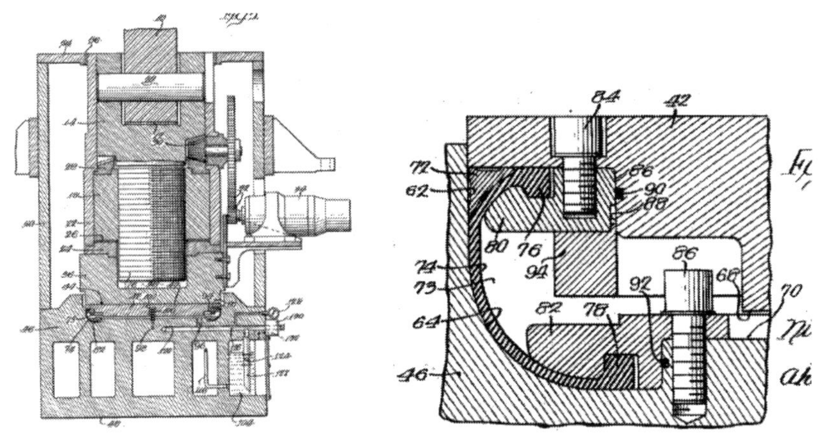

Figure 2. Force overload hydraulic protection system with a hydraulic cushion according to patent US 2936055 [10].

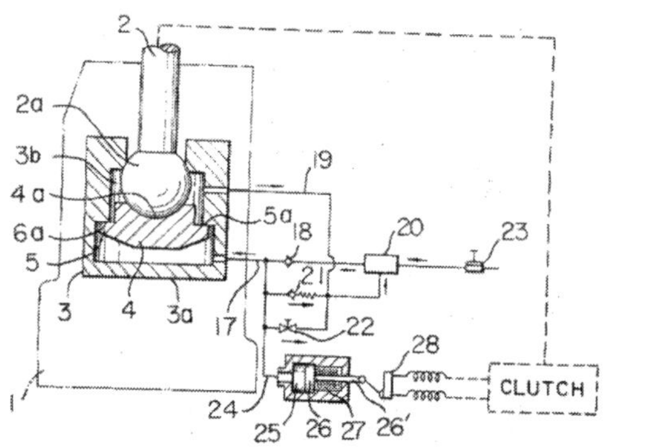

Figure 3. Force overload hydraulic protection system with a hydraulic cushion placed directly between the seat of the rod nut and the slide, according to patent US 3426873 [11].

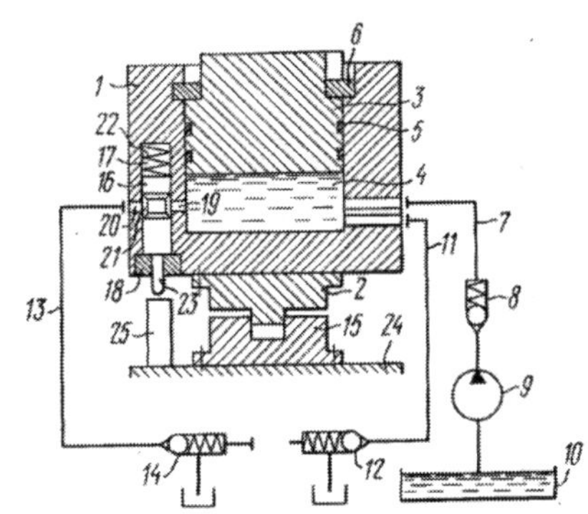

Figure 4. Force overload hydraulic protection system with a hydraulic cushion placed between the seat of the rod nut and the slide, according to patent US 4030336 [12].

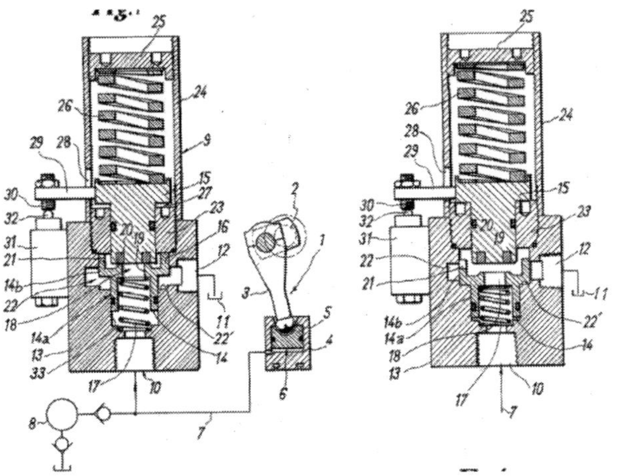

Figure 5. Force overload hydraulic protection system with a hydraulic cushion placed between the seat of the rod nut and the slide, according to patent US 4085669 [16].

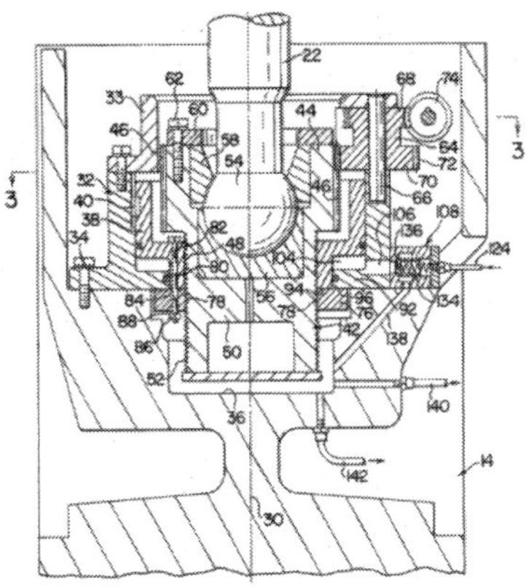

Figure 6. Force overload hydraulic protection system with a hydraulic cushion placed indirectly between the seat of the rod nut and the slide, according to patent US 4166415 [8].

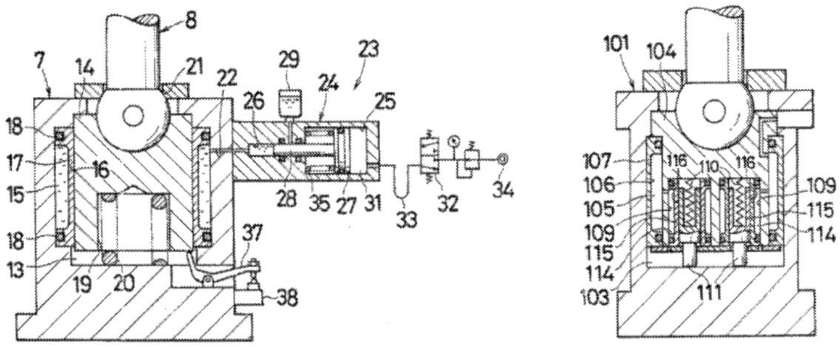

Figure 7. Force overload hydraulic protection system with a hydraulic cushion placed between the seat of the rod nut and the slide, according to patent US 4827839 [17].

Although preferable, "hydraulic cushion" type safety systems do not necessarily disrupt the driving of the slide under overload. Such stopping is included explicitly by the solution proposed by patent US 3426873, figure 3, where the overload causes the clutch to disengage.

In order to use "hydraulic cushion" overload protection systems of the described or similar type in two-crank presses, one such system has to be provided for each rod, the two devices being connected to an adequate general hydraulic circuit. Suggestive examples are offered by patents US 2937733 (figure 8) [13], US 3481171 (figure 9) [14] and US 5638748 (figure 10) [15].

A more particular constructive solution of a force overload hydraulic safety system placed between the slide and the seat of the joint linking it to the rod, designed for forging presses is put forward by patent US 3948077 [18], figure 9. Under overload the tilted slide placed between the press slide and seat of the joint linking it to the rod is displaced, moving the screw that adjusts the position of the slide stroke and of the attached hydraulic piston. Also to be noticed is the particular construction of the rod.

3. "Hydraulic cushion" type protection systems with different locations

Placing the hydraulic cushion between the slide and the seat of the joint linking it to the rod is not the only possible solution. Of the essence is, that between the mobile and stationary parts of the tool a relative motion is generated of opposite direction to the one developed by the crank-rod-slide mechanism. In other words, the protection system needs to induce a relative motion between the supporting face of the machine table and the main shaft axis, thus between the inferior and superior parts of the frame. Translation can be carried out by:

- the rod screw in relation to its body;
- the rod in relation to the main shaft;
- the main shaft in relation to the frame;
- the table in relation to the frame;
- the superior part of the tool in relation to the slide;
- the inferior part of the tool in relation to the table.

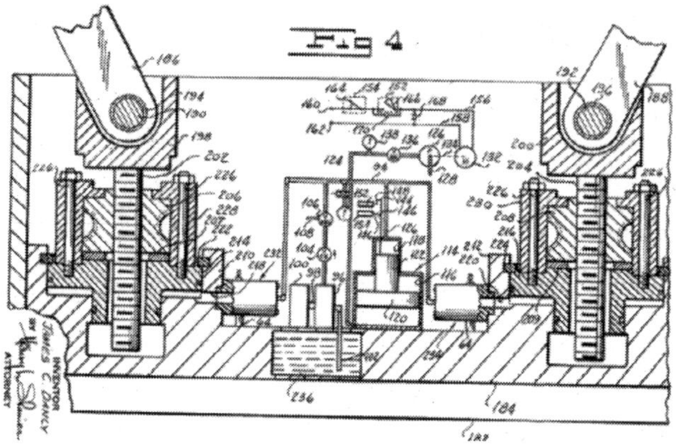

Figure 8. Hydraulic cushion protection system for two-crank mechanical presses according to patent US 2937733 [13].

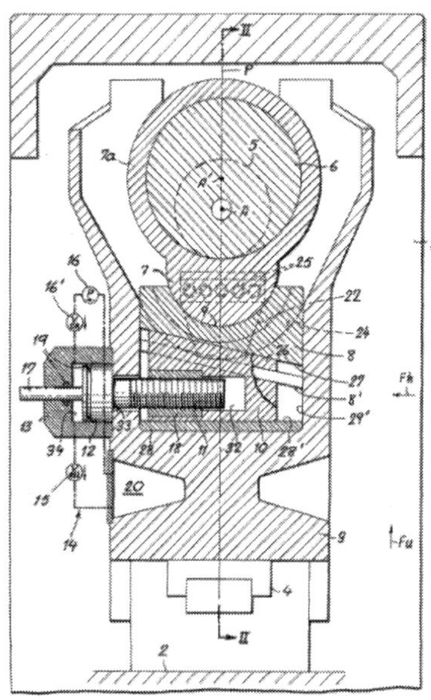

Figure 9. Force overload protection system with a translatable slide and a translatable hydraulic piston according to patent US 3948077 [18]

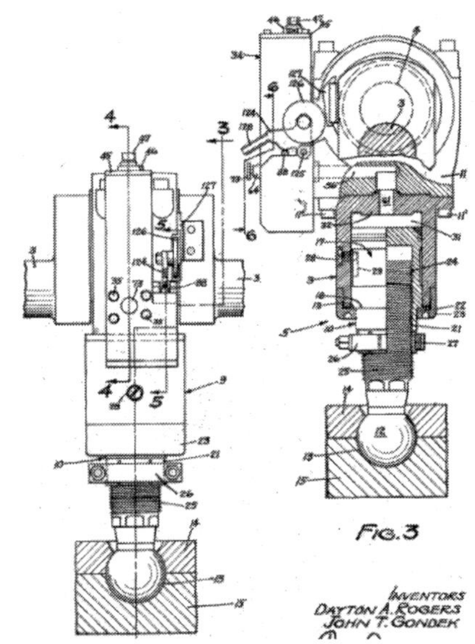

Figure 10. Force overload hydraulic cushion protection system placed between the body of the rod and its screw, according to patent US 2491279 [19].

Positioning the hydraulic cushion between the body of the rod and its screw requires and intermediary assembly assuming the role of a translatable piston. Constructive solutions of this type are put forward, for example, by patents US 2491279 [19] (figure 10) and US 3892143 [20] (figure 11).

To date no patent of a protection system could be identified that entails the translation of the rod in relation to the main shaft. Research in this direction is worth conducting, and this possibility is to be noted as an interesting reserve of technical creativity.

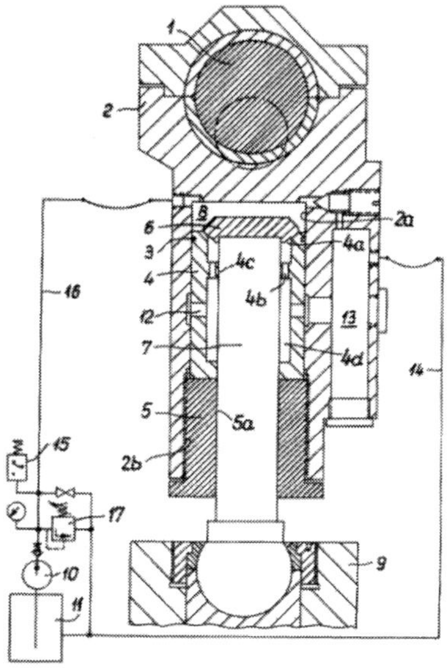

Figure 11. Force overload hydraulic cushion protection system placed between the body of the rod and its screw, according to patent US 3892143 [20].

Placing the hydraulic cushion between the frame and the main shaft entails a supporting slide for the latter, enabling it to translate in relation to the frame. Such a solution is given in patent US 2236154 [21], figure 12.

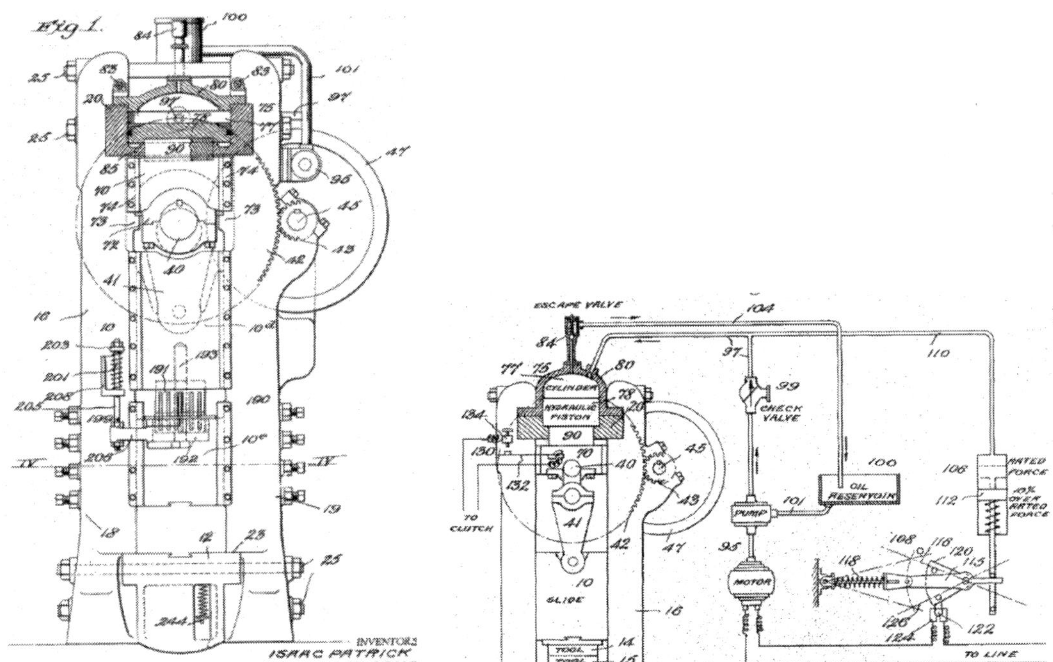

Figure 12. Force overload hydraulic cushion protection system placed between the frame and a main shaft supporting slide according to patent US 2236154 [21].

If the hydraulic cushion type protection system is placed between the table of the press and its frame, the table will behave like a large hydraulic piston. An example in this respect is given by the solution put forward by patent US 2389818 [22], figure 13.

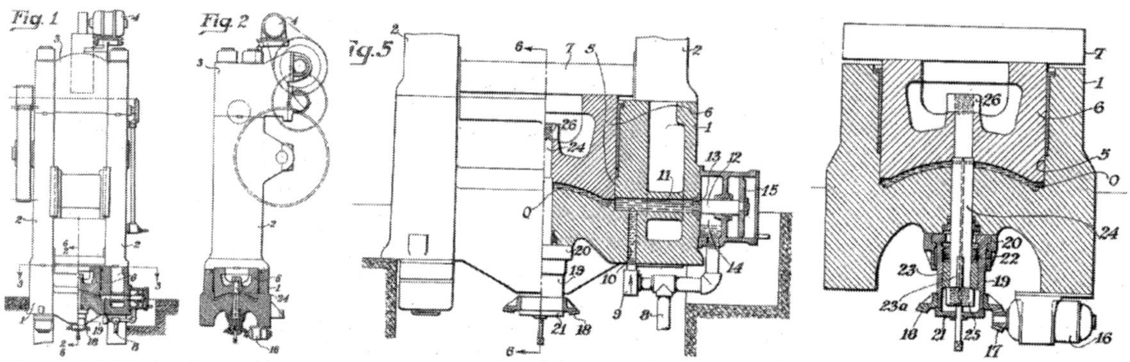

Figure 13. Hydraulic cushion protection system placed between the frame and the table of the press, according to patent US 2389818 [22].

A hydraulic cushion type protection system enabling the translation of the superior part of the tool in relation to the slide, or of the inferior part of the tool in relation to the table can be integrated in either a distancer plate, a "banking plate" if placed on the machine table, under the stationary inferior part of the tool, or even on the superior or inferior part of the tool, respectively. Placing the protection system in one of the parts of the tool is possible, but increases the cost of the tool and no strict correlation with the maximum admitted value of the force for a given press is achieved; such solutions have not been explicitly identified, but the idea is to be noted and can be a source of innovation. The protection system placed in a distancer plate can be found in the current state of technology, an example being patent US 3487772 [23], figure 14. Actually, the patent proposes a module-system of this type, machine-independent, but attachable to any mechanical press as an intermediary element between the tool and the slide or between the tool and the table of the press. Evidently, the solution is applicable also in two-crank mechanical presses, generally in presses with large slides with great surface areas of the superior plate of tool. In such cases two or more independent protection systems of the discussed type can be deployed, ensuring their adequate hydraulic connection.

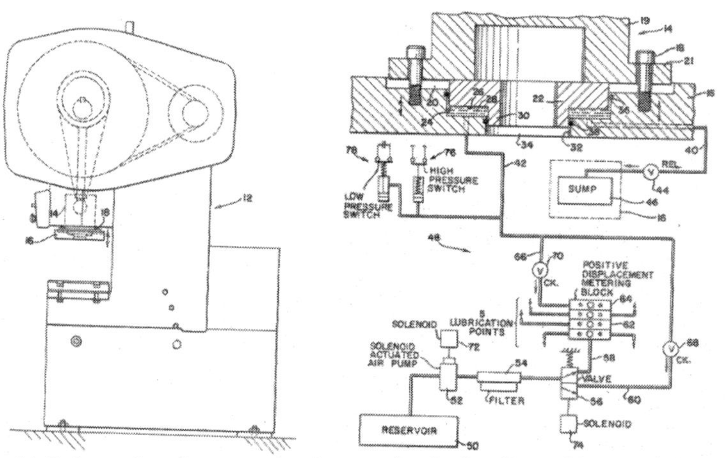

Figure 14. Independent force overload protection hydraulic cushion module- system, attachable to the inferior surface of the slide, according to patent US 3487772 [23].

4. Conclusions

In order to prevent the undesired consequences of possible overload, machine-tools are equipped with various overload protection devices. In presses one of the overloads is by force, when the opposing force at the slide significantly exceeds the nominal force for that the machine was dimensioned and built. In essence, the majority of force overload safety allows the further functioning of the machine, without, however, a useful relative motion between the mobile and the stationary part of the tool.

Reliable and most efficient are "hydraulic cushions" type force overload safety systems, not including destructible parts. In such systems the force at that the protection device can be easily adjusted, and the integration of the hydraulic safety system into the automated control and operation system of the machine is also fairly simple to achieve.

Typically "hydraulic cushion" type protection devices are placed between the slide and the rod, directly or indirectly, as illustrated by the relevant examples given in the paper. Such positioning is, however, not the only alternative. Of essence is that between the mobile and stationary part of the tool a relative motion of opposite direction to the one developed by the crank-rod-slide mechanism occurs. In other words, the protection device needs to induce a relative motion between the supporting surface of the machine table and the main shaft axis, thus between the inferior and superior part of the frame. Constructive solution were identified that ensure the translation of the rod screw in relation to its body, of the main shaft in relation to the frame, of the table in relation to the frame, of the superior part of the tool in relation the slide, and of the inferior part of the tool in relation to the table.

Further possible constructive solutions would be those ensuring the translation of the rod in relation to the main shaft, or of the inferior part of the rod body in relation to its superior part. A study of the current state of technology has not revealed any existing solutions in this sense, which could represent interesting reserves of creativity.

References

[1] Kibbe R R Neely J E White W T and Meyer R O 2009 *Machine Tool Practices* (Pearson)
[2] Cioară R 2008 *Cold Forming Machine Tools* (Brasov: Transilvania University Press)
[3] Cioară R 2005 *Couplings and Brakes for Cold Forming Machine Tools* (Bucharest: Matrix Rom)
[4] Götz B 1996 *Mechanical overload protection device for machines, particularly jaw crushers* (Patent US 5542884)
[5] Richter A 1961 *Safety device for use in eccentric presses and crank presses* (Patent US 3014420)
[6] Beck W D and Thompson T L 1992 *Non-destructive overload apparatus for a mechanical press* (Patent US 5125332)
[7] http://www.aida-global.com/aida-tech/ENG-VOL-2-Hydraulic-Overload-Protection-HOLP.pdf
[8] Spanke E A and L.F. Carrieri L F 1979 Press having overload responsive slide shut height adjusting mechanism (Patent US 4166415)
[9] Pels H 1931 *Hydraulic safety device for power presses and the like* (Patent US 1808606)
[10] Kassnel N J 1960 *Overload protected mechanical press* (Patent US 2936055)
[11] Y. Tezuka Y 1969 *Hydraulic overload release and control for a press machine* (Patent US 3426873)
[12] A.S. Grigorenko et al V V 1977 *Hydraulic protection device for presses* (Patent US 4030336)
[13] Danly J C 1960 *Overload relief assembly for power presses* (Patent US 2937733)
[14] Alexander T W Malolepsy R G and S. Jakso S 1969 *Multiple overload press protection device and adjustable press tonnage device* (Patent US 3481171)
[15] Daniel E 1997 *Hydraulic overload proportional valving system for a mechanical press* (Patent US 5638748)
[16] Yonezawa K 1978 *Overload protector for mechanical press* (Patent US 4085669)
[17] Yonezawa K and Miyajima M 1989 *Hydraulic overload protector for mechanical press* (Patent US 4827839)
[18] Bothe W 1976 *Forging press with independent hammer-position adjustment and overload protector* (Patent US 3948077)
[19] Rogers D A and Gondek J T 1949 Safety overload pitman for power presses (Patent US 2491279)
[20] Kleister H and Sommer W 1975 *Pitman adjustable in length for a press with hydraulic overload safety mechanism* (Patent US 3892143)
[21] Patrik I and Klocke W 1941 *Press combination having means for avoiding jamming* (Patent US 2236154)
[22] Rode F J and Hatch M R 1945 *Press* (Patent US 2389818)
[23] Kraft D H 1970 *Press hydraulic overload system* (Patent US 3487772)

IOP Conf. Series: Materials Science and Engineering **161** (2016) 012007 doi:10.1088/1757-899X/161/1/012007

Wind turbine power curve prediction with consideration of rotational augmentation effects

X Tang, X Huang, S Sun and R Peng

School of Mechanical Engineering, Xiangtan University, 411105 Hunan, China

E-mail: xinzitang@163.com

Abstract. Wind turbine power curve expresses the relationship between the rotor power and the hub wind speed. Wind turbine power curve prediction is of vital importance for power control and wind energy management. To predict power curve, the Blade Element Moment (BEM) method is used in both academic and industrial communities. Due to the limited range of angles of attack measured in wind tunnel testing and the three-dimensional (3D) rotational augmentation effects in rotating turbines, wind turbine power curve prediction remains a challenge especially at high wind speeds. This paper presents an investigation of considering the rotational augmentation effects using characterized lift and drag coefficients from 3D computational fluid dynamics (CFD) simulations coupled in the BEM method. A Matlab code was developed to implement the numerical calculation. The predicted power outputs were compared with the NREL Phase VI wind turbine measurements. The results demonstrate that the coupled method improves the wind turbine power curve prediction.

1. Introduction

Wind energy has been increasingly focused as a widespread green energy source all over the world. For the last decades, wind turbine technology has been undergoing great development. Aerodynamic design is the fundamental process in wind turbine development, as the forces and power produced by the rotor are firstly predicted. The power curve is the rotor power versus the wind speed at the hub height. It is of highly importance for wind turbine development and power production estimation with a candidate site. In both industrial and academic communities, the most popular method for wind turbine design and power prediction is the Blade Element Moment (BEM) based method [1] (also known as Strip theory or Glauert/Wilson method). By dividing the wind turbine blades into sectional blade elements and applying linear momentum conservation to the annular elements, the forces and power are calculated by integration based on the sectional airfoil lift and drag coefficients. The lift and drag coefficients are one of the most important criticalities for the accuracy of power curve prediction in the BEM. These coefficients are often characterized and originally obtained from two-dimensional (2D) airfoil wind tunnel testing. The aerodynamic coefficients of an airfoil are measured over a limited range of angles of attack and are rarely available over the entire range of ±180°. Moreover, the flow pattern in a rotating wind turbine is distinctively different from that in airfoil wind tunnel test. It was pointed out that methods relying on purely 2D flow characterization are not capable of simulating the complex three-dimensional (3D) flows [2]. And the under-prediction of power output using 2D coefficients at high wind speeds for rotating wind turbines is due to the three-dimensional rotational augmentation effects named as "stall-delay" [3].

To build stall-delay models and to represent the lift and drag coefficients, several empirical models have been developed and applied with the BEM method as well as the lifting surface wake theory, such as the Viterna-Corrigan (V-C) model [3] and Du-Selig (D-S) [4]. Snel [5] presented an empirical correction model for lift coefficient but no drag correction was mentioned. Bak [6] developed a correction model based on the pressure distribution, which is a function of span-wise and chord-wise positions. The normal and tangential coefficients are then obtained from integration of the pressure differential model. Hansen [7] proposed a correction model for 3D rotational effects from a quasi-3D Navier-Stokes method. Some researchers tried to find a correct mathematical representation of lift and drag airfoil, i.e. Lanzafame [8] implemented a fifth-order logarithmic polynomial to fit the wind tunnel experimental data in a design code based on the data of the NREL Phase II and VI wind turbine. Some researchers tried using coefficients from 2D CFD and stall-delay model in the BEM method [9]. Other researchers used a disc model with volumetric forces added through source terms in the momentum equations for the simulation of a marine current turbine [10]. These efforts have greatly improved the power performance prediction.

The purpose of this study is to investigate the lift and drag characterization for wind turbine power curve modelling. A Matlab code was developed to implemented the BEM method employing different aerodynamic models for representing the lift and drag coefficients. A 3D computational fluid dynamics (CFD) approach was exploited for characterizing the lift and drag coefficients used in the BEM method for power performance prediction. The discussion was based on the NREL Phase VI wind turbine [11] which is used for validation. A coupled model for representing the lift and drag coefficients covering the pre-stall, stall-develop, post-stall and deep-stall regions was proposed in the BEM method and the results were compared with measurements. The method can be applied to any wind turbines.

2. BEM modelling

2.1. BEM procedure

The analytical aerodynamic approach used for power performance prediction is based on the BEM theory [1]. In the BEM theory, the blade is divided into several sections and each section sweeps an annular area when the rotor rotates. These annuli are separated and no interaction between each other. In other words, the stream tube is decomposed along different radius positions and each annulus has its own momentum balance. By dividing the wind turbine blades into annular blade elements and applying one-dimensional linear momentum conservation to the annular elements, the forces and power are calculated and integrated based on the sectional airfoil lift and drag coefficients, the chords and twist angles of the blade. Based on the standard BEM method, both the Glauert wake model and Prandtl tip-hub loss model were employed. The total power produced by a rotor was then integrated from the root sections to the tip sections in the form of local speed ratio, the effect of the drag was also included.

The aerodynamic coefficients of an airfoil are measured over a limited range of angles of attack and are rarely available over the entire range of $\pm 180°$. Viterna and Corrigan [3] developed a method to do this extrapolation. The Viterna and Corrigan (V-C) model is still widely used for wind turbine power performance prediction nowadays.

Considering the three-dimensional rotational augmentation effects in the wind turbine, Du and Selig [4] tried to build stall delay model. The Du-Selig (D-S) model was used to correct the lift and drag coefficients at high angles of attack where stall-delay exists. The rotor radial locations and the wind speeds are considered in the model.

For a rotating wind turbine blade, the stall is postponed and the lift is augmented. The lift coefficient curve keeps on ascending after the 2D stall angle (observed in 2D wind tunnel testing). From detailed 3D flow analysis, it is obvious that the sectional airfoils in the 3D blade generally experience four stages including pre-stall, stall-develop (up to stall), post-stall and deep-stall. To cover these stages, a coupled model is proposed for the power curve prediction:

$$C_{l,d} = \begin{cases} C_{l,d_{2D}}, \text{ pre-stall} \\ C_{l,d_{3d}}, \text{ stall-develop} \\ C_{l,d_{quasi-FP}}, \text{ post-stall} \\ C_{l,d_{FP}}, \text{ deep-stall} \end{cases} \quad (1)$$

In the linear pre-stall stage, represents the lift and drag coefficient from 2D wind tunnel testing. In the stall-develop stage, represents the derived coefficients from 3D CFD and pressure measurements or the stall-delay models can be employed. In the post-stall region, the turbine can be regarded as a quasi-flat plate, and afterwards the turbine is in deep-stall thus it is considered as a disk and the standard flat plate equations are employed.

A Matlab code was developed to calculate the power coefficients versus different tip speed ratios and the wind turbine power output at different wind speeds. The Glauert wake, Prandtl tip-hub loss and drag effects were considered in the performance calculation. The lift and drag coefficients were interpolated and extrapolated based on the 2D wind tunnel testing, D-S model and V-C model. The 3D coefficients from CFD simulations were also employed and discussed later. The flow chart of the power performance calculation is explained in Figure1.

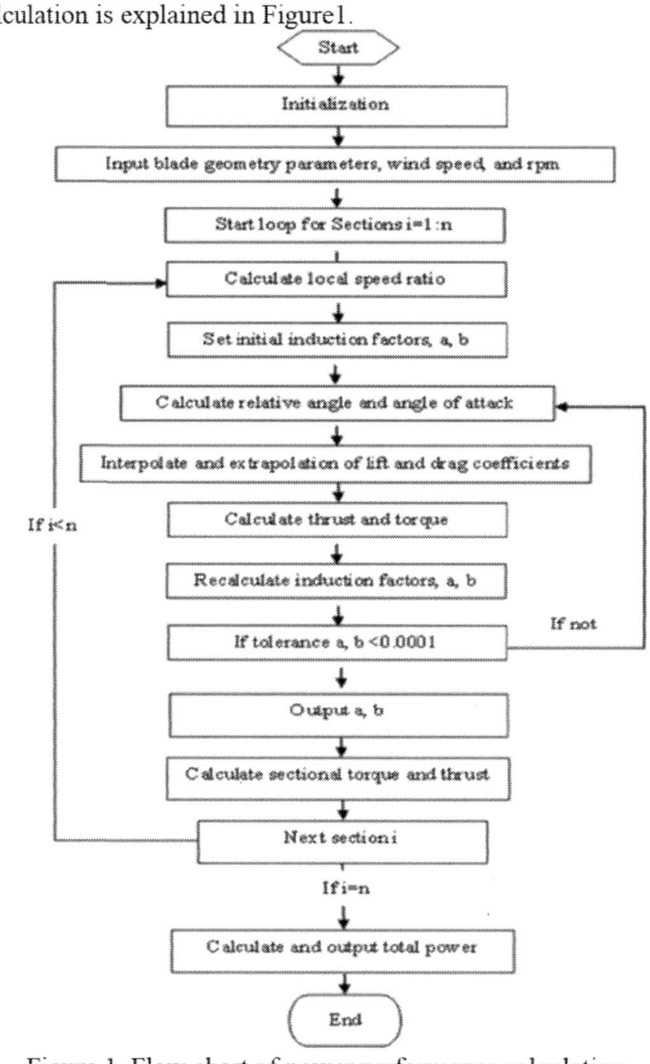

Figure 1. Flow chart of power performance calculation

2.2. BEM validation

To validate the developed BEM code in Matlab, a comparison was conducted with the GH-Blade [12] code and the AeroDyn [13] code. Figure 2 compares the power coefficient of the NREL Phase VI wind turbine predicted by the develop Matlab code with GH-Blade and AeroDyn. Good agreements were obtained thus the BEM method was correctly implemented.

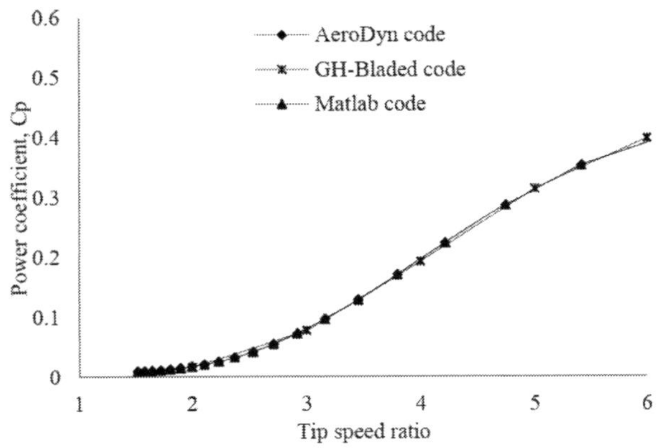

Figure 2. Power coefficient comparison

3. CFD modelling

3.1. CFD approach

The 3D CFD simulations based on the Reynolds-averaged Navier–Stokes (RANS) equations were conducted to analyze the detailed flow pattern along the blade. The turbine and the flow domain are illustrated in Figure 3. Figure 3 depicts the whole flow domain which combines two sub-domains, i.e. the outer domain and the inner domain. The whole domain is 5 times rotor radius in length upstream before the rotor and 15 times rotor radius downstream behind the rotor to allow fully wake development. The outer domain radius is 5 times rotor radius at the inlet and 8 times rotor radius at the outlet. The radius of the inner domain is slightly larger than the rotor radius. The multiple reference frames (MFR) method and sliding mesh method are used to model the rotating wind turbine rotor, i.e. the outer domain is defined as stationary while the inner domain is defined to be rotating with the rotor blades. These boundary conditions are applied: velocity inlet, pressure outlet, far wall, periodic and interface. The blade surface is regarded as non-slipping wall. The two domains interact through mesh interfaces. Periodic boundary was defined to solve a half domain only. These settings are considered to be similar to the real wind tunnel test.

Due to very large size scale difference between the tiny blade surface boundary and the large far field, and also the free-form surface of the turbine blade, the mesh is the most challenging part of wind turbine simulation. Considering the balance between the calculation time and accuracy, a mesh dependency study was performed until there is no obvious change in results. Different mesh scheme and mesh size were tested and evaluated. Finally, a hybrid mesh scheme was applied: a fine tetrahedral mesh was used near blade in the inner domain; a coarser hexahedral mesh was applied to the outer domain. A number of 3.6 million cells in total were generated in ICEM and the minimum Y plus of the first layer near the blade surface is around 1.0 to solve the boundary flow.

A turbulence model dependency study was executed involving the standard k-ε model, the k-ε

IOP Conf. Series: Materials Science and Engineering **161** (2016) 012007 doi:10.1088/1757-899X/161/1/012007

model with enhanced wall treatment, the SST k-ω turbulence model and the SST transition model. Among these models, the SST transition model is capable of simulating turbines accurately but it is very time consuming compared to other models. For this case, the SST k-ω turbulence model provided with reasonably good results within general time limits, thus it was used for all the calculations afterwards. All the calculation was performed in transient mode in FLUENT and a series of time step sizes were examined to secure transient behavior. Again, this is a balance between accuracy and computing time. Providing the monitored results showed no big difference, the setting of 0.001s/step and 1660 steps therefore was then selected considering the balance between accuracy and computing time. All the calculations converged with the residuals under 10^{-5}.

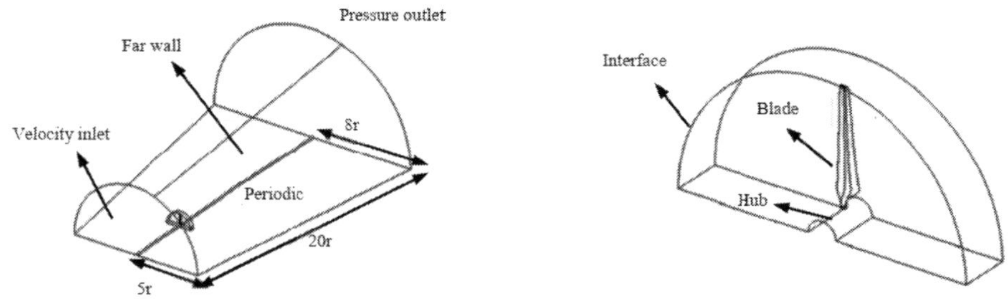

Figure 3. Domain topology and boundary conditions

3.2. CFD validation

Figure 4 plots the calculated pressure distributions compared with measurements for different locations along the blade at wind speed of 7m/s. More pressure distributions were obtained but not presented here due to limited pages. As shown in Figure 4, the results of the CFD simulations well agree with the measurements. Though small quantitative difference exists due to boundary laminar bubble and flow separation, the overall distributions are generally well predicted for all the span locations.

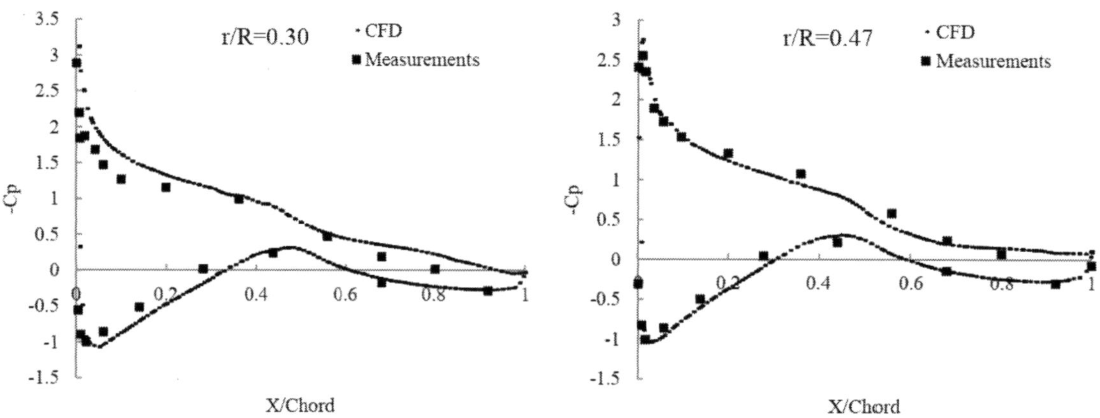

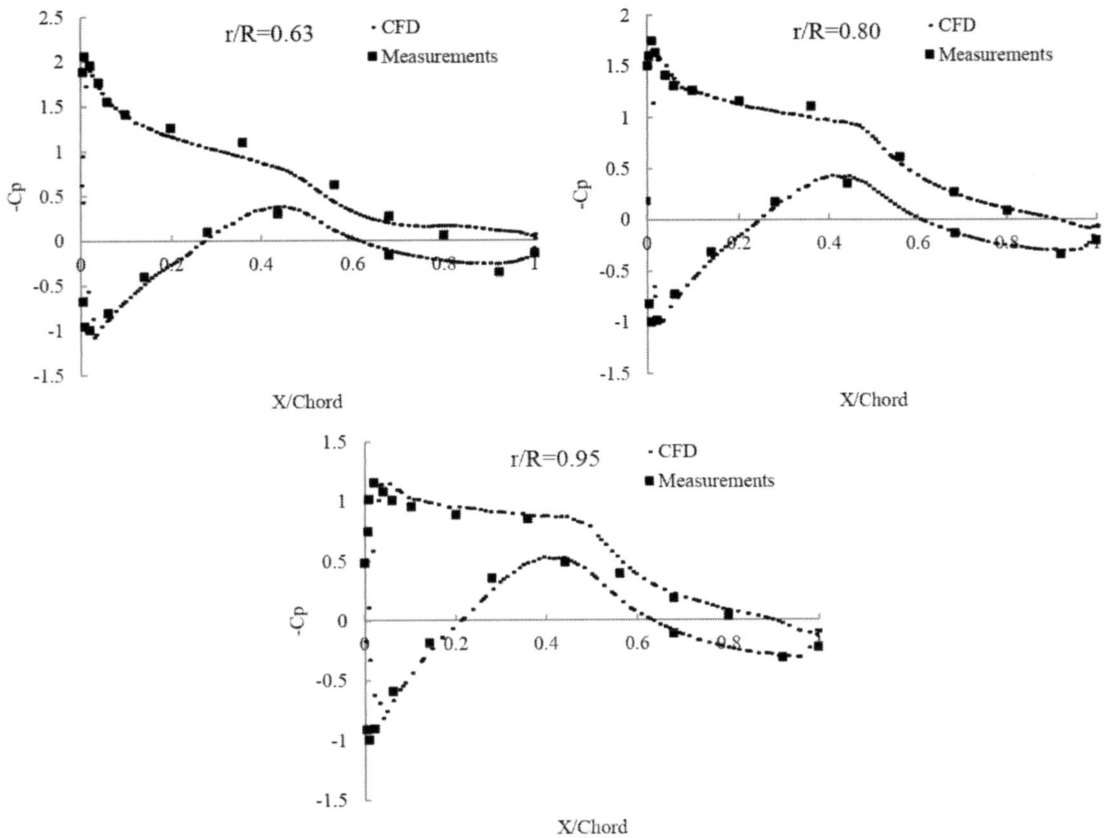

Figure 4. Pressure distributions for different locations along the blade length, wind speed=7m/s.

4. Results and discussion

Figure 5 gives a plot of different series of lift and drag coefficients: (1) 2D coefficients from wind tunnel tests, For the NREL Phase VI wind turbine, the aerodynamic coefficients of S809 from the Delft University of Technology wind tunnel tests at Reynolds number of 1×10^{6} [11] were used, which are named as 2D Cl and Cd; (2) corrected coefficients based on D-S model, named as D-S Cl and Cd; (3) 2D coefficients plus 3D coefficients and corrected coefficients based on V-C model, named as 2D+3D+V-C Cl and Cd; (4) 2D coefficients plus 3D coefficients and coefficients based on modified Flat Plate theory, named as 2D+3D+mFP Cl and Cd .

According to the D-S model from (11) to (16), the D-S Cl and Cd coefficients were calculated based on the 2D wind tunnel coefficients, as presented in Figure 5. Based on the CFD simulations, the inner span locations were regarded as typical stall delay locations other than outer locations. For simplicity, only one radius location r/R=0.3 was implemented, and the coefficients were corrected at the wind speed of 10m/s.

By applying the V-C method, the input aspect ratio was selected as 14 according to the tip chord of 0.358m and radius of 5.029m for the Phase VI turbine blade. Three initial stall angles of 9.21°, 15.23° and 20° were tried, however, none of these initial stall angles produced satisfactory results. With an input of initial stall angle of attack and corresponding lift and drag coefficients, the V-C method showed limited improvement in power prediction. It is noticed that the lift to drag ratios at these two angles of attack do not follow the flat plate theory, and serious "drop" of the lift coefficient curve occurred at these initial stall angles, where the V-C correction starts to work. At the same time, the lift coefficient is claimed to be arising after stall as due to stall-delay in 3D wind turbine flows. Thus it is necessary to solve the problem of the "drop" between the 2D tested aerodynamic coefficients and the V-C corrected coefficients at the angle of attack where the V-C correction model starts to work and keep the corresponding lift to drag ratio been guided by the flat plate theory.

To cope with this "drop", the extracted 3D coefficients derived from the surface pressure of the rotating blades at five span locations based on the CFD calculation were used to "bridge up" this gap, as shown in Figure 5. These 3D coefficients calculated from the blade surface pressure were applied for the angles of attack from 9° to 20°, and there is no big "gap" existed in the 2D+3D+V-C Cl and Cd coefficients.

For the series of 2D+3D+mFP Cl and Cd coefficients, the TUDelft wind tunnel tested lift and drag coefficients are used for angles of attack from 0° to 6.16° for pre-stall region, while for the angles of attack from 6.16° to 20° for stall-develop region, the bridged coefficients from 3D simulations are applied. When the angle of attack is above 20° for post-stall and deep-stall, the coefficients are calculated according to the equations from (2) to (5):

$$C_l = 2 \cdot C_{l\max} \cdot \sin\alpha \cdot \cos\alpha, \quad 20° < \alpha \le 30° \tag{2}$$

$$C_d = C_{d,\max} \cdot \sin^2\alpha, \quad 20° < \alpha \le 30° \tag{3}$$

$$C_l = 2 \cdot \sin\alpha \cdot \cos\alpha, \quad 30° < \alpha \le 90° \tag{4}$$

$$C_d = 2 \cdot \sin^2\alpha, \quad 30° < \alpha \le 90° \tag{5}$$

Here, $C_{l,\max} = C_{l,\alpha=45°}$ and $C_{d,\max} = C_{d,\alpha=90°} = 2$.

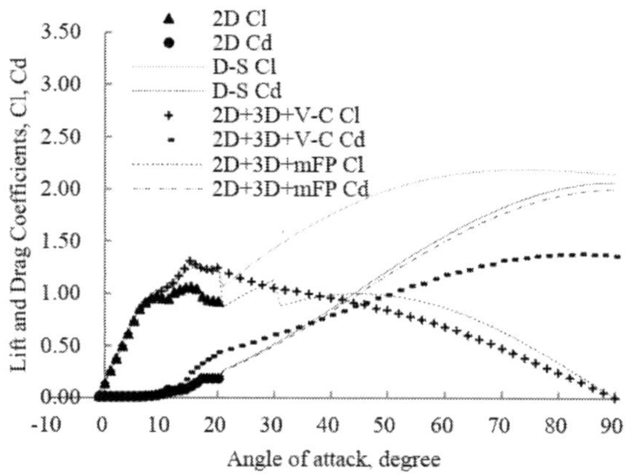

Figure 5. Lift and Drag Coefficients

Providing the lift and drag coefficients, the power curve can be well predicted based on the BEM approach. Figure 6 demonstrates the corresponding power curves for the different series of coefficients in Figure 5. Among these, the "2D+3D+V-C"and the "2D+3D+mFP" curves show excellent agreement in power curve prediction compared with the measurements. It is also demonstrated that the power prediction produced by the"2D+3D+mFP" method for at high wind speeds is improved comparing with the "2D+3D+V-C" method. The "2D+FP" method under-predicts the power comparing to the measurements, which is due to stall-delay in real 3D flows. The "2D+V-C with gap" curve shows a big drop due to the serious "drop" in 2D wind tunnel coefficients and the V-C coefficients as discussed in Figure 5. The "2D+D-S" method has good agreements for low and moderate wind speeds. The predicted power coincides well with the measured power when the wind speed is up to 10m/s. While at higher wind speeds, the prediction is higher than the measurement. Similar results were also reported by Breton [14]. The over-predicted power at high wind speed is mainly caused by the over-corrected lift coefficients at high angles of attack. This is affected by the simplification in the execution of the D-S model. As stated in the D-S model, different radial locations and wind speeds are considered, thus the more the rotor radius locations and wind speeds are calculated, the better prediction is to be obtained. Moreover, the drag coefficients produced by the D-S model are almost the same with those from 2D wind tunnel tests while in fact a higher drag is pronounced in stall-delay. This may also attribute to the over-prediction of the power curve.

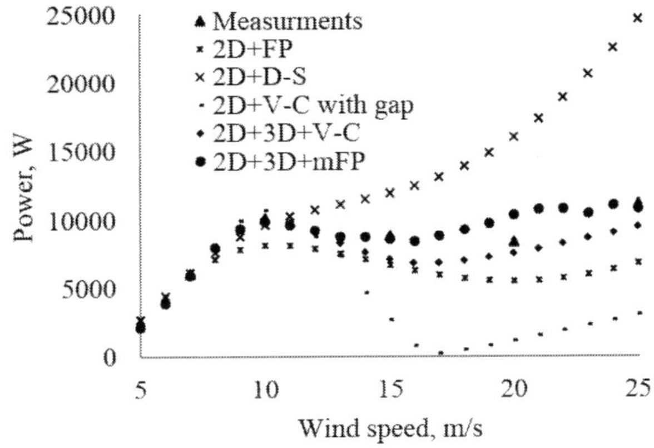

Figure 6. Power curve prediction with different models.

5. Conclusion

The wind turbine power curve prediction was conducted based on the BEM method and the RANS CFD methods. A Matlab code was developed to implemented the BEM method employing different aerodynamic models for representing the lift and drag coefficients. A 3D CFD approach was exploited for characterizing the lift and drag coefficients. A coupled model for representing the lift and drag coefficients covering the pre-stall, stall-develop, post-stall and deep-stall regions was proposed in the BEM method. The lift and drag coefficients derived from the 3D CFD simulations were coupled into the BEM method considering the 3D augmentation effects in the stall-develop stage. The results were compared with the measurements of the NREL Phase VI wind turbine. The proposed coupled model contributes to an improved power prediction especially at moderate and high wind speeds. With advanced turbulence models the results can be further improved. The approach can be applied to any wind turbines for power performance prediction, turbine operation and estimation of wind turbine power production.

Acknowledgment

This research was financially supported by the National Natural Science Foundation of China (Grant No. 51305377).

References

[1] J. Manwell, and J. McGowan, Wind energy explained: theory, design and application, 2nd Edition, John Wiley & Sons Inc, 2009.

[2] R.P.J.O.M. van Rooij, A. Bruining, and J.G.Schepers, " Validation of some rotor stall models by analyses of the IEA Annex XVIII field data," Proceedings of the European Wind Energy Conference, 2003.

[3] J .Tangler, "Wind turbine post-stall airfoil performance characteristics guidelines for blade-element momentum methods," NREL/CP-500-36900, 2004.

[4] Z. Du and M.A.Selig, "A 3-D stall-delay model for horizontal axis wind turbine performance prediction," AIAA-98-0021, 1998.

[5] S.P. Breton, F.N. Coton, and G. Moe, "A study on rotational effects and different stall delay models using a prescribed wake vortex scheme and NREL phase VI experiment data," Wind Energy, Vol.11, pp:459-482, 2008.

[6] C. Bak, J. Johansen, and P.B. Andersen, "Three-dimensional corrections of airfoil characteristics based on pressure distributions," Proceedings of the European Wind Energy Conference, 2006.
[7] P.K. Chaviaropoulos and M.O.L.Hansen, "Investigating three-dimensional and rotational effects on wind turbine blades by means of a quasi-3D Navier Stokes solver," Journal of Fluids Engineering, Vol.122, pp:330-336, 2000.
[8] R. Lanzafame, "BEM theory: How to take into account the radial flow inside of a 1-D numerical code," Renewable Energy, Vol.39, No.1, pp:440-446, 2012.
[9] V. Esfahanian, A.SalavatiPour, I.Harsini b, etc, "Numerical analysis of flow field around NREL Phase II wind turbine by a hybrid CFD/BEM method", Journal of Wind Engineering and Industrial Aerodynamics,Vol.120,pp:29-36,2013.
[10] Q.Guo, L.Zhou, Z.Wang, "Comparison of BEM-CFD and full rotor geometry simulations for the performance and flow field of a marine current turbine", Renewable Energy, Vol.75, pp:640-648, 2015.
[11] M.M.Hand, D.A.Simms, L.J. Fingersh, D.W. Jager, J.R. Cotrell, S. Schreck, and S.M. Larwood, "Unsteady aerodynamics experiment phase VI: wind tunnel test configurations and available data campaigns," NREL/TP-500-29955, 2001.
[12] Garrad Hassan & Partners Ltd, Bladed 4.0 theory manual, 2010.
[13] Moriarty.P.J & Hansen.A.C, "AeroDyn theory manual", Technical report NREL/TP-500-36881, 2005.
[14] S.P. Breton, F.N. Coton, and G.Moe, "A study on rotational effects and different stall delay models using a prescribed wake vortex scheme and NREL phase VI experiment data," Wind Energy, Vol.11, pp:459-482,2008.

IOP Conf. Series: Materials Science and Engineering **161** (2016) 012008 doi:10.1088/1757-899X/161/1/012008

Courses of Action to Optimize Heavy Bearings Cages

V G Szekely

Transilvania University of Brasov, Eroilor 29, 500039 Brasov, Romania

E-mail szekely.valentin.gabriel@unitbv.ro

Abstract: The global expansion in the industrial, economically and technological context determines the need to develop products, technologies, processes and methods which ensure increased performance, lower manufacturing costs and „synchronization of the main costs reported to the elementary values which correspond to utilization". The development trend of the heavy bearing industry and the wide use of bearings determines the necessity of choosing the most appropriate material for a given application in order to meet the cumulative requirements of durability, reliability, strength, etc. Evaluation of commonly known or new materials represents a fundamental criterion, in order to choose the materials based on the cost, machinability and the technological process. In order to ensure the most effective basis for the decision, regarding the heavy bearing cage, in the first stage the functions of the product are established and in a further step a comparative analysis of the materials is made in order to establish the best materials which satisfy the product functions. The decision for selecting the most appropriate material is based largely on the overlapping of the material costs and manufacturing process during which the half-finished material becomes a finished product. The study is orientated towards a creative approach, especially towards innovation and reengineering by using specific techniques and methods applied in inventics. The main target is to find new efficient and reliable constructive and/or technological solutions which are consistent with the concept of sustainable development.

1. Introduction

The continued need for development of existing technological solutions or the need of innovation – invention regarding new higher technical solutions which are determinate by the evolution of the industrial environment at the global level. Thereby the economic efficiency of industrial companies it is an increased fact which determines also an increase of their product's efficiency. Optimization and economic efficiency of the products to ensure minimum consumption during periods of operation and post-operation require the use of specific methods for research and design. In the scientific literature it can be identified a number of methods which are centered on research and design [1] [2] which are using different approaches, techniques and methods of technical creation [3]. It is preferable that the creative work to take into consideration the relevant functions of the product studied; value engineering can be used successfully. This tool, used mainly in industry, ensures the basic functions with minimal costs without neglecting aspects of performance, quality and reliability of products [4]. For example, for the heavy bearing cages a number of key operating functions have been identified. These functions have a direct influence on the material, these being: "ensures resistance", "ensures rigidity", "is reliable", "is sustainable", "is conditioning the operating energy consumption" and "ensures less wear" [5]. It is useful to specify that the functions "is conditioning the operating energy consumption" is ensured implicitly by the reduced mass of the body which depends on the volume and density of the material used. In terms of the production, the function "it is technological" can be expressed by machinability of the material; for example, it can be measured by the cutting force or through another parameter, whether directly or related to the function.

Consistent with the concept of sustainable development of the products [6] is recommended to take into account the CO2 consumption and the degree of material reintegration into the economic or environment circuit. This characteristic appears indirectly from the main function "has an impact on the environment" which is corresponding to the post-use period [5].

2. Important functions and coefficients of the bearing cages

For a certain bearing, the cage selection can be done as a result of a multi-criteria analysis based on the allocation of scores to different functions of the cage. An example of such scores is given in Table 1. It is noted that most of the criteria are dependent upon the nature of the cage material: strength, stiffness, reliability / durability, density, machinability and even the degree of reintegration into the environment. At first glance, the criterion volume doesn't appear to be dependent on the material, but it is dependent to a certain degree of the production technology, technology which it's dependent on the nature of the material. As a result a study of different materials which could be embedded into the design of the heavy bearing cages it is useful and even necessary.

Table 1. Criteria and coefficients of importance.

Criteria	Coefficients of importance
Resistance	0.20
Rigidity	0.10
Reliability / Durability	0.13
Ensures less wear	0.10
Density	0.20
Cost	0.07
Machinability	0.10
Degree of re-integration into the environment	0.05
CO_2	0.05

Within the heavy bearing industry, regardless of the application in which a bearing is incorporated, the cage is mostly made of steel or non-ferrous alloys such as brass or bronze. In certain cases, plastic materials are used, usually polyamides. Plastic cages are used increasingly rare in heavy bearings, primarily due to insufficient mechanical strength. Using plastic cages in the framework of heavy bearings usually leads to a premature failure due to the deterioration of the bearing cage [7].

3. The usage ratio of the material for a brass cage

Due to the anti-friction properties and good strength, durability and preservation of the rolling surface's quality, brass is the main material used in the manufacturing of heavy bearing's cages. Therefore it's justified to choose a cage made of brass as reference for the current study. In Table 2 the dimensions for a heavy bearing and its massive brass cage, used in wind turbine, are given.

Table 2. Main dimensions of the bearing and its cage.

	External diameter	Internal diameter	Width
Bearing	2200 mm	1760 mm	320 mm
Brass cage	2060 mm	1924 mm	247 mm

The blank from which the cage is manufactured, which is a ring-type and the machining allowances can be different depending on the degree of casting precision (examples can be viewed in Table 3). In the case of brass which is a deficient and relatively expensive material it's of interest the usability of the material, defined as the ratio between the volume of the blank and the volume of the cage as a finished product. It can be distinguished between the raw blank obtained by casting and the blank which is a ring-type with rectangular section, obtained by turning, subjected to further machining in order to obtain the pockets for the rolling elements.

Depending on the machining allowance, for the cage with the dimensions given in Table 2, the volume of the raw blank obtained by casting it is shown in Table 4. Also in the same table the values for the degree of material usage are given, for the machining allowance values t_i are shown in Table 3.

In order to determine the degree of material usage the next relationships (1), (2) and (3) are used.

Table 3. Machining allowances depending on the casting precision.

Types of casting	Machining allowances (t_i)
High precision casting (under pressure)	1.4 -1.5 mm
Medium precision casting	2 mm
Low precision casting	2.5 mm

$$\eta_1 = \frac{V_{blankt_i} - V_{ring}}{V_{blankt_i}}; \tag{1}$$

$$\eta_2 = \frac{V_{blankt_i} - V_{cage}}{V_{blankt_i}}; \tag{2}$$

$$\eta_3 = \frac{V_{ring} - V_{cage}}{V_{ring}}, \tag{3}$$

where V_{blankt_i} – is the total volume of the cast blank corresponding to the machining allowance t_i;

η_1 – loss of material due to machining allowance t_i in order to obtain the finished ring;

η_2 – loss of materials in order to obtain the finished cage taking into account the machining allowance t_i and the material milled in order to form the necessary pockets for the rollers;

η_3 – loss of material in order to obtain the pockets for the rollers related to the cage ring without the machining allowances t_i.

The volume of the turned ring is $V_{ring} \cong 105110$ cm^3 and the volume of the cage as a finished product is $V_{cage} \cong 31572$ cm^3.

Table 4. Volume of the blank and the percentage losses of material.

Machining allowances / Name of the characteristics	$t_1 = 1.5$ mm	$t_2 = 2$ mm	$t_3 = 2.5$ mm
$V_{blank\,t_i}$	$\cong 111248$ cm^3	$\cong 113323$ cm^3	$\cong 115412$ cm^3
η_1	5.5%	7.2%	8.9%
η_2	71.62%	72.13%	72.64%
η_3		69.96%	

It is necessary to specify that the design of the cage taken as a reference (see Table 2) consists of two components, namely the base body and a cover, which serves as a retention element for the rollers. In order to determine the degree of material usage the calculated volumes relate to the core body of the cage, without taking into account the volume of the retention ring. The calculation of material losses aren't affected because the retention ring has no machining allowances and the amount of material removed during the process is significantly reduced.

It is preferred that the machining allowance to be small as possible. If the machining allowance would be zero, which is not possible, then the use of the material should be approximately 30% and the loss of material would be 70% (more precisely $\eta_3 = 69.96\%$).

From an economic perspective, as the degree of material usage is higher, with the same amount the number of machining and processing should decrease. As a result the costs of the material and processing will decrease. This provides insofar a more efficient use of the resources and manufacture time. It is therefore justified the identification and use of materials and/or processes to minimize or even eliminate machining allowance for the manufacturing of cages.

Obtaining a high degree of material usage it is however a secondary objective, this being dependent on the production technology adopted. The main priorities remain the functions of the cage, namely strength, durability and the capability to assure less wear of the rolling element.

4. Multi-criteria analysis for choosing the material of the cage

Depending on a particular application, the materials used must fulfill a plurality of functions and a wide range of properties. Selecting the most appropriate material for the heavy bearing cage can be also done based on a multi-criteria analysis. A significant number of materials [8] [9] which can be used to manufacture heavy bearings cages determines the need to select those materials which have similar characteristics or superior to commonly used materials. For the current paper the said materials which have been selected are specified in Table 5, where M1 - Brass (CuZnPb); M2 - Hardened and tempered steel; M3 - Aluminium; M4 - Duralumin; M5 - Hard plastic (thickness 70 mm); M6 – High-Density Polyethylene; M7 - Polypropylene; M8 - Polytetrafluoroethylene; M9 - Polyamide 6.6.

Table5. Properties / Characteristics of materials.

Properties / Characteristics	M. U.	M1	M2	M3	M4	M5	M6	M7	M8	M9
Tensile strength, R_m	[N/mm^2]	200...300	980	70	270	65...100	1...16	17...40	17...28	70
Tensile yield stress, Rp_{02}	[N/mm^2]	290	1030	60	200	1	-	-	-	-
Relative elongation, E_r	[%]	14...20	9	6	5	1	12	10	250	6
Young's modulus, E	[N/mm^2] $\times 10^4$	10.5	21	7	7.1	0.6...1	1.2...1.4	5...19	35...62	32
Hardness	[HB]	60	64...66	200	50	35	55 °Shore	95 °Shore	65 °Shore	15
Density, ρ	[g/cm^3]	8.65	7.8	2.8	2.8	1.4	0.9	0.9	2.1	1.13
Maximum working temperature	[°C]	350	630	250	300	150	70...120	110	300	150
Cost [10] [11]	[€/kg]	4.22	0.49	1.44	1.44	5.5 – 6.66	0.67	0.67	17 - 22	1.18
Machinability		VG	VG	G	G	G	G	G	G	G
Degree of re-integration into the environment		VG	VG	VG	VG	W	G	G	G	G
CO_2		Great	Great	Great	Great	Medium	Medium	Medium	Medium	Medium

M2 material is part of the steel category used in bearings (e.g. RUL1, RUL1V, RUL2, RUL2V and RUL3V), regulated by STAS 1456/1 and STAS 11250 [9].

Various functions of the cage correspond to one or more properties of the material used. Table 6 shows the main characteristics considered in the multi-criteria analysis developed by the author.

Table 6. Correlation between the functions and the characteristics of the multi-criteria analysis.

Function	Material properties / characteristics of the multi-criteria analysis
Resistance	Tensile strength, R_m
Rigidity	Young's modulus, E
Reliability / Durability	Young's modulus, E
Ensures less wear	Hardness; Maximum working temperature
Density / Volume	Density, ρ
Cost	
Machinability	
Degree of re-integration into the environment	
CO_2	

In Table 7 a single set of results are presented from the multi-criteria analysis, scores and weights for all nine materials selected are given. In Figures 1 and 2 the results of the multi-criteria analysis are presented as charts, but downwards and highlighting the most recommended materials. Multi-criteria analysis was repeated several times, allocating different weights and scores in order to identify the materials which are kept into the top of different iterations.

Table 7. Results of the multi-criteria analysis.

Nr crt	Criteriu	Weighted	M1		M2		M3		M4		M5		M6		M7		M8		M9	
			Scor	Weighted value	Scor	Weighted value	Scor	Weighted value	Scor	Weighted value	Scor	Weighted value	Scor	Weighted value	Scor	Weighted value	Scor	Weighted value	Scor	Weighted value
1	Resistance	0.20	4	0.8	5	1	2	0.4	4	0.8	3	0.6	1	0.2	1	0.2	2	0.4	2	0.4
2	Rigidity	0.10	4	0.4	5	0.5	2	0.2	3	0.3	4	0.4	1	0.1	1	0.1	2	0.2	2	0.2
3	Reliability / Durability	0.13	4	0.52	5	0.65	2	0.26	3	0.39	4	0.52	1	0.13	1	0.13	2	0.26	2	0.26
4	Ensures less wear	0.10	5	0.5	1	0.1	3	0.3	4	0.4	4	0.4	3	0.3	3	0.3	5	0.5	4	0.4
5	Density	0.20	2	0.4	2	0.4	3	0.6	3	0.6	5	1	5	1	5	1	4	0.8	5	1
6	Cost	0.07	2	0.14	5	0.35	3	0.21	3	0.21	2	0.14	4	0.28	4	0.28	2	0.14	3	0.21
7	Machinability	0.10	5	0.5	2	0.2	5	0.5	4	0.4	3	0.3	5	0.5	5	0.5	5	0.5	5	0.5
8	Degree of re-integration into the environment	0.05	5	0.25	4	0.2	4	0.2	4	0.2	1	0.05	3	0.15	2	0.1	2	0.1	2	0.1
9	CO2	0.05	3	0.15	2	0.1	1	0.05	1	0.05	2	0.1	5	0.25	5	0.25	4	0.2	5	0.25
Total scor/Weighted value		1.00	34	3.66	31	3.5	25	2.72	29	3.35	28	3.51	28	2.91	27	2.86	28	3.1	30	3.32

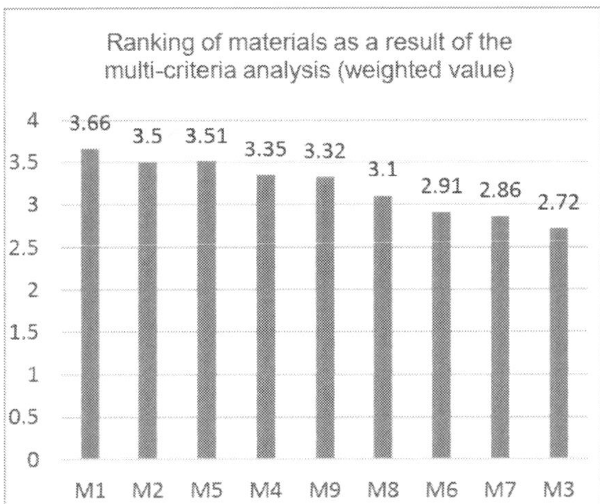

Figure 1. Ordering of materials selected according to weighted score.

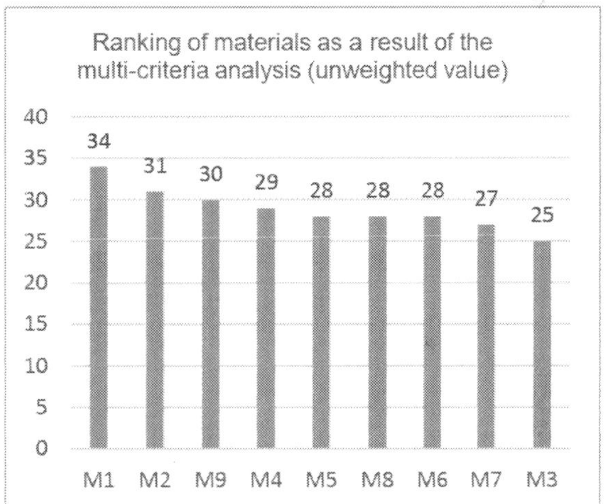

Figure 2. Ordering of materials selected according to unweighted score.

According to the unweighted scores obtained, brass, steel and polyamide P66 occupies the top three places; duralumin is placed on the fourth place. The other materials, except aluminum, are placed at a short distance. According to the weighted scores first places are occupied by brass, hard plastics and steel, while aluminum, polypropylene and polyethylene are occupying disqualifying places. Apart from brass and steel the next materials are noted for further studies: hard plastics, duralumin, polyamide (Polyamide 6.6) and polytetrafluoroethylene. It is necessary to point out that by opting for duralumin the manufacturing technology for a cage does not change. According to the multi-criteria analysis in the case of the presented iteration it can be observed that the materials presented may partially or totally fulfil the requirements.

The approach for optimization of a heavy bearing cage like the type taken as a reference may be summarized only to the change of material, opting for a cheaper one – the effect is to reduce costs, opting for a less dense material – determines the reduction of energy used, or for a material that has a higher degree of machinability. The multi-criteria analysis performed provides solutions for each of these lines of action.

Also it is possible to opt for a combination of materials, course of action which involves primarily a design / redesign effort to identify proper solutions. This is the recommended direction; investing in knowledge it is a sustainable investment and always gives results.

5. Conclusions

A heavy bearing cage satisfies several functions with different coefficients of importance. The current study accomplished by the author aims primarily to obtain higher resistance, to keep or increasing the sustainability of the cage and usage of a material that are cheaper and have low density.

Mainly the material of the cage determines the manufacturing technology. If the manufacturing process involves splintering, then it is necessary to take into account the degree of material usability, especially if the material it is relatively expensive.

Depending on a particular application, the material(s) used must fulfill a plurality of functions and a wide range of properties. Selecting the most appropriate material(s) for the heavy bearing cage can be also done based on a multi-criteria analysis. The current study carried out by the author takes into consideration four metallic materials, four plastic materials and a plastic composite material (hard plastics). From the materials selected the steel, brass, duralumin and hard plastics ensures superior mechanical strength. The same materials also ensure the best reliability and durability properties. Good protection against wear of the rolling surfaces presents brass but also polytetrafluoroethylene, Polyamide (Polyamide 6.6), duralumin and hard plastics, with the exception of steel. Higher costs, but acceptable, has duralumin and polyamide (Polyamide 6.6).

A keen interest it is granted to the material density. Based on this aspect the kinetic energy of the bearing depends, and implicitly the operation energy. Polyamides and hard plastics have low densities, but also polytetrafluoroethylene. Accepted density has also duralumin.

Multi-criteria analysis was repeated several times, allocating different weights and scores in order to identify the materials which are kept into the top of different iterations.

The decision for selecting the most appropriate material is based largely on the overlapping of the material costs and manufacturing process during which the half-finished material becomes a finished product.

The study is orientated towards a creative approach, especially towards innovation and reengineering by using specific techniques and methods applied in inventics. The main target is to find new efficient and reliable constructive and/or technological solutions, consistent with the concept of sustainable development.

References

[1] Georoşeanu G 2011 *Studii şi cercetări de analiza valorii asupra unor grupe de produse din industria de rulmenţi* PhD thesis Transilvania University of Braşov

[2] Mircea C 2009 *Metode noi în proiectare* PhD thesis Polytechnic University of Timişoara

[3] Belous V, Plăhteanu B 2005 *Fundamentele creaţiei tehnice* (Iaşi: Performantica Press)

[4] Crum LW 1971 *Value Engineering: The Organised Search for Value* Ed. Longman (Prentice Hall Press)

[5] Szekely V G 2016 Functions of Heavy Bearing Cages *RECENT* vol 17 no 1(47) (Brasov: Transilvania University Press)

[6] Tureac I, Mărăscu-Klein V, Popescu M şi Cioară R 2006 *Dezvoltarea durabilă a produselor in construcţia de maşini* (Brasov: Transilvania University Press)

[7] *Koyo Ball and Roller Bearings: Failures Causes and Countermeasures* JTEKT CORPORATION Cat.No.B3001E

[8] Ashby M F, Jones D R H 2006 *Engineering Materials 2: An Introduction to Microstructures, Processing and Design* 3th ed. (Oxford: Butterworth-Heinmann)

[9] Mărăscu-Klein V 2000; 2004 *Materiale industriale* vols 1, 2 (Brasov: Transilvania University Press)

[10] *Plastiker*, available at: http://plasticker.de/preise/pms_en.php?show=ok&make=ok&aog=A&kat =Mahlgut

[11] *London Metal Exchange* available at: https://www.lme.com/

IOP Conf. Series: Materials Science and Engineering **161** (2016) 012009 doi:10.1088/1757-899X/161/1/012009

Heavy Bearings Exploitation Energy and Reduction Methods

V G Szekely and R Cioară

Transilvania University of Brasov, Eroilor 29, 500039 Brasov, Romania

E-mail szekely.valentin.gabriel@unitbv.ro

Abstract. The global trend of resource conservation so as "not to compromise the ability of future generation's development" is the fundamental basis of the concept of sustainable development. Concordant with this, the energy efficiency of products is increasingly discussed and frequently taken into account in the design stage. In more cases a product is more appreciated and more attractive as the energy consumption and its associated materials are lower. In the production stage, said consumption advantages primarily the manufacturer, particularly through low consumption thereof. In the operational phase, low energy and materials consumption represents an user advantage and it's a major argument in the decision to purchase and use a particular product. Heavy bearings are frequent products used in wind turbines that are producing non-conventional "clean" energy, as windmills. An enhanced energy efficiency bearing contributes to the enhancement of the overall efficiency of the wind turbines. Based on a suitable mathematical model, this paper identifies and recommends courses of action to reduce the operating energy of heavy bearing through the "cage" - which is the subject of a much larger research - with the highest priority. The identified actions may constitute from a set of requirements for the design stage of the heavy bearing predominantly oriented towards innovation-invention.

1. Introduction

In more cases a product is more appreciated and more attractive as the energy consumption and its associated materials are lower. In the production phase, reduced consumption advantages primarily the producer and it contributes to the reductions of costs. In the operational phase, low energy and materials consumption represents an user advantage and it's a major argument in the decision to purchase and use a particular product.

The energy efficiency of products is increasingly discussed and frequently taken into account at the design stage. Moreover, improving the energy efficiency of products can represent an objective to redesign a product, for its eco-design. The global trend of resource conservation so as not to compromise the ability of future generation's development [1] is the fundamental basis of the concept of sustainable development. Energy efficiency is part of the sustainable development concept [2], and eco-design is one of the specific means. In this context it's necessary to specify that the mission of Transilvania University of Brasov (institution within which the authors of this paper are activating), stated in the first article of the Charter of the University [3], is *"to produce and transfer knowledge to society through: (1) Advanced scientific research, development, innovation and technology transfer in the field of Sustainable Development ..."*. Not incidentally the research institute of the university is called "Pro-DD" (in English "Pro-SD") and gives priority to research for sustainable development.

The quality of the bearings presented in many technical systems largely depends on their energy efficiency. Bearings are effective solutions used from a long time, already of old tradition. Scientific literature regarding bearings is particularly rich: hundreds or even thousands of books, thousands of patents, (tens of) thousands of articles published in specialized magazines and bulletins of professional conferences. Relatively few of them [4] [5] [6] [7] addresses to mathematical models to study bearings either in their entirety or on certain issues. Special mention must be given to doctoral theses in the field; major scientific papers have a higher consistency and originality. Only in Romania have been developed dozens of such work. It seems that everything has been studied until now and that something new can hardly be added.

Heavy bearings are frequent products used in wind turbines which produce unconventional "clean" energy [8]. A bearing with higher energy efficiency contributes to the overall efficiency of the wind turbines, contributing to the increase of the wind turbines efficiency and performance. This paper aims to identify courses of action to reduce the operating energy for heavy bearing, influence of the "cage" – which is the subject of a much larger research – which has higher priority. The identified actions may constitute a set of requirements for the design phase of the heavy bearings, predominantly oriented towards innovation-invention.

2. Movements of the bearing components

Representative components of a bearing are the two rings, inner and outer, intermediate rolling bodies (balls, rollers or needles) and the cage. In heavy bearings are used as rolling elements balls or rollers and the cage is often massive. At many heavy bearings, and not only, the cage is made of brass, a deficient and expensive material.

The model which describes the movements of the bearing components can be simple or complex depending on the multitude of details taken into account. In this case is sufficient a simple model, the purpose of this study is to identify courses of action - especially to redesign and/or develop new constructive solutions - that will contribute to reduce the necessary operating energy of heavy bearings. For this purpose it is sufficient that the model allows the expression of the bearing components revolutions depending on the rotary motion of the moving ring under ideal conditions of contact between the rings and the intermediate rolling bodies.

It is assumed that the components of the bearing do not have dimensional and shape deviations and all intermediate rolling bodies have identical dimensions and that none of the bearing components don't present elastic deformations.

According to the assembly into which a bearing is integrated the outer ring could be fixed and the inner ring would be into a rotary motion, rolling elements and the cage (figure 1a), or the inner ring could be fixed and the outer ring would be into a rotary motion, rolling elements and the cage (figure 1b).

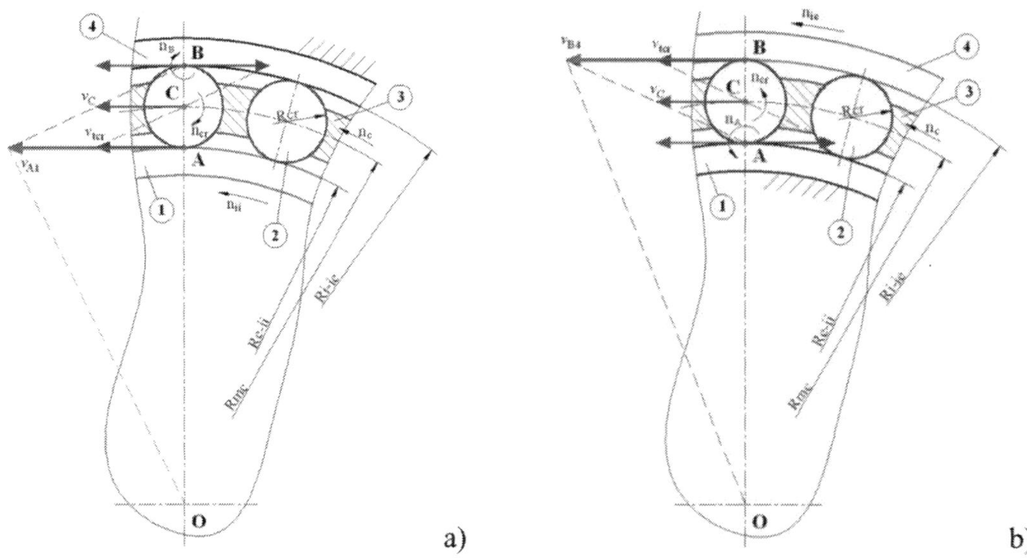

a) b)

Figure 1. Mobile components of a radial bearing and their speed / revolution.

The following measurements are of interest:

n_{ii}, n_{ie} – the revolutions of the inner ring and outer ring;

n_{cr} – the revolution of rolling bodies (balls or rollers) around their own axes of rotation;

n_c – the revolution of the cage which is identical whit the revolution of the rolling element (ball or roller) around the rotation axis of the bearing;

$R_{i\text{-}ii}$ – the inner radius of the bearing inner ring;

$R_{e\text{-}ii}$ – the outer radius of the bearing inner ring, expressed for the theoretical point of contact with the rolling elements (balls or rollers);

$R_{i\text{-}ie}$ – the inner radius of the bearing outer ring, expressed for the theoretical point of contact with the rolling elements (balls or rollers);

$R_{e\text{-}ie}$ – the outer radius of the outer ring;

R_{cr} – radius of the rolling elements (balls or rollers);

R_{mc} – the average radius of the cage;

$R_{i\text{-}c}$, $R_{e\text{-}c}$ – the inner radius of the cage, respectively the outer radius thereof;

B_{ii}, B_{ie} – width of the bearing inner ring and outer ring;

B_c – width of bearing cage.

Obvious,

$$R_{e-ii} + 2R_{cr} = R_{i-ie}. \tag{1}$$

Case A. The inner ring of the bearing is in revolution ($n_{ii} > 0$), while the outer ring is fixed ($n_{ie} = 0$), figure 1a. It is assumed that all intermediate rolling bodies are simultaneously in contact with both rings of the bearing, contact which are considered theoretically ideal and as a result the peripheral speeds of the balls and rings in the contact points are equal.

The following notations are (to review figure 1a):

v_{A1} – the peripheral speed of the inner ring 1 at the point A, point of contact with the rolling element 2;

v_{A2} – cumulated peripheral speed of the body 2 at the point A, point of contact with the inner ring 1;

v_{trc} – peripheral speed of the rolling element 2 exclusively due to its revolution n_{cr};

v_C – velocity of the center C of the rolling element 2 due to the rotation of the cage 3 with the revolution n_c;

n_B – the apparent speed of the rolling element 2 in regard to point B, instantaneous point of rotation of the body 2 in regard to the outer ring 4 which is fixed,

the next relations are obvious:

$$v_{A1} = 2\pi R_{e-ii} n_{ii}; \tag{2}$$

$$v_{A2} = 2\pi(2R_{cr})n_B = 2v_{tcr} = v_{tcr} + v_C; \tag{3}$$

$$v_{tcr} = 2\pi R_{cr} n_{cr}; \tag{4}$$

$$v_C = 2\pi R_{mc} n_c. \tag{5}$$

Using the previous relationships and imposing the condition $v_{A1} = v_{A2}$ the next expressions for revolutions n_{cr} and n_c are obtained:

$$n_{cr} = \frac{1}{2}\frac{R_{e-ii}}{R_{cr}} n_{ii}; \tag{6}$$

$$n_c = \frac{1}{2}\frac{R_{e-ii}}{R_{e-ii} + R_{cr}} n_{ii}. \tag{7}$$

Case B. The inner ring of the bearing is fixed ($n_{ii} = 0$), while the outer ring is in revolution ($n_{ie} > 0$), figure 1b. Using also the next notations:

v_{B4} – the peripheral speed of the outer ring 4 at the point B, point of contact with the rolling element 2;

v_{B2} – cumulated peripheral speed of the body 2 at the point B, point of contact with the outer ring 4;
n_A – the apparent speed of the body 2 in regard to point A, instantaneous point of rotation of the body 2 in regard to the inner ring 1 which is fixed,

the next relations are obvious:

$$v_{B4} = 2\pi\, R_{i-ie} n_{ie}; \tag{8}$$

$$v_{B2} = 2\pi(2R_c) n_A = 2v_{tcr} = v_{tcr} + v_C; \tag{9}$$

Using the relations (4) (5) (which remain valid for the specified case), (8) and (9), and imposing the condition $v_{B4} = v_{B2}$ the next expressions for revolutions n_{cr} and n_c corresponding to case B are obtained:

$$n_{cr} = \frac{1}{2}\frac{R_{i-ie}}{R_{cr}} n_{ie}; \tag{10}$$

$$n_c = \frac{1}{2}\frac{R_{i-ie}}{R_{i-ie} - R_{cr}} n_{ie}. \tag{11}$$

3. Heavy bearing operating/exploitation energy

At any time t_j, the kinetic energy $E_{cj} = E_c(t_j)$ of the bearing is the sum of the kinetic energies of its components in motion of rotation: E_{c-ii} or E_{c-ie} – the kinetic energy of the inner ring, respectively of the outer ring, as the case may be; E_{c-cr} – cumulated kinetic energy of the q rolling elements (balls or rollers); E_{c-c} – the kinetic energy of the cage. Most often the inner ring of the bearing is the ring which is in motion of rotation, hereafter this case is approached. Taking into account the relations (6), (7), (10) and (11) the referred kinetic energies have the next expressions [9]:

$$E_{c-ii} = \frac{1}{2} J_{ii}\omega_{ii}^2 = \pi^3 \rho_{ii} B_{ii}\left(R_{e-ii}^4 - R_{i-ii}^4\right) k_{ii} n_{ii}^2; \tag{12}$$

$$E_{c-c} = \frac{1}{2} J_c \omega_c^2 = \pi^3 \rho_c B_c \left(R_{e-c}^4 - R_{i-c}^4\right) k_c n_c^2 = \frac{1}{4}\pi^3 \rho_c B_c \left(R_{e-c}^4 - R_{i-c}^4\right) k_c \frac{R_{e-ii}^2}{\left(R_{e-ii} + R_{cr}\right)^2} n_{ii}^2; \tag{13}$$

$$E_{c-cr} = q\left(\frac{1}{2} J_{cr}\omega_{cr}^2 + \frac{1}{2} m_{cr} v_C^2\right) = q\frac{\pi^2}{2} R_{e-ii}^2 \left(\frac{J_{cr}}{R_{cr}^2} + m_{cr}\right) n_{ii}^2. \tag{14}$$

where k_{ii} and k_c are coefficients that take into account the particularities of the inner ring geometry and that of the cage in relation to the idealized rectangular section thereof. J_{ii}, J_c and J_{cr} are the moments of inertia in relation to their own rotation axes of the inner ring, cage and respectively of an rolling element, m_{cr} is the mass of an rolling element, ρ_{ii}, ρ_c and ρ_{cr} are the materials densities of the inner ring, cage and rolling elements.

The k_c coefficient takes into account the presence of the q pockets in the cage which are for the rolling elements.

If the rolling element is a full ball, then:

$$m_{cr} = \frac{4}{3}\pi\, \rho_{cr} R_{cr}^3, \tag{15}$$

$$J_{cr} = \frac{2}{5} m_{cr} R_{cr}^2 = \frac{8}{15}\pi\, \rho_{cr} R_{cr}^5, \tag{16}$$

and therefore the relation (14) becomes

$$E_{c-cr} = q\frac{14}{15}\pi^3 \rho_{cr}\, R_{e-ii}^2\, R_{cr}^3\, n_{ii}^2. \tag{17}$$

If the rolling element is a full roller, then:

$$m_{cr} = \rho_{cr} \pi R_{cr}^2 B_{cr}, \tag{18}$$

$$J_{cr} = \frac{1}{2} m_{cr} R_{cr}^2 = \frac{1}{2} \pi \rho_{cr} R_{cr}^4 B_{cr}, \tag{19}$$

and relation (14) becomes

$$E_{c-cr} = q \frac{3}{4} \pi^3 \rho_{cr} R_{e-ii}^2 R_{cr}^2 B_{cr} n_{ii}^2. \tag{20}$$

At any time t_j revolution of the driven ring (the inner ring in the studied case) is $n_{ii} = n_j = n(t_j)$, and the kinetic energy of the bearing is

$$E_c(t_j) = E_c(n_j) = E_{c-ii}(n_j) + E_{c-cr}(n_j) + E_{c-c}(n_j) = K \cdot n_j^2. \tag{21}$$

During its operation, the driven ring of the bearing has a variable speed; very evident in the case of wind turbine where the speed of the port-blade shaft is variable depending on the wind speed, figure 2. The main consequence is that all components of the bearing which are rotating have variable speed.

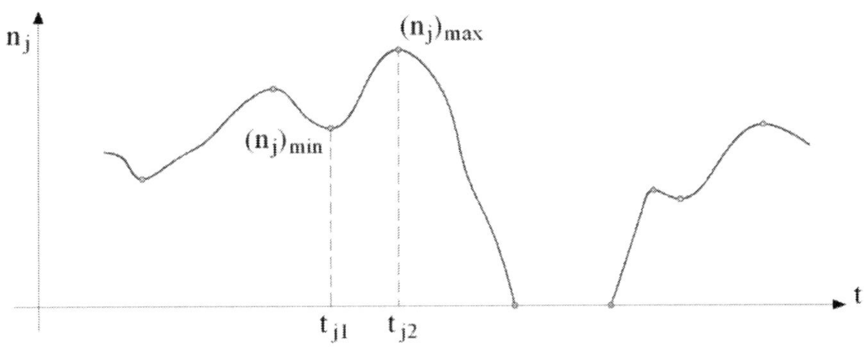

Figure 2. Graph of the time variation of a port-blade shaft speed at a wind power plant.

The operating energy of the bearing is given by the amount of energy required to accelerate the moving parts thereof.

Either $(n_j)_{\min}$ and $(n_j)_{\max}$ two consecutive extreme local values from the port-blade shaft speed chart, corresponding to different moments of time: $(n_j)_{\min} = n(t_{j1})$; $(n_j)_{\max} = n(t_{j2})$, $t_{j1} < t_{j2}$.

Obviously the n_j port-blade shaft speed becomes the speed of the engaged bearing ring. The typical case addressed in the paper is the one with the outer ring fixed and the inner ring rotating, so $n_j = (n_{ii})_j$.

$$E_{c-\exp} = \sum_j \left(E_c(t_{j2}) - E_c(t_{j1})\right) = \sum_j \left(E_c\left((n_j)_{\max}\right) - E_c\left((n_j)_{\min}\right)\right) = \\ = K \cdot \sum_j \left((n_j)_{\max}^2 - (n_j)_{\min}^2\right). \tag{22}$$

Reducing the operating energy of the bearing assumes the necessity to decrease the value of the K factor, the instantaneous n_j speed of the ring engaged depends strictly by the external conditions.

The possibility to reduce the value of the K factor associated to the engaged ring, inner or outer ring, as the case may be, it is quite low.

It is possible to reduce the value of the K factor associated to the rolling elements either by reducing the radius R_{cr} either by using hallow rolling elements [10] [11]. Reducing the radius R_{cr} entail the necessity to increase the number of rolling elements and, to a large extent, this action causes a decrease of the bearing loading capacity. The use of hollow rollers or balls is possible only if their stiffness is not affected, radius R_{cr} must be large enough. Hollow rolling elements have an important influence on the decrease of the K factor due to the subtraction of the mass thereof, and by decreasing the moment of inertia to their own axis of rotation.

The K factor can be reduced by reducing the value which corresponds to the cage. There are two possible courses of action: using a material with low density [12] (the effect would determine a simultaneous reduction, in the same proportions, of the cage mass as well as its moment of inertia) and the development of new constructive solutions with lower masses and/or inertia exclusively due to the cage geometry and not due to their realization of materials with low densities.

Obviously, it is possible to gain cumulative benefits/effects due to the use of less dense material and due to the adoption of favorable constructive solutions.

4. Conclusions

Heavy bearings constitute a frequent product often used in unconventional "clean" energy plants as wind turbines. A higher energy efficacy of the bearing increases the overall wind turbine efficiency.

At any time the kinetic energy of the bearing is the sum of the kinetic energies of its components in motion of rotation: inner or outer ring, as the case may be, the q rolling elements (balls or rollers) and the cage. At wind turbines, but not only, the engaged bearing ring has a variable speed; in consequence all components of the bearing which are engaged have variable speed. Operating energy of the bearing is given at least by the amount of energy required to accelerate the moving parts thereof.

It is unlikely to reduce the kinetic energy by changing the constructive solution of the bearing ring which is rotating.

One way to reduce the kinetic energy, which is evidenced, is to use hollow rolling elements, balls and rollers.

Reducing the kinetic energy of the bearing by reducing the kinetic energy of the cage is rarely addressed. The current paper demonstrates this possibility and identifies, with arguments, two courses of actions: using a material with low density and the development of new constructive solutions whit significantly lower masses and/or inertia compared with those of a reference cage.

References

[1] Harris J M, Wise T A, Gallagher K P and Goodwin N R (eds) 2001 *A survey of sustainable development. Social and Economic Dimensions* (Washington, Covelo, London: Island Press)
[2] Tureac I, Mărăscu-Klein V, Popescu M, Cioară R 2006 *Dezvoltare durabilă a produselor în construcția de mașini* (Brasov: Transilvania University Press)
[3] *Carta Universității Transilvania din Brașov.* Available at: http://www2.unitbv.ro/LinkClick.aspx?fileticket=hm8HJPZxM2M%3d&tabid=61
[4] Bostan V 2014 *Modele matematice în inginerie: probleme de contact. Modelări și simulări numerice în aero-hidrodinamică* (Chișinău: Techical University of Moldavia Press)
[5] Lazovic T, Ristivojevic M and Mitrovic R 2008 Mathematical model of load distribution in roller bearings *FME Trans* vol 36 no 4 p 189-196
[6] Zivkovic A, Zeljkovic M and Tabakovic S 2010 Mathematical model for the roller bearing life determination *AJME* vol 8 no 3 p 108-115
[7] Fritzson D, Fritzson P, Viklund L and Herber J 1994 Object-oriented mathematical modelling – Applied to machine elements *Computer and Structures* vol. 51 no. 3 p 241-253 (Pergamon Press)
[8] Wälzlagertechnik Industrietechnik 1990 (FAG) 501
[9] Deliu Gh 2003 *Mecanica* (Cluj-Napoca: Albastră Publishing House)
[10] Barabaş S A 2010 *Cercetări privind ingineria rulmenților de mari dimensiuni cu scopul reducerii maselor inerțiale și a sporirii eficienței în exploatare* PhD thesis Transilvania University of Brașov
[11] Cioară R et al. 2015 *Bilă pentru rulmenți grei* Patent RO 125679
[12] Csaki I 1999 *Cercetări privind fabricarea coliviilor de rulmenți din materiale compozite ușoare cu parametrii tribologici optimizați, solicitate de industriile de aeronautică și ușoară* PNCD 45 grant C 5076

Studies about the Behavior of the Crash Boxes of a Car Body

B A Constantin[1], D Iozsa[2] and G Fratila

POLITEHNICA University Bucharest, Automotive Department, Bucharest, Romania

E-mail: constantin.bogdan.alexandru@gmail.com[1], daniel_iozsa@yahoo.com[2]

Abstract. A continuous evolution of requirements and standards sheds over the development of new vehicles (for example EuroNCAP ratings) in order to create competition between same market models customer related. The low speed impact protection has to be permanently improved as the damage of the front end structure of the vehicle to be reduced to minimal. As a consequence, a lower damage implies less repair costs and therefore a lower insurance category. The front end structure, including the bumper, responds for the absorption of the kinetic energy created during the impact with maximum efficiency in order to avoid the large deformation of structural components. This is only one of the constraints that the front end structure has to cope with, additionally we can mention the dimensioning of the front end of the vehicle which can affect the packaging, which is mainly influenced by the design, styling and the pedestrian requirements intended to be accomplished by the vehicle. The present paper focuses on the low speed urban impact, offering an overview over the actual state, the load configuration, the applicable regulation, the challenging requirements of a modern front structure, which the modern bumper has to comply with and the finite element simulation of this kind of test.

1. Introduction

Recently, based on the help of advanced development of software and hardware equipment for numerical simulation, the period of time in which a project is finished and a new car is launched on the market has become smaller and smaller. The competition on the automobile market led constructors to seek, apply and improve the latest techniques in the car manufacturing.

The numerical simulation has gained more and more terrain facing the need of cost efficiency and rapidity of the project development. After the manufacturing, a car has to pass in the first place the requirements of the homologation agencies and secondly, the very popular ranking tests (EuroNCAP). Potential problems, which can affect the quality of the product over its life are identified and removed during the project phase.

Using virtual prototyping and numerical simulation, we can improve the performance and the cost of the part before is it actually built. In addition to the numerical test, a physical one is carried out in order to validate that the part meets the requirements. As a consequence, the need to build several sets of physical prototypes of the parts has decreased to a very small number, thus saving time and money.

The advantage of numerical simulation over the physical test consists in observing immediately if one part of the assembly does not comply with the specifications, rather than following an expensive testing procedure and waiting between the test and the post-processing of the results. Thus, we can define the needed adjustments and rerun the simulation until we obtain the desired results. More precisely, while waiting several days for the physical test results for one crash configuration, we can numerically test hundreds of parameters simultaneously while observing in real time the global effects.

If it shows that with the current front bumper design it is impossible to attain the required performances, a geometry change can be proposed [9].

The passive safety load cases can be divided into: high speed crashes for front, side and rear impacts, occupant and pedestrian protection and low speed crashes which are mainly performed for insurance classification purposes. In order to ensure permanent evolution of pedestrian protection requirements significant structural changes of the vehicle front structures are mandatory. All these modifications applied to the front end are in opposition with the performance of the other safety load cases [4]. Therefore, for the automotive engineers it represent a never ending challenge to develop a front end structure that will attend the optimal performance for all the load cases.

The requirements of low speed impacts are imposed by several national and international standards. Some of these are mandatory to be respected in order to homologate the vehicle (CMVSS 215 in US [10] / ECE-R42 in Europe), others to evaluate the insurance level (RCAR).

2. Requirements of modern bumpers

Favorable insurance categories, sufficient deformable zones for pedestrian protection and conflicting low speed legislation aspects require an effective design of bumper systems. The modern bumper system has the function of absorbing the energy of the low speed impact, thus avoiding higher vehicle damage. The energy absorption capacity of the bumper system during a crash is evaluated by the load-displacement response. The area under the load-displacement curve is a measure of the energy absorbed. During a low speed impact, the bumper system has the function of preventing damage to the body in white. Hence, the maximum impact load transmitted through the system has to be limited. The maximal displacement is specified by the vehicle design. The ideal bumper system has a load-displacement response which acts as a step function. The load rapidly reaches the maximum value and remains there throughout the crash.

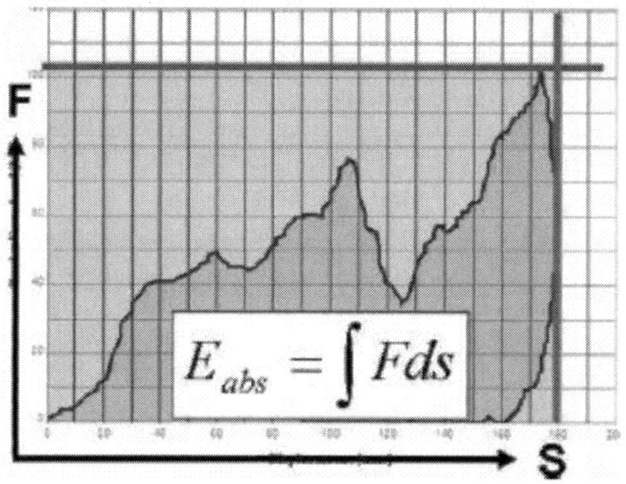

Figure 1. Example of force-displacement curve [8]

The modern screwed bumper systems are composed of a bumper beam and two energy absorbers (e.g. crash box). The old bumper systems were welded to the BIW instead of being screwed. The picture below shows the differences between the two solutions. The advantage of the modern bumper system is the possibility of replacing each part individually in the event of a damage without to pass in the process of cutting and welding. The energy absorbing crash box is placed between the bumper beam and the front rail of the vehicle structure. The crash box has to crush in a specific way, so that the forces and bending moments transmitted to the front rail are limited and the rail does not undergo any significant plastic deformation. As a result the rail does not need to be replaced. In parallel, bot cosmetic damage (e.g. fenders, headlamps, bonnet, etc.) and the damage to the expensive mechanical parts (e.g. radiator, condenser, engine bay components, etc.) have to be kept to the lowest possible [2],[3].

New bumper systems have some disadvantages also: for example, not only does the production complexity increase, but also the cost, weight and vehicle front console. This dimension is caused by the increased distance between the bumper beam and the engine bay components (in order to avoid damage). Additionally, the implementation of deformation elements in the front bumper beam increases the required space for packaging. A third factor responsible for the large front console of the vehicle is the necessary space between the rail and the bumper fascia. Accomplishing the optimal energy management is hard, despite the desire of having a compact front end, required by regular styling reasons [6].

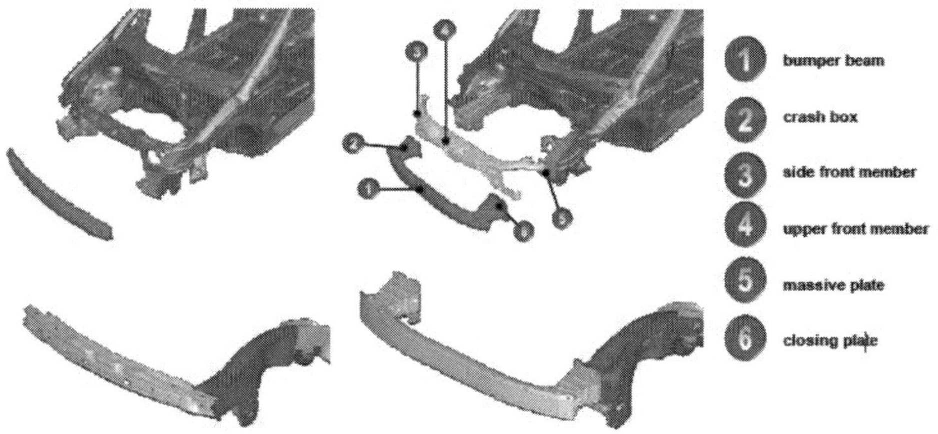

Figure 2. Front end vehicle structure - old/new [8]

As mentioned above, low speed impacts are regulated by an amount of different legal requirements and insurance classification test. It is a challenge to design the most efficient bumper system for all existing load cases simultaneously, because the requirements are in some cases at opposite sides, for example between Europe and North America. While the bumper beam and the crash box should be designed stiff in order to pass the pendulum tests for North American market, the crash box has to be able to crush in the Danner test, without any plastic deformation of the front rail. To solve this problem one alternative used by the global manufacturers is to have different designs and systems for both North American markets and European markets.

Structural changes to the vehicle front end structure to achieve pedestrian protection demands can affect adversely the low speed performance. In order to ensure the required protection for pedestrians, the vehicle must absorb the impact energy by means of a deformable soft structure which has sufficient deformation space. For lower leg protection, an optimized low-density foam in front of the bumper cross member can be integrated. This foam reduces the efficiency of low speed bumper system, so that the barrier intrusion is higher. Hence, without further measures, the vehicle damage and the repair costs would increase, which would intensify the insurance classification.

An effective design of the bumper system must also fulfill certain requirements regarding high speed crashes. During a high speed impact, the crash box has to crush first and the front rail should be able to absorb most of the deformation energy. Due to RCAR test, the crash box is not calibrated in this manner and cannot be efficiently used for high speed crash.

3. Low speed finite element simulation

For researching the low speed impact it was used a finite element model compiled with the explicit solver PAM-CRASH, well known as one the reference software of the automotive industry. This

IOP Conf. Series: Materials Science and Engineering **160** (2016) 012010 doi:10.1088/1757-899X/160/1/012010

model is composed of a simple front beam made from steel with a thickness of 1mm. The interface with the rails of the vehicle structure is chosen to be as the latest solution used, with two crash boxes which have the same characteristics as the front beam. The two crash box elements were designed in several simulations as front rails especially to observe the contrast between the two types of technical solutions and to evaluate the performances.

The striker used is the same as the one described in the Regulation [1], presented also below in the study. It's considered a rigid body [5], guided along the X axis of the vehicle, with a speed of 4km/h and laden with a mass of 1500kg (which can be approximately the mass of a modern vehicle). At the level of the crash boxes there were added TIED contacts in order to define the interface with the crossbeam. At the level of the rail, the crash boxes are considered blocked on all directions, as a hypothesis that deformation may not affect the front structure. The simulation was conducted in three positions: centered, offset and corner.

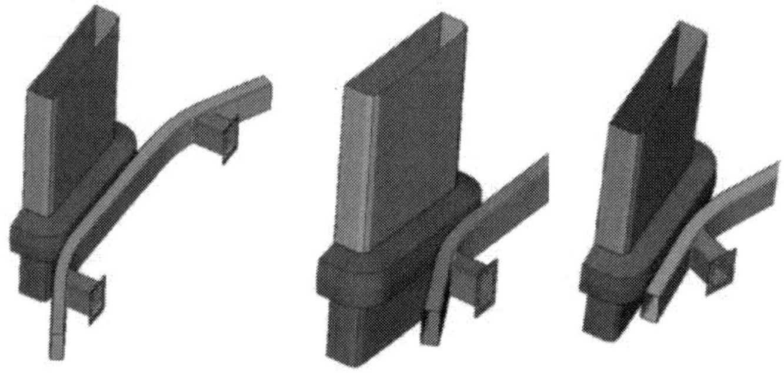

Figure 3. Finite element model – initial condition

The configurations used in the simulations are described in the list below:

1. Classic front end – centered impact
2. Classic front end – offset impact
3. Classic front end – corner impact
4. Modern front end – offset impact / crash box type 1
5. Modern front end – corner impact / crash box type 2
6. Modern front end – centered impact with pedestrian absorber [7]

Below, the results of these simulations will be briefly analysed. It was chosen as a hypothesis a maximum limit of 5% for plastic deformations. (We can consider that above this limit the plastic deformation it becomes visible on the physical parts)

For the 1st impact configuration, we can observe in the figure below that the front beam was deformed well above the limits and we can conclude that the damage cannot be repaired.

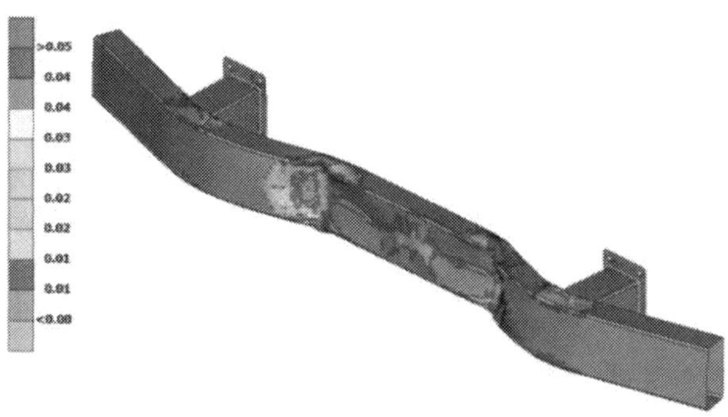

Figure 4. Classic front end – centred impact

The rails, as it were designed in this simulation, were not affected by the impact, so we can conclude that is necessary to design a front absorber in order to reduce the plastic deformation of the crossbeam.

Similar results can be seen in the 2nd and 3rd configuration, the load case being in the offset/corner position. In this case, as clearly appears it's mandatory to develop a crash box structure.

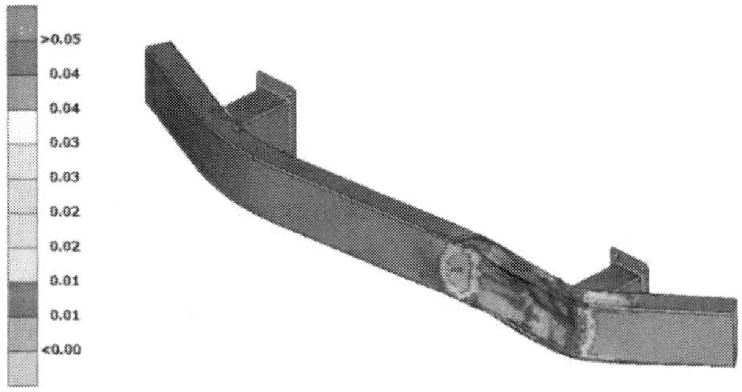

Figure 5. Classic front end – offset impact

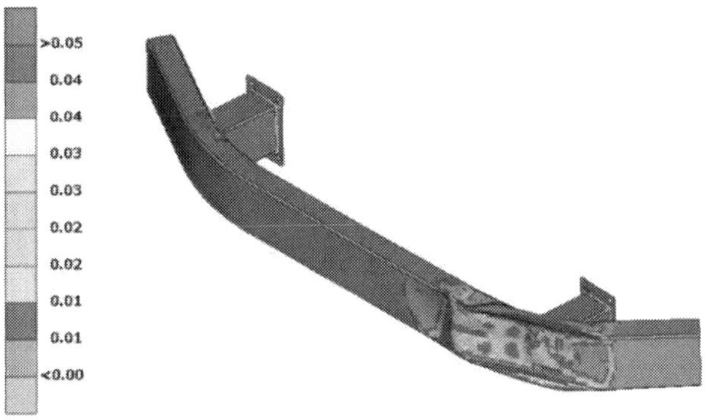

Figure 6. Classic front end – corner impact

The behavior of the crash box was evaluated in two manners described below. In the first it were made cuttings on the part in order to start the deformation. In the second solution, the geometry of the part was reconstructed and also the thicknesses were updated. Many types of technical solutions that are already patented can be applied in order to have the same result. Current trends in bumper design find this kind of part added by the manufacturer in standard equipment of the vehicle [6].

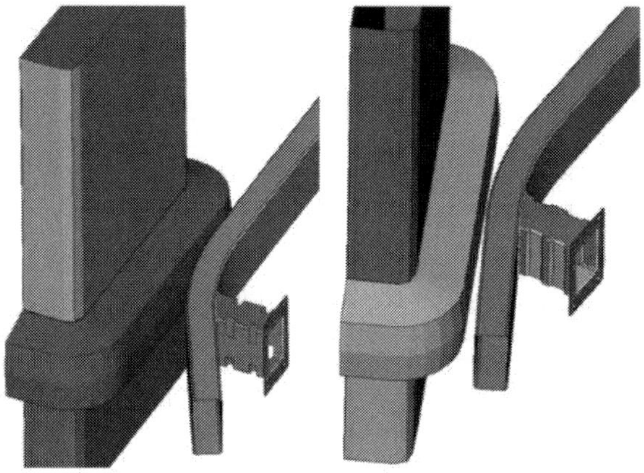

Figure 7. Crash box definition used in simulation

The results obtained are better than the ones of the classic solution, but the plastic deformation still exceeds the requirements imposed.

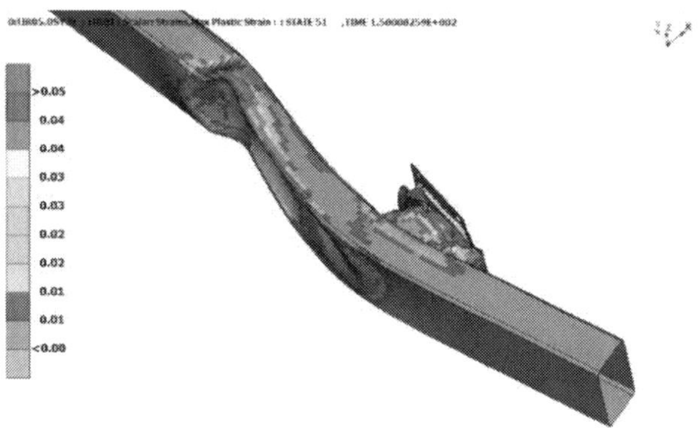

Figure 8. Modern front end – offset impact / crash box 1

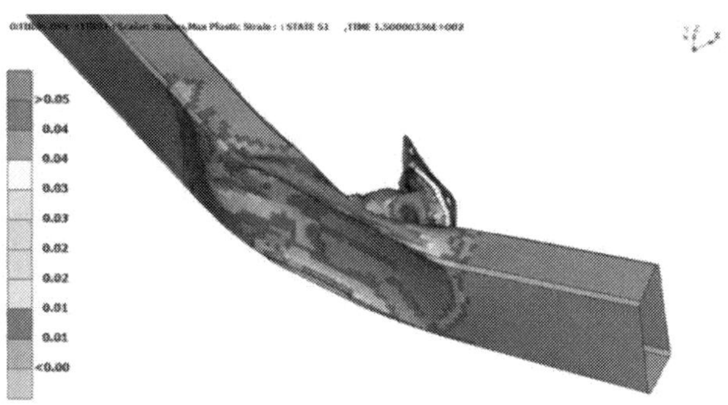

Figure 9. Modern front end – corner impact / crash box 2

In the last case it was designed a pedestrian type of absorber in front of the crossbeam. The material used it was a specific one for this type of application, a PP/PE thermoplastic. The absorber is 100mm long in X axis, with the exterior thickness of 3mm and interior rib thickness of 2.5mm. Additionally, in the centre section it was added a crossbeam support that in modern BIW is welded between the crossbeam and the front end cover, because of the lack of rigidity detected in previous simulation.

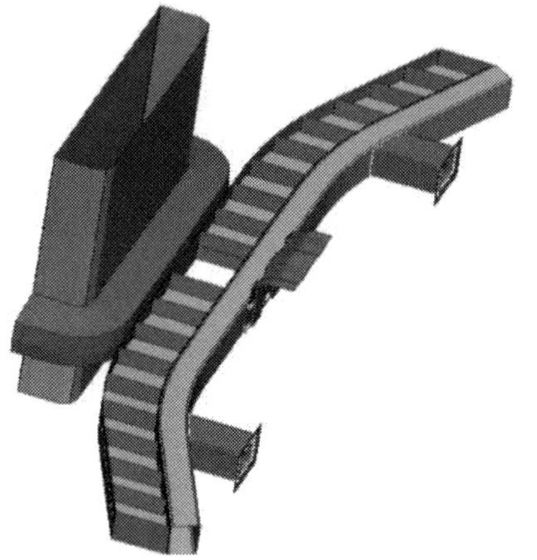

Figure 10. Modern front end – centered impact with pedestrian absorber

In this case it can be seen in the figure below that the energy absorption is made by the front absorber. The deformations at the level of the crossbeam are almost not visible.

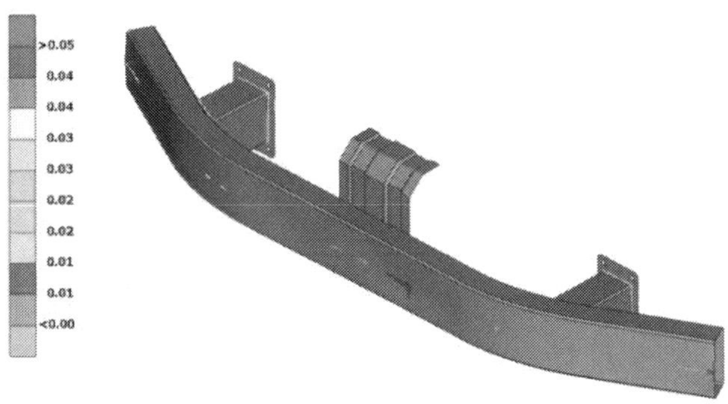

Figure 11. Modern front end – centered impact with pedestrian absorber (front beam plastic deformation)

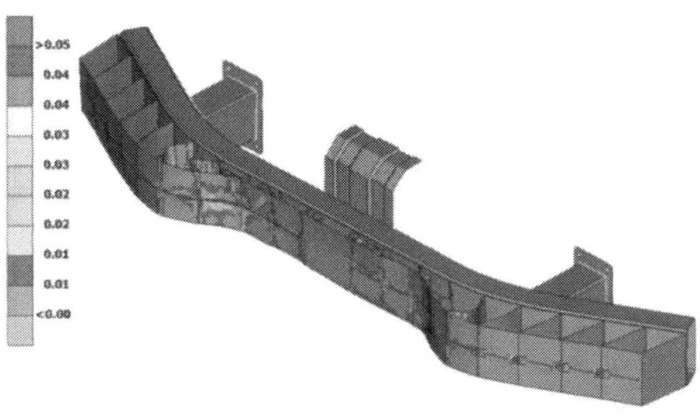

Figure 12. Modern front end – centered impact with pedestrian absorber (front absorber plastic deformation)

Furthermore it was conducted a curve analysis of the contact force for all the simulations launched.

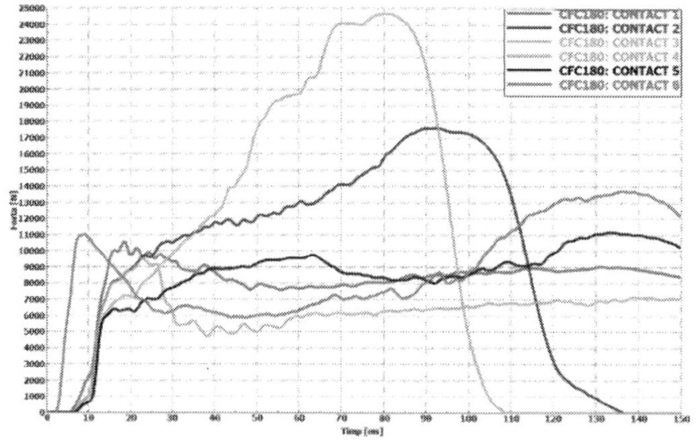

Figure 13. Contact force curves

It is visible that the characteristics are almost the same for the models 1, 4, 5 and 6, the energy absorption being made properly, but in different manners and by different parts. The curves 2 and 3 show that the two crash box that were considered as front rails, were implying the proper manner of absorption, giving a high peak and physically a large deformation of the crossbeam.

This fact can be explained by the internal energy of the each crossbeam during each impact. The first three curves (red, blue and green) show that the part is strongly deformed. The next two curves (black and orange) show that the stresses at the level of the beam were reduced. The final model that represents the modern solution of front end system shows a little influence over the crossbeam, concluding that only the plastic parts were to be damaged, and the cost of the repair being in this case minimal.

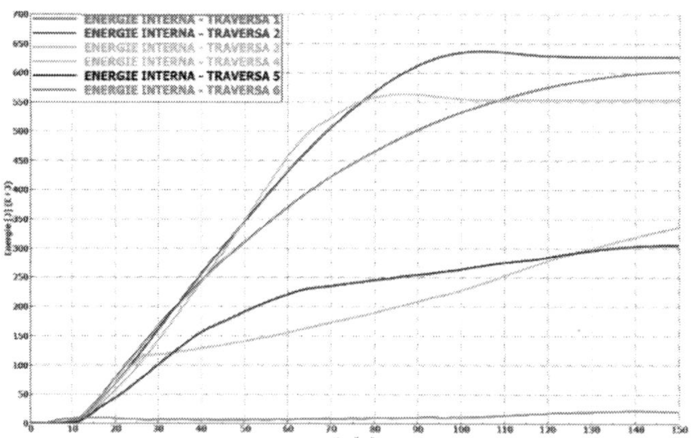

Figure 14. Crossbeam internal energy curves

4. Conclusion

The different low speed urban impact legislations and insurance tests represent a recently and strongly demanding challenge for the automotive industry, and it's known that in the near future it will be more difficult as the severity criteria will increase.

By adding a crash box on each front rail and a plastic or foam absorber on the front cross member it can be reached a compromise in order to comply with low speed impact, high speed impact and pedestrian impact requirements. The level of force transmitted to the front structure can be reduced by approximately 3 times in the case of the vehicle equipped with the parts mentioned earlier. Also, in the majority of the load cases, the worst case scenario is that the front bumper fascia and the front absorber might be replaced. (in typical impact the plastic deformations cannot be observed)

It's mandatory an optimization to be conducted related to front end structure in order to assure all the load cases and styling visions.

The key factor towards to improvement of the front end structure systems performance remains the numerical simulation in upstream phase of the project.

References

[1] ECE R-42 Regulation, European Union, 1980
[2] RCAR bumper test, Research council for automobile repairs, 2010
[3] RCAR Low-speed structural crash test protocol, Issue 2.2, Research council for automobile repairs, July 2011
[4] Front impact absorption system requirements. Part technical specifications. Politecnica University of Catalunya, Barcelona, 2011
[5] Modeling of low speed impact, PhD R. Brach, University of Notre Dame, USA, SAE, 2003
[6] Schuster P., Current trends in bumper design for pedestrian impact, SAE, 2006
[7] Lawrence G.J., Study on feasibility of measures relating to the protection of pedestrians and other vulnerable road users, TRL Ltd, 2004
[8] Ramon-Villalonga L., Enderich Th., Advanced simulation techniques for low speed vehicle impact, Adam Opel GmbH, LS-DYNA Anwenderforum, Frankenthal, 2007
[9] Burnaz E., Giubalca A., Pedestrian crash – Leg/Front bumper analysis, Renault Technologie Roumanie, The annals of Dunarea de jos University of Galati, Fascicle V, Technologies in machine building, 2009
[10] Bumper test and rating protocol, Insurance Institute for Highway Safety, U.S., September 2010

The use of statistical process control to improve the accuracy of turning

Cristian Pisarciuc

Transylvania University of Brasov, Romania

E-mail: pisarciuc.c@unitbv.ro

Abstract. The present work deals with the turning process improvement using means of statistical process control. The approach on improvement is related to the fact that several methods are used in order to achieve quality defined by technical specifications. The experimental data is collected during identical and successive manufacturing processes of turning of an electrical motor shaft. The initial process presents some difficulties because many machined parts are nonconforming as a consequence of reduced precision of turning. The article is using data collected in turning process, presented through histograms and control charts, to improve the accuracy in order to reduce scrap.

1. Introduction

Quality, in manufacturing processes, is defined in concordance with technical specifications. However, there are no two identical products, since all processes are containing multiple sources of variation. Traditionally, in mass production, products quality is assured by postproduction inspection. Each product or batch can be accepted or rejected depending on how much closer is to design specifications. In contrast, Statistical Control Process - SPC use statistical methods to observe the performance of manufacturing process in such way that major variations to be detected before they translate into defective products.

The most recent development in statistical quality control - SQC methods is Motorola's Six Sigma [1]. According to traditional SQC, a process is in control if the variation measured in standard deviations (σ or sigma) is less than one-third the difference between the control limits and the process mean; i.e., the traditional distance between the mean and any control limit fall within the 3-sigma limits. However, Motorola concluded that quality leaders achieved fewer defects by reducing process variations. They estimated that defects would be cut 1000-fold if the process variation could be held to one-sixth the difference between process mean and control limits, or 6σ variation control [2].

The term six sigma (6σ) initiated as a performance measure or a measure of quality. Using six sigma, process goals are set in parts per million in all areas of the production process. In constant development since its origin, six sigma has now evolved into a methodology for improving business efficiency and effectiveness by focusing on productivity, cost reduction, and enhanced quality [3]. The name six sigma comes from the statistical use of the sigma (σ) symbol, which designates standard deviations. The six identifies the number of standard deviations around the mean. A process is six sigma when, the values of collected data for process stability analysis, is at distance 3σ from upper specification limit (USL) and lower specification limit (LSL) and the average absolute data set is at a distance 6σ from LSL to USL (assuming that σ is the standard deviation). Both Statistical Process

Control (SPC) and Statistical Quality Control (SQC) are tools utilized by a six-sigma process. In conclusion, either SPC or SQC are, in the end, methods of reducing the costs associated with production activities.

Sources of variability detected in a process can be classified as:

- Common cause variations that refer to multiple sources of variation that causes constant effects on processes;
- Special cause variations are caused by such factors that affect only some of the processes being most often intermittent and unpredictable.

Identifying these causes is associated to the possibilities of revealing the scrap occurring in manufacturing processes. This is the reason why SPC is used frequently. Motivation is related to the fact that SPC generates savings by reducing the number of interventions in the process, or reducing the waste caused by faulty adjustment operations.

Variation data are monitoring through control diagrams. They are used to highlight common and special causes of variations. The data collection activity and drawing up control charts should have a continuing character in time because when the process does not trigger any detection rule (from control graphics) then the process is stable. The histograms are used as they are the graphical representation of a phenomenon, in this case, process stability. Consequently, effective planning of product quality involves measuring the capability of the production process, which implicates determining the indices Cp (Process Capability) and Cpk (Process Capability Index). In order to assess the process capability the minimal values recommended for Cpk are 1.33 or 1.67 for safety or critical parameter [4].

2. Production process description

Rotor subassembly is the main component of an electric motor, ensuring the transfer of mechanical energy from engine to connected devices. Good quality of the production is vital to the long-lasting evolution and to parameters of the final product, the electric motor. Figure 1 shows a cutaway view through an electric motor highlighting the important parts, shaft being the one that will be analysed in this article.

Figure 1. Cutaway view of an AC motor. Adapted from [5]

The shaft passes to several common machining stages until last one being the drive shaft finishing, more exactly the final processing for bearings mounting diameters (on both drive and non-drive end), finishing the sealing beds and the drive end, where usually is already machined a keyway.

Attaching rotors to CNC machine (Polygim PLG-52) [6] is done using two centres method for faster change. The machines are fitted with a headgear drive mounted in chuck and centering devices of different sizes in order to accommodate all kind of shaft sizes. On tailstock are used ROHM OHM MK5 live centres [7] (revolving), as presented in Figure 2.

For finishing operations of diameters on shaft, Sandvik tool holders are used, model SVJBR 2020K, to support VBMT 16 04 04-PF 4325 [8] cutting inserts (Figure 3).

Figure 2. Shaft mounted before machining

Figure 3. Tool holder and cutting insert used for finishing

3. Initial assessment

The specific shaft model selected for analysis was the one with 15 mm diameter for inner ball bearing ring and 14 mm at drive end. The 15 mm diameter is also the one with the tightest tolerance (+0.009, +0.001).

In order to assess the initial process stability of shaft's finishing, 50 parts were analysed and measured with a micrometre calibrated every five measurements.

In the initial phase, the machining regime for cutting inserts was set-up according to manufacturer recommendations (feed 0.1 mm/rot (0.05-0.2), speed 460 m/min (395-480)) [8], as follows: cutting feed 0.12 mm/rot, cutting speed 425 m/min and spindle speed 3800 rot/min.

The data was collected using an Excel spreadsheet template available from [9]. Using the formulae, already embedded in Excel spreadsheet, the results for statistical control parameters were as follows:

- for 14 mm diameter for drive shaft (+0008, -0.003) (DS), Cpk = 1.168
- for 15 mm ball bearing diameter at drive end (+0.009, +0.001) (DE), Cpk = 0.895
- for 15 mm ball bearing diameter at non-drive end (+0.009, +0.001) (DN), Cpk = 0.862

The first plain conclusion is the fact that process is not adequate from quality point of view, and needs substantial improvements. However, it is interesting to note the importance of tolerance in Cpk's results. The DS diameter of 14 mm having 13 micron tolerance has a better result against DE and DS diameters that have only 8 μm tolerance range, although they were machined with the same cutting conditions and cutting tools. The histograms that illustrate graphically the initial situation are presented in Figures 4, 6, and 8. It can be seen with ease that finishing processes are centred. However, it requires increasing the precision of processing, in order to decrease the number of not conforming parts.

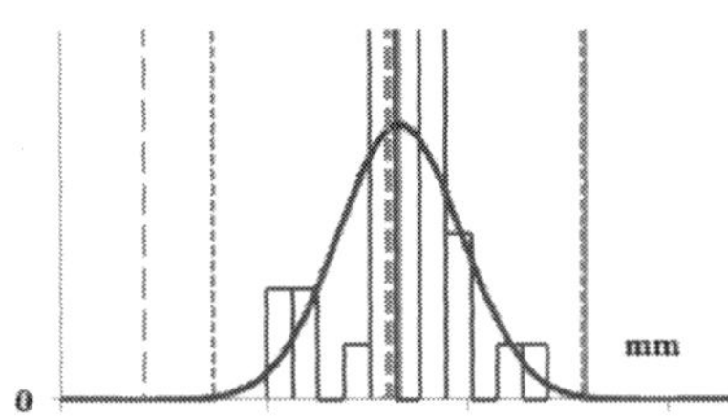

Figure 4. Histogram for DS
ϕ14 (+0.008, -0.003)

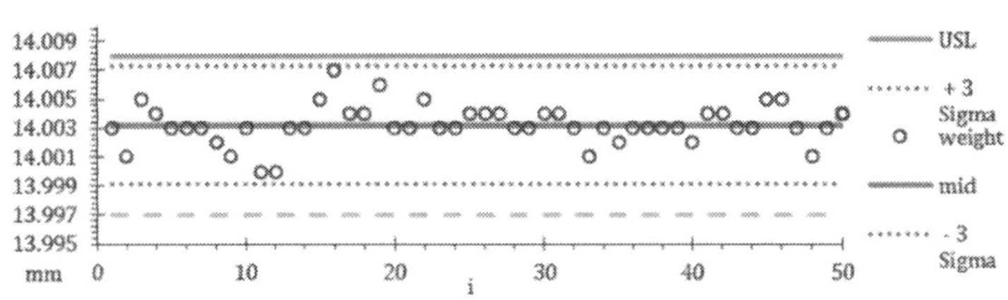

Figure 5. Control diagram for DS
ϕ14 (+0.008, -0.003), initial situation

Analysis of control charts is attesting the fact that there are no unexpected external or disruptive elements on the system. Major influence on the processing capability analysis result is thus internal, most likely processing technology, cutting conditions, or the tools and devices do not meet the process requirements (Figures 5, 7, and 9).

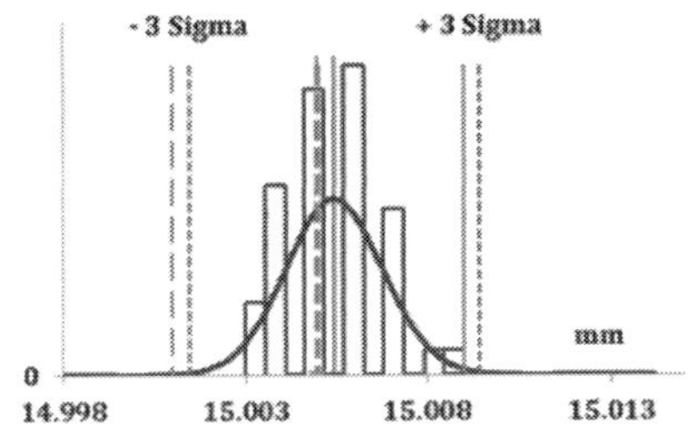

Figure 6. Histogram for DE
ϕ 15 (+0.009, +0.001)

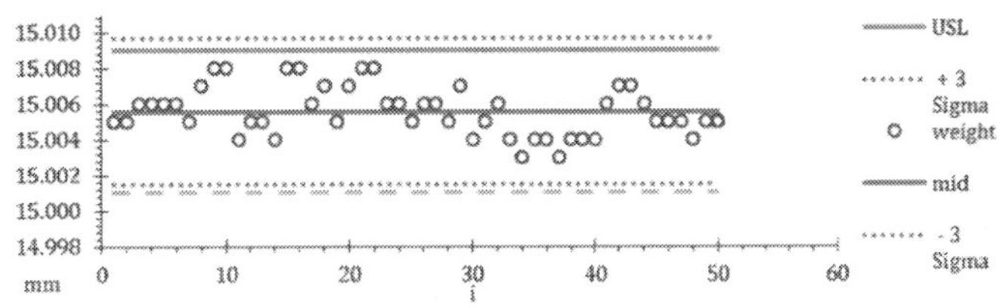

Figure 7. Control diagram for DE
ϕ 15 (+0.009, +0.001), initial situation

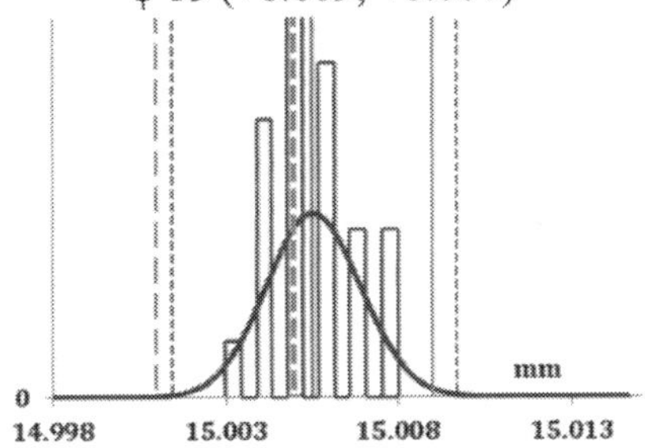

Figure 8. Histogram for DN
ϕ 15 (+0.009, +0.001)

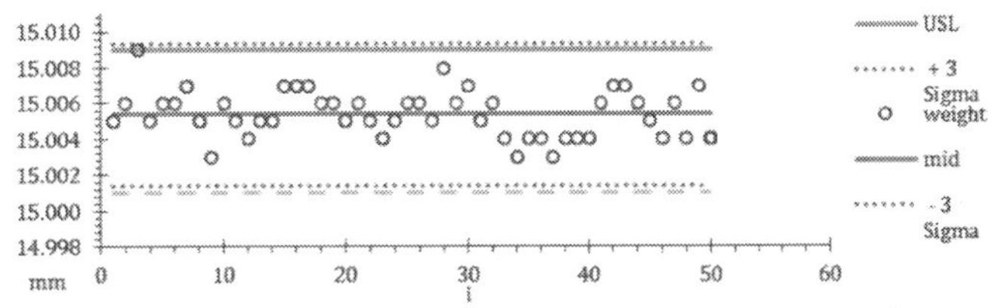

Figure 9. Control diagram for DN
ϕ 15 (+0.009, +0.001), initial situation

4. Experimental research

4.1. Feed turning modifications

The machining parameters for shaft finishing, as mention above, were cutting feed 0.12 mm/rot, cutting speed 425 m/min, and spindle speed 3800 rot/min. Given that the maximum spindle speed for lathe is 4.000 rot/min, raising this speed is not a recommended solution. In the same way, modifying the cutting speed over or under the recommended value of 425 m/min is neither indicated. The only substantial change that could be relevant is related to feed and was decided to test with values of 0.16 mm/rot and 0.08 mm/rot (plus and minus 0.4 mm versus initial parameter). For 0.16 mm/rot the results obtained were:

- for 14 mm diameter for drive shaft (DS), Cpk = 0.79
- for 15 mm ball bearing diameter at drive end (DE), Cpk = 0.72
- for 15 mm ball bearing diameter at non-drive end (DN), Cpk = 0.92

Histograms for feed = 0.16 mm/rot are presented in Figures 10-11.

Straightaway could be noticed that there is a difference between the results for diameter from non-drive end and the drive bearing diameter, the difference that was not seen in the initial tests. Histograms show that the distribution is no longer centred for the two diameters on the drive, and the cutting tip wear has increased significantly here. On non-drive end could be noticed a little higher value (0.86 compared to 0.92), but this is not so relevant, as the machining precision is not influenced.

The results were not as favourable, from process stability point of view, but based on them could be drawn some relevant conclusions about other influences in terms of machining precision. In this stage, only the open keyway could explain the higher wear of cutting inserts. The conclusion can be considered authentic because the two diameters, 14, and 15 mm, are machined by the same tool, in the same cutting conditions but with tools that are mounted in different stations. Consequently, it was discovered a major influence on processing accuracy, more precisely the interrupted turning on drive side of the shaft. Histograms for feed 0.16 mm/rot, shows this effect as the distribution is on the right side, due to wear of cutting tool. Control diagrams presents, more clearly, the wear of cutting insert (Figures 12-13).

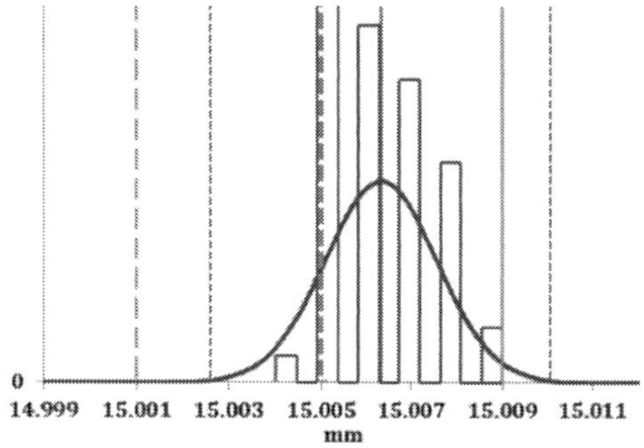

Figure 10. Histogram for DE
ϕ 15, f = 0.16 mm/rot

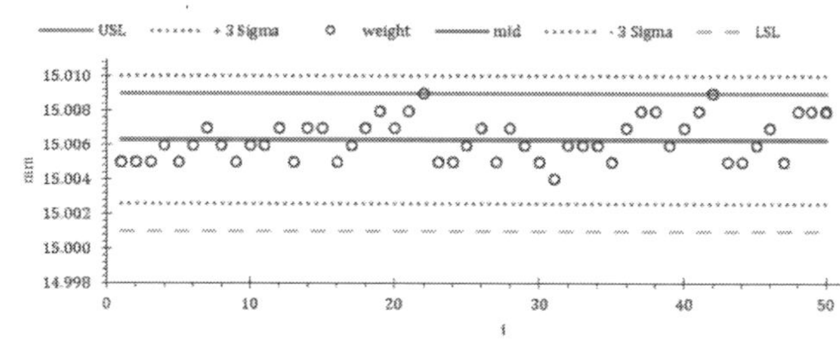

Figure 12. Control diagram for DE
ϕ15, f = 0.16 mm/rot

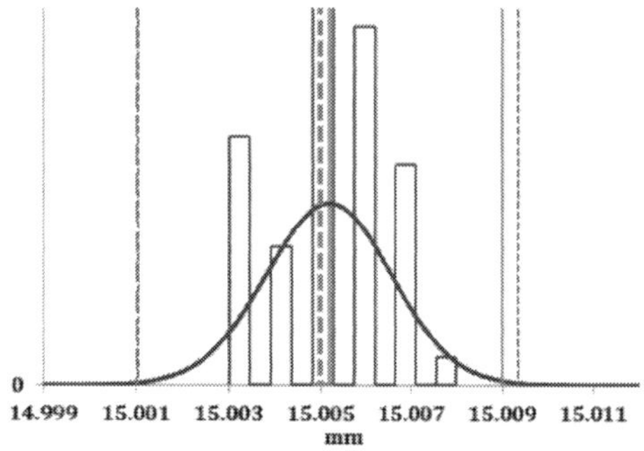

Figure 11. Histogram for DN
ϕ 15, f = 0.16 mm/rot

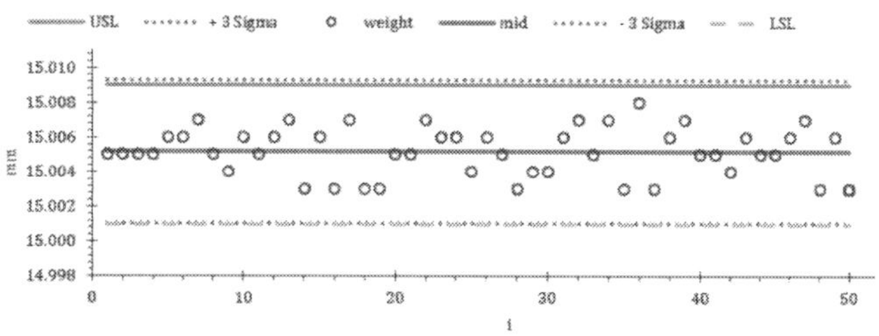

Figure 13. Control diagram for DN
ϕ15, f = 0.16 mm/rot

For feed 0.08 mm/rot the results obtained were:

- for 14 mm diameter for drive shaft (DS), Cpk = 1.10
- for 15 mm ball bearing diameter at drive end (DE), Cpk = 1.03
- for 15 mm ball bearing diameter at non-drive end (DN), Cpk = 1.09

This is the first positive result of the tests, but only for the finishing of 15 mm diameters. The Cpk for drive side of shaft remains the same as the initial situation. The process reaches the natural 3-sigma threshold. Histograms (Figures 14-16) and control charts (Figures 17-19) show a better positioning of the average diameter value on 15.005 mm and 14.003 mm. In order to bypass the problem for 15 mm diameter the only solution is to remake the technological order of machining phases and put the milling of keyway as a last operation. The operation will be done on the same lathe as well.

4.2. Change of cutting inserts

Because the original cutting tools are common carbide turning inserts, in order to maintain the actual productivity, or to raise it, is possible to use cubic boron nitride - CBN inserts. The high hardness of CBN enables it to machine hard materials at high speeds and/or feeds. CBN is chemically inert in ferrous materials and retains its hardness at temperatures over 1000 °C. The variant chosen was Korloy's CBN VBGW160404 [10]. The manufacturer recommends setting to 3200 rot/min the speed of rotation and the feed to 0.18 mm/rot. The process proved to be more stable using these inserts, with 0.3 comparing to the initial situation. Except for some isolated cases, the values have remained at the distance of one micron from the mean value, explicit results being:

- for 14 mm diameter for drive shaft (DS), Cpk = 1.22
- for 15 mm ball bearing diameter at drive end (DE), Cpk = 1.20
- for 15 mm ball bearing diameter at non-drive end (DN), Cpk = 1.28

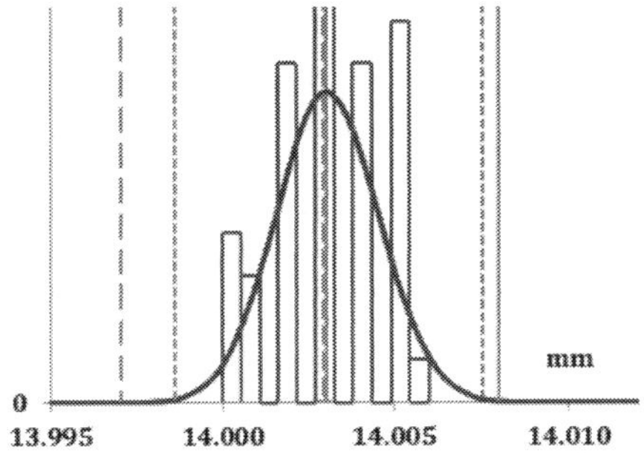

Figure 14. Histogram for DS ϕ14, f = 0.08 mm/rot

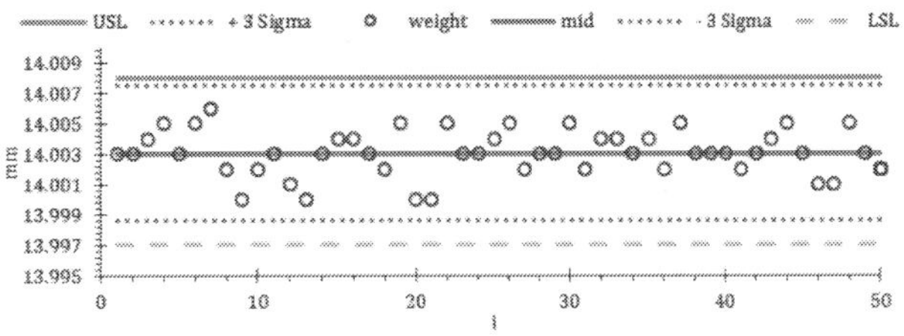

Figure 17. Control diagram for DS ϕ14, f = 0.08 mm/rot

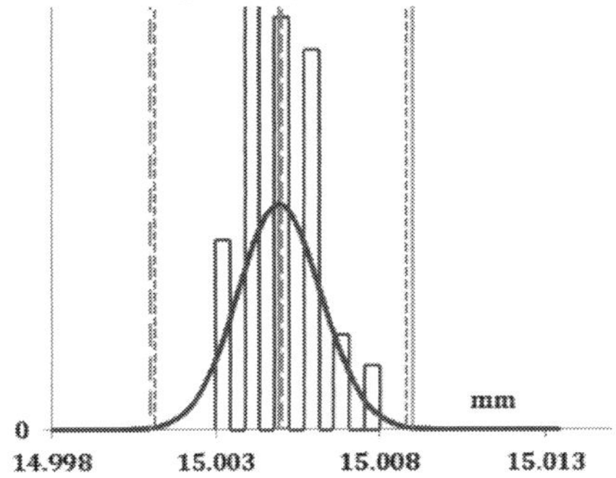

Figure 15. Histogram for DE ϕ 15, f = 0.08 mm/rot

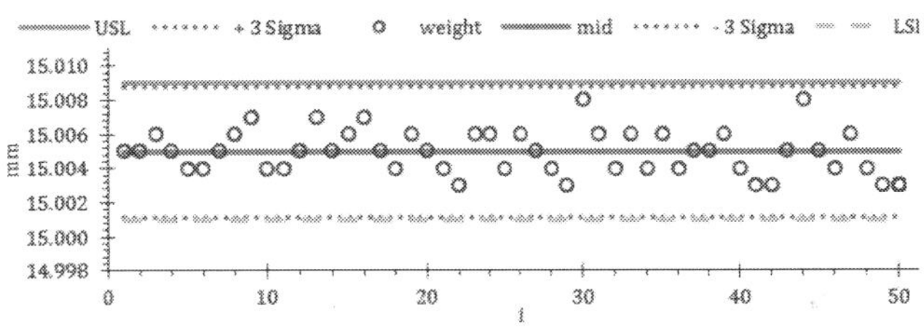

Figure 18. Control diagram for DE ϕ15, f = 0.08 mm/rot

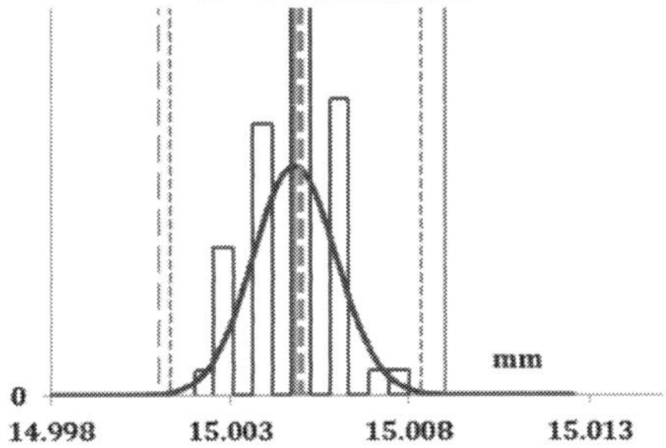

Figure 16. Histogram for DN ϕ 15, f = 0.08 mm/rot

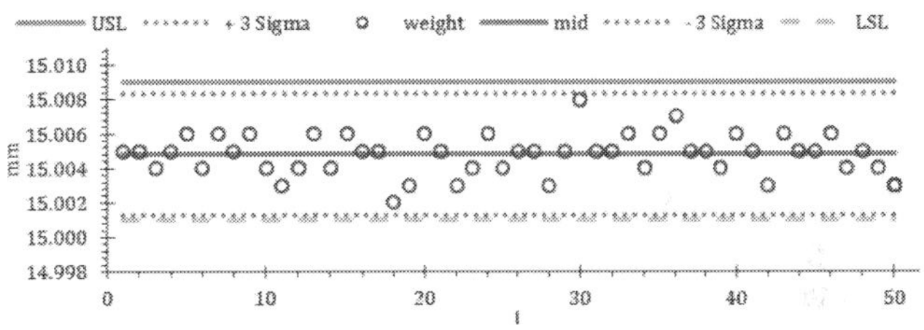

Figure 19. Control diagram for DN ϕ15, f = 0.08 mm/rot

Histograms, in figure 20, and control chart, in figure 21, displays a process more centred, but there are values that appear to be caused by an external factor at non-drive end. The last test demonstrated the influence of an external element in the process. It was discovered that there is great variability on shaft roughing, operation executed on other machine tools. In order to bypass this situation, another set of experiments were performed, more precisely the roughing was performed in a way that the final value for drive end was 17 mm (the machining on CNC lathe is started from this diameter). After this modification, the influence of previous operations was eliminated.

5. Conclusions and further development

First possible solution, in order to overcome the dysfunctionalities identified, involves modifying the technological stages in order to put the milling of the keyway last and lowering the feed to 0.08 mm/rot. This solution does not involve any additional costs because it uses the same cutting inserts but cause a decrease of overall productivity due to longer finishing times.

Other possible solution implies as well modifying the technological phases, putting the keyway machining as a last operation, roughing only to 17 mm diameter, using the VBMT 16 04 04-PF 4325 inserts for machining the drive end using a 0.08 mm/rot feed, and CNB VBGW160404 for machining non drive end. As a result the machining process is conform but with the cost of longer finishing time, and with medium manufacturing costs.

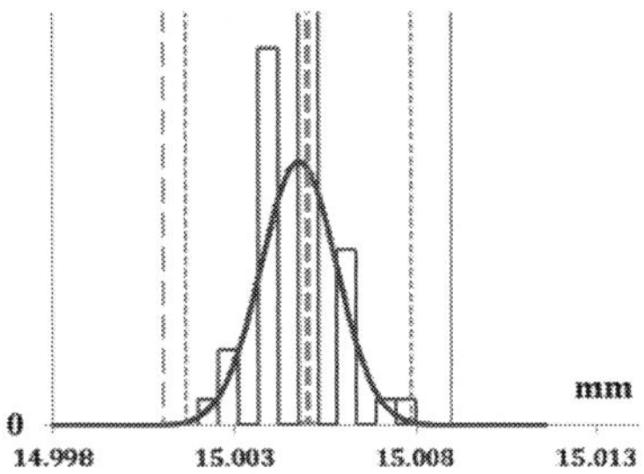

Figure 20. Histogram for DN ϕ 15 (+0.009, +0.001) with CBN insert

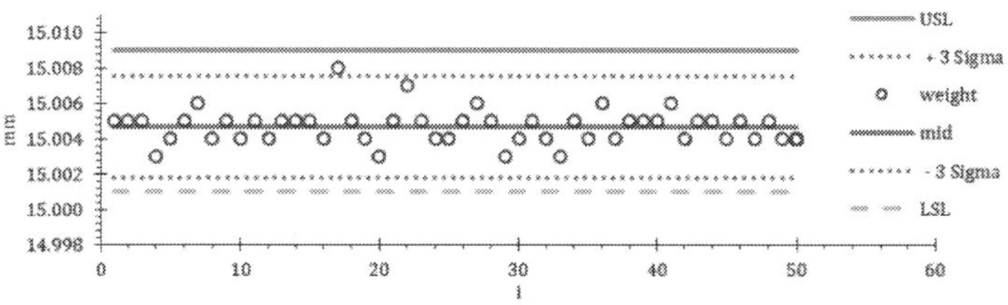

Figure 21. Control diagram for DN ϕ15, ϕ 15 (+0.009, +0.001) with CBN insert

The third solution implies as well modification of manufacturing stages, similar to solution two, and using on both stations CBN VBGW160404 inserts with 0.18 mm/rot feed. As well in this case, process is stable, with lower finishing times but with higher manufacturing costs (a CBN insert price is approximately 40 Euro). Because, in this situation, the speed recommended is 3200 rot/min, lower than the one used for original inserts, in order to speed up the process is possible to implement a balanced roughing operation in one of the variants, Trailing Tool or Simultaneous. In first case, two different tools are used, one for the upper turret and the other for the lower turret. The tool on the lower turret will trail behind the first. In simultaneous turning, the two tools cut together at the same depth during the rough or the finish cut. This option permits to double the feed rate with the same tool load. Machining strategies could be easily implemented using an appropriate CAM software.

Implementation of presented studies proved useful for improving the quality of final products as the number of complaints from customers has decreased considerably, so the purpose of the study to improve the production capability related to customer quality requirements was reached.

References

[1] Harry MJ and Stewart R 1988 *Six Sigma Mechanical Design Tolerancing*. Publication 6s-2-10/88, Motorola University Press, Schaumberg Illinois, USA, Second Edition, p. 1-60

[2] Hinckley CM 1997 Defining the best quality-control systems by design and inspection. *Clinical Chemistry*, ISSN 0009-9147, vol. **43** no. 5 p. 873-879

[3] Plenert G. *Statistical Process Control And Six Sigma*. Available at: www.referenceforbusiness.com/management/Sc-Str/Statistical-Process-Control-and-Six-Sigma.html, Accessed: 12/2015

[4] Douglas M 2004 *Introduction to Statistical Quality Control*. John Wiley & Sons, ISBN 978-0-471-65631-9, New York, USA, p. 776

[5] https://en.wikipedia.org/wiki/Induction_motor, Accessed: 12/2015

[6] http://www.polygim.com.tw/e-pro07.htm, Accessed: 12/2015

[7] http://us.roehm.biz/products/product-overview/?action=showgroups&grpid=7979&cHash=ccfbd93d489c6065df218365eb9d2293, Accessed: 12/2015

[8] www.sandvik.coromant.com/en-gb/products/pages/productdetails.aspx?c=VBMT%2016%2004%2004-PF%204325& m=6434000

[9] http://www.sixsigmablackbelt.de/wp-content/ uploads/ process-capability-excel-template-201501021.xls, Accessed: 05/2015

[10] http://www.korloy.com/newkorloy/eng/file/2015/en/2015%20catalog_EM.pdf, Accessed: 12/2015

IOP Conf. Series: Materials Science and Engineering **161** (2016) 012012 doi:10.1088/1757-899X/161/1/012012

Stiffness selection in synthesis of mechatronic discrete systems

Andrzej Buchacz[1], Damian Gałęziowski[2]
[1,2]Silesian University of Technology, Institute of Engineering Processes Automation and Integrated Manufacturing Systems, The Faculty Of Mechanical Engineering, Konarskiego 18a Street, 44-100 Gliwice, Poland

Emails: [1]andrzej.buchacz@polsl.pl, [2]damian.galeziowski@gmail.com (corresponding)

Abstract. In the paper, the known algorithm of designing of mechatronic discrete systems has been decomposed. As a result, detailed analysis of stiffness selection, during the process of distribution of dynamical characteristics functions, has been done. Based on synthesized one degree of freedom system that utilize piezostack actuator, detailed constrains related to the stiffness and their impact for mechanical, dimensionless and mechatronic parameters, have been investigated. The work extends the known problem of vibration control in discrete mechatronic systems

1. Introduction

Knowing the details related with stiffness selection, vibration isolation in mechatronic discrete systems, by use of synthesis method, can be deeply investigated and significantly improved. Utilization of piezo stack actuators for vibration control is well known [1]. Mathematical model of passive vibration with piezo transducers was shown in [2]. Different types of damping, respectively passive and active type in discrete mechatronic systems was confronted in [3].

Possibility to use negative value elements, in parallel with piezo has been introduced in [4], while using negative capacitance circuit [5]. Based on [4, 5] and works related with negative stiffness [6] authors introduced the problem of usage of negative elements in synthesis of mechatronic discrete systems [7].

The paper is verification and the continuation of Authors [1,3,7] works focused on stiffness selection. The work results in investigations related with stiffness impact for mechanical, dimensionless and mechatronic parameters in the mechatronic one degree of freedom cascade system. Based on the presented example, the paper leads to improvements in the efficiency of physical application of mechatronic systems, that can be easily synthesized and designed for given requirements in form of resonant and anti-resonant frequencies [7], in vibration control area.

2. Synthesis of mechatronic discrete systems

Synthesis of mechatronic systems is well known [1,3,7] and in case of cascade systems, is done based on dynamical characteristics function distribution into continuous fractions. The study on influence of main stiffness selections to all parameters has been done based on 1 degree of freedom (DOF) mechatronic structure, and in 2 DOF of system in mechanical replacement model. Two types of configurations have been in this case applied: *LC* and *LRC*, fig. 1.

The slowness function has been written as:

$$U(s) = H\frac{d_l s^l + d_{l-2}s^{l-2} + ... + d_0}{c_k s^k + c_{k-2}s^{k-2} + ... c_1 s}, \tag{1}$$

where: l – odd degree of numerator, k – degree of denominator, l-k =1, H – any positive real number.

The dimensionless equation of presented in fig. 1. mechatronic structure in *LRC* configuration, and its mechanical replacement models is written as:

$$\begin{bmatrix}1 & 0\\ 0 & \lambda\end{bmatrix}\begin{bmatrix}x_1''\\ x_2''\end{bmatrix} + \begin{bmatrix}0 & 0\\ 0 & 2D_{LRC}\end{bmatrix}\begin{bmatrix}x_1'\\ x_2'\end{bmatrix} + \begin{bmatrix}1+\gamma & -\gamma\\ -1 & 1+\delta\end{bmatrix}\begin{bmatrix}x_1\\ x_2\end{bmatrix} = \begin{bmatrix}0\\ 0\end{bmatrix} \tag{2}$$

where:
λ – nondimensional parameter which describes the relations of inertial elements,
γ, δ – nondimensional parameters which describes the stiffness elements,
$2D_{LRC}$ – dimensionless parameter which describes damping element.

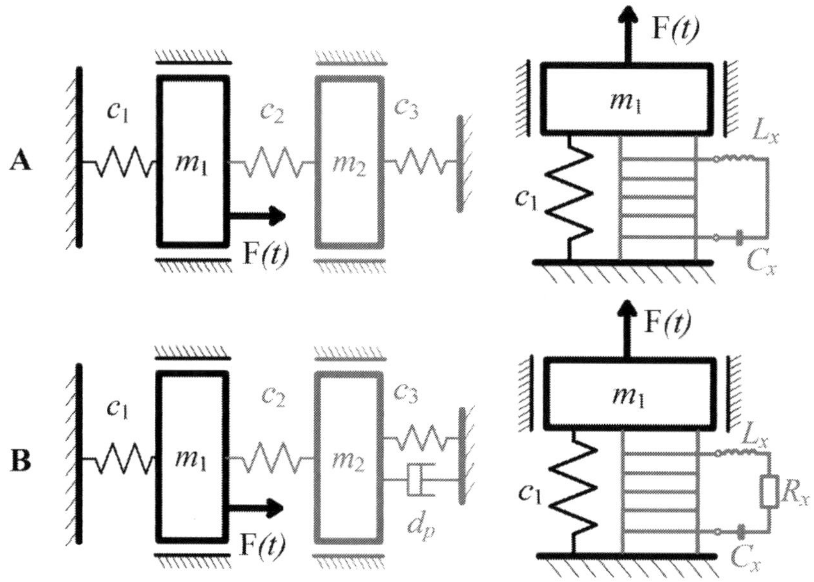

Figure 1. Mechatronic discrete systems in *LC* and *LRC* configuration and their replacement models

3. Mechanical parameters

Stiffness selection, in considered cases (c1), is the first step to determine the structure of designed mechatronic systems. Normally it should be taken from required range:

$$\left(0, H\frac{d_0}{c_1}\right). \tag{3}$$

However it's possible to take it out of the range as well, causing negative value of the stiffness, and therefor impacting also on the other parameters in the system. This selection results in mechanical replacement model (fig. 1). The impact on inertial element m_2 and stiffness elements c_2 and c_3 are shown respectively in the fig. 2 and fig. 3.

For the visualization, requirements in form of poles and zeros have been set in the range of 1000 – 2500 Hz, and for this values following graphs are created.

The value of m_2, is always positive and rapidly increases after taking the c_1 out of required range defined by the equation (2).

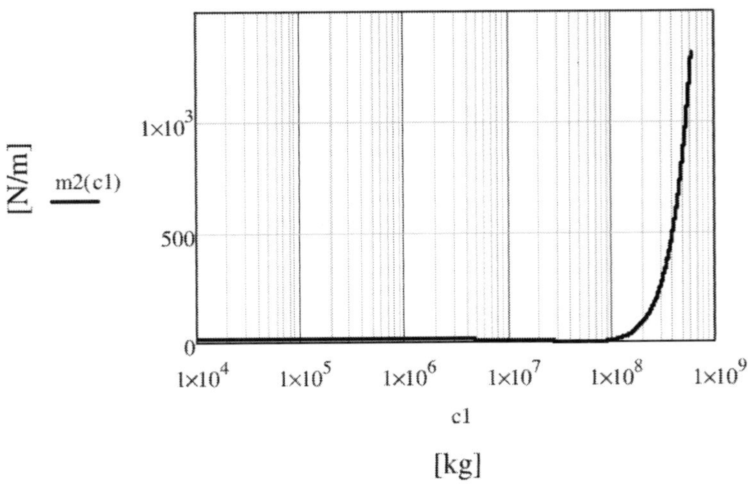

Figure 2. Influence of c_1 selection for m_2

Opposite to inertial element m_2, stiffness elements might have negative value while taking the c_1 above required range. This cause also indirectly the impact on capacitance in external electric network that is connected to piezostack actuator. In this case, negative capacitance might be realized by the system described in [5].

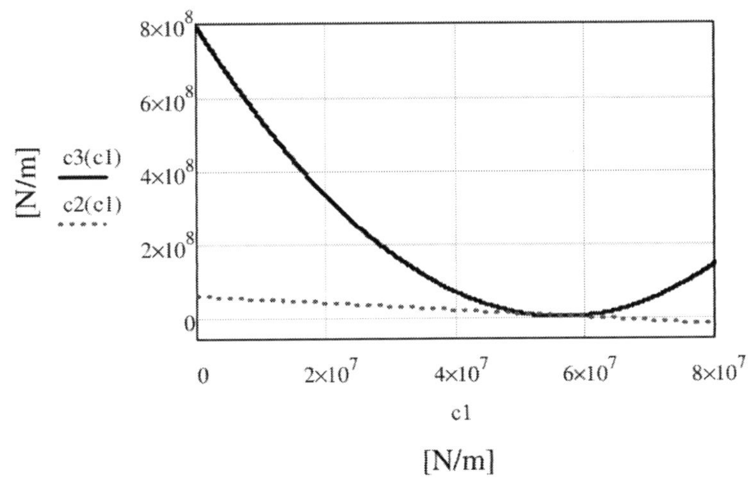

Figure 3. Influence of c_1 selection for remaining in the system stiffness elements c_2 and c_3

4. Dimensionless parameters

After reception of mechanical replacement model, during synthesis process, in next step this model is transformed to dimensional model, so the mechanical and electrical constrains related to piezoelectric effect and external electric network can be combined [7]. The system is analyzed in dimensional time:

$$\tau = \omega_m t, \tag{4}$$

where: ω_m - frequency connected with piezoelement position in the structure.

Influence for dimensionless inertial element is shown in fig. 4. The value of λ is positive, independently from selection of stiffness c_1.

IOP Conf. Series: Materials Science and Engineering **161** (2016) 012012 doi:10.1088/1757-899X/161/1/012012

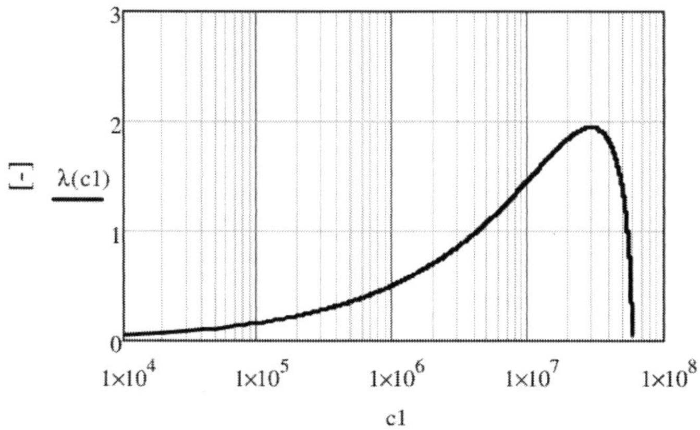

[N/m]

Figure 4. Influence of c_1 selection for parameter λ

In relation to given range (2) it's increasing until upper limit and then above defined ratio decrease. The impact c_1 for parameter γ is shown in fig. 5.

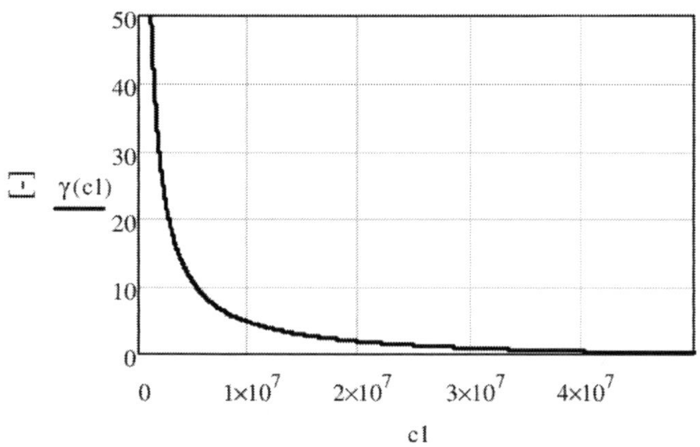

[N/m]

Figure 5. The impact of c_1 for parameter γ

Here selection from or out of the required range is neutral for the value. Similarly to parameter λ, γ which describe stiffness elements in mechanical replacement model is positive, but it's decreasing with increase of the value of c_1.

5. Mechatronic parameters

Retransformation of dimensionless model to mechatronic structure is the last step in synthesis of considered discrete systems process [7]. Following the general piezoelectric equations it's possible to determine parameters of the piezostack actuator, and calculate the values of elements inside the external electric network.

To refer results of investigations to requirements which means to check how the requirements could also affect the values of considered parameters, three different resonant and antiresonant frequencies has been evaluated (from range of 1000 – 2500 Hz). The bigger number in index of investigated parameter, the bigger value of pole and zero is applied.

The impact on selection of c_1 for capacitance in the external electric circuit is shown in fig. 6.

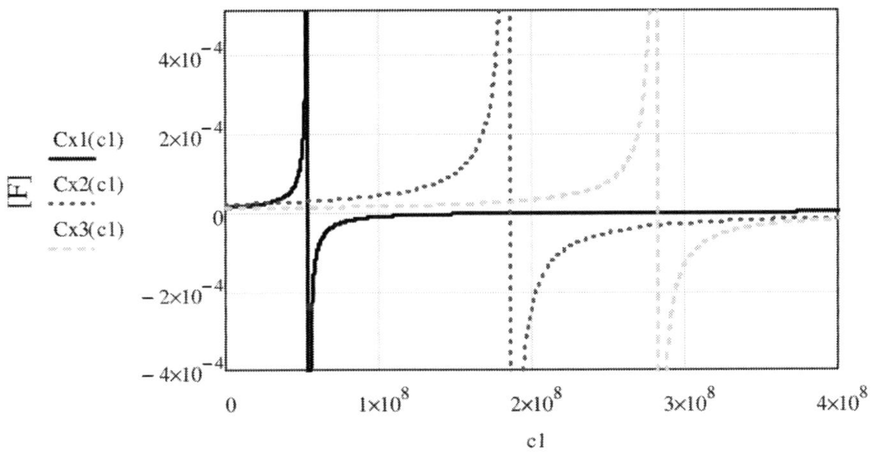

[N/m]

Figure 6. The impact of c_1 for capacitance C_x for different requirements

In all three cases, above the required range of selection of c_1, the value of capacitance in external electric network is negative. The influence of considered stiffness for resistance in electric circuit is shown in fig 7.

In this case, with increase of c_1, the resistance is also increasing. Another parameter that is affecting resistance in electric circuit is selection of parameter d_p, which is in this case proportional to stiffness c_3 and was described inter alia in [7].

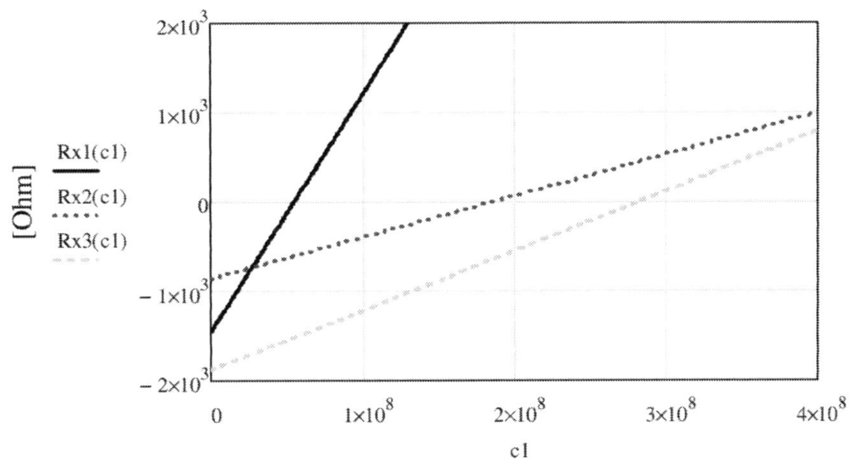

[c1]

Figure 7. Influence of c_1 selection for resistance in external electric network R_x studied for different requirements

Finally, the impact of selection of c_1 for inductance in the final mechatronic system, is shown in fig. 8. Opposite to capacitance and resistance, inductance is decreasing with the increase of c_1.

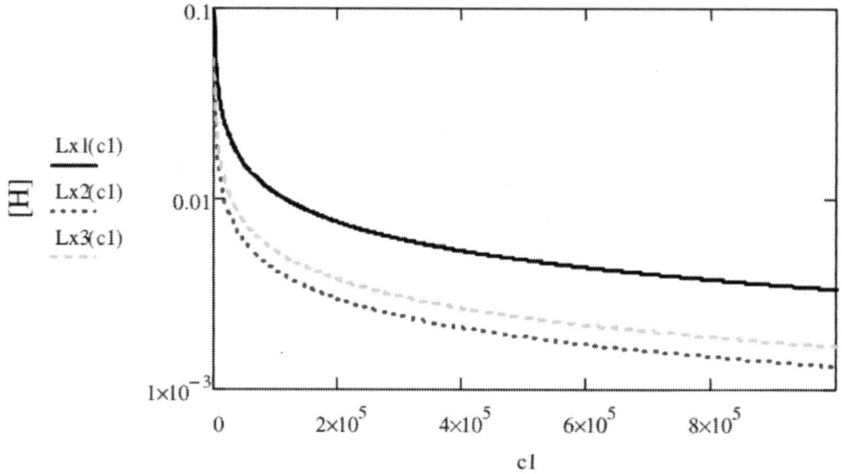

Figure 8. The impact of c_1 for inductance L_x for different requirements

6. Summary

In the paper, study on selection of stiffness elements, which is the key point during synthesis process, has been done. Based on example of cascade system with 1 DOF in mechatronic system, investigations related for verification of the impact for mechanical, dimensionless and mechatronic parameters of the system has been shown. Following the results and presented graphs, increase of the value of c_1 decrease the value of inductance and increase the value of R_x and C_x. Over the upper limit for c_1 selection, described by (3), C_x gets negative value, which is then also increasing with the increase of the value of the stiffness c_1.

It's also possible to select stiffness to obtain different damping functions of the systems: passive damping, where all parameters has positive values and contains passive elements (L_x, R_x), and adaptive-passive (C_x). That leads to negative values of capacitance or resistance elements which are another options for vibration elimination in considered systems that must comply with given requirements in form or resonant and anti-resonant frequencies.

Additionally, detailed graphs related to major constrains have been presented for the extension of possible applications of considered systems in vibration control as dampers that have been built from piezostack actuator and connected *LRC* network with/without negative values elements. Potential applications of the systems can be found in the areas of micro-positioning, optics elements and precise devices.

In further research works, detailed study on stability conditions and physical realization related to available in the market piezoceramic elements and their parameters like piezoelectric capacitance will be studied and constrained.

References

[1] Białas K, Buchacz A, Gałęziowski D 2013 Modelowanie dyskretnych układów mechatronicznych ze względu na funkcję tłumienia, *Modelowanie Inżynierskie*, 16, **47**, pp 31-35,

[2] Buchacz A, Placzek M, Wrobel A 2014 Modelling of passive vibration damping using piezoelectric transducers – the mathematical model, *Maintanace and Reliability,* **16**, I2, pp 301-306.

[3] Białas K, Buchacz A, Gałęziowski D 2015 Passive and active vibration isolation methods in discrete mechatronic systems, *Solid St Phen.*, **220-221**, pp 15-20.

[4] Han X, Neubauer M, Wallaschek J 2004 Improved piezoelectric switch shunt damping technique using negative capacitance, *J. Sound and Vib.*, **332**, I1, 2013, pp7-16.
[5] Fukada E, Date M, Kimura K and others 2004 Sound Isolation by Piezoelectric Polymer Films Connected to Negative Capacitance Circuits, *IEEE Transactions on Dielectrics and Electrical Insulation*, **11**, 2;
[6] Kashdan L, Conner Seepersad C, Haberman M, Wilson PS 2012 Design, fabrication and evaluation of negative stiffness elements using SLS, *Rapid Prot. J.*, **18**, I3, pp 194-200,
[7] Buchacz A, Gałęziowski D 2015 Designing of discrete mechatronic vibration systems with negative value parameters, *Mechanical Systems and Signal Processing*, dx.doi.org/10.1016/j.ymssp.2015.02.003.

Aknowledgment
This work has been conducted as a part of research project PBS2/A6/17/2013 supported by the National Centre for Research and Development in 2013-2016.

The Influence of Machined Surface Microgeometry on Mechanical Hydraulic Removal Mechanism at Ultrasonically Aided EDM Finishing

D Ghiculescu[1,a], N Marinescu[1,b], S Ganatsios [2,c], L Popa[1,d] and G Seritan[1,e]

[1]Polytehnic University of Bucharest, 313 Splaiul Independentei, Bucharest, Romania
[2]Technical Educational Institute of West Macedonia, Koila, Kozani, Greece
[a]daniel.ghiculescu@upb.ro, [e]george.seritan@upb.ro

Abstract. The paper deals with Finite Element Method (FEM) of mechanical-hydraulic component of material removal mechanism at electrical discharge machining aided by ultrasonics (EDM+US) finishing. The influence of two types of crater shapes - produced by commanded and relaxation pulses - is analyzed. Based on FEM results, the ratio between depth and crater diameter is correlated with the consumed power on ultrasonic chain in order to minimize the EDMed surface roughness.

1. Introduction

The process instability at electrical discharge machining (EDM) micromachining and finishing created by narrow working gap is known [1, 2]. A very effective solution to solve this drawback is ultrasonic assistance of electrical discharge machining (EDM+US) due to cyclic acoustic pressure created in the working gap and as a result, ultrasonically induced cavitation within. Therefore a pumping effect of dielectric liquid is created which is able to significantly improve the particles evacuation from the gap [1]. Moreover, ultrasonic assistance produces a major improvement of output technological parameters in terms of machining rate, ultrasonic volumetric relative wear, and surface roughness if the working parameters are optimized [3].

The material removal mechanism at EDM+US has two components: (1) thermal one, specific to pure EDM, but deployed under cyclic variation of dielectric liquid in the gap, and (2) mechanical hydraulic one, specific to ultrasonic shock waves removal. The component (2) is strongly related to surface roughness (Ra) decrease relative to pure EDM at the same input working parameters.

Very few researches reported the decrease of surface roughness as effect of ultrasonic aiding. At wire micro-machining, ultrasonic assistance produces better roughness [4]. Other researchers observed an increase of Ra of 10%, beside the machining rate increase since the removed volume by discharge is grown due to ultrasonic contribution to material removal mechanism [5].

The effective Ra minimization was attained through optimization of the key-parameter, the consumed power on ultrasonic chain (PcUS), which includes electrode-tool or workpiece. If PcUS is too high, Ra could be increased, although the machining rate is greater. But at finishing and micromachining, first objectives to be attained are Ra and volumetric relative wear minimization, i.e. precision increase. Establishing a correlation between machined surface microgeometry shape and optimum value of PcUS represents the main goal of this research.

2. Experimental Data

The decrease of machined surface roughness through ultrasonic aiding at EDM finishing was pointed out through experiments, using both commanded and relaxation pulses. Disk shape samples from X210Cr12 were machined comparatively on Romanian ELER 01 machine with 20 kHz longitudinal ultrasonic oscillations of the tool and without tool oscillations – pure EDM.

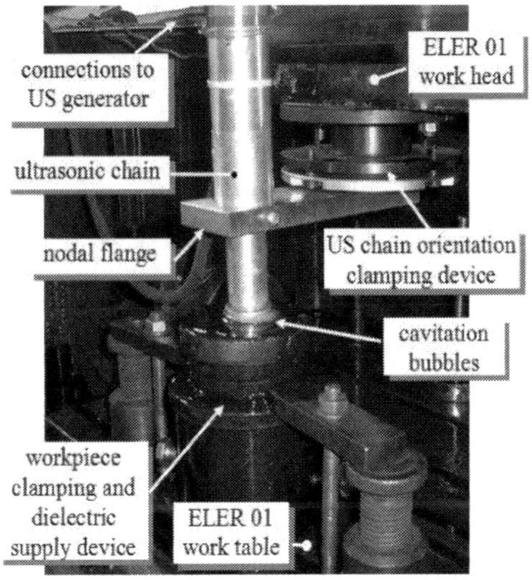

Figure 1. Elements of technological system at ELER 01 tests of EDM+US finishing

The electrode-tool from copper with disk shape of 25 mm diameter, and 3 mm height was used. The ultrasonic (US) chain included: PZT transducer, stepped horn and the tool at its end (figure 1). A nodal flange was used for its clamping. The perpendicularity of ultrasonic longitudinal axis on ELER 01 worktable was obtained with and orientation device based on adjustment by conjugated spherical surfaces [6].

A special generator for finishing and micro- machining, connected to ELER 01 machine was used. At commanded pulses using, the working parameters were: current step, I=1.5 A, positive (tool) polarity; pulse time, ti=12 μs, pause time, t0= 6 μs, injection pressure, phl=0.04 MPa (through workpiece); variation of consumed power on ultrasonic chain in the range, PcUS=0-150W; amplitude, around A=2 μm. At relaxation pulses using, the specific working parameters were: negative polarity, capacitor step, C=68 nF, resistance step (mean for adjusting supply current), R=0.13 kΩ; the ultrasonic parameters were maintained.

Data concerning the machined microtopography obtained with commanded pulses are presented in figures 2-3. The images were taken with the integrated system camera-microscope Neophot 2 – Zeiss. For evaluation of craters profile, the roughness Rz was considered, measured with a Taylor-Hobson surface instrument. The average dimensions of crater diameter and depth were represented comparatively in figures 4, and 5.

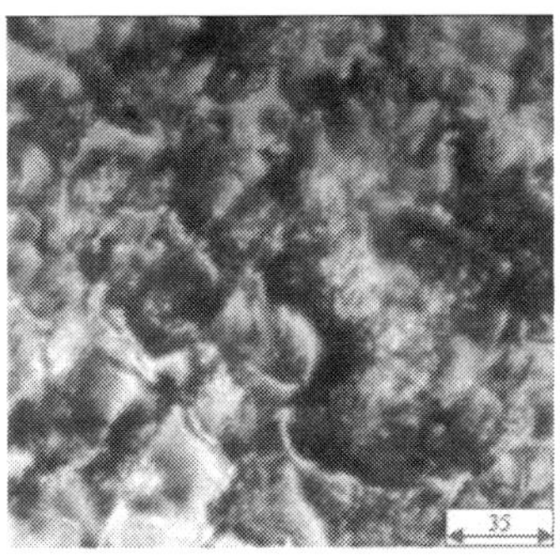

Figure 2. Microtopography at EDM, I=1.5A, polarity +, t_i=12 μs, t_0= 6 μs

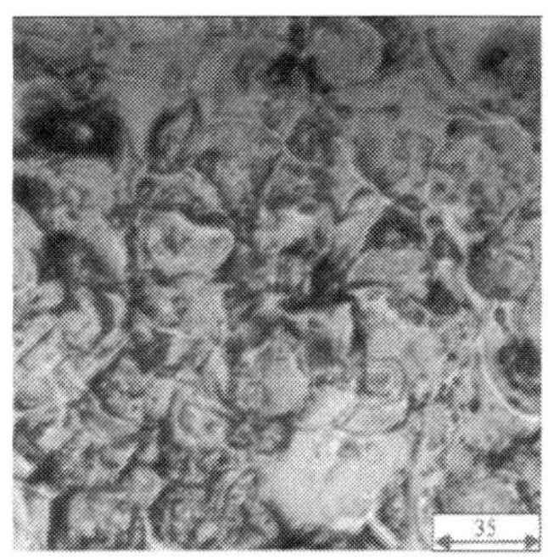

Figure 3. Microtopography at EDM+US, I=1.5A, polarity +, t_i=12 μs, t_0= 6 μs, P_{eUS}=120W

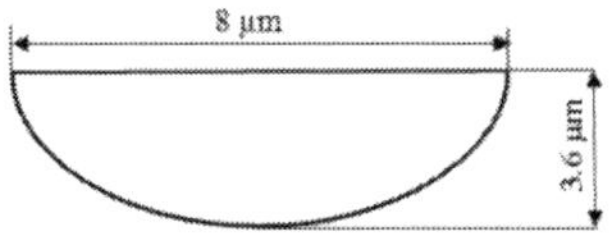

Figure 4. Average dimensions of crater profile at EDM, I=1.5A, polarity +, t_i=12 µs, t_0= 6 µs

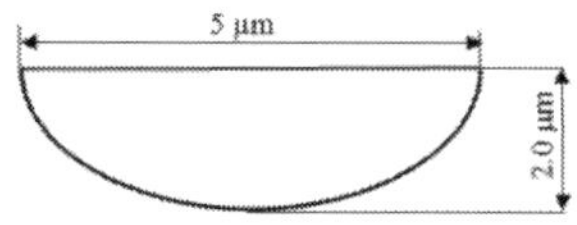

Figure 5. Mean dimensions of crater profile at EDM+US, I=1.5A, polarity +, t_i=12 µs, t_0= 6 µs, P_{cUS}=120W

As it can be noticed, the aspect ratio between depth and diameter of crater was 47.5% in case of EDM and 40% at EDM+US, emphasizing that ultrasonic shock waves modifies the crater profile turning it into a flatter shape. Since this a relative deep kind of crater, some material remains resolidified on crater margins against the case of relaxation pulses, pointed out at FEM modelling.

Data concerning the machined microtopography obtained with relaxation pulses with pure EDM and EDM+US are presented in figures 6, 7, together with their corresponding crater average profile in figures 8, 9.

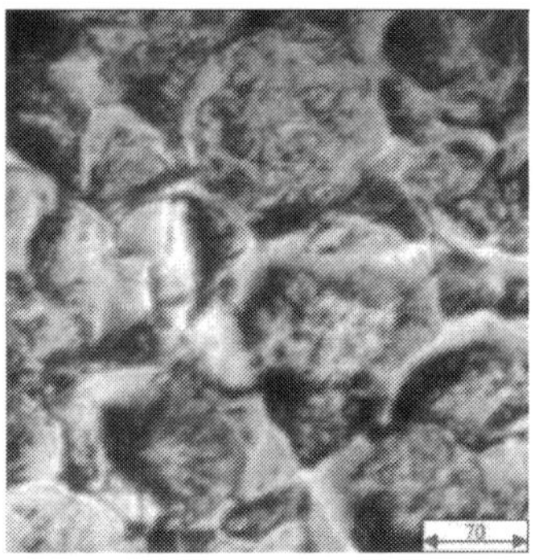

Figure 6. Microtopography at EDM, polarity -, C=68nF, R=0.13kΩ

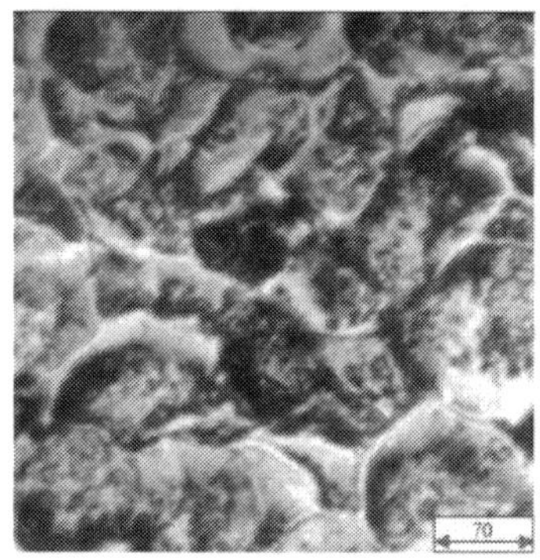

Figure 7. Microtopography at EDM+US, polarity -, C=68nF, R=0.13kΩ, P_{cUS}=90W

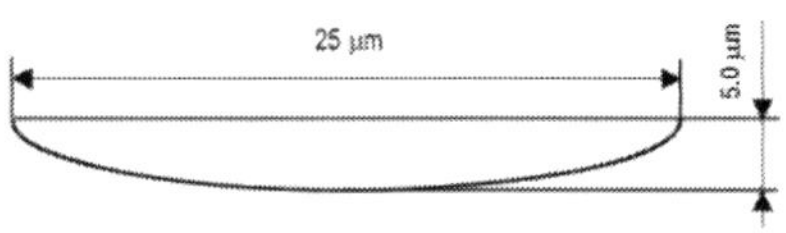

Figure 8. Mean dimensions of crater profile at EDM, polarity -, C=68nF, R=0.13kΩ

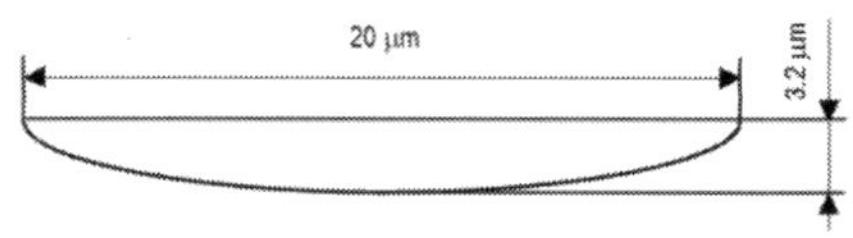

Figure 9. Mean dimensions of crater profile at EDM+US, polarity -, C=68nF, R=0.13kΩ, P_{cUS}=90W

In case of used relaxation pulses, the aspect ratio was 20%. At ultrasonic aiding, the aspect ratio became 16%, which highlights again the trend of reducing this parameter by removing the crater margins. This is a result of shear load produced by ultrasonic cavitation, when PcUS was optimized based on FEM results. This will be finally emphasized in case of both types of pulses.

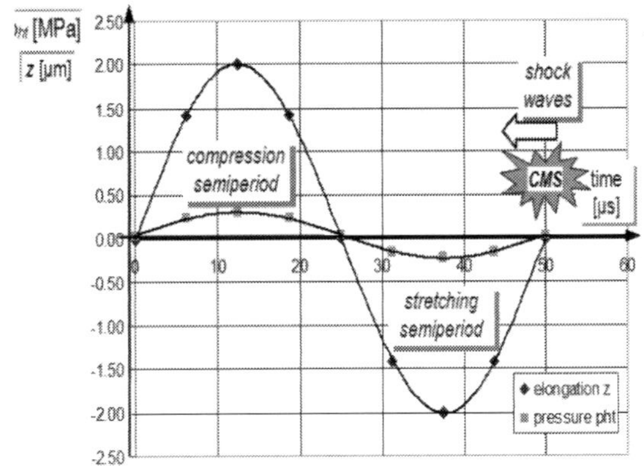

Figure 10. Elongation (z) and total pressure (p_{ht}) in the gap within an ultrasonic period of T_{US}=50 µs (f_{US}=20kHz)

The total hydraulic pressure (pht) variation in the gap is determined by elongation (z) – figure 10, according to relation:

$$p_{ht} = 2\pi \cdot c \cdot \rho \cdot f_{US} \cdot A \sin \omega t + p_{hl} \text{ [Pa]}, \qquad (1)$$

where: c is sound velocity in dielectric liquid [m/s]; ρ - dielectric liquid density [kg/m3]; fUS-ultrasonic frequency [Hz]; A-amplitude of elongation z, [m]; ω=2π fUS [s-1]; phl - local hydraulic pressure [Pa]. In this case, the parameters values were: phl=0.04MPa, ρ=840 kg/m3, K=1.35x109 Pa (K - bulk modulus), c=(K/ρ)1/2=1267.7 m/s, A=2µm, fUS=20kHz. At each final of ultrasonic period TUS, the cumulative microjets stage (CMS) occurs, collective implosion of gas bubbles, generating shock waves with pressure of 100 MPa order of magnitude, oriented along the frontal gap (horizontally) in case of EDM+US finishing tests.

3. Finite Element Modelling

Comsol Multiphysics, Structural Mechanics with Time Dependent variant was used for modelling the mechanical hydraulic component of EDM+US removal mechanism. A 2D parametric model was created for each type of pulse used (figure 11), taking into account the symmetry of machined geometry properties and cavitational phenomena.

▾ Parameters

Name	Expression	Value	Description
rp	15[mm]	0.015 m	workpiece radius
hp	10[mm]	0.01 m	workpiece height
acr	4e-6	4.0E-6	initial crater radius
bcr	3.6e-6	3.6E-6	initial crater depth
rms	0.5e-6	5.0E-7	radius of solidified material
modulE	2.1e11	2.1E11	X210Cr12 Young's modulus
pus	122[MPa]	1.22E8 Pa	ultrasonic shock waves pressure
tus	0.8e-6	8.0E-7	load duration

Figure 11. Example of assigned parameters for the model of hydraulic mechanical removal at EDM+US finishing tests with commanded pulses

The load duration (tUS), produced by CMS is assimilated to time of bubbles shutdown, calculated with Rayleigh relation [7]:

$$t_{US} = 0.915 R_m \sqrt{\frac{\rho}{p_{hl}}} \text{ [s]}, \quad (2)$$

where Rm is maximal radius before contraction, limited by maxium frontal gap, 0.01 mm for these EDM+US finishing tests. The bubbles implosion time was determined around,

tUS=0.8 μs.

The material removal due to ultrasonic contribution is achieved mainly in solid state, by shearing the most sensitive parts of the microgeometry to this load type, the margins of craters, controlled by parameters acr, bcr (e.g. 12.5 μm, and 5 μm at relaxation pulses), and rms. In case of commanded pulses producing deeper craters, the resolidified material amount is greater on crater margins, and therefore rms =0.5 μm against the crater resulted from relaxation pulses using, where rms =0.1 μm.

The boundary conditions were (figure 12): (a) pressure cyclic load (pUS) produced by CMS, applied on one flank of microcraters profile, resulted from previous discharges; (b) fixed constraint on the workpiece inferior surface due to vertical forces applied for clamping - workpiece orientation was on its bottom plane and lateral cylindrical surface through a bushing with inherent clearance.

The workpiece material was D3 (UNS T30403), corresponding to X210Cr12, with properties taken from Comsol library, completed by critical time dependent ones.

The meshing was achieved by 1862 triangular elements with an average quality of 0.967 on 0-1 scale provided by Comsol meshing module, and increasing finesse in the interest zone, where a higher precision is necessary (figure 12. b).

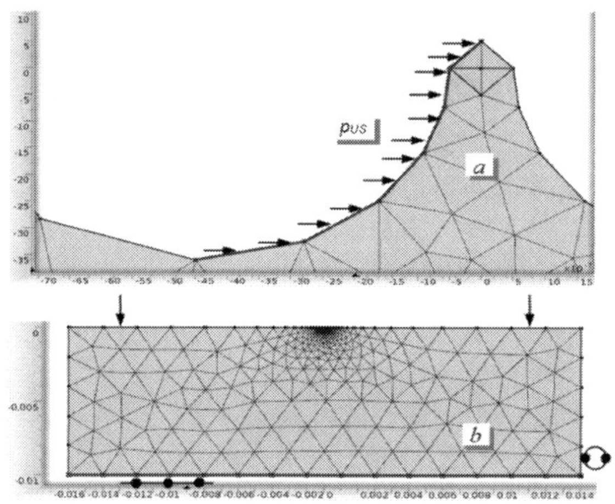

Figure 12. Boundary conditions and meshing: (a) pressure cyclic load on one flank of microcrater profile; (b) fixed constraint at the workpiece bottom with vertical clamping force

In order to underline the volume of material removed by hydraulic mechanical effect of ultrasonic cavitation, the ultimate tensile strength at fatigue pulsing cycle (σ0) was calculated [8]:

$$\sigma_0 = 1.12(40 + 0.16\,\sigma_r), \quad \text{[MPa]} \quad (3)$$

where: σr is the usual ultimate tensile strength. In case of X210Cr12, σr=1500 MPa, and the parameter of interest is σo =313.6 MPa.

4. Results and Discussion

Some significant distributions of Von Mises stress resulted from FEM modelling are presented, affecting the machined surface roughness at EDM+US, using both types of EDM pulses: commanded (figure 13), and relaxation ones (figure 14). The ultrasonic pressure exerted by CMS shock waves (pUS) was varied in the range of 100-125 MPa. Von Mises stress above fatigue ultimate tensile strength (σ0), i.e. the coloured zone, indicated the volume of removed material.

Two zones of machined microgeometry are highlighted from the point of view of material removal: (1) the micropeak zone – when the material is removed from this area, the surface roughness (Ra) is reduced; (2) the microvalley zone – in this case, Ra is increased.

Two distinct behaviours are observed at pUS growing, corresponding to commanded and relaxation pulses, respectively: in figure 13, the material is removed gradually, starting from micropeak to microvalley; in figure 14, the zone of removed material is moved suddenly from micropeak to microvalley. Both threshold pUS values, at which the material removal begins in microvalley, are determined based on FEM results, and presented in figure 15.

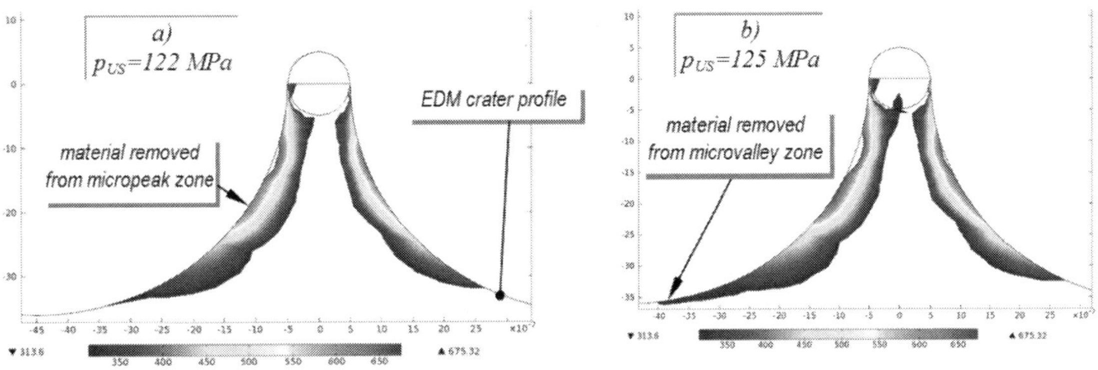

Figure 13. Von Mises stress [MPa] on microgeometry obtained with commanded pulses, higher than ultimate tensile strength of fatigue pulsing cycle, σ_o=313.6 MPa at different ultrasonic pressures, p_{US}

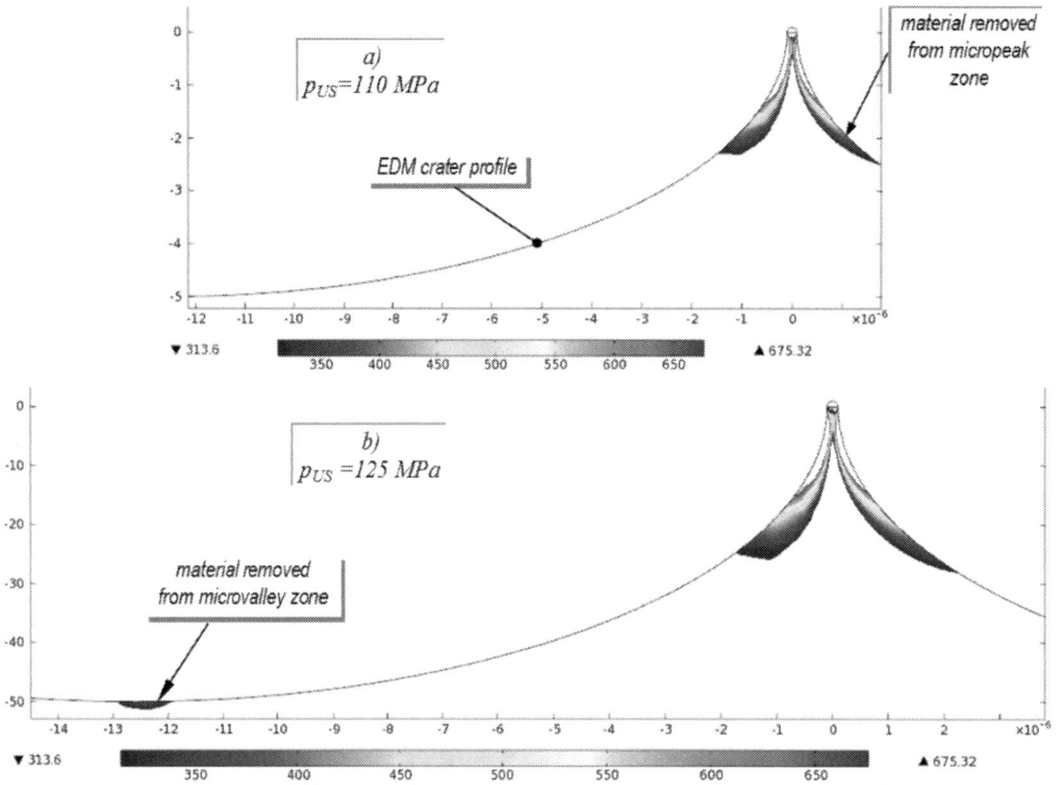

Figure 14. Von Mises stress [MPa] on microgeometry obtained with relaxation pulses, higher than ultimate tensile strength of fatigue pulsing cycle, σ_o=313.6 MPa at different ultrasonic pressures, p_{US}

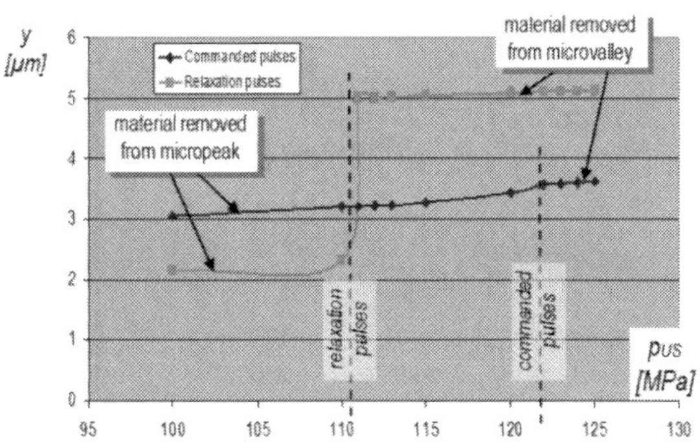

Figure 15. Depth of layer (y) removed as function of ultrasonic pressure (p_{US}) - FEM results

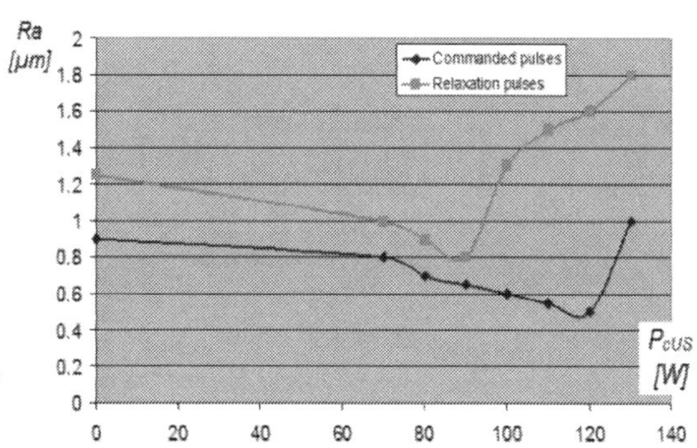

Figure 16. Roughness (Ra) as function of consumed power (P_{cUS}) on ultrasonic chain - experimental data

The thresholds for ultrasonic pressure (pUS) are: 111 MPa at relaxation pulses, and 122 MPa at commanded pulses microgeometry (figure 15). Beyond these values, machined surface roughness increases. These values are in correspondence with the values of consumed power on ultrasonic chain PcUS, which determine the oscillation amplitude (A), affecting first term from relation (1). The experimental data confirmed these assumptions: optimum value of PcUS in terms of Ra minimization is 90 W for relaxation pulses, and 120 W for commanded pulses microgeometry (figure 16).

5. Conclusions

Ultrasonic assistance of EDM finishing is able to reduce a technological key-parameter, the machined surface roughness (Ra) if a critical optimization condition, related to the consumed power on ultrasonic chain (PcUS), which includes the tool, is achieved.

FEM modelling of material removal mechanism at EDM+US finishing reveals that at microgeometry obtained with commanded pulses (deeper craters, high aspect ratio), higher values of ultrasonic pressure cyclic load (pUS) are required to remove the margins of the craters against the case of relaxation pulses microgeometry (flatter clatters, low aspect ratio), where lower values pUS are needed.

This is in agreement with experimental data confirming that the power PcUS is higher at commanded pulses using, in order to reduce the machined surface roughness, i.e. remove the crater borders. In case of relaxation pulses microgeometry, the required power PcUS is lower, the margins to be removed being less robust, and therefore more sensitive to ultrasonic fatigue shear loads.

Acknowledgement

The paper was achieved in Joint Applied Research Project supported by MEN-UEFISCDI, project no. PN-II-PT-PCCA-2013-4-0236, Contract no. 222/2014.

References

[1] B. C. Khatri, P. Rathod, J. B. Valaki, Ultrasonic vibration–assisted electric discharge machining: A research review, Proceedings of the Institution of Mechanical Engineers, Part B, Journal of Engineering Manufacture, March (2015) 1-12.

[2] S. Z. Chavoshi, X. Luo, Hybrid micro-machining processes: A review, Precision Engineering, 41 (2015) 1-23.

[3] D. Ghiculescu, Ultrasonically Aided Electrical Discharge Machining, in: M.P. Jahan (Ed.), Electrical Discharge Machining (EDM), Types, Technologies and Applications, Nova Publishers, 2015, pp. 151-208.

[4] Z.N. Guo, T.C. Lee, T.M. Yue, W.S. Lau, A study of ultrasonic - aided wire electrical discharge machining, Journal of Materials Processing Technology, 63 (1997) 823–828,.

[5] A. Abdullah, M.R. Shabgard, Effect of ultrasonic vibration of tool on electrical discharge machining of cemented tungsten carbide (WC-Co), International Journal of Advanced Manufacturing Technology, October, 38, 11 (2008) 1137-1147.

[6] D. Ghiculescu, N.I. Marinescu, S. Nanu, Equipment for ultrasonic aiding of electrical discharge machining of microslots, Romanian Granted Patent, RO-126191/ 30.05.2012.

[7] I. Anton, Cavitation (in Romanian), Romanian Academy, Bucharest, 1984.

[8] V. Drobota, Resistance of Materials (in Romanian), Didactic and Pedagogical, Bucharest, 1982.

IOP Conf. Series: Materials Science and Engineering **161** (2016) 012014 doi:10.1088/1757-899X/161/1/012014

Contribution on Taguchi's Method Application on the Surface Roughness Analysis in End Milling Process on 7136 Aluminium Alloy

M A ȚÎȚU[1], A B POP[2]

E-mail: [1]mihail.titu@ulbsibiu.ro, [2]bianca.bontiu@gmail.com

Abstract. The resulting surface quality after the cutting process is one of the most important characteristics of product quality and also the most frequent customer requirement. Previous research was focused on the effect investigation of machining parameters: cutting speed [1] and feed per tooth [2] on surface roughness. This paper is in itself a continuation of a previous research [3], in which, with Taguchi's method it was determined the level of influence of the cutting parameters on surface roughness of 7136 aluminium alloy in end milling process. The purpose of this paper is to highlight the importance of Taguchi's method use to analyse the surface roughness of 7136 aluminium alloy in end milling process. To conduct the experiments, three cutting parameters were used: cutting speed, feed per tooth and cutting depth. To analyse the surface quality, the surface roughness Ra (the arithmetic average of the absolute values) was measured. It was determined the recommended configuration regarding the optimum values of each machining parameter and the interactions between them, in order to obtain the better cutting process performance and to reduce the surface roughness sensitivity to uncontrollable factors. Based on a full factorial experiment were confirmed the obtained results by applying the Taguchi's method. Final results are a starting point for further research.

1. Introduction

Both national and international researchers are conducting experiments in almost all fields of research, having the purpose to identify some particular aspects of processes and systems. Literally every experiment is a test. A designed experiment could be defined as a test or series of tests, in which can make changes on the input variables of a process or system, so that through their influence on the process/system, we can identify the reasons for the occurrence of changes which can be seen from the results [4].

Therefore, a mathematical model can be developed which is aimed at achieving these results by the most important input variables and then use this model in the process or system in order to improve taking other decisions. Due to the need of identifying the optimal values of the cutting regime parameters, which affects the end milled surface quality, it was imposed to carry out some research based on experimental plans method. In this research, the construction of experimental plan will be conducted using the Taguchi's method. The argument for this choice is that, G. Taguchi propose a systematic and effective method for conducting the experiment, a method which leads to a proper solution in terms of performance and cost [3].

The proposed method uses an experimental orthogonal matrix which allows to divide the experimental space and also to select an experimental design matrix which contains a large number of variables, with a few experiments. Basically, this method allows to make a smart selection in the experimental field of an experimental parameters subset, which allow to study the entire experimental space with a minimum number of experimental determinations. [5]. this paper is a continuation of a previous research on the application of Taguchi method in the study of the cutting parameters influence over the surface quality, whose results were published in [3]. Similarly identified research was carried out by: [6] which was focused on the optimization of the cutting regime parameters using the Taguchi's method on the turning process of 6061 aluminum alloy, and [7] which has the main objective to optimize the process parameters to obtain a better surface roughness on the turning process of 7075 aluminum alloy. Others research in which the Taguchi's method was investigated are [5], [8] and [9], in which the implementation of this method was analyzed in order to improve the product quality.

2. Research methods

Given the purpose of the Taguchi method - oriented to conduct the experiments, based on this, it was determined the best combinations of the studied parameters. To obtain the expected results, the following steps were followed:

2.1. Process variables and their parameters

1) Identifying the factors involved in the end-milling process of the A7136 aluminum alloy:
- Cutting speed – v (m/min);
- Cutting depth – ap (mm);
- Feed per tooth – fz (mm/tooth).
2) The choice of levels for each machining parameter according to the SECO Tools manufacturer indications and also according to the technological possibilities of the tool, cutting inserts and CNC machine (table 1). The chosen levels reflect the minimum and maximum values of the machining parameters. To carry out the experiments according to Taguchi's method, the minimum and maximum values of each parameter must be used.

Table 1. Cutting process parameters and their values.

A – Cutting speed [m/min]					
495	530	570	610	660	710
B – Cutting depth [mm]					
2	2,5	3	3,5	4	
C – Feed per tooth [mm/tooth]					
0,04	0,06	0,08	0,11	0,14	

$$\begin{pmatrix} 1 & 1 & 1 \\ 1 & 1 & 2 \\ 1 & 2 & 1 \\ 1 & 2 & 2 \\ 2 & 1 & 1 \\ 2 & 1 & 2 \\ 2 & 2 & 1 \\ 2 & 2 & 2 \end{pmatrix}$$

Figure 1. Orthogonal matrix $L_8\,(2^7)$.

3) The column distribution of the parameters with their corresponding levels in the form of a matrix, respecting the rule of settlement. Based on Taguchi's method, were conducted a reduced number of experiments by combining the set factors using the orthogonal tables

(matrices). The orthogonal matrix for the choice of parameter values for each experiment L_8 (2^7) is shown in figure 1.

4) The research experiments according to the Taguchi plan:

2.2. Work Material
This research is focused on the study of arithmetic mean deviation Ra, of the Al7136 end-milled surface. The 7136 aluminum alloy is used in the aircraft industry and many of its properties are superior to other materials, including high strength to weight ratio, high wear resistance, low thermal expansion, corrosion resistance, durability, ductility, and conductivity, which make it a versatile material.

To conduct the 8 tests/experiments according to the Taguchi plan, an extruded block of Al7136 was used which comprises 8 rectangular specimens of 100 mm × 35 mm and 30 mm – figure 2.

Figure 2. The experiment on CNC machining center.

2.3. Cutting tool
The experiments were performed by using 16 mm End milling cutter milling with 100% (16 mm) tool engagement - SECO R217.69-1616.0-09-2AN, holding two indexable cutting inserts with ISO coding XOEX090308FR-E05, H15 - figure 3.

Figure 3. Cutting tool type SECO R217.69-1616.0-09-2AN used to conduct the experiments.

2.4. CNC Machine
The machine used for the milling tests is a HAAS VF2 CNC. Copious amounts of Blasocut BC 35 Kombi SW mineral coolant were provided at the cutting zone throughout the experiment. The pressure of the machine coolant pump was 8 bar. Three hoses of Ø6 diameter, have been used to target the jet.

5) The analysis and the interpretation of experimental data.

2.5. Response

After the machining operations, the surface roughness of each specimen was measured by using a portable surface roughness tester Mitutoyo SURFTEST SJ-210. The covered distance by the measuring device sensor is 5 mm (table 2).

Table 2. The values of 8 experimental setup with the measured roughness R_a values.

No. exp.	v [m/min]	ap [mm]	fz [mm/dinte]	Ra [μm]
1	495	2	0,04	0,070
2	495	2	0,14	0,155
3	660	4	0,04	0,412
4	660	4	0,14	0,108
5	495	4	0,04	0,071
6	495	4	0,14	0,342
7	660	2	0,04	0,074
8	660	2	0,14	0,132

2.6. Results and discussions

Based on the surface roughness measurements conducted in previous research [3], it was determined the configuration regarding the optimum values of each machining parameter and the interactions between them, in order to obtain the better cutting process performance and to reduce the surface roughness sensitivity to the uncontrollable factors.

6) Determination of the most favorable level of each factor to obtain the best possible surface quality: **A2 B1 C1 A1B1 A1C2 B1C1**. In other words, to attain that objective, it is recommended to use a higher cutting speed value at a lower value of cutting depth and feed per tooth. A first finding that emerges here is that the cutting speed has the greatest influence on surface roughness. To confirm this aspect, forwards the research is conducted using a full factorial experiment

3. Research approach using factorial experiment

Using the factorial experiments is recommendable and especially very effective in experiments that containing a several influence factors and it is necessary to study the cumulative effect of these factors on the objective function [5]. At this stage they were used to perform the experiments for all the cutting parameters values (Table 1), resulting a total of 150 cutting regimes. 3 blocks of Al7136 were machined under the same cutting conditions, having the dimensions 500 mm x 101 mm x 24.5 mm. After the end milling process, the Ra measurements were performed – figure 4. Based on the results, an evaluation of the experimental data was carried out that consists of a general appreciation of data homogeneity and particularity, using statistical parameters, such as [5]:

- histograms;
- scatter plot diagrams;
- Response surfaces and contours plot diagrams.

The analysis of the resulting experimental data derived from Ra measurements, was approached by the graphics realization based on the aspects mentioned above, using STATISTICA8 application. By the interpretation of the graphs, are seeking the influences of cutting parameters, that they exert on machined surface roughness. So, using the obtained results, a graphical data evaluation with a histogram distribution was carried out (figure 5). Hence, it appears an approximately normal data distribution and, under the established cutting conditions to conduct the experiment (tool, material and cutting regime) the Ra values are rather low at about 0,5 μm. Using the second parameter statistical processing: scatter plot, we evaluated the orthogonality experimental plan in order to track the dispersion of results to identify the types of relationships between factors and responses.

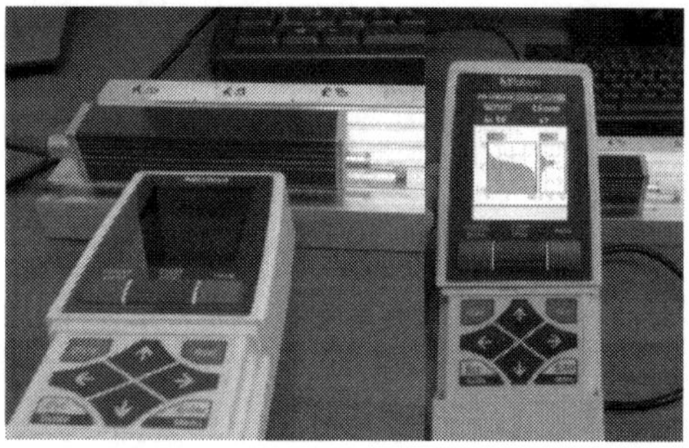

Figure 4. The measurements conducting using Mitutoyo SURFTEST SJ-210.

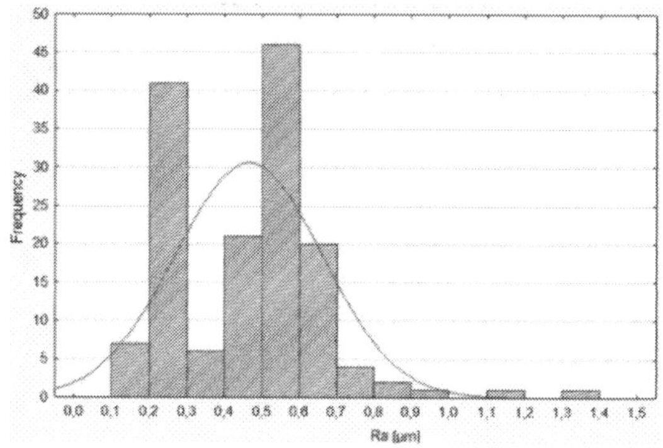

Figure 5. Ra measurements histogram.

Using the second statistical parameter - scatter plot, the orthogonality of the experimental plan was evaluated in order to find the results dispersion to identify the relationships types between factors and responses. By the Ra scatter plot diagram in which the roughness values are analyzed in function by the cutting speed - figure 6, it is derived that the measurements have an increasing trend with increasing this parameter. This finding confirms the issues mentioned in all studies specialized in this direction. But what is specific to this experiment, is the high values distribution of roughness measured in the middle range of speeds (570 m/min - 610 m/min). After the primary analysis on the system behavior of CNC - tool - workpiece, it was supposed that the vibrations were the cause of these increases.

Figure 7 shows that with the increasing of the cutting depth values also there is also a direct correlation in the increase of the roughness distributions. These values may be caused by the same assumption regarding the effect of vibration and chip breaking phenomenon, due to the large loading on tool (high cutting depth and the maximum value of the cutting edge).

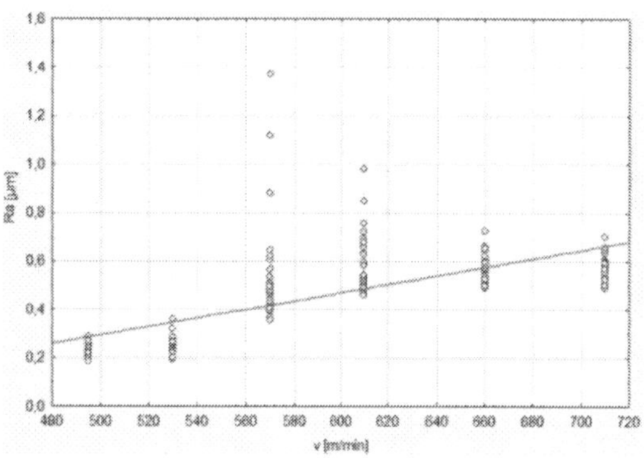

Figure 6. Scatter plot diagram of Ra measurements depending on v [m/mm].

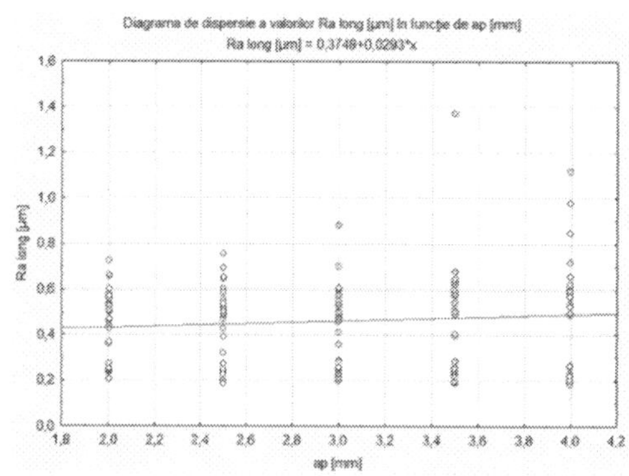

Figure 7. Scatter plot diagram of Ra measurements depending on ap [mm].

Figure 8 shows a slight decrease of the roughness values related to the cutting feed. This trend is directly influenced by the high dispersion of the roughness values at cutting speeds of 570 m/min and 610 m/min. Further, a tridimensional analysis was performed in order to observe the Ra dimensional variation under the machining parameters influence.

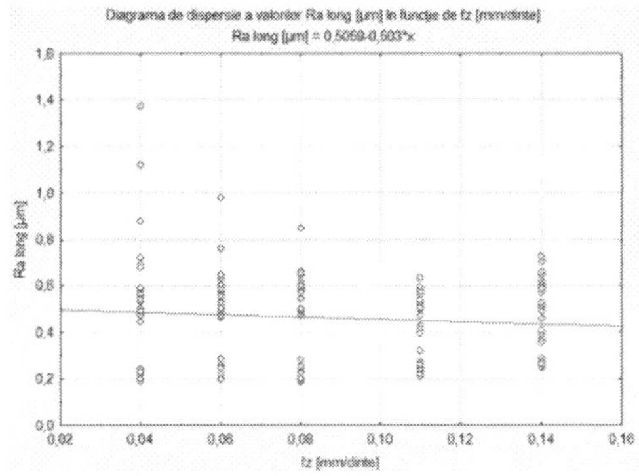

Figure 8. Scatter plot diagram of Ra measurements depending on fz [mm/tooth].

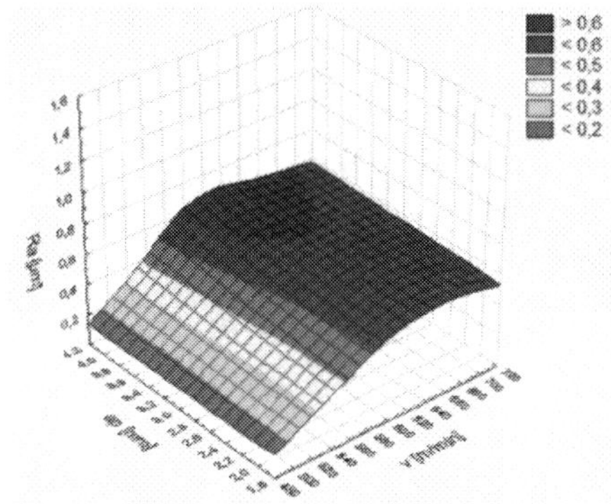

Figure 9. Response surface of Ra under the v [m/min] and ap[mm] influence.

In figure 9, the spatial variation of Ra values depending on the cutting speed and cutting depth influences is presented. It can be observed a fairly large increase of the measurements in the first part of the interval of cutting speeds of 495-570 m/min, on the whole range of cutting depths. However, it appears that the cutting depth has not so great influence on the roughness values at low cutting speeds, but this phenomenon is changing at speeds exceeding 570 m/min.

Figure 10 shows the spatial variation using the contour plots distribution and it can be observed how the roughness values are influenced by the cutting speed and feed per tooth. It is found that the cutting speeds values of 570 to 610 m/min, in combination with a low feed per tooth, have an impact on measured roughness values which increase in whole experimental field. Again, the vibration may be the cause of these increasing. The Ra spatial variation under the cutting depth and feed per tooth influence, is shown in figure 11. After the analysis of this evolution, it appears that the cutting depth related to the cutting feed, influences the surface roughness values of the machined surface, resulting low roughness values. However, at higher feed per tooth values and lower cutting depth values, due to the influence of cutting speeds of 570 to 610 m/min, there is a tendency of roughness increase.

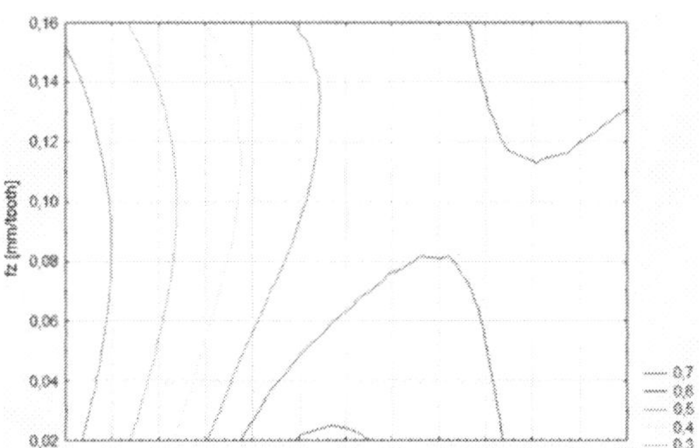

Figure 10. Contur plots of Ra under the v[m/min] and fz [mm/tooth] influence.

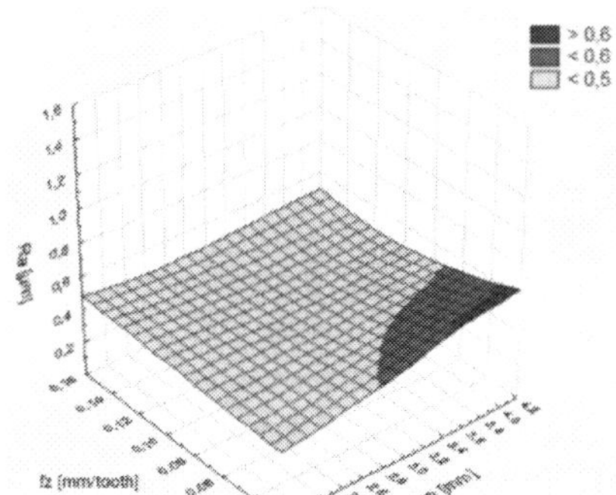

Figure 11. Response surface of Ra under the ap [mm] and fz [mm/tooth] influence.

4. Conclusions

Using the Taguchi's method, the number of experiments has been reduced from 150 to 8.

The optimum configuration composed of each machining parameter, aimed to increase the cutting process performance and to reduce the susceptibility to uncontrollable factors, is: A2 B1 C1 A1B1 A1C2 B1C1.

The histogram shows that the arithmetic mean deviation values of the surface profile, obtained by experiment, and fits the milling process with the studied cutting parameters, in the fine precision class – (1,6 μm).

From the scatter plot diagrams it was found that the Ra measured values, have an increasing trend with the cutting speed increasing. The reason could be the vibrations caused by the resonance regime of the system, tool, material, cutting values and the CNC, and the phenomenon of chip breaking caused by high tool loading.

The conducted tridimensional analysis shows that the cutting speed combined with cutting depth - expresses the strongest influence on the surface roughness.

The cutting parameter with the greatest influence on the surface roughness is the cutting speed. With its increasing the roughness values have also an increasing trend.

5. Future work

Given the dynamic behavior of the system CNC - tool - workpiece, it must be studied the effect of the vibration in this context.

Methods identification that will lead to obtain some mathematical models of others factors – like tool wear, that influence the surface quality

The study of the tool wear evolution in the aluminum machining, based on the varying of the cutting parameters regime.

Expanding the range of current research on other materials and compared to the studied 7136 aluminum alloy.

Increasing the sustainability of economically and in terms of productivity through the use of mathematical modeling of the cutting process.

Expanding the range of current research on other materials and compared to the 7136 Al alloy studied.

6. Acknowledgment

For this paper the logistical support of Universal Alloy Corporation Europe SRL was received.

References

[1] A.B. Bonţiu Pop, M. Lobonţiu, Investigating the Effect of Cutting Speed Variation on Surface Roughness of 7136 Aluminum Alloy in End Milling, Innovative Manufacturing Engineering International Conference, IManE 2015 International Conference, May 21-22, Iaşi, Romania, 2015.

[2] A.B. Bonţiu Pop, M. Lobonţiu, The Influence of Feed per Tooth Variation on Surface Roughness of 7136 Aluminum Alloy in End Milling, Modern technologies in manufacturing, 14-16 October, Cluj Napoca – Romania, 2015.

[3] A.B. Bonţiu Pop, Application of Taguchi's method in the influence study of machining parameters on surface roughness of 7136 aluminum alloy in end milling, Applied Mechanics and Materials: 2015-02-19, ISSN: 1662-7482, Vols. 809-810, pp 123-128, 2015.

[4] Montgomery, D. Design and Analysis of Experiments. Eighth Edition ed. Hoboken: John Wiley & Sons, Inc, 2013.

[5] M. Ţîţu, C. Oprean, Al. Boroiu, Cercetarea experimentală aplicată în creşterea calităţii produselor şi serviciilor, Colecţia Prelucrarea Datelor Experimentale, Editura AGIR, ISBN 978-973-720-362-5, 684 pagini, Bucureşti, 2011.

[6] M. Murthy, K.M. Babu, R.S. Kumar, Optimization of Machinability Parameters of Al6061 using Taguchi Technique. International Journal of Current Engineering and Technology, Special Issue-3, 2014.

[7] N.B. Doddapattar, N. Lakshmana Swamy, An optimization of machinability of aluminium alloy 7075 and cutting tool parameters by using Taguchi Technique. International Journal of Mechanical Engineering And Technology, Vo. 3, Issue 2, pp. 480-493, 2012.

[8] M. Ţîţu, L. Tarnu, S. Ţîţu, Andreea Răulea, Implementing the Taguchi method towards the improvement of product quality, Conferinţa Internaţională MECAHITECH 2015 -"7th International Conference on Innovations, Recent Trends and Challenges in Mechatronics, Mechanical Engineering and New High – Tech Products Development", In: Romanian Review Precision Mechanics, Optics & Mechatronics, Nr. 48/2015, ISBN 2247-7063, ISSN 1584-5982, DOI dx.doi.org/10.17683/rrpmom.issue.48, pag. 128-132, 10-11 Septembrie 2015, Bucureşti, Romania, 2015.

[9] M. Ţîţu, C. Oprean, Cercetarea experimentală şi prelucrarea datelor. Partea I. Colecţia Prelucrarea Datelor Experimentale, Editura Universităţii Lucian Blaga din Sibiu, ISBN (10) 973-739-296-5, ISBN (13) 978-973-739-296-1, ISBN (10) 973-739-297-3, ISBN (13) 978-973-739-297-8, 415 pagini, Sibiu, 2006.

IOP Conf. Series: Materials Science and Engineering **161** (2016) 012015 doi:10.1088/1757-899X/161/1/012015

Regarding to the Variance Analysis of Regression Equation of the Surface Roughness obtained by End Milling process of 7136 Aluminium Alloy

A B POP[1], M A ŢÎŢU[2]

E-mail: [1]bianca.bontiu@gmail.com, [2]mihail.titu@ulbsibiu.ro

Abstract In the metal cutting process, surface quality is intrinsically related to the cutting parameters and to the cutting tool geometry. At the same time, metal cutting processes are closely related to the machining costs. The purpose of this paper is to reduce manufacturing costs and processing time. A study was made, based on the mathematical modelling of the average of the absolute value deviation (Ra) resulting from the end milling process on 7136 aluminium alloy, depending on cutting process parameters. The novel element brought by this paper is the 7136 aluminium alloy type, chosen to conduct the experiments, which is a material developed and patented by Universal Alloy Corporation. This aluminium alloy is used in the aircraft industry to make parts from extruded profiles, and it has not been studied for the proposed research direction. Based on this research, a mathematical model of surface roughness Ra was established according to the cutting parameters studied in a set experimental field. A regression analysis was performed, which identified the quantitative relationships between cutting parameters and the surface roughness. Using the variance analysis ANOVA, the degree of confidence for the achieved results by the regression equation was determined, and the suitability of this equation at every point of the experimental field.

1. Introduction

Previous research was focused on the investigation of the effect of the feed rate on the surface roughness and on the reduction of processing time [1]. This time, the attention will be focused on the mathematical modeling of a regression equation of the surface roughness obtained on the end milled profile.

In the scientific literature, there are studies based on mathematical modeling which can be mentioned: [2] which used the Response Surface Methodology (RSM) as optimization technique, [3] and [4] which used the Taguchi's method, [5] used the regression analysis and [6] which used ANOVA and the regression analysis.

Regression analysis is a statistical method of identifying the quantitative relationships between independent variables (factors) and dependent variables (responses). This method allows the use of experimental data in order to calculate the equation coefficients, an equation that describes the studied system at any point of the experimentally range studied [7].

The technique known as the analysis of variance (ANOVA) uses tests which are based on the variance ratio, to determine if there are significant differences between averages of several data groups and if each data group has a normal distribution. The technique was initially developed and used by R.A. Fisher and his colleagues in England in the early 1920s about ANOVA.

About ANOVA, Fisher says that "it is a cheap and convenient technique for understanding asymmetry" [7]. The variance analysis is a basic tool used to assess the fit of a regression analysis, respectively, the model generated from regression analysis.

ANOVA focuses on estimating various types of variability of responses and estimates made by F-test [7].

To use the ANOVA test for evaluation of the quality match, it is necessary to repeat the experiment at least once in the centre of the experimental range. In these circumstances, using ANOVA, it is possible to divide in two the total variance of a response, one due to the regression model and one due to residual. If the situation in which there exists experimental data obtained from repeated experiments under the same conditions, the residual variance can be divided into the variance related to the model error and related to the variance of replication error [7].

Thus, the first step of the method, is to assume that the experimental plan model was well chosen. Subsequently, the experimental plan model is tested and evaluated to determine if the model and the estimated parameters are significantly different from the experimental data obtained.

After the calculation of the model coefficients, they are evaluated to determine if they were well calculated and if the response value deviation obtained related to the anticipated response value is due to a random variation or due to the uncontrolled influencing factors.

2. Research methods

2.1. Process variables and their parameters

Experiments were conducted based on three machining parameters: cutting speed, feed per tooth and cutting depth.

All of these three parameters takes variables values according to the tool manufacturer recommendation for aluminium machining.

The experiment planning was done respecting the three principles of Montgomery [8]:

- The randomization principle - under which the statistical methods require that the observations (or errors) have random distributions to the parameters. The observation randomization makes this assumption valid.
- The replication principle - involves the repeating of experiment by 3÷7 times for each set of input parameter values. This procedure is necessary to determine the consistency of measurements. In this case, to obtain precise results, a 7x experimental replication was applied.
- The principle of working in "blocks" –It is used to improve the comparison accuracy of the used factors. In this research 7 blocks were used. In the Table 1 is presented this specific working procedure. Under this procedure, I was able to achieve the full factorial experimental plan with 7 blocks.

Table 1. Experimental procedure.

	Levels number					
Cutting speed [m/min]	6 levels					
	495	530	570	610	660	710
	5 levels					
Cutting depth [mm]	2	2,5	3	3,5	4	
Cutting feed [mm/tooth]	5 levels					
	0,04	0,06	0,08	0,11	0,14	
Total experiments	6x5x5=150					
Replications	7					
Total Mesurements	150x7=1050					

2.2. Work Material

The investigations are conducted based on the Al7136- material used in the aircraft industry, which properties are superior to other materials, including high strength to weight ratio, high wear resistance, low thermal expansion, corrosion resistance, durability, ductility, and conductivity, which make it a versatile material.

To conduct the experiment, the specimens have the following dimensions: 500 x 101 x 24,5 mm. On each specimen 50 machining operations were performed for the 50 cutting regimes set based on the combinations of the studied cutting parameters. The machined surface was 50 mm long and 16 mm wide. Thus, the 150 cutting regimes obtained under the combinations, were tested by processing of 21 blocks with dimensions mentioned above, the number of blocks covered the whole number of experiments, including the replications (figure 1).

Figure 1. The Al7136 -21 blocks prepared for machining procedures.

Figure 2. The tool SECO R217.69-1616.0-09-2AN use to conduct the experiments.

2.3. Cutting tool

The experiment was performed using a standard set of tools for aluminium machining - 16 mm End milling cutter milling with 100% tool engagement - SECO R217.69-1616.0-09-2AN, holding two indexable cutting inserts XOEX090308FR-E05, H15 (figure 2). For each of the seven blocks tested from Al7136, out of 21, a set of cutting inserts were used (of 2 pieces each – total 6 cutting inserts).

2.4. CNC Machine

The machine used for the milling tests is a HAAS VF2 CNC. Abundant amounts of Blasocut BC 35 Kombi SW mineral coolant were provided at the cutting zone throughout the experiment. The pressure of the coolant pump of the machine was 8 bar. Three hoses of Ø6 diameter have been used to route the jet. Each specimen was fixed on the CNC table with three clamps, in order to obtain rigidity. So, the specimens are parallel to the table and perpendicular to the main shaft (figure 3). In figure 4 it can be seen, for example, two milled blocks of Al7136.

2.5. Response

After the milling of the 21 blocks of aluminium alloy Al7136, the measurements of the surface roughness were carried out over 5 mm distance using the portable surface roughness tester Mitutoyo SURFTEST SJ-210, with a resolution of 0.002 μm. The delimitation of the performed roughness measurements are indicated in figure 5.

Figure 3. The specimens fixed on the CNC table.

Figure 4. The Al7136 end milled blocks.

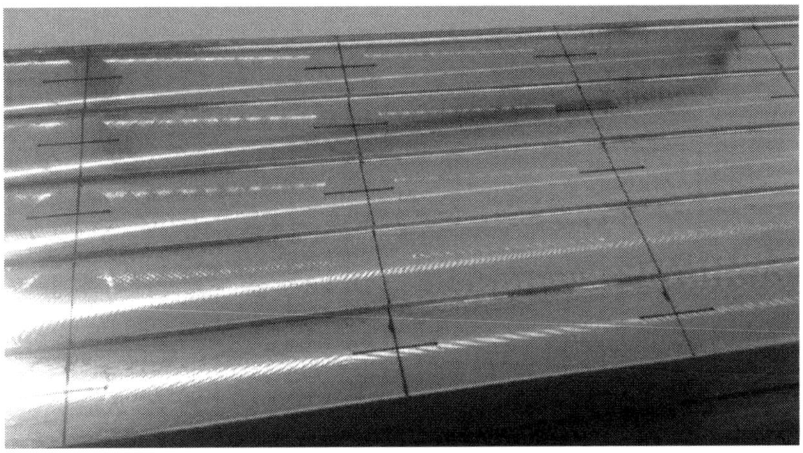

Figure 5. The graphical representation of the Ra measurements area.

2.6. Results

After the measurements performing, the Ra roughness evolution can be analysed according to the studied cutting parameters, as shown below.

Figure 6 shows that with the increasing of cutting speeds between 495 m/min and 660 m/min, the mean roughness values, has an upward trend. Thus, when the cutting speed is 495 m/min, the Ra value is 0,27 μm and for 660 m/min the Ra is 0,57 μm. Regarding the cutting speed of 710 m/min average roughness is 2,3 times higher, having a value of 0,62 μm. When the cutting speed is 570 m/min, the Ra distribution is high enough so that the ratio between the maximum and minimum recorded Ra value is 8 times compared with other cutting speed values for which the ratio is between 2,2 (660 m/min) and 4,99 times (at 495 m/min).

In figure 7, the direct influence exerted on the surface roughness is shown by the combination of cutting feed and feed per tooth at the cutting speed 570 m/min. In this figure, the roughness values obtained at different depths of cut are plotted. The large distribution of the roughness values at cutting speeds of 570 m/min and 610 m/min affects also the maximum roughness values in relation to cutting depth. Given that, the average Ra values are almost linear, it appears that the cutting depth variation does not produce a high fluctuation of roughness values.

The figure 8 shows the experimental data regarding the surface roughness obtained at different feed per tooth values. This indicates that an increase of the feed per tooth values does not the influence significantly the surface quality. This is due to the correlation of feed rate with cutting speed, correlation given by the cutting tools manufacturer.

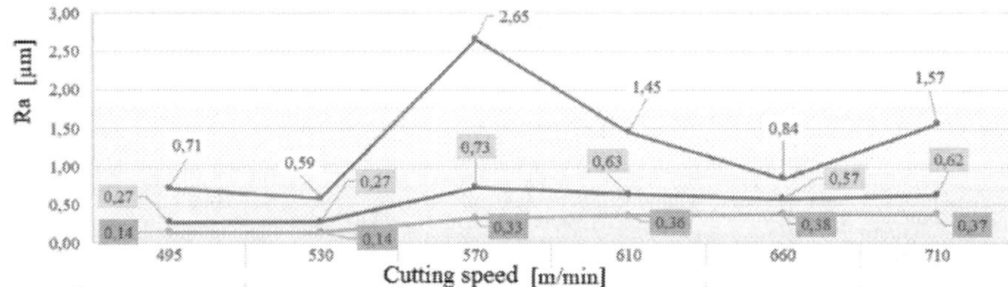

	495	530	570	610	660	710
Ra maximum values	0,71	0,59	2,65	1,45	0,84	1,57
Ra medium values	0,27	0,27	0,73	0,63	0,57	0,62
Ra minimum values	0,14	0,14	0,33	0,36	0,38	0,37

Figure 6. The cutting speed influence on Ra values.

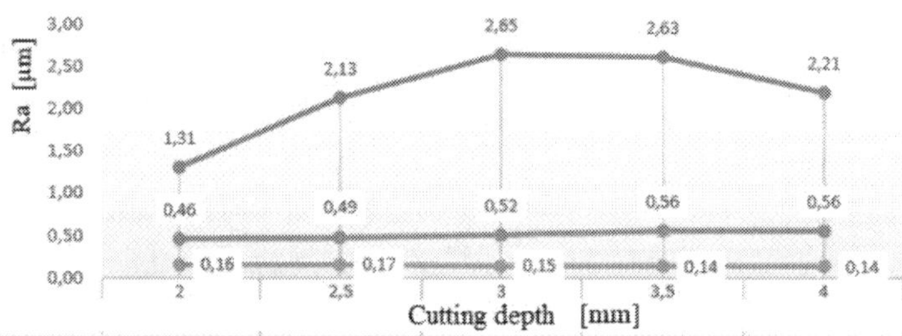

	2	2,5	3	3,5	4
Ra maximum values	1,31	2,13	2,65	2,63	2,21
Ra medium values	0,46	0,49	0,52	0,56	0,56
Ra minimum values	0,16	0,17	0,15	0,14	0,14

Figure 7. The cutting depth influence on Ra values.

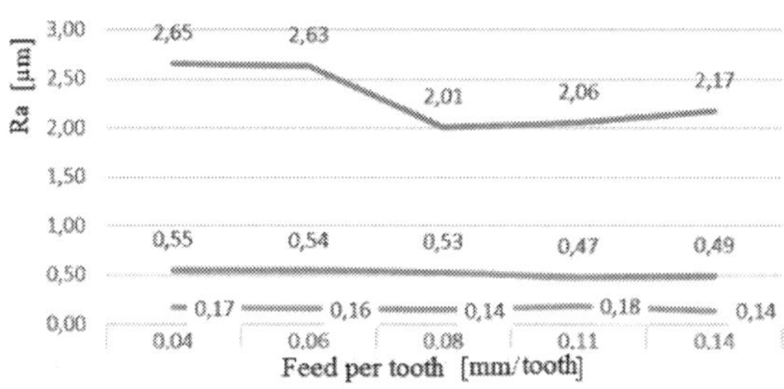

	0.04	0.06	0.08	0.11	0.14
Ra maximum values	2,65	2,63	2,01	2,06	2,17
Ra medium values	0,55	0,54	0,53	0,47	0,49
Ra minimum values	0,17	0,16	0,14	0,18	0,14

Figure 8. The feed per tooth influence exerted on surface roughness.

3. Regression analysis

Using as a starting point the experimental results, the regression analysis was performed in order to obtain a regression equation of the surface roughness according with the studied cutting parameters.

Thus, the regression analysis was carried out using the Methods for Least Squares, a method which can estimate the polynomial equation coefficients which describes the model by minimizing the squared distances between the experimental value and the calculated value. Thus, it was determined the percentage contribution of the cutting parameters influence on the surface roughness, as shown in Table 2.

To facilitate the calculations necessary to achieve the research objective, MINITAB 17 was applied, which is a statistical application and it was determined the sum of squares and the distribution percentage for each factor or factors interaction separately.

Table 2. The cutting parameters influence exerted the surface roughness [%]

Source	FD	SS	%
Regression model	7	199,482	86,12%
Cutting speed A	1	194,322	83,89%
Cutting depth B	1	0,040	0,02%
Feed per tooth C	1	1,157	0,50%
A x B	1	3,093	1,34%
A x C	1	0,316	0,16%
B x C	1	0,366	0,14%
A x B x C	1	0,188	0,08%
Error	827	32,156	13,88%
Fitting error	143	18,735	8,09%
Pure error	684	13,421	5,79%
Total	834	231,638	100%

Thus, by using the calculation procedure predefined on MINITAB 17, the regression coefficients were determined and, therefore, the Ra regression equation is:

$$Ra = 0{,}000287 \cdot A - 5.808 \cdot C + 0.000196 \cdot A \cdot B + 0.01345 \cdot A \cdot C - 0.001582 \cdot A \cdot B \cdot C \quad (1)$$

4. The Variance analysis (ANOVA) of the regression equation

To use the ANOVA test in order to evaluate the match quality, it is necessary to repeat the experiment at least once in the centre of the experimental range. In this research, the experiments were repeated seven times.

In these circumstances, using ANOVA, it is possible to divide the total variance of a response in two, one due to the regression model and one due to the residual. If the situation in which existing experimental data can be used, (from repeated experiments under the same conditions) the residual variance can be divided into the variance of model error and the variance replication error [7].

In the first stage of the method, the model assumes that the experimental plan was well chosen. Subsequently, it is tested and evaluated whether the model and estimated parameters are significantly different from the experimental data obtained.

After calculating the coefficients model, it is assessed whether they were well calculated as well as whether the deviation value of the responses to the responses predicted value is due to random variation or due to the influence of uncontrolled factors.

The first phase of the method assumes that the experimental plan was well chosen. Subsequently, this aspect is tested and then the model and the estimated parameters are evaluated, in order to observe if they are significantly different from the experimental data obtained.

After the calculation of the model coefficients, the deviation of the obtained response against the value of the predicted responses will be evaluated to determine if the deviation is due to the random variation or due to the influence of uncontrolled factors. Therefore, the calculation method is based on the total sum of squares SSTOTAL define as the sum of squares of differences from the average value

[7]. The next phase implies that the absolute values of these estimated variance have to be compared using Fisher test F [7].

F test compares the ratio of two variances and shows the probability that a statistical difference between two variances will exist. In the first step, using the F-test, the variance of the modelled results is compared with the variance of non-modelled results and, it is considered satisfactory if p - the level of confidence is less than 0,05.

In the second step, the model is compared with the replication error and the results are considered satisfactory if p - the confidence level is greater than 0,05 [7].

Each sum of squares can be divided by the number of freedom degrees (fd) to get the average squares (MS) [7]. Going through these steps by using the MINITAB 17, the ANOVA test can be performed for the regression equation Ra. Thus, the Table 3 presents the analysis results. Since the calculated F value is greater than the tabular F value (Fcalc> Ftab) and is always greater than 1, it may be considered that the relation between the modification of the factors levels and the response it is adjusting the relation statistically significant.

Therefore, the mathematical model is acceptable and it is permitted to describe the evolution of the studied phenomenon in the experimentally studied range, with a strong correlation between the studied factors and responses and therefore a strong correlation between the regression equation and the obtained experimental date.

This is confirmed by the value representing 86% of the suitability of the regression model, determined after this analysis. Specifically, the regression equation Ra predicts the chosen experimental plan at any point of the experimental field, with a degree of probability of 86%.

Table 3. ANOVA of regression equation

Source	FD	SS	MS	Fcalc	Ftab
Regression	5	199,38	39,876	1025,09	2,225
A	1	194,322	0,197	5,054	3,853
C	1	1,194	3,685	94,717	3,853
A x B	1	0,569	0,873	22,429	3,853
A x C	1	2,795	2,528	64,997	3,853
A x B x C	1	0,501	0,501	12,871	3,853
Residual	829	32,258	0,039		
Fitting error	145	18,837	0,130	6,628	1,222
Pure error	684	13,421	0,020		
Total	834	231,638	39,915		

5. Conclusion

After the performing of the regression analysis, the quantitative relationships were identified between the studied cutting process parameters (cutting speed, cutting depth and feed per tooth) and the mean deviation of the surface profile.

The percentages of the influence of each cutting parameter were determined (or combination of cutting parameters) exerted on Ra of the machined surface profile, so:

- Cutting speed holds a majority of influence on surface roughness - by 83,89%, depth of cut affects the surface roughness with a rate of 0,02% and the influence on the feed per on Ra is 0,05%;
- Cutting speed with cutting depth influences the surface roughness with 1,34%, cutting speed with feed per tooth, affects the surface roughness with a rate of 0,16%, cutting depth and feed per tooth, influence the surface roughness with a rate of 0,14% and the combination of cutting speed, cutting depth and feed per tooth, affects the surface roughness with a rate of 0,08%.

- To evaluate the experimental data relating to the regression equation, the analysis of variance ANOVA was used, and finally it was found that the significance of the confidence in the obtained results is 85,92%.

6. Future work

- Methods identification that will lead to obtain some mathematical models of others factors – like tool wear, which influence the surface quality.
- The study of the tool wear evolution in the aluminium machining, based on the varying of the cutting parameters regime.
- Expanding the range of current research on other materials and compared to the studied 7136 aluminium alloy.

7. Acknowledgment

For this paper the logistical support of Universal Alloy Corporation Europe, S.R.L was received.

References

[1] A.B. Bonţiu Pop, Investigating the effect of machining parameters on surface roughness of 7136 aluminum alloy in end milling, The International Conference of the Carpathian Euro-Region specialists in industrial systems Ceursis 2014, 10th edition, 11-13 september, 2014, Baia Mare, Romania, pp. 29-35.

[2] A. Kannan, K. Esakkiraja, M. Nataraj, Modeling and Analysis for Cutting Temperature in Turning of Aluminium 6063 Using Response Surface Methodology. Journal of Mechanical and Civil Engineering (IOSR-JMCE), Vol. 9, Issue 4, PP 59-64, Nov. - Dec. 2013.

[3] M. Ţîţu, L. Tarnu, S. Ţîţu, Andreea Răulea, Implementing the Taguchi method towards the improvement of product quality, Conferinţa Internaţională MECAHITECH 2015 - "7th International Conference on Innovations, Recent Trends and Challenges in Mechatronics, Mechanical Engineering and New High–Tech Products Development", In: Romanian Review Precision Mechanics, Optics & Mechatronics, Nr. 48/2015, ISBN 2247-7063, ISSN 1584-5982, DOI dx.doi.org/10.17683/rrpmom.issue.48., pag. 128-132, 10-11 Septembrie 2015, Bucureşti, Romania, 2015.

[4] M. Ţîţu, L. Tarnu, S. Ţîţu, Andreea Răulea, Optimizing and modeling the drying process of the natural gas through the TAGUCHI method, In: 15th International Conference Research and Development in Mechanical Industry RaDMI-2015, 17-20 September 2015, Topola, Serbia, Volumul 2, pag. 71-79, ISBN 978-86-6075-048-0, Septembrie, 2015.

[5] Z.A. Shekh, T.A. Khan, Analysis of Surface Roughness for Turning of Aluminium (6061) Using Regression Analysis. International Journal of Scientific Engineering and Technology, Vol.No.2, Issue No.7, pp: 705-709, 2013.

[6] D. Vakondios, Kyratsis, P., Yaldiz, A. Antoniadis, Influence of milling strategy on the surface roughness in ball end milling of the aluminum alloy Al7075-T6. Measurement, 45(6), 1480-1488, 2012.

[7] Ţîţu, M., Oprean, C. & Boroiu, A. Cercetarea experimentală aplicată în creşterea calităţii produselor şi serviciilor. Bucureşti: Editura AGIR, 2011.

[8] Montgomery, D. Design and Analysis of Experiments. Eighth Edition ed. Hoboken: John Wiley & Sons, Inc, 2013.

Taguchi versus Full Factorial Design to Determine the Influence of Process Parameters on the Impact Forces Produced by Water Jets Used in Sewer Cleaning

N Medan, M Banica

TECH UNIV CLUJ NAPOCA, NORTH UNIV CENTER BAIA MARE, 62A, Dr. Victor Babes Street, 430083 Baia Mare, Romania

E-mail: Nicolae.Medan@cunbm.utcluj.ro

Abstract. The regular cleaning of the materials deposed in sewer networks is realized, especially with equipment that uses high pressure water jets. The functioning of this equipment is dependent on certain process parameters that can vary, causing variations of the impact forces. The impact force directly affects the cleaning of sewer systems. In order to determine the influence of the process parameters on the impact forces produced by water jets the method of research used is the experiment. The research methods used is that Taguchi design and full factorial design. For the experimental determination of the impact forces a stand for generating water jets and a device for measuring the forces of impact are used. The processing of data is carried out using the Software Minitab 17.

1. Introduction

Industrial cleaning is a classic application of water jets technology. In the late 1950s, when reliable high pressure pumps were built, the usage of water jets spread widely in the field of pipes and sewerage cleaning. Phenomena that occur in the cleaning water jets are complex. Adler [1] describes mechanisms occurring at the impact of a jet with a surface. Leach et al [2], Leu et al [3] and Guha et al [4] analyzed pressure distribution along centreline of the water jet. A number of papers have studied the influence of nozzle geometry on water jet [5,6,7].

The aim of the current paper was to determine the influence of the process parameters on the impact forces produced by waterjets. For determine the influence of the process parameters are used two research methods: Taguchi design and Full factorial design. In order to measure the impact forces between the water jet and a flat and rigid surface, we designed and built a stand for generating pressure jets, as well as a device to measure the impact force.

2. Apparatus used and methodology of the measurements

In order to measure the impact forces between the water jet and a flat and rigid surface, we designed and built a stand for generating pressure jets, as well as a device to measure the impact force.

2.1. *Stand to generate pressure jet*

Schematic diagram of the stand to generate pressure jet is shown in figure 1.

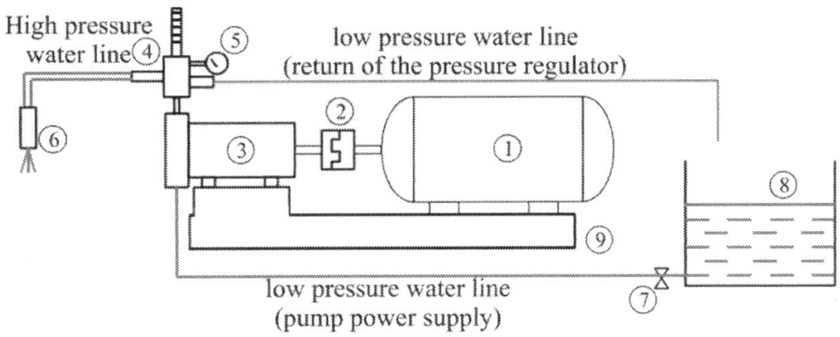

Figure 1. Schematic diagram of the stand to generate pressure jet

Component parts of stand: (1) electric motor (2) flexible coupling; (3) high pressure pump, 4) pressure regulator, 5) pressure gauge, 6) nozzle, 7) tap water, 8) water tank, 9) chassis.

Water coming out of the high pressure pump (3) goes into the pressure regulator (4). Through it adjusts the pressure and flow of water in the path of the high pressure water. This pressure corresponds to the one at the outlet of nozzle.

2.2. *Device for the measurement of the impact force*

In figure 2 is represented the principle diagram of the device for the measurement of the impact force of the water jet and a flat and rigid surface.

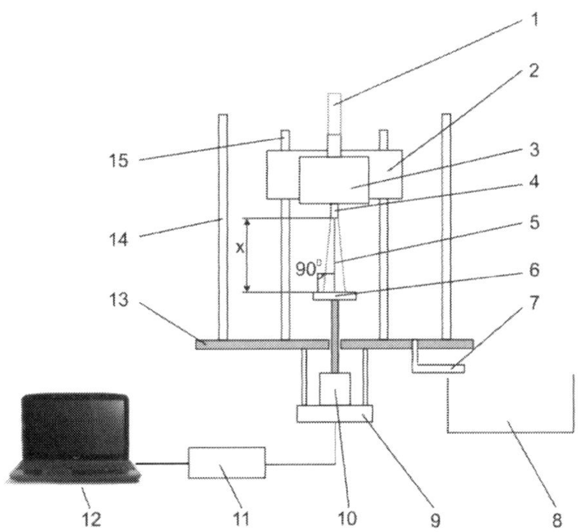

Figure 2. Diagram of the device for the measurement of the impact force of the water jet

In figure 2 is represented the principle diagram of the device for the measurement of the impact force of the water jet and a flat and rigid surface.

Main component parts of the device are: 1) high-pressure water hose, 2) support nozzle, 3) nozzle block, 4) nozzle, 5) water jet, 6) flat and rigid target plate, 7) collection path water, 8) scaled container for measurement of the flow of water jet, 9) piezoelectric sensor mounting, 10) piezoelectric sensor, 11) data acquisition Personal Daq/3000, 12) computer for the processing of data; 13) support plate, 14) acrylic tube, 15) rods for adjusting distance x.

From high pressure water hose (1) come water at a certain pressure p desired. At the outlet of nozzle is generated a water jet (5) that striking target plate (6), who is located at a certain distance x in front of the nozzle. The jet (5) generates an impact force at a time when he meets target plate (6). This force produces axial movement of target plate. This movement is converted into an electric signal by the piezoelectric sensor (10). Electrical signal is collected by data acquisition Personal Daq/3000 (11), which forward data to a computer (12) using DaqView soft processes data actually obtained.

2.3. *Process parameters and geometric configuration of nozzles*

In the water jet cleaning process involved a series of parameters [8]. These parameters can be divided into two major groups, namely: 1) Parameters which shall be defined according to the contact area between the water jet and the surface to be cleaned and 2) the process parameters. In the measurement of the impact forces of a stationary water jet and flat and rigid surface the process parameters involved (figure 3).

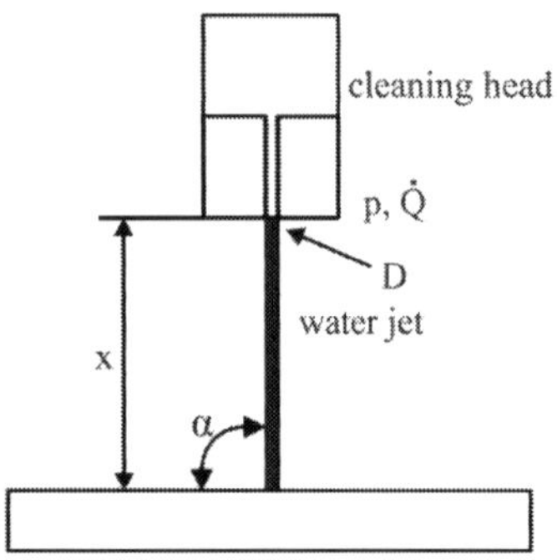

Figure 3. Process parameters

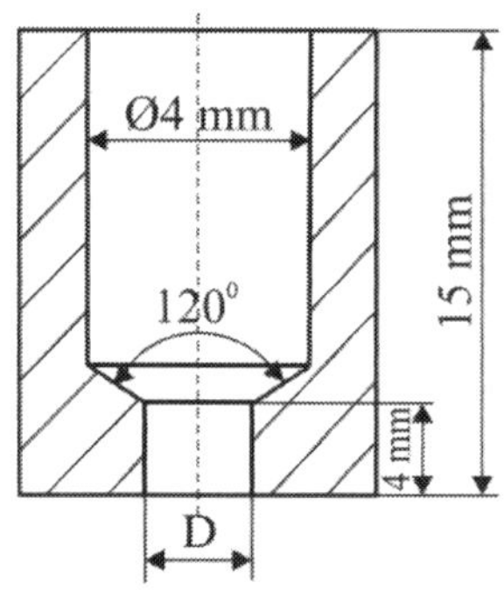

Figure 4. Geometric configuration of the nozzles

They can be divided into two groups, namely hydraulic parameters and the performance parameters. Hydraulic parameters characterized the system high pressure pump-nozzle are represented by work pressure (p), volume flow ($\dot{Q}$) and nozzle diameter (D).

The performance parameters refer to the process of cleaning itself, and are the following:

1) The diameter D of nozzle. The values used are D=1mm, 1.5 mm and 2 mm. The material for nozzles is stainless steel. Geometric configuration of the nozzles used is shown in figure 4.

2) The pressures p used to perform the measurements have the values p=100 bar, p=120 bar, p=140 bar, p=160 bar, p=180 bar and p=200 bar.

3) Cleaning distance x (distance from nozzle to target plate). To perform the measurements distance x has been fixed at the values x=25 mm, x=50 mm, x=75 mm, x=100 mm, x=125 mm, x=150 mm, x=175 mm, x=200 mm.

4) α – impact angle (the angle formed by jet with the target plate); in this work α =60^0, 75^0 and 90^0.

The pressures used to perform the measurements have the values p=100 bar, p=120 bar, p=140 bar, p=160 bar, p=180 bar and p=200 bar.

3. Results

For determine the influence of the process parameters are used two research methods: Taguchi design and Full factorial design.

3.1. *Taguchi method*

In table 1 are presented the values of the used parameters for Taguchi method.

For Taguchi method, each parameter has 2 values, the minim and the maxim, in concordance with the experimental domain.

In table 1 are presented the values of parameters for Taguchi method.

Table 1. The values of the parameters for Taghuci method.

Parameter	Value 1 (minimum)	Value 2 (maximum)
Nozzle diameter D [mm]	1	2
Pressure p [bar]	100	200
Impact angle α [^{0}]	60	90
Distance x [mm]	25	200

Based on the parameters established in accordance with Table 1 the impact forces in Table 2 were determined. For each experiment was performed three measures of impact force and still it has worked with F_{med}, representing the arithmetic average of the three forces measured.

Table 2. The values of impact forces. (Taguchi method).

Experiment no	Parameters				Impact force F [N]			
	Diameter D [m]	Pressure p [bar]	Distance x [m]	Angle α [^{0}]	1	2	3	F_{med}
1	1	100	25	60	8.08	7.69	7.94	7.90
2	1	100	25	90	10.49	10.48	10.36	10.44
3	1	200	200	60	14.79	14.92	14.55	14.75
4	1	200	200	90	19.32	18.70	19.08	19.03
5	1	100	200	60	6.39	6.15	6.40	6.31
6	1	100	200	90	7.84	7.83	8.03	7.90
7	1	200	25	60	17.12	16.98	16.69	16.93
8	1	200	25	90	21.67	21.72	21.83	21.74
9	2	100	200	60	29.06	29.94	29.46	29.48
10	2	100	200	90	36.97	36.71	36.21	36.63
11	2	200	25	60	65.37	65.10	64.80	65.09
12	2	200	25	90	86.00	85.98	86.24	86.07
13	2	100	25	60	30.82	31.31	30.72	30.95
14	2	100	25	90	40.44	40.30	40.18	40.30
15	2	200	200	60	59.95	59.47	60.14	59.85
16	2	200	200	90	76.32	76.30	76.35	76.32

On the basis of the data in table 2 and using the method Taguchi has conducted an analysis of the variance to determine the influence of each parameter and their interactions on the impact force. In Table 3 is presented the analysis of the variance, using Minitab 17.

Table 3. Analysis of Variance (Taguchi method).

Source	DF (degree of freedom)	SS (sum of square)	Contribution
Regression	10	9897,72	99.76%
Diameter	1	6387,61	64.38%
Pressure	1	2253,08	22.71%
Distance	1	53,12	0.54%
Angle	1	281,91	2.84%
Diameter*Pressure	1	757,39	7.63%
Diameter*Distance	1	7,72	0.08%

Table 3. Analysis of Variance (Taguchi method). (continued)

Source	DF (degree of freedom)	SS (sum of square)	Contribution
Diameter*Angle	1	103,67	1.04%
Pressure*Distance	1	7,02	0.07%
Pressure*Angle	1	42,01	0.42%
Distance*Angle	1	4,20	0.04%
Error	5	23,58	0.24%
Total	15	9921,29	100.00%

Based on table 3 (using Taguchi method) results the contribution of the parameters and their interactions on the impact force F:

- nozzle diameter D, with a percentage of influence of 64.38%;
- pressure p, with a percentage of influence of 22.71%;
- interaction between nozzle diameter D and pressure p , with a percentage of influence of 7.63%;
- impact angle α, with a percentage of influence of 2.84%.

The distance parameter and all the other interactions have a value less than 1.04%.

3.2. *Full factorial method*

In table 4 are presented the values of the used parameters for Full factorial method.

Table 4. The values of the parameters for Full factorial method.

Parameter	Values
Nozzle diameter D [mm]	1, 1.5, 2
Pressure p [bar]	100, 120,140,160,180,200
Impact angle α [^{0}]	60, 75, 90
Distance x [mm]	25, 50,75,100,125,150,175,200

Taking into account that there are four parameters, the number of values which they take each of them and using the Full factorial design results a number of 143 experiments. Just like in the case of Taguchi method, for each experiment have carried out a number of three measurements to determine the impact force. As a result there are a number of 432 measurements. With this in mind, for brief of this paper, we dare not to present the table with the values obtained.

Using the Full factorial method has conducted an analysis of the variance to determine the influence of each parameter and their interactions on the impact force. In Table 5 is presented the analysis of the variance, using Minitab 17.

Table 5. Analysis of Variance (Full factorial method).

Source	DF (degree of freedom)	Seq. SS (sum of square)	Contribution
Regression	10	162559	99.14%
Diameter	1	116939	71.32%
Pressure	1	29592	18.05%
Distance	1	389	0.24%
Angle	1	5959	3.63%
Diameter*Pressure	1	7596	4.63%
Diameter*Distance	1	148	0.09%
Diameter*Angle	1	1523	0.93%
Pressure*Distance	1	20	0.01%

Table 5. Analysis of Variance (Full factorial method). (continued)

Source	DF (degree of freedom)	Seq. SS (sum of square)	Contribution
Pressure*Angle	1	385	0.24%
Distance*Angle	1	5	0.00%
Error	421	1406	0.86%
Total	431	163965	100.00%

Based on table 5 (using Full factorial method) results the contribution of the parameters and their interactions on the impact force F:

- nozzle diameter D, with a percentage of influence of 71.32%;
- pressure p, with a percentage of influence of 18.05%;
- interaction between nozzle diameter D and pressure p , with a percentage of influence of 4.63%;
- impact angle α, with a percentage of influence of 3.63%.

The distance parameter and all the other interactions have a value less than 0.93%.

4. Conclusions

In table 6 is presented the contribution of the parameters and their interactions on the impact force F, using Taguchi method and Full factorial method.

Table 6. Contribution of parameters and their interactions (Taguchi versus Full factorial).

Source	Contribution		The difference
	Taguchi method	Full factorial method	
Regression	99.76%	99.14%	0.62%
Diameter	64.38%	71.32%	6.94%
Pressure	22.71%	18.05%	4.66%
Distance	0.54%	0.24%	0.3%
Angle	2.84%	3.63%	0.79%
Diameter*Pressure	7.63%	4.63%	3%
Diameter*Distance	0.08%	0.09%	0.01%
Diameter*Angle	1.04%	0.93%	0.11%
Pressure*Distance	0.07%	0.01%	0.06%
Pressure*Angle	0.42%	0.24%	0.16%
Distance*Angle	0.04%	0.00%	0.04%
Error	0.24%	0.86%	0.62%
Total	100.00%	100.00%	0.00%

Based on the results of experimental work presented in this paper it is found that:

1) Regardless of the method used (Taguchi or full factorial) the classification of parameters and their interactions in the order of importance is the same:

- The most important parameter is the diameter of the nozzle, with a percentage of influence of 64.38% for Taguchi method and 71.32% for Full factorial method;
- The second parameter is the pressure of water jet, with a percentage of 22.71% for Taguchi method and 18.05% for full factorial method;
- The third parameter is the impact angle, with a percentaje of 2.84% for Taguchi method and 3.63% for Full factorial method;
- The distance parameter, with a percentaje of 0.54% for Taguchi and 0.24% for Full factorial method is virtually insignificant;
- Between interactions only diameter and pressure have an important contribution with 6.63% for Taguchi method and 4.63% for Full factorial method;

- All other interactions that have a low importance in both cases (Taguchi and Full factorial methods).

2) In the both cases, for our experimental domain, the distance parameter have an insignifiant value.

3) Although between the two cases there are differences between the calculated values, the order of importance of the parameters and their interactions is the same. But for Taguchi method were carried out 48 measurements while for full factorial were carried out a number of 432 measurements.

The final conclusion: is more convenient to make a research using Taguchi method to determine the influence of the parameters and the interactions between them, because the number of measurements is significantly less than in the case of the full factorial method.

References

[1] Adler W F 1979 The Mechanics of Liquid Impact *Treatise on Materials Science and Technology* 127-183

[2] Leach S J Walker G L 1966 Some Aspects of Rock cutting by High Speed Water Jets *Philosophical Transactions of the Royal Society of London* Series A Vol. 260 295-308

[3] Leu M C *et al* 1998 Mathematical modelling and experimental verification of stationary waterjet cleaning process *Journal of Manufacturing Science and Engineering* 120(3) 571-579

[4] Guha A *et al* 2011 An Experimental and Numerical Study of Water Jet Cleaning Process *Journal of Materials Processing Technology* 610-618

[5] Liu Y *et al* 2010 Development of Impact Test System for Waterjet Descaling Nozzles with LabVIEW, *Web Information Systems and Mining (WISM) International Conference* vol.1 3-7

[6] Annoni M *et al* 2008 Orifice Coefficients Evaluation for Water Jet Application *16th IMECO TC4 Symposium* Florence, Italy

[7] Cotetiu A *et al* 2014 Research about automatic adjustement solution of the advance force at the perffusion drills using fluid elements *Archives of Mining Sciences* Volume 58 Issue 4 1201–1208

[8] Momber A W 2003 Hydroblasting and Coating of Stell structures Oxford Elsevier Ltd

Aspects concerning verification methods and rigidity increment of complex technological systems

M Casian
Technical University of Moldova, 9, Studenţilor str., Chişinău, Republic of Moldova
E-mail: max_casian@yahoo.com

Abstract. Any technological process and technology aims a quality and precise product, something almost impossible without high rigidity machine tools, equipment and components. Therefore, from the design phase, it is very important to create structures and machines with high stiffness characteristics. At the same time, increasing the stiffness should not raise the material costs. Searching this midpoint between high rigidity and minimum expenses leads to investigations and checks in structural components through various methods and techniques and sometimes quite advanced methods. In order to highlight some aspects concerning the significance of the mechanical equipment rigidity, the finite element method and an analytical method based on the use Mathcad software were used, by taking into consideration a subassembly of a grinding machine. Graphical representations were elaborated, offering a more complete image about the stresses and deformations able to affect the considered mechanical subassembly.

1. Introduction

The stiffness criterion of the machine tools together with resistance criterion are ones of the most important criteria to be followed during the process of designing a mechanical equipment. The role of the rigidity is steadily increasing, on the one hand, due to increasing of requirements for precision, and in connection with stopping the growth of the elastic modulus of the material, at the same time with the increasing of the resistance characteristics. In the machine tools building industry, the stiffness criterion has an especially important role. Precision machine tools must be designed more massive than conventional machines for one and the same forces and powers [1]. However, the rigidity should not be regarded as an independent criterion. It affects the work of machine tools by means of machining precision, resistance to vibrations and the durability related to the conditions of contact between parts.

The stiffness of the technological system along a given axis means the ratio of the cutting force on the respective axis and the elastic deformation in the same direction:

$$j = \frac{F}{y},\ N/mm \tag{1}$$

where F – deformation force and y – deformation, appeared from the action of the deformation.

Stiffness of the elastic technological system definitely affects the entire manufacturing process. The deformation of the system is directly proportional related to the size of forces and decisively to the cutting force component.

Thus, the permanent monitoring of this force variation during the cutting process could highlight the continuous deformation of the system. In order to optimize the grinding process and increase productivity, it is necessary to impose some major requirements aiming increasing system rigidity and keeping constant deformation, reducing the time needed to charge the system until adjusted parameters are equal to the real ones and also to reduce the time needed to detente phase.
Besides static stiffness, we can also mention the dynamic stiffness. Dynamic stiffness is very important in the correct study of the processing precision, especially when considering variations in cutting forces, these forces are compared to their nominal values.
In addition, there is a variation in the time (no inconsiderable):

$$\frac{dF_y}{d\tau} \neq 0 \tag{2}$$

where dF_y is the variation of the force exerted along the y axis and $d\tau$- a short time interval.
In the study of dynamic stiffness, there is the dynamic component ΔFd that generates a dependence of the machining accuracy on the dynamic characteristic of elastic technological system, evaluated just by the dynamic stiffness and dynamic failure of this system.
The value of system rigidity significantly influences manufacturing accuracy, especially in case of high specific values of the cutting parameters (such is the case of roughing operations). It is evident that only some of the errors that occur during machining process are due to elastic deformations. It is correct that these errors must be only a fraction of the tolerance that has to be ensured. This condition can be expressed as:

$$y_{din} = y_{stat} \cdot \chi \leq y_{adm} \tag{3}$$

where y_{din} - deformation of system taking into account the dynamic stiffness, y_{stat}- static deformation of system, χ - dynamic coefficient, y_{adm} - the allowable deformation of system.
Cutting machines are complex systems, whose constructive solutions depend on the direction of change of the cutting forces and the positions of the nodes in them can need a variety of joint surfaces, respectively, and various stiffness values could be necessary. Therefore, the experimental tests must be as close as possible concordant to the most realistic typically processing.
Over the years, the researchers were interested to obtain scientific information and to establish mathematical models able to highlight the significance of the concept of the rigidity for the behaviour of the mechanical components and to establish conditions in which a higher rigidity could be obtained. Thus, Boca developed [1] a research concerning the influence of the rigidity of the universal lathe on the machining accuracy. He proposed some mathematical models able to highlight the influence of the elastic deformation on the machined surface shape in transversal sections and the change of the tool desired trajectory under the action of the cutting forces. Another research subject addressed in [1] referred to the shape error of the machined surface generated by the deformation of the technological system along a radial direction.
Kudoiarov et al. took into consideration the possibilities of diminishing the elastic deformations of the main shaft as a machine tool subassembly during the turning process [3]. A software able to determine the influence of turning conditions on the machined surface errors was proposed, by considering the technological system rigidity. The authors concluded that in order to obtain manufactured parts characterized by a high machining accuracy, information concerning the lathe rigidity must be known and used.
The objective of the research presented in this paper was to apply the finite element method and of an analytical method in order to study some aspects and identify optimized solutions for ensuring a high rigidity of the cutting machine tool components.

2. Verification methods

The research of the static rigidity is performed by experimental methods applied directly on the machine tools and developing spatial mechanical (physical) models or by using theoretical analyses. The total rigidity of the machine tool could be evaluated by measurements developed directly on the machine tool. The balance of the elastic deformations, namely the total amount of movements recorded by the measuring instruments on the workpiece, could be achieved by taking into consideration the individual movements of the support system elements (Figure1).

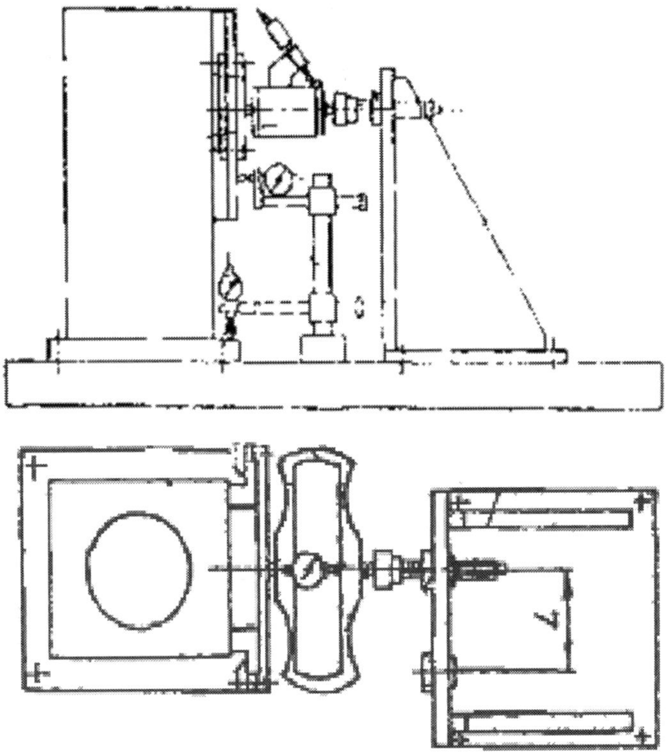

Figure 1. Stand for checking the static rigidity of the physical model

The balance of elastic displacements could be done in the following order:
1) Evaluation of the stiffness of individual elements;
2) Estimation of the rigidity effect of some elements on the total stiffness of the machine tool and subsequently the stiffness increasing of the weak elements;
3) Testing of hypotheses that have been put into the calculation and simplifying/ clarifying the calculation [2].

Depending on each purpose, the movements balance can be done at various levels of detail. In order to obtain an overall distribution of elastic displacements in nodes and to determine the main directions of increasing the rigidity, the overall balance of elastic movements is determined. At the same time, the measuring devices/instruments must be placed so that they ensure the possibility of determining the components of movement of the force application point compared to movements of the main nodes of the machine tool.

If it is necessary to improve the positions of machine nodes (this meaning an increase of machine tool rigidity), to take into consideration the information concerning elements rigidity and the verification of hypotheses initially formulated, the analysis of balance of elastic displacements could be developed in detail. At the same time, the displacements components of forces application points compared to own deformations in elements connections [3] should be considered.

3. Grinding elastic system - verification

Another methods of checking the deformations are represented by effort method and displacement method used in the strength of materials and applied in order to establish dimensions and to test structures, bars and solid bodies; these methods could be considered as precise methods (figure 2). In order to highlight the deformation of parts that directly support the cutting forces, we will study the system consisting of the motor shaft and abrasive wheel (figures 2-3). We will use in our analysis two methods: finite element method (by means of SolidWorks 2010, Simulation 2010) and analytical method (by means of Mathcad14) software [4], respectively.

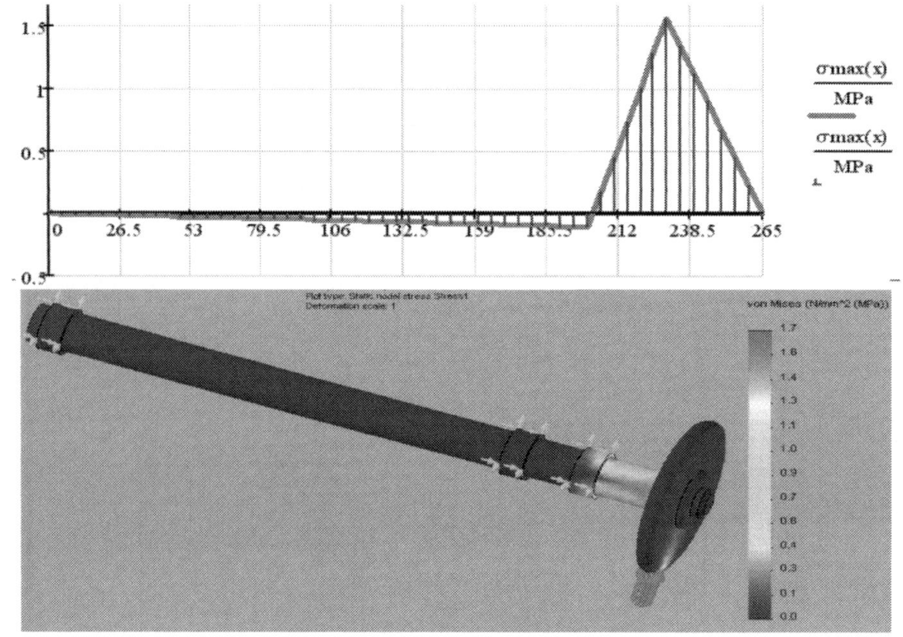

Figure 2. Stress distribution

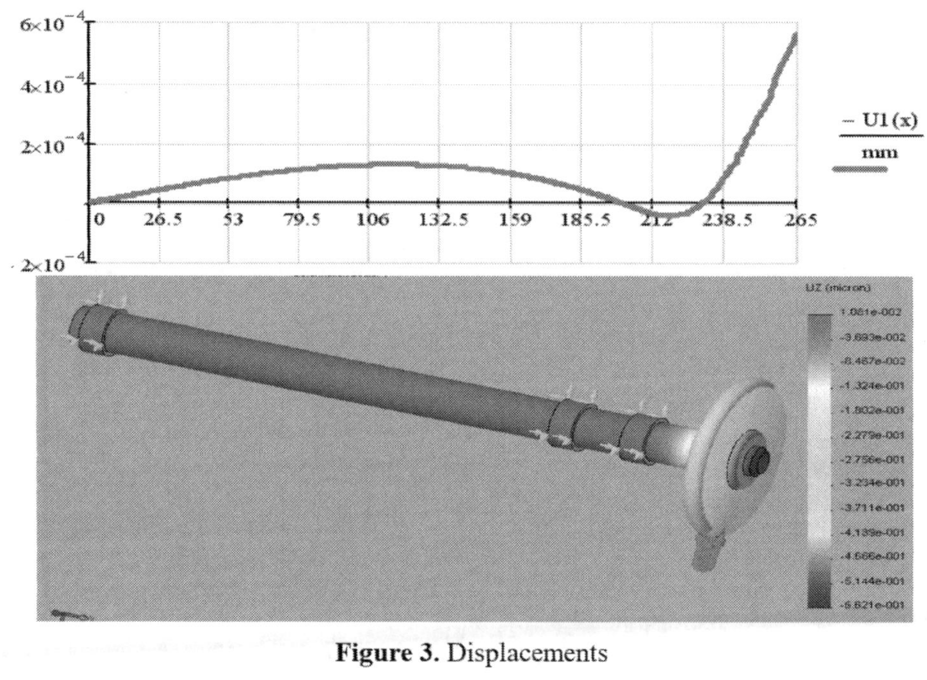

Figure 3. Displacements

The approximate solutions in analytic form are less used. Most often numerical solutions using computers were obtained (figure 4). The existence of a huge number of computer programs and of personal computers have made the numerical methods are very easy to be used obtaining results in a short time. One may show that the results could be offered in a very suggestive and easy to interpret form especially, namely in graphical form (figures 4-5).
The advantage of finite element method (FEM) are numerous and important. When planning a new design, we can model and study the behaviour of the structure in various load environments. Therefore, based on the obtained results, the model can be modified before the execution of the final drawings. Once the CAD model is developed, the design of structure can be analysed in detail applying FEM.

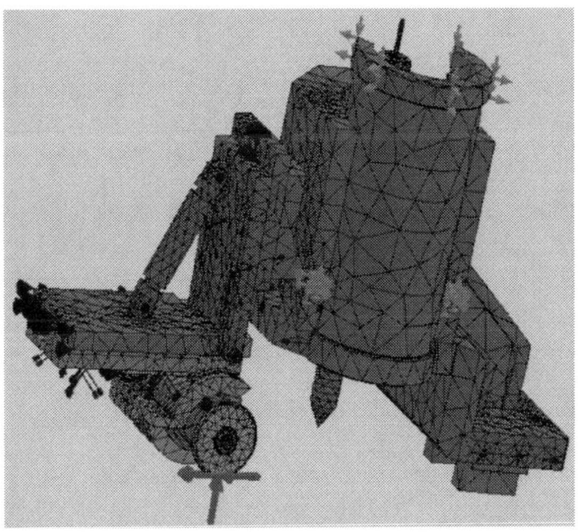

Figure 4. Meshing of CAE model

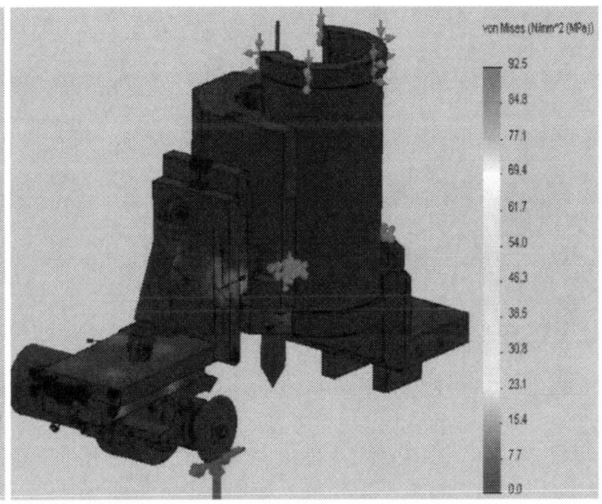

Figure 5. Stress (von Mises)

Using FEM, one can save time and money by reducing the number of required prototypes. In the case of an existing product with certain problems during use, or when improvements are required, the design model could be analysed using FEM, in order to accelerate the design changes and also to reduce design costs.

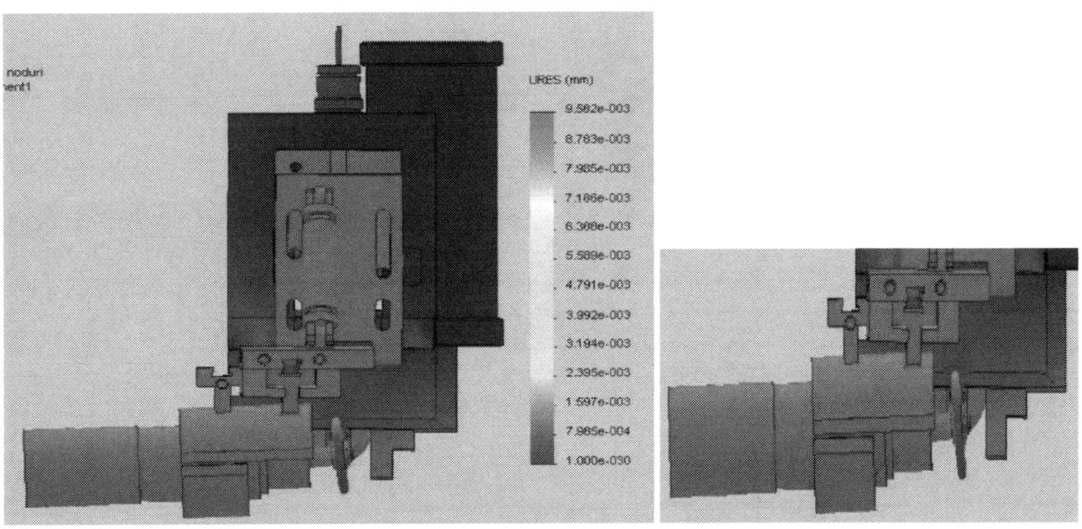

Figure 6. Total deformations of system (CAE model)

Usually the values obtained by means of CAE model (figure 6) and amounts obtained by the classic resistance calculation are very close to each other, and we can say with certainty that a calculation developed by means of FEM is no worse than a classic calculation (even sometimes it has superior advantages, if handled quite well).

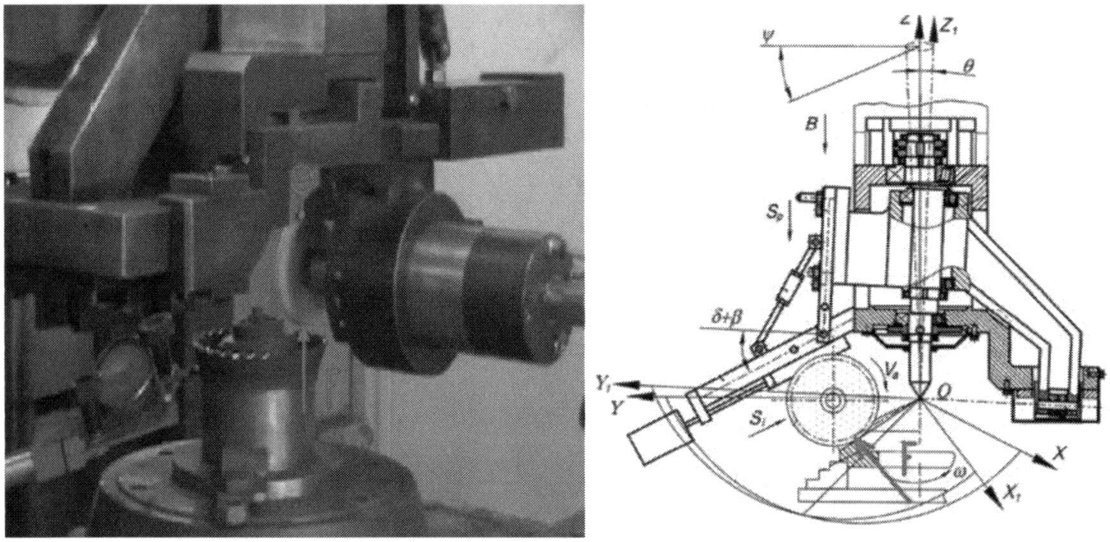

Figure 7. Determination of static displacements on the machine tool

Experimental methods for measuring the machine tool stiffness are limited to measuring the forces that act in loading nodes and corresponding elastic deformations. Loading nodes in machine tool is done by means of certain loading devices. Elastic deformations could be measured by using standard dial gauges (figure 7) [5]. Loading devices must ensure the possibility of generating in machine tool forces characterized by values and directions corresponding to the points where it can be considered as acting the cutting forces. The constructive solutions of loading devices and used parts, of methods of installation on the machine tool can be very various.

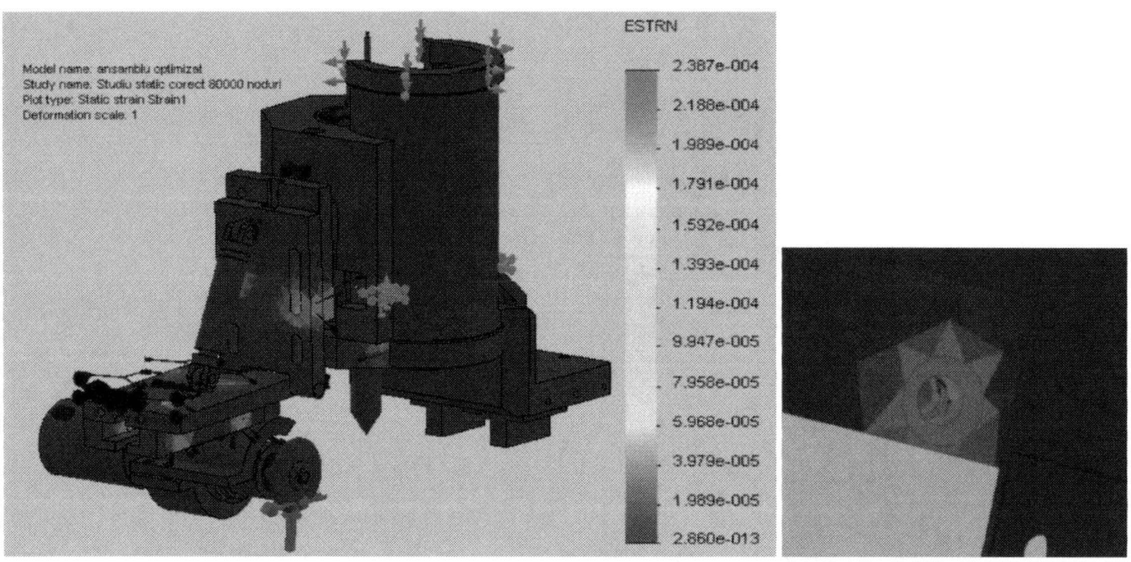

Figure 8. Equivalent strain for entire system and for the place of bolt

Complex universal devices allow to measure the joint stiffness for direct and indirect loading. For example, direct loading refers to exterior turning, and the indirect – to interior turning [6]. Direct and indirect loading give the possibility to evaluate not only the rigidity of nodes, but sometimes also to determine the value of clearances.
The safety verification of the system must be made taking into account the distribution of equivalent strains. For example, in figure 8, we can see the highest values of deformations [4] and thus we have information about the zones where the deformation are too high.
If the values corresponding to the deformations are analysed, it can say that they are less than offset values for the entire technological system and this demonstrates that the decrease of stiffness factor must be considered in places of parts and assemblies contacts.

4. Appendices

The importance of knowing stiffness values of technological systems components is limited in evaluation of quality and accuracy of the finished product. Often these values are low and, as a result, technological problems occur and the product is characterized by a low quality. If high values of the stiffness are necessary, the cost of technologic systems is high. Therefore, it is very important to find a middle value and this solution is almost impossible without application of verification methods and techniques. These methods and techniques should be studied and applied both individual and combined, in order to better understanding the system stiffness which is a feature quite complex.

Thus, the rigidity increase of the machine tools can be achieved [7] by:

- Increasing the intrinsic rigidity of machine tool parts, by selecting the most advantageous sections shapes and edges, frame structures, additional supports, materials with high value of modulus of elasticity;
- Reducing the number of joints in nodes and between them;
- Selection of adequate clearances, application of preload and compensators that automatically establish the clearances;
- Improving the quality (surface roughness) of the contact surfaces;
- Careful assembly of parts and the machine tool itself.

References

[1] M. Boca, *Cercetări teoretice şi experimentale privind eroarea de prelucrare cauzată de rigiditatea sistemului tehnologic la strunjire*, Doctoral thesis, Iaşi, 2011.

[2] Iu. V. Kirilin, *Ispitania i isledovania metalorejushih stankov: metodicheskie ucaznia k laboratornim rabotam*, UlGTU, Ulianovsk, 2012.

[3] R. G. Kudoiarov, R. R. Basharov, O. D. Pirkova, *Method of determining and increasing the technological rigidity of lathes* (in Russian), Vestnik Ufimskovo Gosudarstvennovo Aviatsionnnovo Technicheskovo Universiteta, vol. 18, no.3 (64), pp. 87-92, 2014.

[4] A. A. Alyamovskii, A. A. Sobachkin, Ye. V. Odintsov, A. I. Kharitonovich, N. B. Ponomarev, SolidWorks 2007/2008. *Computer modeling in engineering practice* (in Russian), BKHV-Peterburg, 2008.

[5] Bostan I., Mazuru S., Olevschi A., *Procedeu de prelucrare a dinţilor angrenajului precesional.* MD Patent 3532 (2006).

[6] Ia. N. Otenii, *Isledovania jestkosti tehnologhicheskoi sistemi: metodicheskie ucaznia c laboratornim rabotam*, RPK „Politehnic" , Volgograd, 2004.

[7] Ryzhov E.V., Sagarda A.A., Il'inskii V.B., Chelovetskii I.Kh., *Kachestvo poverkhnosti pri almazno-abrazivnoi obrabotke [Surface quality when diamond abrasive machining]*. Kiev, Naukova dumka Publ., 1979.

IOP Conf. Series: Materials Science and Engineering **161** (2016) 012018 doi:10.1088/1757-899X/161/1/012018

Optimization of the shell compliance by the thickness changing using PSO

M Szczepanik, A Poteralski and M Kalinowski

Institute of Computational and Mechanical Engineering, Faculty of Mechanical Engineering, Silesian University of Technology, Poland

Email: miroslaw.szczepanik@polsl.pl

Abstract. The paper is devoted to an application of the particle swarm methods and the finite element method to the optimization of 2-D structures (plane stress/strain, bending plates and shells). The shape, topology and material or thickness of the structures are optimised for the stress, strain and volume criteria. The numerical examples demonstrate that the method based on particle swarm computation is an effective technique for solving computer aided optimal design.

1. Introduction

Reinforced structures are often used in practice because they are resistant, stiff and stable. A typical area of application of such structures is an aircraft and automotive industry where light, stiff and highly resistant structures are required. Many car and aircraft elements are made as thin panels reinforced by stiffeners or by changing thickness. Reinforced structures are frequently subjected to static or dynamic loads and it is important to get an information about their response. If the response is not satisfactory, it must be improved in order to satisfy proper requirements. These improvements can be achieved, for instance, by the optimization process [1]. Optimal choice of number of stiffeners and their locations or thickness arrangement in a structure decides about the effectiveness of the reinforcement. In the paper the reinforced structures are subjected to static loads and analyzed by the FEM. The aim of optimization is to find the optimal thickness of the structure by minimize maximal displacements with constrains imposed on volume of the structure.

Strength of structures [2][3] with an arbitrary geometry, material properties and boundary conditions can be obtained by carrying out laboratory tests but they are usually very expensive and time consuming. In order to reduce costs and time, computer simulations are performed instead of experimental investigations. As a result, static or dynamic quantities of interest like displacements, velocities, accelerations, forces, stresses, i.e. can be determined. The most versatile methods of analysis of structures subjected to arbitrary static and time dependent boundary conditions are the finite element method (FEM) or boundary element method (BEM).

In the present paper, FEM with swarm method in optimization of statically loaded reinforced structures is presented. The additional comparisons of the effectiveness of particle swarm optimizer (PSO) with the effectiveness of the evolutionary algorithms (EA) and the artificial immune systems (AIS) are presented in the papers [4][5][6].

2. The particle swarm optimiser

The particle swarm algorithms, similarly to the evolutionary and immune algorithms, are developed on the basis of the mechanisms discovered in the nature. The swarm algorithms are based on the models of the animals social behaviours: moving and living in the groups. The animals relocate in the three-dimensional space in order to change their stay place, the feeding ground, to find the good place for reproduction or to evading predators. We can distinguish many species of the insects living in swarms, fishes swimming in the shoals, birds flying in flocks or animals living in herds.

A simulation of the bird flocking was published by Raynolds [7]. He assumed that this kind of the coordinated motion is possible only when three basic rules are fulfilled: collision avoidance, velocity matching of the neighbours and flock centring. The computer implementation of these three rules showed very realistic flocking behaviour flaying in the three dimensional space, splitting before obstacle and rejoining again after missing it. The similar observations concerned the fish shoals. Further observations and simulations of the birds and fishes behaviour gave in effect more accurate and more precise formulated conclusions [7]. The results of this biological examination where used by Kennedy and Eberhart [8], who proposed Particle Swarm Optimiser – PSO. This algorithm realizes directed motion of the particles in n-dimensional space to search for solution for n-variable optimization problem. PSO works in an iterative way. The location of one individual (particle) is determined on the basis of its earlier experience and experience of whole group (swarm). Moreover, the ability to memorize and, in consequence, returning to the areas with convenient properties, known earlier, enables adaptation of the particles to the life environment. The optimization process using PSO is based on finding the better and better locations in the search-space (in the natural environment that are for example hatching or feeding grounds).

The algorithm with continuous representation of design variables and constant constriction coefficient (constricted continuous PSO) has been used in presented research. In this approach each particle oscillates in the search space between its previous best position and the best position of its neighbours, with expectation to find new best locations on its trajectory. When the swarm is rather small (swarm consists of several or tens particles) it can be assumed that all the particles stay in neighbourhood with currently considered one. In this case we can assume the global neighbourhood version and the best location found by swarm so far is taken into account – current position of the swarm leader (figure 1).

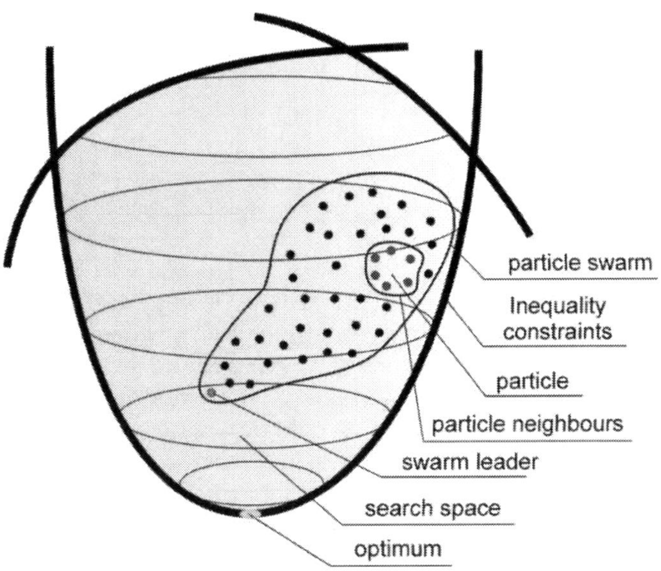

Figure 1. The idea of the particle swarm

The position of the *i*-th particle is changed by stochastic velocity v_i, which is dependent on the particle distance from its earlier best position and position of the swarm leader. This approach is given by the following equations:

$$v_{ij}(k+1) = wv_{ij}(k) + \phi_{1j}(k)\left[q_{ij}(k) - d_{ij}(k)\right] + \phi_{2j}(k)\left[\hat{q}_{ij}(k) - d_{ij}(k)\right] \quad (1)$$

$$d_{ij}(k+1) = d_{ij}(k) + v_{ij}(k+1), \quad i = 1,2,...,m\ ;\ j = 1,2,...,n \quad (2)$$

where:
$\phi_{1j}(k) = c_1 r_{1j}(k);\ \phi_{2j}(k) = c_2 r_{2j}(k)$,
m – number of the particles,
n – number of design variables (problem dimension),
w – inertia weight,
c_1, c_2 – acceleration coefficients,
r_1, r_2 – random numbers with uniform distribution [0,1],
$d_i(k)$ – position of the i-th particle in k-th iteration step,
$v_i(k)$ – velocity of the i-th particle in k-th iteration step,
$q_i(k)$ – the best found position of the i-th particle found so far,
$\hat{q}_i(k)$ – the best position found so far by swarm – the position of the swarm leader,
k – iteration step.

The velocity of *i*-th particle is determine by three components of the sum in Eqn 1. The first component $wv_i(k)$ plays the role of the constraint to avoid excessive oscillation in the search space. The inertia weight w controls the influence of particle velocity from the previous step on the current one. In this way this factor controls the exploration and exploitation. Higher value of inertia weight facilitates the global searching, and lower – the local searching. The inertia weight plays the role of the constraint applied for the velocities to avoid particles dispersion and guaranteeing convergence of the optimization process. The second component $\phi_1(k)[q_i(k) - d_i(k)]$ realizes the cognitive aspect. This component represents the particle distance from its best position found earlier. It is related to the natural inclination of the individuals (particles) to the environments where they had the best experiences (the best value of the fitness function). The third component $\phi_2(k)[\hat{q}_i(k) - d_i(k)]$ represents the particle distance from the position of the swarm leader. It refers to the natural inclination of the individuals to follow the other which achieved a success.

The flowchart of the particle swarm optimiser is presented in figure 2. At the beginning of the algorithm the particle swarm of assumed size is created randomly. Starting positions and velocities of the particles are created randomly. In the next step the objective function values are evaluated for each particle. The data necessary for calculation of the objective function are obtained in consequance of solving the boundary value problem by means of the FEM (described in the next section). After obtaining information about fitness of all the particles, the best positions of the particles are updated and the swarm leader is chosen. Then the particles velocities are modified by means of the Eqn 1 and particles positions are modified according to the Eqn 2. The process is iteratively repeated until the stop condition is fulfilled. The stop condition is typically expressed as the maximum number of iterations. The general effect is that each particle oscillates in the search space between its previous best position (position with the best fitness function value) and the best position of its best neighbour (relatively swarm leader), hopefully finding new best positions (solutions) on its trajectory, what in whole swarm sense leads to the optimal solution.

The modified version of PSO algorithm with additional procedure has been used in presented research and was originally proposed by the author. The additional procedure consists of two stages. First the clones of the swarm leader are created. The clones substitute the particles with the worst fitness. Next the clones parameters are mutated with the declared probability by adding random numbers with uniform distribution from the range of the design parameters variation

$$\hat{q}_{ij}(k+1) = \hat{q}_{ij}(k) + rand\left[h_j^{\min},\ h_j^{\max}\right] \tag{3}$$

where $rand\left[h_j^{\min},\ h_j^{\max}\right]$ - function which returns the random numbers with uniform distribution from the range of the design parameters variation $\left[h_j^{\min},\ h_j^{\max}\right]$. The prescribed PSO parameters are: number of particles = 30, inertia weight w = 0.73, acceleration coefficients c_1, c_2=1.47, number of the clones = 5, probability of mutation = 50%.

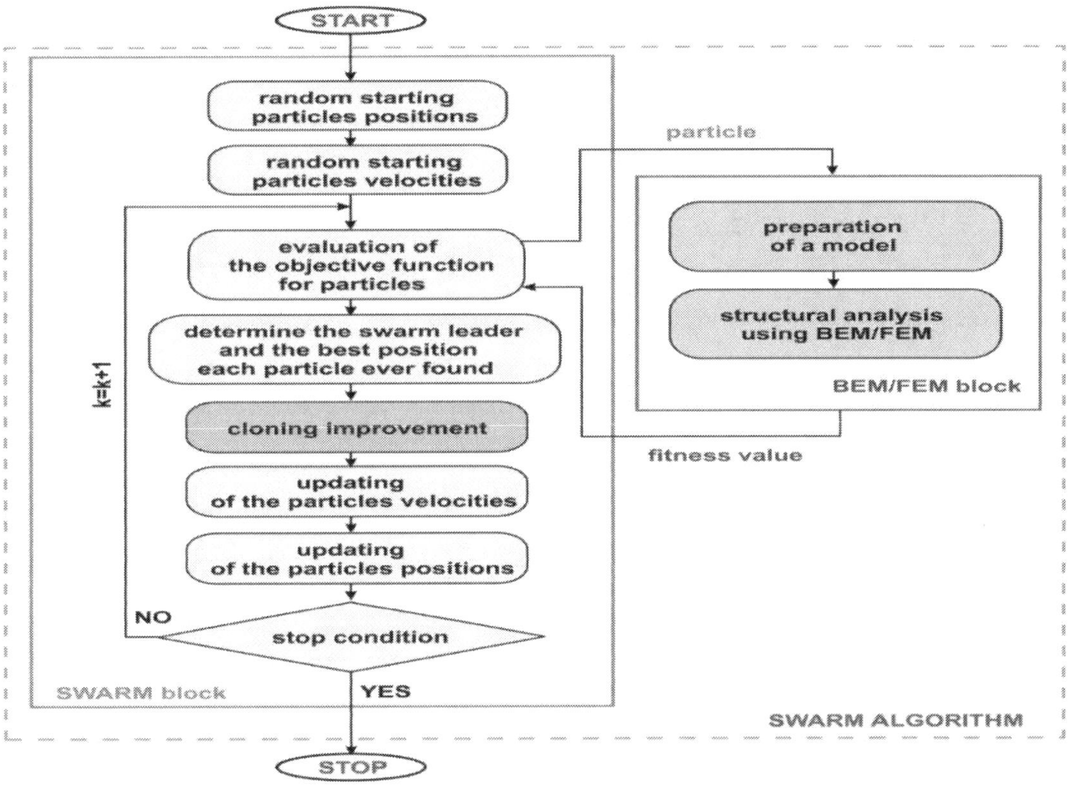

Figure 2. The flowchart of the swarm algorithm

3. Formulation of the problem

Consider a structure (plate in plane stress/strain, bending plate or shell) which, at the beginning of an iterative process, occupies a domain $\Omega_0\left(in\ E^d,\ d=2\ or\ 3\right)$, bounded by a boundary Γ_0. The domain Ω_0 is filled by a homogeneous and isotropic material of a Young's modulus E_0 and a Poisson ratio ν. The thickness of the structure g_0 is also constant at the beginning of the particle swarm process. The 2-D structures are considered in the framework of theory of elasticity. During the particle swarm process the domain Ω_t, its boundary Γ_t and the field of Young's modulus $E(\boldsymbol{x})=E_t, \boldsymbol{x}\in\Omega_t$ or the thickness $g(\boldsymbol{x})=g_t$ can change for each generation t (for t=0, E_0=const, g_0=const). The particle swarm process proceeds in an environment in which the structure fitness is describing by minimize of the maximal displacements

$$J = \max |u| \tag{4}$$

with a constraint imposed on the volume of the structure $V = |\Omega| \le V_{max}$.

In order to solve the formulated problem FE models of the structures are considered [9]. The structure is divided into finite elements $\Omega_e, e = 1,2,...,R$, and node displacements are calculated by solving a system of linear algebraic equation

$$\mathbf{KU=F} \tag{5}$$

where **U** is a column matrix of unknown displacements, **F** is a known column matrix of acting forces and **K** is a known global stiffness matrix of the structure whose elements are given as follows:

$$\mathbf{K}^e\mathbf{=B}^T\mathbf{DBV}^e \tag{6}$$

where **D** and **B** are the known elasticity and geometrical matrices, respectively, $\mathbf{V}^e$ represents the volume of the finite element.

The distribution of thickness $E(x), x \in \Omega_t$ in the structure is describing by a surface $W(x), x \in H^2$ (for plate in plane stress/strain, bending plate) or a hipersurface $W(x), x \in H^3$ (for shell). The surface (hipersurface) $W(x)$ is stretched under $H^d \subset E^d$, $(d = 2, 3)$ and the domain Ω_t is included in H^d, i.e. $(\Omega_t \subseteq H^d)$.

The shape of the surface (hipersurface) $W(x)$ is controlled by parameters of particle h_j, j=1,…,N, which create a particle

$$par = \langle h_1, h_2, ..., h_j, ..., h_N \rangle, \qquad h^{min} \le h_j \le h^{max} \tag{7}$$

where h^{min} - the minimum value of the parameter of particle and h^{max} - the maximum value of the parameter of particle.

Parameters of particles are values of the function $W(x)$ in interpolation nodes x_j, i.e. $h_j = W(x_j)$, j=1,2,…,N.

The assignation of of thickness to each finite element $\Omega_e, e = 1,2,...,R$ is adequately performed by the mapping:

$$g_e = W(x_e), x_e \in \Omega_e, e = 1,2,...,R \tag{8}$$

It means that each finite element can have different material. When the value of thickness for the e-th finite element is included in:

- the interval $0 \le g_e < g_{min}$, the finite element is eliminated and the void is created,
- the interval $g_{min} \le g_e < g_{max}$, the finite element remains (figure 4).

The idea of particle swarm generation for 2-D structure in the figure 3 is presented.

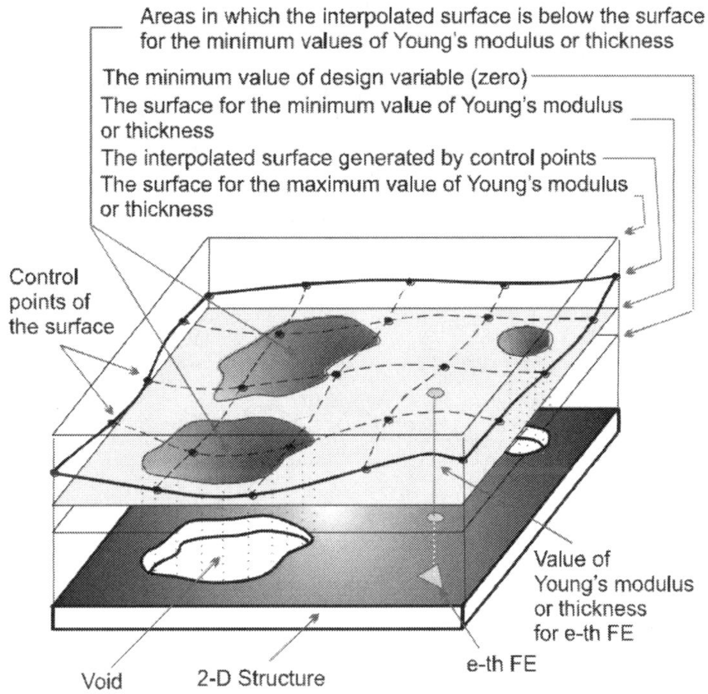

Figure 3. Illustration of the idea of particle swarm generation for 2-D structure

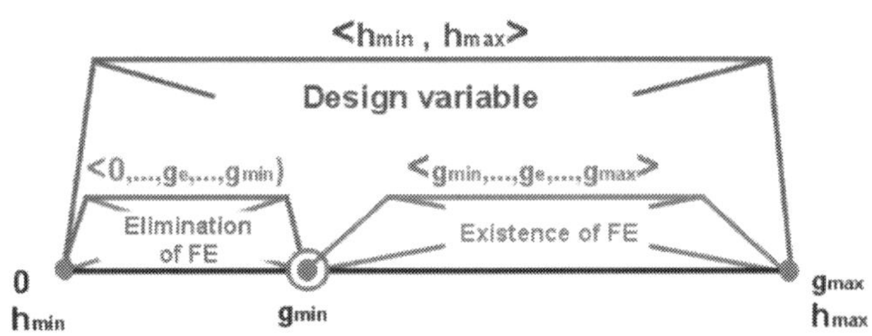

Figure 4. Requirements for elimination and existence of the finite elements

4. Examples of particle swarm optimization of structures

The bracket of the car bumper is considered. The geometry of the structure is shown in the figure 5. The considered construction is stiffly supported as is presented in the figure 6. and is loaded with a pressure 0.02 MPa. It is made from the polypropylene with aiddition of the glass fibre (E = 6 000 MPa, ν=0.35). The thickness of the shell construction is equal 3 mm and in the region around the hole for the tow hook is equal 6 mm. Calculation parameters of optimization of the bracket of the car bumper in the table 1 are presented. Distribution of control points of interpolation hipersurface for the bracket of the car bumper in the figure 7 is presented. Map of displacements of the bracket of the car bumper in the figure 8 and 9 are presented.

Figure 5. Geometry of the bracket of the car bumper

Table 1. Calculation parameters of optimization of the bracket of the car bumper

pressure c^0 [MPa]	Number of design variables		Number of particles	V_{max} [cm^3]
0.02 MPa	31		20	530
material	Young module E	Poisson ratio	range of parameters of particle [mm]	Thickness of the bracket outside optimization area [mm]
polypropylene with addition of the glass fibre	6 000 MPa	0.35	3.0 ÷ 8.0	3.0

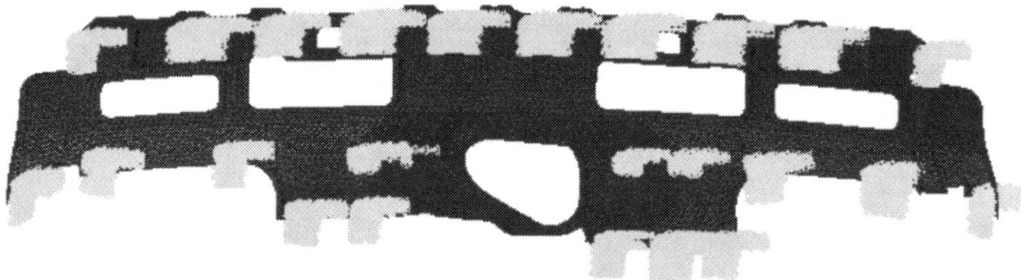

Figure 6. Boundary condition of the bracket of the car bumper

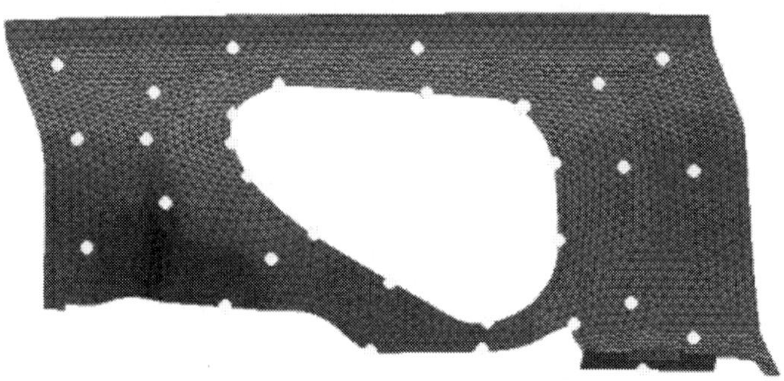

Figure 7. Distribution of control points of interpolation hipersurface for the bracket of the car bumper

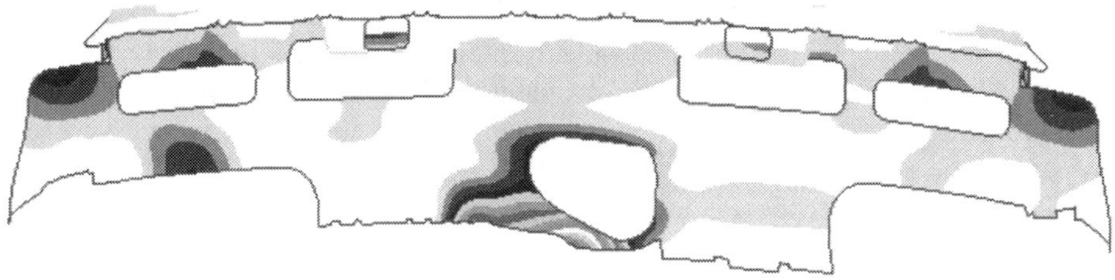

Figure 8. Map of displacements of the bracket of the car bumper

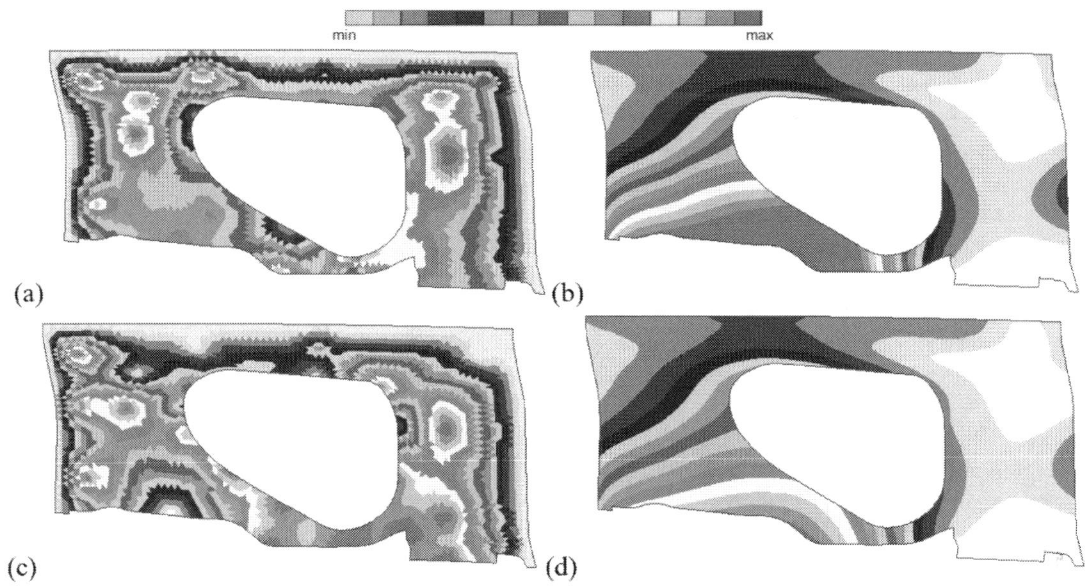

Figure 9. Results of optimization for the bracket of the car bumper
(a), (b) the best solution in the first iteration; (c), (d) the best solution after optimization process
(a), (c) map of thickness; (b), (d) map of displacements

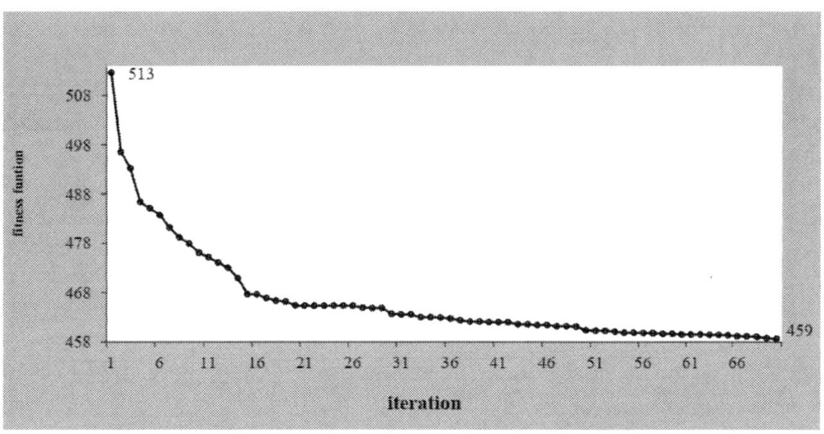

Figure 10. History of the fitness function for the best particle

IOP Conf. Series: Materials Science and Engineering **161** (2016) 012018 doi:10.1088/1757-899X/161/1/012018

5. Conclusions

An effective tool of particle swarm optimization of reinforced structures has been presented. Using this approach shape, topology and thickness optimization are performed simultaneously [10][11]. The important feature of this approach is a strong probability of finding the global optimal solutions received by implementation of the particle swarm algorithm and its universality consisting in application of this approach to the optimization of the 2-D structures [12]. The particle swarm algorithms [13][14], similarly to the evolutionary [15] and immune algorithms [16][17][18], are developed on the basis of the mechanisms discovered in the nature. Comparison between AIS, PSO and EA in the paper [4][5][19] is presented.

References

[1] Bendsøe M P 1989 Optimal shape design as a material distribution problem. Struct. Optim.

[2] Poteralski A Szczepanik M and Burczyński T 2015 Immune Optimal Design of 2-D and 3-D Structures International Conference on Artificial Intelligence and Soft Computing (ICAISC) Lecture Notes on Artificial Intelligence vol 9120 Springer pp 471-482

[3] Poteralski A Szczepanik M and Burczyński T 2015 Swarm and immune computing of dynamically loaded reinforced structures International Conference on Artificial Intelligence and Soft Computing (ICAISC) Lecture Notes on Artificial Intelligence vol 9120 Springer pp 483-494

[4] Poteralski A Szczepanik M Dziatkiewicz G and Kuś W 2013 Comparison between PSO and AIS on the basis of identification of material constants in piezoelectrics Springer-Verlag Berlin Heidelberg ICAISC 2013 Part II LNAI 7895 pp 569–581

[5] Poteralski A Szczepanik M Beluch W and Burczyński T 2014 Optimization of composite structures using bio-inspired methods Artificial intelligence and soft computing ICAISC 2014 vol 8468 pp 385-395

[6] Xie Y M and Steven G P 1997 Evolutionary Structural Optimization London Springer

[7] Reynolds C W 1987 Flocks herds and schools A distributed behavioral model Computer Graphics vol 21 pp 25–34

[8] Kennedy J and Eberhart R C 1995 Particle Swarm Optimization. Proceedings of IEEE Int. Conf. on Neural Networks Piscataway NJ pp 1942-1948

[9] Poteralski A 2015 Data processing in immune optimization of the structure Beyond Databases Architectures and Structures (BDAS) Communications in Computer and Information Science vol 521 Springer pp 309-319

[10] Szczepanik M and Burczyński T 2012 Swarm optimization of stiffeners locations in 2-D structures Bulletin of the Polish Academy of Sciences Technical Sciences vol 60 no 2 pp 241-246

[11] Szczepanik M Poteralski A Długosz A Kuś W and Burczyński T 2013 Bio-inspired optimization of thermomechanical structures Springer-Verlag Berlin Heidelberg ICAISC 2013 Part II LNAI 7895 pp 79-90

[12] Zienkiewicz O C and Taylor R L 2000 The Finite Element Method. Butterworth Heinemann Oxford

[13] Burczyński T and Szczepanik M 2013 Intelligent optimal design of spatial structures Computer & Structures Elsevier

[14] Yamada T et al. 2010 A topology optimization method based on the level set method incorporating a fictitious interface energy Comput. Methods Appl. Mech. Engrg.

[15] Burczyński T and Osyczka A 2003 Evolutionary Methods in Mechanics Proc. IUTAM Symposium Kluwer Dordrecht

[16] Poteralski A Szczepanik M Ptaszny J Kuś W and Burczyński T 2013 Hybrid artificial immune system in identification of room acoustic properties Inverse Problems in Science and Engineering Taylor & Francis

[17] Poteralski A 2014 Optimization of mechanical structures using artificial immune algorithm

Beyond Databases Architectures, and Structures Communications in Computer and Information Science vol 424 pp 280-289

[18] Szczepanik M Poteralski A Ptaszny J and Burczyński T 2012 Hybrid Particle Swarm Optimizer and Its Application in Identification of Room Acoustic Properties Swarm and Evolutionary Computation - International Symposia, SIDE 2012 and EC 2012, Held in Conjunction with ICAISC 2012 Zakopane Poland Proceedings. Lecture Notes in Computer Science vol 7269 Springer

[19] Burczyński T Poteralski A and Szczepanik M 2015 Immune and Swarm Optimization of Structures Advances in Evolutionary and Deterministic Methods for Design Optimization and Control in Engineering and Sciences Chapter no 19

IOP Conf. Series: Materials Science and Engineering **161** (2016) 012019 doi:10.1088/1757-899X/161/1/012019

Increase Productivity and Cost Optimization in CNC Manufacturing

MUSCA Gavril, MIHALACHE Andrei, TABACARU Lucian

"Gheorghe Asachi" Technical University of Iasi, Department of Machine Manufacturing Technology, Blvd. D. Mangeron, No.59A, Iasi-700050, Romania

musca@tcm.tuiasi.ro, andrei.mihalache@yahoo.com, lucian.tabacaru@yahoo.com

Abstract. The advantage of the technological assisted design consists in easy modification of the machining technologies for obtaining machine alternation, tool changing, working parameters variation or the modification of loads to which the tools are subjected. By determining tool movement inside machining and by using tool related moving speeds needed for both positioning and manufacturing we are able to compute the required machining time for each component of the machining operation in progress. The present study describes a cost optimization model for machining operations which uses the following components: machine and its operator related cost, set-up and adjustment, unproductive costs (idle state), direct and indirect costs. By using manufacturing technologies assisted design procedures we may obtain various variants for the technological model by modifying the machining strategy, tooling, working regimes or the machine-tool that are used. Simulating those variants allows us to compare and establish the optimal manufacturing variant as well as the most productive one.

1. Introduction.

CNC mechanical machining performance requires special adapted methods for designing technologies and data management inside manufacturing industries. Finding optimum variants of machining calls for advanced knowledge in terms of working times and their costs. There are various computing mathematical relations given for the determination of the working times for simple machining if the machining related parameters are well known. But for mechanical parts which have an intricate geometry the working times are very hard to compute manually thus making their findings only possible by CAM means. Those software allow, achieve and simulate manufacturing processes in order to check the correctness of the project before its implementation.

One of the advantages of the assisted design of the machining technologies is that it allows the simulation of the working environment.

The simulation of an operation or the entire machining process consists of:

- Determination of the tool's trajectory and collision avoidance between the tool and the blank, avoidance of impacting the machine or its devices during manufacturing;
- Determination of the machining time for each operation as well as for the entire part. For machining operation's simple components there are mathematical relations which give us

exact operation times but for most of the machining the sheer number of the movements and components is way bigger and difficult to model. Using simulation procedures we can determine the total number of movements and their components which gives us exact or very close to reality working times specific values and those of auxiliary movements.

The simulation process provides important data regarding the determination of the machine's capabilities, the determination of the machining time and cost and the establishment of the optimal working variant.

Conception of the NC driven machining processes leads most of the times to different solicitations of the tools that are used inside manufacturing operations. There are differences between tools that are used for roughing as opposed to finishing or solicitation related differences appear depending on the machined surfaces geometries. Different machining times for various tools lead to different values for the expected tool life. For example, by expressing this parameter in time dependent units or in the number of the machined parts we get different estimated times for the CNC setup. But because setup times are usually higher they may influence the overall productivity of the manufacturing process as well as the machining related cost. The results obtained up till now by the authors led them to the conclusion that there is room for improvement in terms of optimization of the CNC machines adjustment times by modifying the values of the working regimes related parameters.

2. CAM model analyses and their significance in identifying the machining cost components.

The advantage of the technological assisted design consists among others in being able to easily modify the machining technology in order to obtain several machining variants which will allow the achievement of some practical requirements such as machine alternation, tool changing, working parameters variation, the modification of loads to which the tools are subjected, etc.

By adjusting the machining related parameters for different working regimes we may determine working time values as well as the auxiliary times for each tool and programmed NC machining operation which lead automatically to tool life expectancy related to the number of the machined parts. This case is especially useful when dealing with intense working regimes which use reduced working times and the related working regimes parameters also diminished in case of tools that are hardly solicited. By simulating manufacturing processes we can determine correspondent adjustment times.

One of the main advantages that make manufacturing processes simulation so necessary is tool trajectory generation which includes imposed machining movements, tool withdrawal necessary for new crossings as well as tool change. By determining tool movement inside machining and by using tool related moving speeds needed for both positioning and manufacturing we are able to compute the required machining time for each component of the machining operation in progress. After setting up the CNC we add the required times needed for starting and stopping the machine and tool change related times thus by means of simulation being able to compute the total machining time for a given part.

The above mentioned elements are obtained by means of simulation procedures provided by NX CAM software which emphasis the importance of simulation in terms of geometrical verifications and a good knowledge of the CNC's programming productivity. There have been generated CAM models for different parts using NX CAM. For each distinct part there multiple simulation sessions were carried out by shifting CNCs, tools, machining variables or the strategy regarding the removal of the working related additions [2], [6].

The simulations help at verifying the CAM machining model, its precision and quality and give us data regarding times for each machining apart as well as for the entire process.

An example for a machining model is that presented in fig. 1.

At this moment we feel that the simulation of manufacturing on machine-tool devices driven by NC must be completed with a distinct module which may allow exact machining costs preview.

Machining simulation gave us in this case complete data regarding the precision, productivity and manufacturing costs offering criteria for the optimization of the manufacturing processes on NC driven machine-tools devices.

3. Proposed algorithm for the determination of the machining cost.

The CNC machining cost depends on a multitude of elements such as CNC equipment cost, of the tools that are to be used in the process, the maneuvering and surveillance of the machining system, the quantity and cost of consumables, energy cost, etc.

The influence of cutting tools over manufacturing is a known issue in the industry. This study proposes a computational model which depends upon a newly introduced inside the manufacturing process of a cutting tool as shown in [4], [5], and [7]. This model uses the cutting time related cost as well as the one of the tool and the fine-tuning of the CNC equipment.

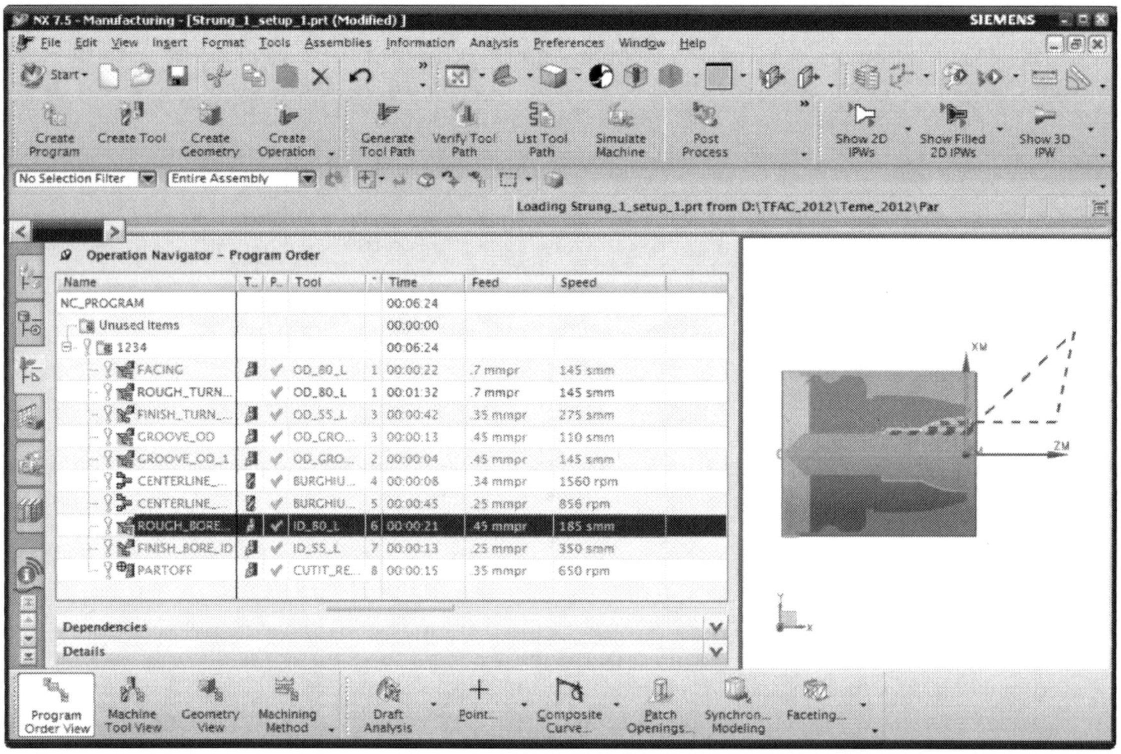

Figure 1. General view over a model prepared for manufacturing (NX CAM).

The analysis of spending share related to electrical energy consumption [1] uses a cost dependent computational model based on manufacturing experiments. The model is accompanied by an industry extracted particular case which estimates resulted savings and costs. The authors demonstrate that there are no electric energy consumption savings because energy related cost in CNC machining are already low but highlighting the fact that substantial savings are recorded in case of a high production volume by means of machining parameters optimization.

The present study describes a cost model for machining operations which uses the following components: machine and its operator related cost, set-up and adjustment, unproductive costs (idle state), direct and indirect cost imposed by the manufacturing process.

According to [1], [3], and [8] cost components related to energy consumption rates may be grouped as follows:

- Direct cost with the energy that has been consumed during the manufacturing process;
- Indirect cost regarding the supplementary energy consumed during the working process such as general illumination, heating, etc.;

- External cost dependent on the environment, or given by advertising and commercial presentation of the product.

The authors present cost components and their conclusion is that 38.7 % are represented by direct cost, 34.7% is tool related cost and the energy related cost is low peaking a 0.45% from the overall. We thus estimate that energy cost isn't high enough to arouse any special interest or to consider it as a radical saving component inside CNC manufacturing processes [1]. Nevertheless savings can be made at the real cost level by decreasing working times. In case of a high production volume by optimizing the machining parameters we can safely assume that energy related savings and costs may become important factors.

The proposed algorithm ensures the cost computation per one part.

The total machining cost is composed from blank related cost, CNC equipment related expenses, and its operator related cost or the tools that are used (1):

$$C = C_{blank} + C_{machine} + C_{operation} + C_{tool} \tag{1}$$

The cost related to the machine is represented by the electrical energy amortization one consumed during a single part manufacturing (2):

$$C_{machine} = \frac{P_{equipment\_aquisistion}}{t_{amortization} \times t_{working\_time}} \times t_{program} + P_{electric\_energy} \times t_{program} \times P_{per\_kW} \tag{2}$$

Where P signifies "*price*" and t "*time*".

The cost per operation is represented by the operator remuneration and machine human adjuster (3):

$$C_{operation} = (t_{program} + f \times t_{program}) \times \frac{salary}{\min} \tag{3}$$

Where f stands for a fraction of the $t_{program}$ needed for the adjustments and the programming activities.

The last component of the cost computational algorithm (4) is represented by the tools that are used to machine the part and results as a sum of the tooling cutting heads related cost with regards of the manufacturing time for each tool, tool durability and the number of cutting edges it has.

$$C_{tool} = C_{cutting\_head} \times \frac{t_{tool}}{T_{economic} \times N_{no.\_of\_cutting\_edges}} \tag{4}$$

Where $T_{economic}$ stands for an economical tool life.

4. Different costs for various CNC machining.

The machining cost computational algorithm was successfully tested in MS Excel. It uses modules in order to determine cost components for each machining operation as well as total cost.

Computing manufacturing steps related cost has been carried out by considering NX CAM program component which resulted as shown in figure 2. For each machining operation we know the manufacturing regime related parameters, the tool code and its location on the machine and the working time.

For each machining it has been determined the cost driven by the cutting tools. By knowing the tool's price, its lifespan and the number of repositioning of the cutting component we have been able to determine the price per minute of each tool based on which we could compute the tool imposed cost per working time for each machining operation.

Based on the total time we can determine machine imposed cost as well as the ones for the operator and electrical energy consumption. The spreadsheet also depicts graphical representations of these costs in order to highlight the significance of cost components which also makes it easier to interpret the simulation results.

The cost driven by the machine-tool uses as reference the machine cost for one minute of utilization. Based on the machine's electrical power we can determine the consumed electrical energy related cost per minute. In order to simplify the computation we appreciate that the consumed electrical power equals the nominal power. In such a situation the value of the electrical energy cost is approximately 1/6th of machine cost, but realistically this ratio has a much smaller value considering either idle state or the fact that finishing requires reduced electrical power.

For the computation of the salary cost for each working minute of the operator we consider both direct and indirect salary related expenses. As results from the shown elements cost computation is being made in Euro's and RON's.

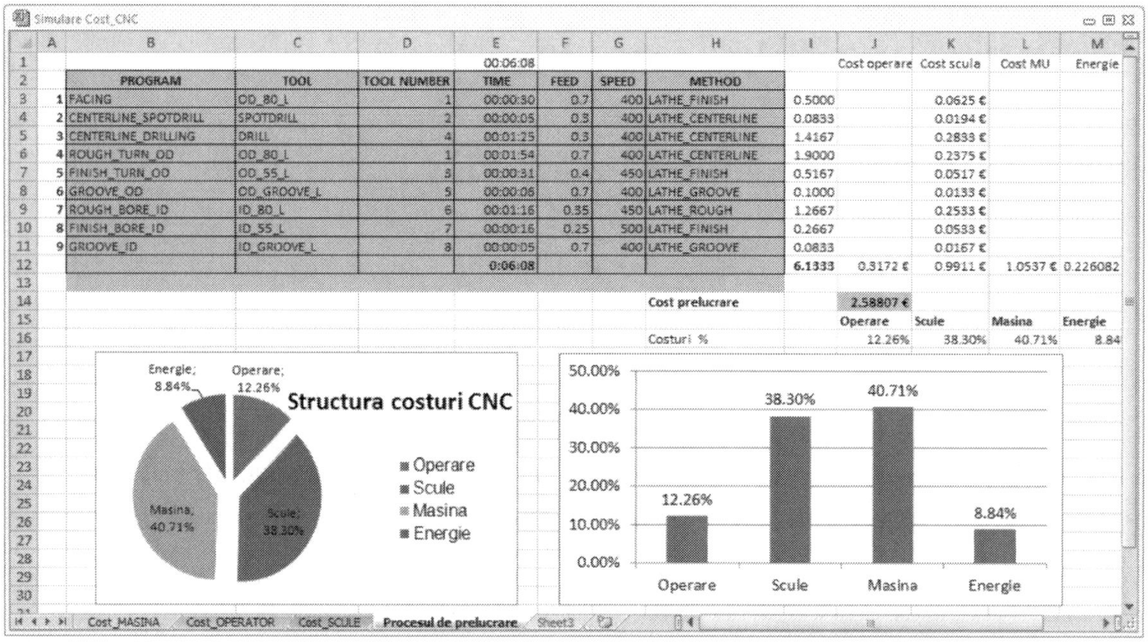

	PROGRAM	TOOL	TOOL NUMBER	TIME	FEED	SPEED	METHOD		Cost operare	Cost scula	Cost MU	Energie
				00:06:08								
1	FACING	OD_80_L	1	00:00:30	0.7	400	LATHE_FINISH	0.5000		0.0625 €		
2	CENTERLINE_SPOTDRILL	SPOTDRILL	2	00:00:05	0.3	400	LATHE_CENTERLINE	0.0833		0.0194 €		
3	CENTERLINE_DRILLING	DRILL	4	00:01:25	0.3	400	LATHE_CENTERLINE	1.4167		0.2833 €		
4	ROUGH_TURN_OD	OD_80_L	1	00:01:54	0.7	400	LATHE_CENTERLINE	1.9000		0.2375 €		
5	FINISH_TURN_OD	OD_55_L	3	00:00:31	0.4	450	LATHE_FINISH	0.5167		0.0517 €		
6	GROOVE_OD	OD_GROOVE_L	5	00:00:06	0.7	400	LATHE_GROOVE	0.1000		0.0133 €		
7	ROUGH_BORE_ID	ID_80_L	6	00:01:16	0.35	450	LATHE_ROUGH	1.2667		0.2533 €		
8	FINISH_BORE_ID	ID_55_L	7	00:00:16	0.25	500	LATHE_FINISH	0.2667		0.0533 €		
9	GROOVE_ID	ID_GROOVE_L	8	00:00:05	0.7	400	LATHE_GROOVE	0.0833		0.0167 €		
				0:06:08				6.1333	0.3172 €	0.9911 €	1.0537 €	0.226082
							Cost prelucrare		2.58807 €			
									Operare	Scule	Masina	Energie
							Costuri %		12.26%	38.30%	40.71%	8.84

Figure 2. Obtained results using NC machining cost computational algorithm.

The cost per one minute of use of the cutting tool utilizes structural elements from a database for each tool, producer or supplier, data related to the number of cutting edges and facets, tool durability levels and price.

By knowing the tool type and the one of the blank's material we have determined values for the manufacturing regime parameters which helps defining the NX CAM working steps and thus providing machining times.

5. Conclusions.

The simulation of the manufacturing process on NC driven machine-tool equipment's allows the visualization of the tool's trajectory, the possibility to check the correctness of the machine's programming as well as the avoidance of the cutting tool with the blank, the machine itself or its devices.

The tool movement during machining consists of rapid approaching moves, feeds and rapid withdrawals for which we have to determine the execution times. By summing them up we obtain the required times for each machining operation as well as the total machining time.

The simulation of the manufacturing process that has been approached in the current study has led to the necessity to develop a machining cost computational algorithm. It has been implemented in MS Excel and it uses the following modules in order to identify the necessary costs for the machine-tool equipment: operation, tools, energy and blank material.

For a given part for which we knew its geometry and manufacturing process structure there have been simulated variants characterized by different values for used tool life spans (15 to 45 minutes). Thus we had an increase of the working times and a decrease of cost. Identifying cost components weights such as 42% for the machine-tool, 38% tooling, 12% operation and 8% energy allowed us to propose a strategy for the optimization of cost and machining operation variants.

Analyzing data we saw that the most significant cost component is represented by the CNC machine having an almost 42% weight in the overall process. Very close come the tools used in the process with almost a 38% share. We conclude that by using expensive but performant tools we can assure the quality of manufacturing processes.

The CNC operator salary related cost is about 12% in weight which represents nearly a quarter of the CNC machine cost. That being said we can safely assume that expenses made by four operators cover the acquisition cost of a CNC machine thus explaining why companies focus on a reduced number of highly qualified operators.

The total weight for energy related cost is about 8% but in reality may have smaller shares. When increasing tooling cost we expect a significant decrease of CNC machine related cost, operation cost and energy cost.

Because of the tool lifespan of 15 to 45 minutes, manufacturing regimes are less intense which leads to the possibility of increasing machining times and decreasing productivity. By analyzing data from the diagram it results that by increasing the working time by 32.49% we may achieve an economy up to 10.33% for the manufacturing cost.

References:

[1] Anderberg, S., Kara, S., Energy and cost efficiency in CNC machining, The 7th CIRP Conference on Sustainable Manufacturing, 2009

[2] Anderberg, Staffan, Methods for improving performance of process planning for CNC machining - An approach based on surveys and analytical models, thesis, *Department of Materials and Manufacturing Technology, Chalmers University of Technology, Göteborg,* Sweden, 2012

[3] Bouaziz, Z., and al., Cost estimation system of dies manufacturing based on the complex machining features, *Int J Adv Manuf Technol* (2006) 28: 262–271, DOI 10.1007/s00170 004-2179-3

[4] Curran, R., and al., Modelling of aircraft manufacturing cost at the concept stage, *in Int J Adv Manuf Technol* (2006) 31: 407–420 DOI 10.1007/s00170-005-0205-8

[5] Darie, V., Balan, R., Blajina, O., Breazu, A., Determination of the economic efficiency of investments to the use of CNC machining centers, *U.P.B. Sci. Bull., Series D,* Vol. 74, *Iss. 3,* p.*141*-152, 2012, ISSN 1454-2358

[6] Mohammadjafari, M., Ahmed, S., Zawiah, S., Dawal, M., Zayandehroodi, H., A Review of Literature on Reducing Time and Cost of New Product in SMEs through Project Management by E-Collaboration, *Proceedings of the 2011 International Conference on Industrial Engineering and Operations Management Kuala Lumpur*, Malaysia, p.125-130, January 22 – 24, 2011

[7] Mohammadjafari, M., Dawal, S.Z.M., Ahmed, S., Zayandehroodi, H., Toward a Theoretical Concept of E-Collaboration through Project Management in SMEs for Reducing Time and Cost in New Product: A Review, *Journal of Applied Sciences* 11 (1): 174-182, 2011, ISSN 1812-5654 / DOI 10.3923/jas.2011.174.182

[8] Selvaraj,P., and al., An integrated approach to design for manufacturing and assembly based on reduction of product development time and cost, *Int J Adv Manuf Technol* (2009) 42:13 29, DOI 10.1007/s00170-008-1580-8

IOP Conf. Series: Materials Science and Engineering **161** (2016) 012020 doi:10.1088/1757-899X/161/1/012020

Mechanical device for determining the stiffness and the viscous friction coefficient of shock absorber elements modelled by bond graph

R Ibănescu[1] and M Ibănescu[2]

[1]Department of Theoretical Mechanics, "Gheorghe Asachi" Technical University of Iaşi, Bd. Dimitrie Mangeron nr. 67, 700050, Iaşi, Romania
[2]Department of Structural Mechanics, "Gheorghe Asachi" Technical University of Iaşi, Bd. Dimitrie Mangeron nr. 67, 700050, Iaşi, Romania

ribanesc@yahoo.com

Abstract. The present paper presents a mechanical device for the assessment of the fundamental parameters of a shock absorber: the spring stiffness and the viscous friction coefficient, without disassembling the absorber. The device produces an oscillatory motion of the shock absorber and can measure its amplitude and angular velocities. The dynamic model of the system, consisting of the mechanical device and the shock absorber, is performed by using the bond-graph method. Based on this model, the motion equations are obtained, which by integration lead to the motion law. The two previously mentioned parameters are determined by using this law and the measured values of two amplitudes and of their corresponding angular velocities. They result as solutions of a system of two non-linear algebraic equations.

1. Introduction

The shock absorbers are used in various applications, especially in the automotive industry. The main parts of a shock absorber are a spring and a dumper. Consequently, the most important parameters of such a device are the spring stiffness and the viscous friction coefficient. These parameters are usually determined separately for the spring [1], [2], [3], and for the dumper [4], [5], [6], [7], by using appropriate theoretical methods or devices. There are papers which study both parameters [8], [9], [10]. Some papers present the two parameters assessment by using the force provided by the force sensor located at the upper end of the shock absorber. This force is the sum of the force developed by the spring and by the dumper and cannot be used for determining these parameters. The paper presents a mechanical device which can provide both parameters by exciting the shock absorber with an inertial mechanism. The mechanism consists of two identical eccentric masses rotating with the same constant angular velocity in opposite directions. The mechanism is fixed at the upper end of the shock absorber, which is in a vertical position. The force of inertia of these two masses induces an oscillatory vertical motion of the shock absorber. The motion equation of the whole device is determined by using the bond graph method [11], [12], [13]. It is necessary to measure two values of the amplitude at corresponding angular velocities, in order to determine the spring stiffness and the viscous friction coefficient of the dumper. Finally, the required parameters of the shock absorber are obtained by resolving an algebraic system of two equations.

The device is very useful in industry, because it provides a very affordable method for determining the parameters of shock absorbers.

2. The mechanical device

The mechanical device consists in a box (1), having mass M (Fig.1). It can slide on two parallel columns (3) and it is supported by a shock absorber, which is made of a spring of stiffness k_e and a dumper, whose viscous friction coefficient is γ. The shock absorber is fixed according to the code [14]. Additional masses can be added to the box (1), when a pre-load is required.

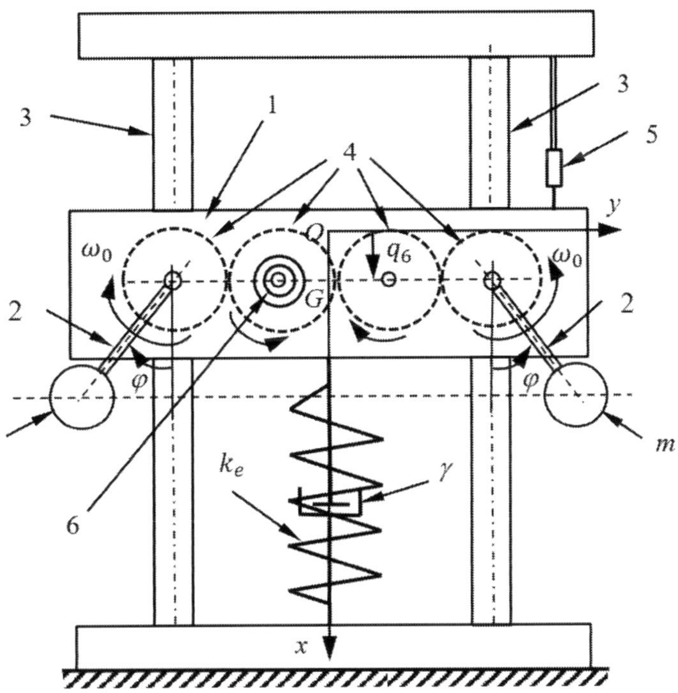

Figure 1. The mechanical device

A mechanical system, consisting of four identical gear wheels (4), driven by an electric motor with variable angular velocity, is located inside the box (1). The motor set in rotational motion two rods (2), l in length. The rods, which have a negligible mass, rotate with the same, constant angular velocity ω_o, in opposite directions, starting from the vertical position ($\varphi = 0$). The rods angle of rotation is $\varphi = \omega_o t$. Each rod it is provided with a mass m at its end. When the rods rotate, the box (1) has a vertical oscillatory motion, induced by the inertial force of the two masses. For the assessment of spring stiffness and of viscous friction coefficient, it will be shown that is necessary to measure the amplitude of the oscillatory motion for mass (1) and the angular velocity for bars (2). The amplitude measurement is performed by using a displacement sensor (5) of type F38000204, with a travel of 102.87mm and a linearity of 0.1%. For the angular velocity measurement, a sensor, resistant to vibrations and shocks, of type TQG19E (30g for vibration and 100g for shocks) is recommended. The electric motor and the sensor for angular velocity measurement are located on the axis of one of wheels (4) and they are both denoted by (6) in figure 1.

3. The bond-graph model of the mechanical device

The bond-graph model of the mechanical device is pictured in Fig.2. It contains two sources of flow, corresponding to the rods angular velocities, ω_o. The inertial force of masses *m,* which induces the system motion, is oriented in the positive direction of *Ox* axis and its velocity is $\omega_o l sin(\omega_o t)$. The transition from the angular velocity, ω_o, to the velocity of the inertial force is modeled by introducing a transformer defined by the parameter $lsin(\omega_o t)$. Then, a 0-junction, which distributes the inertial force power to masses *m* and *M,* is introduced. A 1-junction is associated to each mass. A capacitor, corresponding to the spring, a dissipative element corresponding to the dumper and an inertial element corresponding to mass *M* are attached to 1-junction for mass *M.*

As a consequence of the causal assignment, two elements in integral causality (the spring and mass *M*), and two elements in derivative causality (the two masses *m*) result.

The bond-graph model leads to the following system of differential-algebraic equations:

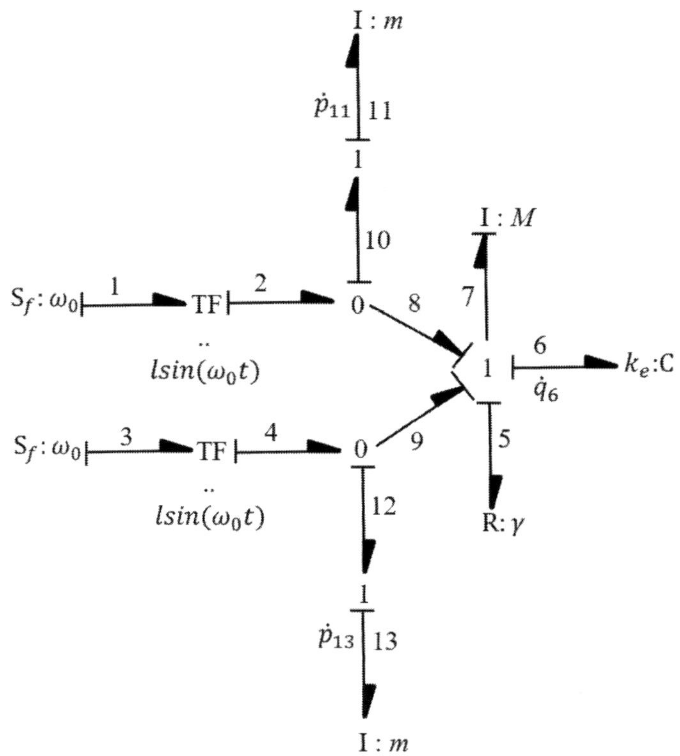

Figure 2. The bond graph model

$$\dot{p}_7 = -k_e q_6 + \gamma \frac{p_7}{M} + \dot{p}_{11} + \dot{p}_{13} \quad (1)$$

$$\dot{q}_6 = \frac{p_7}{M} \quad (2)$$

$$p_{11} = m\omega_0 lsin(\omega_0 t) - m\frac{p_7}{M} \quad (3)$$

$$p_{13} = m\omega_0 lsin(\omega_0 t) - m\frac{p_7}{M} \quad (4)$$

For finding the motion equation of mass *M* in terms of displacement variable q_6, equations (2), (3) and (4) are differentiated and introduced in equation (1).

If variable q_6 is denoted by x, the following motion equation of mass M results:

$$\ddot{x}(M+2m)+\gamma\dot{x}+k_e x=2m\omega_0^2 l cos(\omega_0 t) \quad (5)$$

which for integration is written in the form:

$$\ddot{x}+\gamma(M+2m)^{-1}\dot{x}+k_e(M+2m)^{-1}x=2m\omega_0^2 l(M+2m)^{-1}cos(\omega_0 t) \quad (6)$$

4. Stationary response assessment

The solution of the homogeneous equation:

$$\ddot{x}+\gamma(M+2m)^{-1}\dot{x}+k_e(M+2m)^{-1}x=0 \quad (7)$$

is a non-periodic motion or an oscillatory motion but, in both circumstances, this component of the general solution tends to 0.

A particular solution of equation (6) is chosen of the following shape:

$$x_p=Dsin(\omega_0 t)+Ecos(\omega_0 t) \quad (8)$$

that is the stationary response of mass M motion.

C and D are constants, which are determined by imposing relation (8) as solution of equation (6).

The two constants have the values:

$$D=2ml\omega_0^2\left(\frac{\gamma}{M+2m}\right)\left((M+2m)\left[\left(\frac{k_e}{M+2m}-\omega_0^2\right)^2+\left(\frac{\gamma\omega_0}{M+2m}\right)^2\right]\right)^{-1} \quad (9)$$

$$E=2ml\omega_0^2\left(\frac{k_e}{M+2m}-\omega_0^2\right)\left((M+2m)\left[\left(\frac{k_e}{M+2m}-\omega_0^2\right)^2+\left(\frac{\gamma\omega_0}{M+2m}\right)^2\right]\right)^{-1} \quad (10)$$

The final shape of the stationary response is:

$$x_p=\sqrt{D^2+E^2}\sin(\omega_0 t+\theta)=A\sin(\omega_0 t+\theta) \quad (11)$$

where

$$A=\sqrt{D^2+E^2} \text{ and } \tan(\theta)=\frac{E}{D} \quad (12)$$

The amplitude of the stationary response has the form:

$$A=(2ml\omega_0^2)(M+2m)^{-1}\left[\left(\frac{k_e}{M+2m}-\omega_0^2\right)^2+\left(\frac{\gamma\omega_0}{M+2m}\right)^2\right]^{-1/2} \quad (13)$$

The amplitude of the stationary response depends on ω_o, but

$$\lim_{\omega_0\to\infty} A=\frac{2ml}{M+2m} \quad (14)$$

which demonstrates that the amplitude depends no more on the spring stiffness and on the dumper viscous friction coefficient, when ω_0 has great values.

5. Experimental assessment of spring stiffness and dumper viscous friction coefficient

The experimental assessment of these two parameters is based on the measuring of two amplitudes A_1 and A_2, corresponding to the angular velocities ω_1 and ω_2, both much lower than the limit value, given by equation (14).

The following system of equations, whose unknowns are k_e and γ, is obtained:

$$A_1=\frac{2ml}{M+2m}\omega_1^2\left[\left(\frac{k_e}{M+2m}-\omega_1^2\right)^2+\left(\frac{\gamma\omega_1}{M+2m}\right)^2\right]^{-1/2} \quad (15)$$

$$A_2 = \frac{2ml}{M+2m}\omega_2^2\left[\left(\frac{k_e}{M+2m} - \omega_2^2\right)^2 + \left(\frac{\gamma\omega_2}{M+2m}\right)^2\right]^{-1/2} \tag{16}$$

In order to resolve the system, the following notations are performed:

$$L = \frac{2ml}{M+2m},\ P = \frac{k_e}{M+2m},\ \text{and } Q = \frac{\gamma}{M+2m} \tag{17}$$

The new shape of the system is:

$$A_1 = L\omega_1^2[(P - \omega_1^2)^2 + Q^2\omega_1^2]^{-1/2} \tag{18}$$

$$A_2 = L\omega_2^2[(P - \omega_2^2)^2 + Q^2\omega_2^2]^{-1/2} \tag{19}$$

and the unknowns become P and Q.
Unknown P expression is:

$$P = \omega_1\omega_2\left[1 + \frac{L^2}{\omega_2^2-\omega_1^2}\left(\frac{\omega_1^2}{A_1^2} - \frac{\omega_2^2}{A_2^2}\right)\right]^{1/2} \tag{20}$$

and, based on relations (17), it results:

$$k_e = (M + 2m)\omega_1\omega_2\left[1 + \frac{L^2}{\omega_2^2-\omega_1^2}\left(\frac{\omega_1^2}{A_1^2} - \frac{\omega_2^2}{A_2^2}\right)\right]^{1/2} \tag{21}$$

In the same manner, unknown Q expression is:

$$Q = \left(\frac{L^2}{A_1^2}\omega_1^2 - \frac{(P-\omega_1^2)^2}{\omega_1^2}\right)^{1/2} \tag{22}$$

and, based on the same relations (17), γ results:

$$\gamma = (M + 2m)\omega_1\left[\frac{L^2}{A_1^2} - \left(\frac{P}{\omega_1^2} - 1\right)^2\right]^{1/2} \tag{23}$$

6. Conclusions

The shock absorbers have as main components a spring and a dumper. The parameters of these two elements are: the spring stiffness and the dumper viscous friction coefficient. The magnitude of these parameters is of a great interest for two main reasons. The first one is the possibility of checking if they coincide with the designed values. The other one is the necessity of using their values in the mathematical model of the mechanical device, the shock absorber belongs to.

The mechanical device presented in the work permits the assessment of the stiffness of a spring and the viscous friction coefficient of a dumper, as elements of the shock absorber, without disassembling it.

The mathematical model of the mechanical device is constructed by using the bond-graph method. This mathematical model provides the relations for computing the shock absorber parameters, based on the measurements of two oscillatory motion amplitudes and of the corresponding angular velocities. They both have to be much lower than a limit computed value.

References

[1] Cleveland J, Manne S, Bocek D and Hansma P 1993 A nondestructive method for determining the spring constant of cantilevers for scanning force microscopy *Rev. Sci. Instrum* Vol. **64** 2 pp 403-405

[2] Li X, Chen J and Lan F 2015 Dynamic simulation and determination of stiffness of two springs, *Proc. of SAE-China Congress* (Shanghai China: Springer) pp 389-402

[3] Forrester M 2001 *Stiffness Model of a Die Spring* Thesis submitted to the Faculty of the Virginia Polytechnic Institute & State University for the degree of Master of Science in Mechanical Engineering Blacksburg Virginia https://theses.lib.vt.edu/theses/available/etd-05062002-

004428/unrestricted/Etd.pdf

[4] Świder J, Zbilski 2012 A Viscous friction measurement technique in robot joint with the use of surrogate mass *Pomiary Automatyka Robotyka* **12** pp 121-128

[5] Jiuhong J, Jianye D, Yu W, Honxing H 2008 Design method for fluid viscous dampers, *Arch. Appl. Mech.* **78** 9 pp 737-746

[6] Skagerstrand H 2014 *Shock Absorber Modelling* Master's thesis Chalmers University of Technology Göteborg Sweden http://publications.lib.chalmers.se/records/fulltext/211571/211571.pdf

[7] Cui Y, Kurfess T R and Messman M 2010 Testing and modeling of nonlinear properties of shock absorbers for vehicle dynamics studies *Proc. of the World Congress on Engineering and Computer Science 2010* ed S I Ao, L Gelman *et al* (San Francisco Newswood Limited) **II** pp 949-954

[8] Rao M D and Gruenberg S 2002 Measurement of equivalent stiffness and damping of shock absorbers, *Experimental Techniques* **26** 2 pp 39-42

[9] Li S, Lu Y and Li L 2012 Dynamical test and modeling for hydraulic shock absorber on heavy vehicle under harmonic and random loadings, *Research Journal of Applied Sciences, Engineering and Technology* **4** 13 pp 1903-1910

[10] Konieczny Ł 2016 Analysis of simplifications applied in vibration damping modelling for a passive car shock absorber *Shock and Vibration* http://dx.doi.org/10.1155/2016/6182847

[11] Borutzky W 2010 *Bond Graph Methodology – Development and Analysis of Multidisciplinary Dynamic System Models* (London: Springer-Verlag) pp 89-174

[12] Ibănescu R and Păstrăvanu O 2001 *Bond-Graph Language in Modeling and Simulation of Physical-Technical Systems* ("Gheorghe Asachi" Publishing House, Iaşi, Romania) pp 195-322

[13] Karnopp D, Margolis D and Rosenberg R 2012 *System Dynamics: Modeling , Simulation and Control of Mechatronic Systems* (New Jersy: John Willey & Sons Inc) pp 162-202

[14] Indian Standard IS 12648

Selection of active elements in system reduction of vibration

K. Bialas

Silesian University of Technology University, Faculty of Mechanical Engineering, Institute of Engineering Processes Automation and Integrated Manufacturing Systems, Konarskiego 18A, 44-100 Gliwice, Poland

katarzyna.bialas@polsl.pl

Abstract. This work presents non-classical method of design of mechatronic systems. The purpose of this paper is also introduces synthesis of mechatronic system understand as design of mechatronic systems. The synthesis may be applied to modify the already existing systems in order to achieve a desired result. The system was consisted from mechanical and electrical elements. Electrical elements were used as subsystem reducing unwanted vibration of mechanical system. Electrical elements can be realized in the form of coils with movable core. The system was modelled in Matlab Simulink.

1. Introduction

The phenomenon of vibration is often observed in everyday life. Some types of vibration are used in the operation of machines and devices, but most of them are of a harmful nature. This fact is related to the influence they exert on real objects causing their malfunction. An additional issue is the negative impact vibration has on the human body. For this reason the issue of the reduction of unrequired vibration is of great significance. That is why many research institutes are investigating methods of preventing the improper application or exploitation/operation of newly built machinery or methods of adjusting already existing and operating machines to specified requirements. There are many method of reduction of vibration, one of them is active reduction. The method of active reduction of vibration is characterised by the fact that vibration is compensated by vibration from additional external sources. Active execution systems may utilise mechanical, pneumatic, hydraulic, electromagnetic or electro-dynamic elements. Active systems may supply or absorb energy, in a specific manner, from any area of equipment. The application of active methods of reducing vibration enables the elimination of limitations present in passive systems [4-6]. In this paper is presented active subsystem realized by using electrical elements and selection value of those elements.

2. Active elements in reduction of vibration

The model of the research is system of two degrees of freedom (figure 1). In order to reduce vibration added actuator to each inertial element (figure 2). Actuators generate electrodynamic force F_{E1} and F_{E2} counteracted to a mechanical excitation applied to the body with a mass m_1. Diagram of the system modeling in Matlab Simulink is shown in figure 3.

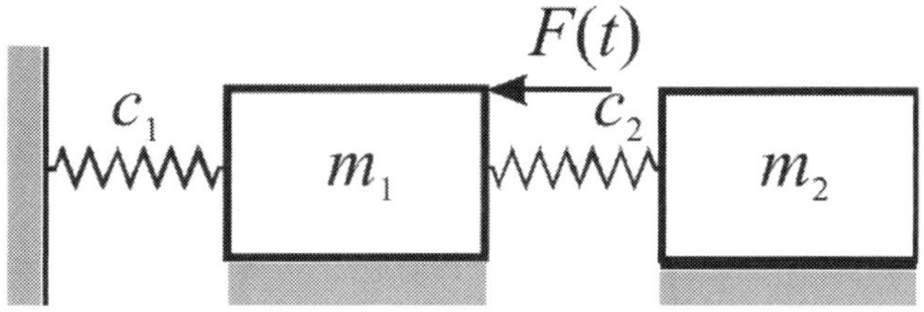

Figure 1. System of two degrees of freedom

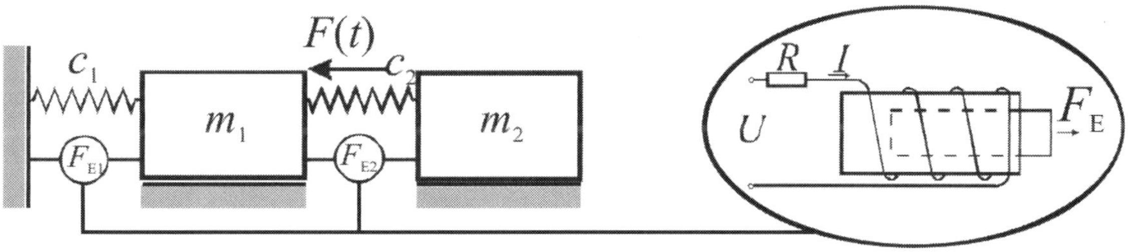

Figure 2. A model of the system with electric elements

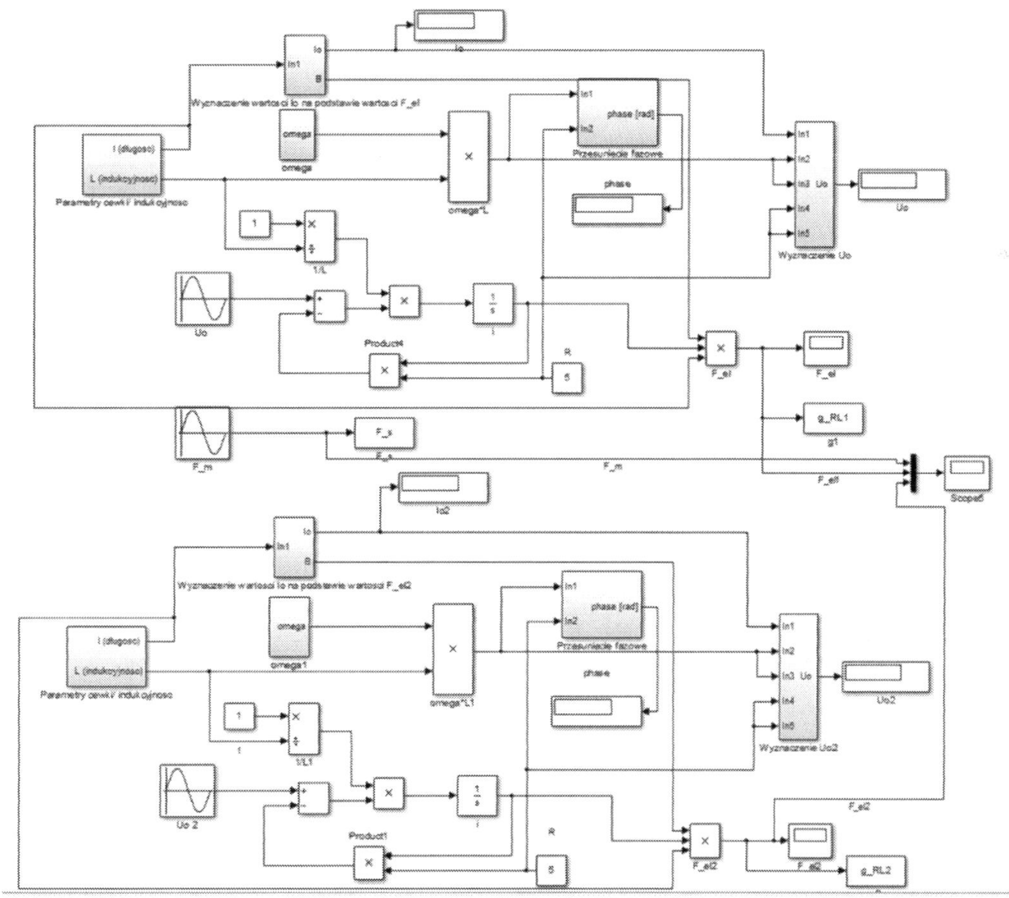

Figure 3. System in Matlab Simulink

In order to determine the value of electrodynamic force it is necessary to make use of the dependence presented below (1) [7,8]:

$$F_E = BIL_C \tag{1}$$

where:

F_E – force acting on the actuator, B – magnetic flux density, I – current in the conductor, L_C – length of the conductor.

Input to the program, were the values of the electrodynamic forces F_{E1} and F_{E2}. For each of the actuators there are separate layout generate active forces (figure 4,5).

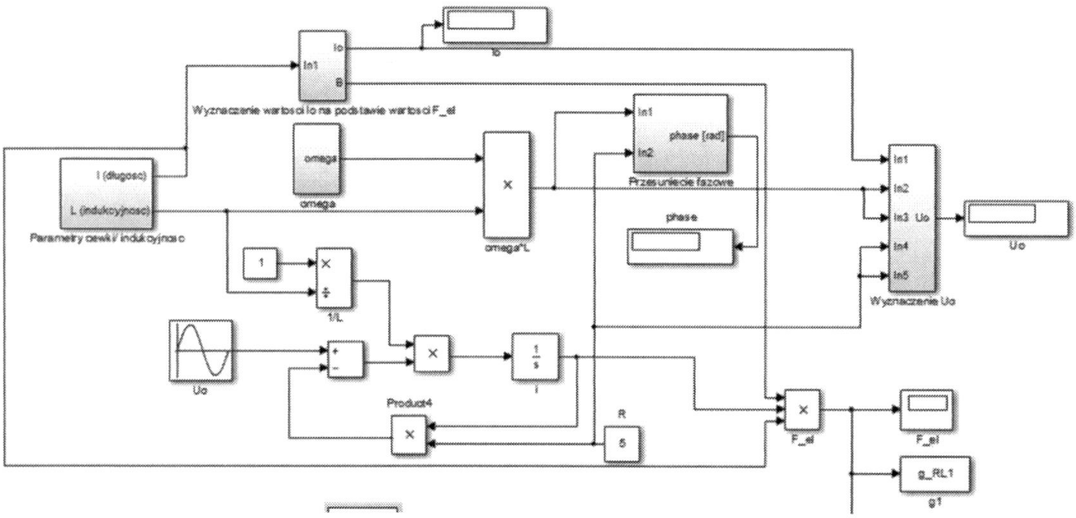

Figure 4. First subsystem (inertial element m_1)

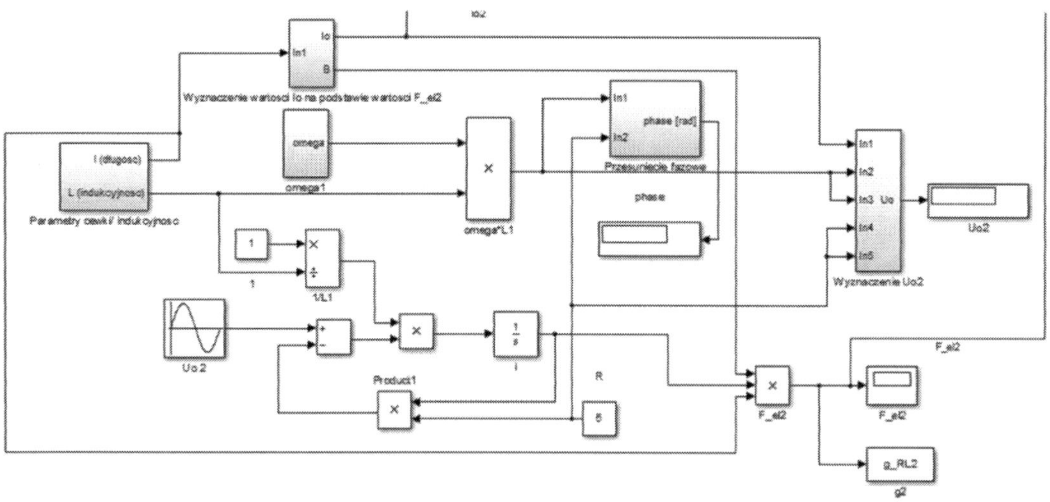

Figure 5. Second subsystem (inertial element m_2)

The value of the active forces of the FE1 and FE2 to be get to reduce unwanted vibration is entered in the subsystem (F_el) shown in figure 6.

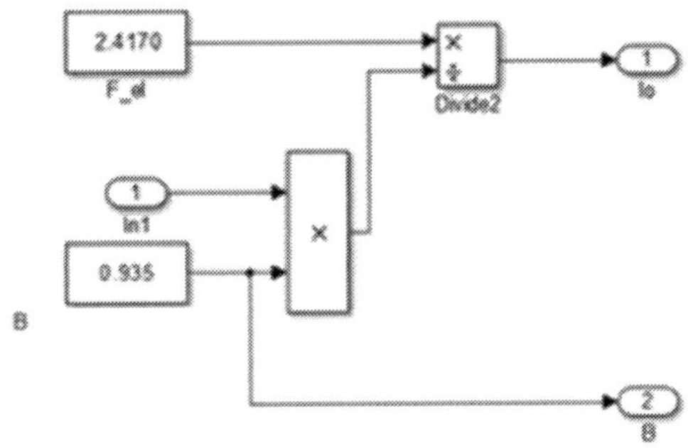

Figure 6. Subsystem – value of active force

This subsystem (figure 6) is used to designate the current that must flow in the RL to generate the desired electrodynamic force. Team blocks maps converted into equation (1). Input 1 passes for subsystem length guide. It is also the value of magnetic induction. Assumes constant value of magnetic induction, in order to simplify the model number.

A parameter that allows you to control the obtained value of the forces is voltage. Subsystem implementing this task is shown in figure 7.

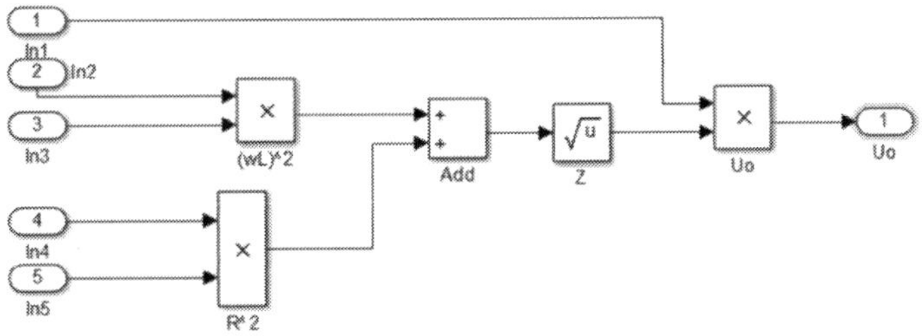

Figure 7. Subsystem – value of voltage

The entrance to the subsystem are the following:
- 1-designated current,
- 2.3-the product of the resonance frequency and the inductance,
- 4.5-resistance of the resistor.

To generate the strength of electrodynamic force was in phase with the mechanical excitation – the subsystem is in figure 8.

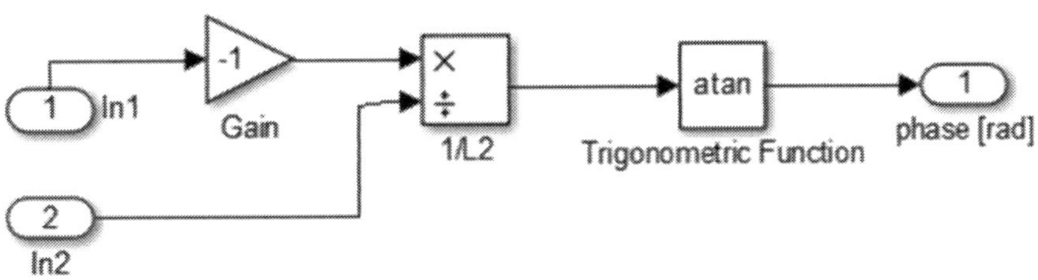

Figure 8. Subsystem – phase of electrodynamic force

Block U_0 introduces to the perimeter of the correct control voltage – figure 9.

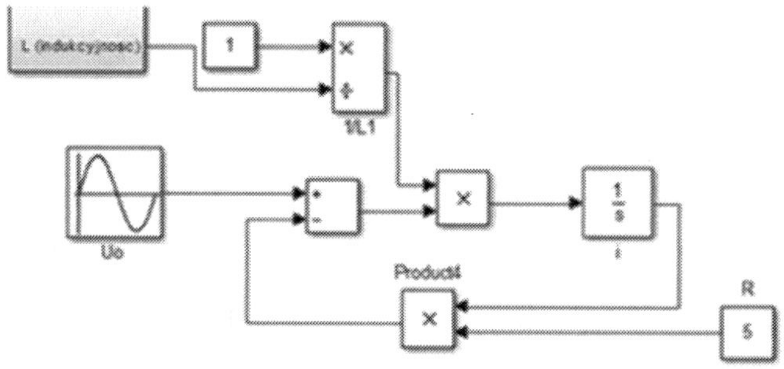

Figure 9. Subsystem – voltage.

The output signal is the strength of electrodynamic force calculated with equation (1) (figure 10).

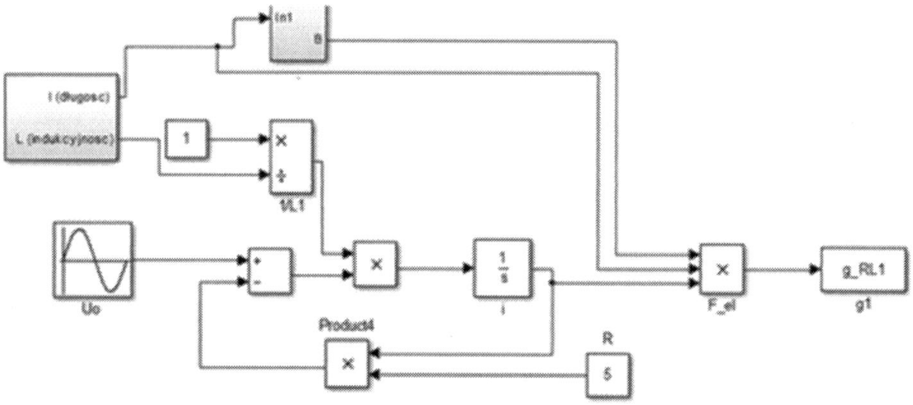

Figure 10. The designation of the active electrodynamic force

Analysis of passive and active vibration reduction is based on an analysis of the data obtained on the basis of the generated graphs of frequency-phase. On the developed charts compares the results obtained for the active and passive system. Blue color marked amplitude value of vibration for a passive system. Red dotted line indicated the amplitude of vibration for the active vibration reduction (figure 11,12).

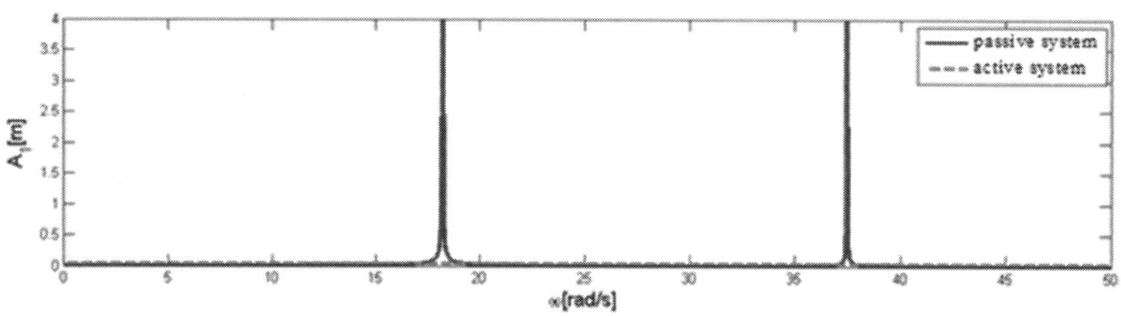

Figure 11. Diagram of amplitude A_1 of the system with electrical elements

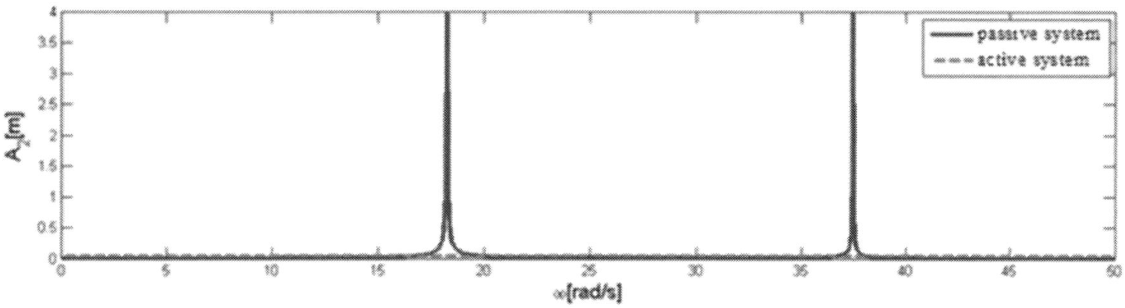

Figure 12. Diagram of amplitude A_2 of the system with electrical elements

3. Summary

The modelling and analysis of mechanical systems with the active reduction of vibration implemented with electric elements, presented in the work, constitute a new approach to the problem of designing such systems. In order to design a system containing active elements it is necessary to synthesize or identify a passive system which does not contain vibration damping elements and next, depending on the structure of the passive system and on excitations affecting it, select a structure containing active elements from the sequence of structures. Use of software Matlab Simulink allows fast modification of parameters without modeling a new system. Developed blocks can be used in other more complex systems.

Acknowledgement

This work has been conducted as a part of the research project No. PBS2/A6/17/2013 supported by the National Centre for Research and Development

References

[1] Bialas K. 2013 *Electrical Elements in Reduction of Mechanical Vibrations* (Applied Mechanics and Materials) 371 pp 657-661.
[2] Bialas K. 2012 *Mechanical and electrical elements in reduction of vibrations* (Journal of Vibroengineering) 14 (1) pp 123-128.
[3] Białas K., Buchacz A., Dzitkowski T. 2009 *Synthesis of vibrating active mechanical systems with dumping in view of polar graphs and structural numbers* (Gliwice Monograph 230, Silesian University of Technology Press), (in Polish).
[4] Buchacz A., Płaczek M. 2010 *Selection of Parameters of External Electric Circuit for Control of Dynamic Flexibility of a Mechatronic System* (Solid State Phenomena) 164 pp 323-326.
[5] Engel Z., Kowal J. 1995 *Vibro-accoustic processes control* (Kraków AGH Press) (in Polish).
[6] Michałowski S. 1994 *Active systems in machines construction* (Cracow Publication by Cracow University of Technology, Monograph 171) (in Polish).
[7] Onwubolu G. C. 2005 *Mechatronics Principles and Applications* (Elsevier Butterworth-Heinemann)
[8] Preumont A. 2006 *Mechatronics Dynamics of Electromechanical and Piezoelectric Systems* (Published by Springer. The Netherlands)

IOP Conf. Series: Materials Science and Engineering **161** (2016) 012022 doi:10.1088/1757-899X/161/1/012022

Optimal management of reconfigurable manufacturing system modeling with Petri nets developed three-dimensional - RPD3D

F Teodor, V Marinescu and A Epureanu

"Dunărea de Jos" University of Galaţi, Faculty of Mechanics, Manufacturing Engineering Department

E-mail: theo.florin@gmail.com

Abstract: Modeling of reconfigurable manufacturing systems would have done using existing Petri net types, but the complexity and dynamics of the new manufacturing system, mainly data reconfiguration feature, required looking for a more compact representation with many variables that to model as accurately not only the normal operation of the production system but can capture and model and reconfiguration process.

Thus, it was necessary to create a new class of Petri nets, called RPD3D (Developed Petri nets with three dimensional) showing the name of both lineage (new class derived from Petri nets developed, created in 2000 by Prof. Dr. Ing Vasile Marinescu in his doctoral thesis) [1], but the most important of the new features defining (transformation from one 2D model into a 3D model).The idea was to introduce the classical model of a Petri third dimension to be able to overlay multiple levels (layers) formed in 2D or 3D Petri nets that interact with each other (receiving or giving commands to enable or disable the various modules together simulating the operation of reconfigurable manufacturing systems). The aim is to present a new type of Petri nets called RPD3D - Developed Petri three-dimensional model used for optimal control and simulation of reconfigurable manufacturing systems manufacture of products such systems.

1. Introduction

The necessity of new manufacturing systems has emerged as a feedback to the globalization of the economy, generating complex applications for producers (a variety of products that meet the quality and low price special).

The concept of reconfigurable manufacturing system involves including the capability to change the physical structure of the production system (repositioning resources, changing their functionality, etc.) to process new pieces of the same family of parts or new families.

The concept of reconfigurable manufacturing system (RMS) and composition of resources (Reconfigurable Manufacturing Machines - RMT, transportation systems parts and semi-finished industrial robots, storage, etc.) have been proposed in 1999 at the Center for Systems Research Reconfigurable Manufacturing (ERC / RMS) at the University of Michigan by Prof. dr. Koren Yoram known in the scientific world as the "father of RMS".

He defined as the RMS modern manufacturing systems, it has precisely the required production, just when it is needed. *This means that the manufacturing system adapts in real time to the point of manufacturing requirements by reconfiguring processes that can be grouped into several classes, as follows:*

R1 – Initial setup of all elements RMS - manufacturing workshop (building the list of resources available and the choice of which will be included in a synchronization RMS on technical characteristics of the resource and technological operation to be performed);

R2 – Reconfiguring all the factors RMS by opting for a technological flow or another (FT - order ot achieve technological operations and relations of dependency between them);

R3 – RMS reconfiguration to ensure system stability (capacity calculation positions RPD3D variables associated RMS Petri net so this is viable, stable and without jamming);

R4 – RMS manufacture reconfiguration during simulation by changing states of RES resources (either an RES becomes available at some point, or becomes unavailable because of failure or the need to use it in another RMS);

R5 – RMS reconfiguration under changing production specifications (changing the number of copies made of a particular product, in or out of the manufacture of a product);

RMS's aim is therefore to achieve an overall manufacturing orders or family of parts in a number of samples while maintaining optimal configurations the available resources according to the chosen criteria (minimum production time, maximum productivity) in terms of system stability.

Reconfigurability's RMS must be based on economic factors that will ensure that the benefits obtained outweigh the costs and production losses reconfiguration for this operation.

We assume that this condition is satisfied in all cases and suppose that reconfiguration time is as small (eg R3 and R5 time should not exceed the time necessary to manufacture a one product). Also, the reconfigurability to be atomic - is carried out in a single operation to reconfigure then be re-evaluated RMS's condition.

2. The idea

Modeling of RMS would have done using existing Petri net types, but the complexity and dynamics of the new manufacturing system, mainly data reconfiguration feature, required looking for a more compact representation with many variables that to model as accurately not only the normal operation of the production system but can capture and model and reconfiguration process.

If any products to manufacture enough to have a workshop with some machine tools, industrial robots, intermediate storage of parts and tools to make several technological operations, then we can modularize the whole process by associating each resource and generic RPD3D each technological operations, they differ only by working parameters and status. These can be predefined generic RPD3D are editable and are stored as modules, parameter values so that individualized product or its manufacturing process technology being stored in the database.

Shaping extremely simple to manufacture a product (5-6 technological operations on 2-3 machine tools supplied by an industrial robot and equipped with deposits of semi-finished and finished parts, plus inter-conditionings and error correction subnets) with a classical Petri net we get a model that will contain dozens of positions and transitions and hundreds of links between these elements will have a development model prevailing in the vertical relationship between the two dimensions may have values of $6 \div 20$, the network obtained is hardly noticed as a whole.

For a more complex product manufacturing (tens of technological operations, dozens of machine tools, industrial robots, elements of the transport and storage, several members of the product family, the whole assembly being able to reconfigure dozens of other manufacturing systems) we get a classical Petri network of thousands of elements linked by tens of thousands of links. Such a network would be represented in the form factor represented and viewed almost impossible, but adding a third dimension Petri net can obtain a compact representation, even beautiful, which will also resolve the difficult problem of tracking a link type feedback (reset) of one of the last transition to one of the top positions commonly used type of connection.

In nature the deoxyribonucleic acid has a 3D structure of double helix just to save space, not influence function as essentially a 'ladder' right is functionally identical to the one of a double helix.

2.1 Analysis of the state of resources

During operation of RMS, a resource can be in one of the following six states: Ready (R), Work (W), Damaged (D), Pause (P), Error correction (E) or IntoRMS - built into RMS (I) - the resource is ready but requires time to be entered in RMS. During operation of RMS, a resource can be in one of the following six states: Ready (R), Work (W), Damaged (D), Pause (P), Error correction (E) or IntoRMS - built into RMS (I) - the resource is ready but requires time to be entered in RMS.

The transition between a state and the other resources can be automated (P to W, W to R, E to I, E to W) or may be subject to state one of the six markers (ON - marking is active, thus allowing change of status OFF - the mark is inactive and blocks the change of state (see figure 1).

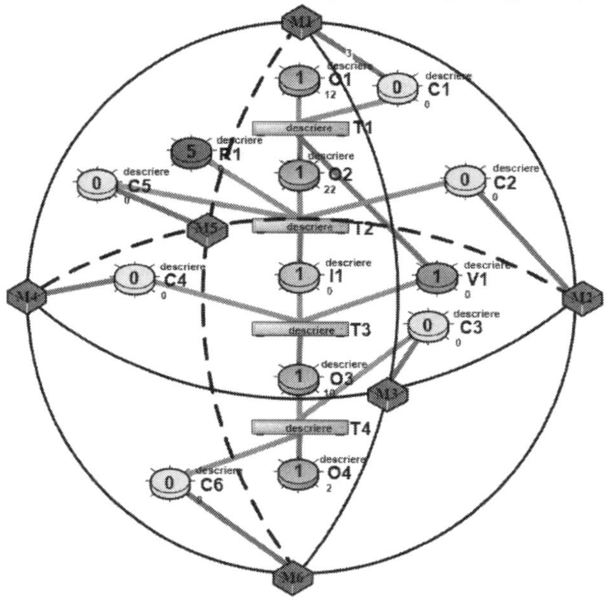

Figure1. RPD3D network that models the operation of RMS module

3. Developed three-dimensional Petri RPD3D - presenting the model

To create and simulate the operation of RMS with RPD3D author created an application called SODRMS (System to Optimal Driving for RMS) based on the NetBeans8 IDE development platform in tandem with a MySQL database and graphics were made with Java3D interface.

Note that all programs used and related documentation (NetBeans, MySQL, Java3D) are available free on the Internet, so long as they are not used for commercial purposes, so they are easily available to anyone who would like to create work environment for application SODRMS. Like any Petri net, RPD3D consists of two main elements: position (represented by spheres) and transitions (represented by parallelepiped), connected by different thickness oriented segments called links.

As a RP-D, a position can shape operations (processing, transport, handling, etc.),fixed resources (processing centers, AGV's, robots), variable resources (pallets, parts, buffers) or conditions conduct required operations. A technological process can be divided into different phases, the shape and the position that the operation can be replaced by a subnet Petri net model.

In this model, operations positions are called O-positions. As in the case of Petri networks developed from which the one or more chips in a position corresponding resource means that the resource is available, the number indicating the amount of chips.

Resources can be divided into two classes, resources whose number is fixed by design planning such as robots and conveyors, AGV sites (positions corresponding to this resource type are called R-positions, and variable resources (pallets, pieces or tasks to be processed) as in sharing resources (these are called V-positions). Marking of variable resources must be determined so that the system does not block or run unloaded or loaded partial.

The new model is maintained and intermediate positions, called I-positions that can model a process involving a variable number of resources such as buffer storage operations or deposits, or to ease of maintenance of properties behavior patterns processing systems.

Thus, to enable ordering of subsystems activities are maintained and positions that shape the ordering of necessary control information, called C-positions.

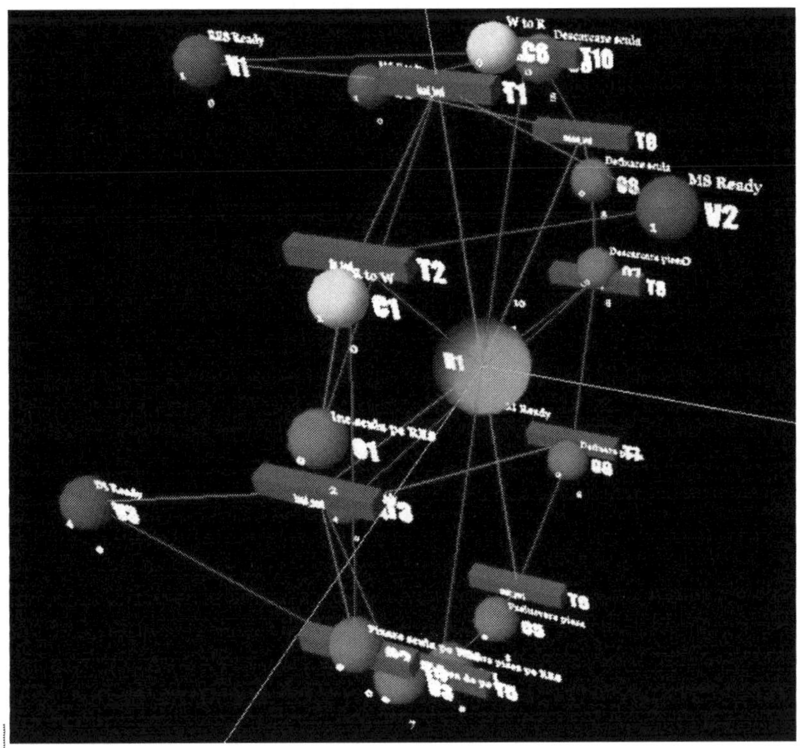

Figure 2. RPD3D graphical representation

Usually they enter a position at the entrance to the first transition control subsystem for controlling its activities and start counting the number of cycles controlled and controlling position at the exit end of the last transition of operations to control and count the cycles of activities for each subsystem type of machined parts in the system. In this model are transitions that represent the beginning or end of an operation - O-transitions and transitions that trigger subnets for error handling are E-transitions.

As a novelty, the colors used to represent positions and transitions were differentiated for easy viewing in 3D space and understanding RPD3D operation: operational positions - red (O), fixed resources positions - green (R), position variable resources - blue (V), intermediate positions - Gray (I), control positions - yellow (C), operational transitions - blue (TO), transitions, input / output of subnets - yellow (TE) (see figure 2).

Each position RPD3D may contain one or more marks (or chips). Representation in classical RP was achieved by points that were added or removed from the circle symbolizing the position, but in the transition to three-dimensional representation, marking positions to be adapted and opted for a numerical representation placed on the outside of the sphere that symbolizes the position.

The graphical representation of the elements is a ball-type position (diameter 0,5 - 1 - 2 or 3 units), and the type of transition elements is a parallelepiped with dimensions of length and 0.5 x 3 units section 0.5 units - 3 being colored according to the convention above.

Positions and transitions have some metadata displayed graphically, the others being stored and handled only by RMS related MySQL database. Four position description are shown (top right), titled (right) mark (middle range) and execution time (bottom right) and the transition name (right) and description (middle).

4. Presentation tables in the database for the main elements RPD3D

Table POZ – type stores data items in a module position RPD3D how a resource can be associated to a workshop production or technological operations of a manufacturing flow of a product, data like: *ID position* - number generated by the database, incremental, used as the primary key of the table (sorting, unique records), *name position* - like O25, *description* (interpretation) position - remained formulating grammatical descriptions used in the definition of RP-D model [1], *marking* the original position, *the coordinate of the center* position, *scale* position - which is the position size (0.5 X, 1X, 2X and 3X), *the maximum allowable execution modeled operation* position (T_{max}) in seconds, *visibility* -1 (ON) and 0 (OFF) and *timestamp* - when inserting position as a 10-character numeric sequences; The insertion type Timestamp field value is automatically calculated and can be used to reconfigure the RMS (you can compare the current time with TSTP and find out bond or 3D object is more recent or older. Ex: 1256953732 = Sat Jul 23 02:16:57 2005

Table TRZ – stores data items in a module type transitions RPD3D, how that can be associated with a resource or a workshop production of a technological operations manufacturing flow of a product, data like: *ID transition* - number generated by the database, incremental, used as the primary key of the table (sorting, unique records), *the name of the transition* - the type TOnnn, TCnnn, where nnn is a number of 8 characters, *the description* (interpretation) transition - remained formulation grammatical descriptions used in the definition of RP-D model [1], *the coordinate* of the center the transition, *initial error code* of the transition, *scale transition* - the size of which is the transition (0.5X, 1X, 2X and 3X), *visibility* transition-1 (ON) and 0 (OFF), *timestamp*- timestamp of transition – transition time of insertion as a 10-character numeric sequences;

Can thus be defined 10^5 error codes, enough for any module RPD3D each error code with a description and a way of resolving the error, even if for most of these error codes, solving resource is switching state D (Damaged) and excluding it from the RMS's analysis.

Links between elements are oriented (sense) and are carrying information about the manufacturing system reconfiguration.

Table LEG – stores data items in a module type connected RPD3D how that can be associated with a resource or a workshop production of a technological operations manufacturing flow of a product, data like: *ID link* - number generated by the database, incremental, used as the primary key of the table (sorting, unique records), *the name* of the connection, *name position*, *name transition*, *meaning* the link (0-direct, 1-reverse), *the weight of the link* (probability of extinction of that links to a possible reconfiguration), *description of the link* (optional, in particular for the weights 1 and 2), *TTB_LEG* - TimeToBuild - information, in numerical form (seconds), which is the length of materialization or selling link in the process of reconfiguration, directly proportional to the weight of the link, *visibility of link* - 1 (ON) and 0 (OFF), *timestamp link* - hold when inserting links as a 10-character numeric sequences; useful to compare the connections between them by their age (maintenance required reconfiguration process).

The weight three is given arcs of the elements constituting modules RPD3D, weight 2 is given RPD3D arcs between modules on the same level and the share of one of the modules is given arcs RPD3D located on different levels.

The link thickness indicates durability link, weakest link indicating a higher probability of extinction of that links to a possible reconfiguration of RMS, the share of the three strongest and the weakest one. Data links between modules (share 1 and 2) will be noted in Table CON.

In addition, in the RPD3D model, easy to make the connections between the different modules of the RMS, placed on the same level or on different levels, it adopted a third type of element, called connector, represented by a red cube with sides equal to the diameter of the sphere which is a position connectors are arranged on a sphere circumscribing RPD3D compact (obtained after compaction operation). The based connections are six, are placed in the points where the axes intersect the sphere which includes RPD3D compact and are referred to as M1 (y +), M2 (X +), M3 (z +), M4 (x-), M5 (z-), M6 (y-). Outside the module, the six connectors are connected by arcs of weight 1 or 2 other connectors and inside the module, they are connected by arcs of 3 weight control positions which they convey different commands given from higher hierarchical levels (ie : beginning or end of OT, transfer the number of finished parts made, changing the state of RES). Connectors does not affect the operation or its associated RPD3D analysis but were only meant to send it to and from the upper levels commands will result in the reconfiguration or RMS's operation.

Table CON – connector type stores data items in RMS: *ID connector* - number generated by the database, incremental, used as the primary key of the table (sorting, unique records), *RPD3D module ID* which includes, *name connector* - type M1 (means M1 connector module 23 associated with the RMS), *description connector*–optional, fields *X_CON, Y_CON and Z_CON* are the coordinates of the connector relative to the center of the sphere circumscribed, *PLI* - internal liaison position (inside the module ex: "C2"), *PLE* - external link position (another connector module ex: "M1_MOD_23"), *weight* - leg. between RMS modules are on different levels are value 1 and leg. between modules of the same level are value 2, *STARE_CON* (active / inactive) that transmits the activation signal to another module or not, *visibility Connector* – (1 ON and 0 OFF), *visibility of link* -1 ON and 0 OFF, *timestamp*- hold when inserting links as a 10-character numeric sequences; useful to compare the connections between them by their age (maintenance required reconfiguration process) şi *timestamp position* - when inserting the connector as a 10-character numeric sequences;

As a novelty, it will be able to use the zoom in levels - the number of items displayed on the page (visible) will be chosen depending on the zoom level. This feature uses zoom levels of visibility of items in the RPD3D, switching from one zoom level to another leading to enable or disable the visibility of certain elements.

Also, all the new element was introduced feature scale (representing the same type of item 0.5 of normal size or 2x or 3x).A greater representation of an item RPD3D suggests greater importance given that item (robots, other important resources).

Thus: which models resource positions (fixed or variable) will be represented at 2x or 3x scale positions and operational transitions (those of processing, assembling, etc.) at 1x scale, control and intermediate positions will be represented at the scale 0.5 x.

Since we started from the premise that RPD3D RMS modeling the operation's resources look the same regardless of size, for leaders or materials involved, which leads to a standardization of these networks, they can be built, it will create the basic forms RPD3D resources, the rest can be created when it will be needed.

5. Conclusions

Just as DNA is essential to the identity of any body, so RPD3D is suitable mathematical model to describe the process of reconfiguration of reconfigurable manufacturing system, the combination and interaction of the modules RPD3D inspired by way of combining DNA elements whose sequence determines the diversity of individuals of the same species.

Taking on the comparison of genetic identity and modeling of manufacturing systems can be said that the method of differentiation of individuals of the same species by a minor modification of the combination of the DNA could be used to create a procedure to enable the manufacture of entire families of parts per same RMS adjustments (reconfiguration) minor. This can be compared with the reconfigurability of RMS temporal evolution of a species nearing RMS sites more than any other previous production systems can integrate into manufacturing of bioengineering features.

References:

[1] Marinescu V 2000 *Research on the flexible management of metalworking processes cold* PhD Thesis (University Lower Danube Galati)

[2] Reuven K 2007 *Design principles of reconfigurable machines* Int J AdvManufTechnol **34** pp 430–439

[3] Bi Z M, Sherman Y, Lang T, Verner M and Orban P 2007 *Development of reconfigurable machines* Int J AdvManufTechnol

[4] Puttegowda K 2003 *Context Switching in a Run-Time Reconfigurable System* The Journal of Supercomputing **26** pp 239-257

[5] Toguyeni A, Dangoumau N and Lee E *A Synthesis Approach for Reconfigurable Manufacturing Systems Design Based on Petri Nets* LAGIS - CNRS UMR 8146 (EcoleCentrale de Lille)

The parameters impact on cutting forces and determination of regression function

T D Popovici[1,2]

[1] Department of Manufacturing Engineering, The faculty of Engineering and Management of Technological Systems, University Politehnica Bucharest, 313 Splaiul Independentei, Sector 6, Bucharest, Romania

[2] Department of CN Programming and Technology Research, Romanian Research & Development Institute for Gas Turbine, 220 D Iuliu Maniu Bd., Sector 6, Bucharest, Romania

E-mail: tabydana@yahoo.com

Abstract: The aim of this paper is to present results of certain experiments related to material workability characteristics, taking into consideration its high quality features and large usability for special manufacturing. It is highly used in various industrial processes, such as aeronautics, petrochemical industry, chemical industry, etc. It was examined the influence of milling parameters upon the three dimensional cutting force and after it was determinate regression function.

1. Introduction

Many problems regarding the pieces manufacturing posed challenges for many researchers which contributed and are still contributing to determine the most efficient approach in describing the complex process of turning. The entire spectrum of the actual researches regarding the workability represents the multiple directions of this complex knowledge. After a slight examination, an expert is tempted to define the turning workability related to its field of activity [1, 2,7]. Although many researches were focused on the material workability, at the time of speaking there is no clear definition and no appropriate index widely accepted. In terms of methods regarding the material machining, is well known that there are many manufacturing processes, but with no strict constraints according to the material. Machining restrictions depending on the material were defined by many facts [1, 3, 4, 5]; they are strictly related to the effective time of machining, which has to be as short as possible, decreased costs and possibilities of standardisation [1]. In this paper approaches the 41MoCr11 workability using an experimental method (using defined milling conditions), following the material workability markers as the cutting efforts which appear in the machining process, and after we determinate the regression function with specific program.

2. Paper contents

The aim of the experiment is to determine machinability of 41MoCr11 material in terms of forces that appear during cutting, knowing the processing system parameters, used at the plane milling operation and determine the influence of each parameter of cutting regime on these efforts.

Because "the basis of machinability is any relationship of dependence between the quantities involved in processing" [3,11], so that the resulting relationship 1,

$$\bar{Y} = \Gamma(\bar{x}_1, \bar{x}_2, \dots \bar{x}_j, \dots, \bar{x}_k) \tag{1}$$

Y is the dependent variable and X_n are independent variables.

In this case the relations (2,3,4) used are:

$$F_x = \Gamma(v, f_z, a_e, a_p), N \quad (2)$$
$$F_y = \Gamma(v, f_z, a_e, a_p), N \quad (3)$$
$$F_z = \Gamma(v, f_z, a_e, a_p), N \quad (4)$$

Where:

Fx, Fy, Fz – cutting force on X, Y and Z [N]
a_p – depth of cut [mm]
a_e – redial depth [mm]
f_z – feed per tooth [mm/tooth]
n – cutting speed [m/min]

3. Experimental Details

All the experiments of this paper were made with machine-tool MCV 300. This is CNC milling machine with 5 axes. For the tests was used a milling cutting tool, with coated carbide, with z = 3 and 18 mm diameter, made by Taegu Tec Company [7,11,12]. Flat milling was chosen. The machining method is dry type cutting. It consists of two activities:

- the mechanical activity, including the machine tool, cutting tools – milling cutters, work piece made of 41MoCr11
- the data acquisition, including the KISTLER measuring table (dynamometer), an electronic amplifier and the computer having a data acquisition system (figure 2).

Table 1 Initial characteristic of piece.

Material		Piece	Hardness HRC	
			Interval	Value
MP4	Steel 41MoCr11		31,78 - 33,00	32,4

Table 2. Chemical composition.

Element	C[%]	Mn[%]	P[%]	S[%]	Si[%]	Cr[%]	Ni[%]	Mo[%]	Fe[%]
Conform STAS 8185	0.38-0.45	0.4-0.8	Max. 0.035	Max. 0.035	0.17-0.37	0.9-1.30	Max. 0.3	0.15-0.30	**Rest**

Experimental	0.40	0.67	<0.025	<0.025	0.29	1.04	<0.025	0.23	**Rest**

After this analysis, with special spectrometer it can conclusion that material is 41MoCr11 steel (see value compared with STAS 8185); and in figure1 we can see a uniformity dispersion of elements, a sorbite structure with fine grain in ASTME112 standard.

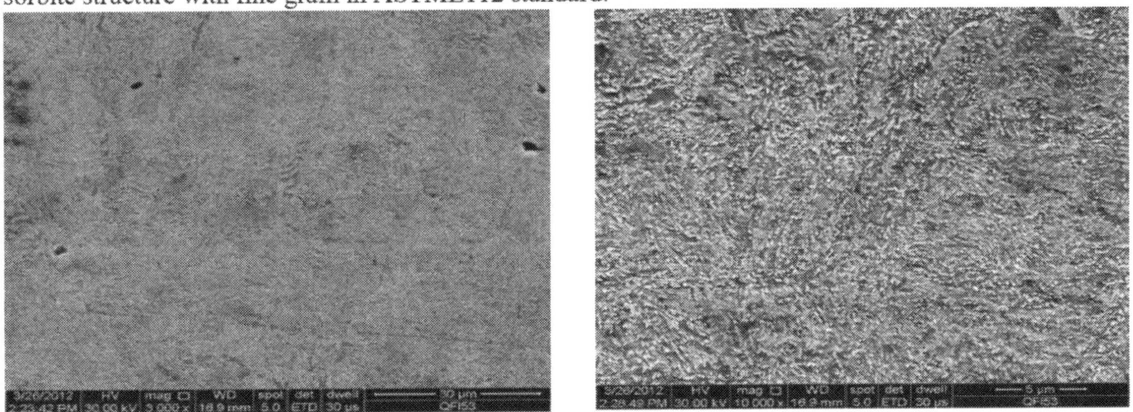

X 3 000 X 10 000

Figure1 Metalographic structure of 41MoCr11 material.

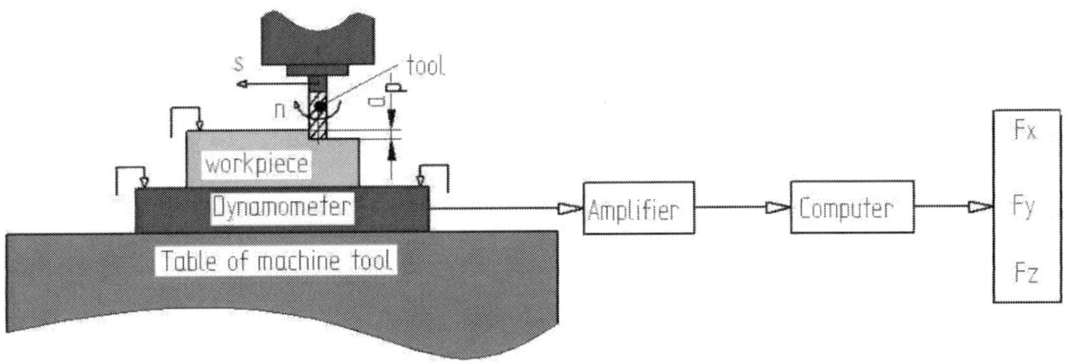

Figure 2 Experimental set-up

Taking into account the recommendations of the supplier tool [6,9,13], the working rate was established in experimental program. Using Response Surface theory, the experimental program was coded (table 3).

Table 3 The experimental program.

Independent variables	**Code**		
	X_1	X_2	X_3
	-1	0	1
depth of cut, a_p: 1 to 2.5 mm	1	1,58	2,5
radial depth a_e: 0,5 to 2,5 mm	0,5	1,12	2,5
feed rate f_z, mm / tooth : 0,01 to 0,1 mm / tooth	0.01	0.032	0.1
cutting speed: 200 to 300 m/min	200	245	300

The registration data were done by the DynoWare program. In figure 3 (a, b, c,) graph forces and torque (Fx, Fy, Fz, Mz) for only first processing are presented.

(a) (b)

(c)

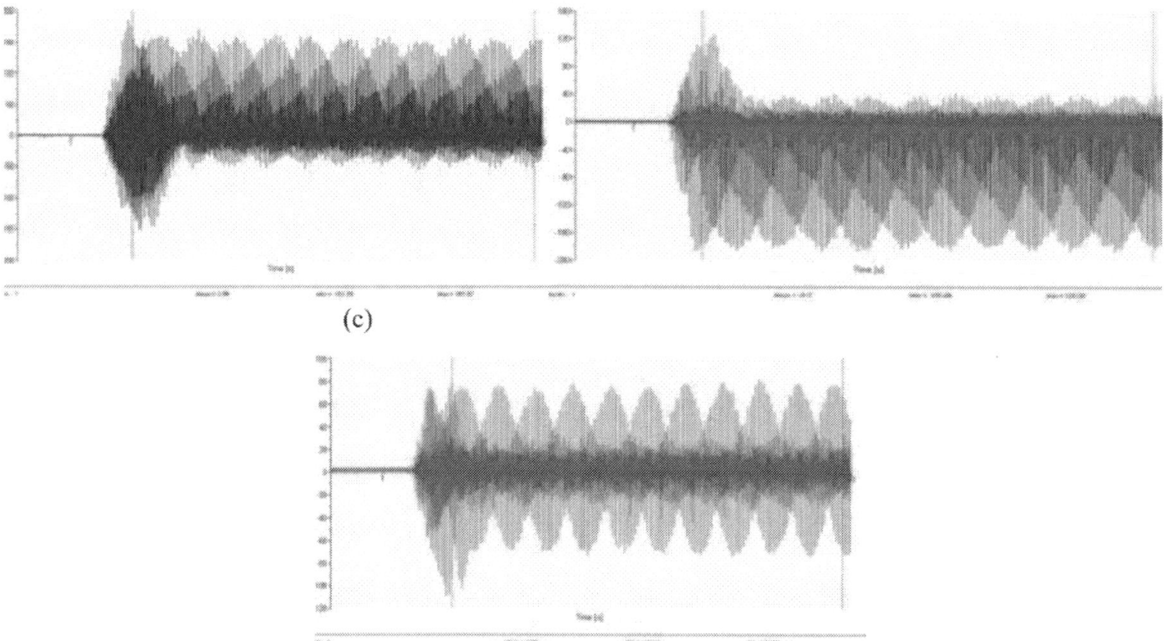

a)$F_x(a_p,a_e,f_z,v)$=167,57 N
b)$F_y(a_p,a_e,f_z,v)$=185,49 N
c)$F_z(a_p,a_e,f_z,v)$=101,90 N

Figure 3 The graph of cutting forces F_x, F_y, F_z for first experience

Cutting regime parameters in this case are: a_p= 1mm; a_e = 0.5 mm; f_z= 0.01mm/tooth; v= 200 m/min. The experiments revealed the next values of cutting forces on the three directions: F_x € [167,57;886,14]N, F_y € [185,49; 992,07]N, F_z € [101,9, 627,2]N. To determine the total cutting force we used the formula 5.

$$F = \sqrt{F_x^2 + F_y^2 + F_z^2}, N \quad (5)$$

The values obtained for total cutting force are in the interval, F € [269.94, 1470.65] N. The variation of cutting force depending on the parameter is shown in Figure 3, 4, 5 and 6.

Influence of depth of cut on the cutting force; study of the obtained diagrams reveals that, with the increase of the depth of cut (see figure 3), the cutting force increase on the first interval [13,7,9] and decrease on the second interval [11,10, 8]. Influence of radial depth on the cutting force in figure 4 the obtained diagrams reveals that cutting force increase with the increase of radial depth. This increase of cutting force is approximately constant. Influence of feed rate on the cutting efforts; study of the obtained diagram reveals that cutting force F have same trend with the increase of the feed rate (see figure 5). In this case we see that the influence of feed rate is very high. Influence of cutting speed on the cutting efforts; study of the diagram (see figure 6) reveals that force F have same trend with the easy increases of the spindle speed. After study all obtained diagrams that cutting force increase when the feed rate increases. This influence is very high.

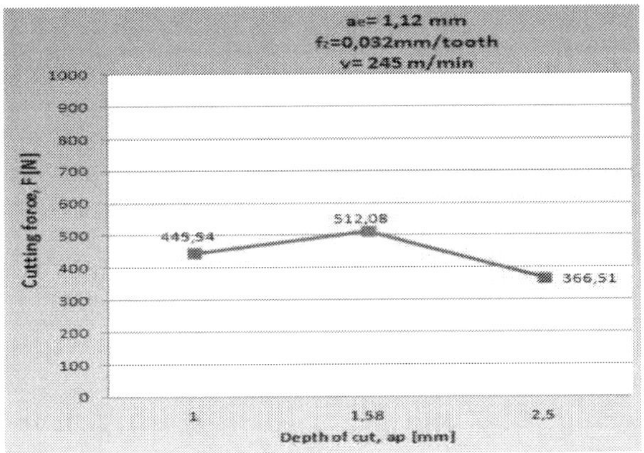

Figure 4 Variation of cutting force with depth of cut

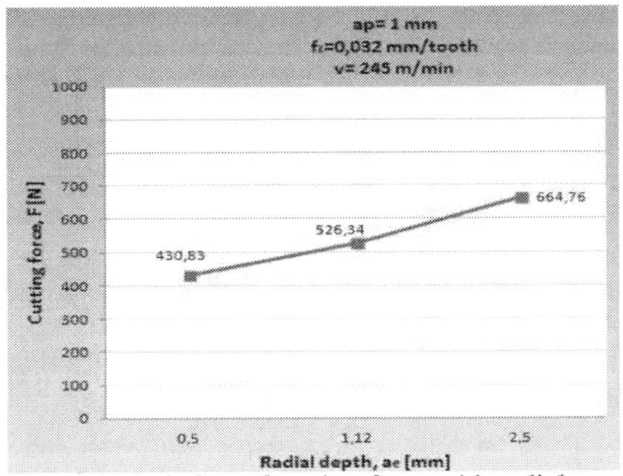

Figure 5 Variation of cutting force with radial

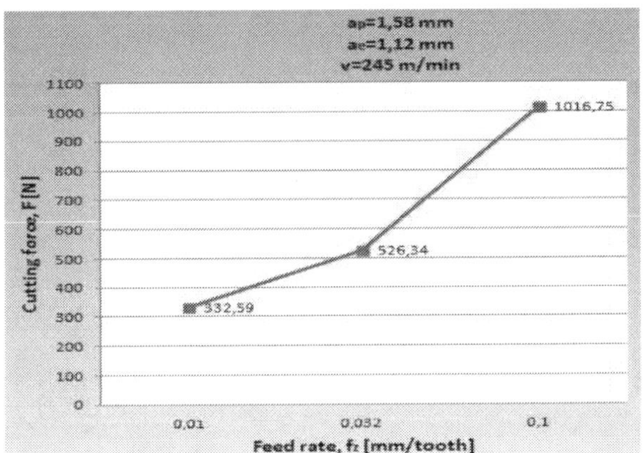

Figure 6 Variation of cutting force with feed rate

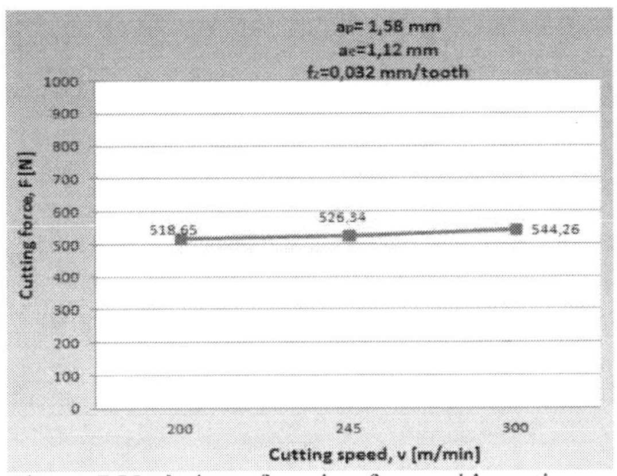

Figure 7 Variation of cutting force with cutting

The cutting speed hasn't important influence of the cutting force. Regression functions for cutting force F_x, F_y and F_z. The regression function for cutting force Fx, Fy and Fz were determinate as multivariate regression functions for each material processed form:

$$\bar{Y} = A_0 X_1^{A_1} X_2^{A_2} X_3^{A_3} X_4^{A_4} \Leftrightarrow y = a_0 + \sum a_j z_j, y = \ln Y, z_j = Xj, J = \overline{1,4} \qquad (6)$$

Practically, the experimental data obtained and with DataFit 9.0 program were calculated ecuation (7,8,9) the coefficients A0, Aj, j = (1,4).

$$F_x = e^{5.120750561 60716} * X_1^{0.145937380699512} * X_2^{0.0937555117427452} * X_3^{0.64671035677042} * X_4^{0.512564541459054} \qquad (7)$$

$$F_y = e^{6.18615740615678} * X_1^{0.0429273773906433} * X_2^{0.49241326670457} * X_3^{0.470860523927147} * X_4^{0.225044060032545} \qquad (8)$$

$$F_z = e^{3.84961631600464} * X_1^{0.158047134562595} * X_2^{0.386177602152276} * X_3^{0.553671988766012} * X_4^{0.584742532633664} \qquad (9)$$

4. Conclusions

This work allows the following conclusions:

- the variation depth of cut have nonlinear influence for cutting force and require a higher degree of attention and is necessary more study for another intervals.
- the cutting force variation is direct proportional with radial depth variation.
- the increase of feed rate have the greatest influence on cutting force.
- the increase of cutting speed haven't influence or have easy influence for the variation of cutting force.
- to optimize cutting processing on the 41MoCr11 is necessary to use all the cutting parameters.

Research Directions - develop research methods cutting machinability of this material in order to standardize some of them, determination of machinability data for this material using different tools geometry, coating, composition and processing methods, and the development of data banks workability, machinability, continuous updating of data in accordance with the new tool materials with improved characteristics precision, stiffness and power of machine tools.

5. References

[1]C Picoş *et al*, 1981 Turning workability of ferrous alloys, *Ed. Tehnică*, Buchares

[2] T Grămescu et al, 2000 Material workability, *Ed. Tehnica-Info*, Chişinău,

[3] M Gheorghe, 1984 Research on the machinability of malleable cast ironremoving, *PhD Thesis*, Bucharest

[4] N Andrei, 2010 Simulation of integrated CAD-CAM-CAE for partsof the composition of turbo engines. Experimentation IQMD integrated, Bucharest

[5] M Sime, 2011.Contributions regarding workability and working rate optimisation on drilling of high alloy steel, *Doctoral thesis*, Bucharest,

[6] Metal Working Cutting Tools, Taegu Tec, Member IMC Group, 2011.

[7] D L Martin, A N Tabenkin, F G Parsons, 1995 Precision Spindle and Bearing Error Analysis, International Journal of Machine Tools & Manufacture,**187-193** vol. 35 No 2

[8] N Lynagh, H Rahnejat, M Ebrahimi, R Aini, 2000 Bearing induced vibrations in precision high speed routing spindles, *International Journal of Machine Tools & Manufacture,* **561-577** vol. 40

[9] D M. Marin, D Marin, R.C Varban, Dynamical behaviour analysis of three-axis milling centres, Scientific Bulletin, series D: **Mechanical Engineering**, vol. **71.**

[10] S Aspiewak and T Nickel, Vibration based preload estimation in machine tool spindles, *International Journal of Machine Tools & Manufacture,* **567-588** vol. 41

[11] Iliescu M Vladareanu L Spanu 2010 Modelling and Controlling of Machining Forces when Milling Polymeric Composites, *Materiale Plastice* , **231-235**, WOS:000281051300022, SN 0025-5289 .

[12] F.T Jay 1996 Strain field analysis and sensor design for monitoring machine tool spindle bearing force, *International Journal of Machine Tools & Manufacture*, **203- 216** vol. 36 No 2

[13] Sandru O I, Vladareanu L, Sandru A, 2008 A new method of approaching the problems of optimal control, Proceedings of the 10th WSEAS *International Conference on Mathematical and Computational Methods in Science and Engineering,* **390- 393,** WOS: 000262436800083.

Tolerance Analysis of 3d-MJM parts according to IT grade

K Kitsakis[1,2,4a], J Kechagias[1], N Vaxevanidis[3] and D Giagkopoulos[2]

[1]Department of Mechanical Engineering, Technological Educational Institute of Thessaly, GR 41110, Larissa, Greece

[2]Department of Mechanical Engineering, University of Western Macedonia, GR 50100, Kozani, Greece

[3]Department of Mechanical Engineering Educators, School of Pedagogical and Technological Education, GR 14121, N. Heraklion Attikis, Greece

E-mail: [a]kitsakis@teithessaly.gr,

Abstract. In this study we investigate the dimensional accuracy of 3D printed objects and define the International Tolerance (IT) Grade for the linear sizes of parallel opposite surfaces and the diameter of a cylinder. Eight (8) exact parts were printed by an Objet Eden250 3D printer using the Multi Jet Modelling (MJM) process. The printed objects were measured and their dimensions analysed to define the IT grade of the specific process. MJM is an Additive Manufacturing (AM) or 3D Printing manufacturing process that can be used in place of CNC to save weeks of lead time, allows us to create parts with a superior strength to weight ratio and can significantly improve the efficiency of the completed system. MJM process is used for a great variety of applications on concept models to model performance parts as functional prototypes, jigs and fixtures, manufacturing tooling and sometimes even production parts. The result of this investigation is that the MJM 3D printing process can produce objects with IT11 grade or better, which classify it to the Fits area of IT grades.

1. Introduction

According to the American Society for Testing Materials (ASTM), Additive Manufacturing is "A process of joining materials to make objects from 3D model data, usually layer upon layer, as opposed to subtractive manufacturing technologies" [1] [2]. It is a key technology for the future, with strong impact on the addition of business value for manufacturers, over the next decade. Its history begins in 1983 when Chuck Hull invented a process called Stereolithography, enabling a 3D object to be printed from CAD data and took almost 20 years before the first consumer devices began to appear. 3D printing is an Additive Manufacturing process invented and patented by the Massachusetts Institute of Technology in 1993 [3]. In 2009 the technology patent for the material extrusion process called Fuse Deposition Modeling (FDM) expired, so the technology became much cheaper and available to a much broader cross section of producers.

The 3D printing process begins with a representation of an object as a 3D model in computer aided design (CAD) based software. That model can be created directly in the software, or it can be input into the software, through the use of a laser scanning device, that will take a physical object and bring it into the system. Once that design is created, the Standard Tessellation Language (STL) file is generated by the software. STL is the most popular file format for Additive Manufacturing. To tessellate something, as in Standard Tessellation language, means to break it down into a series of polygons -and to be more specific in triangles- to represent not just its external structure, but its internal structure, too. Once that file is created, the system slices it into many different layers and passes that information to the Additive Manufacturing device [4].

The Additive Manufacturing system itself creates the object, by definition layer by layer, until we have a finished object. Very often, post production is required. That might be the removal of dust or other material, it might require some machining or a process called sintering where we're closing voids, or some sort of infiltration process where we're filling voids within the object itself with other materials. The goal is to create a physical object from a 3D model.

Although traditional-subtractive manufacturing has its advantages and is preferred when producing very high volumes of similar objects, when we need a diversity of different materials to choose from and when we want to create large parts, 3D printing is preferred when there is high design complexity, the need for rapid production and the necessity for restriction of the material waste [5].

According to American Society for Testing and Materials group (ASTM) the Additive Manufacturing processes which are used today are [6]:

- Vat Photo polymerization
  - Stereolithography (SLA)
  - Digital Light Processing (DLP)
- Material Jetting
  - Multi-jet Modeling (MJM)
- Material extrusion
  - Fused Deposition Modeling (FDM)
  - Fused Filament Fabrication (FFF)
- Powder Bed Fusion
  - Electron beam melting (EBM)
  - Selective Laser Sintering (SLS)
  - Selective Heat Sintering (SHS)
  - Direct Metal Laser Sintering (DMLS)
- Binder Jetting
  - Powder Bed and Inkjet Head 3D printing (PBIH)
  - Plater-Based 3D printing (PP)
- Sheet Lamination
  - Ultrasonic Consolidation (UC)
  - Laminated Object Manufacturing (LOM)
- Direct Energy Deposition
  - Laser Metal Deposition (LMD)

In Multi-Jet Modeling (MJM) technology a print head selectively deposits material on the platform. These droplets are most often comprised of photopolymers with second materials, such as wax, used to support structures during the building process. At the next step, an ultraviolet light solidifies the photopolymer material to shape the cured parts. At the final pass, the planerizer levels the material to create a nice flat surface (Figure 1). When the parts have been printed, a post-building process is required to remove the support material. The advantages of this 3D printing technology are the high accuracy and surface finish, the capability to use multiple materials and colors on the same part and the hands-free removal of support material. The disadvantages are the limited range of materials and the relatively slow build process [6] [1].

Various implementations for this technology can be found in the industries of industrial and consumer products and, more specific, industries of automotive, medical, aerospace and defense. Applications of this technology are Rapid Prototyping (RP), Rapid Tooling (RT) and Direct Digital Manufacturing (DDM)[7] [8]. The parts produced for these sectors need to be firm and fit the prospective functionality, but 3D printing for RP, RT and DDM is a relatively new technology and in the literature there are only a few studies related to dimensional accuracy and categorization, [9] [10]. In this paper we studied the coherence and repeatability of linear and circular dimensions of parts produced by a 3D printer using the Multi-jet Modeling (MJM) process. We classified the above dimensions and categorized their quality according the International Tolerance Grade.

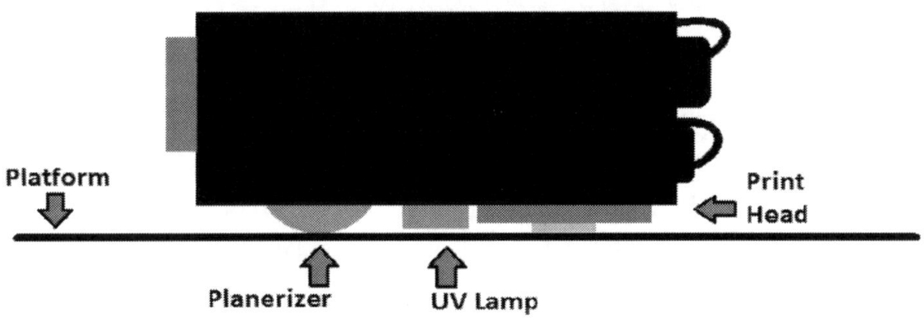

Figure 1. Multi-Jet Modeling printer parts.

2. Scope

The most important applications for MJM 3D printing technology are any situation in which there is a need for a high fidelity mock up, with complex objects, moving parts, constant service finish, fine resolution, and accuracy. The most crucial aspect for ensuring the dimensional repeatability of the printed parts is the dimensional accuracy which represents the degree of agreement between designed prescription and the manufactured product dimensions. The current dimensioning and tolerance standards evaluate the dimensional accuracy of a component part through its size and shape [11].

The variation in size of the manufactured products is very crucial for the applications that this technology is applied for. We examined different linear dimensions in the X, Y and Z axis and circular dimensions at the top and the bottom of a vertical cylinder to categorize the produced samples according to the ISO 286 International Tolerance Grades table reference.

3. Experimental Work

For our experiment we designed an L like 3D shape combined with a cylinder near one of its corner, but not tangential to the edges, as shown in Figure 2. This shape gave us the opportunity to examine external linear dimensions (length, width and height of the L shape) and outer circular dimensions (diameter of the cylinder). The dimensions have been chosen so that they belong to different Basic Size ranges of the International Tolerance Grades Table.

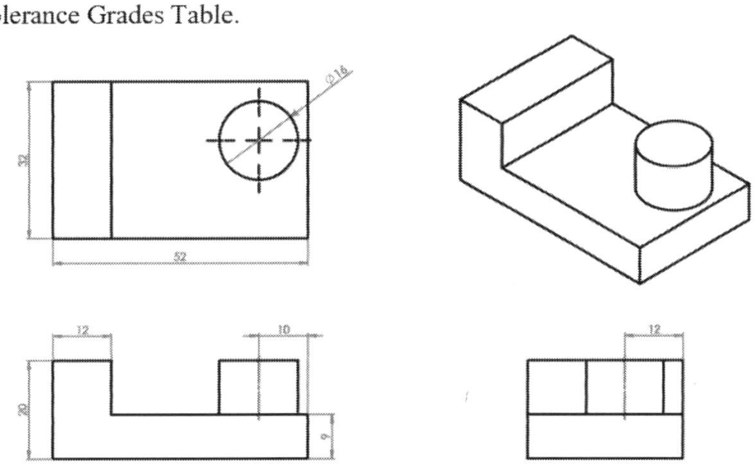

Figure 2. Designed part dimensions in mm.

Eight independent parts (Figure 3) were produced from the same STL file using an Objet Eden250 3D printer. We selected High-speed printing mode (30-micron) and glossy surface at the printing process to have more sleek surfaces and only at the bottom of the parts the FullCure705 support was printed to ground them. For all parts Veroblack Opaque material was used.

Figure 3. The eight printed parts.

The Eden250 printer (Figure 4) is an ultra-thin-layer, high-resolution 3D printer for rapid prototyping and rapid manufacturing. Some technical specifications of the printer are: Net build size 250mm X 250mm X 200mm, Build resolution at X-axis: 600dpi, at Y-axis: 300dpi, and Z-axis: 1600dpi, Horizontal build layers down to 16 micron at High Quality and down to 30 micron at High speed printing mode, 0.1-0.2mm typical accuracy varying according to geometry, part orientation and print size, Materials supported: FullCure 720 Model Transparent, VeroWhite Opaque, VeroBlue Opaque, VeroBlack Opaque and Support type FullCure705 Support [12].

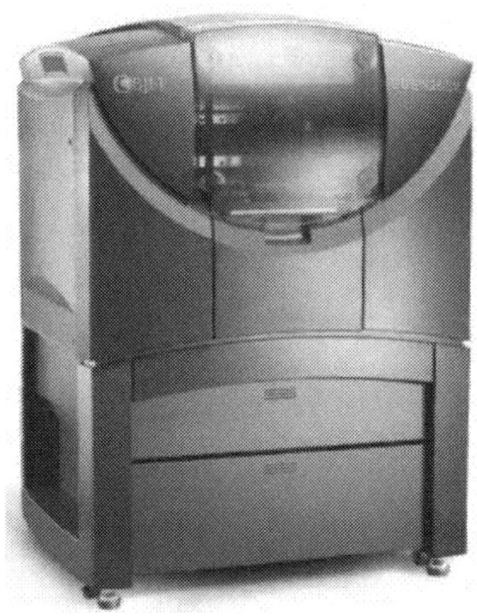

Figure 4. OBJET EDEN250 MJM 3D printer.

The printed parts were measured by three micrometers, with measuring ranges of 0-25mm, 25mm-50mm and 50mm-75mm each. We measured:

- The height (Z-Print Axis) at ten measurement points on the upper top surfaces (rectangle and circle surface) and at nine measurement points on the lower top surface (points HA1-HA10, HB1-HB9 - Figure 5)
- The length (Y-Print Axis) at six measurement points on the upper rectangle and at six measurement points on the lower rectangle (LA1-LA6, LB1-LB6 - Figure 6)
- The width (X-Print Axis) at ten measurement points (W1-W10 - Figure 7) and the dimeter at four measurement points (D1-D4 - Figure 8).

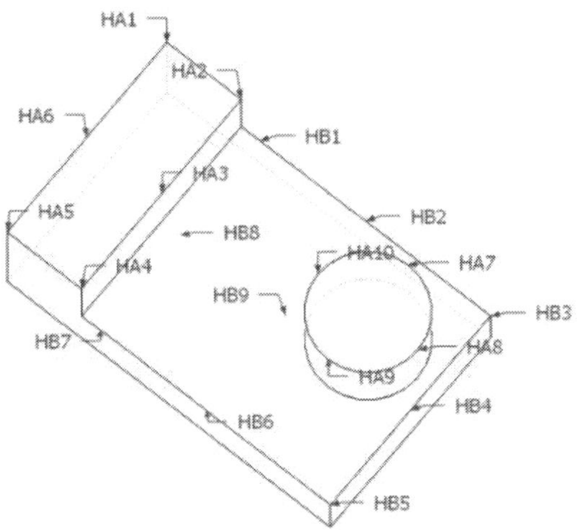

Figure 5. Measured Height points.

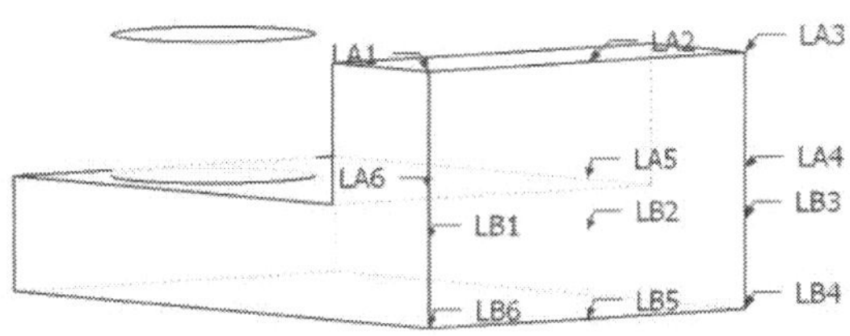

Figure 6. Measured Length points.

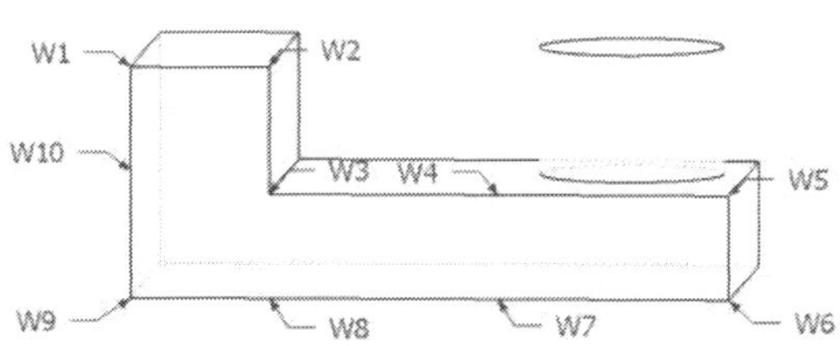

Figure 7. Measured Width points.

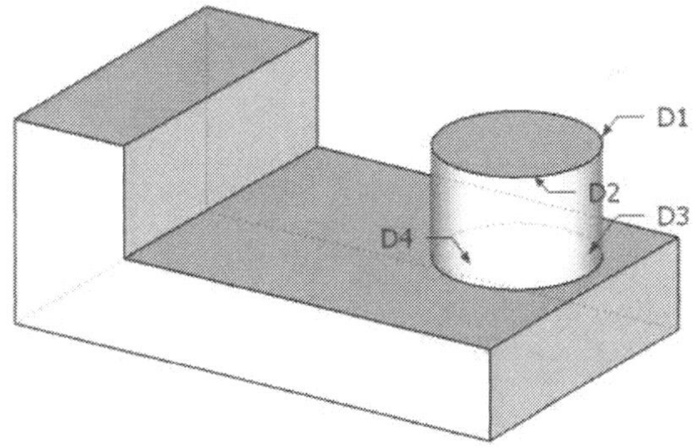

Figure 8. Measured Diameter points.

4. Results and Analysis

In Chart 1 and Chart 2 the values of height or Z-Print Axis is presented and although the mean is very close to the designed dimension, the variation of errors (±3σ) sometimes extends the printer's nominal accuracy of 0.2mm given by the constructor. The measurement points 7-10 of HA, which correspond to the top face of the cylinder have even better median values, but larger variation values too. We used the High Speed print mode and probably at High Quality print mode this phenomenon could be eliminated, so further investigation has to be done with another experiment.

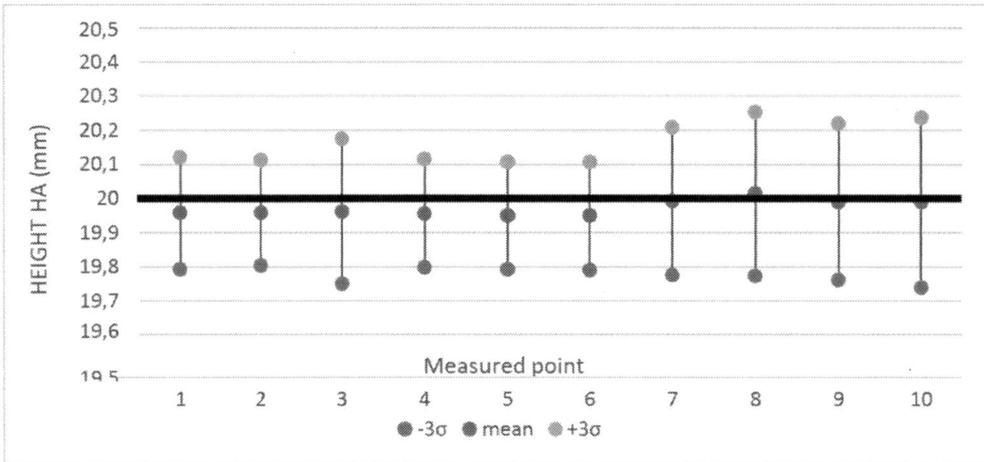

Chart 1. Variations of Height HA.

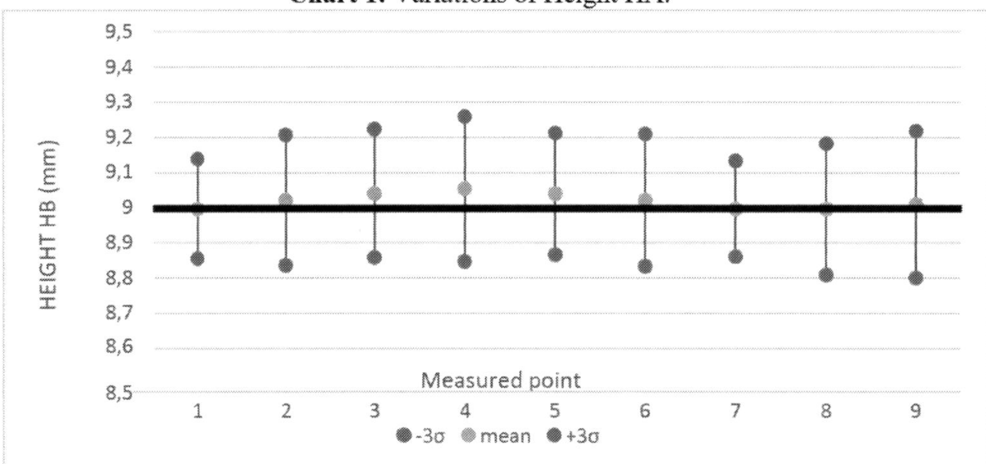

Chart 2. Variations of Height HB.

By examining the Y-Print Axis with dimensions LA and LB, values of which are represented in Chart 3 and Chart 4, we note that length is oversized and the variation of errors is very small, excluding measurement points 4, 5 and 6 of LB, which are at the lower layer.

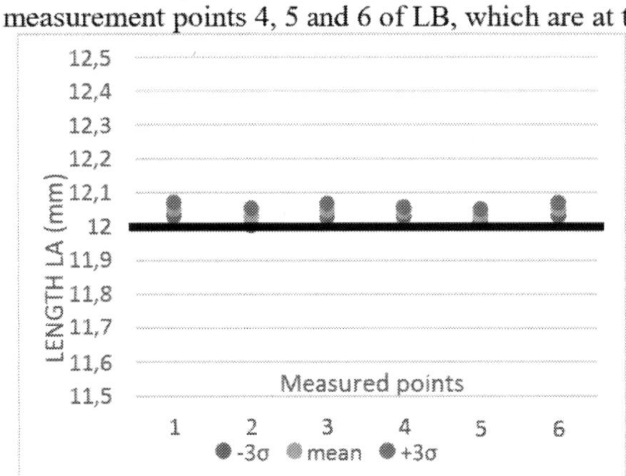

Chart 3. Variations of Length LA.

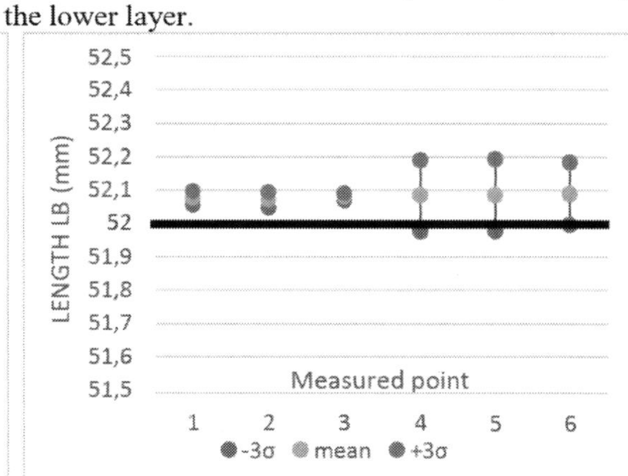

Chart 4. Variations of Length LB.

This observation occurs also in Chart 5, for measurement points W6 – W9 of width dimension, which represents the X-Print Axis. The oversizing of the base layer dimensions is thought to be caused by the support material layer which is printed prior to the main printing process of the part. The Support material is less compact and causes partial diffusion into the lower levels of the printed parts.

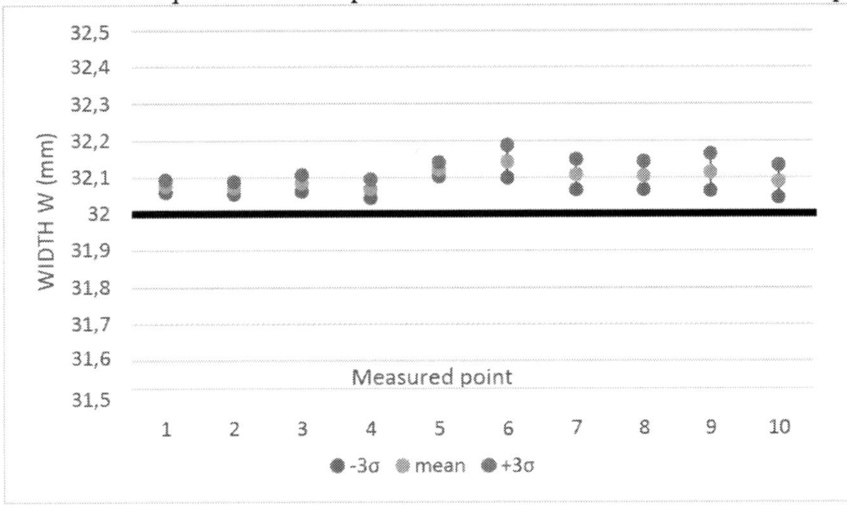

Chart 5. Variations of Width W.

The medians and the variations in diameter of a cylinder at four different measurement points, two on the top and two on the bottom are represented in Chart 6. From the results we ascertain that the bottom of the cylinder is undersized, while the top face is oversized and with bigger variation. This finding confirms that geometry is a very basic aspect considering the variation of accuracy and more investigation about the geometry error is required.

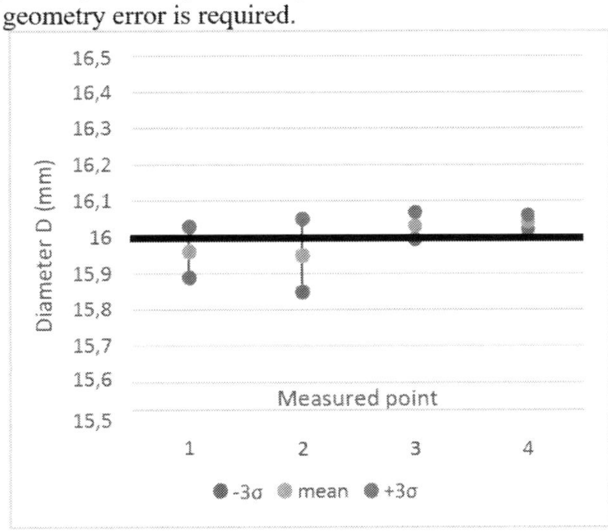

Chart 6. Variations of Diameter D.

The International Tolerance Grade (ITG) specifies tolerances with associated manufacturing processes for a given dimension. It indicates how precise an industrial process is and according to ISO 286-1:2010 for sizes up to 3150mm there are twenty values, IT01, IT0, IT1..IT18. The lower the IT grade number is, the higher is the precision of a machining process. For measuring tools IT01-IT7 is required. IT7- IT11 are grades for Fits and IT12-IT18 for Large Manufacturing Tolerances [13][14].

The process Tolerance T can be calculated by the following function:

$$T = 10^{\frac{ITG-1}{5}}(0.45\sqrt[3]{D} + 0.001D)$$

where ITG is the IT Grade category value (for IT5-IT18),
and D is the Geometrical mean dimension in mm:

$$D=\sqrt{D_{min}D_{max}}$$

where Dmin and Dmax are the limits of the dimension range.

There are twenty-one ranges for sizes up to 3150mm and the values of standard tolerance grades according to IT grades and Nominal size are shown in Table 4 [15].
To specify the IT grade for the size of two parallel opposite surfaces we examined the shaft through all LA values for one dimension (Y-Axis) and W values 1,2,3,10 for the second dimension (X-Axis). The size of cylinder type was examined with all the values of D. Table 1 - Table 3 show the minimum and maximum values for the examined measurement points and the difference of these in μm.

Table 1. Tolerance of the Y-Axis.

LA	Min mm	Max mm	(Max-Min) X 1000 μm/mm
1	12,04	12,06	**20 μm**
2	12,02	12,04	**20 μm**
3	12,04	12,055	15 μm
4	12,04	12,05	10 μm
5	12,02	12,04	**20 μm**
6	12,04	12,06	**20 μm**

Table 2. Tolerance of the Y-Axis.

W	Min mm	Max mm	(Max-Min) X 1000 μm/mm
1	32,07	32,085	15 μm
2	32,06	32,08	20 μm
3	32,07	32,09	20 μm
10	32,07	32,11	**40 μm**

Table 3. Tolerance of the cylinder.

D	Min mm	Max mm	(Max-Min) X 1000 μm/mm
1	15,93	16,00	70 μm
2	15,91	16,00	**90 μm**
3	16,01	16,05	40 μm
4	16,03	16,05	20 μm

For every size, we located its value at the first column of Table 4 and with the maximum value of the above tables, we assigned the corresponding IT grade. Analytically:

- For LA size, Y-Print Axis, with the nominal value of 12 mm and maximum difference of 20 μm we assign the IT8 grade.
- For W size, X-Print Axis, with the nominal value of 32 mm and maximum difference of 40 μm we assign the IT9 grade
- For D size, cylinder type, with the nominal value of 16 mm and maximum difference of 90 μm we assign the IT11 grade.

All three sizes are categorized to IT11 grade at maximum which means that they belong to the Fits area of IT grades.

Table 4. Values of standard tolerance for nominal sizes up to 120 mm.

Nominal size mm		Standard tolerance grades																			
		IT01	IT0	IT1	IT2	IT3	IT4	IT5	IT6	IT7	IT8	IT9	IT10	IT11	IT12	IT13	IT14	IT15	IT16	IT17	IT18
Above	Up to and inclu-ding	Standard tolerance values µm											mm								
—	3	0,3	0,5	0,8	1,2	2	3	4	6	10	14	25	40	60	0,1	0,14	0,25	0,4	0,6	1	1,4
3	6	0,4	0,6	1	1,5	2,5	4	5	8	12	18	30	48	75	0,12	0,18	0,3	0,48	0,75	1,2	1,8
6	10	0,4	0,6	1	1,5	2,5	4	6	9	15	22	36	58	90	0,15	0,22	0,36	0,58	0,9	1,5	2,2
10	18	0,5	0,8	1,2	2	3	5	8	11	18	27	43	70	110	0,18	0,27	0,43	0,7	1,1	1,8	2,7
18	30	0,6	1	1,5	2,5	4	6	9	13	21	33	52	84	130	0,21	0,33	0,52	0,84	1,3	2,1	3,3
30	50	0,6	1	1,5	2,5	4	7	11	16	25	39	62	100	160	0,25	0,39	0,62	1	1,6	2,5	3,9
50	80	0,8	1,2	2	3	5	8	13	19	30	46	74	120	190	0,3	0,46	0,74	1,2	1,9	3	4,6
80	120	1	1,5	2,5	4	6	10	15	22	35	54	87	140	220	0,35	0,54	0,87	1,4	2,2	3,5	5,4

5. Conclusions

From the experimental values and the analysis that followed we made some conclusions, which are summarized below:

- Dimensions in the X and Y axis are always oversized.
- The accuracy of the base of the produced parts is poorer in both X and Y axis.
- The accuracy in Z-axis is smaller than the other two axis and it depends on both the geometrical shape, rectangle or cylinder, and the height of layer.
- The cylinder diameter is undersized at the top, oversized at the base and with less accuracy at the top level.
- The X, Y and cylinder size belong to the Fits area of IT grades.

This is a preliminary study on the IT grade specification for the MJM 3D-print process. Further study examining the tolerance of holes and channels in 3D printed objects by this technology is required.

References

[1] Cotteleer M and Joyce J 2014 3D opportunity: Additive manufacturing paths to performance, innovation, and growth *Deloitte University Press*

[2] Kechagias J 2007 Investigation of LOM process quality using design of experiments approach *Rapid Prototyping Journal* 13 316-323

[3] Chandler D 2011 MIT research continues to push the boundaries of the burgeoning technology of 3-D printing *MIT News*

[4] Jacobs P and Reid D 1992 *Rapid prototyping & manufacturing* (Dearborn, MI: Society of Manufacturing Engineers in cooperation with the Computer and Automated Systems Association of SME)

[5] Yan X and Gu P 1996 A review of rapid prototyping technologies and systems *Computer-Aided Design* 28 307-318

[6] Cotteleer M, Holdowsky J and Mahto M 2014 The 3D opportunity primer: The basics of additive manufacturing *Deloitte University Press*

[7] Kruth J, Leu M and Nakagawa T 1998 Progress in Additive Manufacturing and Rapid Prototyping *CIRP Annals - Manufacturing Technology* 47 525-540

[8] Kechagias J, Iakovakis V, Katsanos M and Maropoulos S 2008 EDM electrode manufacture using rapid tooling: a review *J Mater Sci* 43 2522-2535
[9] Dimitrov D, van Wijck W, Schreve K and de Beer N 2006 Investigating the achievable accuracy of three dimensional printing *Rapid Prototyping Journal* 12 42-52
[10] Dimitrov D, Schreve K and de Beer N 2006 Advances in three dimensional printing – state of the art and future perspectives *Rapid Prototyping Journal* 12 136-147
[11] Kechagias J, Stavropoulos P, Koutsomichalis A and Vaxevanidis N 2014 Dimensional Accuracy Optimization of Prototypes produced by PolyJet Direct 3D Printing Technology *Advances in Engineering Mechanics and Materials* (International Conference on Industrial Engineering - INDE '14) pp 61-65
[12] Anon 2010 Documentation | Stratasys *Stratasys.com*
[13] Conway H 1966 *Engineering tolerances* (London: Pitman)
[14] Farmer L 1999 *Dimensioning and tolerancing for function and economic manufacture* (Gladesville, N.S.W.: Blueprint Publications)
[15] Kverneland K 1996 *Metric standards for worldwide manufacturing* (New York: ASME Press)

IOP Conf. Series: Materials Science and Engineering **161** (2016) 012025 doi:10.1088/1757-899X/161/1/012025

A Study of the dimensional accuracy obtained by low cost 3D printing for possible application in medicine

K Kitsakis[1,2,4a], P Alabey[1], J Kechagias[1], and N Vaxevanidis[3]

[1]Department of Mechanical Engineering, Technological Educational Institute of Thessaly, Larissa, GR 41110, Greece

[2]Department of Mechanical Engineering, University of Western Macedonia, Kozani, Greece

[3]Department of Mechanical Engineering Educators, School of Pedagogical and Technological Education, GR 14121 N. Heraklion Attikis, Greece

E-mail: [a]kitsakis@teithessaly.gr,

Abstract. Low cost 3D printing' is a terminology that referred to the fused filament fabrication (FFF) technique, which constructs physical prototypes, by depositing material layer by layer using a thermal nozzle head. Nowadays, 3D printing is widely used in medical applications such as tissue engineering as well as supporting tool in diagnosis and treatment in Neurosurgery, Orthopedic and Dental-Cranio-Maxillo-Facial surgery. 3D CAD medical models are usually obtained by MRI or CT scans and then are sent to a 3D printer for physical model creation. The present paper is focused on a brief overview of benefits and limitations of 3D printing applications in the field of medicine as well as on a dimensional accuracy study of low-cost 3D printing technique.

1. INTRODUCTION

Three-dimensional Printing (3-DP) and/or Rapid Prototyping (RP) describes the creation of 3-D physical objects manufactured layer by layer. Since the 80's many RP methods have been developed and commercialized. Despite their differences, the majority of them uses the same methodology to construct a physical prototype [1-5]: Creation of a 3-D digital model; STL surface model creation; cross sections creation; generation of vector or raster data for machine tool control; creation of the physical model layer by layer; and finally post processing of the final physical model [6].

Although RP machines may use different kind of materials to construct a model (liquid, solid, powder) the procedure presented above is generally the same. The use of such a technology is becoming increasingly popular in the medical sector, especially in surgical planning and simulation. The need for surgeons to have a closer look at the patient's problem with noninvasive methods is imperative. Many of the processes initially developed for industrial reasons are now to be modified in order to serve medical science.

For the construction of a bio-model the main difference to an industrial model lays in the way the 3D data is obtained. In this case, a computed tomography scan (CT scan) of the patient is used in order to create the virtual model. Depending on the conditions under which the CT scan is performed, the data may be altered. For example, movement of the patient during the examination may cause variations in the recorded dimensions. This may lead to dimensional discrepancies both in the virtual and the physical model that will be created. Therefore, it is essential to be really thorough during the CT scan to make sure that the data will be as accurate as possible.

The second step is the image segmentation that converts tomography into the 3D model. For intensity images (i.e. those represented by point-wise intensity levels) the most popular approaches are: the threshold technique; the edge-based method; the region-based method; and the connectivity-preserving relaxation-based segmentation method [5].

In the third step the physical model is created by using a 3D printing machine. Not all RP methods are suitable for the construction of a bio-model. The most suitable method is selected depending on the size, the materials and mainly the use of the bio model. The materials used are usually liquids, powders or solids such as resins, thermoplastics, polycarbonate, hydroxyapatite, metals etc. [7-9]. After the construction, post-processing takes place. All the support structures are removed and the bio-model is prepared in order to be touched by bare hands.

In the last few years various medical applications of RP have been developed [10]. This area is highly challenging because advances in this area require always a multidisciplinary approach and include a minimum of tasks related to medical imaging and 3D modeling, medical treatment and actual RP technology.

Prosthetic devices and implants have been developed a while ago using 3D printing. Recently, implants with optimized geometries have been fabricated directly in high-strength "final" materials using advanced processes such as laser powder forming.

Plastic surgery and reconstruction prosthetics are also very promising areas for 3D printing technologies. Breast and similar prosthetic devices are being investigated, as well as customized mandible substitutes and anatomically accurate prostheses for cranio-maxillo-facial reconstruction [11].

3D Printing can be used by surgeons to assist them in planning and analyzing complex operations, especially craniofacial and maxillofacial surgeries.

The transparency of the model and recent developments in color resins allow distinct visualization of tumors or other anomalies within the surrounding tissue or bone. High-cost applications in comparison to CNC modeling techniques have been a restraint and therefore rapid prototyping has been used in the past only in cases where the benefits obtained justify the extra expense. However, the advent of lower- cost RP systems and full color capability in recent years, in addition to the growing need for awareness in the medical community is likely to result in a wider use of such models [12]

Nowadays, RP applications are involved in many manufacturing procedures and there is a need for educating individuals belonging to various scientific backgrounds. Moreover, many types of prototypes, like bio-models, can be used to educate medical students in the terms of the technologies and the methods used to construct models, moulds and finally implants. Typical applications are illustrated in Figs 1-4.

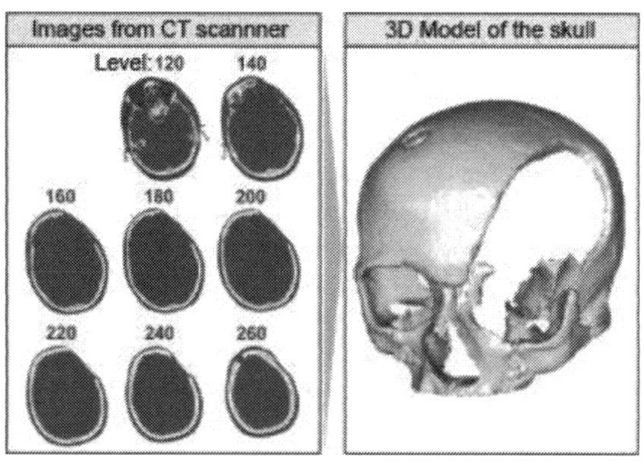

Figure 1. Images from CT SCANNER and 3D digital model [13].

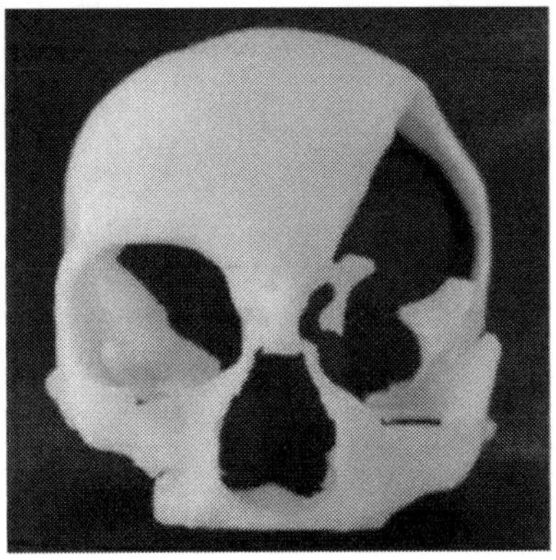

Figure 2. 3D printed skull [13].

2. LOW COST 3-D PRINTING

Low-cost 3-D printing is frequently referred to the application of semi-professional printers. A typical equipment of this category is the Wanhao Duplicator 4X; a 3D printer operating under the Fused Filament Fabrication (FFF) technique, which is one the most common 3D-Printing technique today.
The Duplicator 4X is a 3D printer with dual extrusion up for 1.75 mm filament (Fig. 3). It has a protective cover around the printer to ensure a stable temperature during printing. The layer alignment results in a 100 μm resolution. The drive system was designed to give the optimum positioning accuracy. The entire system has a sturdy 3/8″ diameter linear shafting. The bearings, which are made of sintered stainless steel, are self-aligned. The Z-axis construction consists of a stepper driven, cantilevered stage. The speed of the Z axis ranges from 150 mm/minute to 1.000 mm/minute.

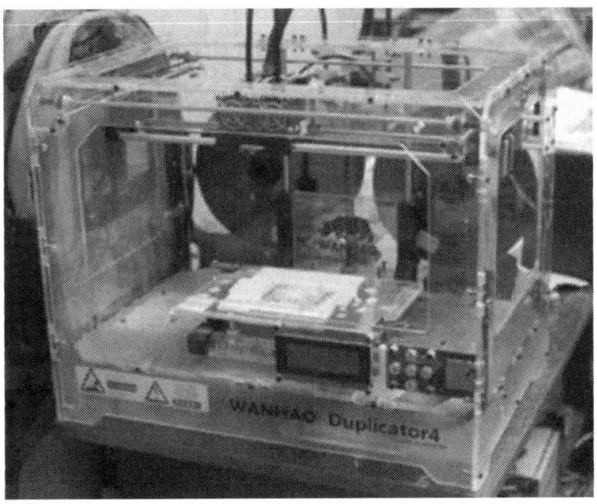

Figure 3. The Duplicator 4X 3D Printer.

Dimensional accuracy of the parts produced by using rapid prototyping technologies and the relevant factors affecting it, are major issues of investigation in the literature. The main parameters influencing dimensional accuracy were found to be layer thickness, infill, number of shells, printing and movement speed.

An investigation of the dimensional accuracy of parts produced by Wanhao Duplicator 4X 3D Printer is presented by performing nine (9) printing operations. In order to achieve randomized experiments the L9 Taguchi orthogonal array was used. The selected part geometry has been prepared in STL format. The part designed and its specifications are presented in Fig. 4.

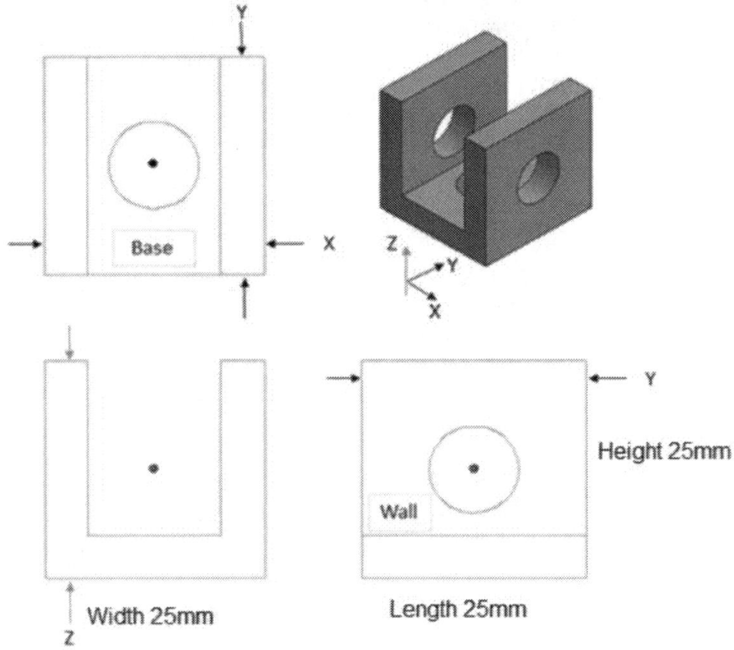

Figure 4. STL part specifications.

The process parameters used were the Printing Material (PM), the Infill Rate (IR), the Number of Shells (NS) and the Layer Height (LH). The Printing Material is defined by the control factors Acrylonitrile Butadiene Styrene (ABS) and Polylactic Acid (PLA). For the Infill Rate, three levels equal to 20%, 50% and 70% of the solid model were selected. The Number of Shells stands for the number of outlines printed on each layer of the object and is defined by three levels equal to 1, 2 and 3. Finally, for the Layer Height three values for the thickness of each layer were considered, 0.1, 0.2 and 0.3 mm. The process parameters and their respective levels are summarized in Table 1.

Table 1. Parameter Design.

No	Process Parameter	1	2	3
1	Printing Material (PM)	ABS	PLA	-
2	Infill Rate (IR)	20%	50%	70%
3	Number of Shells (NS)	1	2	3
4	Layer Height (LH)	0.1 mm	0.2 mm	0.3 mm

Average linear dimensions in X, Y and Z direction for each experiment have been measured with a digital caliper with an accuracy of 0.01 mm; see Figure 5 and Table 2).

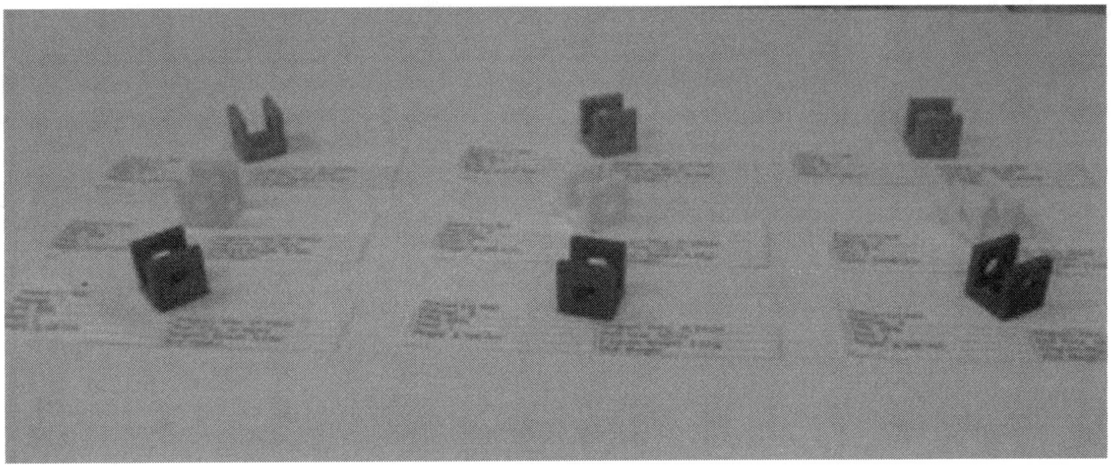

Figure 5. The 9 test parts.

Table 2. Measurements in X, Y, and Z directions.

No of Exp.	PM	IR	NS	LH		X	Y	Z
1	1	20	1	0.1		24,84	24,69	25,085
2	1	50	2	0.2		24,94	24,685	25,145
3	1	70	3	0.3		25,11	24,79	24,9875
4	2	50	3	0.1		24,97	24,74	25,19
5	2	70	1	0.2		24,87	24,6825	25,2
6	2	20	2	0.3		25,00	24,895	25,08
7	1	70	2	0.1		24,96	24,555	25,12
8	1	20	3	0.2		25,05	24,745	25,075
9	1	50	1	0.3		24,77	24,57	24,5
					average	24,94	24,71	25,04
					Max	25,11	24,90	25,20
					Min	24,77	24,56	24,50
					Range	0,34	0,34	0,70
					σ^2	0,089168	0,088	0,364675

3. CONCLUSIONS AND FUTURE CHALLENGES

Since the need for bio-models and implants is imperative and the role of RP technologies is essential in their construction there are many promising studies that might help to overcome all the arising issues. In the present study a dimensional accuracy investigation of low-cost 3D printing was carried out and the results were tabulated on Table 2. The physical models manufactured by this low- cost 3D printing can be used for medical application after optimization of the process in order to achieve dimensional accuracy superior to the accuracy of the CT scan data.

REFERENCES

[1] Gibson I, Rosen D and Stucker B 2010 *Additive manufacturing technologies* (New York: Springer)

[2] Kechagias J, Anagnostopoulos V, Zervos S and Chryssolouris G 1997 Estimation of build times in Rapid Prototyping processes. *In Proceedings of the 6th European Conference on Rapid Prototyping and Manufacturing*, P.M. Dickens, ed., University of Nottingham, pp 137–148.

[3] Hieu L, Zlatov N, Vander Sloten J, Bohez E, Khanh L, Binh P, Oris P and Toshev Y 2005 Medical rapid prototyping applications and methods *Assembly Automation* 25 284-292

[4] Giannatsis J and Dedoussis V 2007 Additive fabrication technologies applied to medicine and health care: a review *The International Journal of Advanced Manufacturing Technology* 40 116-127

[5] Alabey P, Pappas M, Kechagias J and Maropoulos S, 2010 Medical Rapid Prototyping and Manufacturing: Status and Outlook, *In Proceedings of the ASME 2010 10th Biennial Conference on Engineering Systems Design and Analysis, ESDA2010*, Vol. 1, pp 739-745

[6] Chua C K, Vu T T, Leong K F, Lim C S, 2009 An interactive multimedia approach to enhance learning of rapid prototyping, *In Proceedings of the 4th Innovative Developments in Design and Manufacturing: Advanced Research in Virtual and Rapid Prototyping (VR@P4)*, P.J. da Silva Bartolo et al., eds., CRC Press, pp.497-503.

[7] Wilson C, de Bruijn J, van Blitterswijk C, Verbout A and Dhert W 2003 Design and fabrication of standardized hydroxyapatite scaffolds with a defined macro-architecture by rapid prototyping for bone-tissue-engineering research *Journal of Biomedical Materials Research* 68A 123-132

[8] Ang T, Sultana F, Hutmacher D, Wong Y, Fuh J, Mo X, Loh H, Burdet E and Teoh S 2002 Fabrication of 3D chitosan–hydroxyapatite scaffolds using a robotic dispensing system *Materials Science and Engineering: C* 20 35-42

[9] Seitz H, Rieder W, Irsen S, Leukers B and Tille C 2005 Three-dimensional printing of porous ceramic scaffolds for bone tissue engineering *Journal of Biomedical Materials Research Part B: Applied Biomaterials* 74B 782-788

[10] Winder J and Bibb R 2005 Medical Rapid Prototyping Technologies: State of the Art and Current Limitations for Application in Oral and Maxillofacial Surgery *Journal of Oral and Maxillofacial Surgery* 63 1006-1015

[11] Limpanuphap S and Derby B 2002 Manufacture of biomaterials by a novel printing process *Journal of Materials Science: Materials in Medicine* 13 1163-1166

[12] Ulbrich C, Hermini H and Zavaglia C 2009 The use of technologies as Rapid Prototyping and scanner inspection in surgical planning to medical application *Innovative Developments in Design and Manufacturing* 161-165

[13] Maravelakis E, David K, Antoniadis A, Manios A, Bilalis N and Papaharilaou Y 2008 Reverse engineering techniques for cranioplasty: a case study *Journal of Medical Engineering & Technology* 32 115-121

Artificial immune algorithm implementation for optimized multi-axis sculptured surface CNC machining

N A Fountas[1], J D Kechagias[2] and N M Vaxevanidis[1]

[1]Laboratory of Manufacturing Processes & Machine Tools (LMProMaT)-Department of Mechanical Engineering Educators, School of Pedagogical and Technological Education (ASPETE), GR 141 21, N. Heraklion, Athens, Greece

[2]Department of Mechanical Engineering, Technological Educational Institute of Thessaly, Larissa, Greece

N.M. Vaxevanidis: vaxev@aspete.gr

Abstract. This paper presents the results obtained by the implementation of an artificial immune algorithm to optimize standard multi-axis tool-paths applied to machine free-form surfaces. The investigation for its applicability was based on a full factorial experimental design addressing the two additional axes for tool inclination as independent variables whilst a multi-objective response was formulated by taking into consideration surface deviation and tool path time; objectives assessed directly from computer-aided manufacturing environment A standard sculptured part was developed by scratch considering its benchmark specifications and a cutting-edge surface machining tool-path was applied to study the effects of the pattern formulated when dynamically inclining a toroidal end-mill and guiding it towards the feed direction under fixed lead and tilt inclination angles. The results obtained form the series of the experiments were used for the fitness function creation the algorithm was about to sequentially evaluate. It was found that the artificial immune algorithm employed has the ability of attaining optimal values for inclination angles facilitating thus the complexity of such manufacturing process and ensuring full potentials in multi-axis machining modelling operations for producing enhanced CNC manufacturing programs. Results suggested that the proposed algorithm implementation may reduce the mean experimental objective value to 51.5%

1. Introduction

Standard tool-paths implemented to machining modeling are generated by NC programmers considering technical attributes such as cutting tool type/material; raw material properties and machine tool specifications; as well as the applicable range for machining parameters like axial/radial cutting engagement; tool inclination angles -in the case of 5-axis machining-; maximum tool-path discretization step; etc. Such parameters are still user dependent, hence; left in experience or trial-and-error simulations to decide upon the most convenient outcome for actual deployment. Should the requirements for high accuracy; quality and productivity rates are considered, intelligent optimization methods for sculptured surface machining tool-paths is inevitable. Sculptured surface machining refers to several research directions aiming at providing tool-paths for high- precision machining whilst, simultaneously, maintaining high productivity. In the past few years research was focused on creating new tool-path generation approaches for 3- and 5-axis surface machining [1-6], capable of operating under a constant interpolation error [7-9] or maintaining equal scallops [10]. Even though such tool-paths come up with highly sophisticated features for seamless integration and production facilitation, no significant aid for machining parameter specifications is provided whereas their efficiency is away from being optimum given the manual selection for their settings. Till date, modern soft computing and artificial intelligence tools are employed to formulate efficient optimization modules to address complex sculptured surface milling problems. Li et al. [11], suggested a back-propagation neural network to optimize a multi-objective optimization problem involving the cutting conditions. To deal with the problem of generating poor surface roughness when it comes to tool-path planning for flank surface milling; Chu et al. [12], developed a spline-constrained tool-path optimization module. In their work, the coefficients of spline's equations are determined by applying a particle swarm optimization scheme. Hsieh and Chu [13] also addressed 5-axis flank milling problems and a particle swarm optimization algorithm was implemented to produce the cutter locations so that the machining error is reduced. Other research efforts have also presented results concerning sculptured surface machining optimization problems using particle swarm optimization (PSO) technique [14]. Other artificial intelligence approaches that have already been implemented to address complex machining optimization problems include artificial immune systems [15] and hybrid modules incorporating more than one intelligent function [16] mandatory to present the method/methods used in research, in order to solve the problem/problems.

2. Tool-path exploitation for optimization via *Clonal-G* artificial immune algorithm

Artificial immune systems (AIS) mimic the mechanism of interaction among antigens and antibodies technically modeled via the parameter of affinity measure. These artificial entities are formulated mainly by employing immune functions such as clonal selection modules; immune nets or negative selection algorithms. In the case of 5-axis CNC machining tool-paths, the two inclination angles are operated regarding their predetermined constant values whilst the part's curvature manipulates the tool's trajectory which is of a standard pattern under a given cutting strategy. Tool inclination angles for 5-axis CNC programming ought to be constrained just as the rest of machining parameters; i. e. feed-rate; cutting depth; step-over, cutting speed; etc. Therefore a search domain for an intelligent algorithm would be subjected to satisfy such constraints and this is achieved if exploitation is conducted within lower and upper limits. Search domain constraints may constitute the antigen population in an artificial immune system whereas the population of candidate solutions is expressed as a vector of a magnitude depended on the number of variables. In this study a x-vector represents two variables; x(1) and x(2) illustrating lead inclination and tilt inclination angle respectively. The algorithm implemented in *Mathworks®* *MATLAB R2016a* [17] and operated under the following settings: Number of generations 50; Population size 20; Mutation probability 0.01; Hypermutation control 0.8 and Number of clones per candidate 10.

3. Experimental design for exploiting multi-axis CNC tool-paths

The sculptured surface tool-path optimization problem was determined by examining the responses of surface deviation and tool-path time against the independent variables; lead angle and tilt angle holding 4 level each; under a full factorial design of sculptured surface machining simulation experiments. Thereby, 16 experiments were performed in *SIEMENS® NX10* manufacturing environment, where a 50x75 mm benchmark sculptured surface was utilized as it has been mentioned to conduct tool-path planning tests and summon the outputs for exploiting the simplified response's variations. Table 1 shows the parameters and the levels of the experiment resulting to 16 simulation runs.

Table 1. Full factorial design of experiments for tool-path simulations.

Tool-path parameters	Levels			
Lead angle [*deg*]	10	20	30	40
Tilt angle [*deg*]	10	20	30	40

The cutting strategy employed was a zig-zag pattern under "dual-axis on drive" swept style with intermediate X-Y tool passes to eliminate the scallops among the subsequent radial cutting intervals and fit the local curvatures of the surface whilst simultaneously varying primary and secondary tool inclination axes, thus; lead and tilt angles. Note that this special tool-path is for simultaneous 5-axis finish machining. Therefore a roughing tool-path was first applied to remove the main material volume with a 6mm flat-end mill and leave a uniform iso-scallop remaining volume to be finished. When it comes to multi-axis surface machining toroidal cutters are usually implemented owing to their efficiency and high-performance against ball-end mills or flat-end mills. Fig. 1(a) depicts the benchmark part and Fig. 1(b) the finishing surface tool-path applied.

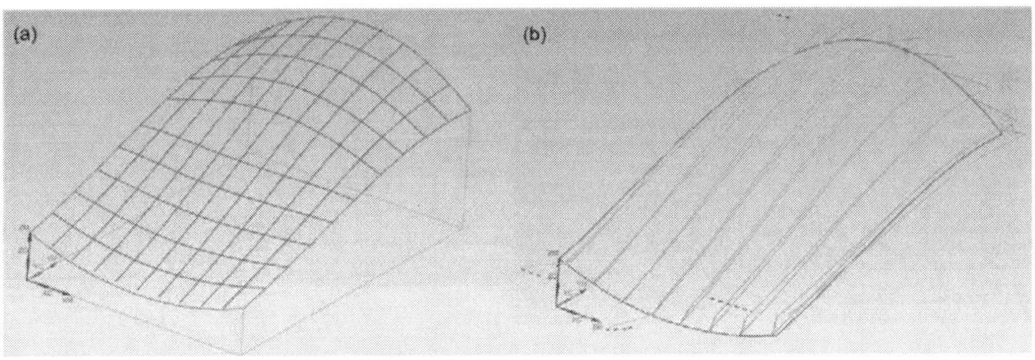

Figure 1. Test part used: (a) 3D part; (b) "dual axis on drive" finishing tool-path for 5-axis CNC machining.

A Ø10mm toroidal end-mill with 2mm corner radius was determined whilst the cutting tolerance was set to 0.1mm to facilitate the virtual surface topography in terms of the scallops left on the machined contour. The test benchmark surface was divided to 10 zones towards X direction and 10 zones towards Y direction thus; ending up with a set of 100; equally spaced discrete points which were measured normal to surface using software tools found in *SIEMENS® NX10* environment. For each of the 16 experimental runs; the surface deviation between the ideal and the simulated surface was the mean of the 100 measurements on all 16 3D machined model outputs.

4. Results and discussion

According the results obtained from the experimental process a regression equation based on the full quadratic model representation was finally generated to play the role of a fitness function to be evaluated by the artificial immune algorithm. The regression equation is given in Eq. 1.

$$ObjFun = 0.27 - 0.035x(1) + 0.1045x(2) + 88\times10^{-3}x(1)^2 - 1387\times10^{-3}x(2)^2 - 3\times10^{-5}x(1)x(2)\_(1)$$

Results were analyzed using *Stat-Ease MINITAB® 17* statistical package. The two inputs (inclination angles) are set as a common vector x = [x(1), x(2)]. The search range for both inclination angles was that of the experimental design; [10, 40]. 10 evaluations were conducted to validate the optimum inclination angle set whilst corresponded convergence diagrams were examined to check the final fitness value. The algorithm exhibited quite good repeatability attaining almost identical outcome to all series of evaluations referring to both fitness value and result for inclination angles. Fig. 2(a) illustrates the best run's response surface the regression model produced which is actually a depiction of the two inclination angles' variation on the response of the common objective for surface deviation and tool-path time. The initial population to be evolved towards the optimum solution is also observed. Fig. 2(b) illustrates the final population's convergence pattern whilst Fig. 3(c) gives the convergence plot for the mean fitness values attained by the artificial immune algorithm as well as the evolution diagram representing the convergence of the candidate solutions to the final point which is deemed the optimum one for the two parameters.

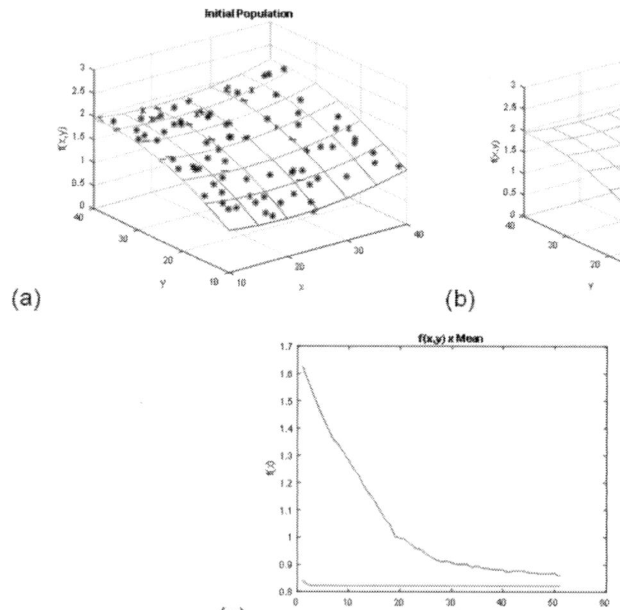

Figure 2. Best result attained: (a) Initial population; (b) optimum solution region; (c) Convergence plot.

Table 2 presents the experimental results for surface deviation; tool-path time and their common; simplified expression taking into account the values for the independent variables as well as the normalization of outputs which is an essential feature when the inherent bias owing to different magnitudes of outputs is considered. The fitness values of algorithmic validation runs as well as the optimum final points for the independent variables obtained; are given in Table 3.

Table 2. Parameter values and experimental results.

Run no.	Lead angle *[deg]*	Tilt angle *[deg]*	Surface deviation *[mm]*	Tool-path time *[min]*	Common Objective *(normalized)*
1	10	10	1.09589	1.66666	1.216038835
2	10	20	1.27824	2.41666	1.489268966
3	10	30	1.29802	3.13333	1.632982428
4	10	40	1.41588	4.16666	1.934120402
5	20	10	0.53805	1.53333	0.724298206
6	20	20	1.17077	2.08333	1.343401943
7	20	30	1.50669	2.65	1.724054743
8	20	40	1.87719	3.53333	2.184555878
9	30	10	0.51195	1.63333	0.727234138
10	30	20	1.26855	2.05	1.424594364
11	30	30	1.90303	2.63333	2.077249593
12	30	40	1.74334	2.91666	1.972289256
13	40	10	1.04981	1.85	1.201811564
14	40	20	1.77516	2.16666	1.902796516
15	40	30	2.26025	2.51666	2.39626539
16	40	40	1.55603	2.93333	1.811538531

Table 3. Genetic algorithm validation evaluations and optimum final points.

Validation run	ObjFun	x(1)	x(2)
1	0.821972072	20.31	10
2	0.82749375	17.5	10
3	0.823223438	21.25	10
4	0.822052574	20.44	10
5	0.821957714	20.28	10
6	0.843675	25	10
7	0.843675	25	10
8	0.821972072	20.31	10
9	0.843675	25	10
10	0.822756268	21.01	10

5. Conclusions and future work

The methodology presented in this work constitutes an efficient optimization framework in the special case of sculptured surface CNC tool-path optimization. Its efficiency lies heavily on the stochastic exploration of the problem's regions via the immune algorithm. Even if the problem mentioned is a well-established research field, no significant efforts have yet been made to provide optimization infrastructure using machining modeling environment and intelligent systems. Such an approach presents originality on both the application domain of artificial intelligence tools and the philosophy underpinning the problem statement. From production engineering perspectives the methodology dramatically reduces evaluation costs since no physical experimentation is required and main work is conducted using latest generation tool-path planning software. Finally the immune algorithm implementation led to the reduction of mean value for the common objective studied; which was 51.5%.

References

[1] Masood A, Siddiqui R, Pinto M, Rehman H and Khan MA 2015 Tool Path Generation for Complex Surface Machining, Using Point Cloud Data. *Proc CIRP* **26** 397 – 402

[2] Taejung K and Sarma SE 2002 Tool path generation along directions of maximum kinematic performance; a first cut at machine-optimal paths. *Comput. Aided Des.* **34** 453–468

[3] Chen T and Ye TA 2002 tool path generation strategy for sculptured surface machining. *J. Mater. Process. Technol.* **127** 369–373

[4] Park S 2004 Sculptured surface machining using previous triangular mesh slicing. *Comput. Aided Des.* **36** (3) 279-288

[5] Ding S, Mannan MA, Poo AN, Yang DCH and Han Z 2003 Adaptive iso-planar tool path generation for machining of free-form surfaces. *Comput. Aided Des.* **35** 141-153

[6] Han Z, Yang DCH and Chuang JJ 2001 Isophote-based ruled surface approximation of free-form surfaces and its application in NC machining. *Int. J. Prod. Res.* **39** 1911-1930

[7] Tsai MS, Nien HW and Yau H-T 2008 Development of an integrated look-ahead dynamics-based NURBS interpolator for high precision machinery. *Comput. Aided Des.* **40** 554-566

[8] Chu CH, Wu PH and Lei WT 2012 Tool path planning for 5-axis flank milling of ruled surfaces considering CNC linear interpolation. *J. Intell. Manuf.* **23** 471-480

[9] Lin Z, Fu J, Shen H and Gan W 2014 An accurate surface error optimization for five-axis machining of freeform surfaces. *Int. J. Adv. Manuf. Technol.* **71** (5-8) 1175-1185

[10] Lee E 2003 Contour offset approach to spiral tool-path generation with constant scallop height. *Comput. Aided Des.* **35** (6) 511–518

[11] Li L, Liu F, Chen B and Li BC 2015 Multi-objective optimization of cutting parameters in sculptured parts machining based on neural network. *J. Intell. Manuf.* **26** (5) 891-898

[12] Chu C-H, Hsieh H-T, Lee C-H and Yan C 2015 Spline-constrained tool-path planning in five-axis flank machining of rules surfaces. *Int. J. Adv. Manuf. Technol.* **80** (9) 2097-2104

[13] Hsieh HT and Chu CH 2012 Improving optimization of tool path planning in 5-axis flank milling using advanced PSO algorithms. Robot. Comput. Integr. Manuf. **29** (3) 3-11

[14] Bharathi-Raja S and Baskar N 2012 Application of particle swarm optimization technique for achieving desired milled surface roughness in minimum machining time. *Exp. Syst. Appl.* **39** (5) 5982-5989

[15] Ulker E, Turanalp ME and Halkaci HS 2009 An artificial immune system approach to CNC toolpath generation. *J. Intell. Manuf.* **20** 67-77

[16] Oysu C and Bingul Z 2009 Application of heuristic and hybrid-GASA algorithms to tool-path optimization problem for minimizing air time during machining. *Eng. Appl. Artif. Intell.* **22** (3) 389-396

[17] De Castro LN and Von Zuben FJ 2001 Immune and neural network models: Theoretical and empirical comparison. *Int. J. Comput. Intell. Appl.***1** 239-257

IOP Conf. Series: Materials Science and Engineering **161** (2016) 012027 doi:10.1088/1757-899X/161/1/012027

Ring-tool profiling – graphical method in CATIA based on Generating trajectories theorem

G Frumuşanu, V Teodor and N Oancea
"Dunărea de Jos" University, Manufacturing Engineering Department,
Domnească Street 111, 800201 – Galaţi, Romania

E-mail: virgil.teodor@ugal.ro

Abstract. Machining of threads having high dimensions and multiple starts by turning is a challenging problem. An alternative possibility is to machine them by milling. The most productive milling solution is when using tools with inner active surface, namely ring tools. In the case of threads with multiple starts, the reciprocal enwrapped profile of the ring tool is considerably different to the shape of the thread axial (normal) section. In this paper, we suggest a methodology to profile the generator ring tool, based on a complementary theorem from enwrapped surfaces field. At the same time, a graphical algorithm aiming to find the ring tool profile, developed in CATIA graphical environment has been applied in the concrete case of a trapezoidal thread. The graphical profiling solution is presented in comparison to an analytical solution, in order to test the results precision. The graphical profiling method proves to be rigorous, easy to apply and highly intuitive.

1. Introduction

The helical surfaces can be generated with tools delimited either by an outer revolution surface (disc-tool, end mill) [1, 2], or by an inner revolution surface (ring-tool) [3]. The ring-tool for worms machining do have the big advantage of being much more productive, relative to the other above mentioned types of tools.

In all cases, the tool profiling methods currently used are lying on the fundamental theorems of the enwrapped surfaces (Olivier, Gohman, Nikolaev) [1, 2], or on complementary theorems (Minimum distance method, Family of substitutive circles method, Tangents method) [4].

The present paper proposes a modality for finding the ring-tool profile on the base of a new expression for the enwrapping theorems – the Plane generating trajectories method [5]. The novelty in using this method is the enabling of an easy interpretation in graphical form, under CATIA designing environment. Hereby, a dedicated algorithm to profile the ring-tool has been developed together with a graphical method in CATIA, having the same purpose. The method might be later extended for also profiling other types of tools – e.g. the tangential ring-tool.

2. Generating helical surfaces with a ring-tool

The helical surfaces generation by using a ring-tool is a highly productive process, currently used for machining long screws (e.g. the driving screws from machine tools construction) in large series.

The tools have relative high diameters, their active surface being an inner revolution surface. The ring-tool axis is disjoint to the helical surface axis, being placed on a position normal to the tangent at the helix corresponding to exterior diameter of the helical surface.

The generating process kinematics involves the following motions:

- *I* – worked piece rotation around its own axis, $\vec{V}$;
- *II* – worked piece translation along $\vec{V}$ axis, in correlation with the rotation motion, the ensemble formed by the two motions giving a helical motion having the same axis and parameter as the helicoid to generate;
- *III* – cutting motion, executed by the ring-tool.

The formal expression of the above presented kinematics can be obtained after defining the following reference systems (see figure 1):

- *XYZ*, meaning a system associated to the helical surface (the worked piece) and having *Z* axis overlaid to surface axis, and
- $X_1Y_1Z_1$ – system attached to the ring-tool axis, having Z_1 and X_1 axis overlaid to ring-tool axis $\vec{A}$, respectively to *X* axis of the surface system.

The distance between *Z* and Z_1 axis, measured along X (X_1), is denoted by *a* and represents a constructive parameter. The angle α between *Z* and Z_1 axis can be found by imposing to $\vec{A}$ axis the condition of being normal to the helix of D_{ex} diameter, from the helical surface to be generated (see figure 2):

$$\alpha = \arctan\left({2p}/{D_{ex}} \right). \tag{1}$$

The problem needing to be solved is that in which, by starting from the helical surface shape, one intends to find the profile of its generator ring-tool.

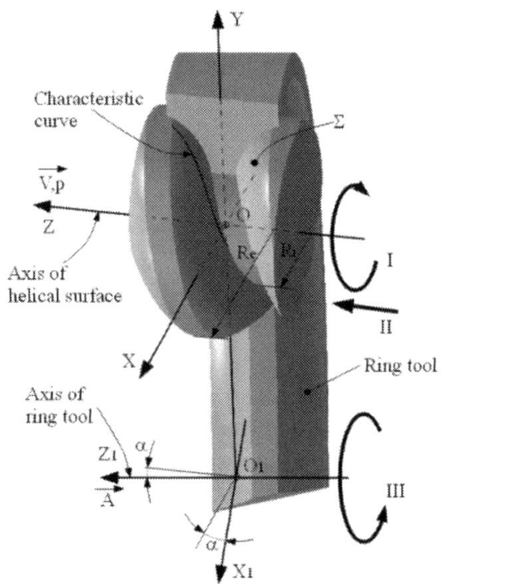

Figure 1. Ring-tool inner surface. Reference systems.

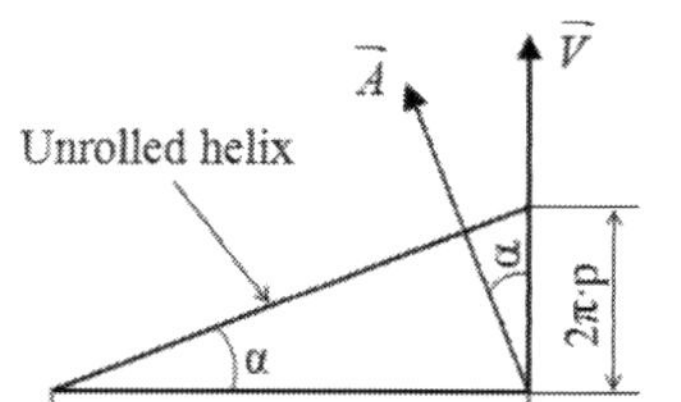

Figure 2. The relative position between $\vec{A}$ and $\vec{V}$ axis.

3. Ring-tool profiling algorithm

3.1. Plane generating trajectories method

The method, if applied in the addressed problem, starts from the equations of the helical surface to be generated, having the general form:

$$\Sigma \mid X = X(u,\varphi); Y = Y(u,\varphi); Z = Z(u,\varphi). \tag{2}$$

with u and φ – independent variables.

The equations of Σ surface are then transferred from XYZ into $X_1Y_1Z_1$ system through the transform:

$$X_1 = \alpha[X - \delta], \tag{3}$$

with

$$\alpha = \begin{pmatrix} 1 & 0 & 0 \\ 0 & \cos\alpha & \sin\alpha \\ 0 & -\sin\alpha & \cos\alpha \end{pmatrix}, \quad \delta = \begin{pmatrix} a \\ 0 \\ 0 \end{pmatrix}, \tag{4}$$

a being the above mentioned constructive parameter.

In principle, the equations of the helical surface (2), after applying the transform (3), become:

$$\Sigma \,\Big|\, X_1 = X_1(u,\varphi); Y_1 = Y_1(u,\varphi); Z_1 = Z_1(u,\varphi). \tag{5}$$

By sectioning Σ surface with $Z_1 = H$ plane (with H – arbitrary variable), C_{Σ_H} curve results, having, in principle, the equations:

$$C_{\Sigma_H} \,\Big|\, X_1 = X_1(u); Y_1 = Y_1(u); Z_1 = H, \tag{6}$$

see also figure 3.

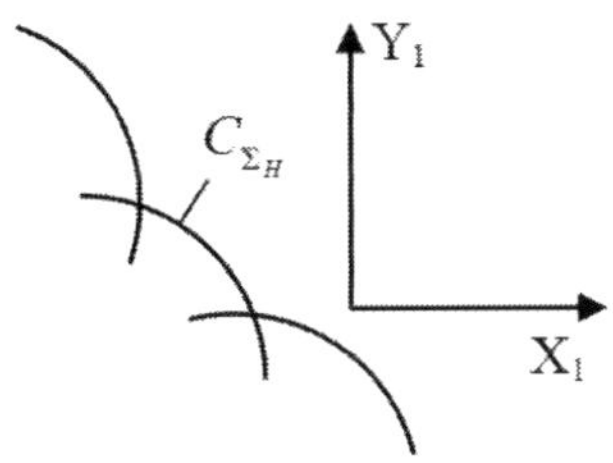

Figure 3. $\left(C_{\Sigma_H}\right)_v$ family of trajectories from $Z_1 = H$ plane.

3.2. The family of plane generating trajectories

It is defined as the family described by the curve belonging to the helical surface and placed in a plane section, normal to the future ring-tool axis, which has a rotation motion of v angular parameter around ring-tool axis [6]:

$$X_1 = \omega_3^T(v)\begin{pmatrix} X_1(u) \\ Y_1(u) \\ H \end{pmatrix}, \tag{7}$$

where ω_3^T is the matrix of the co-ordinates transform corresponding to rotation around Z_1 axis. After developing, we obtain:

$$\left(C_{\Sigma_H}\right)_v \left| \begin{aligned} & X_1 = X_1(u)\cdot\cos v - Y_1(u)\cdot\sin v; \\ & Y_1 = X_1(u)\cdot\sin v + Y_1(u)\cdot\cos v; \\ & Z_1 = H. \end{aligned} \right. \tag{8}$$

3.3. The enveloping condition

The enveloping condition can be deducted by starting from the specific theorem of the Family of generating trajectories method. According to this, the inner revolution surface reciprocal enwrapped to the family of generating trajectories is the locus of that points owning to C_{Σ_H} curves, from different plane sections H, for which the normal to the curve does intersect the ring-tool axis.

The normal in the current point of C_{Σ_H} curve (6) is then defined and its direction parameters are calculated:

$$\vec{N}_{C_{\Sigma H}} = \begin{vmatrix} \vec{i} & \vec{j} & \vec{k} \\ \dot{X}_{1_u} & \dot{Y}_{1_u} & 0 \\ 0 & 0 & 1 \end{vmatrix} = \dot{Y}_{1_u} \cdot \vec{i} - \dot{X}_{1_u} \cdot \vec{j}. \tag{9}$$

The equations of the normal can be now written as:

$$N_{C_{\Sigma H}} \begin{vmatrix} X_1 = X_1(u) + \lambda \cdot \dot{Y}_{1_u}; \\ Y_1 = Y_1(u) - \lambda \cdot \dot{X}_{1_u}; \\ Z_1 = H, \end{vmatrix} \tag{10}$$

with λ – scalar variable parameter.

The equation of normals family (9) result from (10) and (7):

$$\begin{pmatrix} X_1 \\ Y_1 \\ Z_1 \end{pmatrix} = \begin{pmatrix} \cos v & -\sin v & 0 \\ \sin v & \cos v & 0 \\ 0 & 0 & 1 \end{pmatrix} \cdot \begin{pmatrix} X_1(u) + \lambda \cdot \dot{Y}_{1_u} \\ Y_1(u) - \lambda \cdot \dot{X}_{1_u} \\ H \end{pmatrix}, \tag{11}$$

or, after developing:

$$\left(N_{C_{\Sigma H}}\right)_v \begin{vmatrix} X_1 = \left[X_1(u) + \lambda \cdot \dot{Y}_{1_u}\right]\cos v - \left[Y_1(u) - \lambda \cdot \dot{X}_{1_u}\right]\sin v; \\ Y_1 = \left[X_1(u) + \lambda \cdot \dot{Y}_{1_u}\right]\sin v + \left[Y_1(u) - \lambda \cdot \dot{X}_{1_u}\right]\cos v; \\ Z_1 = H, \end{vmatrix} \tag{12}$$

with v being the family parameter.

Into $X_1Y_1Z_1$ reference system, associated to ring-tool, $\vec{A}$ axis has the equations:

$$A \big| X_1 = 0;\ Y_1 = 0;\ Z_1 = H, \tag{13}$$

with H variable. Hereby, the condition meaning that the family normals intersect the ring-tool axis is:

$$\begin{cases} \left[X_1(u) + \lambda \cdot \dot{Y}_{1_u}\right]\cos v - \left[Y_1(u) - \lambda \cdot \dot{X}_{1_u}\right]\sin v = 0; \\ \left[X_1(u) + \lambda \cdot \dot{Y}_{1_u}\right]\sin v + \left[Y_1(u) - \lambda \cdot \dot{X}_{1_u}\right]\cos v = 0. \end{cases} \tag{14}$$

By eliminating the parameter λ from equations system (14), the specific enveloping condition is obtained as:

$$X_1(u) \cdot \dot{X}_{1_u} + Y_1(u) \cdot \dot{Y}_{1_u} = 0. \tag{15}$$

3.4. The profile of the ring-tool axial section

The equations ensemble (8) giving the plain trajectories family together with the condition (15), when H is arbitrarily varying, represent the enveloping surface, namely the inner revolution surface of the future ring-tool, which has, in principle, the equations:

$$S \big| X_1 = X_1(v,H); Y_1 = Y_1(v,H); Z_1 = H. \tag{16}$$

The ring-tool axial section results in the form (see also figure 4):

$$SA \begin{cases} R = \sqrt{[X_1(v,H)]^2 + [Y_1(v,H)]^2}, \\ Z_1 = H. \end{cases} \tag{17}$$

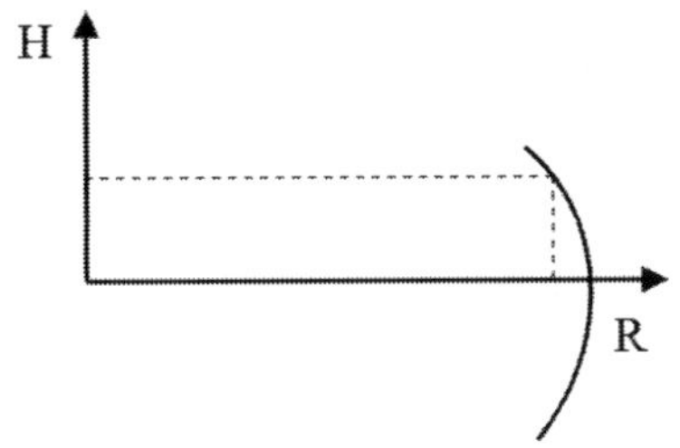

Figure 4. Ring-tool axial section.

4. Numerical application

A numerical application of the profiling algorithm from above is further presented. It addresses the profiling of the ring-tool for generating a trapezoidal thread (figure 5), such like the ones used at driving screws, into machine tools kinematical chains.

The screw axial profile is referred to a system *XYZ*, having its *Z* axis overlaid to the axis of the thread to be generated, while *X* is the symmetry axis of the axial profile gap, which is formed by three straight segments:

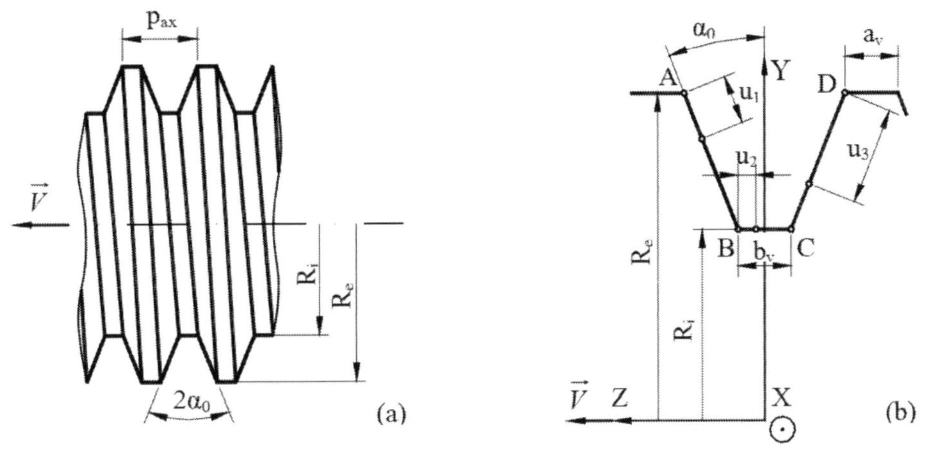

Figure 5. The trapezoidal thread: (a) – model, (b) – axial profile.

$$\overline{AB}\big|\, X=0;\ Y=R_e-u_1\cdot\cos\alpha_0;\ Z=b_v/2+\left(u_{1\max}-u_1\right)\sin\alpha_0, \tag{18}$$

with u_1 – variable, $u_{1min}=0$, $u_{1max}=(R_e-R_i)/\cos\alpha_0$,

$$\overline{BC}\big|\, X=0;\ Y=R_i;\ Z=b_v/2-u_2, \tag{19}$$

with u_2 – variable, $u_{2min}=-b_v/2$, $u_{1max}=b_v/2$,

$$\overline{CD}\big|\, X=0;\ Y=R_e-u_3\cdot\cos\alpha_0;\ Z=-b_v/2+\left(u_3-u_{3\max}\right)\sin\alpha_0, \tag{20}$$

with u_3 – variable, $u_{3min}=0$, $u_{3max}=(R_e-R_i)/\cos\alpha_0$.

The composite helical surface – right hand worm, of $p=p_{ax}/2\pi$ helical parameter, is generated through the helical motion of $\vec{V}$ axis and φ angular parameter:

$$\Sigma_{AB}\left|\begin{aligned}&X=-\left(R_e-u_1\cdot\cos\alpha_0\right)\sin\varphi;\\&Y=\left(R_e-u_1\cdot\cos\alpha_0\right)\cos\varphi;\\&Z=b_v/2+\left(u_{1\max}-u_1\right)\sin\alpha_0+p\cdot\varphi;\end{aligned}\right. \tag{21}$$

$$\Sigma_{BC}\left|\begin{array}{l} X=-R_i\cdot\sin\varphi; \\ Y=R_i\cdot\cos\varphi; \\ Z=b_v/2-u_2+p\cdot\varphi; \end{array}\right. \tag{22}$$

$$\Sigma_{CD}\left|\begin{array}{l} X=-(R_e-u_3\cdot\cos\alpha_0)\sin\varphi; \\ Y=(R_e-u_3\cdot\cos\alpha_0)\cos\varphi; \\ Z=-b_v/2+(u_3-u_{3\max})\sin\alpha_0+p\cdot\varphi. \end{array}\right. \tag{23}$$

After applying the transform (3), the helical surface equations are referred to $X_1Y_1Z_1$ (see figure 1):

$$\Sigma_{AB}\left|\begin{array}{l} X_1=-(R_e-u_1\cdot\cos\alpha_0)\sin\varphi\cdot\cos\alpha-[b_v/2+(u_{1\max}-u_1)\sin\alpha_0+p\cdot\varphi]\sin\alpha; \\ Y_1=(R_e-u_1\cdot\cos\alpha_0)\cos\varphi+a; \\ Z_1=-(R_e-u_1\cdot\cos\alpha_0)\sin\varphi\cdot\sin\alpha+[b_v/2+(u_{1\max}-u_1)\sin\alpha_0+p\cdot\varphi]\cos\alpha; \end{array}\right. \tag{24}$$

$$\Sigma_{BC}\left|\begin{array}{l} X_1=-R_i\cdot\sin\varphi\cdot\cos\alpha-(b_v/2-u_2+p\cdot\varphi)\sin\alpha; \\ Y_1=R_i\cdot\cos\varphi+a; \\ Z_1=-R_i\cdot\sin\varphi\cdot\sin\alpha+(b_v/2-u_2+p\cdot\varphi)\cos\alpha; \end{array}\right. \tag{25}$$

$$\Sigma_{CD}\left|\begin{array}{l} X_1=-(R_e-u_3\cdot\cos\alpha_0)\sin\varphi\cdot\cos\alpha-[-b_v/2+(u_3-u_{3\max})\sin\alpha_0+p\cdot\varphi]\sin\alpha; \\ Y_1=(R_e-u_3\cdot\cos\alpha_0)\cos\varphi+a; \\ Z_1=-(R_e-u_3\cdot\cos\alpha_0)\sin\varphi\cdot\sin\alpha+[-b_v/2+(u_3-u_{3\max})\sin\alpha_0+p\cdot\varphi]\cos\alpha. \end{array}\right. \tag{26}$$

Regarding the condition (15) specific form for the addressed application, the partial derivatives expressions for each of the three profile segments result from relations (24) – (26):

$$\overline{AB}\left|\begin{array}{l} \dot{X}_{1_{u1}}=\cos\alpha_0\cdot\sin\varphi\cdot\cos\alpha+\sin\alpha_0\cdot\sin\alpha; \\ \dot{Y}_{1_{u1}}=-\cos\alpha_0\cdot\cos\varphi; \end{array}\right. \tag{27}$$

$$\overline{BC}\left|\begin{array}{l} \dot{X}_{1_{u2}}=\sin\alpha; \\ \dot{Y}_{1_{u2}}=0; \end{array}\right. \tag{28}$$

$$\overline{CD}\left|\begin{array}{l} \dot{X}_{1_{u3}}=\cos\alpha_0\cdot\sin\varphi\cdot\cos\alpha-\sin\alpha_0\cdot\sin\alpha; \\ \dot{Y}_{1_{u3}}=-\cos\alpha_0\cdot\cos\varphi. \end{array}\right. \tag{29}$$

The equations (24) – (26) together with condition (15) in its specific form, when H is varying, give the characteristic curves at the contact between the composite helical surface and the ring-tool one.

We further present a numerical example in the concrete case when the profile to be generated is defined by the following values of its geometric parameters: R_e = 45 mm, R_i = 40 mm, p_{ax} = 16 mm, a_v = 4 mm, b_v = 4 mm, α = 20°, a = 225 mm, thread number of starts k = 1.

The characteristic curve has been found through 11 points, for the tool flank that generates CD flank of the trapezoidal thread. These points' co-ordinates have been determined with the help of a MatLab application, dedicated to this purpose and developed on the base of the suggested profiling algorithm, and they are presented in table 1. Table 2 shows the co-ordinates of the points that define the axial section of the ring-tool, corresponding to the same flank, in the form (17). In figure 6, one can see the results of the addressed numerical application, in graphical form.

Note In the points B and C, onto the generated profile there are discontinuities that should be separately approached for tool profiling.

Table 1. The characteristic curve

Point no.	X_1 [mm]	Y_1 [mm]	Z_1 [mm]
1	1.3417	269.9745	-3.80
2	1.2200	269.4775	-3.62
3	1.0998	268.9806	-3.44
4	0.9795	268.4839	-3.26
5	0.8592	267.9873	-3.08
6	0.7405	267.4909	-2.90
7	0.6219	266.9946	-2.72
8	0.5035	266.4985	-2.54
9	0.3861	266.0025	-2.36
10	0.2691	265.5067	-2.18
11	0.1525	265.0111	-2.00

Table 2. The axial section

R [mm]	H [mm]
269.9778	3.80
269.4802	3.62
268.9829	3.44
268.4857	3.26
267.9887	3.08
267.4919	2.90
266.9953	2.72
266.4989	2.54
266.0028	2.36
265.5068	2.18
265.0112	2.00

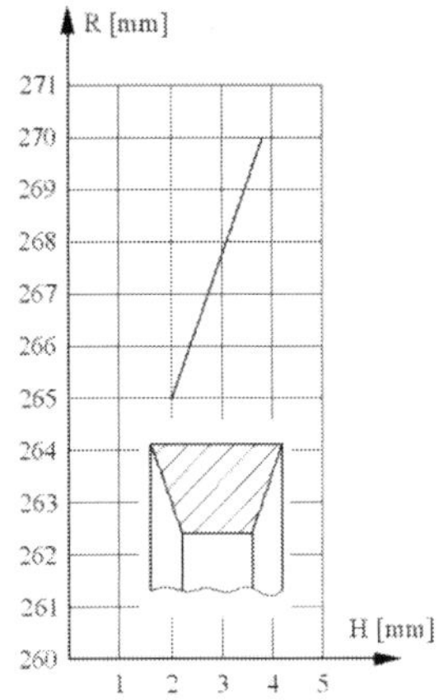

Figure 6. Ring-tool profile.

5. Graphical method in CATIA

A graphical algorithm has also been developed in order to solve the same profiling problem, in the graphical design environment CATIA.

The algorithm starts by creating the two reference systems mentioned in section 2. The helical surface's generatrix is then drawn into *XOY* plane. The helical surface is virtually generated by applying *SWEEP* command, after defining as director curve the helix of $\vec{V}$ axis and p helical parameter and setting as translation direction the axis $\vec{V}$. Into the same reference system, a line tangent to the helix is drawn, following to be used for defining the ring-tool axis.

Into the tool reference system – $X_1Y_1Z_1$, a plane is generated by the method "*through point and line*", the definition point and line being the origin of tool reference system and the above mentioned tangent-line, respectively. The axis $\vec{A}$ is generated as a line normal to this plane, and which passes by the origin of $X_1Y_1Z_1$ system. For this line, we set the option "*length type: infinite*". The other two axis of the reference system have been defined as lines perpendicular to a curve, namely $\vec{A}$, also passing by the origin of $X_1Y_1Z_1$ system, and having as support their corresponding planes (Z_1X_1 plane for X_1 axis and Z_1Y_1 plane for Y_1 axis).

A plane parallel with the plane X_1Y_1 is generated at offset at H value. It is obtained the intersection curve between the helical surface and this plane ("*INTERSECTION*" command). The obtained curve is an in-plane curve and belongs to the helical surface.

On this curve it is generated a point, using the option "*on curve*" and with "*distance to reference*" on type "*ratio of curve length*". Initially, the value of this ratio is arbitrary (figure 7).

A point is generated as intersection between the tool's axis and the plane $Z_1 = H$. A line is drawn by this point and the one from the intersection curve ("*LINE*" command, type "*point-point*", figure 8).

Consequently, from the point belonging to the intersection curve it is generated a line tangent to the curve and the angle between this line and the previously drawn line is measured.

Changing the value of "*ratio of curve length*" parameter, the angle's value is also changed. Tracking the angle's value is possible to determine the ratio for which the angle is 90°. In this position, the tangent is perpendicular to the line which link a point from the curve with a point from the tool's axis. This perpendicularly means that the line is normal to the curve and, from construction, passes through a point from the tool's axis. So, the enveloping condition is accomplished and the point from the intersection curve belongs to the characteristic curve. Resuming this process for various values of H a set of points from the characteristic curve are determined.

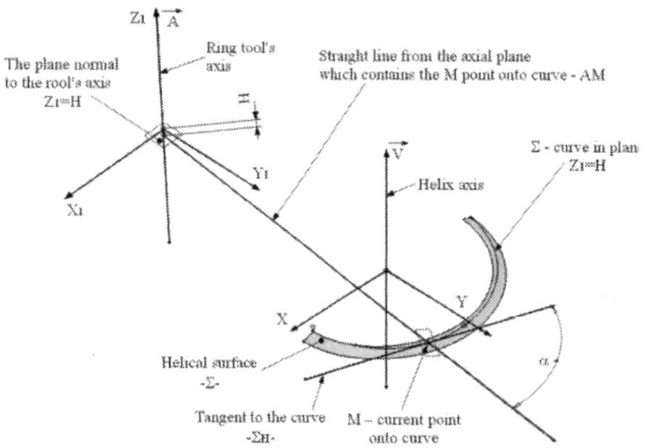

Figure 7. Current point onto $\sum_H$: tangent line to $\sum_H$ and straight line between current point and tool's axis — AM, $(\alpha \neq \pi/2)$.

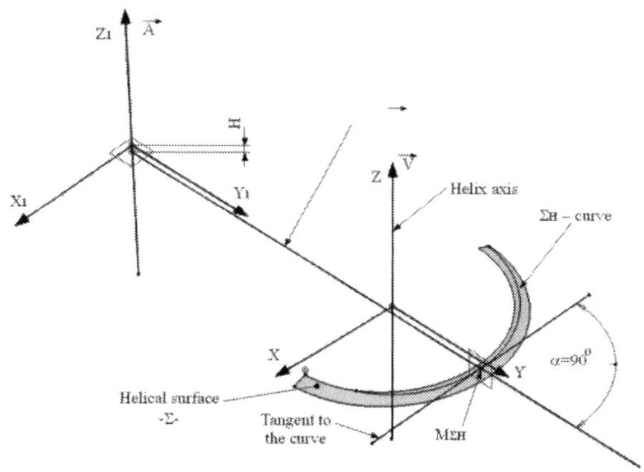

Figure 8. Point onto characteristic curve: normal to $\sum_H$ intersecting the tool's axis — AM_N, $(\alpha = \pi/2)$.

The results such obtained are very close to the ones analytically determined. For example, instead of the point having the co-ordinates (0.6219; 266.9946), from Z_1 = -2.72 plane, analytically found, it was determined, by applying the graphical method, the point (0.6230; 266.9960), from the same plane. The distance between the two points is about $1.78 \cdot 10^{-3}$ mm, meaning a very good concordance between the results delivered by the two methods.

6. Conclusion

In this paper, an alternative method for profiling the ring-tools for machining by milling the helical surfaces is presented. The method lies on a complementary theorem of the enwrapped surfaces – the Plane generating trajectories theorem, which has been implemented in CATIA graphical design environment. Method application is sampled in a concrete case – profiling of a ring tool used for machining a screw trapezoidal thread.

In order to validate the method application, the same profiling problem has been solved analytically too. Between the results separately obtained in the two cases there is a very good concordance. Beyond its remarkable precision, the newly developed method proves to be easy to apply and has a pregnant intuitive character.

References

[1] Litvin F L 1984 *Theory of gearing* (Washington: NASA Scientific and Technical Information Division)
[2] Radzevich S P 2008 *Kinematics geometry of surface machining* (London: CRC Press)
[3] Berbinschi S, Teodor V and Oancea N 2012 A study on helical surface generated by the primary peripheral surfaces of ring tool *Int. J. Adv. Manu. Tech.* **61** p 15-24
[4] Oancea N 2004 *Surface generation by enwrapping, vol. II. Complementary theorems* (Galaţi: „Dunărea de Jos" Publishing House)
[5] Teodor V 2010 *Contributions to the elaboration of a method for profiling tools – Tools which generate by enwrapping* (Beau Bassin: Lambert Academic Publishing)
[6] Baroiu N, Teodor V and Oancea N 2015 A new form of the in-plane trajectories theorem. Generation with rotary cutters *Bulletin of the Polytechnic Institute of Iaşi* **3** p 29-36

Acknowledgement

This work was supported by a grant of the Romanian National Authority for Scientific Research and Innovation, CNCS – UEFISCDI, project number PN-II-RU-TE-2014-4-0031.

The Rack-Gear Tool Generation Modelling. Non-Analytical Method Developed in CATIA, Using the Relative Generating Trajectories Method

V G Teodor, N Baroiu, F Susac and N Oancea

"Dunărea de Jos" University of Galaţi, Faculty of Engineering, Department of Manufacturing Engineering, Galaţi, Romania

Email: virgil.teodor@ugal.ro

Abstract. The modelling of a curl of surfaces associated with a pair of rolling centrodes, when it is known the profile of the rack-gear's teeth profile, by direct measuring, as a coordinate matrix, has as goal the determining of the generating quality for an imposed kinematics of the relative motion of tool regarding the blank. In this way, it is possible to determine the generating geometrical error, as a base of the total error. The generation modelling allows highlighting the potential errors of the generating tool, in order to correct its profile, previously to use the tool in machining process. A method developed in CATIA is proposed, based on a new method, namely the method of "relative generating trajectories". They are presented the analytical foundation, as so as some application for knows models of rack-gear type tools used on Maag teething machines.

1. Introduction

The method of relative generating trajectories, as complementary method for study of enwrapping surfaces, by rolling method, was presented [1]. A variant of this method is presented in this paper. The variant includes a CATIA developed application, regarding the modelling of generation with rack-gear tool for an ordered curl of surfaces (profiles) associated with a pair of circular centrodes. These centrodes rolls on the rectilinear centrode, associated with the rack-gear tool.

The issue presented in this paper was approached in others forms by Baicu [2], using the 3D modelling, as so as by Berbinschi and all [3], based on the "minimum distance" complementary theorem [5].

Also, the generation modelling is used in order to identify the optimal cutting scheme at generation with hob mill of the involutes teeth by Antoniadis [4].

At the same time, Teodor [6] developed an analytical form of the trajectories method, applied also for generation modelling.

A variant of the "relative generating trajectories" method is presented in this paper, as complementary method for study of the enwrapping surfaces associated with a pair of rolling centrodes.

Obviously, the complementary theorems are alternative ways for express the fundamental theorem of Willis [7].

The graphical method proposed in this paper, applied for modelling of surfaces generated with the rack-gear, is based on the "relative generating trajectories" method. This method determines the trajectories of points belong to the tool's cutting edge in their relative motion regarding the blank.

The graphical expression is based on the specific algorithm, developed for a tool with profile known in discrete form, as an ordered cloud of points. This cloud of points is obtained by measuring of the generating tool.

The graphical method is characterised by simplicity of application because use the CATIA capabilities. The method is very rigorous and intuitive at the same time.

The obtained results for presented applications are compared with results obtained by an analytical method, in order to prove the graphical method's quality.

2. The generation modelling using rack-gear tool

The rolling centrodes, reference systems and profile of the rack-gear tool are presented in figure 1.

They are defined the reference systems:

xy is the global reference system;

XY — the reference system joined with the blank and the C_1 centrode;

$\xi\eta$ — relative reference system joined with the C_2 centrode. The generating rack-gear's profile is joined with the C_2 centrode.

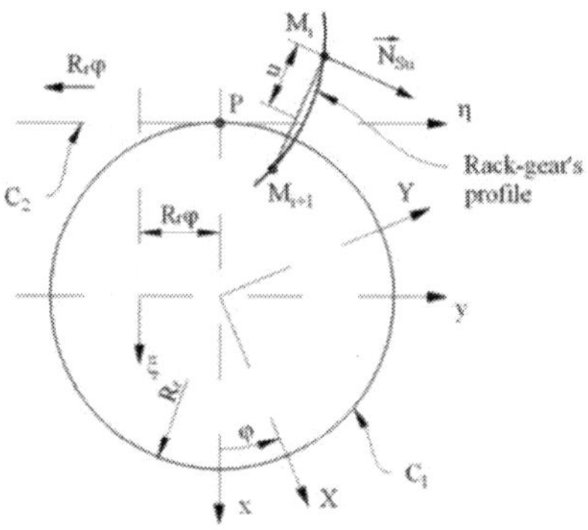

Figure 1. Centrodes and reference systems associated; the profile of rack-gear.

The tool's profile is defined in the $\xi\eta$ reference system, in a discrete form,

$$S = \begin{pmatrix} \xi_1 & \eta_1 \\ \vdots & \vdots \\ \xi_i & \eta_i \\ \vdots & \vdots \\ \xi_n & \eta_n \end{pmatrix}, \ i = 1\ldots n. \tag{1}$$

For M_iM_{i+1} elementary segment which replace the rack-gear tool, with equations:

$$S_u \left| \begin{aligned} \xi &= \xi + u \cdot \cos\beta_i; \\ \eta &= \eta + u \cdot \sin\beta_i, \end{aligned} \right. \tag{2}$$

with,

$$\tan\beta_i = \frac{\Delta\eta}{\Delta\xi}; \ \Delta\xi = \xi_{i+1} - \xi_i; \ \Delta\eta = \eta_{i+1} - \eta_i. \tag{3}$$

For $i = 1 \ldots n$, with n big enough, the profile of the rack-gear tool can be described very rigorous.

The $\vec{N}_{Su}$ normal versor to the elementary profile M_iM_{i+1},

$$\vec{N}_{Su} = \sin\beta_i \cdot \vec{i} + \cos\beta_i \cdot \vec{j} \tag{4}$$

is defined.

In this way, the parametrical equations of the normal to the elementary profile, in the point $M_i(\xi\eta)$, are given by

$$\vec{N}_{Su} \begin{vmatrix} \xi = \xi_i + u \cdot \cos\beta_i + \lambda \cdot \sin\beta_i; \\ \eta = \eta_i + u \cdot \sin\beta_i + \lambda \cdot \cos\beta_i, \end{vmatrix} \tag{5}$$

withλ scalar parameter. The direction of $\vec{N}_{Su}$ versor is given by (5).

In the relative motion between the rack-gear tool and the blank,

$$X = \omega_3(\varphi)[\xi + a];\ a = \begin{pmatrix} -R_r \\ -R_r \cdot \varphi \end{pmatrix}, \tag{6}$$

the normal (6) describes the family:

$$\left(\vec{N}_{Su}\right)_\varphi \begin{vmatrix} X = [\xi_i + u \cdot \cos\beta_i + \lambda \cdot \sin\beta_i - R_r] \cdot \cos\varphi + [\eta_i + u \cdot \sin\beta_i + \lambda \cdot \cos\beta_i - R_r \cdot \varphi] \cdot \sin\varphi; \\ Y = -[\xi_i + u \cdot \cos\beta_i + \lambda \cdot \sin\beta_i - R_r] \cdot \sin\varphi + [\eta_i + u \cdot \sin\beta_i + \lambda \cdot \cos\beta_i - R_r \cdot \varphi] \cdot \cos\varphi. \end{vmatrix} \tag{7}$$

In equations (7), u and φ are variable parameters.

For $\lambda = 0$, the equations (7) represents the trajectory family of a point from the rack-gear's profile, in the relative motion regarding the blank.

If the goal is to determine the S profile's enwrapping, it is necessary that the normals family $\left(\vec{N}_{Su}\right)_\varphi$ to pass through the gearing pole, P, for various rolling positions [7]. The different rolling positions are obtained for various values of the φ parameter.

If the gearing pole, P, is defined in the XY reference system,

$$P \begin{vmatrix} X_P = -R_r \cdot \cos\varphi; \\ Y_P = R_r \cdot \sin\varphi, \end{vmatrix} \tag{8}$$

then, from (7) and (8) results the equations assembly:

$$\begin{aligned} &[\xi_i + u \cdot \cos\beta_i - R_r] \cdot \cos\varphi + [\eta_i + u \cdot \sin\beta_i - R_r \cdot \varphi] \cdot \sin\varphi + \lambda \cdot \sin(\varphi + \beta_i) = -R_r \cdot \cos\varphi; \\ &-[\xi_i + u \cdot \cos\beta_i - R_r] \cdot \sin\varphi + [\eta_i + u \cdot \sin\beta_i - R_r \cdot \varphi] \cdot \cos\varphi + \lambda \cdot \cos(\varphi + \beta_i) = R_r \cdot \sin\varphi. \end{aligned} \tag{9}$$

The λ variable scalar is eliminated between the two conditions (9) and the enwrapping condition specifically for the relative generating trajectories is determined. This condition establish the blank's flank as enveloping for the trajectories family for points belong to the rack gear tool's flank, regarding the XY reference system of the blank,

$$\varphi = \frac{(\xi_i + u \cdot \cos\beta_i) \cdot \cos\beta_i - (\eta_i + u \cdot \sin\beta_i) \cdot \sin\beta_i}{-R_r \cdot \sin\beta_i}. \tag{10}$$

The equations assembly:

$$\begin{aligned} X &= (\xi_i + u \cdot \cos\beta_i - R_r) \cdot \cos\varphi + (\eta_i - u \cdot \sin\beta_i - R_r \cdot \varphi) \cdot \sin\varphi; \\ Y &= -(\xi_i + u \cdot \cos\beta_i - R_r) \cdot \sin\varphi + (\eta_i - u \cdot \sin\beta_i - R_r \cdot \varphi) \cdot \cos\varphi, \end{aligned} \tag{11}$$

and the condition (10), determine the enwrapped profile — the analytical model of the blank's generated surface.

3. Application

The involute flank generated with a rack-gear tool is presented in figure 2. In the same figure are presented the pair of rolling centrodes and the reference systems.

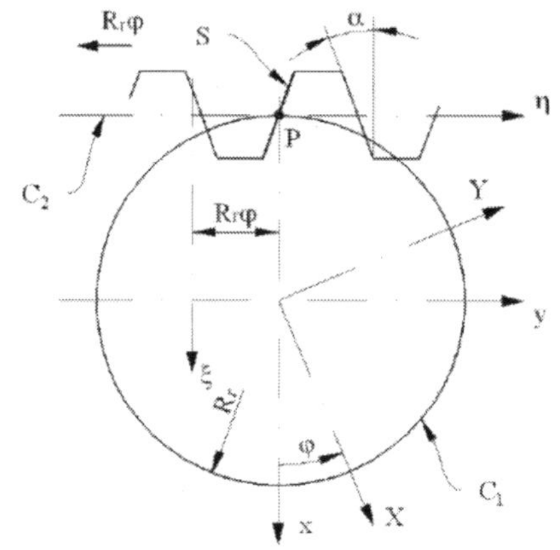

Figure 2. Rack-gear; reference systems.

The *S* profile of the generating rack-gear was obtained in discrete form by measuring with a profile projector, see figure 3.

Figure 3. The real rack-gear flank's measuring.

In table 1 are presented the point's coordinates measured on the rack-gear's profile.

Table 1. Coordinates measured on rack-gear profile.

Nr. crt.	X [mm]	Y [mm]
1	0.000	0.000
2	0.150	0.348
3	0.139	0.312
4	0.302	0.723
5	0.690	1.722
6	1.040	2.724
7	1.357	3.721
8	1.708	4.725
9	2.032	5.721
10	2.380	6.724
11	2.742	7.720
12	3.086	8.725
13	3.430	9.721
14	3.783	10.726
15	4.148	11.726
16	4.511	12.722
17	4.660	13.250

The equations assembly (11) and (10) determines the generating profile for the input data: $R_r = 50$ mm, rolling circle radius; $m = 5$ mm, modulus and $\alpha = 20°$.

In figure 4 is presented the generated profile of an involute tooth, the analytical modelled profile.

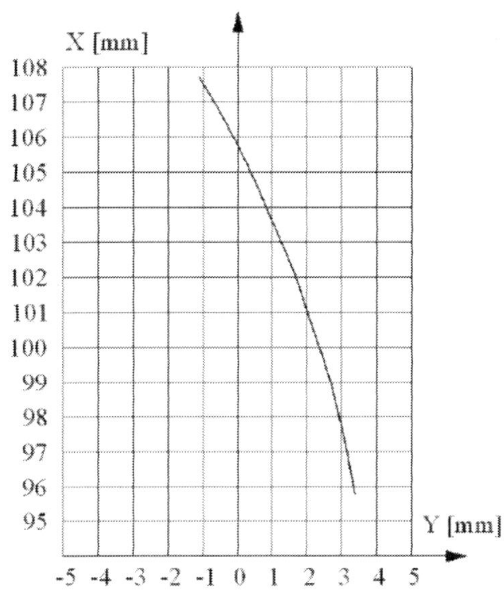

Figure 4. The modelled profile

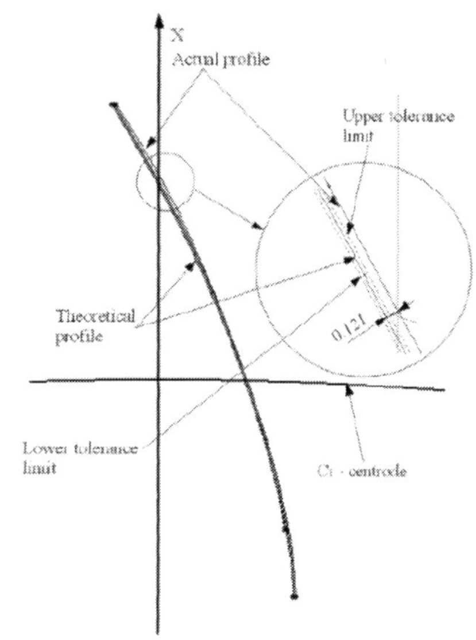

Figure 5. The generated profile

A graphical solution was developed in CATIA design environment.

According to this method, in DMU Kinematics module, the generating process was simulating by rolling for the tooth's flank, using the rack-gear tool.

The rack-gear measured flank was modelled using a spline curve which has as control points those points with coordinates measured using the profile projector.

In the same file was drawn a normal line to the tool's profile (to the spline curve) which was constrained to pass through one of the control points.

By simulating the mechanism function, keeping as fixed element the rack-gear, is found the position when the previously drawn normal line pass through the gearing pole. The intersection point between this normal and the modelled profile will be a point from the generated involute profile.

This last step is repeated for each of the measured points, obtaining points onto the generated profile.

Due to the fact that the deviation of the rack-gear tool's profile exceeds the allowed limits we can expect that the deviations of the tooth's profile exceed the allowed limits too.

In figure 5 are presented the limits obtained for the given input data.

4. Conclusions

The relative generating trajectories method is a complementary method for study of the enwrapping surfaces.

The presented applications refers to conjugated profiles associated with a pair of rolling centrodes — the particular case of the generation with rack-gear regarding the generation starting from the discrete form of the tool's profile.

The graphical solution developed in CATIA use the capabilities of this design environment. The analytical results and those obtained with the graphical algorithm are identical from the technical point of view.

The graphical method is simple, intuitive and easy to apply.

The generating trajectories may highlight the problem linked with the profile's interference at machining.

References

[1] Teodor V, Păunoiu V, Berbinschi S, Baroiu N and Oancea N 2015 The method of "in-plane generating trajectories" for tools which generate by enveloping – application in CATIA *Journal of Machine Engineering* **15** p 69-80

[2] Baicu I et al 2002 Modelling of surfaces generation *The Annals of "Dunărea de Jos" University of Galați* **5** p 48-53

[3] Berbinschi S et al 2001 A graphical method developed in CATIA design environment for the modelling of generation by enveloping *The Annals of "Dunărea de Jos" University of Galați* **5** p 25-30

[4] Dimitrou V, Antoniadis A 2008 *CAD-based simulation of the hobbing process for the manufacturing of spur and helical gears* (London: Springer-Verlag) p 347-357

[5] Oancea N 2004 *Surfaces generation by enwrapping* (Galați: "Dunărea de Jos" University Publishing House)

[6] Teodor V 2010 *Contributions to the elaboration of a method for profiling tools. Tools which generate by enwrapping* (Beau Bassin: Lambert Academic Publishing)

[7] Litvin F L 1984 *Theory of gearing* (Washington: NASA Scientific and Technical Information Division)

Acknowledgments

This work was supported by a grant of the Romanian National Authority for Scientific Research and Innovation, CNCS-UEFISCDI, project number PN-II-RU-TE-2014-4-0031.

A Case Study of Reverse Engineering Integrated in an Automated Design Process

R Pescaru[1] P Kyratsis[2] and G Oancea[1]

[1]Department of Manufacturing Engineering, Transilvania University of Brasov, Brasov, Romania
[2]Department of Mechanical Engineering and Industrial Design, Western Macedonia University of Applied Sciences, Kila Kozanis, Greece

E-mail: gh.oancea@unitbv.ro

Abstract. This paper presents a design methodology which automates the generation of curves extracted from the point clouds that have been obtained by digitizing the physical objects. The methodology is described on a product belonging to the industry of consumables, respectively a footwear type product that has a complex shape with many curves. The final result is the automated generation of wrapping curves, surfaces and solids according to the characteristics of the customer's foot, and to the preferences for the chosen model, which leads to the development of customized products.

1. Introduction

It is well known that the current trends on the product markets refer to shortening the launch cycle of products on the market [1]. This cycle takes into account the conceiving, designing, prototyping and manufacturing of the product. Nevertheless, if the aim is to develop new products starting only from the physical model, new technologies allow major flexibility and bring high profits. Despite the fact that in the beginning, the Reverse Engineering (RE) technique has not been considered a time saving technology that could shorten the time before the output of new products [1] on the market, it has gradually been proved to be an extremely useful connection between designing and manufacturing [2]. The RE technique can be used in the product quality evaluation process or in the redesigning process of products for which no documentation is available [2].

The development of the product digitization technology has led to the emergence of a various range of 3D scanners. Scanners have been divided in two categories: with contact or without (non-contact) [3]. The choice of the scanner type is made according to the object surfaces; a scanner that has contact is thus preferred for surfaces that contain edges, spikes or holes, while scanners without contact are chosen for surfaces that contain details. However, the scanners without contact have proved to be more useful due to their portability, high speed of data capturing and high precision.

The variety of shapes in the point clouds obtained from the scanning process, led to the emergence and development of dedicated RE software systems such as: Geomagic [4], Rapid Form [5], CATIA [6] and so on. These software systems allow the conversion of point clouds into parametric surfaces by taking into account the following steps: processing the point clouds, designing the curves and generating the surfaces.

However, the parametric surface obtained by means of dedicated RE software systems represent "a lot of patches" [7] which is the virtual model of the physical product, a model that is frozen [8], that does not allow their automatic modification. Even if the generation of curves/surfaces over point cloud is performed automatically, using the dedicated functions, the development of a customized model cannot be implicitly performed [9]. In terms of designing, a customized model actually means a product which automatically updates once the parameters have been established at the beginning, thus leading to create a model entirely adapted to the customer needs. In order to outline the customized products the authors have used [10] the RE technique for assemblies that contain parts with simple configuration. Based on customer needs have developed a series of algorithms to automate the creation of products dedicated to human [11].

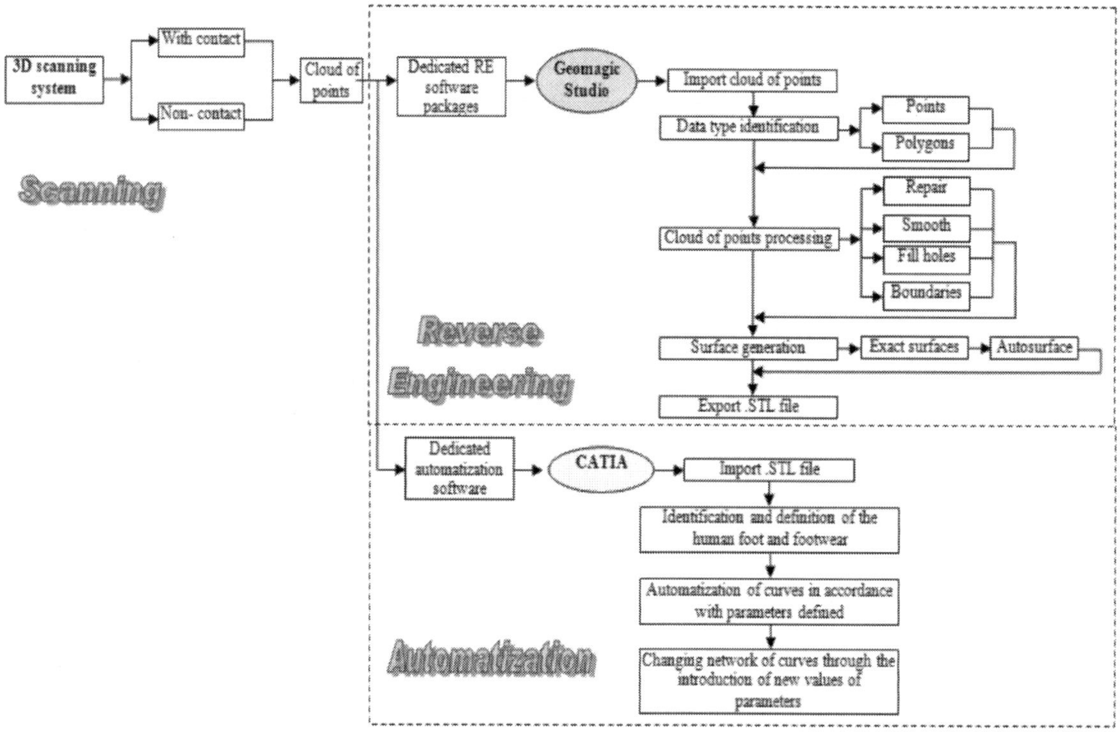

Figure 1. Stages of design methodology

The design methodology proposed by the authors (figure 1) starts with the specific stages of RE technique, namely: digitization of the product, processing the point clouds, conceiving of the curves over the finished point cloud. Its originality consists in the curves automation by introducing the parameters corresponding to the human component and to the product. These parameters stand at the basis of defining the characteristic points of the product and they allow the creation of new curves. All these curves will be entirely parametric, but above all they will be customized. The paper presents how the methodology is used for a complex customizable product.

2. Data Acquisition

The digitization stage of the product was achieved using a solution from Steinbichler Company with the non - contact scanner named COMET L3D with blue light [12].

The dedicated software package of this scanner is Comet Plus and it provides a series of specific tools for object digitization, alignment the point clouds, processing the point clouds, creating and editing of the mesh. The result of scanning process (figure 2) is a file in STL format.

3. Processing of Cloud Points

The Geomagic Studio software dedicated to the RE technique is used to process the point cloud. Firstly, in order to obtain a quality of the point cloud, the types of data that make it up have to be identified [13]: ordered points, unordered points or polygons. In this case the point cloud consists of polygons (figure 3). In order to achieve a high quality mesh, the point cloud is being analyzed (e.g. non-manifold edges, self-intersections, highly creased edges, spikes, small tunnels, and small holes) and the issues occurring during the object scanning are removed.

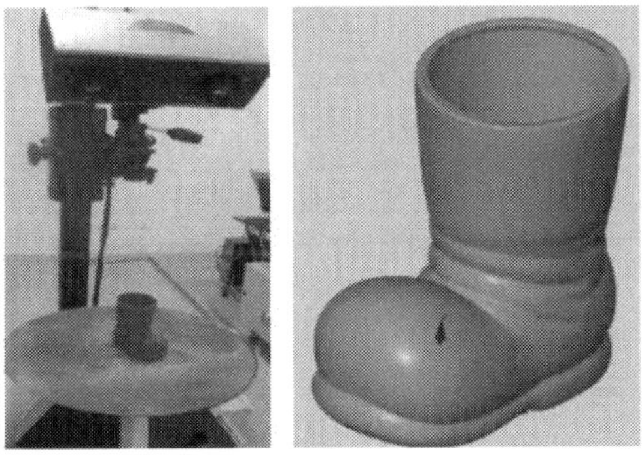

Figure 2. COMET L3D and scanning result.

Figure 3. Default cloud of points.

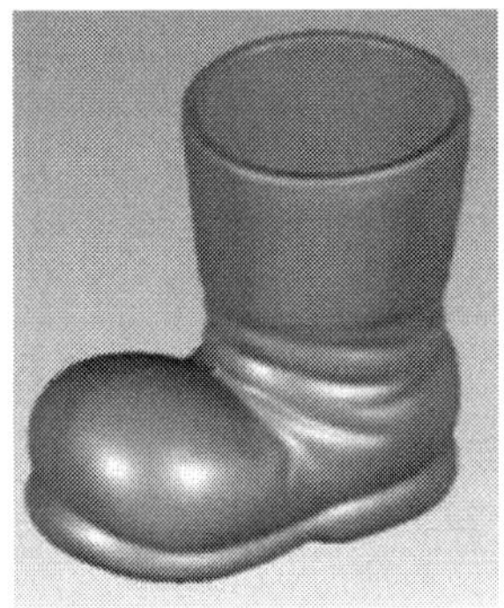

Figure 4. Patch surfaces

Decimation (reducing the number of polygons) is performed in order to facilitate working with point cloud. The following operations: cleaning the disconnected triangles of point cloud, noise reductions, spike removal, filling of holes, are done automatically, using the predefined functions. If the result of the cleaning operations and the editing of point cloud are satisfactory, the type patch surfaces are performed (figure 4). The obtained result is saved using the STL extension in order to import them into the CATIA environment.

4. Automated Product Design

The idea of automating the resulted curves over the point cloud is the first step to be taken in order to obtain customized products. This stage can be done in CAD environment such as CATIA, PRO-Engineer, Solid WORKS, by means of dedicated tools. The CATIA software package contains these specific tools in the Shape module [14].

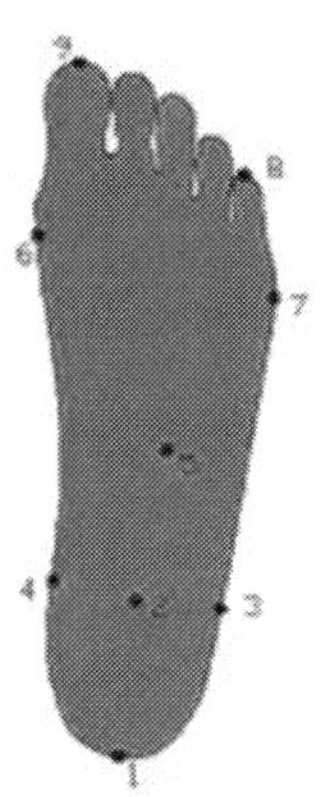

Figure 5. Anthropometric points of human foot [15]

Table 1. Anatomical points of human foot.

Point name	*Number of point*
Anterior ankle point	1
Heel centre	2
Center of lateral malleolus	3
Center of medial malleolus	4
Bending point of foot	5
1st Metatarsophalangeal Joint	6
5th Metatarsophalangeal Joint	7
Extreme anterior point of the 5th toe	8
Extreme anterior point of the foot	9

The tools of the Digitized Shape Editor toolbar allow the creation of sections (in blue on the point cloud) and curves (in white on the point cloud). Both sections and curves can be achieved by means of automatic and/or manual method. The sections are generally obtained by intersecting the point cloud with working planes, while the curves emphasize details of the object.

Firstly, before obtaining the sections over the point cloud, the working planes must be defined using the functions on the Wireframe and Design toolbar. These tools allow the defining of points, lines, working plans, circles, spline curves, projection of points on the curves, intersection of curves, using various methods.

4.1. Human Foot Parameters

The anthropometry applied in shoe industry targets the measuring of the human foot dimensions. The dimensions are measured between landmarks known as anthropometrical points or characteristic points of the human foot [15].

In figure 5 and table 1 [16] are presented the anthropometrical points needed to define the main dimensions used to characterize the human foot, respectively for designing the footwear. The parameters of the human foot have been defined in accordance with the anthropometrical points shown in figure 6 and table 2.

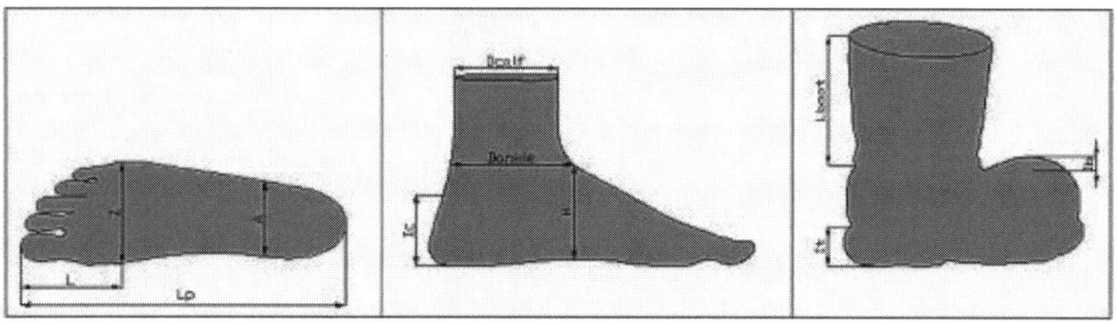

Figure 6. Parameters of human foot and of boot

Table 2. Characteristic parameters of human foot and of product

Parameters of human foot
A – specific length for narrowing of foot
Lp –length of foot
Z – specific length for width of foot
L –length from top of foot up to section where width of foot is known
H –height of foot
Ic –height of heel of foot
Dankle –diameter specific for ankle of foot
Dcalf –diameter specific for calf of foot
Parameters of product
It –height of boot heel
Ib –height of boot toe cap
Lboot –height of boot

4.2. Identification of Wrapping Curves

The process of automatization starts with identifying the base curve (which is obtained by intersecting the point cloud with the default XOY plane). On the curve shown in figure 7 are defined the following: the origin of the own coordinate system (the point corresponding to the narrow part of the foot), the support points of the object (defined in accordance with the point origin and parameters of the foot) and the auxiliary points (which allow keeping the curve shape). Each point, previously defined, will have a working plane that will become the basis for the other points.

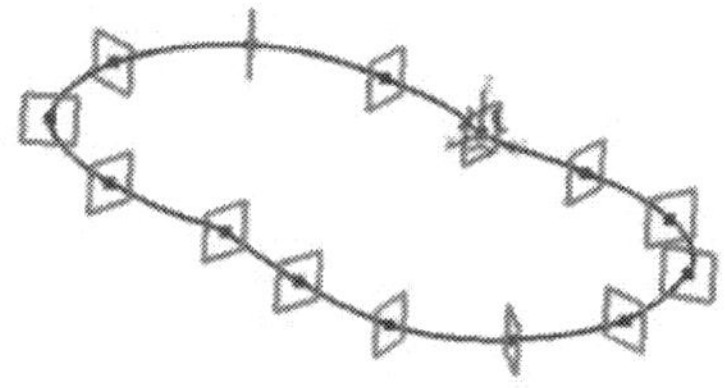

Figure 7. Necessary points

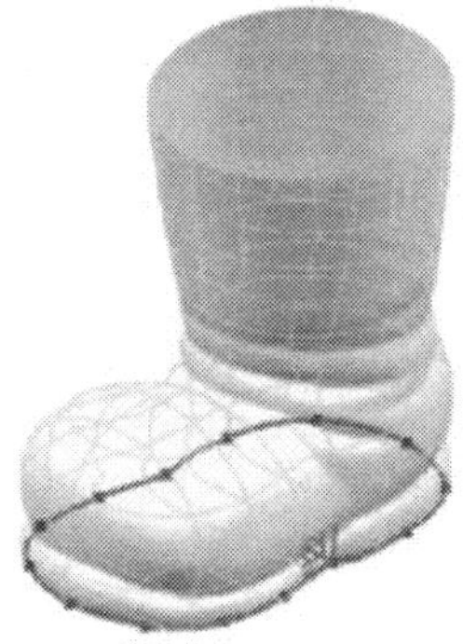

Figure 8. Network of wrapping curves

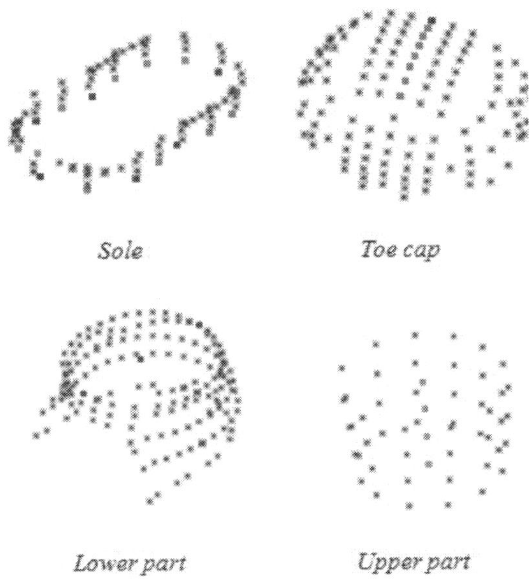

Figure 9. Parametric points of boot components

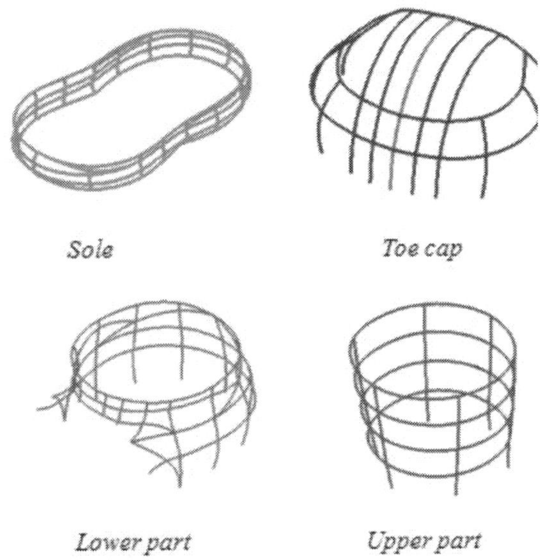

Figure 10. Automated network of curves

The previously defined working planes are used to intersect the point cloud in order to obtain the sections and the curves that define the object. The result of this operation (figure 8) is a network of wrapping curves.

4.3. Automated Generating of Wrapping Curves

The points that represent the basis for defining the wrapping curves are introduced according to: a point of origin (default or chosen by the user), a series of lengths corresponding to the human foot and to the approached product type. All these points are gathered from the generated curves, on the finished point cloud (figure 8), using manual or automatic methods.

The automatization stage begins by defining the previously presented parameters (section 4.1) and continues with calculating the point coordinates, defining them by means of parameters and formulas. In order to keep a better control over the product shape, the product has been divided in components

[17]. For this product the following four components are used (figure 9 and figure 10): (1) sole, (2) toe cap, (3) lower part and (4) upper part of the product.

The first component, the boot sole, contains the base curve of the wrapping curves network and is made up of the supporting points, respectively of the auxiliary points. The supporting points are defined, to the following parameters: A –specific length for foot narrowing; Lp –length of foot; Z – specific length for foot width; L –length from the top of the foot up to the section where the foot width is known. In figure 9 are shown all the points necessary for the automated design of the boot sole.

In what concerns the second component, the toe cap, the operation starts by defining the points on the middle curve of the toe cap. The reference point for this component is coloured in blue (figure 9) and is defined according to the origin point and to the parameters that are necessary for the automated design of the component: H –height of foot and Ib –height of boot toe cap.

For the automated modelling of the third component, the lower part, of the boot are defined three points related to the point corresponding to the leg bending centre and to the parameters H –height of foot and Dankle - specific diameter for foot ankle. To keep a symmetrical distance from the diameter of the ankle, a lot of supplementary points have been defined to allow maintaining the boot shape up to the heel height. The points that were automatically generated for the lower part in figure 9 are shown.

The upper part of the boot is influenced by two parameters, namely: Lboot –height of boot and Dcalf –specific diameter for foot calf. The upper part component is divided into many levels along the boot height. A new centre has been defined for each level (colour magenta as shown in figure 9). This centre was defined in relation to the point considered to be the boot centre, and also related to the following parameters: Lboot, Dcalf and Dankle (figure 9).

If it is considered the previously defined points to be control vertices then the curves network can be defined by means of the Spline curves (figure 10). In figure 11 are shown two footwear products of the boot type, the first is obtained using the default parameter values and the second is automatically generated with modified values.

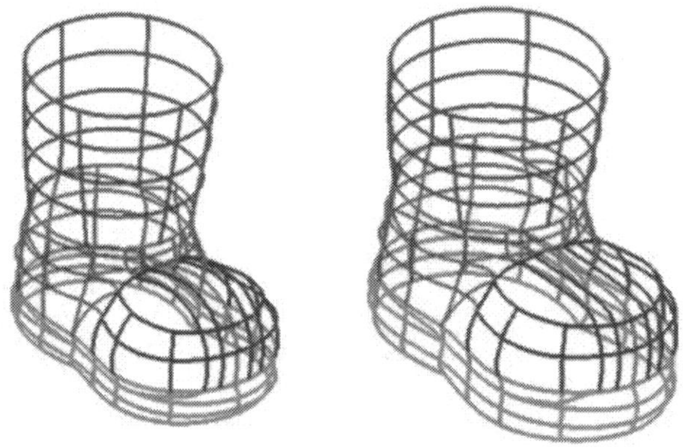

Figure 11. 3D wireframe model obtained using automatically generated curve networks

4.4. Automated Generating of Surfaces

The surfaces for each component of the footwear type product were developed, based on the wrapping curves network. Due to the fact that the surfaces which define the product are not of the primitive type, there were used complex type surfaces Loft and Fill type. Another factor that led to the choice of these types of surfaces is that only they allow automatization, once their parameter values have been modified. Figure 12 shows the surfaces based on wrapping curves that were generated for each boot component.

If the result is satisfactory, a specific thickness is applied throughout the entire product, resulting in a customized solid. The final result is presented in the figure 13.

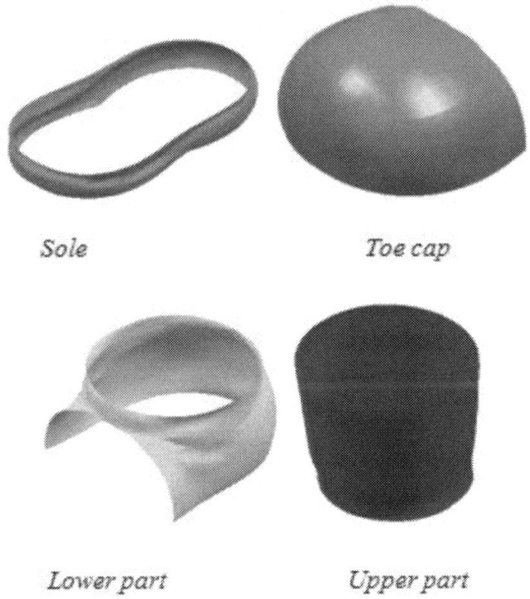

Figure 12. Boot components obtained using automatically generated curve networks

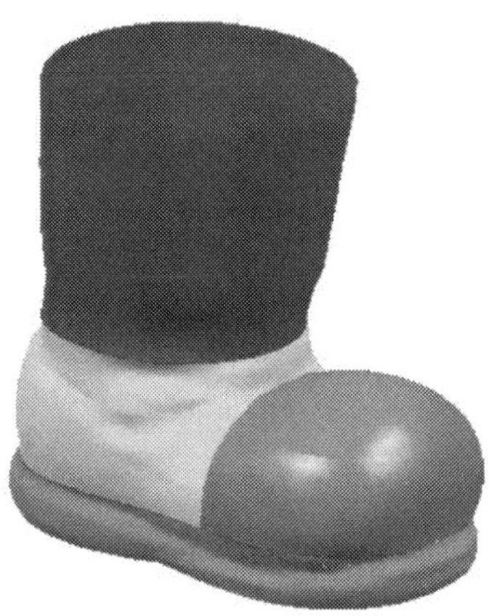

Figure 13. Final parametric boot

5. Conclusion
In this paper, the authors have presented a methodology and associated algorithm used for the automated design of products with high level of customization for which no documentation is available. The algorithm underlying that the methodology contains the stages through which an existent object passes, in order to have its automated design: digitization, processing point cloud, designing the wrapping curves over the point cloud, the automation in correlation with parameters of the human component and product. For the studied footwear product a network of Spline curves is generated and based on it, networks of surfaces / solids are designed, the final result being a parametric CAD model.

Acknowledgment
This work supported by the Sectoral Operational Programme Human Resources Development (SOP HRD), financed from the European Social Fund and by the Romanian Government under the project number POSDRU/89/1.5/S/59323 and we hereby acknowledge the structural founds project PRO-DD (POS-CCE, O.2.2.1., ID 123, SMIS 2637, ctr. No 11/2009) for providing the infrastructure.

References

[1] Draghici G 1991, Ingineria integrata a produselor [Integrated product engineering], Ed. Eurobit, Timisoara, Romania.
[2] Raja V and Fernandes K 2008, Reverse Engineering an Industrial Perspective, Springer Series in Advanced Manufacturing, UK.
[3] Gibson I, Rosen D and Stucker B 2010, Additive Manufacturing Technologies. Rapid Prototyping to Direct Digital Manufacturing, Springer
[4] Geomagic Available: http://www.geomagic.com/en/products/design/overview. Accessed 12 April 2015
[5] Rapid Form Available: http://www.rapidform.com/products/xor/overview/. Accessed 12 January 2015
[6] CATIA Available: http://www.3ds.com/products-services/catia/solutions-by-industry/ Accessed 1May 2015
[7] Chang K and Chen C 2011, 3D Shape engineering and design parameterization, *Computer - Aided Design and Applications*. **8**(5) 681-692.
[8] Durupt A, Remy S and Ducellier G 2010, Knowledge based Reverse Engineering – An approach for Reverse Engineering of a mechanical part. *ASME Journal of Computing and Information Science in Engineering*. **10**(4) 044501-1-044501-4.
[9] Durupt A, Remy S and Ducellier G and Eynard B 2008, From a 3D point cloud to an engineering CAD model: a knowledge-product-based approach for reverse engineering. *Virtual and Physical Prototyping*, **3**(2) 51-59.
[10] Pescaru R and Oancea G 2012, CAD modeling of part assemblies using reverse engineering technique, *Advanced Materials Research*, **591-593** 7-10.
[11] Wang C 2010, An analysis and evaluation of fitness for shoe lasts and human feet. *Computers in industry*, **61**(6) 532-540.
[12] Scanner COMET L3D Available: http://www.steinbichler.com/products/3d-scanning/comet-l3d, Accessed 15 February 2015
[13] Geomagic (2012) Tutorial Geomagic Studio 2012.
[14] Catia Documentation Available: http://www.catiadesign.org/_doc/v5r14/. Accessed 15 June 2014
[15] Albu A 2011, Contribuţii la elaborarea constructivă şi tehnologică a calapoadelor în industria de încălţăminte [Contributions to the constructive and technological designing of the last like in the footwear industry] PhD Thesis, University of Oradea, Romania
[16] Mălureanu G and Mihai A 2003, Bazele proiectării încălţămintei [Footwear design basics]. Ed. Performantica, Iasi, Romania
[17] Zhang D, Wang J and Yang Y 2014, Design 3D garments for scanned human bodies, *Journal of Mechanical Science and Technology*, **28** (7) 2479-2487

Profiling procedure for disk cutter to generate the male rotor, screw compressors component, using the "Substitute Family Circle" - graphic method in AUTOCAD environment

CL Popa[1] and V Popa[2]

[1] "Dunărea de Jos" University, 47 Domnească Street, Galați, Romania
Faculty of Engineering, Mechanical Engineering Department

Email: lpopa@ugal.ro

Abstract. This paper proposes a profiling method for the tool which generates the helical groove of male rotor, screw compressor component. The method is based on a complementary theorem of surfaces enveloping - "Substitute Family Circles Method". The specific theorem of family circles of substitution has been applied using AUTOCAD graphics design environment facility. The frontal view of the male rotor, screw compressor component, has been determinate knowing the transverse profile of female rotor, and using this theorem of "Substitute Family Circle". The three-dimensional model of the rotor makes possible to apply the same theorem, leading to the surface of revolution enveloping the helical surface.
An application will be also presented to determine the axial profile of the disk cutter, numeric and graphics, following the proposed algorithm.

1. Introduction

The helical surface generating, as constituent surfaces of machine bodies (gears, active elements of helical pumps, screw compressors), is done using tools bounded with revolution surfaces, disk cutter type tool. A distinctive problem in these types of tools is to determine the shape of revolution surface that must be conjugated to the helical surface. The mathematical solution to this problem is known as Theorem I Olivier, or Gohman Theorem [1].

Lyukshin [2] has also developed a vectorial solution to the problem, based on decomposition of helical motion in several rotation motions - Nikolaev Theorem. This solution represents the most often form used for solving analytically the problem of profiling the primary surface of the disk cutter, due the ease expressing of the enveloping condition - the condition of coplanarity of three vectors: helix axis, tool axis disk and a vector that connects two points of those axes.

Also have been elaborated specific complementary theorems as: "Substitute Family Circles" and "Minimum Distance Method" as analytical theorems for approaching this issue [3].

Graphic design software products in AutoCAD or CATIA environments allowed approaching the problem graphically [4,5,6]. The graphic solving is very rigorous and intuitive.

2. Screw Compressor Rotors Front View Profile [7]

In the following we are proposing a graphical method developed in AutoCAD, based on the theorem of "Substitute Family Circles", for profiling the disk cutter which generates the male rotor, screw compressor component, first of all for a particular case, but with prospect to generalizing the method to the other helical surfaces (Figure 1).

The mobiles systems $X_2Y_2Z_2$ and $X_1Y_1Z_1$ will be defined, being attached to, respectively, the male rotor and female rotor, in their transverse section, while they will rotate with the global systems $x_1y_1z_1$ and $x_0y_0z_0$.

The female rotor profile is made up of a series of arcs:

$\overset{\frown}{A_2B_2}$ arc, radius r;

$$A_2B_2 = \begin{vmatrix} X_2 = r_2 - r\cos\theta_1; \\ Y_2 = -r\sin\theta_1; \\ Z_2 = 0; \end{vmatrix} \qquad \theta_{1\min} = 0;\ \theta_{1\max} = \left| ar\cos\frac{r + r_0}{2r_2} \right|; \tag{1}$$

$\overset{\frown}{B_2C_2}$ arc, radius r_0;

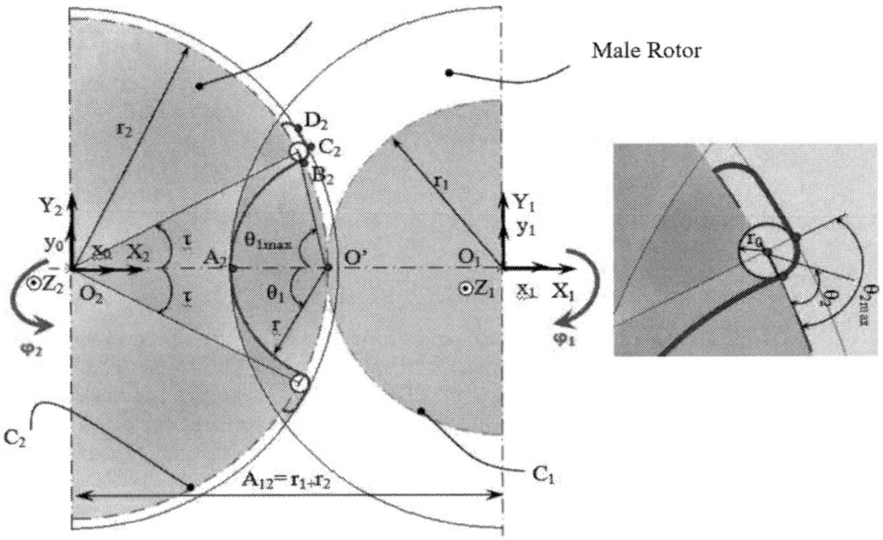

Figure1.The frontal geometry of rotors, screw compressor component [7];
C_1 and C_2 are centrodes in rolling motion, associated to transverse sections of the rotors

$$B_2C_2 = \begin{vmatrix} X_2 = r_2 cos\,\tau + r_0 cos(\theta_{\max 1} - \theta_2); \\ Y_2 = -r_2 sin\,\tau - r_0 sin(\theta_{\max 1} - \theta_2); \\ Z_2 = 0; \end{vmatrix} \tag{2}$$

$$\theta_{2\min} = 0;\ \theta_{2\max} = (\pi - \theta_{1\max}); \tag{3}$$

$\overset{\frown}{C_2D_2}$ arc, radius r_2+r_0;
Knowing the absolute movements:

$$\begin{vmatrix} x_2 = \omega_3^T(\varphi_2)\cdot X_2; \\ x_1 = \omega_3^T(-\varphi_1)\cdot X_1; \end{vmatrix} \tag{4}$$

and the relative position of global systems,

$$x_1 = x_2 + \left\| \begin{matrix} A_{12} \\ 0 \end{matrix} \right\|, \tag{5}$$

it will be determine the relative motion

$$X_1 = \omega_3(-\varphi_1)\left[\omega_3^T \cdot X_2 + \left\| \begin{matrix} A_{12} \\ 0 \end{matrix} \right\|\right]. \tag{6}$$

This motion will determine the envelopes of profiles made up of arcs $\overset{\frown}{A_2B_2}, \overset{\frown}{B_2C_2}, \overset{\frown}{C_2D_2}$, representing the female rotor profile. One circle belongs to the "Substitute Circles Family" (Figure 2) is described by equations as follow:

$$C_i \left| \begin{array}{l} X = -R_{r_1} \sin \varphi_1 + r_i \sin \beta_i; \\ Y = R_{r_1} \cos \varphi_1 - r_i \cos \beta_i, \end{array} \right. \tag{7}$$

r_i , β_i si φ_1 variable.

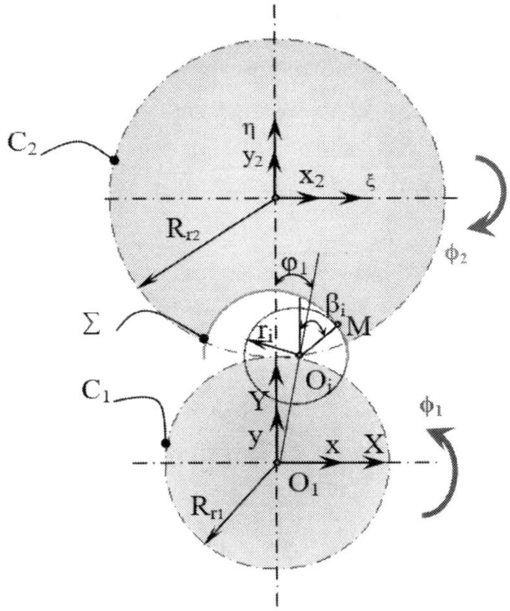

Figure 2.Circle of substitute circles family; generated profile $\sum$; reference systems

The circle, center O_i, fulfils the tangency conditions to $\sum$ profile. The tangency condition between circle C_i and $\sum$ profile means: conditions of contact points (Equation 8) and common tangent in contact point (Equation 9):

$$\left| \begin{array}{l} X(u) = R_{r_1} \sin \varphi_1 - r_i \sin \beta_i; \\ Y(u) = R_{r_1} \cos \varphi_1 + r_i \cos \beta_i, \end{array} \right. \tag{8}$$

$$\begin{array}{l} \dot{X}_u^k = -r_i \cos \beta_i; \\ \dot{Y}_u^k = -r_i \sin \beta_i, \end{array} \tag{9}$$

The equations assembly (8) and (9) allows us to determinate the next condition, as follow:

$$\dot{Y}_u \sin \varphi_1 - \dot{x}(u) \cos \varphi_1 = \frac{x(u)\dot{X}_u + y(u)\dot{Y}_u}{R_{r_1}}, \tag{10}$$

and

$$tg\beta_i = \frac{\dot{Y}_u}{\dot{X}_u} \quad , \tag{11}$$

meaning the enveloping condition.

The transposition in rolling motion of the substitute family circles leads to envelope shape:

$$\begin{array}{l} \xi = -r_i \cos(\varphi_1 + \varphi_2 + \beta_i) - R_{r_2} \cos \varphi_2; \\ \eta = -r_i \sin(\varphi_1 + \varphi_2 + \beta_i) + R_{r_2} \sin \varphi_2, \end{array} \tag{12}$$

$$R_{r_1}\varphi_1 = R_{r_2}\varphi_2 \,. \qquad (13)$$

Equation 13 represents the rolling condition of the two centrodes; τ, r_0, r, r_1, r_2 are known sizes.
It will propose, in the following, a graphical solution to the problem, using the "Substitute Family Circles" method.

3. The male rotor profiling

The substitute family circles method [3] assumes the next definition for the *family circles*: family circles belonging to a compound profile, linked of a couple of centrodes in rolling motion, is the family circles having centres on the associated centrode and in tangency with the compound profile, at the same time, as shown in Figure 3. $\overset{\frown}{A_2B_2C_2}$ is made up of some arcs with centres on centrode C_2 and the substitute family circles linked with centrode C_2 is reduced of two circles of radius r_1 and centre O_1', respectively, radius r_0 and centre O_0'.

The substitute family circles for $\overset{\frown}{C_2D_2}$ is made up of circles of radius r_0 and centres on centrode C_2 - circles c_1, c_2.

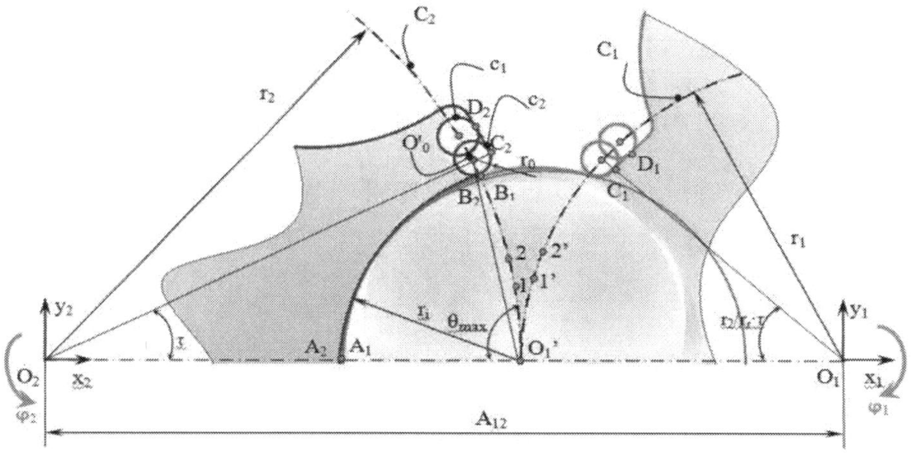

Figure 3.The substitute family circles

In this particular case, the transposed of substitute family circles on centrode C_1, to whom the arcs radius r, belong, is made up of arcs, having centres on centrode C_1, in O_1', for $\overset{\frown}{A_1B_1}$ and in O_2', for $\overset{\frown}{BC_1}$, as shown in Figure 4. O_2' will be determinate through condition that arc $\overset{\frown}{O_1'O_0'}$ measured on centrode C_2 to be equal to arc $\overset{\frown}{O_1'O_2'}$ measured on centrode C_1.

$\overset{\frown}{A_1B_1}$ belongs to circle radius „r" centre in O'_1,

$\overset{\frown}{BC_1}$ belongs to circle radius "r_0" centre in O'_2.

$\overset{\frown}{B_2B}$ belonging to female rotor profile, will be determinate using point B, which is initially superposed to point B_2, in the relative motion of the two centrodes.

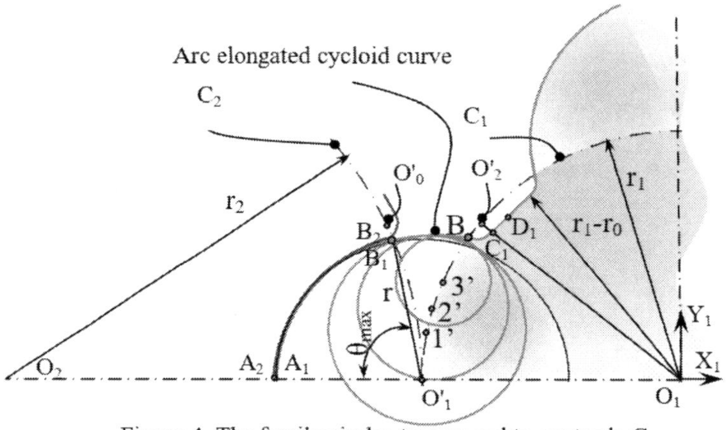

Figure 4. The family circles transposed to centrode C_1

The successive positions of the two centrodes, in the rolling process, and the position of point B, in relation to centrode C_1 attached to male rotor, as well, are represented in Figure 5.

Will be considere a large number of points, on arc $\overset{\frown}{O_1'B}$ angularly equidistant, thus arc $\overset{\frown}{O'_1 1}$ belonging to centrode C_1, to be equal to arc $\overset{\frown}{O_1' 1'}$ belonging to centrode C_2, and so on.

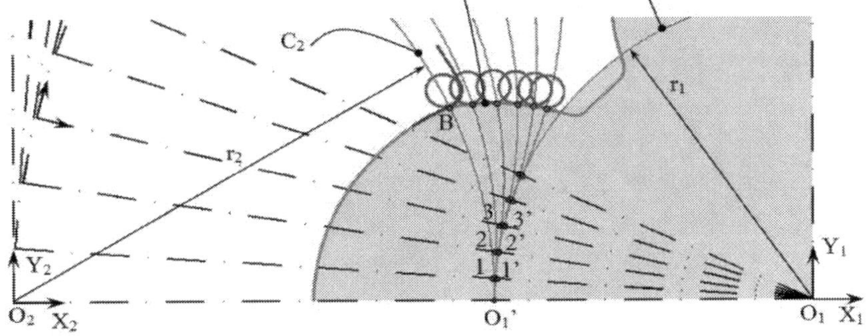

Figure 5. The relative position of centrodes C_1 and C_2 in the rolling process and the trajectory of point B toward the system and X_1 Y_1

The male rotor profile is described by the reference system, whose X_1' axis represents the symmetry axis of the gap between the two successive lobes belonging to male rotor, Figure 6.

Coordinate transformation between the two systems is described by Equation 14:

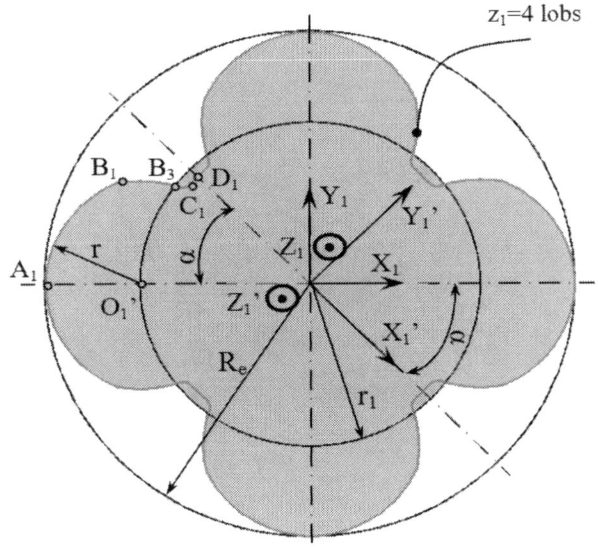

Figure 6. The new reference system X_1'Y_1' of female rotor ($\alpha=45^0$)

$$\begin{Vmatrix} X_1' \\ Y_1' \end{Vmatrix} = \begin{Vmatrix} \cos\alpha & -\sin\alpha \\ \sin\alpha & \cos\alpha \end{Vmatrix} \cdot \begin{Vmatrix} X_1 \\ Y_1 \end{Vmatrix} \quad (14)$$

4. The disk cutter for male rotor

Will be make the solid model of male rotor, Figure 7, reported to the system $X_1'Y_1'Z_1'$, for a rotor left screw, pitch 188.67mm.

The substitution family circles method stipulates that the contact between a helical surface, cylindrical and constant pitch and a revolution surface (the primary peripherical surface of disk cutter) is based on the specific theorem [3], as follows:

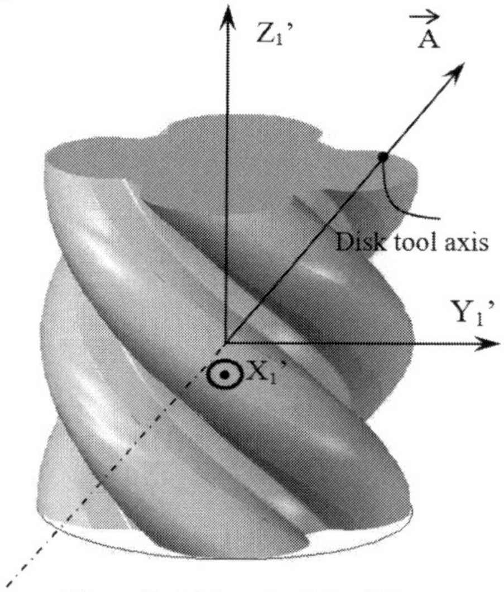

Figure 7. Male rotor 3D solid

The revolution surface (disk cutter), mutually envelope with one helical surface, cylindrical and constant pitch, is the locus made up of family circles, which have the common axis identical with the axis of the future disc cutter; those family circles are tangents to the intersection curves between the helical surface and the orthogonal planes to the disc cutter axis, in transverses planes with the future revolution surfaces axis (Figure 8).

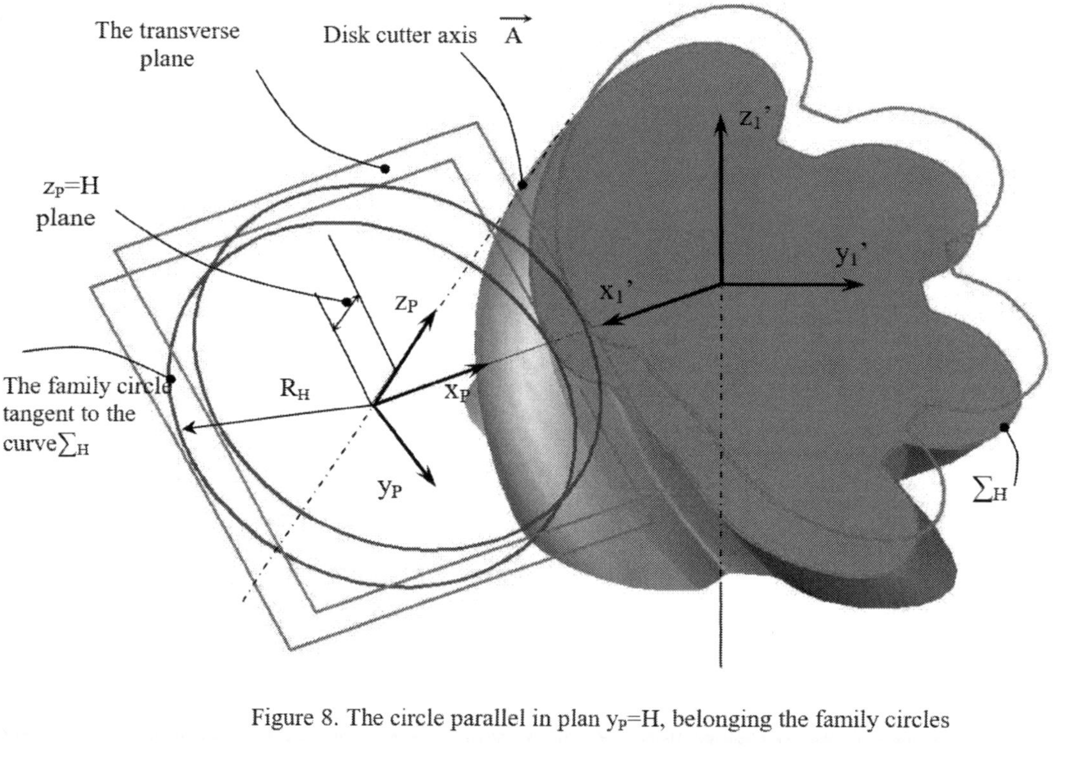

Figure 8. The circle parallel in plan y_P=H, belonging the family circles

The coordinates ensemble (H,R_H) defines the disk cutter axial section, mutually envelope with the helical groove belonging to the generated rotor, for different size of H parameter.

5. Profiling disk cutter generating helical groove of male rotor, using graphical procedure in AUTOCAD environment

Different sizes of H parameter and the coordinates ensemble (H, R_H) determine the axial section of the future disk cutter, which is mutually envelope with the helical grove of male rotor (Figure 9). The centres of the circles radius R_H belong to the tool axis, Z_p and are in tangency with the curves $\sum_H$ – planes curves of the surfaces to generate.

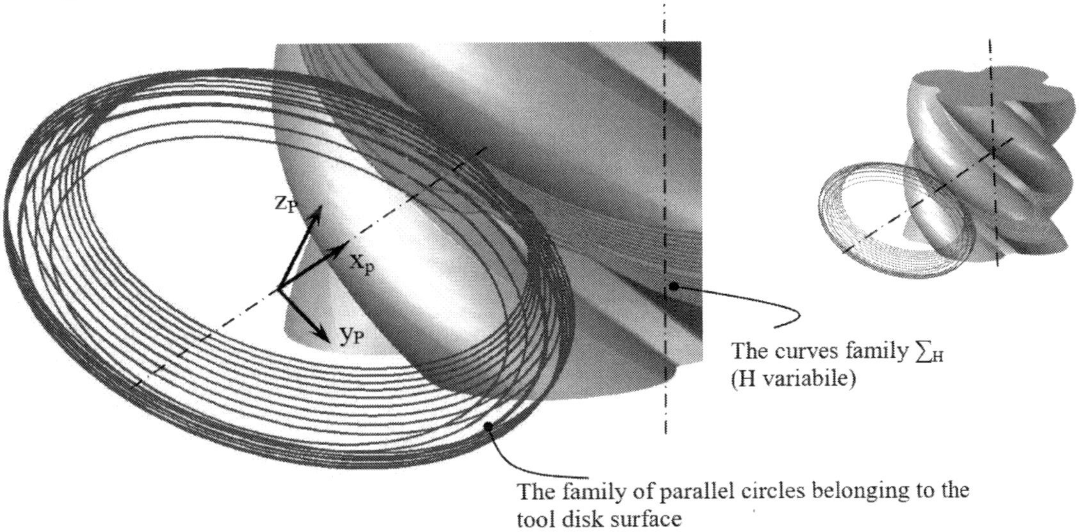

Figure 9. The male rotor solid and "the family of parallel circles" of the revolution surface (disk cutter)

Consider the male rotor screw compressor component, having dimensions as follow: r_1=32mm; r_2=48mm; Re_1=50mm; Re_2=50mm; i=6/4; r=22mm; r_0=1.8mm [7], $\alpha=45^0$, helical pitch P_E=188.67mm; the axis $\vec{A}$, of the future disk cutter will be determinate being in tangency to the counterclokwise helix.

5.1. AUTOCAD application description

It will be considere the 3D solid of the male rotor, shown in Figure 7.

The succession of commands it will be as follow:

-UCS command for rotating the coordinates system with the complement of twist angle of the helix;
-SLICE command for determination the transverse plan;
-SECTION command for definition the intersection surface by the transverse plan;
-UCS command for the new coordinates system in the transverse plan;
-EXPLODE the section defined previously;
-CIRCLE command for determination the circles radius minimal and tangent to the curve.
-Repeat the entire sequence above.

Table1 describes the coordinates of the axial section of the disk tool which generates the helical grove of the male rotor.

Table1. The axial section coordinates of the disk cutter

R[mm]	38.936	41.749	43.768	47.791	48.006	..	39.993	39.252	38.560	38.207
H[mm]	3.300	4.300	5.300	6.300	8.000	..	17.000	18.000	19.000	20.000

Figure 10 describes the shape of revolution surface of the disk tool.

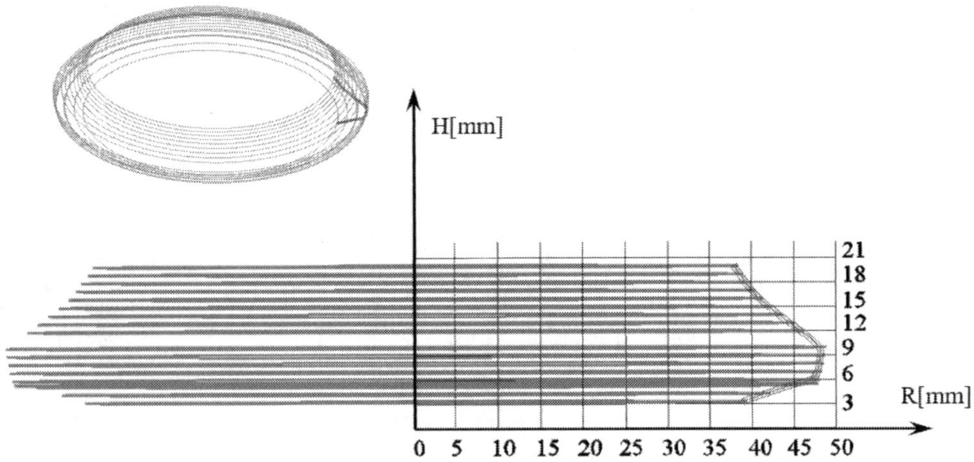

Figure10. The axial section shape of the disk cutter

6. Conclusions

The paper consists in presentation in AUTOCAD environment, regarding a graphical algorithm to profiling the disk cutter which generates the male rotor screw compressor component.

The family circles method has been used, for both, the determination of male rotor transverse profile and for the disk cutter solid generating of helical groove.

The graphical method to express the substitute family circles allows an accurate and fast approach of the mutually envelope profiles, which are in rolling process; this method is also appropriate to shape the revolution surfaces who generate helical surfaces.

For the presented example, the solve of the problem was made up first manually, using AUTOCAD commands. In the future, it can be written a specific code AUTOLISP to allow the automation, after the parameterization of the tool and surfaces dimensions.

The graphical interpretation of points is made with o polyline edited with the option Spline. Once the increasing of number of points, the polyline is more accurate with the theoretical shape of the axial section of the tool.

References

[1] Litvin FL 1984 *Theory of gearing, Refrences Publication 1212 NASA, Scientific and Technical Information Division, Washiongton DC*

[2] Lyukshin VS 1968 *Theorya vintovich poverhnostey v proektirovanyi rejuschich instrumentov (Theory of helical surfaces for designing cutting tools)*), Mashinostroenye, Moskva

[3] Oancea N 2004 *Generarea suprafeţelor prin înfăşurare (Surfaces generation trough winding)*, Vol. I-II, Galati University Press, ISBN 973-627-170-4

[4] Theodor V 2013 and all *Profiling of Rack Gear Tool for Generation of the helical surfaces* Proceedings of the ICAM Engineering and Technologies, NEV TECH, Sweeden, pp.63-72, ISBN 978-91-7501-892-8

[5] Baicu I 2002 and all *Profilarea sculelor prin modelare solidă (Solid modeling for profiling tools)* ed.Tehnica Info, Chisinau, ISBN 9975-63-172-X

[6] Popa CL 2015 *Graphical Method in CAD Environment for Profiling End Mill Tool*, Applied Mechanics and Materials Volumes 809-810 pp. 787-792, Iasi, ISSN 1662-7482

[7] Amosov P E et al 1977 *Vintovîe kompresornîe mashinîi - Spravochnik (Screw Compression Machines - Handbook),* Mashinstroienie, Leningrad

A study on using pre-forming blank in single point incremental forming process by finite element analysis

K I Abass

The University of Mustansiriyah, Mechanical Department, Baghdad, Iraq

E-Mail: khalil1969a@yahoo.com

Abstract: Single Point Incremental Forming process (SPIF) is a forming technique of sheet material based on layered manufacturing principles. The edges of sheet material are clamped while the forming tool is moved along the tool path. The CNC milling machine is used to manufacturing the product. SPIF involves extensive plastic deformation and the description of the process is more complicated by highly nonlinear boundary conditions, namely contact and frictional effects have been accomplished. However, due to the complex nature of these models, numerical approaches dominated by Finite Element Analysis (FEA) are now in widespread use. The paper presents the data and main results of a study on effect of using pre-forming blank in SPIF through FEA. The considered SPIF has been studied under certain process conditions referring to the test work piece, tool, etc., applying ANSYS 11. The results show that the simulation model can predict an ideal profile of processing track, the behaviour of contact tool-workpiece, the product accuracy by evaluation its thickness, surface strain and the stress distribution along the deformed blank section during the deformation stages.

1. Introduction

Single Point Incremental Forming (SPIF) is a displacement controlled process performed on a CNC machine. A clamped blank is deformed by the movement of the tool that follows a prescribed tool path [1], a sketch of SPIF is presented in figure 1.

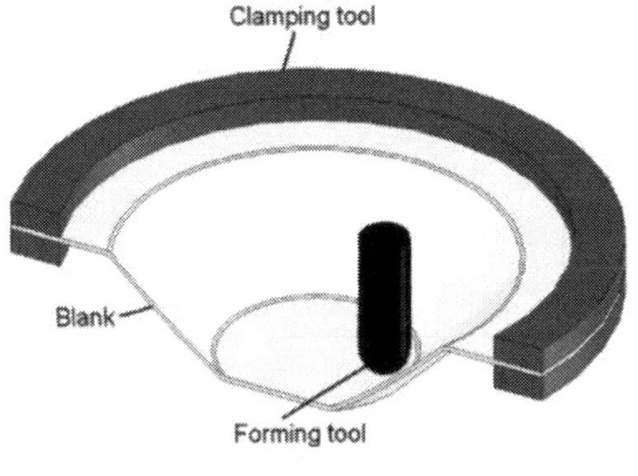

Figure 1. The basic component of SPIF process [1].

The complicated geometry information is resolved into a series of two-dimensional layers, and then the plastic deformation is carried out layer-by-layer through the CNC controlled movements of a simple hemispherical forming tool to get the desired part [2]. SPIF is a flexible sheet metal forming process with a high potential for small quantities production and for rapid prototyping applications. With this process it is possible to form sheet metal parts without specialized dies [3].

Advantages of the technology are high process flexibility, relatively low hardware costs and enhanced formability. SPIF enables production of even complex shapes without costly die sets. A die-less forming process leads to a significant lower time-to-market. Applications for which SPIF would be especially useful include prototyping and small-lot production for automobile, aerospace and biomedical industries [4].

Some of SPIF limitations are a lack of tolerance predictability due to springback, wall angle limitations, the material thinning and the surface quality. Some methods applied aimed to improve the formability. These were warm forming [5], laser assisted heating [6] and electric heating through the tool [7]. A hybrid process concentrates on integrating Asymmetric Incremental Sheet Forming (AISF) and stretch forming [8]. The effect of using this new hybrid process is shorter production time, compared to AISF, especially when the areas to be processed are identified in advance. Also, there are improvement is thickness distribution with a reduced amount of maximum thinning.

Another innovative method is to cut slots and tabs in which the area to be formed is inside the blank [9]. The effects were measured by forming a simple and a complex part, with and without cutouts and with and without backing plates. Compared to using conventional blanks, the parts formed using this new method were more accurate when unsupported and less accurate than when using a stiff cutout supporting plate. The results showed low effects in case of partial cutouts and higher accuracy when using stiff backup plate. Adding a backing plate, using a kinematic supporting tool and modifying the final part of the tool path have also been researched in terms of effects on the geometrical accuracy [10]. The results showed lower sheet bending near to the initial tool contact location in case of adding the backing plates, reduced springback in case of using the kinematic support tool and no pillow effect at the sheet center in case of extending the tool path. The overall effect of these features shows improvement in the general accuracy and in the thickness distributions of the final product. A produced heating unit which includes dedicated laser optics and laser was integrated into the forming machine. This shows that temperature control and reduction of friction and wear are mandatory in SPIF process [11].

The self-capabilities of the process to rectify inaccuracies when different steps of ISF are carried out on both surfaces of the part was studied [12]. At the second repeated step, the process shows increased accuracy. The subsequent steps do not bring any significant improvement to the accuracy. In case of using a back-drawing strategy, the deviation was reduced by 15-25%.

Hot incremental sheet forming process consists of supplying a continuous current in order to generate local heating, enabling higher formability than cold forming. This process was investigated [13] and the results showed a different grain distribution due to electrical heating and induced strain, dependent on the material properties. By increasing the wall angle of inclination, a lower surface accuracy results with higher energy consumption. A new way of using Two-Point Incremental Forming (TPIF) with partial die was proposed [14]. Based on the results, the analytical model is able to handle the deformation mechanics of SPIF and TPIF with partial die while the neck formation is eliminated in TPIF. The incrementally produced parts by TPIF with partial die have a more accurate geometry than the parts produced by SPIF. This is the effect of smaller elastic recovery upon unloading. The effects of friction on the surface finish, forming load, material deformation and formability are studied using a newly developed Oblique Roller Ball (ORB) tool [15]. The experimental results show better surface finish, lower forming load and higher formability and considerable through-the-thickness-shear along the tool moving direction, which can be considered as a secondary effect in the SPIF process.

2. Development of FE model

An FE model of SPIF process simulation has been developed. The model includes the effect of contact between the tool set - forming tool, die and blank holder - and work piece, as well as the elastic-plastic material behavior of the work piece. To develop the FE model, the effect of using pre forming product was studied with constant features of forming tool, fixture and blank, as illustrated in the figure 2.

Symmetric elements are used to model blanks that are rotationally symmetric about an axis. The deformed blanks are subjected to the loads from the forming tool and supporting die. A two dimensional analysis of a sector of the deforming blank is carried out in order to yield the complete stress and strain distributions.

A quadrilateral mapped mesh is used for blank. The boundary conditions - displacement, loading and the contact - are represented in the figure 2 and the forming tool motion was specified in profile with constant speed.

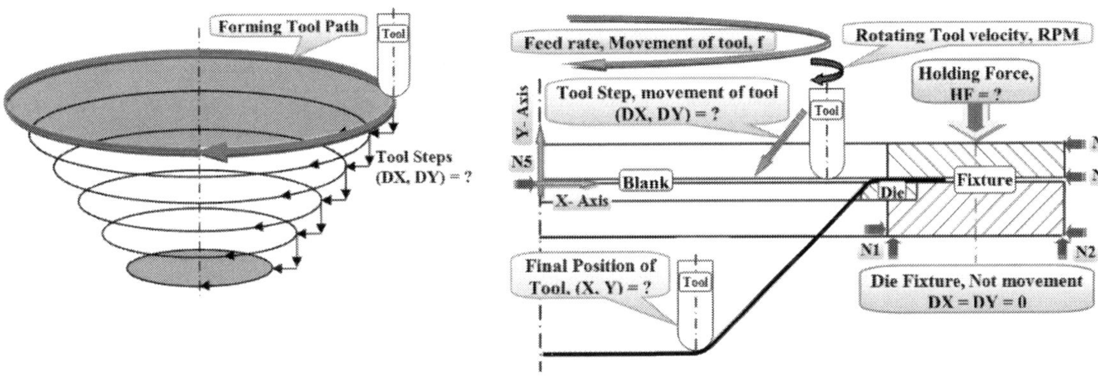

Figure 2. Boundary conditions in the SPIF process.

For the model, the blank type is of 226 mm outside diameter (D_b) and 0.9 mm thickness (t), the friction coefficient (μ) is 0.05, forming tool profile diameter (D_p) of 10 mm, and all six studies are with the same die profile, D_d of 5 mm.

A pure Aluminum Al1050 has been used, with the specific mechanical properties resulted from Stress-Strain curve of a tensile test as shown in the table 1.

Table 1. Material properties of the blank.

Variable	Density, ρ, g/m^3	Young's modulus, E, GPa	Poisson's ratio, ν	Yield stress, σ_y, MPa	Tangent modules, Eτ, GPa
Value	2700	75	0.3	80	0.5

The element PLANE42-2D to represent the tool set and the element V15C0106-2D has been chosen to represent the workpiece, and both have translations in the X and Y directions.

The elements representing the contact work piece - tool set, which include the forming tool, die and blank holder, are CONTAC169 and CONTAC171-2D Point-to-Surface Contact, like a Rigid-to-flexible contact. The real constant set has been used for each contact surface. A non-linear analysis, convergence criteria, incremental load and specified load steps is applied. The convergence tolerance was based on minimization of the force residual.

3. Results and discussion

Six predicted AutoCAD models to produce dome profile on depths 10, 20, 30, 40 and 50 mm are represented in the figure 3. Using these products as a primary part to start production a new profile by SPIF process depends on the forming tool path movement, 50 mm depth and with 45° degree.

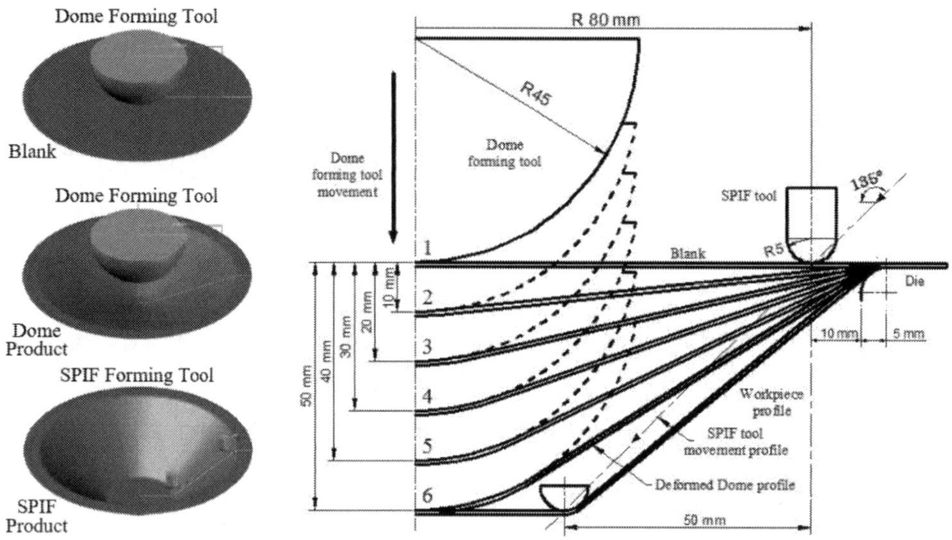

Figure 3. Typical Dome profiles and SPIF forming tool movement.

Depending on the information detailed in the figure 3, the FEM models of the dome products are presented. The evaluation of the strains along the deformed dome section is illustrated in the figure 4.

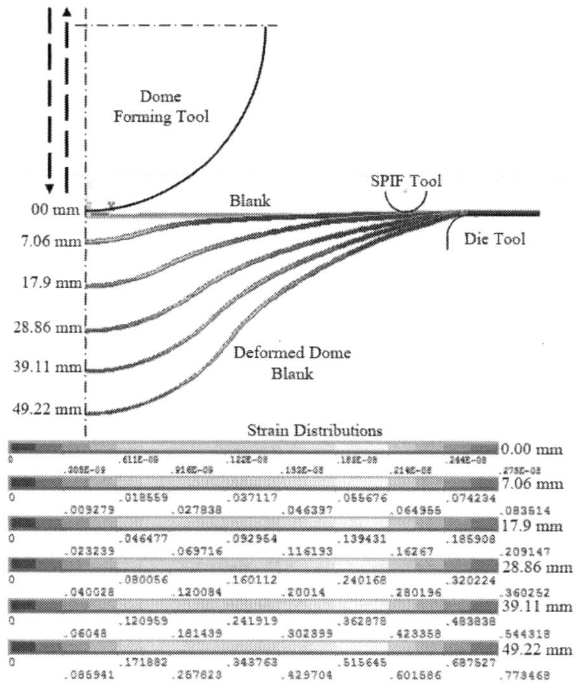

Figure 4. Stress distributions stages of deformed dome parts.

The springback clearly appeared after removing the forming tool effect, where, the AutoCAD predicted models depth, 10, 20, 30, 40 and 50 mm becomes 7.06, 17.9, 28.86, 39.11 and 49.22 mm depth, respectively. The sprinback is reduced with increasing the depth of the deformation, where, the increasing of stresses by the forming tool causes increasing the strains, especially under the forming tool profile.

Thickness distribution and strain field across the formed part are necessary to any intermediate formability evaluation. In this section it was determined the thickness variation and localized strain across the formed part. The evaluation of the thickness along the deformed section of dome after removing the tool path effect is represented in the figure 5, it is clear that the maximum thinning is under the forming tool profile and the increases in the thinning with increases the depth of the deformation to get maximum value, 0.415 mm when produce the 50 mm dome depth. Also, there is little reducing in the thickness on the region affected by the die radius.

The lowest thickness values for 0, 10, 20, 30, 40 and 50 mm dome depths are 0.9, 0.84, 0.74, 0.63, 0.52 and 0.415 mm depth and 5 mm distance from the blank center. The location of the maximum thinning depended on the depth of the deformation. In the area where high reduction in thickness happens, a neck results indicating failure of the product at this stage.

The evaluation of the thickness distribution of the SPIF deformed blanks is illustrated in the figure 6 for a 50 mm forming tool stroke when the part is completely drawn. The blank section is divided into four regions, as follows:

- fixed region, 100 to 113 mm from blank center that is not affected by the forming tool movement;
- backing plate radius effect region, 90 to 100 mm from blank center, minimum reducing is 0.8 mm when using 50 mm dome depth;
- forming tool movement effect region, 30 to 90 mm from blank center, clear improvement in thickness distribution when using the dome product deeper (max. pre deforming product);
- non direct contact region, 0 to 30 mm from blank center, where the thinning percentage of the thickness decreases with increasing the dome product depth used, as is illustrated in the figure 7.

Strain distribution that results from simulation model represents strain value for each node in numerical mode of SPIF and also strain values of all nodes at each step size of forming tool, resulting a huge number of results.

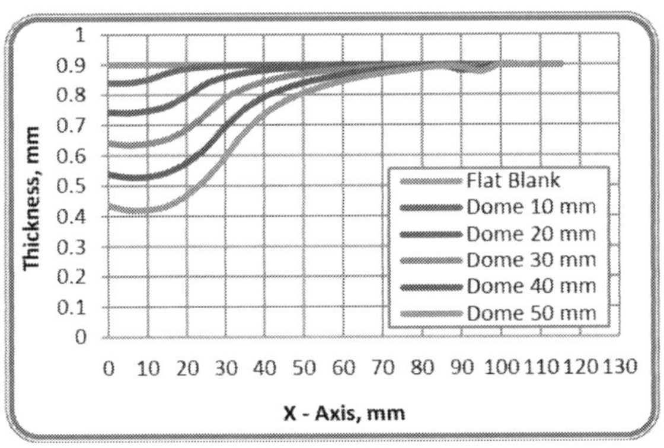

Figure 5. Thickness distribution of deformed dome parts.

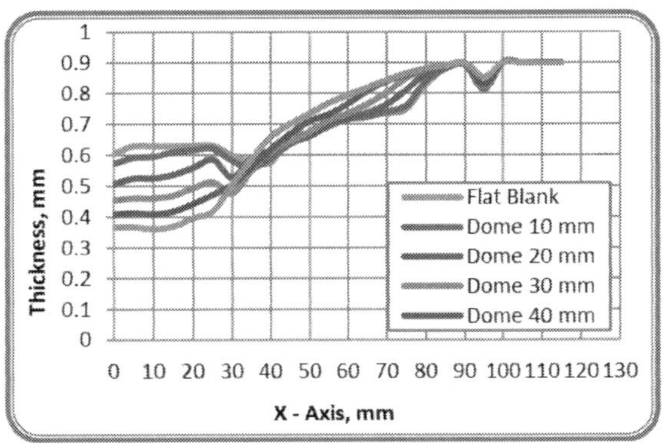

Figure 6. Thickness distribution of deformed SPIF parts.

The effect of using pre deformed profile on strain distribution at upper surface of SPIF workpiece is represented in the figure 8, for forming tool stroke at 50 mm depth and 45° angle, i.e., when the part is completely drawn and removed from the die.

It is also concluded that the peak values of strains are concentrated between 20 to 40 mm distance from the blank center and they increase toward the blank center. This means that the maximum values of strain will be concentrated on the places where the forming tool touches and moves, as well as, the places at the center of the blank, even there is no direct contact with the forming tool 0 to 20 mm from the blank center. No forming operation occurs under blank holder, the strain distribution is more uniform and the strain values were reasonable at 40 to 90 mm distance from blank center.

The distributions of upper surface strain of the 50 mm dome depth was more uniform and had more reasonable values than the others.

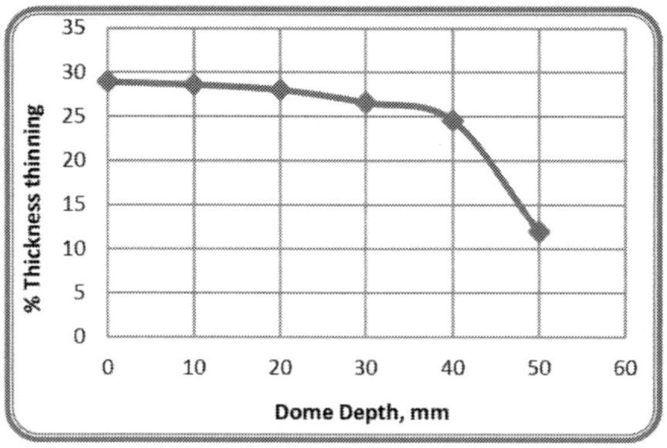

Figure 7. Thickness thinning of deformed blank.

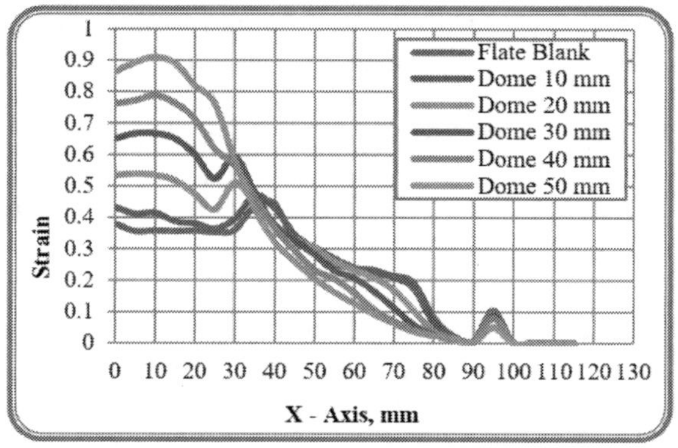

Figure 8. Strain distribution of deformed SPIF parts.

Successive stages that used pre forming product to produce the SPIF part profile are shown in the figure 9. It is clear that the forming tool stroke influences greatly the stress values, in which these values increase with stroke. The peak height appears to be a good indicator to assess whether a forming operation will be successful or not. The severe of the stress distribution is located at the end of stroke and the occurrence of a localized neck is obvious. Fracture was observed along a sharp corner. This was predicted by a high value of the stress.

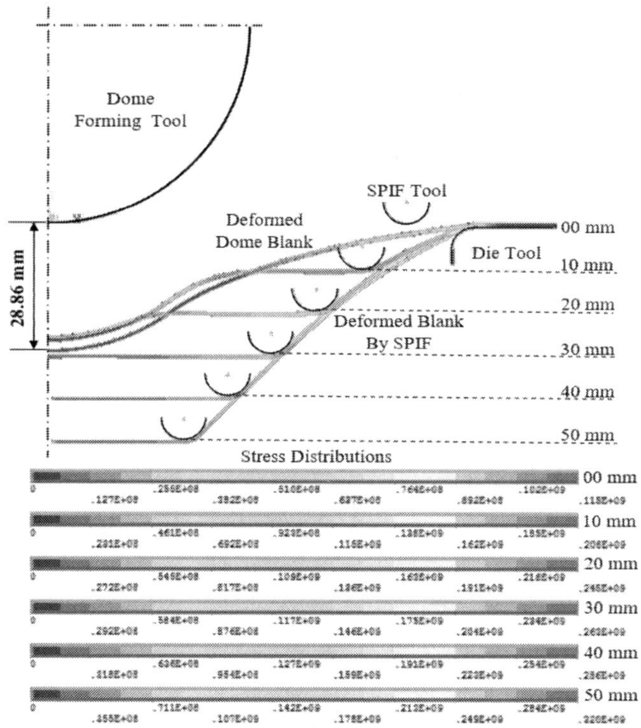

Figure 9. Stress distributions stages of deformed SPIF.

Depending on the dome product depth, the deformation of the SPIF process can be divided into: a) the primary stages, are reorientation the blank shape depending on the product profile, with elastic deformation follows the forming tool position; and b) the secondary stages, starts when the forming tool gets to the final depth of the dome product until the required depth of the SPIF product. Both depend on the SPIF tool path.

4. Conclusions

The simulation model of Single Point Incremental Forming, SPIF by FEM with pre forming product can predict an ideal profile of processing track, and springback error of SPIF is effectively eliminated. The use of proper simulation system (prediction model) and high accuracy procedure when entering as many variables as possible in the simulation model produce high level prediction of the results.

The high values of stress and strain distributions along the section of the product profile appear to be good indicators to assess whether a product will be successful or not.

The sheet thickness distributions play an important role in the forming of defects, and using pre forming product in SPIF process showed an improvement in thickness distribution along the deformed section of the product profile, especially when using pre forming product depth equal to the SPIF product depth requirement.

References

[1] Hadoush A and Boogaard A H van den 2009 Substructuring in the implicit simulation of single point incremental sheet forming, the incrementally updated approach *Int. J. Mater. Form.* **2** pp 181-189

[2] Singh R P and Goyal G 2014 FEA analysis to study the influence of various forming parameters on spring back occurs in single point incremental forming *Int. J. Eng. Research and Applications* (IJERA) ISSN:2248-9622 pp 33-37

[3] Benmessaoud R 2014 A tool path generation method for the multi-pass incremental forming process investigation *Int. J. Adv. Research in Comp. Sci. and Soft. Eng.* **4** pp 1035-1044

[4] Dittrich M A, Gutowski T G, Cao J and Roth J T 2012 Exergy analysis of incremental sheet forming *Prod. Eng. Res. Devel.* **6** pp 169-177

[5] Ambrogio G, Filice L and Manco GL 2008 Warm incremental forming of magnesium alloy AZ31 *CIRP Annals - Manufacturing Tehcnology* **57** 1 pp 257-260

[6] Duflou, J R *et al* 2008 Improved SPIF performance through dynamic local heating *Int. J. Mach. Tools and Manufacture* **48** 5 pp 543-549

[7] Guoqiang F, Fentao S, Xiangguo M, Lin G and Guoquan T 2010 Electric hot incremental forming of Ti-6Al-4V Titanium sheet *Int. J. Adv. Manuf. Technology* **49** 9 pp 941-947

[8] Araghi B T, Manco G L, Bambach M and Hirt G 2009 Investigation into a new hybrid forming process: Incremental sheet forming combined with stretch forming *CIRP Annals - Manufacturing Technology* **58** pp 225-228

[9] Allwood J M *et al* 2010 The effect of partially cut-out blanks on geometric accuracy in incremental sheet forming *J. Mate. Processing Technology* **210** pp 1501-1510

[10] Essa K and Hartley P 2011 An assessment of various process strategies for improving precision in single point incremental forming *Int. J. Mater. Form.* **4** pp 401-412

[11] Göttmann A *et al* 2011 Laser-assisted asymmetric incremental sheet forming of titanium sheet metal parts *Prod. Eng. Res. Devel.* **5** pp 263-271

[12] Ambrogio G and Filice L 2012 On the use of Back-drawing Incremental Forming (BIF) to improve geometrical accuracy in sheet metal parts *Int. J. Mate. Form.* **5** pp 269-274

[13] Ambrogio G, Filice L and Gagliardi F 2012 Formability of lightweight alloys by hot incremental sheet forming *Materials and Design* **34** pp 501-508

[14] Silva M B and Martins P A F 2013 Two-Point Incremental Forming with partial die: theory and experimentation *JMEPEG* **22** pp 1018-1027

[15] Lu B, Fang Y and Xu D K 2014 Mechanism investigation of friction-related effects in single point incremental forming using a developed oblique roller-ball tool *Int. J. Mach. Tool and Manufacture* **85** pp 14-29

IOP Conf. Series: Materials Science and Engineering **161** (2016) 012032 doi:10.1088/1757-899X/161/1/012032

Kinematic Structures for Processing of Surfaces with a Circle Directrix and a Straight Line Generatrix (Part IV)

R Cioară

Department of Engineering and Industrial Management, Transilvania University, Brasov 500036, Romania

E-mail: cioarar@unitbv.ro

Abstract. A body with a given geometry can be obtained by various manufacturing procedures, entailing, however, different surface quality, dimensional precision and (most importantly) productivity. Each body is characterised by one or more surfaces, simple or complex, as the case may be. It is known that a surface is characterised by a directrix and a generatrix, as well as by the modality of its physical achievement by means of the machines and tools uses for manufacturing. The directrix is obtained as a result of a primary motion and possibly one or more secondary motions. The generatrix is obtained as a result of one secondary motion, rarely more. If the generatrix is materialized on a tool, then a secondary motion is not necessary. In addition one or more auxiliary motions are needed. Other generation modalities not excluded, from the theory of surface generation on machine-tools it is known that the directrix, as well as the generatrix can be: materialized; generated by copying; kinematically generated as the trajectory of a point; kinematically generated as the envelope of a family of curves; generated by rolling; or programmed. Typically in literature generation "by copying" and "programmed generation" of the directrix and generatrix are not addressed distinctively, both being assumed as of the same nature. The recent evolution of industrial electronics and implicitly of machine-tools has determined a clear differentiation between the kinematics and construction of machine-tools (still) using generation by copying from a master and the kinematics of NC machine-tools. The paper is one of kinematic synthesis, oriented towards innovation-invention, and presents by means of examples, not necessarily known or typical, all six cases of generation of surfaces characterised by a straight line generatrix and a circle directrix obtained kinematically as the envelope of a family of curves.

1. Introduction

It is the destination of a machine-tool, as asserted in various definitions [1] [2], to manufacture components of pre-defined future technical systems. Manufacturing a part entails deliberate action, consisting of a sequence of certain steps in order to obtain a body of the desired geometry and characteristics in accordance with a known model or a "mental" image, described or graphically represented. The geometry of the part corresponds to one or several of its functions and is obtained by manufacturing or fabrication deploying other dedicated bodies or assemblies of bodies (tools, instruments, devices, equipment, and machines). The characteristics of a part, other than its geometry, can be very diverse and relate to its (mechanical, corrosion, etc.) strength, hardness, aspect and so on.

A body with a given geometry can be obtained by various manufacturing procedures, entailing, however, different surface quality, dimensional precision and (most importantly) productivity. Among other attributes, each of the manufactured parts is characterised by one or more surfaces, simple or complex, as the case may be. It is known [3] [4] [5] that a surface is characterised by a directrix and a generatrix, as well as by at least one modality of its physical achievement by means of the machines and tools uses for manufacturing.

Other generation modalities not excluded, from the theory of surface generation [3] on machine-tools it is known that the directrix (D), as well as the generatrix (G) can be: materialized (M; m); generated by copying (Co; co); kinematically generated as the trajectory of a point (Ci_{tp}; ci_{tp}); kinematically generated as the envelope of a family of curves (Ci_{fc}; ci_{fc}); generated by rolling (R; r); programmed (P; p). Consequently the set of possibilities of achieving a surface characterised by a certain directrix and a certain generatrix is given by the set of possible combinations of the modalities of obtaining the respective directrices and generatrices [5] [6]. For example, Ci_{tp}&m means D(Ci_{tp})&G(co), that is a modality of surface generation where the directrix D is obtained kinematically as the trajectory of a point (Ci_{tp}), and the generatrix G is obtained by copying (co). Parts of these combinations are found in practice and are described by literature, as illustrations of known modalities of surface generation [1] [3] [4] [7]. While any directrix-generatrix combination is theoretically possible, only relatively few are also practically efficient. The systematic approach explores the entire set of generation possibilities, applied or not (yet) in practice, hence being oriented towards discovery and even inventics [8] [9]. The paper is limited to case of generation with a circle directrix kinematically generated as the envelope of a family of curves.

Typically in literature [1] [3] generation "by copying" and "programmed generation" of the directrix and generatrix are not addressed distinctively, both being assumed as of the same nature. The recent evolution of industrial electronics and implicitly of machine-tools has determined a clear differentiation between the kinematics and construction of machine-tools (still) using generation by copying from a master and the kinematics of NC machine-tools.

Each motion is the result of the action (of the final effector element) of a generation or auxiliary linkage, as the as the case may be. Depending on the way they are connected, linkages can be independent or partially overlapping. The directrix is obtained as a result of a primary motion and possibly one or more secondary motions. The generatrix is obtained as a result of one secondary motion, rarely more. If the generatrix is materialized on a tool, then a secondary motion is not necessary. In addition one or more auxiliary motions are needed.

Over time three modalities of obtaining a material body have been identified and utilised: a) by material removal (in order to obtain a certain body the excess matter is removed from a body of initially greater volume; b) by material redistribution (the matter of a body of initially the same volume as the targeted volume is redistributed; or the matter of a body of initially greater volume is redistributed, thus obtaining the desired body plus an excess that is removed), and c) by adding material (matter is added to a body of initially smaller volume; or matter is added to fill an empty cavity the form and volume of that are identical to those of the desired body). Various processes of cutting, plastic forming and non-conventional technologies are utilized [7].

The examples given in this paper refer exclusively to the modality of "material removal".

2. Generation with a circle directrix obtained kinematically as the envelope of a family of curves and a straight line materialised directrix

Milling with the active cylindrical or tilted part of various mills, typically slab mills or side-and-face cutters, surfaces generation is conducted with a materialised generatrix and a directrix obtained as the envelope of a family of curves (family of identical epicycloids, prolonged, described by the tips of all tool teeth). The directrix can be a circle, when the presence of a circular feed is required, carried out either by the workpiece, the necessary motions are illustrated in figure 1, or the tool (the case of "planetary milling" [2]).

The main motion I, conducted by the tool, and the circular feed II, conducted by the workpiece, are both rotations. The directions of motions I and II are recommended to be opposing, so that, inter alia the relative speed in the tangent point (the cutting speed) is obtained as the sum of the peripheral speeds of tool and workpiece. The generation feed motion II participates in the forming of the directrix. As the generatrix is materialized, this machining does not require a feed linkage for obtaining the generatrix. The translation motions III, IV and V are auxiliary positioning motions.

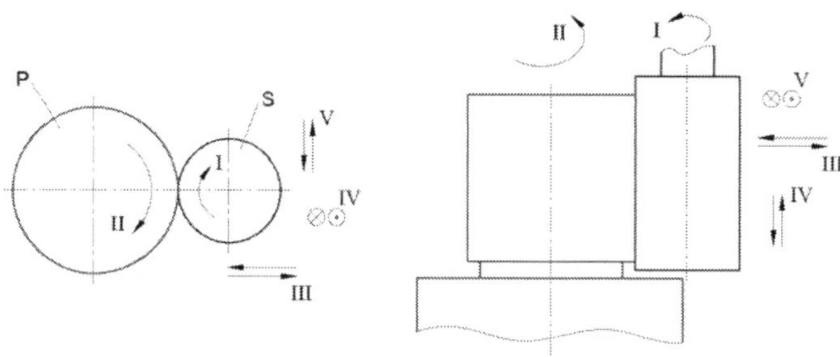

Figure 1. Machining schematic for milling with a materialised generatrix of a cylindrical surface (example for case Ci_{fc}&m).

The generated surface can also be an „interior" one, not just „exterior".

In the case of planetary milling, motion II – circular feed, as all other motions, is carried out by the tool. This is the case most often when the workpiece is large and cannot or is inefficient to be actuated to conduct any generating or auxiliary motion.

Kinematic structure of machines for such processing is known [2].

3. Generation with a circular directrix obtained kinematically as the envelope of a family of curves and a straight line generatrix obtained by copying

A kinematic directrix is obtained most often as the envelope of a family of curves by milling and grinding, with a variety of directrix curves. The "circular directrix" case is a usual one. In order to obtain a straight line directrix by copying required are a master Sb (a materialised „programme"), on that the surface to be copied is tilted in relation to the axis of the workpiece by the same inclination as that of the conical generatrix of workpiece P, and a feeler Pp of the same profile as the generatrix of the tool S. Figure 2 illustrates a machining schematic for milling a surface described by a circle directrix obtained kinematically as the envelope of a family of curves, and a straight line generatrix obtained by copying.

Motion II is a circular feed carried out by the workpiece in order to obtain the whole circular directrix. Together with the tool's axial feed III, a continuous motion and of small value compared to motion II, also the translation III' is carried out by feeler Pp. Motions III and III' are simultaneous and identical. Motion III' of the feeler determines motion IV' of the feeler, a translation perpendicular to motion IV conducted in its plane. Motion IV' is transmitted by the same ratio as motion IV towards the axis of the tool. In this case the simultaneity and proportionality of motions III and IV ensure the obtaining of the desired straight line generatrix.

The master Sb can be replaced by a ruler, in which case the feeler Pp is replaced by a shoe.

Figure 3 shows a kinematical diagram of principle compatible with the considered machining schematic. This reveals certain particularities worth mentioning:

- motion II, the workpiece's circular feed can be carried out towards either the left or the right. This is ensured by direction inverter INV_2;
- motion III (implicitly also motions III', IV' and IV) can be discrete, what is ensured by the presence of intermitter device INT_3. In the absence of this, motion III is continuous;
- motion III (implicitly also motions III', IV' and IV) can be conducted in both directions, a possibility ensured by inverter INV_3;
- even if motion III is carried out, feeler Pp could not be actuated, in which case clutch $C_{3'}$ is disengaged. This facility is not offered by the machining schematic in figure 3;
- also clutches $C_{4'}$ and C_4 can be provided for the possible decoupling of motions IV' and IV, or only of motion IV, respectively;

- motion IV can be identical to motion IV' or can differ from this by a non-dimensional proportionality factor i_{C4}, ensured, for example, by a lever system. This possibility is not captured in the schematic in figure 2;
- the auxiliary motions (V and VI) can be actuated mechanically or (only) manually.

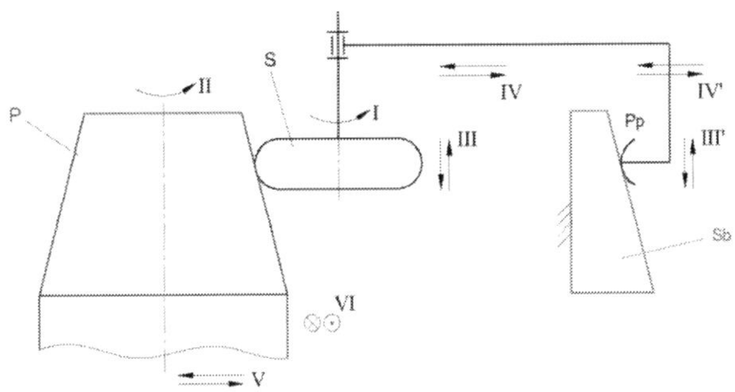

Figure 2. Machining schematic for milling a conical surface, the straight line generatrix being obtained by copying (example for case Ci_{fc}&co).

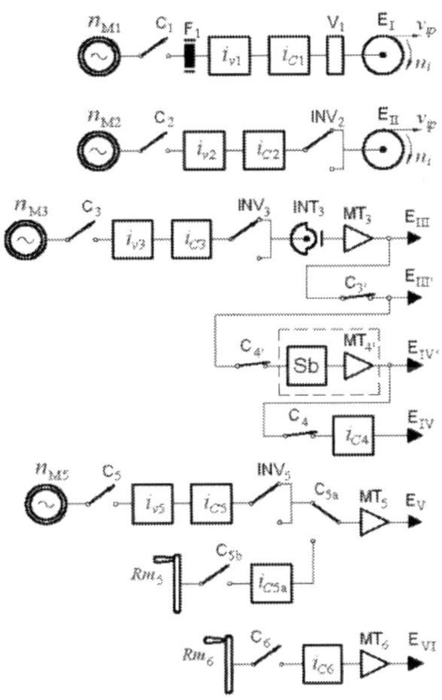

Figure 3. A kinematic diagram of principle for the case Ci_{fc}&co presented in figure 2.

4. Generation with a circle directrix obtained kinematically as the envelope of a family of curves and a straight line generatrix obtained kinematically as the trajectory of a point

A possible example for the considered case is the finishing of an exterior cylindrical surface by grinding with the peripheral part of a cylindrical grinding wheel, figure 4. The tool S, the grinding wheel, carries out the main motion I and the radial feed IV, a positioning feed needed to ensure grinding at the imposed dimension. The workpiece P carries out the circular feed II, necessary for obtaining (together with motion I) of the circle directrix, and axial feed III, continuous and of reduced speed by the axis of the part, necessary for obtaining the straight line generatrix of the workpiece. Figure 5 shows the kinematic diagram of principle corresponding to the considered case, minimal, and of strictly mechanical structure.

In many classical circular grinding machines of general use, the adjustment mechanism i_{v1} is absent from the main linkage, or is very simple. Typically there is also no clutch C_1.

In the circular feed linkage the adjustment function represented by adjustment mechanism i_{v2} is typically transferred to the actuating electric motor M2, which is a DC motor allowing continuous speed adjustment. The presence of clutch C_2 is not imperative.

In order to achieve the axial feed motion III, a "mechanical" actuation as well as a manual one was provided by means of the same transformation mechanism MT_3. It has to be mentioned though, that in circular grinding machines the "mechanical" axial feed linkage has most often a hydraulic structure, the final element being a linear hydraulic motor, while for the manual axial feed a pinion-rack transformation mechanism is used, the two actuation modes being thus kinematically disjunct.

In order to obtain radial positioning motion IV, in the schematic in figure 5 only manual actuation is provided. In practice, however, often a second linkage is encountered for this motion, independent from the first one and of hydraulic structure.

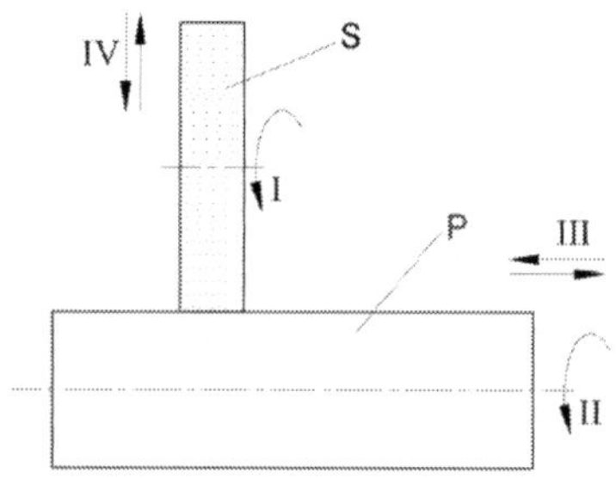

Figure 4. Machining schematic for cylindrical grinding with the peripheral part of a grinding wheel (example for case $Ci_{fc}\&ci_{tp}$).

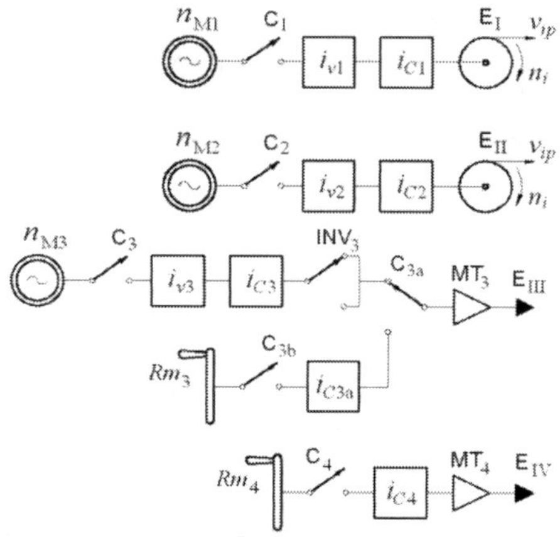

Figure 5. A kinematic diagram of principle for the case $Ci_{fc}\&ci_{tp}$ presented in figure 4.

5. Generation with a circle directrix and a straight line generatrix, both obtained kinematically as the envelope of a family of curves

This case can be exemplified by the machining schematic in figure 6, namely the milling of a surface with a circle directrix by means of a slab mill or the cylindrical part of a side-and-face cutter; the tool S is positioned tangentially to the cylindrical surface of workpiece P and carries out main motion I, a rotation, and axial feed III along the workpiece axis, the latter being the motion that determines the achievement of the generatrix. The workpiece P carries out the circular feed II that determines the achievement of the directrix. To be satisfied is the condition that the axial feed (given by motion III) of the tool is small in relation to the circular feed of the workpiece, thus that a sufficiently small translation (of the order of tenths of millimeter) of the tool S along the direction of motion III corresponds to one revolution of workpiece P. The machining schematic does not necessarily require perpendicularity between the axes of the tool and the workpiece.

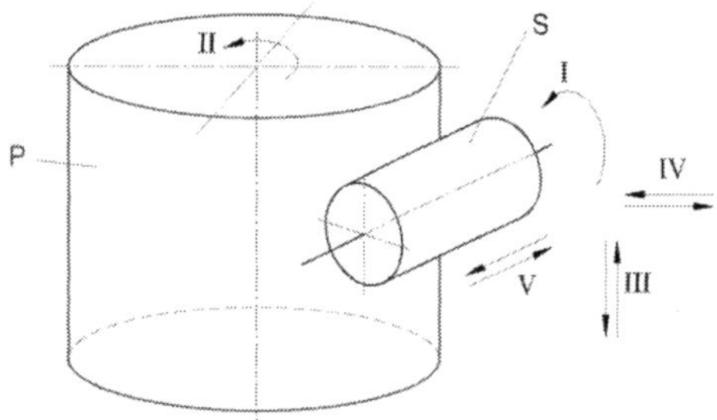

Figure 6. Machining schematic for the milling of cylindrical surfaces, the straight line generatrix being obtained as the envelope of a family of curves (example for the case $Ci_{fc}\&ci_{fc}$).

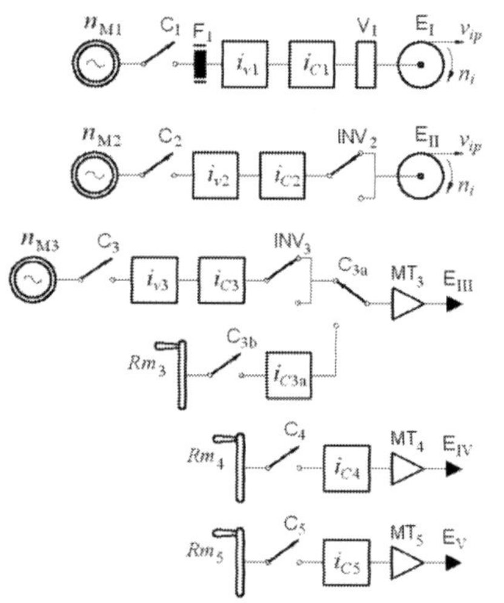

Figure 7. A kinematic diagram of principle for the case $Ci_{fc}\&ci_{fc}$ presented in figure 6.

Figure 7 presents a kinematic diagram of principle that satisfies all requirements for achieving the machining schematic of figure 6. The main linkage for the rotation I of the tool in a single direction includes a clutch C_1, a brake F_1, a speed control mechanism of the final shaft (that achieves an adjustable final transmission ratio i_{v1}), a flywheel V_1 and, most probably, a motion transmission with a constant total transmission ratio i_{C1}. The function of flywheel V_1 can be assumed by another part of the main linkage, for example by one of the gears.

The linkage for the circular feed motion II of the workpiece includes an inverter mechanism INV_2, so that the revolution of the workpiece can be carried out to either the left or the right. The output speed is small and consequently the presence of a brake in the structure of this linkage is not required. In figure 7 only a mechanical actuation is indicated, but the possibility of manual actuation is also provided frequently.

The linkage for the axial feed motion III is provided with both mechanical and manual actuation, and it is by means of a clutch C_{3a} that the branch of the two is selected, by that at a given time motion is transmitted to the final effector E_{III} (a translation slide). Motion III can be carried out by either the tool or the workpiece.

In figure 7 distinctive, manually actuated linkages of identical structure are provided for auxiliary motions IV and V.

6. Generation with a circle directrix obtained kinematically as the envelope of a family of curves and a straight line generatrix obtained by rolling

It is known that rolling entails a mobile curve (rolling curve) that rolls without slippage over a fixed curve, called base. If the rolling curve does not carry out all motions entailed by rolling, than both rolling curve and base curve are mobile. The base and the rolling curve need to be mutually envelopable.

The base and the rolling curve can be most diverse, some most difficult or even impossible to generate kinematically. In most technical applications, the base and the rolling curve are easily generated plane curves, typically a straight line or a circle. The base and the rolling curve cannot be simultaneously both straight lines, but can be simultaneously both circles. Consequently, most technical applications of machining by rolling a resumed to the rolling of a circle over a straight line, of a circle over another circle (at its interior or exterior, as the case may be), or of a straight line over a circle. In the particular case where the generatrix to be obtained is a straight line, then it is taken as the base, and the rolling curve will be a circle (or arc of circle) materialised as a generatrix on the tool used for machining. This situation is illustrated in the machining schematic presented in figure 8. Figure 9 presents a kinematic diagram of principle that ensures the specified necessary motions.

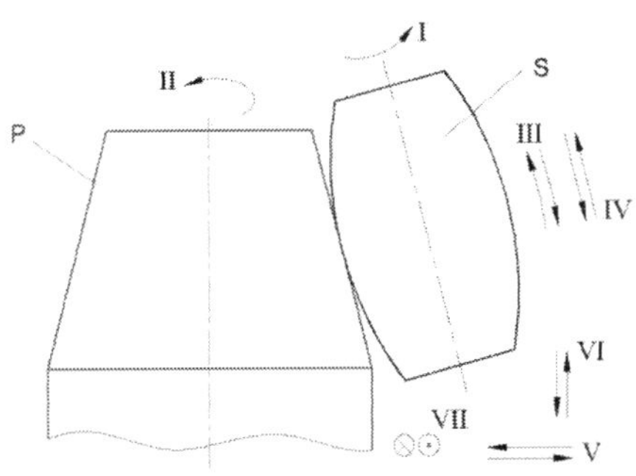

Figure 8. Machining schematic for milling a conical surface, example for the case Ci_{fc}&r.

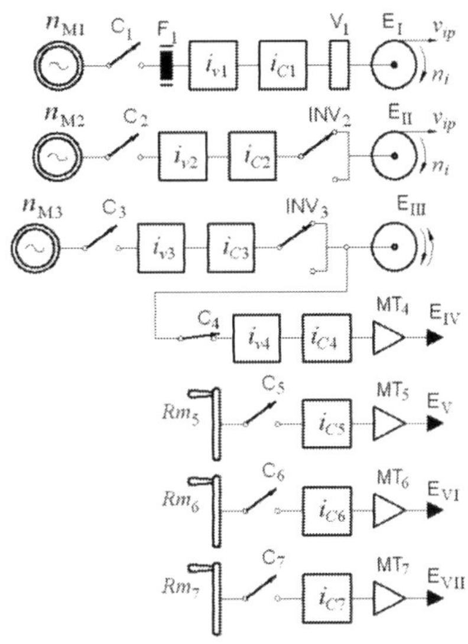

Figure 9. A kinematic diagram of principle for the case Ci_{fc}&r presented in figure 8.

The tool S carries out both the main motion I, a rotation motion, and motions III and IV that together ensure the rolling without slipping of the tool's circle profile (its materialised generatrix) over the straight line generatrix of the machined workpiece P. Workpiece P carries out the circular feed II either as climb or conventional milling (in the same or opposite direction as the feed). Consequently the structure of the linkage needs to include an inverter mechanism. The tool S or the workpiece P carry out all or only one/some of the auxiliary motions V, VI and VII, which are positioning motions by the directions of the coordinate axes.

The main motion I, which is the revolution of the tool by its own axis, and the circular feed II of the workpiece are independent. The rotation III and translation IV have a rigid kinematic link and contribute together to the generation by rolling of the workpiece generatrix.

For the auxiliary motions V, VI and VII that ensure positioning along the directions of the coordinate axes linkages of minimal kinematic structures are provided, manually actuated from handwheels R_{m5}, R_{m6} and R_{m7}.

7. Generation with a circle directrix obtained kinematically as the envelope of a family of curves and a programmed straight line generatrix

As already mentioned, the directrix of a surface generated by milling is typically obtained as the envelope of a family of curves, most often a family of significantly prolonged cycloids. Consequently a schematic of milling machining can be used as an example in this case too.

Any of the machining schematics with a circle directrix obtained kinematically as the envelope of a family of curves and a straight line generatrix achieved kinematically (by copying, as the trajectory of a point, as the envelope of a family of curves or by rolling) can be converted into a machining schematic with a "programmed" straight line generatrix. Figure 10, however, exemplifies a slightly different schematic from the in figure 6, the generated surface being conical.

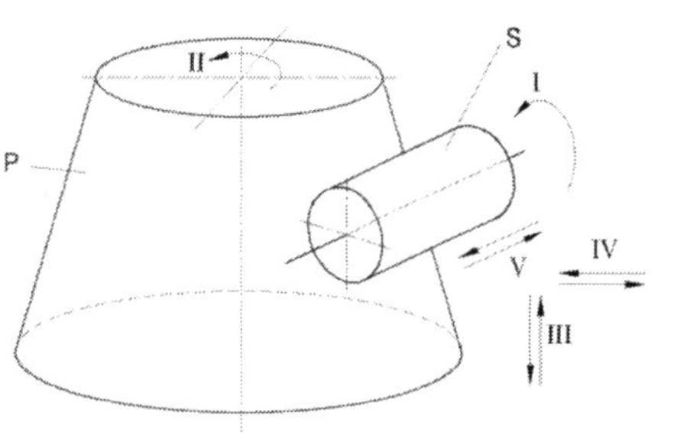

Figure 10. Machining schematic for milling a conical surface, example for the case Ci_{fc}&p.

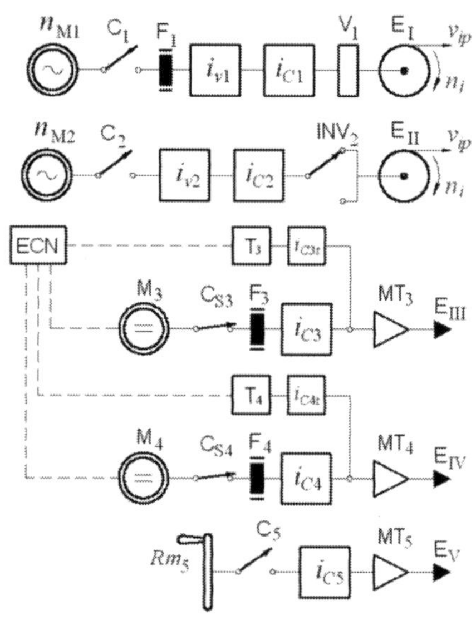

Figure 11. A kinematic diagram of principle for the case Ci_{fc}&p presented in figure 10.

Because the generatrix is "programmed", only motions III and IV, translations, need to be numerically controlled, as shown in the kinematic diagram of principle in figure 11. The main motion I and the circular feed motion of the workpiece II contribute together to obtaining the (continuously variably diameter) circle directrix described as the envelope of a family of curves.

Outside the machining time, motions III and IV can also have the role of auxiliary positioning motions. Only motion V, the tangential feed of the tool is exclusively an auxiliary positioning motion.

8. Conclusions

A body with a given geometry can be obtained by various manufacturing procedures, entailing, however, different surface quality, dimensional precision and (most importantly) productivity. Each body is characterised by one or more surfaces, simple or complex, as the case may be. A surface is characterised by a directrix and a generatrix, as well as by the modality of its physical achievement by means of the machines and tools uses for manufacturing.

At present six modalities of achieving the directrices and generatrices are known and deployed: materialized; generated by copying; kinematically generated as the trajectory of a point; kinematically generated as the envelope of a family of curves; generated by rolling; and programmed. Typically in literature generation "by copying" and "programmed generation" of the directrix and generatrix are not addressed distinctively, both being assumed as of the same nature. The recent evolution of industrial electronics and implicitly of machine-tools has determined a clear differentiation between the kinematics and construction of machine-tools (still) using generation by copying from a master and the kinematics of NC machine-tools.

The set of possibilities of achieving a surface characterised by a certain directrix and a certain generatrix is given by the set of possible combinations of the modalities of obtaining the respective directrices and generatrices. Parts of these combinations are found in practice and are described by literature, as illustrations of known modalities of surface generation. Any directrix-generatrix combination is theoretically possible, however only relatively few are also practically efficient. The systematic approach explores the entire set of generation possibilities, applied or not (yet) in practice, hence being oriented towards discovery and even inventics.

At least one machining schematic corresponds to each combination of directrix and generatrix, depending on their nature and on the modality of their obtaining. Many of these schematics are known and frequently deployed in practice, while others – possibly just as many ore more, the reserve of creativity being difficult to assess – are unknown. Revealing and analysing as many as possible of these is a contribution to the development of knowledge and can lead towards identifying new efficient machining schematics. Several kinematic diagrams of principle correspond to each machining schematic, as well as an undefined number of details kinematic diagrams of machine-tools capable of generating surface according to the respective machining schematic.

This paper represents a kinematic synthesis, oriented towards innovation and invention, and presents by means of examples – not necessarily known or usual – all six cases of generation of surfaces characterised by a straight line generatrix and a circle directrix obtained kinematically as the envelope of a family of curves. The obvious connection is emphasized between the machining schematic and the kinematic diagram of principle of a machine-tool, as a starting point in the kinematic synthesis of machine-tools with special destinations, either particular or more general. The subject is too rarely encountered in the landscape of current scientific research, and any new information represents a contribution to the development of knowledge in the field of machine-tool kinematic synthesis and beyond. In fact, the evolution of human society is in essence based on knowledge and seems to be aimed at developing and expanding knowledge, including its treasuring.

References

[1] Botez E 1977 *Machine-tools* (vol. I, *Theory*) (Bucharest: Tehnică)
[2] Cioară R 2005 *Secialisated and Special Machine-tools* (Brasov: Transilvania University Press)
[3] Botez E 1966 *Fundamentals of Surfaces Generation on Machine-tools* (Bucharest: Tehnică)
[4] Sandu I Gh 2008 *Surfaces Generation. Treaty* (Bucharest: Romanian Academy Press)
[5] Cioară R 2011 *Proc. of ModTech (Chisinau)* pp 225–228
[6] Cioară R 2011 *Proc. of the 9th Int. Conf. CHER 21 (Sozopol)* vol. 9 ed T Tashev et al (Sofia: Heron Press) p 196-200
[7] Weck M 1988 *Werkzeugmaschinen* (Düsseldorf: VDI Verlag)
[8] Belous V 1992 *The Inventics* (Iasi: "Gh. Asachi" Press)
[9] Stănciulescu T D Belous V and Moraru I 1998 *Treaty of Creatology* (Iasi: Performantica)

IOP Conf. Series: Materials Science and Engineering **161** (2016) 012033 doi:10.1088/1757-899X/161/1/012033

The impact of temperature changing on surface roughness of FFF process

D Chaidas[1], K Kitsakis[3], J Kechagias[1], S Maropoulos[2]

[1]Mechanical Engineering Department, Technological Educational Institute of Thessaly, Larissa, Greece
[2]Mechanical Engineering and Industrial Design Department, Technological Educational Institute of West Macedonia, Kozani, Greece
[3]Department of Mechanical Engineering, University of Western Macedonia, Kozani, Greece

Email: maropou@teiwm.gr

Abstract. The current study investigates the surface roughness of models produced by a 3D printer. All models were produced by addition of solid material, a process called fused filament fabrication (FFF): initial extrusion into plastic filament, second extrusion and trace-binding during the 3D printing process. A low cost 3D printer Ultimaker was used to print these items. Polylactic acid (PLA) was used as main polymer material for printing. The temperature was parameter under direct variations in order to examine if there was an influence on roughness of 3d printed models. The surface roughness parameters were: the average mean surface roughness (Ra, μm), the surface roughness depth (Rz, μm), the total height of the roughness profile (Rt, μm) and the arithmetic mean width of profile elements (Rsm, μm). The examination showed conditionality: as temperature was increased the surface roughness parameters were further decreased.

1. Introduction

Fused filament fabrication (FFF), can also be found as fused deposition modeling (FDM) that creates computer aided design (CAD) models to be 3D printed from Polylactic Acid (PLA) [1], Acrylonitrile Butadiene Styrene (ABS) or similar substrates.

This process is very easy and cheaper than other methods of rapid prototyping thus can be implemented in many applications: Architecture, Product Design, Biomedicine, Aerospace, Education [2] etc.

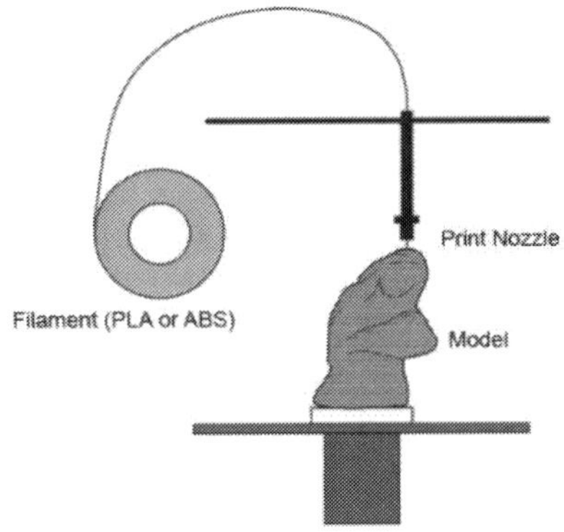

Figure 1. Fused Filament Fabrication process

The FFF 3D printers are very popular (appliance also in home 3D printing) during last years on the market. These machines deposit filament of a certain material (mainly thermoplastics) into thin layers which are cooled very fast (with cold fan) thus the next layer can be added into the previous layer in figure 1. Thermoplastic polymers are plastics which become semi-liquid above a specific temperature and return to a solid state when cooling down. The 3D model is made by positioning of the nozzle in relation to the Plexiglas print bed, while depositing fused PLA (or ABS) onto the built part [3]. All models were printed on this print bed. This Plexiglas bed may be stationary or may move along one the X, Y, Z axes. In this research a 3D printer with a stationary print bed was used. There are various printers with different technical specifications such as: quality of print (Layer resolution), build volume, print speed. The main difference of SLA (Stereolithography) 3D printers is that SLA printers are used for high detailed printed objects. One is able to decide which parameters can change to take the best print in FFF process [4]. In this research Cura software was chosen.

Cura (created by the makers of the Ultimaker printer) is a free software that prepares the model that is designed in the CAD program (Autocad in this case). This software converts STL (or OBJ) files (slice the model) into G-Code. The 3D printer must handle the file that is designed in order for the model to be translated into a suitable format.

Dividing the model into thin layers is called slicing. Cura is suitable software for Ultimaker printer for performing slicing. By this software lots of settings can be changed such as layer height or layer thickness. All CAM (Computer Aided Manufacturing) software need an STL file in order to turn a 3D model into a machine (3D printer) friendly format. This format is called G-Code and Cura converts STL files to G-Code.

2. 3D Printer

All 3D printers are composed of electromechanic components (hardware) and software applications that are developed by engineering companies [5]. Home edition 3D printers have easier functions than rapid prototyping machines. These are suitable for small offices (home 3D printing) and use less energy. Manufactured for printing small plastic objects, home edition 3D printing technology is a low cost procedure and there is continuous development so the price of these machines will be reduced compared to expensive rapid prototyping machines.

For measures in this research popular 3D printer called Ultimaker was used (figure 2). The Ultimaker Original (by Ultimaker Ltd) is an open source 3D printer and it is based on the Rep Rap (Replicating Rapid-prototyper) technique. This model has been awarded as the fastest and most accurate 3D printer in 2012. It can print complex drawings that have been designed in CAD software. Ultimaker supplies PLA (Polylactic Acid) and ABS (Acrylonitrile Butadiene Styrene) as materials for printing.

IOP Conf. Series: Materials Science and Engineering **161** (2016) 012033 doi:10.1088/1757-899X/161/1/012033

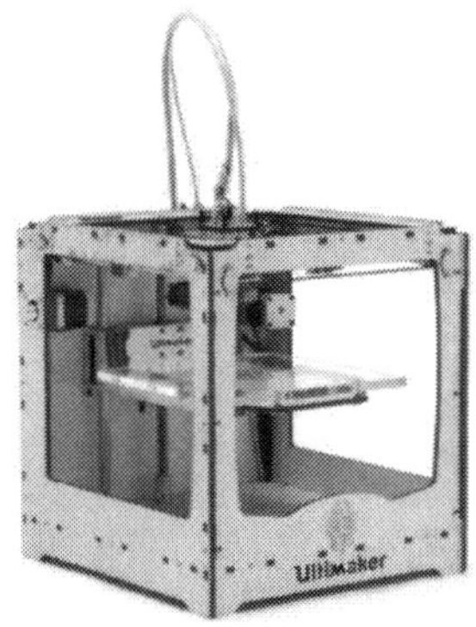

Figure 2. Ultimaker 3d Printer

3. Technical details

Some technical specifications of the printer are given: Layer resolution up to 20 µm, Build volume 21 cm × 21 cm × 20.5 cm, Print speed 30-300 mm/s, Travel speed 30-350 mm/s, Recommended filament diameter 2.85 mm, Nozzle diameter 0.4 mm, Printing technology Fused filament fabrication (FFF), Stand-alone SD-card printing, Frame dimension X Y Z: 35.7 cm × 34.2 cm × 38.8 cm, Operation nozzle temperature: 180-260 °C, Software: Cura - Official Ultimaker. PLA is a biodegradable polyester and non-toxic material suitable for many applications [6]. There are many applications such as medical usage, injection molding, packaging and 3D printing within past years. Its smell is very pleasant during printing. With this material you can get very good quality of prints in any color.

PLA or Polylactic acid derived from renewable resources like corn starch, tapioca roots or sugarcane. It is a very ecological thermoplastic polyester because it naturally degrades in outdoor environment conditions. Different parameters in RepRap additive manufacturing system for PLA models could be selected [7].

For this research a 3D model was designed (figure 3) by Autocad software and it was extracted in STL format. It was converted by Cura software to G-Code and the layer was built with Ultimaker 3D printer. Suitable settings have been put into software changing only two parameters (filament temperature and shell thickness) each time. The printing settings were: layer thickness (height - mm) - 0.2, shell thickness (mm): 3-2-1, fill density (%) - 20, print speed (mm/sec) - 100, printing temperature (°C): 210-220-230.

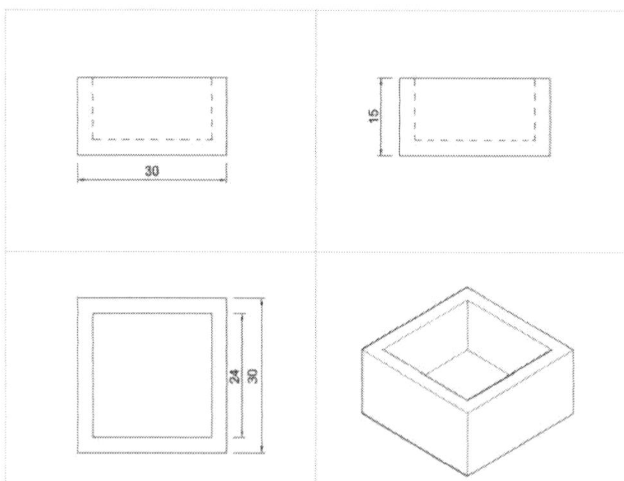

Figure 3. 3D model

Figure 4. (a) 1st item (3 mm, 210 °C), (b) 2nd item (3 mm, 220 °C), (c) 3rd item(3 mm, 230 °C), (d) 4th item (2 mm, 210 °C), (e) 5th item (2 mm, 220 °C), (f) 6th item (2 mm, 230 °C), (g) 7th item (1 mm, 210 °C),(h) 8th item (1 mm, 220 °C), (i) 9th item (1 mm, 230 °C)

Nine items were printed (figure 4), shell thickness:

- D1 = 1 mm;
- D2 = 2 mm;
- D3 = 3 mm;

and three different temperature printing parameters were used:

- T1= 210 °C;
- T2= 220 °C;
- T3= 230 °C .

All models were measured on each surface roughness by a device, which is a user friendly instrument for surface roughness measurement Mitutoyo Surftest SJ-210, designed as a handheld tool. By this device surface roughness waveforms can be viewed on a color LCD screen as well.

The surface texture parameters measured during this examination were the following:

- Ra (μm): the arithmetic mean surface roughness (arithmetical mean of the sums of all profile values). Ra is by far the most commonly used parameter in surface finish measurement and for general quality control. Despite its inherent limitations, it is easy to measure and offers a good overall description of the height characteristics of a surface profile [8].

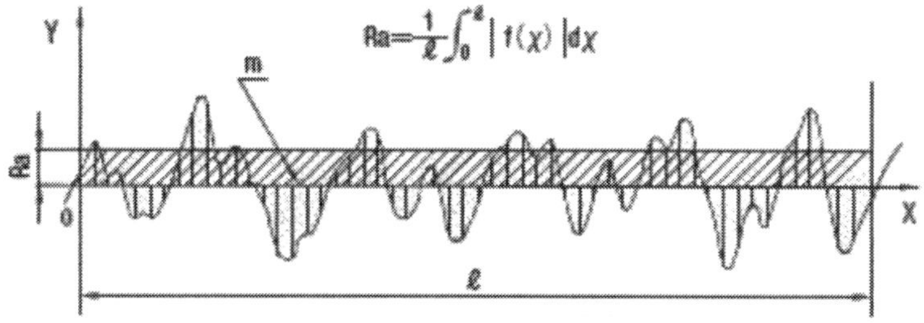

Figure 5. Surface texture parameters

- Rz (μm): surface roughness depth. It is the arithmetic mean value of the single roughness depths of consecutive sampling lengths.

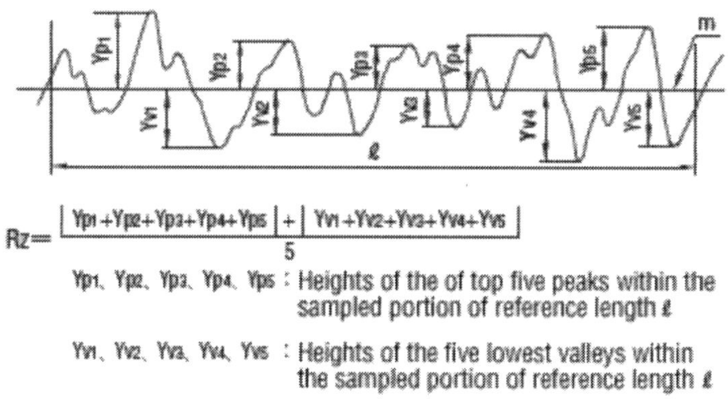

Figure 6. Surface texture parameters

- Rt (μm): total height of the roughness profile, i.e., the vertical distance between the highest peak and the lowest valley along the assessment length of the profile. This parameter is very sensitive to the high peaks or deep scratches.

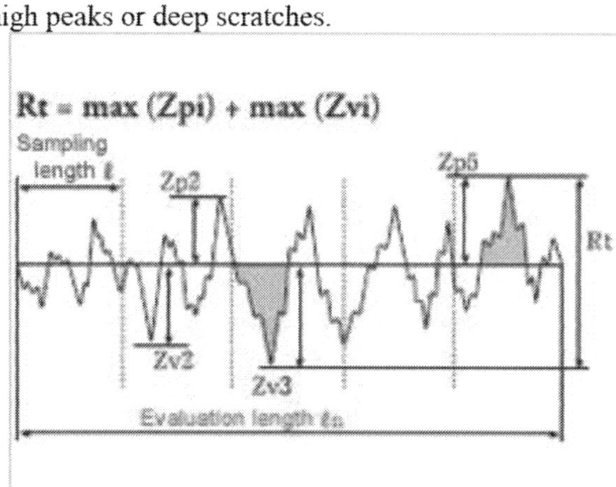

Figure 7. Surface texture parameters

- Rsm (µm): the arithmetic mean width of profile elements. It is the arithmetic mean value of the widths of the profile elements of the roughness profile, where a profile element is a peak and valley in the roughness profile.

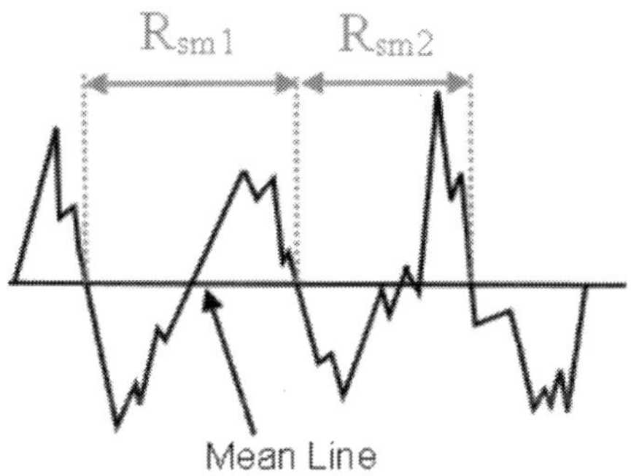

Figure 8. Surface texture parameters

The obtained results are in the following tables:

Table 1. Surface roughness measurements (1mm shell)

R (mean)	Model 1 (T_1, D_1)	Model 2 (T_2, D_2)	Model 3 (T_3, D_3)
R_a (µm)	17.00	14.66	14.08
R_z (µm)	83.47	75.96	63.98
R_t (µm)	123.80	103.90	68.86
R_{sm} (µm)	229.73	224.10	203.10

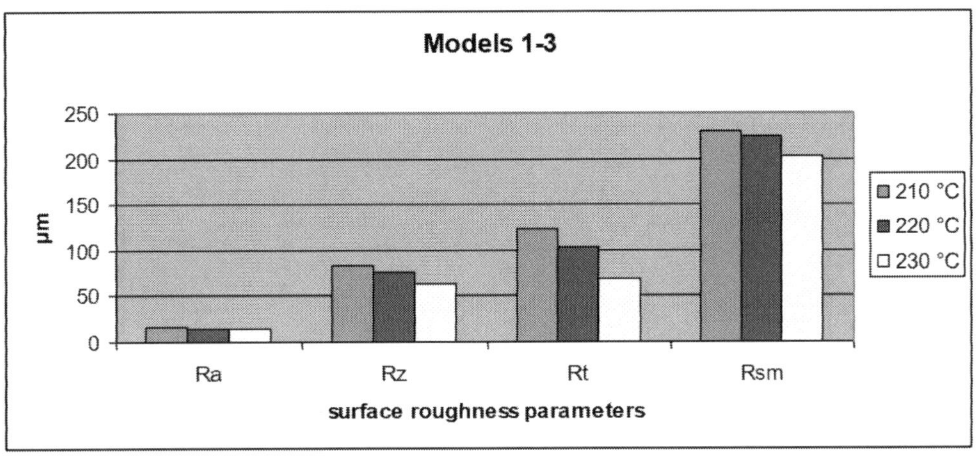

Table 2. Surface roughness measurements (2mm shell)

R (mean)	Model 4 (T_1, D_1)	Model 5 (T_2, D_2)	Model 6 (T_3, D_3)
R_a (µm)	16.48	13.04	12.84
R_z (µm)	85.74	62.20	62.26
R_t (µm)	85.74	83.07	69.22
R_{sm} (µm)	321	205.46	196.73

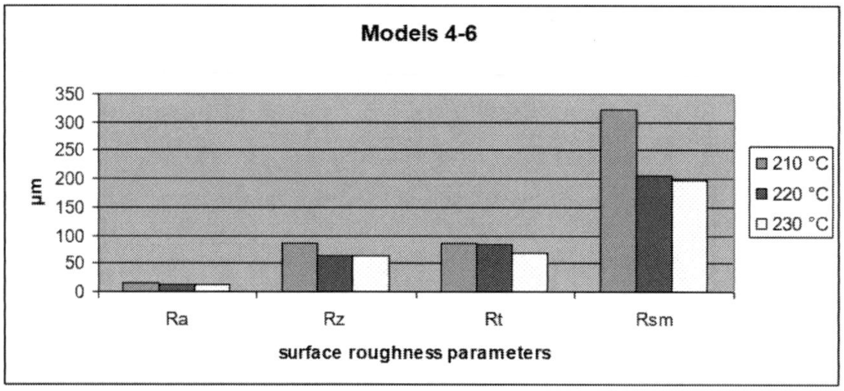

Table 3. Surface roughness measurements (3mm shell)

R (mean)	Model 7 (T_1, D_1)	Model 8 (T_2, D_2)	Model 9 (T_3, D_3)
R_a (µm)	21.77	19.49	12.98
R_z (µm)	128.48	98.21	57.84
R_t (µm)	74.39	145.60	65.62
R_{sm} (µm)	204.3	248.50	198.5

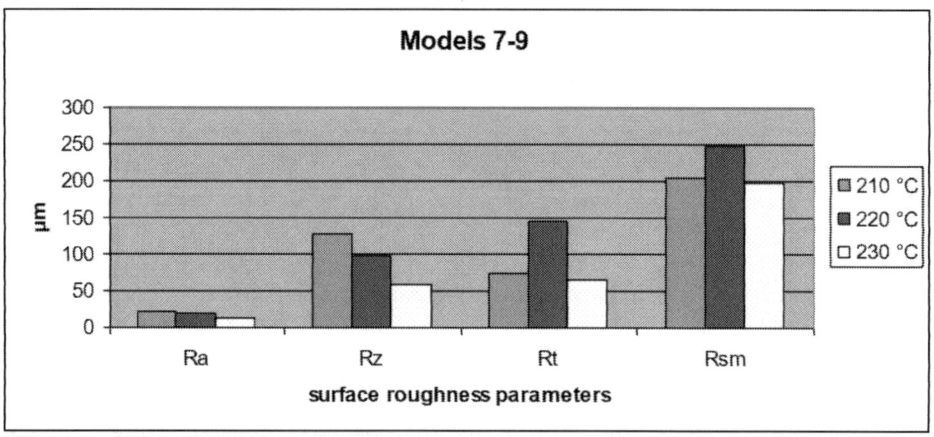

IOP Conf. Series: Materials Science and Engineering **161** (2016) 012033 doi:10.1088/1757-899X/161/1/012033

4. Conclusion

The examination of the influence of the temperature factor to the surface roughness was made. The results showed that all surface roughness parameters Ra, Rz, Rt, Rsm decreased (smoother surfaces when R parameters tend to zero) as the temperature increased from 210 °C to 230 °C.

References

[1] Skliutas E, Malinauskas M 2015 Fused filament fabrication of biodegradable polylactic acid threedimensional microstructures, *Conference: Open Readings*, Vilnius, **58**

[2] Schelly C, Anzalone G C, Wijnen B, Pearce J M 2015 Open-source 3-D printing Technologies for education: Bringing Additive Manufacturing to the Classroom *Journal of Visual Languages & Computing* vol 28 pp 226-237

[3] Jonusauskas L, Skliutas E, Malinauskas S B M 2015 Custom on demand 3D printing of functional microstructures, Lithuanian journal of physics, vol 55 no 4

[4] Moza Z, Kitsakis K, Kechagias J, Mastorakis N 2015 Optimizing Dimensional Accuracy of Fused Filament Fabrication using Taguchi Design, *14th International Conference on Instrumentation, Measurement, Circuits and Systems*, Salerno, Italy

[5] Romero L, Guerrero R, Espinosa M M , Jimenez M, Dominguez I A, Dominguez M 2014 Additive manufacturing with RepRap methodology: current situation and future prospects *25th Annual International Solid Freeform Fabrication (SFF) Symposium*, Univertity of Texas. Austin

[6] Martin O, Averous L 2001 Poly(lactic acid): plasticization and properties of biodegradable multiphase systems *Polymer* vol 42 issue 14 pp. 6209-6219

[7] Ciurana J, Sereno L, Valles E 2013 *Selecting Process Parameters in RepRap Additive Manufacturing System for PLA Scaffolds Manufacture* Procedia CIRP vol 5 pp 152-157

[8] Barclift M W, Williams C B 2012 Examining variability in the mechanical properties of parts manufactured via polyjet direct 3D printing *International Solid Freeform Fabrication Symposium* August 6 - 8, Austin, Texas

The dimensional design of machining technologies

A Toca[1], I Stingaci[2] and I Rusica[3]

[1] Associate professor, Technical University of Moldova, Chisinau, RM
[2] Senior lecturer, Technical University of Moldova, Chisinau, RM
[3] Associate professor, Technical University of Moldova, Chisinau, RM

E-mail: a_toca@mail.utm.md

Abstract. In the paper we analyze the mutual influence of constructive and technological dimensional links on conditions of formation of the machining accuracy sizes. It is shown, that the formation the sizes from technological locating datum surface demands higher accuracy of the technological sizes, but in this case, the machining allowances are more stable. At the formation the sizes by the means of transitions from technological locating datum surface to adjustment surface, the accuracy sizes is achieved without necessity of increase of an accuracy of the technological sizes, but thus, it is observed a growth of tolerances of the machining allowances and decreasing of the accuracy if some surface will not be machined. The dimensional optimality is not characterized only by the number of technological sizes, but it is necessary to take into account the growth (sometimes complicated) of the technological sizes accuracy (a case of formation of the sizes as closing link of dimensional chains).

1. Introduction.

Designing of technological process of a detail's manufacturing is very responsible stage of works. From quality of designing of technological processes, from their depth of study in many cases respects efficiency of the manufacture, size of initial costs, and also costs connected with elimination of mistakes depends. The dimensional analysis takes a special place during designing technology, concerning the stage of designing, and also the stage of technological designing [1-4].

The nature of the constructive dimensional links is defined by the designer who takes into account the technological features of machine tools, but not in detriment of their functionality. Therefore, carrying out of the dimensional analysis on a joint of these two phases it is very important with the purpose of improvement of dimensional adaptability to manufacturing of made details.

One of most complicated problems during the elaboration of the manufacturing process is the synthesis of dimensional structure. It is necessary to execute not only dimensional analysis of a developed manufacturing process but also to achieve on this basis the optimum dimensional structure.

The dimensional analysis of the manufacturing process is a methodology which offers the possibility: to establish these links at the elaboration stage of the manufacturing process, to establish the accuracy of machining, to choose correctly the machine tool's accuracy after the accuracy of machining, to establish the accuracy norms for the technological device, permits to trace out the narrow places, makes the manufacturing processes to be well-balanced. At the same time, the tackling of the dimensional analysis problems is done on the special positions, especially regarding the mechanism of the accuracy assurance [2, 3]. It treats, first of all, the effect of compensation of machining errors, which allows with the greater reliability to evaluate the machining accuracy considering the phenomenon of errors compensation.

IOP Conf. Series: Materials Science and Engineering **161** (2016) 012034 doi:10.1088/1757-899X/161/1/012034

2. The optimality of the machining technologies.

One of criteria of optimality of the technologies has a structural nature. The manufacturing process is considered optimum, if in the structure of all technological dimensional chains, the number of the technological sizes is minimum [2]. This condition is respected if for each constructive size there corresponds within the technological dimensional chain the unique technological size (figure 1), and in dimensional chains for machining allowances, each machining allowance is determined by two technological sizes or one technological size and one size on blank (figure 2). The last has two technological sizes formed at the other phase of a manufacturing process.

The minimum possible number of the technological sizes $N_{T_{MIN}}$ is determined from the relation $N_{T_{MIN}} = N_C + 2 \cdot N_{Ad}$, where N_C - number of the constructive sizes, N_{Ad} - number of machining

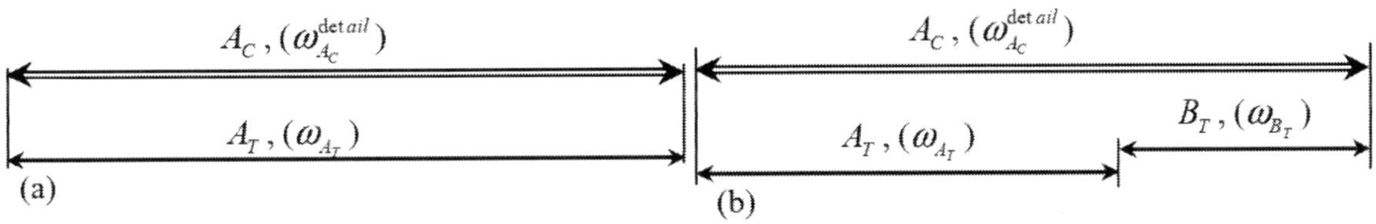

Figure 1. The constructive size is submitted: (a) - only by the technological size, $\omega_{A_C}^{\det ail} = \omega_{A_T}$; (b) - by the technological size and one (it is possible more) in already existing (historical) size, $\omega_{A_C}^{\det ail} = \omega_{A_T} + \omega_{B_T}$

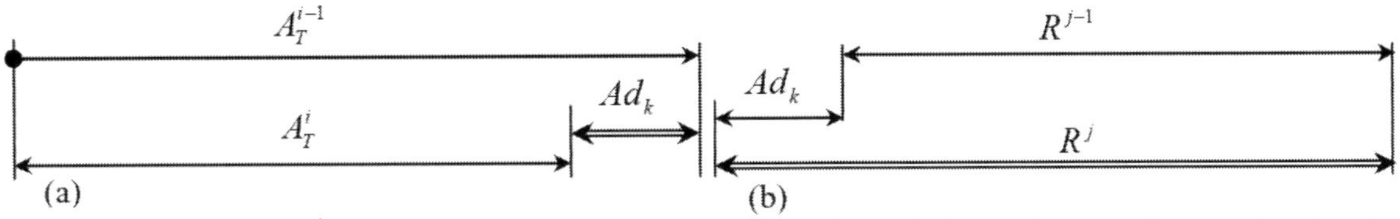

Figure 2. Two opportunities of transformation of the technological sizes: (a) - it is direct and under the control over technological base, $\omega_{Ad_k} = \omega_{Ad_T^{i-1}} + \omega_{A_T^i}$; (b) - it is not direct, as consequence of direct formation of other size, $\omega_{R^j} = \omega_{R^{j-1}} + \omega_{Ad_k}$.

allowances. The optimum manufacturing process becomes ideal, if each surface is processed only once, thus providing the final constructive size. It is necessary to note, that the dimensional optimality is not characterized only by the number N_T, but it is necessary to take into account the growth (sometimes complicated) of the technological sizes accuracy (a case of formation of the sizes as closing link of dimensional chains).

3. Interaction of the constructive and technological dimensional links.

During machining, the technological dimensional links are subordinated to constructive dimensional links, in a degree necessary for deriving on a detail the real sizes which are shaped within the limits of the appropriate tolerances. Thus it is insufficiently clear, the mechanism of formation of dimensional technological chain, the correlation of structures of dimensional constructive and technological links, the mechanism to form the favorable conditions of size's accuracy assurance.

The setting-up size is formed as a closing link of the dimensional adjustment chain, the components and their structure depend on the way of adjustment. The adjustment size goes into, alongside with others, as a making link into the dimensional chain, closing size which is a technological size. Technological dimensional links reflect the technological opportunities of machine tools. In narrow sense, these technological opportunities are reduced to conditions of formation the linear, circular etc. distances - the sizes between a processed surface and technological locating datum surface or between two processed surfaces.

At its turn the technological sizes are links of a dimensional chain with the closing size – the derivative constructive size. The constructive sizes with the attributes (nominal, limit deviations) are present only on a drawing of a detail. The identical sizes on drawings of the blank during its transformation to a detail are derivatives constructive sizes and most frequently used as the closing sizes, substituted by the technological sizes.

It represents an interest to evaluate the conditions of accuracy sizes formation depending on structures of dimensional constructive and technological links. There were analyzed examples of details machining with different constructive dimensional links (figure 3) with a formation of the sizes from technological locating datum surface and from adjustment surfaces.

The formation of the sizes from technological locating datum surface is submitted on figure 4.

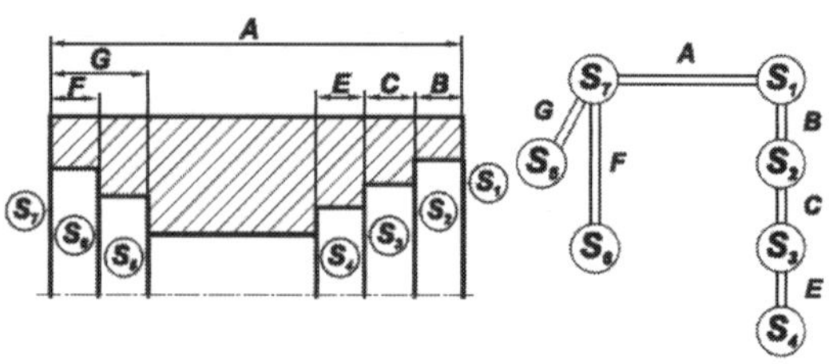

Figure 3. Example of constructive variants of mashined details and the appropriate structures of the dimensional technological links

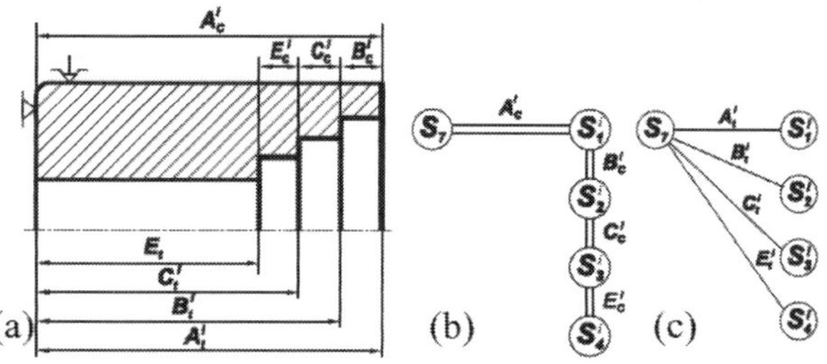

Figure 4. Formation of the sizes from technological locating datum surface. (a) - the operational sketch, (b) - constructive dimensional links, (c) - technological dimensional links

Drawing up of dimensional chains consists: in the choice of the derivative constructive size from the scheme of constructive links as closing; in definition of that technological size by means of which it will be assured (from the scheme of technological dimensional links); in the choice of the other technological sizes already generated to close the dimensional chain. At a stage i of machining the derivative constructive size A_c^i coincides with the technological size A_t^i, therefore the dimensional chain contains only two links and the technological accuracy of the machine tool is transferred to a detail lost-free (figure 5 a). For the size B_c^i (as for the sizes C_c^i and E_c^i with identical dimensional links) the dimensional chain powers up three links (figure 5 b). In a ratio of tolerances ω it is possible to write: $\omega_{A_t^i} \leq \omega_{A_c^i}$; $\omega_{B_t^i} + \omega_{A_t^i} \leq \omega_{B_c^i}$; $\omega_{C_t^i} + \omega_{B_t^i} \leq \omega_{C_c^i}$; $\omega_{E_t^i} + \omega_{C_t^i} \leq \omega_{E_c^i}$

Obtained inequalities specify necessity to establish the accuracy parameters $\omega_{A_t^i}, \omega_{B_t^i}, \omega_{C_t^i}$ and $\omega_{E_t^i}$ of the technological sizes A_t^i, B_t^i, C_t^i and E_t^i, and at the same time to reach the accuracy of the derivative constructive sizes it is necessary to execute the technological sizes with the greater accuracy.

At repeated machining all surfaces at a stage $i+1$ under the same scheme inequalities such as (1) will be fair. From processed surfaces S_1, S_2, S_3 and S_4 the machining allowances are deleted and their tolerances identified by the accuracy of the same technological sizes at two sequential stages (figure 6). After machining of surface S_1 the size B_c possess the value B_c^{int} with tolerance $\omega_{B_c^{\text{int}}} = \omega_{A_t^{i+1}} + \omega_{B_t^i}$. It is visible, that the accuracy increased by value $\omega_{A_t^i} - \omega_{A_t^{i+1}}$ (figure 7). Thus the sizes C_c and E_c, do not vary. At machining the surfaces S_1 and S_2 the tolerance of the size B_c^{i+1} possess the value $\omega_{B_c^{i+1}} = \omega_{A_t^{i+1}} + \omega_{B_t^{i+1}}$ also the accuracy of the size is a little improved C_c - $\omega_{C_c^{\text{int}}} = \omega_{B_t^{i+1}} + \omega_{C_t^i}$. In conclusion, for the sequential machining of surfaces we shall receive:

S_1 : $\omega_{A_t^{i+1}} \le \omega_{A_c^{i+1}}$

$\omega_{Ad_1^{i+1}} = (\omega_{A_t^i} + \omega_{A_t^{i+1}})$,

$\omega_{B_c^{\text{int}}} = \omega_{A_t^{i+1}} + \omega_{B_t^i}$

S_1, S_2: $\omega_{B_t^{i+1}} + \omega_{A_t^{i+1}} \le \omega_{B_c^{i+1}}$

$\omega_{Ad_2^{i+1}} = (\omega_{B_t^i} + \omega_{B_t^{i+1}})$,

$\omega_{C_c^{\text{int}}} = \omega_{B_t^{i+1}} + \omega_{C_t^i}$

$S_1 \dots S_3$: $\omega_{C_t^{i+1}} + \omega_{B_t^{i+1}} \le \omega_{C_c^{i+1}}$

$\omega_{Ad_3^{i+1}} = (\omega_{C_t^i} + \omega_{C_t^{i+1}})$

$\omega_{E_c^{\text{int}}} = \omega_{C_t^{i+1}} + \omega_{E_t^i}$

$S_1 \dots S_4$: $\omega_{E_t^{i+1}} + \omega_{C_t^{i+1}} \le \omega_{E_c^{i+1}}$

$\omega_{Ad_4^{i+1}} = (\omega_{E_t^i} + \omega_{E_t^{i+1}})$

$\omega_{Ad_4^{i+1}} = (\omega_{E_t^i} + \omega_{E_t^{i+1}})$

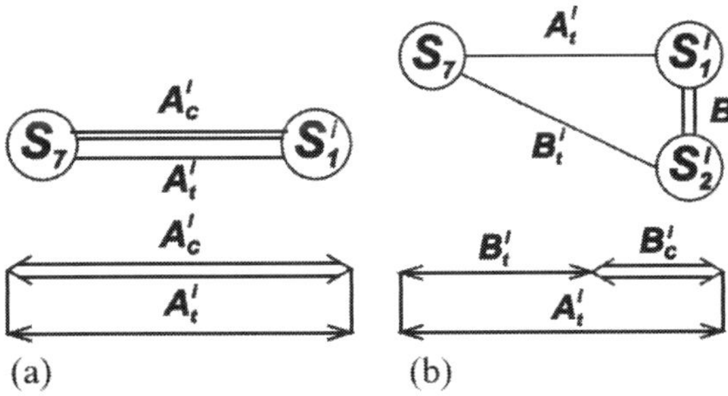

Figure 5. Dimensional chains for the sizes A_c^i (a) and B_c^i (b)

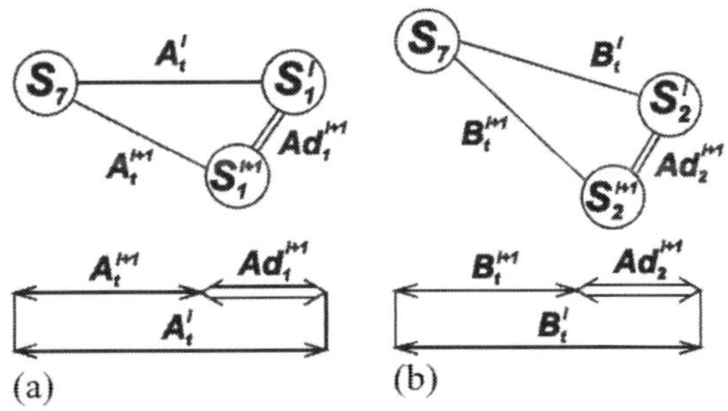

Figure 6. The dimensional chains for intermediate machining allowances

The machining of detail with formation the sizes from adjustment surfaces is submitted on figure 8. The adjustment surfaces for the sizes B_c, C_c and E_c correspond sequentially to a surface S_1, S_2 and S_3. It is visible, that schemes of constructive and technological dimensional links completely coincide. As a consequence, the dimensional chain for all derivative constructive sizes are powered up with only two links and a technological accuracy of the machine tool are transferred to a detail lost-free: $\omega_{A_t^i} \le \omega_{A_c^i}$; $\omega_{B_t^i} \le \omega_{B_c^i}$; $\omega_{C_t^i} \le \omega_{C_c^i}$; $\omega_{E_t^i} \le \omega_{E_c^i}$.

For a machining stage $i+1$, under the same scheme inequalities are identical and fair. At machining a surface S_1 the machining allowances Ad_1 is deleted, the size A_c passes in a condition A_c^{i+1} (figure 6 a), the size B_c passes in an intermediate condition B_c^{int}. The machining of the surface S_2 with the elimination of the machining allowances Ad_2 provokes the transition of the size B_c from the condition B_c^{int} to a condition B_c^{i+1}, and the size C_c passes in an intermediate condition C_c^{int}.

The solving of the dimensional chains taking in the consideration the sequential machining of

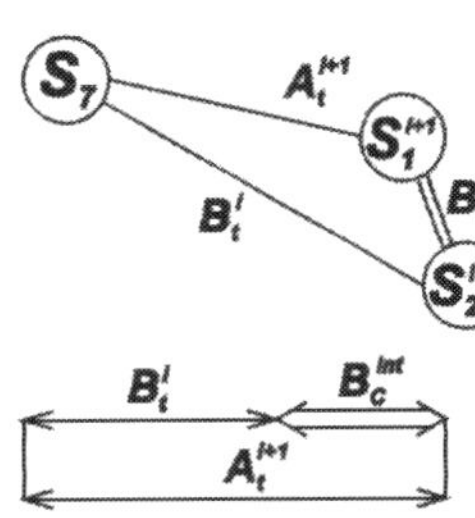

Figure 7. The dimensional chains for intermediate sizes

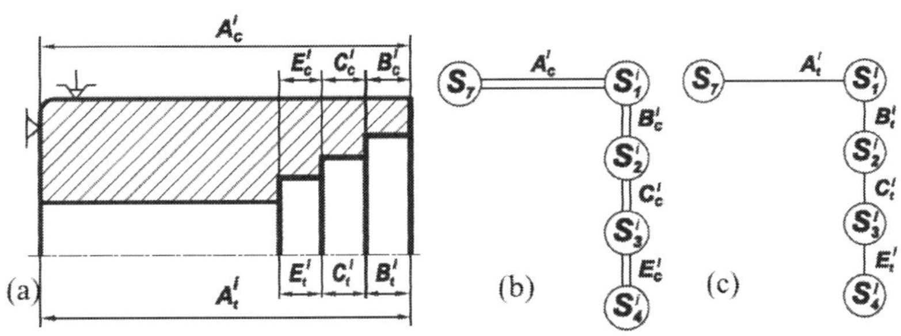

Figure 8. Formation of the sizes from adjustment surfaces. (a) - the operational sketch, (b) - constructive dimensional link, (c) - technological dimensional links

surfaces we shall receive:

S_1:

$$\omega_{A_t^{i+1}} \le \omega_{A_c^{i+1}}$$
$$\omega_{Ad_1^{i+1}} = (\omega_{A_t^i} + \omega_{A_t^{i+1}})$$
$$\omega_{B_c^{\text{int}}} = \omega_{B_t^i} + \omega_{Ad_1^{i+1}}$$
$$\omega_{B_c^{\text{int}}} = \omega_{B_t^i} + (\omega_{A_t^i} + \omega_{A_t^{i+1}})$$

S_1, S_2:

$$\omega_{B_t^{i+1}} \le \omega_{B_c^{i+1}}$$
$$\omega_{Ad_2^{i+1}} = \omega_{B_t^{i+1}} + \omega_{B_c^{\text{int}}}$$
$$\omega_{Ad_2^{i+1}} = (\omega_{B_t^i} + \omega_{B_t^{i+1}}) + (\omega_{A_t^i} + \omega_{A_t^{i+1}})$$
$$\omega_{Ad_2^{i+1}} = (\omega_{B_t^i} + \omega_{B_t^{i+1}}) + \omega_{Ad_1^{i+1}}$$
$$\omega_{C_c^{\text{int}}} = \omega_{C_t^i} + (\omega_{B_t^i} + \omega_{B_t^{i+1}}) + (\omega_{A_t^i} + \omega_{A_t^{i+1}})$$

$S_1 \ldots S_3$:

$$\omega_{C_t^{i+1}} \le \omega_{C_c^{i+1}}$$
$$\omega_{Ad_3^{i+1}} = \omega_{C_t^{i+1}} + \omega_{C_c^{\text{int}}}$$
$$\omega_{Ad_3^{i+1}} = (\omega_{C_t^i} + \omega_{C_t^{i+1}}) + (\omega_{B_t^i} + \omega_{B_t^{i+1}}) + (\omega_{A_t^i} + \omega_{A_t^{i+1}})$$
$$\omega_{Ad_3^{i+1}} = (\omega_{C_t^i} + \omega_{C_t^{i+1}}) + \omega_{Ad_2^{i+1}}$$
$$\omega_{E_c^{\text{int}}} = \omega_{E_t^i} + \omega_{Ad_3^{i+1}}$$
$$\omega_{E_c^{\text{int}}} = \omega_{E_t^i} + (\omega_{C_t^i} + \omega_{C_t^{i+1}}) + (\omega_{B_t^i} + \omega_{B_t^{i+1}}) + (\omega_{A_t^i} + \omega_{A_t^{i+1}})$$

$S_1, \ldots S_4$:

$$\omega_{E_t^{i+1}} \le \omega_{E_c^{i+1}}$$
$$\omega_{Ad_4^{i+1}} = \omega_{E_t^i} + \omega_{E_c^{\text{int}}}$$
$$\omega_{Ad_4^{i+1}} = (\omega_{E_t^i} + \omega_{E_t^{i+1}}) + (\omega_{C_t^i} + \omega_{C_t^{i+1}}) + (\omega_{B_t^i} + \omega_{B_t^{i+1}}) + (\omega_{A_t^i} + \omega_{A_t^{i+1}})$$
$$\omega_{Ad_4^{i+1}} = (\omega_{E_t^i} + \omega_{E_t^{i+1}}) + \omega_{Ad_3^{i+1}}$$

As the surfaces are formed sequentially from each other, the phenomenon of error's compensation operates. Then it is possible to write: $\omega_{A_t^i} \le \omega_{A_c^i}$; $\omega_{B_t^i} \le \omega_{B_c^i} + 2\omega_{1^i2^i}^{comp}$; $\omega_{C_t^i} \le \omega_{C_c^i} + 2\omega_{2^i3^i}^{comp}$; $\omega_{E_t^i} \le \omega_{E_c^i} + 2\omega_{3^i4^i}^{comp}$. For a machining stage $i+1$, under the same scheme inequalities are identical and fair. The solving of the dimensional chains taking in the consideration the sequential machining of surfaces and the phenomenon of error's compensation we shall receive:

S_1:

$$\omega_{A_t^{i+1}} \le \omega_{A_c^{i+1}}$$
$$\omega_{Ad_1^{i+1}} = (\omega_{A_t^i} + \omega_{A_t^{i+1}}) - 2\omega_{1^i1^{i+1}}^{comp}$$
$$\omega_{B_c^{\text{int}}} = \omega_{B_t^i} + \omega_{Ad_1^{i+1}}$$
$$\omega_{B_c^{\text{int}}} = \omega_{B_t^i} - 2\omega_{1^i2^i}^{cmp} + (\omega_{A_t^i} + \omega_{A_t^{i+1}}) - 2\omega_{1^i1^{i+1}}^{comp}$$

S_1, S_2:

$$\omega_{B_t^{i+1}} \le \omega_{B_c^{i+1}} + 2\omega_{1^{i+1}2^{i+1}}^{comp}$$
$$\omega_{Ad_2^{i+1}} = (\omega_{B_t^i} + \omega_{B_t^{i+1}}) - 2\omega_{2^i2^{i+1}}^{comp} + (\omega_{A_t^i} + \omega_{A_t^{i+1}}) - 2\omega_{1^i1^{i+1}}^{comp}$$
$$\omega_{Ad_2^{i+1}} = (\omega_{B_t^i} + \omega_{B_t^{i+1}}) - 2\omega_{2^i2^{i+1}}^{comp} + \omega_{Ad_1^{i+1}}$$
$$\omega_{C_c^{\text{int}}} = \omega_{C_t^i} - 2\omega_{2^i3^i}^{cmp} + (\omega_{B_t^i} + \omega_{B_t^{i+1}}) - 2\omega_{2^i2^{i+1}}^{cmp} + (\omega_{A_t^i} + \omega_{A_t^{i+1}}) - 2\omega_{1^i1^{i+1}}^{cmp}$$

$$S_1 \dots S_3: \quad \omega_{C_t^{i+1}} \le \omega_{C_c^{i+1}} + 2\omega^{comp}_{2^{i+1}3^{i+1}}$$

$$\omega_{Ad_3^{i+1}} = (\omega_{C_t^i} + \omega_{C_t^{i+1}}) - 2\omega^{comp}_{3^i 3^{i+1}} + (\omega_{B_t^i} + \omega_{B_t^{i+1}}) - 2\omega^{comp}_{2^i 2^{i+1}} +$$
$$+ (\omega_{A_t^i} + \omega_{A_t^{i+1}}) - 2\omega^{comp}_{1^i 1^{i+1}}$$

$$\omega_{Ad_3^{i+1}} = (\omega_{C_t^i} + \omega_{C_t^{i+1}}) - 2\omega^{comp}_{3^i 3^{i+1}} + \omega_{Ad_2^{i+1}}$$

$$\omega_{E_c^{\text{int}}} = \omega_{E_t^i} - 2\omega^{comp}_{3^i 4^i} + (\omega_{C_t^i} + \omega_{C_t^{i+1}}) - 2\omega^{comp}_{3^i 3^{i+1}} +$$
$$+ (\omega_{B_t^i} + \omega_{B_t^{i+1}}) - 2\omega^{comp}_{2^i 2^{i+1}} + (\omega_{A_t^i} + \omega_{A_t^{i+1}}) - 2\omega^{comp}_{1^i 1^{i+1}}$$

$$S_1, \dots S_4: \quad \omega_{E_t^{i+1}} \le \omega_{E_c^{i+1}} + 2\omega^{comp}_{3^{i+1}4^{i+1}}$$

$$\omega_{Ad_4^{i+1}} = (\omega_{E_t^i} + \omega_{E_t^{i+1}}) - 2\omega^{comp}_{4^i 4^{i+1}} + (\omega_{C_t^i} + \omega_{C_t^{i+1}}) - 2\omega^{comp}_{3^i 3^{i+1}} +$$
$$+ (\omega_{B_t^i} + \omega_{B_t^{i+1}}) - 2\omega^{comp}_{2^i 2^{i+1}} + (\omega_{A_t^i} + \omega_{A_t^{i+1}}) - 2\omega^{comp}_{3^i 3^{i+1}}$$

$$\omega_{Ad_4^{i+1}} = (\omega_{E_t^i} + \omega_{E_t^{i+1}}) - 2\omega^{comp}_{4^i 4^{i+1}} + \omega_{Ad_3^{i+1}}$$

4. Conclusions.

The schemes of machining with a formation of sizes from technological locating datum surface are characterized by unfavorable conditions of accuracy formation for constructive sizes, but the machining allowances are more stable.

The schemes of machining with substitution of technological locating datum surface by adjustment surfaces are characterized by favorable conditions for accuracy formation of the constructive sizes as dimensional links power up only on two elements. The machining allowances are instable because their tolerance depends on multiples change of bases, and lowering of the accuracy if the line-up of repeated machining is broken off. The machining allowances also depends on the number of operation elements.

To develop of optimal dimensional technologies the following optimality criteria to be considered: the number of the technological sizes is minimum and equal to the number of design sizes; in the dimensional chains in which the machining allowance is the closing link the number of units should be equal to 3; the amount of machining allowances should be minimal, and the value of each of them shall not result in additional operation element; the number of operation elements to assure the accuracy of each surface should be as low as possible.

References

[1] Kuznetsov P M 2014 Operativnaâ razrabotka sistem tehnologičeskogo proektirovaniâ v mašinostroitel'nom proizvodstve. *Tehnologiâ mašinostroeniâ.* Moskva. **5**, 40 – 43

[2] Mokrushin Y A and Shamin V Y 2013 Necessity of improving technological process planning in machine building. *University proceedings. Volga region. Engineering sciences. Machine sciences and building.* **1**, 123 – 130

[3] Rahmilevich E G 2013 Avtomatizaciâ rasčetov količestvennyh pokazatelei tehnologičnosti s ispol'zovaniem konstryktorsko - tehnologičeskoi parametrizacii. *Tehnologiâ mašinostroeniâ.* Moskva. **1**, 53 – 56

[4] González Contreras F 2012. An improved tolerance charting technique using an analysis of setup capability. *The International Journal of Advanced Manufacturing Technology*, Volume **62**, Issue **9 – 12**, Springer, 1205 - 1218

Research regarding stiffness optimization of wires used for joints actuation from an elephant's trunk robotic arm

C Ciofu and G Stan

Vasile Alecsandri University of Bacau, Department of Industrial Engineering, Calea Marasesti Street, No.157, 600115 Bacau, Romania

E-mail: blgcatalina@yahoo.com

Abstract. Elephant's trunk robotic arms driven by wires and pulley mechanisms have issues with wires stiffness because of the entailed elastic deformations that is causing errors of positioning. Static and dynamic loads from each joint of the robotic arm affect the stiffness of driving wires and precision positioning. The influence of wires elastic deformation on precision positioning decreases with the increasing of wires stiffness by using different pre-tensioning devices. In this paper, we analyze the variation of driving wires stiffness particularly to each wire driven joint. We obtain optimum wires stiffness variation by using an analytical method that highlights the efficiency of pre-tensioning mechanism. The analysis of driving wires stiffness is necessary for taking appropriate optimization measures of robotic arm dynamic behavior and, thus, for decreasing positioning errors of the elephant's trunk robotic arm with inner actuation through wires/cables.

1. Introduction

Robotic structures with joints actuated by wire pulleys have solely the possibility of wires pulling, not pushing. Therefore, the actuating wires are constantly in a tension that increases elastic deformations. Wire pulleys are used for actuating specific robotic arms, flexible positioning and high workspace with respect to their size. The structural forces and joints torques moments which change while robotic arms are operating, lead also to tension forces and affect wire stiffness. Elephant's trunk robotic arms with wire-driven joints and pulley mechanism have stiffness issues of the structure because of wires elastic deformations that affect positioning precision. The influence of elastic deformations on positioning precision is generally reduced by different devices along each actuating wire of joints to optimize tension distribution and, thus, to increase the robotic arm stiffness [1, 2, 3].

In this paper, we consider a robotic arm with 5 degrees of freedom, serial connected joints and the performed motions similar to the elephant's trunk. In the analysis of wires elastic deformations, we consider no device for automatically control of driving wires stiffness. We obtain variation of wires stiffness depending on the particular wire tensions from each actuated joint. The tensile forces applied to the actuating wires comprises of the total loads given by: moments of resistance, frictions between driving wires and surfaces from joints interior, and rotated masses by each joint. The optimum wires stiffness is analytically determined for driving wire used to actuate each of five joints from the elephant's trunk robotic arm.

We use an analytical method to optimize wires stiffness by pre-tensions. The stiffness of wires used for joints driving mechanism from elephant's trunk robotic arms influence dynamic behavior of the arm structure and has been of interest for different research personnel. Some research teams have suggested stiffness-oriented tension resolution algorithms or wire tension state sensing components to optimize positioning precision [4, 5]. However, all these control devices of wires stiffness make the robotic arm to be heavier, robust and, thus, inappropriate for manipulation in narrow spaces.

Inner actuation with wires and pulley mechanism is advantageous because of miniaturization possibility of robotic arms diameter to achieve thinner robotic structure for manipulation in narrow spaces. For the specific robotic arm structure, we suggest stiffness optimization by pre-tensioning the actuating wires. The obtained results of wire stiffness influence on elastic deformations are necessary for taking suitable measures of dynamic behavior optimization and prove method viability.

2. Tension analysis of driving wires from an elephant's trunk robotic arm

The particular structure of elephant's trunk robotic arm enables joints actuation by pulley mechanisms with tensioned wires. Driving system consists of AC servomotors to actuate the joints in a closed loop control. Elastic deformations of driving wires causes positioning errors of elephant's trunk robotic arm that are outside of the feedback control system. Therefore, tension analysis of driving wires is necessary to evaluate variation of elastic deformations that also affects stiffness of wires and, of motion transmission system. The Hooke's law given in equation (1) defines elastic deformations Δl_i of driving wires.

$$\Delta l_i = (T_i \cdot l_i)/(E \cdot A_w) \tag{1}$$

We noted: T_i - tensile load of the wire that drives joints i, l_i - length of the wire that drives joints i before loading, E - Young modulus of the driving wires made of stainless steel with Teflon coating, şi A_w - cross sectional area of driving wire with diameter of 0.38 *mm*.

From Hooke's law results that we can decrease elastic deformation either by decreasing the length of wire or of tensile loads, either by increasing wires cross sectional area. We consider in the structure designing stage a minimum length of driving wires according to the dimensional limitations of structure components. Diameter size of driving wire is restricted to the lowest possible value to achieve minimum arm diameter and flexible manipulation in very narrow spaces.

Moreover, from the variation law of wires stiffness to bending given by parameter R_f in equation (2) results that simultaneously with the increase of wires diameter d_w we have an increase of wires stiffness too. However, robotic arm structure entails that driving wire diameter allowing the winding on circular surfaces from structure interior and this requirement unfits the increase of driving wire diameter.

$$R_f = E \cdot I = E \cdot (\pi \cdot d_w^4)/64 \tag{2}$$

In equation (2) we noted I as the inertia moment of wire cross section about neutral axis and d_w as the driving wire diameter.

With all the mentioned above requirements we consider optimization of driving wires stiffness using a pre-tensioning device for reducing tensile loads T_i that are causing elastic deformations of driving wires along their length on different sections l_i, l_i^f şi l_i'' between points A, B, C and D from figure 1 and figure 2. The actuation of joints by pulley mechanical with wires for motion transmission supposes several tensile loading T_i of driving wires at the lower ends. Tensile loads T_i' at the upper ends of wires that drives joints i are smaller than tensile loads T_i, equation (3). Parameter η defines the coefficient of efficiency for each transmission mechanism.

$$T_i' = \eta^{-1} \cdot T_i \tag{3}$$

Stiffness optimization of active wires, used for driving the joints from elephant's trunk robotic arm, consists in using a pre-tensioning mechanism, which reduces elastic deformations so that positioning errors are at minimum values while operating. After applying a pre-tension load $T_{0,i}$ to the active wire of joint i there is an additional elastic deformation $\Delta l_{s,i}$ as shown in figure 2. While the robotic arm is operating, an extra tension load $T_{a,i}$ is causing elastic deformation $\Delta l_{a,i}$ besides pre-tensioning load.

In addition, efficiency coefficient of the transmission mechanism (wire-pulley) affects tension loads from the two ends of driving wire as given in equation (4) for pre-tension loads and in equation (5) for extra tension loads.

$$T'_{0,i} = \eta^{-1} \cdot T_{0,i} \tag{4}$$

$$T'_{a,i} = \eta^{-1} \cdot T_{a,i} \tag{5}$$

From equations (4-5) we find elastic deformations caused by pre-tension loads and extra loads mentioned above, equations (6-7).

$$\Delta l'_{s,i} = \eta^{-1} \cdot \Delta l_{s,i} \tag{6}$$

$$\Delta l'_{a,i} = \eta^{-1} \cdot \Delta l_{a,i} \tag{7}$$

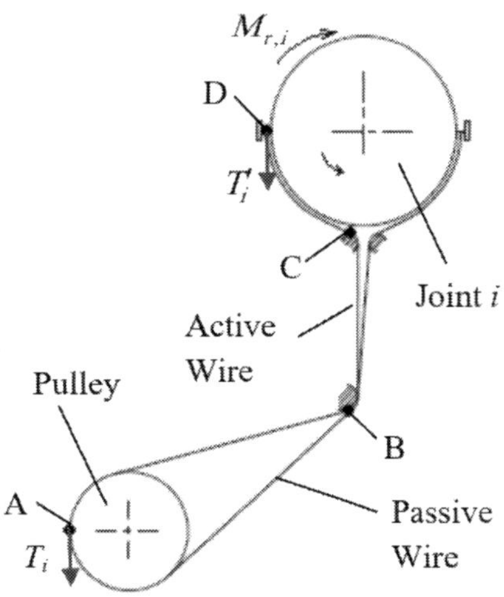

Figure 1. Tension loads of active wire that drives joint i.

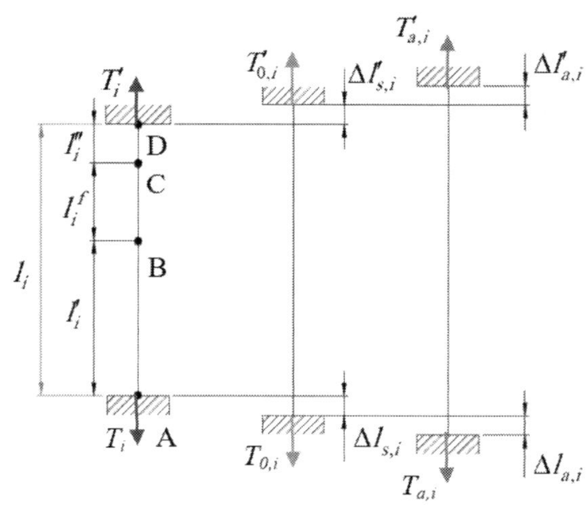

Figure 2. Elastic deformations at active wire ends.

The values of tension loads T_i are known from features dimensioning of transmission mechanism as given in table 1. We have specific features of the driving wire, E=252523·10[6] *N/m* and A_w=0,114 mm^2, and using equation (1) we obtain maximum elastic deformations caused by tension loads T_i as given in table 1.

Table 1. Specific features of the driving wire *i*

i	T_i (N)	l_i (mm)	Δl_i (mm)
1	4	200	0.000358
2	3	250	0.000336
3	2	300	0.000269
4	1	350	0.000156
5	0.02	400	0.000036

Elastic deformations Δl_i of wires that actuate the five joints from elephant's trunk robotic arm vary as shown in the diagram from figure 3.

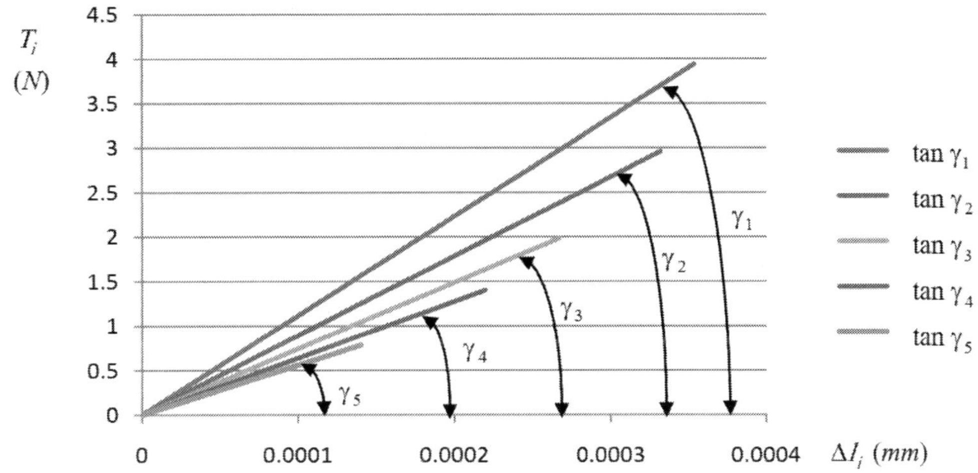

Figure 3. Elastic deformations before pre-tensioning of active wires.

The ratio between tension loads T_i and elastic deformations Δl_i define stiffness of the wire that actuates joint i, given as $\tan\gamma_i$. From the variation diagram of elastic deformation, figure 3, we find that wire stiffness decrease as we go from the first joint to the fifth joint because the length of driving wires l_i increase and the manipulated masses decrease, tension loads T_i respectively.

3. Stiffness optimization of driving wires by pre-tensioning

Stiffness optimization of wires that actuates five joints from an elephant's trunk robotic arm by pre-tensioning aims to reduce elastic deformations, caused while robotic arm operates, and systematic positioning errors respectively. The method consists in analytical determination of the optimum value given for pre-tension loads $T_{0,i}$ to achieve optimum stiffness of the wire that actuates joint i, figure 4. Pre-tension loads $T_{0,i}$ and $T'_{0,i}$ from the lower and the upper ends of active wires, figure 4.a and figure 4.b respectively, cause additional elastic deformations $\Delta l_{s,i}$ and $\Delta l'_{s,i}$ given by equation (8) and equation (9) respectively.

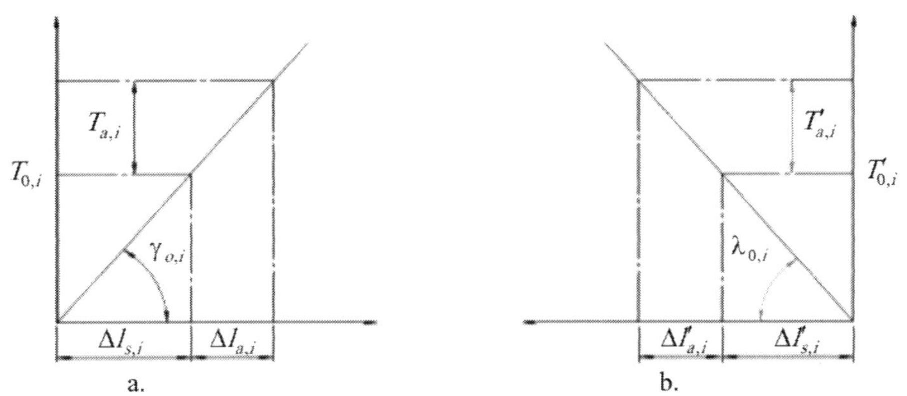

Figure 4. Elastic deformations of the wire that drives joint i located at:
a. Lower wire end, b. Upper wire end.

From equations (8-9) we find that the stiffness $\tan\gamma_{o,i}$ of active wire at lower end, equation (10), is equal to the stiffness $\tan\lambda_{o,i}$ of active wire at upper end, equation (11).

$$\Delta l_{s,i} = (T_{o,i} \cdot L_i)/(E \cdot A_w) \tag{8}$$

$$\Delta l'_{s,i} = (T'_{o,i} \cdot L_i)/(E \cdot A_w) \tag{9}$$

$$\tan\gamma_{o,i} = T_{0,i}/\Delta l_{s,i} \tag{10}$$

$$\tan\lambda_{o,i} = T'_{0,i}/\Delta l'_{s,i} \tag{11}$$

Transposition of the two diagrams from figure 4 until to the overlapping of pre-tension loads $T_{0,i}$ and $T'_{0,i}$ comes out in diagram from figure 5, which shows elastic deformations of the wire that drives joint i. If the active wire is pre-tensioned with load $T_{0,i}$ and joint i is rotating under tension load T_i, then it lengthens additionally with elastic deformation $\Delta l_{a,i}$ and the pulley-joint assembly contracts with the same length because of pre-tension load decrease at an amount given by load $T_{r,i}$.

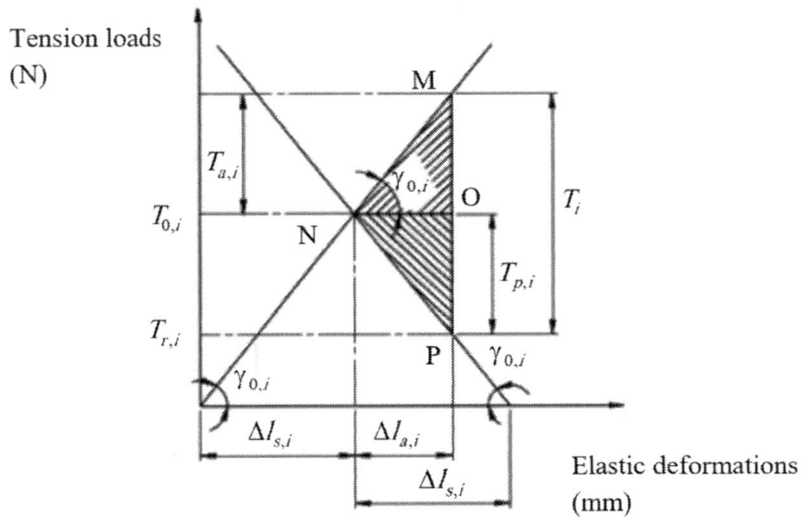

Figure 5. Elastic deformations diagram of active wire that actuates joint i.

When the elephant's trunk robotic arm operates, the total tension load of active wire from joint i becomes T_t, equation (12).

$$T_t = T_{0,i} + T_{a,i} = T'_{r,i} + T_i \tag{12}$$

As given by equations (8-11), we have same variation of wire stiffness at its two ends given in figure 5 by angle $\gamma_{0,i}$ and, thus, it follows the equality $T_{a,i} = T_{p,i} = T_{0,i} + T_{r,i}$. Assuming that we know tension load T_i applied at lower end of driving wire and that it is the sum between extra-tension load $T_{a,i}$ and tension load $T_{p,i}$ then, as a consequence of pre-tensioning of driving wire, we find the equality $T_{a,i} = T_i/2$, shown in diagram from figure 5. The material elasticity of wire used for actuating the joints from elephant's trunk robotic arm allows the extension and contraction with the same amount $\Delta l_{a,i}$ while operating and finally the driving wire remains with elastic deformation $\Delta l_{s,i}$. In case of pre-tension load $T_{0,i}$ absence, elastic deformations from driving wire decrease its stiffness at the value given by tangent of angle γ_i as shown in diagram from figure 6.

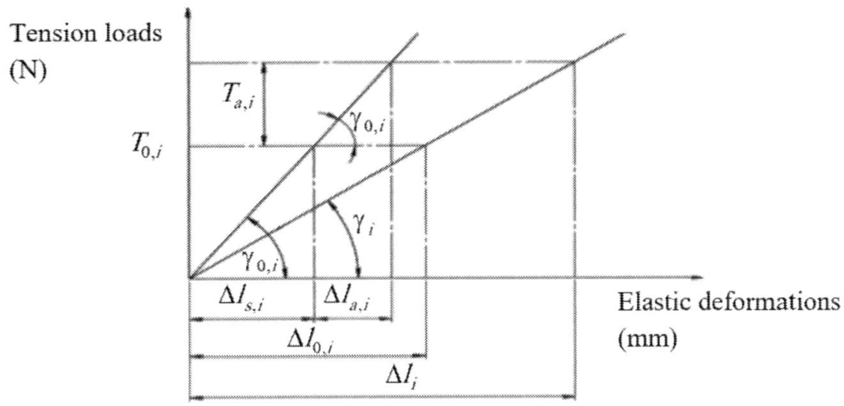

Figure 6. Variation of elastic deformation before and after pre-tensioning.

In diagram from figure 6, $\Delta l_{0,i}$ defines elastic deformation of active wire with stiffness $tan\gamma_i$. From diagram we find stiffness variation of active wire before pre-tensioning, depending upon pre-tension load $T_{0,i}$, that is $\tan\gamma_i = T_{0,i}/\Delta l_{0,i}$. From figure 5 we find the equality $T_{0,i}/\Delta l_{s,i} = T_{a,i}/\Delta l_{a,i}$ that is equivalent with $\Delta l_{s,i}/\Delta l_{a,i} = T_{0,i}/T_{a,i}$. Thus, using equation (10) we determine analytically the stiffness optimization of pre-tensioning on active wire, equation (13).

$$\delta = 100\cdot(\tan\gamma_i - \tan\gamma_{0,i})/\tan\gamma_i = 100\cdot(1-\Delta l_{s,i}/\Delta l_{0,i}) \qquad (13)$$

From figure 5 we know that $\Delta l_{s,i} > \Delta l_{a,i}$, and from figure 6 we find the elastic deformations dependence $\Delta l_{0,i} >> 2\cdot\Delta l_{a,i}$ that determine amount of quantity $\delta > 100\cdot(1-0.5\cdot\Delta l_{s,i}/\Delta l_{a,i})$. With these assumptions and with ratio $\Delta l_{s,i}/\Delta l_{a,i} \cong 1$, we obtain that quantity δ is higher than 50%.

4. Conclusion

In this paper, we highlighted the need of stiffness optimization of active wires that actuate joints from an elephant's trunk robotic arm through structure interior. Tension loads, manipulated masses of each joint, inherent resistance moments and friction loads, cause elastic deformations of active wires. Length variation of driving wires directly affects the stiffness of motion transmission system and, implicitly, the precision positioning. The total elastic deformations cumulated because of the mentioned loads require compensation methods to reduce them with pre-tensioning mechanism or by electronic adjustments made in programming algorithm of the robotic arm. We determined by an analytical method that a pre-tension mechanism used for optimizing the active wire stiffness goes to minimum 50% stiffness increase. The analytically obtained results are necessary for feature experimental determinations of positioning precision from elephant's trunk robotic arm.

References

[1] Direkwatana C, Suthakorn J and Wilasrusmee C 2011 Development of Wire-Driven Laparoscopic Surgical Robotic System *Int. Conf. on Rob. and Biomi.* 485-490

[2] Yeo S H, Yang G and Lim W B 2013 Design and analysis of cable-driven manipulators with variable stiffness *Mechanism and Machine Theory* **69** 230-244

[3] Ning K and Worgotter F 2012 Control System Development for a Novel Wire-Driven Hyper-Redundant Chain Robot 3D-Trunk *IEEE/ASME Trans. on Mech.* **17** 949-959

[4] Dehghani M and Moosavian S A A 2014 Dynamics Modeling of a Continuum Robotic Arm with a Contact Point in Planar Grasp *Journal of Robotics* 13

[5] He B, Wang Z, Li Q, Xie H and Shen R 2013 An Analytic Method for the Kinematics and Dynamics of a Multiple-Backbone Continuum Robot *Int. J. of Advanced Robotic Sys.* **10** 84

A locus problem solved by using a mechanism with three dyads and two leading elements

I Popescu[1], L Sass[2] and A E Romanescu[2]

[1] Departament of Mechanical Engineering, Academy of Technical Sciences, Bucuresti, 26, Bd. Dacia, 010413, Roumanie
[2] Faculty of Mechanical Engineering, University of Craiova, 107, Calea Bucuresti, 200478, Roumanie

E-mail: ludmila_sass@yahoo.com

Abstract. In Geometry there are many types of loci, solved by means of classic geometrical considerations and yielding lines and arcs of circles or conics. Yet more complicated locus can be solved by means of the Theory of Mechanisms. Our research starts from a locus and provides a solution based on the Theory of Mechanisms, finding the equivalent mechanism. The structural and cinematic analysis of the mechanism is made, determining the trajectory of a point representing the locus which presents interest. The mechanism has three dyads and two leading elements, for which the movements were correlated by means of a coefficient q. For various values of q different loci were obtained, similar for close values of q but different for significantly different values of q.

1. Introduction

Many mechanisms generating geometric curves were built in an empiric manner along time, by ingenuity and intuition. Afterward, mathematicians like Pafnuti Cebîşev, Evans, Paucellier, Cardan, Reuleaux and many others obtained mechanisms with generating properties by using geometric considerations, relying on the properties of curves and the Theory of Mechanisms. A significant number of contributions are authored by I. I. Artobolevskii [1], who determined techniques to apply some geometric considerations. They enabled him to create many mechanisms which generate mathematic curves [2].

Cinematic possibilities of mechanisms with two and three leading elements, with linearly correlated movements are studied in [3]. Many examples of such mechanisms and examples of plotted curves are provided.

Scale models of mechanisms which move such as to describe loci of some points with aesthetic properties are presented in [4].

Many examples of mechanisms used to generate different curves as geometric curves are presented in [5]. The types of curves fall into well known categories like ellipses, parables, lemniscates etc. Imposed loci are used as starting points in [6] in order to define various mathematic curves. Generating mechanisms which determine such curves are determined. When rotating, they generate aesthetic surfaces.

In [7] graphical constructions of some ordinary mathematic curves are given.

IOP Conf. Series: Materials Science and Engineering **161** (2016) 012036 doi:10.1088/1757-899X/161/1/012036

2. The locus problem

Considering the system of axis HOB from figure 1, a parallel to the ordinate is plotted such as to include H. We search for the locus corresponding to the intersection point of two lines which move with their heads following the axes OH, OB and HD.

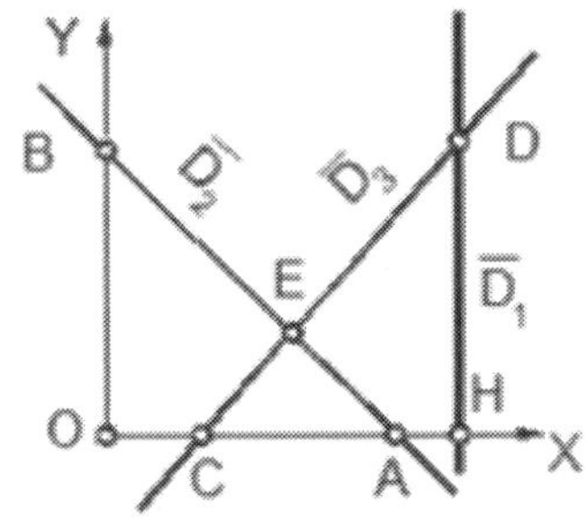

Figure 1. Geometric construction

3. Study of mechanism

Coulisses are introduced at the ends in order to force the heads of the lines to move always along the axes, figure 2. In order to find the locus of the point E which represents the intersection between the mentioned lines, two coulisses will be introduced. They move along the lines and are related by means of a rotation couple.

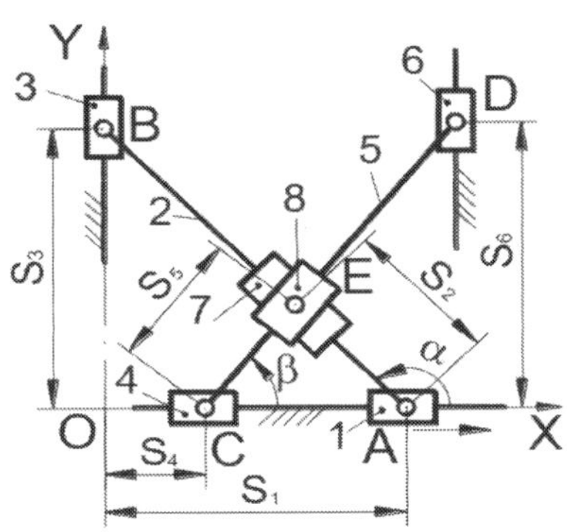

Figure 2. Equivalent mechanism of type P-P-RRP-RRP-PRP

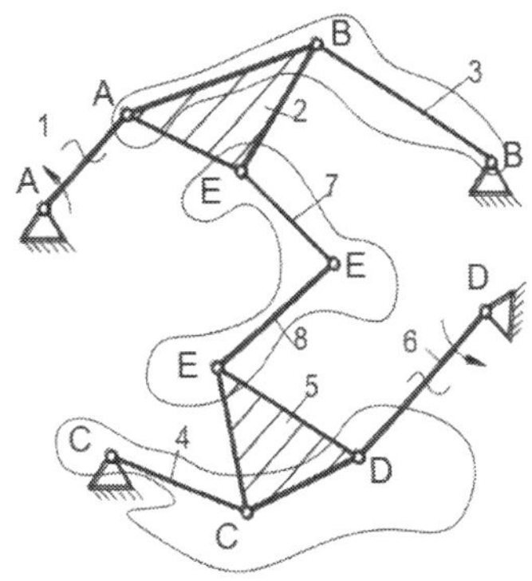

Figure 3. The mechanism structure

The mechanism's degree of mobility is: M=3n-2C5-C4=3 8-2 11=2. Consequently there are two leading elements (S_1 and S_6), figure 1 reveals that both lines can perform movements independently to each other.

The mechanism's structure is given in figure 3. The mechanism has two leading elements (S_1 and S_6) and three dyads. Its type is P-P-RRP-RRP-PRP.

The following equations can be written by using the method of contours:

$$\begin{cases} x_B = S_1 + AB\cos\alpha = 0 \\ y_B = AB\sin\alpha = S_3 \end{cases} \quad (1)$$

$$\begin{cases} x_D = S_4 + CD\cos\beta = const. \\ y_D = CD\sin\beta = S_6 \end{cases} \quad (2)$$

$$\begin{cases} x_E = S_4 + S_5 \cos\beta = S_1 + S_2 \cos\alpha \\ y_E = S_5 \sin\beta = S_2 \sin\alpha \end{cases} \tag{3}$$

From equations (1) one can determine S_3 and α. S_4 and β can be determined from equations (2). The next step is to calculate S_2, S_5 and the coordinates of the tracing point E from equations (3).

4. Results

The values used as initial data are: AB=68, CD=70, x_D=56. The movements of the leading elements are considered to be linearly correlated through the equation (4):

$$S_6 = q \cdot S_1 \tag{4}$$

Different loci are expected to be obtained for different values of the coefficient q.

It should be mentioned that the length AB imposes limits over S_1, which for this case varies within the range – 68 ... 68. Similarly the limits for the variation of S_6 [-70, 70] are imposed by the length CD. These quantities appear in the above equations, as having known values. For certain values of "q", the values for S_6 can fall outside the range – 70… + 70. In this case the mechanism can no longer operate.

Therefore the program includes special sequences to detect these cases and prompts the user to provide other values for "q".

For q= - 0.6 we obtained the position of the mechanism depicted by figure 4.

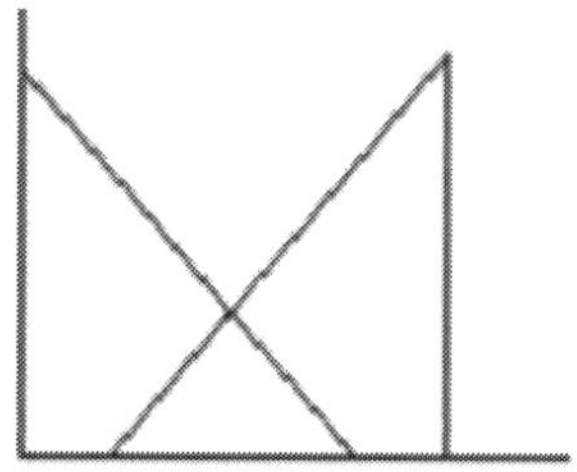

Figure 4. Position of the mechanism for q= -0.6

For q=1 there were obtained the locus from figure 5 and the values of the races from figure 6.

This is a curve not known from Geometry. It has a turning point in a specific area and exhibits other properties to be studied with methods from Differential Geometry. Such curve can be used, for example, in technical applications - e.g. for tracing operations or for trajectories imposed by certain points from equipment in the textile industry.

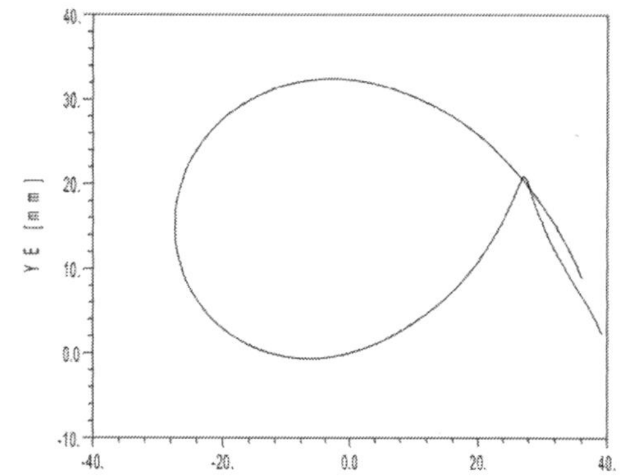

Figure 5. Locus for q=1

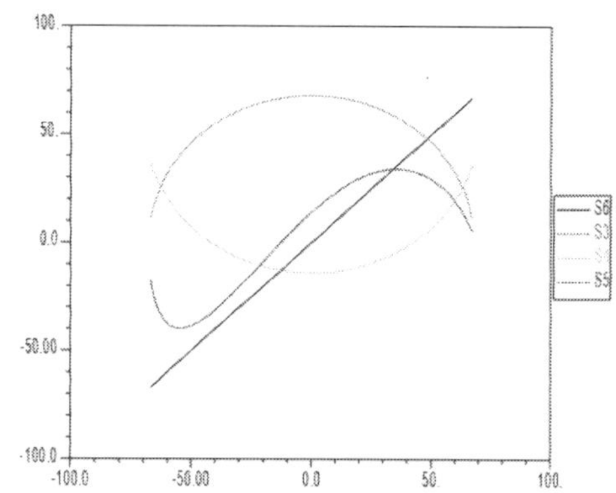

Figure 6. Values of races for q=1

The locus is an open curve, similar to the letter α . The following conclusions can be drawn based on figure 6:

- the curve described by S_6 is linear;
- the curve described by S_5 is similar to a cosine ;
- the curves for S_3 and S_4 are similar, with opposite curvatures and present symmetries.

For q=0.6 there was obtained the locus from figure 7 and the values of races from figure 8.

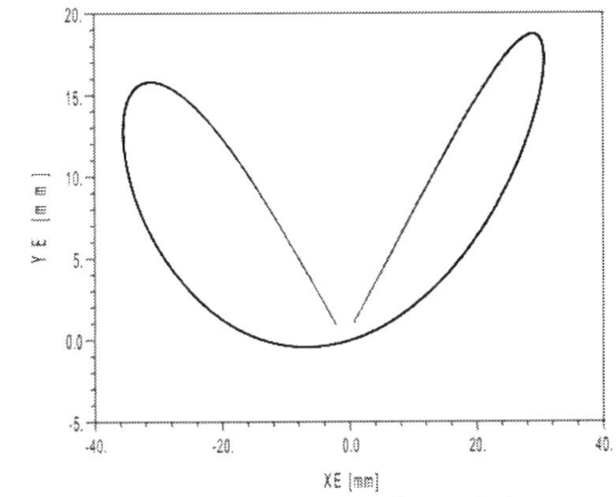

Figure 7. Locus for q=0.6

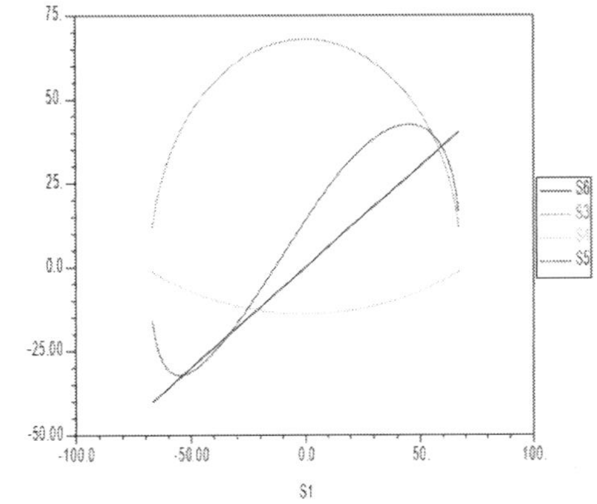

Figure 8. Values of races for q=0.6

In this case the locus is a curve with a shape of flower with two petals. Races similar to those described above were obtained, except for that corresponding to S_4, which is opposite to that corresponding to S_3 now. Different sizes characterize the new curves.

The locus obtained for q= - 1 is depicted by figure 9 whilst the values of races are presented in figure 10.

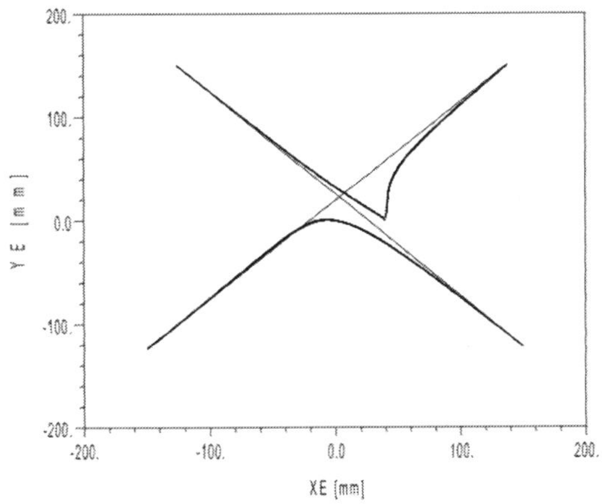

Figure 9. Locus for q=–1

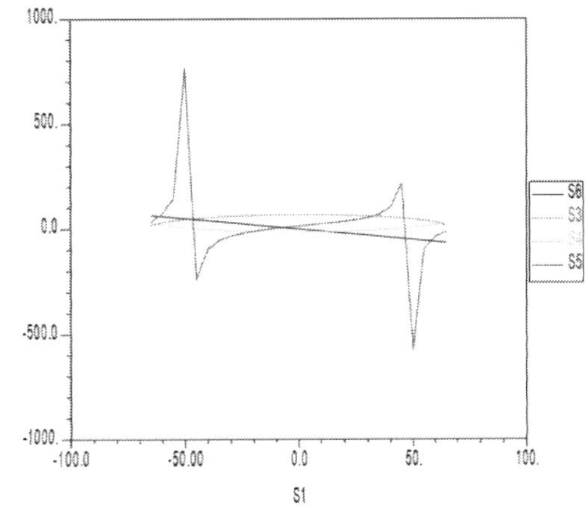

Figure 10. Values of races for q=–1

In this case the locus is an interesting curve, with rectilinear portions, similar to a hyperbola with asymptotes, not known from Geometry. The influence of the coefficient "q" is noticeable.

Different loci, obtained for various values of q are depicted by figures 11-15.

The curves are different to each other. For close values of q, similar curves are obtained. On the other side, when the values of c are significantly different, completely different curves are obtained.

All the curves are open. Some of them have complex shapes, whilst others are simple branches, similar to lines. When changing the movement sense of the element 6, when q<0, curves similar to those corresponding to the initial sense are obtained, but their positioning with respect to the system of axes is different.

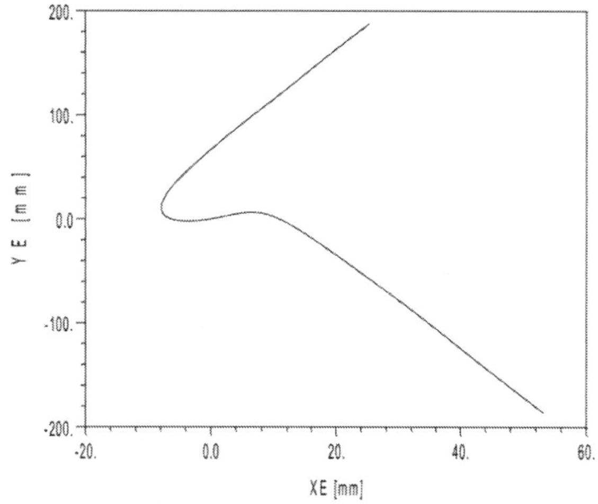

Figure 11. Locus for q=5

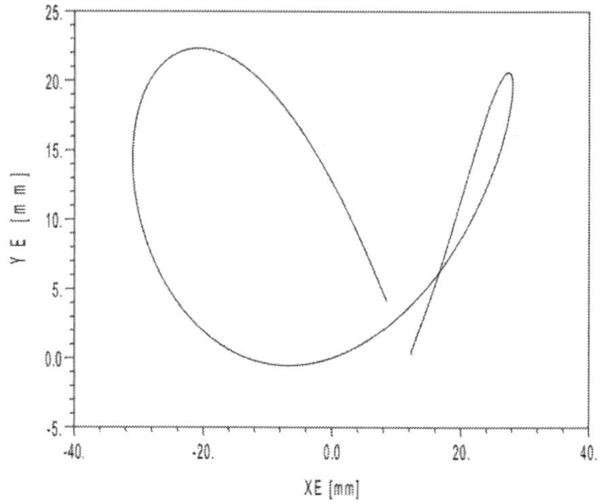

Figure 12. Locus for q=0.8

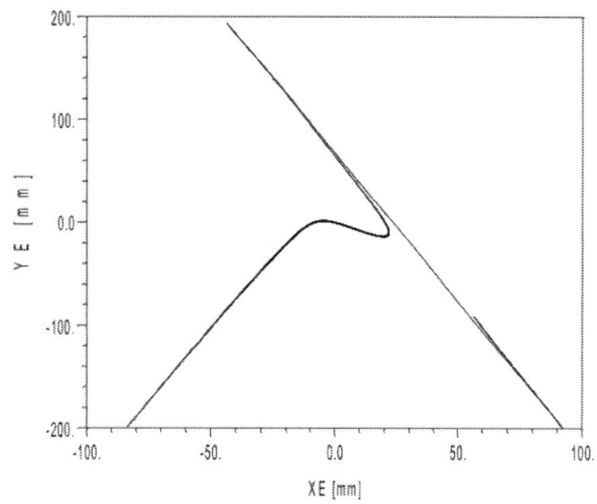

Figure 13. Locus for q= -3

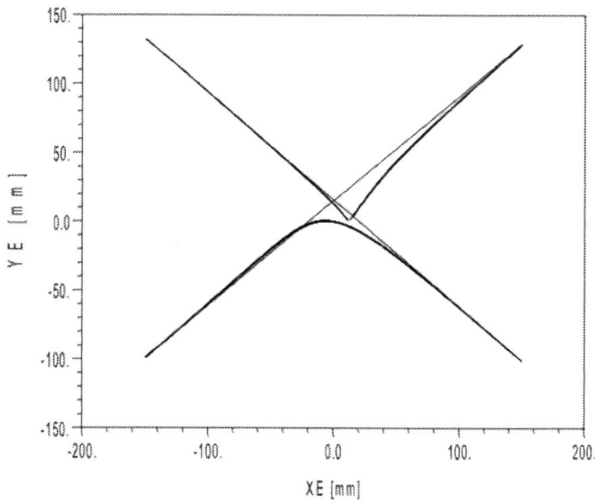

Figure 14. Locus for q= -0.8

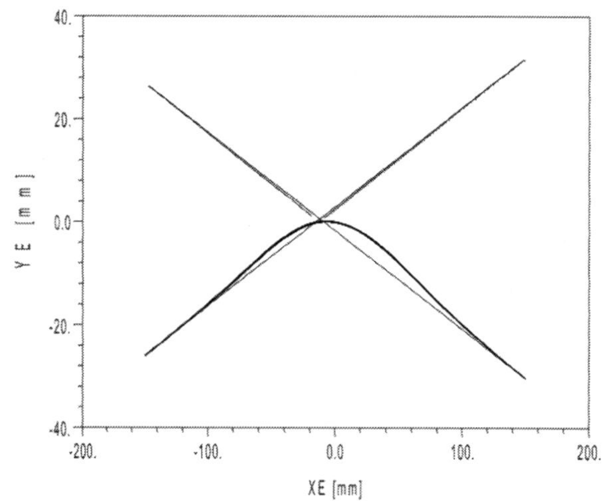

Figure 15. Locus for q= -0.2

Figures 16-18 present successive positions of the mechanism for different values of q.

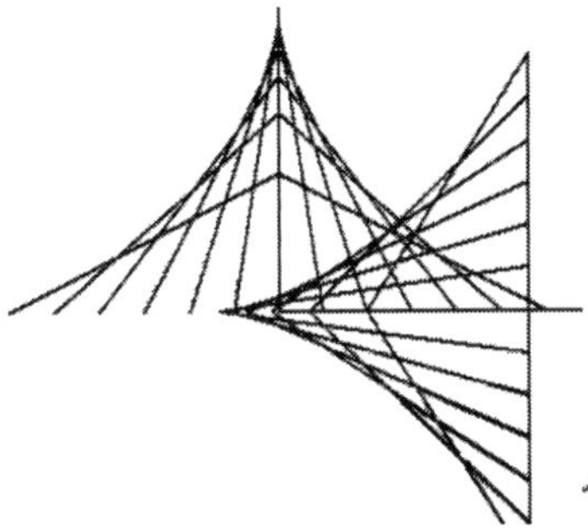

Figure 16. Successive positions of the mechanism for q=1

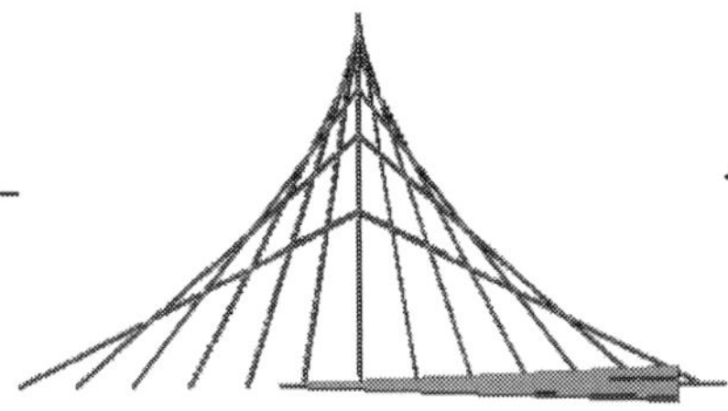

Figure 17. Successive positions of the mechanism for q= 0.05

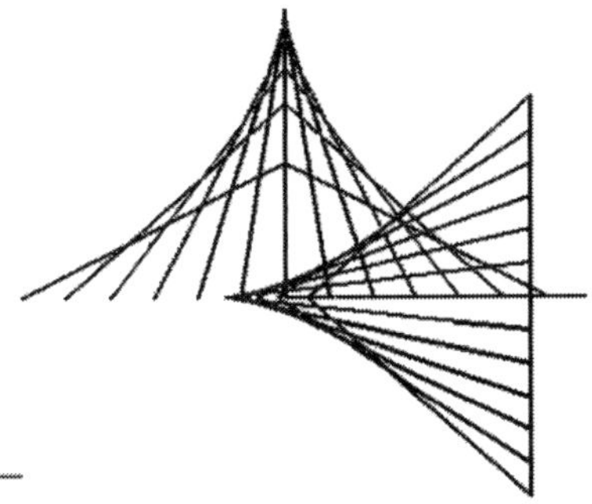

Figure 18. Successive positions of the mechanism for q=0.8

Some remarks must be done:
- the lines CD and AB fall out the system of axes toward negative values;
- for small values of q, the line CD is moving inside a narrow domain;
- the race of the point C toward the left limit of the system of axes is smaller for high values of the race of the point D, because the difference between the line HC and the arc of circle with the radius CD and the center at the left limit of C is small;
- for q<0 identical successive positions to those corresponding to q>0 are obtained for the same absolute values of q.

5. Conclusions

The problem of locus used as starting point was made equivalent to a problem of cinematic synthesis and analysis (positions and trajectories) of an equivalent mechanism. The mechanism was analyzed and the searched locus was determined. It consists of the trajectory of the point representing the intersection of two lines. These trajectories are not known from Mathematics. Their usability in technique for example can be related to the design of some equipment in the textile industry.

Because the movement of both lines were independent, they were correlated by means of a coefficient q in order to prevent the random character of their movement. For various values of q, various loci were obtained. For certain domains corresponding to the variation of q, the obtained curves are similar, but have different sizes. For significantly different values of q, different curves were obtained. All the curves are open and their shapes are either complex, or simple arcs.

References

[1] Artobolevskii I I, 1959, *Teoria mehanizmov dlia vosproizvedenia ploschih crivâh*. Izd. Academii Nauk, Moskva
[2] Artobolevskii I I, 1979, *Mehanizmî v sovremennoi tehnike*, Vol. I-V, Izd. Nauka, Moskva
[3] Popescu I, 2015, *Curbe de bielă generate de mecanisme cu două sau trei elemente conducătoare*. Editura Sitech, Craiova
[4] Online at: *http://graphicalcommunication.skola.edu.mt/syllabus/linkages-and-mechanisms*, accessed: 10.01.2016
[5] Online at: *http://www.joshuanava.biz/geometric/loci.html*, accessed: 11.01.2016
[6] Popescu I, 2012, *Structura şi cinematica unor mecanisme generatoare de curbe şi suprafeţe estetice*, Editura SITECH, Craiova
[7] Online at: *http://www.engineeringdrawing.org/2012/05/points-locus-problem-2-9.html*, accessed: 10.12.2015

A locus problem solved through methods from the Theory of Mechanisms

I Popescu[1], A E Romanescu[2] and L Sass[2]

[1] Departament of Mechanical Engineering, Academy of Technical Sciences, Bucuresti, 26, Bd. Dacia, 010413, Roumanie
[2] Faculty of Mechanical Engineering, University of Craiova, 107, Calea Bucuresti, 200478, Roumanie

E-mail: romanescualinaelena@yahoo.com

Abstract. The specialty literature provides many mathematic problems dealing with graphical or analytical solutions for loci. Many mathematic curves are actually loci of particular points. In this paper with present the results of our researches related to solving a problem of locus by using methods from the Theory of Mechanisms. The locus studied in this paper is placed on the side of an anti-parallelogram. The equivalent mechanism for which we made a synthesis is of type R-RRR-RRP-PRP (it has a leading element and three dyads). The locus consists in a curve with a "banana" shape, distorted at one at its heads, modifying its sizes when changing the length of the mechanism's crank while preserving its shape. Two solutions were found for the analyzed problem consisting in identical but displaced curves.

1. Introduction

Many mathematic curves were studied since ancient times. Many old statues and bas-reliefs, as well as remaining ornaments carved in stone or jewels come to certify the above. In time mathematic studies addressed those curves, determining their geometric properties, graphic methods to plot them and other details. Other curves were derived as a consequence and in time they became so common that many studies were focused on them. In general the mathematicians limited their efforts to the determination of curves' properties, in few cases even conceiving graphical methods to plot them. The books of Geometry contain many (usually simple) loci. For example in [1] some examples of classical loci are given. There were obtained ordinary geometrical elements like lines and circles, secants, tangents, perpendicular lines etc. Usually mathematic details are also accompanying them. Mechanisms are approached in [2]. They generate ellipses, parables, hyperboles, cissoids and pedals of curves. The mechanisms' synthesis is accomplished by starting from loci and from the properties of the corresponding curves.

A conicograph mechanism obtained when starting from a locus problem is presented in [3]. The obtained mechanism has 7 elements and 10 couples of class V. It consists in a leading element with rotation movement, two dyads of type RPR and a dyad of type PRP. Examples related to plotting of ellipses, parables and hyperboles are provided. In [4] Robert Ferréol, Jacques Mandonnet and Alain Esculier give examples of various curves which represent loci of certain points. Also are provided animations of the lines generating curves through certain points. Patents relying on loci can be found too. An example in this sense is [5], where a new profile for the gears teeth are provided.

Many mathematic problems for geometric loci are solved graphically or analytical in [6]. Simple problems are used as start points and more difficult problems are afterward approached. Movements of some mechanisms used to plot geometric curves are treated in [7].

2. The locus problem

We consider the anti-parallelogram ABCD and a point E on CD. From E the line EF is plotted (F belongs to the line AF perpendicular on AD). The locus of the point G is to be determined, where G represents the intersection of the lines BC and EF, figure 1.

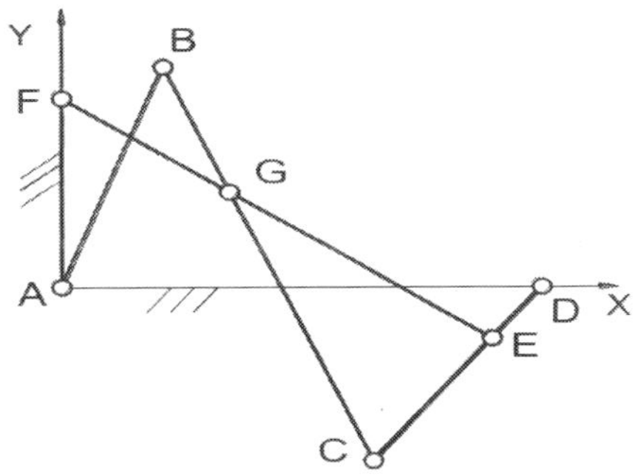

Figure 1. Geometric construction

3. Study of the equivalent mechanism

The articulated anti-parallelogram mechanism ABCD was built firstly,fFigure 2.

A point E is belonging to CD is considered and a line EF is plotted. F belongs to the line AF, which becomes the ordinate while AD represents the abscissa. In order to force the movement of F along a fixed line, while preserving constant values for the lengths AB, BC, CD, AD, DE and FE, a coulisse and a rotation couple are placed in F. In order to find the point where BC intersects EF, two coulisses are placed in G (they are related by means of a rotation couple). In this way we obtained an equivalent mechanism of the locus problem.

The structural schema of the mechanism from figure 2 is depicted by figure 3. One can see that the mechanism is of type R-RRR-RRP-PRP (it has a leading element and three dyads).

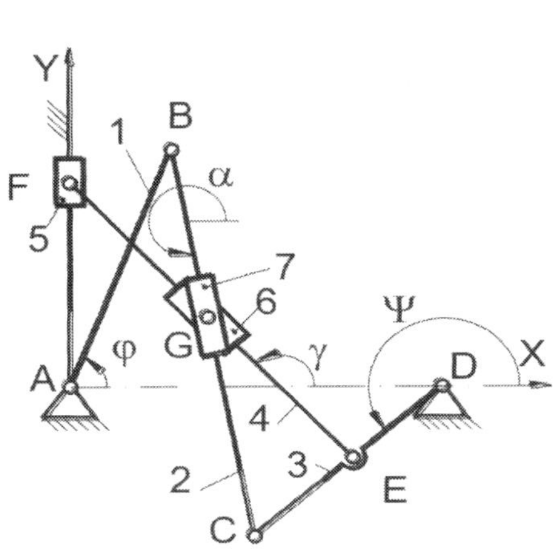

Figure 2. Equivalent mechanism of type R – RRR – RRP - PRP

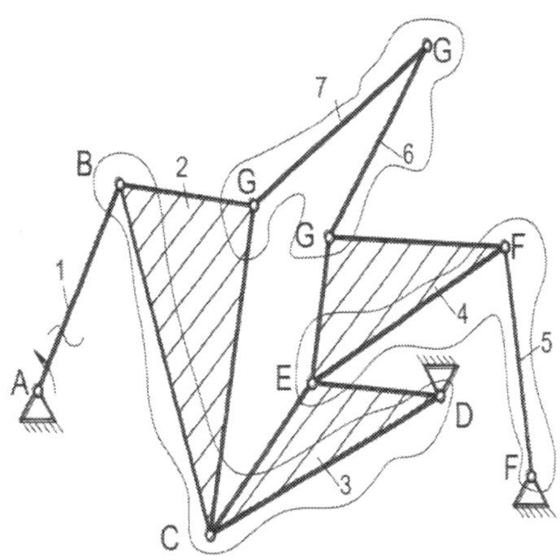

Figure 3. The mechanism structure

4. Relations for calculus

Starting from figure 2 we can write the following equations by using the method of contours for the cinematic analysis of mechanisms:

$$\begin{cases} x_C = AB\cos\varphi + BC\cos\alpha = x_D + DC\cos\psi \\ y_C = AB\sin\varphi + BC\sin\alpha = DC\sin\psi \end{cases} \quad (1)$$

$$\begin{cases} x_E = x_D + DE\cos\psi \\ y_E = DE\sin\psi \end{cases} \quad (2)$$

$$\begin{cases} x_F = x_E + EF\cos\gamma = 0 \\ y_F = y_E + EF\sin\gamma \end{cases} \quad (3)$$

$$\begin{cases} x_G = AB\cos\varphi + BG\cos\alpha = x_E + EG\cos\gamma \\ y_G = AB\sin\varphi + BG\sin\alpha = y_E + EG\sin\gamma \end{cases} \quad (4)$$

By using equation (1) we can solve the Kinematics of the contour ABCD. The lengths of elements are known and we must determine the angles α and ψ. Equation (2) can be used to determine the position of E on CD providing that the length ED is known. x_F and y_F are obtained from equation (3).

From the equation (3) of x_F we can get $\cos\gamma$. In order to obtain y_F we must calculate $\sin\gamma$. A square root is obtained, with the signs $\pm$. Therefore two positions for the curve to be determined as locus are obtained. Further on, from equation (4) we can determine the variable lengths BG and EG and finally the coordinates of the point G.

5. Results

The following values were used as initial data for the mechanism's elements: AB=22; BC=80; DC=AB; x_D=BC; AD=BC; DE=DC/2; EF=95.

Figure 4 depicts the cinematic chain ABCD and the trajectory of the center of the rod BC, which falls into the category "lemniscate of Bernoulli", represented here in order to see a specific locus of this cinematic chain.

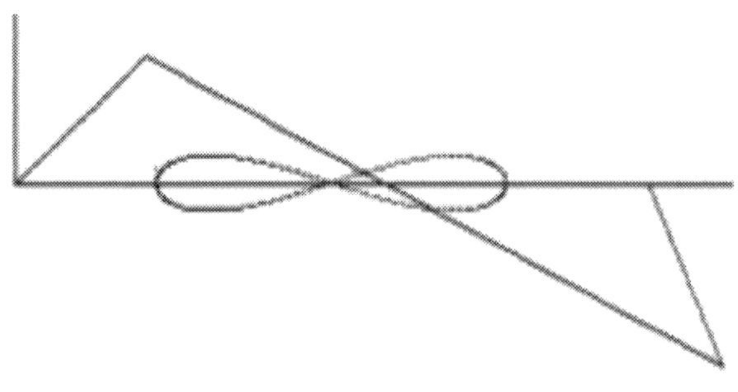

Figure 4. Cinematic chain ABCD and the trajectory of the center of the rod BC

For q=1 there were obtained the locus from figure 5 and the values of the races from figure 6.

This is a curve not known from Geometry. It has a turning point in a specific area and exhibits other properties to be studied with methods from Differential Geometry.

Such curve can be used, for example, in technical applications - e.g. for tracing operations or for trajectories imposed by certain points from equipment in the textile industry.

Figure 5 depicts the successive positions of this cinematic chain. The successive positions of the entire mechanism are given in figure 6.

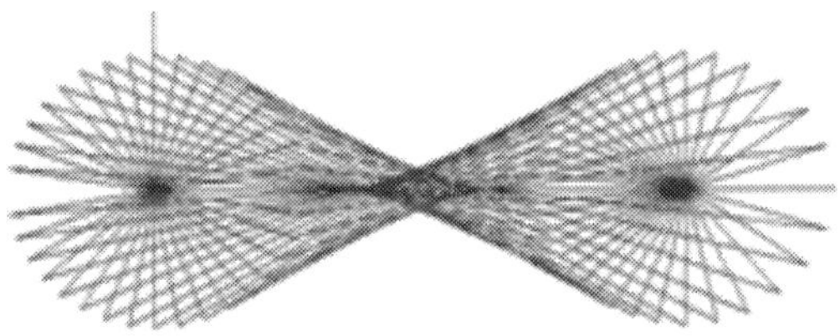

Figure 5. Successive positions of this cinematic chain ABCD

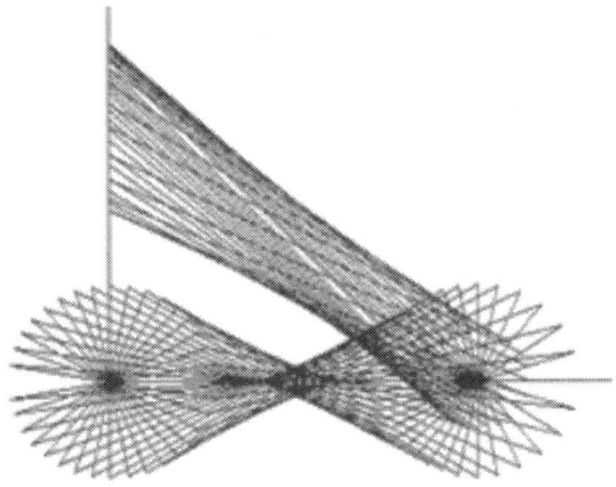

Figure 6. Successive positions of the mechanism

Figure 7 depicts the mechanism in a certain position, along with the trajectories of the centers of the rods BC and EF. One can see that the trajectory of the center of the EF rod is curve resembling to an ellipse.

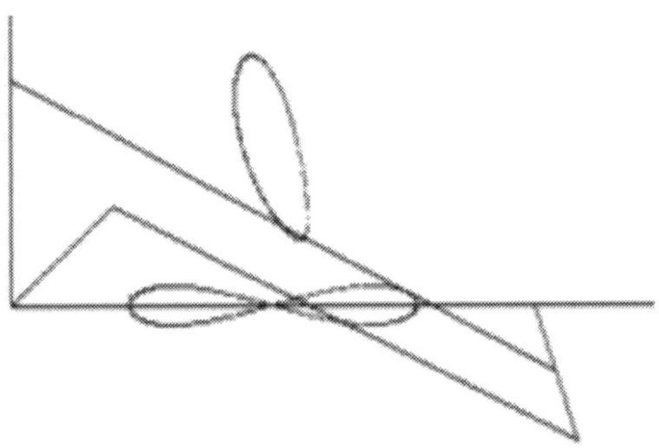

Figure 7. The mechanism in a certain position and the trajectories of the centers of the rods BC and EF

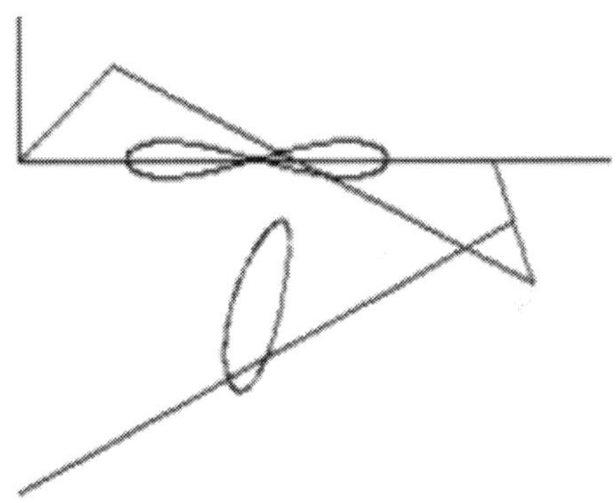

Figure 8. The mechanism in the second position

Because the algebraic system provides two solution, we also considered the alternative with opposite sign (-). We obtained figure 8, where the trajectory of the center of the rod E is similar to that corresponding to the positive sign. Its position now is under the abscissa. The images from figure 7 and figure 8 provide information on the positions of the curves obtained as locus for G, when comparing to the trajectory of the center of the parallelogram's rod.

The curve of intersection between the rods – that is the locus to be determined, corresponding to the point G, is represented by figure 9. It is a rod curve of "banana" shape, with a small distortion at its upper head. Mathematically, the tracing point falls outside the rods (the coulisse 6 can exceed the length EF and the coulisse 7 can exceed the length BC – possible trajectories of coulisses being along the rods' extensions). When the studies are concerned only with the mechanism, disregarding the locus problem, restrictions can be imposed such as to prevent the intersection point to fall outside the rod's lengths. In this case the crank AB can no longer describe full rotations.

When considering the solution of the algebraic system with the sign (-), for EF=95 one can obtain a curve as that from figure 9 and figure 10, placed now overhead the abscissa, figure 10.

Figure 11 provides the variations of the intersection point's coordinates for the curve from figure 9, (that is for the locus). One can notice a certain symmetry of these curves along some sub-ranges, reaching the limit values for φ=50°. The curves are continuous, without any jumps or discontinuities.

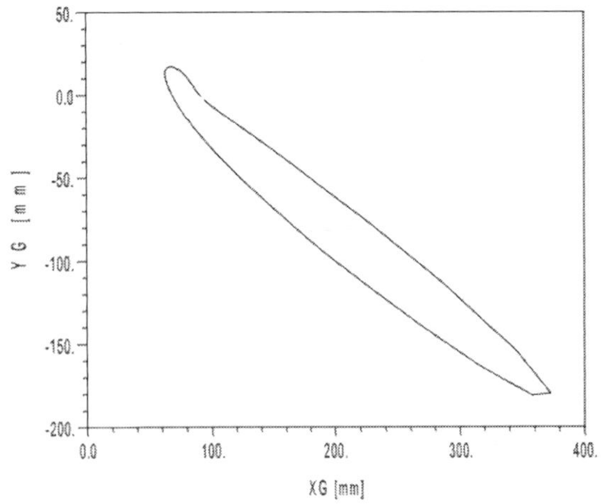

Figure 9. Locus of point G, solution for (+)

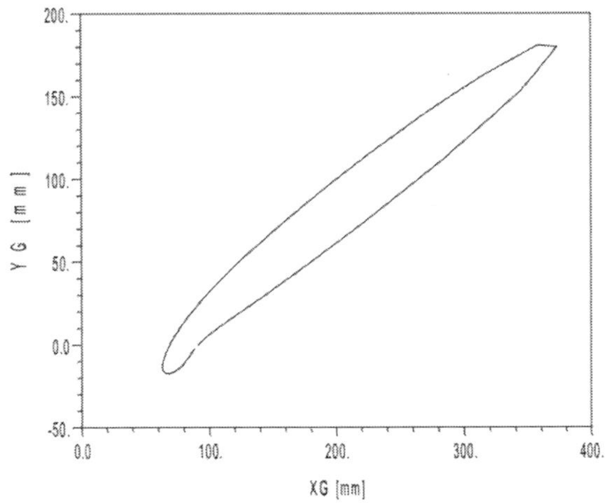

Figure 10. Locus of point G, solution for (-)

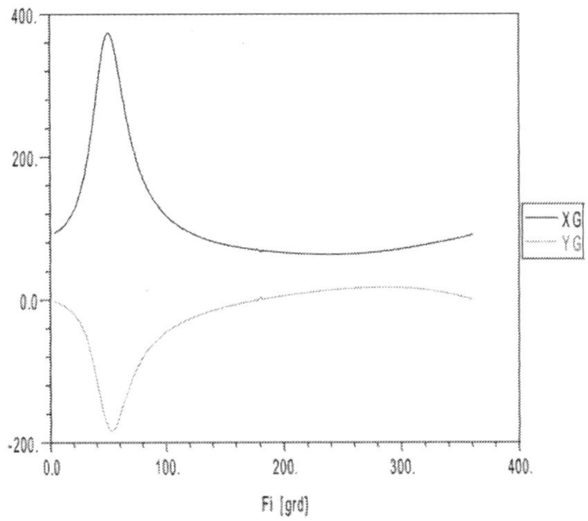

Figure 11. Variations of the intersection point's coordinates for the solution (+)

The curves described above are associated to the data considered in the initial stage for the mechanism. Modifications of data result into similar curves, with different sizes.

Figure 12 provides examples for them. When evaluating visually these figures, knowing that the figures are resized in order to fit their place in page, one should consider that the length of the abscissa from figure 12, figure 13, and figure 14 is identical to that from figure 9.

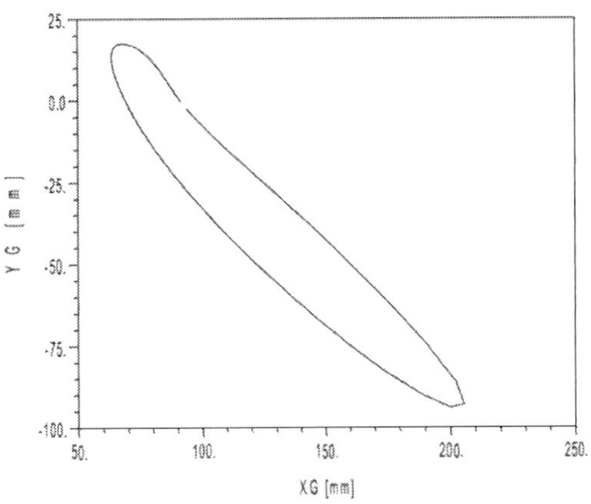

Figure 12. Curve obtained for element EF=97

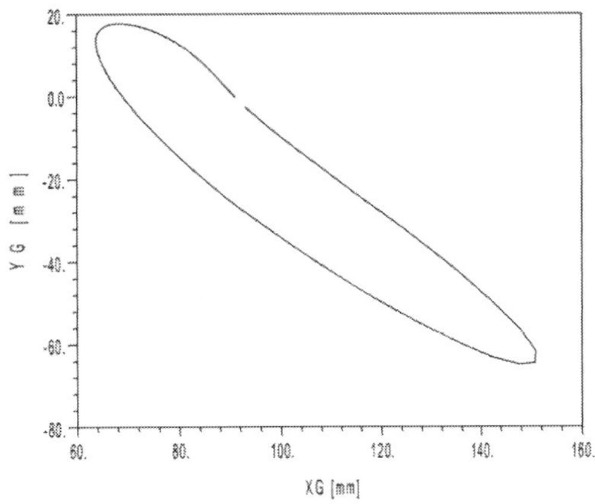

Figure 13. Curve obtained for element EF=100

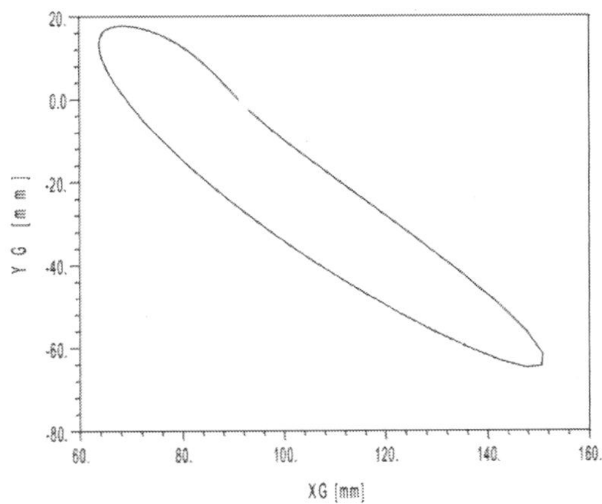

Figure 14. Curve obtained for element EF=120

6. Conclusions

A locus problem was solved by using the method of mechanisms' theory. A mechanism with three dyads of type R-RRR-RRP-PRP was obtained. The points from the rods describe curves like lemniscate or ellipses and the locus was found to be a curve with a shape of banana, with a distortion at its upper head.

When resizing the mechanism, similar curves were obtained. They have different sizes. The second solution of the involved algebraic system provides the same curves, shifted with respect to the first solution.

References

[1] Online at *http://www.bymath.net/studyguide/geo/sec/geo10.htm*, accessed: 20.12.2015

[2] Popescu I, Luca L, Cherciu M, 2013, *Structura şi cinematica mecanismelor. Aplicaţii*. Editura Sitech, Craiova

[3] Popescu I, Luca L, Mitsi S, 2011, *Geometria, structura şi cinematica unor mecanisme*. Editura Sitech, Craiova

[4] Online at *http://www.mathcurve.com/courbes2d/hyperbole/hyperbole.shtml*, accessed: 01.11.2015.

[5] Online at *http://www.google.com/patents/EP0252156A1?cl=fr*, accessed: 15.11.2015

[6] Online at *http://www.matheureka.net/Q169.htm*, accessed: 01.12.2015

[7] Nava J. *Arts. Geometric drawing. How to draw loci for crank mecahanism*, available at: *http://www.joshuanava.biz/geometric/loci.html*, accessed: 01.12.2015

IOP Conf. Series: Materials Science and Engineering **161** (2016) 012038 doi:10.1088/1757-899X/161/1/012038

Packaging design for the Greek olive oil industry

A Kouveli[1], D Tzetzis[1] and P Kyratsis[2]

[1] International Hellenic University, School of Economics, Business Administration and Legal Studies, 14km Thessaloniki - N. Moudania, Thermi, GR57001, Greece
[2] Western Macedonia University of Applied Sciences, Department of Mechanical Engineering and Industrial Design, Kila Kozani, Greece, GR50100, Greece

E-mail: pkyratsis@teiwm.gr

Abstract. Packaging is one of the most important elements that characterize product design. It does not only refer to the appearance of the product, that creates the first impression to the user, but it is a whole communication code. This has been identified by the enterprises involved and their competition is high. Nowadays, the most successful of them pay a great deal of attention on every day's trends concerning packaging design and try to apply these trends upon their own designs. The present paper describes the packaging design in today's industries and the existing trends. It will therefore reveal what makes a package design successful, from the company's perspective and how the company studies such factors in order to achieve the best result. Those factors and the research results are applied on the olive oil packaging industry, producing a final packaging design solution.

1. Introduction
Packaging design is a very important sector for the today's industries. A package is the outer part of a product and therefore the first element that a user is going to visually observe. Due to the fact that nowadays the image plays a significant role in every activity of the individual's life, from the internet world to the outdoors advertising, the success of a business depends mainly upon a well-designed package.

The competition between the industries is increased day by day and packaging sales on the emerging markets are expected to continue to show strong growth. Therefore, the companies are always informed about the general trends in packaging design in order to achieve the highest profit.

The conduction of this research with the subsequent proposal will produce results that will be helpful for the industries and the designers themselves in order to create a profitable package design. The result of an efficient packaging will therefore induce an achievement in the overall product design sector of the industry.

Consequently, the market in olive oil packaging design sector is going to be examined and general evidence about the design, the materials and the lines of the package are going to be pointed out as well. Finally, a combination between the theoretical background in general packaging design and the olive oil packaging design is going to lead to a proposed design solution.

2. Theoretical Background
The process for the design of a package is similar enough with the general process of the product design. In general terms, planning, developing and completing are the three basic phases for the

packaging design. Planning is a more organizational phase and therefore it consists of the initial research about what is going to be designed (development of a strategy, organizing of groups etc.), while development concerns the strong identity that the company gives to the package. Finally, the completing phase is an overall procedure that the industry should mind the final colors, lines, materials of the design, and the general image of the package and communicate this image to the client in the best way [1].

A basic packaging attribute is to hold a strong identity. The enrichment of the company is made though branding. Branding is a very important issue for company's good sells: top industries do always include packaging design on innovative positioning and solid, consistent brand strategy. This strategy is nowadays the most important marketing tool and packaging is its complementary power. Therefore, branding is the driving force for the connection between the customer and the image of the product. Many categories of consumer products are dominated by a particular shape and structural trend over another and there is a reason about this fact. Trends are usually leaded by function and driven by competition and price sensitivity. An example shows exactly this statement: it is impossible to pack milk in hand tag, whilst it is unaffordable to sell it in a crystal carafe [2].

Generally, the memory footprint upon the client and the bond with the brand gets stronger when the memory touch points are activated. Therefore, bonding with a brand requires a multi – sensory experience. When this happens, an optimized match between perception and reality and a substantial creation of a brand platform for product extension will be added as important dimensions to the business' organizational base. [3]

Therefore, in order to design an effective package a company should be always informed and even better, forecast the latest trends [4]. This procedure happens with the marketing strategy of the homonym organizational field. Some general actualities that do characterize this packaging design era are mentioned bellow:

- Hand drawn logos and labeling: warm graphics that look homemade or friendly.
- Hyperfunctional Forms: more than one functions in a package.
- Masterchef: consumer to 'cook like a chef'.
- ASAP – As simple as possible: pure and distinctive design.
- Intelligent Packaging: gaining digital access with the package to exclusive content.
- Geometry: graphic design's philosophy by making geometric patterns.
- Kraft Paper: varying of thickness to package a product, eco – friendly product.
- Ornately old – fashioned: idealization of the past in a modernized manner.
- 'Authentic' Design Aesthetics: products on the shelf with more authentic and intimate look.
- E-tailing volumes: online food retail driven by consumers' requirements.

From all of the products that exist in the packaging design market, food products are of top design oddity amongst the other product categories not only due to the fact that the user's experience with a food product involves all five senses, but also because of their expiration date. The user therefore is to reject a product if it smells bad, or if it is not tasteful. This is a reason that the design of a package of a food product is quite peculiar. Olive oil belongs to the food products' industry and its top design difficulty is to face its liquid form [5]. An industry player must realize the fact that it should include packaging design in the whole management and designing field, and prioritize it as a high importance process. The phases for an exemplary managing of this process are three, viz planning – developing – creating, and the most important elements are to be always informed about the today's trends and/or forecast them were it is possible in order to have an appropriate perspective about the company's competitors and the strategies that can make the business stand out in the market.

3. Olive Oil Use and Market Trends

3.1. The use and quality of olive oil

The olive oil is a product with recognized value from the ancient years. It was used as a consumable product but it had also a symbolic value. The olive trees and the olives were used as a sign of peace and purity. Nowadays, it is cultivated mostly in Europe, whereas Greece is on the top three countries with the highest production.

The use of olive oil is high not only in the Mediterranean areas but also worldwide. The Mediterranean nutrition, which is sustained by wheat, olives and grapes, traditionally has become more and more popular. The olive oil is the central element on this type of diet and it provides several heart health benefits, such as favourable effects on cholesterol oxidation, whilst it exerts antiflammatory, antithrombotic and antihypertensive effects. The olive oil's freshness and flavour depends not only on the stakeholders and the production system but also on the material that is going to be used for packaging. The olive oil in general diminishes over time; however a wise choice of materials prolongs the freshness of the product. Therefore, glass is a well-researched packaging material for olive oil. Dark glass offers better protection from light contrary to the clear glass, which is though more acceptable and preferred by the customer. Consequently, if brand decides to use clear glass, it is better to cover the bottle with a label or UV blocker or design a cardboard case for protection [6].

Concerning the other materials, the aluminium has not yet been studied extensively by the scientific community, whilst the plastic permits the air and humidity to pass through the bottle. Stainless steel is the best material for olive oil packaging, though it costs a lot and it is less suitable for commercial packaging and customer's preferences. Tinplate cans are not clear yet concerning their suitability, as half of the researchers agree on its being good for storage, while others do not. Finally, coated paperboard is highly effective but not acceptable from the client, while the bag–in–box packaging needs more research for olive oil storing, because it may use plastics that can influence the shelf life of the product [7].

3.2. The today's olive oil market

The olive oil is a product that is cultivated in 23 countries worldwide and Greece is the world's third largest producer of olive oil (11% of total volume production), coming after Spain (40%) and Italy (14%). Greek olive oil is also of superior quality since 80% of production is of extra virgin grade, compared with its two biggest competitors, Spain and Italy, for which the relative percentage is 30 % and 65% respectively. Despite the fact that it has the highest annual consumption in the world (16kg), its exports are limited (only 1/3 of its overall production). According to the International Olive Council, Greece is the world's leading exporter for extra virgin olive oil. Greek olive oil mainly goes to EU (90%) while the 10% goes to Canada, Australia, Japan and United States [8].

Concerning the numerical part of the exports, the 80% of the exports of Greek olive oil in EU for the year 2012 were bulk, whilst only the 10% was exported by Greek brands [9]. Today the percentage that reaches this stage of labelling/branding, has increased to 27 % of the Greek olive oil production, which is still a low percentage, as the corresponding percentages for Spain and Italy, which are the biggest competitors for Greece in olive oil production, are 50% and 55 % for Spain and 75% and 97% for Italy respectively [10].

Though Greece possesses the highest percentage in extra virgin grade, the Greek olive oil has not managed to gain one of the first eighteen positions neither in the rankings that have been exported by the non –government organization World's Best Olive Oils, nor to the New York International Olive Oil Competition for the year 2015. WBOO collects the top olive oils from all worldwide's competitions, some of which to be granted by IOC, with Spain and Portugal to be on the top positions and Greece on lower ones [11].

In the NYIOOC competition, Spain got the first positions, whilst Australia and Italy were also on the top 10. Nineteen Greek olive oils managed to gain the 'gold' and 'silver' award , and two of them, were on the top thirty [12]. However, it should be referred that Greece has possibilities to ascend, due to the fact that more and more producers show interest in the sector: the country had the biggest number in entries (168) in NYIOOC for the year 2015 and also more than any other year.

3.3. Trends in EU consumption

Taking into account the fact that EU holds the ¾ of world's olive oil production, it asserts the largest percentage on sales. Also, the popularity of the healthy Mediterranean diet increases, especially for the olive oil, which is shown by the fact that the demand for the product has been doubled in other countries, except for the three main producers during the last 20 years.

A negative part on the whole statistical analysis is that the olive oil production in Greece (and Italy) has been reduced. Indeed, in 1990 the production was higher than the year 2015 contrary to EU in general and specifically the other EU countries that their production increased dramatically from 1990 and on. A quite positive element for Greece though is that its olive oil production has remained almost stable through years, with a downtrend the last decade.

Finally, the evidence show a downtrend in olive oil production and the 2012 to 2020 projections forecast that the exports are going to slightly fall. However, the new data show that there are positive elements about the development of the Greek olive oil agricultural sector. The percentage of the olive oil producers that have labelled their olive oil has been increased by 20% in the last two years. Therefore, the interest for the Greek olive oil production is higher, something that is further underlined by the 168 olive oil brands which participated in the competition that was organized by the NYOOC [13].

4. The Proposed Packaging Design Concept

4.1. Concept Development

The proposed olive oil packaging design concept is a derivative of the aforementioned analysis. Accordingly, the ideas themselves but also the colour palettes and the lines were made up and designed by taking into account all of the above. Therefore, the final concept is named *'The olive and its core'* and it concerns two characters that are complementary to each other. The olive itself needs the core in order to mature.

The narrative has generated a series of products with an integrated design language. Each product has its complementary part; either it is the package itself or the product that is protected by the pack, exactly as the core is protected by the olive. The concept is playful and pure in order to be stamped in the customer's mind and the design simple and definite: the shapes of the packages appear like the shape of the olive (figure 1).

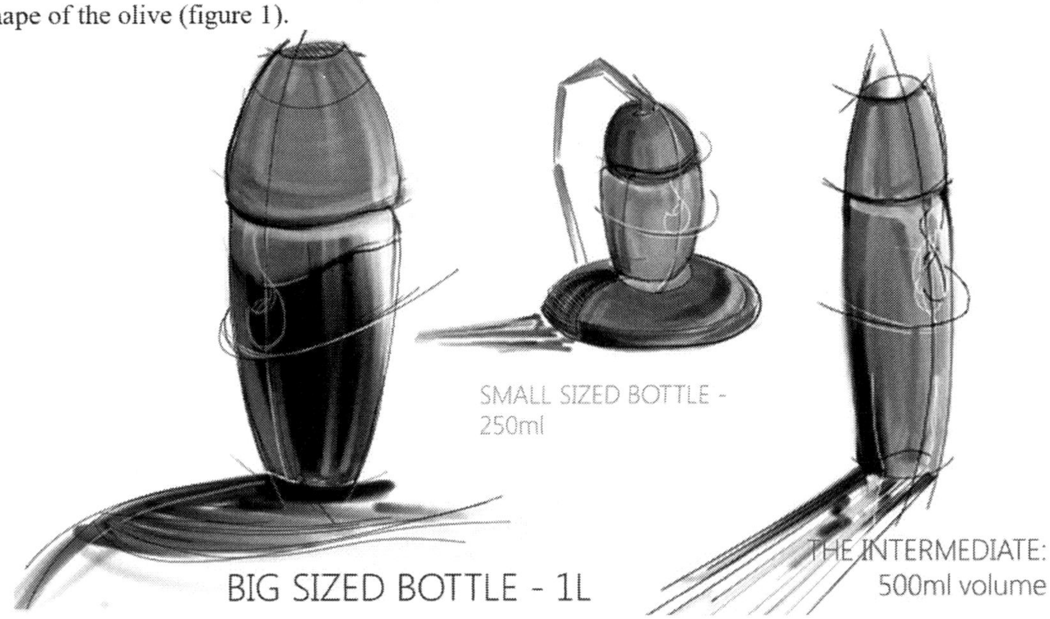

Figure 1. The proposed design concept

4.2. Brand Name/Typography

This particular brand name brings in mind the main purpose that the package serves: to illustrate verbally the image of an element related with the olive oil, viz the olive on this case. Due to the fact that the brand in general represents the story of a company and the image is the most important element in brand smash framework, the creation of such a hinted 'story' is appropriate. If the name was something completely different from the olives, then the most probable result would be the confusion of the user and his consequent detachment from the product itself.

Complementary to the brand name, the logo of the product is an abstractive and simple illustration of the olive and its core. It therefore shows simply and directly the concept of the packaging. The color of the logo can vary depending on the purpose and especially the element that is to be enhanced. However, a pallet of earth colors was chosen to be used ad hoc, because the olive tree is a product related with the earth. It should be mentioned though that the olive oil is an extremely luxurious product by itself. Thus in order to accentuate the gold olive oil's color and not absorb its value, neutral colors were chosen for the logo on the bottles (not the overall packaging system). The blue and red that are used traditionally by the most well-known companies, were set aside, as they can't be harmonized with this case (figure 2).

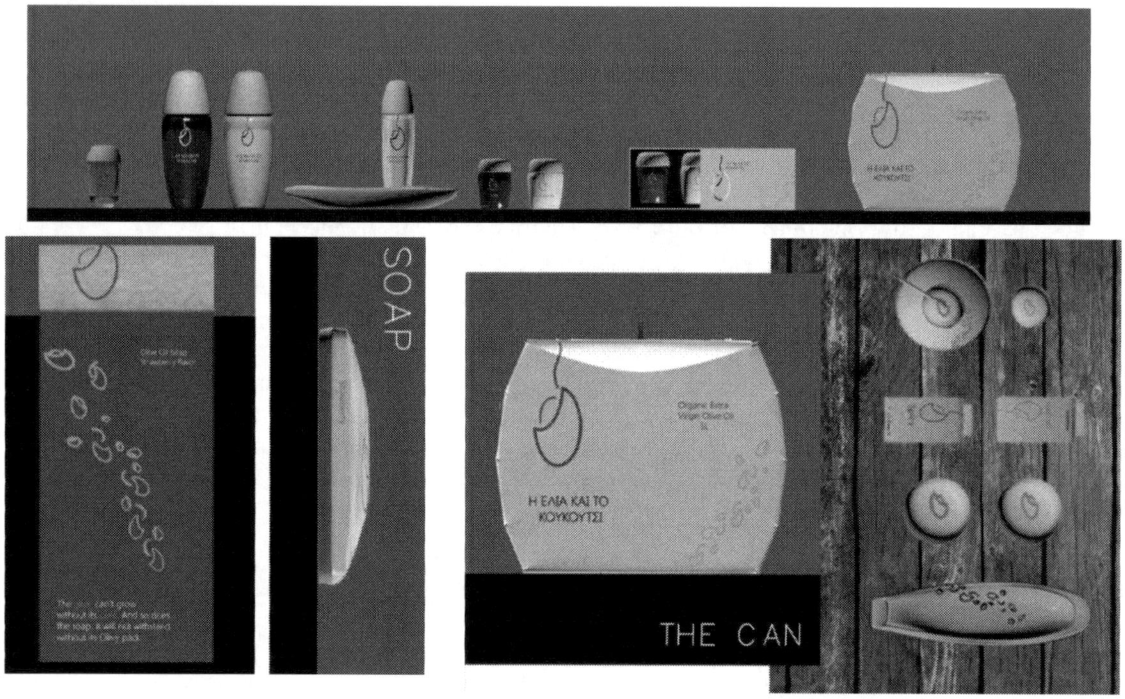

Figure 2. The series of the proposed olive oil products

4.3. Visibility – Materials

Concerning the visibility part, the designed bottles are followed by the mild and symmetric curves of the olive's shape. These forms are not the typical ones for the designing of an olive oil bottle, but are connected a lot with the olive oil due to their general shape and the story behind - logo. Its difference from the traditional forms and the immediate reminding of the olive oil are elements that can make the user differentiate the product on the shelf. Furthermore, the name is light enough, not at all austere, contrary to other brands and the curvy cute shapes create a positive aura to be used for food consumption.

It should also be mentioned that the mild curves of the shaping of the bottles are those that will lead to a more overall positive attitude. On the other hand, the angular shapes inspire intense taste sensations, an element that is contradictory to the mild flavor of the olive oil, as it functions as an accompanying element to the cooked or raw food (figure 3).

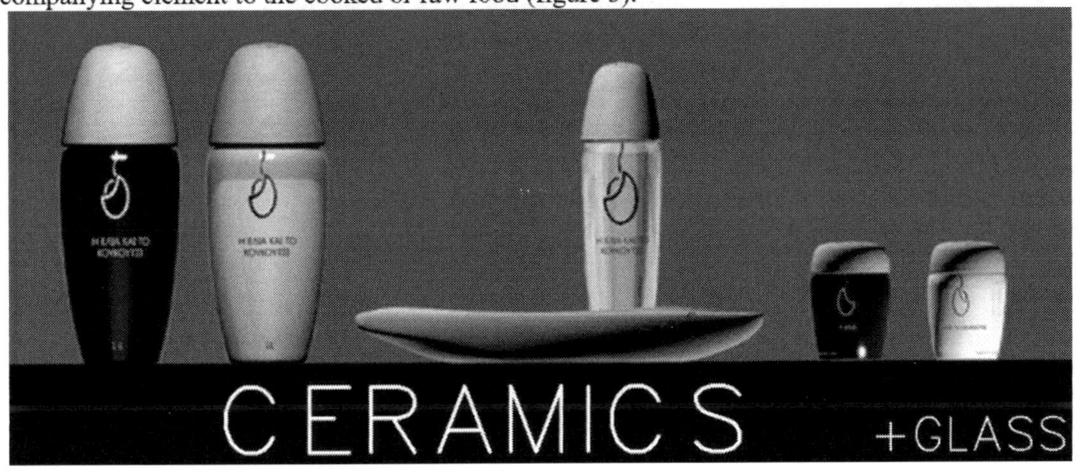

Figure 3. The materials and shapes of the packaging system

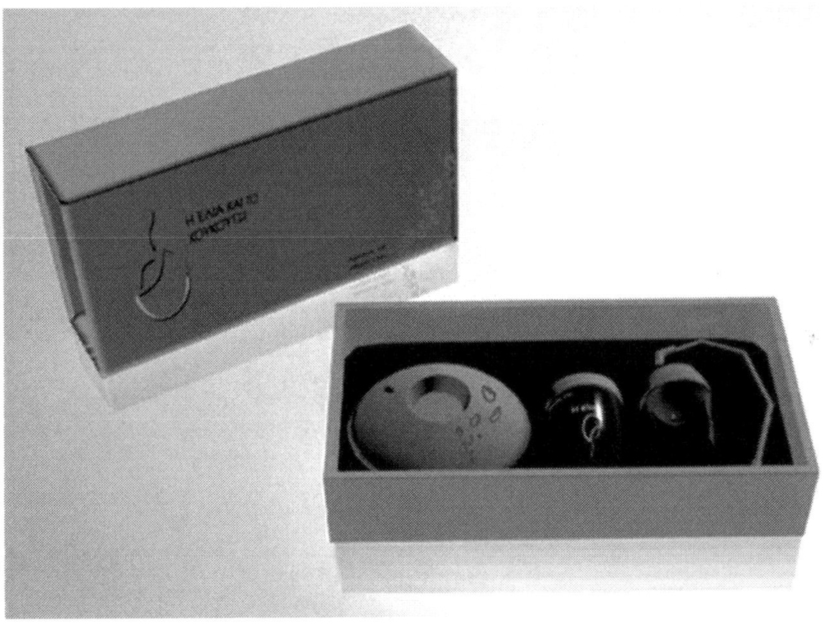

Figure 4. The proposed case

The contrast on the color of the packages and the logo are alert points for the client together with the materials combination, leading also to a cognitive mapping of the brand. The materials that are used for the final packaging design are those that have been studied above and are appropriate for the conservation of the olive oil, viz the ceramic, the wood and the glass for the bottles and stainless steel for the can and the cases (figure 3). In case of the clear glass, cases have been designed not only for better protection but also to fit more than one product in one pack (figure 4). Also, the craft paper is used for the smallest pieces, e.g. soap packages and individual olive oil proportions (figure 5).

5. Conclusions

The expansion of the market nowadays has created risks in the business sector, so that companies are closing down or being dissolved every day. However, such a failure is happening due to the fact that most industries have not figured out the key factors that can increase their profit. Most of them neglect

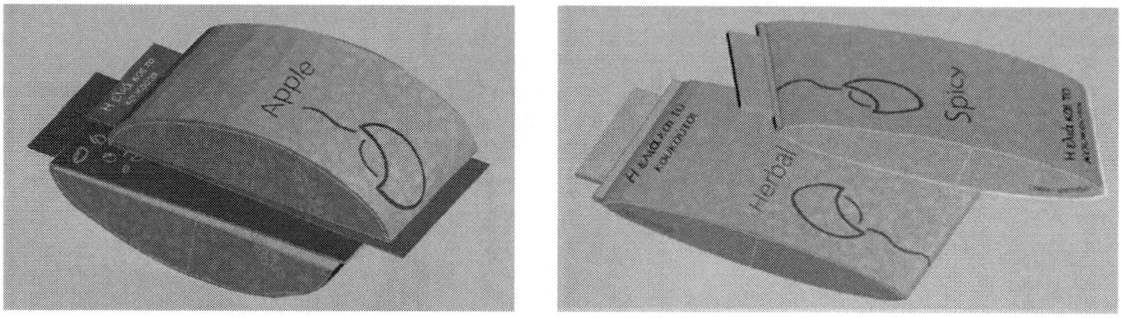

Figure 5. The proposed packaging for soaps and olive oil proportions

first of all the packaging design of a product, whilst others emphasize insignificant elements, do not follow a holistic strategic plan and set aside the characteristics that can improve them in the competitive market.

The present paper stressed the importance of packaging design in the olive oil industry. The market trends were presented and a case study was proposed at a conceptual level. In addition, the large margins for profitability offered to the Greek olive oil industry were revealed.

References

[1] Capsule 2008 *Design Matters: Packaging, an essential primer for today's competitive market* Rockport Publishers Inc.

[2] Dupuis St and Silva J 2011 *Package Design Workbook: The Art and Science of Successful Packaging* Rockport Publishers Inc.

[3] Lindstrom M 2005 *Brand Sense: Build Powerful Brands through Touch, Taste, Smell, Sight and Sound* Free Press, New York

[4] Chapman Available at: http://www.packagingoftheworld.com/2014/10/5-emerging-packaging-design-trends-for.html, Accessed: 03/05/2016

[5] Desmet PMA, Fenko A, Labbe D, Martin N and Schifferstein HNJ 2013 Influence of Package Design on the Dynamics of multisensory and emotional food experience *J. Food Quality and Preference* **27** 18-25

[6] Piscopo A and Poiana M 2012 *Packaging and Storage of olive oil, included in the book Olive Germplasm – The Olive Cultivation, Table Olive and Olive Oil Industry in Italy* edited by Muzzalupo I, ISBN:978-953-51-0883-2

[7] Wang S, Li X, Wang, Rodriguez R and Flynn D 2014 *Packaging influences on olive oil quality: A review of the literature* UC Davis Olive Center

[8] Olive Oil Available at: https://www.nbg.gr/greek/the-group/press-office/e-spot/reports/Documents/Olive%20Oil_2015.pdf. Accessed: 04/04/2016

[9] IOC Available at: http://www.internationaloliveoil.org/estaticos/view/136-country-profiles, Accessed: 05/04/2016

[10] SEVITEL Available: http://www.oliveoil.gr/component/content/article/1-Site/45-Το-ελληνικό-ελαιόλαδο-με-ορίζοντα-το-2020. Accessed: 02/05/2016

[11] WBOO Available at: http://www.worldsbestoliveoils.org/worlds-best-olive-oils.html, Accessed: 04/05/2016

[12] NYIOOC Available at: http://www.bestoliveoils.com/pairing/, Accessed: 20/04/2016

[13] EU Commission Available at: http://ec.europa.eu/agriculture/olive-oil/economic-analysis_en.pdf, Accessed: 01/05/2016

Experimental investigation and modelling of surface roughness and resultant cutting force in hard turning of AISI H13 Steel

M Boy[1], N Yaşar[1] and İ Çiftçi[2]

[1] Karabük University, TOBB Technical Sciences Vocational College, 78050, Turkey

[2] Karabük University, Department of Manufacturing Engineering, 78050, Turkey

Email: mboy@karabuk.edu.tr

Abstract. In recent years, turning of hardened steels has replaced grinding for finishing operations. This process is compared to grinding operations; hard turning has higher material removal rates, the possibility of greater process flexibility, lower equipment costs, and shorter setup time. CBN or ceramic cutting tools are widely used hard part machining. For successful application of hard turning, selection of suitable cutting parameters for a given cutting tool is an important step. For this purpose, an experimental investigation was conducted to determine the effects of cutting tool edge geometry, feed rate and cutting speed on surface roughness and resultant cutting force in hard turning of AISI H13 steel with ceramic cutting tools. Machining experiments were conducted in a CNC lathe based on Taguchi experimental design (L_{16}) in different levels of cutting parameters. In the experiments, a Kistler 9257 B, three cutting force components (Fc, Ff and Fr) piezoelectric dynamometer was used to measure cutting forces. Surface roughness measurements were performed by using a Mahrsurf PS1 device. For statistical analysis, analysis of variance has been performed and mathematical model have been developed for surface roughness and resultant cutting forces. The analysis of variance results showed that the cutting edge geometry, cutting speed and feed rate were the most significant factors on resultant cutting force while the cutting edge geometry and feed rate were the most significant factor for the surface roughness. The regression analysis was applied to predict the outcomes of the experiment. The predicted values and measured values were very close to each other. Afterwards a confirmation tests were performed to make a comparison between the predicted results and the measured results. According to the confirmation test results, measured values are within the 95% confidence interval.

1. Introduction

The hard turning process is gradually replacing the traditional grinding in the automotive and tool and die industries. The traditional method of machining hardened materials included heat treatment, rough turning, and then grinding process. As a result of the advances in machine tools and cutting tool technology, hard turning becomes an effective manufacturing process to produce parts with precision and surface quality. Hard turning provides a number of potential benefits to grinding machining like as higher material removal rates, more flexibility, lower energy consumption, elimination coolants, shorter set up time in complex parts process. The hard turning is generally performed without a coolant using ceramics and cubic boron nitride (CBN) cutting tools due to the required tool material hardness.

Especially, advanced alumina-based ceramic cutting tool is widely used as machining of various types of steel, cast iron, non-ferrous metals, and refractory nickel-based alloys [1-5].

Traditionally, a wide range of tool steels is used in producing dies and molds. AISI H13 hardened steel exhibits high hardness, toughness, high temperature strength, high resistance to thermal shock, thermal fatigue and thermal softening. Due to its mechanical properties and aggressive cutting conditions, hardened AISI H13 steel is recognized to be a difficult-to-cut material [6]. Many researchers have conducted various studies on cutting edge geometry, cutting force, surface roughness, cutting tool material, tool wear in hard turning of various hardened steels. Özel et al. used Taguchi's L16 orthogonal array to determine the effects of the cutting tool edge geometry, workpiece hardness, feed rate and cutting speed on surface roughness and resultant forces in finish hard turning of AISI H13 steel using CBN tools experimental results of their study indicated that the effect of cutting edge geometry on the surface roughness was remarkably significant. The cutting forces were influenced not only by cutting conditions but the cutting edge geometry and workpiece surface hardness also had a significant effect [7]. Yalçın studied the performance of CBN and the mixed ceramic tools in turning soft and hard H13 tool steel with different cutting parameters. The interaction of the cutting parameters on the surface roughness and cutting forces was determined by using ANOVA. The results showed that the surface roughness in hard turning was lower with the CBN than with the ceramic tool [8]. Thiele and Melkote investigated the effects of tool cutting edge geometry and workpiece hardness on the surface roughness and cutting forces in the finish hard turning of AISI 52100 steel. They found that increasing the edge hone radius tends to increase the average surface roughness and the effect of edge hone on the surface roughness decreased with an increase in workpiece hardness. Also, it was revealed that the cutting forces were influenced by the workpiece hardness and the cutting edge geometry. [9]. De Oliveira et al. studied the performance of CBN and whiskers reinforced ceramics in continuous and interrupted hard turning with constant cutting conditions. Their results showed that, in continuous turning, the longest tool life is achieved using PCBN, but similar tool life is attained in interrupted turning using both PCBN and ceramic. In terms of roughness, the PCBN tools showed better results for continuous and interrupted surfaces [10]. Aslan et al. conducted an optimization study by machining a hardened AISI 4140 grade (63 HRC) steel on a lathe by using Al_2O_3 + TiCN coated ceramic inserts. The relationship between the parameters and the responses were determined using multiple linear regression analysis. They found that the VB value decreased as the cutting speed and the depth of cut increased; however, it first decreased and then increased as the feed rate increased. On the other hand, the surface roughness decreased as the cutting speed increased. In contrast surface roughness increased when the feed rate increased [11]. Fnides et al., investigated the effect of three cutting parameters on surface roughness in turning of X38CrMoV5-1 hardened steel treated at 50 HRC using mixed ceramic tool. Their results revealed that the effect of feed rate on surface roughness is more significant than cutting speed, whereas the depth of cut is not significant [12]. Aouici et al. investigated the effects of cutting speed, feed rate, workpiece hardness and depth of cut on surface roughness and cutting force components in hard turning. Mathematical models were developed for surface roughness and cutting force components using the response surface methodology (RSM). Results showed that the cutting force components were influenced principally by depth of cut and workpiece hardness; however, both feed rate and workpiece hardness had statistical significance on surface roughness [13]. Bensouilah et al. adopted ANOVA and RSM to effect of cutting parameters on the surface roughness and cutting force components during hard turning of AISI D3 cold work tool steel with ceramic inserts. The results revealed that the minimum surface roughness obtained with the coated CC6050 ceramic insert. However, the uncoated ceramic insert was useful in reducing the machining force [14]. Günay and Yücel investigated optimizing the cutting conditions for the average surface roughness (Ra) in machining of high-alloy white cast iron (Ni-Hard) at two different hardness levels (50 HRC and 62 HRC). ANNOVA results showed that feed rate is the most significant variable for Ni-Hard with 62 HRC while the cutting speed is the most significant for Ni-Hard with 50 HRC [15]. In a further work, they investigated the effect of cutting conditions for cutting force and surface roughness. They found that depth of cut and feed rate is the most significant factor on surface roughness and cutting force for both the ceramic and CBN cutting tool.

Also they obtained the smallest surface roughness with CBN insert during machining of Ni-Hard with 62 HRC [16]. Xiong et al. investigated the effect of cutting parameters on the tool life and tool wear mechanism of conventional cemented carbide for machining of AISI H13 hardened steel. The results showed that WC–5TiC–10Co ultrafine cemented carbide possessed higher hardness values and transverse rupture strength, and exhibited better cutting performance than conventional inserts under the same cutting conditions [17].

This study aims at investigating the effects of cutting tool edge geometry, feed rate and cutting speed on surface roughness and resultant cutting force in hard turning of AISI H13 steel with ceramic cutting tools of different edge preparations. Taguchi's design and analysis of experiment process has been used to achieve this purpose. L16 orthogonal array was used in the design of experiment. Furthermore, analysis of variance is used to determine the statistical significance of the cutting parameters. Finally, confirmation tests were carried out using the optimal cutting conditions which were determined by Taguchi optimization method. Also, a mathematical model was developed to estimate the resultant cutting force and average surface roughness (Ra) with regression analysis by using experimental results.

2. Experimental

2.1. Experimental Conditions and Equipment's

The hard turning tests were performed on hot work tool steel (AISI H13). The cylindrical AISI H13 specimens utilized in these experiments had a diameter of 45 mm and length of 200 mm. The specimens were through-hardened and tempered to obtain the hardness value of 55 HRC. The chemical composition of AISI H13 steel is given Table 1. Mixed ceramic (Al_2O_3+TiC) inserts with two different types of edge preparations were investigated in this study. The cutting tools used were commercial grade ceramic inserts produced by Kyocera with the geometry of DNGA 150404 S/T02025 PT600M. PT600M is a special coating called "Megacoat". These inserts were clamped mechanically on a rigid tool holder with an ISO designation of DDJNR 2525M-1504. The inserts were positioned into a tool holder characterized by a negative rake angle of γ=−6°, clearance angle of α=−6°, and an approach angle of χ=93°. Cutting forces were measured with a Kistler 9257 B type piezoelectric dynamometer and an associated 5019 B130 charge amplifier connected to PC employing Kistler Dynoware software. During the hard turning tests, the cutting forces known as the primary cutting force Fc, the feed force Ff, and the radial force Fp were recorded. Longitudinal turning was conducted on CNC lathe (Johnford TC35) at a constant depth of cut at 0.1 mm. The length of cut for each test was 20 mm in the axial direction. Average surface roughness (Ra) value which is one of the most important machinability criteria was measured according to ISO 4287 standard. Ra measurements were performed by using a Mahrsurf PS1 surface roughness gauge with a cut-off length of 0.8 mm and sampling length of 5 mm. Ra values were calculated by averaging three roughness values obtained from three different positions of machined surface. The flowchart for modelling of the hard turning parameters is shown in figure 1.

Table 1. Chemical composition of AISI H13 tool steel (%).

C	Si	Mn	P	S	Cr	Mo	V
0.38	0.97	0.34	0.02	0.002	5	1.34	0.93

2.2. Design of experiment and analysis

A there-factor four-level factorial design was used to determine the effects of the tool edge geometry, feed rate and cutting speed on surface roughness and resultant forces in the hard turning of AISI H13 steel. Four different levels of cutting speed (Vc) and feed rate (f) and two different levels of tool edge geometry (TEG) are chosen as the control factors and their levels were determined as shown in Table 2.

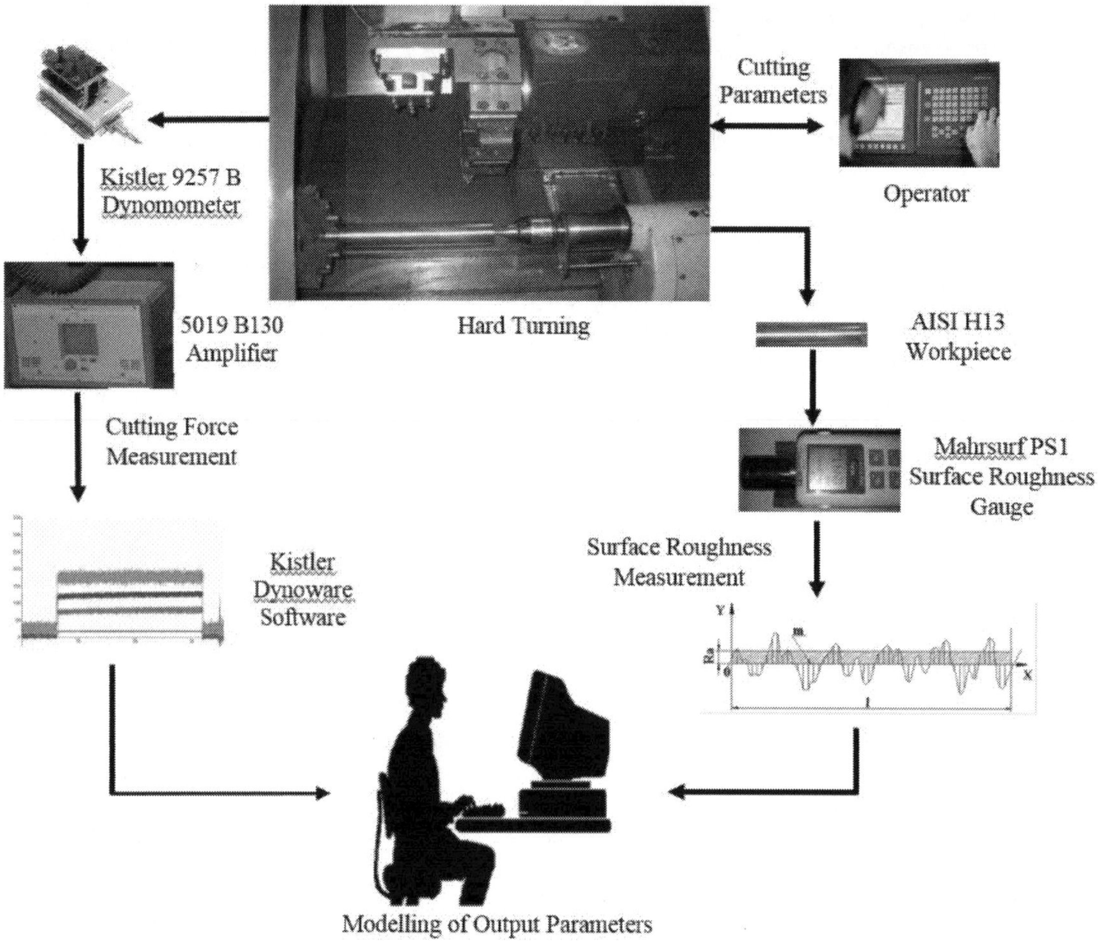

Figure 1. The flowchart for modelling of hard turning parameters.

The parameter levels were chosen within the intervals recommended by the cutting tool manufacturer. The L_{16} mixed orthogonal array of the Taguchi method was used to determine the optimum cutting conditions and analyze the effects of the machining parameters. ANOVA was applied with 95% confidence level to determine the significance level of the variables on average surface roughness (Ra) and resultant cutting force (F). The purpose of ANOVA is to determine which cutting parameters significantly affect the performance characteristics of independents variables. The ANOVA test was performed to evaluate the statistical significances of the fitted regression model and factors involved therein for the response factors namely F and Ra. From the experimental data the effects of three predictor factors (cutting speed, feed rate and cutting tool geometry) upon two response variables were analyzed by using MINITAB statistical software. Cutting conditions have been optimized separately for Fc and Ra. Generally, the smaller-the-better, the higher the- better, and the nominal-the-best quality characteristics use in the analysis of the S/N ratio. The aim of this study was to minimize surface roughness and resultant cutting force. Therefore, the-smaller-the-better performance characteristic for Ra and F were applied in order to obtain the optimal cutting conditions. S/N ratio (η) is defined as follows:

$$\frac{S}{N(\eta)} = -10 \cdot \log\left[\frac{1}{n} \cdot \sum_{i=1}^{o} Y_i^2\right] \tag{2.1}$$

where Y_i is the observed data at the experiment and n is the number of experiments. The S/N ratios of Ra and F were calculated using equation (2.1).

Table 2. Factors and their levels.

Factors	Level 1	Level 2	Level 3	Level 4
A: Cutting speed, Vc (m/min)	60	80	100	120
B: Feed rate, f (mm/rev)	0.04	0.06	0.08	0.10
C: Tool Edge Geometry, (TEG)	1[a]	2[b]		

a: Chamfered tool edge geometry
b: Chamfered + honed tool edge geometry

3. Experimental Results and Discussion

3.1. Analysis of the Signal-to-Noise (S/N) Ratio

Average surface roughness and resultant cutting force occurred during machining of the AISI H13 steel were measured after the experiments performed according to the L_{16} orthogonal array. S/N ratios of the Ra and F data obtained from the experimental results, which will be used to determine the optimal levels of each variable, were calculated according to the equation (2.1). The experimental results and S/N ratios calculated according to Taguchi's "the-smaller-the-better" quality characteristic were given in Table 3.

Analysis of the effect of each control factor on the surface roughness and resultant cutting force was performed with a S/N response table. The response tables of S/N for Ra and F are shown in Table 4. This table, which is made by using the Taguchi method shows the optimum levels of control factors for the optimum surface roughness and resultant cutting force values.

The level values of control factors for Ra and F given in Table 4 are shown in graph forms in figures 2 and 3. Optimum machining parameters of the control factors for minimizing the surface roughness and resultant cutting force can be easily determined from these graphs. The best level for each control factor was found according to the highest S/N ratio in the levels of that control factor. According to this, the levels and S/N ratios for the factors giving the best Ra value were specified as factor A1, B1 and C2.

Table 3. Experimental results and corresponding S/N ratios.

Run	Control factors			Experimental results		Signal-to-noise ratio (S/N)	
	A- (f) (mm/rev)	B-(Vc) (m/min)	C- (TEG)	R_a (μm)	F (N)	R_a	F
1	0.04	60	1	0.523	255.7	5.62997	-46.1967
2	0.04	80	1	0.552	244	5.16122	-45.6350
3	0.04	100	2	0.446	173.1	7.02305	-45.1267
4	0.04	120	2	0.449	167.6	6.95507	-44.7015
5	0.06	60	1	0.509	305.5	5.86564	-48.0791
6	0.06	80	1	0.535	299.6	5.43292	-47.8703
7	0.06	100	2	0.579	227.8	4.74643	-47.1056
8	0.06	120	2	0.480	222.3	6.37518	-47.1045
9	0.08	60	2	0.529	311.4	5.53089	-49.3525
10	0.08	80	2	0.563	284.9	4.98983	-48.4393
11	0.08	100	1	0.736	344.3	2.66244	-48.9713
12	0.08	120	1	0.601	331.4	4.42251	-48.6357
13	0.10	60	2	0.741	344.1	2.60364	-50.2740
14	0.10	80	2	0.775	335.4	2.21397	-49.9645
15	0.10	100	1	0.955	399.5	0.39993	-50.4816
16	0.10	120	1	0.985	385.7	0.13128	-50.0525

In other words, an optimum Ra value was obtained at a feed rate (A1) 0.04 mm/rev, at cutting speed (B1) 60 m/min and with chamfer plus hone edge (C2) (figure 2). Therefore, by considering the S/N ratios in Table 4, the optimum cutting conditions for the resultant cutting force were A1, B4 and C2. The smallest surface roughness and resultant cutting force and their S/N ratio that can be obtained according to these levels were calculated by using equation (3.1) and (3.2). In these equations, η_G is the S/N ratio calculated at the optimum levels (dB), $\bar{\eta}_G$ is the average S/N ratios of all variables (dB), $\bar{A}o, \bar{B}o, \bar{C}o$ are the mean S/N ratios when the factors A, B and C are at optimum levels (dB), and Ra_{cal} is the calculated surface roughness value. At the end of the machining tests, the average values of the surface roughness and resultant cutting force were calculated to be 0.427 μm and 173.62 N respectively. Similarly, average values of S/N ratio for surface roughness and resultant cutting force were calculated to be 7.387 dB and -44.792 dB respectively.

$$\eta_G = \bar{\eta}_G + (\bar{A}_O - \bar{\eta}_G) + (\bar{B}_O - \bar{\eta}_G) + (\bar{C}_O - \bar{\eta}_G) \tag{3.1}$$

$$Ra_{cal} = 10^{-\eta_G/20} \tag{3.2}$$

Table 4. Response table for S/N ratios of surface roughness and resultant cutting force.

	Surface roughness			Resultant cutting force		
Level	A	B	C	A	B	C
1	6.192	4.908	3.713	-45.41	-48.48	-48.24
2	5.605	4.449	5.055	-47.54	-47.98	-47.76
3	4.401	3.708		-48.85	-47.92	
4	1.337	4.471		-50.19	-47.62	
Δ	4.855	1.200	1.344	4.78	0.85	0.48

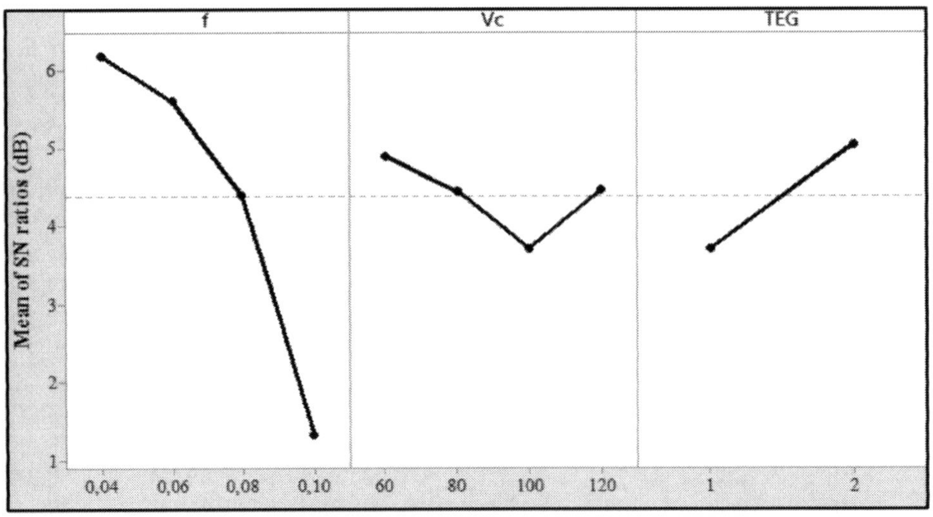

Figure 2. Effect of process parameters on average S/N ratio for Ra

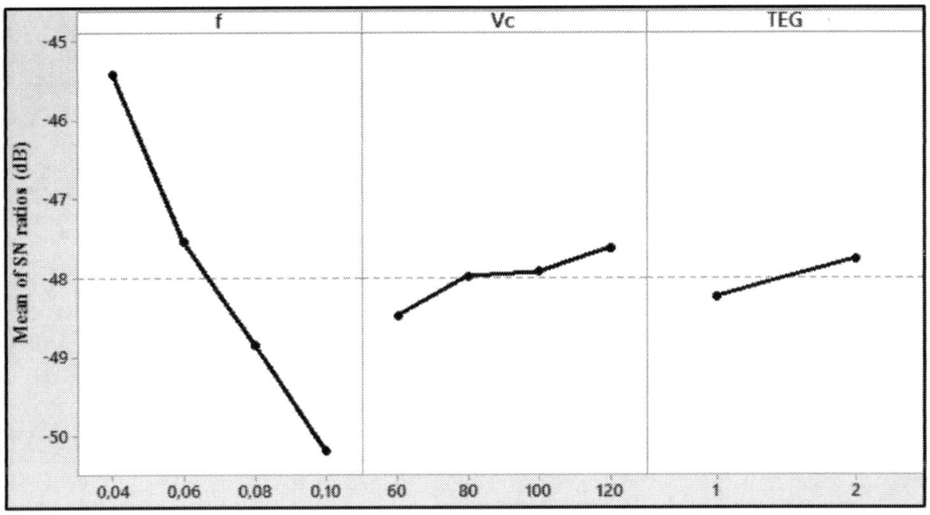

Figure 3. Effect of process parameters on average S/N ratio for F.

3.2. ANOVA method

The analysis of variance (ANOVA) was used to investigate which design parameters significantly affect the surface roughness and resultant cutting force. This analysis was carried out for a significance level of a = 0.05, i.e. for a confidence level of 95%. Table 5 show the P-values, that is, the realized significance levels, associated with the F-tests for each term of variation. The terms with a P-value less than 0.05 are considered to have a statistically significant contribution to the performance measures and the obtained models are considered to be statistically significant. It shows that that the conditions chosen in the model have significant effects on the responses. The ANOVA results for surface roughness are shown in Table 5. This table also shows the degrees of freedom (DF), sum of squares (SS), mean square (MS), F-values (F) and probability (P) in addition to the percentage contribution ratio (PCR) of each factor. F ratios and their PCR were taken into consideration to identify the significance level of the variables. Table 5 indicates that the most effective variable on the Ra value is the feed rate with 80.7% of PCR. It is well known that the theoretical geometrical surface roughness is primarily a function of the feed for a given nose radius and changes with the square of the feed rate value. It is once again shown that feed rate has important effect on surface roughness in hard turning. The other variables having effect on Ra are cutting tool edge geometry with 10.33 %. The cutting speed had a small influence of 4.27%.

Table 5. ANOVA for S/N ratio of surface roughness and resultant cutting force.

Source	DF	SS	MS	F-Ratio	P	PCR (%)
Surface roughness						
A	3	56.177	18.7257	45.95	0.000	80.7
B	3	2.972	0.9906	2.43	0.140	4.27
C	1	7.199	7.1987	17.66	0.003	10.33
Error	8	3.26	0.4075			4.7
Total	15	69.608				100
Resultant cutting force						
A	3	49.7034	16.5678	536.85	0.000	94.9
B	3	1.4984	0.4995	16.18	0.001	2.9
C	1	0.9281	0.9281	30.07	0.001	1.8
Error	8	0.2469	0.0309			0.04
Total	15	52.3768				100

Table 5 showed that the main effects of cutting speed, feed rate and edge geometry are all significant on the resultant cutting force. Table 5 shows that the most effective variable on the resultant cutting force is the feed rate with 94.9% of PCR. The other variables having effect on resultant cutting force are cutting speed with 2.9% and tool edge geometry with 1.8%. The error ratio was calculated as 0.04% and it is the smallest ratio.

3.3 Regression equations
The correlation between the cutting conditions and the machining output factors were performed by regression analysis. Regression analysis are used for the modeling and analyzing of several variables where there is relationship between a dependent variable and one or more independent variables. The predictive equations which were obtained by the linear regression model of surface roughness and resultant cutting force are given in equation (3.3) and (3.4). Coefficient of determination (R^2) value for the Ra and F were calculated as 80,34 % and 98.7 %, respectively.

$$R_a = 0.256 + 5.982xf + 0.001162xVc - 0.1043xTEG \tag{3.3}$$

$$F = 147.7 + 2241xf - 0.3346xVc - 11.93xTEG \tag{3.4}$$

The comparison of experimental results and predicted values of machining output factors calculated from equation (3.3) and (3.4) are illustrated in figure 4. Evaluating the figures and R^2 values, the predicted models found for average surface roughness and resultant cutting force are satisfactory. The above mathematical model can be used to predict the values of the surface roughness parameters and resultant cutting force components the limits of the factors studied. The results of comparison were demonstrated to predict the values of surface roughness parameters and resultant cutting force close to those readings recorded experimentally with a 95% confidence interval.

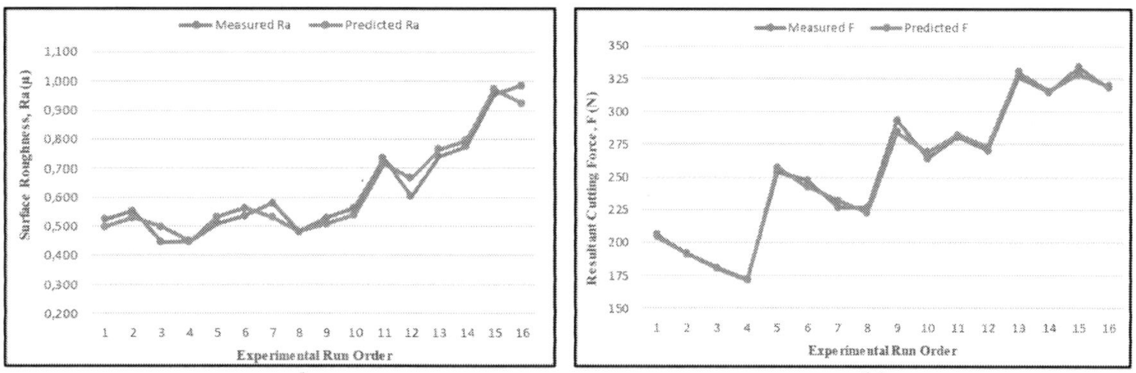

a) b)

Figure 4. Comparison between measured and predicted values; a) Ra, b) F.

3.4. Confirmation tests
Confirmation tests are a crucial step recommended by Taguchi to verify experimental conclusions. The purpose of the confirmation experiment is to validate the conclusions drawn during the analyzing. For this purpose, the following equations were used in the specification of the confidence interval (Cl) for estimated surface roughness. Confirmation test was required in the present case study because the optimum combination of parameters and their levels i.e. A1B1C2 did not correspond to any experiment of the orthogonal array. Optimum combination of parameters and their levels i.e. A1B4C2 for resultant cutting force corresponded to experiment of the orthogonal array (OA). Therefore, it may be noted that if the optimal combination of parameters and their levels coincidently match with one of the experiments in the OA, then no confirmation test is required.

$$CI = \sqrt{F_{0,05}(1, \upsilon_e)V_e\left(\frac{1}{n_{eff}} + \frac{1}{r}\right)} \tag{3.5}$$

$$n_{eff} = \frac{N}{1+\upsilon_T} \tag{3.6}$$

Table 6. Comparison between confirmatory test results and calculated values

Confirmatory	Experiment results	Calculated values		Differences	
Ra_{mea}(μm)	η_{mea}(dB)	Ra_{cal}(μm)	η_{cal}(dB)	$\lvert Ra_{mea} - Ra_{cal} \rvert$	$\lvert \eta_{mea} - \eta_{cal} \rvert$
0.479	6,393	0.427	7.387	0.052	0.994

Here, υ_e is the error degree of freedom, V_e is the error variance, n_{eff} is the repeating number of the experiments (equation 3.5) N is the total number of the experiments, υ_T is the variable's degree of freedom and r is the number of confirmation tests (equation 3.6). Using equations (3.5) and (3.6), confidence value of ±1.472 dB was obtained for surface roughness. Table 6 shows the differences between the values obtained by confirmatory tests and the values calculated by equations (3.5) and (3.6) of the S/N ratios. It is seen that the difference of 0.994 dB is under 5% confidence interval of 1.472 dB for surface roughness. Therefore, the optimal control factor settings for all the cutting conditions were confirmed as confident.

4. Conclusion

In this study, the Taguchi method was used to obtain optimum cutting conditions (feed rate, cutting speed and cutting tool edge geometry) in hard turning of AISI H13 steel. Experimental results were evaluated using ANOVA. The results can be drawn as follows:

- According to the results of statistical analysis, it was found that the feed rate and tool edge geometry was the most significant factor on surface roughness with a PCR of 80.7% and 10.33% respectively. The feed rate was the most significant parameters for the resultant cutting force with a percentage contribution of 94.9%. The other variables having effect on resultant cutting force were cutting speed with a percentage contribution of 2.9% and tool edge geometry with a percentage contribution of 1.8%.
- The optimum levels of the control factors for minimizing the surface roughness and resultant cutting force using S/N ratios were determined. The optimal conditions for surface roughness was observed at A1-B1-C2 (i.e., feed rate = 0.04 mm/rev, cutting speed = 60 m/min and chamfer+honed edges). The optimal conditions for resultant cutting force was observed at the same levels A1-B4-C2 (i.e., feed rate = 0.04 mm/rev, cutting speed = 120 m/min and chamfer+honed edge).
- Cutting edge geometry, feed rate and cutting speed are all found to be effective on resultant cutting force.
- Developed regression models demonstrated a very good relationship with high correlation coefficients (Ra = 0.803 and F = 0.987) between the measured and predicted values for surface roughness and resultant cutting force.
- Confirmation tests at optimal conditions were carried out. According to the confirmation test results, measured values are within the 95% confidence interval.

References

[1] Tönshoff H K, Arendt C and Ben Amor R 2000 CIRP Ann **49** 547-566
[2] Davim J P 2011 *Machining of Hard Materials* (London:Springer) pp 1-31
[3] Grzesik W 2008 *Advanced Machining Processes of Metallic Materials: Theory, Modelling and*

Application, (Oxfort:Elsevier) pp 245-259
[4] König W, Berktold A and Koch K F 1993 CIRP Ann **42(1)** 39-43
[5] Klocke F, Brinksmeier E and Weinert K 2005 CIRP Ann **54(2)** 22-45
[6] Çiçek A, Kara F, Kıvak T and Ekici E 2013 Int J Refract Met H **41** 461–469
[7] Özel T, Hsu T K and Zeren E 2005 Int J Adv Manuf Technol **25** 262–269
[8] Yalçın B 2015 T Can Soc Mech Eng **39(2)** 323
[9] Thiele J D and Melkote S 1999 J Mater Process Tech **94** 216-226
[10] Oliveira A J, Diniz A E and Ursolino D J 2009 J Mater Process Tech **209(12–13)** 5262-5270.
[11] Aslan E, Camuşcu N and Birgören B 2007 Journal Mater Des **28** 1618–1622
[12] Fnides B, Yallese M A, Mabrouki T and Rigal J F 2009 Mechanika **3(77)** 68-73
[13] Aouici H, Yallese M A, Chaoui K, Mabrouki T and Rigal J F 2012 Measurement **45** 344–353
[14] Bensouilah H, Aouici H, Meddour I, Yallese M A, Mabrouki T and Girardin F 2016 Measurement **82** 1–18
[15] Günay M and Yücel E 2013 Measurement **46** 913–919
[16] Yücel E and Günay M 2013 Proc. Inst. Mech. Eng. C J. Mech. Eng. Sci. **227(10)** 2280-2290
[17] Xiong J, Guo Z, Yang M, Wan W and Dong G et al 2013 Ceram Int **39** 337–346

The application of artificial intelligence in the optimal design of mechanical systems

A Poteralski and M Szczepanik

Institute of Computational and Mechanical Engineering, Faculty of Mechanical Engineering, Silesian University of Technology

Email: arkadiusz.poteralski@polsl.pl

Abstract. The paper is devoted to new computational techniques in mechanical optimization where one tries to study, model, analyze and optimize very complex phenomena, for which more precise scientific tools of the past were incapable of giving low cost and complete solution. Soft computing methods differ from conventional (hard) computing in that, unlike hard computing, they are tolerant of imprecision, uncertainty, partial truth and approximation. The paper deals with an application of the bio-inspired methods, like the evolutionary algorithms (EA), the artificial immune systems (AIS) and the particle swarm optimizers (PSO) to optimization problems. Structures considered in this work are analyzed by the finite element method (FEM), the boundary element method (BEM) and by the method of fundamental solutions (MFS). The bio-inspired methods are applied to optimize shape, topology and material properties of 2D, 3D and coupled 2D/3D structures, to optimize the termomechanical structures, to optimize parameters of composites structures modeled by the FEM, to optimize the elastic vibrating systems to identify the material constants for piezoelectric materials modeled by the BEM and to identify parameters in acoustics problem modeled by the MFS.

1. Introduction

In the present paper, the application of bio-inspired methods in optimization of several structures is presented [1][7]. The evolutionary algorithms (EA), the artificial immune systems (AIS) and the particle swarm optimizers (PSO) are used to optimize shape, topology and material properties of 2D [16] and 3D structure, to optimize of termomechanical structures, to optimize of composites structures, to optimize of elastic vibrating systems and to identify in acoustics problem. Structures considered in this work are analyzed by the finite element method (FEM), the boundary element method (BEM) and by the method of fundamental solutions (MFS).

2. Optimization methods

Soft computing techniques resemble human reasoning more closely than traditional techniques, which are largely based on conventional logical systems or rely heavily on the mathematical capabilities of a computer. These computing techniques are often used to complement each other in applications. It should be pointed out that simplicity and complexity of systems are relative, and certainly, most successful mathematical modeling of the past have also been challenging and very significant. Unlike hard computing schemes, which strive for exactness and for full truth, soft computing techniques exploit the given tolerance of imprecision, partial truth, and uncertainty for a particular problem.

Another common contrast comes from the observation that inductive reasoning plays a larger role in soft computing than in hard computing.

Three important areas of soft computing methods, namely:

- Evolutionary Computation (EC),
- Artificial Immune Systems (AIS),
- Particle Swarm Methods (PSM),

are presented in the paper.

2.1. Evolutionary Computation (EC)

Evolutionary algorithms [5] are algorithms searching the space of solutions and they are based on the analogy to the biological evolution of species. Like in biology, the term of an individual is used, and it represents a single solution. Evolutionary algorithms operate on populations of individuals, so while the algorithm works, all the time we deal with a set of problem solutions. An individual consists of chromosomes. Usually it is assumed that an individual has one chromosome. Chromosomes consist of genes which are equivalents of design variables in optimisation problems. The adaptation is computed using fitness function. All the genes of an individual decide about the fitness function value.

In the first step, an initial population of individuals is created. Usually, the values of the genes of particular individuals are randomly generated. In the next step, the fitness function value for each individuals is computed. Then, evolutionary operators change genes of the parent population individuals, they are then selected for the offspring population, which becomes a parent population and the algorithm is continuing iteratively till the end of the computation. The termination condition of the computation can be formulated as the maximum number of iterations.

In evolutionary algorithms the floating-point representation is applied, which means that genes included in chromosomes are floating-point numbers. Usually the variation of the gene value is limited.

Evolutionary operators change gene values like the biological mechanisms of mutation and crossover. Different kinds of operators are presented in publications, and the basic ones are:

- uniform mutation,
- mutation with Gaussian distribution,
- boundary mutation,
- simple crossover,
- arithmetical crossover.

An important element of an evolutionary algorithm is the mechanism of selection. The probability of the individual's survival depends on the value of the fitness function. Ranking selection is performed in a few steps. First, the individuals are classified according to the value of the fitness function, then a rank value is attributed to each individual. It depends on the individual's number and the rank function. The best individuals obtain the highest rank value, the worst obtain the lowest one. In the final step individuals for the offspring generation are drawn, but the probability of drawing particular individuals is closely related to their rank value.

2.2. Artificial Immune Systems (AIS)

The artificial immune systems (AIS) are developed on the basis of a mechanism discovered in biological immune systems [17]. An immune system is a complex system which contains distributed groups of specialized cells and organs. The main purpose of the immune system is to recognize and destroy pathogens - funguses, viruses, bacteria and improper functioning cells. The lymphocytes cells play a very important role in the immune system. The lymphocytes are divided into several groups of cells. There are two main groups B and T cells, both contains some subgroups (like B-T dependent or

B-T independent). The B cells contain antibodies, which could neutralize pathogens and are also used to recognize pathogens.
The artificial immune systems [3] take only a few elements from the biological immune systems. The most frequently used are the mutation of the B cells, proliferation, memory cells, and recognition by using the B and T cells. The artificial immune systems have been used to optimization problems in classification and also computer viruses recognition. The cloning algorithm presented by de Castro [3] uses some mechanisms similar to biological immune systems to global optimization problems. The unknown global optimum is the searched pathogen. The memory cells contain design variables and proliferate during the optimization process. The B cells created from memory cells undergo mutation. The B cells evaluate and better ones exchange memory cells. In Wierzchoń [17] version of Clonalg the crowding mechanism is used - the diverse between memory cells is forced. A new memory cell is randomly created and substitutes the old one, if two memory cells have similar design variables. The crowding mechanism allows finding not only the global optimum but also other local ones. The presented approach is based on the Wierzchoń [17] algorithm, but the mutation operator is changed. The Gaussian mutation is used instead of the nonuniform mutation in the presented approach.
In the first stage of a flowchart of an artificial immune system the memory cells are created randomly. They proliferate and mutate creating B cells. The number of nc clones created by each memory cell is determined by the memory cells objective function value. The objective functions for B cells are evaluated. The selection process exchanges some memory cells for better B cells. The selection is performed on the basis of the geometrical distance between each memory cell and B cells (measured by using design variables). The crowding mechanism removes similar memory cells. The similarity is also determined as the geometrical distance between memory cells. The process is iteratively repeated until the stop condition is fulfilled. The stop condition can be expressed as the maximum number of iterations [10].

2.3. Particle Swarm Methods (PSM)
The particle swarm algorithms [4][13], similarly to the evolutionary and immune algorithms, are developed on the basis of the mechanisms discovered in the nature. The swarm algorithms are based on the models of the animals social behaviours: moving and living in the groups. The animals relocate in the three-dimensional space in order to change their stay place, the feeding ground, to find the good place for reproduction or to evading predators. We can distinguish many species of the insects living in swarms, fishes swimming in the shoals, birds flying in flocks or animals living in herds.
A simulation of the bird flocking was published in [13]. They assumed that this kind of the coordinated motion is possible only when three basic rules are fulfilled: collision avoidance, velocity matching of the neighbours and flock centring. The computer implementation of these three rules showed very realistic flocking behaviour flying in the three dimensional space, splitting before obstacle and rejoining again after missing it. The similar observations concerned the fish shoals. Further observations and simulations of the birds and fishes behaviour gave in effect more accurate and more precise formulated conclusions [13]. The results of this biological examination where used by Kennedy and Eberhart [4], who proposed Particle Swarm Optimiser – PSO. This algorithm realizes directed motion of the particles in n-dimensional space to search for solution for n-variable optimisation problem. PSO works in an iterative way. The location of one individual (particle) is determined on the basis of its earlier experience and experience of whole group (swarm). Moreover, the ability to memorize and, in consequence, returning to the areas with convenient properties, known earlier, enables adaptation of the particles to the life environment. The optimisation process using PSO is based on finding the better and better locations in the search-space (in the natural environment that are for example hatching or feeding grounds).
The algorithm with continuous representation of design variables and constant constriction coefficient (constricted continuous PSO) has been used in presented research. In this approach each particle oscillates in the search space between its previous best position and the best position of its neighbours, with expectation to find new best locations on its trajectory. When the swarm is rather small (swarm

consists of several or tens particles) it can be assumed that all the particles stay in neighbourhood with currently considered one. In this case we can assume the global neighbourhood version and the best location found by swarm so far is taken into account – current position of the swarm leader.
At the beginning of the algorithm the particle swarm of assumed size is created randomly. Starting positions and velocities of the particles are created randomly. The objective function values are evaluated for each particle. In the next step the best positions of the particles are updated and the swarm leader is chosen. Then the particles velocities are modified and particles positions are also modified. The process is iteratively repeated until the stop condition is fulfilled. The stop condition is typically expressed as the maximum number of iterations.
The general effect is that each particle oscillates in the search space between its previous best position (position with the best fitness function value) and the best position of its best neighbour (relatively swarm leader), hopefully finding new best positions (solutions) on its trajectory, what in whole swarm sense leads to the optimal solution.

3. Overview of the scientific research

Three optimization algorithms described above were applied to the following tasks which are authors' research i.e.:

- optimization of shell and shell solid structures due to the shape, topology and material properties,
- optimization of termomechanical structures,
- identification and optimization of composites structures,
- optimization of elastic vibrating systems,
- identification in acoustics problem.

The first task concerns the optimization of shell and shell solid structures (figure 1a). At the beginning of optimization process the material is homogeneous. During this process material properties can be changed (density of material). Finally after optimization heterogeneous material is obtained [8]. As a result of changing of the material properties a part of the finite elements could be eliminated, resulting in a change of the outer boundary of shell or shell-solid structure (shape optimization) and were generated new internal boundaries as a holes (topology optimization). The optimization was performed using the artificial immune system, the particle swarm optimizer and the evolutionary algorithm. In the table 1 input data are presented i.e.: total load Q, constraint on maximal displacement u, thickness of 2D structure and range of changing of densities ρ_e.

Table 1. The input data to the optimization task of a shell-solid structure

shell thickness [mm]	u [mm]	Q [kN]	range of ρ_e [g/cm^3]
10.0	1.2	1.08	$0 \le \rho_e < 2.36$ elimination $2.36 \le \rho_e \le 7.86$ existence

Result after optimization process in the form of the map of displacements after deformation in the figure 1b is presented.
Examples of application the optimization of shell and solid structures due to the shape, topology and material properties has been published in several articles [2][12].

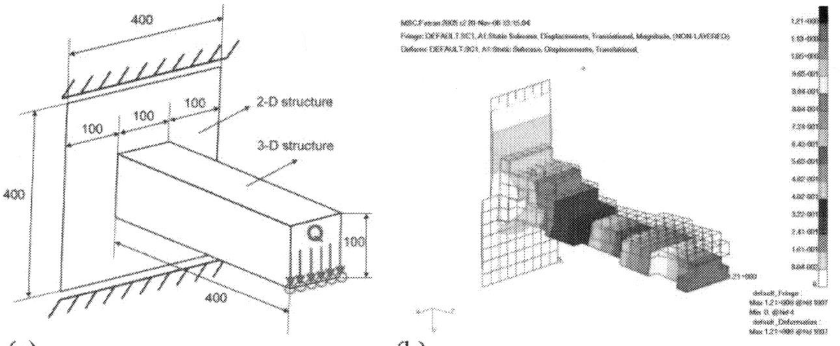

(a) (b)

Figure 1. (a) Geometry and scheme of loading, (b) The map of displacements after deformation for optimal construction

The next task concerns the optimization of thermoelasticity problem (coupled field) [14]. Considered bodies were subjected to simultaneous impact of thermal and mechanical fields (figure 2a). The coupled fields are a special case of coexistence phenomena of different physical nature. For a coupled problems, a set of variables describing physical phenomenon data cannot be reduced and replaced by a description of a single physical phenomenon and the phenomenon occurring in specific areas they cannot be separated. For coupled fields there are two classes of problems: coupling occurs at the interface areas by boundary conditions (eg. interaction fluid - solid), areas where there are phenomena overlap partially or completely, so that coupling takes place by the equations describing different physical phenomena (eg. thermoelasticity, piezoelectricity, electromagnetism). As an optimization example of coupled structures, thermo-mechanical systems are considered, for which the optimum shape of radiators modelled using Bezier curve (figure 2b) for several objective function is searched (minimization of volume, temperature and the maximum value of equivalent stresses). The optimization was performed using the artificial immune system and the particle swarm optimizer. In the table 2 input data are presented i.e.: pressure, heat flux, ambient temperature and heat convection coefficient.

Table 2. Boundary conditions values

Pressure	5000Pa
Heat flux	1000W/m^2
Ambient temperature	25 °C
Heat convection coefficient	2W/m^{2}K

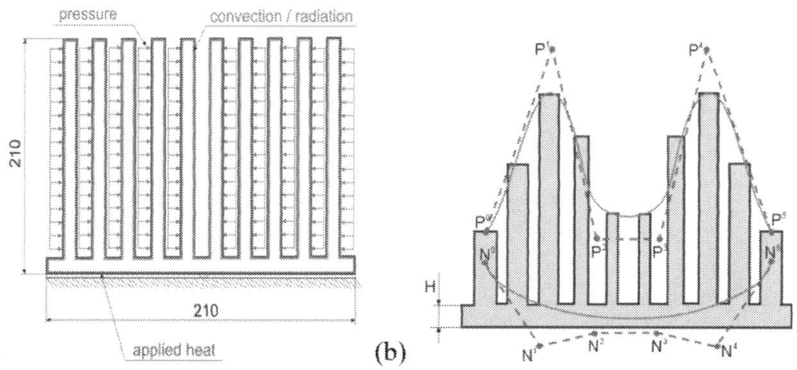

(a) (b)

Figure 2. (a) Geometry and scheme of loading, (b) Design parameters

Geometry after optimization process for three different optimization criteria in the figure 3 is presented.

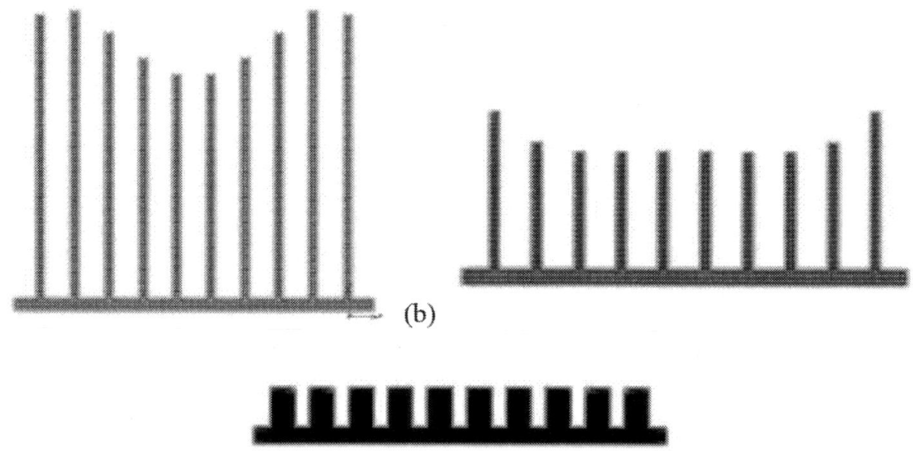

(a) (b)

(c)

Figure 3. Geometry after optimization process: (a) minimization of temperature, (b) minimization of volume, (c) minimization of equivalent stresses

Examples of application the optimization of termomechanical structures has been published in the paper [14].

Another problem is the optimization and identification of certain parameters for composite systems, which have high strength-to-weight ratio compared with traditional materials. The laminates, being multilayered, fibre reinforced composites, have especially superior properties. Searched: the number and order of layers and arrangement of layers and their thickness allowing for optimal strength properties. Identification of elastic constants material in a multi-layer laminate of different sequences of layers were also performed. The optimization was performed using the artificial immune system and the particle swarm optimizer.
A symmetric hybrid laminate plate made of two materials is considered (figure 4b). The external plies of the laminate are made of material M_e, the core of the plate is made of the material M_i .
The properties of materials are:

- material M_e (graphite-epoxy, T300/5280): E_1=181GPa, E_2=10.3GPa, G_{12}=7.17GPa, ν_{12}=0.28, ρ=1600kg/m^3;
- material M_i (glass-epoxy, Scotchply 1002): E_1=38.6GPa, E_2=8.27GPa, G_{12}=4.14GPa, ν_{12}=0.26, ρ=1800 kg/m^3.

where: E_1 – axial Young's modulus, E_2 - transverse Young's modulus, G_{12} - axial-transverse shear modulus, ν_{12} - axial-transverse Poisson ratio.

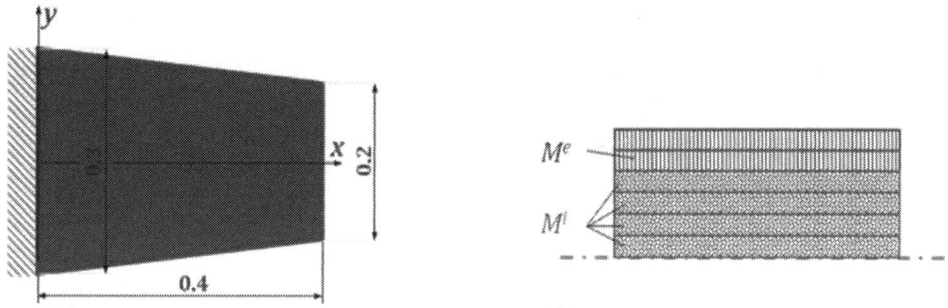

(a) (b)

Figure 4. The hybrid laminate plate: (a) dimensions and bearing; (b) location of materials (for 12-plies case)

The aim of the optimization is to find the optimum ply angles of the hybrid laminate for the given number and thicknesses of the laminas (stacking sequence of plies). It is assumed that the number of laminas made of particular materials is constant. Results in the form of ply angles of the hybrid laminate for four variants in the table 3 are presented.
The AIS and PSO are employed to solve the optimization problem.

Table 3. Optimization results

Variant	Plies no.	Stacking sequence (ply angles)	max($\|\omega_{cl}-\omega_{ex}\|$) [Hz]
continuous	12	(-48.3/-49.9/50.3/50.2/50/50.4)s	64.864
	24	(49.1/-48.9/48.9/49.1/49.2/49.4/49.2/-49.2/49.2/49.2/49.1/-48)s	65.718
5°	12	(50/-50/-50/-50/-55/-55)s	64.633
	24	(-50/50/50/-50/-50/50/50/-50/-50/-50/50/-45)s	65.613
15°	12	(45/-45/-60/-60/-60/-60)s	63.02
	24	(45/45/45/-45/45/-60/75/-60/60/-75/75/-60)s	63.750
45°	12	(45/-45/-45/-90/-90/-90)s	60.663
	24	(-45/45/45/-45/-45/-90/45/45/45/-90/-90/45)s	63.525

Examples of application the identification of certain parameters for composite structures has been published in paper [9].

Next problem is devoted to reinforced structures considered in this work are dynamically loaded and analyzed by the coupled boundary and finite element method (BEM/FEM).
The task of analysis was based on the determination of the fields of displacements, accelerations and the boundary forces (stresses) in shell structures, reinforced with stiffeners dynamically loaded and optimization of the structural form of these structures for the optimization criteria, built on designated fields. The optimal positions of stiffeners are searched in order to maximize stiffness of the plate subjected to the sinusoidal load, the Heaviside load and for three kinds of supports. The optimization was performed using the artificial immune system and the particle swarm optimizer.
The optimization of the reinforced rectangular plate shown in figure 5 is presented. The plate is dynamically loaded and different kinds of load and support are considered (figure 6).

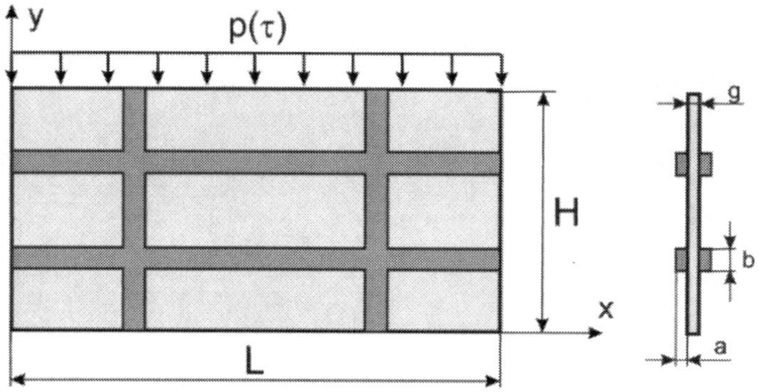

Figure 5. Reinforced rectangular plate

Three different supports are considered: a) support A – the plate is fixed on the left and right edge, b) support B – the plate is fixed on two supports, each of 0.5 cm long, c) support C – the plate is supported on the bottom.

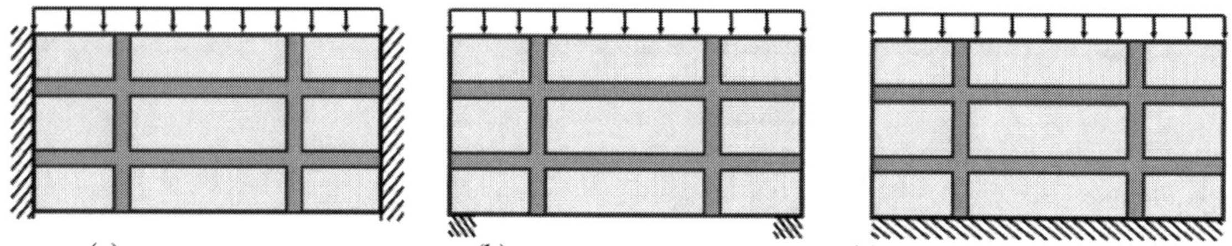

(a) (b) (c)

Figure 6. Types of supports: (a) support A, (b) support B, (c) support C.

The plate is reinforced by the frame structure composed of 4 straight beams of square cross-section (2a×b). The length and the height of the plate is L=10 cm and H=5 cm, respectively. The thickness of the plate is g=0.25 cm, the dimensions of cross-section of beams are 2a=0.5 cm and b=0.5 cm.

The material of the plate in plane stress and the frame is aluminum, for which the values of mechanical properties are: modulus of elasticity E=70 GPa, Poisson's ratio ν=0.34 and density ρ=2700 kg/m3. The material is homogeneous, isotropic and linear elastic.

The uniformly distributed load is applied at the upper edge of the plate. Two kinds of time dependent loads are considered (figure 7): a) the sinusoidal load $p(\tau)=p_o\sin(2\pi\tau/T)$ with the period of time T=20π μs, and b) the Heaviside load $p(\tau)=p_oH(\tau)$. The amplitude of the load in both cases is p_o=10 MPa.

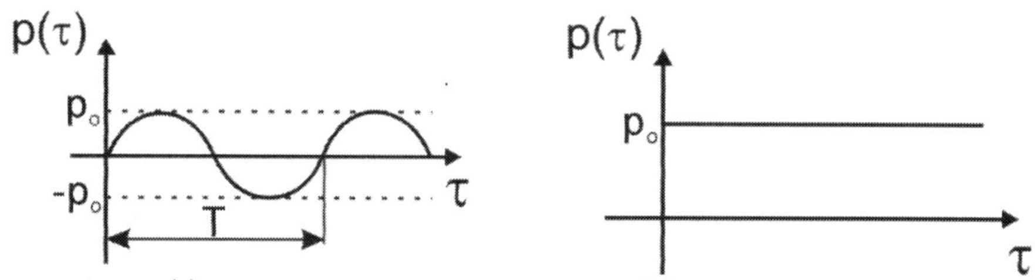

Figure 7. Dynamic loadings: (a) sinusoidal, (b) Heaviside

Results after optimization problem in the table 4 are presented.

Table 4. Values of design variables, J and R

Load	Support	Design variables [cm]				J_o [10^{-4}cm]	J [10^{-4}cm]	R [%]
		X1	X2	Y1	Y2			
AIS								
Sinus.	A	4.75	2.86	0.88	2.75	89	76	15
	B	4.75	1.81	0.57	2.75	92	73	21
	C	1.20	1.82	0.50	2.75	82	62	24
Heav.	A	0.50	4.75	0.50	4.50	112	91	19
	B	4.75	1.41	0.50	4.50	211	149	29
	C	0.50	2.20	1.70	2.80	49	42	14
PSO								
Sinus.	A	4.75	2.86	0.88	2.75	89	76	15
	B	4.75	1.81	0.57	2.75	92	73	21
	C	1.20	1.82	0.50	2.75	82	62	24
Heav.	A	0.50	4.75	0.50	4.50	112	91	19
	B	4.75	1.41	0.50	4.50	211	149	29
	C	0.50	2.20	1.70	2.80	49	42	14

More information of the optimization of dynamically loaded reinforced structures in paper [11] is presented.

The last problem, which uses a bio-inspired algorithms is identification of complex values of the impedances of room walls for the acoustic problem. The acoustic field control and design require the determination of structure parameters, e.g. the acoustic absorption of building materials. The pressure measurements are simulated by the method of fundamental solutions MFS (meshless method). The optimization was performed using the artificial immune system and the particle swarm optimizer.
Complex values of the impedances $Z_1 \div Z_4$ of room walls are identified. The geometry and other parameters of the analysed 2-D model of a room are presented in Figure 8. The same structure was considered by Dutilleux et al. The dimensions of the room model are: a = 3.4 m and b = 2.5 m. The acoustic medium is air at the temperature 20°C. The analysis is performed for the frequency equal to 100 Hz. Eight sensors are located at points with coordinates related to the wave length λ and the room dimensions. The number of both the boundary points and the sources is equal to 54. The sources are located at a circle of radius r = 2.5 m, centered at the geometric centre of the rectangle (Figure 8).
The AIS and its hybrid version HAIS [6], EA and PSO are employed to solve the optimization problem (table 5).

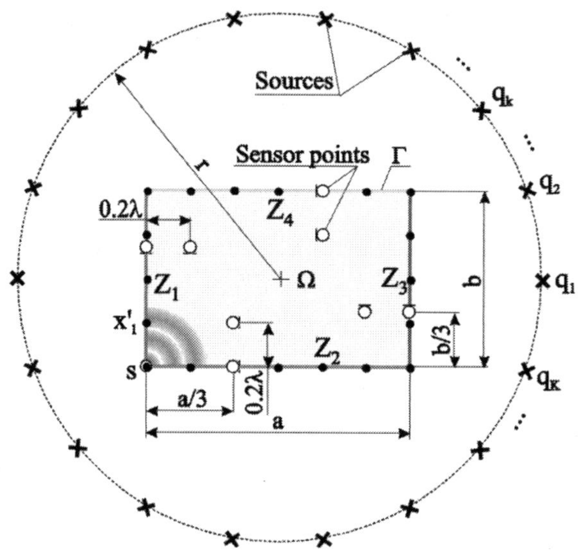

Figure 8. Scheme of the problem solved by the MFS and the HAIS

Table 5. Reference values of genes and results obtained by using different methods

Results	Reference values	(8,56)-ES, Dutilleux et al.	AIS, EA, PSO	HAIS
p_1	4920	4706	4833	4920
p_2	1590	1591	1484	1590
p_3	2390	2487	2362	2390
p_4	3720	3756	3678	3720
p_5	3400	3400	3370	3400
p_6	50	84	49	50
p_7	3500	3568	3720	3500
p_8	2200	2174	2256	2200

Examples of application identification in acoustics problem has been published in several articles [6][15].

4. Conclusions

In the paper the formulation and application of the finite element method, boundary element method, method of fundamental solutions and the bio-inspired methods to optimization of several structures is presented. The bio-inspired methods can be simply implemented because it needs only the values of objective functions. The important feature of this approaches is the strong probability of finding the global optimal solutions. The described approaches are free from limitations connected with classic gradient optimization methods.

References

[1] Burczyński T, Poteralski A and Szczepanik M Immune and Swarm Optimization of Structures, Advances in Evolutionary and Deterministic Methods for Design, Optimization and Control in Engineering and Sciences, Chapter No. 19, 2015

[2] Burczyński T and Szczepanik M Intelligent optimal design of spatial structures. Computer & Structures, Elsevier 2013

[3] Castro L N and Timmis J Artificial Immune Systems as a Novel Soft Computing Paradigm, Soft Computing, 7(8):526-544, 2003
[4] Kennedy J and Eberhart R C 1995 Particle Swarm Optimisation. Proceedings of IEEE Int. Conf. on Neural Networks, Piscataway, NJ, pp 1942-1948.
[5] Michalewicz Z 1996 *Genetic algorithms + data structures = evolutionary algorithms* Springer-Verlag Berlin
[6] Poteralski A Szczepanik M Ptaszny J Kuś W and Burczyński T 2013 Hybrid artificial immune system in identification of room acoustic properties Inverse Problems in Science and Engineering Taylor & Francis
[7] Poteralski A Szczepanik M Dziatkiewicz G Kuś W 2013 Comparison between PSO and AIS on the basis of identification of material constants in piezoelectrics Springer-Verlag Berlin Heidelberg 2013 ICAISC 2013 Part II LNAI 7895 pp 569–581
[8] Poteralski A 2014 Optimization of mechanical structures using artificial immune algorithm, Beyond Databases Architectures and Structures Communications in Computer and Information Science vol 424 pp 280-289
[9] Poteralski A Szczepanik M Beluch W and Burczyński T 2014 Optimization of composite structures using bio-inspired methods Artificial intelligence and soft computing ICAISC 2014 vol 8468 pp 385-395
[10] Poteralski A Szczepanik M and Burczyński T 2015 Immune Optimal Design of 2-D and 3-D Structures International Conference on Artificial Intelligence and Soft Computing (ICAISC) Lecture Notes on Artificial Intelligence vol 9120 Springer pp 471-482
[11] Poteralski A Szczepanik M and Burczyński T 2015 Swarm and immune computing of dynamically loaded reinforced structures International Conference on Artificial Intelligence and Soft Computing (ICAISC) Lecture Notes on Artificial Intelligence 9120 Springer pp 483-494.
[12] Poteralski A 2015 Data processing in immune optimization of the structure, Beyond Databases Architectures and Structures (BDAS) Communications in Computer and Information Science vol 521 Springer pp 309-319
[13] Reynolds C W 1987 Flocks, herds, and schools, A distributed behavioral model Computer Graphics 21 pp 25–34
[14] Szczepanik M Poteralski A Długosz A Kuś W and Burczyński T 2013 Bio-inspired optimization of thermomechanical structures Springer-Verlag Berlin Heidelberg ICAISC 2013 Part II LNAI 7895 pp 79-90
[15] Szczepanik M Poteralski A Ptaszny J Burczyński T 2012 Hybrid Particle Swarm Optimizer and Its Application in Identification of Room Acoustic Properties Swarm and Evolutionary Computation - International Symposia ICAISC 2012 Proceedings. Lecture Notes in Computer Science 7269
[16] Szczepanik M and Burczyński T 2012 Swarm optimization of stiffeners locations in 2-D structures Bulletin of the Polish Academy of Sciences Technical Sciences vol 60 no 2 pp 241-246
[17] Wierzchoń S T 2001 Artificial Immune Systems theory and applications EXIT (in Polish).

Numerical analysis of eccentric orifice plate using ANSYS Fluent software

D Zahariea

Department of Fluid Mechanics, Fluid Machinery and Fluid Power Systems, Technical University "Gheorghe Asachi" Iași, România

E-mail: dzahariea@yahoo.com

Abstract. In this paper the eccentric orifice plate is qualitative analysed as compared with the classical concentric orifice plate from the point of view of sedimentation tendency of solid particles in the fluid whose flow rate is measured. For this purpose, the numerical streamlines pattern will be compared for both orifice plates. The numerical analysis has been performed using ANSYS Fluent software. The methodology of CFD analysis is presented: creating the 3D solid model, fluid domain extraction, meshing, boundary condition, turbulence model, solving algorithm, convergence criterion, results and validation. Analysing the numerical streamlines, for the concentric orifice plate can be clearly observed two circumferential regions of separated flows, upstream and downstream of the orifice plate. The bottom part of these regions are the place where the solid particles could sediment. On the other hand, for the eccentric orifice plate, the streamlines pattern suggest that no sedimentation will occur because at the bottom area of the pipe there are no separated flows.

1. Introduction

One of the most important, common and long history of use methods for fluid flow measurement is the differential pressure device method (head-loss device method), for which the difference in pressure Δp measured at two points, upstream and downstream sides of the measurement device, is proportional with the rate of fluid flow.

The flow measurement device, which in this case, is a restriction device causing the obstruction to the fluid flow, will generate a pressure drop which will be measured with two pressure taps positioned upstream and downstream of the restriction device. The most common taps installed configurations are: flange pressure taps, corner pressure taps and vena contracta taps (D and $D/2$ pressure taps), [1]. There are many restriction devices such as nozzles, Venturi nozzles and Venturi tubes, but by far, the most commonly used restriction device for flow measurement in pipes is the orifice plate (orifice meter) which is designed usually as: squared-edged orifice plate, squared-edged with conical back-cut orifice plates, sharp-edged orifice plates, conical entrance orifice plates, quarter-circle entrance orifice plate, segmental orifice plate and eccentric orifice plates, [1].

In order to obtain the desired accuracy some requirements should be accomplished: the circularity of both the orifice and the pipe, the plate flatness, the roughness of the upstream plate surface, the sharpness of the orifice edge, the thickness of both the plate and the orifice, the pressure taps, the pipe roughness, the eccentricity (the distance between the pipe axis and the orifice axis). There are many results referring to orifice plates, from which, in this context, should be mentioned: the comparative numerical analysis of different orifice plates [2], the numerical analysis of orifice plate working under non-standard conditions [3], as well as the comparative numerical analysis of single hole and multi hole orifice plates [4]. However, all this research is about the conventional concentric orifice plate. Despite the fact that for conventional orifice plates, the pipe and the orifice should be concentric, there are some exceptional circumstances for which the eccentricity should be an effective design option. The eccentric orifice plates are recommended for flow measurement of fluids with entrained gas, liquid or sediments for which the sedimentation of such inclusions can thus be avoided, [5].

The eccentricity of the orifice is defined by the difference between the pipe axis and the orifice axis, and for the particular case analyzed in this paper, can be express through the formula:

$$e = \frac{D}{2}(1 - \beta) \qquad (1)$$

where D is the pipe diameter, $\beta = d/D$ is the diameter ratio, and d is the orifice diameter.

This paper discusses the numerical analysis of an eccentric orifice plate used for flow measurement of liquids using ANSYS Fluent software, as compared with the conventional concentric orifice meter. The eccentric orifice plate has the same orifice diameter as the concentric orifice plate.

2. Geometry

Both orifice meters are characterized by the same diameter ratio $\beta = d/D = 0.5$, where d=50 mm is the orifice diameter and D=100 mm is the pipe diameter, as well as by the same pipe Reynolds number $Re = 10^5$. Considering the fluid as water with density ρ=998.2 kg/m^3 and dynamic viscosity μ=0.001003 Pa·s, the fluid average velocity at the pipe inlet will result as V=1.0048 m/s and the mass flow rate will be Q_m=7.8775 kg/s. The eccentricity computed using the equation (1), is equal with 25 mm.

The orifice plate is located into a horizontal pipe with total length of 35D, between two pipe sections, the upstream pipe section with length L_1=15D and the downstream pipe section with L_2=20D, figure 1.

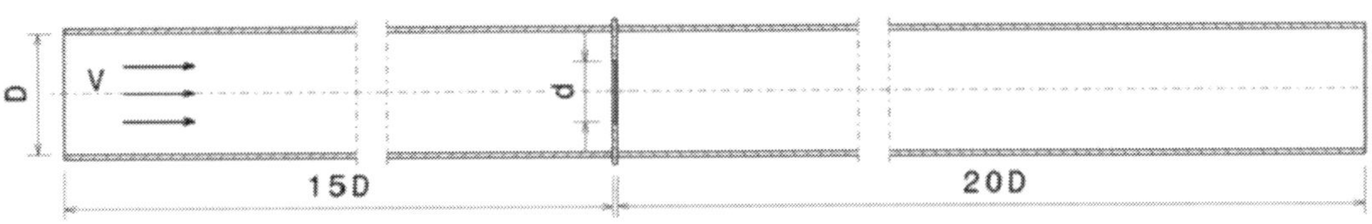

Figure 1. Geometry of the computational domain.

Because of the longitudinal symmetry of the geometry, a longitudinal symmetry cutting plane has been used, thus only one half of the pipe geometry will be modeled. The 3D model has been created using CATIA. The 3D geometric model has been transferred in ANSYS using the STP file format. The solid domain geometry of the orifice meters is presented in figure 2 for the concentric orifice plate, and in figure 3 for the eccentric orifice plate.

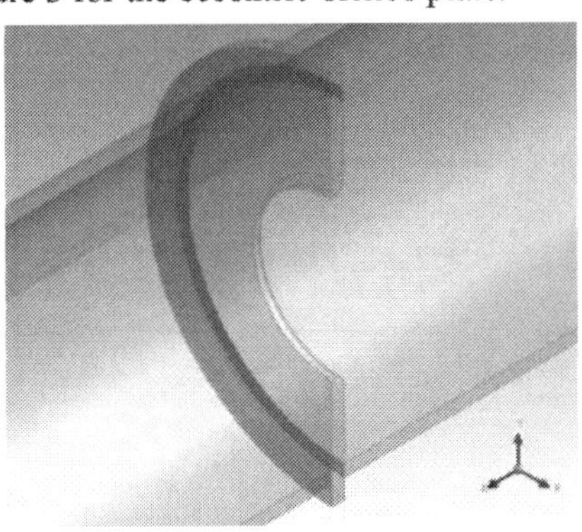

Figure 2. Concentric orifice plate-solid domain.

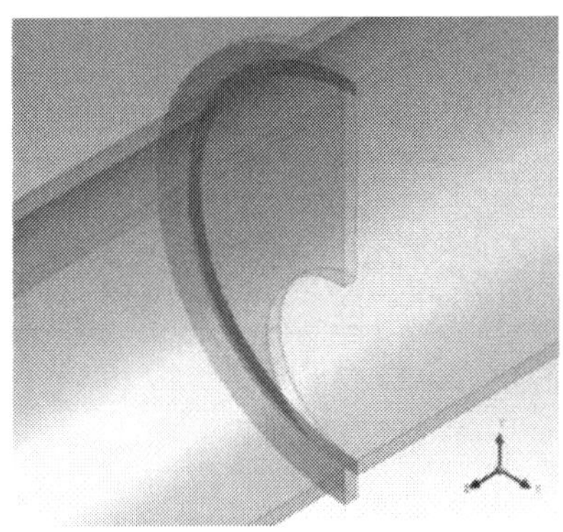

Figure 3. Eccentric orifice plate-solid domain.

3. Mesh

Using fluid flow domain extraction techniques, the fluid flow domain has been obtained. The fluid flow domains from the orifice plate zone are presented in figure 4 for the concentric orifice plate, and in figure 5 for the eccentric orifice plate.

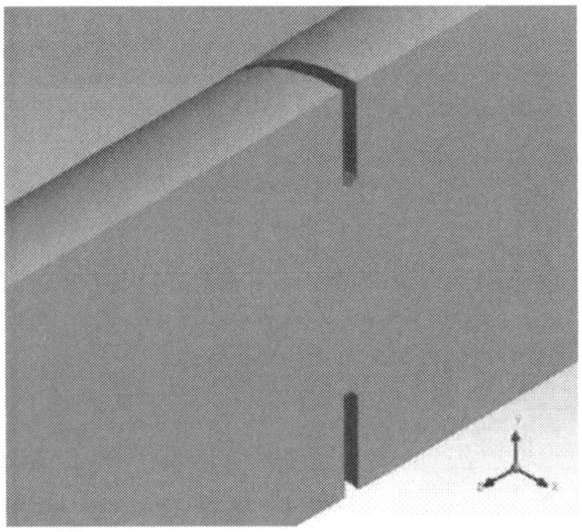

Figure 4. Concentric orifice plate-fluid domain.

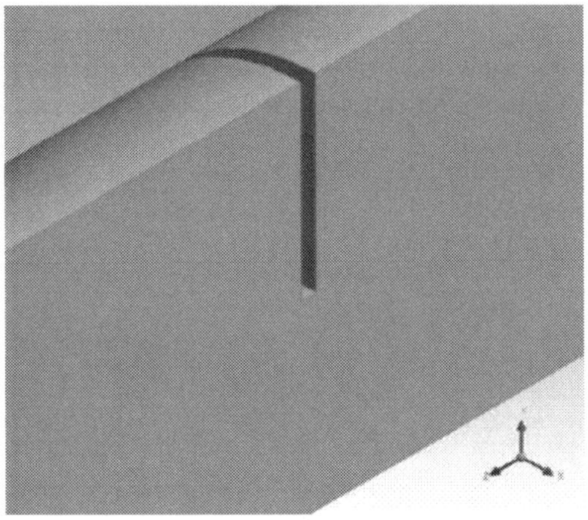

Figure 5. Eccentric orifice plate-fluid domain.

The mesh obtained for the concentric orifice plate is presented in figure 6.

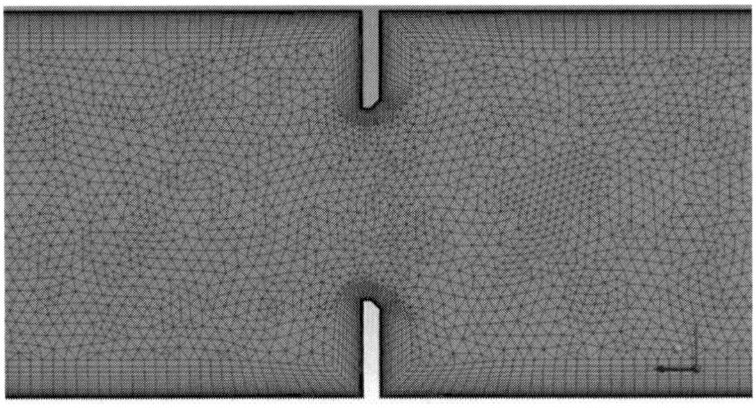

a) symmetry plane detail

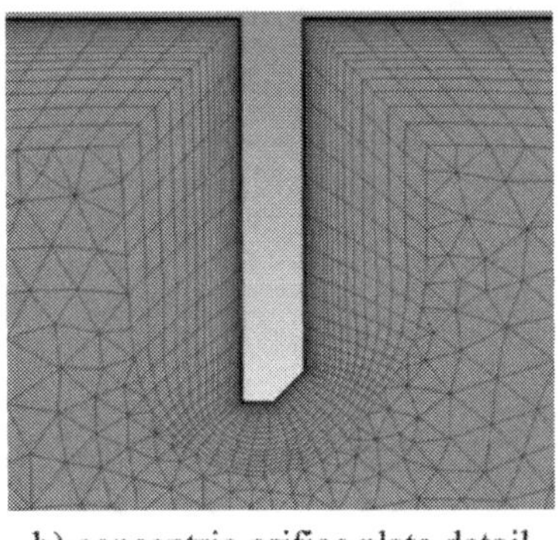

b) concentric orifice plate detail

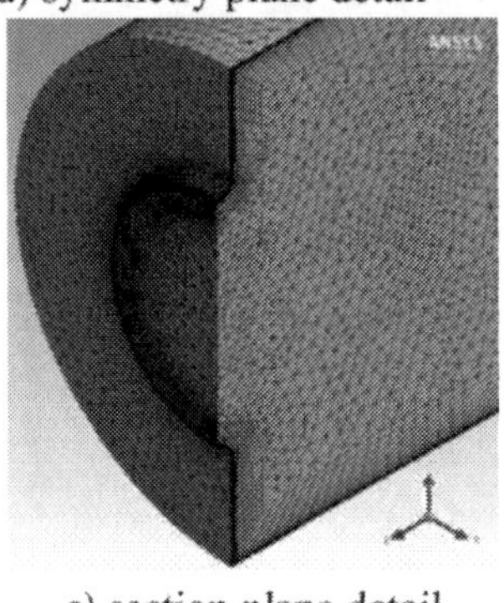

c) section plane detail

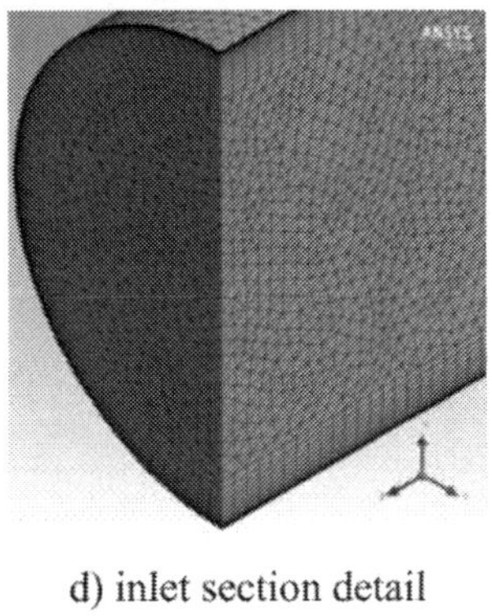

d) inlet section detail

Figure 6. Concentric orifice plate - mesh.

The mesh obtained for the eccentric orifice plate is presented in figure 7.

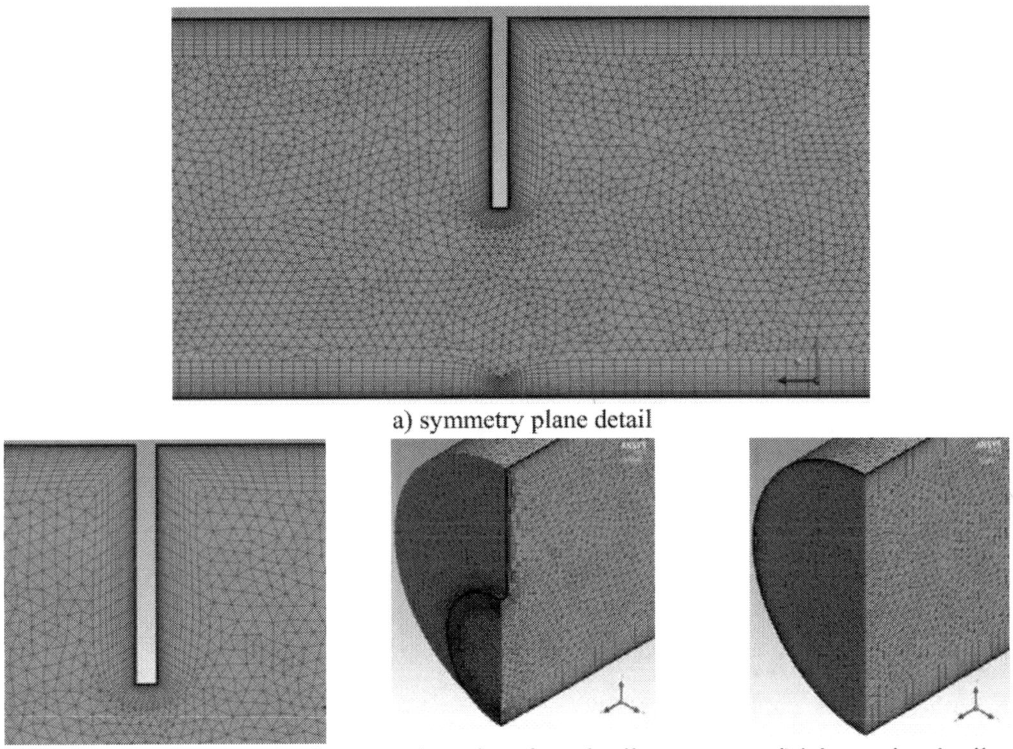

a) symmetry plane detail

b) eccentric orifice plate detail c) section plane detail d) inlet section detail

Figure 7. Eccentric orifice plate - mesh.

Both the mesh for concentric orifice plate, as well for the eccentric orifice plate has been obtained using Mesh application from ANSYS Workbench. In order to proper resolve the boundary layer along the fluid domain walls, as well as around the orifice plate an inflation mesh property has been defined with 25 layers and the first layer thickness adjusted for the dimensionless wall distance $y^{+} \cong 1$. The mesh principal parameters are presented in table 1.

Table 1. Mesh characteristics.

Orifice plate	Nodes	Elements	Skewness
Concentric	2496422	6650615	0.896
Eccentric	2455446	6550947	0.916

4. CFD analysis

The CFD analysis has been performed using Fluent software for solving the Reynolds-Averaged Navier-Stokes equations with the following main settings: $k - \omega$ SST turbulence model; steady pressure based solver with SIMPLE pressure-velocity coupling algorithm with second order upwind spatial discretization for momentum, turbulent kinetic energy and specific dissipation rate.

The boundary conditions of the problem are: the velocity V=1.0048 m/s on the inlet section; the atmospheric pressure on the outlet section; no-slip boundary condition at the flow domain walls, as well at the orifice plate walls; zero normal velocity and zero normal gradients of all variables at the symmetry plane; the effect of gravity has been considered.

The convergence criterion of the steady state solution obtained by the CFD analysis has been defined as a sum of three convergence check criteria: the residual error for continuity, for x-, y- and z-velocity, for turbulence kinetic energy k and for specific rate of dissipation ω should be under 10^{-4}; the pressure at the inlet section, as one of the relevant variable of the analysis, should reach a steady solution; the relative mass flow rate imbalance between the inlet and the outlet sections should be less than 1% of the smallest mass flow through the flow domain. The convergence has been obtained after 880 iterations for concentric orifice plate analysis, and after 1283 iterations for the eccentric orifice plate analysis. At the convergence point, the relative mass flow rate imbalance is 3.12e-04% for concentric orifice plate and 5.82e-04% for the eccentric orifice plate.

The computations have been parallelized performed on 32 cores of a SUPERMICRO workstation with 4 Intel XEON E5-4640 2.4 GHz processors, 256 GB DDR4 RAM, 2 TB HDD and NVIDIA Quadro K4000 3 GB GDDR5 video card.

5. Results

The streamlines, both 3D and 2D, are presented in figure 8 for the concentric orifice plate, and in figure 9 for the eccentric orifice plate.

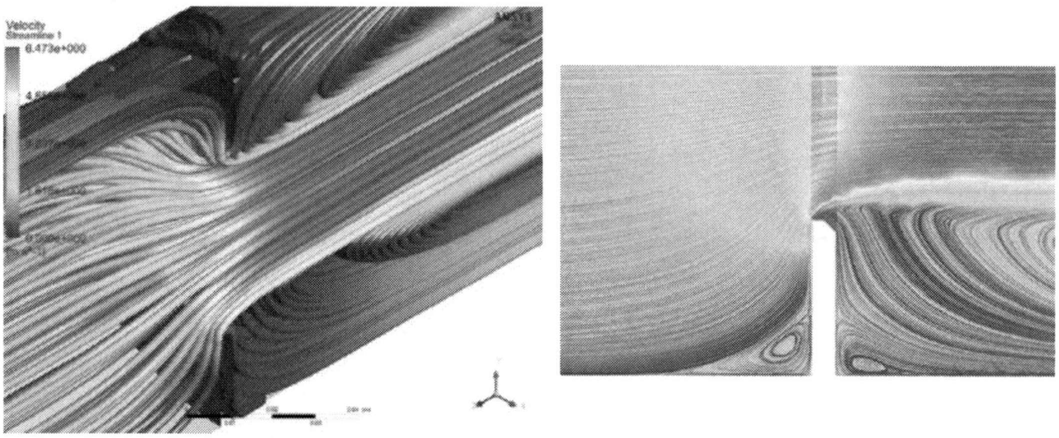

a) 3D b) 2D

Figure 8. Concentric orifice plate - streamlines.

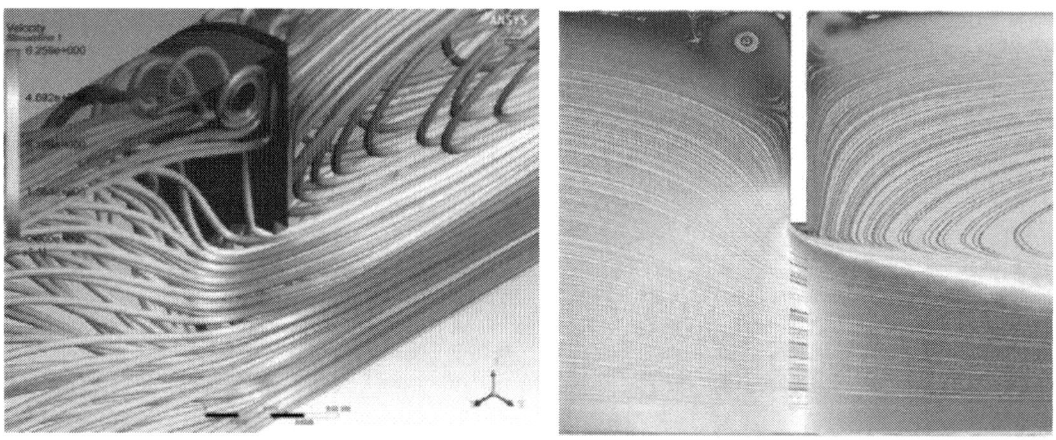

a) 3D b) 2D

Figure 9. Eccentric orifice plate - streamlines.

6. Validation

The validation of the CFD analysis has been done by comparing the CFD results with the results provided by the standards ISO 5167, [1] and ISO TR 15377, [5] in terms of discharge coefficient.

The discharge coefficient is defined as, [1]:

$$C = \frac{4Q_m\sqrt{1-\beta^4}}{\varepsilon\pi d^2\sqrt{2\rho\Delta p}} \quad (2)$$

where ε=1 is the expansion factor.

The validation results are presented in table 2.

Table 2. Validation results.

Orifice plate	Pressure tappings	Discharge coefficient, C		Relative error
		Standard	CFD	
Concentric	D and D/2	0.6062	0.6084	0.36%
	Flange, ±25.4 mm	0.6062	0.6117	0.90%
	Corner, ±3 mm	0.6069	0.6707	10.50%
Eccentric	Corner, ±3 mm	0.6270	0.6887	9.83%

7. Conclusions

The pressure drop across the orifice plate computed by CFD analysis is in very good agreement with the standard results, but starting from a certain distance from the orifice plate. In the close vicinity of the orifice plate, the errors are increased, and for this reason, further mesh improvement are required. However, the qualitative results obtained related with the streamlines pattern, confirm that the separated flow areas at the bottom part of the pipe, upstream and downstream of the concentric orifice plate where the solid particles could sediment, are not present for the eccentric orifice plate, and as consequence, for the eccentric orifice plate the sedimentation process will be avoided.

Acknowledgments

The author would like to acknowledge the computational resources offered by the Computer-Aided Fluid Engineering Laboratory from the Department of Fluid Mechanics, Fluid Machinery and Fluid Power Systems, Technical University "Gheorghe Asachi" Iaşi, Romania. The Computer-Aided Fluid Engineering Laboratory has been equipped with hardware and software resources with the financial support of the grant ENERED, POSCCE-A2-O2.2.1-2009-4, ID 911.

References

[1] ISO 5167 2003 Measurement of fluid flow by means of pressure differential devices inserted in circular cross-section conduits running full

[2] Sridevi T, Sekhar D and Subrahmanyam V 2014 Comparison of Flow Analysis Through a Different Geometry of Flowmeters using Fluent Software *Int. J. of Research in Eng. and Technol.* **3(8)** 141-149

[3] Karthik G S, Kumar K J Y and Seshadri V 2015 Prediction of Performance Characteristics of Orifice Plate Assembly for Non-Standard Conditions Using CFD, *Int. J. of Eng. and Technical Research* **3(5)** 162-167

[4] Malatesch B, Ganesha T and Math M C 2014 CFD Analysis and Comparison of Fluid Flow Through a Single Hole and Multi Hole Orifice Plate, *Int. J. of Research in Advent Technol.* **2(8)** 6-15

[5] ISO/TR 15377 2007 Measurement of fluid flow by means of pressure-differential devices – Guidelines for specification of orifice plates, nozzles and Venturi tubes beyond the scope of ISO 5167

Requirements statement in product design

L Slătineanu[1], O Dodun[1], M Coteaţă[1], I Coman[1], V Manole[1], C V Gika[1]

[1]Machine Manufacturing Technology Department, "Gheorghe Asachi" Technical University of Iaşi, RO

E-mail: oanad@tcm.tuiasi.ro

Abstract. One of the activities corresponding to the product design and development refers to clarifying the requirements necessary to be fulfilled by the final product. There are distinct opinions concerning the ways in which these requirements could be formulated and applied. The paper presents the results of a succinct analysis aiming to highlight the developing of product requirements in the case of applying the value analysis method and the axiomatic design method, respectively. The way in which the requirements are formulated in the case of a patent application is also considered. Results of the study were applied in the cases of designing some mechanical products.

1. Introduction

In the life cycle of a product, there is a stage when its design is necessary. There are different ways of defining the product design; here, we will consider that the product design includes the activities developed after a certain product idea is well contoured and up to the moment when all the drawings are definitively achieved. Similar aspects could be considered about the technological process design, in such a case the activities being finalized by technological documentation.

In order to meet the objectives for which the product is designed, the substage of adequate requirements formulation could be taken into consideration. There are also distinct opinions of researchers about the ways in which the requirements specific to a certain product could be formulated and developed.

Thus, a certain guide could be constituted by the way in which the product functions are established in accordance with *the value analyses method*. Some decades ago, Lawrence Miles, Jerry Leftow, and Harry Erlicher, when they were working for General Electric, defined the value analysis as a method of functional systemic analysis of the product, in order to diminish the manufacturing costs, by removing the costs which were not contributing to the product use. Over the time, the value analysis developed gradually and new techniques were identified in order to solve various problems.

Leber et al. investigated the usefulness of two methods of innovative management, namely the value analysis and the conjoint analysis that could be applied in order to design a new product [4]. They concluded that an objectively determined value of a product can have a significant role in obtaining a product of interest for customer or user.

Crow formulated a set of considerations concerning the possibilities of using a function analysis system technique as an analysis tool [1]. He proposed a methodology of applying a value improvement process, by using the functional analysis system technique as a communications vehicle and a tool for addressing the stage of the creative phase in the activities of product design.

On the other hand, in the last decades, the validity of the principles corresponding to the so-called *axiomatic design* was proved. This design method considers that there are two fundamental axioms, namely the independence axiom, which needs the independence of functional requirements necessary to be met by the product, and the axiom of information, which stipulates the necessity of minimizing the quantity of information used during the design activities [6]. When applying the axiomatic design method, after the customer needs are defined, the functional requirements must be formulated.
Thompson (2013) showed that a method of structuring the process of formulating the requirements could be considered [7]. She appreciated that a system of classifying the functional requirements could be used, in order to diminish the probability of excluding one or more key stakeholders during the design activity.
Thompson developed also a set of considerations concerning the classification of procedural errors possible to appear in the process of defining the functional requirements in accordance with the axiomatic design theory. She mentioned five classes of such procedural errors met in the definition of the functional requirements both by novice and expert designers [8].
Within this paper, the problem of designing certain mechanical devices/products/processes was formulated and the ways in which the requirements specific to these devices were investigated; the paper presents the results of this investigation.

Table 1. Functions corresponding to the device tap elaborated in accordance with the requests specific to value analysis method.

Function code	Function name	Function category if the importance is considered	Function category if the possibilities of measuring functions are considered
A	Adaptable to the most of classical sink faucets	main	objective
B	Acted by foot	main	objective
C	Needs an acting force achievable by foot	secondary	objective
D	Including a connection between the tap and the acting by foot subassembly	secondary	objective
E	Ensuring protection of the components again the corrosive action of the environment	secondary	objective
F	Ensuring force for closing the tap	secondary	objective
G	Ensuring ease use and maintenance	secondary	objective
H	Characterized by an aesthetic aspect	secondary	subjective
I	Characterized by low volume	secondary	objective

2. Aspects specific to establishing the product functions in the case of using the value analysis method

The so called nomenclature of product functions must include only the necessary functions. Two rules must be applied when the product functions must be defined [5]:
1) A function is considered as useful if it adds value of usage to the product and if it could exist in an independent way in relation to the other functions;

2) In order to establish if a certain function is not useful for the product, the rule of function removal could be applied. If a product function is removed and the product utility is not affected, this means that the function is not useful.

If the possibilities to measure a function are considered, one can notice that a function could be *objective*, when it can presents measurable dimensions by means of one or many well determined measure units and *subjective*, respectively, when it is not measurable and it could express a certain affective state.

If the importance of a function is taken into consideration, one can mention that there are also *main functions,* that define the objectives that must be met by the product, and *secondary* or *auxiliary functions,* that ensure the conditions for fulfilling the main functions and that indirectly contribute to the achieving the product total value of using the product.

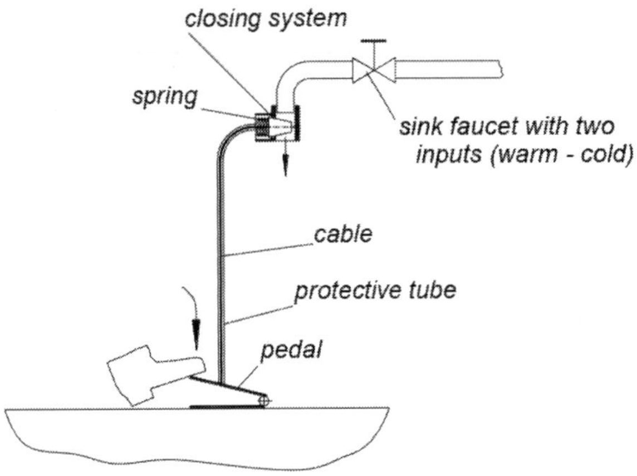

Figure 1. Schematic representation of a water saving faucet identified by considering the product functions and the rules valid in the case of applying the values analysis method.

There is the risk of developing confusions between the *functions* and *social needs*, between functions and *functional characteristics*, between functions and *the technical solutions for achieving the functions*, between the *main functions* and *the secondary or auxiliary functions* [5].

The above mentioned rules were applied in the case of a faucet with water saving. It is known that usually, the faucet is maintained open even when the water is not used and this could mean a waste water. If the faucet itself or a device added to the classical tap could be acted by foot only when the water is necessary, a certain water saving could be achieved.

The product identified functions corresponding to this faucet or device faucet were placed in the table 1, where the qualifier objective/ subjective was also included.

On the base of the functions included in the table 1, the water saving faucet whose schematic representation is presented in Figure 1 was identified. One can see that a way in which an additional device could be added to the common faucet was identified, in order to save the water consumption by acting a pedal only when the water must really flow.

3. Aspects specific to formulation of the functional requirements in the case of applying the axiomatic design method

In the case of using the axiomatic design method, four domains could be initially identified: customer domain, where the customer needs are highlighted, functional domain, where functional requirements are mentioned, physical domain, where the design parameters are specified and process domain, in which the process variables must be showed (Fig. 3).

The errors met frequently when defining the functional requirements could be the following [7]: a) not all the stakeholders are identified; b) One or more classes of stakeholders are wrongly emphasized or excluded; c) One or more categories of requirements are wrongly mentioned or excluded.

In the present paper section, the problem of identifying a technological solution for obtaining threads with variable pitch necessary in a doctoral research is taken into consideration. One can notice that the researcher could be considered as customer. The customer need could be the design and developing a technological solution adaptable on one of the universal machine tools found in the research laboratory, in order to study the possibilities of obtaining threaded surfaces with variable pitch by cutting.

In accordance with this customer need, the main functional requirement could be formulated in the following way: design a technological solution able to allow obtaining and studying the threaded surfaces with variable pitches and distinct characteristics on a universal machine tool found in the research laboratory.

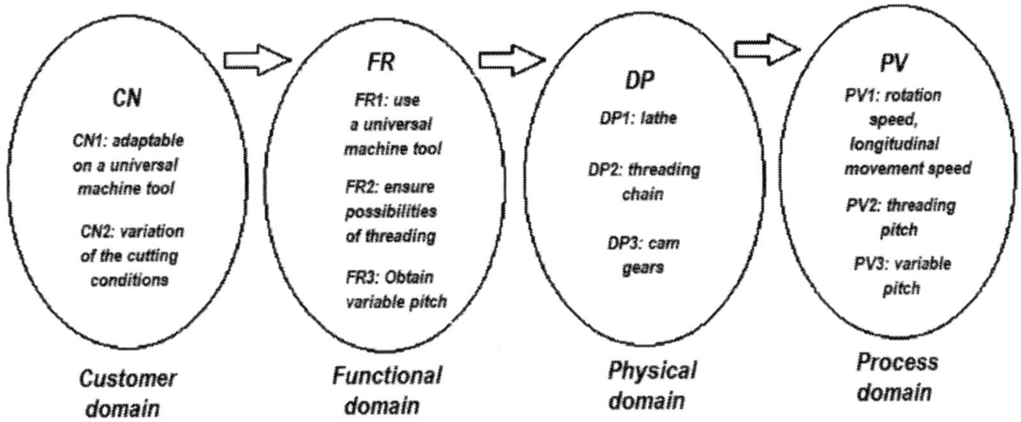

Figure 2. Domains specific to using the axiomatic design method in the case of a technological solution for variable pitch threading.

When applying the axiomatic design, the analysis of the main functional requirement (functional requirements of zero level, FR0) is continued by identifying the functional requirements of first level, of second level etc.

In the case of the desired technological solution, as functional requirements of first level, the following aspects could be taken into consideration:

FR1: Use a universal machine tool among the available machine-tools found in research laboratory;

FR2: Ensure possibilities of threading;

FR3: Ensure possibilities of machining threaded surfaces with variable pitches;

If the problem of identifying functional requirements of second order is formulated, one can take into consideration:

FR1.1: Use a rotation movement of workpiece;

FR1.2: Use distinct values for the rotation speed of the workpiece / test sample, in accordance with the type of workpiece material;

FR2.1: combine the workpiece rotation movement with a translation movement of the treading tool, characterized by a variable speed;

FR2.2: ensure possibility to change the speed of the longitudinal rectilinear movement of the cutting tool;

FR2.3: use a tool able to generate the thread profile;

FR3.1: ensure changing of the leading screw rotation speed just during the threading process;

FR3.2: ensure obtaining a certain variation of the lead screw rotation speed;
In the subsequently stage of applying the axiomatic design, *the design parameters* must be identified and eventually highlighted by means of a matrix. In the case of the approached problem, some of the considered design parameters could be:
DP1.1: main shaft of the lathe;
DP1.2: speed gear box of the lathe;
DP2.1: threading chain of the lathe;
DP2.2: gear train and feed gear box
DP2.3: threading tool
DP3.1: pair of cam-gears in the train gear;
DP3.2: profile of the cam gears.
The functional requirements and design parameters corresponding to the desired technological solution were included in table 2; one must mention that some simplifying assumptions were taken into consideration, in order to ensure only an illustration of the way in which the functional requirements could be formulated when the axiomatic design method is used.
By considering also other subsequently stages of applying the axiomatic design method, the solution schematically presented in Figure 4 was found as meting the mentioned functional requirements. One can see that a pair of cam gears could be placed in the kinematic chain of the lathe, in order to change the rotation speed of the lathe lead screw and, in this way, to obtain a threaded surface with variable pitch. From the point of view of axiomatic design principles, the solution presented in Figure 4 could be affected by an improvement, since there are two functional requirements (FR3.1 and FR3.2) which are met by the same design parameter (DP3.1); this is the so called case of *coupled design*.

Table 2. Functional requirements and design parameters corresponding to use of axiomatic design method in the case of a technological solution for obtaining threaded surfaces with variable pitch.

Line no. 1	**Design parameters**				**Design parameters**						
2					**Design parameters of first order**						
3					**DP1**		**DP2**			**DP3**	
4	**Functional requirements**				**Design parameters of second order**						
5					**DP1.1**	**DP1.2**	**DP2.1**	**DP2.2**	**DP2.3**	**DP3.1**	**DP3.2**
6 Col. no. 1	2	3	4	5	6	7	8	9	10	11	12
7	Functional requirements of first level	FR1	Functional requirements of second level	FR1.1	X						
8				FR1.2		X					
9		FR2		FR2.1			X				
10				FR2.2				X			
11				FR2.3					X		
12		FR3		FR3.1						X	
13				FR3.2						X	X

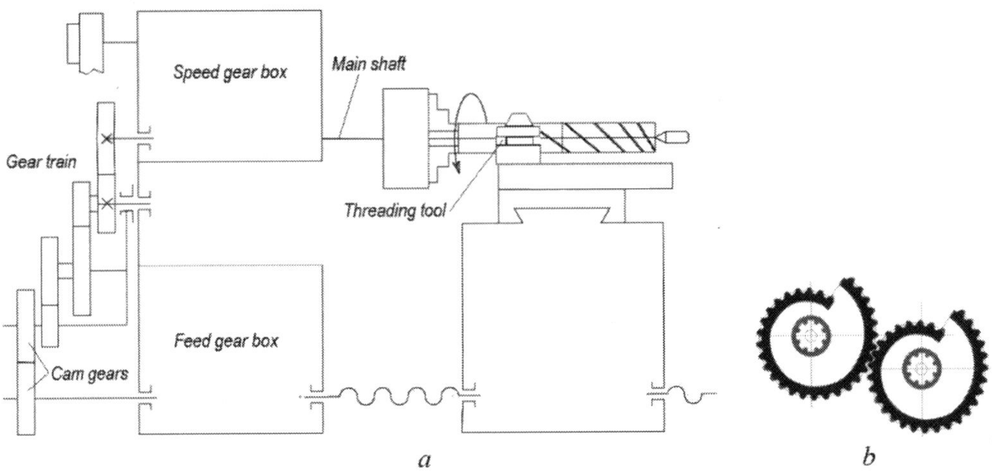

Figure 3. Schematic representation of the solution proposed for obtaining threaded surfaces with variable pitch (a) and cam gears which can be used in order to reach this objective (b).

Establishing the functions in the case of patent application

Another example of formulation the functions specific to designing an equipment for studying the machinability by drilling under constant force feed could be discussed in connection with the elaboration of a patent application. It is known that the so called *claims* could highlight the innovative solutions by considering functions able to ensure novelty, industrial application and inventiveness to the proposed technical solution. In fact, within a patent application, adequate answers to new or improved functions could be elaborated.

Table 3. Formulation of the functions leading to an improved solution for the equipment of testing the machinability by drilling under constant force feed.

Function code	Function name	Function category if the importance is considered	Solution (subassembly) proposed
A	Eliminating the stage of diminishing the error of coaxiality between the drill axis and the rotation axis of the test piece	Objective	Vice attached to the slide
B	Diminishing the possible influence of the friction forces in the guides of the slide supporting the test piece	Objective	Roller guide
C	Diminishing and balancing the force exerted by the slide and other parts during the drilling test	Objective	Balancing weight acting by means of pulley

An example concerning this aspect is presented in table 3 and the result could be seen in Figure 4.

Essentially, in the case of the test of drilling under constant force feed, the feed of drill into the test piece is materialized by means of a weight 1 of known value; by measuring the depth of the hole achieved in a certain testing time in the test piece material, one could obtain information concerning the machinability by drilling under constant force feed of the analyzed material. As a result of documentation activity [2, 3] and some experimental researches, one noticed that a high duration is necessary to ensure a certain coincidence between the axis of test sample found in rotation movement and the axis of drilling tool found in vice. In order to diminish the duration of this sequence (considered as a necessary function), one used a feed movement of the test sample (clamped in a vice), from up to down, to the drilling tool rotated by a motor placed on the machine tool table. In order to diminish the possibility of influencing the test results by the friction forces developed between the guide and the slide able to achieve the feed movement (this being an additional function – request), rollers guide were proposed to be used. Finally, a possibility to balance the weight of the slide should be find, in order to ensure a certain drilling force (this being the function to be met) and a pulley and a supplementary weight 2 were included in the test equipment.
The functions specific to the desired equipment (only those able to improve the known solution) are mentioned in the table 3.

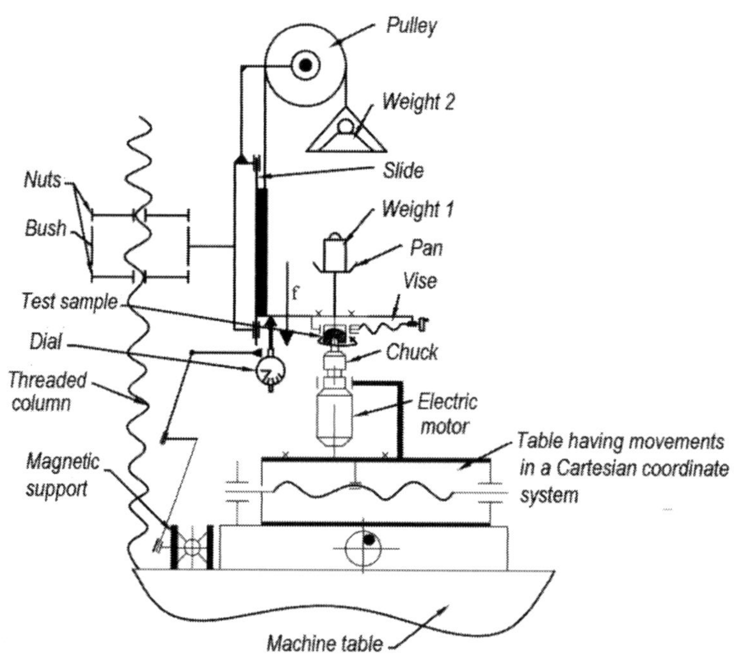

Figure 4. Schematic representation of the improved equipment for evaluation of the machinability by drilling under constant force feed.

4. Discussion and conclusions

Both in the case of applying the value analysis method and the axiomatic design method, the problem of defining the functions or requirements that must be met by the final technical solution is taken into consideration. Each of the above mentioned methods has proper rules and principles of defining *the product functions* – in the case of value analysis method and *the functional requirements*, in the case of axiomatic design method, respectively. There are distinct classifications and characteristics of the functions that must be met by the final technical solution in the case of each of the discussed methods, but there are also common aspects.

In the present paper, the principles valid for the value analysis method were applied in the case of a water waste faucet and, respectively, the principles valid for the axiomatic design method were applied in order to identify and design a technological solution for obtaining threaded surfaces with variable pitch and adaptable on a universal machine tool. One can notice that the both methods are very useful in a clearer defining of the functions or of the requirements specific to the product to be designed and this is the main benefit of the applying the above mentioned methods. A second benefit could derive from the possibility of identifying new or improved solutions for the addressed problem and this fact was appreciated as corresponding to the proposed solutions for the desired products.

Another way of functions formulation is taken into consideration in the case of patent application elaboration, when the researcher must focus his preoccupations only on the functions able to add improvements or new advantages. An application of this aspect was considered for a device for evaluation of the materials machinability under constant force feed.

In the future, there is the intention to deeply study the other substages of applying the three methods, in order to improve the possibilities of their applying in the processes of products and processes design.

References

[1] Crow K (adapted by) 2002, Value analysis and function analysis system technique, 2002. Available at http://www.npd-solutions.com/va.html, accessed: 18.02.2016

[2] Gherman L, Coteaţă M, Slătineanu L 2011 *Maşinostroenie i tehnosfera XXI veka. Sbornik trudov XVIII mejdunarodnoi naucino-tehnicesoi konferenţii Doneţk.* Tom **4** 58

[3] Gherman L, Slătineanu L, Coteaţă M 2011 *Patent RO128618-A*

[4] Leber M, Bastič M, Mavrič M, Ivanišević A 2014 Value Analysis as an Integral Part of New Product Development, Procedia Engineering **69** 90

[5] Stan F publishing year not mentioned available: http://elth.ucv.ro/student1/Cursuri/Stan%20(Ivan)%20Felicia/Analiza%20si%20ingineria%20valorii/Seminarii/Proiect%20AIV.pdf accessed: 5.04.2016

[6] Suh N P 2001 *Axiomatic Design: Advances and Applications* (Oxford: University Press)

[7] Thompson M K 2013 *CIRP Ann-Manuf Techn* **62** 115

[8] Thompson M K 2013, Proceedings of ICAD2013 The Seventh International Conference on Axiomatic Design, Worcester – June 27-28 ICAD-2013-16

Formwork application optimization by using augmented reality

R Diaconu, R Petruse and P D Brindasu
Lucian Blaga" University of Sibiu, Faculty of Engineering, 4 Emil Cioran Street, Sibiu, Romania
e-mail raluca.diaconu@ulbsibiu.ro

Abstract. By using the PLM (Product Lifecycle Management) principle on the formwork case study, after determining the functions and the technical solutions, the application must be made as optimum as possible in order to assure productivity and provide the necessary information as quick as possible. The concept is to create a complex management for the formwork based on augmented reality. By taking into account the development rate of the information, augmented reality is tending to be one of the widest (in term of domain) visualization instrument. Also used in the construction domain, augmented reality can be applied also for the case of formwork design and management. The application of the solution will be retrieved in the construction of the product, its transportation and deposit. The usage of this concept will help reduce, even eliminate human or technical errors and can offer a precise state of a specific required formwork from the stock.

1.Formwork

1.1. Introduction

One of the main problems in the formwork domain is management. The paper deals with the storage and inventory aspects of this field.

The aim is to obtain a stock – pile that offers the possibility to select a certain formwork, to obtain all the necessary information regarding the desired product (location on site/ in the deposit, number of panels available, state: clean or prepared for cleaning), to transport it and to prepare the falsework for delivery on site. Also, for the construction of a certain element, after the selection of the panels, one can virtually assemble the modules to create the pouring form and to design the mounting stages that the site worker must execute. By using the A.R. you can simulate the position and correct the errors that can appear, in this way at the building yard the assembling process will ensue normally. In order to group the formwork panels we first need to determine the selection method. Taking into account the existing formwork panels (suppliers), the principle selection factors are:

- shape
- dimensions
- assembling method
- material.

1.2. Formwork shape/form

In the construction domain, buildings are erected not only in the traditional shapes and forms. Architects exceeded this barrier and different regular and irregular shaped institutions can be seen all over the world. By putting together all the two – dimensional and 3D forms, it results that for each concrete building a specific casting mould was used (whether it is a known shape or not).

Figure 1 is an exemplification of such buildings.

IOP Conf. Series: Materials Science and Engineering **161** (2016) 012043 doi:10.1088/1757-899X/161/1/012043

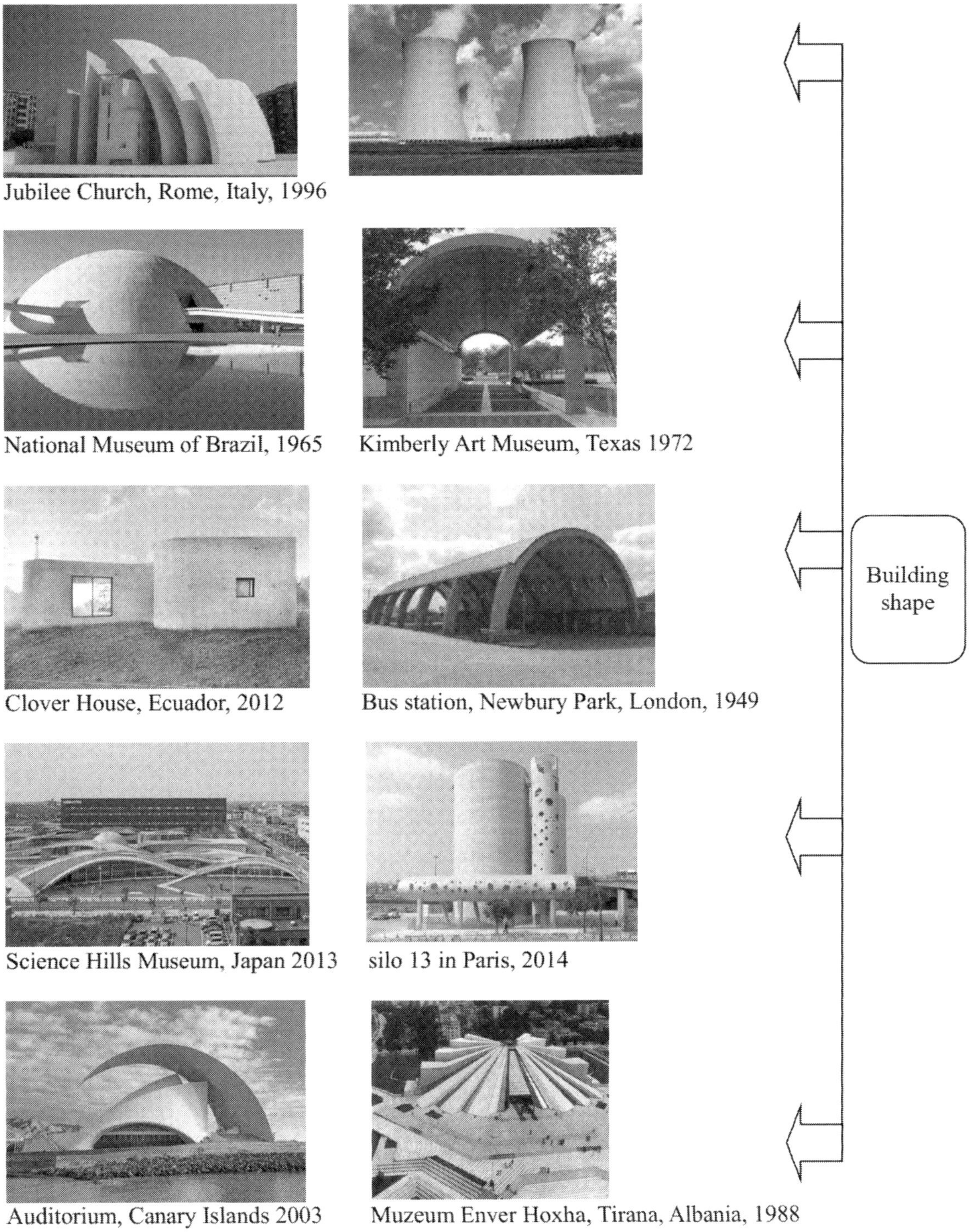

Figure 1. Different shaped buildings [16]

By analyzing the previous affinity diagram and the canonical form of the functions that determine the buildings' shape we can use the unique representation and use that formula in order to obtain the necessary formwork.

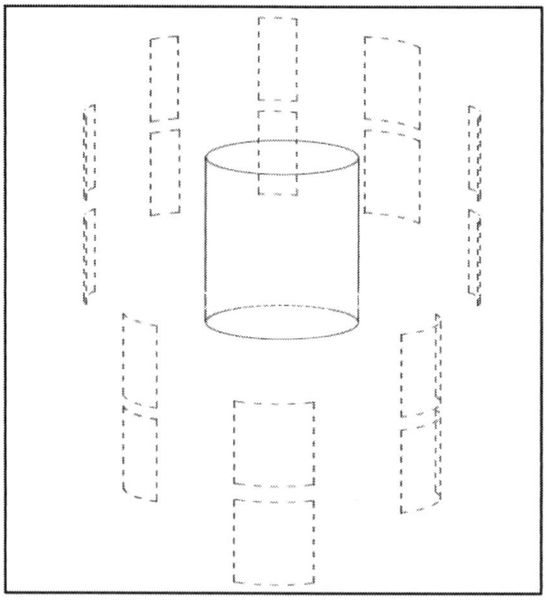

Figure 2. Cylinder wall – formwork components

Because every complex shape is formed of simple parts, for the modular panels it is essential to define the line's origin, radius, second and third point (if necessary). By knowing all these, one can easily determine the canonical form of the geometric figure.
Hence, each modular component can be reduced to a curve or segment (the case when the segments' ends represent the radius of the circumscribed circle).
After assembling the parts the desired formwork can be used for pouring. Such an example is represented in Figure 2, where a cylinder wall is exploded into different modular parts – formwork panels. For each modular panel we define the radius, length and corresponding height (function of the supplier) and we can obtain the different curved panels that we need.
When the height of the panels is not enough one can easily combine the existing modules and reach the desired dimensions by maintaining the curved shape constant to the upper and lower panels.
For a proper storage of the modular panels we can simply write the radius dimension first, the length, the height and if necessary the coordinates of another point of the panel (the origin will be considered the left hand lower corner of the panel of the pouring position) and generate a simple bar code.
For example, for a middle panel from the cylindrical formwork column we will introduce as input data the radius R, L, H, and then 0, 0, H and L,0, H. If the panel has 2.5m radius, with a height of 5m and 1.8 m long, then the bar code would be:

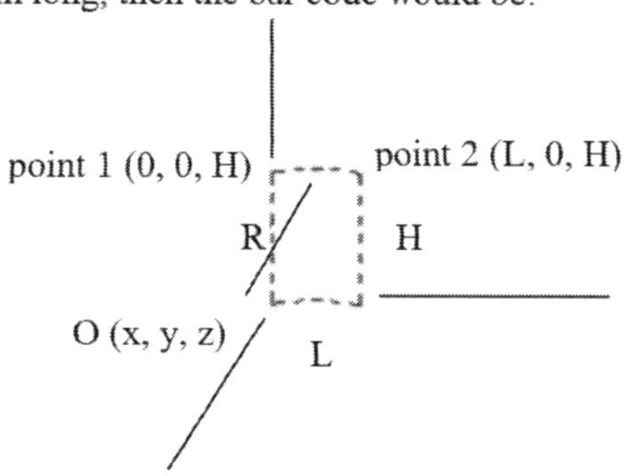

2.5-1.8-5-0,0,5-1.8,0,5
TEC-IT.COM

Figure 3. Bard code generation

On line bar code reader can easily determine the code numbers (the type code used initially was Code –

128) and we can simply determine the necessary parameters as follows:

2.5 - 1.8 - 5 - 0,0,5 - 1.8,0,5

radius length height first point coordinates second point coordinate

1.3. Formwork dimensions

Function of the supplier that we work with, we can easily define the available dimensions of the straight and circular panels.

In order to have the desired productivity other aspects regarding the product are mandatory (pressure resistance, material, connection type, estimated assembly time).

As a case study, we selected 9 formwork suppliers to work with, known world-wide, in order to exemplify the storage functionality.

Table 1 is a classification of each supplier according to their existing products.

Table 1. Formwork characteristics.

	Supplier							
Characteristic	Neru	RMD	Ulma	Doka	Noe	Meva	Hunebeck	Peri
Pressure[a]	<50-70	<60	<60-80	<60		<60	<60	<60
Min. radius[b]	2.5	2.5	2.25/1.125/2	3.5	2.75	2.5	4	1
Height[b]	2.5/3/3.5 3.65/3.95 4/4.95	3/1.5 0.75	3/2/1.5	0.7/1.2/ 2.4/3/3.6 4.8	3/1.5/0.75	3/1.5	1.5/2/3	0.6/1.2/1.8 2.4/3/3.6
Length[b]	<4	2.4/1.2/0.6	3/2/1.5	2.5	2.5/1.27/ 0.71	2.5/1.25	2.5/1.28	2.5/1.28/ 0.85
Connection type	'u'clamp (double edge)	trapeze key bolt options/ multi clamp	tapped vice	double vice and tapped vice	tapped vice	rigid clamp, clippers	bolt clamp, wailing clamp, tapped vice	tapped vice
Material	plywood + wood/ steel	plywood + steel	plywood + aluminium	plywood + Steel/ aluminium	plywood + steel	plywood + steel	plywood + steel	plywood + steel

[a] unit measure 'kN/m^2'.
[b] unit measure 'm'.

Taking into account that all buildings elements are modular ones (foundation elements: sleeves, foundation girders, columns, walls, girders, beams), the modules are realized in order to assure the pouring height and width as optimum as possible. Hence it results an order/variation of all the subsets formworks that fulfill the requirements

1.4. Formwork assembling methods

Function of the producer, there are different type of clamps that connect the panels together.

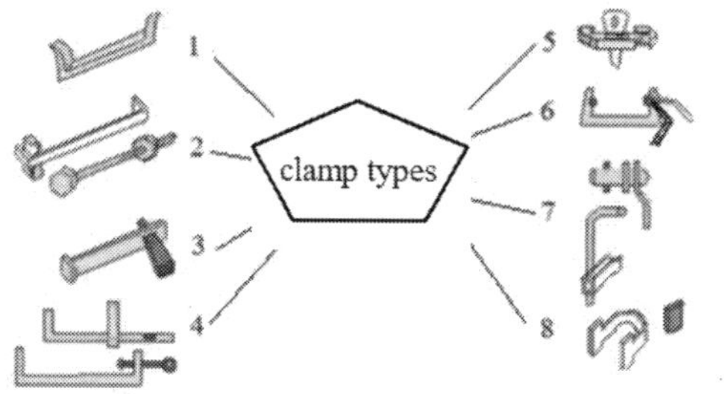

1. with elastic clamp
2. with rigid clamp and wedge/purlin
3. with bolt and wedge or screw
4. with double vice and wedge/purlin
5. with tapped vice
6. with rivet key
7. with clippers
8. with double wedge/purlin

Figure 4. Formwork coupling

1.5. Formwork material

Now-a-days there are 5 known materials: timber, steel, aluminum, composite and fabric. In the following figure (Figure 5) one can see the different shapes and materials that the formwork has.

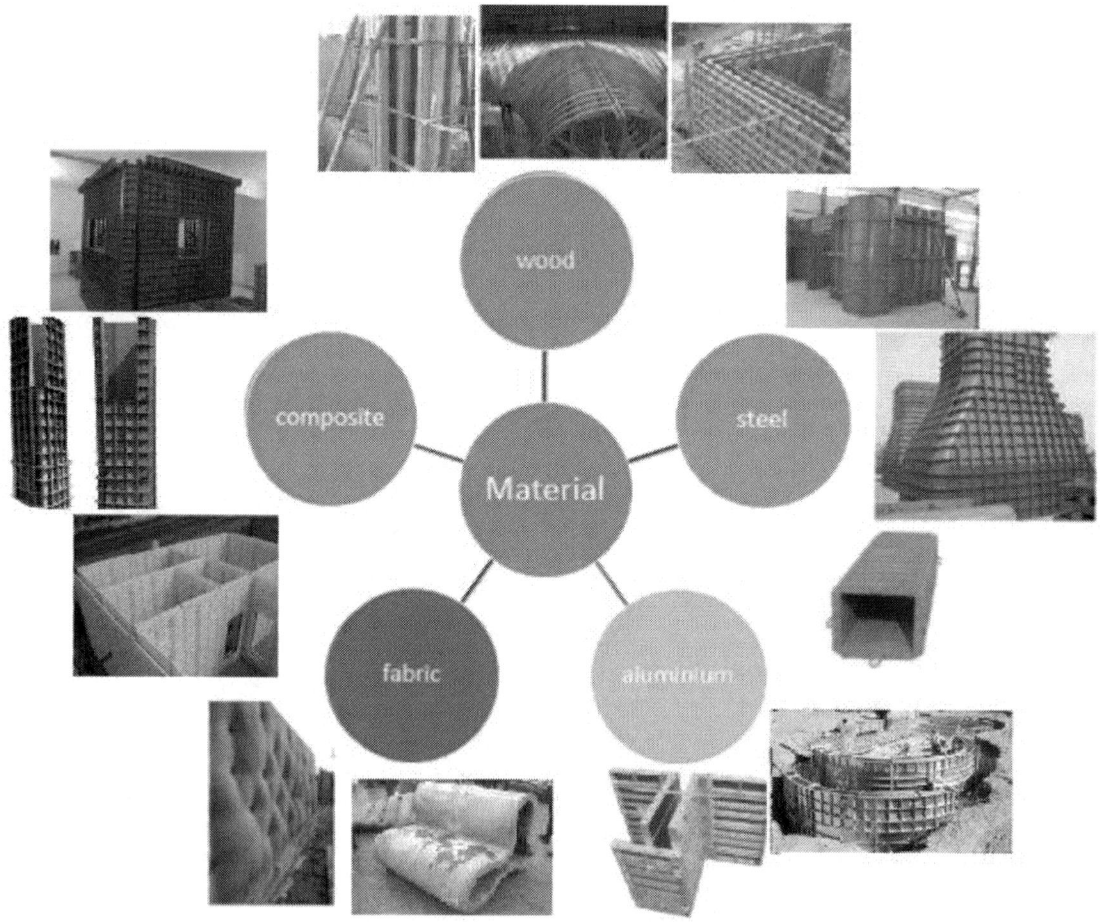

Figure 5. Formwork material [17]

1.6. Formwork selection

A "Lean construction" represents [1] an improved system that eliminates all the "waste" factors from the planning and execution domain. For the case of formwork storage, the following aspects need to be evaluated: transportation time, unnecessary movement, inventory, rework, maintenance planning.

IOP Conf. Series: Materials Science and Engineering **161** (2016) 012043 doi:10.1088/1757-899X/161/1/012043

The formwork (as stated before) is one of the important aspects concerning productivity and on site scheduling.

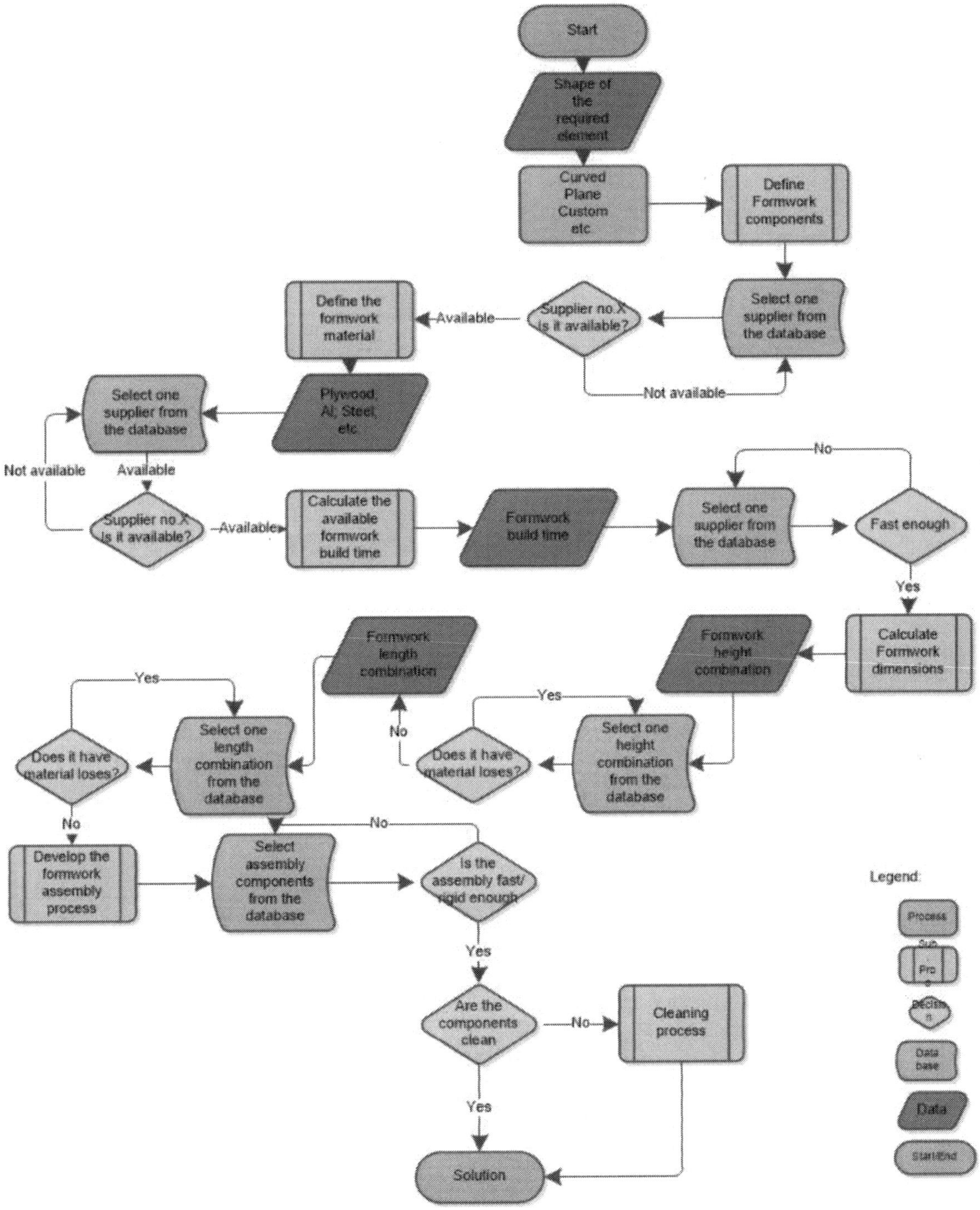

Figure 6. Selection algorithm

Taking into account the discussed selection parameters, an algorithm was conducted in order to determine the necessary type of formwork.

The applied algorithm is a decision tree type. The idea is to determine the "leaf" element (the main selection aspects) and to obtain a method for a decision making. Because many consumers (from the construction execution field) do not analyse the selection process thorough, problems can appear during the pouring process concerning the quality of the resulted element (e.g.: porous surfaces, reworks, formwork failure) or additional costs (auxiliary tie-rods for extra resistance, rent payments for unnecessary formwork panels) that can easily be avoided.
The simple path that we must work on is:

select shape -> material -> dimensions -> number of modules + type -> assembly method

In order to have the expected results, for each decision leaf time is of the essence.
Hence for each branch other subdivisions must be created, obtaining Figure 6.
After having the necessary components established, by scanning a bar code of those specific modules and auxiliary items, we can see in the inventory management program if that type of form is on stock, the number of pieces available or its current position, days of usage or time till it is storage back. The bar code is available for each shape – dimension – supplier – material.
Having all the data known we can assemble the mounting kit. The obtained group will now have a new code that will assure a correct mounting on site. Also by using Augmented Reality we can visualize if the selected panels assemble in the correct manner.

2. Augmented reality

The concept of augmented reality is a tool that can easily guide and minimize errors before executing works. This technique blends in a 3D space both the elements of the real world and virtual one creating a simulation of what the virtual product could look like in a certain (real) environment.
In the construction domain because of the new technologies and materials the traditional ways of execution and design have also changed. Augmented reality offers the possibility to answer all the questions regarding the site – design correlations such as: position on the terrain, possible excavations, execution works or even dimensioning precision. Recent studies [2] show that the main construction domain of focus for the Augmented Reality applications is the building sector followed by infrastructures and highways. The purpose of using such technology can be diverse [2]. From the monetarization one to error detection, model evaluation, update and validation to infrastructure tasks [3] such as: site dimensioning, excavations, geometry, supervision, planning or control [4].
The usage of Augmented Reality systems could be classified as stationary or mobile. The table below shows the tendency for each construction domain mentioned above taking into account the latest researches.

Table 2. Construction domain leaning

Construction domain	Article	Purpose	Usage type
Buildings	"Management of building supported on virtual interactive models" [5]	Offers and idea about the stages of the construction in a 4D application (function of time)	mobile
Infrastructure	"Potential of Mobile Augmented Reality for Infrastructure Field Tasks" [6]	Offers real time information about the site characteristics (network, tunnels, water drainage)	mobile
Highway	"Development of methodology and virtual system for optimized simulation of road design data" [7]	Simulates road design taking into account the natural real environment	stationary
	"Evaluation of image-based modelling and laser scanning accuracy for emerging automated performance monitoring" [8]	An A.R. system that can recognize images for design input data	mobile + stationary

For our case study we used a combined system: mobile + stationary. By using a camera (mobile system) we can scan the QR code from the formwork 2D drawing and see the step by step assembly. By the help of the stationary device (the A.R. program) we can see on the laptop the steps a worker has to perform in order to put together the modular components.
The case study represents a concrete wall having the dimensions L x H x g: 4.5m x 3.4m x 0.5m. By applying the selection chart we determined the optimum modular panels to work with. Figure 7 is a representation of the panel position.
The panels type 1 have the dimensions L x H: 0.9 m x 2.7 m, panel 2: L x H: 0.9 m x 0.9 m and panel no. 3 L x H: 2.7 m x 0.75 m.
When reading the technical documentation and drawings, by using the QR code, the assembling steps will ensure the correct position and mounting of the three panel types.

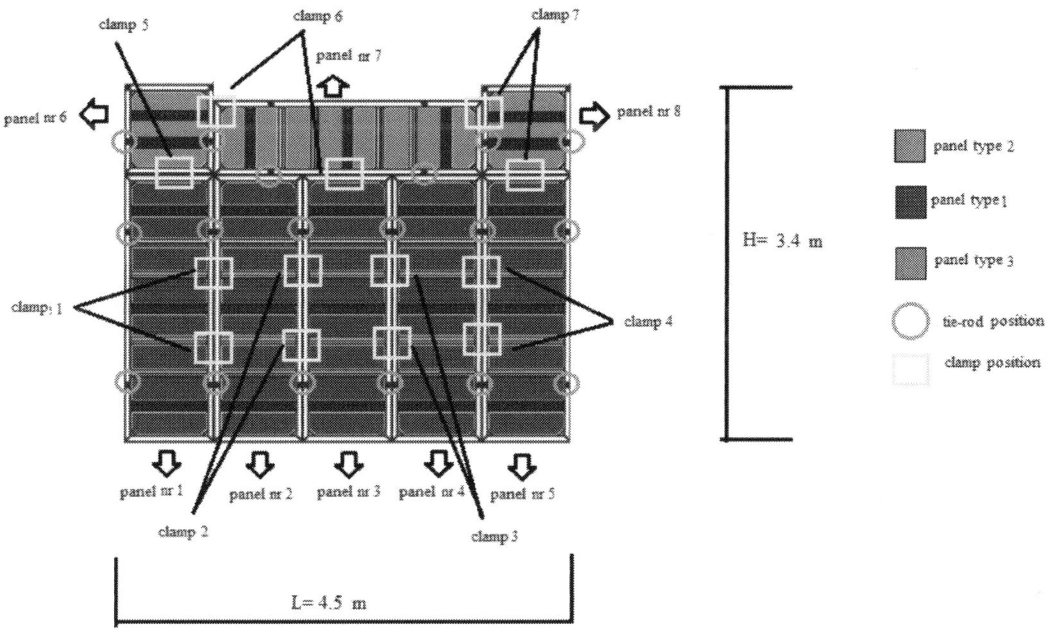

Figure 7. Formwork sketch

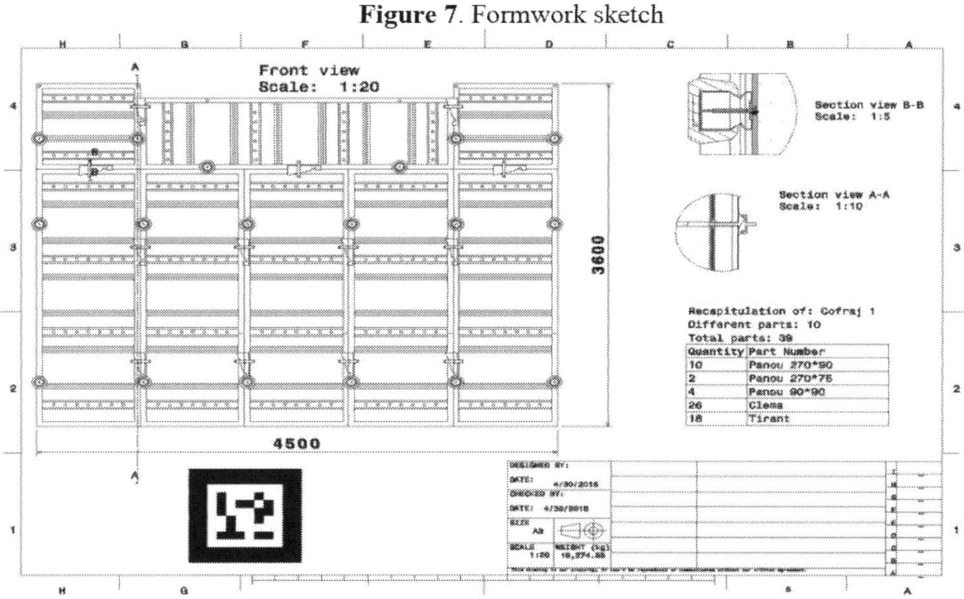

Figure 8. Formwork plan

IOP Conf. Series: Materials Science and Engineering **161** (2016) 012043 doi:10.1088/1757-899X/161/1/012043

Figure 8 is the execution plan of the wall formwork.

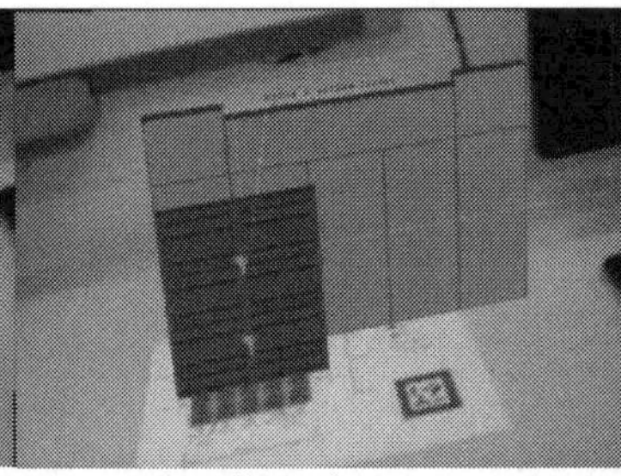
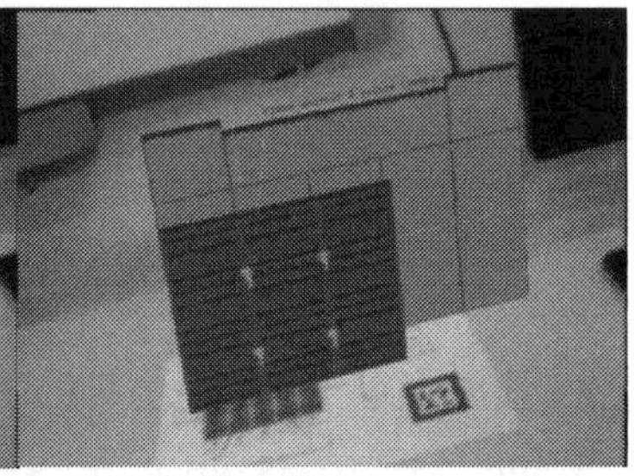

Step 1. Step 3. Step 5.

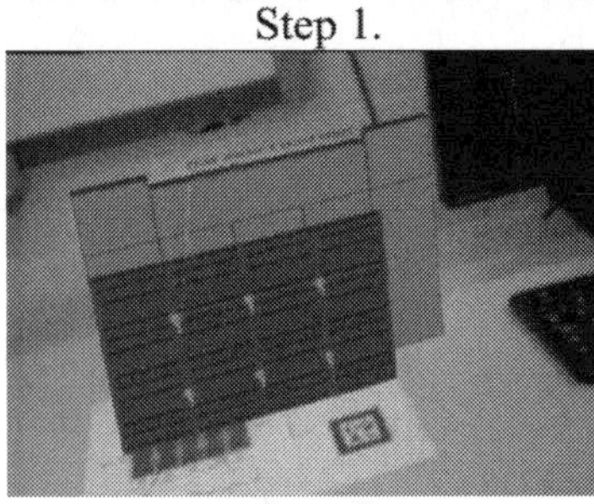
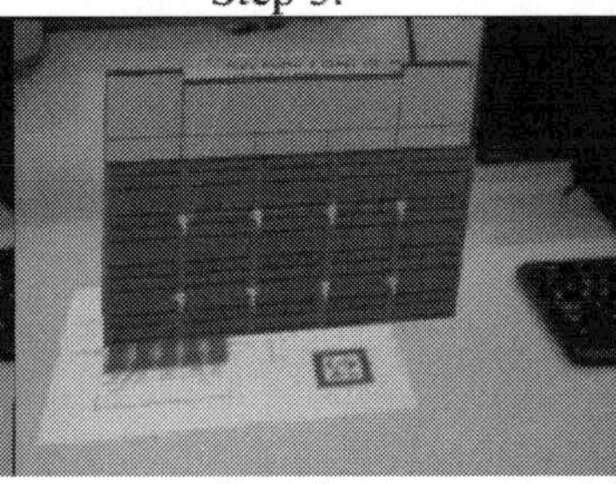
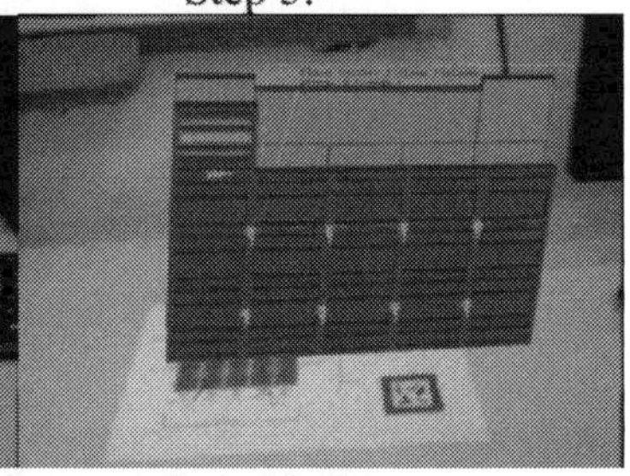

Step 7. Step 9. Step 11.

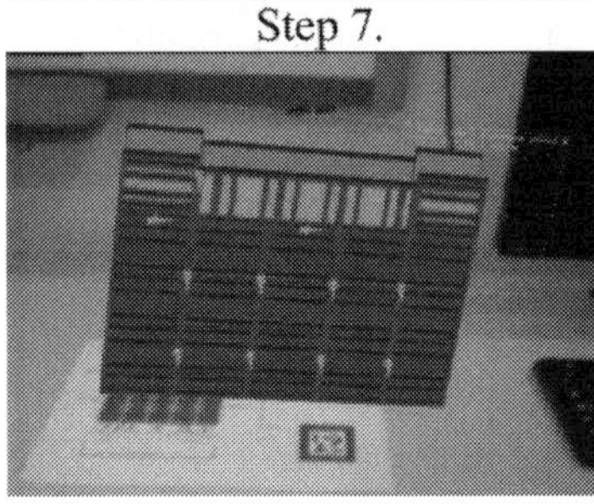

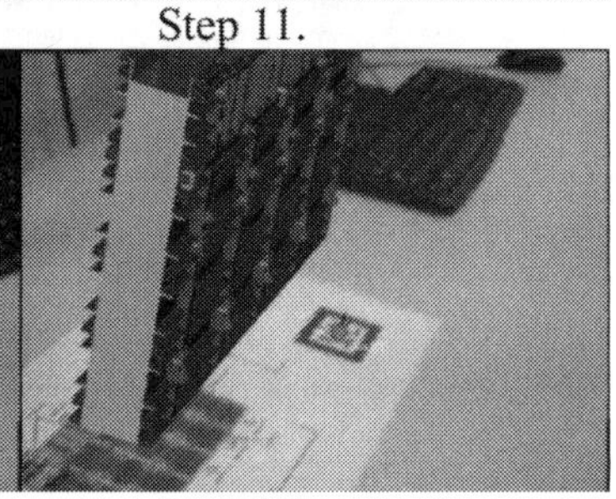

Step 14. Step 15. Step 16.

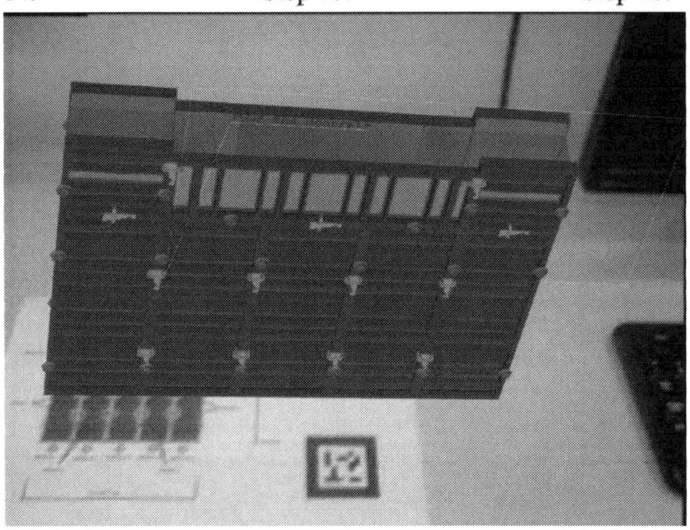

Step 17.

Figure 9. Augmented reality for formwork assembly

In order to apply the A.R. after designing the form in Catia v5 we transformed the product and its parts into a simplified format called .wrl (virtual reality modeling language). Using this format, we could

virtually see the assembling steps.
In order to execute this mounting operation 17 steps need to be performed:

panel 1 ->	panel 2 ->	clamps ->	panel 3 ->	clamps ->	panel 4 ->	clamps ->	panel 5->	clamps ->
step 1	step 2	step 3	step 4	step 5	step 6	step 7	step 8	step 9

panel 6 ->	clamps ->	panel 7->	clamps ->	panel 8 ->	clamps ->	tie-rods ->	concrete pouring
step 10	step 11	step 12	step 13	step 14	step 15	step 16	step 17

Also, for the clamps and tie-rods another code can be defined in order to see the section and the corresponding position. Figure 10 and figure 11 represent a detailed representation of the two components.

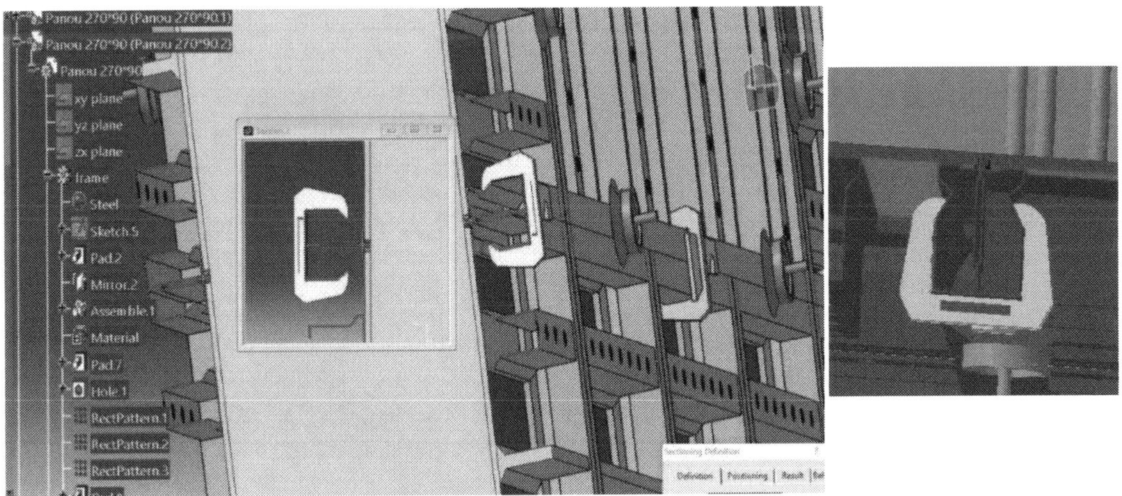

Figure 10. Formwork clamp 2D and 3d detail

Concerning the tie-rods, they connect the two panels that form the formwork.

Figure 11. Formwork tie-rod position

3. Conclusions

The storage assures the proper selection of the necessary form and offers the proper information about the components.

IOP Conf. Series: Materials Science and Engineering **161** (2016) 012043 doi:10.1088/1757-899X/161/1/012043

The usage of Augmented Reality not only helps avoiding errors but also increases labor productivity. By selecting the optimum solution for the panel, the 3D visualization helps the workers assemble the formwork precise and correct.

References

[1] Ansah R, Sorooshian S, Mustafa S 2015 Lean Construction Techniques A Framework towards Elimination of Waste in Construction Industry *Proceeding of the 2015 International Conference on Operations Excellence and Service Engineering (IEOM 2015)* Orlando Florida USA

[2] Rankohi S, Waugh L 2013 *Review and analysis of augmented reality literature for construction Visualization in Engineering* p 1-9

[3] Behzadan A H, Dong S and Kamat V R 2015 Augmented reality visualization: A review of civil infrastructure system applications *Advance Engineering Informatics* **29** 252-267D

[4] Sahin, Togay A 2016 Augmented reality applications in product design process *Global Journal on humanities and Social Sciences* **3** 115-125

[5] Sampaio A, Gomes A R and Santos J P 2012 Management of building supported on virtual interactive models *Journalof Information Technology in Construction* **17** 121-133

[6] A. Hammad, J. Garrett, H. Karimi 2002 Potential of Mobile Augmented Reality for Infrastructure Field Tasks *Applications of Advanced Technologies in Transportation* 425-432

[7] Kang L S, Moon S H, Dawood N and Kang M S 2010 Development of methodology and virtual system for optimized simulation of road design data *Journal of Automation in Construction* **19** 1000–1015

[8] Golparvar-Fard M, Bohnb J, Teizerb T et al. 2011 Evaluation of image-based modeling and laser scanning accuracy for emerging automated performance monitoring *Journal of Automation in Construction* **20** 1143–1155

[9] Chi H-L, Kang S-C and Wang X 2013 Research trends and opportunities of augmented reality applications in architecture, engineering and construction *Automation in Construction* **33** 116-122

[10] Akhavian R and Behzadan A H 2012 An integrated data collection and analysis framework for remote monitoring and planning of construction operations *Advance Engineering Informatics* **26** issue 4 749-761

[11] Behzadan A H, Dong S and Kamat V R 2015 Augmented reality visualization: A review of civil infrastructure system applications *Advance Engineering Informatics* **29** 252-267

[12] Agarwal S 2016 Review on application of augmented reality in civil engineering *International Conference on Inter disciplinary Research in Engineering and Technology*

[13] Chien-Ho K and Jiun-De K 2015 Making formwork construction lean *Journal of Civil Engineering and Management* **21** issue 4 444 – 458

[14] Abdelhamid T S, El-Gafy M and Mitropoulos P 2009 Selection of Roof Casting Formwork Systems for the Bird Island Project: Case Study *Practice Periodical on Structural Design and Construction* **14** issue 4

[15] Schall G, Mendez E et.al 2009 Handheld Augmented Reality for underground infrastructure visualization *Personal and Ubiquitous Computing* **13** issue 4 281-291

[16] Images Figure 1 available at: http://www.architecturaldigest.com/; http://dels.nas.edu/; http://www.designboom.com/ sq.wikipedia.org; http://design-milk.com/

[17] Images Figure 5 available at: http://studio-tm.com/; http://www.tradekorea.com/;

[18] Formwork suppliers' sites available at: http://www.peri.com/en; www.huennebeck.ro; www.mevaformwork.com; www.noe.de/en/homepage.html; www.doka.com/en/; www.ulmaconstruction.com/en/formwork; www.rmdkwikform.com; www.neru.com.tr/en/index.php; www.mesaimalat.com/

IOP Conf. Series: Materials Science and Engineering **161** (2016) 012044 doi:10.1088/1757-899X/161/1/012044

CAD - CAM Procedures Used for Rapid Prototyping of Prosthetic Hip Joint Bone

Luminita I. POPA [1], Vasile N. POPA [2]

[1] Department of Automation and Information Technologies,Transilvania University , Mihai Viteazul street no.5 , Braşov , România

[2] Regional Department for Defense Resources Management Studies, Mihai Viteazul street no.130, Brasov, Romania, Associate Professor

Email: [1] mluminita2001@yahoo.com

Abstract. The article addresses the issue of rapid prototyping CAD/ CAM procedures, based on CT imaging, for custom implants dedicated to hip arthroplasty and the correlation study to be achieved between femoral canal shape, valued by modern imaging methods, and the prosthesis form. A set of CT images is transformed into a digital model using one of several software packages available for conversion. The purpose of research is to obtain prosthesis with compatible characteristics as close to the physiological, with an optimal adjustment of the prosthesis to the bone in which it is implanted, allowing the recovery of the patient physically, mentally and socially.

1. CAD-CAM Procedures

In the literature, CAD (Computer Aided Design) / CAM (Computer Aided Manufacturing) is an acronym that means using computer design and manufacturing.

This innovative technology that uses digital computers to conduct various design and manufacturing functions tends to fully integrate these activities which, traditionally, have been regarded as two distinct and separate functions. Overall, CAD /CAM develop future enterprise advanced technology, full assisted by computer. The computing system consists of optimized for design hardware and software. Within CAD activity, based on executing drawings, were made 3D models of shields. Based on the overall design, taking into account the existing links between the components, it has been established asynchronous motor assembly. CAM includes indirect applications in which the computer has the support role for manufacturing operations. CNC (Computer Numerical Control) machine tools are designed to change the shape and size of objects in terms of productivity, dimensional accuracy and quality of machined surface. I followed the issues such as energy consumption, process complexity and costs. The emergence and integration of CAD/CAM systems within the production process, enabled engineers to create 3D CAD project, to generate trajectories in the generic language of CAM system and then to perform conversions in multiple languages. The CAD and CAM activities are closely interdependent, because CAD is underlying CAM activities. CAD tools are defined based on three concepts: geometric modeling, computer

graphics and design tools. CAM tools consist of the intersection of three fields used in manufacturing: CAD tools, network concepts and machine tools.
Computer- Aided Manufacturing – CAM is defined as the use of a computer system in planning, management and control of the operations of a plant, by any direct or indirect interface between the computer and production resources.
For implant complex digital prototype realization were taken the following steps:
1. Modeling assembly components;
2. Setting the base assembly (fixed element - "anchored" in assembly)
3. "Assemble" of all components of virtual 3D mode
4. Analysis of functionally prototype obtained

After obtaining the digital prototype, we can start mechanics simulation. Analysis of efforts and deformations can be achieved by using FEA / CFD (Finite Element Analysis / Computer Fluid Dynamics). Implant model generation was performed starting from the 3D surface of the femur obtained from the three-dimensional reconstruction and visualization model of pelvis bones. There were necessary radiographic image processing (extraction important parameters in arthroplasty) and processing of CTs and MRIs Magnetic Resonance Imaging (stacks of 2D), in order to reconstruct 3D surface which is representing the hip bones. Using 3D software application, it was obtained the femur interior, upon which was built the model prosthesis. Based on the obtained models, implants prototypes were made, using rapid prototyping device and were performed mechanical tests on prototypes.

Figure 1 below shows the normal anatomy of the pelvic bone and femur.

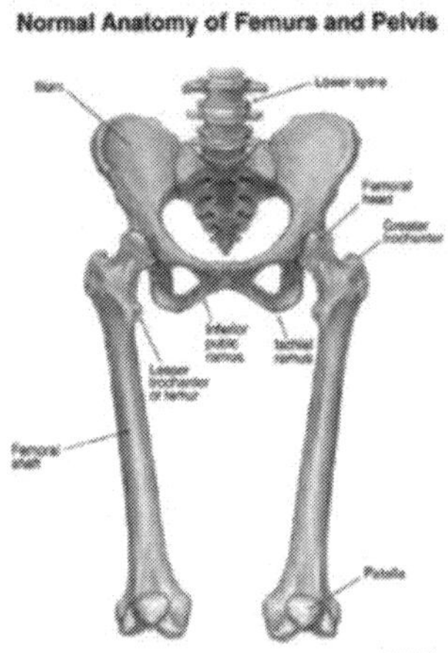

Figure 1. The normal anatomy of the pelvic bone and femur.

Inserting the prosthesis in hip arthroplasty involves placing the body and neck of prosthesis in the place of bone marrow at the level of femoral body and neck and femoral head replacement with prosthesis head .
Three-dimensional reconstruction and visualization have included the following steps:

- The possibility of charging a stack of CT images in DICOM or BMP format
- Viewing stack of 2D images (one image at a time)
- Reconstruction of three-dimensional surface representing femur and viewing it (with the possibility of displacement and rotation of the pattern in the scene);
- Editing stack of 2D images and viewing its effects over three-dimensional rebuilt surface;
- The possibility of radiodensity change of the tissue to be rebuilt;
- Exporting stack of 2D images (DICOM format) in PNG format;

- Exporting the reconstructed surface in a format recognized by 3D software (for the next module of implant generation)

Implant generation comprises the following components:

- Loading polygon surface representing the femur;
- Getting femur interior;
- Prosthesis construction, starting from the femoral bone interior.

The figure 2, below shows the construction of bones from a CT data set of some body parts.

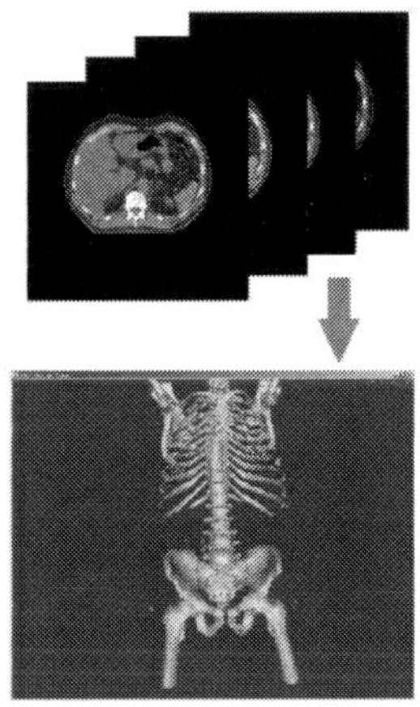

Figure 2. Bones reconfiguration from a CT data set.

2. Implant assembly modeling

Many researchers have found that a geometric adaptation shape of cementless implant rod shape to the inner contour of the proximal femur is essential to optimize the load taking over and for a better mechanical stability. Based on femoral bone obtained from the 3D reconstruction and visualization software (based on tomographic images) we have passed in modeling of customized implant.

When is designing the femoral component we can use specialized programs for modeling suite SolidWorks, CATIA, SolidEdge, Pro Engineer, etc (figure 3,4,5).

Within the created assemblies, are defined various components through their degrees of freedom including geometric information and relative positioning of parameters (figure 6,7,8).

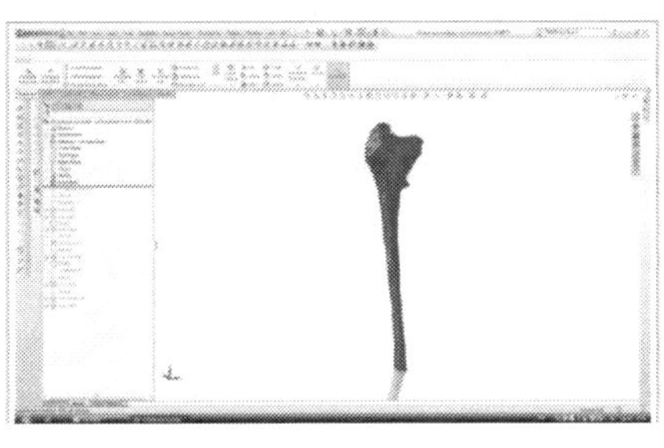

Figure 3. Femoral bone view in SolidWorks

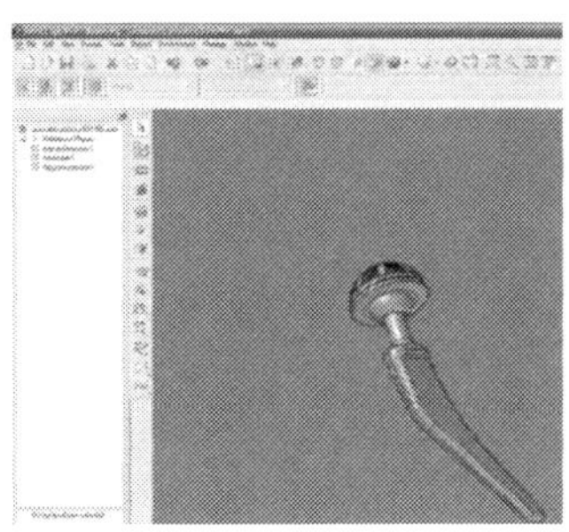

Figure 4. CATIA view implant assembly

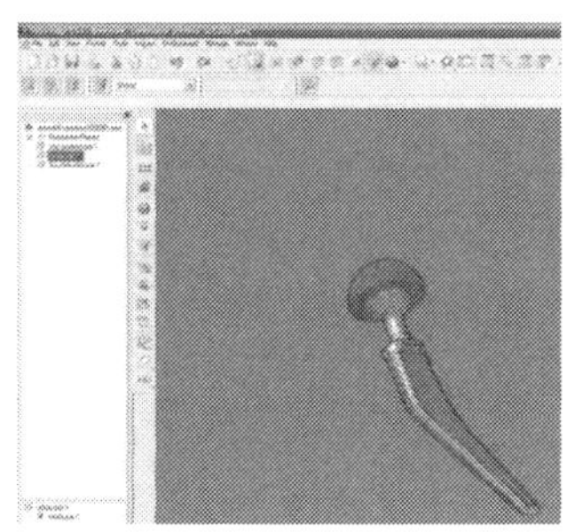

Figure 5. SolidEdge view implant assembly

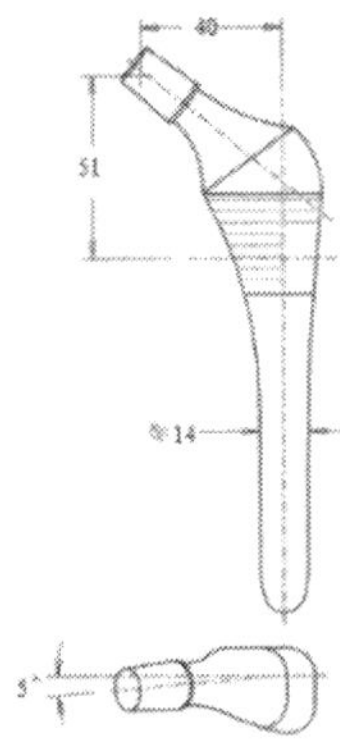

Figure 6. 2D representation of hip prosthesis

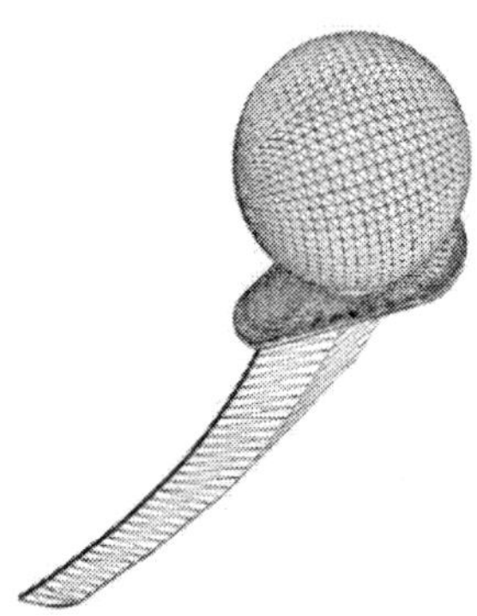

Figure 7. 3D modelling surfaces

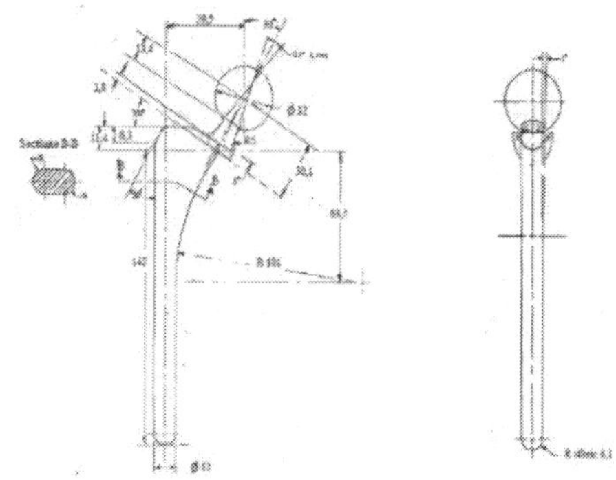

Figure 8. Implant assembly drawings

Important roles in the optimization of physical – mechanical parameters have design, analysis and test for different loads of newly created implants applications. Starting with the design phase is envisaged performing finite element analysis (FEA Finite Elements Analysis) in the orthopedic area of interest in order to obtain its loads status. Mechanical failures that occur in vivo are due to wear or fatigue request.

Implant assembly components may be made of:

- Metals (pure or alloy);
- Titan Ti6Al4V and stainless steel or 316 stainless steel;
- Chromium is the major element in stainless steel having composition with minimum concentration of 11%.
- Alloys type Co-Cr -Mo;
- Ceramic materials, at ceramic materials manufacturing are used: alumina (Al2O3) or zirconia (ZrO2).

In Figure 9, bellow was modeled the fastening assembly of femur prosthesis.

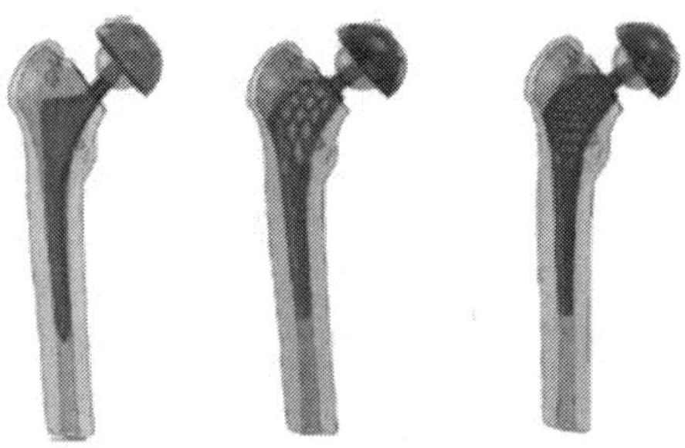

Figure 9. Fastening assembly of femur prosthesis modeling

-using finite element method, we have determined the request forces appearing in support position and while walking, joint femoral loading, outside different requests influence over all assembly elements.

-determining the safety factor for every constituent assembly to determine possible overloads which could lead to implant failure;

-realization of comparative study for Medin customized femoral rod requests to establish achievement effort opportunity of dedicated implant;

Movement across the bone-prosthesis are shown in Figure 10.

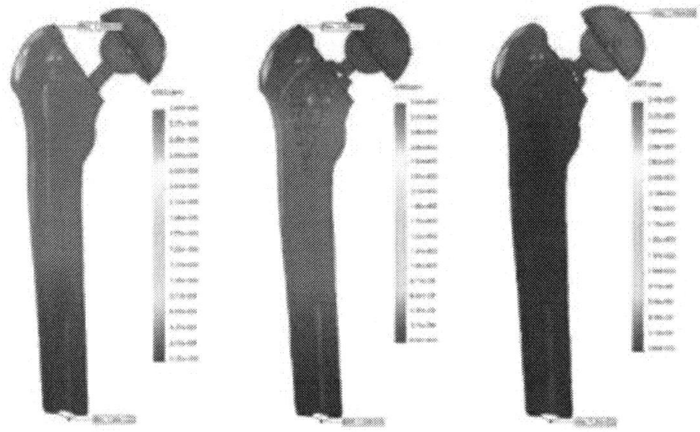

Figure 10. Movement across the bone-prosthesis

-using finite element method, we can obtain animations which are illustrating the evolution of von Mises efforts, during joint request for a better choice of taking the load (Figure 11).

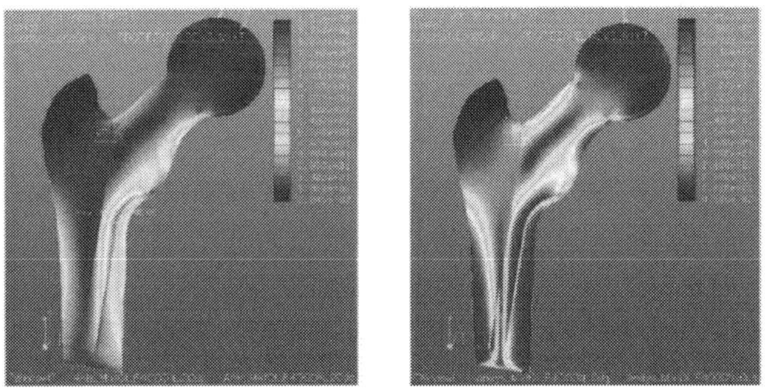

a) **b)**

Figure 11 Comparisons between von Mises stress states of a hip prosthesis - main section

a) operating under 20 degree angle; **b)** operating under 0 degrees angle;

3. Rapid prototyping techniques

Rapid prototyping technologies are defined as a set of technologies that enable the manufacturing of physical models directly from three-dimensional models or virtual prototyping, through programs such as CAD, CAM, which are transferred to a computer responsible for its materialization.

Rapid Prototyping techniques have as final goal realization in of prototypes in a short time.

The advantages of these processes are the following:

- Low energy consumption;
- Little loss of material;
- High speed manufacturing;
- It can be parts or products with complex geometries.

Rapid prototyping is the technology used in obtaining prototypes based on a 3D CAD model, which is digitally processed in order to recognize the data format by specialized machines through the following processes:

- SLA (Stereo lithography) - polymers solidifying at the impact of a light ray;
- LOM (Laminated Object Manufacturing) - obtained by layering part;

- FDM (Fused *Deposition* Modeling) - process for deposition of material;
- SLS (Selective laser *sintering)* / SLM (Selective laser melting) of metal powders;
- DMLS (Direct metal laser sintering) we can get some small unique tools;
- Printing three-dimensional (3D Printing).

"Rapid Prototyping Technologies" have grown in importance in the process of arthroplasty due to the fact that within these technologies, the manufacturing times are relatively low from the stage of conception to final completion of the prostheses, the costs of assimilation and manufacture of individualized prostheses have reduced very much.

Rapid prototyping technologies are applied in making parts by casting metal alloys in three directions:

- When designing parts, patterns and molds;
- Making models;
- Making molds.

3.1. The realization of a hip prosthesis

Parts of implant bone are particularly demanding because it is subject to complex applications. Hip prosthesis is made of high alloy steel with Cr-Ni-Mo-Mn. The hip prosthesis should approach the most of the natural shape of the bone forming the joint. In order to achieve three-dimensional model we have used SolidWorks software. The three-dimensional design of bone obtained after requests, may be subject to real bone; The most conclusive mechanical tests is fatigue test. The same program can be used for the processing by milling on a controlled numerically machine with unique prostheses, custom made for each patient.

Forging in mold is recommended to obtain prostheses in series; this process ensures maximum durability of the part in terms of fatigue. The shape of the mold cavity can be processed by milling using the same coordinates of contours obtained by sectioning and applying the necessary correction range.

Following the heat treatment, the mold cavity will be finished by Electrical Discharge Machining (EDM), with an electrode obtained by electroplating after the model made by SLS.

The easiest way to rebuild bone structure of a patient is to use Computed Tomography (CT) images which are already exist from previous treatments of the patient. A set of CT images can be converted to a digital model using one of several available SolidWorks software packages available for conversion.

Input Data to this software is usually in the form of DICOM files and output data is predominantly STL (Standard Tessellation Language), which can be directly used in most of Rapid Prototyping technologies to produce real models. The reconstruction of the patient bone structure can be individualized with minimal effort and in complete safety thanks to handling CT files type DICOM and transforming them into STL files and / or CAD that can be 3D processed. Rapid prototyping technology is the technology used to obtaining prototyping based on a 3D CAD model, which is digitally processed in order to recognize the data format by specialized machines.

The STL files are required for Rapid Prototyping machines. Digital prototype is used to define the assembly of 3D entities.

The advantages of digital prototype are:

- Removing the need for physical prototype;
- Reducing the execution costs;
- Increasing the product quality;
- Reducing the time required to launch the product on the market;
- Information being accessible to design teams takes place much faster.

4. CAM Technologies for individualized prostheses

In the stage of physical achievement of prostheses (CAM) there are used the following technologies:

1. CAM technologies processing through the removal of material, starting from a semi-finished

IOP Conf. Series: Materials Science and Engineering **161** (2016) 012044 doi:10.1088/1757-899X/161/1/012044

product (a large amount of raw material) and the excess material is removed using conventional methods (turning, milling, grinding, etc.), or by using unconventional methods (EDM - Electrical Discharge Machining, laser machining, ultrasound, etc.).
2.CAM processing technologies by redistribution of material. According to these technologies, a fair amount of raw material is redistributing in the required prosthesis form, using solid forming (forging, stamping, drawing, extrusion, etc.) or redistribution in liquid or semi (molding, injection molding, etc.).
3. Rapid Manufacturing Technologies or Rapid Prototyping (RP), performing the part by adding material as needed and where needed.
4. Modern technologies, which are based on the implementation of modular elements with simple geometries, easily assembled into a solid body with complex geometry.

5. Stereo lithography (SLA) procedures

Modern lithographic procedures were used initially in electronic components technology, which exploits the property of photosensitive substances, which were initially liquid monomers, polymerizing under a radiation of a certain wavelength (typically ultraviolet radiation). If the monomer layer is selectively covered by means of a cliché or the used radiation is concentrated as a spot, the light-sensitive substance will polymerize selectively; such a configuration results "in plan". The configuration thickness depends on the depth of penetration of used radiation. Unexposed monomer is removed by washing. Stereo lithography is defined as polymers solidifying at the impact of a light ray. Stereo lithography (SLA) was patented in 1982 by Chuck HULL and uses the principle of superposition of successive shaped in plan layers and polymerized under the action of ultraviolet radiation emitted by a laser. It was the first method of prototyping technique. The polymerization is a reaction where smaller molecules (monomers) are combining chemically (by main valences) to form a macromolecule, branched approximately in the same composition. SLA cannot be imposed except by using the computer. The desired model is designed in three dimensions using 3D software modeling. The work principle of stereo lithography is resulting in the figure 12.The stereo lithography installation is shown in figure 13.
A photosensitive resin in the form of liquid monomer is put in machine tank car. At the liquid surface is focused a beam originating from a laser generator using ultraviolet radiation. The beam can be scanned with a computer controlled mirrors system on the liquid surface contour plan in desired coordinates X - Y; The plan represents a cross-sectional through the work part. The result will be polymerization of monomers after the programmed contour. To obtain three-dimensional form, the model support will receive a Z- axis movement, whose value is a direct function of the thickness of the laser beam polymerized layer to obtain a compact and rigid form. Thus, the model dips in tank as it is formed.

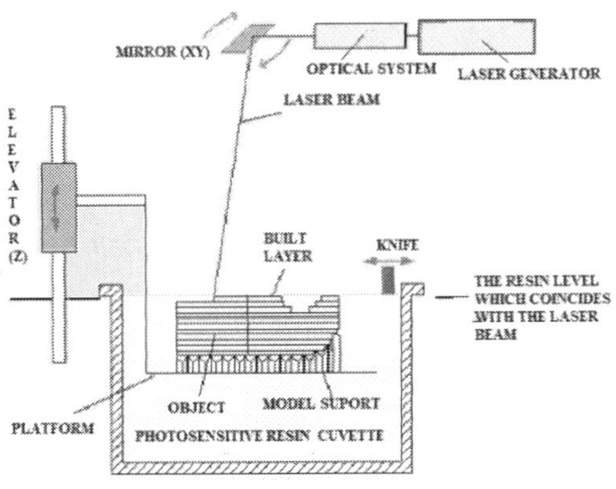

Figure 12 The work principle of stereo lithography

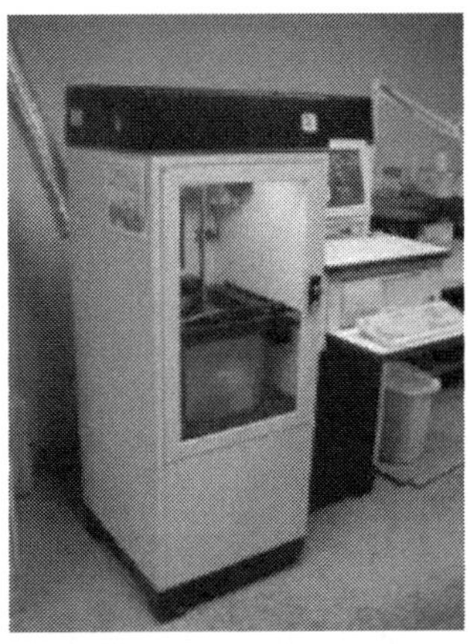

Figure 13 The stereo lithography installation

SLA is the fastest current process for obtaining the photopolimerlizabile resin complicated models, with the possibility of change or adapt quickly the shape, which is impossible without a CAD system.

5.1. The use of photosensitive substances

The photo polymerizable resin is used consisting mainly of a photo initiator and a mixture of various monomers. Photo initiators selectively absorb light radiation, in the range of ultraviolet and visible range, ie between 250-450 nm; They convert light energy into chemical energy that primes the polymerization reaction. The monomer mixture is containing various molecules belonging to the family of acrylic and methacrylic esters. The design model is oriented so as to obtain a secure basis on which to "build" model, with maximum stability and avoiding detachment of parts of the model that could float freely in the liquid.

6. Selective Laser Sintering (SLS)

A similar process with the same advantages but using powdered metallic materials (SLS - Selective Laser Sintering) allows not only getting wax models, polycarbonate or nylon but even finished parts of the metal or polymer. Selective laser sintering principle (SLS) of using as raw material polymer powders, is shown in the figure 14. A selective laser sintering system is shown in the figure 15.

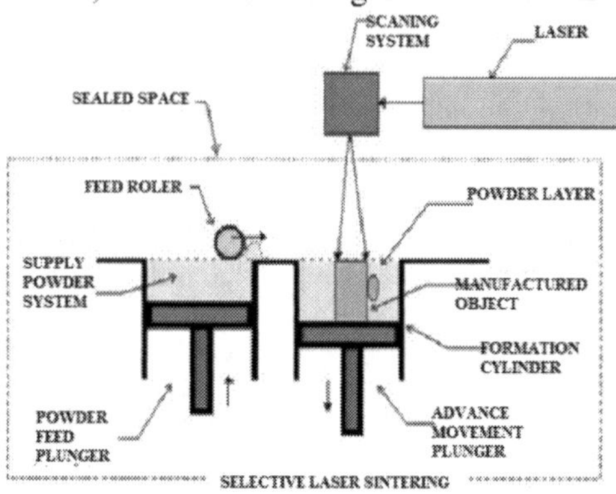

Figure 14 Selective laser sintering principle (SLS)

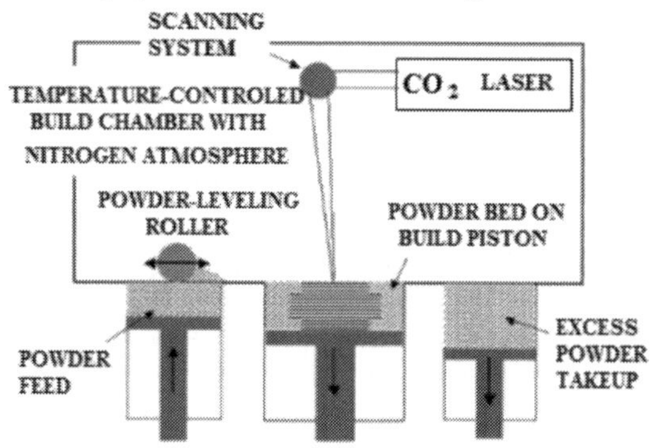

Figure 15 The selective laser sintering system

6.1. The metallic powders used in Selective Laser Sintering (SLS)

The metallic powders that can be used in Selective Laser Sintering include a variety of materials including implantable stainless steel, cobalt or nickel superalloys, titanium alloys and pure titanium. All these materials have excellent mechanical and physical qualities, better than cast or forged materials, so are recommended for *biomedical implants*. The biocompatibility properties, corrosion resistance and low specific gravity, combined with the infinite possibilities of 3D geometric modeling recommend this technological process as a tool of the future micro-mechanical engineering, mechatronics and robotics. The parts obtained by this process can be further processed as desired, if deemed necessary, by any known mechanical process: milling, turning, drilling, countersinking, reaming, tapping, slotting, grinding, broaching, lapping, honing and over finishing.

Also, the parts can be further processed by sandblasting, polishing and welding, Electrical Discharge Machining and coating.

The process of rapid prototyping by laser direct sintering unfolds in five stages:

- Creating virtual 3D-CAD model and exporting of CAD file in STL format;
- Part orientation and positioning, supports generation and STL file checking;
- STL 3D file conversion into 2D files and preparation for prototyping machine;
- Making physical model by melting successive layers;
- Post processing of physical model.

Sintering achievement of two models of customized hip implants is shown in figure 16.

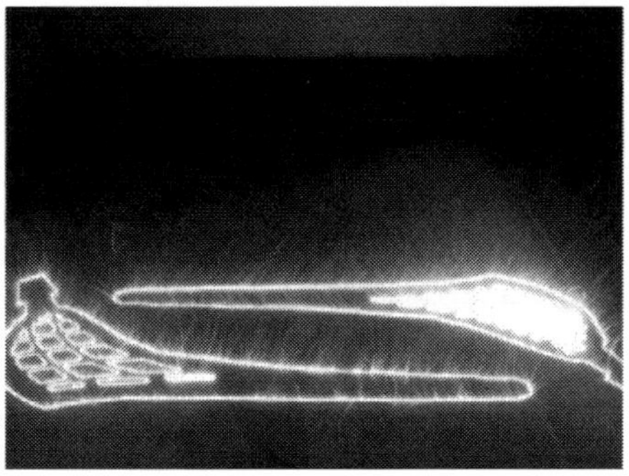

Figure 16. Sintering achievement of two models of customized hip implants

Whatever rapid prototyping method of hip prosthetic works, the test pieces obtained from new material will be tested before (to determine its characteristics) and after manufacturing, performing mechanical (tensile, compression, shear, etc.) and physical tests (corrosion) on real models according ISO 7206-4 (2002), ISO 7206-8 (2002).Different types of prosthetic hip implant are presented in figure 17

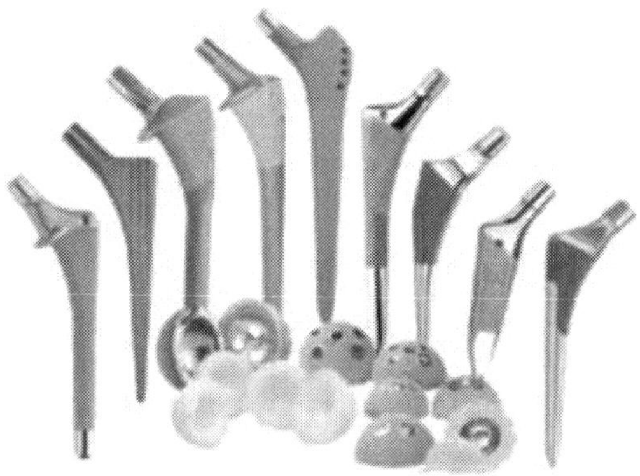

Figure 17. Different types of prosthetic hip implant

7. Conclusions

The doctor and engineer always have the real picture of the patient's anatomy and its data can be manipulated without it bothering the patient at all. The final physical version of hip prosthesis can be tested on static and dynamic machines mechanical test to confirm the safety of designed solution.

Through advanced methods of CAD /CAM computer we can create the necessary elements for customized prosthetics which is tailored to each individual. The use of CAD CAM procedures increases the efficiency of therapeutic methods, leading to more efficient health services, by decreasing the number of required intervention, and increasing creativity in healthcare, because prosthetic implants are adapting to the characteristics of each individual. In the article it was presented briefly the possibility of obtaining customized prosthesis by sintering and stereolithography, using computer generated 3D models.

The flow chart bellow (figure 18) was made by the authors as a synthesis of all CAD / CAM procedures needed to achieve physical hip implant.

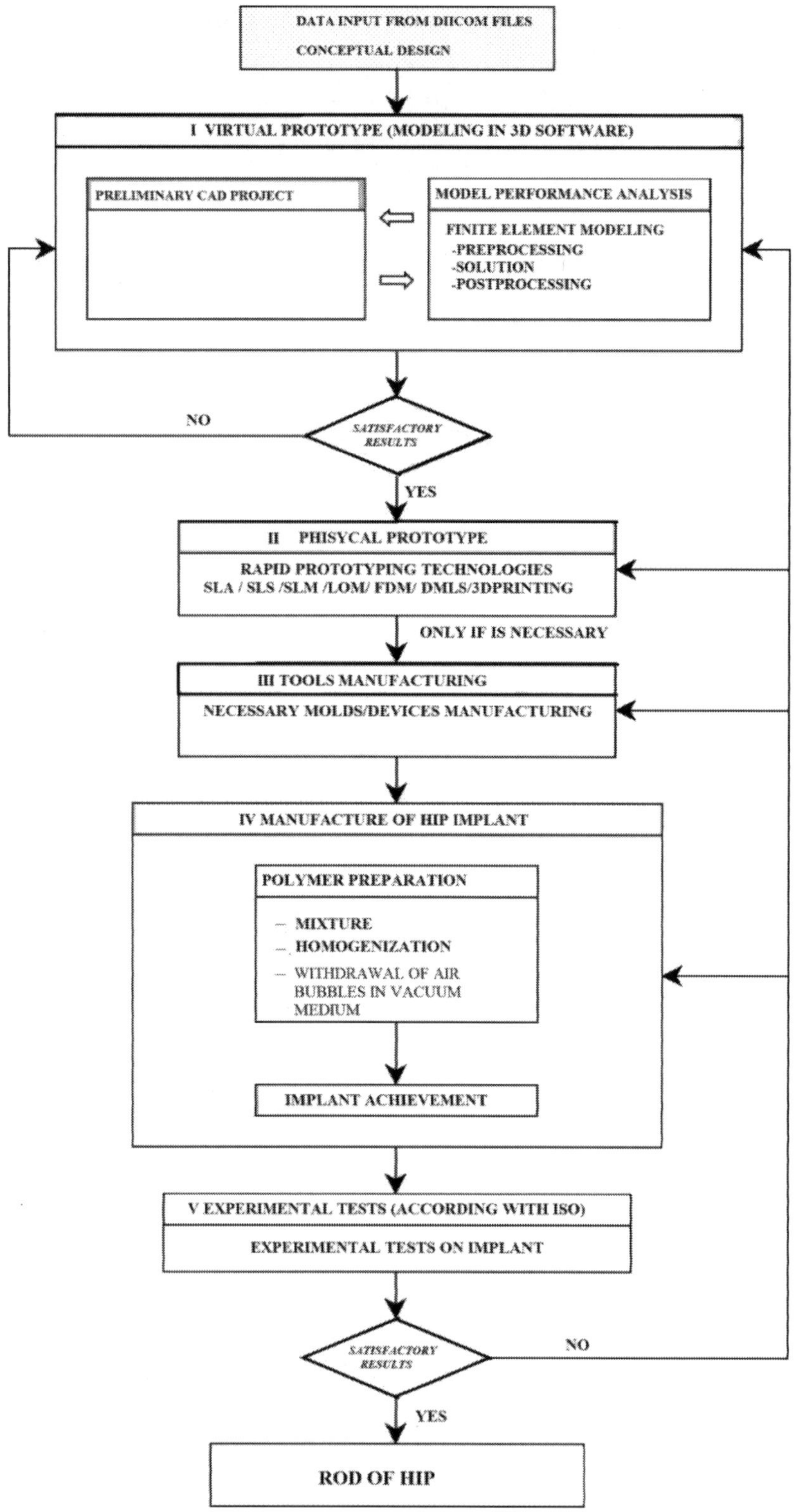

Figure 18. The synthesis of all procedures CAD / CAM needed to achieve physical hip implant
Different types of prosthetic hip implant

References

[1] Gheorghe I Gh *et al.* 2010 *Microtehnologii avansate prin prototipare rapida utilizând sinterizarea directa cu laser* (Bucureşti: Editura CEFIN)

[2] ISO 7206-4:2002: Implants for surgery - Partial and total hip prostheses, Part 4-Determination of endurance properties of stemmed femural components.

[3] ISO 7206-8:2002: Implants for surgery - Partial and total hip prostheses, Part 8-Endurance performances of stemmed femural components.

[4] Hua, J *et al. The role of 3D Image Reconstruction and Rapid Prototyping Models in Total Hip Arthroplasty (*http://www.materialise.com/materialise/view/en/422029)

[5] Sariali E *et al.* 2009,Three dimensional hip anatomy in osteoarthritis, analysis of the femural offset. *J Arthroplasty*; 24/990—7.

IOP Conf. Series: Materials Science and Engineering **161** (2016) 012045 doi:10.1088/1757-899X/161/1/012045

Random lattice structures. Modelling, manufacture and FEA of their mechanical response

G Maliaris[1], I T Sarafis[2], T Lazaridis[2], A Varoutoglou[2] and G Tsakataras[2]

[1]Laboratory for Special Engineering, Department of Electrical & Computer Engineering, School of Engineering, Democritus University of Thrace, GR67100, Xanthi, Greece

Email: gmaliari@ee.duth.gr

[2]Mechanical Engineering Department, Technological Educational Institute of East Macedonia-Thrace, Agios Loukas, GR65404, Kavala, Greece

Abstract. The implementation of lightweight structures in various applications, especially in Aerospace/ Automotive industries and Orthopaedics, has become a necessity due to their exceptional mechanical properties with respect to reduced weight. In this work we present a Voronoi tessellation based algorithm, which has been developed for modelling stochastic lattice structures. With the proposed algorithm, is possible to generate CAD geometry with controllable structural parameters, such as porosity, cell number and strut thickness. The digital structures were transformed into physical objects through the combination of 3D printing technics and investment casting. This process was applied to check the mechanical behaviour of generated digital models. Until now, the only way to materialize such structures into physical objects, was feasible through 3D printing methods such as Selective Laser Sintering/ Melting (SLS/ SLM). Investment casting possesses numerous advantages against SLS or SLA, with the major one being the material variety. On the other hand, several trials are required in order to calibrate the process parameters to have successful castings, which is the major drawback of investment casting. The manufactured specimens were subjected to compression tests, where their mechanical response was registered in the form of compressive load – displacement curves. Also, a finite element model was developed, using the specimens' CAD data and compression test parameters. The FE assisted calculation of specimen plastic deformation is identical with the one of the physical object, which validates the conclusions drawn from the simulation results. As it was observed, strut contact is initiated when specimen deformation is approximately 5mm. Although FE calculated compressive force follows the same trend for the first 3mm of compression, then diverges because of the elasto-plastic FE model type definition and the occurred remeshing steps.

1. Introduction

Biomimicry, man effort to imitate nature strategies in solving human problems has found many applications during man evolution. In this respect, humans managed to imitate the foam-like structures, widely used by nature, by manufacturing metal foams and honeycomb structures using various production methods [1]. Nowadays, metal foams are used for a broad range of applications in aerospace/ automotive industries and recently in orthopaedics. They combine unique properties such as high strength to weight ratios as well as energy absorption characteristics [2]. In the previous mentioned examples two different cellular arrangements are introduced; ordered and stochastic one. The geometry of random lattices (metal foams) consists of irregular, non-periodic cells which form a network of interconnecting walls or struts. On the other hand, regular lattice structures (honeycomb) are formed by the repetitive placement, at the same distance in 3D space, of a representative cell.

Understanding mechanical behaviour is the key parameter for the proper adoption of cellular materials to various applications. Experimental testing combined with finite element analysis has proved to be a solid method for material characterization. In the case of cellular geometries with stochastic arrangement, experimental testing is an easy task in contrast to FEA where geometry definition, an essential step towards seamlessly FE model setup, imposes difficulties due to random structure nature. Various solutions have been proposed by researchers in order to simplify the effort required for the representation of cellular geometry. The most accurate method depends on the reconstruction of the real foam geometry using micro CT scans. Although μCT has gained popularity nowadays, it involves the operation of expensive infrastructure which normally generates huge amount of data that require a time consuming process to translate raw data into geometry. On the other hand, methods that rely on mathematical approximations of the lattice geometry using statistical analysis of pivotal geometric data represent the majority of the published work on this scientific field. Initial efforts for foam modelling relied on previously submitted articles on foam froth geometry approximation. In these investigations the Kelvin cell was used which partitions space with minimum surface area for identical cells [3, 4]. Although it is proven that Kelvin tetrakaidecahedron cannot accurately represent foam geometry, because the natural variations of cell size and shape are not considered [5], researchers kept on using it upon recently [6], mainly as a reference structure. A promising method, recently employed by many researchers, is the generation of irregular cells applying the Voronoi tessellation technique. This technic is used to partition 3D space into random polyhedra. The geometric features of the generated polyhedra are later used in lattice structure modelling.

The mechanical behaviour of cellular solids has been thoroughly investigated by various researchers the last 15 years. The elastic response of open cell foams was studied by [6, 7, 8]. In their investigations researchers considered the effect of various parameters such as relative density, cell irregularity, cell shape and strut cross sectional variations. It was found that relative density and cell irregularity are the dominant parameters affecting the elastic moduli. Also, as energy absorbers, foams are frequently subjected to dynamic loadings or impacts. Therefore, it is quite significant to analyse the large plastic deformation behaviour considering also dynamic phenomena such as rate dependent material properties. Large plastic deformation regime of cellular materials consists of two regions; extended stress plateau and densification region [2]. In order to capture the densification effect through FE analysis, self-contact between struts should be implemented in the developed FE models. The later requirement increases the computation effort by a large margin, so simplification of the strut FE modelling is a necessary step towards efficient computational time management. Two dimensional FE models [9, 10] and beam-based strut realization [6, 10, 11] are the main methods which are commonly applied.

The geometrical characterization of stochastic cellular structures is achieved through the determination of geometric parameters such as cell irregularity, strut cross sectional shape, strut equivalent radius, porosity. Up to recently it was not possible to prescribe these parameters, since they are coupled to the applied processing method, thus they follow physical laws and remain uncontrolled parameters. The last years, the emergence of various additive manufacturing methods, with enhanced accuracy, rendered feasible the materialization of stochastic and ordered lattice structures. The mechanical behaviour of 3D – printed ordinate structures using the DMLS (Direct Metal Laser Sintering) or SLM (Selective Laser Melting) additive technologies where investigated by several researchers [12 – 14]. Relationship between unit cell type and porosity and the fatigue behaviour of selective laser melted meta-biomaterials was studied by Amin Yavari et al. The failure mechanisms of additively manufactured porous biomaterials considering porosity and unit cell type was investigated in [13]. The effects of build orientation and heat treatment on the microstructure and mechanical properties of selective laser melted Ti6Al4V lattice structures was explored in [14]. The above mentioned research papers were focused on metallic lattice structures which were modelled using specific unit cells.

In the present study the random lattice structures were created applying the voronoi tessellation technic. The Voronoi tessellation (or Voronoi diagram) is a space partitioning method which creates polyhedral (cells), when applied in 3D space. The struts are modelled as solid cylinders with constant radius. With the proposed algorithm is possible to control various geometric characteristics such as porosity and cell mean volume. The modelled geometry is then 3D printed with a photosensitive castable resin. The final parts are manufactured by investment casting. The specific manufacturing method permits the use of various materials such as steel alloys, aluminium alloys, titanium alloys, gold, silver, platinum, etc.. The casted parts are subjected to compressive experiments which are coupled to FEA simulations in order to investigate their mechanical behaviour. The use of a self-contact algorithm is considered in the FEA models so that the interaction between the struts can be detected, which is crucial for capturing material behaviour in the densification area.

2. Solid modelling

The Voronoi tessellation (or Voronoi diagram) is a space partitioning method which creates polyhedra, when applied to a 3D geometry (a control volume) that share a common characteristic. The distance between the seed point of a polyhedron and any point included in the specific polyhedron is the minimum one with respect to other seed points. The tessellation is applied to N randomly distributed seed points into a control volume, which is usually a simple geometry such as cube, cylinder, sphere, etc.. The employed seed point distribution method is of a great interest since it defines the statistical distribution of pivotal geometric characteristics. In the next step, the neighbouring points are connected with lines which are then bisected by their perpendicular planes. For a given seed point, its Voronoi cell is defined by the smallest polyhedron that contains that point. The above mentioned procedure is presented in figure 1 for 10 seed points.

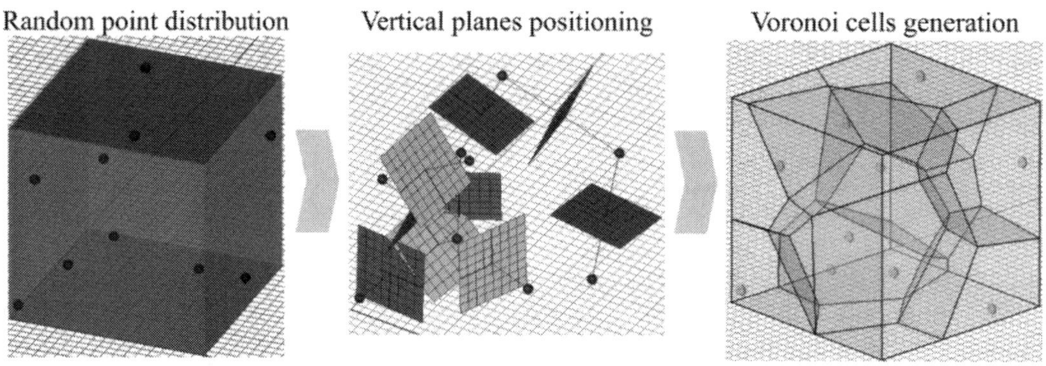

Figure 1: The Voronoi tessellation procedure

The edges of the Voronoi cells form the skeleton which can be appropriately dressed to represent the struts of the lattice structure. The majority of the researchers use the vertices of the cells in order to define beam elements during the FE model definition [4, 6, 10, 15, 16, 17]. Although this simplification leads to geometries that do not capture the cross section variation along strut length and mass concentration at the junctions, is a preferred method of strut modelling in order to limit the required computational effort.

In this work the extracted Voronoi cell edges are "dressed" with solid cylinders with constant radius, to appropriately capture strut deformation during compression, after contact has been established. This approximation is more realistic than the definition of beam elements and provides more insight for the determination of the structure failure mode. After the properly positioning of the cylinders in 3D space, another algorithm is applied that joins the cylinders together. During that procedure, some geometrical errors occur which are overcome by applying a wrap function on the whole geometry. The latter function applies a wrapping surface on the selected entities and can easily fix topology errors such as small tunnels, intersecting triangles, entrapped surfaces, etc. The result is a watertight mesh with clean geometry. A sample lattice structure created with the above mentioned methodology is presented in figure 2.

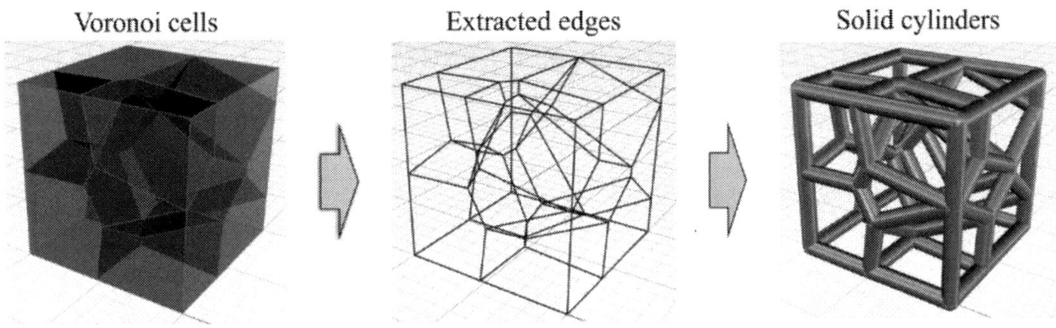

Figure 2: Solid modelling procedure

3. Lattice structure manufacturing

The developed solid modelling algorithm was used to create lattice structures with various characteristics such as porosity and strut radius. The manufacturing route of the digital lattices starts with the 3D printing of the geometry with a photosensitive castable resin. The specific type of photosensitive resins have gained great attention, especially from goldsmiths, which is attributed to the fact that have provided a mean for producing intricate jewellery designs using a well-established production method such as investment casting. The modelled stochastic lattices were printed using two different 3D printers, a consumer level printer (Formlabs Form1+) and a professional one (Asiga FreeForm Pro75). Both utilize the SLA technic (Stereolithography) but use different light sources. The Form1+ a UV laser, where Asiga a high definition UV DLP projector. The printed samples were transformed to metal lattices by investment casting using the Orotig Speedcast 220MJ – Ti casting machine. The specific machine has automated casting cycles, where the metal is melted by the electric arc in presence of Argon-gas. When the casting takes place, the molten metal is pushed down into the flask by a controlled Argon-gas overpressure. This make possible to obtain very compact, oxidation-free casting extremely accurate in details and quite precise and fitting even in very diminutive sections, thus is the perfect candidate for manufacturing the stochastic lattices.

The initial trials revealed certain constrains of the casting procedure which relate with the geometric characteristics of the modelled structures. The cell size and the strut radius should be large enough to allow continuous metal flow and also removal of the investment material with little effort. As it is observed in figure 3, the casting was unsuccessful as both geometric constrains did not have appropriate values.

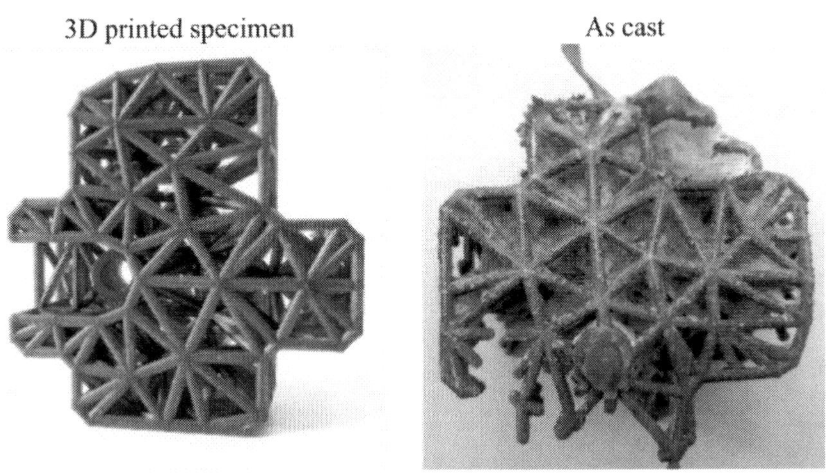

Resin: Asiga SuperCAST, Printing layer height: 10μm
Casting material: Steel 304
Figure 3: Unsuccessful casting due to inappropriate cell size and strut radius.

After conducting many trials, it was concluded that minimum cell size should be larger than 4mm, as well strut radius should be no less than 0.1mm. Based on these dimensional constrains, new stochastic lattices were modelled 3D printed and manufactured. The results are illustrated in the following figure:

Resin: Asiga SuperCAST, Printing layer height: 10μm
Casting material: Steel 304
Figure 4: Successful casting after proper modification of geometrical constrains.

The number of manufactured specimens, the casting material, as well as geometric properties of modelled structures are provided in table 1.

Table 1. Manufactured specimens.

Casting material	Shape	Dimensions [mm]	# of specimens	Porosity [%]	Strut radius [mm]
Aluminium A1100	Cylinder	ØXH 31X31	3	76.7	0.7
Aluminium A1100	Cube	31X31X31	3	76.7	0.7
Stainless steel 304	Cube	31X31X31	1	87.4	0.5

As it is observed, usable specimens with fine geometric features were possible to manufacture with stainless steel 304. The melting temperature of aluminium A1100 was not high enough, thus it was possible to fill the mold up to a point for specimens that had a strut radius of 0.5mm. On the other hand, that was not observed for specimens with larger radius.

4. Mechanical tests

The compressive experiments were conducted using an INSTRON 8801 servohydraulic machine. Its maximum axial load capacity is ±100kN. Control of the 8801 is achieved with the use of a controller and a PC, equipped with appropriate software that provides complete control of the system. During the test, load or displacement control can be achieved, in order to meet the required target. The testing is procedure programmed accordingly with the use of an appropriate application for static tests, which accompanies the testing machine (Bluehill). All specimens were slightly prestressed (<10N) in order to eliminate any planarity errors. Thereafter, a constant compression displacement rate was applied to the specimen, equal to 2mm/min. The compression force was continuously monitored through the installed load cell. Similarly, the position of the load piston was registered. All tests were carried out until displacement was equal to 30% of the specimen's height.

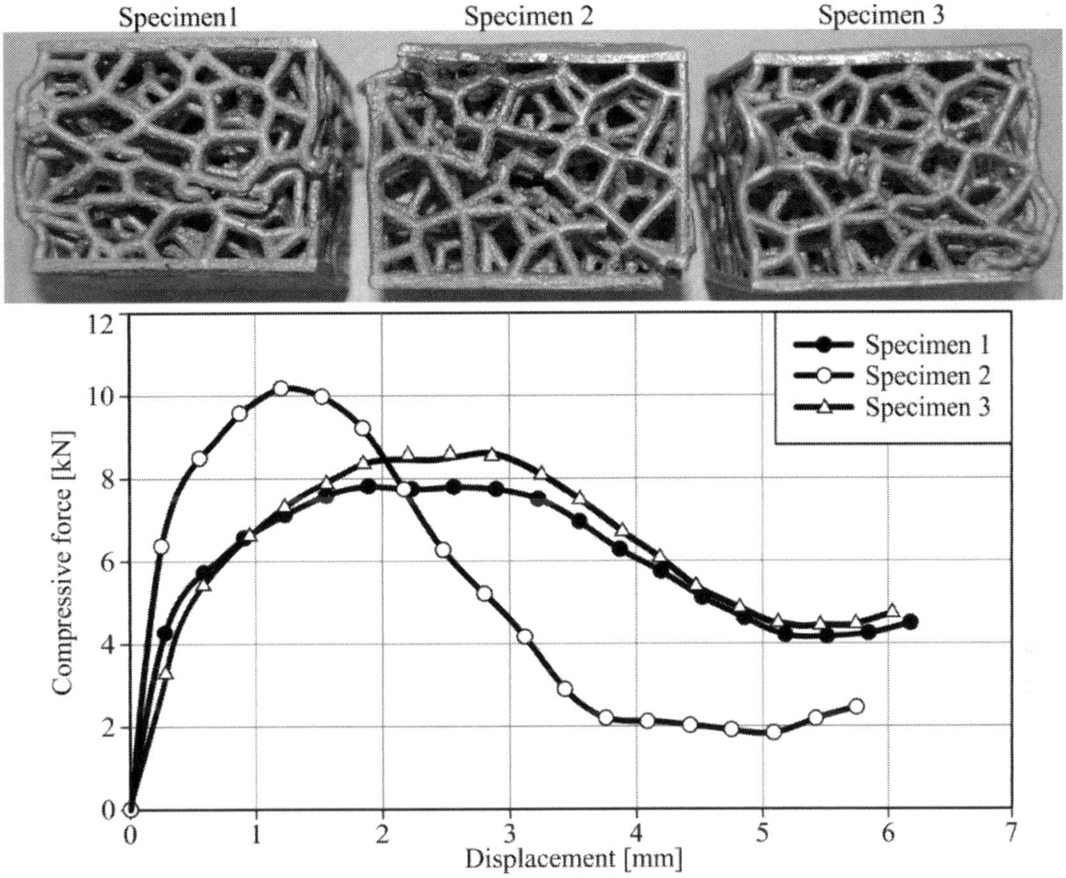

Material: Aluminum A1100, Cube shaped specimens
Porosity: 76,7%, Strut radius: 0.7mm

Figure 5: Experimental results for cube shaped aluminium stochastic lattice specimens.

The depicted mechanical behaviour, as it is presented in the form of compression force – displacement curves in figure 5, was unexpected. Two of the three specimens (one and three) exhibit similar behaviour, as their curves present small differences. Specimen #2 shows a different behaviour, indicating a deviation on the experimental parameters. The common observation between all specimens is the small (in specimens one and three) or completely absent (in specimen two) of the expected plateau region. The main characteristic of this region is the structure capability to retain a constant force value while deforming for a significant displacement range. Specimens one and three seem to develop such a region during compression for a displacement range approximately between 1.8 and 2.8mm, which is quite limited. Specimen two does not seem to develop this region. Rather than developing a plateau region, a premature collapse is engaged, which is also unexpected for a ductile material as aluminium, as it is also shown in the diagram of figure 5. Although metallographic analysis is necessary to investigate microstructural characteristics, the immediate quenching into water, which was applied to remove investment material, seems to be the cause of the limited material elongation. The same behavior was observed also for the cylindrical aluminum specimens. (45°)

On the other hand, the mechanical behavior of the stainless steel specimen, as it is presented in figure 6, did not demonstrate any surprises. The plateau region starts when the specimen's displacement reaches 2mm and extends beyond the illustrated displacement range, since there isn't an abrupt change in force value that would indicate a transition to the densification area. A minor change of the curve's tangent is observed when the compression displacement lies between 5 and 7mm. As it will be presented in the next section, contact between struts is initiated in the specific displacement region, causing the slight increment of the registered force value.

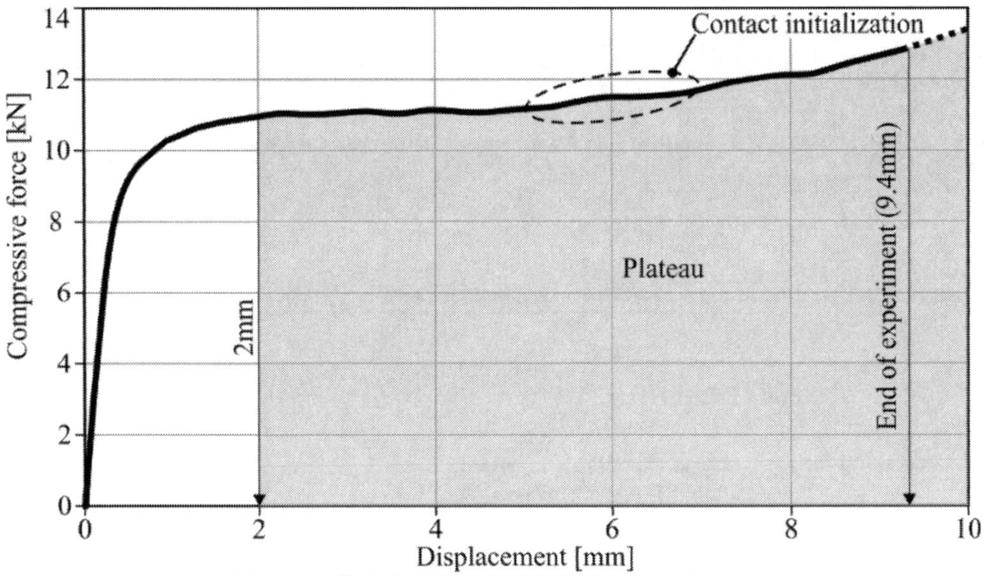

Material: Stainless Steel 304, Cube shaped specimens
Porosity: 87.1%, Strut radius: 0.5mm

Figure 6: Experimental results for cube shaped stainless steel stochastic lattice specimens.

5. FE assisted calculations

The finite element analysis of the compression response of the generated foam geometry was conducted using Deform 3D FEA software. The simulation of the compression of the modeled stochastic lattice structure involves large unconstrained plastic deformation under variable strain rates and localized phenomena due to the variable cross sectional geometry of the foam. The rigid-viscoplastic formulation, in which the elasticity effects are completely neglected, is applied to this analysis. This is a valid assumption since the plastic strains are typically orders of magnitude greater than the elastic ones. In order to represent the evolution of the deformed geometry, the updated Lagrangian description of motion is used, i.e. the mesh follows the deformation of the solid. However, one of the main limitations of Lagrangian formulations for large plastic deformation problems is the excessive distortion of the elements that is caused, leading to unacceptably small time steps and numerical instability of the solution. A viable way to overcome this difficulty is to simulate compression by continuously remeshing (or rezoning) the plastic workpiece (foam).When certain conditions are met, such as zero or negative element Jacobians in one or more elements, a remeshing of the deformed area that triggered the event is necessary and an automatic mesh generation method is used to facilitate the convergence. Also, an adaptive meshing is applied in order to resolve steep gradients that may be present in the solution. Weighting factors can be individually assigned for specifying the relative mesh density to regions of high curvature, steep strain or strain rate. In this way, mesh adaptability can maintain a high-quality mesh in severely deformed zones.

The stochastic lattice structure was modelled as an elasto-plastic object and meshed using 4-node tetrahedral elements. Due to the high total surface area of the modelled specimen and its complicated geometry, a total of nearly 900,000 elements were created in order to represent the imported volume with acceptable accuracy. A box-shaped simple rigid object in contact with the lattice was included in the finite element model setup to apply the compressive load by defining the displacement towards the specimen with constant but adjustable velocity, according to the strain rate desired. The opposite specimen side was simply supported by defining a z=0 displacement restriction. In order to assist solver convergence, the compressive speed was set equal to 0.5mm/s, which is higher than the one used during the experiments, but the resultant strain rate had a negligible effect on the compressive force. Also, an appropriate self-contact detection algorithm was used, to account for the added reaction force caused by structure's struts interaction. The generated structure was deformed up to approximately 50% of its size in the direction of compression.

The determination of the rate dependent material properties is of a great importance since high strain rates can differentiate the material mechanical response by a great extent. Several models have been proposed that try to approximate the viscoplastic behavior of materials over a wide range of strain rates and temperatures. The Johnson – Cook model, Zerilli – Armstrong and Norton – Hoff are characteristic examples of such models. Another method is the definition of the rate dependent material properties in table format. This method is simpler and can represent material behaviours not possible with the previously mentioned ones. The drawback is the table definition which is rather time consuming and difficult to edit. The developed FE model as well as the defined material properties for stainless steel 304 are presented in figure 7.

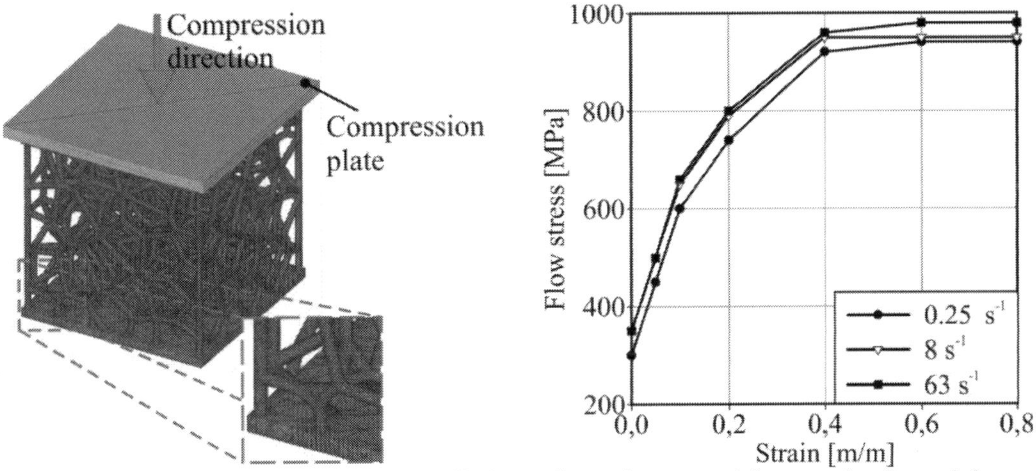

Figure 7: Developed FEA model and applied rate dependent material properties for stainless steel 304.

The simulation completed after solving 3100 steps, with a variable stroke (displacement) per step. A total of 13 remeshing steps were necessary to correct poorly formed elements during simulation. It took almost 7 days on an i7 3770K PC equipped with 32Gb of RAM. The saved data files had a size that reached 60Gb.

5.1. Simulation results

It was of high interest to check whether the calculated deformation could accurately capture the deformation of the specimen during experimental testing. As it is pictured in figure 8, the two geometries are almost identical.

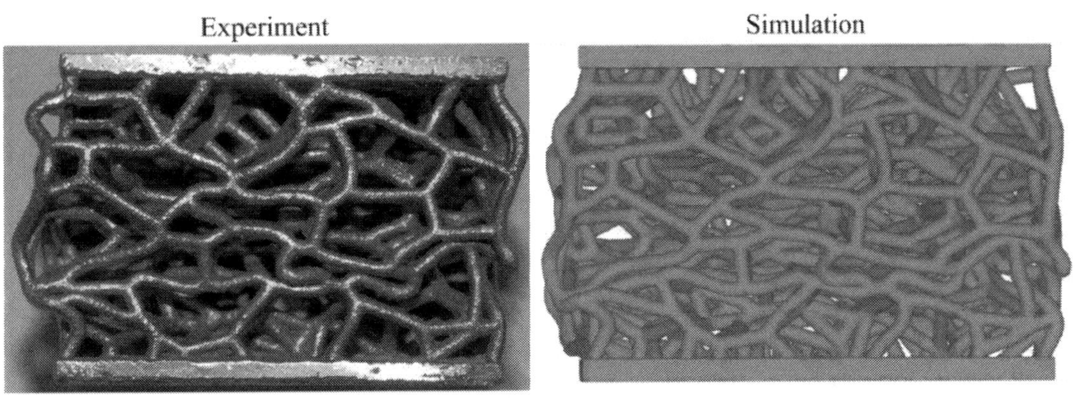

Material: Stainless Steel 304, Cube shaped specimens
Porosity: 87.1%, Strut radius: 0.5mm
Figure 8: Specimen deformation as depicted by experiment (right) and simulation (left).

The above mentioned result verifies that FEA, with proper model setup and material properties, is possible to deliver accurate results even in cases where geometry is very complex. On the other hand, the calculated compressive force presents a slight different behaviour. As it is presented in figure 9, up to 3.5mm both curves coincide. After this point, the calculated compressive force curve does not converge anymore with the measured one, although both of them exhibit an increasing trend. At the end of the experimental curve, the deviation between the two curves reaches 6.5%. This deviation can be attributed to a number of parameters such as:

- Defined material properties. The material data used for the FE model setup were provided with the FEA software. So it is not possible to describe the behaviour of the material used for castings, with an accuracy of 100%.
- Remeshing steps. During simulation it is necessary to rebuild the mesh at areas where element distortion exceeds quality criteria. After remeshing, an interpolation routine is applied to transfer the solution data from the old mesh to the new one. This process smooths the transferred solution data, as it can be observed at the various remeshing steps of figure 9, where the compressive force decreases abruptly, and it recovers after 50 to 60 solution steps.
- Elastoplastic material model. Force recovery is also caused by the elastoplastic material model. Due to remeshing, elastic deformation is lost, so in the next simulation steps the calculated deformation of the lattice structure lies within its elastic range. Thus, it is necessary to deform the lattice structure to regain its elastic portion, before any plastic deformation occurs. That procedure affects the compressive force calculation up to the point where material enters the plastic deformation region.

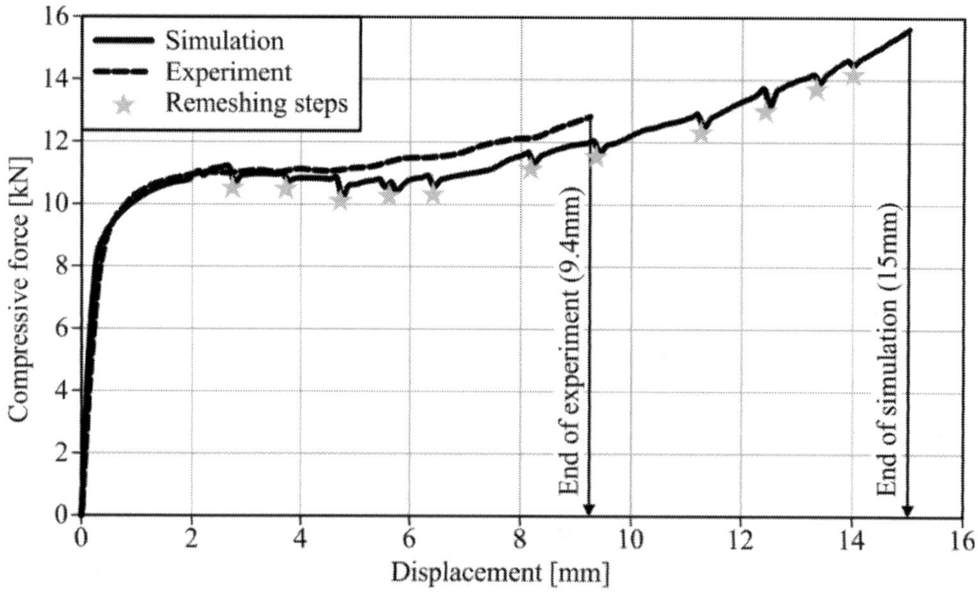

Material: Stainless Steel 304, Cube shaped specimens
Porosity: 87.1%, Strut radius: 0.5mm
Figure 9: Measured and FEA calculated compression force.

Another interesting simulation outcome is the progression of the number of contact sites. In figure 6 it was identified that contact is initiated between the displacement range of 5 to 7mm. As it was previously mentioned, a contact algorithm was defined during FE model setup. In figure 10, the contacted nodes as well as the contact area for 3 displacement values (5, 6, 7mm) are presented. As compressive displacement reaches 5mm, the first nodes begin to interact, but the effect on the compressive force value is negligible. At 6mm, the number of contact nodes is multiplied, also the total contact area becomes 1.77mm^2, and the interaction between the ligaments starts to compensate on force calculation. At 7mm the same trend is observed, thus it is verified that contact is established in the specific displacement range.

Displacement:	5.00mm	6.00mm	7.00mm
Contact area:	$0.27mm^2$	$1.77mm^2$	$3.39mm^2$

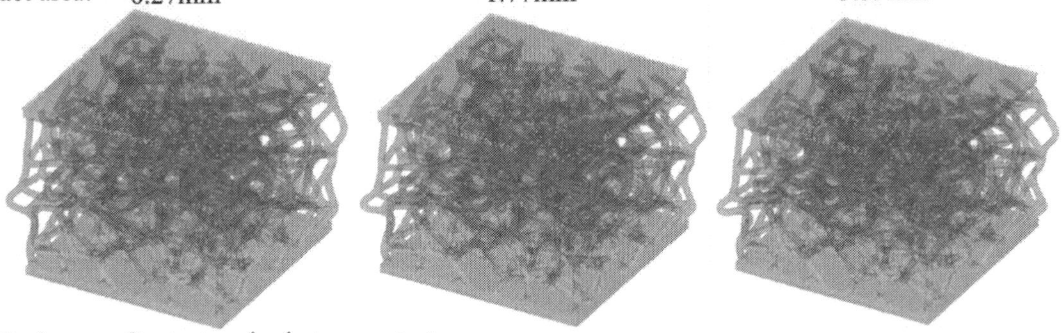

Red spots: Contact nodes between struts

Green spots: Contact nodes between specimen upper surface and compression plate

Figure 10: Evolution of contact area.

6. Conclusions

This work had an initial target to materialize stochastic lattice structures with a production method that permits the use of several materials. That was accomplished with the aid of 3D printing and investment casting. After several trials, the design and process parameters were optimized and successful castings were conducted. That success is a huge step towards manufacturing of random lattice structures with various materials, since the only alternative manufacturing method is 3D printing with metal powder (SLM or SLS), which imposes restrictions on material availability. Furthermore, the conduction of compressive material tests showed that the manufactured specimens exhibit very high load to weight ratio, a key parameter of metal foams. Also, the results of the FEA of the experimental procedure verified the FE model setup and proved that is possible to design stochastic structures based on FEA results before manufacturing.

7. References

[1]. Banhart J 2000 Manufacturing routes for metallic foams *JOM* **52** 22-27

[2]. Gibson L J and Ashby M F 1997 *Cellular Solids – Structure and Properties* (Cambridge: Cambridge University Press)

[3]. Li K, Gao X L and Subhash G 2006 Effects of cell shape and strut cross-sectional area variations on the elastic properties of three-dimensional open-cell foams *J. Mech. Phys. Solids* **54** 783–806

[4]. Alkhader M and Vural M 2008 *Mechanical response of cellular solids: Role of cellular topology and microstructural irregularity* Int. J. Eng. Sci. **46** 1035–1051

[5]. Raj S V 2011 Microstructural characterization of metal foams: An examination of the applicability of the theoretical models for modelling foams *Mat. Sci. Eng. A-Struct.* **528** 5289–5295

[6]. Jang W Y, Kyriakides S and Kraynik A M 2010 On the compressive strength of open-cell metal foams with Kelvin and random cell structures *Int. J. of Solids Struct.* **47** 2872–2883

[7]. Gong L and Kyriakides S 2005 Compressive response of open cell foams Part II: Initiation and evolution of crushing International *Int. J. of Solids Struct.* **42** 1381–1399

[8]. Jang W Y, Kraynik A M and Kyriakides S 2008 On the microstructure of open-cell foams and its effect on elastic properties *Int. J. of Solids Struct.* **45** 1845–1875

[9]. Borovinšek M and Ren Z 2008 Computational modelling of irregular open-cell foam behaviour under impact loading *Materialwiss. Werkst.* **39** 114–120.

[10]. Alsayednoor J, Harrison P and Guo Z 2013 Large strain compressive response of 2-D periodic representative volume element for random foam microstructures *Mech. Mater.* **66** 7–20.

[11]. Michailidis N 2011 Strain rate dependent compression response of Ni-foam investigated by experimental and FEM simulation methods *Mat. Sci. Eng. A-Struct.* **528** 4204–4208
[12]. Ahmadi S, Yavari S, Wauthle R, Pouran B, Schrooten J, Weinans H and Zadpoor A 2015 Additively Manufactured Open-Cell Porous Biomaterials Made from Six Different Space-Filling Unit Cells: The Mechanical and Morphological Properties *Materials* **8** 1871–1896
[13]. Kadkhodapour J, Montazerian H, Darabi A C, Anaraki A P, Ahmadi S M, Zadpoor A A and Schmauder S 2015 Failure mechanisms of additively manufactured porous biomaterials: Effects of porosity and type of unit cell *J Mech Behav Biomed Mater* **50** 180–191
[14]. Wauthle R, Vrancken B, Beynaerts B, Jorissen K, Schrooten J, Kruth J -P and Van Humbeeck J 2015 Effects of build orientation and heat treatment on the microstructure and mechanical properties of selective laser melted Ti6Al4V lattice structures *Addit. Manuf.* **5** 77–84
[15]. Gaitanaros S and Kyriakides S 2015 On the effect of relative density on the crushing and energy absorption of open-cell foams under impact *Int. J. Impact Eng.* **82** 3–13
[16]. Gaitanaros S, Kyriakides S and Kraynik A M 2012 On the crushing response of random open-cell foams. International *Int. J. of Solids Struct.* **49** 2733–2743
[17]. Sotomayor O E and Tippur H V 2014 Role of cell regularity and relative density on elastoplastic compression response of 3-D open-cell foam core sandwich structure generated using Voronoi diagrams *Acta Mater.* **78** 301–313

IOP Conf. Series: Materials Science and Engineering **161** (2016) 012046 doi:10.1088/1757-899X/161/1/012046

Biomimicry in Product Design through Materials Selection and Computer Aided Engineering

G Alexandridis[1], D Tzetzis[1] and P Kyratsis[2]

[1] International Hellenic University, 14km Thessaloniki - N. Moudania, Thermi, 57001, Greece
[2] Western Macedonia University of Applied Sciences, Department of Mechanical Engineering and Industrial Design, Kila Kozani, Greece, GR50100, Greece

E-mail: pkyratsis@teiwm.gr

Abstract. The aim of this study is to demonstrate a 7-step methodology that describes the way nature can act as a source of inspiration for the design and the development of a product. Furthermore, it suggests special computerized tools and methods for the product optimization regarding its environmental impact i.e. material selection, production methods. For validation purposes, a garden chaise lounge that imitates the form of a scorpion was developed as a result for the case study and the presentation of the current methodology.

1. Introduction

Nature is an inexhaustible source of inspiration. Since their existence, people used to observe and study nature in order to find solutions to several of their daily problems. This practice of imitating the models, systems, and elements of nature in order to solve complex human problems is known as biomimicry or biomimetics [1]. Despite the fact that biomimicry has actually a history of millions of years, namely during the whole timeline of human evolution, it was not until the recent years that it has begun to be approached in a systematic way and with a more intense interest in an attempt for more sustainable product designs.

2. Sustainability and Life Cycle Assessment

Sustainable development is the one that meets the needs of the present without compromising the ability of future generations to meet their own needs [2, 3]. Sustainability is based on three interrelated factors (economic, environmental and social), which are often found to conflict with one another. The goal of sustainable development is to achieve a complete interaction and balance between these factors in order to satisfy basic social and economic needs, while maintaining all the natural resources for future generations and improving the well-being of the environment and ecosystem on which life depends [4, 5, 6, 7].

Under the light of eliminating the negative environmental impact and comply with the three principles of sustainability, the Life-Cycle Assessment (LCA) method was used for the environmental assessment of the a chaise lounge. LCA is a method that quantitatively analyses and evaluates the environmental impact of a product throughout its entire lifecycle - from cradle to grave with all the intermediate stages. Environmental impact is associated with air pollutant emissions and consumption of resources, both of which contribute to a wide range of impacts, such as climate change, stratospheric ozone depletion, tropospheric ozone creation, eutrophication, acidification, toxicological stress on human health and ecosystems, the depletion of resources, water use, land use and noise.

LCA is a decision making tool that calculates indicators of potential environmental impacts and provides alternative choices for potential improvements of the product under examination, identifying opportunities for pollution prevention and reductions in resource consumption. In the current research, the LCA method was implemented using the Solidworks Sustainability™ LCA tool.

3. The Proposed Methodology

The main idea behind this project was the development of a methodology that would suggest a way of how to take advantage of a special characteristic or mechanism from a biological system and adapt it in the development of a product in order to provide a solution to a customer need. There are two distinct orientations in biomimicry as a design process [8]: a) solution driven, where an interesting biological phenomenon or mechanism inspires the search for potential applications, adapting it in the development of a new product, b) problem driven, where the identification of a given problem stimulates the search for biological mechanisms that could help to the solution of the problem.

The current methodology falls in the category of solution driven methodologies and its objective is oriented on product design with special emphasis on sustainability and eco-friendliness. That was accomplished with the aid of state of the art CAD and material selection systems. The proposed methodological framework comprises 7 steps as presented in the figure 1.

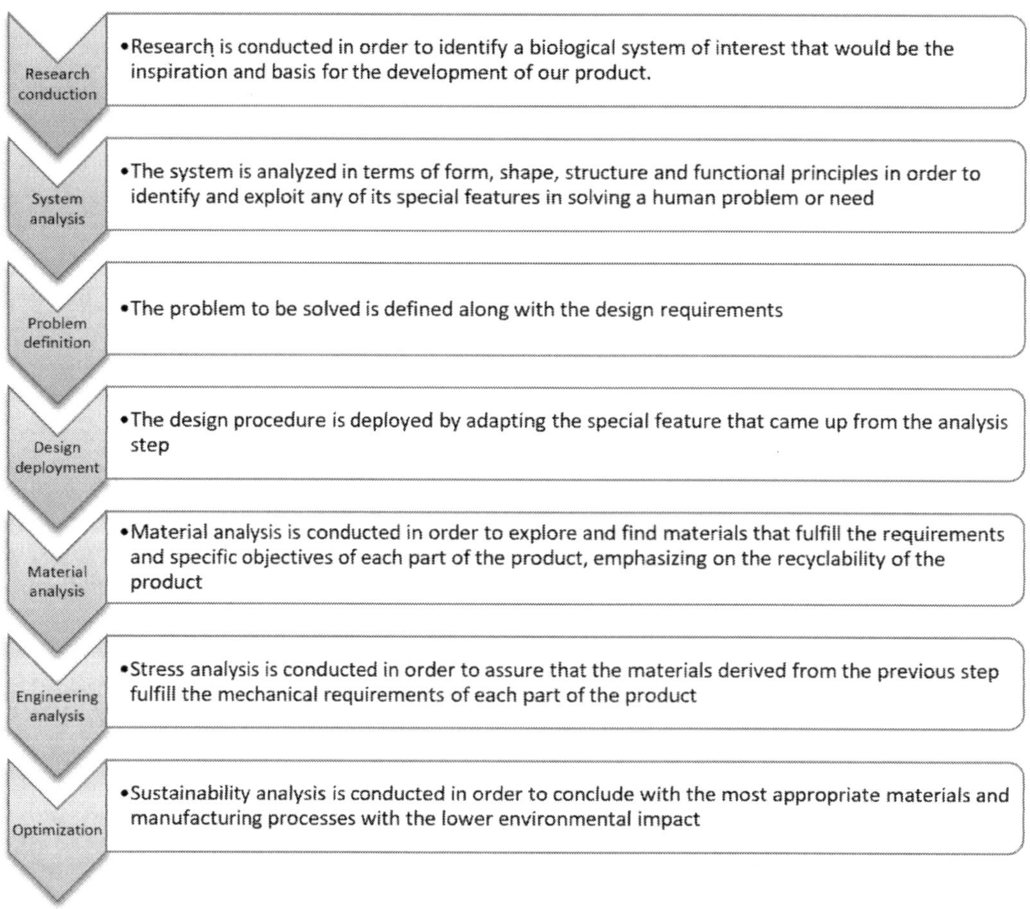

Figure 1. The 7 steps of the proposed methodology

- ***Research conduction:*** the initial step of this project was to identify a biological system with a special feature or shape that would be the inspiration for the product development. In this case, the scorpion was selected.
- ***System analysis:*** the next step was to understand and analyze the morphology of the scorpion in terms of form, shape, structure and function. Scorpions have eight legs with their most recognizable characteristics being the pair of grasping pedipalps and the narrow, segmented tail, carried in a forward curve over the back, ending with a venomous stinger [9]. Through the process of abstraction and simplification a series of rough sketches were made in order to serve as the guideline in an attempt to translate its features into geometrical principles and identify any of its characteristics that could be accommodated into a daily product.
- ***Problem definition:*** focusing on the shape of the tail, the idea was born that it could be exploited in solving the problem of providing adequate shadow while lying under the hot sun. The concept that came up was to design a chaise lounge that would take advantage of the curved shape of the scorpion's tail by adapting an umbrella at the tip of its backrest, where the stinger is supposed to be. In addition to the first idea, another one came up to make an adaptable slot at the tip of the backrest that could make it possible to adapt not only just an umbrella but also a mosquito net in order to provide further protection from the mosquitos' attack or a source of light for someone who enjoys reading, while relaxing in the evening. In addition, a series of design requirements were considered (i.e. outdoor use, lightweight, modular system, minimal design, appealing shape).
- ***Design deployment:*** For the development of the design the following decisions were taken into account: user population (includes everyone except young children), product size (95% of the potential customers). Based on these design parameters together with the shape of the scorpion, the scaffold of the chaise lounge's frame was developed.
- ***Material analysis:*** engineering materials are classified into six broad families: metals, polymers, elastomers, ceramics, glasses, and hybrids. The members of a family have certain features in common (e.g. similar properties, processing routes, etc.) and often similar applications. Furthermore, each material has particular attributes (density, strength, cost, etc.). A design demands a certain profile of these (e.g. high strength, low density, modest cost, etc.). Hence, the aim is not depicting a material but a certain profile of properties. Mechanically speaking, the frame of the chaise lounge is considered a panel loaded in bending, while the leg is considered a beam loaded in bending. Hence, both of them must be strong and stiff enough to carry compressive loads and bending moments without failure, while at the same time being as light as possible.

  The translation of the design requirements of the two components is summarized in
  Table **1**. In order to select the materials, that not only meet the design requirements previously described, but those that also maximize performance, two material property charts were examined with the use of the CES EduPack™ piece of software i.e. a chart of Young's modulus against density and a chart of fracture toughness against price (Figure 2). The materials that fulfilled the design requirements for each component were:
    - For the frame: polypropylene–PP (index 0.00269), polyethylene terephthalate-PET (index 0.00112) and Polyvinylchloride-PVC (index 0.001).
    - For the legs: Cast Al-alloys (index 0.00332), Age-hardening wrought Al-alloys (index (0.00319), Non age-hardening wrought Al-alloys (0.00311).
- ***Engineering Analysis:*** this step was to assure that the chosen materials meet the mechanical requirements of the design. That was accomplished using the Finite Element Method (FEM – figure 3). For the aim of this study, each component of the chaise lounge was examined separately. Thus, stress analysis was carried out for all the 3 materials derived from the material selection phase. The materials examined for the frame were: a) Polypropylene (PP), b) Polyethylene terephthalate (PET) and c) Polyvinylchloride (PVC). The materials examined

for the legs were: a) Cast Aluminum alloys, b) Age-hardening wrought Aluminum alloys and c) Non age-hardening wrought Aluminum alloys.

Table 1. The design requirements of the chaise lounge's components

	FRAME	LEG
FUNCTION	Light and stiff frame	Light and stiff beam
OBJECTIVE	Minimize the mass, ***m***, of the frame	Minimize the mass, ***m***, of the leg
CONSTRAINTS	a. Length, ***L*** (specified) b. Bending stiffness, ***S*** (specified) c. Fracture Toughness, $K_{IC} > 4.5$ MPa m$^{1/2}$ d. Cost, $C_m < 5$ €/kg e. Recyclable	a. Length, ***L*** (specified) b. Bending stiffness, ***S*** (specified) c. Fracture Toughness, $K_{IC} > 15$ MPa m$^{1/2}$ d. Cost, $C_m < 5$ €/kg e. Recyclable
FREE VARIABLES	Frame thickness, ***t***	Section Area, ***A***

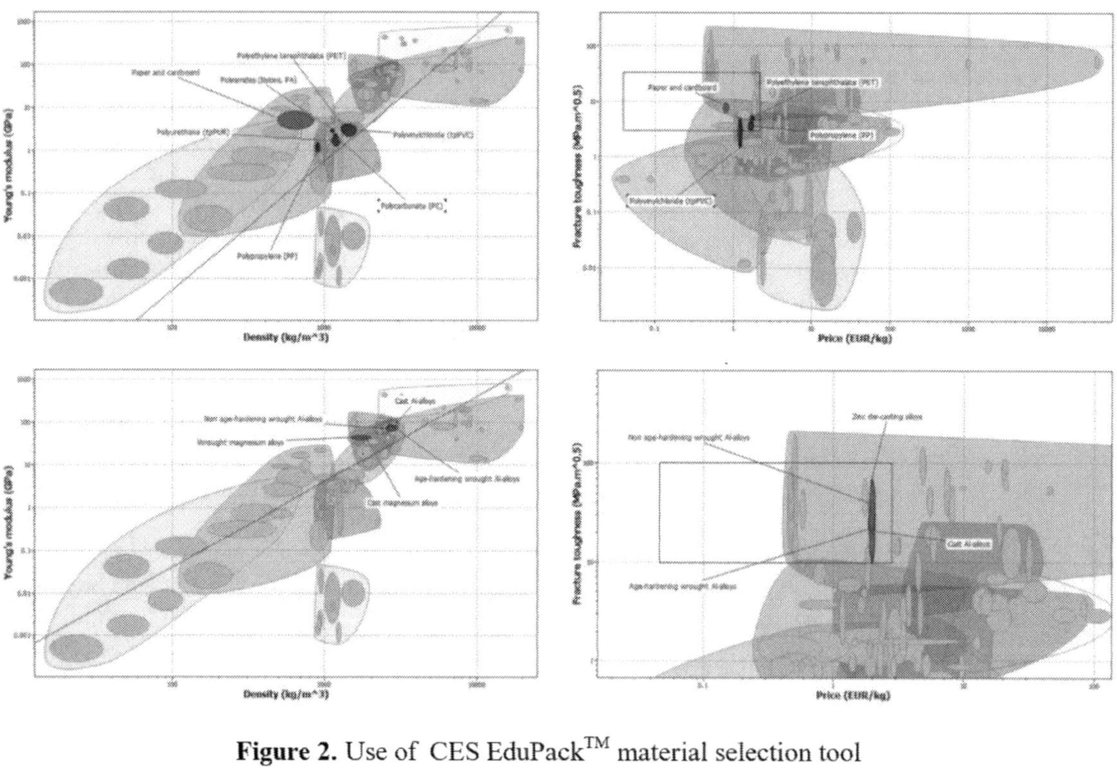

Figure 2. Use of CES EduPack™ material selection tool

- ***Environmental Impact Assessment:*** In this last step the materials derived from the material selection phase along with their corresponding manufacturing processes were evaluated and compared regarding their environmental impact during the entire lifecycle of the product. The objective of this phase was to finally conclude and select those materials that attribute to the design the optimum characteristics based on both the aspect of their mechanical characteristics as well as their sustainability and eco-friendliness. The evaluation of the environmental impact was achieved with the use of the Solidworks Sustainability™ tool, which employs the Life Cycle Assessment (LCA) method.

The data used in the Sustainability^{TM} module are based on the GaBi^{TM} software (PE International's database). The analysis is conducted in two steps. In the first step, the software gathers specific environmental impacts of each component of the life cycle inventory and in the second step, the impact of the four environmental indicators were calculated (Carbon footprint, Energy consumption, Air acidification, Water Eutrophication).
For the analysis of the present study the comparison was made assuming that the place of production and use is Europe, the mode of transportation is by truck for an average of 1900 km and the product's lifetime is 10 years. The manufacturing process selected for the production of the frame is injection molding and for the production of the legs is die casting. Additionally, in the case of the leg three different types of materials were selected out of a wide range that fall into the categories of the cast aluminum alloys derived from the material selection phase. The materials selected for the performance of the analysis were: a) 201.0-T6, b) 356.0-T6, c) C355.0-T61.
The following figures 4 to 7 depict the results of the evaluation procedure of each material and manufacturing process based on the four environmental indicators for each component separately.

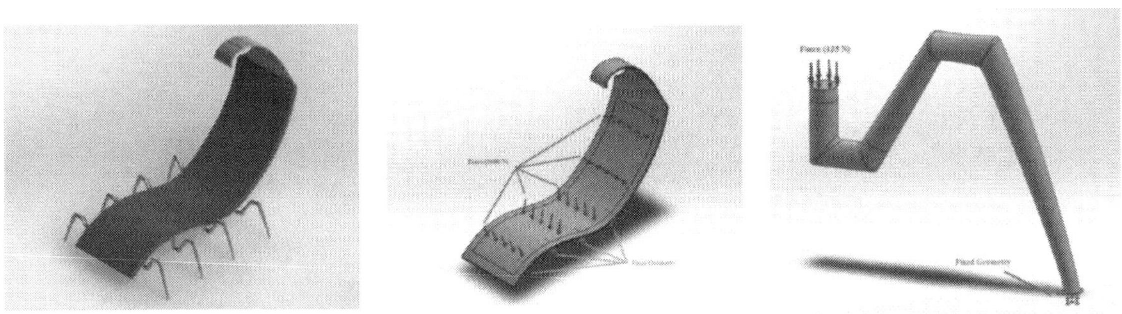

Frame	Mass (kg)	Von Mises Stress Max (MPa)	URES: Resultant Displacement Max (mm)
PP	17.030	9.27	115.17
PET	27.171	20.01	87.35
PVC	24.875	9.78	55.96

Legs	Mass (kg)	Von Mises Stress Max (MPa)	URES: Resultant Displacement Max (mm)
Cast-Al	0.836	160.68	3.52
Age-hard Al	0.836	160.68	3.92
Non Age-hard Al	0.836	160.68	4.36

Figure 3. FEM implementation

Comparing the results in the charts (figures 4 to 7) and in figure 3, polypropylene (PP) is the most appropriate material for the production of the frame since it meets the mechanical requirements of the product and additionally it will lead to a frame with the lowest weight. At the same time, it has the lowest environmental impact. As far as the production of the legs is concerned, the three materials present either no or very small differences on the aspect of their mechanical characteristics. Thus, the choice was made on the basis of their environmental impact. The material with the lowest impact is the Cast Aluminum alloy 356.0-T6 and that was the final selection for the production of the legs.

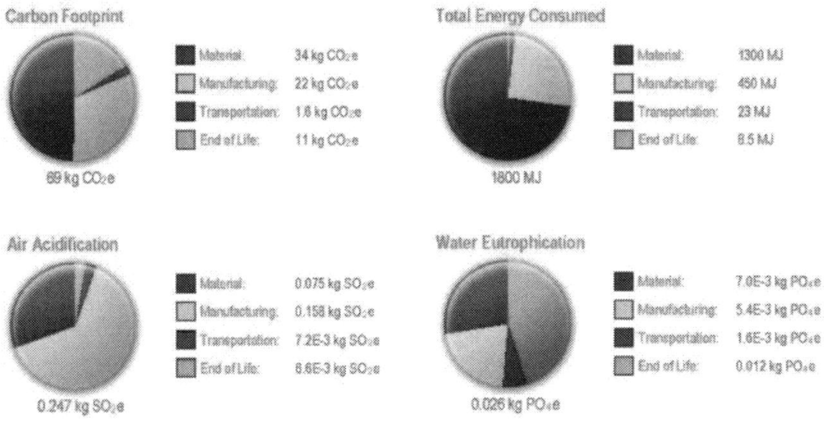

Impact of the PP frame material on the environmental factors

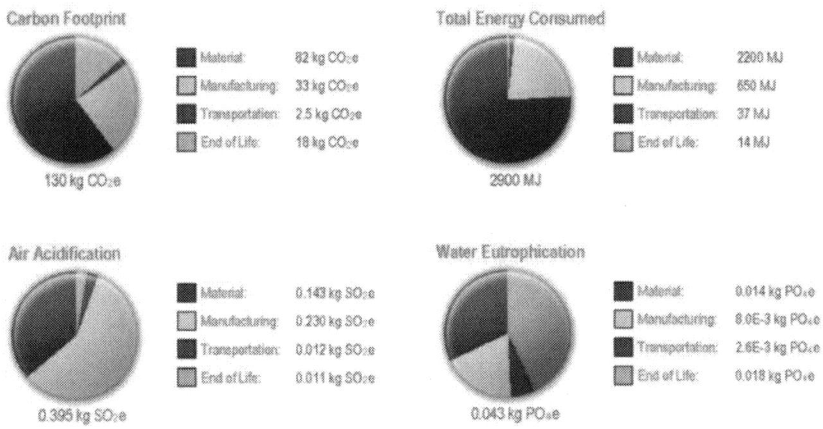

Impact of the PET frame material on the environmental factors

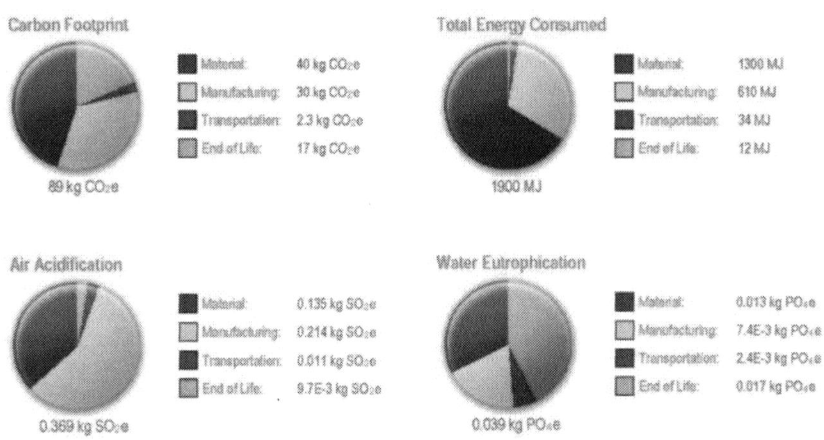

Impact of the PVC frame material on the environmental factors

Figure 4. Frame's materials impact on the environmental factors

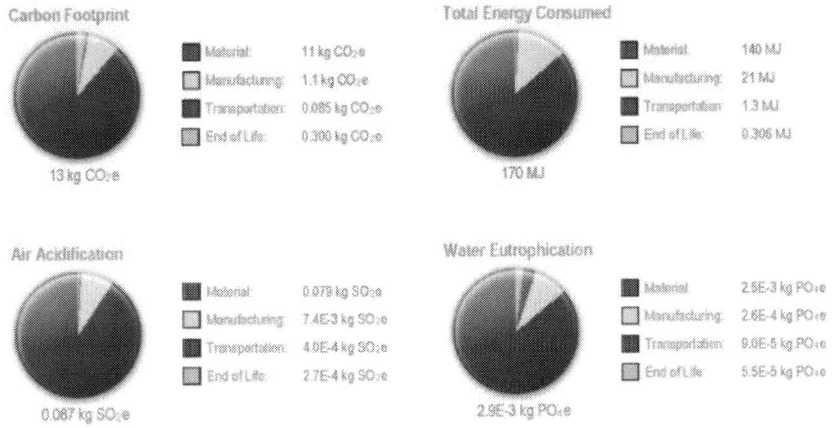

Impact of the 201.0-T6 leg material on the environmental factors

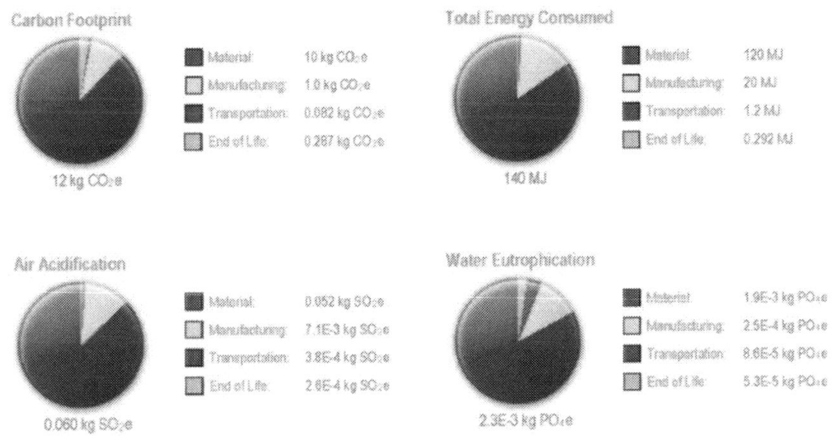

Impact of the 356.0-T6 leg material on the environmental factors

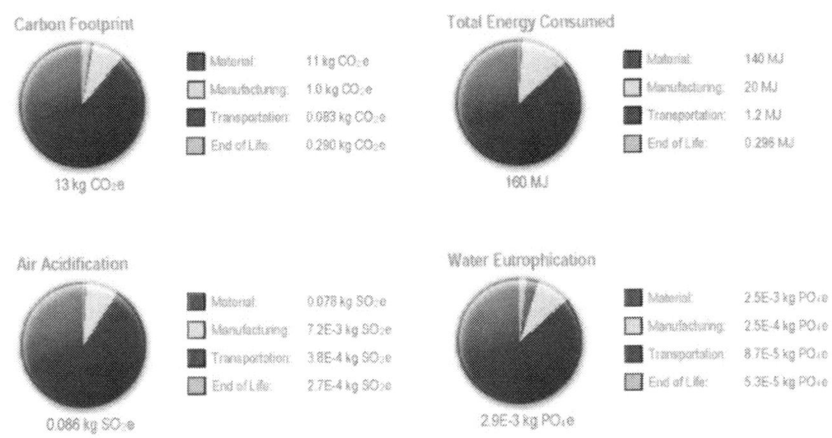

Impact of the C355.0-T61 leg material on the environmental factors

Figure 5. Leg's materials impact on the environmental factors

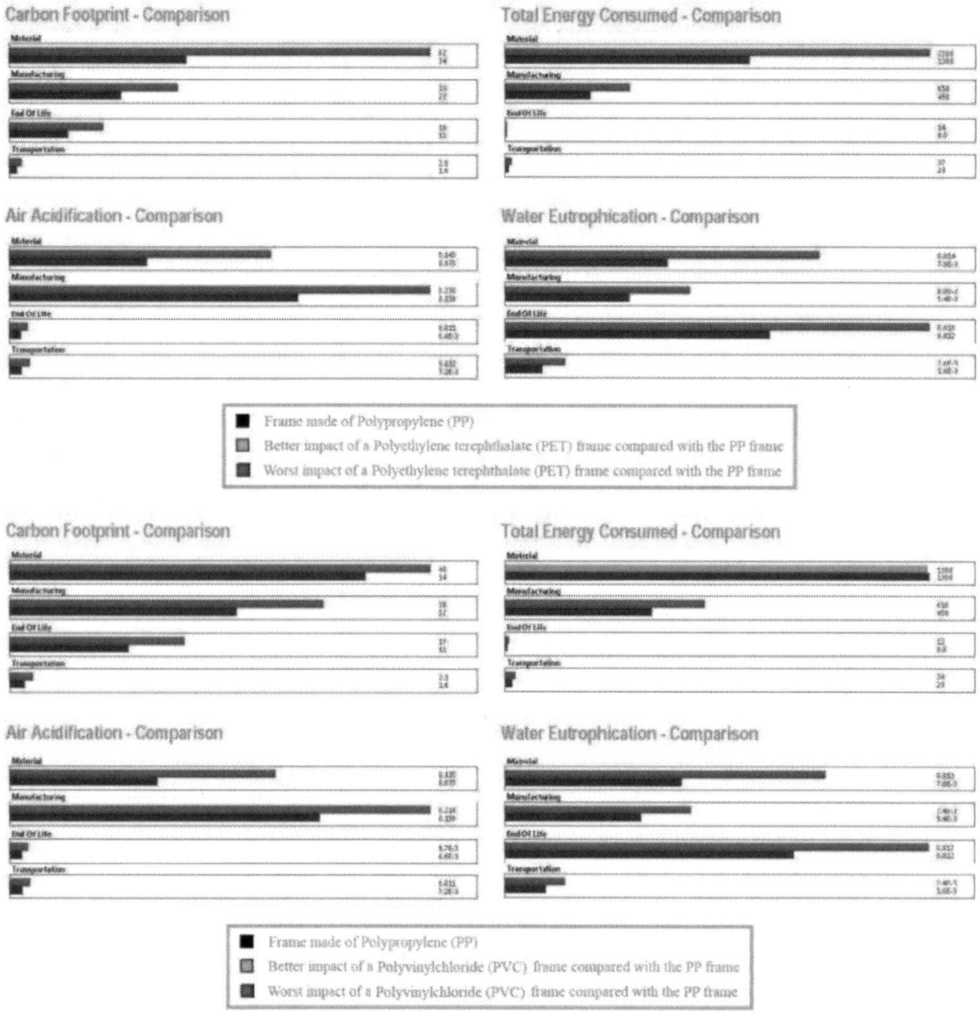

Figure 6. Comparison of the environmental factors per lifecycle stage for the frame

4. Conclusion

The aim of this study was to propose a methodology that can be used in product design and which combines biomimicry with sustainable development. As a result, more eco-friendly and green products are to be produced. The aim of this methodology is to provide a roadmap for a potential product designer in order to exploit nature as a source of inspiration for the design and development of a product, while at the same time, optimizing this product in terms of its environmental impact utilizing state of art computerized tools.

It should be mentioned that no matter how well-structured a methodology is, it is not a panacea to the success of a product, since there are many other factors affecting it, two of the most prominent being the creativity and the technical expertise.

Figure 7. Comparison of the environmental factors per lifecycle stage for the legs

References

[1] Benyus J 1997 *Biomimicry: Innovations Inspired by Nature* William Morrow & Co

[2] Brundtland G, Khalid M, Agnelli S, Al-Athel S, Chidzero B, Fadika L et al 1987 *Our common future – The World Commission on Environment and Development* Oxford University Press

[3] Pedersen Zari M 2007 Biomimetic approaches to Architectural Design for increased sustainability. *The SB07 NZ Sustainable Building Conference*

[4] Adams W 2006 The future of sustainability: Re-thinking environment and development in the twenty-first century *In Report of the IUCN renowned thinkers meeting* **29**

[5] Available at: http://www.solidworks.com/sustainability/products/2795_ENU_HTML.htm Accessed: 04/04/2016

[6] Versos CA, and Coelho DA 2011 Biologically inspired design: methods and validation. *Industrial Design-New Frontiers*, pp 101-120

[7] Coelho D and Versos C 2011 A comparative analysis of six bionic design methods. *International Journal of Design Engineering* **4(2)**, pp 114-131

[8] Goel A, McAdams D and Stone R 2014 Biologically Inspired Design *Computational Methods and Tools*, Springer-Verlag

[9] Available at: https://en.wikipedia.org/wiki/Scorpion Accessed: 04/04/2016

IOP Conf. Series: Materials Science and Engineering **161** (2016) 012047 doi:10.1088/1757-899X/161/1/012047

A Reverse Engineering method for 3d parametric modeling of geometries based on processing of laser scanning data and a sweep technique

E Varitis

Department of Mechanical Engineering and Industrial Design, Technical Educational Institute of West Macedonia Koila, Kozani 50100, Greece

Email: varitisem@gmail.com

Abstract. A Reverse Engineering (RE) method for parametric modelling is presented in this paper. According to this method laser scanning data are processed by means of an algorithm and a parametric geometry is produced. The algorithm generates a spline used as a driving curve for a 2D profile, both approximated from the point cloud data, with the final geometry being produced through with a sweep based technique. The method was applied to digitize a commercial product, a bottle, and the geometry was reconstructed at high accuracy and surface quality. Finally the results of the proposed method were compared with auto surfacing from *.stl files and with surfaces generated by means of sweep commands without converting curves to splines.

1. Introduction

For decades reverse engineering methods have been used for direct and accurate geometry reconstruction, without contacting the real object. Reverse engineering is suitable for remodeling objects with complicated geometry e.g. archaeological objects (1) or biomedical applications (2,3,4). In this circumstances the obtained point cloud could, after smoothing and local operations processes, produce surface geometry with high accuracy.

The instant remodeling method of reverse engineering is also very useful for geometry reconstruction of specimens used in industry or industrial design consumer products. The electronic file in these circumstances could be used for mold creation (5) and G code generation or for better design optimization as well as changing product dimensions. Changes in dimensions length and smooth surfaces are great advances or surface representation with parametric modeling.

On the other hand modeling of a real part in a parametric CAD software like Autodesk Inventor, SolidWorks, Creo parametric, could be a complicated and time consuming processes. In some cases is necessary for the user to have expert knowledge of advanced CAD modeling techniques like sweep, loft or boundary. Moreover even if the user has the appropriate knowledge it is likely that he will have to try many of these techniques in order to find the most appropriate one.

Finally point cloud processing software's produce small surfaces with non-uniform area curvature and continuity with local inaccuracies, when auto generating surfaces. Such surface generation are not suitable for applications such as G code production. The more suitable process of converting areas of point cloud to planar or b spline surfaces, requires surface modeling techniques with trimming and knitting functions, in order to produce the final object. This is in some circumstances not accurately enough done automatically, as shown in the present paper.

All the aforementioned reasons necessitate parametric modeling of object surface with automatic functions able to circumvent these problems, especially for applications involving industrial use.

In this paper an automatic method is presented for parametric generation of object surface through laser scanning data. The processes is based on an algorithm which converts point cloud data to a sweep based feature when the application has the suitable form.

2. Proposed method

According to the proposed method, employing data derived from a laser scanner, two 2D curves are produced. A 2D profile and a guide curve. The algorithm approaches the 2D profile with an ellipse and the guide curve with a spline curve.

Initially a plane is defined where the 2D profile will be produced and the ellipse's axis is modelled based on the min and max distances of the points in this area. The ellipse adjusts to the length of the two axes driven by the guide curve.

The guide curve represents a spline approximation with a max. of four polynomial points , a start point an endpoint and if necessary, two polynomial points in between to maintain a simple formulation of the curve. A point on the major axis of ellipse is considered as the start point for guide curve generation, whereas the endpoint of the spline is projected in the far opposite area of the scanned object. The shape of the spline is generated and controlled from the algorithm described in previous papers (6,7) and verified in papers (8,9). The accuracy of the sketch profile and guide curve are controlled from distance measurements of point cloud and curves.

Once the curves are imported into a parametric CAD Software, a Sweep function is used to generate the desired surface.

3. Application of the proposed method

The proposed method was applied to a commercial bottle shown in Figure 1. This product, modelled in a parametric CAD software with sweep based techniques, represents a characteristic example of geometries for which the proposed method could provide accurate results.

Figure 1. Specimen.

IOP Conf. Series: Materials Science and Engineering **161** (2016) 012047 doi:10.1088/1757-899X/161/1/012047

The specimen was measured by a laser scanner. The result of the measurement, after an automated smoothing processes, is shown in figure 2. The proposed method produced two 2D curves, an ellipse and a spline, which are shown in figure 3 in relation to the point cloud.

In a following step, the ellipse, curve and point cloud were imported into a parametric CAD software and a surface was generated with sweep based technique. In figure 4 the resulting surface and zebra strips are shown. The shape of strips are optimized to geometry smoothness and continuity

Finally the distance deviation between point cloud and surface was measured and is shown in figure 6. The overage deviation is less than 0,03mm and the maximum is below 0,1mm.

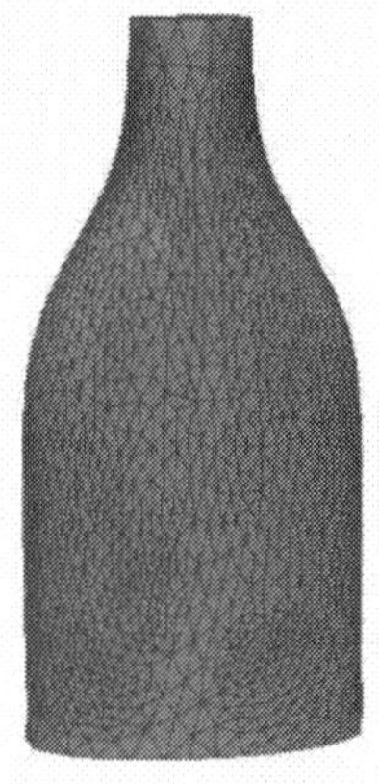

Figure 2. Polygons.

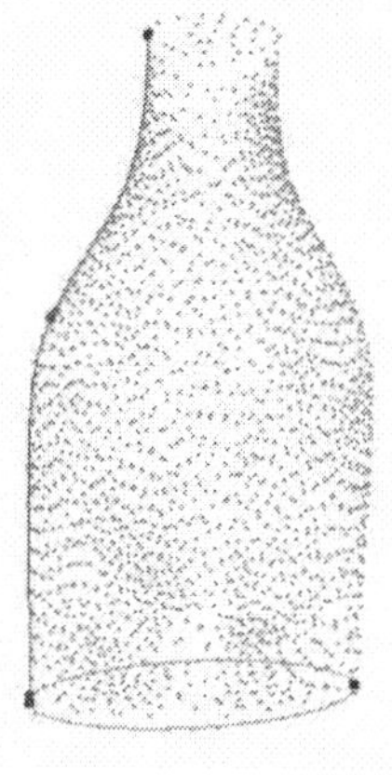

Figure 3. Spline (in red) and ellipse (in blue) used to generate the final surface

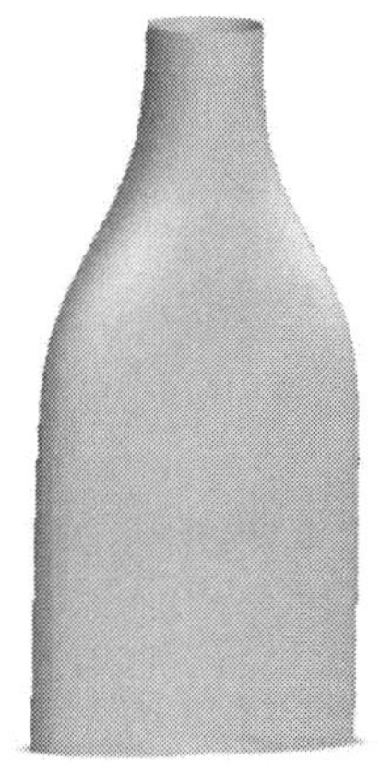

Figure 4. Sweep surface resulting from the proposed method.

Figure 5. Zebra strips of the generating surface.

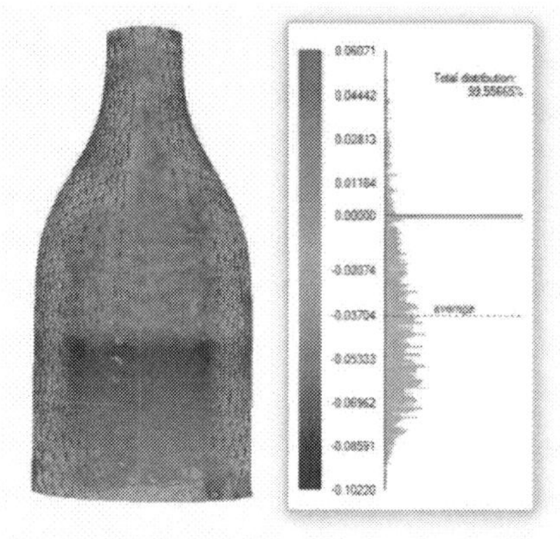

Figure 6. The overage deviation is less than 0,04 mm

To provide a reference to the above technique, the object's surface was also created with an automated process provided by the point cloud processing software. The produced surface is shown in figure 7 and the corresponding zebra strips in figure 8. It is obvious that the surface is not smooth enough.

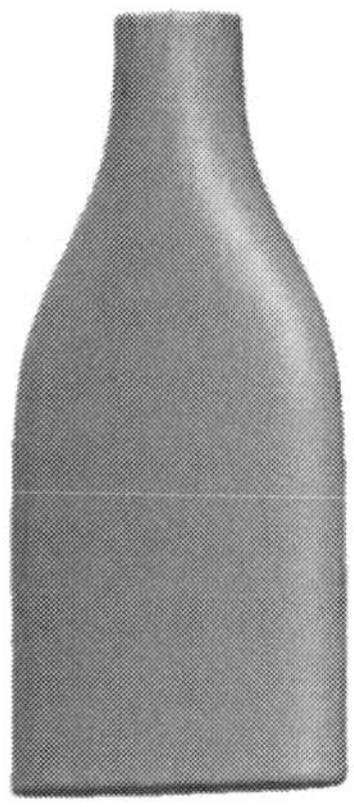

Figure 7. auto creation surface

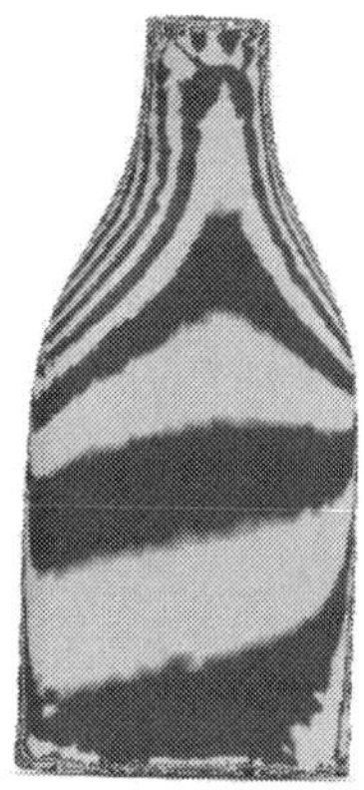

Figure 8. Zebra stripes of the auto creation surface.

The third method used to generate the surface, was based on a guide curve, produced within the point cloud processing software, through the projection of points in a reference plane. The produced curve consist of small line segments. The geometry of the object for the final method is produced once again with a sweep technique, as shown in figure 9, whereas the corresponding zebra strips are demonstrated in figure 10. In this case, similar to the one above, the produced zebra strips are not satisfactory.

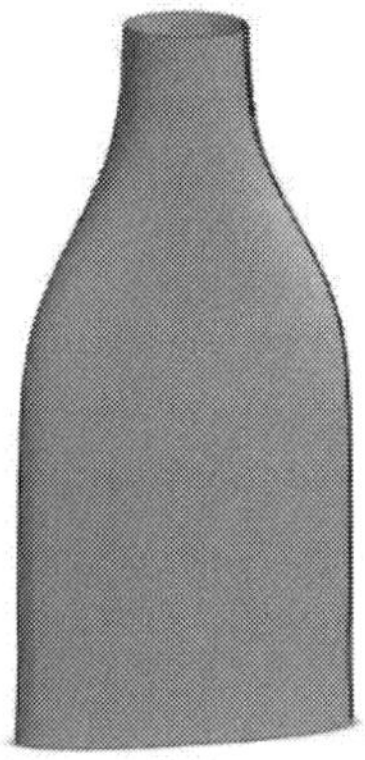

Figure 9. Surface creating with sweep command but the guide curve does not be spline

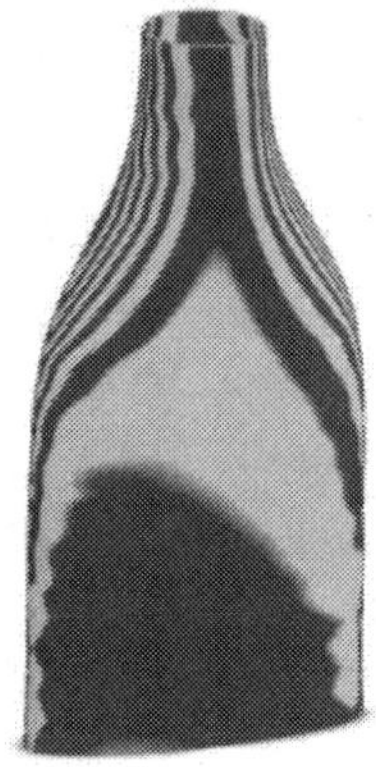

Figure 10. Zebra stripes of the generating geometry.

4. Conclusions

In this paper a method for reverse engineering an object with parametric surface characteristics is presented. More specifically, any industrial type specimens or consumer products can be reconstructed with sweep based techniques. The proposed method automatically extracts a 2D sketch profile of ellipse and a spline curve (used as a guiding profile for the sweep function) consisting of a maximum of four polynomial points to maintain a simple formulation at a smooth enough representation. The final geometry is produced instantly and in easily, without the need of experienced user involvement or any manual intervention (e.g. curve design). The final geometry is superimposed onto the initial point cloud data and the overage deviation was below 0,04mm. Moreover the geometry of the specimen is produced with two alternative methods and compared to the aforementioned one. An auto surface creation inside the point cloud processing software and a method where the guide curve was produced by projecting points onto a reference plane. The Zebra strips are investigated and the results of the proposed methodology compared favorably to the lattertwo scenarios.

References

[1] Bouzakis K.-D, Pantermalis D., Efstathiou K., Varitis E, Michailidis N. and Mavroudis I 2005 Archaeological objects solid geometry determination based on computer tomographies, digital photography and appopriate processing of the related data, 2nd International Conference on Manufacturing Engineerign 675-94

[2] Tsouknidas A., Maropoulos S., Savvakis S and Michailidis N. 2011 FEM assisted evaluation of PMMA and Ti6AL4V as materials for cranioplasty resulting mechanical behaviour and the neurocranian protection. Bio-Medical Materils and Engineering **21,** 139-47

[3] Tsouknidas A., Savvakis S., Asaniotis Y., Anagnostidis K., Lontos A., Michailidis N.: The effect of kyphoplasty parameters on the dynamic load transfer within the lumbar spine considering the response of a bio-realistic spine segment. Clinical Biomechanics, 2013, vol. 28, issue 9-10, pp. 949-55.

[4] Tsouknidas A.: The effect of pedicle screw implantation depth and angle on the loading and stiffness of a spinal fusion assembly. Bio-Medical Materials and Engineering, 2015 vol. 25, issue 4, pp 425-433.

[5] Bouzakis K.D., Maliaris G., Tsouknidas A.: A FEM supported semi-solid high pressure die casting process optimization based on rheological properties by isothermal compression tests

at thixo temperatures extracted. Computational Materials Science, 2012, vol. 59, pp. 133-139.

[6] Mansour G, Bouzakis K-D, Sagris D, Varitis E and Pappa M 2006 Reverse engineerign of parts using 3D measuring device (CMM) by means of splines, Acta Technica napocensis, Applied mathematics and mechanics **49**, 575-82

[7] Mitsi S, Bouzakis K.-D, Sagris D. and Mansour G 2008 Deternmination of optimum robor base location considering discrete and-effector positions by means of hybrid generic algorihm, Robotics and Computer-Intergrated Manufacturing, **24**, 50-59

[8] D. Sagris, S. Mitsi, K.-D. Bouzakis, G. Mansour 2011 Spatial RRR robot manipulator optimum geometric design by means of a hybrid algorithm, The Romanian review precision mechanics, Optics & Mechatronics, No. **39**, pp. 141-144.

[9] S. Mitsi, K.-D. Bouzakis, G. Mansour, D. Sagris, G. Maliaris 2005 Off line programming of an industrial robot for manufacturing, International Journal of Advanced Manufacturing Technology, Vol.**26**, pp.262-267.

Special features of raising of operational data of machine components based of functionally oriented coatings

A Mikhaylov, M Petrov, E Sydorova and D Mikhaylov

Donetsk National Technical University, 58 Artiom, 83001, Donetsk

Email: tm@mech.dgtu.donetsk.ua

Abstract. The paper examines different rational approaches to raise operational data of machine components based on functionally oriented coatings. A structural model has been developed and principles of implementation of the functionally oriented coatings are given. The paper classifies the coatings and gives examples of technological support of the implementation of the functionally oriented coatings.

1. Introduction

Provision of features of machine components based on application of special coatings is one of the promising ways to raise their quality [1, 2, 3, and 4]. The existing coating types allow to solve many problems related to the machine component quality raise, namely: provide protection of the component surface layer from corrosive attacks, oxidation processes, salt and erosion damages, acid and alkaline affects; reduce damageability of the main material; improve the component heat resistance and its parameters; reduce radiation affect made on the main material; solve design and esthetic problems and the entire complex of other issues.

Different materials and their combinations, substance compositions and structures are used to make coatings for the machine components. Metal, intermetallic, composite and com-position and complex and combined coatings are used. The machine component surface coatings can be single-and-multilayer.

Many different methods are used to make machine component coatings. The most promising methods of coating making are chemical deposition (HT-CVD), physical deposition (PVD), the electrolytic method, gas-thermal sputtering, electric-spark alloying and welding.

Application of different types of the coatings allows raising the machine component operational data and solves the entire complex of functional tasks. However, the type of service of the components with the coatings is often connected with complex types of the coating damages, namely:

- the coating uneven wear;
- the coating sector corrosive and erosion and other attacks;
- influence of uneven specific contact load on the coating surface;
- complex impact of different operational functions on the coating (operational function structure different data) and other operational features.

The goal of the paper is to research the raise of the machine component quality based on creation of specific features of every component surface and the feature structure of the entire component under the influence of many uneven operational functions at the expense of the application of the functionally oriented coatings (FOC) and provision of their features [5, 6].

To reach the goal the paper solves the following tasks: analyzes the features of break-ages of the unit functional components under the operation; develops the general approach and principles of the FOC implementation; develops the FOC classification; offers technological provision and researches specific features of implementation of the component function-ally oriented characteristics.

2. The general approach and principles of implementation of the functionally oriented coatings of the machine components

Many operational functions influence the component under its operation. Each surface is affected by uneven loads and every zone of influence of the specific load has certain topological parameters with the operating functions having the certain structure. To provide the component functionally oriented characteristics under the conditions the correspondence among the operational functions F, technological influence TI and product characteristics C is to be defined (Figure 1). There are the certain links [5] among the operational functions F, the technological influence TI and the product characteristics C, namely: similarity, correspondence identity, analogy, equivalence, and adequacy.

Figure 1. shows the model of interconnection of the system objects: the operational functions F, the technological influence TI and the unit characteristic C. The implementation technological scheme FOC is implemented on the interconnection basis [5]. It should be mentioned that the structure of the interconnections among the model components has the closed form that allows defining the parameters of the technological influence and the unit features depending on the characteristics of operational function impact on the unit functional components. All the parameters are defined on the basis of the group of the special principles of element and technological influence orientation [5]:

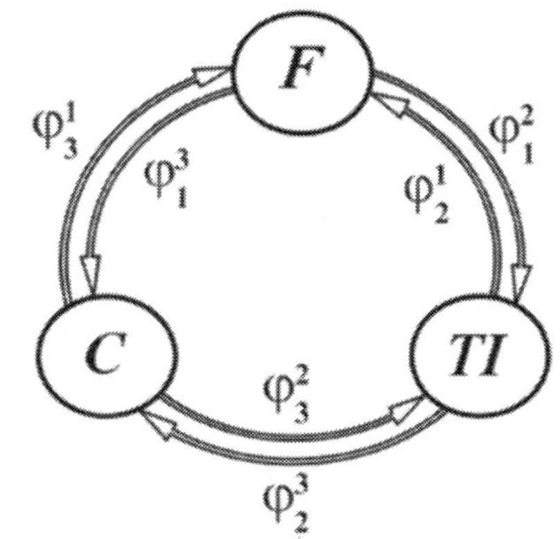

Figure 1. The model of interconnections φ_i^j of the system objects: the operational functions *F*, the technological influence *TI*, and the unit characteristic *C*.

1. Functional correspondence of the features of the action of the elementary function at each functional component of the unit to the characteristics of implementation of the technological influence and the parameters of provision of the necessary features in the functional component.
2. Topological correspondence of the geometrical parameters of the unit functional component, in which the elementary function acts under the operation, to the geometrical parameters of the zone of the technological influence on the unit and the geometrical parameters of the feature zone.
3. Quantitative correspondence of the multitude of the operational components in which many different elementary functions act under the operation to the multitude of implementation of different technological influence and the multitude of the components providing the necessary features in the unit functional components.
4. Adequate dependence of the spatial features of action of the elementary function under the operation, technological influence and operational characteristics in the space of every unit functional component.
5. Adequate dependence of the temporal features of action of the elementary function under the operation, temporal and spatial features of implementation of the technological influence and temporal operational characteristics in every functional component of the unit at every level of the technology depth.
6. Structural correspondence of the multitude of the elementary functions, implementation of the multitude of the technological influence and multitude of characteristics in the unit functional

elements out of the condition of provision of the given, required and limit features of the entire unit.

7. Adequate structural and functional correspondence of every functional component to the given, required or maximum potential of the general characteristics of the entire unit in the space and time.
8. Adequate structural and functional correspondence of characteristics in the neighbourhood of every functional component to the given, required or maximum potential of the general characteristics of the entire unit in the space and time.

It is to be mentioned that the process of the given principles of technological impact and unit characteristics implementation can be made on the basis of an iteration approach [5]. The iteration approach allows fulfilling of the FOC synthesis through the implementation of the given principles with taking into account of successive processes and multiple repeated recurrent processes made at the expense of feedback couplings.

The FOC can have the following characteristics (Figure 2):

1. Changeable characteristics (evenly and unevenly changeable, functionally changeable, gradient and others) (Figure 2, a) within every surface of a product and/or unit structure surface.
2. Step-function characteristics (in one direction, two directions, wide, middle, small steps and others) (Figure 2, b) within every surface of a product and/or unit structure surface.
3. Zonal characteristics (zone traditional sizes, macro-zonal, micro-zonal, with change-able characteristics according to the zones and others) (Figure 2, c) within every surface of a product and/or unit structure surface.
4. Spotted characteristics (big, average, small size spots with different or similar characteristics and others) (Figure 2, d) within every surface of a product and/or unit structure surface.
5. Special characteristics within every surface of a product and/or unit structure surface.

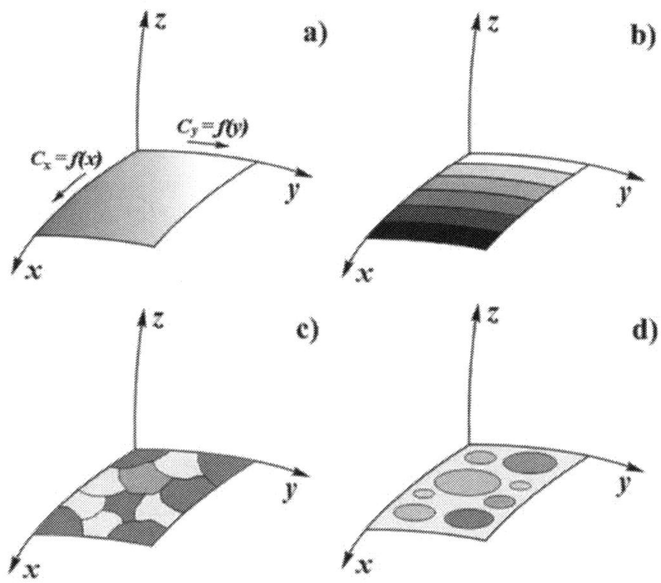

Figure 2. Hypothetical alternatives of single-layer FOC.

The machine component FOC provide fulfillment of the following characteristics of operation under the influence of the changeable parameters of the function on the surface and unequal structural components of the multitude of functions on the unit:

1. Complete adaptation of the unit under manufacturing to the characteristics of its operation in the machine or the technological scheme.
2. Maximum operational potential of the unit or multiple of it in the machine (the multiplicity degree is defined by a designer).
3. The stated wear of every coating on the unit surface at the given period of its operation.
4. Coordinated simultaneous complete wear of all types of coatings on the unit surface at the given period of its operation.
5. Qualitatively new set of the product characteristics under operation and other features.

IOP Conf. Series: Materials Science and Engineering **161** (2016) 012048 doi:10.1088/1757-899X/161/1/012048

3. Classification of the functionally oriented coatings

It can be noted that the coatings used for mechanical engineering products can have different physical and mechanical characteristics (FMC) the parameters of which are formed within the limits peculiar for traditional and super- strong coatings. The coatings can have changeable thickness on the product surface which is defined depending on the operational function action. At the same time the coatings made of the materials with different FMC can be used for the unit different surfaces. Besides, the coatings on the unit surfaces can be single-layer and multilayer (Figure 3).

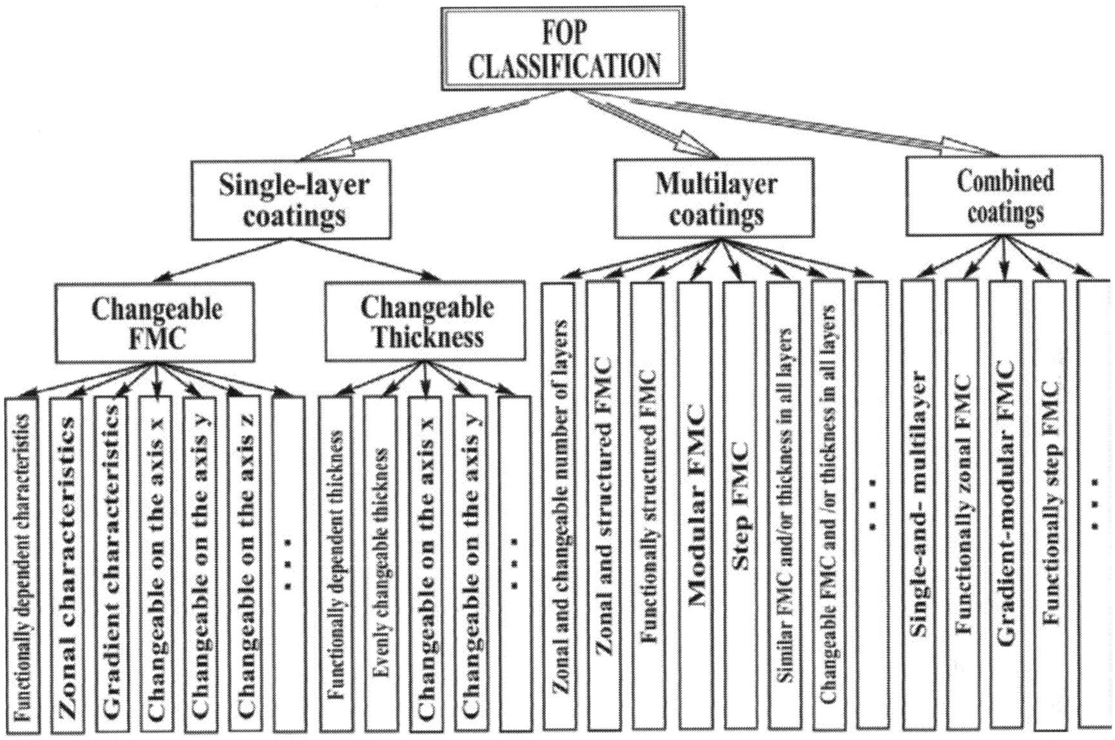

Figure 3. Classification of the FOC for machine components.

It is to be mentioned that the thickness of the single-layer and multilayer coatings (unit fibers of the multilayer coating) can be different in general and it depends on the functional purpose of the coating and the method(s) of its depositing.

The FOC can be used for different machine components. In this case they can be classified as the scheme shown in Figure 3. Here the FOC are grouped as follows: the single-layer coatings, the multilayer coatings and the combined coatings.

The single-layer coatings can have the changeable FMC with the changeable thickness and have the following characteristics: functionally dependent, zonal, gradient, and changeable on the axis x, changeable on the axis y, changeable on the axis z and others.

To provide the coating simultaneous complete wear if there is uneven wear on the unit functional components in some number of cases, the single-layer coatings with the changeable thickness with the following parameters can be used: functionally dependent thickness, evenly dependent thickness, changeable on the axis x, changeable on the axis y, other characteristics.

To raise the unit resource the multilayer coatings which can have the following characteristics can be used: zonal changeable number of layers, zonal structured FMC, functionally structured FMC, modular FMC, step FMC, similar FMC and/or thickness in all layers, changeable FMC and/or thickness in all layers, and other characteristics.

In a number of cases it is reasonable to use the combined coatings for the units. They have the following characteristics: combined FMC of the single-layer and multilayer coatings, functionally zonal FMC, gradient and modular FMC, functionally step FMC and other characteristics.

Figure 4 shows some unit FOC: Figure 4, a – the single- layer coating of constant thickness and characteristics; Figure 4, b – the single-layer coating of constant thickness and changeable characteristics; Figure 4, c – the single-layer coating of variably changeable thickness and constant characteristics; Figure 4, d – the single-layer coating of constantly changeable thickness and constant characteristics; Figure 4, e – the zonal multilayer coating with the layers of similar thickness; Figure 4, f – the zonal multilayer coating with the layers of different thickness.

The given mathematical structural and functional symbol models of the single-layer and multilayer coatings and the alternatives of the functionally oriented coatings allow making synthesis of the unit given or required characteristics. It should be mentioned that the synthesis of the unit necessary characteristics can be done on the basis of the morphological approach with the application of morphological matrixes and methods of morphological synthesis of the coating alternatives. This approach allows generating of new solutions as to non-traditional characteristics of the coatings.

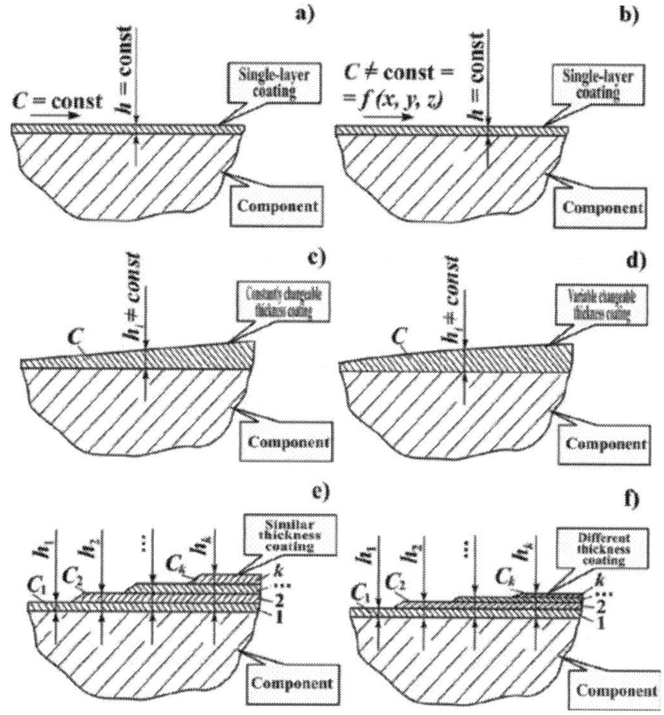

Figure 4. Some alternatives of the unit FOC:
a – the single-layer coating of the constant thickness and characteristics; b – the single-layer coating of the constant thickness and changeable characteristics; c – the single-layer coating with the variably changeable thickness and constant characteristics; d – the single-layer coating of the constantly changeable thickness and constant characteristics; e –the zonal multilayer coating with the layers of similar thickness; f – the zonal multilayer coating with the layers of different thickness.

4. Technological support and features of implementation of the unit functionally oriented characteristics

To implement the FOC technological support, namely technological plants that make coatings on the basis of the group of special principles of orientation of technological impacts and the unit feature provision, is necessary.

To implement some methods of the FOC spraying special plants were constructed at Donetsk National Technical University:

1. The plant for vacuum ion and plasma spraying of the FOC made on the basis of the modernized system ННВ 6.6-И1.
2. The experimental detonation plant DEPLA-5.
3. The spark plant to form the FOC. It is aimed at implementing of spark alloying of the surface layer of the unit with different functionally oriented characteristics of the sprayed layer.

IOP Conf. Series: Materials Science and Engineering **161** (2016) 012048 doi:10.1088/1757-899X/161/1/012048

Figure 5 shows the plant for vacuum ion and plasma spraying of the FOC based on the modernized system ННВ 6.6-И1. It allows making coatings for different machine components.

Figure 5. The plant to spray functionally oriented vacuum ion and plasma coatings.

Let us look into the process of making the FOC for the gas-and turbine engine compressor blade with the application of the method of vacuum ion and plasma spraying. Figure 6 shows the single-layer coating destruction scheme and consequent forming of the multilayer FOC at 6 stages – one at each. The single layer destruction scheme shows forming of the en-tire zone of the surface complete destruction which starts with the zone 1-a1-b-1I which is spread to the zone 2-1-a1-b-1I-2I-a2, then to the zone 3-2-1-a1-b-1I-2I-3I-a3 and so on up to the zone 5-4-3-2-1-a1-b-1I-2I-3I-4I-5I-an. Under the destruction of the corresponding zone of the single layer coating the wear of the blade main material takes place here. According to the characteristics of the surface destruction under the blade operation formation of the multilayer coating with topological orientation of every layer on the surface of the blade fluted land is offered. It is to be done in accordance with the special principle of the surface orientation de-scribed by the expression (2). In this case the FOC is formed with the application of the special shielding matrixes.

The FOC is technologically formed as a result of consequent spraying of the single-layer coating at every operation stage with taking into account the topological orientation of every layer surface (Figure 6). First coating № 1 with its topological orientation on the surface according to the wear scheme is sprayed, then coating № 2 and the next layers. The coating is sprayed on the entire surface of the blade fluted land at the end of the stage. In this case every coating characteristic can be defined in accordance with the first principle.

The multilayer coatings formed in accordance with the first and the second principles are characterized by one-time complete destruction at the given period of operation of the ГТД compressor blades. It allows avoiding of destruction of the blade fluted land main material for a long period of time and at the same time raising technical and economic indices of their restoration.

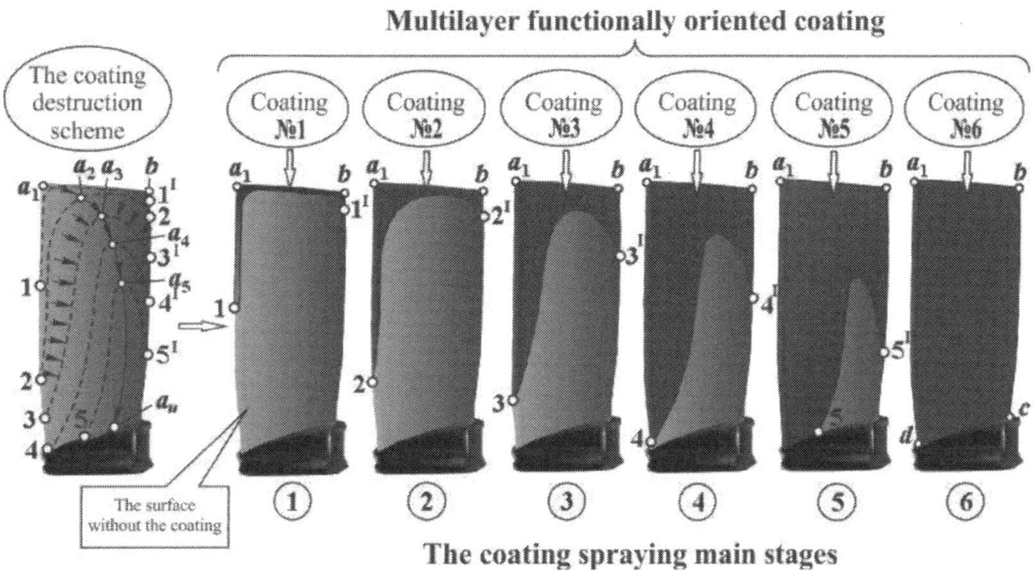

Figure 6. Schemes of destruction of the single-layer coating and forming of the multilayer FOC.

Figure 7 shows the experimental detonation plant DEPLA-5 used for the FOC spraying. The formation of the functionally oriented characteristics of the coatings takes place at the expense of several dispensers for different types of powder material installed into the plant, specified individual modes of the plant unit cycle and application of different technological methods of the powder material spraying.
The plant body is filled with a mixture of gases able to detonate within the plant operation unit cycle. The corresponding dispensers send a portion of the powder material at the specified time. The powder material composition is changed in accordance with the characteristics of the FOC structure formation and then the igniter (firing light) is actuated. The gas burning occurring in the plant body is transformed into the detonation. The detonation wave is a complex of a shock wave and a gas chemical reaction zone. The detonation wave moves along the plant body at a supersonic speed.

Figure 7. Experimental detonation plant *DEPLA-5*.

The detonation products in the plant body carry the powder material mixture particles along, heat them and carry them to the component machined surface where the coating is formed. The powder material welding, collision and evaporation take place in the two-phase flow that is formed in the plant body. The detonation processes are repeated in a cycle when the coatings are sprayed and according to frequency of the plant operation, for example 15 shots per second.
Figure 8 shows: a – the scheme of the spark plant to form the FOC, b- the zone and elements to form the coatings, c- the scheme of the FOC with changeable characteristics.
The offered plant (Figure 8, a) operates similar to a 3D-printer that produces the characteristic specified "design". It makes the spark alloying of the component surface layer with functionally oriented characteristics. Figure 8 shows: 1 – a cylindrical component (cathode); 2 – a tool block system; 3, 4, 5, 6 – tool blocks; 7, 8, 9, 10 – exchangeable alloying tools (electrodes-anodes) made of different materials – molybdenum, tungsten, vanadium and chromium; 12 – a reel (solenoid); 13 – a spring (resilient member); 14 an electrode position regulator that provides the specified amplitude of its vertical alternate motion; 15 – the guides in the tool block system; 16 – electric arc among the electrodes during the spark alloying; 17 – a unit zone (spot) of the alloying metal in the surface formed coating (with the diameter of r); 18 – a unit surface layer made of alloying metals with the multitude zones of their inclusions.
It is to be noted that with the help of the plant (Figure 8, a) the mechanism of formation of the coating alloying layer formation is made as a result of both carrying and interaction of the electrode material and heat impact of electrical spark. The specified characteristics of the coating surface layer can be implemented at the expense of both the anode separate materials and a combination of the material mixture of four anodes.
The operation of the offered plant (Figure 8, a) is based on several schemes. Figure 9 shows the alternatives of the tool tracing schemes as to the component (rimier). It shows the following: Figure 9, a – the motion B1 is periodical, the motion П is alternate; Figure 9, b – the motion B2 is swinging, the motion П is alternate; Figure 9 c – the motion B3 is constant and swinging, the motion П is alternate.

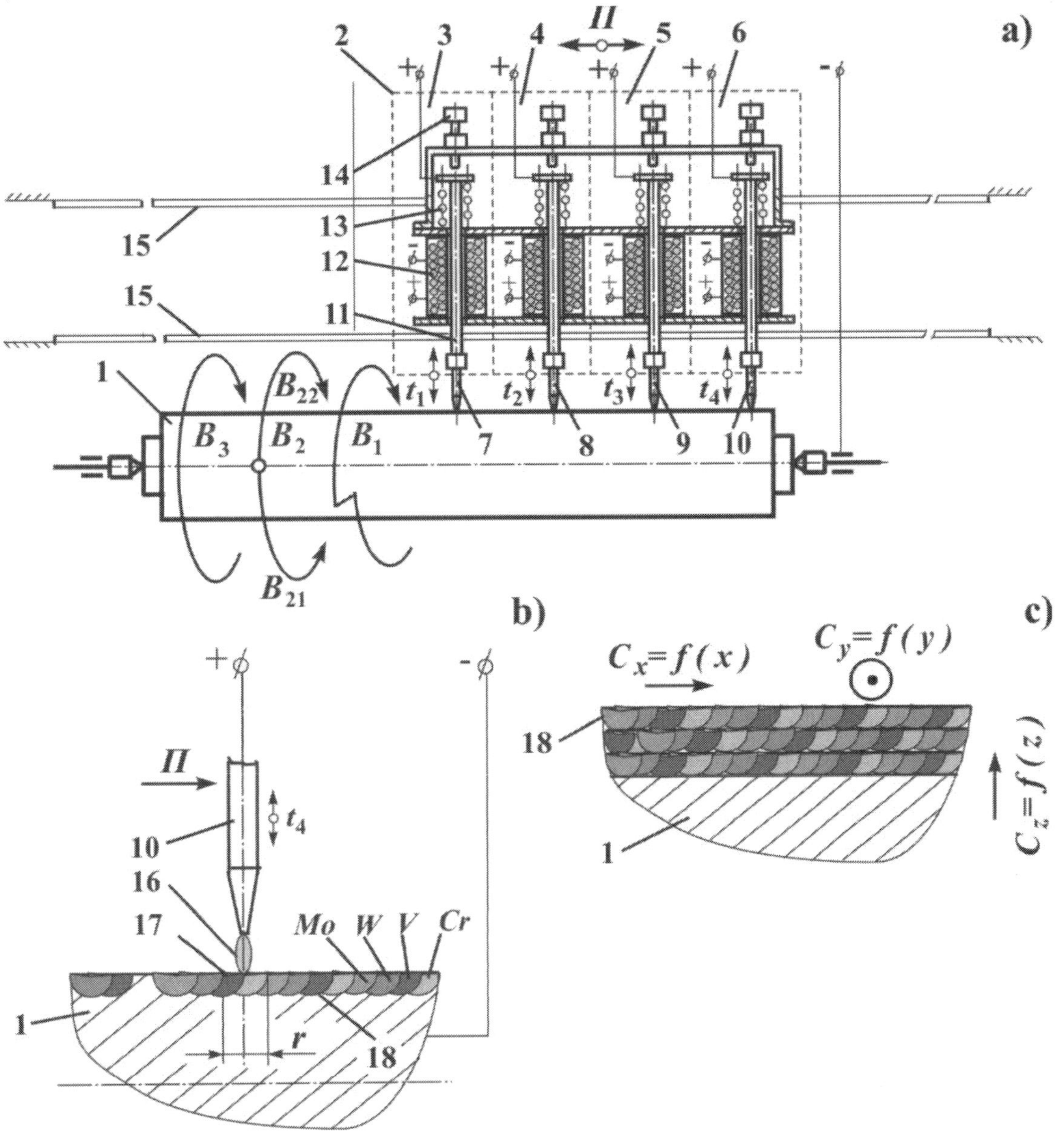

Figure 8. The FOC spark spraying; a – the plant scheme, b- the FOC spraying zone and components, c- the FOC scheme with changeable characteristics.

5. Conclusion

The paper under discussion shows the research in improvement of the machine components quality based on special characteristics of every component surface and the entire component characteristic structure under the operation of the multitude of uneven operational functions at the expense of the FOC. The research allowed the following:

1. Analysis of the features of destruction of the unit functional components under the operation. It was stated that in a number of cases the machine components were under the influence of the varying operational functions that led to the resource reduction.

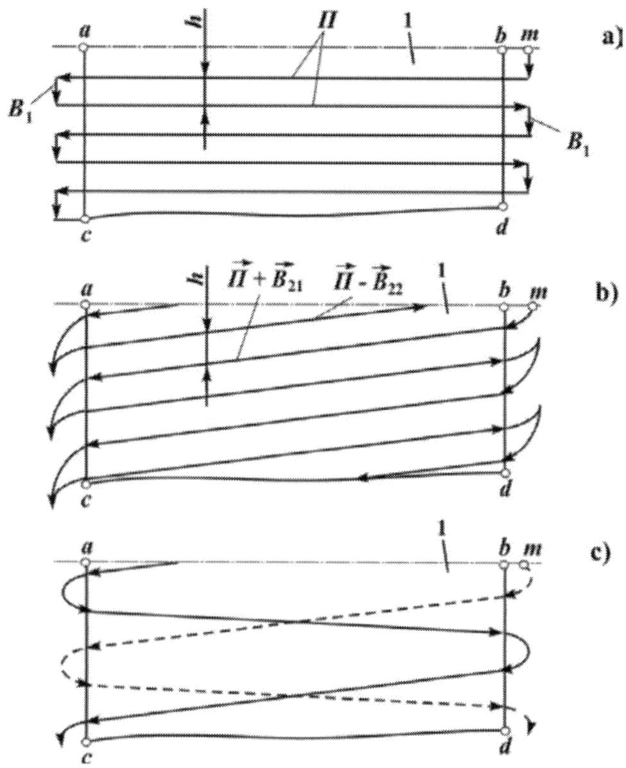

Figure 9. Alternatives of the tracing schemes of the tool as to the component: a – the motion B_1 is periodical, the motion Π is alternate; b – the motion B_2 is swinging, the motion Π is alternate; c – the motion B_3 is constant and swinging, the motion Π is alternate

2. The development of the general approach to the FOC spraying on the machine component surfaces. It is based on the group of special principles of the FOC making.
3. The FOC classification to sort different alternatives of the component coatings and characteristics.
4. Development of the special technological support to spray the FOC on the machine components.
5. The research in characteristics of the implementation of the functionally oriented characteristics of the components based on the FOC. It was stated that the FOC provided the qualitatively new set of the machine components.

It is to be noted that to spray the FOC a lot of methods could be applied and a complex of different facilities could be used. It depends on the design features of the certain component and the material it is made of, the FOC necessary parameters, the technological approach as to the coating spraying and a number of other conditions.

References

[1] A.F. Puryakov 2003 *Teoreticheskiye osnovy tekhnologii plazmennogo napyleniya.* [Technological Foundations of Plasma Spraying] (Moscow: MGTU named after N. Bauman) p 360 [in Russian]

[2] S.N. Grigoriev 2009 *Metody povysheniya stoikosti rezhishchego instrumenta* [Methods to Raise Tool Cutting Durability] (Moscow: Mashinostroyenie) p 368 [in Russian]

[3] G.L. Amitan, I.A. Baisupov, Y.M. Baron 1988 *Spravochnik po elektrokhimicheskim I electrofizicheskim metodam obrabotki* [Reference Book in Electric and Chemical and Electric and Physical Methods of Machining] (Leningrad: Mashinostroyenie) p 719 [in Russian]

[4] K.A. Yushchenko, Y.S. Borisov, Y.S. Kuznetsov and V.D. Korzh 2007 *Inzheniriya poverkhni* [The Surface Engineering] (Kiev: Naukova Dumka) p 558 [in Ukrainian]

[5] A. N. Mikhailov 2009 *Osnovy sinteza funktsionalno orientirovannykh tekhnologiy* [Basics for the Synthesis of Functionally Oriented Technologies] (Donetsk: DonNTU) [in Russian]

[6] D.A. Mikhailov, A.P. Nedashkowski and T.G. Ivchenko 2014 *Tekhnologicheskiye osobennosti vosstanovleniya lopatok compressora ГТД s primeneiem funktsionalno orientirovannykh pokrytiy* [Technological Characteristics of Restoration of Blades of the Compressor ГТД with the Application of Functionally Oriented Coatings] *Progresivno tekhnologii I systemy mashinobuduvannya* vol 47 (Donetsk: DonNTU) p 213-224 [in Russian].

IOP Conf. Series: Materials Science and Engineering **161** (2016) 012049 doi:10.1088/1757-899X/161/1/012049

An Energy-Critical Plane Based Fatigue Damage Approach for the Life Prediction of Metal Alloys

N Pitatzis[1] and G Savaidis[1]

[1]Department of Mechanical Engineering, Aristotle University of Thessaloniki, GR-54124 Thessaloniki/Greece

E-mail: pitatzis@auth.gr

Abstract. This paper presents a new energy-critical plane based fatigue damage approach for the assessment of the fatigue life under uniaxial and multiaxial proportional and non-proportional fatigue loading. The proposed approximate method, based on Farahani's multiaxial fatigue damage model, takes into account the critical plane orientations during a loading cycle and the values of the respective damage parameters on them. The uniqueness of the proposed method lies on the fact that it considers a weighted contribution of each critical plane orientation to the material damage. The relative weighting factors depend on the declination of each critical plane with respect to the critical plane, where the damage parameters exhibit their maximum values during a fatigue loading cycle. Herein, several low, mid and high-cycle fatigue loading cases are being investigated. The induced elastic-plastic stress-strain states are approximated by means of respective finite element analyses (FEA). Several experimental fatigue data derived from uniaxial and multiaxial fatigue tests on StE460 steel alloy thin-walled hourglass-type specimens have been used to verify the model's calculation accuracy. Comparison of experimental and calculated fatigue lives confirm remarkable fatigue life calculation accuracy in all cases examined.

1. Introduction

The need for lightweight, safe and financially optimized engineering systems performing under fatigue loading implies an in-depth investigation of the physical phenomena related to the loss of structural integrity, and the establishment of respective mathematical methods for the approximate calculation of structural durability. During the last three decades the research interest on Fatigue has been focused on the multiaxial case [1, 2, 3, 4, 5] as a one-way path to attain lighter and even safer structures. This is due to the fact that in the case of multiaxial fatigue, neither the physical phenomena related to the damage of the material have been thoroughly investigated, nor the former mathematical material constitutive models were proven to lead to results of satisfactory accuracy [6, 7, 8, 9]. The effort on multiaxial fatigue could be distinguished into two major research fields, namely the pursuit for the establishment of new, reliable material models to approximate the elastic-plastic multiaxial stress-strain state [1, 10, 11, 12] and the challenge for the constitution of fatigue damage approach methods in order to attain credible approximation of fatigue life [13, 14, 15, 16, 17, 18].

This work focuses on the second field of fatigue life approach and incorporates a proposed multiaxial fatigue criterion established on energy-critical plane basis. The assessment of multiaxial fatigue life can be achieved via three potential paths; stress, strain and energy-based approaches. It has been referred [16] that for all the above-mentioned fatigue life assessment methods, when the damage parameters are being considered on the critical planes, then the respective approximate methods lead to the most satisfactory results. Furthermore, energy based approaches are proven to be the less parameter-demanding methods [17, 18, 19] compared to stress and strain based approaches so as for contiguous fatigue life approximation accuracy to be achieved. These facts constitute important incentives for further research towards energy-critical plane based fatigue assessment methods, as disclosed herein.

2. The proposed energy-critical plane fatigue damage approach

The significance of using energy-based fatigue damage parameters considered on critical planes lies on the fact that these planes and the energy conveyed on them are being directly related to the physical causes of materials' failure due to fatigue [20, 21]. According to Taylor [22], the deformation of solid materials under external loading results from the dislocation movement detected on atomic scale due to shear stresses acting on crystal lattice planes with the minimum shear strength. Accordingly, it is proved [23, 24] that the initiation of material consistency degradation under fatigue loading is being detected on plane orientations where shear stresses and strains attain their maximum values. Figure 1 shows a thin-walled hourglass specimen subjected to combined axial-torsional loading with time varying axial force F and torsional moment M_T, as well as the local plane stress on the highlighted finite free surface of the specimen on a specific moment, the corresponding critical plane orientation where shear stress, τ_{yz}, reaches its maximum value, τ_{max}, the perpendicular to the critical plane normal stress component, σ_n, responsible for the propagation of the failure and the corresponding maximum shear and normal strain, $\gamma_{max} / 2$ and ε_n respectively.

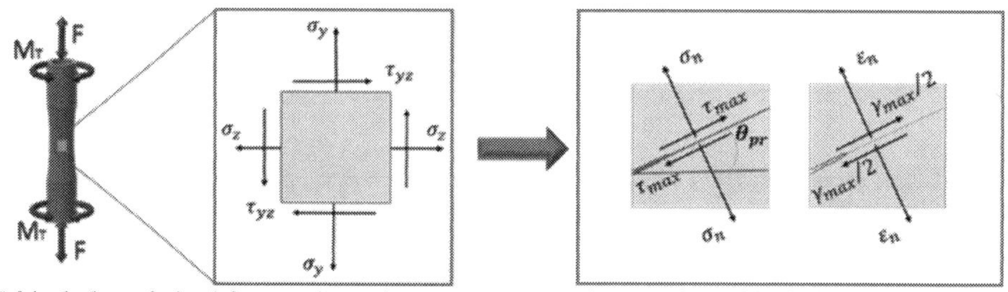

Figure1. Critical plane derived from a given plane stress state and respective critical plane orientation angle, θ_{pr}.

The magnitude of τ_{max} and σ_n and the extend of non-proportionality in conjunction with the microstructure of the material [25] are the major parameters related to the materials' durability under fatigue loading. The critical plane orientation could be defined according to the stress transformation formulas of Mohr, applicable to both uniaxial and multiaxial stress states [17]. When combined fatigue loading components are being applied, the orientation of the critical plane alters by time. In the case of non-proportional fatigue this alternation becomes prominent compared to the corresponding critical plane orientation changes under proportional loading with the same nominal stress amplitudes as shown in figure 3.

Taking all the above into account, the proposed fatigue life approximate method is being implemented according to the following steps:

1st step: Calculation of local elastic-plastic strains and stresses.

IOP Conf. Series: Materials Science and Engineering **161** (2016) 012049 doi:10.1088/1757-899X/161/1/012049

The stress-strain state is being approximately calculated by means of respective elastic-plastic FEA on the external free surface of smooth, thin-walled, hourglass-type specimens made of StE460 steel alloy. The von Mises yield criterion [26] in conjunction with the non-linear kinematic hardening model of Chaboche [27] are being applied. The geometrical parameters of the specimen and the mechanical properties of the material under investigation are being included in Table 1. The geometrical characteristics of the specimen and a representative finite element mesh configuration are shown in figure 2.

Axisymmetric solid elements (SOLID273, [28]) have been applied to model the axisymmetric solid structure. The elements are being defined by eight nodes on a master plane and the rest of the nodes are being automatically defined in the circumferential direction based on the eight master plane nodes. Each element node has three degrees of freedom, namely the translations on the x, y and z directions. A total number of 1065 elements, with a minimum length of 0.1mm detected on the z axis, progressively increasing as the distance from the xz plane increases to reach the maximum length of 5.5mm, have been used for the creation of the finite element mesh (figure 2).

Table 1. Geometrical data of the thin-walled, hourglass-type specimens and mechanical properties of steel StE460 [29, 30].

Geometrical Parameters					
$D_o(mm)$	$D_i(mm)$		$D_h(mm)$	$R(mm)$	$H(mm)$
50	41		36	130	220
Mechanical Properties					
$R_m[MPa]$	$R_{p0.2}\ [MPa]$	$E\ [MPa]$	v	$K'[MPa]$	n'
643	500	208500	0.3	1115	0.161
ε_f'	$\sigma_f'\ [MPa]$	γ_f'	$\tau_f'\ [MPa]$	c	b
0.281	969.6	1.87	585	-0.493	-0.086

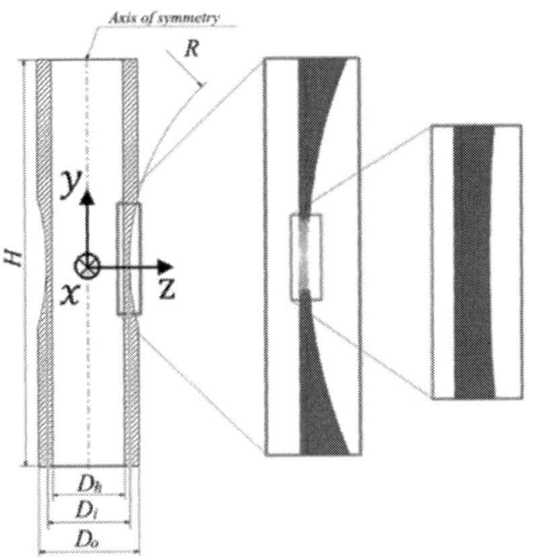

Figure 2. Thin-walled hourglass type specimen and typical finite element mesh configuration.

The normal stress-normal strain and the shear stress-shear strain stabilized hystereses derived from FEA after five (5) fatigue cycles are being used for the assessment of the respective fatigue life according to the proposed fatigue life assessment method. Significant engineering parameters used in the framework of the present work are the nominal normal and shear stresses, ***S*** and ***T*** respectively, which are being defined according to Ref. [31], as follows:

$$S = \frac{F}{\pi((D_i^{\ 2} - D_h^{\ 2})/4)} \tag{1}$$

$$T = \frac{2M_T}{\pi((D_i^{\ 3} - D_h^{\ 3})/8)} \tag{2}$$

2nd step: Calculation of fatigue damage parameters.

Two fatigue damage parameters, Π_I and Π_{II}, have been implemented, namely the product of the maximum shear stress and the respective maximum shear strain on the critical plane, relative to the critical plane orientation, θ_{pr}:

$$\Pi_I = \left[\tau_{max} \cdot \left(\frac{\gamma_{max}}{2}\right)\right](\theta_{pr}), \tag{3}$$

and the product of the normal to the critical plane stress and the respective normal strain relative to the critical plane orientation, θ_{pr} (figure 1),

$$\Pi_{II} = [\sigma_n \cdot \varepsilon_n](\theta_{pr}). \tag{4}$$

Given the stabilized stress-strain hysteresis loops derived from the FEA according to the first step, the calculation of the before-mentioned parameters becomes feasible for a full fatigue cycle. The bandwidth of possible critical plane orientations ($-45° \leq \theta_{pr} \leq 45°$) is being divided into orientation segments of 5 degrees ($\Delta\theta_{pr} = 5°$) and the sum of each damage parameter, Π_I and Π_{II} is being calculated for each 5-degree segment and for a full loading cycle. A typical distribution of $\sum_T \Pi_I(\Delta\theta_{pr})$ and $\sum_T \Pi_{II}(\Delta\theta_{pr})$ during a fatigue cycle under proportional and non-proportional fatigue, where combined axial and torsional loading components are being applied, are shown in the qualitative bar diagrams of figure 3a and 3b, respectively, where the significant difference between the two distributions is prominent.

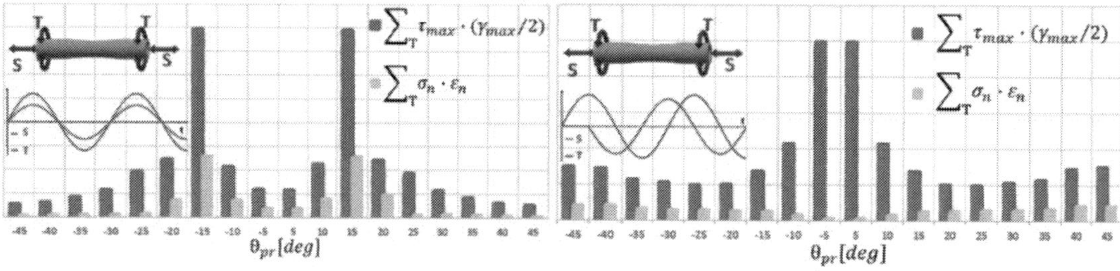

a. *b.*

Figure 3. Qualitative bar diagrams of $\sum_T \Pi_I(\Delta\theta_{pr})$ and $\sum_T \Pi_{II}(\Delta\theta_{pr})$ distribution during a full loading cycle under (a) proportional and (b) non-proportional fatigue, where combined sinusoidal axial and torsional loading components with the same nominal stress amplitudes ($S_{a,prop} = S_{a,non-prop}$ and $T_{a,prop} = T_{a,non-prop}$) and phase differences of $\Delta\varphi = 0°$ and $\Delta\varphi = 90°$ respectively are being applied.

3rd step: Introduction of fatigue damage weighting factors.

Taking into consideration the maximum shear stress-critical plane approach [13, 15], the 5-degree critical plane orientation bandwidth, where $\sum_T \boldsymbol{\Pi}_{II}(\boldsymbol{\Delta\theta}_{pr})$ attains its maximum value, is the one where the fatigue crack is more likely to be induced. According to the proposed fatigue life approximate method, the contribution to the material damage of $\sum_T \boldsymbol{\Pi}_{II}(\boldsymbol{\Delta\theta}_{pr})$ parameter on the rest 5-degree segment orientations is being decreased -by the introduction of a relative damage weighting factor- as the declination of each critical plane with respect to the critical plane where the damage parameters exhibit their maximum values increases. The physical explanation of this assessment lies on the fact, that the normal stress on other critical planes than the one where the crack is expected to be induced, could not contribute to the crack opening-closure -and accordingly to fatigue damage- at the same extend, because of the declination between the orientation of the theoretical crack propagation plane and the several critical plane orientations that significantly deviate from the critical plane orientation 5-degree bandwidth where $\sum_T \boldsymbol{\Pi}_{II}(\boldsymbol{\Delta\theta}_{pr}) = \boldsymbol{max}$, as shown in figure 4. It is plausible that this contribution lessens for critical planes with larger declination with respect to the orientation, where $\sum_T \boldsymbol{\Pi}_{II}(\boldsymbol{\Delta\theta}_{pr}) = \boldsymbol{max}$. The following weighting factor is proposed to describe this assumption mathematically:

$$W_{f\sigma,i} = 1 - i\left(\frac{1}{9}\right),\ i = 1, \ldots, 8 \tag{5}$$

Each weighting factor $\boldsymbol{W}_{f\sigma,i}$ acts as a multiplier on each respective sum, $\sum_T \boldsymbol{\Pi}_I(\boldsymbol{\Delta\theta}_{pr})$ as depicted in figure 4.

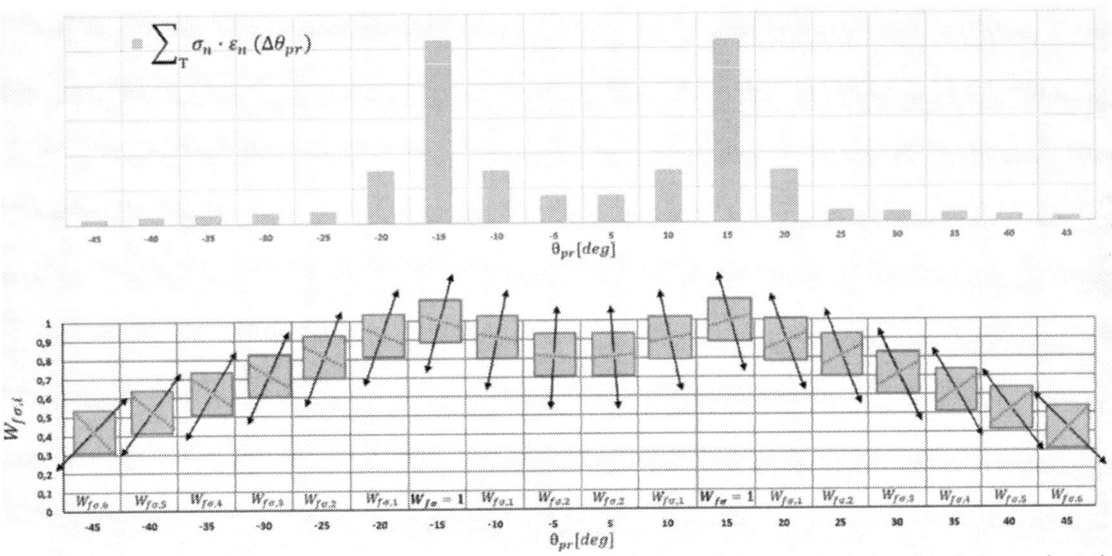

Figure 4. Qualitative bar diagram of $\sum_T \Pi_{II}(\Delta\theta_{pr})$ distribution during a fatigue cycle (up), critical plane orientations susceptible to crack initiation, where $W_{f\sigma,i} = 1$, and $W_{f\sigma,i}$ values distribution for each 5-degree critical plane orientation segment (down).

Furthermore, a second fatigue damage weighing factor, $\lambda_{\sigma_{n,max,i}}$ is being introduced, which equals the ratio of the maximum value of σ_n identified on each 5-degree critical plane orientation segment over the stress that corresponds to strain value of 0.2%, considering the stabilized stress-strain curve of the material under cyclic fatigue loading. Considering this, $\lambda_{\sigma_{n,max,i}}$ is calculated for each 5-degree critical plane orientation bandwidth according to the equation:

$$\lambda_{\sigma_{n,max,i}} = \frac{\sigma_{n,max,i}}{\sigma_{\varepsilon=0.2\%}}, i = 1, \ldots, 9 \tag{6}$$

Each one of the nine weighting factors, $\lambda_{\sigma_{n,max,i}}$, acts as a multiplier on each respective sum, $\sum_T \Pi_I(\Delta\theta_{pr})$. Eqs. (5) and (6) have been derived from several parametrical fatigue life assessment analyses, and after comparative studies with corresponding uniaxial and multiaxial fatigue test data [24], that have been accomplished in order to attain the most reliable fatigue life assessment.

4th step: Fatigue life assessment.

Taking into account the above mentioned fatigue damage weighting factors and the concept of Farahani (Fa) [17, 18] the following equation for the approximate calculation of fatigue life is being proposed:

$$\frac{(1+\frac{\sigma_n^m}{\sigma'_f})}{(\tau'_f \gamma'_f)}(\Sigma\Pi_I(\Delta\theta_{pr,\tau,1}))+\lambda_{\sigma_{n,max,1}} W_{f\sigma,1}\frac{1}{(\sigma'_f \varepsilon'_f)}(\Sigma\Pi_{II}(\Delta\theta_{pr,\sigma n,1}))+\ldots+\frac{(1+\frac{\sigma_n^m}{\sigma'_f})}{(\tau'_f \gamma'_f)}(\Sigma\Pi_I(\Delta\theta_{pr,\tau,9}))+$$
$$\lambda_{\sigma_{n,max,9}} W_{f\sigma,9}\frac{1}{(\sigma'_f \varepsilon'_f)}(\Sigma\Pi_{II}(\Delta\theta_{pr,\sigma n,9})) = (\frac{\sigma'_f}{E}(N_f)^b + \varepsilon'_f(N_f)^c) + (\frac{\tau'_f}{G}(N_f)^b + \gamma'_f(N_f)^c \tag{7}$$

It should be noted that when $W_{f\sigma,i}$ and $\lambda_{\sigma_{n,max,i}}$ take unit values, the expression (7) yields Farahani's equation [17]. The qualitative effect of the implementation of $W_{f\sigma,i}$ and $\lambda_{\sigma_{n,max,i}}$ to the fatigue life assessment according to Eq. (7) is depicted in figure 5. It should be noted that the Wöhler curves shown in figure 5 are qualitative and indicative and are only used to illuminate on the effect of the proposed fatigue damage weighting factors to the approximate fatigue life calculations. Nonetheless an indicative fatigue limit (parallel to the N_f axis segment of the indicative curves) is being used, given that the material under study (StE460 steel alloy) exhibits fatigue limit [31].

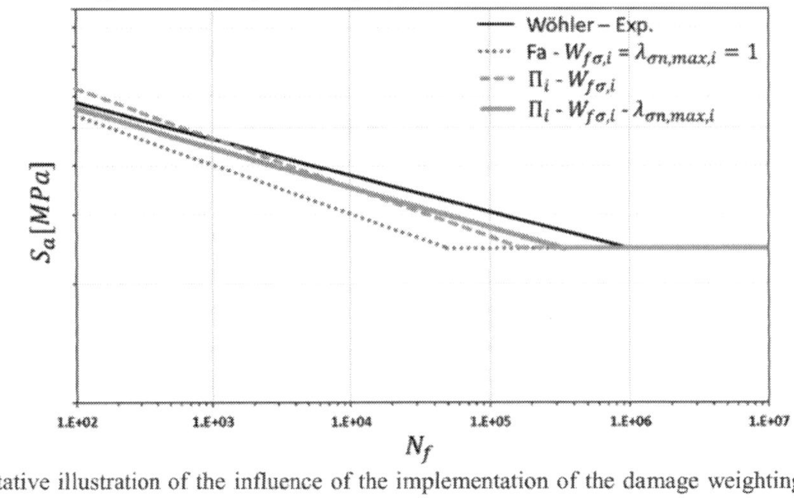

Figure 5. Qualitative illustration of the influence of the implementation of the damage weighting factors $W_{f\sigma,i}$ and $\lambda_{\sigma_{n,max,i}}$ on the Wöhler curve position in comparison to the Wöhler curve calculated using Farahani's concept.

IOP Conf. Series: Materials Science and Engineering **161** (2016) 012049 doi:10.1088/1757-899X/161/1/012049

Specifically, the introduction of solely the proposed weighting factor, $W_{f\sigma,i}$, (dashed green curve in figure 5) leads to the assessment of longer fatigue life, compared to that calculated if $W_{f\sigma,i}$ is not considered (red dotted line = Farahani's concept). Practically, the use of the weighting factor, $W_{f\sigma,i}$, leads to parallel shift of the Wöhler curve calculated using Farahani's model to higher fatigue lives. Consequently, the contribution of $W_{f\sigma,i}$ to the fatigue life calculation eliminates the reported disadvantage of Farahani's model to underestimate fatigue life in the regime of high cycle fatigue [19].

$\lambda_{\sigma_{n,max,i}}$ yields the rotation of the Wöhler curve counter-clockwise as qualitatively shown in figure 5. It acts as corrective factor that leads to more reliable fatigue life assessment in the of low cycle fatigue regime. Thereafter, the introduction of both fatigue damage weighting factors, $W_{f\sigma,i}$ and $\lambda_{\sigma_{n,max,i}}$, to the fatigue life calculation, as proposed herein, yields the thick green-coloured Wöhler curve shown in figure 5.

3. Evaluation of the proposed method and future perspectives

The here proposed energy-critical plane based fatigue damage approach for the life prediction of metal alloys has been evaluated through comparison with corresponding experimental results. In particular, four fatigue loading categories have been examined; uniaxial tension-compression, pure torsion, multiaxial proportional and non-proportional cyclic loading with combined tension – torsion components [24]. Table 2 gives a summarizing overview of the theoretical and experimental results and corresponding % logarithmic errors, $E_{er}(\%)$, where

$$E_{er}(\%) = \log\left(\frac{N_{f,predicted}}{N_{f,experimental}}\right)\ (\%) \tag{8}$$

for the loading cases investigated are included in Table 2.

Table 2. Loading data for all cyclic fatigue cases examined, experimental and calculated fatigue lives according to the proposed and Farahani's method, and respective % logarithmic errors.

Loading	S_a [MPa]	T_a [MPa]	[illegible]	$N_f - EXP$	$N_f - \Pi_i$	$N_f - Fa$	$E_{er}(\%) - \Pi_i$	$E_{er}(\%) - Fa$
Tension-Compression	338.8	-	-	50100	42556	22123	-7.08	-35.49
	385	-	-	7690	6921	4614	-4.57	-22.18
	450.2	-	-	1630	1888	912	6.38	-25.21
Pure Torsion	-	204	-	38250	31112	31112	-8.97	-8.97
	-	212.7	-	23000	19182	19182	-7.88	-7.88
	-	257.5	-	1820	2433	2433	12.60	12.60
Tension-Torsion Proportional	172.2	137.4	0	130300	103714	75801	-9.91	-23.52
	216.5	147.3	0	521000	410502	325587	-10.35	-20.41
	244.2	147.3	0	31100	24880	17416	-9.69	-25.18
Tension-Torsion non-Proportional	215	149.2	90	574600	295885	133585	-28.82	-63.36
	295	193.4	90	51900	32178	16089	-20.76	-50.86
	318.8	184.2	90	39670	29752	15074	-12.49	-42.02
	391.5	230.9	90	6570	5626	2811	-6.73	-36.87
	480.9	269.3	90	540	480	210	-5.06	-40.89

Figures 6 to 9 contain comparative illustration of the results. Therein, the calculation results obtained from the proposed model are illustrated with triangle shaped symbols. The respective fatigue lives derived from Farahani's model, i.e. Eq. (7) considering $W_{f\sigma,i} = \lambda_{\sigma_{n,max,i}} = 1$, are represented with circle spots. Square symbols are used for the corresponding experimental data. In all fatigue loading cases studied, the approximate fatigue life results attained from the proposed method (Π_i) led to fatigue life prediction of better accuracy compared to the respective accuracy derived from Eq. (7) if no fatigue damage weighting factors are being considered (Farahani concept). The characteristic coincidence between results derived from the proposed method and the method based on the concept of Farahani is also distinguishable in the case of pure torsion (figure 7), due to the theoretically zero value of the perpendicular to the critical plane normal stress component, σ_n.

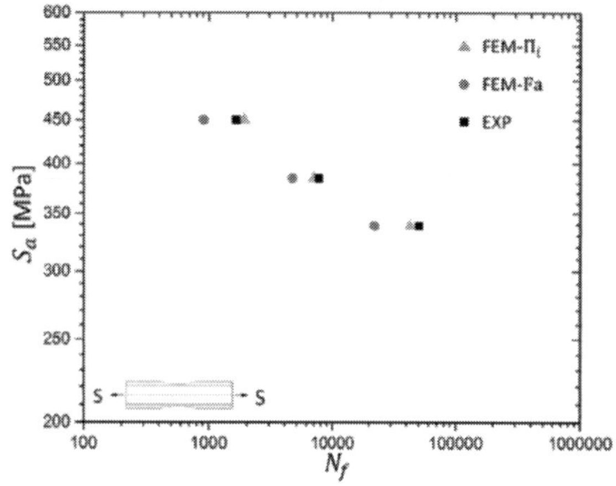

Figure 6. Experimental (EXP) and corresponding calculated fatigue lives according to the proposed method (FEM-Π_i) and Farahani's model (FEM-Fa) for constant amplitude tension-compression cyclic fatigue loading.

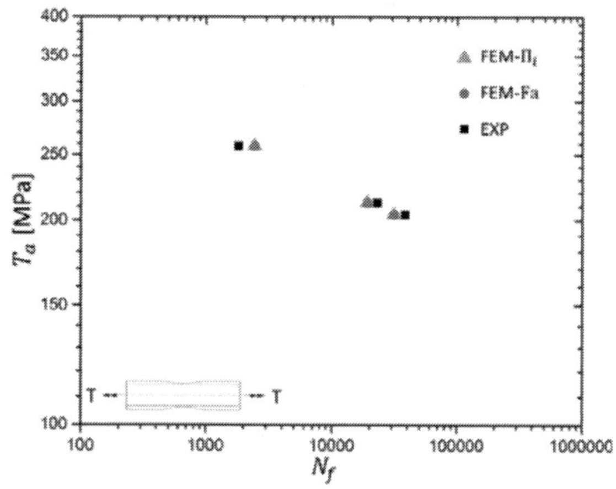

Figure 7. Experimental (EXP) and corresponding calculated fatigue lives according to the proposed method (FEM-Π_i) and Farahani's model (FEM-Fa) for pure torsion cyclic loading.

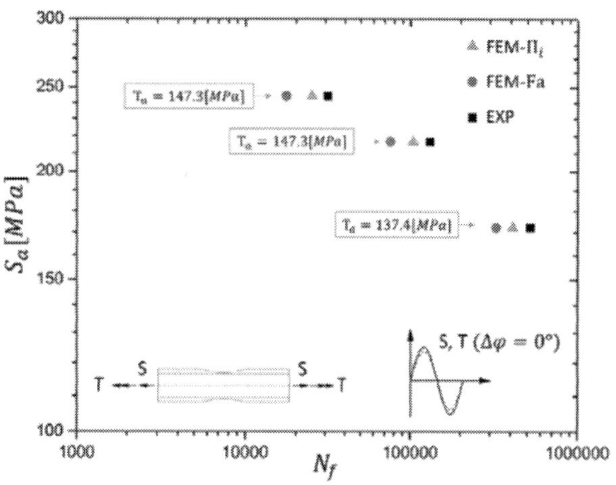

Figure 8. Experimental (EXP) and corresponding predicted fatigue lives according to the proposed method (FEM-Π_i) and Farahani's model (FEM-Fa) for proportional tension-torsion cyclic loading.

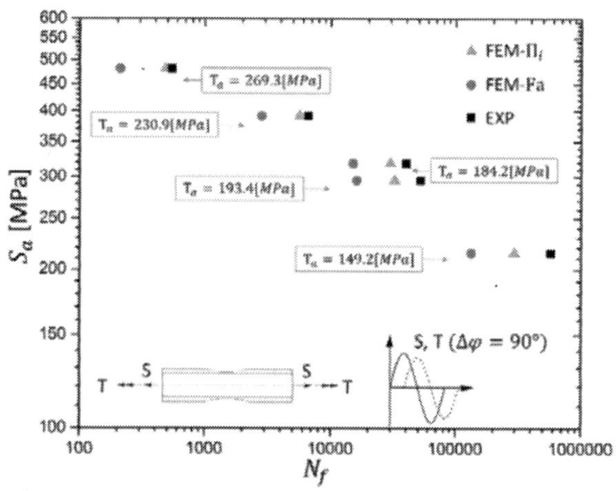

Figure 9. Experimental (EXP) and corresponding predicted fatigue lives according to the proposed method (FEM-Π_i) and Farahani's model (FEM-Fa) for non-proportional tension-torsion cyclic loading.

The first significant advantage of the proposed method lies in the fact that it introduces a weighted contribution to fatigue damage of the normal stress and strain that act perpendicular to the critical plane. This is achieved by mapping the damage parameters Π_I and Π_{II} through a full fatigue cycle not only as far as their values is concerned, but also taking into account their corresponding values as a function of the critical plane orientation angle, θ_{pr}. The combined implementation of the proposed damage weighting factors $W_{f\sigma,i}$ and $\lambda_{\sigma_{n,max,i}}$ aims at the mathematical description of respective physical phenomena related to the loss of structural integrity due to fatigue. In particular:

- $W_{f\sigma,i}$ covers the influence of the critical plane orientation shift (figure 3 and 4) to fatigue life. Herein, this influence is not regarded only as a function of the accumulated damage induced by the effect of σ_n and ε_n to the crack propagation on several critical plane orientations during a loading cycle, but also as a function of the divergence between each critical plane orientation and the critical plane where the crack due to fatigue is more likely to be induced.
- $\lambda_{\sigma_{n,max,i}}$ describes the material's sensitivity to micro crack propagation due to fatigue under the influence of σ_n and ε_n, related to the strain limit where plastic deformation becomes prominent under cyclic loading ($\sigma_{\varepsilon=0.2\%}$). According to the proposed concept the material experiences structural integrity loss under the influence of σ_n and ε_n that is not just proportional to the value of the relative damage parameter $\Pi_{II} = [\sigma_n \cdot \varepsilon_n](\theta_{pr})$ but is also proportionally affected by the ratio of σ_n over the stress value that corresponds to the barrier over which significant plastic strain occurs under cyclic loading ($\sigma_{\varepsilon=0.2\%}$). Therefore, the incorporation of fatigue damage weighting factor $\lambda_{\sigma_{n,max,i}}$ acts as a sensitivity parameter that induces more damage per fatigue cycle in the regime of low cycle fatigue and as damage alleviator in the regime of high cycle fatigue, leading to fatigue life approximations of better accuracy.

Another worth-mentioning advantage of the proposed method is that its implementation requires the same minimum -compared to corresponding multiaxial fatigue damage models [19]- number of material parameters as in the case of Farahani's model, while, in parallel, the calculation of the proposed weighting factors' values does not require additional material parameters.

Nonetheless it should be pinpointed that further evaluation of the proposed model's reliability on other materials and additional loading cases is also required. Accordingly, and taking into account that StE460 belongs to the category of cyclic softening materials, the investigation of the approximation accuracy of the proposed method in the case of cyclic hardening materials, for instance Al5083, is a scope of great significance. Presently, the authors work on the extension of the model's applicability to notched components, where stress concentration, stress gradients and surface treatment are also to be taken into account.

4. Conclusions

A reliable, easily applicable energy-critical plane based fatigue damage approach for the life prediction of metal alloys has been developed based on the model of Farahani [17, 18]. In all fatigue loading cases examined, it was proven that the proposed method yields approximate fatigue lives of remarkable accuracy. The advantage of the proposed method lies on the fact that, all the material parameters needed for its implementation are available in the bibliography [31] and the proposed damage weighting factors can be analytically approximated for other materials according to the procedure followed here, in the case of StE460. In the framework of the present work the finite element method has been applied to approximate the stress-strain elastic-plastic state. Significant incentives for relative future work are the perspective to upgrade the proposed model to become applicable on notched components as well as its combination with analytical methods for the approximate calculation of the stress-strain state under multiaxial fatigue, as an alternative to the finite element method.

References

[1] M. Hoffmann and T. Seeger, A Generalized Method for Estimating Multiaxial Elastic–Plastic Notch Stresses and Strains, *J. Eng Mater Tech.* **107** (1985) 250-254.
[2] R. Yuuki, K. Akita and N. Kishi, The effect of biaxial stress state and changes of state on fatigue crack growth behavior, *International Journal of Fatigue.* **12** (1989) 93-103.
[3] C. C. Chu, Multiaxial fatigue life prediction method in the ground vehicle industry, *International Journal of Fatigue.* **19** (1997) 325-330.
[4] B. Li, L. Reis and M. De Freitas, Simulation of cyclic stress/strain evolutions for multiaxial fatigue life prediction, *International Journal of Fatigue.* **28** (2006) 451-458.
[5] A. K. Marmi, A. M. Habraken and L. Duchene, Multiaxial fatigue damage modelling at macro scale of Ti–6Al–4V alloy, *International Journal of Fatigue.* **31** (2009) 2031-2040.
[6] C. A. Gonçalves, J. A. Araújo and E. N. Mamiya, Multiaxial fatigue: A stress based criterion for hard metals, *International Journal of Fatigue.* **27** (2005) 177-187.
[7] D. F. Socie, Critical plane approaches for multiaxial fatigue damage assessment, *Advances in Multiaxial Fatigue, STP 1191, Philadelphia, ASTM.* (1993) 7-36.
[8] C. H. Wang and M. W. Brown, Life prediction techniques for variable amplitude multiaxial fatigue—Part 1: Theories, *J. Eng. Mater. Technol.* **118** (1996) 367-370.
[9] T. Itoh, T. Nakata, M. Sakane and M. Ohnami, Nonproportional low cycle fatigue of 6061 aluminum alloy under 14 strain paths, *European Structural Integrity Society.* **25** (1999) 41-54.
[10] R. Döring, J. Hoffmeyer, T. Seeger and M. Vormwald, A plasticity model for calculating stress–strain sequences under multiaxial nonproportional cyclic loading, *Computational Materials Science.* **28** (2003) 587-596.
[11] I. V. Papadopoulos, et al., A comparative study of multiaxial high-cycle fatigue criteria for metals, *International Journal of Fatigue.* **19** (1997) 219-235.
[12] T. Palin-Luc and S. Lasserre, High cycle multiaxial fatigue energy criterion taking into account the volume distribution of stresses, Elsevier, *European Structural Integrity Society.* (1999) 115-129.
[13] A. Fatemi and D. F. Socie, A critical plane approach to multiaxial fatigue damage including out-of phase loading, *Fatigue & Fracture of Engineering Materials & Structures.* **11** (1998) 149-165.
[14] S. E. Onome, et al., A new distortion energy-based equivalent stress for multiaxial fatigue life prediction, *International Journal of Non-Linear Mechanics.* **47** (2012) 29-39.
[15] D. L. McDiarmid, A shear stress based critical-plane criterion of multiaxial fatigue failure for design and life prediction, *Fatigue Fract. Eng. Mater. Struct.* **17** (1995) 1495-1484.
[16] D. F. Socie and G.B. Marquis, Multiaxial Fatigue. Ed. SAE International, USA. (2000).
[17] A. Varvani-Farahani, A new energy-critical plane parameter for fatigue life assessment of various metallic materials subjected to in-phase and out-of-phase multiaxial fatigue loading conditions, *International Journal of Fatigue.* **22** (2000) 295-305.
[18] A. Varvani-Farahani, Silicon MEMS components: a fatigue life assessment approach, *Microsystem Technologies.* **11** (2005) 129-134.
[19] Y. Y. Wang and W.X. Yao, Evaluation and comparison of several multiaxial fatigue criteria, *International Journal of Fatigue.* **26** (2004) 17-25.
[20] T. Itoh, M. Kameoka and Y. Obataya, A new model for describing a stable cyclic stress-strain relationship under non-proportional loading based on activation state of slip systems, *Fatigue Fract. Eng. Meter. Struct.* **27** (2004) 957-967.
[21] S. B. Lee, A criterion for fully reversed out-of-phase torsion and bending, ASTM STP 853, Philadelphia, 1985.
[22] G. I. Taylor, Plastic strain in metals, *J. Inst. Metals.* **62** (1938) 307-324.
[23] N. Jayaraman and M. M. Ditmars, Torsional and biaxial (tension-torsion) fatigue damage mechanisms in Waspaloy at room temperatures, *International Journal of Fatigue.* **11** (1989) 309-318.
[24] Y. Jiang, O. Hertel and M. Vormwald, An experimental evaluation of three critical plane

multiaxial fatigue criteria, *International Journal of Fatigue*, **29** (2007) 1490-1502.

[25] Y. Xue, D. L. McDowell, M. F. Horstemeyer, M. H. Dale and J. B. Jordon, Microstructure-based multistage fatigue modeling of aluminum alloy 7075-T651, *Engineering Fracture Mechanics*. **74** (2007) 2810-2823.

[26] R. von Mises, Mechanik der festen körper im plastisch deformablen zustand, *Math. Phys.* 1 (1913) 582-592.

[27] J. L. Chaboche, Constitutive equations for cyclic plasticity and cyclic viscoplasticity, *International Journal of Plasticity*. **5** (1989) 247-302.

[28] ANSYS Theory Reference, ANSYS Documentation, Axisymmetric Shells (11.4.1.). ANSYS Inc. (2015).

[29] J. Hoffmeyer, R. Döring, T. Seeger and M. Vormwald, Deformation behavior, short crack growth and fatigue lives under multiaxial non-proportional loading, *International Journal of Fatigue*. **28** (2006) 508-520.

[30] C. M. Sonsino, Influence of load and deformation-controlled multiaxial tests on fatigue life to crack initiation, *International journal of Fatigue*. **23** (2001) 159-167.

[31] FKM (Forschungskuratorium Maschinenbau), Analytical Strength Assessment of Components, VDMA Publ., 6th revised edition (2012).

IOP Conf. Series: Materials Science and Engineering **161** (2016) 012050 doi:10.1088/1757-899X/161/1/012050

Experimental Research on the Elastic Deformation Mode of S235JR Rolled Steel Fastened between the Centers of a Universal Lathe

LLTabacaru[1], E Axinte[2] and G Musca[3]

[1,2,3] "Gheorghe Asachi" Technical University of Iasi, Faculty of Machine Manufacturing and Industrial Management, Iasi, Romania

E-mail: ltabacaru@tcm.tuiasi.ro

Abstract. Elastic deformations of the technological system occur during the mechanical treatment of a blank, regardless of the manner in which it is fastened. The elastic deformation of the blank is significant especially when machining shaft-like parts. The purpose of our research is to compare the mathematical model of blank deformation to the experimental model when the blank, which is a part belonging to the shaft class, is fastened between centers.

1. Introduction

The technological system may suffer elastic deformations under the action of exterior forces. These deformations occur on the direction of action of the forces. System rigidity is defined as its capacity to resist deformation. The forces that act in the technological system are the cutting forces. The Py force is usually used to determine system rigidity. The proper approach would be to calculate the resultant of the three forces first and then the rigidity [1].

The rigidity of a system is calculated as the component/deformation ratio

$$R = \frac{P_y}{y}\,[\text{daN/mm}]$$

The force/deformation ratio was determined experimentally, by entering different values of the Py cutting force into the technological system and then measuring the deformation triggered by them. Figure 2 shows the deformation law applying to regular machine tools where the y_0 represents plastic deformation is actually the weight shift in view of the processing of the clearances occurring between the assembly components.

The surface between the curves is the
mechanical work meant to resist the friction forces between components on clearance processing, on unevenness processing.

Further to repeated loading on the same component, the surface between the two curves decreases and the mechanical work diminishes. This surface shape is called rigidity hysteresis.

$$R_{med} = \frac{AB}{OB} = tg\alpha \text{ - mean rigidity}$$

Further to repeated loading on the same component, the surface between the two curves decreases and the mechanical work diminishes. This surface shape is called rigidity hysteresis:

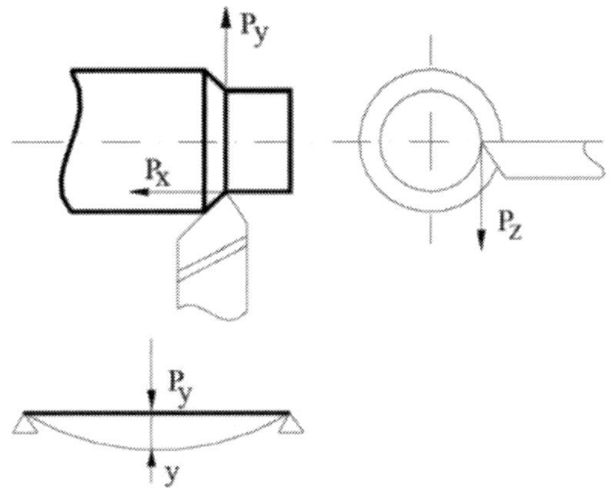

Figure 1. Elastic deformation caused by the $P_y$force

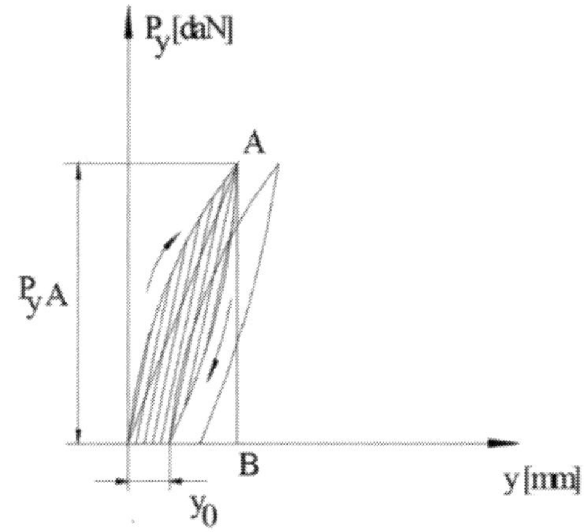

Figure 2. Graphical representation of load-deformation.

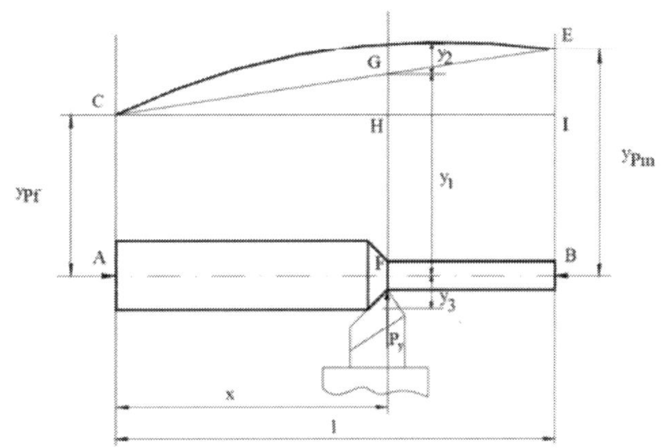

Figure 3. Pattern used to determine errors caused by elastic deformation when the blank is fastened between centers.

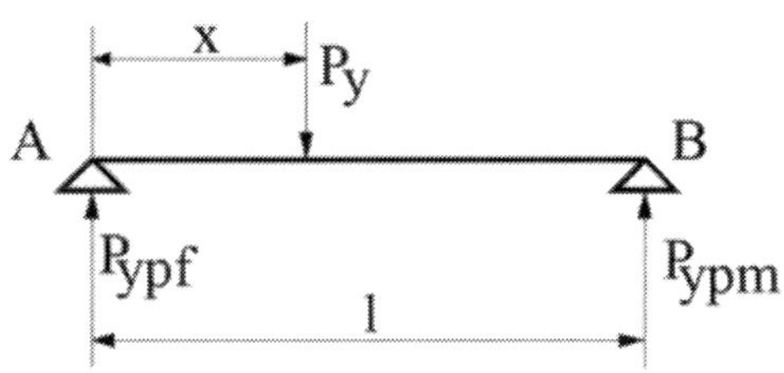

Figure 4. Py force distribution on supports

2. Mathematical model

The work pattern shown in figure 3 [1] is used to determine the processing error caused by the elastic deformation of the technological system when the blank is placed between centers (figure 13).

The following technological system deformations occur under the influence of the Py force:

- headstock shift;
- tailstock shift;
- part shift in section x caused by headstock and tailstock shift
- part arrow in section x;
- tool and tool support shift under the action of P_y;

$$t_{pr} - t_r = t_{rem} = y_1 + y_2 + y_3 \ ; \ y_1 = FH + GH = y_{Pf} + \frac{x}{l} \cdot \left(y_{pm} - y_{pf} \right)$$

$$\Delta CGH \sim \Delta CEI$$

$$\frac{GH}{CH} = \frac{EI}{CI} \ GH = \frac{x}{l} \cdot \left(y_{pm} - y_{pf} \right)$$

The problem arising here is to determine the extent to which the Py force acts on the tailstock and the extent to which the same force acts on the headstock (figure 4).

$$P_{y_{pf}} = P_y \cdot \frac{l-x}{l}$$

$$P_{y_{pm}} = P_y \cdot \frac{x}{l}$$

$$y_{pt} = \frac{P_{y_{pf}}}{R_{pf}} = \frac{P_y}{R_{pf}} \cdot \frac{l-x}{l};$$

$$y_{pm} = \frac{P_{y_{pm}}}{R_{pm}} = \frac{P_y}{R_{pm}} \cdot \frac{x}{l};$$

$$y_1 = \frac{P_y}{R_{pf}} \cdot \frac{l-x}{l} + \frac{x}{l}\left(\frac{P_y}{R_{pm}} \cdot \frac{x}{l} - \frac{P_y}{R_{pf}} \cdot \frac{l-x}{l}\right);$$

$$y_2 = \underbrace{\frac{P_y}{R_{pf}}\left(\frac{l-x}{l} - \frac{x}{l} \cdot \frac{l-x}{l}\right)} + \frac{P_y}{R_{pm}} \cdot \frac{x^2}{l^2};$$

$$y_1 = \frac{l}{R_{pf}}\left(\frac{l-x}{l}\right)^2 + \frac{P_y}{R_{pm}} \cdot \left(\frac{x}{l}\right)^2$$

$$y_2 = \frac{P_y}{3EI} \frac{x^2(l-x)^2}{l} \text{ - arrow ratio}$$

$$y_3 = \frac{P_y}{R_{sc}}$$

$$t_{pr} = t_r + \frac{P_y}{R_{pf}}\left(\frac{l-x}{l}\right)^2 + \frac{P_y}{R_{pm}}\left(\frac{x}{l}\right)^2 + \frac{P_y}{3EI} \cdot \frac{x^2(l-x)^2}{l} + \frac{P_y}{R_{sc}} \text{ if } P_y = c \cdot k \cdot t_r;\ t_r = \frac{t_{pr}}{1+A}$$

$$t_{rem} = t_{pr} - t_r = t_{pr}\left[1 - \frac{1}{1+A}\right]$$

When a batch of parts is processed, the prescribed depth ranges from to with a variation which causes an error on the y direction.

$$\Delta_y = t_{rem_{max}} - t_{rem_{min}} = \left[t_{pr_{ax}} - t_{pr_{min}}\right] \cdot \left[1 - \frac{1}{1+A}\right]$$

on the diameter:

$$\Delta D = 2\Delta_y$$

Assuming that the machine tool is very rigid, which means that tailstock, headstock and tool shifts are absent, y1 and y3 are null and hence the processing error is the result of the elastic deformation of the part. The shape of the part is shown in figure 5.

When the part is very rigid, i.e. it does not undergo any elastic deformation, and assuming that the tool and the support are very rigid, the shape of the part is influenced solely by y_1.

The processing error will only be determined by y_1(figure 5, 6).

$$t_{rem} = y_1 = \frac{P_y}{R_{pf}}\left(\frac{l-x}{l}\right)^2 + \frac{P_y}{R_{pm}}\left(\frac{x}{l}\right)^2$$

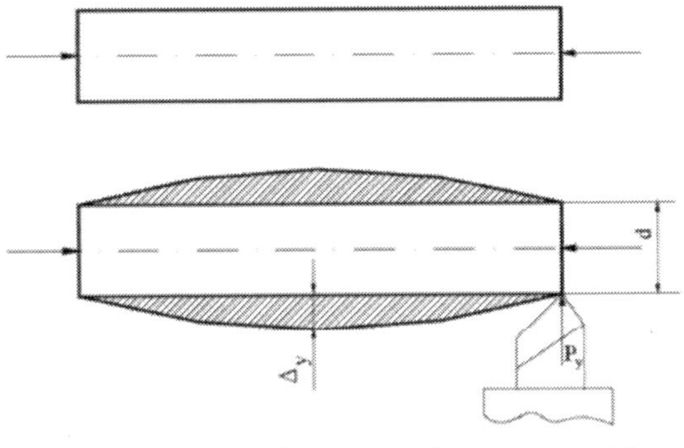

Figure 5. Shape of the part when the machine tool is very rigid

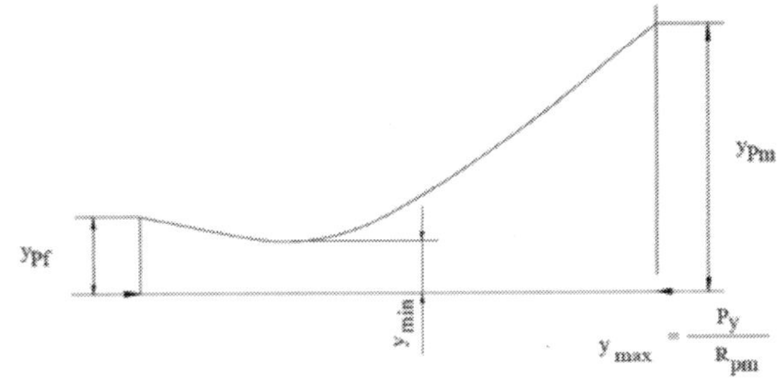

Figure 6. Processing error caused solely by y_1.

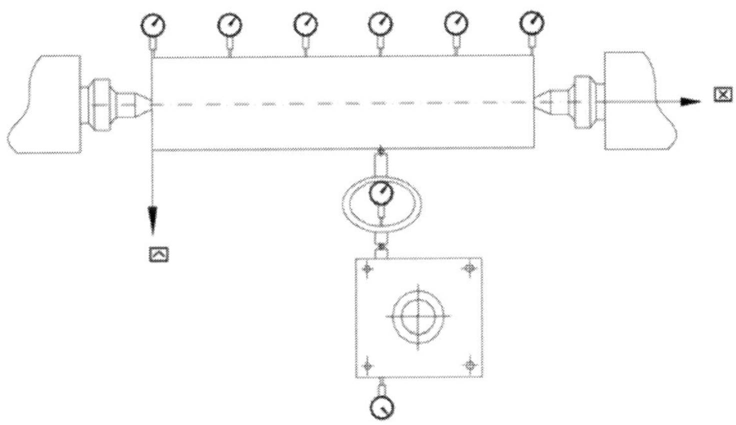

Figure 7. Experimental pattern used to assess the static elastic deformation of various components subject to the $F_y$force

A shape error also occurs, which is a quadric.

$$t_{rem} = y_{\max} - y_{\min}$$

- In order to determiney_{min}we will vanish $t'_{rem} = 0$;

$$t'_{rem} = \frac{P_y}{R_{pf}} \cdot \frac{2x}{l^2} - \frac{P_y}{R_{pf}} \cdot \frac{2l}{l^2} + \frac{P_y}{R_{pm}} \cdot \frac{2x}{l^2} = 0$$

$$\frac{x}{l}\left[\frac{1}{R_{pf} \cdot l} + \frac{1}{R_{pm} \cdot l}\right] = \frac{1}{R_{pf} \cdot l} \Rightarrow \frac{\mathrm{x}}{1} = \frac{R_{pm}}{R_{pf} + R_{pm}}$$

By replacing in t_{rem}we determine y_{min}:

$$y_{\min} = \frac{P_y}{R_{pf}}\left[1 - \frac{R_{pm}}{R_{pf} + R_{pm}}\right]^2 + \frac{P_y}{R_{pm}}\left(\frac{R_{pm}}{R_{pf} + R_{pm}}\right)^2 \cdot \left[\frac{R_{pf} + R_{pm} - R_{pm}}{R_{pm} + R_{pm}}\right]^2$$

$$y_{\min} = P_y \cdot \frac{R_{pf}}{\left(R_{pf} + R_{pm}\right)^2} + P_y \cdot \frac{R_{pm}}{\left(R_{pf} + R_{pm}\right)^2} = \frac{P_y}{R_{pf} + R_{pm}}$$

3. Experimental model

In this paper we aim at determining the processing errors caused by the elastic deformations of the technological system when placing parts between centers.
For this experiment we used a 1000 mm long and 40 mm wide S 235 JR metal bar, which observed the SR EN 10025 requirements, and a SN 250 machine tool.

The method consists of the static simulation of the forces generated during the turning process; the elastic deformations of the various components are determined after the force applied on the work piece has been increased, which should provide data on the rigidity of the turning system figure 7.

Indeed, from the viewpoint of the various methods of work piece fastening, it may be interesting to assess the system's resistance to deformation caused by the cutting forces. Several simplifying hypotheses were considered for the case described in figure 7, namely that the Fy force.

As shown in figure 7, a dynamometer is placed between the tool holder and the work piece in order to read at any time the force entered into the system, by the rotation of one of the dynamometer nuts. We used dial gauges placed every 170 mm and at the ends of the blank, in order to read the elastic deformation of the work piece subject to different Fy load forces, which materialize various cutting forces. The dynamometer touched the work piece and a parallelepiped piece replacing the turning tool, by means of two metal balls, with the aim of ensuring the best possible transmission of the radial force.

Table 1. Elastic deformation read in various points on the part for various Fy force values applied at x = 170 mm

Exp. no.	Force Fy[daN] (increments) applied to x=170 mm	Elastic deformation [mm]											
		Rotating center - headstock		x=170		x=340		x=510		x=680		Rotating center - tailstock	
		C	D	C	D	C	D	C	D	C	D	C	D
1	0 (0)	0.0	0.0	0.0	0.004	0.0	0.01	0	0.007	0	0.01	0.0	0.006
2	18.18 (20)	0.09	0.01	0.05	0.11	0.09	0.21	0.09	0.165	0.065	0.07	0.14	0.18
3	36.36 (40)	0.36	0.31	0.12	0.18	0.21	0.31	0.18	0.27	0.14	0.05	0.29	0.375
4	54.54 (60)	0.53	0.49	0.255	0.24	0.30	0.40	0.275	0.36	0.21	0.12	0.44	0.52
5	73.05 (80)	0.62	0.65	0.33	0.31	0.405	0.50	0.38	0.46	0.285	0.19	0.60	0.72
6	91.57 (100)	0.80	0.79	0.41	0.38	0.50	0.58	0.48	0.54	0.36	0.26	0.67	0.80
7	109.19 (120)	0.98	0.93	0.50	0.45	0.60	0.68	0.58	0.64	0.39	0.34	0.83	0.89
8	127.27 (140)	1.10	1.11	0.57	0.58	0.75	0.78	0.71	0.73	0.41	0.42	1.03	1.05
9	145.45 (160)	1.18	1.18	0.60	0.60	0.86	0.86	0.80	0.80	0.415	0.415	1.20	1.20

$$t_{rem} = y_{\max} - y_{\min} = \frac{P_y}{R_{pm}} - \frac{P_y}{R_{pf} + R_{pm}}$$

In order to determine the highest value of the Fy experimental force that may be used during the experimental research, we considered hard turning using a regular cutting tool. The work piece was

Table 2. Elastic deformation read in various points on the part for various Fy force values applied at x = 340 mm.

Exp. no.	Force Fy[daN] applied at x=340 mm	Elastic deformation [mm]											
		Rotating center - headstock		x=170		x=340		x=510		x=680		Rotating center - tailstock	
		C	D	C	D	C	D	C	D	C	D	C	D
1	0	0.0	0.009	0.0	0.007	0.0	0.004	0	0.0035	0.0	0.002	0.0	0.0
2	18.18	0.12	0.29	0.09	0.13	0.14	0.21	0.11	0.15	0.06	0.09	0.0	0.0
3	36.36	0.30	0.48	0.19	0.23	0.31	0.37	0.23	0.28	0.12	0.15	0.09	0.15
4	54,54	0.51	0.67	0.28	0.32	0.48	0.54	0.35	0.39	0.185	0.215	0.24	0.40
5	73.05	0.72	0.85	0.42	0.43	0.65	0.71	0.48	0.52	0.25	0.28	0.52	0.70
6	91.57	0.91	1.00	0.52	0.525	0.82	0.87	0.61	0.64	0.315	0.345	0.84	0.98
7	109.19	1.10	1.17	0.62	0.63	0.97	1.02	0.74	0.77	0.38	0.40	1.19	1.25
8	127.27	1.30	1.32	0.69	0.70	1.16	1.18	0.89	0.89	0.44	0.475	1.42	1.43
9	145.45	1.50	1.50	0.79	0.79	1.33	1.33	1.01	1.01	0.55	0.55	1.50	1.50

Table 3. Elastic deformation read in various points on the part for various Fy forcevalues applied at x = 510 mm.

Exp. no.	Force Fy[daN] applied at x=510 mm	Elastic deformation [mm]											
		Rotating center - headstock		x=170		x=340		x=510		x=680		Rotating center - tailstock	
		C	D	C	D	C	D	C	D	C	D	C	D
1	0	0.0	0.005	0.0	0.004	0.0	0.004	0.0	0.002	0.0	0.001	0.0	0.22
2	18.18	0.10	0.22	0.11	0.14	0.15	0.19	0.135	0.17	0.08	0.11	0.22	0.66
3	36.36	0.27	0.41	0.21	0.24	0.33	0.43	0.29	0.34	0.18	0.21	0.55	1.00
4	54.54	0.45	0.59	0.29	0.32	0.49	0.54	0.44	0.48	0.26	0.29	0.91	1.20
5	73.05	0.64	0.73	0.39	0.42	0.65	0.69	0.60	0.65	0.35	0.38	1.20	1.30
6	91.57	0.82	0.90	0.48	0.59	0.82	0.86	0.75	0.80	0.425	0.46	1.37	1.75

7	109.19	0.97	1.25	0.58	0.595	0.98	1.03	0.91	0.96	0.51	0.54	1.61	1.99
8	127.27	1.16	1.25	0.68	0.71	1.16	1.19	1.07	1.11	0.60	0.62	2.01	2.28
9	145.45	1.31	1.31	0.80	0.80	1.33	1.33	1.25	1.25	0.73	0.73	2.58	2.58

madeof carbon steel and used the following mode of operation: cutting depth, t= 4mm and s=0.4mm/rot. In this case, the Fy force may be calculated using the following relation [1,2,3]:

Table 4. Elastic deformation read in various points on the part for various Fy force values applied at x = 680 mm.

Exp. no.	Force Fy[daN] applied at x=680 mm	Elastic deformation [mm]											
		Rotating center - headstock		x=170		x=340		x=510		x=680		Rotating center - tailstock	
		C	D	C	D	C	D	C	D	C	D	C	D
1	0	0	0.001	0	-0.001	0	0.001	0	-0.002	0	0.002	0	0.003
2	18.18	0.008	0.12	0.12	0.004	0.11	0.008	0.13	0.18	0.07	0.08	0.24	0.30
3	36.36	0.22	0.30	0.14	0.15	0.25	0.28	0.22	0.25	0.13	0.15	0.59	0.75
4	54.54	0.32	0.43	0.21	0.35	0.37	0.50	0.41	0.42	0.24	0.28	0.94	1.02
5	73.05	0.47	0.55	0.28	0.48	0.51	0.64	0.55	0.59	0.33	0.39	1.33	1.45
6	91.57	0.63	0.70	0.34	0.37	0.63	0.59	0.73	0.76	0.41	0.49	1.66	1.73
7	109.19	0.71	0.82	0.41	0.42	0.75	0.76	0.89	0.93	0.50	0.59	2.04	2.10
8	127.27	0.83	0.90	0.49	0.53	0.89	0.90	1.05	1.10	0.61	0.72	2.61	2.69
9	145.45	0.95	0.95	0.55	0.55	0.94	0.94	1.22	1.22	1.24	1.24	2.98	2.98

$$F_y=C_{Fy}t^x s^y HB^n$$

whereC_{Fy} is a coefficient the value of which depends on the cutting tool type, on the material the cutting tool and the work piece are made of C_{Fy} =0.027 and HB is work piece material hardness HB=150- S 235 JR. The exponents are x=0.9, y=0.75, n=2.0. When the cutting depth is t=4 mm and the cutting feed is s= 0.4 mm/rot,Fy =0.027 40.9 0.40.75 1502= 105daN for S 235 JR.

On the other hand, the forces generated in the work piece during the experimental research should not exceed the maximum elastic deformation values applying to the work piece material.

A Fy force of variable value, ranging from 0 to 145.45 daN, was entered into the system by means of the dynamometer, and then the F_y force was decreased to 0 daN; the deformation values shown in tables 1-5 were read by the dial gauges for each F_y force value [4, 5].

Table 5. Elastic deformation read in various points on the part for various Fy force values applied at x = 850 mm.

Exp. no.	Force Fy[daN] applied at x=850 mm	Elastic deformation [mm]											
		Rotating center - headstock		x=170		x=340		x=510		x=680		Rotating center - tailstock	
		C	D	C	D	C	D	C	D	C	D	C	D
1	0	0	0.005	0	0.004	0	0.01	0	0.003	0	0.001	0	0.02
2	18.18	0.12	0.20	0.005	0.19	0.009	0.18	0.15	0.28	0.12	0.30	0.035	0.11
3	36.36	0.19	0.34	0.29	0.30	0.15	0.24	0.24	0.37	0.21	0.39	0.075	0.17
4	54.54	0.24	0.40	0.60	0.62	0.21	0.30	0.32	0.445	0.30	0.47	0.11	0.22
5	73.05	0.31	0.48	0.82	0.84	0.28	0.38	0.40	0.52	0.39	0.56	0.16	0.30
6	91.57	0.36	0.54	0.83	0.85	0.35	0.44	0.485	0.60	0.48	0.65	0.205	0.39
7	109.19	0.42	0.59	0.84	0.86	0.42	0.51	0.56	0.68	0.54	0.72	0.25	0.44
8	127.27	0.51	0.60	0.85	0.88	0.51	0.55	0.62	0.76	0.78	0.79	0.37	0.47
9	145.45	0.64	0.64	0.90	0.90	0.63	0.63	0.87	0.87	0.87	0.87	0.54	0.54

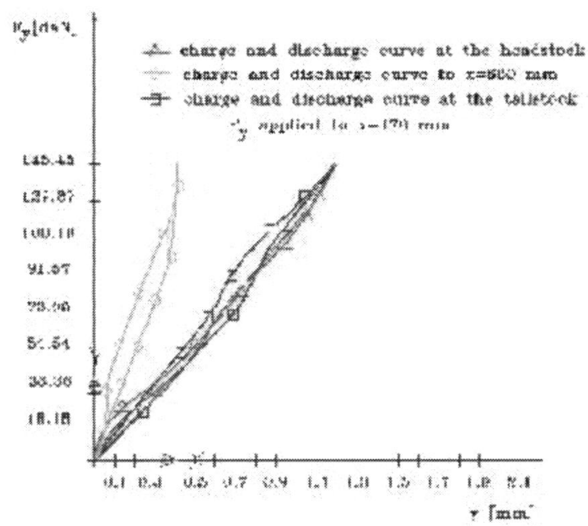

Figure 8. Graphical representation of the elastic deformation variation depending on different loading and unloading values of the technological system at the x = 170 mm distance

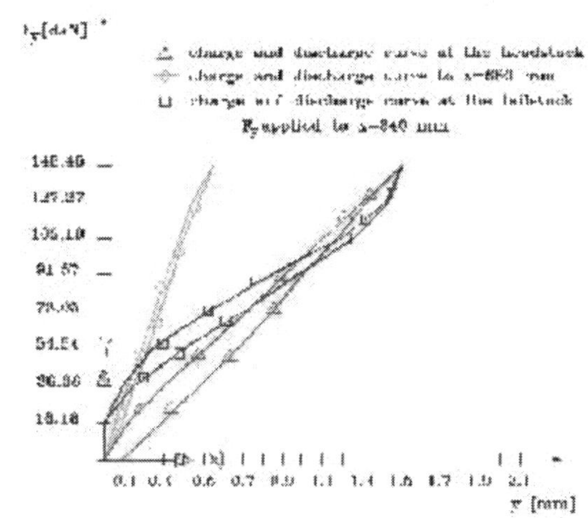

Figure 9. Graphical representation of the elastic deformation variation depending on different loading and unloading values of the technological system at the distance 340 mm

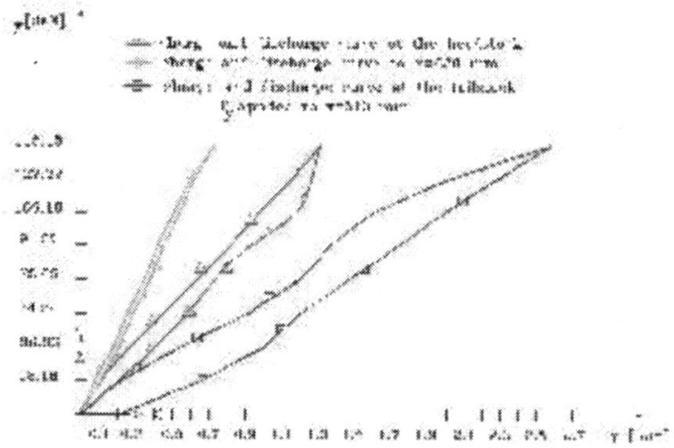

Figure 10. Graphical representation of the elastic deformation variation depending on different loading and unloading values of the technological system at the x = 510 mm distance

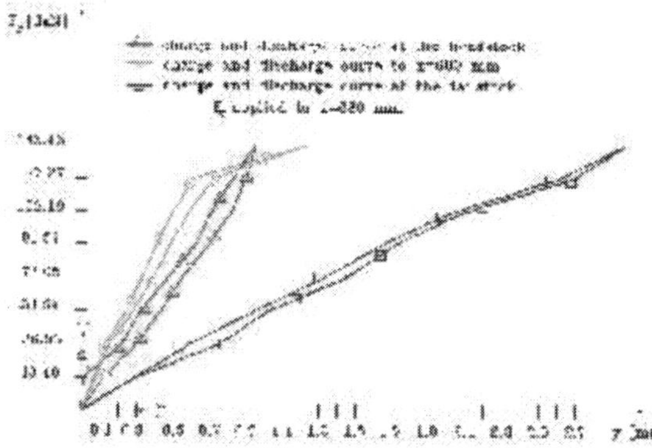

Figure 11. Graphical representation of the elastic deformation variation depending on different loading and unloading values of the technological system at the x = 680 mm distance

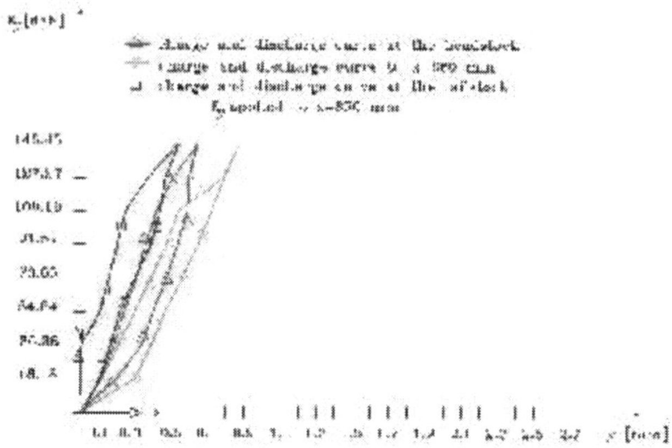

Figure 12. Graphical representation of the elastic deformation variation depending on different loading and unloading values of the technological system at the x = 850 mm distance

Figure 13. Picture taken during the elastic work piece deformation measuring process

4. Conclusion

Further to the analysis of the tables containing the experimental values and of the graphical representations in figure 8-12, we may conclude that:

1. a cutting force of Fy = 145.45 daN applied at the x=680 mm distance from the left end of the work piece causes a maximum deformation value in the tailstock area of y = 2.98 mm. When the same cutting force value is applied in the same work piece area at a x=510 mm distance, the following work piece deformation value is read: y= 2.58 mm;
2. in figure 11, the unloading curve read in the tailstock area for the Fy force applied at the x=510 mm distance leads of a maximum plastic deformation value of y=0.3 mm; this value proves that the technological system components shifted to resist the friction forces occurring in the system and to absorb the system clearances;
3. our research confirmed that, among the components in direct contact with the work piece, the universal support had the highest static rigidity;
4. the elastic deviation of the work piece is one of the factors that may impede upon the accuracy of the processing during the turning process. The extent of this deviation significantly depends on the static rigidity of the components of the processing systems;
5. our research fully confirms the mathematical model, when the blank is fastened between centers.

Appendix 1. Dynamometer calibration

Pull			Compression		
F	Gauge value	E	F	Gauge value	E
daN	ind.	ind./daN	daN	ind.	ind./daN
50	55.0	1.100	50	54.9	1.100
60	66.0	1.100	60	66.0	1.105
80	88.0	1.095	80	88.1	1.095
100	109.9	1.092	100	110.0	1.116
125	137.2	1.104	125	137.9	1.092
150	164.8	1.100	150	165.2	1.106
200	219.8	1.104	200	220.5	1.106
250	275.0	1.098	250	275.8	1.104
300	329.9	1.102	300	331.0	1.106
350	385.0	1.100	350	386.3	1.114
400	440.0	1.106	400	442.0	1.118
450	495.3	1.112	450	497.9	1.114
500	550.9		500	553.6	

References

[1] Benardos P G Mosialos S Vosniakos G C 2006 *Prediction of workpiece elastic deflection under cutting forces in turningRobotics and Computer-Integrated Manufacturing***22**ISSN 0736-5845pp 55-514

[2] Picos C Pruteanu O Bohosevici C and others 1992 *Proiectarea tehnologiilor de prelucrare mecanică prin strunjire*Manual Designvol 1 University Publishing ISBN 5-362-00970-2 (Chişinău)

[3] Tabacaru L and Pruteanu OV 2007*Conceptia si managementul tehnologiilor de fabricatie* Junimea Publishing ISBN 978-973-37-1210-7 (Iasi)

[4] Tabacaru L and Pruteanu O V 2010*Managementul tehnologiilor de fabricatie* Politehnium Publishing ISBN 978-973-621-294-9 (Iasi)

[5] Topal E S and Cogun C 2005 A cutting force induced error elimination method for turning operations Official Journal of Technological Processes of the materials vol **170** no 1-2 ISSN 0924-0136pp 190-203

TiO_2 nanostructured surfaces for biomedical applications developed by electrochemical anodization

G Strnad[1], **C Petrovan**[2], **O Russu**[3], **L Jakab-Farkas**[4]

[1] Faculty of Engineering, Petru Maior University, Tirgu Mures, RO
[2] Faculty of Dental Medicine, University of Medicine and Pharmacy, Tirgu Mures, RO
[3] Faculty of Medicine, University of Medicine and Pharmacy, Tirgu Mures, RO
[4] Faculty of Technical and Human Sciences, University Sapientia, Tirgu Mures, RO

E-mail: gabriela.strnad@ing.upm.ro

Abstract. Present research demonstrates the formation of self-ordered nanostructured oxide layer on the surface of two phase Ti6Al4V alloy by using electrochemical anodization in H_3PO_4/HF electrolytes. Our results show that the ordered oxide nanotubes grow on large areas on the samples surface, on both phases of (α+β) Ti6Al4V titanium alloy. We developed nanotubes of 70 nm (internal diameter) using 0.3 wt% HF and of 80 nm using 0.5 wt% HF additions to 1M H_3PO_4, at an anodization potential of 20 V, and an anodization time of 2 hours. We show that anodization potential has a strong influence on nanostructures morphology. Our results show that nanotubes' internal diameter is ~30 nm at 10 V potential, ~40 nm at 15 V potential, and ~70-80 nm at 20 V potential in anodization process performed in 1M H_3PO_4 + 0.5 wt% HF, 2 hours. The thickness of the developed nanostructured oxide layer is in 200-250 nm range.

1. Introduction

Titanium based implants are of great importance in craniofacial surgery and orthopaedics, providing excellent outcomes in terms of functionality. These devices utilize the unique biocompatibility properties of titanium and titanium alloys, which facilitate the apposition of bone directly onto the metal surface, a phenomena known as osseointegration. The modification of the implant surface at the micro level has been shown to have a major effect on the rate and degree of osseointegration. Recent studies demonstrate that surface modification at nano scale level in order to produce an ordered nanostructured surface, promotes further enhanced bone apposition during the early stages of osseointegration when compared with those seen on micro-rough surfaces [1-9].

Nowadays the use of commercially pure titanium (CpTi) is dominant in dental implants, while the use of Ti6Al4V (Ti grade 5) alloy is of major importance in orthopaedics, due to its increased mechanical properties [10].

Electrochemical anodization of titanium and titanium based alloys can shape the morphology, structure and chemical composition of oxide layers to enhance the surface properties [11, 12]. Several recent studies have shown that nanostructured (nanoporous or nanotubular) surfaces can be developed by optimizing the process parameters of anodization [13-16]. These nanostructures combine very well with osseous tissue and can be the perfect basis for osteoblasts integration during the process of bone regeneration.

The mechanism of nanotubes formation is a very complex one and occurs in interdependent conditions. In a simple approach, the formation of nanotubes can be viewed as being governed by the competition between anodic oxide formation and chemical dissolution of the oxide as soluble complexes. Briefly explained, the nanotubular layer develops in three stages: first is compact oxide formation, second is the initial porous structure formation, and third is self-organized nanotubes growth [17]. In recent studies Kulkarni, Krasicka, and Lee emphasize that a proper combination of electrolyte composition, anodization potential, anodization time, and a perfect balance between potentiodynamic stage and potentiostatic stage are required in order to develop uniform highly-organized nanotubular oxide structures on titanium [1, 11, 15]. Moreover, the formation of self-ordered nanostructures on two phase titanium alloys is a more demanding process, due to the selective dissolution of the less stable phase and/or different reactions rate of the different phases of the alloy.

Phosphate-fluoride solutions proved to be very effective on TiO_2 nanotubular structures development. As during the electrochemical anodization process a competition between oxide formation and dissolution exists, in H_3PO_4/HF electrolytes hydrofluoric acid is the key factor which causes nanoporous or nanotubular oxide layer development.

In this context our major research direction is enhancement of surfaces of biomedical titanium based implants, both at micro and nano level, in order to improve their osseointegration. Our recently reported research led to successful development, using sand blasting and acid etching (SLA) processes, of micro-rough SLA surfaces with an average surface roughness R_a of 1-2 μm, surface topography with micropits less than 10 μm (these micropits are essential for a proper osseontegration), large valleys of ~50 μm and different size peaks, and enhanced hydrophilic properties [18-20].

Presently, the major goal of our current research work is to develop an effective and efficient procedure to prepare modified implant surfaces with uniform, self-organized nanotubular oxide layers. The objective of the present paper was to develop ordered TiO_2 nanotubes on the surface of two phase Ti6Al4V alloy for biomedical implants, using electrochemical anodization procedure. Our results show that the goal was accomplished; our nanotubular structures for medical implants are fully comparable with the latest results reported by several research teams working in the field of nanostructured titanium based modified surfaces [1, 2, 8, 13, 16].

2. Methodology

For the present set of experiments we prepared ∅16 x 3 mm disc samples of Ti6Al4V titanium alloy. The roughness of samples was measured using a Surtronic 25 (Taylor Hobson) roughness tester. The weight of samples was measured using Kern ARJ 220-4M laboratory balance with a reproducibility of 0.1 mg. Before the anodization all samples were subjected to a cleaning process in distilled water followed by ethanol, and drying.

The anodization process was performed in an electrochemical cell having two electrode configuration. The samples were connected to the anode of the cell, while the cathode was in form of ∅16 x 3 mm disc, made of high purity copper (99.99%) rod. For all experiments the distance anode-cathode was 15 mm. We used a direct current stabilized power supply, and voltmeter and ammeter for electrical parameters control and registration. 1M H_3PO_4 with different concentration of HF was used as electrolyte in anodization experiments. All electrolytes were prepared from reagent grade chemicals (Chemical Company) and DI water. Experiments were carried out at room temperature.

The electrochemical anodization process consisted in an initial potentiodynamic stage, with a sweep rate of 0,1 V/s until the end potential was reached, followed by a potentiostatic stage by holding the potential at the end potential (10 V, 15 V, and 20 V). The total duration for the anodization process was 2 hours for all experiments. After anodization, the samples were rinsed with deionized water, cleaned in ethanol and dried.

The surface morphology was evaluated using scanning electron microscopy, performed in a JSM 5200 (JEOL) scanning electron microscope, operated at 25 kV. For a proper interpretation of the results, the micrographs were collected at magnifications of 1500X, 10000X, 20000X, 35000X, and 100000X using secondary electrons and backscattered electrons. The measurements of nanostructures developed on the modified surfaces were made on SEM micrographs using a graphical image processing software (Gimp).

3. Results and discussion

The main parameters of anodization processes performed on Ti6Al4V samples in order to develop ordered TiO_2 nanostructures are listed in Table 1. The table shows also the results of the experiments regarding the diameters of the nanotubes developed on the modified surfaces.

Table 1. Anodization process parameters and results

	Process parameters		Results			
	Electrolyte	Anodization	Nanostructures dimensions			
	1M H_3PO_4 + HF	potential	Internal diameter D_i [nm]		External diameter D_e [nm]	
Sample	[% wt] HF	U [V]	range	mainly	range	mainly
60	0.5	20	60-100	70	100-160	120
61	0.5	20	50-100	70	90-160	120
63	0.5	20	50-100	80	90-160	120
62	0.3	20	40-90	70	70-130	100
65	0.5	15	30-50	40	60-100	70
64	0.5	10	25-45	30	50-90	60
70	chemical etch in 0.5 wt% HF for 30 s (sample for assessment of the initial surfaces)					

Ti6Al4V is a two phase (α+β) alloy, with 6 wt% aluminum stabilizing the α phase (hcp), and 4 wt% vanadium stabilizing the β phase (bcc). Figures 1a and 2a show the SEM micrograph of the surface after etching in 0,5 wt% HF for 30 seconds. On the surface there are regions formed by β phase (bcc) – white areas, and α phase (hcp) – black regions.

After anodization process the surface of the samples is covered by a nanostructured layer, consisting in ordered nanotubular structures. From figure 1b, collected at low magnification (1500X), appear that a continuous TiO_2 layer is formed on the surface of the anodized surface. This layer is interrupted by 1-5 µm micro craters, where, at this magnification, it is not clear if there is or not a nanostructured development of an oxide layer.

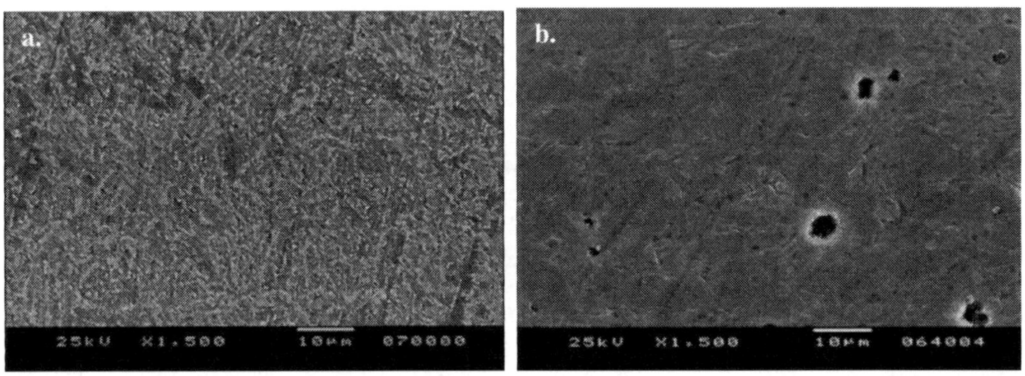

Figure 1. SEM micrographs of initial Ti6Al4V samples surface (a) and anodized surface (b) showing the uniform development of nanostructured oxide layer – 1500X magnification (a – sample 70, etched in 0.5 wt% HF for 30 s; b – sample 64 anodized in 1M H_3PO_4 + 0.5 wt% HF, U=10 V, 2 hours)

At higher magnification (10000X), figure 2b, compared with 2a, shows that the nanotubes are developed on both phases of the titanium alloy, even if there are some differences in nanotubes dimensions, depending on which phase they grew. The SEM micrograph presented in figure 2b shows more uniform nanotubes development on β phase. This phase seems to be more favorable for an ordered nanostructured development of the oxide layer. An explanation for this can be the beneficial influence of vanadium content. More research in order to elucidate the causes of these differences, that occur, probably, from the initial microstructure and/or the processes involved in the initial potentiodynamic stage of anodization, is planned in the nearest future.

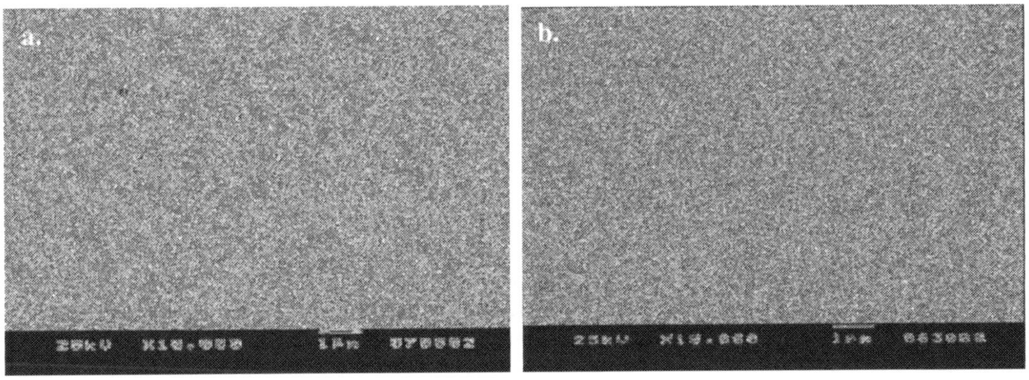

Figure 2. SEM micrographs of initial Ti6Al4V samples surface (a) and anodized surface (b) showing the uniform development of nanostructured oxide layer – 10000X magnification (a – sample 70, etched in 0.5 wt% HF for 30 s; b – sample 64 anodized in 1M H_3PO_4 + 0.5 wt% HF, U=10 V, 2 hours)

As regard the *influence of electrolyte composition* our experiments aimed to study its effect on nanostructures morphology in phosphate-fluoride solution containing 1M H_3PO_4 and different additions of HF (see Table 1). Figure 3 shows the morphology of the surfaces of the samples anodized using the same anodization potential (20 V) and modifying the content in hydrofluoric acid. Our results show titanium oxide development as nanotubular structures on all samples.

Comparing the nanotubes dimensions measurements (made by using Gimp graphical image processing software and presented in Table 1) we found that they only slightly differ when using 0.3 wt% HF (figure 3a) and 0.5 wt% HF (figure 3b) additions to 1M H_3PO_4.

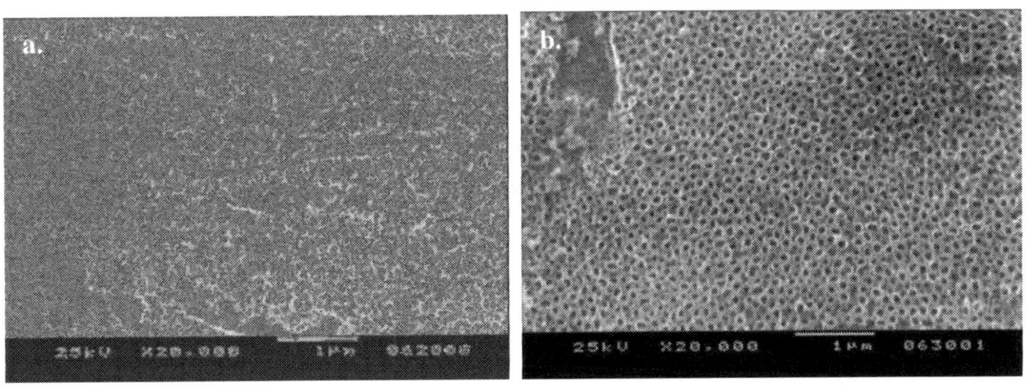

Figure 3. SEM micrographs of anodized Ti6Al4V surfaces showing the development of self-ordered nanotubular oxide layers and the influence of electrolyte composition on surface morphology (a – sample 62 anodized in 1M H_3PO_4 + 0.3 wt% HF, U=10 V, 2 hours – 20000X; b – sample 63 anodized in 1M H_3PO_4 + 0.5 wt% HF, U=20 V, 2 hours – 20000X)

On sample 62, developed at U=20 V, in 1M H_3PO_4 + 0.3 wt% HF the nanostructures (figure 3a) present internal diameters (D_i) ranging from 40-90 nm and external diameters (D_e) of 70-130 nm. The majority of nanotubes have D_i=70 nm and D_e=100 nm.

Sample 63, developed at U=20 V, in 1M H_3PO_4 + 0.5 wt% HF, presents nanotubes (figure 3b) with internal diameters in 50-100 nm range and external diameters of 90-160 nm. The majority of nanotubes have D_i=80 nm and D_e=120 nm.

By analyzing the surfaces' morphology at lower magnifications (SEM not shown here) we found that the oxide coverage of the surface is better in the case of the 1M H_3PO_4 + 0,5 wt% HF electrolyte, compared with 1M H_3PO_4 + 0,3 wt% HF electrolyte. This is explainable by the fact that on the surface is growing a compact oxide layer when using only phosphoric acid solution. The addition of hydrofluoric acid is the key factor in achieving the necessary conditions of the complex process of nanotubes development by anodizing process in phosphate-fluoride solutions.

Other parameter of anodization process which is reported to have an important influence on surface morphology is *anodization potential*. In order to elucidate its effect we performed our experiments in 1M H_3PO_4 + 0.5 wt% HF electrolyte, which we found to be the most beneficial in terms of Ti6Al4V surface coverage, using different anodization potentials (10 V, 15 V, and 20 V) for potentiostatic stage of our anodization process (see Table 1). The SEM micrographs presented in figures 4 and 5 show the development of ordered TiO_2 nanotubes on all surfaces. The results of nanostructures measurements (made by using Gimp graphical image processing software and presented in Table 1) show a strong influence of anodization potential on nanotubes diameters.

At U=10 V the nanostructures (figure 4a) present internal diameters ranging from 25-45 nm and external diameters of 50-90 nm. The majority of nanostructures have D_i=30 nm and D_e=60 nm.

By anodization at U=15 V the nanostructures have slightly bigger dimensions (figure 4b). Internal diameters are ranging from 30-50 nm and external diameters from 60-100 nm, with the majority of nanostructures having D_i=40 nm and D_e=70 nm.

Using a higher anodization potential of U=20 V, the dimensions of nanostructures are clearly bigger compared to those developed at 15V and 10V. As figures 5a (collected at a magnification of 35000X) and 5b (collected at 100000X magnification) show, internal diameters are in 50-100 nm range and external diameters in 90-160 nm range. The majority of nanostructures have D_i=80 nm and D_e=120 nm.

The thickness of nanostructured oxide layer developed in 1M H_3PO_4 + 0.5 wt% HF electrolyte at U=20 V and an anodization time of 2 hours is in 200-250 nm range (figure 6).

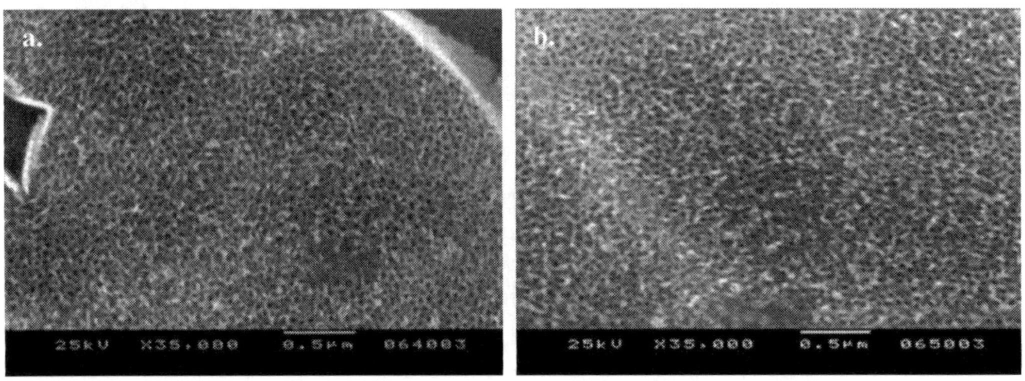

Figure 4. SEM micrographs of Ti6Al4V surfaces, anodized in 1M H_3PO_4 + 0.5 wt% HF, 2 hours, showing the development of ordered nanostructured oxide layers and the influence of anodization potential on surface morphology (a – sample 64 anodized at U=10 V - 35000X; b – sample 65 anodized at U=15 V - 35000X)

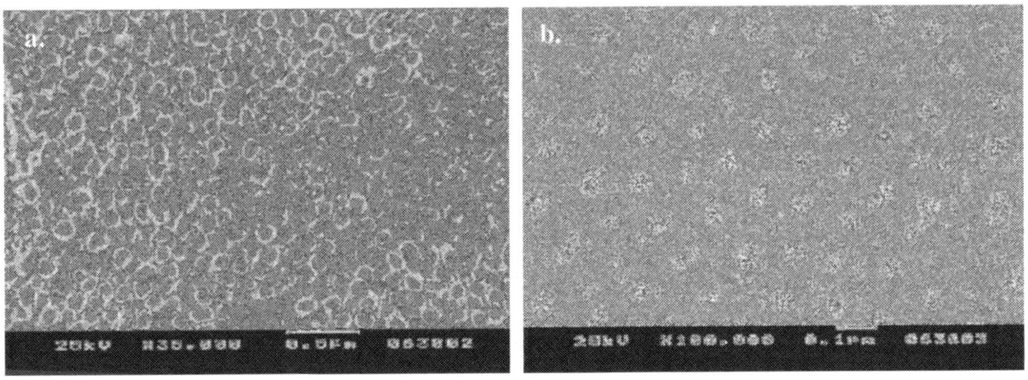

Figure 5. SEM micrographs of Ti6Al4V surface, anodized in 1M H_3PO_4 + 0.5 wt% HF, at U=20 V, 2 hours, sample 63, showing the development of well defined, highly ordered nanostructured oxide layer (a – 35000X magnification; b – 100000X magnification)

Figure 6. SEM micrographs of Ti6Al4V surface, anodized in 1M H_3PO_4 + 0.5 wt% HF, U=20 V, 2 hours, showing the development of ordered nanotubular oxide layer and its thickness (sample 63 – 35000X)

Our results show the best nanotubes separation in the case of the modified surfaces developed at 20 V anodization potential. All samples developed by anodization in 1M H_3PO_4+0.5 wt% HF electrolyte, at U=20 V, applied with a potential ramp of 0.1 V/s, and then maintained constant until reaching the total time of the process of 2 hours, presents self-ordered TiO_2 nanotubes, with high uniformity on large areas of the surface. By our present research we demonstrate the possibility of formation of a nanostructured, ordered oxide layer on the surface of two phase Ti6Al4V (α+β) alloy.

4. Conclusions

In the present set of experiments we have successfully developed ordered nanostructured TiO_2 layer on the surface of Ti6Al4V alloy for biomedical applications (orthopaedic and dental implants) by using electrochemical anodization in H_3PO_4/HF electrolytes.

Our results show that the ordered oxide nanotubes grow on large areas on the samples surface, on both phases of (α+β) Ti6Al4V titanium alloy, with some differences in size (diameter and height) and uniformity, depending, probably, on the initial microstructure and/or the processes that occur in the initial potentiodynamic stage of anodization. More research in order to elucidate the causes of these differences is planned in the nearest future.

Regarding the *influence of electrolyte composition on surface morphology* we found that in phosphate based solutions (1M H_3PO_4) an addition of 0.5 wt% HF provides the best conditions for nanotubes development on entire surface of Ti6Al4V alloy.

Our results on *the influence of anodization potential on surface morphology* show the ordered nanostructures development in 1M H_3PO_4 + 0.5 wt% HF electrolyte at 10 V, 15 V, and 20 V. The potential has an important effect on nanotubes diameter. At 10 V and 15 V potentials we measured nanotubes internal diameter of ~30-40 (±10) nm and by using 20 V potential the nanotubes internal diameter was ~70-80 (±20) nm. In our anodization cell, at 20 V potential and a process time of 2 hours, the thickness of the developed nanostructured oxide layer is 200-250 nm.

Based on present results, future research is planned by us in order to further analyze and elucidate the complex and interdependent influence of surface microstructure, surface preparation prior to anodization, potential applying ramp in the potentiodynamic stage of the anodization, current in the potentiostatic stage of the anodization, and anodization time on the uniform development of highly ordered nanotubular oxide structures on two phase titanium alloy Ti6Al4V using electrochemical anodization in H_3PO_4/HF electrolytes.

Acknowledgments
The Ti6Al4V titanium alloy bar for substrates preparation was supplied by SC Procam SRL, Tirgu Mures. The authors thank Anca Mazare and Biro Dominic for their helpful discussions, Andrei Serban Gaz and Nicolae Chirila for their help with chemical solutions formulation and preparation, and Petrica Morar for his help with electrical set-up of anodization cell. Dissemination of the results of present research work is partially supported by Petru Maior University research centre Advanced Technologies of Design and Assisted Manufacturing (TAPFA).

References
[1] Kulkarni M et al 2015 Titanium nanostructures for biomedical applications *Nanotechnology* **26** 062002
[2] Kulkarni M et al 2014 Biomaterial surface modification of titanium and titanium alloys for medical applications (*Nanomedicine*) chapter 5 pp 111-135
[3] Minagar S et al 2013 Cell response of anodized nanotubes of titanium and titanium alloys *Journal of Biomedical Materials Research A* **9999A** 1-14
[4] Mazare A et al 2012 Changing bioperformance of TiO_2 amorphous nanotubes as an effect of inducing cristallinity *Biochemistry* **87** 124-131
[5] Narendrakumar K et al 2015 Adherence of oral streptococci to nanostructured titanium surfaces *Dental Materials* **31** 1460-1468
[6] Mazare A 2016 Corrosion, antibacterial activity and haemocompatibility of TiO_2 nanotubes as a function of their annealing temperature *Corrosion Science* **103** 215–222
[7] Ross A and Webster T 2013 Anodizing color coded anodized Ti6Al4V medical devices for increasing bone cell functions *International Journal of Nanomedicine* **8** 109-117
[8] Beltrán-Partida E et al 2015 Improved osteoblast and chondrocyte adhesion and viability by surface-modified Ti6Al4V alloy with anodized TiO_2 nanotubes using a super-oxidative solution *Materials* **8** 867-883
[9] Karazisis D et al 2016 The role of well-defined nanotopography of titanium implants on osseointegration: cellular and molecular events in vivo *International Journal of Nanomedicine* **11** 1367-1381
[10] Wang W and Poh C K 2013 Titanium alloys in orthopaedics (*Titanium alloys – advances in properties control*) chapter 1 pp 1-20
[11] Krasicka-Cydzik E 2012 Anodic layer formation on titanium and its alloys for biomedical applications (*Titanium alloys – towards achieving enhanced properties for diversified application*) chapter 8 pp 175-200
[12] Brammer K et al 2011 Biomaterials and biotechnology schemes utilizing TiO_2 nanotube arrays

(*Biomaterials science and engineering*) chapter 9 pp 194-210

[13] Kulkarni M et al 2016 Influence of anodization parameters on morphology of TiO_2 nanostructured surfaces *Advanced Material Letters* **7(1)** 23-28

[14] Mazare A et al 2014 Electrochemical behavior in simulated body fluid of TiO_2 nanotubes on TiAlNB alloy elaborated in various anodizing electrolite *Surf. Interface Anal.* **46** 186-192

[15] Lee K et al 2014 One-Dimensional Titanium Dioxide Nanomaterials: Nanotubes *Chemical Reviews* **114** 9385–9454

[16] Al-Mobarak N A and Al-Swayih A A 2014 Development of titanium surgery implants for improving osseointegration through formation of a titanium nanotube layer *Int. J. Electrochem. Sci.* **9** 32–45

[17] Macak J M et al 2007 TiO_2 nanotubes: Self-organized electrochemical formation, properties and applications *Current Opinion in Solid State and Materials Science* **11** 3–18

[18] Strnad G and Chirila N 2015 Corrosion rate of sand blasted and acid etched Ti6Al4V for dental implants *Procedia Technology* **19** 909-915

[19] Strnad G et al 2015 Effect of surface preparation and passivation treatment on surface topography of Ti6Al4V for dental implants *Applied Mechanics and Materials* **809-810** 513-518

[20] Strnad G et al 2016 Contact angle measurment on medical implant titanium based biomaterials *Procedia Technology* **22** 946-953

IOP Conf. Series: Materials Science and Engineering **161** (2016) 012052 doi:10.1088/1757-899X/161/1/012052

Effect of stacking sequence on the coefficients of mutual influence of composite laminates

I Dupir (Hudişteanu), N Ţăranu and A Axinte

Technical University "Gheorghe Asachi" of Iasi, , 67 Prof. D. Mangeron Blvd., 700050, Iasi, Romania

E-mail: iulianahudisteanu@ce.tuiasi.ro

Abstract. Fiber reinforced polymeric (FRP) composites are nowadays widely used in engineering applications due to their outstanding features, such as high specific strength and specific stiffness as well as good corrosion resistance. A major advantage of fibrous polymeric composites is that their anisotropy can be controlled through suitable choice of the influencing parameters. The unidirectional fiber reinforced composites provide much higher longitudinal mechanical properties compared to the transverse ones. Therefore, composite laminates are formed by stacking two or more laminas, with different fiber orientations, as to respond to complex states of stresses. These laminates experience the effect of axial-shear coupling, which is caused by applying normal or shear stresses, implying shear or normal strains, respectively. The normal-shear coupling is expressed by the coefficients of mutual influence. They are engineering constants of primary interest for composite laminates, since the mismatch of the material properties between adjacent layers can produce interlaminar stresses and/or plies delamination. The paper presents the variation of the in-plane and flexural coefficients of mutual influence for three types of multi-layered composites, with different stacking sequences. The results are obtained using the Classical Lamination Theory (CLT) and are illustrated graphically in terms of fiber orientations, for asymmetric, antisymmetric and symmetric laminates. Conclusions are formulated on the variation of these coefficients, caused by the stacking sequence.

1. Introduction

One of the most important advantages of laminated composites is their potential to orient the laminas in different suitable directions or to choose the convenient stacking sequence configurations, such that the structural response to the complex state of stresses can be improved [1].

The *coefficients of mutual influence* or the *shear-extension coupling coefficients* describe the structural response of the multi-layered composites with shear or normal strains, caused by normal or shear stresses, respectively.

Depending on the applied stresses, the different mechanical strains induced at the interface of the adjacent laminas, can be high enough to cause failure [2]. The stacking sequence, which is an important parameter, firstly studied in [3-4], has a major influence on the failure modes at the free edges such as: *delamination* and *transverse cracks through the thickness of the laminate*, caused by interlaminar stresses. The mismatch of the material properties between adjacent plies, quantified by the Poisson's ratio or by the coefficients of mutual influence, is still an actual concern in the design of composite laminates.

The delamination, which may occur in multi-layered composites due to the mismatch of the material or due to the residual stresses induced by the manufacturing process, is frequently studied through the suitable experimental tests or numerical modelling analysis using FEM [5, 6].

Changing the stacking sequences, while maintaining the same orientation angles in composite laminates, influences the bending response and the coupling effects, which may occur, but it does not influence the in-plane response [7].

The effect of the stacking sequence on the in-plane and flexural coefficients of mutual influence is studied in this paper, through the suitable choice of three lay-up configurations as illustrated in figure 1: a general asymmetric laminate $[\pm\theta/(\theta)_2]$, an antisymmetric angle-ply laminate $[(\pm\theta)_2]$ and a symmetric angle-ply laminate $[\pm\theta]_s$.

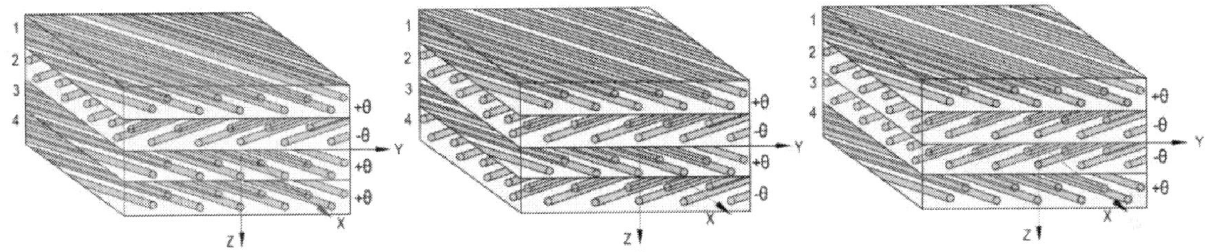

a. b. c.

Figure 1. Stacking sequences of the analysed laminates: a. $[\pm\theta/(\theta)_2]$ general asymmetric laminate; b. $[(\pm\theta)_2]$ antisymmetric angle-ply laminate; c. $[\pm\theta]_s$ symmetric angle-ply laminate.

Special particularities of the variation of the coefficients of mutual influence are discussed, depending on the chosen stacking sequences.

2. Evaluation of the coefficients of mutual influence

The constitutive relations of the multi-layered composites expressed by the force-deformation and moment-deformation equations are:

$$\begin{Bmatrix} N_x \\ N_y \\ N_{xy} \end{Bmatrix} = \begin{bmatrix} A_{11} & A_{12} & A_{16} \\ A_{12} & A_{22} & A_{26} \\ A_{16} & A_{26} & A_{66} \end{bmatrix} \begin{Bmatrix} \varepsilon_x^0 \\ \varepsilon_y^0 \\ \gamma_{xy}^0 \end{Bmatrix} + \begin{bmatrix} B_{11} & B_{12} & B_{16} \\ B_{12} & B_{22} & B_{26} \\ B_{16} & B_{26} & B_{66} \end{bmatrix} \begin{Bmatrix} k_x \\ k_y \\ k_{xy} \end{Bmatrix} \quad (1.1)$$

$$\begin{Bmatrix} M_x \\ M_y \\ M_{xy} \end{Bmatrix} = \begin{bmatrix} B_{11} & B_{12} & B_{16} \\ B_{12} & B_{22} & B_{26} \\ B_{16} & B_{26} & B_{66} \end{bmatrix} \begin{Bmatrix} \varepsilon_x^0 \\ \varepsilon_y^0 \\ \gamma_{xy}^0 \end{Bmatrix} + \begin{bmatrix} D_{11} & D_{12} & D_{16} \\ D_{12} & D_{22} & D_{26} \\ D_{16} & D_{26} & D_{66} \end{bmatrix} \begin{Bmatrix} k_x \\ k_y \\ k_{xy} \end{Bmatrix} \quad (1.2)$$

where: $(\varepsilon_x^0, \varepsilon_y^0, \gamma_{xy}^0)$ are the laminate mid-plane strains; (k_x, k_y, k_{xy}) are the laminate curvatures; $[A_{ij}]$, $[B_{ij}]$, $[D_{ij}]$ represent the terms of the *extensional stiffness matrix* $[A]$, *bending-stretching coupling matrix* $[B]$ and *bending stiffness matrix* $[D]$, respectively.

The effect of the axial-shear coupling are expressed by the coefficients of mutual influence that have been introduced by Lekhnitskii [8] and defined in [9-13]:

$\eta_{i,ij}$ represents the *in-plane coefficient of mutual influence of the first kind* that describes the stretching in the i direction, induced by the shear stress in the ij plane;

$\eta_{ij,i}$ represents the *in-plane coefficient of mutual influence of the second kind* that describes the shearing in the *ij* plane, induced by normal stress in the *i* direction;

$\eta^{f}_{i,ij}$ represents the *flexural coefficient of mutual influence of the first kind* that describes the bending with respect to the *i* direction, induced by twisting in the *ij* plane;

$\eta^{f}_{ij,i}$ represents the *flexural coefficient of mutual influence of the second kind* that describes the twisting in the *ij* plane, induced by bending moment with respect to the *i* direction.

In the present paper the relations for the evaluation of the in-plane and flexural coefficients of mutual influence, valid for any type of general composite laminate, are developed; however, they are also available for special orthotropic laminates or other particular multi-layered composites. The equations are developed, starting from the reports formulated by *Nettles* [14], where the in-plane engineering constants of non-symmetric composite laminates are written in terms of the fully-populated [*A*], [*B*], [*D*] stiffness matrices, after solving a Cramer system. Therefore, the expressions for the in-plane and flexural coefficients of mutual influence are presented, both for the general case of composite laminates and for a particular case of symmetric angle-ply laminates, when [*B*] = *0*.

2.1. *In-plane coefficients of mutual influence*

If laminate bending-stretching coupling does not occur, the in-plane coefficients of mutual influence are characterized by the extensional stiffness matrix through the terms A_{16} and A_{26}. Otherwise, almost all the terms of the assembled stiffness matrix are implied in the evaluation of the in-plane structural response, produced by the axial-shear coupling.

The in-plane coefficients of mutual influence are defined according to equation (2):

$$\eta_{x,xy} = \varepsilon_x^0 / \gamma_{xy}^0,\ \eta_{y,xy} = \varepsilon_y^0 / \gamma_{xy}^0,\ \text{when } \tau_{xy} \neq 0,\ \sigma_x = 0,\ \sigma_y = 0;$$

$$\eta_{xy,x} = \gamma_{xy}^0 / \varepsilon_x^0,\ \text{when } \sigma_x \neq 0,\ \sigma_y = 0,\ \tau_{xy} = 0; \qquad (2)$$

$$\eta_{xy,y} = \gamma_{xy}^0 / \varepsilon_y^0,\ \text{when } \sigma_y \neq 0,\ \sigma_x = 0,\ \tau_{xy} = 0.$$

2.1.1. *General composite laminates*

The in-plane coefficients of mutual influence for any general case of composite laminate are determined with equations (3)-(6), as following:

$$\eta_{x,xy} = \frac{\begin{vmatrix} A_{12} & A_{16} & B_{11} & B_{12} & B_{16} \\ A_{22} & A_{26} & B_{12} & B_{22} & B_{26} \\ B_{12} & B_{16} & D_{11} & D_{12} & D_{16} \\ B_{22} & B_{26} & D_{12} & D_{22} & D_{26} \\ B_{26} & B_{66} & D_{16} & D_{26} & D_{66} \end{vmatrix}}{\begin{vmatrix} A_{11} & A_{12} & B_{11} & B_{12} & B_{16} \\ A_{12} & A_{22} & B_{12} & B_{22} & B_{26} \\ B_{11} & B_{12} & D_{11} & D_{12} & D_{16} \\ B_{12} & B_{22} & D_{12} & D_{22} & D_{26} \\ B_{16} & B_{26} & D_{16} & D_{26} & D_{66} \end{vmatrix}} \ (3), \qquad \eta_{y,xy} = -\frac{\begin{vmatrix} A_{11} & A_{16} & B_{11} & B_{12} & B_{16} \\ A_{12} & A_{26} & B_{12} & B_{22} & B_{26} \\ B_{11} & B_{16} & D_{11} & D_{12} & D_{16} \\ B_{12} & B_{26} & D_{12} & D_{22} & D_{26} \\ B_{16} & B_{66} & D_{16} & D_{26} & D_{66} \end{vmatrix}}{\begin{vmatrix} A_{11} & A_{12} & B_{11} & B_{12} & B_{16} \\ A_{12} & A_{22} & B_{12} & B_{22} & B_{26} \\ B_{11} & B_{12} & D_{11} & D_{12} & D_{16} \\ B_{12} & B_{22} & D_{12} & D_{22} & D_{26} \\ B_{16} & B_{26} & D_{16} & D_{26} & D_{66} \end{vmatrix}} \ (4)$$

$$\eta_{xy,x} = \frac{\begin{vmatrix} A_{12} & A_{22} & B_{12} & B_{22} & B_{26} \\ A_{16} & A_{26} & B_{16} & B_{26} & B_{66} \\ B_{11} & B_{12} & D_{11} & D_{12} & D_{16} \\ B_{12} & B_{22} & D_{12} & D_{22} & D_{26} \\ B_{16} & B_{26} & D_{16} & D_{26} & D_{66} \end{vmatrix}}{\begin{vmatrix} A_{22} & A_{26} & B_{12} & B_{22} & B_{26} \\ A_{26} & A_{66} & B_{16} & B_{26} & B_{66} \\ B_{12} & B_{16} & D_{11} & D_{12} & D_{16} \\ B_{22} & B_{26} & D_{12} & D_{22} & D_{26} \\ B_{26} & B_{66} & D_{16} & D_{26} & D_{66} \end{vmatrix}} \quad (5), \qquad \eta_{xy,y} = -\frac{\begin{vmatrix} A_{11} & A_{12} & B_{11} & B_{12} & B_{16} \\ A_{16} & A_{26} & B_{16} & B_{26} & B_{66} \\ B_{11} & B_{12} & D_{11} & D_{12} & D_{16} \\ B_{12} & B_{22} & D_{12} & D_{22} & D_{26} \\ B_{16} & B_{26} & D_{16} & D_{26} & D_{66} \end{vmatrix}}{\begin{vmatrix} A_{11} & A_{16} & B_{11} & B_{12} & B_{16} \\ A_{16} & A_{66} & B_{16} & B_{26} & B_{66} \\ B_{11} & B_{16} & D_{11} & D_{12} & D_{16} \\ B_{12} & B_{26} & D_{12} & D_{22} & D_{26} \\ B_{16} & B_{66} & D_{16} & D_{26} & D_{66} \end{vmatrix}} \quad (6)$$

2.1.2. Particular case of symmetric angle-ply laminates

The direct and inverse form of force-deformation constitutive relation for symmetric composite laminates are presented in equations (7.1) and (7.2), as following:

$$\begin{pmatrix} N_x \\ N_y \\ N_{xy} \end{pmatrix} = \begin{bmatrix} A_{11} & A_{12} & A_{16} \\ A_{12} & A_{22} & A_{26} \\ A_{16} & A_{26} & A_{66} \end{bmatrix} \begin{Bmatrix} \varepsilon_x^0 \\ \varepsilon_y^0 \\ \gamma_{xy}^0 \end{Bmatrix} \quad (7.1) \qquad \begin{pmatrix} \varepsilon_x^0 \\ \varepsilon_y^0 \\ \gamma_{xy}^0 \end{pmatrix} = \begin{bmatrix} A'_{11} & A'_{12} & A'_{16} \\ A'_{12} & A'_{22} & A'_{26} \\ A'_{16} & A'_{26} & A'_{66} \end{bmatrix} \begin{pmatrix} N_x \\ N_y \\ N_{xy} \end{pmatrix} \quad (7.2)$$

For the particular case of symmetric angle-ply laminates, the in-plane coefficients of mutual influence can be established with equations (8) – (11):

$$\eta_{x,xy} = \frac{A'_{16}}{A'_{66}} \quad (8), \qquad \eta_{y,xy} = \frac{A'_{26}}{A'_{66}} \quad (9), \qquad \eta_{xy,x} = \frac{A'_{16}}{A'_{11}} \quad (10), \qquad \eta_{xy,y} = \frac{A'_{26}}{A'_{22}} \quad (11)$$

2.2. Flexural coefficients of mutual influence

The assessment of the bending structural response of the general composite laminates, caused by the normal-shear coupling, can be realized with the implication of all terms of the stiffness matrices.

In case of symmetric laminates, the flexural coefficients of mutual influence are expressed through the terms D_{16} and D_{26} of the bending stiffness matrix.

The flexural coefficients of mutual influence are defined according to equation (12), as following:

$$\eta^f_{x,xy} = k_x / k_{xy}, \ \eta^f_{y,xy} = k_y / k_{xy}, \text{ when } \tau^f_{xy} \neq 0, \ \sigma^f_x = 0, \ \sigma^f_y = 0;$$
$$\eta^f_{xy,x} = k_{xy} / k_x, \text{ when } \sigma^f_x \neq 0, \ \sigma^f_y = 0, \ \tau^f_{xy} = 0; \quad (12)$$
$$\eta^f_{xy,y} = k_{xy} / k_y, \text{ when } \sigma^f_y \neq 0, \ \sigma^f_x = 0, \ \tau^f_{xy} = 0.$$

2.2.1. General composite laminates

The flexural coefficients of mutual influence of general composite laminates are determined with equations (13) – (16), such as:

$$\eta^{f}_{x,xy}=\frac{\begin{vmatrix}A_{11}&A_{12}&A_{16}&B_{12}&B_{16}\\A_{12}&A_{22}&A_{26}&B_{22}&B_{26}\\A_{16}&A_{26}&A_{66}&B_{26}&B_{66}\\B_{11}&B_{12}&B_{16}&D_{12}&D_{16}\\B_{12}&B_{22}&B_{26}&D_{22}&D_{26}\end{vmatrix}}{\begin{vmatrix}A_{11}&A_{12}&A_{16}&B_{11}&B_{12}\\A_{12}&A_{22}&A_{26}&B_{12}&B_{22}\\A_{16}&A_{26}&A_{66}&B_{16}&B_{26}\\B_{11}&B_{12}&B_{16}&D_{11}&D_{12}\\B_{12}&B_{22}&B_{26}&D_{12}&D_{22}\end{vmatrix}}\ (13),\qquad \eta^{f}_{y,xy}=-\frac{\begin{vmatrix}A_{11}&A_{12}&A_{16}&B_{11}&B_{16}\\A_{12}&A_{22}&A_{26}&B_{12}&B_{26}\\A_{16}&A_{26}&A_{66}&B_{16}&B_{66}\\B_{11}&B_{12}&B_{16}&D_{11}&D_{16}\\B_{12}&B_{22}&B_{26}&D_{12}&D_{26}\end{vmatrix}}{\begin{vmatrix}A_{11}&A_{12}&A_{16}&B_{11}&B_{12}\\A_{12}&A_{22}&A_{26}&B_{12}&B_{22}\\A_{16}&A_{26}&A_{66}&B_{16}&B_{26}\\B_{11}&B_{12}&B_{16}&D_{11}&D_{12}\\B_{12}&B_{22}&B_{26}&D_{12}&D_{22}\end{vmatrix}}\ (14)$$

$$\eta^{f}_{xy,x}=\frac{\begin{vmatrix}A_{11}&A_{12}&A_{16}&B_{11}&B_{12}\\A_{12}&A_{22}&A_{26}&B_{12}&B_{22}\\A_{16}&A_{26}&A_{66}&B_{16}&B_{26}\\B_{12}&B_{22}&B_{26}&D_{12}&D_{22}\\B_{16}&B_{26}&B_{66}&D_{16}&D_{26}\end{vmatrix}}{\begin{vmatrix}A_{11}&A_{12}&A_{16}&B_{12}&B_{16}\\A_{12}&A_{22}&A_{26}&B_{22}&B_{26}\\A_{16}&A_{26}&A_{66}&B_{26}&B_{66}\\B_{12}&B_{22}&B_{26}&D_{22}&D_{26}\\B_{16}&B_{26}&B_{66}&D_{26}&D_{66}\end{vmatrix}}\ (15),\qquad \eta^{f}_{xy,y}=-\frac{\begin{vmatrix}A_{11}&A_{12}&A_{16}&B_{11}&B_{12}\\A_{12}&A_{22}&A_{26}&B_{12}&B_{22}\\A_{16}&A_{26}&A_{66}&B_{16}&B_{26}\\B_{11}&B_{12}&B_{16}&D_{11}&D_{12}\\B_{16}&B_{26}&B_{66}&D_{16}&D_{26}\end{vmatrix}}{\begin{vmatrix}A_{11}&A_{12}&A_{16}&B_{11}&B_{16}\\A_{12}&A_{22}&A_{26}&B_{12}&B_{26}\\A_{16}&A_{26}&A_{66}&B_{16}&B_{66}\\B_{11}&B_{12}&B_{16}&D_{11}&D_{16}\\B_{16}&B_{26}&B_{66}&D_{16}&D_{66}\end{vmatrix}}\ (16)$$

2.2.2. Particular case of symmetric angle-ply laminates

The direct and inverse form of moment-deformation constitutive relations are performed by equations (17.1) and (17.2):

$$\begin{Bmatrix}M_x\\M_y\\M_{xy}\end{Bmatrix}=\begin{bmatrix}D_{11}&D_{12}&D_{16}\\D_{12}&D_{22}&D_{26}\\D_{16}&D_{26}&D_{66}\end{bmatrix}\begin{Bmatrix}k_x\\k_y\\k_{xy}\end{Bmatrix}\ (17.1)\qquad \begin{pmatrix}k_x\\k_y\\k_{xy}\end{pmatrix}=\begin{bmatrix}D'_{11}&D'_{12}&D'_{16}\\D'_{12}&D'_{22}&D'_{26}\\D'_{16}&D'_{26}&D'_{66}\end{bmatrix}\begin{pmatrix}M_x\\M_y\\M_{xy}\end{pmatrix}\ (17.2)$$

The evaluation of the flexural coefficients of mutual influence for symmetric angle-ply laminates is simplified, based on the equations (18) – (21):

$$\eta^{f}_{x,xy}=\frac{D'_{16}}{D'_{66}}\ (18),\qquad \eta^{f}_{y,xy}=\frac{D'_{26}}{D'_{66}}\ (19),\qquad \eta^{f}_{xy,x}=\frac{D'_{16}}{D'_{11}}\ (20),\qquad \eta^{f}_{xy,y}=\frac{D'_{26}}{D'_{22}}\ (21)$$

3. Results

The variations of the coefficients of mutual influence are represented graphically with respect to the fiber orientation angles, for the three analyzed composite laminates. The plies of the multi-layered composites are realized alternatively of S glass or E glass fibers, embedded in an epoxy resin matrix. The mechanical characteristics of the composite material constituents are illustrated in table 1.

Table 1. Mechanical properties of constituent materials [15]

Constituents of composite plies	Longitudinal Young's modulus (GPa)	Poisson's ratio
S glass fibres	85.5	0.22
E glass fibres	72.4	0.22
Epoxy resin matrix	4.1	0.40

The results are centralized such that three distinct comparative analysis can be observed, depending on the stacking sequence, on the in-plane or bending structural response and function of the fiber volume fractions for the considered lay-ups.

The differences and the particularities of the studied asymmetric, antisymmetric and symmetric composite laminates, in terms of in-plane and flexural coefficients of mutual influence of first kind and second kind are illustrated in figure 2.

Figure 3 and figure 4 show a comparative analysis between the in-plane and flexural coefficients of mutual influence, both for the general composite and for the symmetric angle-ply laminate.

In figures 5-7, the variation of the in-plane and flexural coefficients of mutual influence is presented for every studied configuration with non-zero results, for suggestive fiber volume fractions.

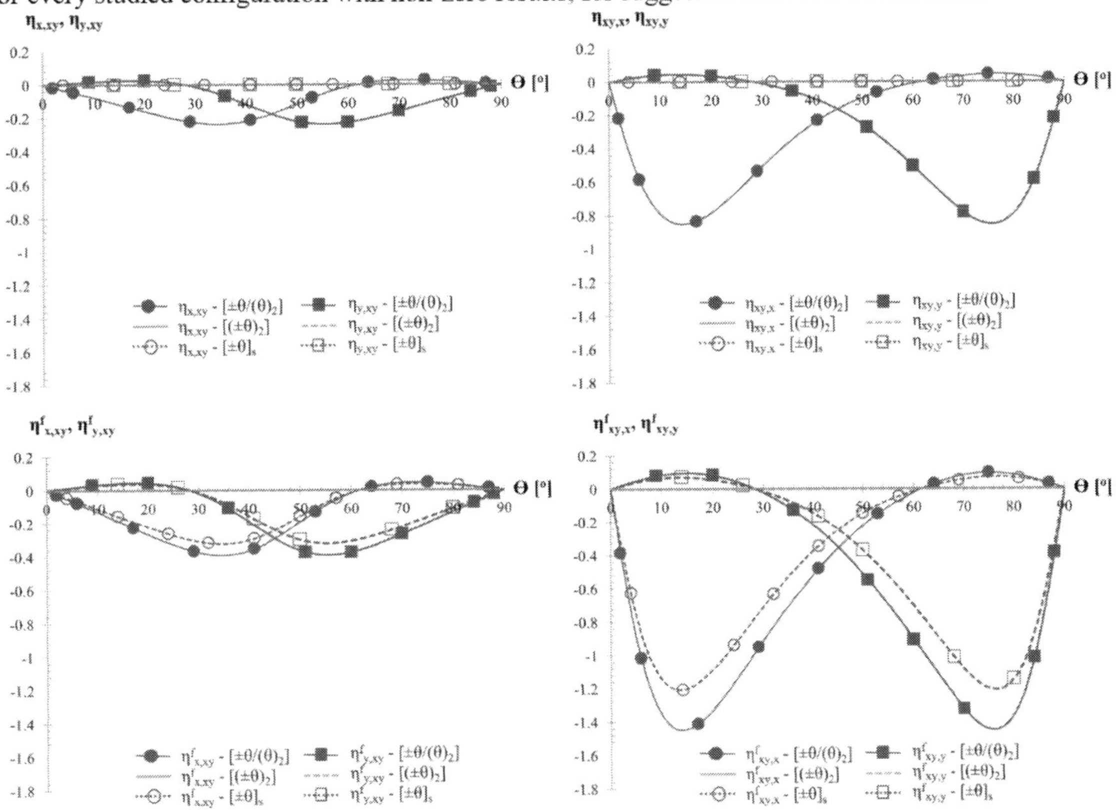

Figure 2. In-plane and flexural coefficients of mutual influence for the analysed laminates.

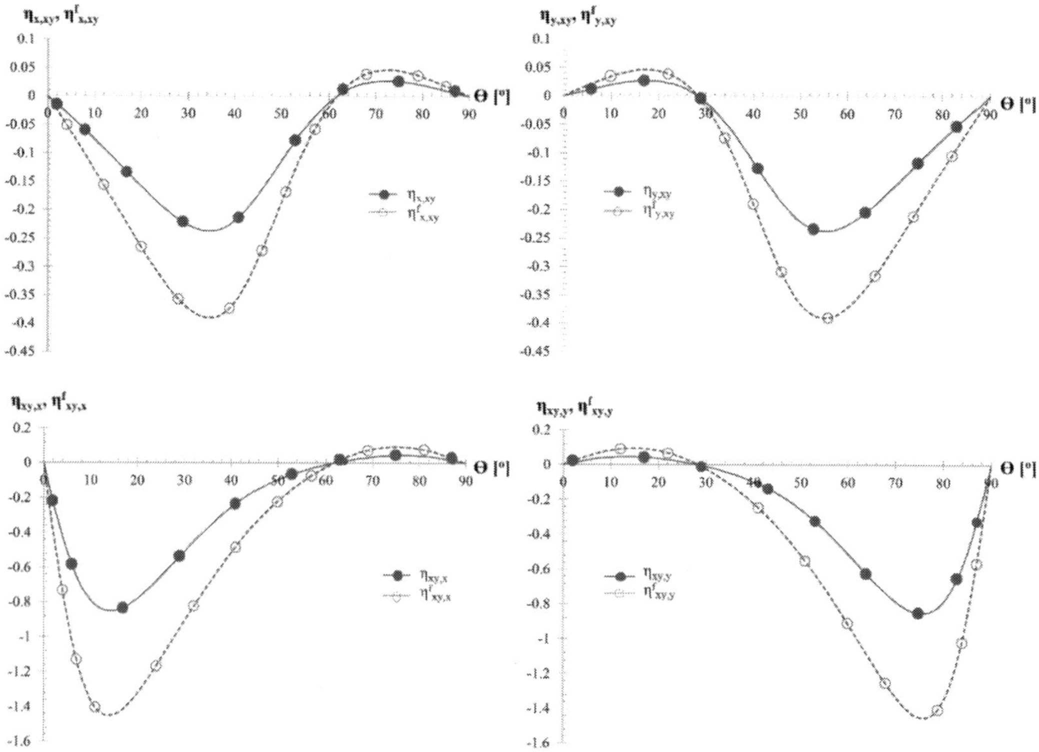

Figure 3. In-plane and flexural coefficients of mutual influence for general asymmetric laminates.

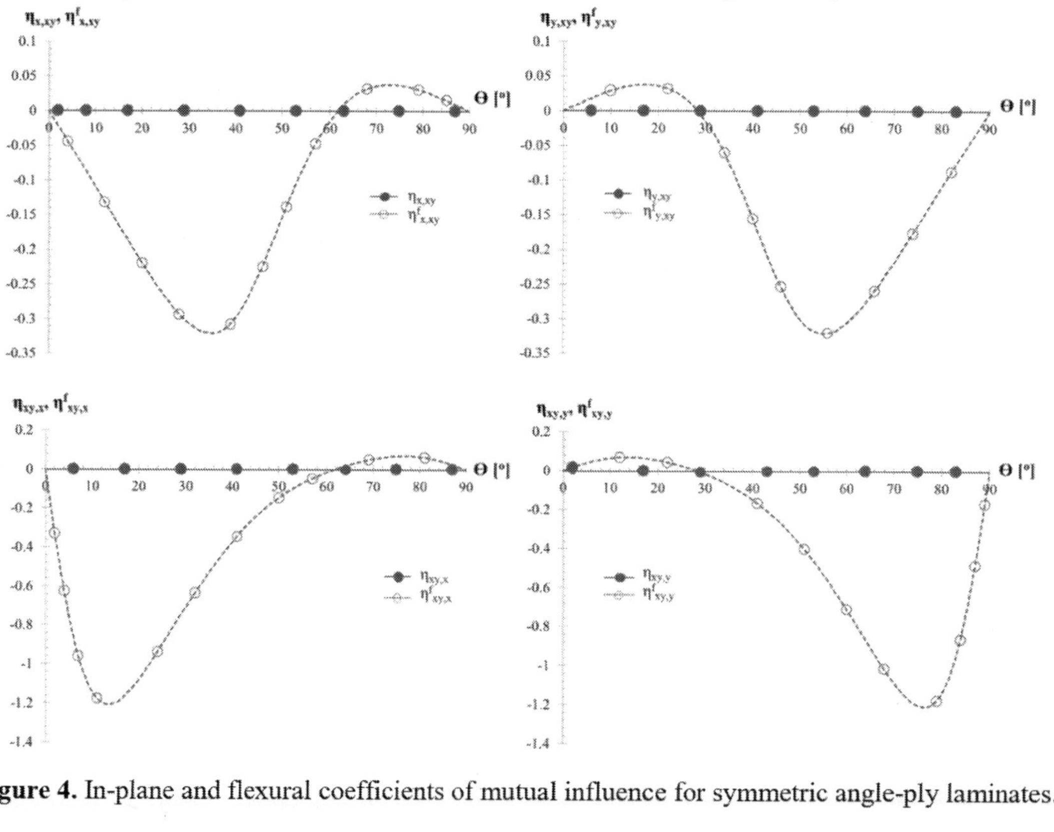

Figure 4. In-plane and flexural coefficients of mutual influence for symmetric angle-ply laminates.

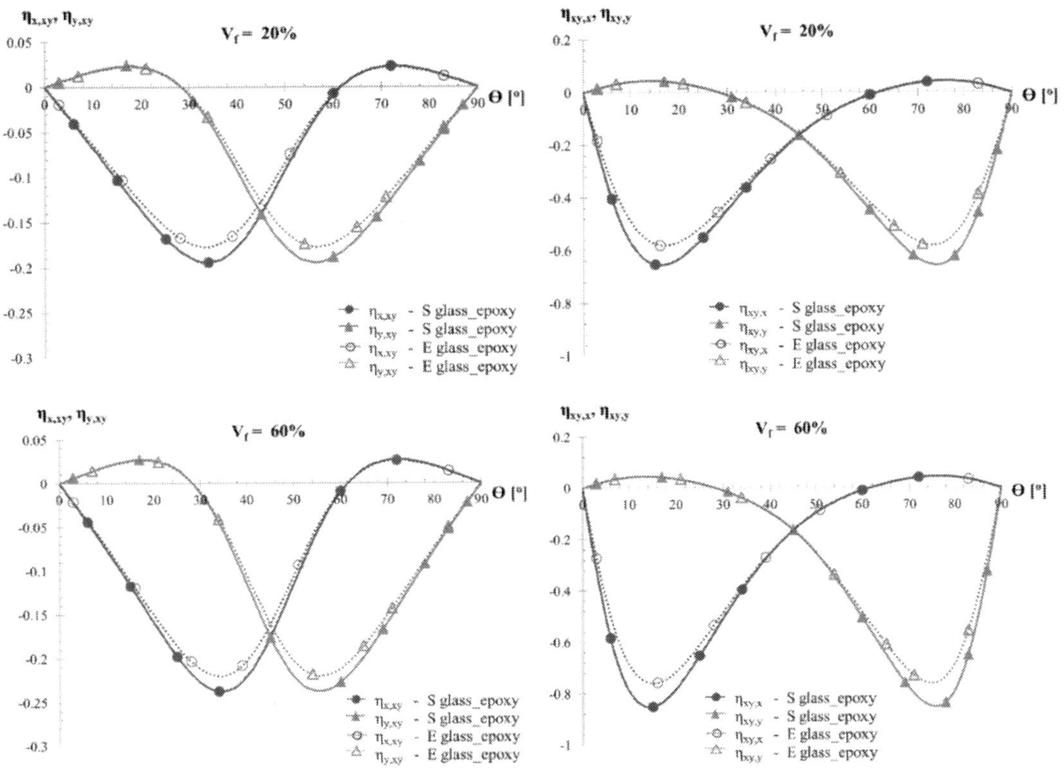

Figure 5. In-plane coefficients of mutual influence for general asymmetric laminates.

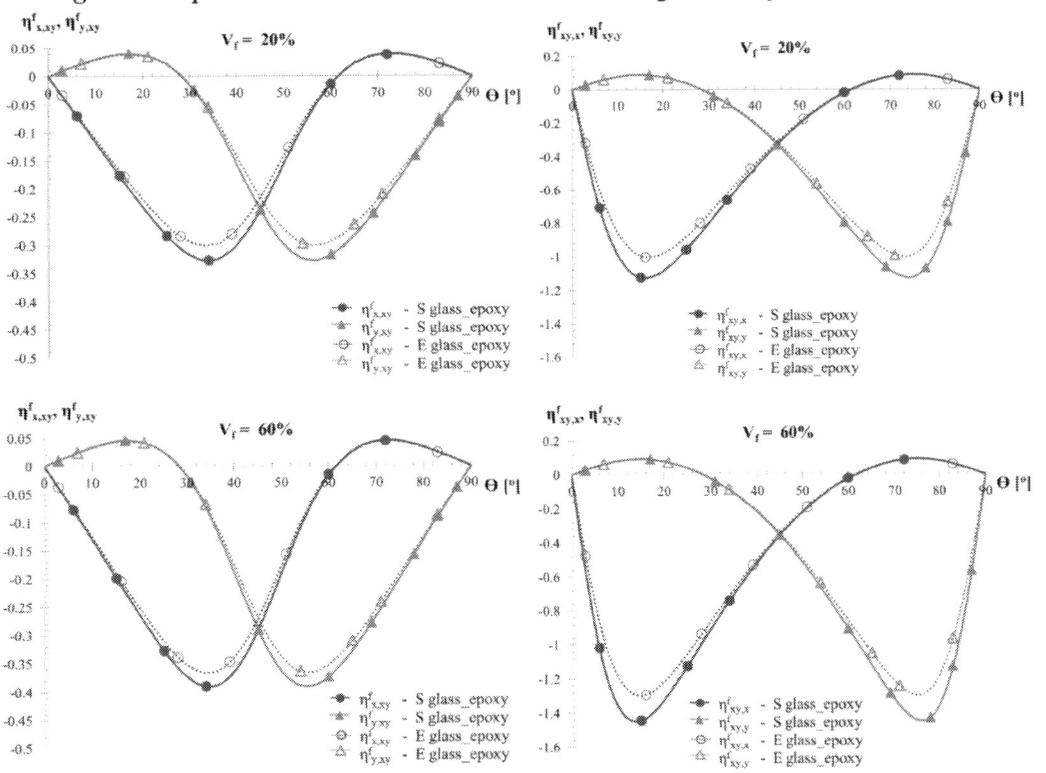

Figure 6. Flexural coefficients of mutual influence for general laminates.

IOP Conf. Series: Materials Science and Engineering **161** (2016) 012052 doi:10.1088/1757-899X/161/1/012052

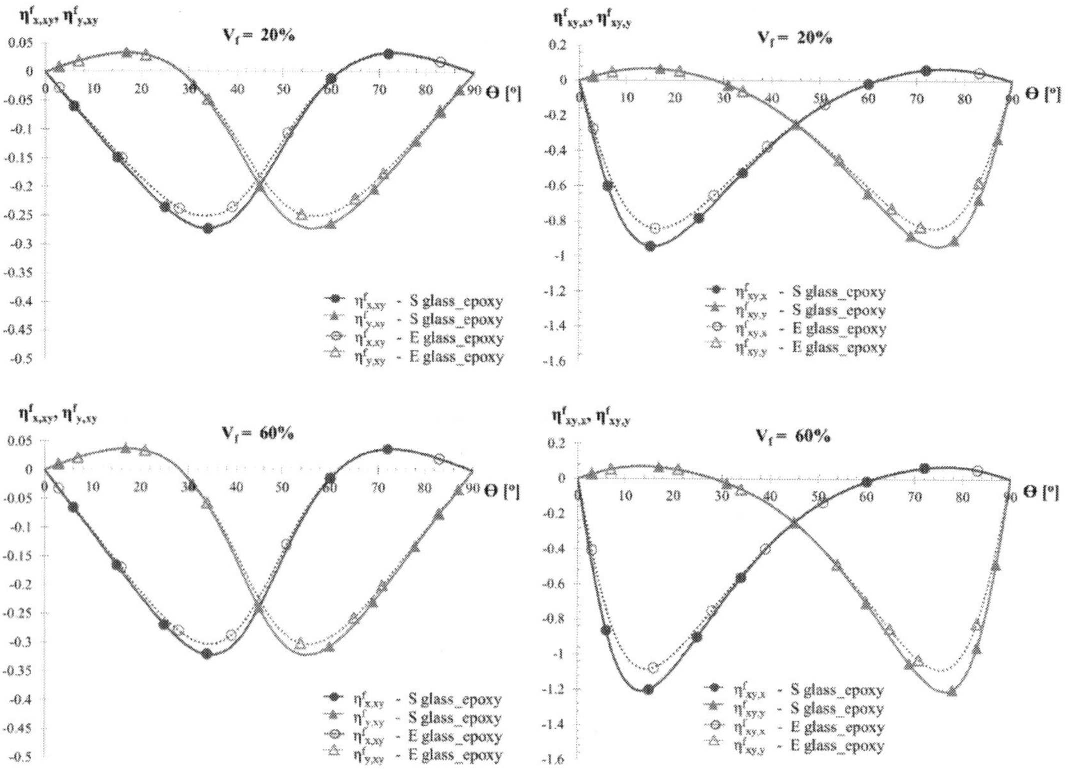

Figure 7. Flexural coefficients of mutual influence for symmetric angle-ply laminates.

4. Discussion

The general asymmetric $[\pm\theta/(\theta)_2]$ composite laminate presents nonzero values for both in-plane and flexural coefficients of mutual influence, having the flexural engineering constants greater than the in-plane ones.

The analyzed antisymmetric $[(\pm\theta)_2]$ angle-ply laminate has all coefficients of mutual influence equal to zero, since the angle-ply laminate is a special orthotropic laminate. All antisymmetric special orthotropic laminates have zero terms for A_{16}, A_{26}, D_{16}, D_{26}, which implies no in-plane and flexural coefficients of mutual influence. However, this is not a general rule for every antisymmetric laminate.

The symmetric $[\pm\theta]_S$ angle-ply laminate has zero in-plane coefficients of mutual influence and nonzero flexural coefficients of mutual influence. The results are confirmed by the relations which express that the terms $A_{16} = A_{26} = 0$, while $D_{16}, D_{26} \neq 0$, valid for symmetric angle-ply laminates.

Comparison between the flexural shear-extension coupling coefficients shows greater values in case of the general laminate than for the symmetric angle-ply laminate. Moreover, the coefficients of influence of first kind are smaller than the ones of second kind. As to clearly show the differences, in case of the studied configuration of laminates and composite materials constituents', the maximum general values for the coefficients of influence of first kind range from 0.05 to -0.40, while in case of the coefficients of mutual influence of second kind the values extend between 0.05 to -1.50.

In the case of the in-plane and flexural coefficients of mutual influence for general laminate or for the flexural coefficients of mutual influence for symmetric angle-ply laminate, the analysis has been run for fiber volume fractions of 20% and 60%. Increases of the coefficients' values with fiber volume fraction are noticed, until V_f reaches 60%.

5. Conclusions

The stacking sequence of the composite laminate has a major importance on the variation of the flexural coefficients of mutual influence, rather than the in-plane ones.

The special orthotropic laminate configurations are generally the recommended solutions for engineering applications, due to their reductions of the stiffness matrix terms and consequently for the simplification of the laminate analysis. The symmetric special orthotropic laminates have zero in-plane coefficients of mutual influence, but nonzero values for the flexural ones. The antisymmetric angle-ply laminate is a good alternative when both in-plane and flexural coefficients of mutual influence have to be zero, due to the complex loading actions.

Therefore, the configuration lay-up of a multi-layered element can be rationally designed, such as the differences in the coefficients of mutual influence between the adjacent plies to be as small as possible, to avoid failure caused by interlaminar stresses and/or ply delamination at the free edges.

References

[1] Peters S T 1998 *Handbook of composites Second edition* Cambridge University Press (Great Britain: Chapman & Hall)

[2] Bailie J A, Ley R P and Pasricha A 1997 *A summary and review of composite laminate design guidelines* NASA Contract NAS1-19347 Northrop Grumman Military Aircraft Systems Division (El Segundo California)

[3] Pagano N J and Pipes R B 1971 *The influence of stacking sequence on laminate strength* (*J Comp Mat* Vol **5**) p 50-57

[4] Pipes R B and Pagano N J 1970 *Interlaminar stresses in composite laminates under uniform axial extension* (*J Comp Mat* Vol **4**) p 538-548

[5] Chermoshentseva A S, Pokrovskiy A M and Bokhoeva L A 2016 *The behavior of delaminations in composite materials – experimental results* (*IOP Conf Series: Materials Science and Engineering* Vol **116**) p 1-5

[6] Kovalovs A, Rucevskis S, Kulakov V and Aniskevich A 2016 *Delamination detection in carbon fibre reinforced composites using electrical resistance measurement* (*IOP Conf Series: Materials Science and Engineering* Vol **111**) p 1-6

[7] Dupir (Hudişteanu) I and Ţăranu N 2015 *The influence of the stacking sequence on stress and strain distributions for quasi-isotropic laminates* (*The Bulletin of the Polytechnic Institute of Jassy Construction. Architecture Section* LXI **2**) p 97-110

[8] Lekhnitskii S G 1982 *Theory of elasticity of an anisotropic body* Mir Publishers (Moscow)

[9] Reddy J N 2004 *Mechanics of laminated composite plates and shells Theory and analysis Second edition* CRC Press (United States of America: Boca Raton)

[10] Jones R M 1999 *Mechanics of composite materials Second edition* Taylor & Francis Inc. (Philadelphia)

[11] Cristescu N D, Craciun E-D and Soós E 2004 *Mechanics of elastic composites* CRC Series (Modern Mechanics and Mathematics) Chapman & Hall/CRC (United States of America)

[12] Voyiadjis G Z and Kattan P I 2005 *Mechanics of composite materials with Matlab* Springer (Netherlands)

[13] Herakovich C T 1998 *Mechanics of fibrous composites* John Wiley & Sons Inc. (University of Virginia: United States of America)

[14] Nettles A T 1994 *Basic mechanics of laminated composite plates Second edition* NASA Reference Publication (1351)

[15] Taranu N, Bejan L, Cozmanciuc R and Hohan R 2013 *Elements of composite materials* Politehnium Publishing House (Jassy)

Durability influence depending on the thickness and micro-hardness of AlTiN and TiN thin layers deposed on cutting inserts

A Bădănac[1], O Lupescu [1], D Paraschiv[1], C Ungureanu[1] and A Rotaru[2]

[1] Technical University "Gheorghe Asachi" of Iasi-Romania, Department of Machine Manufacturing Technology, Blvd. Mangeron, No. 59A, 700050, Iasi, Romania

[2] Technical University "Gheorghe Asachi" of Iasi-Romania, Department of Mechanical Engineering, Blvd. Mangeron, No. 59A, 700050, Iasi, Romania

E-mail: badanac_ana@yahoo.com

Abstract. During the exploitation process of cutting tools occurs phenomens as wear which leads to deterioration of functional parameters, performance and partial or total loss of the functioning capacity. The wear resistance of cutting tools depends by the nature of the material and by the working conditions (the quality of surface processing, lubrication quality, temperature). For increased the durability of cutting tools and increase the wear resistance are taken action which aim, improving the surface quality. For increase the surfaces quality are known different deposit methods in vacuum of materials by coating technology PVD as: aluminum-titanium nitride (AlTiN) and titanium nitride (TiN). The coatings realized by vacuum deposition have gained special attention because of their unique physical and chemical properties for example excellent resistance to oxidation at high temperatures. The most important characteristics which are checked in the deposition process of thin layers are: thickness and microhardness. The thickness and microhardness influence significantly the cutting tool durability. In this paper the authors carried out researches regarding the coatings of AlTiN and TiN in thin layers, on surface of some cutting inserts. It was studied the durability influence depending on the thickness and microhardness of AlTiN and TiN thin layers.

1. Introduction

During the technological process of cutting, is obtained modification of the form of cutting tools by removing surplus material in the form of chips, in order to obtain surfaces with certain configurations, in a determined field of tolerance with a required roughness. The form modification is obtained for the blank. The processability factors (the mechanical forces occurring during processing, the formation and evacuation of chips, generation and transfer of heat, as well as the tool wear) adversely affect the life of the tool, reliability and machining process duration, as well as the quality of the final product. For strengthening surfaces of cutting tools are used coatings with different materials with better properties than the base material of the tool. Deposition of some thin layers with materials ultra-hard on cutting tools it led to an increased durability and to reducing wear. Reducing the cutting tool wear it is important for achieve productivity high enough. The coatings of titanium aluminum nitride and titanium nitride are very tough and increase the cutting tools sustainability, [1,2]. The tools coated with aluminum-titanium nitride and titanium nitride can process at high speeds and provides the possibility to process a high number of parts in a smaller interval than the uncoated tools. The coatings are resistant to all types of wear as well as: abrasive wear, the wear of settlement surface and clearance, crater wear, the flanks wear and the forming on cutting edge some deposition. The occurrence of wear on cutting inserts it occurs depending on the thickness and microhardness of AlTiN and TiN layers deposited on cutting inserts. As the layers of TiN and TiN deposited on cutting inserts are thicker and tougher with that much wears occurs harder, [3,4].

2. Method and Equipment's used

In order to carry out thickness tests of the deposited layers was used devices BAQ kaloMax and RMA 5, available in the S.C. RULMENŢI S.A. BÂRLAD company. The BAQ kaloMax device is a grinder type spherical cap realized to determine the thickness of coatings and the layers system. The result of the grinding process depends on the both sides of friction system, of the sample and of the kaloMax device. The KaloMax device parameters which influence the result are: diamond abrasive paste, the number of revolutions, grinder period, diameter and the normal force exercised by the ball. The functioning of BAQ kaloMax device relies on sectioning the deposited layer, in measuring the spherical cap size obtained by abrasion by means of a steel ball which slide over the covered cutting inserts after previously it was deposited on insert some diamond paste. The upper part of vise jaws is amended by approaching and distancing depending on the shape of the sample. The clamping system is mounted on a table movable longitudinally and transversely. Ball is engaged in advance by an electric motor, [5,6]. After a certain period of time the steel ball achieved through diamond paste, a spherical cap. For determining the thickness of the layer deposited on the sample, was measured from its surface, the diameter of the cap and the diameter of the boundary between the layer and the base material. These diameters were measured by reflected light microscope, RMA 5. Both equipment's BAQ kaloMax respectively RMA 5 are shown in figure 1.

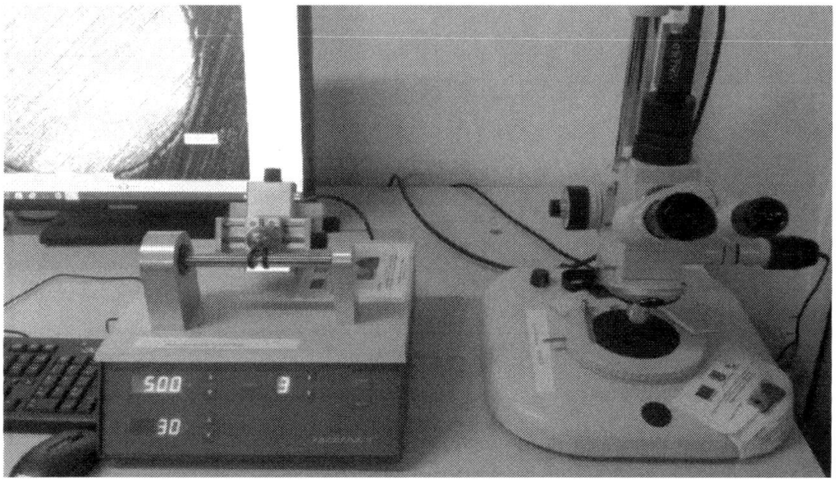

a) b)

Figure 1. The BAQ kaloMax and RMA 5 devices
a) BAQ kaloMax; b) RMA 5

The usual method to achieve a microhardness value is to measure the depth or indentation left by an indenter of a specific shape, with a specific force applied for a specific time. MV-100A/102A, figure 2, existing in the S.C. RULMENTI S.A. BÂRLAD company is specifically designed to test microhardness of small samples or metal, metal foils, thin layers reinforcement and layers of galvanized. The test device MV-100A / 102A exert some testing forces on the sample, using weights and a lever mechanism (amplifier force). The device engine drive the lever to release the corresponding weights for testing force selected. Then the weights released press the indenter down achieving a form on the sample at the right time. After indenter left the sample to return to the original position, is adjusted the turret at 40x, to measure the diagonals. The resulting values after measuring diagonals, resulting from the indentation are inserted in the integrated computer of device with that obtains a value of the microhardness, depending on the piece used.

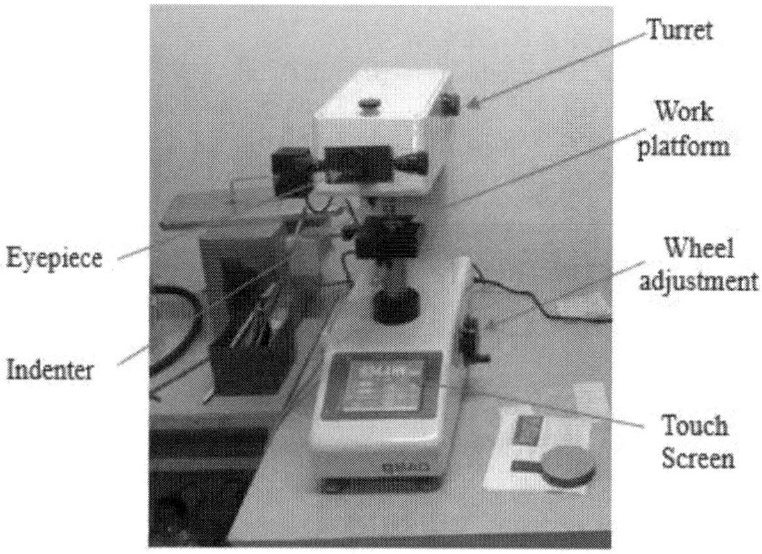

Figure 2. Device for testing microhardness MV-100A/102A

The test was carried out in two stages. First, the diamond indenter is driven into the surface of the test material by applying a known assay force. Secondly, the length of the diagonal measurement of the resulting the indentation is introduced into computer integrated, from where resulted the value of microhardness at Vickers scale (HV).

3. Results

For the achievement of experimental research were used some cutting inserts type SPUN 120312. After the cutting inserts were coated with aluminium-titanium (AlTiN) thin layer and with titanium (TiN) thin layer with these were performed several operations to obtain radiuses and chamfers on some different work-pieces. The cutting inserts coated with aluminium-titanium (AlTiN) thin layer are shown in figure 3 respectively those coated with titanium (TiN) thin layer are shown in figure 4.

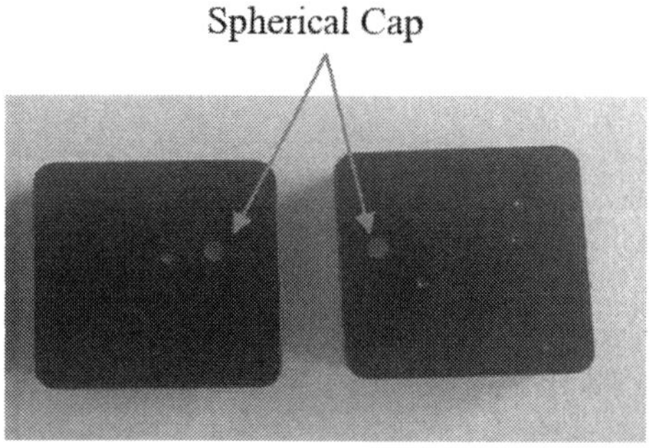

Figure 3. The cutting inserts coated with AlTiN thin layer

Figure 4. The cutting inserts coated with TiN thin layer

The thickness values of cutting inserts type SPUN 120312 coated with AlTiN resulted after using the ball-cratering method by using the BAQ kaloMax device are shown in figure 5 and for cutting inserts type SPUN 120312 coated with TiN are shown in figure 7. In figure 6 are presented the measurement values for microhardness of cutting inserts type SPUN 120312 coated with AlTiN, respectively in figure 8 are presented the measurement values for cutting inserts type SPUN 120312 coated with TiN. The measurement for microhardness were completed using the equipment MV-100A/102A.

Figure 5. Thickness measurements of cutting inserts type SPUN 120312 coated with AlTiN

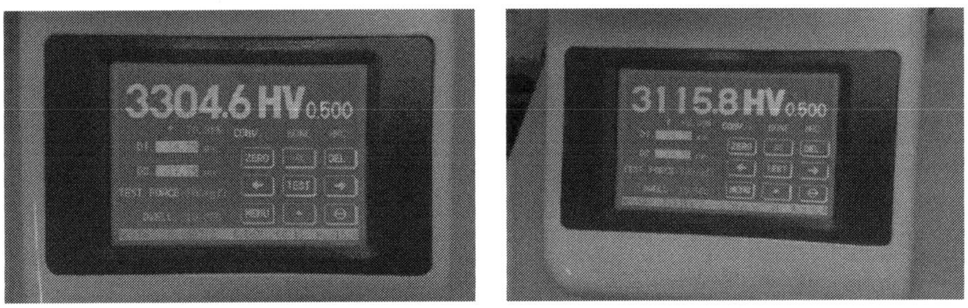

Figure 6. Microhardness measurements of cutting inserts type SPUN 120312 coated with AlTiN

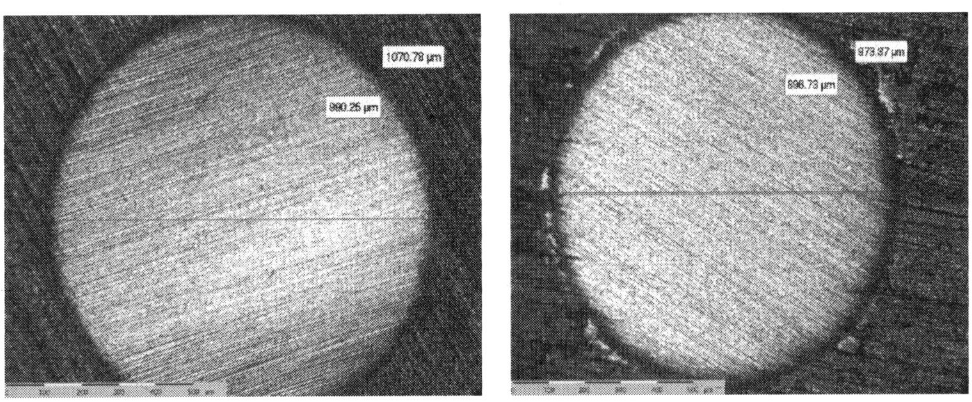

Figure 7. Thickness measurements of cutting inserts type SPUN 120312 coated with TiN

Figure 8. Microhardness measurements of cutting inserts type SPUN 120312 coated with TiN

The increased durability obtained, due to the deposition of AlTiN and TiN thin-layer on the cutting inserts and the layer thickness, respectively layer microhardness of the both types of cutting inserts used are listed in table 1. The data indicate getting the highest values of durability for the cutting inserts type SPUN 120312 coated with aluminium-titanium with the the biggest microhardness, that were placed at the highest height in vacuum chamber, case in which was processed the highest volume of manufactured parts.

Table 1. The values obtained for microhardness, durability and thickness layer

Type of Layer Deposited	Revolutions per minute	Layer thickness [μm]	Ball diameter [mm]	Layer microhardness [HV]	Durability tested [pcs]
AlTiN	500 1/min	0.97	30	3304.6	700
AlTiN	500 1/min	0.82	30	3115.8	500
TiN	500 1/min	1.38	30	2347.1	400
TiN	500 1/min	1.19	30	2153.3	350

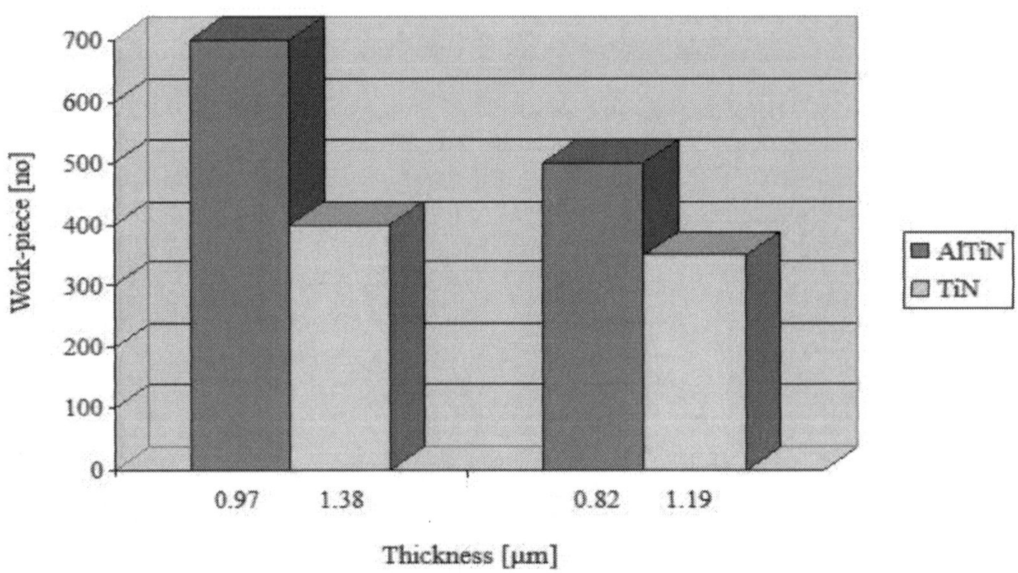

Figure 9. Graphical representation of processed parts number reported to thickness with the type cutting insert SPUN 120312

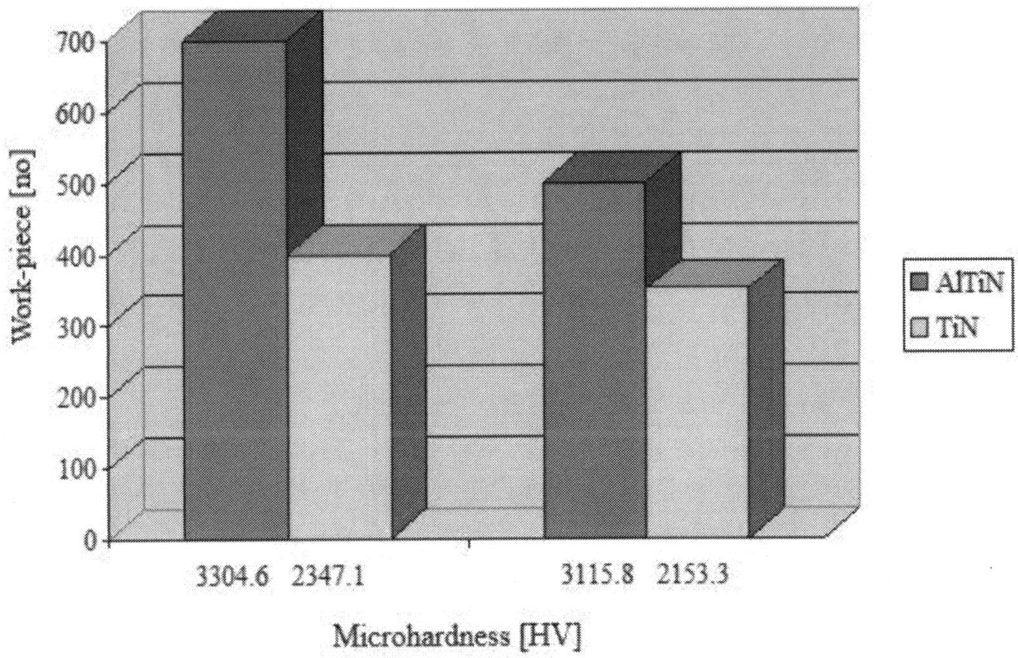

Figure 10. Graphical representation of processed parts number reported to microhardness with the type cutting insert SPUN 120312

In the graphical representations is rendered the manufactured parts volume, processed with the cutting inserts for: layer thickness in figure 9 and layer microhardness in figure 10.

4. Conclusion

Regarding the volume of processed parts with cutting inserts coated with AlTiN and TiN it is noted a direct proportionality with increasing the time as how the cutting inserts keep their qualities of cutting but, with a lower productivity in the case of using cutting inserts coated with TiN. The results obtained opens new opportunities for continuing the studies in this domain, future researches following: diversification of the qualitative types of cutting inserts coated with thin layers of TiN and AlTiN and the study of their behaviour in the processes of machining; analyse of thin layers of TiN and AlTiN influence on the microhardness increase and the durability of the cutting inserts.

5. References

[1] Enache S, Tănase I, Străjescu E, 1988, *The cutting tools technology*, vol. II, (Bucharest: Technique).

[2] Paraschiv D, et. all, 2010, *Technologies for high-performance for increase the parts durability*, (Iasi: Junimea).

[3] Martinu L et al., 2009, *Handbook of Deposition Technologies for Films and Coatings*, Third Edition, Peter M Martin, Ed., Elsevier, 417.

[4] Brogren M, Harding G. L, Karmhag R, Ribbing C, Gunnar G, Niklasson A., Stenmark L, 2000, *Titanium – aluminum – nitride coatings for satellite temperature control*. Thin Solid Films 370, pp.268 – 277.

[5] Teer BC-2, Jan. 2009, *Ball Crater Device Specification and User Guide*, Teer Coatings Ltd.

[6] Gee M G, et. all, 2003, *Progress towards standardisation of ball cratering, Wear*, vol. 255, pp. 1-13.

IOP Conf. Series: Materials Science and Engineering **161** (2016) 012054 doi:10.1088/1757-899X/161/1/012054

Influence of the Metal Volume Fraction on the maximum deflection and impact load of GLARE plates subjected to low velocity impact

GSE Bikakis, A Savaidis, P Zalimidis, S Tsitos

Department of Mechanical Engineering Educators, School of Pedagogical and Technological Education, N. Heraklion, GR 141 21, Athens, Greece

E-mail: bikakis.george@yahoo.com

Abstract. Fiber-metal laminates are hybrid composite materials, consisting of alternating metal layers bonded to fiber-reinforced prepreg layers. GLARE (GLAss REinforced) belongs to this new family of materials. GLARE is the most successful fiber-metal laminate up to now and is currently being used for the construction of primary aerospace structures, such as the fuselage of the Airbus A380 air plane. Impact properties are very important in aerospace structures, since impact damage is caused by various sources, such as maintenance damage from dropped tools, collision between service cars or cargo and the structure, bird strikes and hail. The principal objective of this article is to evaluate the influence of the Metal Volume Fraction (MVF) on the low velocity impact response of GLARE fiber-metal laminates. Previously published differential equations of motion are employed for this purpose. The low velocity impact behavior of various circular GLARE plates is predicted and characteristic values of impact variables, which represent the impact phenomenon, are evaluated versus the corresponding MVF of the examined GLARE material grades. The considered GLARE plates are subjected to low velocity impact under identical impact conditions. A strong effect of the MVF on the maximum impact load and a significant effect on the maximum plate deflection of GLARE plates has been found.

1. Introduction

Fiber-metal laminates are hybrid composite materials, consisting of alternating metal layers bonded to fiber-reinforced prepreg layers. ARALL (Aramid Reinforced ALuminum Laminates), CARALL (CArbon Reinforced ALuminum Laminates) and GLARE (GLAss REinforced) belong to this new family of materials. GLARE is the most successful fiber-metal laminate up to now and is currently being used for the construction of primary aerospace structures, such as the fuselage of the Airbus A380 air plane. Further applications have also been considered: aircraft cargo floors of Boeing 777, aircraft engine cowlings, bonded GLARE patch repair, aircraft stiffeners with a wide variety of shapes, cargo containers, seamless GLARE tubes.

Impact properties are very important in aerospace structures, since impact damage is caused by various sources, such as maintenance damage from dropped tools, collision between service cars or cargo and the structure, bird strikes and hail [1]. Much work has been published on the subject of impact, concerning conventional composite materials, which are frequently used in aerospace structures. Analytical, numerical and experimental impact studies have been considered. Recently, due to the importance of understanding the response of GLARE structures to impact loading, many researchers have studied impact phenomena concerning GLARE. In these studies, the low and high velocity impact, and the ballistic impact response of GLARE are analyzed.

This article focuses on predicting the influence of Metal Volume Fraction (MVF) on the response of thin circular clamped GLARE plates, subjected to low velocity impact. Low velocity impacts represent damage from service trucks, cargo containers and dropped tools during maintenance [1]. A typical test method, employed by researchers in order to study the low velocity impact behavior of GLARE, is to apply impact loads on circular clamped plates using a drop weight impact tester, equipped with a hemispherical impactor [1, 2]. Consequently, prediction of the response of circular GLARE plates to low velocity impact has great practical importance.

This article deals with the theoretical prediction of the dynamic response of thin circular clamped GLARE plates, subjected to low velocity impact by a lateral hemispherical impactor, striking at their center. In the work of Guocai et al. this problem was studied experimentally [2]. Caprino et al. [3, 4] developed a mechanistic model to predict the response of square fibreglass-aluminum laminates under low velocity impact, which requires the implementation of several experimental tests. Hoo Fatt et al. [5] employed a spring-mass model to predict the ballistic response of clamped square GLARE panels. In two articles by Tsamasphyros and Bikakis, the differential equations of motion, corresponding to the impact of a hemispherical impactor striking on a circular clamped GLARE plate, were derived using a spring-mass model as well [6] and solved [7].

The principal objective of this article is to evaluate and compare the influence of the MVF on the maximum impact deflection and the maximum impact load of thin circular clamped fiber-metal laminates consisting of different standard GLARE grades. Previously published equations are employed for this purpose. The low velocity impact behavior of GLARE 2A-3/2-0.4, GLARE 2A-4/3-0.238, GLARE 3-3/2-0.4, GLARE 4-3/2-0.317, GLARE 5-3/2-0.233, GLARE 2A-4/3-0.256, GLARE 4-3/2-0.342 and GLARE 5-3/2-0.283 circular plates of equal total thickness and of equal areal weight, for various diameters, is predicted under identical impact conditions. Then, the effect of the MVF of these GLARE grades on the maximum plate deflection and the maximum impact load is compared and analyzed, in connection with the thickness and the areal weight of the plates. The presented results shed light to the impact response of GLARE fiber-metal laminates and will help engineers and researchers to understand the behavior of the examined GLARE grades. The derived conclusions can be used for the design of GLARE structures when their exposure to low velocity impact is expected. To the authors' knowledge, no other similar impact study concerning GLARE fiber metal laminates has been published in the current scientific literature.

2. Analytical equations

We consider a thin clamped circular GLARE plate with radius α and thickness h. The plate is struck at its center by an impactor with large mass M_0 and initial kinetic energy E_k. The impactor has a hemispherical tip with radius R. The plate consists of alternating layers of aluminum and glass-epoxy. The aspect ratio α/h is assumed high, so that shear deformation and local indentation are negligible. The plate is clamped along its boundary. As the impactor progresses, the load P due to the global deflection of the plate and the corresponding central plate deflection w_o increase, while the velocity of the impactor decreases, until the plate reaches its maximum deformation. At this point, due to the elastic strain energy of the prepreg layers, which have been stretched during the loading stage, the plate starts to move toward the opposite direction, until the impact load P becomes equal to zero. Using a spring-mass model, the differential equations of motion corresponding to this physical phenomenon were derived in reference [6] and solved in reference [7]. Here, the basic employed equations from reference [7] for our calculations are presented. These equations have been validated by comparisons with experimental data [6, 7]. For large plate deflections, the impact load during loading is given by:

$$P_L(w_o) = K_p w_o + K_{el} w_o^3 \tag{1}$$

Equation (1) is valid for $0 \le w_o \le w_o^{max}$ where w_o^{max} is the maximum central plate deflection due to the impact. The corresponding unloading impact load is given by:

$$P_U(w_o) = K_p(2w_o - w_o^{max}) + K_{el}w_o^3 \tag{2}$$

Equation (2) is valid for $w_o^f \le w_o \le w_o^{max}$, where w_o^f is the final deflection of zero impact load. It is noted that the final deflection is equal to the permanent dent depth caused by the impact on the GLARE plate [6]. Analytical expressions to calculate w_o^{max} and w_o^f can be found in reference [7].

The coefficients K_p and K_{el} are as follows [6-8]:

$$K_p = 0.576N_x + 0.576N_y + 0.734N_{xy} \tag{3}$$

$$K_{el} = [0.62A_{11} + 0.62A_{22} + 0.412(A_{12} + 2A_{66})]\frac{1}{\alpha^2} \tag{4}$$

The in-plane forces N_x, N_y and N_{xy} of the aluminum layers are [6-8]:

$$N_x = N_y = m\sigma_o t_{Al}, \qquad N_{xy} = m\frac{\sigma_o}{\sqrt{3}}t_{Al} \tag{5}$$

where m is the number of aluminum layers, σ_o is the yield stress of aluminum and t_{AL} is the thickness of each aluminum layer. A_{ij} are the extensional stiffnesses of the laminate.

The low velocity impact is simulated in three stages. During the first stage, the impactor starts to deform the plate until internal damage due to delamination occurs. For the first impact stage, the time interval Δt_1, in order for the impactor to move from the initial position $w_o^1 = 0$ to an arbitrary position $w_o^2 \le w_o^d$, is calculated numerically by the following integration:

$$\Delta t_1 = \int_0^{w_o^2} \frac{dw_o}{\sqrt{v^2 - \frac{K_p w_o^2}{M_o + m_e} - \frac{K_{el} w_o^4}{2(M_o + m_e)}}} \tag{6}$$

where w_o^d is the delamination position calculated using the pertinent analytical equation of reference [7]. The initial velocity v of the impactor is equal to $\sqrt{2E_k / M_o}$ and m_e is the effective plate mass calculated as indicated in [6, 7, 9].

The second impact stage starts after delamination and continues up to the position of maximum plate deformation w_o^{max}. For the second impact stage, the time interval Δt_2, in order for the impactor to move from the initial position $w_o^1 = w_o^d$ to an arbitrary position $w_o^2 \le w_o^{max}$, is calculated numerically by the following integration:

IOP Conf. Series: Materials Science and Engineering **161** (2016) 012054 doi:10.1088/1757-899X/161/1/012054

$$\Delta t_2 = \int_{w_o^d}^{w_o^2} \frac{dw_o}{\sqrt{v^2 - \frac{2E_{del}}{M_o + m_e} - \frac{K_P w_o^2}{M_o + m_e} - \frac{K_{el} w_o^4}{2(M_o + m_e)}}}. \tag{7}$$

E_{del} is the energy required for the sudden delamination among glass-epoxy layers, and can be calculated as indicated in references [5-7, 9-12].

The third impact stage starts from the position of maximum plate deformation w_o^{max} and is completed when the impact load becomes equal to zero.

3. Results

We consider plates with identical total thickness equal to 1.7 mm consisting of the following standard grades: GLARE 2A-3/2-0.4, GLARE 2A-4/3-0.238, GLARE 3-3/2-0.4, GLARE 4-3/2-0.317 and GLARE 5-3/2-0.233. We also consider plates with identical areal weight equal to 4.23x10^{-5} N/mm^2 consisting of the following standard grades: GLARE 2A-3/2-0.4, GLARE 2A-4/3-0.256, GLARE 3-3/2-0.4, GLARE 4-3/2-0.342 and GLARE 5-3/2-0.283.

The examined GLARE plates consist of alternating S2-glass UD fiber prepreg layers and 2024-T3 aluminum layers. Each prepreg ply has a thickness of 0.125 mm. All aluminum layers of each GLARE plate have equal thickness. We apply the equations presented in the previous section in order to calculate the low velocity impact response of the aforementioned standard GLARE grades. The material properties considered for our calculations are from references [1, 5, 13]. The prepreg tensile failure strain, aluminum and prepreg weight density, have been determined from our correspondence with the manufacturer of GLARE. We have analyzed circular plates with 50 mm, 65 mm and 80 mm radiuses for each of the considered different GLARE configurations.

We consider that the impactor has a mass of 6.29 Kg, which is the same as in the experiments of [2], and initial kinetic energy equal to 13 J in all examined impact cases. At this energy level, the velocity of the impactor is low yielding no sensitivity of the material properties related to strain rates effects [4, 14]. Furthermore, we have calculated using the analytical model of reference [8] the first failure energy of all examined GLARE plates under static lateral indentation and verified that it is greater than 13 J in all cases. As explained in reference [6], this verification yields that the considered initial kinetic energy of the impactor does not cause first failure of the glass-epoxy layers due to tensile fracture.

In figures 1-3 the % deviation from the average maximum plate deflection due to impact versus the MVF, of GLARE plates with 50 mm, 65 mm and 80 mm radius is depicted. The plates have equal areal weight. MVF is defined as the ratio of the total thickness of aluminum layers divided by the laminate thickness, and is expressed as % percentage. A significant effect of the MVF on the maximum plate deflection can be observed in these figures. It can also be observed that for increasing MVF, the maximum plate deflection increases, with the exception of GLARE 2A-4/3-0.256 which has MVF equal to 58. This trend is not valid for all GLARE plates with equal thickness. In figure 4 the % deviation from the average maximum plate deflection due to impact versus the MVF, of GLARE plates with 80 mm radius is depicted. The plates have equal thickness.

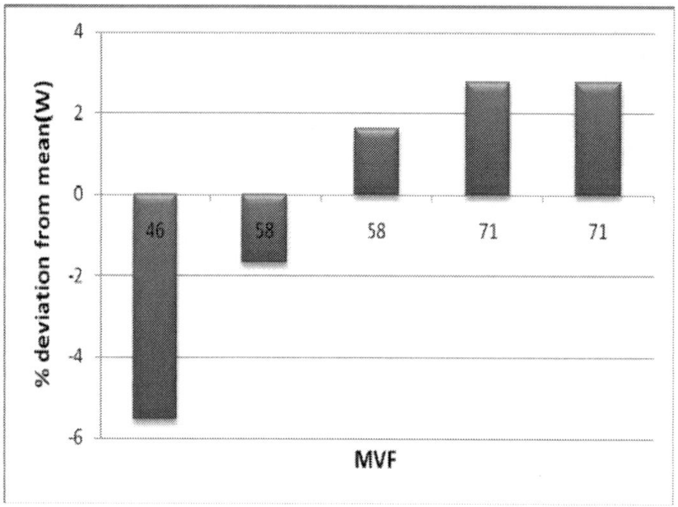

Figure 1. Maximum deflection vs MVF of plates with equal areal weight and 50 mm radius

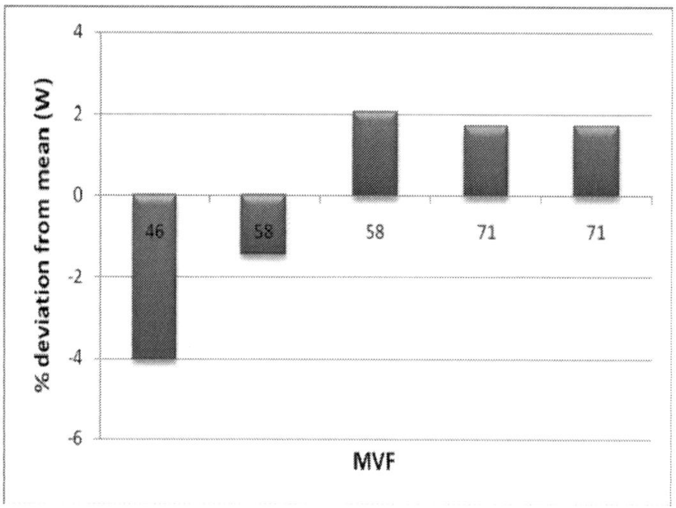

Figure 2. Maximum deflection vs MVF of plates with equal areal weight and 65 mm radius

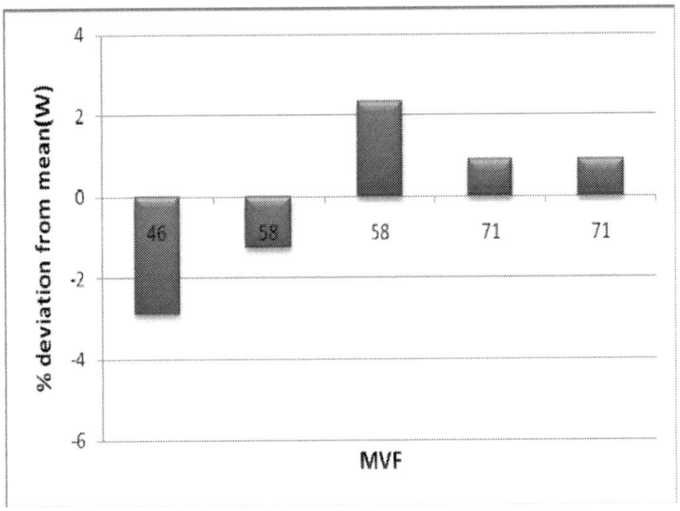

Figure 3. Maximum deflection vs MVF of plates with equal areal weight and 80 mm radius

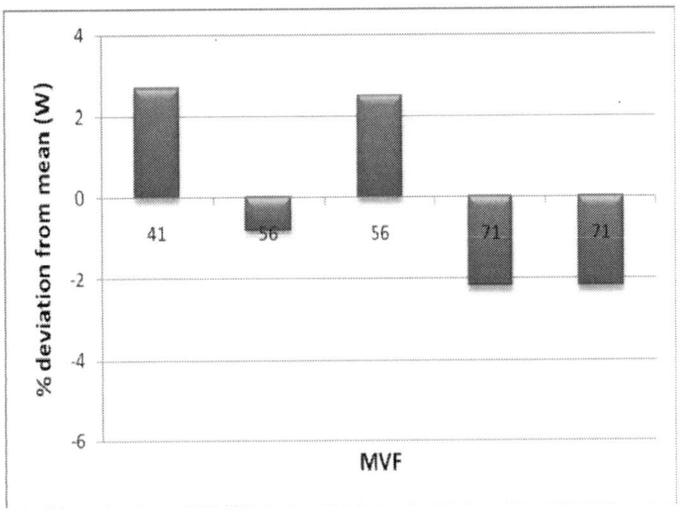

Figure 4. Maximum deflection vs MVF of plates with equal thickness and 80 mm radius

In figures 5-7 the % deviation from the average maximum impact load versus the MVF, of GLARE plates with 50 mm, 65 mm and 80 mm radius is depicted. The plates have equal areal weight. A strong effect of the MVF on the maximum impact load can be observed in these figures. It can also be observed that for increasing MVF, the maximum impact load increases, with the exception of GLARE 2A-4/3-0.256, which has MVF equal to 58. This trend is also valid for GLARE plates with equal thickness.

In figure 8 the % deviation from the average maximum impact load versus the MVF, of GLARE plates with 65 mm radius is depicted. The plates have equal thickness.

Based on our results, it is found that MVF can be used as a qualitative design parameter in order to alter the maximum impact load of GLARE plates with equal thickness or equal areal weight in a consistent manner, provided that the basic lay-up, number of aluminum and prepreg layers, does not change, as it changes in the case of GLARE 2A-4/3. It is also found that MVF can be used as a qualitative design parameter in order to alter the maximum plate deflection due to impact of GLARE plates with equal areal weight in a consistent manner, under the same provision.

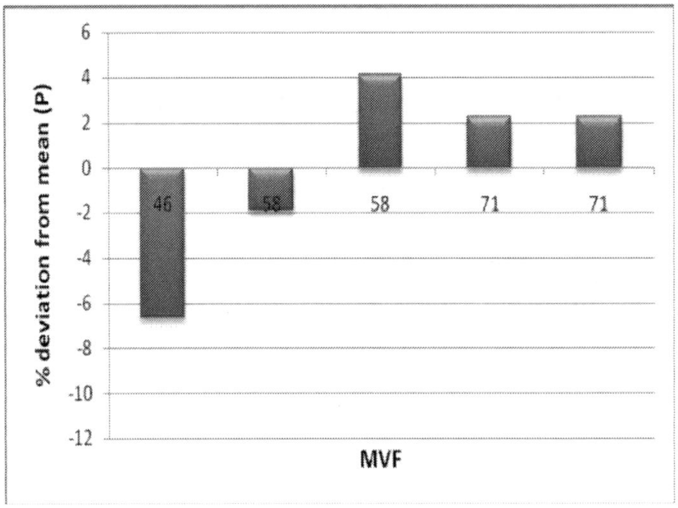

Figure 5. Maximum impact load vs MVF of plates with equal areal weight and 50 mm radius

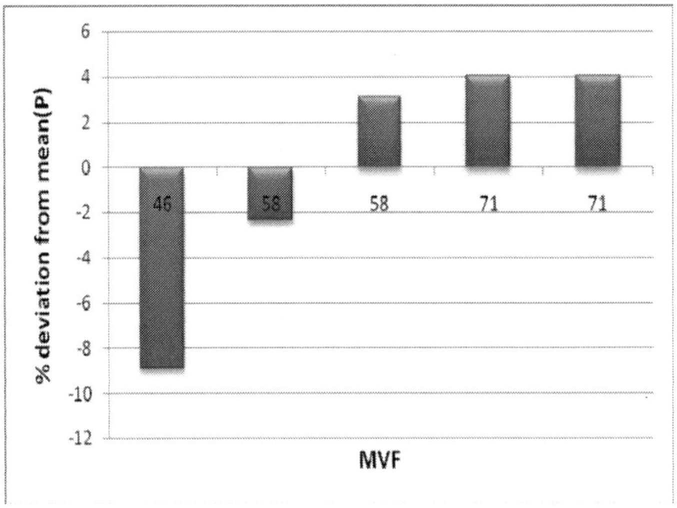

Figure 6. Maximum impact load vs MVF of plates with equal areal weight and 65 mm radius

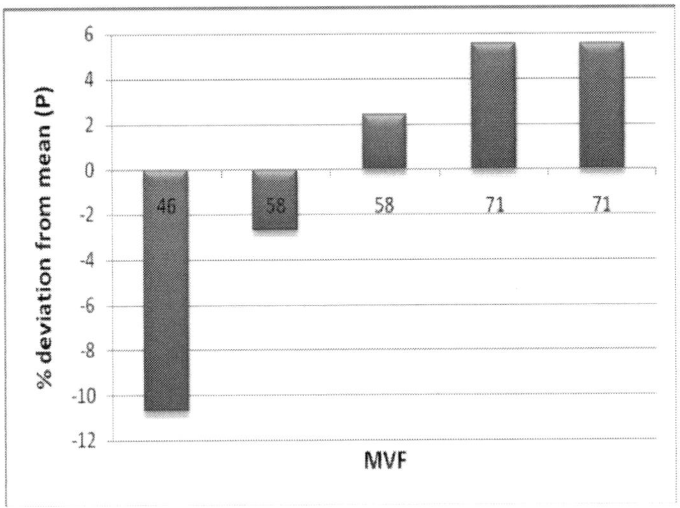

Figure 7. Maximum impact load vs MVF of plates with equal areal weight and 80 mm radius

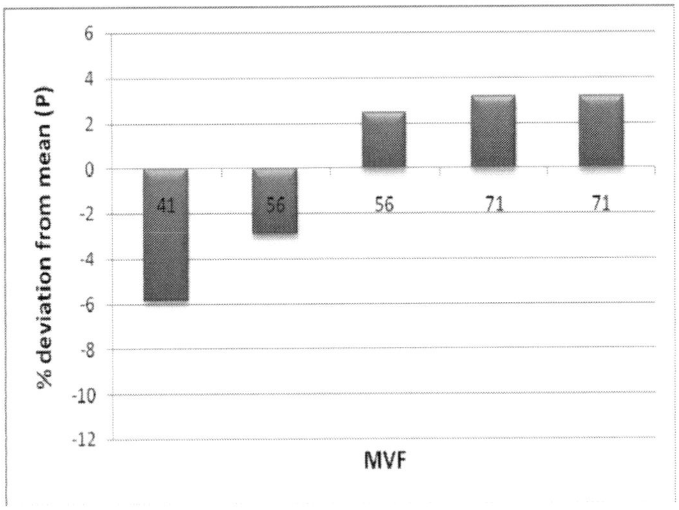

Figure 8. Maximum impact load vs MVF of plates with equal thickness and 65 mm radius

4. Conclusions

In this article, the influence of the MVF on the low velocity impact response of GLARE fiber-metal laminates is studied. Previously published differential equations of motion are employed for this purpose. The considered GLARE plates have equal thickness or equal areal weight and are subjected to low velocity impact under identical impact conditions.

A strong effect of the MVF on the maximum impact load and a significant effect on the maximum plate deflection of the GLARE plates has been found. Considering the same basic lay-up, when the MVF increases, the maximum impact load increases whereas the maximum plate deflection increases only when the plates have equal areal weight.

To the authors' knowledge, no other similar impact study concerning GLARE fiber metal laminates has been published in the current scientific literature. The influence of the MVF on other characteristic values of impact variables is an option for future research.

References

[1] Vlot A 1996 *Int. J. Impact Eng.* **18** 291-307
[2] Guocai W, Jenn-Ming Y, Hahn H T 2007 *J. Mater. Sci.* **42** 948-957
[3] Caprino G, Spataro G, Del Luongo S 2004 *Composites Part* A **35** 605-616
[4] Caprino G, Lopresto V, Iaccarino P 2007 *Composites Part* A **38** 290-300
[5] Hoo Fatt M S, Lin C, Revilock D M Jr et al 2003 *Comp. Struct.* **61** 73-88
[6] Tsamasphyros G J and Bikakis G S 2011 *J. Reinf. Plast. Compos.* **30** 978-987
[7] Tsamasphyros G J and Bikakis G S 2013 *Aerosp. Sci. Technol.* **29** 28-36
[8] Tsamasphyros G J and Bikakis G S 2009 *Adv. Compos. Lett.* **18** 11-19
[9] Tsamasphyros G J and Bikakis G S 2014 *Mech. Adv. Mater. Struc.* **21** 39-46
[10] Lin C, Hoo Fatt M S 2006 *J. Comp. Mater.* **40** 1801-1840
[11] Moriniere F D, Alderliesten R C, Sadighi M et al 2013 *Comp. Struct.* **100** 89-103
[12] Moriniere F D, Alderliesten R C, Benedictus R 2013 *Mech. Mater.* **66** 59-68
[13] Alderliesten R 2005 *Fatigue crack propagation and delamination growth in GLARE* (Delft : Delft University Press)
[14] Belingardi G and Vadori R 2002 *Int. J. Impact Eng.* **27** 213-229

IOP Conf. Series: Materials Science and Engineering **161** (2016) 012055 doi:10.1088/1757-899X/161/1/012055

Influence of the Metal Volume Fraction on the permanent dent depth and energy absorption of GLARE plates subjected to low velocity impact

GSE Bikakis, A Savaidis, P Zalimidis, S Tsitos

Department of Mechanical Engineering Educators, School of Pedagogical and Technological Education, N. Heraklion, GR 141 21, Athens, Greece

E-mail: bikakis.george@yahoo.com

Abstract. Fiber-metal laminates are hybrid composite materials, consisting of alternating metal layers bonded to fiber-reinforced prepreg layers. GLARE (GLAss REinforced) belongs to this new family of materials. GLARE is the most successful fiber-metal laminate up to now and is currently being used for the construction of primary aerospace structures, such as the fuselage of the Airbus A380 air plane. Impact properties are very important in aerospace structures, since impact damage is caused by various sources, such as maintenance damage from dropped tools, collision between service cars or cargo and the structure, bird strikes and hail. The principal objective of this article is to evaluate the influence of the Metal Volume Fraction (MVF) on the low velocity impact response of GLARE fiber-metal laminates. Previously published differential equations of motion are employed for this purpose. The low velocity impact behavior of various circular GLARE plates is predicted and characteristic values of impact variables, which represent the impact phenomenon, are evaluated versus the corresponding MVF of the examined GLARE material grades. The considered GLARE plates are subjected to low velocity impact under identical impact conditions. A strong effect of the MVF on the permanent dent depth and energy absorption of GLARE plates has been found.

1. Introduction

Fiber-metal laminates are hybrid composite materials, consisting of alternating metal layers bonded to fiber-reinforced prepreg layers. ARALL (Aramid Reinforced ALuminum Laminates), CARALL (CArbon Reinforced ALuminum Laminates) and GLARE (GLAss REinforced) belong to this new family of materials. GLARE is the most successful fiber-metal laminate up to now and is currently being used for the construction of primary aerospace structures, such as the fuselage of the Airbus A380 air plane. Further applications have also been considered: aircraft cargo floors of Boeing 777, aircraft engine cowlings, bonded GLARE patch repair, aircraft stiffeners with a wide variety of shapes, cargo containers, seamless GLARE tubes.

Impact properties are very important in aerospace structures, since impact damage is caused by various sources, such as maintenance damage from dropped tools, collision between service cars or cargo and the structure, bird strikes and hail [1]. Much work has been published on the subject of impact, concerning conventional composite materials, which are frequently used in aerospace structures. Analytical, numerical and experimental impact studies have been considered. Recently, due to the importance of understanding the response of GLARE structures to impact loading, many researchers have studied impact phenomena concerning GLARE. In these studies, the low and high velocity impact, and the ballistic impact response of GLARE are analyzed.

This article focuses on predicting the influence of Metal Volume Fraction (MVF) on the response of thin circular clamped GLARE plates, subjected to low velocity impact. Low velocity impacts represent damage from service trucks, cargo containers and dropped tools during maintenance [1]. A typical test method, employed by researchers in order to study the low velocity impact behavior of GLARE, is to apply impact loads on circular clamped plates using a drop weight impact tester, equipped with a hemispherical impactor [1, 2]. Consequently, prediction of the response of circular GLARE plates to low velocity impact has great practical importance.

This article deals with the theoretical prediction of the dynamic response of thin circular clamped GLARE plates, subjected to low velocity impact by a lateral hemispherical impactor, striking at their center. In the work of Guocai et al. this problem was studied experimentally [2]. Caprino et al. [3, 4] developed a mechanistic model to predict the response of square fibreglass-aluminum laminates under low velocity impact, which requires the implementation of several experimental tests. Hoo Fatt et al. [5] employed a spring-mass model to predict the ballistic response of clamped square GLARE panels. In two articles by Tsamasphyros and Bikakis, the differential equations of motion, corresponding to the impact of a hemispherical impactor striking on a circular clamped GLARE plate, were derived using a spring-mass model as well [6] and solved [7].

The principal objective of this article is to evaluate and compare the influence of the MVF on the permanent dent depth and impact energy absorption of thin circular clamped fiber-metal laminates consisting of different standard GLARE grades. Previously published equations are employed for this purpose. The low velocity impact behavior of GLARE 2A-3/2-0.4, GLARE 2A-4/3-0.238, GLARE 3-3/2-0.4, GLARE 4-3/2-0.317, GLARE 5-3/2-0.233, GLARE 2A-4/3-0.256, GLARE 4-3/2-0.342 and GLARE 5-3/2-0.283 circular plates of equal total thickness and of equal areal weight, for various diameters, is predicted under identical impact conditions. Then, the effect of the MVF of these GLARE grades on the permanent dent depth and impact energy absorption is compared and analyzed, in connection with the thickness and the areal weight of the plates. The presented results shed light to the impact response of GLARE fiber-metal laminates and will help engineers and researchers to understand the behavior of the examined GLARE grades. The derived conclusions can be used for the design of GLARE structures when their exposure to low velocity impact is expected. To the authors' knowledge, no other similar impact study concerning GLARE fiber metal laminates has been published in the current scientific literature.

2. Analytical equations

We consider a thin clamped circular GLARE plate with radius α and thickness h. The plate is struck at its center by an impactor with large mass M_0 and initial kinetic energy E_k. The impactor has a hemispherical tip with radius R. The plate consists of alternating layers of aluminum and glass-epoxy. The aspect ratio α/h is assumed high, so that shear deformation and local indentation are negligible. The plate is clamped along its boundary. As the impactor progresses, the load P due to the global deflection of the plate and the corresponding central plate deflection w_o increase, while the velocity of the impactor decreases, until the plate reaches its maximum deformation. At this point, due to the elastic strain energy of the prepreg layers, which have been stretched during the loading stage, the plate starts to move toward the opposite direction, until the impact load P becomes equal to zero. Using a spring-mass model, the differential equations of motion corresponding to this physical phenomenon were derived in reference [6] and solved in reference [7]. Here, the basic employed equations from reference [7] for our calculations are presented. These equations have been validated by comparisons with experimental data [6, 7]. For large plate deflections, the impact load during loading is given by:

$$P_L(w_o) = K_p w_o + K_{el} w_o^{\ 3} \tag{1}$$

Equation (1) is valid for $0 \le w_o \le w_o^{max}$ where w_o^{max} is the maximum central plate deflection due to the impact. The corresponding unloading impact load is given by:

$$P_U(w_o) = K_p\left(2w_o - w_o^{max}\right) + K_{el}w_o^{\,3} \tag{2}$$

Equation (2) is valid for $w_o^f \le w_o \le w_o^{max}$, where w_o^f is the final deflection of zero impact load. It is noted that the final deflection is equal to the permanent dent depth caused by the impact on the GLARE plate [6]. Analytical expressions to calculate w_o^{max} and w_o^f can be found in reference [7].

The coefficients K_p and K_{el} are as follows [6-8]:

$$K_p = 0.576N_x + 0.576N_y + 0.734N_{xy} \tag{3}$$

$$K_{el} = \left[0.62A_{11} + 0.62A_{22} + 0.412\left(A_{12} + 2A_{66}\right)\right]\frac{1}{\alpha^2} \tag{4}$$

The in-plane forces N_x, N_y and N_{xy} of the aluminum layers are [6-8]:

$$N_x = N_y = m\sigma_o t_{Al}\,, \qquad N_{xy} = m\frac{\sigma_o}{\sqrt{3}}t_{Al} \tag{5}$$

where m is the number of aluminum layers, σ_o is the yield stress of aluminum and t_{AL} is the thickness of each aluminum layer. A_{ij} are the extensional stiffnesses of the laminate.

The low velocity impact is simulated in three stages. During the first stage, the impactor starts to deform the plate until internal damage due to delamination occurs. For the first impact stage, the time interval Δt_1, in order for the impactor to move from the initial position $w_o^1 = 0$ to an arbitrary position $w_o^2 \le w_o^d$, is calculated numerically by the following integration:

$$\Delta t_1 = \int_0^{w_o^2} \frac{dw_o}{\sqrt{v^2 - \dfrac{K_P w_o^{\,2}}{M_o + m_e} - \dfrac{K_{el}w_o^{\,4}}{2(M_o + m_e)}}} \tag{6}$$

where w_o^d is the delamination position calculated using the pertinent analytical equation of reference [7]. The initial velocity v of the impactor is equal to $\sqrt{2E_k / M_o}$ and m_e is the effective plate mass calculated as indicated in [6, 7, 9].

The corresponding kinetic energy is:

$$E_k = \frac{1}{2}M_o\left(v^2 - \frac{K_P w_o^{\,2}}{M_o + m_e} - \frac{K_{el}w_o^{\,4}}{2(M_o + m_e)}\right) \tag{7}$$

The second impact stage starts after delamination and continues up to the position of maximum plate deformation w_o^{max}. For the second impact stage, the time interval Δt_2, in order for the impactor to move from the initial position $w_o^1 = w_o^d$ to an arbitrary position $w_o^2 \le w_o^{max}$, is calculated numerically by the following integration:

$$\Delta t_2 = \int_{w_o^d}^{w_o^2} \frac{dw_o}{\sqrt{v^2 - \dfrac{2E_{del}}{M_o + m_e} - \dfrac{K_P w_o{}^2}{M_o + m_e} - \dfrac{K_{el} w_o{}^4}{2(M_o + m_e)}}} \tag{8}$$

E_{del} is the energy required for the sudden delamination among glass-epoxy layers, and can be calculated as indicated in references [5-7, 9-12].

The kinetic energy during the second stage is given by:

$$E_k = \frac{1}{2} M_o \left(v^2 - \frac{2E_{del}}{M_o + m_e} - \frac{K_P w_o{}^2}{M_o + m_e} - \frac{K_{el} w_o{}^4}{2(M_o + m_e)} \right) \tag{9}$$

The third impact stage starts from the position of maximum plate deformation w_o^{max} and is completed when the impact load becomes equal to zero. For this impact stage, the time interval Δt_3 , in order for the impactor to move from the initial position $w_o^1 = w_o^{max}$ to an arbitrary position $w_o^2 \geq w_o^f$, is calculated numerically by the following integration:

$$\Delta t_3 = \int_{w_o^{max}}^{w_o^2} \frac{-dw_o}{\sqrt{\dfrac{2K_P w_o (w_o^{max} - w_o) + 0.5K_{el} (w_o^{max\,4} - w_0{}^4)}{M_o + m_e}}} \tag{10}$$

The kinetic energy during the third stage is given by:

$$E_k = \frac{1}{2} M_o \left(\frac{2K_P w_o (w_o^{max} - w_o) + 0.5K_{el} (w_o^{max\,4} - w_o{}^4)}{M_o + m_e} \right) \tag{11}$$

3. Results

We consider plates with identical total thickness equal to 1.7 mm consisting of the following standard grades: GLARE 2A-3/2-0.4, GLARE 2A-4/3-0.238, GLARE 3-3/2-0.4, GLARE 4-3/2-0.317 and GLARE 5-3/2-0.233. We also consider plates with identical areal weight equal to 4.23x10^{-5} N/mm^2 consisting of the following standard grades: GLARE 2A-3/2-0.4, GLARE 2A-4/3-0.256, GLARE 3-3/2-0.4, GLARE 4-3/2-0.342 and GLARE 5-3/2-0.283.

The examined GLARE plates consist of alternating S2-glass UD fiber prepreg layers and 2024-T3 aluminum layers. Each prepreg ply has a thickness of 0.125 mm. All aluminum layers of each GLARE plate have equal thickness. We apply the equations presented in the previous section in order to calculate the low velocity impact response of the aforementioned standard GLARE grades. The material properties considered for our calculations are from references [1, 5, 13]. The prepreg tensile failure strain, aluminum and prepreg weight density, have been determined from our correspondence with the manufacturer of GLARE. We have analyzed circular plates with 50 mm, 65 mm and 80 mm radiuses for each of the considered different GLARE configurations.

We consider that the impactor has a mass of 6.29 Kg, which is the same as in the experiments of [2], and initial kinetic energy equal to 13 J in all examined impact cases. At this energy level, the velocity of the impactor is low yielding no sensitivity of the material properties related to strain rates effects [4, 14]. Furthermore, we have calculated using the analytical model of reference [8] the first failure energy of all examined GLARE plates under static lateral indentation and verified that it is greater than 13 J in all cases. As explained in reference [6], this verification yields that the considered initial kinetic energy of the impactor does not cause first failure of the glass-epoxy layers due to tensile fracture.

In figures 1-3 the % deviation from the average permanent dent depth due to impact versus the MVF, of GLARE plates with 50 mm, 65 mm and 80 mm radius is depicted. The plates have equal areal weight. MVF is defined as the ratio of the total thickness of aluminum layers divided by the laminate thickness, and is expressed as % percentage. A strong effect of the MVF on the permanent dent depth can be observed in these figures. It can also be observed that for increasing MVF, the permanent dent depth increases. This trend is also valid for GLARE plates with equal thickness, with the exception of GLARE 2A-4/3-0.238, which has MVF equal to 56.

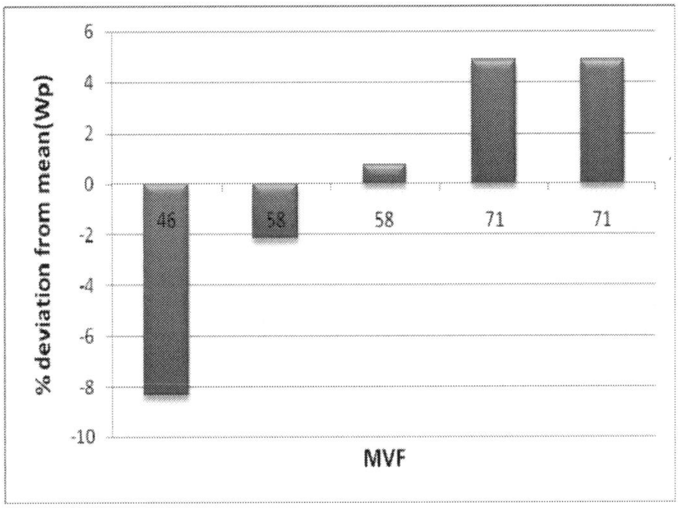

Figure 1. Permanent dent depth vs MVF of plates with equal areal weight and 50 mm radius

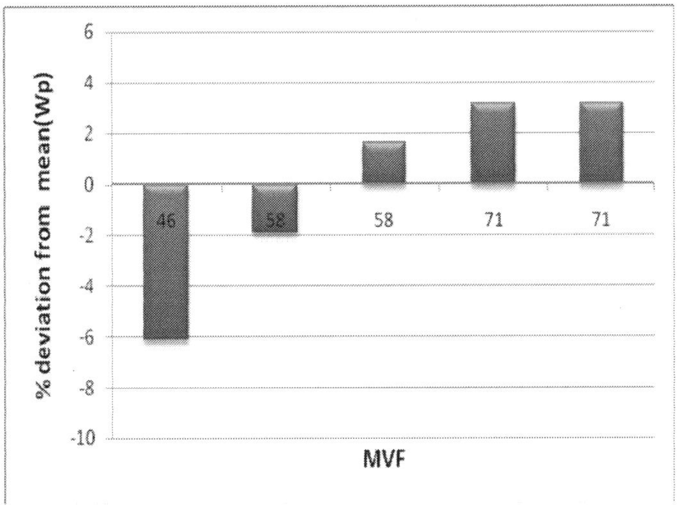

Figure 2. Permanent dent depth vs MVF of plates with equal areal weight and 65 mm radius

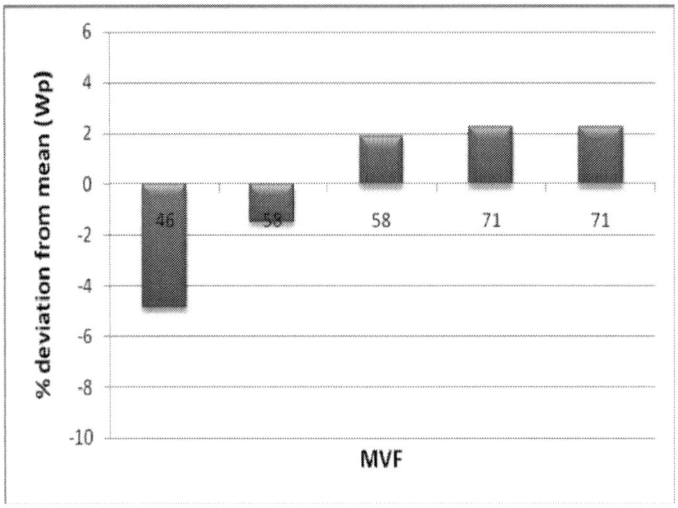

Figure 3. Permanent dent depth vs MVF of plates with equal areal weight and 80 mm radius

In figures 4-6 the % deviation from the average permanently absorbed impact energy versus the MVF, of GLARE plates with 50 mm, 65 mm and 80 mm radius is depicted. The plates have equal thickness. A strong effect of the MVF on the permanently absorbed impact energy can be observed in these figures. It can also be observed that for increasing MVF, the permanently absorbed impact energy decreases, with the exception of GLARE 2A-4/3-0.238, which has MVF equal to 56.

Based on our results, it is found that MVF can be used as a qualitative design parameter in order to alter the permanent dent depth and the absorbed impact energy of GLARE plates with equal thickness or equal areal weight in a consistent manner, provided that the basic lay-up, number of aluminum and prepreg layers, does not change, as it changes in the case of GLARE 2A-4/3.

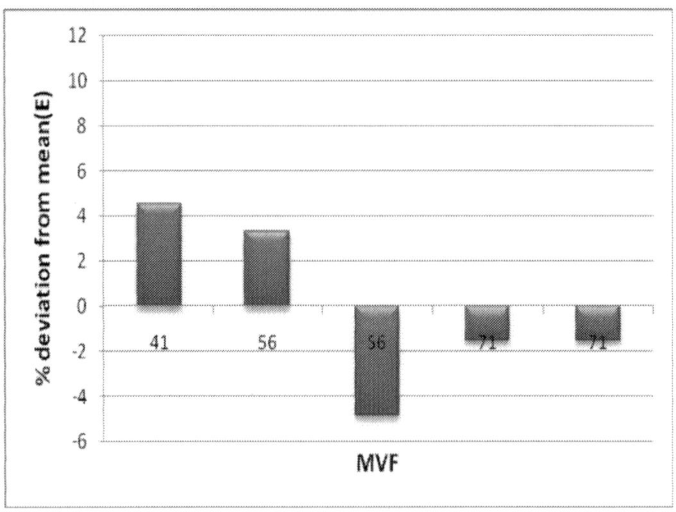

Figure 4. Absorbed energy vs MVF of plates with equal thickness and 50 mm radius

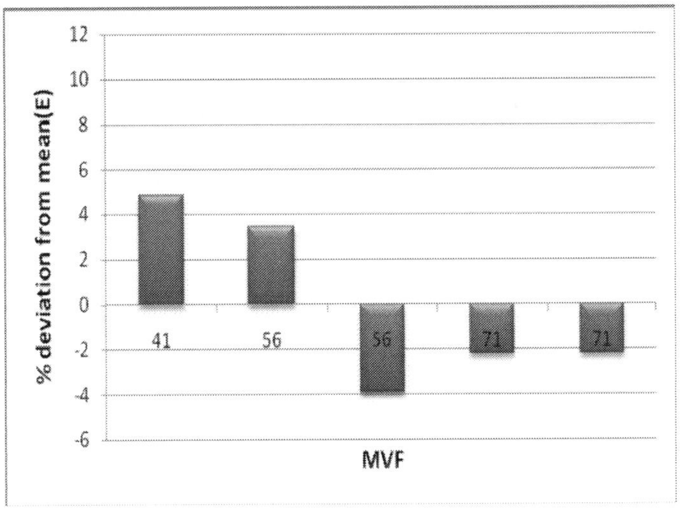

Figure 5. Absorbed energy vs MVF of plates with equal thickness and 65 mm radius

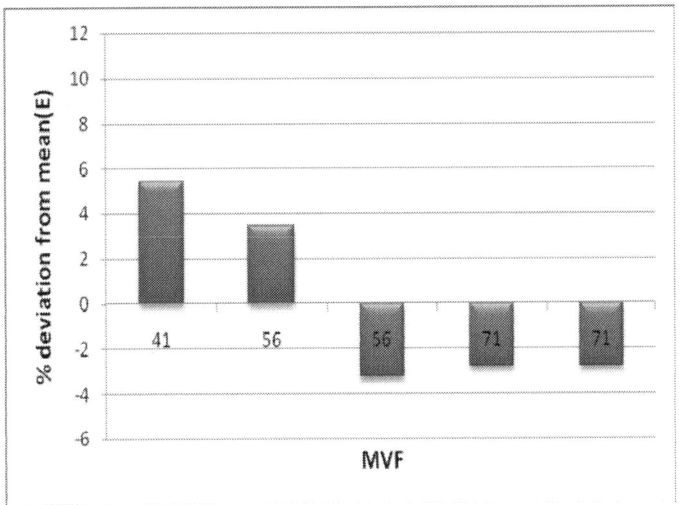

Figure 6. Absorbed energy vs MVF of plates with equal thickness and 80 mm radius

4. Conclusions

In this article, the influence of the MVF on the low velocity impact response of GLARE fiber-metal laminates is studied. Previously published differential equations of motion are employed for this purpose. The considered GLARE plates have equal thickness or equal areal weight and are subjected to low velocity impact under identical impact conditions.

A strong effect of the MVF on the permanent dent depth and energy absorption of the GLARE plates has been found. Considering the same basic lay-up, when the MVF increases, the permanent dent increases whereas the permanently absorbed impact energy decreases.

To the authors' knowledge, no other similar impact study concerning GLARE fiber metal laminates has been published in the current scientific literature. The influence of the MVF on other characteristic values of impact variables is an option for future research.

References

[1] Vlot A 1996 *Int. J. Impact Eng.* **18** 291-307
[2] Guocai W, Jenn-Ming Y, Hahn H T 2007 *J. Mater. Sci.* **42** 948-957
[3] Caprino G, Spataro G, Del Luongo S 2004 *Composites Part* A **35** 605-616
[4] Caprino G, Lopresto V, Iaccarino P 2007 *Composites Part* A **38** 290-300
[5] Hoo Fatt M S, Lin C, Revilock D M Jr et al 2003 *Comp. Struct.* **61** 73-88
[6] Tsamasphyros G J and Bikakis G S 2011 *J. Reinf. Plast. Compos.* **30** 978-987
[7] Tsamasphyros G J and Bikakis G S 2013 *Aerosp. Sci. Technol.* **29** 28-36
[8] Tsamasphyros G J and Bikakis G S 2009 *Adv. Compos. Lett.* **18** 11-19
[9] Tsamasphyros G J and Bikakis G S 2014 *Mech. Adv. Mater. Struc.* **21** 39-46
[10] Lin C, Hoo Fatt M S 2006 *J. Comp. Mater.* **40** 1801-1840
[11] Moriniere F D, Alderliesten R C, Sadighi M et al 2013 *Comp. Struct.* **100** 89-103
[12] Moriniere F D, Alderliesten R C, Benedictus R 2013 *Mech. Mater.* **66** 59-68
[13] Alderliesten R 2005 *Fatigue crack propagation and delamination growth in GLARE* (Delft : Delft University Press)
[14] Belingardi G and Vadori R 2002 *Int. J. Impact Eng.* **27** 213-229

IOP Conf. Series: Materials Science and Engineering **161** (2016) 012056 doi:10.1088/1757-899X/161/1/012056

Method and device for dynamic modelling of rubbery materials applied to human soft tissues. Part I: determination of mechanical characteristics and dynamic model proposal

S Alaci[1], M C Ciornei[2], F C Ciornei[1], C Filote[1] and D Amarandei[1]

[1] "Stefan cel Mare" University of Suceava, Romania
[2]The University of Medicine and Pharmacy "Carol Davila" Bucuresti, Romania

E-mail: alaci@fim.usv.ro

Abstract. The paper aims to study the dynamic behaviour human soft tissues, based on the similarity noticed between the tensile stress-strain curves, for soft biological tissues and rubbery materials. Elastic materials, like rubber and steel, were tested for different types of loadings. In order to establish a dynamical model, the force-elongation dependency of the rubber wire is required. A rubber wire was loaded axially by increasing loads and the deformations up to more than 100% strains were registered. The experimental points from force-displacement plot were interpolated and the function proposed was chosen.

1. Introduction

Development of cutting edge technology and engineering focussed to new requirements imposed to working materials. For materials producers the challenges expand day by day, [1]. Time and again, the needs imposed to new materials are opposing, being dictated by the extreme conditions under which different devices must work, [2]. As a result to the large variety of imposed characteristics, new materials occurred - composite materials, metal carbides etc., [3]. Also, at the same time, researchers reconsidered classical materials, [4]. One of the most widespread materials and employed in versatile applications is beyond doubt, rubber. Without rubber, the existence of automotive industry would be unthinkable, [5]. Rubber is also met in everyday life, from sealing techniques to irrigations systems, [6]. The broad range of mechanical characteristic variations recommended rubber, for the last period, for applications in peak technologies, military industry, space engineering, [7], bioengineering, [8], medical equipment, [9]. Finally, one can say that rubber found a broad practicability as alternative for soft human tissues. Employment of rubber as replacement for soft human tissues supposes a very good knowledge of mechanical characteristics of live tissue, [10] to be substituted, finding the formula of the artificial materials with the characteristics closest to the living tissue, [11], [12] and not least, precautions that, in time, the properties of the artificial material will remain unchanged. The last constraint is especially important considering that the substitution of original tissue with an artificial one is made regularly by surgery. If the characteristics of the artificial material alter in time, obviously, the necessity of repeated surgery is obvious. As emphasised above, understanding the mechanical characteristics of rubber is particularly important. This is an extremely difficult task considering the polymeric structure of rubber. An isotropic and non-homogenous behaviour is expected, [13]. Additionally, as outlined in scientific literature, rubber can be characterised by means of rheological properties.

2. Experimental force-deformation tests

The stress-strain characteristic curve of a materials subjected to a certain test offers a wealth of information. The shape of this curve depends essentially by the test type. The simplest examples presented are the traction test of a steel probe and the contact test for elastic materials.

In the case of traction test, for stresses below a definite value, proportionality between strains and stresses is observed but for the Hertzian contact, the force depends on the deformation raised to half-cube power.

Next, the elastic characteristics obtained for rubber bodies of different shapes are presented. The following situations were considered:

- Contact test between a steel ball and a rubber block
- Compression test for a rubber block
- Traction test for filiform rubber bodies

The first three tests were performed on GADALBINI QUASAR 600 equipment, [14]. The first two types of tests required the design of a special loading device. Figure 1 presents the grips of the equipment, the loading device together with the ball and rubber block.

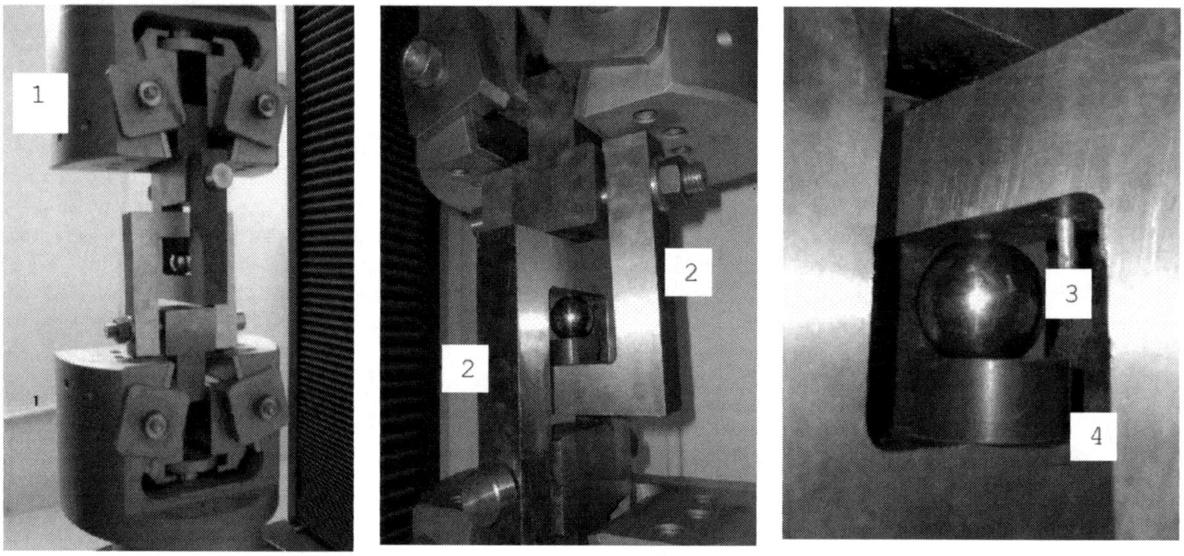

Figure 1. Device for contact tests of a cylindrical rubber block
1 - grips; 2 - loading device; 3 - $\phi 40$ bearing ball; 4 - rubber block

In figure 2 there are presented the load-displacement curves for a cylindrical rubber body compressed by steel ball. The curves present similar shapes to the ones given in technical literature, [10]. One can notice the hysteresis phenomenon and the fact that there is no remnant deformation. In order to interpolate the points from the loading and unloading curves, exponential function can be used. The precision in obtaining the parameters of interpolation curves is essentially depending on the accuracy of identifying the point of contact initiations. This aspect was highlighted for metallic bodies by Oliver and Pharr, [15-16] and Prchlik, [17].

To exemplify the above considerations, in figure 2 there is presented the time variation for the loading curve of a contact between a bearing ball and a plane disc made of mild steel. It is clearly that the transition from horizontal variation to the actual loading curve is made via a transition region where it is difficult to appreciate what actually happens: the contact initiation takes place or there are deformations of the loading system. In figure 3 there are presented comparatively the time dependences of loading force with ball (1) and without ball (2). In both situations it is observed the transition region. The different shapes of the curves in the transition regions don't conduct to a conclusion concerning the point of contact beginning.

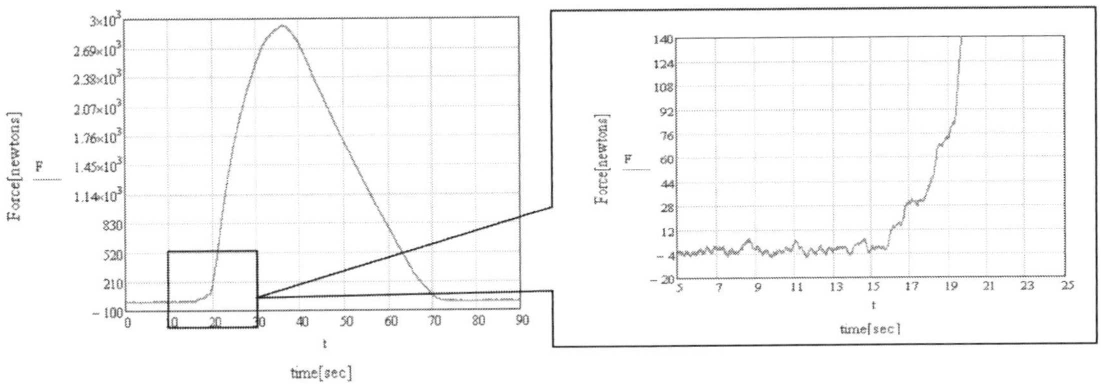

Figure 2. Difficulty encountered in identifying the point of contact initiation

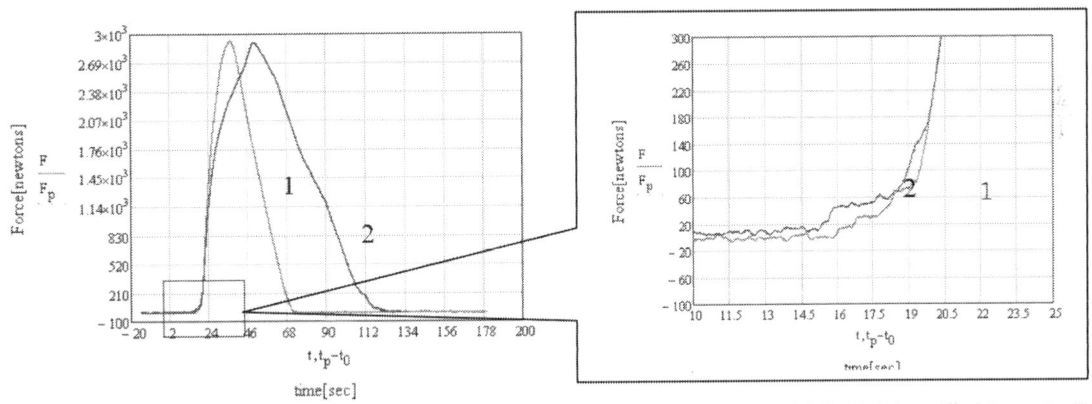

Figure 3. Comparison between loading force variation for the cases: with ball (1) and without ball (2)

Figure 4. Compression of a rubber plate between two steel discs

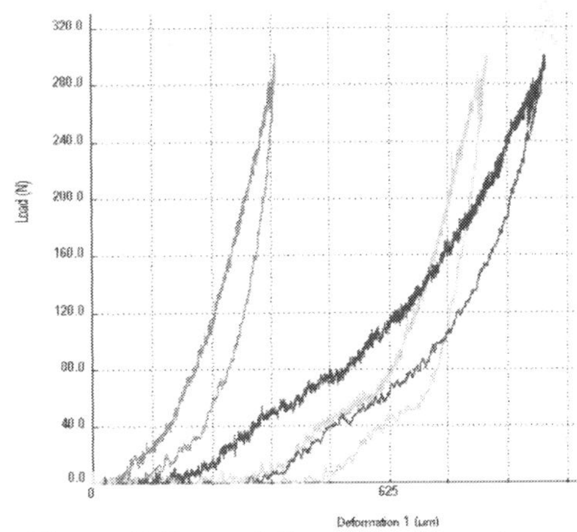

Figure 5. Force-deformation dependency for the compression test of rubber plate – three different compositions

For performing the compression test, in the loading mechanism was assembled a device consisting of two steel discs and a rubber plate. The loading force variation versus deformation is presented in figure 5. The deformations of the rubber probe are in the large deformations domain.

3. Experimental traction tests for rubber string

The last test considered was of a filiform rubber probe under traction. There were considered cords made from the same type of rubber with different diameters of the cross-section: 2 mm, 2.5 mm and 3 mm. They were loaded using weights and measuring after each loading augmentation, the distance between two marks set on the un-stretched wire. Figure 6 shows two curves corresponding to the traction tests performed on two cords from the same material, with the same cross-section but with different initial lengths, $l_{10} = 0.505m$ (1), $l = 0.755m$ (2).

To compare the effect of force upon the dimensions of loaded cord, the force versus strain variations were represented. For the first cord, one could notice that after surpassing *100%* strain, the strain increases very much and the creep occurs. For the second rope, the forces were lower and produced strains below *100%*. Interpolation points lie down on a curve having a shape similar to the curves presented by Reeves, [18], for the force-deformation variation for a human tendon.

The points from this graph were considered in finding the interpolation curve. Numerous expressions may be chosen for the interpolation curve. Assuming that the expression of the interpolation curve will be further used in structuring a dynamic model, it is aimed that the interpolation function should contain few parameters as possible but as much as necessary to accomplish a good quality interpolation. To this reason, the expression proposed by Chacon, [19], was selected for the interpolation function:

$$f(x) = -k\left[x[2 - tanh(x/a)] + bx^3\right] \quad (1)$$

where k, a and b are constants found from the condition of minimization of the function

$$\phi(k,a,b) = \sum_{k=1}^{n}[f(x_k) - F_k]^2 \quad (2)$$

In figure 7 there are presented the experimental point from curve (2), figure 6 and the interpolation curve.

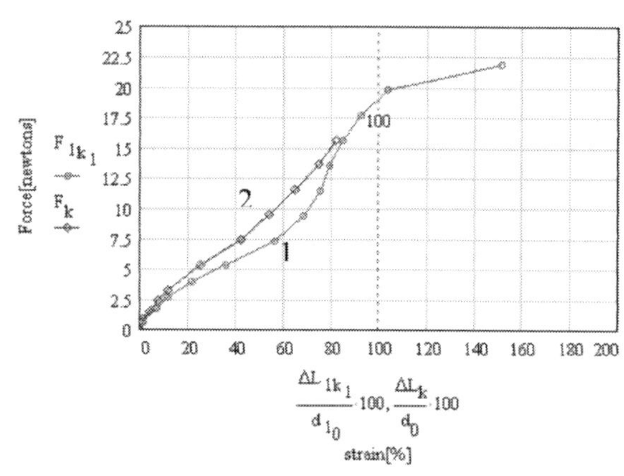

Figure 6. Loading curves for two cords with the same cross section but different lengths

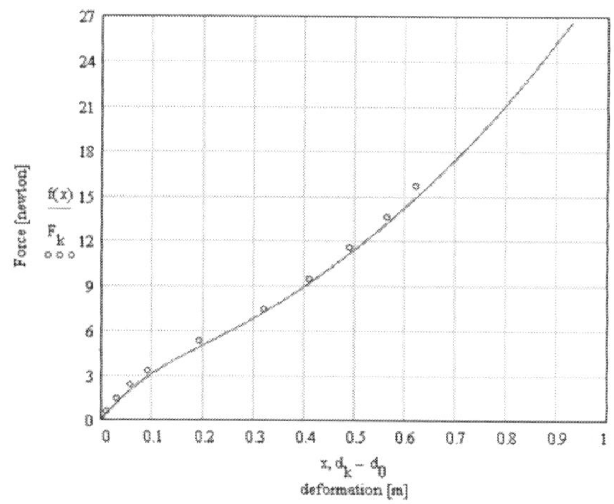

Figure 7. Experimental points and the interpolation curve

From figure 6 and figure 7 it is easily observed the non-proportional relationship between forces and strains that will have immediate consequences upon the type of differential equation selected to describe the behaviour of the system containing the rubber body. The differential equation will be a nonlinear one. It is obvious that from the structure of differential equation should not be neglected the damping factor - a consequence of internal friction from rubber string. This aspect is evidenced in the second part of the paper.

4. Conclusions

The aim of the paper is to propose a dynamical model capable to describe the dynamical behaviour of rubber bodies when employed as human soft tissues substitutes. Taking into account that the main feature required to these materials is to present physical and mechanical characteristics closer to the original tissue, the first part of the work presents the experimental results obtained on original test rigs with the purpose of plotting the force-deformation curves for diverse loading types and body models.

There are presented the characteristic curves obtained at the compression of a steel ball on a rubber block, the compression curve of a rubber plate and the traction curve of a filiform rubber body.

Comparisons with the correspondent curves presented in technical literature traced for the soft human tissues revealed that the best similarity exist for the traction curves of filiform bodies. For this type of test there were made several experiments and each time it was concluded that there is an initial region where the deformations increase more rapidly than the force, followed by a zone where the other way happens.

Proposition of a dynamical model assumes the existence of an analytical relation between force and deformation. Using for the interpolation function an expression proposed in scientific literature, a relationship between force and deformation was obtained and it will be further employed in construction of dynamical model and presented in the second part of the paper.

5. References

[1] Vahlas C editor, accessed 12.03.2016 Materials Science and Engineering in Europe: Challenges and Opportunities *Materials Science and Engineering Expert Committee (MatSEE)* http://www.esf.org/fileadmin/Public_documents/Publications/MatSEEC_ChallengesOpportunities.pdf

[2] Millar D I A 2012 *Energetic materials at extreme conditions* (Springer)

[3] Ball P 1999 *Made to Measure: New Materials for the 21st Century* (Princeton University Press)

[4] Brandes U, Stich S and Wender M 2013 *Design by Use: The Everyday Metamorphosis of Things (*Birkhauser)

[5] Pacejka H 2012 *Tire and Vehicle Dynamics,* Third Edition (Butterworth-Heinemann)

[6] Harp S L 2015 *A World History of Rubber: Empire, Industry, and the Everyday* (Wiley-Blackwell)

[7] Conference Proceedings Toulouse 2007 *Polymers in Defence and Aerospace Applications* (Smithers Rapra Technology Ltd)

[8] Sabu T and Ranimol S 2010 *Rubber Nanocomposites: Preparation, Properties and Applications* (John Willey& Sons Asia)

[9] Gad S C and Gad-McDonald S 2016 *Biomaterials, Medical Devices, and Combination Products: Biocompatibility Testing and Safety Assessment* (CRC Press)

[10] Derby B and Akhtar R editors 2015 *Mechanical Properties of Aging Soft Tissues* (Springer)

[11] Nobuyuki Nakajima 2000 *Science and Practice of Rubber Mixing* (Rapra Technology Limited)

[12] Wei Sun, Sacks M S and Scott M J 2005 Effects of Boundary Conditions on the Estimation of the Planar Biaxial Mechanical Properties of Soft Tissues *Journal of Biomechanical Engineering* **127(4)** 709-15

[13] Öchsner A and Altenbach H 2015 *Mechanical and Materials Engineering of Modern Structure and Component Design* (Springer)

[14] ***http://www.iu.hio.no/bygglab/Strekklab/utstyr/manualer/QUASAR_250-V87C.pdf

[15] Oliver W C and Pharr G M 2004 Measurement of hardness and elastic modulus by instrumented indentation: Advances in understanding and refinements to methodology *Journal Materials Research* **19 (1)** 3-20

[16] Oliver W C and Pharr G M 1992 An improved technique for determining hardness and elastic modulus using load and displacement sensing indentation experiments, *Journal Materials Research* **7 (6)** 1564-83
[17] Prchlik L 2004 Input error sensitivity of hardness and elastic modulus evaluated from indentation load-displacement records by Oliver and Pharr method *Journal of Materials Science* **39** 1185–93
[18] Reeves N D, Maganaris C N, Ferretti G and Narici M V 2005 Influence of 90-day simulated microgravity on human tendon mechanical properties and the effect of resistive countermeasures *Journal Applied Physiology* **98** 2278 -86
[19] Chacon R, Suarez P, Sanchez-Bajo F and Muniz P 1997 Nonlinear and chaotic oscillations of an india-rubber band *Physical Review E* **56(5)** 5321-26

Acknowledgments

The infrastructure used for this work was partially supported by the project "Integrated Center for research, development and innovation in Advanced Materials, Nanotechnologies, and Distributed Systems for fabrication and control", Contract No. 671/09.04.2015, Sectoral Operational Program for Increase of the Economic Competitiveness co-funded from the European Regional Development Fund.

IOP Conf. Series: Materials Science and Engineering **160** (2016) 012057 doi:10.1088/1757-899X/160/1/012057

Method and device for dynamic modelling of rubbery materials applied to human soft tissues. Part II: device and experimental results

S Alaci[1], M C Ciornei[2], F C Ciornei[1], C Filote[1] and I C Romånu[1]

[1] "Stefan cel Mare" University of Suceava, Romania
[2] The University of Medicine and Pharmacy "Carol Davila" Bucuresti, Romania

E-mail: alaci@fim.usv.ro

Abstract. The paper presents the experimental results obtained on an experimental device where a horizontal rubber wire is stretched by a transversal oscillating force - that is a body with an acceleration sensor attached, placed at the middle of it that oscillates freely. A nonlinear model was proposed for the experimental test rig, the differential equation of motion was offered and a series of curves were traced and compared to the experimental ones. One can conclude that the theoretical model certifies very well the behaviour of the real model. An open problem remains the manner of adopting the parameters characteristic to the dissipative element of the system.

1. Introduction

Analyzing the shape of traction force-deformation curves corresponding to human tissues, [1], [2], it was noticed a good agreement between the experimental curve stress-strain of a human tendon and the same curve traced for a rubber cord during a traction test. This fact suggested the idea of modelling soft human tissues as a fascicle of wires made of rubber. Besides elastic characteristics, the rubber also presents rheological properties that together with inertial characteristics will state the dynamical behaviour of the systems where the rubber elements will operate.

The second section of the paper presents the experimental results obtained on a test rig where rubber cords are used. The device is designed by authors and constructed in the laboratory.

In the third section, a dynamic model is proposed and the differential equation, its integration and preliminary conclusions are deduced.

Conclusions are presented in the last section and it is emphasised that the theoretical model proposed describes accurately the actual behaviour of the tissue.

2. Device and experimental results

Based on the observations from the first part of the paper, a device was built, with two rubber cords (1) were pre-stressed with the same traction force. On the midst of these cords was attached a mass (2) which can vibrate in vertical plane, figure 1. On the body of the mass is fixed an acceleration sensor (3). The body is (2) is moved downwards with controlled amplitude, with the aid of limiting part (4). When the body is set free, it oscillates and the data offered by the acceleration sensor [3], are saved in digital form and processed afterwards.

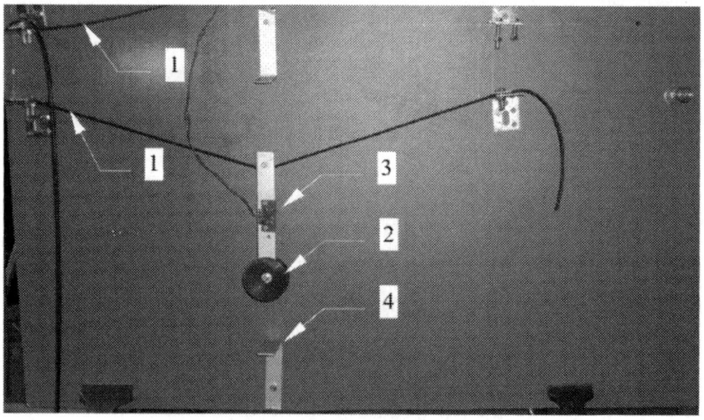

Figure 1. Experimental set-up

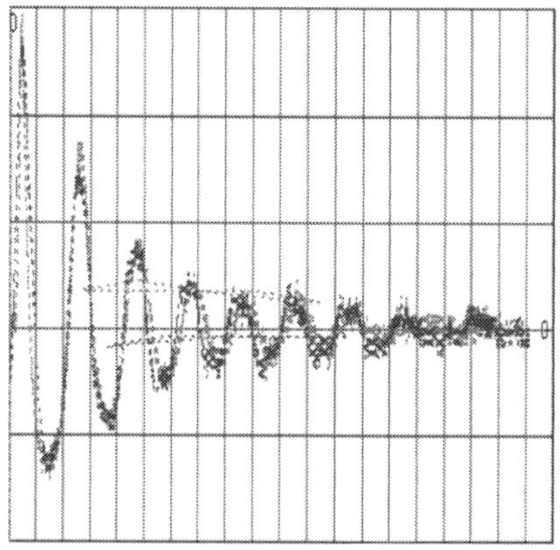

Figure 2. Accelerations for two successive tests

During the tests it was noticed that it is practically impossible to ensure identical behaviours to both wires. The main causes are the related to ensuring the same pre-stretching force for the two cords and mounting the suspended body precisely at the middle of the wire. As a consequence, after recording the signal generated by the acceleration sensor one could perceive the "beat" phenomenon produced by the different values of periods of oscillation resultant from the two wires. Thus, it was abandoned the two wires variant and a single one was used, without pre-stretching.

In figure 2 there are presented the results of two successive, identical launchings of the mobile body, delayed by a period of several tens of minutes, in order to complete an image upon system's repeatability behaviour. As it can be observed from figure 2, except for a restrain on horizontal, the signals are quasi-identical. Also form figure 2 it can be noticed an evident signal damping and the presence of a noise overlapped on the body's vibration. To eliminate the noise, the numerical filtering was performed using computer software. In figure 3 there are presented the initial signal and the filtered one, [4], [5].

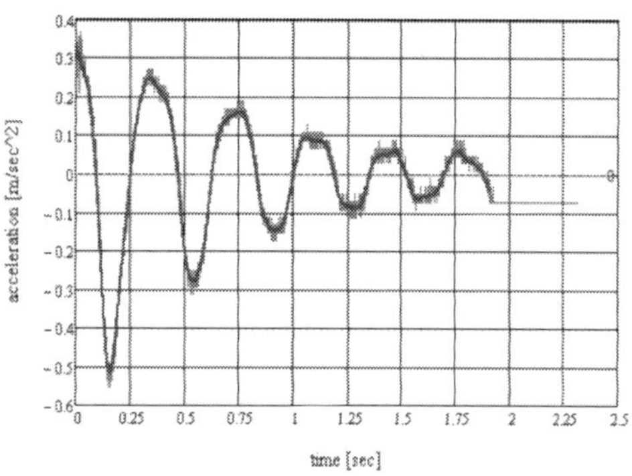

Figure 3. Original signal and filtered one

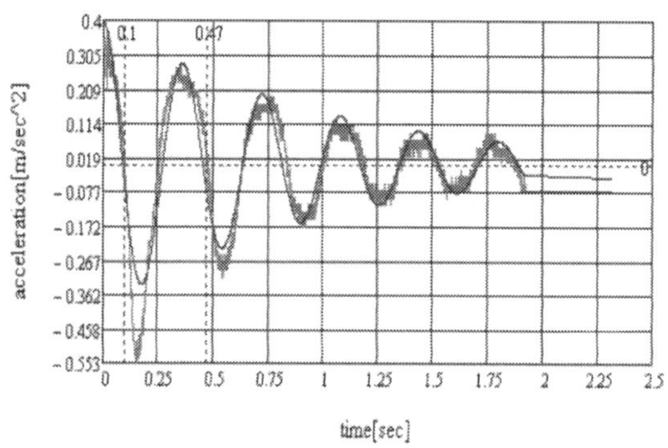

Figure 4. Original signal and signal interpolated with a curve characteristic to a linear system

From figure 3, besides the important damping, other two conclusions are drawn: a) the motion is a quasi-periodical one since the time intervals between two instants when the acceleration passes through zero in the same direction are equal; b) the asymmetry of the signal with respect to equilibrium position, fact that directs to the conclusion that the dynamical system proposed should be a non-linear one. In figure 4 there are presented the experimental data and the results of interpolation of these data with a curve of type $y(t) = A\exp(\mu t)\cos(\omega t + \varphi_0)$, characteristic to the model of a dynamic system having the behaviour described by a second order homogenous differential equation with constant coefficients. The parameters of interpolation curve were established using the subroutine *genfit* of MATHCAD14, [6]. The asymmetrical aspect of acceleration plot is noticeable in figure 4.

After numerical integration of filtered signal of acceleration, the velocity and vertical displacement of the middle point of rubber wire, where the weight is fixed. Using these data, the curve from phase diagram and the hysteresis curve were traced and the graphs are presented in figure 5 and figure 6, respectively.

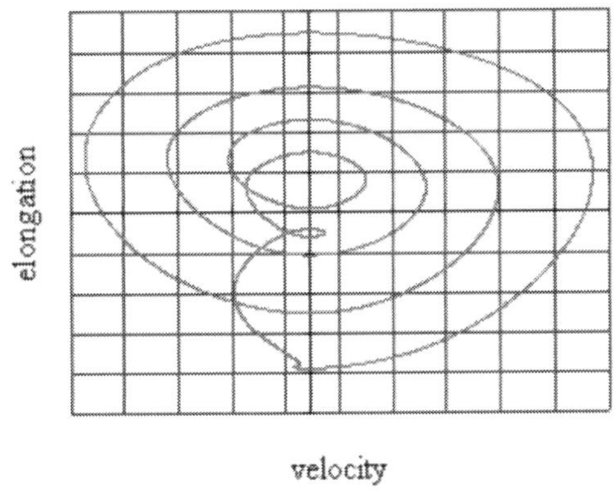

Figure 5. Phase diagram

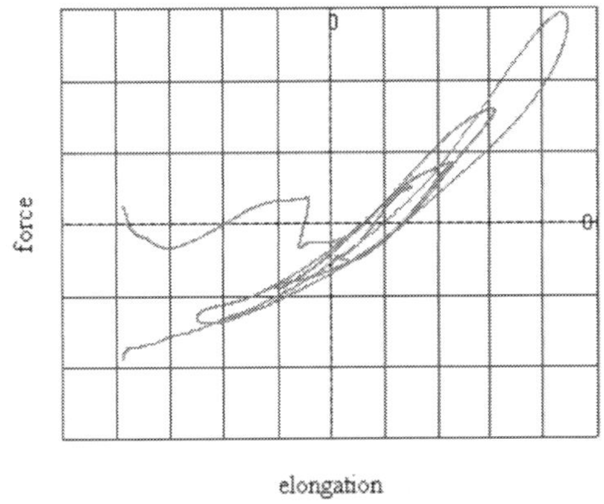

Figure 6. Hysteresis loop

3. Theoretical model proposition

In the subsequent section is presented the theoretical model proposed to describe the behaviour of the suspended body motion.

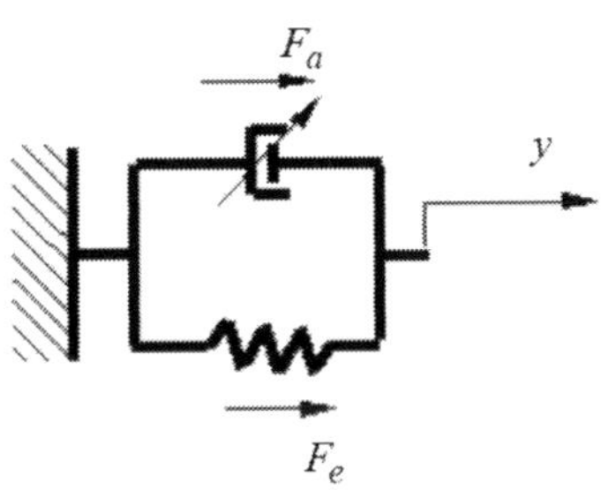

Figure 7. Model for the rubber cord

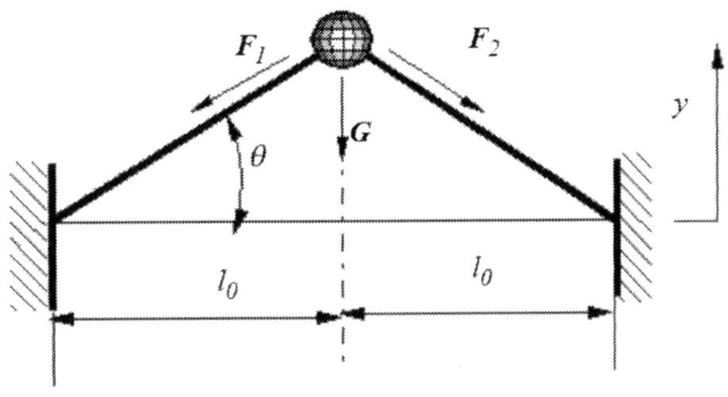

Figure 8. Forces acting upon the body

The rubber wire is modelled as an elastic element linked in parallel with a damper. The damper has a variable characteristic. The idea of using such a type of damper is due to Hunt and Crossley, [7]. Their challenge to model the damped contact of two elastic balls using a linear model conducted to a hysteresis curve that didn't close in origin as a loop but remained opened. This fact contradicts the problem stating that at the instants of contact initiation and contact detachment the interaction forces between the balls should be zero. To overcome this drawback, Hunt and Crossley showed that the damping force is required to be proportional both to deformation velocity and to deformation raised at power $3/2$ (power equal to the exponent of deformation from the force expression in the case of Hertzian contact, [8]).

The expression of elastic force F_e from the stretched rubber wire was deduced by interpolation of experimental data with a curve having the form, [9], in the first part of the paper, namely:

$$f(x) = -k\left[x[2 - tanh(x/a)] + bx^3\right] \quad (1)$$

where k, a and β are constants and x is the strain (specific elongation). Denoting by y the vertical displacement of the mass, then the elongation of the wire is:

$$\Delta l = \sqrt{l_0^2 + y^2} - l_0 \quad (2)$$

where l_0 is the initial, unstretched length of the wire and l is the current length. Therefore, the strain is:

$$x = \frac{\Delta l}{l_0} = \frac{\sqrt{l_0^2 + y^2}}{l_0} - 1 \quad (3)$$

Considering that in relation (2) the displacement y is a function of time t, the derivation with respect to time of relation (2) leads to the expression of velocity of deformation of the rubber cord:

$$v(t) = \frac{y(t)\dot{y}(t)}{\sqrt{l_0^2 + y(t)^2}} \,. \quad (4)$$

For the damping force, the general form is proposed:

$$F_a(t) = \mu[v(t)]^\alpha[\Delta l]^\beta \quad (5)$$

where μ, α and β are constants. The vector equation of motion of the body is (figure 8):

$$M\boldsymbol{a} = \boldsymbol{G} + \boldsymbol{F}_1 + \boldsymbol{F}_2 \quad (6)$$

where M and $\boldsymbol{G}$ are the mass and the weight of the body, respectively $\boldsymbol{F}_1$, $\boldsymbol{F}_2$ are the tractions from the two wire branches, each one equal in module with the sum between an elastic force and damping force (according to the model from figure 7):

$$F = F_e + F_a = f\left(\frac{\sqrt{l_0^2 + y(t)^2} - l_o}{l_0}\right) + \mu\left[\frac{y(t)\dot{y}(t)}{\sqrt{l_0^2 + y(t)^2}}\right]^\alpha \left[\sqrt{l_0^2 + y(t)^2} - l_o\right]^\beta \quad (7)$$

The projection of relation (6) on vertical direction gives the scalar equation:

$$Ma(t) = -Mg - 2(F_e + F_a)\sin\theta \,, \quad (8)$$

where:

$$\sin\theta = \frac{y}{\sqrt{l_0^2 + y^2}} \quad (9)$$

The motion equation takes the final form:

$$M\ddot{y}(t)=-g-\left\{f\left(\frac{\sqrt{l_0^2+y(t)^2}-l_o}{l_o}\right)+\mu\left[\frac{y(t)\dot{y}(t)}{\sqrt{l_0^2+y(t)^2}}\right]^{\alpha}\left[\sqrt{l_0^2+y(t)^2}-l_o\right]^{\beta}\right\}\frac{y(t)}{\sqrt{l_0^2+y(t)^2}} \tag{10}$$

The equation (10) is a second order nonlinear differential equation, [10]. To integrate this equation, applying a numerical procedure is necessary. The authors used the Runge-Kutta, [11], method of fourth order. It is obvious that for obtaining a tangible solution, knowing the parameters μ, α and β that characterize the damping force, is required. In a first stage, the characteristic parameters were chosen: $\mu=0.7$, $\alpha=1$ and $\beta=0$, corresponding to a linear viscous damper.

Concerning the characteristics of rubber wire, the parameters from equation (1) are the ones found experimentally in the first part of the paper: $\kappa=13.11$; $a=0.328$; $b=0.174$; $l_0=0.25$. The mass of the body attaché to the wire is $M=0.22\,Kg$. In figure 9 and figure 10 there are presented the variations of displacement and velocity, respectively, for the theoretical model. With these variations, using the relation (10), the acceleration can be found directly, the numerical derivation being unnecessary.

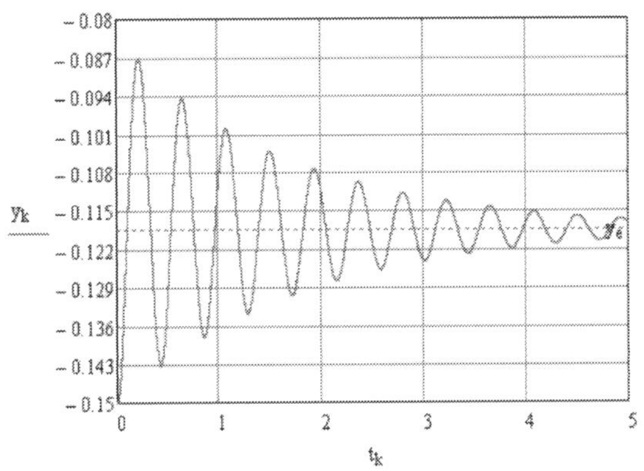

Figure 9. Displacement of theoretical model

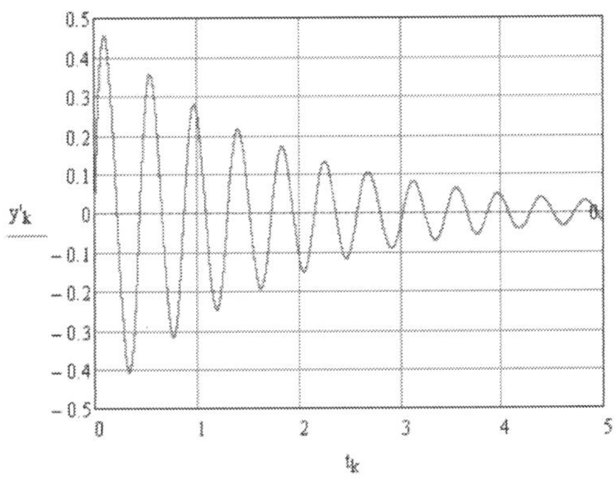

Figure 10. Velocity of theoretical model

In figure 10, the displacement y_e represents the value around which the oscillatory motion performs. This value is coincident to the equilibrium elongation value, when the body is acted by its own weight and by the traction forces from cords. To find this value, it is considered the equilibrium case when the damping force is not present in the structure of the traction force, and therefore it is necessary:

$$2f\left(\frac{\sqrt{l_0^2+y(t)^2}-l_o}{l_o}\right)\frac{y(t)}{\sqrt{l_0^2+y(t)^2}}=-Mg \tag{11}$$

Equation (11) is transcendent and was solved using the Newton-Raphson algorithm and gives the elongation value for the equilibrium state of the body. In figure 11 and figure 12 there are presented the phase diagram and the hysteresis loops. The variation of force from the elastic wire was also represented in figure 12.

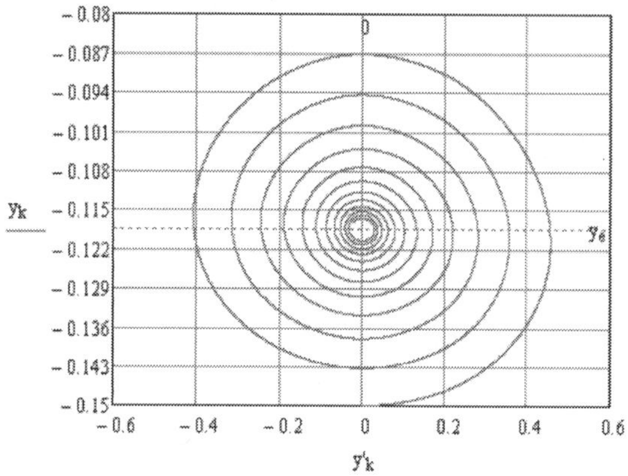

Figure 11. Phase diagram

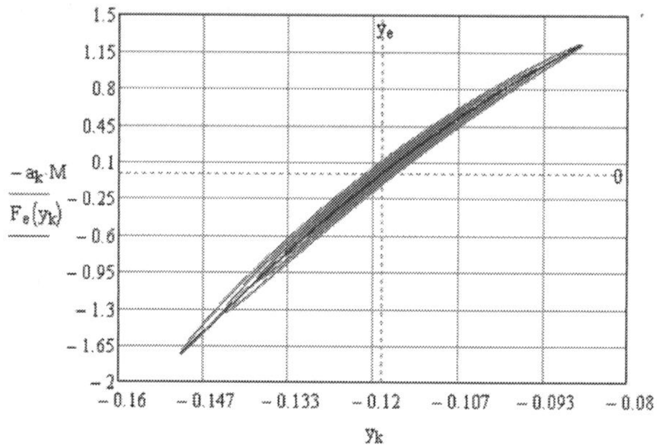

Figure 12. Hysteresis loops

Comparing the curves from the phase space and the hysteresis loops for the theoretical model, with the ones obtained experimentally, it can be affirmed that the proposed model can describe very well the behaviour of the actual system. The manner of determining the values μ , α and β that describe as accurate as possible the dissipative component of the system is the aim of future research.

4. Conclusions

The paper proposes a test rig and methodology for characterizing mechanical properties of rubber when used as substitute for soft human tissues. Initially, different tests applied to rubber bodies of diverse shapes and tracing the force-deformation diagram characteristic to each type of loading are considered. A comparison with the results from literature shows a good agreement between the experimental curve and the theoretical one for the traction of a filiform rubber body. For such type of body, the experimental points from the traction curve were established and afterwards, the data were interpolated with a function having a form proposed in technical literature.

Next, the paper presents a test rig, consisting of a vibratory system, made of an horizontal rubber cable, having attached at its midst a mass of small dimensions. An acceleration transducer is attached to the mobile body, used for acquiring experimental data. The repeatability of the results was tested in a first stage. Then, the experimental data were filtered and integrated in order to obtain, for body oscillatory motion, the time dependency of velocity and elongation. These data were used in plotting the experimental curves for phase space and hysteresis loops that attested the nonlinear character of the dynamical system considered and the occurrence of a significant damping.

Finally, a theoretical model was proposed for simulation of dynamic experimental system, and the equation of motion was deduced that proved to have a strong non-linear character. The equation of motion was integrated numerically, and following the integrating process, the time dependencies of velocity and elongation were plotted, together with the phase space and hysteresis loops. A qualitative comparison showed that the proposed theoretical describes very well the dynamical behaviour of the experimental dynamic system. For future research directions, the manner to deduce from experimental data the optimum values of the parameters characterizing the dissipative component of the system is aimed.

5. References

[1] Derby B and Akhtar R editors 2015 *Mechanical Properties of Aging Soft Tissues* (Springer)
[2] Reeves N D, Maganaris C N, Ferretti G and Narici M V 2005 Influence of 90-day simulated microgravity on human tendon mechanical properties and the effect of resistive

countermeasures *Journal Applied Physiology* **98** 2278 –86

[3] Sinclair I 2001 *Sensors and Transducers* 3rd edition (Newnes)

[4] Miroslav D, Lutovac D, Tošić V and Evans B L 2001 *Filter Design for Signal Processing Using MATLAB and Mathematica* (Prentice Hall)

[5] Mitra S K 1999 *Digital Signal Processing: A computerBased Approach* (McGraw-Hill New York)

[6] Mathcad 14 User's Guide 2007 http://home.iitk.ac.in/~santanum/EE360_Web/Mathcad%20Users%20Guide.pdf

[7] Hunt K H and Crossley F R E 1975 Coefficient of Restitution Interpreted as Damping in Vibroimpact *ASME Journal of. Applied Mechanics* 440–45

[8] Johnson K L 1985 *Contact Mechanics* (Cambridge University Press)

[9] Chacon R, Suarez P, Sanchez-Bajo F and Muniz P 1997 Nonlinear and chaotic oscillations of an india-rubber band *Physical Review E* **56(5)** 5321-26

[10] Jordan D W and Smith P 2007 *Nonlinear differential equation. Problems and Solutions* A Sourcebook for Scientists and Engineers Oxford Texts in Applied and Engineering Mathematics (Book 11) (Oxford University Press)

[11] Lambert J D 1993 *Numerical Methods for Ordinary Differential Systems* (John Wiley & Sons)

Acknowledgments

The infrastructure used for this work was partially supported by the project "Integrated Center for research, development and innovation in Advanced Materials, Nanotechnologies, and Distributed Systems for fabrication and control", Contract No. 671/09.04.2015, Sectoral Operational Program for Increase of the Economic Competitiveness co-funded from the European Regional Development Fund

IOP Conf. Series: Materials Science and Engineering **161** (2016) 012058 doi:10.1088/1757-899X/161/1/012058

The electron beam welding of dissimilar materials – case study

A Munteanu

"Gheorghe Asachi" Technical University of Iasi, Department of Machine Tools, Blvd. D. Mangeron, 59 A, 700050 Iasi, Romania

adycypmunteanu@yahoo.com

Abstract. The modalities to realize the welding workpieces are multiple. The electron beam welding is one of them. One can weld two different types of materials that give the possibility to reduce the cost of workpiece, if the active part is realised of rich materials welded on components with inferior phisico-mecanical characteristics. The procedure provides great flexibility to the product designs through efficient use of each type of material. So this aspects lead to the necessity to join dissimilar metals. Different tables are given in the specific literature regarding the possible combination. Conflicts may arise by the compromises required for to the optimum heat control of the two dissimilar materials used. But nowadays, more and more frequently are meet the welding of dissimilar metals, thus, the objective of this article is to provide information regarding the particular case of welding between stainless steel and copper without the filler material use.

1. Introduction

The procedures mostly described (traditional procedure or un-conventional methods) for obtaining quality welds mentioned in the literature has been between identical metals or at least welds of metals of similar compositions and properties. There are many applications, however, in which the joints are made from metals of different compositions because in the industrial practice must be resolve the mechanical wear problem or a high-temperature situation, or other conditions in which different properties are required from different parts of the same weldment. This aspects lead to the necessity to join dissimilar metals, because in production, time is money, imposed both of the economical and the functional constraints. A successful weld between dissimilar metals is one that is as strong as the weaker of the two metals being joined, but possessing sufficient tensile strength and ductility so that the joint will not fail in the weld, tells the specialists.

A classification of the most commonly used welding processes for dissimilar materials was realized by Kah [1] and Avery [2]: as fusion weld (shielded metal arc, gas metal arc, gas tungsten arc), low dilution weld (electron beam welding, laser beam and plasma arc) and non fusion weld (friction welding, explosion welding). Avery considered that all fusion welds are dissimilar metal welds because the metals being joined have a wrought structure and the welds have a cast structure. Usually dissimilar metal join are made by fusion, the other two are more often used for high productivity and for special applications [2].

In the electron beam welding case (a low dilution welding method [1, 2]), the welding materials behavior is generally good and can be combined a large numbers of materials but in terms of quality. The main applications of electron beam welding returns to steels, but an important place belongs to dissimilar joints. Tables are given in the specific literature regarding the possible combinations (table 1), so that one can avoid unwanted combination because it forming the intermediary components.

In the electron beam welding process the heat is generated at the impact of high energy electron beam the workpieces. Electrons emitted by a electronic gun are accelerated at a rate that approaches 30% ... 70% of the speed of light and, through their high kinetic energy in contact with the workpiece, releases the heat necessary to achieve the weld. Thus, this process is characterized by high power density of the order of 108-109 W / cm2 that reduces heat loss by conduction and allows controlling the penetration depth of the weld. Also, because of this, the specific areas of the electron beam joint have a temperature gradient much higher than in other welding processes.

Table 1. Workpieces weldment with electron beam (after [23])

	Cr	Cu	Fe	Mg	Mn	Mo	Nb
Cr	-	2	2	5	2	1	5
Cu	2	-	2	5	1	3	2
Fe	2	2	-	3	2	2	5
Mg	5	5	3	-	5	3	4
Mn	2	1	2	5	-	3	5
Mo	1	3	2	3	3	-	1
Nb	5	2	5	4	5	1	

1 = Good weldment – wished combinations; 2 = Acceptable weldment –can be obtained complex structures; 3 = Used with precaution – insufficient data in the specific literature; 4 = Used with great precaution – doesn't exist in the specific literature; 5 = Unwanted combination – are forming intermediary components.

Regarding the welding between stainless steel and copper the researchers were interested to approach this theme lately. Findik and Durgutlu studied this type of joint in explosion welding case [3, 4]. Chen used laser welding in 2015 like a successfully method and obtained a mixed zone of this two materials [5].

Magnabosco [6], Weit et al [7] using electron beam welding obtains a joint with a heterogeneous fusion zone between copper and ASI 304L stainless steel, with a chemical composition characterized by gradual variations in concentrations of iron and copper [8, 9].

The literature [10] mentioned that beam offset can influence the joint Cu-Fe and Shung Guo studied in detailed the diffusion behavior of Fe/Cr, cooling rate and the joint strength. So Cu-Fe alloys are completely miscible in the stable liquid state at high temperatures and do not form brittle intermetallic compounds [3, 9, 10]. Zhang welded stainless steel with copper alloy using copper filler wire [11] and was investigate the effects of process parameters on the tensile strength of the joints, and the optimization of these process parameters.

Even if there are study that investigated dissimilar copper-steel welds, the specific information are not fully established. The present work was focused on the morphological analysis of dissimilar electron beam welded joint between copper and austenitic stainless steel in order to evaluate their potential in industrial application.

2. Materials and experimental conditions

The materials used for the experiment were a copper workpiece and a stainless steel one. It was used an austenitic stainless steel with the following composition: C=0,063%; Mn=1,18%; Si=0,71%; P=0,040%; S=0,017%; Cr=18,307%; Ni=9,93%; Cu=0,328%; Mo= 0,251%; As<0,003%; Al=0,006%; Co=0,162%; Nb=0,055%; Ti=0,017%; V=0,088%; W=0,417%; Pb<0,004%; Sn=0,014%; Mg=0,021%; Zr=0,014%; Bi=0,009%; Ca=0,002%; Ce=0,018%; Ta<0,008%; B=0,002%; Zn=0,025%; La=0,003%, Fe<68,321%.

Generally, these stainless steels have electron beam welding good behavior. Depending on the alloying elements ("chrome equivalent" or "nickel equivalent") may draw conclusions on the peculiarities of welding this material. There are three main categories of stainless steel - ferritic, martensitic and austenitic. Ferritic stainless steels are magnetic and have low carbon content. Its include chromium as the main component, generally in a proportion of 13% ... 17%. Martensitic stainless steels are magnetic, with a typical content of 12% chromium and carbon content of the medium. Non-magnetic steels are austenitic and, in addition to chromium, in the usual concentrations of 18%, containing nickel, which increases the corrosion resistance. In case of stainless steel used for the experimental tests, the percentage of chromium show that it is an austenitic stainless steel (Cr =

18.307% Ni = 9.93%) and a percentage of 68% Fe states that it is a austenitic stainless steel with high carbon grade, which ensures a high corrosion resistance and can be used at high temperatures.

Between two stainless steel materials of the above composition was used a copper material (figure 1). The sandwich structure was preferred other conventional welding because lately the industrial practice uses this type of overlap.

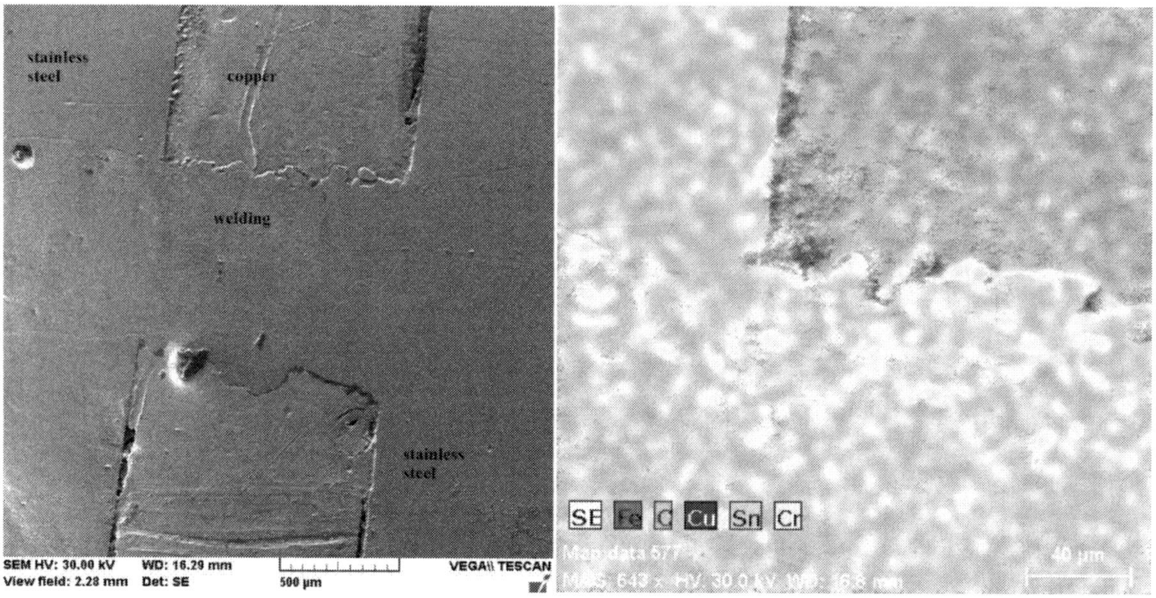

Figure 1 Micrographs of the sandwich joint- stainless steel and copper

The equipment used in the experiments is a welding equipment electron beam ELA 60/60 (AFE) equipped with a vacuum and working chamber. The few technical characteristics of the equipment are given in the following:- Accelerating voltage - 60 kV; - Welding current, mA - welding regime from 5 to 1000, tracing regime from 2-5; - The current focus, mA 400-1000; - Gun pressure, Pa (mm Hg) - 10.3 10.7 * (8 * 10-5)

Working conditions for the research work are presented in table 2, it was realized a joint with a single welding pass but with a current intensity value that ensure the optimal penetration depth.

Table 2. Working conditions

Parameters	Conditions
Acceleration voltage (kV)	60
Curent beam (mA)	30
Focal distance (mA)	744
Gun specimen distance (mm)	150
Working chamber vacuum (Pa)	$10{,}7*10^{-2}$ Pa
Electron beam gun vacuum (Pa)	$10{,}7*10^{-3}$ Pa
Welding speed (m/min)	0.23
Welding pass	1

The EBW (electron beam welding) technology provides strong joints and offers many benefits and has advantages in many applications because of the degree of concentration of the welding energy. But the problems arise between welds of the dissimilar metals related to the transition zone between the metals and the intermetallic compounds formed in this transition zone. So, this transition zone was analyzed with a scanning electron microscope (SEM). To perform this study a Vega Tescan, scanning electron microscope (model VEGA II LMH) was used.

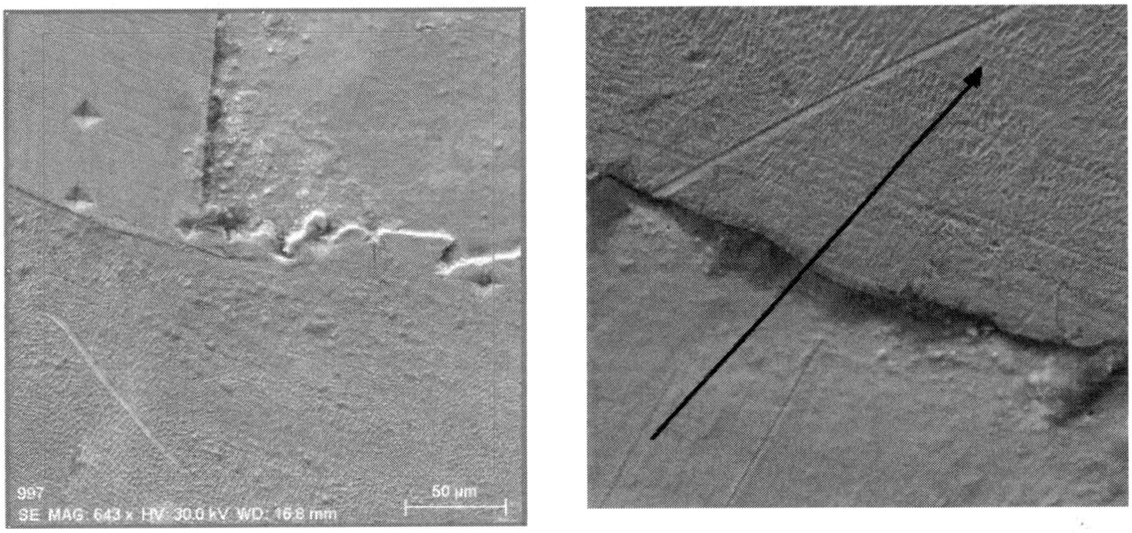

Figure 2 SEM micrographs of the analyzed area

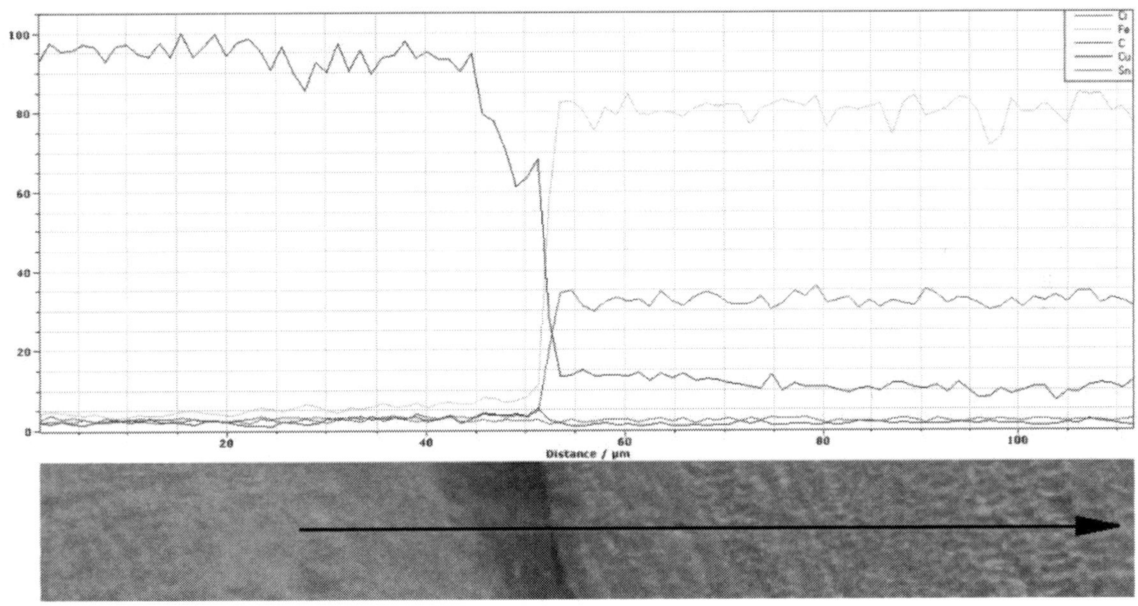

Figure 3 Chemical composition of the melting zone

The welding of two dissimilar metals with different melting temperature or thermal conductivity is complicated by the fact that one material will melt prior to the other one, but using a high-power density heating source such as electron beam permits to reach the fusion of both metals.

The results showed welds with poor mixing of materials. The copper have high thermal conductivity than stainless steel which leads to different temperature gradient. Because the melting point is different the initial presumption was that the melting zone and mixing zone will be larger so that the two materials will merge. But the stainless steel it passed like a nail through copper and the mixing zone is very small. Even if the most researchers put in fusion zone this type of welding, the microstructure studied with scanning electron microscopy emphasized the researcher's point of view, who study this process like low dilution welding process. The higher density of copper to iron must have as result higher mixing area, but with the effect of gravity, it was performed a riveting.

3. Conclusion

In the electron beam welding case of austenitic steel, the crack phenomena may occur in the hot molten metal, by placing at the grain boundary of phosphorus, silicon and niobium impurities, due to precipitation of chromium carbides. Even if the electron beam welding could be categorized as ideal fusion welding process sometimes the melting zone changes may occur that have the effect on the shape and the structure on the weld like imperfections. In the paper research case metallographic inspection of the joint did not reveal defects such as microcracks or porosity. The two studied materials have large differences in their melting temperature and copper have high thermal conductivity, but with the gravity help the stainless steel acts as a permanent joint like rivet.

4. Acknowledgements

The author gratefully acknowledges the support provided by lector Nicanor Cimpoesu (SIM, Gheorghe Asachi Technical University of Iasi) regarding the SEM microscopy.

References

[1] P. Kah, M. Shrestha, J. Martikainen, Trends in Joining Dissimilar Metals by Welding, Applied Mechanics and Materials 440 (2014) 269-276.
[2] R. E. Avery, Pay attention to dissimilar-metal welds, Chemical Engineering Progress, 1991.
[3] F. Findik, Recent developments in explosive welding [J]. Materials & Design (2011) 32(3):1081-1093.
[4] A. Durgutlu, B. Gulenc, F. Findik. Examination ofcopper/stainless steel joints formed by explosive welding. Materials and Design (2005), 26: 497–507.
[5] S H.Chen, Huang J H., Xia J., Zhao X K., Lin S B. Influence of processing parameters on the characteristics of stainless steel/copper laser welding. Journal of Materials Processing Technology, 2015, 222:43–51.
[6] I. Magnabosco, P. Ferro, F. Bonollo, L. Arnberg, An investigation of fusion zone microstructures in electron beam welding of cooper-stainless steel, Materials Science and Engineering A 424 (2006) 163-173
[7] T.A. Mai, A.C. Spowage, Characterisation of dissimilar joints in laser welding of steel–kovar, copper–steel and copper–aluminium, Materials Science and Engineering: A, Volume 374, Issues 1–2, (2004) 224–233.
[8] P.S. Wei, Y.K. Kuo, J.S. Ku, Fusion Zone Shapes in Electron-Beam Welding Dissimilar Metals, Journal of Heat Transfer 122(3), (2000) 626-631.
[9] Welding guide for the joining of dissimilar metals, 2006, available at: www.esabsp.esab.net/.../Handbooks/Welding%20Con, accessed: 21.11.2015.
[10] Shun Guo, Qi Zhou, Jian Kong, Yong Peng, Yan Xiang, TianYuan Luo, KeHong Wang, Jun Zhu, Effect of beam offset on the characteristics of copper/304stainless steel electron beam welding, Vaccum (2016), doi: 10.1016/j.vacuum.2016.03.034.
[11] Zhang B.G, Zhao J, Xiao-Peng L I, et al. Electron beam welding of 304 stainless steel to QCr0.8 copper alloy with copper filler wire. Transactions of Nonferrous Metals Society of China, 2014, 24(12):4059–4066.
[12] J. C. Tan, S. A. Westgate , T. W. Clyne, Resistance welding of thin stainless steel sandwich sheets with fibrous metallic cores: experimental and numerical studies, Science and Technology of Welding and Joining (2007) Volume 12, Issue 6.
[13] G. Madhusudhan Reddy, K. Srinivasa Rao, Microstructure and mechanical properties of similar and dissimilar stainless steel electron beam and friction welds. International Journal Advance Manufacturing Technologie (2009) 45: 875-888.

IOP Conf. Series: Materials Science and Engineering **161** (2016) 012059 doi:10.1088/1757-899X/161/1/012059

The Study of Some Cracks on the Panels of a Minibus Body Using Modal Analysis

D Iozsa [1]and D A Micu[2]

[1] Associate Professor, POLITEHNICA University of Bucharest, ROMANIA

[2] Lecturer, POLITEHNICA University of Bucharest, ROMANIA

E-mail: daniel_iozsa@yahoo.com, dan.alexandru.micu85@gmail.com

Abstract. The body vibrations caused by the disturbing forces which acts on this one cause many problems connected, generaly, with the passengers confort (mechanical vibrations, noises) and also connected with the body reliability. The paper presents the study of some craks appeared on the roof panels of the body of a minibus and the study of some attempts to improve the behaviour. The study was made using the modal analysis in two steps. First was studied the behaviour of the hole structure and than was studied only the panel where cracks appear. The analysis of the first four eigenmodes confirmed cause of the cracks. Finally, it was offered an improved solution.

1. Introduction

The study presented in this paper started after the observation of some cracks which are appearing after a fairly short running distance, on the first panel of the roof of a minibus (figure 1). The producer noticed that the cracks appear especially at romanian users [3].

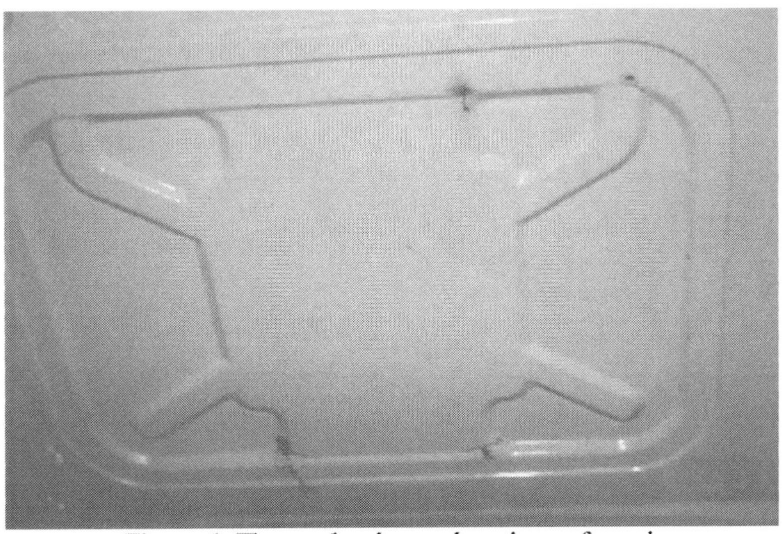

Figure 1. The cracks observed on the roof panel

The main causes could be the frequencies and the amplitude of the vibrations which excite the body structure. Areas of maximum stresses appear in the sensible zones with less rigidity. If the amplitude of the vibration is higher, cracks could be generated in these areas after a relatively small number of cycles.

Modal analysis offers many informations about the vibrational behaviour of a structure. The study of the vibration eigenmodes of the structure gives information about the resonance frequencies, the shape of the structure during vibration and the distribution of stresses for each eigenmodes. It is important to study that eigenmodes of the structure which are excited in the frequencies domain of the disturbing forces and to observe the areas where the structure has less stiffness, with peaks of stresses, areas where could be generated cracks. To perform such analysis is very useful a deformable model of finite elements [1], [2].

2. Modal analysis of the body structure

In the first stage of the analysis were studied the eigenmodes of the whole structure of the body using a simplified model. This model was made by beams and shells finite elements, and the calculus was performed with ADINA, a software specialized in the linear and nonlinear analysis of the structures.

Many of the first modes of the body stucture were excited in the frequencies domain due to running on the road, showing deformations of the roof panels. The first mode was manifested on the first panel, the panel with cracks (figure 2) [3].

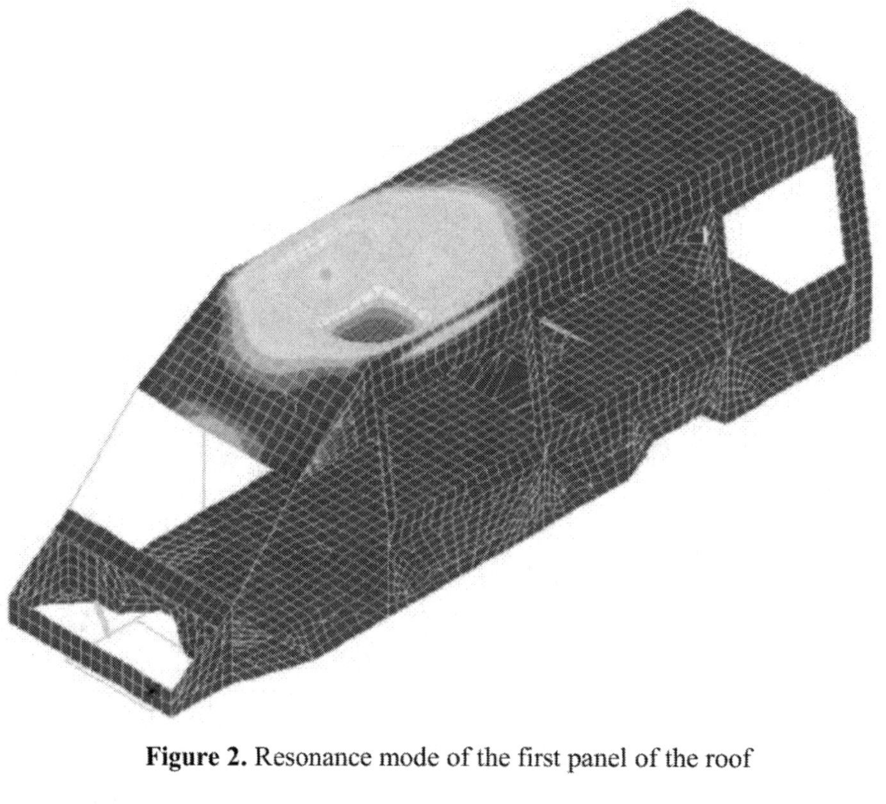

Figure 2. Resonance mode of the first panel of the roof

Vibrations of the first panel of the roof were manifested, also, in other eigenmodes determined by modal analysis. This shows that the body has a low stiffness and a higher vibration sensibility on this first panel of the roof, which can be excited by the vibrations generated by the running of the minibus on the road. In this stage, the panels where modeled like flat panels without ribs.

In the next step of the study was made a detaliated modeling of the first panel of the roof in order to observe the causes which are determining the appearance of the cracks, in the areas showed in the figure 1. The researching of eigenmodes was performed using the Modal analysis modulus of ANSYS, an well known software in the automotive industry.

The geometry of the panel presented in figure 3 shows that were modeled detailed all the ribs made by stamping on the surface of the panel. The finite elements model was made using only shell elements. It was used a relatively high mesh density, so the level of trust for the results obtained is also great.

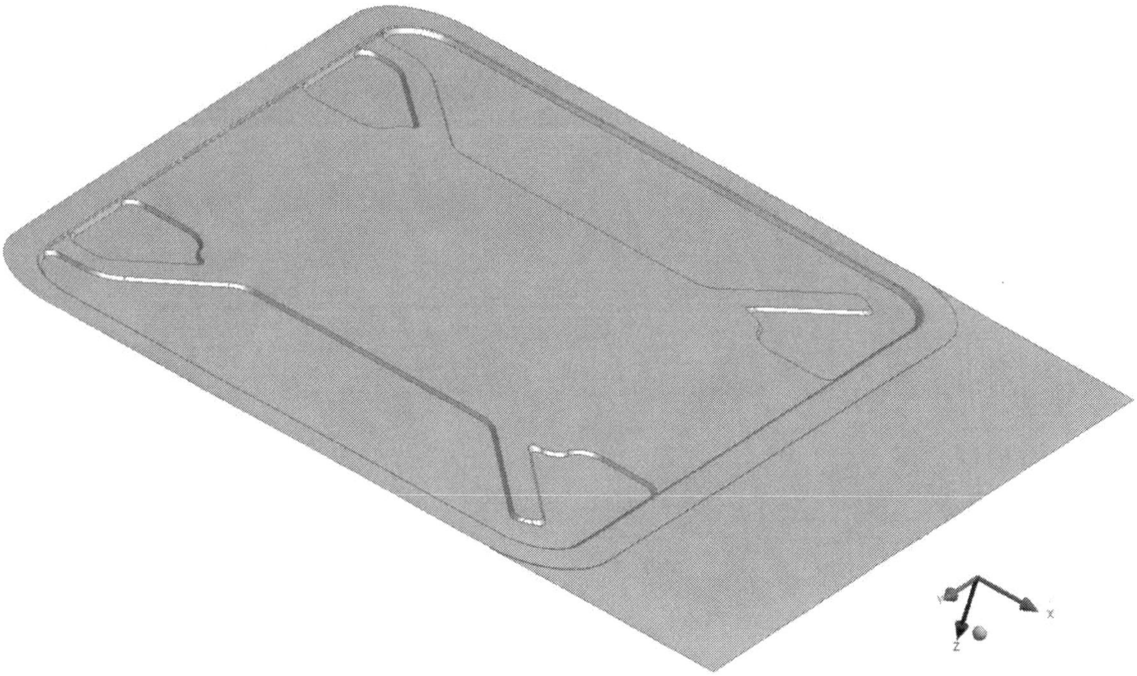

Figure 3. The geometry of the first panel of the roof

The model is fixed on the boundary lines in the same way that the panel is fixed on the beam structure of the minibus body.

The modal analysis of this panel shows that the first three modes are excited at low frequencies which could be caused by rolling on the road or by the operation of some subassemblies (Table 1). The fourth mode is excited at 43.14 Hz.

The displacements of the panel in these four modes are presented in figure 4.

Table 1. Resonance frequency of the first modes

Mode No.	Frequecy
Mode 1	14.05 Hz
Mode 2	26.985 Hz
Mode 3	27.976 Hz
Mode 4	43.14 Hz

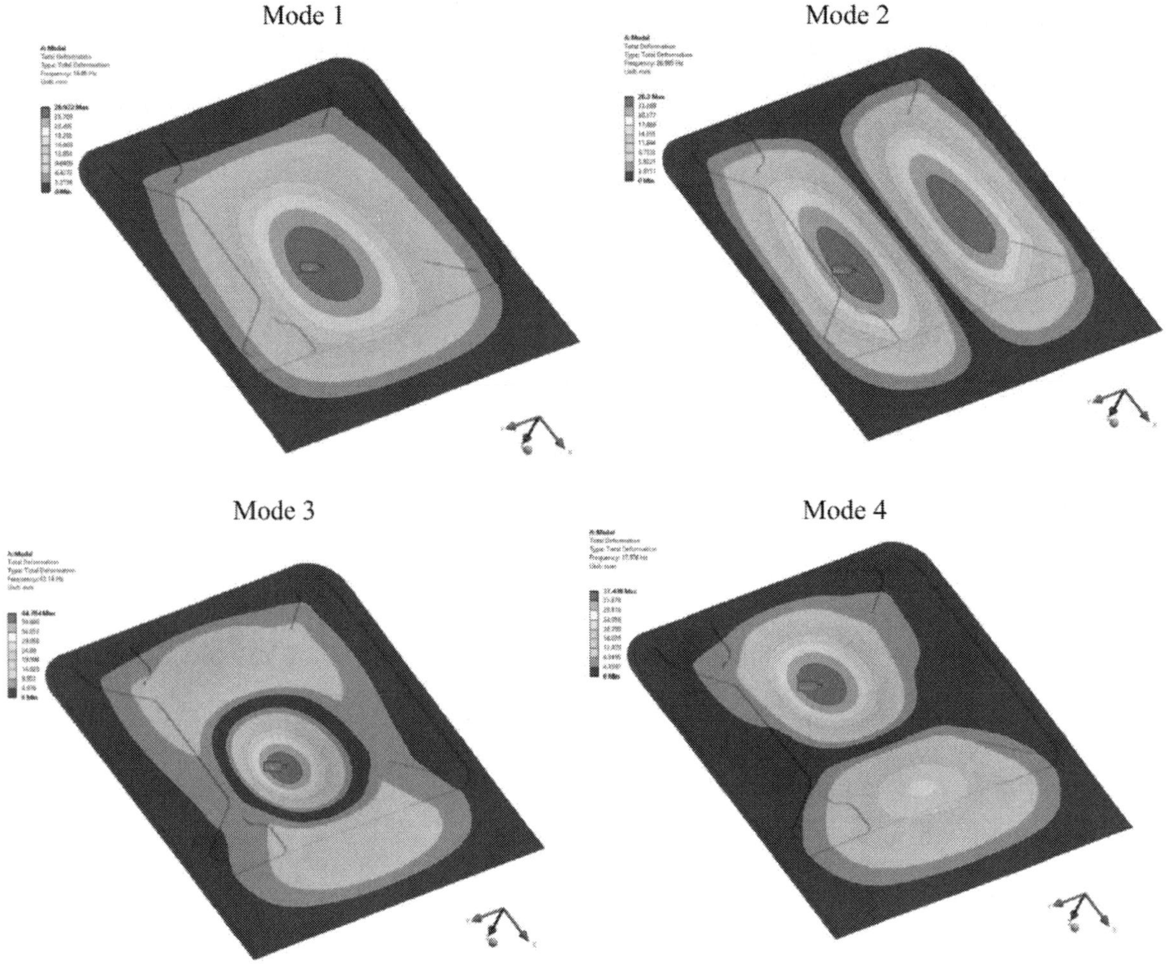

Figure 4. The displacement of the panel in the first four modes

Typicaly for a rectangular plate, the first and the second mode have one, respectively two peaks of amplitude [4]. The third mode is a single peak mode of the central zone of the panel and the fourth mode is a two peaks mode. The shapes of the last two modes are influenced by the shape of the ribs stamped on the panel.

The study of the equivalent stresses in the first three eigenmodes highlights that peaks of maximum stresses are developed in the same places where the cracks appear (figure 5). It was observed that the position of these peaks of stress is influenced by the shape of the ribs stamped on the panel. These peaks are placed in the blend area beetwen the central ribs and the edge rib.

The fourth eigenmode with a higher frequencie, has a peak of maximum stress in a different place.

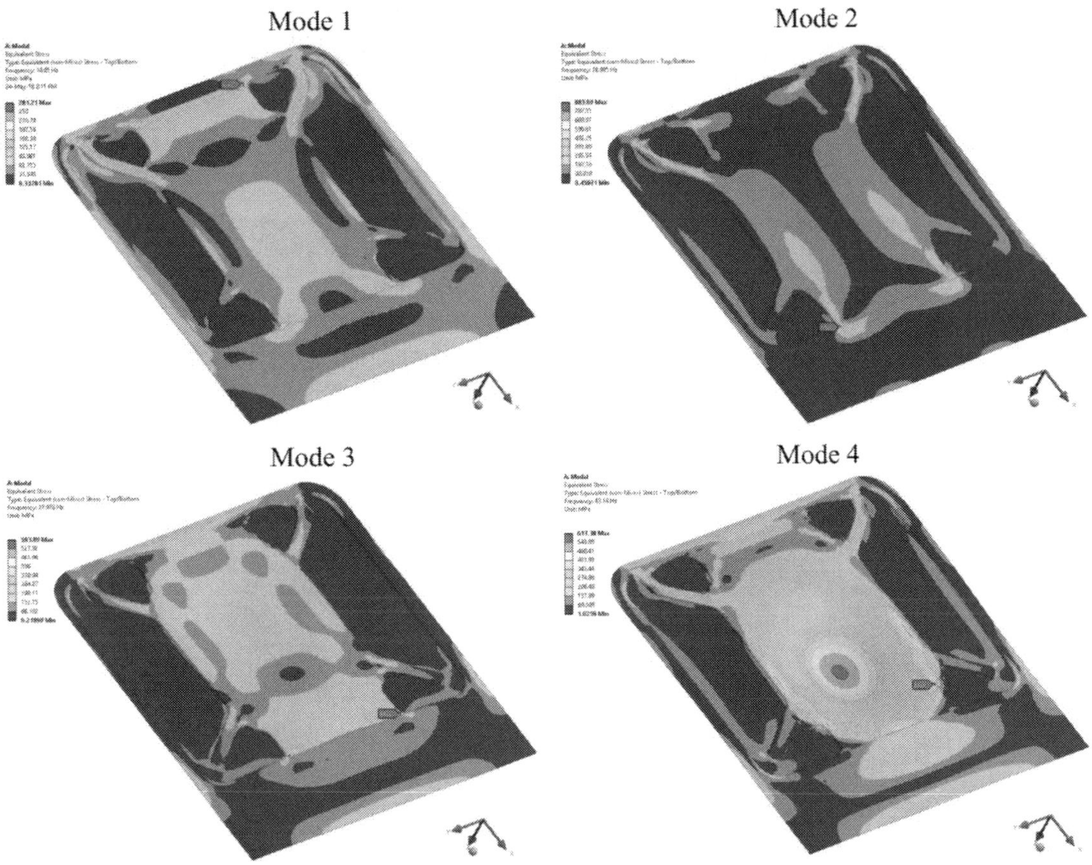

Figure 5. The equivalent stresses in the first four modes

One solution for this problem could be modifying the shape of the stamped ribs in order to increase the rigidity of the panel. So, there were studied some different shapes of the stamped ribs and better results were obtained with a X-shape ribs (figure 6).

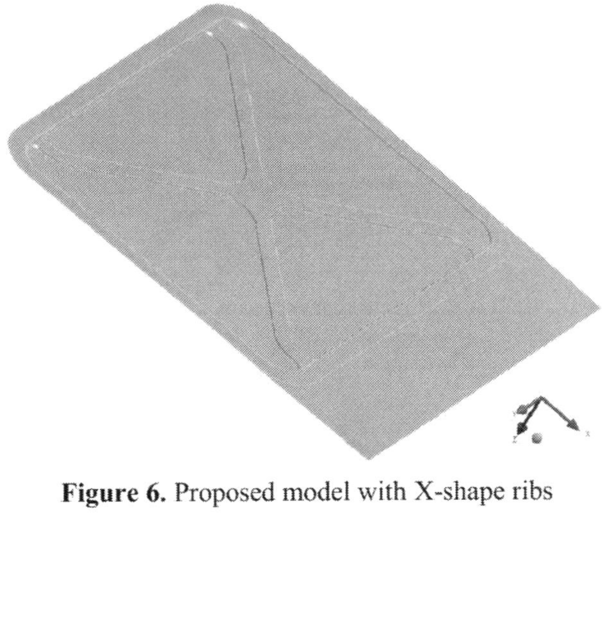

Figure 6. Proposed model with X-shape ribs

The modal anlysis of this model showed that the first mode is also a single peak of amplitude, excited at a higher frequency of 46.16 Hz, which is not caused by the running of the vehicle on the road. The second mode is a two peaks mode and is excited at 76.51 Hz. The position of the peaks was influenced by the X-shape ribs. The first and the second modes are presented in figure 7.

The proposed panel is harder with 23 mg but will have a better rigidity.

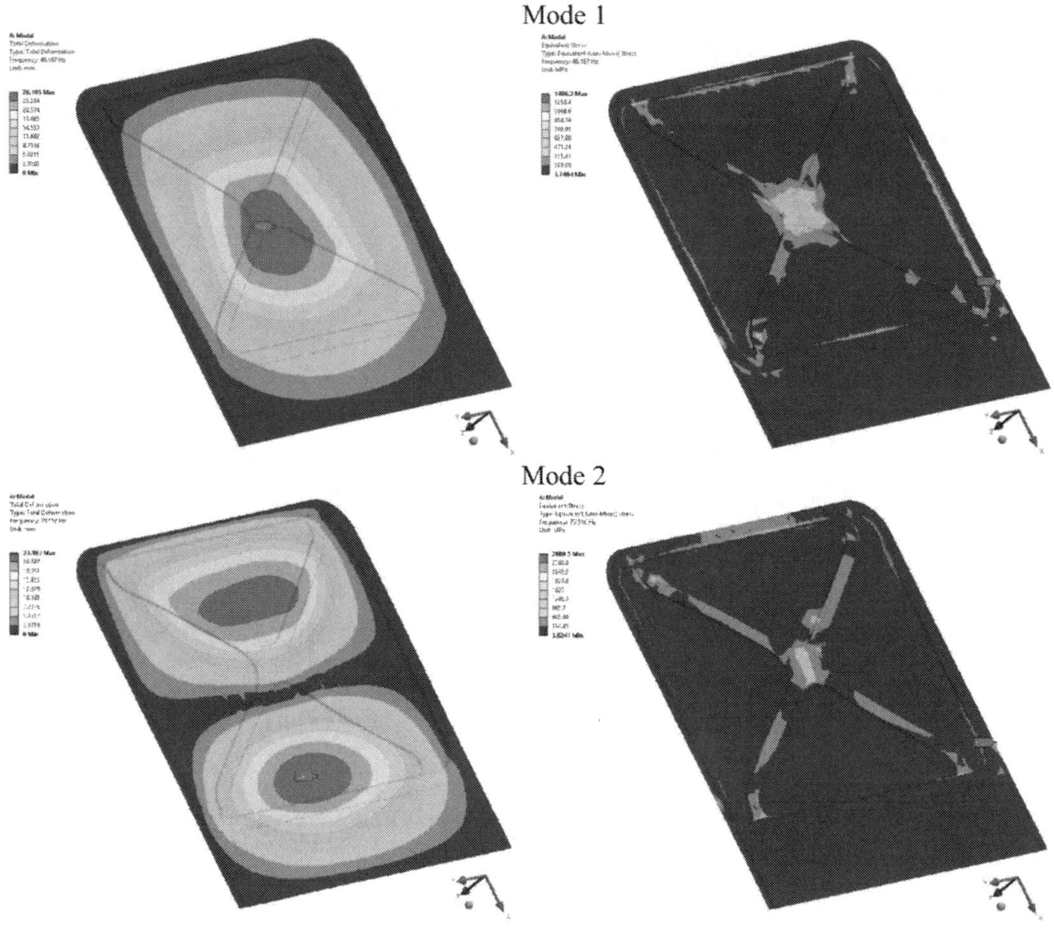

Figure 7. The mode shape and the equivalent stresses in the first two modes of the proposed model

3. Conclusions

The producer of the minibus observed that these craks are appearing only at minibuses of users from Romania, which indicates the well known problems of roads from this country. The vibrational loads which act on the minibus is much higher, favoring the appearance of cracks. The body of these minibuses have a raised roof which is less rigid and more sensitive to the vibrations.

The cracks are caused by the reduced rigidity of the structure in the area of the first panel of the roof. The shape of the stamped sheet metal, the ribs and the blends of the surfaces do not stiffen enough this panel.

References

[1] Zhi-Fang Fu, Jimin He, Modal Analysis, Butterworth-Heinemann, 2001
[2] Zaveri K, Modal Analysis of Large Structures, Brüel&Kjær, 1985
[3] Iozsa D et al, Study of the Dynamic Behaviour of the Car Body using Modal Analysis, ESFA 2003 Conference, Bucharest, 2003
[4] Jiu Hui Wu, A.Q. Liu, H.L.Chen, Exact Solutions for Free-Vibration Analysis of Rectangular Plates Using Bessel Function, Journal of Applied Mechanics, Vol.74, November 2007

Nondestructive testing of advanced materials using sensors with metamaterials

Steigmann Rozina[1], Danila Narcis Andrei[1,2], Iftimie Nicoleta[1], Tugui Catalin-Andrei[1,3], Novy Frantisek [4], Fintova Stanislava [5,6]

Vizureanu Petrica[3], Savin Adriana[1]

[1] Nondestructive Testing Department, National Institute of R&D for Technical Physics, Iasi, Romania
[2]Faculty of Mechanical Engineering, Technical University Gh.Asachi Iasi, Romania
[3]Faculty of Science and Materials Engineering, Technical University Gh.Asachi Iasi, Romania
[4]Faculty of Mechanical Engineering, University of Žilina, Žilina, Slovak Republic
[5] Institute of Physics of Materials AS CR v.v.i., Brno, Czech Republic
[6] Materials Research Centre, Brno University of Technology, Brno, Czech Republic

Email: steigmann@phys-iasi.ro

Abstract. This work presents a method for nondestructive evaluation (NDE) of advanced materials that makes use of the images in near field and the concentration of flux using the phenomenon of spatial resolution. The method allows the detection of flaws as crack, non-adhesion of coating, degradation or presence delamination stresses correlated with the response of electromagnetic sensor.

1. Introduction

Electromagnetic nondestructive evaluation (eNDE) is based on the modification produced by the variation of electrical properties of the examined material due to small inclusions, mechanical /thermal fatigue or of structural components [1]. The term of discontinuity can be defined by the flaw induced in the material during exploitation as well as an imperfection appeared during the technological process. The dimensions and the location of discontinuity, evaluated by noninvasive tests, considered as flaw, are specified in the product standard. The importance of small flaws detection cannot be exaggerated, being known that these can develop due to fatigue or corrosion during exploitation, leading to catastrophic failures. The discontinuities can appear during the fabrication stage and during the coating of metals when the ceramic coatings aims to increase the capacity of high-temperature resistance and protection from thermal oxidation of the substrate. Few coating flaws cannot influence the use of a component, but, per assembly, can decrease the resistance of the finite product. Ceramic materials are optimal for thermal barrier coating (TBC) for their high melting point and possibility to be realized by insular deposition. TBC are essential in aerospace application and mechanical components and systems, for their capabilities of lowering metal surface temperatures. The inspection of equipment from industrial installations represent an efficient way to foreseen and reduce possible risks, but, the use of new materials presenting special properties in extreme environment conditions is nowadays the ideal solution to reach this desideratum. From this reason, the NDE is a domain in continuous developing.

For the material electrical conductive and the flaws possible to appear at the surface or immediately under its surface, one of the NDE possibility is represented by eddy current method.

For the detection of depth cracks with small opening is difficult to apply high frequency electromagnetic methods using "classical" sensors because the standard penetration depth is comparable with crack depth. The detection of surface or subsurface material discontinuities in flat conductive materials is usually performed by electromagnetic method with the sensor scanning the surface to be inspected. For the better localization of discontinuity position, new type sensor with metamaterials (MM) lens is used.

Metamaterials [2], [3] has attracted interest the scientific community, due its possible applications in the optical domain and nondestructive evaluation of materials by construction of performant sensors (perfect lenses, controllable reflection and transmission devices, and electromagnetic absorbers) [4-6]. A generic NDE system is presented below.

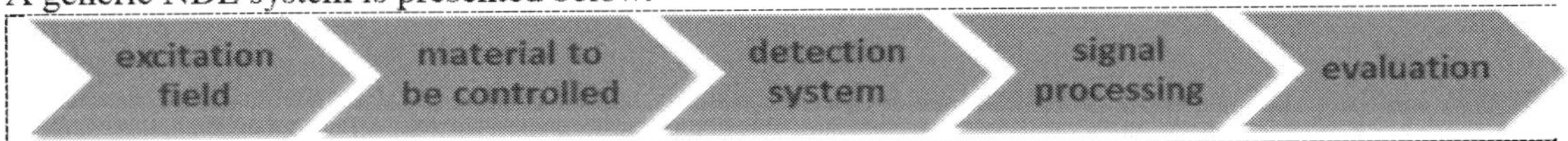

Figure 1. The generic NDE system

The efficient acquisition of data and the displaying of interpretable results of materials properties and their quality has led to development of equipment able to evaluate components/systems without damaging [7].

This paper presents a new type of electromagnetic transducer for detection of micrometric cracks, opened at the surface or immediately subsurface of a material with roughness smaller than the depth of crack.

2. Studied samples; experimental set-up

Let us a steel block, relatively thick, made of advanced materials such as:

a) *steel plates AISI 4340*, low alloy martensitic steel were studied. These present microscopic cracks due to the hydrogen attack. Such steels, when are heat treated to obtain very high strength and hardness, are prone to present microscopic hydrogen cracks due to its high carbon content (0,38-0,43 %) and addition of Mn, Ni, Cr and Mo. The combined high carbon and alloy contents of this steel can lead to the formation of martensite when it is cooled too fast from high temperature to below the transformation temperature. The carbon content is sufficiently high to form hard martensite that may be brittle. When exposed to rich hydrogen content environments by the manufacturing process or by the service conditions, frequently exhibits what it is called hydrogen induced cracking (HIC). One of the faces of the sample has been polished, with R_a=1 μm, determined using Taylor Hobson roughness and precision ultra-surface finish software V5.

HIC were obtained for the samples according to the test conditions of the NACE TM 0284 standard [8], without external loads.

b) *steel plates AISI 316L*, coated with ZrO_2 and ZrO_2 and 20 % Y_2O_3. The optimization of the substrate surface topography preferably is based on more complete characterization in order to achieve a reasonable balance between the level of induced delamination stresses and the mechanical bonding, minimization of the defect size. Yttria Stabilized Zirconia (YSZ) ceramics are considered Thermal Barrier Coating (TBC) materials due to their thermal conductivity (~2.0 $m^{-1}.K^{-1}$ at 1100 °C), refractory, chemical inertness, and compatible thermal expansion coefficient (10.1 × 10^{-6} K^{-1} at 873 K) with metallic support. Laminar structures of YSZ TBC layers deposited on stainless steels are typically porous and the pore size and character depends on the process parameters. The AISI 316L steel is low carbon version of AISI 316 (0.018 %C), which may be susceptible to intergranular corrosion (IGC) in certain corrosive media after it is welded or otherwise heated at temperatures between 430 and 860 °C.

In order to detect flaws as cracks, non-adhesion of coating, etc. high frequency electromagnetic methods must be used, so that the standard penetration depth shall be comparable dimensions of flaws. For detection the possible discontinuities is use an electromagnetic sensor with MM lens, absolute send-receiver-type [9].

In electromagnetic NDE [10, 11], the frequency is chosen so that h≈3δ where, h is the depth of the flaw and $\delta = \sqrt{\dfrac{1}{\pi f \sigma \mu}}$ represents the standard penetration depth, $\mu = \mu_r \mu_0 = \mu_r \cdot 4\pi \cdot 10^{-7}$ H/m.

The sensor with MM lens [12, 13] used in electromagnetic nondestructive evaluation (eNDE) [10, 11] is an absolute receiver-type sensor. The lens is made using two conical Swiss rolls (CSR) [7] having a large basis face to face and is focusing the electromagnetic field and also the evanescent waves (Figure 2a). A CSR consists of a number of spiral wound layers of an insulated conductor on a conical mandrel.

The geometrical parameters of CSR are 20 mm base diameter, 3.2 mm top diameter, 20° aperture angle, 55 mm height and 3 turns made from 18 µm thickness copper foil adhesiveless laminated with 12 µm thickness polyimide foil (LONGLITE™200 produced by Rogers Corporation), in order to decrease the losses at high frequencies. The operation frequencies depending both by the constitutive parameters of MM as well as by the polarization of the incident electromagnetic field (TE_z or TM_z). The focal distance of this lens is $f=l$, where l represents the height of the CSR. The frequency dependence of the effective magnetic permeability of the CSR was determined measuring the S parameters (S_{11} and S_{21}) and applying the effective medium method using a 4395A Network/Spectrum/Impedance Analyzer coupled with an Agilent 87511A S Parameters Test kit. CSRs are tuned at 105 MHz frequency [9].

MM lenses are lenses that, at working frequency, have either $\varepsilon_{eff} = -1$ and then electric evanescent modes can be manipulated, either $\mu_{eff} = -1$ and then the lens can focus magnetic evanescent modes [15]. As shown in [14] the sensor with MM lens functioning in the range of frequencies such that μ_{eff} are a maxim value. For a resonance frequency of 105 MHz, the relative magnetic permeability is 24. The incident field is generated by a one-turn rectangular coil, having 35 × 70 mm, using a Cu wire with 1 mm diameter. In focal plan of the MM lens is placed one turn circular coil with 1mm diameter and 0.1mm wire. During the measurement, the sensor is maintained in fix position, the relative displacement sensor – object to be controlled is made with a XY motorized stage Newmark, which assures the displacement in plan with ±10 µm precision, at the scanning of area with established steps in both directions. Once the level of noise was estimated, the optimal signal processing method is started. The scanning, measurements, acquisition and the processing of data were commanded by a PC using programs developed in Matlab 2014b.

The scheme to connection of sensor with MM lens with equipment is presented in figure 2b.

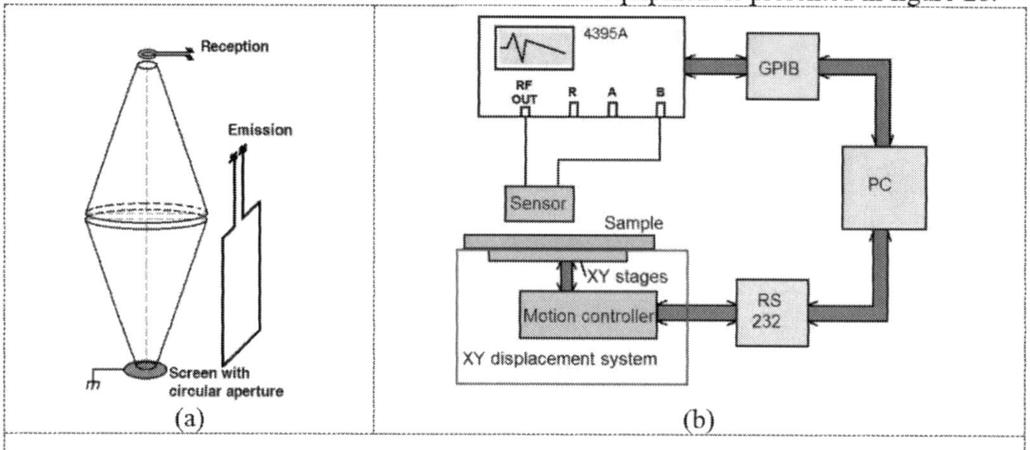

Figure 2. Scheme of : Sensor with MM lens (a); connection of sensor with MM lens with equipment (b)

During the scanning of the selected region of sample the presence of discontinuities disturb the propagation field and the sensor detect the signal. The image delivered by the sensor is amplified.

The method presents the advantage of giving in one representation both the information about amplitude and phase obtained from the sensor.

The sensor is connected to Network/Spectrum/Impedance Analyzer 4395A Agilent coupled through IEEE 488 interface switch to a PC which also commands the displacement system. The physical realization of this is presented in Figure 3a and the connection with equipment in Figure 3b.

(a) (b)

Figure 3. Electromagnetic sensor with MM lens: physical realization (a); experimental set-up (b)

3. Results

3.1. *The samples AISI 4340 steel* attacked with hydrogen are one of the faces polished, with R_a=1μm, determined using Taylor Hobson rugosimeter and precision ultra-surface finish software V5. Presence of the cracks appears due to hydrogen attack and these are invisible with naked eyes. Using a microscope BX51 Olympus USA, functioning in reflection regime, with magnification 200X this cracks is identified and marked as direction. This sample was scanned on 100x40μm^2 with 1 μm scanning step.

In Figure 4a is presented the microscopic image of crack identified and in Figure 4b the amplitude of electromotive force induced in the reception coil of the electromagnetic transducer at the scanning of the corresponding zone from the sample. For the measurements, the amplitude represented in each measurement point is the arithmetic average of three measurements.

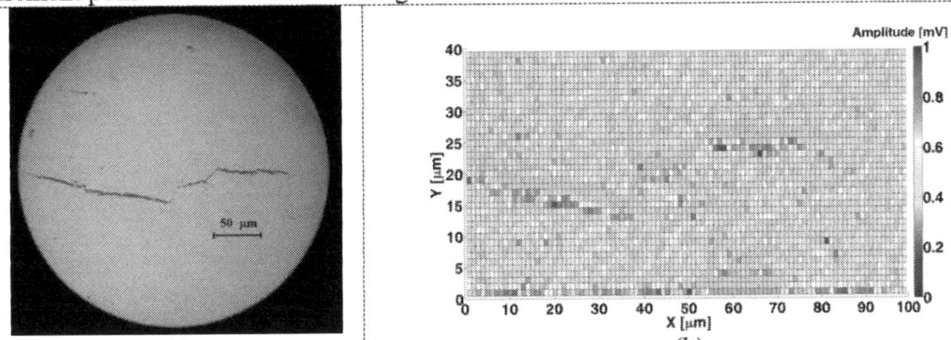

(a) (b)

Figure 4. Comparison between optical images (200X magnification) (a) and electromagnetic images (b) for the crack

IOP Conf. Series: Materials Science and Engineering **161** (2016) 012060 doi:10.1088/1757-899X/161/1/012060

3.2. *Steel plates AISI 316L* - for the characterizing the sample, the SEM method has been used, so SEM images (Figure 5a) emphasize that the 316L steels substrate is compact, with some inclusions and/or pores between it and the ZrO_2 layer.

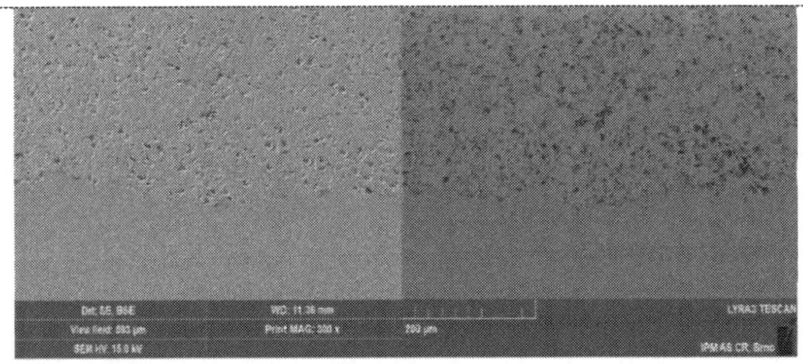

(a)

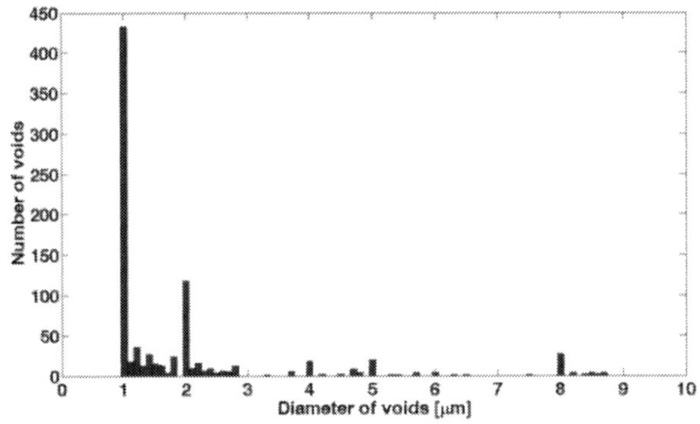

(b)

Figure 5. The SEM images (a)and voids counting (b) for specimen with 0.2 mm thick monolithic coating ZrO_2 with addition of 20 % Y_2O_3

In particular, the results obtained [16] show that for the specimens, with 0.2 mm thick monolithic coating of ZrO_2 with addition of 20 % Y_2O_3, the 316L substrate is compact, some little inclusions and/or pores between the support and the ZrO_2 layer. Using a procedure for image processing was determined that pores created about 3 up-to 12% of the analyzed layers. Pores are distributed random and nonhomogeneous. EDX analysis shows the interface between the steel and the zirconia layer, a higher degree of oxidation takes place, more scattered within the zirconia doped with yttria. Nondestructive identification of the last type of oxidation is of interest in this paper. For this type of specimen the ceramic top-coating (nonconductive and nonmagnetic) become a material which create a probe lift-off. The response of the sensor depends only on coating thickness and the existence or lack of adhesion. On the other hand, the residual stresses in the TBC affected its performance. The amplitude of the signals received by the sensor is presented in Figure 6.

Examining Figure 6, it can be observed that the values of electromotive force induced in the reception coil of sensor are usually affected by several parameters, as lift-off, inspection frequency, material composition, and the presence of inhomogeneity on or near the object surface. Monolithic coating zirconia, with addition of 20% yttria having different thickness does not presents non adherence zones. The amplitude of the voltage induced in the reception coil has relatively constant value, excepting the regions where the variations appear due probable to oxide agglomerates. The surface of metallic substrate was blasted to create perfectly clean and rough enough surface for good adhesion of deposited ceramic coating.

The eNDE is non-invasive method, interpretation of results requires considerable knowledge of the electromagnetic properties of the samples as well as the technique itself.

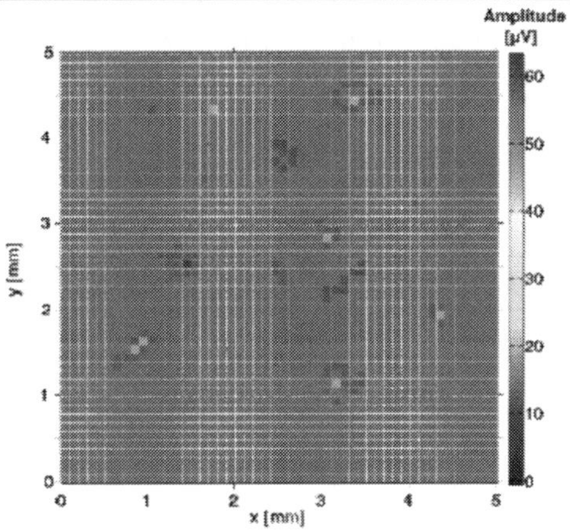

Figure 6. The amplitude of the voltage induced in the reception coil of the sensor at the scanning of the sample.

New types of materials or of materials already studied in special environmental conditions need other adaptable methods for characterization and evaluation. The challenge for the development of new types of electromagnetic sensors consists in the obtaining of good detection sensitivity for a minimum 3/1 signal to noise ratio (signal strength relative to background noise) as well as good spatial resolution. The signal to noise ratio has been determined measuring the signal in the reception coil over the tested samples and in free space. The analysis of the results show the possibility to apply this method to determine the eventual dislocations of coating layer from the substrate layer and can be improved further to complete the existent methods (infrared thermography, ultrasounds, etc).

4. Conclusions

In order to detect the intensification of evanescent waves, a sensor with MM lens is used at normal incidence, in the near field. Using plane electromagnetic waves, TMz polarized; evanescent wave can appear in cracks, or which are diffracted by the crack's edges. Using this type of sensor, the evanescent waves can be manipulated and the electromagnetic images of the detected cracks are obtained. These images are in good concordance with the images obtained by optical microscopy.

New nanocomposites, with nanometer - sized second phase particles dispersed in ceramic matrix and/or at grain boundaries have shown an ability to withstand thermal cycling at temperatures significantly higher than simple layer YSZ-coatings. They also have significant better strength and creep resistance than other coatings. Hence the evaluation of the surface structure and possible delamination at the interface of deposited layers for this type of zirconia coating on stainless steel is important.

Further tests on a larger number of specimens with different coating aspects of the surface / number of layers are needed to establish the accuracy of the results and also the correlation between the localized small-sized discontinuities and the resulting MM sensor indication.

5. References

[1] Georgiou GA, 2009 *Mater. Sci. Eng.* **3** 76
[2] Veselago V G 1968 *Sov. Phys. Usp.* **10** 509
[3] Smith D R, Padilla W J, Vier D C, Nemat-Nasser S C, Schultz S 2000 *Phys. Rev. Lett.* **4(18)**:4184

[4] Pendry J 2004 *Contemp Phys.* **45** 191
[5] Smith D R, Schultz S, Markoš P, Soukoulis CM 2002 *Phys. Rev. B.***65**(19) 195104
[6] Savin A, Bruma A, Steigmann R, Iftimie N, Faktorova D 2015 *Appl. Sci.* **5**(4) 1412
[7] Haimbaugh R E 2015 Practical induction heat treating. (ASM nternational)
[8] NACE TM0284-2011, Standard Test Method - Evaluation of Pipeline and Pressure Vessel Steels for Resistance to Hydrogen-Induced Cracking, National Association of Corrosion Engineers.
[9] Savin A, Steigmann R, Bruma A and Šturm R 2015 *Sensors* **15** (7) 15903
[10] Grimberg R, Savin A and Steigmann R 2012 *NDT & E Int.* **46** 70
[11] Grimberg R and Savin A 2011 Patent RO126245-A0
[12] Engheta N and Ziolkowski R W 2006 *Electromagnetic Metamaterials: Physics and Engineering Explorations*, (Hoboken: Wiley)
[13] Pendry J B 2000 *Phys. Rev. Lett.* **85** 3966
[14] Grimberg R, Savin A, Steigmann R, Serghiac B and Bruma A 2011 *INSIGHT* **53** 132
[15] Pendry J B, Holden A J, Robbins D J, Stewart W J 1999 *IEEE T Microw Theory* **47**(11) 2075
[16] Faktorova D, Novy Fr, Savin A, Iftimie N, Turchenko V and Craus ML 2016 Stud Appl Electromag 41 (TBP)

Acknowledgments
This work was supported by a grant of the Ministry of National Education, CNCS – UEFISCDI, project number PN-II-ID-PCE-2012-4-0437 and Nucleus Program PN 16 37 01 01.

IOP Conf. Series: Materials Science and Engineering **161** (2016) 012061 doi:10.1088/1757-899X/161/1/012061

Behaviour of a ZnO thin film as MSG for biosensing material in sub-wavelength regime

Iftimie N[1], Steigmann R[1,2], Danila N A[1], Iacomi F[2], Faktorova D[3] and Savin A[1]

[1] Nondestructive Testing Department, National Institute of R&D for Technical Physics, Iasi, Romania
[2]Faculty of Physics, Alexandru Ioan Cuza University, Iasi, Romania
[3]Faculty of Electrical Engineering, University of Žilina, Žilina, Slovak Republic

Email: niftimie@phys-iasi.ro

Abstract. Zinc oxide nanostructured materials, such as films and nanoparticles, could provide a suitable platform for development of high performance biosensing material due to their unique fundamental material properties. In this study, the enzyme biosensing consisting of a zinc oxide (ZnO) nanoparticles were grown on SiO_2/Si substrates by vacuum thermal evaporation method and their sensing characteristics are examined in air and investigated. The film morphology is characterized by X-ray diffraction (XRD) the film crystalline quality and by scanning electron microscopy (SEM). Also, the interest in surface waves appeared due to evanescent waves in the metallic strip grating structure (MSG-Ag/ZnO/SiO_2/Si) in sub-wavelength regime. Before testing the sensor with metamaterials (MMs) lens in the sub-wavelength regime, a simulation of the evanescent wave's formation has been performed at the edge of Ag strips, with thicknesses in the range of micrometers.

1. Introduction

The stability of ZnO nanostructures under physiological conditions, being important for sensing and biosensing applications, depends on the crystal quality of the nanostructures but can be assumed as sufficient for structures grown by thermal evaporation methods [1]. Nanostructures made of ZnO are easy and reliable to produce in a wide manner of different forms and structures [2]. ZnO is very versatile and due to its high coupling coefficient, many applications exist, like Surface Acoustic Wave (SAW) devices [3], bulk acoustic wave devices, gas sensors [4-5], infrared detectors, tactile sensor arrays and enzyme biosensors [6]. The evidence of plasmons apparition in structures of wires with micrometric diameters has been shown for Al at frequency of 8.2 GHz [7].

The possibility to excite plasmons at lower or equivalent frequencies in complex structures containing silver metallic strip gratings (MSGs) was investigated. A MSG interface supports the well-known surface plasmon-polariton (SPP) modes, yielded by the coupling of the EM field with the coherent oscillations of the free electrons in the metal. Their ability to store and propagate the EM energy at subwavelength scales is essential for many applications such as biosensing, photonic circuits and optical data storage.

In this paper, in order to confirm the availability of ZnO nanostructures for convenient high-efficiency biosensors, the glucose sensing performance via the electrical properties to be measured in air was investigated. Also, the interest in surface waves appeared due to evanescent waves in the MSG in sub-wavelength regime. The experimental test is making with the sensor with metamaterials lens [8,9] in the sub-wavelength regime, and a simulation of the evanescent wave's formation has been performed at the edge of Ag strips, with thicknesses in the range of micrometers. In this work, high quality ZnO films were grown on SiO_2/Si substrates by vacuum thermal evaporation method is characterized by X-ray diffraction the film crystalline quality and by Scanning Electron Microscopy (SEM) the film morphology. This method has many advantages such as being a low cost, simple, high yield, 10^{-5} Torr pressure, deposition rate of 0.05 mm/s and a deposition time of 30 s. The high electrochemical response can be attributed to the unique structural properties of our sensor electrode like the high surface to volume ratios of ZnO nanoparticles, which can provide a favorable microenvironment for the immobilization of uricase enzyme and retain the good enzymatic activities which in turn enhances the sensitivity of sensor electrode for the analyte, as demonstrated by the detection of uric acid in the absence of a mediator.

2. Experimental section

The morphology of the nanostructure significantly affects its electrochemical properties, so numerous ZnO nanostructures have been investigated for application in enzyme biosensor. The physical, chemical and surface properties of the desired support determine the method of uricase enzyme immobilization, the nature of the immobilized sensing molecules, and the overall biosensor performance. While the redox capabilities of enzymes are not typically enhanced because of their insulated redox centers, specific ZnO nanostructures can facilitate direct electron transfer (DET) between enzyme electroactive sites and external electrodes [10-14]. The synthesis of ZnO nanostructures by different techniques has promoted the fabrication of enzyme biosensors.

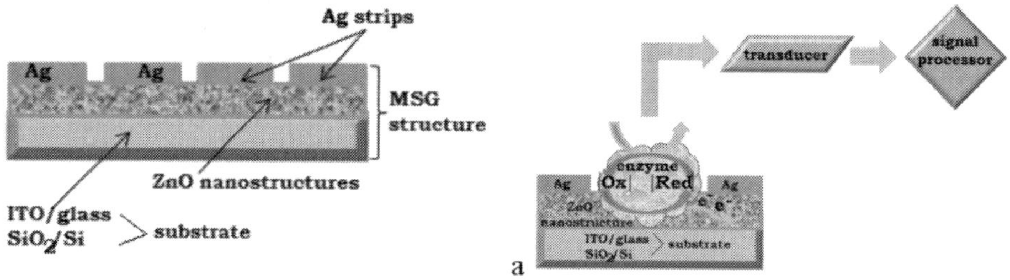

Figure 1. a) MSG structure; b) Working principles of a ZnO nanostructure-based enzyme biosensing.

To sensing mechanism for enzyme biosensing (e.g. uric acid and ascorbic acid) designed in this purpose, the voltage applied across the two electrodes causes a current to flow via electron tunneling through the potential barriers between nanoparticles. Detecting uric acid in physiological fluids is important for diagnosing disorders associated with altered purine metabolism [15,16]. Equation (1) describe the uricase catalyzing reaction:

$$\begin{aligned} &C_5H_4N_4O_3(\text{uric acid}) + O_2 + 2H_2O \\ &\xrightarrow{uricase} C_5H_6N_4O_3(\text{allantoin}) + CO_2 + H_2O_2 \end{aligned} \quad (1)$$

A reagentless uric acid biosensor based on uricase functionalized ZnO nanostructures exhibits good thermal stability and anti-interference capability. When uric acid is oxidized in the presence of uricase it is turned into allantoin along with carbon dioxide and hydrogen peroxide. Due to the presence of water (H-OH), it is a high probability that allantoin will accept a proton from (H-OH) converting it to allantoinium ion, which in turn will interact with the ZnO nanostructures and produce a potential change at the electrode. As the concentration of ions changes in surrounding the ZnO nanostructures and the electrode potential will change. The potentiometric responses of the sensor electrodes were studied in uric acid solutions made in buffer (PBS pH 7.0).

3. Structural characterization

The XRD patterns of the obtained structures were carried out on a diffractometer using CuKα radiation (λ =1.5406 Å) (Figure 2). The samples were analyzed in the range of 2θ=5°÷80° with a scanning angle rate of 0.02 and a 2 s/step count time. The experimental XRD patterns were identified using Crystallographica Search-Match programme. The XRD peaks for ZnO crystalline phases are visible. The reflection peaks, at (100), (002), (101), (110), (103) and (112) were indicative of the hexagonal wurtzite ZnO nanostructure.The ZnO film is polycrystalline in nature having a preferred grain growth orientation along (002) and (101) planes which correspond with the peaks at 34.4 and 36.2°. The weakness of the peaks is related to the thickness of the thin films. This is due the nature of the source material and it is assumed that only nanoparticles migration from the source to the substrate takes place.

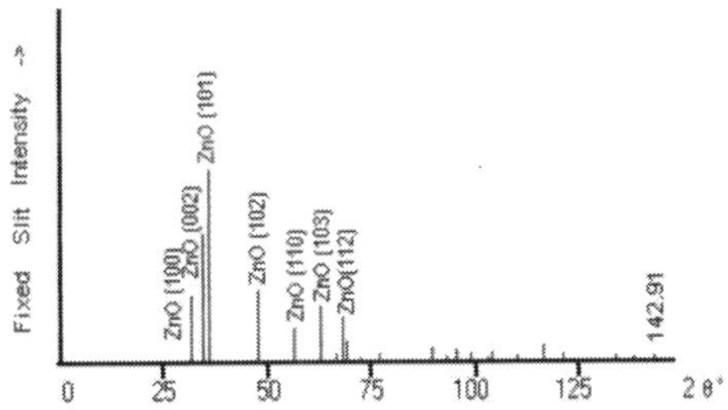

Figure 2. XRD ZnO thin film deposited on SiO_2/Si substrate.

For structural characterization of Ag/ZnO/substrate nanostructures scanning electron microscopy, SEM, investigations were performed. The SEM showed that the structure of the films of ZnO is columnar (Figure 3 a). It can be shown also that of ZnO nanostructured without bound enzyme had a uniform film, while ZnO nanostructured with bound enzyme had many globular structures. These observations confirmed immobilization of enzyme. The surfaces are without inclusions and defects, fact that make them appropriate for radio frequencies applications.

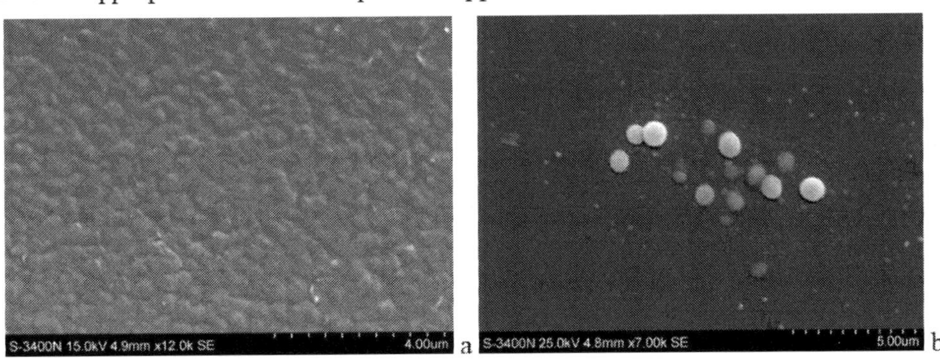

Figure 3. The SEM images of cholesterol oxidase bound ZnO nanostructures: (a) without enzyme; (b) with immobilized uricase

4. Evanescent waves. FDTD simulation

Due to experimental difficulties in obtaining a perfect lens, the manipulation of evanescent modes can be made with an electromagnetic sensor with MMs lenses that have, at the operation frequency, either μ_{eff} = -1, and the lens can focus magnetic evanescent modes [18-20,9,21-23]. The spatial resolution of the system was verified according to [20] and the analysis of data obtained show that the realization of MM lenses in the radiofrequency range is possible using CSR, whose distortions are minimal and whose calculation are based on Fourier optic principles [24].

The sensor and experimental set-up is presented [25,17]. Generation and detection of evanescent waves in slits are made using the electromagnetic sensor lens. The rectangular frame used for the generation of TMz polarized waves is identical with rectangular frame used for the simulation model and the lens is made from a CSR [8,17]. The working frequency was 474 MHz. The simulation was make using XFDTD 6.3 software produced by REMCOM [26]. In Figure 4 is shown the result of simulation with XFDTD. The E_y component, which counts in our case, is displayed. In [27] it has been presented the behavior of the field with air in the slits, it can be shown that, for uricase between the strips, the amplitude of the electric field has the same behavior as in [22] but the amplitude decreasing due to electrical permittivity high of the uricase. Thus, the symmetrical maxima appear in the middle of the slits, decreasing to the minimum value, on the strips edges. Inside the strips, other pair of maxima appears, followed by the decreasing to middle of the strip.

The Fig. 6 shows the dependency of voltage amplitude induced in the reception coil of sensor at the scanning of the MSG taken into study, the image showing that the type of sensor allows the correct emphasizing of extremely thick conductive strips and eventual interruptions.

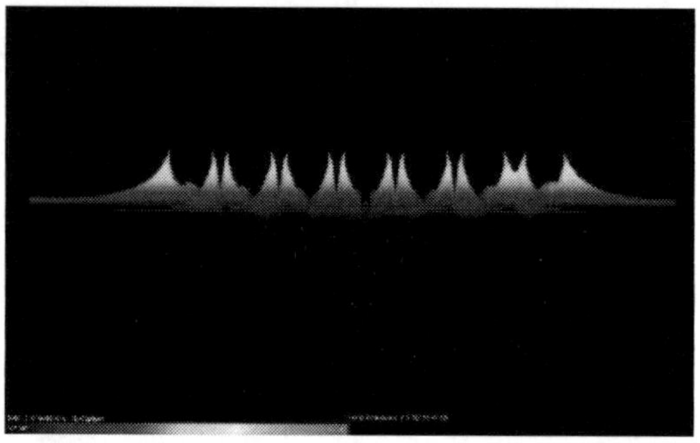

Figure 4. Numerical results for electric-field amplitude distribution near the strips; the field values are normalized to the amplitude of the incident field.

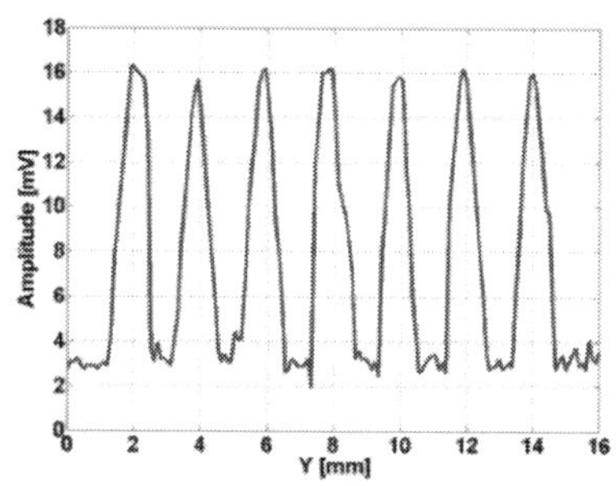

Figure 5. Amplitude of voltage induced in the reception coil at the scanning of silver strip grating.

The existence of a single evanescent mode, theoretically foreseen, is experimentally confirmed by the existence of a local maximum in the middle zone of the slits, followed by an accentuated decreasing on the edge of the strips. The results are in good concordance with theoretical estimations, that confirms good adhesion of silver on $ZnO/SiO_2/Si$, also good alignments of strips.

5. Experimental set-up and biosensing properties

ZnO nanostructured have excellent prospects for interfacing biological recognition events with electronic signal transduction as a new generation of biosensors in which active enzymes sites are coupled directly with a nanostructured ZnO electrode resulting in direct electron transfer between the enzyme and nanostructured ZnO with improved biosensing properties.

The initial experimental dates were carried out using a multimeter device because of its versatility in measuring resistance, voltage and ampere. The multimeter probes were immersed into 5 ml sodium phosphate buffer (0.02 M, pH 7.0) in a beaker. The reaction was started by adding 0.5 ml of enzyme solution. Based on the current produced during the chemical reaction the number of molecules of uricase and consequently the amount of uricase present in the sample could be determined. The current was measured at varying concentrations of uricase to obtain a full working range. The sensing mechanism of most electrochemical uric acid sensors is based on an enzymatic reaction catalyzed by uricase as described in Figure 6.

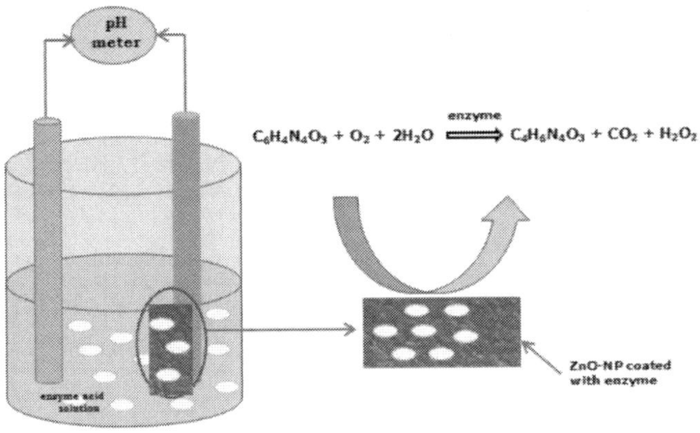

Figure 6. Schematic diagram of the uric acid sensing setup using ZnO-NP coated with uricase as working electrode showing the possible electrochemical reaction near the working electrode.

To study effect of uricase concentration, the concentration was varied from 100 to 1300 mg/dl and amperometric measurements were made after 0.5 ml of uric acid added. The probe was immersed into the solutions and the displayed ampere was noted against the concentration of the solution.

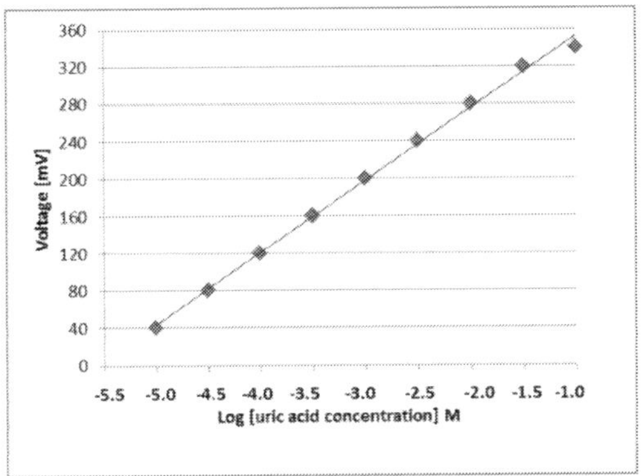

Figure 7. The calibration curve of uric acid biosensor based on $ZnO/SiO_2/Si$ nanostructures for linear concentration range of 5.2 to 20.8 mM

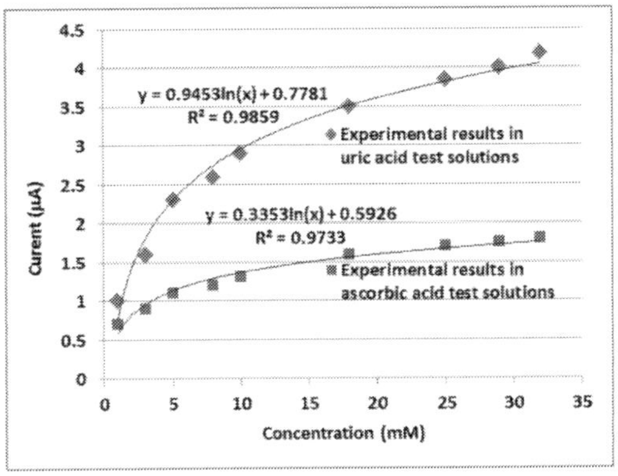

Figure 8. Variation in the current at a fixed potential of 0.04 V with changing uricase concentration to $Ag/ZnO/SiO_2/Si$ architecture biosensing

Figure 7 shows the relation between the output potential response and uricase concentration for the element biosensing. It is clearly seen from the graph that the response current increases as the concentration of uricase increases and saturated at high concentration of uricase which suggests the saturation of active sites of the enzymes at those uric acid levels. The selectivity study suggests that the presence of interferants, like ascorbic acid, have a negligible effect on the performance of the $Ag/ZnO/SiO_2/Si$ architecture toward sensing of uricase (Figure 8). The calibration curves were obtained in the range of 100-1300 mg/dl (2.6-33.8 mM), with coefficient of determination was estimated to be $R^2 = 0.9859$. Under optimized conditions, the steady-state current showed a linear dynamic range of 200-800 mg/dl (5.2-20.8 mM).

6. Conclusions

A simple fabrication technique for a biosensing electrochemical uric acid sensor based on ZnO nanostructures deposited on SiO_2/Si substrate by vacuum thermal evaporation was performed. The Ag strip grating thickness of µm order is comparable with the depth penetration of electromagnetic wave of

474 MHz, so that evanescent waves appear at the edge of strips and their manipulation can improve the resolution power of the sensor.

The experimental measurement was made with a sensor with metamaterials lens. It has been proved that sensors with metamaterials lens can be used in amplitude evaluation of the evanescent waves formed at the edge of strips of $Ag/ZnO/SiO_2/Si$ structure. The calibration curves were obtained in the range of 2.6 to 33.8 mM, with coefficient of determination was estimated to be $R^2 = 0.9859$. By results obtained it is shown that the element biosensing structure investigated is rapid and sensitive for the detection of uricase. One can concluded that due to the simple synthesis and electrode fabrication, good sensitivity, low detection limit and fast response, the ZnO nanostructured opens a way for the fabrication of highly efficient cholesterol biosensors.

References

[1] Zhou J, Xhu N S and Wang Z L 2006 Dissolving behavior and stability of ZnO wires in biofluids: a study on biodegradability and biocompatibility of ZnO nanostructures *Adv. Mater.* **18** 2432-2435

[2] Wang Z L 2004 Zinc oxide nanostructures: growth, properties and applications *J. Phys. Condens. Matter.* **16** 829-858

[3] Fu Q, Deng J, Zhou D, Gong S, Zhong S and Luo W 2012 *IMCS 2012 – The 14th International Meeting on Chemical Sensors* 905-908

[4] Rambu A P, Tiron V, Nica V and Iftimie N 2013 Functional properties of ZnO films prepared by thermal oxidation of metallic films *J. Appl. Phys.* **113** 234506-6

[5] Iftimie N, Iacomi F and Rezlescu N 2008 High performance gas sensing materials based on nanostructured zinc oxide films *Journal of Optoelectronics and Advanced Materials* **10** 1810–1813

[6] Faktorova D 2010 Dielectric spectroscopy of glucose turnover in cancerous tissue model *In: Acta Technica CSAV* **55** (1) 29-37

[7] Pendry J B, Holden A J, Stewart W J and Youngs I 1996 Extremely Low Frequency Plasmons in Metallic Mesostructures *Phys. Rev. Lett.* **76** 4773-4776

[8] Grimberg R, Savin A, Steigmann R, Serghiac B and Bruma A 2011 Electromagnetic non-destructive evaluation using metamaterials *INSIGHT* **53** (3) 132-137

[9] Savin A, Steigmann R, Bruma A and Sturm R 2015 An electromagnetic sensors with a metamaterial lens for nondestructive evaluation of composite materials *Sensors* **15** 15903-15920

[10] Zhao Y, Yan X, Kang Z, Lin P, Fang X, Lei Y, Ma S and Zhang Y 2013 Highly sensitive uric acid biosensor based on individual zinc oxide micro/nanowires *Microchim. Acta.* **180** (9) 759–766

[11] Liu J, Guo C, Li C M, Li Y, Chi Q, Huang X, Liao L and Yu T 2009 Carbon-decorated ZnO nanowire array: a novel platform for direct electrochemistry of enzymes and biosensing applications *Electrochem. Commun.* **11** 202–205

[12] Dai Z, Shao G, Hong J, Bao J and Shen J 2009 Immobilization and direct electrochemistry of glucose oxidase on a tetragonal pyramid-shaped porous ZnO nanostructure for a glucose biosensor *Biosens. Bioelectron.* **24** (1) 286–1291

[13] Kavitha T, Gopalan A I, Lee K P and Park S Y. 2012 Glucose sensing, photocatalytic and antibacterial properties of graphene–ZnO nanoparticle hybrids *Carbon* **50** (8) 2994–3000

[14] Jafari M, Khodadadi A A, Mortazavi Y and Ghorchian H 2012 *IMCS 2012—The 14th International Meeting on Chemical Sensors* 687–689

[15] Usman Ali S M, Alvi N H, Ibupoto Z, Nur O, Willander M and Danielsson B 2011 Selective potentiometric determination of uric acid with uricase immobilized on ZnO nanowires *Sensors and Actuators B* **152** 241–247

[16] Zhang Y, Kang Z, Yan X and Liao Q 2015 ZnO nanostructures in enzyme biosensors, *Sci. China. Mater.* **58** 60–76

[17] Steigmann R, Iftimie N and Savin A 2015 Characterization of the Thin Films Structures in Subwavelength Regime as Biosensing Materials *Applied Mechanics and Materials: Advanced Research in Aerospace, Robotics, Manufacturing Systems, Mechanical Engineering and Bioengineering* **772** 62-66
[18] Pendry J B, Holden A J, Stewart W J and Youngs I 1996 Extremely Low Frequency Plasmons in Metallic Mesostructures *Phys. Rev. Lett.* **76** 4773-4776
[19] Pendry J B, Holden A J, Robbins D J and Stewart W J 1999 Magnetism from conductors and enhanced non-linear phenomena *IEEE Trans. Microw. Theory Tech.* **47** 47-58
[20] Gribic A and Eleftheriades G V 2003 Growing evanescent waves in negative-refractive-index transmission-line media *Applied Physics Letters* **82**(12) 1815-17
[21] Grimberg R and Tian G Y 2012 High-frequency electromagnetic non-destructive evaluation for high spatial resolution, using metamaterials *Proc. R. Soc. A*, **468** 3080–3099
[22] Savin A, Steigmann R and Bruma A 2014 Metallic Strip Gratings in the Sub-Subwavelength Regime *Sensors* **14**(7) 11786-11804
[23] Savin A and Steigmann R, 2014 *patent RO129801-A0/2014*, Electromagnetic sensor for assessing integrity of braids of single-layer printed wiring on flexible support and of layered meso-structures
[24] Born M and Wolf E 1999 *Principle of Optics* 7th ed. (Cambridge: Cambridge University Press)
[25] Grimberg R 2013 Electromagnetic metamaterials *Mater. Sci. Eng. B* **178** 1285–1295
[26] Lagendijk A, Van Tiggelen B and Wiersma D S 2009 Fifty years of Anderson localization *Physics Today* **62**(8) 24-29
[27] Iftimie N, Tascu S, Salaoru I, Steigmann R, Savin A, Irimia M and Iacomi F 2015 The evanescent waves in metallic strip gratings and complex structures in subwavelength regime *Materials Today: Proceedings*, **2**(10) 3486-3452

Acknowledgments
This paper is supported by Romanian Ministry of National Education under project PN-II-ID-PCE-2012-4-0437 and Nucleus Program PN 16 37 01 01.

The analysis of the distribution of unitary stresses for the universal plowshare in tiller seeder combos (UPTSC)

D Chiorescu [1], E Chiorescu [1], O Dodun [2], V Crăciun [3]
[1]„Ion Ionescu de la Brad" University of Agricultural Sciences and Veterinary Medicine, 3, Mihail Sadoveanu Alley, Iaşi, 700490, Romania
[2]„Gheorghe Asachi" Tehnical University, 59 A, Dimitrie Mangeron Blvd, Iași, 700050, Romania
[3]„Gheorghe Asachi" Tehnical University, 43, Dimitrie Mangeron Blvd, Iași, 700050, Romania

E-mail: dan_chiorescu@yahoo.com

Abstract: The sustainable development of agriculture is an important component of economic and social progress of the mankind aiming especially at promoting environmentally friendly systems and technologies. Thus, the implementation of sustainable agriculture also requires some high performance farming aggregates such as tiller seeder combos. Their most stressed active working part is the plowshare which has an important part in cutting the soil. For this reason, we consider that theoretical and experimental research is needed for the tear to which this working part is subjected to. This paper analyses the behavior of the universal plowshare, component part of UPTSC, using the Finite Element Method (FEM) and the Ansys software program. With the help of FEM, we analyzed the universal plowshare in the material structure during the soil cutting process, highlighting the deformation degree and the stress field in the working part. In the first stage, we identified a representative set of problems concerning the soil cutting process, for which we designed the solutions through numerical simulations. In the processing stage, we designed a 3D model which respects entirely the geometric shape of the active element in Cartesian coordinates. In order to simulate the soil cutting process in accordance with the real conditions, the compilations are done for various refinement degrees of the discretization network in finite elements. In the same stage we introduced the constraints represented by: the fixation of the plowshare support, direction, as well as the action of the cohesion and shear strength. Using the Explicit Dynamics module of the Ansys software, which allows studying the plowshare behavior, we analyzed in real conditions, the normal and the shear stresses as well as the deformation, for various soil types and various soil states. Considering the data on the existent stresses, following the FEM analysis of the working part, we determined the wear and suggested the safety coefficients for this case.

Introduction

For humankind on the whole, agriculture has a major contribution to the sustainable development of the economy, by providing food to people, thus ensuring their life on the planet.

The increase in productivity and economic efficiency in agriculture has been made possible by the continuous development of the mechanized agriculture technologies, specifically the use of ever bigger and heavier tractors and machines, leading to the intensification and expansion of some physical degradation processes, as well as to the anthropogenic destructuring and compaction of the soil [1].

The Earth population has tripled in the last century, making it necessary for the world economy to grow over 20 times, the industrial production increasing over 50 times, leading to a 30 times higher fuel consumption.

According to the latest reports an area of 1ha of productive land is lost every six seconds due to various forms of degradation, so the future expansion of agriculture on a "horizontal" plane is no longer possible [1].

Considering all these aspects, there is a growing interest worldwide in the conservative agriculture system which allows the rational management of natural resources with direct implications in preventing and/or minimizing soil degradation and restoring its productive capability. Thus, now more than ever, there is necessary to adopt measures on a global level in order to stop environmental degradation and allow a rational and safe future development of society [1].

Using the Finite Element Method (FEM) for analysis is one of the most recent and advantageous calculus and simulation methods, being a must in engineering, manufacturing and high performance research.

The Finite Element Method FEM analyses the behavior of the working part (the universal plowshare), ensuring the validaty and the accuracy of the used finite element program, its results being comparable to the experimental ones, as well as highlighting the main changes taking place in the material structure during deformation: the thinning degree of the material, the stress field, the orientation, the sense and the velocity of the deformations.

Simulating the working process of the universal plowshare, as well as its behavior when altering the soil, we can predict the deformation and the wear of the working part, we can analyze the stress and determine the working strength and study the possible cracks that may occur. FEM has a wide range of uses, solving a multitude of problems that may occur in engineering.

Material and Method

In order to study the behavior of the universal plowshare using the finite element method, we designed a tridimensional model which respects entirely the geometry of the active element, using the drawing software CATIA and saving the file with the .igs extension so that it could be imported. We imported the drawing representing the universal plowshare into the Explicit Dynamics module.

The ANSYS software package has a materials library in which their main properties are predefined. In this case there are the physical-mechanical properties of the material used for the plowshare, OLC 45 according to the Romanian STAS 880 – 80 and European standard SR EN 10083-2:1995 with the symbol 1C45 (Table 1).

Table 1 Mechanical and physical characteristics of the 1C45 steel

Steel brand	Tensile yield strength $Rp_{0,2}$ [MPa]	Ultimate tensile strength R_m [MPa]	Poisson Coefficient n_{med}	A [%]
1C45	410	800	0.3	14

To study the behavior of the universal plowshare, we chose size 8 with the following constructive dimensions presented in table 2, where B is the plowshare width and g its thickness.

Table 2 Main dimensions of the universal plowshare

B [mm]	220
g [mm]	6

Since the simulation of the soil displacement process done by the universal plowshare fitted on a tiller seeder combo (UPTSC) is tridimensional, a system of coordinates made of three axes (Ox, Oy and Oz) is introduced, the direction of the motion being represented by the Ox axis.

The motion of the plowshare takes place on the Ox alignment, the motion sense being opposite to the axes system, as showed (Figure 1). The Oz orientation refers to the angle of the working part in relation to the field surface.

We are making a tridimensional study on the unitary and tangential stresses according to the direction of the plowshare motion during the soil displacement process Ox.

After introducing the mechanical characteristics of the material, to design the simulations set we need to choose the optimum discretization for the working part (the plowshare) and the soil constraints from the whole system, according to the importance it has in the working process [2]. Thus, for the universal plowshare we choose the 1 mm discretization.

We are doing the simulation for three types cernoziom non -carbonated soil textures: sandy, sandy - loam, sandy - clay, according to the soil state, taking into consideration soil cohesion, the internal coefficient of friction between the soil particles and the external coefficient of friction between the soil and the working part [3].

Following simulation we notice that the area of maximum deformation, and dangerous as regard the wear and tear, is located in the centre of the plowshare as shown (Figure 2).

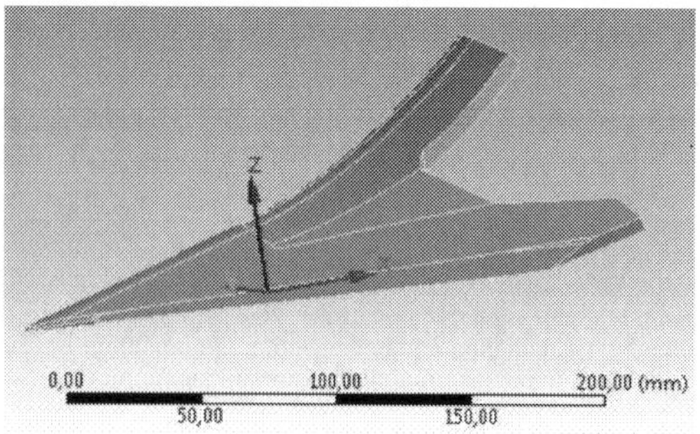

Figure 1 The axes system and the direction of the plowshare motion

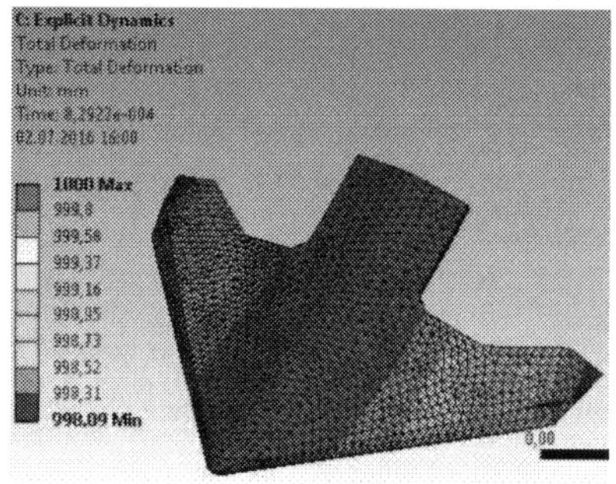

Figure 2 Total deformation for the plowshare

A representative set of trials with finite elements was done on the solid model during simulation in accordance with the type of soil, in Cartesian coordinates and same conditions [4, 5].

We notice that for a sandy clay soil the working part that is 120 mm deep at an angle of 3^0, the maximum normal stress is 615.91 MPa (Figure 3) and the shear stress is 320.36 MPa (Figure 4), observing a dangerous area of the red color, possibly leading to cracks occurring in the plowshare.

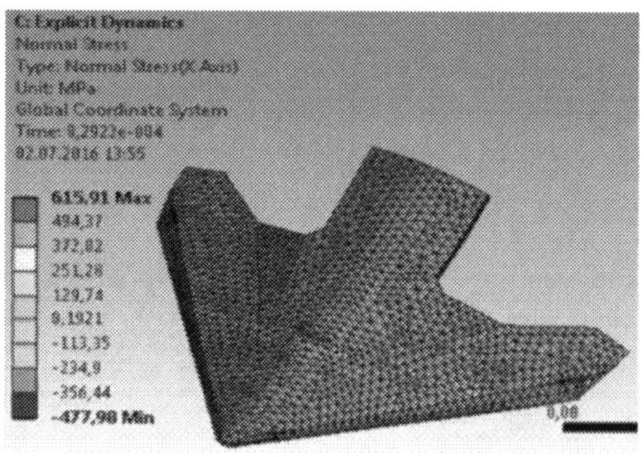

Figure 3 The normal stress on the motion direction OX for the sandy clay texture

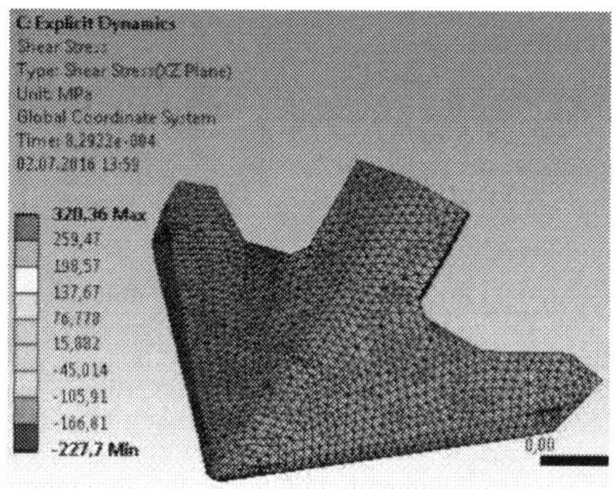

Figure 4 The shear for the sandy clay texture (XZ)

The loading of the plowshare was done using directly the pressure corresponding to the working depth, according to the type of texture: P_{max} = 1.00 MPa, P_{min}= 0.5 MPa for the sandy clay texture; P_{max} = 0.50 MPa, P_{min} = 0.30 MPa for the sandy loam texture; P_{max} = 0.25 MPa, P_{min} = 0.10 MPa for the sandy texture.

It can be observed that for a sandy loam soil in the same working conditions as for the previous trial, the maximum normal stress is 305.52 MPa (Figure 5) and the shear stress is 158.91 MPa (Figure 6).

For the sandy soil texture, in which the plowshare is at a depth of 120 mm and 3^0 angle as in the previous cases, the maximum normal stress has the value of 158.55 MPa (Figure 7), while the maximum shear stress is 82.469 MPa (Figure 8).

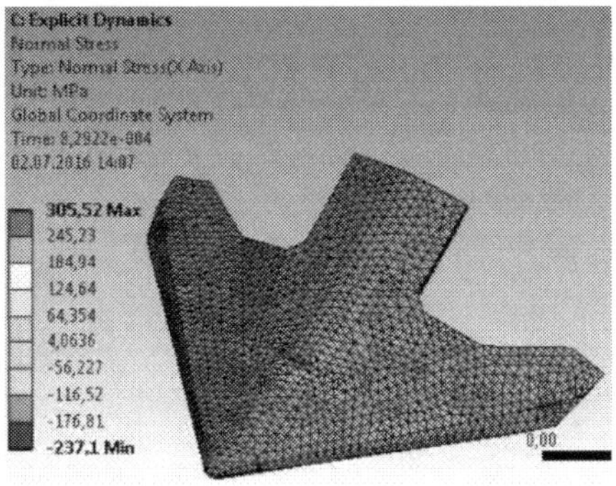

Figure 5 The normal stress on the motion direction OX for the sandy loam texture

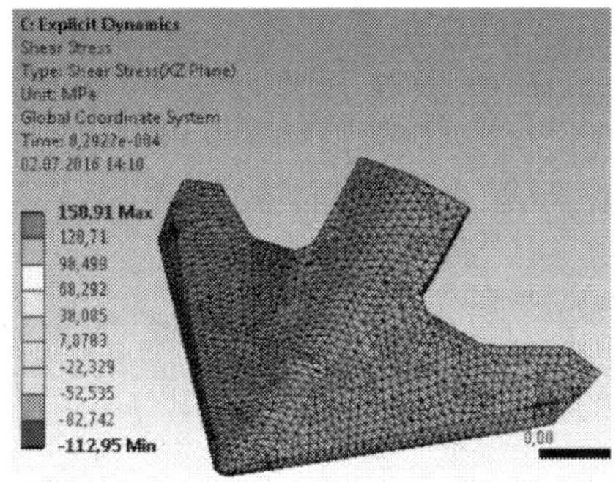

Figure 6 The shear for the sandy loam texture (XZ)

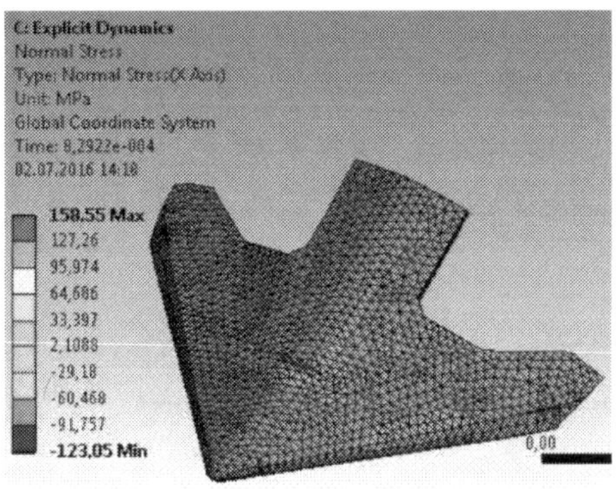

Figure 7 The normal stress on the motion direction OX for the sandy texture

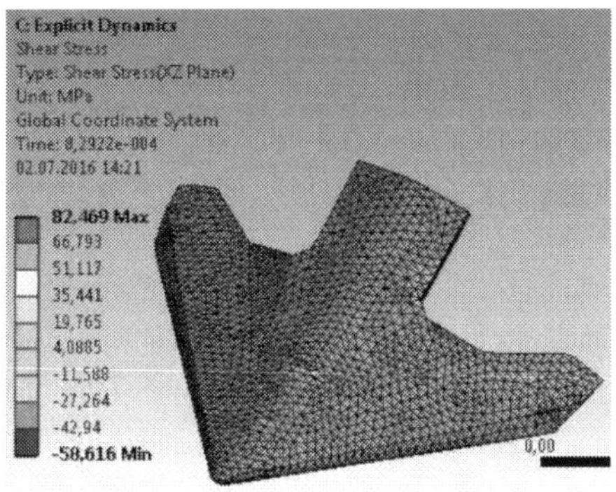

Figure 8 The shear for the sandy texture (XZ)

The distribution of the equivalent Von Mises stresses on the model subjected to the pressure according to the textures type is depicted (Figure 9).

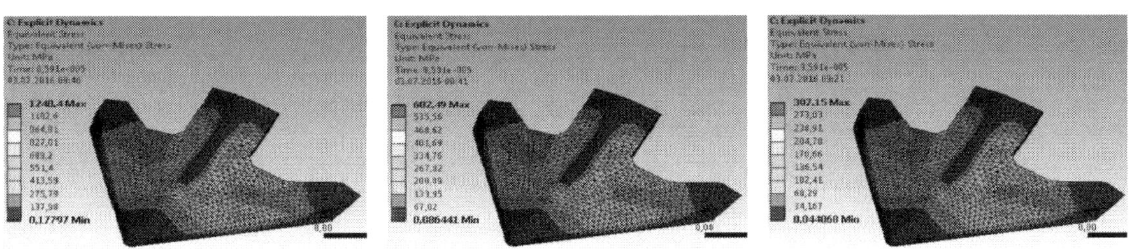

Sandy clay texture | Sandy loam texture | Sandy texture

Figure 9 The distribution of the equivalent Von Mises stresses for the universal plowshare

We can conclude after the simulation that the soils of the sandy-clay texture can take over high yield or shear stresses due to the binding strengths between the particles or between the particles and the aggregates, thus leading to small deformations in these types of texture. In the case of the sandy texture, big deformations occur under the action of small value stresses. For the sandy-loam texture, one can notice after the simulation that the stresses taken over have average values.

Table 3. The maximum and minimum values of the normal unitary and shear stresses for the minimum pressure

Texture type	P_{min} [MPa]	Maximum normal stress [MPa]	Minimum normal stress [MPa]	Maximum shear stress [MPa]	Minimum shear stress [MPa]
Sandy	0.1	64.761	-49.27	33.182	-21.927
Sandy loam	0.3	189.16	-148.52	97.592	-68.149
Sandy clay	0.5	305.52	-237.1	158.91	-112,95

Table 4. The maximum and minimum values of the normal unitary and shear stresses for the maximum pressure

Texture type	P_{max} [MPa]	Maximum normal stress [MPa]	Minimum normal stress [MPa]	Maximum shear stress [MPa]	Minimum shear stress [MPa]
Sandy	0.25	158.55	-123.05	82.469	-58.616
Sandy loam	0.5	305.52	-237.1	158.91	-112,95
Sandy clay	1	615.91	-477.98	320.36	-227.7

In relation to the maximum stress σ_{max} and the minimum one σ_{min} determined by using the plowshare model with the finite elements (Table 3) and (Table 4), we calculated the wear and tear determining the wear safety coefficients for different types of textures [6]. We determine the average value of the stresses σ_m and the stress range σ_a as follows:

$$\sigma_m = \frac{\sigma_{max}+\sigma_{min}}{2}; \ \sigma_a = \frac{\sigma_{max}-\sigma_{min}}{2} \quad (1)$$

The asymmetry coefficient of the cycle is given by the ratio between σ_{min} and σ_{max}:

$$R = \frac{\sigma_{min}}{\sigma_{max}} \quad (2)$$

The wear and tear determination starts from the wear limit for the alternating symmetric cycle which is calculated using the formula [7]:

$$\sigma_{-1}=0.25(\sigma_R+\sigma_C)+50 \text{ [MPa]} \quad (3)$$

The maximum range of the cycle after Soderberg is determined according to the yield limit σ_c, while after Goodman it is in relation to the shear strength:

$$\sigma_{aS} = \sigma_{-1}\left(1-\frac{\sigma_m}{\sigma_c}\right); \quad \sigma_{aG} = \sigma_{-1}\left(1-\frac{\sigma_m}{\sigma_R}\right) \quad (4)$$

The safety coefficient for variable stresses after the Soderberg criterion and after the Goodman criterion is determined using the following formulae:

$$C = \frac{1}{\frac{\sigma_a}{\sigma_{-1}}+\frac{\sigma_m}{\sigma_c}}; \quad C = \frac{1}{\frac{\sigma_a}{\sigma_{-1}}+\frac{\sigma_m}{\sigma_R}} \quad (5)$$

Following the determinations, the following safety coefficients dependent on the texture type were reached (Table 5):

Table 5. The safety coefficients after Soderberg and Goodman criteria

Texture type	The safety coefficient after Soderberg		The safety coefficient after Goodman	
	P_{min}	P_{max}	P_{min}	P_{max}
Sandy	0.89	0.92	0.88	0.90
Sandy loam	0.92	0.95	0.90	0.92
Sandy clay	0.95	1.05	0.92	0.96

Conclusions

A representative set of six problems regarding the process of soil displacement using the universal plowshare, was elaborated, differentiated by the value of minimum and maximum pressure.

All six problems concerning the soil displacement process were solved through numerical simulations, using the Ansys software package, employing the same refinement degree of the discretization network into finite elements in Cartesian coordinates.

The maximum deformation is also defined as being dangerous and is observed in the central area of the universal plowshare where its wear occurs during the working process.

The experiments were realised around the Western Romanian Steppes where the soil is non – carbonated cernoziom in which the predominant textures are : sandy, sandy loam and sandy clay.

Moreover the improvement of the productivity is made by the implementation of new hydroameliorative methods and their protection.

For the analyzed texture types of non-carbonated cernoziom, the wear value of the working part is higher in the middle of the plowshare where the angle of inclination changes.

It transpires from determinations that, although the universal plowshare is resistant to static stresses, it does not cope well with the wear stresses at the oscillating loading cycle to which it is subjected in the soil.

The safety coefficient according to Goodman criterion is subunitary for all three types of analyzed textures (sandy C=0.90; sandy loam C=0.92; sandy clay C=0.96), meaning that the plowshare is resistant to stresses.

According to the Soderberg criterion, the coefficient is subunitary for the sandy texture (C=0.92) and the sandy loam texture (C=0.95). For the sandy clay texture, the coefficient is over one unit (C=1.05) which means that there is the possibility of cracks occurring, which, due to variable stresses, can lead to irreversible deformations or the breaking of the working part if the maximum pressure reaches values over 1 MPa. This phenomenon is caused by the stresses inside the working part that are over the yield limit or the shear limit of the material of which is made, fact confirmed by finite element simulations.

For this reason, specific actions are required such as increasing the durability of these plowshares through various methods. Based on the conclusions achieved, we are planning to continue the research in order to compare the experimental results with the simulated ones.

References

[1] Jităreanu G, Tenu I, et.all. 2007 Technologies and Machines for Mechanizing Soil Tillage in Order to Employ the Sustainable Agriculture Concept Eds. "Ion Ionescu de la Brad"

[2] Chiorescu E, Prepelita D, Popescu S, 2009 The local storage coefficient of layered soil, *Envrironmental Engeneering and Management Journal*, Vol.**8**, nr.3,pp. 447-452, ISSN 1582-9596- Iaşi

[3] Chiorescu E., Popescu Şt., Toma D., 2008 Equations for alluvial soil's storage coefficients, Envrironmental Engeneering and Management Journal, Vol.**7**, nr. 6, ISSN 1582-9596, pp. 809-813, Iasi

[4] Bentaher H., Ibrahmi A., Hamza E., Hbaieb M., Kantchev G., Maalej A., et al, 2013 Finite element simulation of moldboard–soil interaction, *Soil Till*, Res. **134**, pp. 11–16.

[5] Bernacki H., Haman J, Kanafojski CZ ., 1972 Agricultural Machines, Theory and Construction, vol. I, *Naţional Technical Information Service*, Springfield, Ilinois, USA

[6] Jafari R., 2008 Sensitivity analysis of factors affecting on finite element analysis in soil-tool interaction. Proceeding F-and-B Proceedings of the 1st WSEAS International Conference on Finite Differences-Finite Elements-Finite Volumes-Boundary Elements, pp. 122–127

[7] Kosmas C., Gerontidis S., Marathianou M., Detsis B., Zafiriou T., Nan Muysen W., Govers G., Quine T., Vanoost K., 2001 The effects of tillage displaced soil on soil properties and wheat biomass, *Soil Till*, Res. **58**, pp. 31–44

Optimization of the materials distribution in composite systems

A Poteralski and M Szczepanik

Institute of Computational and Mechanical Engineering, Faculty of Mechanical Engineering, Silesian University of Technology

Email: arkadiusz.poteralski@polsl.pl

Abstract. The optimization of structures in macro scale is widely used nowadays. The goal of the paper is to apply optimization techniques to obtain better performance on the micro level. The presented methods open new possibilities. The structures build with the use of materials with optimal microstructure can obtain the best performance. The microstructure can be optimized taking into account loads of the macro structure. The optimization of microstructure is not easy currently, but in future, in applications where performance of the structure is very important, the presented approach may be used with success. A bio-inspired method based on the artificial immune system (AIS) is used to solve the optimization problem. Immune computing provides a great probability of finding the global optimum. The optimal topology is generated by the level set approach. Optimization (identification) of the topology and the distribution of the mass density of microstructure considered to two materials by the minimization of the fitness function which depended on the coefficients of stiffness matrices. The paper presents methodology, algorithm of optimization and numerical examples.

1. Introduction

The paper is devoted to immune optimization (identifiation) in multiscale problems. The structure modelled as a macrostructure with a local periodic microstructure is considered. The multiscale analysis is performed with the use of the homogenization method. The objective function evaluation with the use of the homogenization algorithm is considered.

The computational homogenization is the one of the numerical techniques which enables multiscale analysis of the structures. The detailed description of the computational homogenization in which the structures with a local periodicity are considered is presented in the paper[14]. In the local periodicity there are areas of structure with the same microstructure. The microstructures can also be built from lower scale locally periodic microstructures. The analysis of the structure taking into account the local periodicity of microstructures is the goal of the computational homogenization. Analysis in a few scales which allows to use models with at least a few orders of degrees of freedom lower than model created in one scale is the main advantage of the computational homogenization.

The material parameters for each integration point in finite elements depend on the solution of a representative volume element (RVE) in the lower scale. The RVE is a model of the microstructure which in most cases modelled as a cube or a square. Using this method voids, inclusions and other properties of microstructure can be included in the model. The boundary value problem for RVE is solved using the numerical method like FEM. The periodic displacements boundary conditions are taken into account. The strains from the higher level are prescribed as additional boundary conditions. The RVE for each integration point of the higher level model must be created and stored for the next iteration steps if the nonlinear problem with plasticity is considered. The transfer of information both form lower to higher and higher to lower scales is needed in most cases. In the linear problem the one way transfer of results from lower to higher scales is possible. The material parameters for the higher scale are obtained on the basis of solving a few direct problems for RVE in the lower scale. The homogenized material parameters depend on average stress values in RVE obtained after applying average strains to RVE. The strain-stress relation obtained using RVE is used in the higher level model. The average strains are strains in the integration point from the higher level.

The present researches are based on the application of the artificial immune system and the finite element method to the topology optimization (identification) of microstructure RVE. This work is an extension of previous researches of Burczyński, Poteralski and Kuś concerning such optimization problems [1][2]. Recently, immunology methods have found various applications in mechanics [9][10][11].

2. Artificial immune systems

The artificial immune systems [3][4] are developed on the basis of a mechanism discovered in biological immune systems. The artificial immune systems take only a few elements from the biological immune systems. The most frequently used are the proliferation, mutation of the B-cells, memory cells, and recognition by using the B and T-cells. This algorithm is based on the Wierzchoń method [8]. Difference is that the mutation operator is changed. Instead of the nonuniform mutation [5][6][7] the Gaussian mutation is used in the presented approach. At the beginning of the AIS the memory cells are created randomly. They proliferate and mutate creating B-cells. The number of clones nc created by each memory cell is determined by the memory cells objective function value. The objective functions for B-cells are evaluated. The selection process exchanges some memory cells for better B-cells. The selection is performed on the basis of the geometrical distance between each memory cell and B-cells (measured by using design variables).

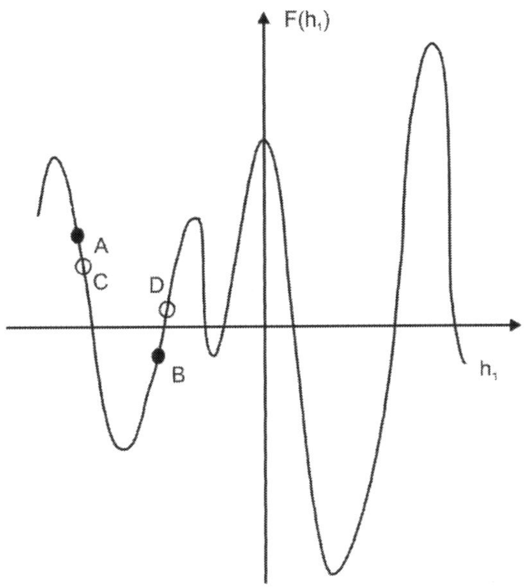

Figure 1. The idea of the selection mechanism

In the selection stage two memory cells "A" and "C" are compared and the better one (memory cell "C" - because the objective function is better) goes to next iteration.

The crowding mechanism removes similar memory cells. The similarity is also determined as the geometrical distance between memory cells. The process is iteratively repeated until the stop condition is fulfilled.

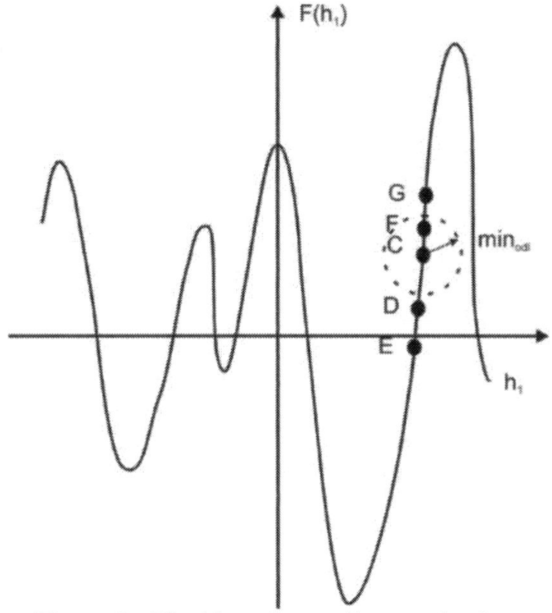

Figure 2. The idea of crowding mechanism

During the crowding mechanism for memory cell "C" only memory cell "F" is in the similarity area. The better one (memory cell "C") stay in population and a worse (memory cell "F") is eliminated. Finally new memory cell is generated in randomly way.
The stop condition can be expressed as the maximum number of iterations.

3. The multiscale modeling

The dependences between two or more scales is possible using the multiscale modeling (figure 3).

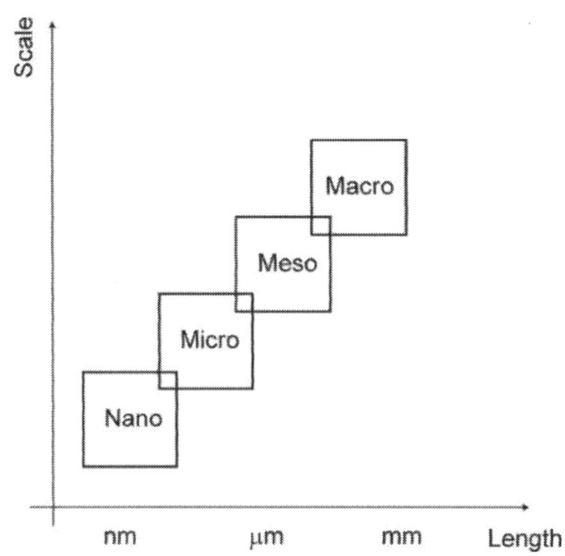

Figure 3. Mulsticale modeling

Computational homogenization [12] is the one of the methods which is used in the multiscale modelling [12]. Instead of homogenous material the heterogeneous material is introduced. (figure 4).

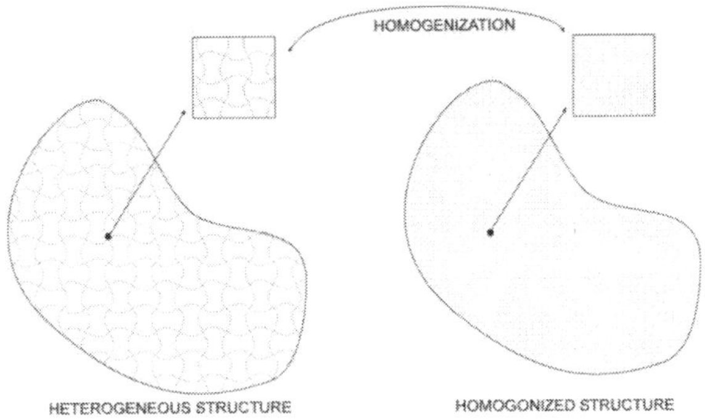

(a) (b)

Figure 4. Homogenization of material: (a) heterogeneous structure, (b) structure after homogenization

When the microstructure is periodic the homogenization method is very useful. On the basis of a numerical solution to the boundary value problem performed in each scale is obtained using the influence between scales in the computational homogenization. Analysis of the structure in the two-scale in figure 5 is presented.

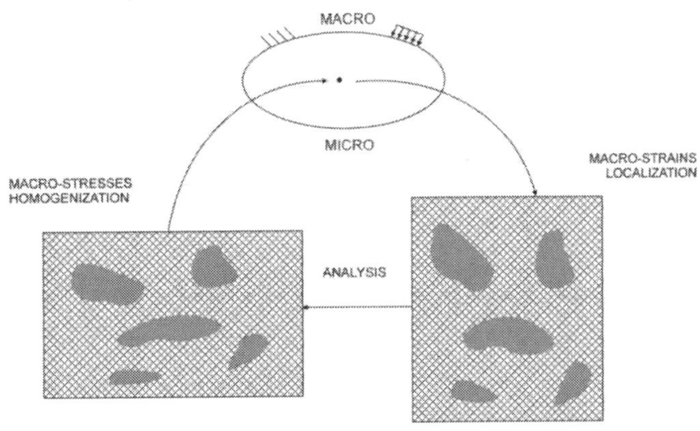

Figure 5. Two-scale computational homogenization

Using the Gauss integration point the boundary value problem for RVE could be solved. The strain values are transferred to the micromodel during the localization stage. The displacements, traction and periodic boundary conditions are applied to the microstructure. The stresses obtained after boundary value problem analysis are used to obtain homogenized average values which are transferred after the homogenized stage to the higher scale. The relationship between strains and stresses for an orthotropic elastic material are expressed as follows:

$$\sigma = C\varepsilon \quad (1)$$

$$\varepsilon = S\sigma \quad (2)$$

where:

$$\sigma = [\sigma_{11}, \sigma_{22}, \sigma_{33}, \sigma_{12}, \sigma_{13}, \sigma_{23}]^T \quad (3)$$

$$\varepsilon = [\varepsilon_{11}, \varepsilon_{22}, \varepsilon_{33}, \varepsilon_{12}, \varepsilon_{13}, \varepsilon_{23}]^T \tag{4}$$

are vectors of stresses and strains.
C and S are the stiffness and compliance matrices, of the orthotropic linear elastic material, respectively. They can be written as

$$C = \begin{bmatrix} c_{11} & c_{12} & c_{13} & 0 & 0 & 0 \\ & c_{22} & c_{23} & 0 & 0 & 0 \\ & & c_{33} & 0 & 0 & 0 \\ & & & c_{44} & 0 & 0 \\ & sym. & & & c_{55} & 0 \\ & & & & & c_{66} \end{bmatrix} \tag{5}$$

$$S = C^{-1} = \begin{bmatrix} s_{11} & s_{12} & s_{13} & 0 & 0 & 0 \\ & s_{22} & s_{23} & 0 & 0 & 0 \\ & & s_{33} & 0 & 0 & 0 \\ & & & s_{44} & 0 & 0 \\ & sym. & & & s_{55} & 0 \\ & & & & & s_{66} \end{bmatrix} \tag{6}$$

where:

$$\begin{array}{lll} s_{11} = \dfrac{1}{E_1} & s_{22} = \dfrac{1}{E_2} & s_{33} = \dfrac{1}{E_3} \\ s_{44} = \dfrac{1}{G_{12}} & s_{55} = \dfrac{1}{G_{23}} & s_{66} = \dfrac{1}{G_{13}} \\ s_{12} = \dfrac{-\nu_{21}}{E_2} & s_{13} = \dfrac{-\nu_{31}}{E_3} & s_{21} = \dfrac{-\nu_{12}}{E_1} \\ s_{23} = \dfrac{-\nu_{32}}{E_3} & s_{31} = \dfrac{-\nu_{13}}{E_1} & s_{32} = \dfrac{-\nu_{23}}{E_2} \end{array} \tag{7}$$

where E_i is Young's modulus along the axis i, G_{ij} is the shear modulus in the direction j on the plane whose normal is in the direction i, and ν_{kl} is Poisson's ratio that corresponds to contraction in the direction j when an extension is applied in the direction k.
The 9 variables are independent in the fully orthotropic elastic material [13][14] due to symmetry of the stiffness and compliance matrices.
The material coefficients can be obtained once for each microstructure in the case of linear problems. For each microstructure the six analyses should be performed to obtain the 9 independent orthotropic material coefficients. The average stresses and strains for RVE are defined as follows

$$\varepsilon_{av} = \frac{1}{|\Omega_{RVE}|} \int_{\Omega_{RVE}} \varepsilon d\Omega_{RVE} \tag{8}$$

$$\sigma_{av} = \frac{1}{|\Omega_{RVE}|} \int_{\Omega_{RVE}} \sigma d\Omega_{RVE} \tag{9}$$

where Ω_{RVE} is the area of RVE.
The constitutive relation between them has the form

$$\sigma_{av} = C^h \varepsilon_{av} \tag{10}$$

where C^h is the stiffness tensor of the equivalent homogenous material that fulfils the elastic deformation characteristic for the heterogeneous material. A detailed description of the algorithm of computational homogenization is presented in figure 6.

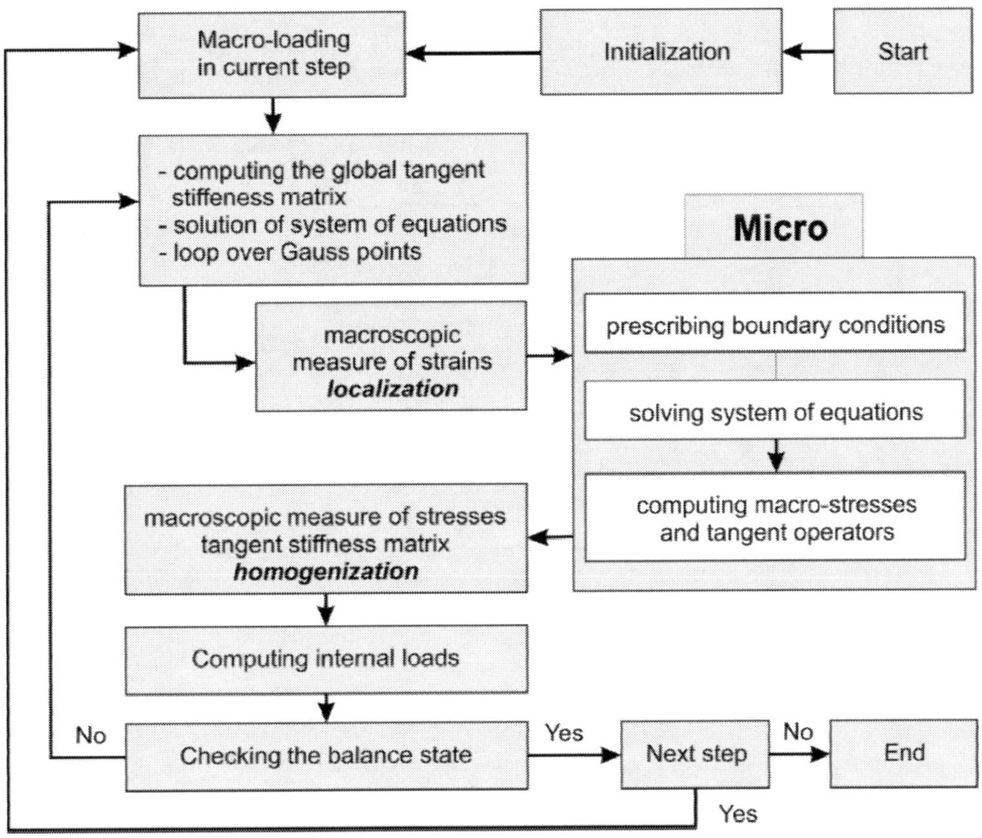

Figure 6. Computational homogenization algorithm

4. Topology optimization and parameterization

The distribution of mass density $\rho(X)$, $(X) \in \Omega_t$ in the structure is described by a hyper surface $W_\rho(X), (X) \in H^3$. The hyper surface $W_\alpha(X), \alpha = \rho$ is stretched under $H^d \subset E^d, (d=3)$ and the domain Ω_t is included in H^d, i.e. $(\Omega_t \subseteq H^d)$.

The shape of the hyper surface $W_\alpha(X), \alpha = \rho$ is controlled by parameters d_j, j=1,2,...,G, which create a B-cell receptor

$$B - cell = \langle d_1, d_2, ..., d_j, ..., d_G \rangle \tag{11}$$

$$d_j^{min} \le d_j \le d_j^{max} \tag{12}$$

where d_j^{min}, d_j^{max} - are minimum and maximum values of the parameters of B-cell receptor, respectively. Parameters of B-cell receptor are the values of the function $W_\alpha(X)$, $\alpha = \rho$ in the control points $(X)_j$ of the hyper surface, i.e. $d_j = W_\alpha[(X)_j]$, $j = 0,1,2,...,G$.

The finite element method is applied in analysis of the structure. The domain Ω of the structure is discretized using the finite elements, $\Omega = \bigcup_{e=1}^{E} \Omega_e$.

The assignation of the mass density to each finite element $\Omega_e, e = 1,2,...,E$ is performed by the mappings:

$$\rho_e = W_\rho\left[(X)_e\right], (X)_e \in \Omega_e, e = 1,2,...,E \tag{13}$$

It means that each finite element can have different mass density.

When the value of the mass density for the e-th finite element is included in the interval $0 \le \rho_e < \rho_{\min}$ the finite element is eliminated and the void is created, the interval $\rho_{\min} \le \rho_e < \rho_{\max}$ the finite element remains. In the last step the Young's modulus for the e-th finite element is evaluated using the following equation

$$E_e = E_{\max}\left(\frac{\rho_e}{\rho_{\max}}\right) \tag{14}$$

where

$E_{\max}$, $\rho_{\max}$ - Young's modulus and mass density for the same material, respectively.

The key stage in the structural optimization is parameterization. The optimization process is not effective when the number of design variables is very large. A connection between number of finite element and design variables (parameters of B-cell receptor) leads to poor results. The better results can be obtained when the hyper surface of mass density distribution is interpolated by suitable number of values given in control points $(X)_j$. Using this approach, on the one hand, the number of design variables should be small and on the other hand should provide the good interpolation. Two different types of the interpolation procedures were applied. First the interpolation bases on neighbourhood of elements and second multinomial interpolation.

5. Numerical example

Numerical example is considered, i.e. the identification of the topology and the distribution of the mass density of microstructure considered to two materials by the minimization of the fitness function (15). The procedure described in Chapter 4 has been slightly modified, as follows: when the value of the mass density for the e-th finite element is included in the interval $0 \le \rho_e < \rho_{\min}$ the finite element remains and has a value of mass density for material number 1 (table 3), when the value of the mass density for the e-th finite element is included in the interval $\rho_{\min} \le \rho_e < \rho_{\max}$ the finite element also remains and has a value of mass density for material number 2 (table 3). The structures are considered in the framework of the theory of elasticity. The results of the examples are obtained by using an identification method based on the artificial immune system with the parameters included in table 1. Periodic boundary conditions are applied to the microstructure presented in the figure 7.

Table 1. The parameters of the artificial immune system

the number of memory cells	5
the number of the clones	10
crowding factor	50%
Gaussian mutation	50%

The immune process proceeds in the environment in which the structure fitness is described by the minimization of the fitness function:

$$J(\mathbf{x}) = \sum_{i=1}^{n} \sum_{j=1}^{n} \left\| S_{ij}(\mathbf{x}) - S_{ij_{ref}} \right\| \tag{15}$$

where:
S_{ij} , S_{ijref} – coefficients of compliance matrices (5).

The results of the identification process in the form of distribution of map of densities are presented in the figure 9. Geometry of the microstructure RVE in the figure 7 is presented (a, b, c =1).

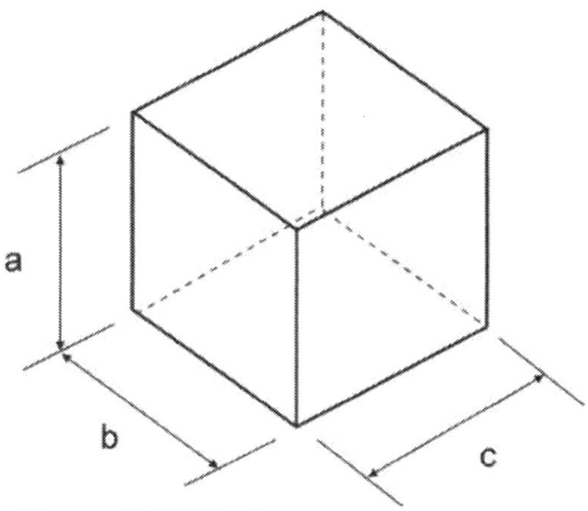

Figure 7. RVE microstructure - geometry

Material data for optimize structure in the table 2 are presented.

Table 2. Material data

	material no 1 - glass	
Poisson ratio 0.203076	Mass density (ρ_{max}) 2.5 g/cm^3	Young modulus 69150.9 MPa
	material no 2 - epoxy	
Poisson ratio 0.175124	Mass density (ρ_{max}) 0.55 g/cm^3	Young modulus 3818.47 MPa

Information about distribution of two materials in the microstructure in the table 3 is presented.

Table 3. Condition of distribution for two different materials

$\rho_e<50\%\rho_{max}$	material no 2
$\rho_e\geq50\%\rho_{max}$	material no 1

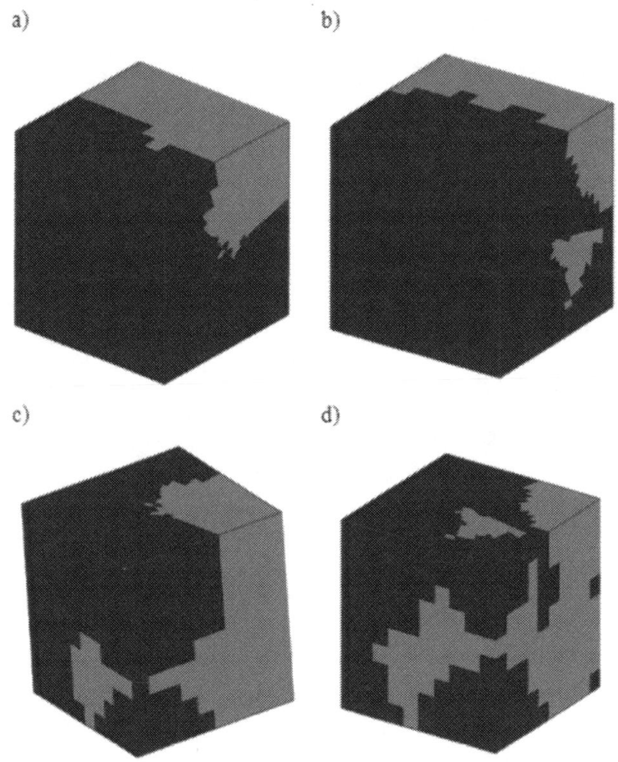

Figure 8. The results after identification process: (a), (c) the distribution of materials (Young's modules) for two different positions for the reference structure, (c) the distribution of materials (Young's modules) for two different positions for the best structure

The results of the identification for the best solution after 86 iterations, where the color red corresponds to material no. 1, and the color blue corresponds to material no. 2 are presented on the figure 8 and the table 4.

Table 4. Results of optimization

Number of iterations	Number of fitness function evaluations	Value of fitness function
86	3010	897.106018
Coefficients of compliance matrices	Reference values	Values after identifications
S_{11}	9018.59	9012.28
S_{12}	1925.6	2081.16
S_{22}	18037.4	18041.2
S_{23}	3227.03	3434.94
S_{33}	22071.2	22069.1
S_{31}	1921.72	1969.77
S_{44}	4708.77	4681.95
S_{55}	5989.53	6212.67
S_{66}	5989.53	6212.67

6. Conclusions

The paper describes the author's method of simultaneously optimization (identification) of the shape, topology and material properties of 3D structures for multiscale problem. During the optimization process the structure, which in the initial phase consisted of a homogenous material, as a result of changing material properties (change of material density) became heterogeneous material (optimization of material properties). In order to control a small number of design variables density function has been introduced describing the distribution of the material densities in a two or three-dimensional region in which the optimized mechanical structure is "submerged" (parameterization method). In this approach, design variables have values of interpolation function in appropriate distributed control points (interpolation nodes). This allows that the number of design variables is equal to the number of control points of the interpolation hyper surface and this number doesn't depend of discretization of the structure. The procedure described in this paper was used for the multiscale problem. For the multiscale modeling method of computational homogenization is used. The multiscale modeling allows one to take into account the dependences between two or more scales. The most important stage of a computational homogenization method was to develop a model of the considered structure on a lower scale, with a requirement that the size of this fragment of the structure was much smaller than the characteristic dimension of the body in a higher scale. A fragment of the structure, occurring during the computational homogenization procedure is called RVE (Representative Volume Element). Use of an artificial immune system allowed for optimal distribution of the two materials for the microstructure RVE for the objective function dependent on the parameters of the stiffness matrix. Numerical example is given and good results are obtained (figure 9) - the value of the fitness function is to close to zero. Comparison between AIS, PSO and EA in another optimization problem in the paper [15][16][17][18] is presented.

References

[1] Poteralski A Szczepanik M Dziatkiewicz G Kuś W and Burczyński T 2011 Immune identification of piezoelectric material constants using BEM Inverse problems in science and engineering vol 19 issue 1 Taylor & Francis

[2] Poteralski A Szczepanik M Ptaszny J Kuś W and Burczyński T 2013 Hybrid artificial immune system in identification of room acoustic properties Inverse Problems in Science and Engineering Taylor & Francis

[3] Ptak M and Ptak W 2000 Basics of Immunology, Jagiellonian University Press, Cracow (in Polish)

[4] Balthrop J, Esponda F Forrest S and Glickman M 2002 Coverage and Generalization in an Artificial Immune System Proceedings of the Genetic and Evolutionary Computation Conference GECCO 2002 pp 3-10 Morgan Kaufmann New York

[5] de Castro L N and Timmis J 2003 Artificial Immune Systems as a Novel Soft Computing Paradigm Soft Computing 7(8) pp 526-544

[6] de Castro L N and Von Zuben F J 2001 Immune and Neural Network Models: Theoretical and Empirical Comparisons, International Journal of Computational Intelligence and Applications (IJCIA) 1(3) pp 239-257

[7] de Castro L N and Von Zuben F J 2002 Learning and Optimization Using the Clonal Selection Principle IEEE Transactions on Evolutionary Computation Special Issue on Artificial Immune Systems vol. 6 n. 3 pp 239-251

[8] Wierzchoń S T 2001 Artificial Immune Systems, theory and applications EXIT (in Polish)

[9] Poteralski A Szczepanik M and Burczyński T 2015 Immune Optimal Design of 2-D and 3-D Structures International Conference on Artificial Intelligence and Soft Computing (ICAISC) Lecture Notes on Artificial Intelligence vol 9120 Springer pp 471-482

[10] Poteralski A Szczepanik M and Burczyński T 2015 Swarm and immune computing of dynamically loaded reinforced structures International Conference on Artificial Intelligence and Soft Computing (ICAISC) Lecture Notes on Artificial Intelligence 9120 Springer pp

483-494

[11] Burczyński T Poteralski A and Szczepanik M 2015 Immune and Swarm Optimization of Structures, Advances in Evolutionary and Deterministic Methods for Design, Optimization and Control in Engineering and Sciences Chapter no 19

[12] Madej Ł Mrozek A Kuś W Burczyński T and Pietrzyk M 2008 Concurrent and upscaling methods in multi scale modelling – case studies *Computer Methods in Material Science*, **8**, 1, AGH Krakow

[13] Bąk R and Burczyński T 2009 *Computational Strength of Materials* (in Polish *Wytrzymałość materiałów z elementami ujęcia komputerowego*) WNT Warszawa

[14] Schneider R Faust G Hindenlang U and Helwig P 2009 Inhomogeneous, orthotropic material model for the cortical structure of long bones modeled on the basis of clinical CT or density data *Comput. Methods Appl. Mech. Engrg.* **1298** pp 2167-2174

[15] Poteralski A Szczepanik M Dziatkiewicz G and Kuś W 2013 Comparison between PSO and AIS on the basis of identification of material constants in piezoelectrics Springer-Verlag Berlin Heidelberg ICAISC 2013 Part II LNAI 7895 pp 569–581

[16] Poteralski A Szczepanik M Beluch W and Burczyński T 2014 Optimization of composite structures using bio-inspired methods, Artificial intelligence and soft computing ICAISC 2014 vol 8468 pp 385-395

[17] Szczepanik M Poteralski A Długosz A Kuś W and Burczyński T 2013 Bio-inspired optimization of thermomechanical structures Springer-Verlag Berlin Heidelberg ICAISC 2013 Part II LNAI 7895 pp 79-90

[18] Szczepanik M Poteralski A Ptaszny J and Burczyński T 2012 Hybrid Particle Swarm Optimizer and Its Application in Identification of Room Acoustic Properties Swarm and Evolutionary Computation - International Symposia SIDE 2012 and EC 2012 Held in Conjunction with ICAISC 2012 Zakopane, Proceedings. Lecture Notes in Computer Science 7269 Springer

IOP Conf. Series: Materials Science and Engineering **161** (2016) 012064 doi:10.1088/1757-899X/161/1/012064

Experimental measurements of composite materials from the bearing structure of railway vehicles

I SEBEŞAN[1], L M BABICI[2], and D FLOROIU[2]

[1]Polytechnic University of Bucharest, Faculty of Transports, Rolling Stock Department, 313 Splaiul Independenţei, zip code 0600042, Bucharest, Romania

[2]Romanian Railway Authority -RRA, 393 Calea Griviţei, zip code 010719, Bucharest, Romania

E-mail: ioan_sebesan@yahoo.com

Abstract. Composite materials belong to the category of "new materials", artificially created to meet the demands of science and technology.On the one hand, these are a perfect replacement of ferrous and nonferrous materials, and a perfect solution for a whole series of technical problems in various industrial branches, on the other hand.The composite materials represent fiber arrangements - continuous or not - from a resistant material (reinforcing elements), which are coated with a matrix, the mechanical strength of which is much lower.The Matrix maintains the desired geometrical arrangement of fibers and transmits stresses that the piece is subjected to.The present paper proposes to analyze behavioral patterns during the tests of specimens made of composite materials.

1. Introduction

The composite materials represent a class of engineering materials with special technical and scientific interest. They are not only a perfect replacement of ferrous and nonferrous materials but also the perfect solution for solving a whole series of technical problems in different industrial branches. Some technical problems are difficult or sometimes impossible to solve using traditional materials.

A composite material is a combination of two or more different materials from a chemical point of view, with an interface between them [7].

The constituent materials maintain their separate identity in a composite, though their combination generates all the features and characteristics of the various constituent materials in part.One of the items is called a matrix and is defined as forming a continuous phase.

IOP Conf. Series: Materials Science and Engineering **161** (2016) 012064 doi:10.1088/1757-899X/161/1/012064

The other main element called armature is added to the matrix to improve or modify its properties. Reinforcement is a discontinuous phase, equally distributed in the whole volume of the matrix [2].

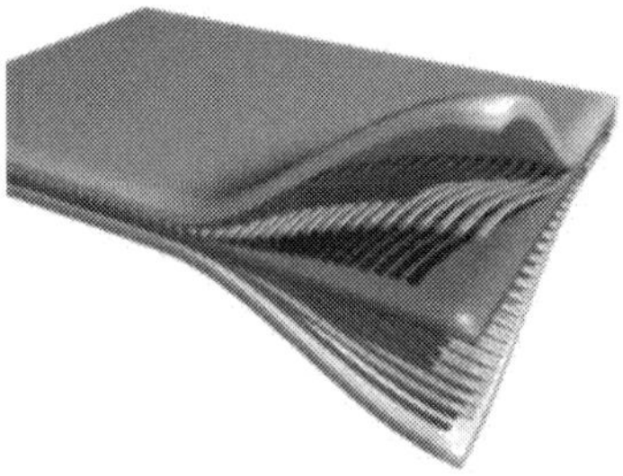

Figure 1. Layers composite material [2]

Figure 2. Kawasaki bogie- uses CFRP (Carbon Fiber Reinforced Plastic) [10]

The composite materials represent a class of engineering materials with special technical and scientific interest. They are not only a perfect replacement of ferrous and nonferrous materials but also the perfect solution for solving a whole series of technical problems in different industrial branches. Some technical problems are difficult or sometimes impossible to solve using traditional materials.

A composite material is a combination of two or more different materials from a chemical point of view, with an interface between them [9].

The constituent materials maintain their separate identity in a composite, though their combination generates all the features and characteristics of the various constituent materials in part.

One of the items is called a matrix and is defined as forming a continuous phase.

The other main element called armature is added to the matrix to improve or modify its properties. Reinforcement is a discontinuous phase, equally distributed in the whole volume of the matrix [2].

The composite materials are part of "new materials" and are specifically created to respond to particular requirements with regard to:
- Mechanical strength and stiffness;
- Resistance to corrosion;
- Resistance to action of chemical agents;
- Low weight;
- Dimensional stability;
- To withstand varying loads, shock and wear;
- Insulating and aesthetic properties [1].

For the construction of railway rolling stock, composite materials were inserted out of the need to reduce masses and to obtain sound and thermal insulation, better braking performance and traction, etc.

The main requirements for composite materials are to successfully replace the traditional materials and to make improvements as much as possible.

From the point of view of the masses, the composite materials represent a big plus contributing to a significant reduction of loads and thus leading to improved performance of railway vehicles.

The noise emissions during traffic can be reduced through the use of composite materials with superior insulation properties [8].

For example, the elastomers based thermoplastic polyester offers a successful combination for many parts and components used in isolation railway vehicles.
In the case of rail vehicles, composite materials are used in construction: fuel tanks, fan blades, bodies, frames, chairs, floors, dashboards, radiators, doors, friction pads and brake blocks, cisterns, parts for wagons, elements of decoration and furniture subway trains, thermal and electrical insulation, and so on.
The classification of composite materials is usually done according to a series of criteria among which the most important are:

a) Depending on the nature of the matrix:
   - the metal matrix composite (Al, Cu, Ni, Mg, Al alloys, Cu);
   - organic matrix composites (polymers);
   - ceramics matrix composite (silicon carbide, silicon nitride).

b) After the geometric configuration of the reinforcement material:
   - composites with discontinuous fibers (short fiber, mono or multifilaments);
   - composites with continuous fibers;
   - composites with large particles (graphite, oxides, nitrides, carbides and alloys)
   - composites microparticles (reinforcing material represents 1 ... 15%, and an average particle diameter does not exceed 0.1 μm),
   - lamellars and multi-layers composite

c) After the structure and method of production

Design of composite materials is very different than that of the conventional products of synthetic materials [7].

A „gelcoat" is an in-mold coating used in plastics material reinforced with fibers and molded polymers.
Maxguard gelcoat offers superior performance for critical use in environments with added value of high protection in combination with special cosmetic products and exceptional properties for application in the process.
PAFS assemblies and subassemblies are used for:
• inner cladding of departments and halls of trains and passenger motorcars;
• interior and exterior cladding booths traction of passenger motorcars, interior walls, ceilings, booths traction and passenger wagons, luggage, and so on.

PAFS assemblies and subassemblies must be resistant to environmental conditions, heat, excessive humidity, detergents, influence of solar radiation, and must also be fireproof. They are designed so that the number of wear parts to be minimal, with these arranged to be interchangeable and accessible.

They are designed so that the number of wear/scuffing parts to be minimal, with these arranged to be interchangeable and accessible.

PAFS assemblies and subassemblies of the railway wagons are made from materials such as synthetic resins whose physical-chemical and mechanical properties must meet specific quality and
current standards, so the finished product and panels match the characteristics required by European standards.
PAFS are builts on the molds of the same material with a high degree for processing active surfaces by laminating manually the successive layers of fiberglass impregnated with polyester resins. The molds are obtained on models made of wood or PVC depending on the technological possibilities of execution.
Products can then be painted with paints that do not affect fireproofing classification.
The products have the role of inner lining and coating or of support mechanisms and various installations.
PAFS assemblies and subassemblies, plated with HPL, must be resistant to environmental conditions, heat, excessive humidity, detergents, influence of solar radiation, and must also be fireproof. They are

designed so that the number of scuffing piece/slice to be minimal, with these arranged to be interchangeable and accessible.

2. Experimental results

Using mechanical tests qualitative data are obtained of the behavior of materials in the corresponding strain conditions during these tests and the values of physical or conventional values called mechanical characteristics, which can be used as quantitative parameters for expressing mechanical properties.
The tests were performed at Romanian Railway Authority-Romanian Railway Notified Body- Laboratory department.

2.1. Determination of tensile strength at full load, in the direction parallel and perpendicular to the direction of lamination

In order to highlight the particularities of the mechanically stressed composite material's reinforcements tensile stress was used.
The test was conducted in accordance with [3].
Tensile testing was carried out on specimens made of the composite material, having the shape and dimensions according to [5].
For testing 10 rectangular specimens were used, type 2 with dimensions (250 x 25 x 5) mm, 5 taken in the longitudinal direction and 5 taken in the sectional direction of the material lamination.
In the case of the specimens the width b and the thickness h were measured. The samples underwent tensile testing, and the speed of displacement of mobile jaw was of 2 mm / min. There was a breaking force F. Maximum tensile stress was calculated using the formula:

$$\sigma = F / bh$$

The test results for determining the tensile stress at maximum load for resin type Hetron F645 TFE are found in Chart 1 and 2 and test mode samples are shown in fig. no. 3- 6:

Figure 3. Test specimens during the tensile test

Figure 4. Test specimens during the tensile test

Figure 5. Test specimens caught in the jaws of the machine during the tensile test

Figure 6. Test specimens caught in the jaws of the machine during the tensile test

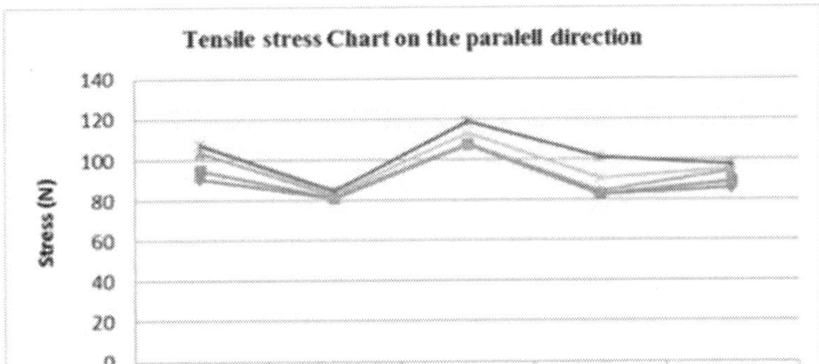

	2012	2013	2014	2015	2016
70	90.7	80.96	107.1	82.8	85.99
80	94.5	81.03	107.2	82.9	88.95
90	103.6	82.13	108	84.3	93.96
100	107.2	83.04	113.3	90.7	95.19
110	107.3	84.63	118.7	101.1	97.23

Chart 1- Tensile stress

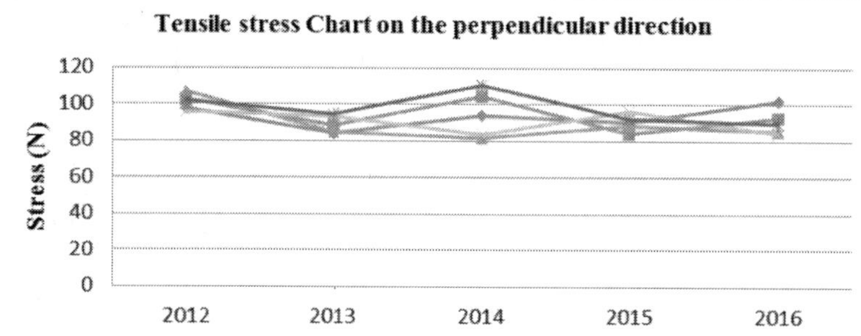

	2012	2013	2014	2015	2016
	97.5	84.05	93.7	90.5	102.12
	102.8	88.64	104.7	84	92.77
	106.8	84.94	81.5	88.7	85.23
	95.6	93.16	83.3	96.5	83.96
	101.7	94.63	110.6	92.3	89.98

Chart 2- Tensile stress

2.2. Determination of the maximum bending stress conducted on the parallel direction and perpendicular direction of lamination

To highlight the particularities of the composite material, a bending stress test was used.
The test was conducted in accordance with [4].
For testing 10 rectangular specimens were used with dimensions (80 x 10 x 4) mm, 5 taken in the longitudinal direction and 5 taken in the sectional direction to the rolling direction of the material. For the test method A was applied (method of 3 points).
For the specimens the width b and the thickness h were measured. The samples were placed one by one on the fixed bearing of the machine and the requested bending test was done with a moving speed of the mobile jaw of 10 mm / minute. The maximum bending force F was checked. The bending stress at maximum load was calculated using the formula:

$$\sigma f = 3FL / 2bh^2$$

where $L = 16 \times h_{medium} = 68$mm (represents the distance between the two supports).

The test results for determining the bending stress at maximum load for resin type Hetron F645 TFE are found in Chart 3 and 4 and the test mode samples are shown in fig. no. 7 - 8:

Figure 7. Test specimens caught in the jaws of the machine during the bending test

Figure 7. Test specimens caught in the jaws of the machine during the bending test

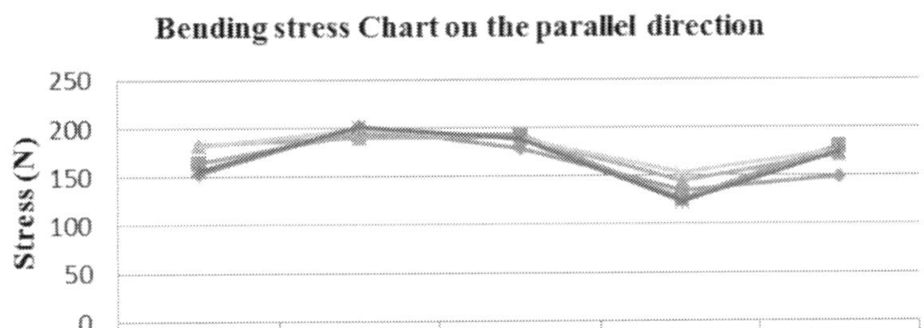

	2012	2013	2014	2015	2016
—◆—	153.5	202.08	178.7	134.5	148.33
—■—	164.4	194.87	192.4	125.8	179.76
—▲—	183.4	189.41	190.2	144.1	172.04
—×—	181.1	198.22	188	153.5	175.79
—✱—	156	201.61	188.3	122.7	174.11

Chart 3- Bending stress on the parallel direction

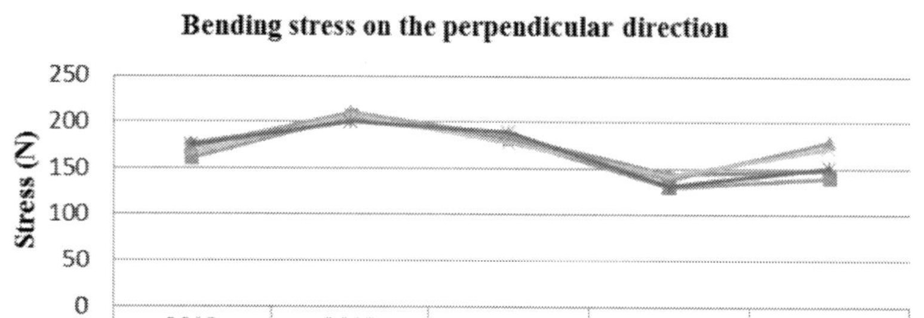

	2012	2013	2014	2015	2016
	175.9	203.28	184.8	143.9	147.81
	160.6	205.51	181.5	130.4	139.42
	171.9	210.52	185.4	138.7	179.13
	166.2	207.59	177.4	142.4	172.45
	174.9	199.83	189.3	131.6	150.5

Chart 4- Bending stress on the perpendicular direction

2.3. Determination of Charpy impact strength, on samples without notch (on the parallel and perpendicular direction to the rolling direction) and C-notch specimens (on the parallel and perpendicular direction to the direction of lamination)

To highlight the particularities of the composite materials breaking behaviour, a bending impact test was used.
The test was conducted in accordance with [5].
For testing, 40 rectangular specimens were used with dimensions (80 x 10 x 4) mm, 20 taken in the longitudinal direction (10 with notch type C and 10 without notch) and 20 taken in the sectional direction (10 with notch type C and 10 without notch) to the rolling direction of the material.
The width **h** and the thickness **b** for specimens without notch respectively thickness $\mathbf{b_k}$ to notch specimens was measured.
An energy shock for breaking specimens $\mathbf{A_n}$ without notch, respectively $\mathbf{A_k}$ - for notch specimens was recorded.
Charpy impact strength was calculated using the formula:

$$\mathbf{a_n = A_n/hb} \text{ sau } \mathbf{a_k = A_k/hb_k},$$

depending on the type of the specimen.

For specimens that were used to determine the Charpy impact strength on specimens with the type C notch, notch execution was carried out at AFER with a broaching machine having the Channel V profile with a depth of 2 mm.
The test results for the determining the Charpy impact strength (specimens without notch) for resin type Hetron F645 TFE are found in Chart 5 and 6 and the test mode samples are shown in fig. no. 9 - 10:

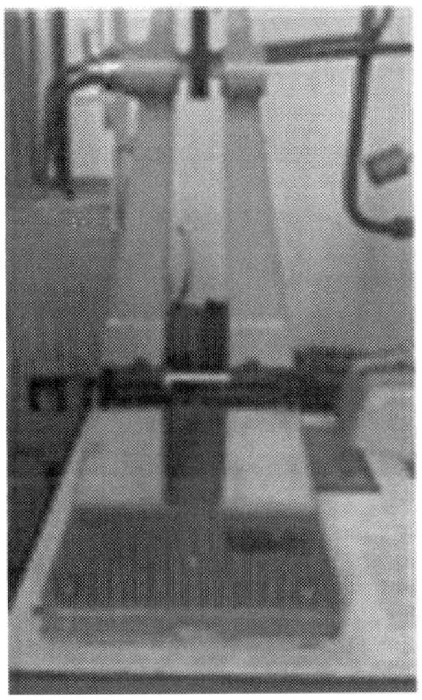

Figure 9. Charpy machine and test specimens without notch

Figure 10. Charpy machine and test specimens with notch

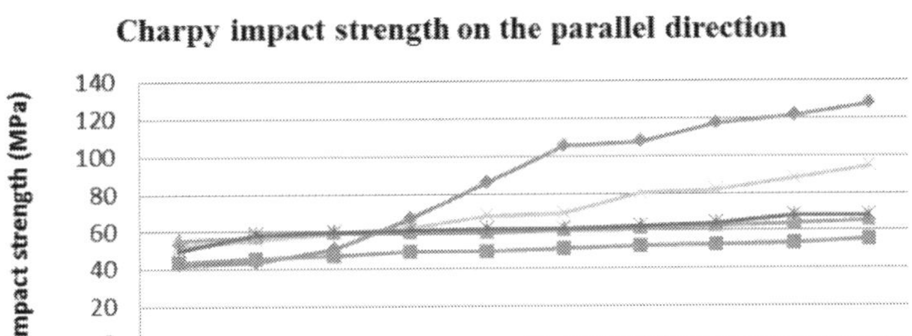

2012	42.3	43.6	50.3	66.6	86	105.6	107.8	117.1	121.5	127.3
2013	43.26	45.66	46.86	48.85	49.27	50.47	51.72	52.87	53.28	55.34
2014	55.7	56.8	59.1	58.6	59.1	60.2	61.2	62.5	63.6	64.8
2015	51.9	55.3	58.6	61.4	68	69.3	80.3	82	88.3	94.5
2016	50.01	58.03	59.4	60.19	61.03	61.25	62.14	63.5	67.95	68.01

Chart 5 - Test specimens without notch on the parallel direction

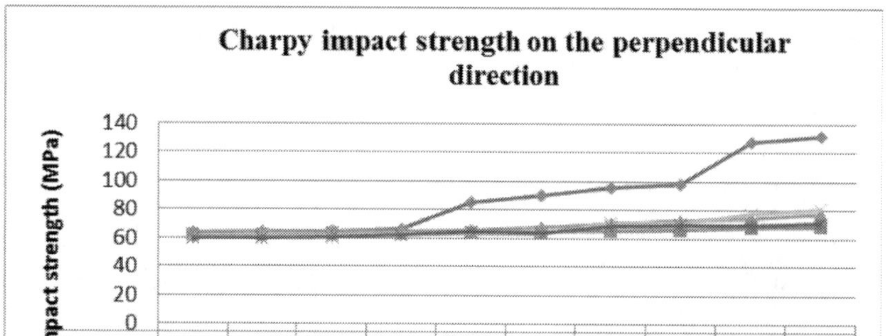

2012	60.5	62.2	64.8	66.5	85	90.4	95.7	98.4	127.5	131.4
2013	61.86	62.52	62.91	63.24	63.75	64.58	65.52	66.02	67.85	68.86
2014	64	64.5	64.9	65.1	65.5	67.8	70.2	72.5	74.4	77.9
2015	61.7	61.9	62.5	62.8	62.9	63.4	70.9	71.6	78.1	81
2016	60	60	60.47	62.39	64.59	64.16	69.16	69.3	69.99	71.2

Chart 6 - Test specimens without notch on the perpendicular direction

The test results for the determining the Charpy impact strength (specimens with C notch) for resin type Hetron F645 TFE are found in Chart 7 and 8:

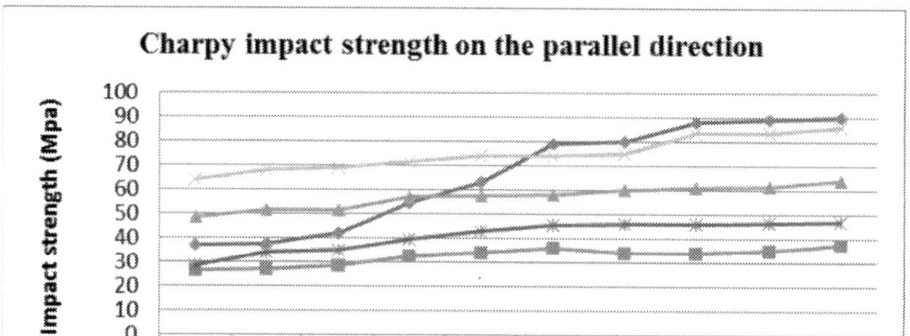

2012	37	37.6	41.9	54.6	63.2	79	80.2	87.9	89.3	90.3
2013	26.47	27.14	28.42	32.46	33.91	36.25	33.9	34.12	35.29	37.41
2014	48.8	51.5	51.8	57.3	57.7	57.9	60.2	61	61.5	64.2
2015	64.2	68.1	69	71.5	74	74.2	75.1	83.7	83.8	86
2016	28.5	34.2	35.2	39.5	43.2	45.6	46.1	46.3	46.7	47

Chart 7- Test specimens with notch on the parallel direction

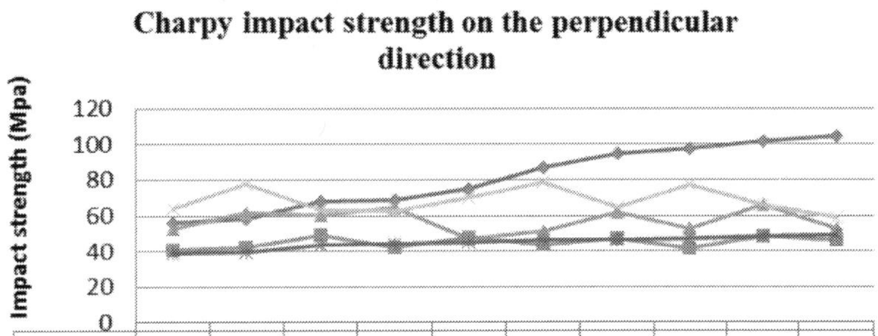

2012	56.1	58.5	67.9	68.5	75.1	86.7	94.4	97.5	101.8	104.2
2013	40.41	41.9	49.39	42.31	47.89	43.58	47.06	41.18	48.53	46.65
2014	52.9	61.7	60.3	64.7	47	51.4	61.7	52.9	66.1	52.9
2015	63.7	77.7	63.1	62.2	69.9	78.3	64.8	77.2	66	58.7
2016	38.5	39.7	43.4	44.5	45	46.1	46.7	47.4	48.8	49.3

Chart 8- Test specimens with notch on the perpendicular direction

3. Conclusion

The continuously evolving rail transport has led to the development of new technologies that meet market requirements. Thus were created new types of materials with new properties, capable of raising the performance of railway vehicles.

The use of composite materials has an important meaning, firstly by significantly reducing the masses, which leads to lower consumption of energy or fuels and increasing the amount of cargo or number of passengers.

This paper presents an experimental study on composite materials, respectively PAFS assemblies and subassemblies and the testing purpose was the approval / certification of products and framing outcomes within the limits of standards method.

For the verification of the manufacturing technology composite materials, the specimens taken from the composite material were subjected to mechanical tests for certification and underwent real life use.

The products tested were consistent with technical requirements for use in rail vehicles bearing /carrying capacity structure structure.

We can conclude from charts 1 and 2 that during the period of 5 years the results for the tensile stress in both cases of application (the direction perpendicular or parallel to the lamination direction) are very close, the difference being less than 2 %. Also, these results are about 25% higher than the minimum required.

We can conclude from charts 3 and 4 that during those five years all breaks were initiated by requesting stretching external fibers. It is noted that the bending stress results in both cases of application (the direction perpendicular or parallel to the lamination direction) are very close, the difference being less than 2.5%. Also, these results are about 40% higher than the minimum required.

In 2015, it was observed that immediately after the start of bending tests on the surface of specimens, small cracks appeared in resin, in the middle of them, even if the test was held in elastic material specimens. As the test continued, the cracks grew in size and number. Meanwhile the loading force of the specimen decreased. After a while, the resin in the middle area of the specimen lost its rigidity being

destroyed by fissures and the specimen was no longer resisting. The application force stabilized. The test specimens did not break, keeping their integrity due to the layered structure of fiberglass.
In conclusion, it was found that the composite material studied can withstand greater demands than those that it will likely have to undergo and can be used with confidence in the support structure of railway vehicles.

ACKNOWLEDGMENTS
The study results carried out by the present paper will be developed in the researches included in European railway projects NETIRAIL şi SPARTACUS as well as the contract PN II PCCA Contract 192/2012 "Solutions to improve the dynamic performance and the crashworthiness of traction railway vehicles to align with the requirements of European regulations", contract financed by UEFISCDI.

REFERENCES

[1] *Sebeşan I., Băiaşu D.,* Mechanical models in the dynamics of the railway vehicles, Romanian Academy, ISBN 978-973-27-2443-9, Bucharest 2014
[2] *Cruceanu, C.,* The sustainable development of transport itself, 2015
[3] *EN ISO 527-4,* Determination of tensile properties of plastics. Part 4: Test conditions for isotropic and orthotropic fibre- reinforced plastic composite
[4] *EN ISO 14125,* Fibre- reinforced plastic composite. Determination of flexural properties
[5] *EN ISO 179-1,Plastics Determination of Charpy impact proprierties*
[6] *Sebeşan I.,* The dynamics of railway vehicles MatrixRom Publishing, ISBN 978-973-755-725-4, Bucharest 2011
[7] Zecheru Gh., Drăghici Gh., Elements of science and materials engineering, vol. 1 şi 2, Publishing ILEX and Publisher University of Ploiesti, 2001
[8]Stanciu A., Cotoros D., Simulation of Mechanical Properties for Fibre Reinforced Composite Materials, Theoretical and Experimental Aspects of Continuum Mechanics, WSEAS Cambridge, UK, Febr.2008, ISBN 978-960-6766-38-1, ISSN 1790-2769.
[9] Almoreanu E., Chiriţă R., Pillars and plates from composite materials, Technical Publishing, ISBN 973-31-1105-8, Bucharest 1997
[10] http://global.kawasaki.com/en/mobility/rail/bogie/efwing.html- TechnoBox efWing, October, 2014

IOP Conf. Series: Materials Science and Engineering **161** (2016) 012065 doi:10.1088/1757-899X/161/1/012065

Fatigue Design of Leaf Springs for New Generation Trucks

E Giannakis[1], M Malikoutsakis[1] and G Savaidis[1]

[1]Department of Mechanical Engineering, Aristotle University of Thessaloniki, GR-54124 Thessaloniki/Greece

E-mail: efstratg@auth.gr

Abstract. Taking as example high-performance parabolic leaves made of the high-strength steel 52CrMoV4 used for the suspensión of heavy truck axles, the present paper focuses on the effect of the manufacturing process (heat treatment and stress peening) on their performance under fatigue loading. Investigations of the microstructure and the mechanical properties of the core and the surface of selected prototype specimens reveal the degradation of the mechanical surface properties due to the applied manufacturing process. The results of these investigations have been used as input for analytical fatigue life calculations and the theoretical assessment of every individual life-influencing factor (material, heat treatment, surface treatment, loading). Wöhler (S-N) curves for various probabilities of survival were experimentally determined and compared to the calculated ones. The agreement between Experiment and Theory is satisfactory.

1. Introduction

Nowadays, leaf springs constitute the most effective suspension way of light- to heavy-duty commercial vehicles. Their outmost advantage is that they group together all main operational actions (i.e. suspension, lateral support and guidance) into a single component, rendering a large number of additional components redundant. On the other hand, this creates the need for a detailed and accurate design and a consequent reliable manufacture of components able to withstand these actions. Previously published papers [1,2] have already covered some of the most important design aspects of leaf springs, namely the strength and fatigue performance for a set of specifically determined fatigue-related properties, as well as the design aspects concerning the kinematic behavior (guidance to suspension coupling). An exemplary overview of a leaf-spring-based suspension system for the front axle of a heavy-duty truck is illustrated in figure 1. Therein, the various components constituting the suspension, guidance and braking sub-systems are clearly indicated, as well as the main load components, which act upon operation.

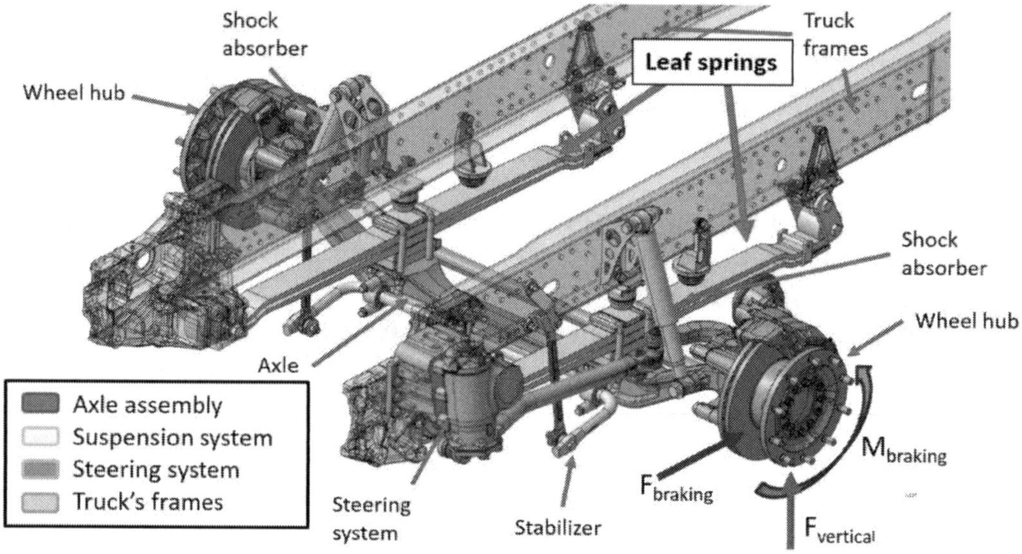

Figure 1: Front leaf spring based suspension system featuring a 2-leaf spring

Rather than examining the overall design process (alike the previous contributions [1,2]), the present paper attempts to go deeper into the indispensable design input data, by taking a closer look to the above-stated set of strength- and fatigue-related properties. Hereby, these very properties and basic technological factors, which act together as fundamental perquisites for the design itself, are approximated through some basic material and surface properties related to the specific manufacturing processes used during the production of high-performance leaf springs.

More specifically, the present study focuses on the influence of the manufacturing process on the microstructure, and the mechanical and surface properties dominating the resultant fatigue life of parabolic leaf spring specimens made from the typical leaf spring steel 52CrMoV4 (Material number 1.7701 acc. to EN 10089: 2002 [3]). This process involves a series of different procedures, namely the transformation of the raw material microstructure through quenching and tempering, the stress-shot-peening of the failure-critical surface and finally the normalization of the induced residual stresses over the aforementioned surface through a "controlled" plastification procedure.

The inspection of samples (leaf spring samples produced in serial conditions) under the Optical Microscope has revealed the microstructure and the surface decarburization state owed to the heat-treatment process. Vickers' (HV) micro-hardness profiles were measured to verify the above findings (extent of the decarburized zone), while Rockwell C (HRC) macro-hardness measurements over the surface and core of the samples were used to evaluate the fundamental strength properties resulting from the manufacturing process altogether. Additionally, surface roughness profiles were determined to quantify the influence of the technological parameter concerning the roughness caused by the Stress-Shot-Peening (SSP) process on the overall fatigue behavior.

Using these carefully resolved technological parameters in conjunction with the material's mean stress sensitivity approximation formula after Schütz [4], theoretical fatigue life calculations for parabolic leaf spring specimens made of the 52CrMoV4 high-strength spring steel have been performed based on the FKM guideline [5] along with some inevitably arbitrary assumptions (but in line with the accumulated experience [1,2]). The influence of every individual factor was successfully identified and assessed in respect to the total fatigue life estimation, while uniaxial cyclic, constant-amplitude, three-point bending tests were conducted on a series of specimens, in order to determine the fundamental stress-life curves for various probabilities of survival and validate the accuracy of the respective theoretical calculations. Finally, a limited set of variable amplitude loading tests following the operational loading spectra after Grubisic and Fischer [6] have been used to check the accuracy of the developed theoretical stress-life curve in conjunction to the Miner's elementary rule, assuming a damage sum of D=1 (100%).

2. Production stages of leaf springs

The raw material 52CrMoV4 used for leaf spring production comes into flat bars (geometry and dimensions acc. To EN 10092:1 [7]) with an original bainitic-perlitic dominant microstructure. These bars are initially heated up to >900° C and rolled in order to achieve the desired thickness distribution along their length (it should be mentioned at this point that due to the similarity of the function which describes the desired thickness distribution, to the mathematical parabolic function ($f(x)=a/x^2$, a=const.), the leaf springs of this type have acquired the characteristic designation "parabolic"). Other areas that need geometrical forming, i.e. steps and eyes (see also [1,2]), are created at this stage as well, using forging moulds and specialized forming machines (for eye forming). Each leaf, which is still up to this point almost a straight bar, is then inserted into another forging press that will give it its characteristic elliptical shape shown in figure 2.

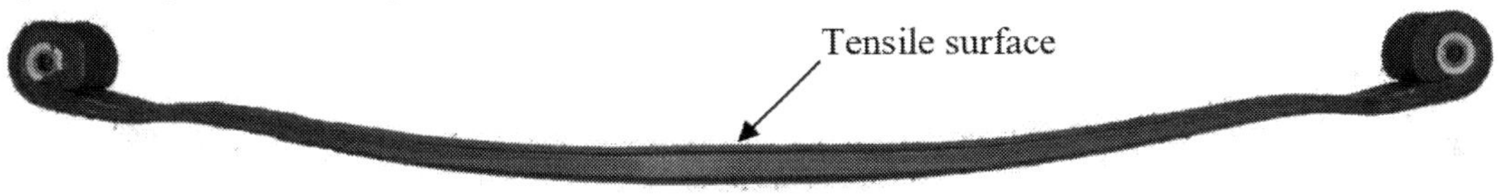

Figure 2: Unloaded mono-leaf spring illustrating its characteristic elliptical shape

The next phase comprises of the quenching of the leaf in a temperature-controlled oil bath in order to achieve a cooling gradient that will transform the microstructure to almost a 100% martensitic one over both the surface and core. The leaf is then tempered in a controlled environment (the whole procedure is commonly referred to as martensitic quenching and tempering) to relax the internal stresses induced by all previous actions (severe hot-rolling, forming and quenching) and regain some ductility. The side effect of this mandatory annealing is the formation of a decarburized zone on the surface of the leaf, which is both unfavourable, but also somewhat desired (up to a certain extent) in terms of fatigue. In order to understand this ambiguous behaviour, one must go further than the obvious reduction in hardness and consequent lower material strength (lower ultimate tensile strength R_m) owed to the expulsion of carbon from the surface, and realize that the impelled increase of the ductility can offer an enhanced crack resistance in fatigue loading. Therefore, it becomes evident that the sequential heat-treatments should be balanced to achieve an optimum behaviour overall.

The next production stage of each leaf includes the bending of the leaf near the material's nominal tensile strength (~1600-1750MPa) using a designated actuator, its insertion to a "cassette" that will hold this bent position in place and the shot-peening of the tensile surface of the leaf (upper surface in Figure 2) with shots of certain hardness and diameter. The goal of this "stress-shot-peening" process is the entrapment of favourable compressive residual stresses that will decrease significantly the maximum effective stress (the stress that the material on the surface of the leaf is subjected to during operation) and hence increase the fatigue life. Common compressive residual stress values induced by this procedure range from 550 (quite bad quality) to 1200MPa (superior quality), which demonstrate the importance of this stage [1,2].

Again, an unfortunate side effect of the later procedure is the formation of increased roughness, which has the opposite outcome. Regarding this point, it should be mentioned that studies on the intensity, coverage and number of passes of the shot peening process [8] have shown that the roughness can be brought back to its original state or even achieve a better roughness profile, but in the expense of the total product cost (more than one passes). Thus, such methods (multiple passes) have not been adopted yet by the leaf spring industry.

Finally, the last stage involves the "pre-setting" of the leaf spring after its assembly (in case of multi-leaf springs), which is the bending of the spring to the extent of going over the material's yield stress at the tensile surfaces of all leaves, in order to evenly redistribute the above stated residual stresses. This action also serves the avoidance of the suspension's settling and shagging after only a small portion of its operational life, since a relaxation and redistribution of stresses is otherwise inevitable.

3. Material microstructure and mechanical properties & Surface inspection

Extracting material samples from both the tensile surface and core of the specimens has enabled the inspection of the influenced areas by the heat- and surface- treatments. The metallurgical inspection under the Optical Microscope (OM) has revealed the almost 100% martensitic microstructure at all areas (core and surface), as well as the extent of the anticipated decarburized zone to a depth of about 150μm from the surface. Vicker's micro-hardness measurements at the same area have verified this finding as illustrated in figure 3, which presents an exemplary area near the surface, along with the micro-hardness diagram over the depth (distance from surface). The decarburized zone becomes evident through editing the OM photo to offer a better view of the decarburization intensity (white areas indicate the absence of carbon).

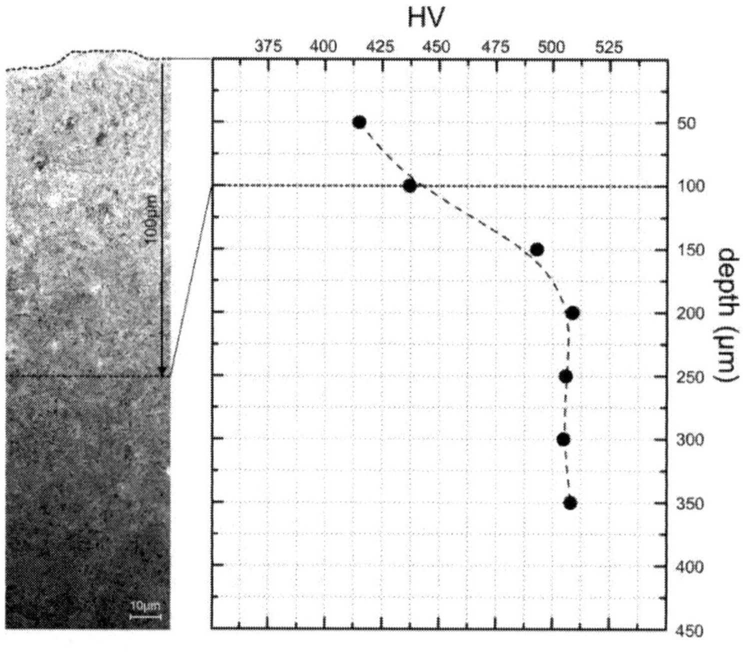

Figure 3: Decarburized zone over the depth near the surface and Vicker's micro-hardness profile

As expected, the decarburization intensity varies from an almost complete (adjacent to the surface) to a partial one, which gradually fades away along the depth. The micro-hardness profile illustrates the respective behaviour, following an increasing function (from surface to core) up to the maximum of 500-530HV. These values coincide with the measured core hardness and comply with the almost martensitic microstructure of the core of the specimen, as shown in figure 4. The later figure is displayed hereby in order to additionally verify the uniformity of the material after quenching and tempering along the whole thickness of the leaf and to illustrate the solid quality of the microstructural grid (evenly distributed phase areas).

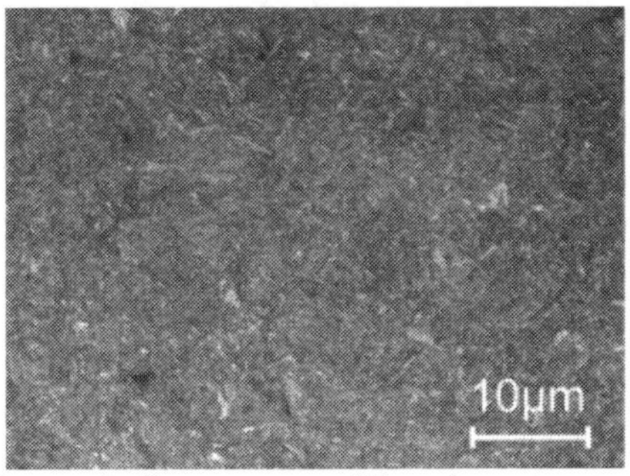

Figure 4: Core martensitic microstructure

Rockwell C (HRC) measurements over the surface and core of the samples present an almost uniform behaviour in hardness, with a tendency of achieving a slightly higher value on the surface (core hardness ranges form 47.4 to 48.7HRC while surface hardness has an average value of 49.5HRC). According to the hardness-to-R_m conversion standard [9], the surface hardness corresponds to a $R_m \approx 1728$ MPa, representing the failure-critical area of the leaf spring and is therefore used to evaluate the respective material properties.

Finally, the surface roughness profile is measured in various locations to acquire the strength related roughness parameter R_z (peak-to-valley height) to be used in the fatigue life calculations. An exemplary measurement of the roughness profile is illustrated in figure 5, indicating the characteristic R_z value of 32μm.

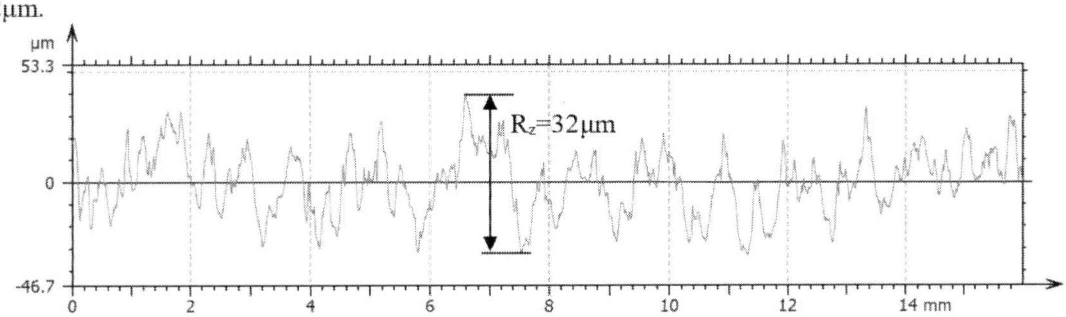

Figure 5: Measured roughness profile

4. Fatigue life theoretical calculation

The primary target of the theoretical fatigue-life calculation is to develop a reliable stress-life curve for a constant stress ratio ($R=\sigma_{min}/\sigma_{max}$) or a better-suited in this case constant mean stress σ_m, and constant amplitude loading (Wöhler curve). This can be achieved by determining at least two points of interest that a line can pass through. These two points are obviously the well-established fatigue limit ($N_D=10^6$ cycles) on the one side and another one that corresponds to a different characteristic behavior, i.e. the knee point that corresponds to the initiation of the low cycle fatigue regime. Regarding the first point, a combination of the instructions found in the FKM guideline [5] with the option to include the Mean Stress Sensitivity factors established by Schütz [4] (instead of the FKM guideline proposed ones), can lead to a reliable estimation. On the other hand, no clear guidance is given anywhere for the determination of any other point of the fatigue-life curve, hence some improvisation is required.

Taking into consideration the expected stress-strain behavior of such a material (adjacent values of R_m and yield stress as well as limited ductility) one can speculate that the yield stress will be around the 90% of the R_m. Therefore this value ($0.9 \cdot R_m$) can be selected as characteristic and paired to an arbitrarily chosen fatigue life that respects the limited ductility assumed, i.e. $N=10^3$ cycles. The later selection, although seeming completely arbitrary and unreliable, is actually in accordance with the NASTRAN FE solver assumptions [10] used for many years now, while it has been additionally ascertained by own experience by its well-suited conformity to a vast number of experiments. Moreover, it can be easily realized that the selection of this point inside the area of $N=10^3$ to $N=10^4$ cycles has, in all cases, a minor effect on the fatigue life curve slope.

Having drawn the line passing through these two points (line 1 in figure 6), the FKM guideline instructions are sequentially applied, starting by applying the material related factors (loading mode, roughness and residual stresses) and afterwards the load-level-related ones (mean stress sensitivity factor). This means that the whole line is first shifted upwards to account for bending (original fatigue limit is given for tension-compression) by a factor of 1.016 (line 2 in figure 6) and then downwards to rationalize the measured roughness (hereby a factor of 0.691 for R_z=32μm – line 3 in figure 6). The maximum factor of 1.3 according to the FKM guideline is then selected to multiply the stresses and consider the effect of the favorable compressive residual stresses (line 4 in figure 6), since the guideline includes only the effect of normal stress-peening (not the combined stress-shot-peening). Finally, the mean stress sensitivity factor of M=0.5 is used to translate the curve to the constant mean stress applied during the constant mean stress – constant amplitude experimental procedure, which conforms well to both the FKM guideline and Schütz's proposal (line 5 in figure 6).

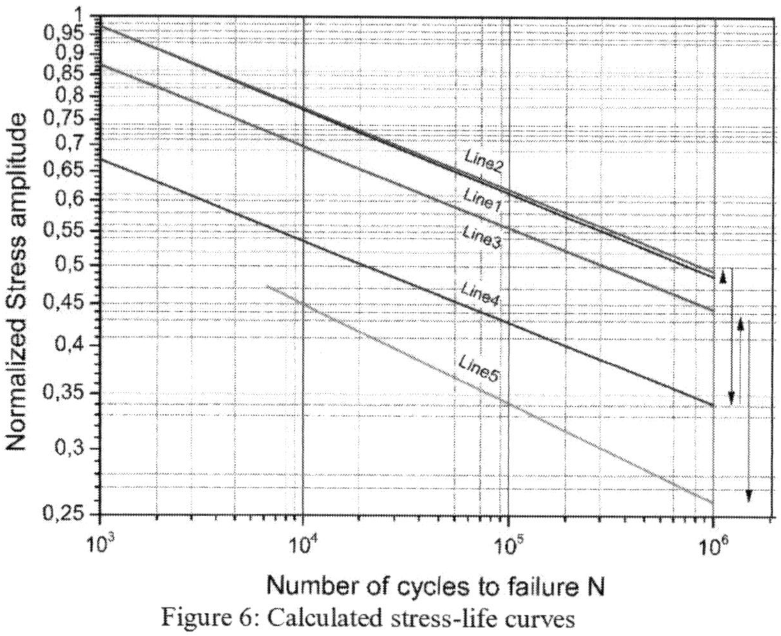

Figure 6: Calculated stress-life curves

It should be emphasized at this point that in accordance with the FKM guideline, the resulting theoretical fatigue-life curve corresponds to a Probability of Survival P_s=97.5% and hence it should be compared to experimental curves having explicitly the same P_s as well.

It is also evident that the curve corresponding to the material's fatigue life (for R=-1 or σ_m=0, line 4 in figure 6) can be used to create the respective curves for any mean stress desired by simply using the same mean stress sensitivity factors. This means that having the stress distribution for various variable amplitude loadings for a given geometry, one can calculate the effective mean stresses and stress amplitudes at any given point and thus calculate the corresponding anticipated damage for each load cycle $D=1/N_i$ (where N_i is the fatigue life calculated from transforming line 4 to a new line 5 in Fig. 6, which corresponds to the calculated mean stress at the point examined and the respective stress amplitude). The damage sums for each load cycle can then be calculated using the Miner's elementary rule at each point and plotted against the position of the aforementioned point. This way, a damage sum plot is used to visualize the fatigue life of the component (hereby a mono-leaf spring) versus a characteristic position (hereby the length). Figure 7a shows the load sequence (extracted from the spectrum of Grubisic and Fischer [6]) used to calculate the characteristic stress plots illustrated in figure

7b (normalized to the mean stress). These plots represent the stresses obtained by FE Analysis of a given (already optimized) geometry for some of the main characteristic combination of load component values and are sequentially used to calculate the respective damage sums shown in figure 7c.

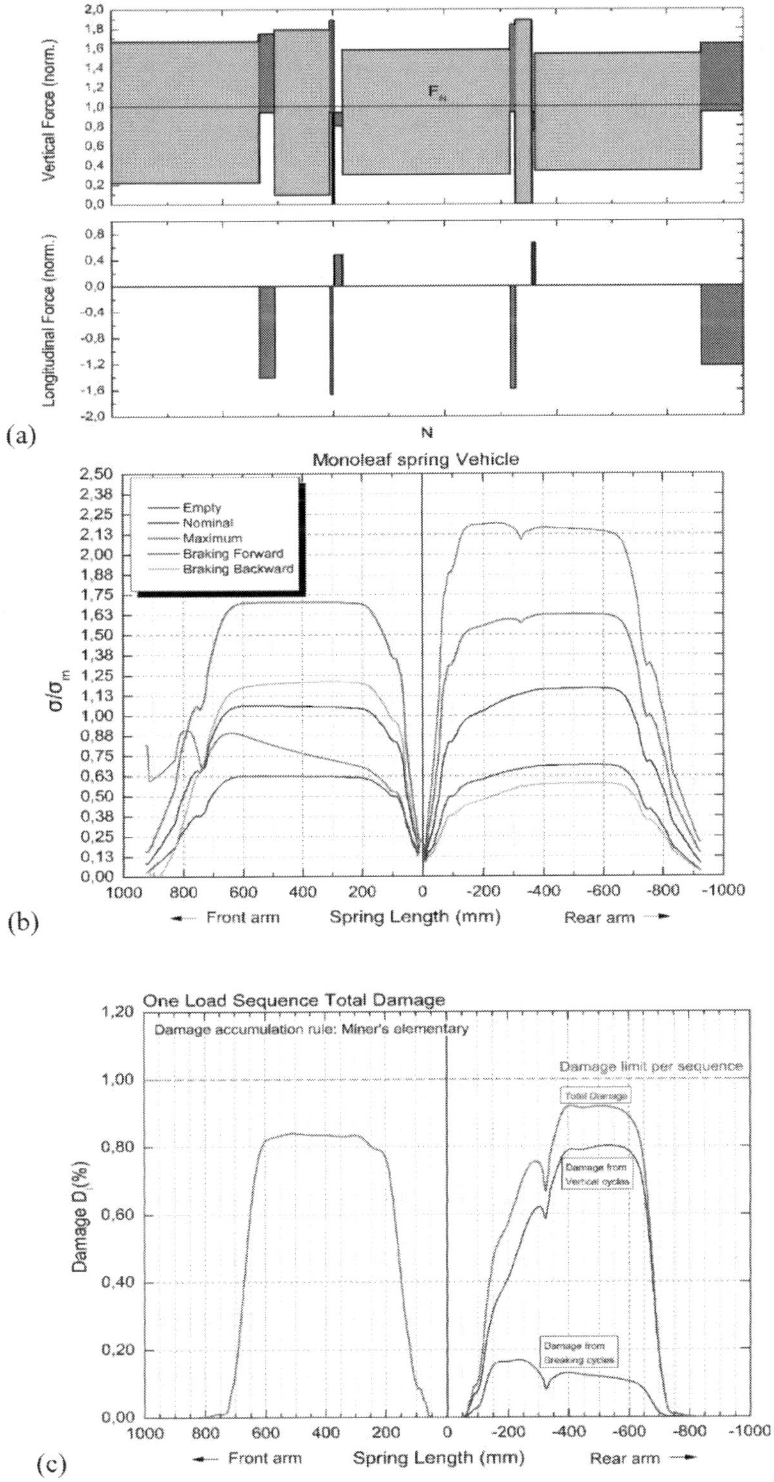

Figure 7: (a) Load sequence extracted from the spectrum after Grubisic and Fischer [6]; (b) Stress plots versus the length of the leaf for characteristic operational actions and (c) One load sequence damage sums calculated along the length of the leaf

The damage limit line in figure 7c represents the (normalized to itself) damage allowed for each sequence in order for the leaf spring to achieve the operational life requirement. It can be seen that the damage plot is below (around 90%) of this limit, which means that the design of this particular leaf spring is expected to present a higher operational fatigue life by around 10% compared to the requirement.

5. Experimental validation

The experimental validation consists of (a) the verification of the fundamental theoretical fatigue life curve with a set of constant mean stress, constant amplitude loading tests, and (b) the experimental determination of the operational fatigue life under variable amplitude loading. Figure 8 presents the (normalized to the mean stress) experimentally determined curve from a set of 10 specimens, along with the respective scatter-band for various probabilities of survival. As shown below, the tests (round points in figure 8) have been performed in three different load levels and can be considered adequate for a reliable statistical analysis (at least in the finite-life region), since the conventional scatter-band produced is quite narrow ($T_N=N_{Ps=90\%}/N_{Ps=10\%}=1:2.08$).

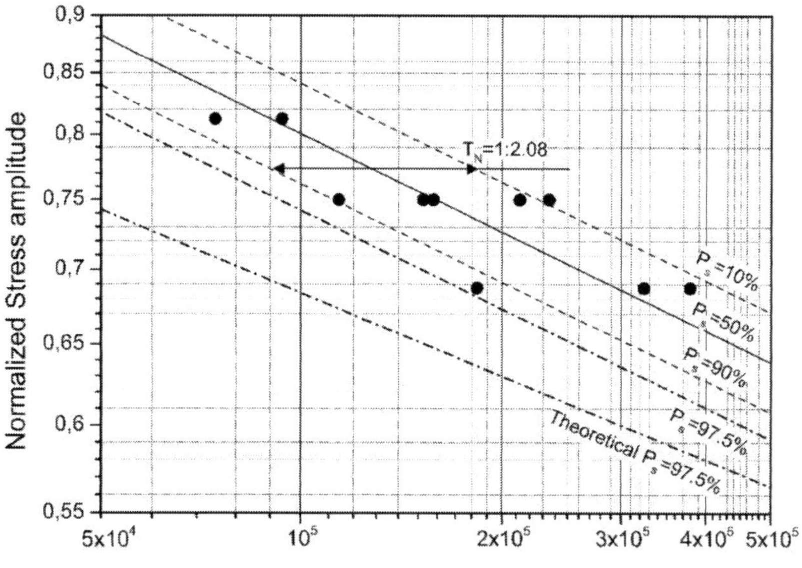

Figure 8: Experimental stress life curve for various probabilities of survival, conventional scatter-band and theoretically determined curve

The theoretical stress-life curve is plotted too in the above figure versus the experimental curve with a P_s=97.5%, showing a satisfactory agreement. It is noticeable that the theoretical curve comes very close to the experimental one near the conventional fatigue limit ($N=10^6$ cycles), while larger deviations are observed towards the low cycle fatigue regime. Nevertheless, this approximation seems to be reliable enough to proceed with the assessment of the operational fatigue life under variable amplitude loading using directly the theoretical curve as is.

The operational fatigue life tests using the load sequence shown in figure 7a have been conducted using the in-house built bi-axial fatigue test rig for leaf springs. An overview of this test rig is illustrated in figure 9. The test rig consists of a vertical actuator bringing the vertical loads and two horizontal actuators bringing the longitudinal loads (during braking actions). The test rig is built in such a way to test simultaneously two complete leaf springs and receive all original serial parts that contribute to the suspension system (original vehicle frames and peripheral eye bushings, buffers clamping devices, etc.). This way a simulation of the vehicle conditions is approached as best as possible.

Figure 9: Bi-axial test rig

In order for the reader to be easier acquainted with this test rig, it should be clarified that the leaf springs are assembled upside down, meaning that the road (where the loads originate from) is in the plane of the horizontal actuators, parallel to the vehicle's frames.

The preliminary testing of only two specimens resulted to fatigue lives of 98.3% and 127.6% of the requirement respectively, which means that the theoretical calculations have the potential to successfully describe the fatigue behavior of such components.

6. Summary and Conclusions

The leaf spring production stages have been described and each expected outcome, in terms of influencing the strength and fatigue behavior, discussed in a stepwise manner. The weight of the present study has been laid upon the understanding of the material state and properties of the failure-critical tensile surfaces of leaf springs. A theoretical methodology for calculating the expected fatigue life under both constant and variable (operational) loadings has been presented, based solely on fundamental material data. The FKM guideline has been utilized along with some relatively reliable assumptions to complete the stress-life curve, along with the recommendations after Schütz regarding the mean stress sensitivity factors and the Miner's elementary damage accumulation rule.

An optimized design of a mono-leaf spring has been used for the stress calculations through FE analysis, as well as tested in constant and variable amplitude loading. The comparison between the theoretical and experimental curves (for constant amplitude) and fatigue lives (for variable amplitude – operational loading) seems to produce a satisfactory agreement. Although these first results are encouraging, further testing (with both constant and variable amplitudes) with different leaf spring designs and materials is required to come to a safe conclusion.

References

[1] S. Karditsas, G. Savaidis, M. Malikoutsakis, Advanced leaf spring design and analysis with respect to vehicle kinematics and durability, Int. J. Struct. Integrity. 6 (2015) 243-258.

[2] M. Malikoutsakis, G. Savaidis, A. Savaidis, C. Ertelt, F. Schwaiger, Design, Analysis and Multi-Disciplinary Optimization of High-Performance Front Leaf Springs, Theoretical and Applied Fracture Mechanics, Vol. 83, (2016), 42–50.

[3] EN 10089:2002, Hot rolled steels for quenched and tempered springs. Technical delivery conditions.

[4] W. Schütz, Über eine Beziehung zwischen der Lebensdauer bei konstanter und veränderlicher Beanspruchungsamplitude und ihre Anwendbarkeit auf die Bemessung von Flugzeugbauteilen. Zeitschrift für Flugwissenschaften, 15, 407-419, 1967.

[5] FKM-Guideline. Analytical Strength Assessment. 5th Edition, VDMA Publishing, 2003.

[6] V. Grubisic, Determination of load spectra for design and testing, Int. J. Vehicle Design. 15 (1994) 8-26.

[7] EN 10092-1:2003, Hot roller spring steel flat bars – Part 1: Flat bars – Dimensions and tolerances on shape and dimensions.

[8] St. Baiker (Editor). Shot peening – a dynamic application and its future, Metal Finishing News Publishing, 2006.

[9] ASTM A 370 – 03a: Standard Test Methods and Definitions for Mechanical Testing of Steel Products.

[10] MSC Software, MSC Nastran Embedded Fatigue Documentation.

Shape memory composite antennas for space applications

Loredana Santo, Fabrizio Quadrini, Denise Bellisario

Industrial Engineering Dep., University of Rome Tor Vergata, via del Politecnico 1, 00133 Rome, Italy

loredana.santo@uniroma2.it

Abstract. Future space missions will require large space infrastructures in order to achieve scientific and technological objectives characterized by an intrinsic complexity. In this study, the development of shape memory composite structures for aerospace applications is described. In particular, the structure of a small-scale self-deployable mast has been prototyped as a proof of concept for its feasibility. The mast structure is made by interlocking two shape memory polymer composite (SMPC) strips, each one made of two layers of carbon fiber fabric with a shape memory (SM) epoxy resin interlayer. A complete deployment of the SMC structure was achieved. The versatility of this technology has been also demonstrated in previous studies, in which small scale deploying solar panels were fabricated. Obtained results are very promising in terms of manufacturing technology, and shape recovery of manufactured parts.

1. Introduction

Deployable structures are largely used for space applications and infrastructures. Generally, solar panels, antennas, radars and masts of the satellites are deployable to save space during launch. At the same time, it is very important that such structures are as light as possible and they require good stability in the deployed configuration to support structural loads and to resist external disturbances. Most conventional deployable structures are highly complex structures making use of numerous mechanical components such as linkages, hinges, motors and energy storage devices [1]. However, the large-sized mechanisms and complex assembly processes of these kinds of devices could involve problems like heavy weights, complexity of mechanisms and high costs [2]. Some of these structures as the able deployable articulated mast (ADAM) and the large deployable antennas (LDAs) use complex linkage mechanisms resulting in a large structural mass in order to obtain a good performance during deployment [3,4]. For example, deployable masts have been studied as telescopic booms which deploy via a series concentric prismatic joints [5]. Recently, masts with box bellows are constructed from flat rectangular panels joined by hinges [6], but generally the use of articulated trusses is very widespread [7,8]. However, recent advances in smart materials and structures that are easy to control, light weight, inexpensive, and capable of continuous deformations have opened new possibilities to design compact deployable structures. Shape memory polymers (SMPs) are polymeric smart materials that have the ability of recovering from a deformed shape to their original shape induced through the application of an external stimulus. SMP-based bending actuators and hinges for foldable and deployable structures have been studied and developed in the last few years [9-12]. Instead, SMPCs combine structural properties of continuous-fibre polymer-matrix composites with functional behaviour of shape memory polymers. In previous studies, the authors have developed and prototyped some self-deploying configuration structures by using SMPCs. A first result was obtained in the production of composite laminates with shape memory properties. Composite sheets consisted of two layers of carbon fibre fabric with a shape memory epoxy resin interlayer. Carbon fibre prepregs can be also used: an optimal adhesion between the different layers was achieved thanks to the compatibility of the prepreg matrix and the shape memory material [13]. SMC sandwiches, having a SM foam core and fibre reinforced skins, were also produced [14]. Among several applications it can be cited truss structures, solar reflectors, space station's modules, antennas and solar sails [15]. The shape recovery of a bent thin sheet of such composites was also tested for the first time in microgravity conditions during the BION-M1 mission of the Soyuz spacecraft (20th of April 2013) [16]. Interesting results were achieved in terms of shape recovery and the effect of microgravity in the case of small samples is found negligible if compared with recovery forces. Instead, microgravity strongly affects performances of heating devices and it has to be taken into account for designing of self-deployable structures in future space applications. Recently, also other studies on SMPCs application with higher strength and elastic modulus have been studied to fabricate a deployable truss which can unfold and has a low stiffness after being deployed [1]. Moreover, conceptual design of small-scale structures was prototyped with the aim to define several configurations which are able to self-deploy. In particular, the SMC prototypes were manufactured to represent two examples of self-deployable devices for de-orbiting systems: a composite cross without sail and a dual sail structure with composite frames and kapton sails [17].

In the current study, the acquired knowledge about SMPCs' technology is used to develop and prototype a different SMPC deployable structure for space applications. In fact this technology offers a new foldable/deployable lightweight structure that can be scaled to large antenna dishes or similar deployable structures. In particular, a prototype of space deployable articulated mast is developed. Deployable masts have been used in space extensively for deploying and support space operations. The basic idea is to mould the composite structure in the elongated (opened) configuration and subsequently shift to a shortened (closed) shape in the memory step. The composite structure would recover the opened configuration by heating. Obtained results are very promising in terms of manufacturing technology, and shape recovery of manufactured parts. Thus, as a result of heating, initial configuration can be successfully recovered without failures. In laboratory tests, composite heating was provided by a hot air gun. Commercially available materials have been used in the experimentation for both composite layers and SMP interlayer.

2. Materials and methods

2.1. Materials

Commercially available materials have been used for prototyping SMPC structures. The thermosetting carbon fibre reinforced (CFR) prepregs utilized are Hexcel Carbon/Epoxy prepreg composite. HexPly® M49/42%/200T2X2/CHS-3K is a woven high strength carbon epoxy prepreg, whereby M49 is the resin type; 42% is the resin content by weight; 200T2X2 is the reinforcement reference and CHS-3K refers to high strength carbon fiber. This kind of prepreg is a 0/90 fabric typically used for aeronautical applications: the epoxy matrix shows high stiffness and strength without any shape memory behavior. The SMP interlayer was an uncured epoxy resin (3M Scotchkote 206 N) which was available as a green powder, and was a one-part, heat curable, thermosetting epoxy coating.

2.2. SMPC production

Two shape memory composite strips were produced by using a sandwich structure in a laboratory scale procedure. These two strips properly combined and memorized are going to form the mast structure. The prepreg sheet was cut into four strips of 15x400 mm^2. Two prepreg strips were used as the outer skins of the sandwich structure and between them an interlayer SMP of about 0.1 mm thick was deposited (Figure 1a). The behavior of the SMPC structures was relatively complex because of the presence of two epoxy systems with different glass transitions temperatures. It has been observed that their transition range were comparable [13] and accordingly 150°C has been considered as an appropriate temperature for material forming whereas higher temperatures could lead to degradation. Therefore, the two sandwich strips were placed between two sheets of thermoplastic film (used for mold-release) and finally formed by hot pressing in a hydro-pneumatic press (by ATS FAAR). The applied pressure was 1.3 MPa, the moulding temperature 150°C, and the holding time 15 min.

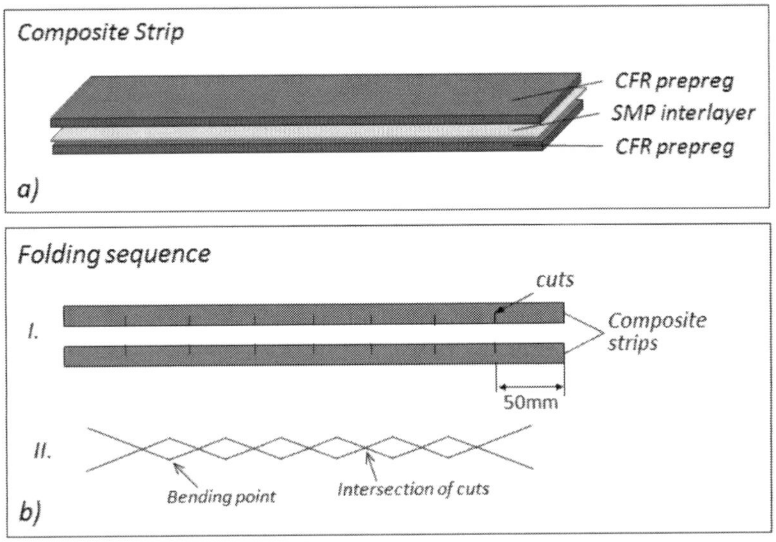

Figure 1: a) Scheme of the composite strip; b) Scheme of the sequence of combination and folding of the composite strips.

2.3. Memory and recovery step

The initial shape of the strips after forming was a flat configuration. The two sandwiches were combined and shaped as showed in Figure 1b. On each sandwich structure, a horizontal cut of 7 mm was made every 50 mm, in order to obtain 7 cuts on each strip.

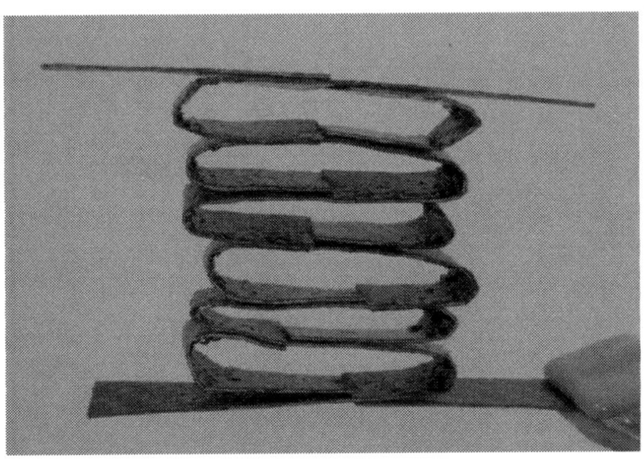

Figure 2: Final configuration of the mast after memorizing step.

During memory step, the two strips were heated and combined intersecting two homologous cuts and bending of about 180° the centre portion of strip between two consecutive cuts. This operation has been performed for six times, in order to fold and close the structure. A hot gun was used to provide 150 °C for the memory phases. The final memorized configuration was showed in Figure 2. After manufacturing, the SMPC mast was tested with several memory-recovery cycles. Shape recovery was obtained using a hot gun. The aspect of the reduction of the total length and so the folding sequence is fundamental for the correct deployment that is also function of the heating source. In particular, heating must promote the correct sequence of deployment avoiding the uncontrolled recovery of parts.

3. Results and discussion

The main design criteria in deployable structures are the compactness of the stored configuration, dimensional tolerances, light weight, ease of deployment, durability and endurance. In this study, it has been assessed the possibility to optimize the design and the configuration deployment of shape memory mast. The SMPC mast was memorized with a configuration which reduces the initial length of 7 times. Thus, it is possible to deploy a mast of an initial length (memorized) of 57.3 mm to a final length of 400 mm. The deployment of the mast has been achieved only by exploiting the shape recovery properties of the structure and no telescopic booms, switching mechanisms or rigid joints are used. The memory-recovery process is very easy and rapid both in the memory and recovery phases. After identifying the optimal folding design and memorizing the SMPC structures, it was possible to deploy the SMPC mast by heating with a hot air gun. In order to test the performances of the SMC prototypes, simple memory-recovery cycles have been carried out. In particular the initial configuration and the phases of the recovery are showed in Figure 3. The SMC mast folded configuration recovered its initial flat shape (Figure 4).

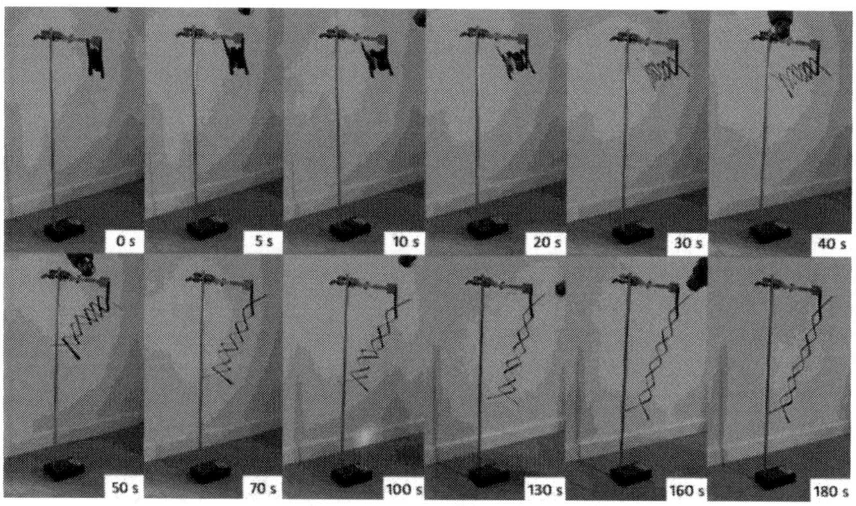

Figure 3: Recovery sequence of the composite mast.

During tests the composite mast was fixed on a metallic support. The SMC mast deployed smoothly without problems achieving the final shape (the elongated one), demonstrating multiple folding and related deployment of the two strips without issues. About 200 s were necessary for a full recovery of the SMC mast, due to the greater extension of the structure and to the localized heat source used.

Accordingly, the SMPC mast unfolds correctly six times for single strip, demonstrating the deployment of a mast with a reduced length of seven times.

Figure 4: Deployed SMPC mast.

An important achievement was that no damages or evident cracks in the bending zones, or the surrounding areas, were found.

In the laboratory tests, a localized heat source as a hot air gun was chosen to memorize and recovery the SMC shape due to the limited sizes of the laboratory prototypes. In final applications, higher sizes are expected and local heating could be easily embedded to control shape recovery. Embedding the heaters could also help in differential deployment along different unfolding axis.

Starting from these results, the possibility to scale-up the prototypes could be furtherly investigated. Additionally, in order to deepen the present work, some aspects should be more thoroughly investigated, such as: the damage during and after the memory stage, the residual stiffness of the SMC, the final geometry to evaluate possible defects, etc.

4. Conclusion

In this study, a SMPC mast was prototyped in a laboratory scale process and its memory-recovery properties were assessed. SMPC structures can be easily produced by placing a SMP interlayer between composite layers. In particular, the proposed SMPC mast prototype was designed and produced so that the structure can deploy stretching six-fold compared to the initial memorized size. The experimental results showed that SMPC mast can successfully self-deploy following the desired design constraints and recovering the original flatness without noticeable damages. Obtained results are very promising in terms of manufacturing technology for producing SMPC slender structures, of compactness of the stored configuration and of ease of deployment. The presented results are expected to represent a first step to understand and study the SMPC mast structure for space applications. Further studies will be performed to quantitatively evaluate the rigidity of the deployed configuration, the durability and endurance of the system, the amount of recovery and of damages.

References

[1] Zhang, R., Guo, X., Liu, Y. & Leng, J. (2014). "Theoretical analysis and experiments of a space deployable truss structure". Compos. Struct. 112, 226–230

[2] Huang, H., Deng, Z., Li, B. (2012). "Mobile Assemblies of Large Deployable Mechanisms". J. Space Eng. 5, 1–14

[3] Gross, D. & Messner, D. (1999). "The able deployable articulated mast–enabling technology for

the shuttle radar topography mission". In Proc. of the 33rd Aerospace Mechanisms Symp. Pasadena

[4] Meguro, A., Shintate, K., Usui, M. & Tsujihata, A. (2009). "In-orbit deployment characteristics of large deployable antenna reflector onboard Engineering Test Satellite VIII". Acta Astronaut. 65, 1306–1316

[5] D. S. H. Chae, D. J. Rohweller, M. W. Thomson, (1994).US Patent 5135795

[6] F. Jensen, and S. Pellegrino, (2001). "Arm development – review of existing technologies, Tech. Rep. CUED/D-STRUCT/TR198", Department of Engineering, University of Cambridge, Cambridge, UK.

[7] H. Guo, R. Liu, and Z. Deng, (2008). "Preliminary design method study for the space deployable articulated mast," Proc. 1st Int. Conf. on Intelligent Robotics and Applications, Vuhan, China, pp. 322-331

[8] L. R. Adams, (1988). "The X-beam as a deployable boom for the space station," Proc. 22nd Aerospace Mechanisms Sym., California, US, , pp. 59-66

[9] Lee, Y., Lee, H., Hwang, T., Lee, J. & Cho, M. (2015). "Sequential Folding using Light-activated Polystyrene Sheet". Sci. Rep. 5, 16544

[10] Xiao, X. et al. (2015). "Shape memory polymers with high and low temperature resistant properties". Sci. Rep. 5, 14137

[11] Mao, Y. et al. (2015). "Sequential Self-Folding Structures by 3D Printed Digital Shape", Memory Polymers. Sci. Rep., 5, 13616

[12] Felton, S., Tolley, M., Demaine, E., Rus, D. & Wood, R. (2014)."A method for building self-folding machines Science", 345, 644–646

[13] Quadrini, F. (2014). "Polymer matrix composites with shape memory properties." Advanced Materials Research, 783-786, pp.2509-2516.

[14] Santo, L., F. Quadrini, and L. De Chiffre (2013). "Forming of shape memory composite structures." Key Engineering Materials, 554-557, pp.1930-1937.

[15] Accettura, A., L. Santo, and F. Quadrini (2012). "Space mission opportunities using shape memory composites." Proceedings of "63rd International Astronautical Congress" IAC 2012.

[16] Loredana Santo, Fabrizio Quadrini, Pier Luigi Ganga, Valfredo Zolesi (2015). Mission BION-M1: Results of Ribes/Foam2 experiment on shape memory polymer foams and composites, Aerospace Science and Technology, 40, pp. 109-114.

[17] Santo L., Quadrini F., Bellisario D., Accettura A.G, "Conceptual prototypes of composite structures for aerospace", Proceedings of the 2016 Manufacturing Science and Engineering Conference, MSEC2016, June 27 to July 1, 2016, Virginia Tech, Blacksburg, VA, USA.

Fatigue testing of reinforced-concrete steel bars

S Maropoulos[1],D Fasnakis, Ch Voulgaraki, S Papanikolaou, A Maropoulos[2], A Antonatos[3]

[1]Department of Mechanical Engineering and Industrial Design, Technical Educational Institute of West Macedonia Koila, Kozani 50100, Greece

[2]Department of Mechanical Engineering, Aristoteles University of Thessaloniki, Thessaloniki, Greece

[3]Testing Research & Standards Centre, Public Power Co, Kantza, Greece.

E-mail: maropou@teiwm.gr

Abstract. A number of low-cycle fatigue tests were conducted on reinforced-concrete steel bars of various diameters to study their behaviour under axial loading according to EN 10080 and EN 1421-3. Scanning electron microscopy was used to study the specimen fracture surfaces. The problems faced during testing are presented and a specimen preparation method is described that will aid researchers on fatigue testing to obtain accurate test results and save on material and time.

1. Introduction

The fatigue strength of reinforced concrete bars has been the subject of research for many years. Usually the fatigue characteristics of reinforced-concrete steel bars are studied with axial fatigue tests using conventional testing machines and specimens in the as-delivered state, usually ribbed rods. The advantages of axial tests include low costs at a relatively high testing frequency (up to 200 Hz) and the possibility of defining test conditions precisely. A very significant disadvantage is related to problems connected with clamping bars in testing machines, where local stresses in bar clamping areas are responsible for the premature failure of the bars. It is also difficult to avoid the effect of secondary stresses caused by the non-rectilinearity of specimens or their misalignment in testing machine jaws [1]. A number of studies [2-4] contain information about various methods to prepare the grip parts of

*Corresponding author

specimens usually connected with using an added layer between a specimen and a testing machine jaw and /or slight machining of specimen ends. Some of these include the use of leather belts wrapped around the specimen grip part, low melting point metallic fillings, resins and aluminium pads. However, little information is available concerning the effectiveness of the measures mentioned.

Bars and ribbed rods made of B500SP steel are used for reinforcing concrete elements and structures designed in accordance with the specifications in EN 10080 [5] for C class steels of a yield point of 500 MPa. The requirements for fatigue strength tests are described in EN 10080 and EN 1421-3 [6] and are summarised below:

- Maximum stress, σmax=300 MPa,
- Stress amplitude, σa=150 MPa,
- Load cycle frequency, 1-200 Hz,
- Acceptance criterion, failure-free after $2 \cdot 10^6$ load change cycles

2. Experimental

High cycle fatigue tests were conducted on reinforced-concrete steel 8, 12 and 20 mm diameter bars (Figure 1) to study their behaviour under axial tension loading according to EN 10080 and EN 1421-3 using a 100kN capacity Instron machine (Figure 2) until cracking or reaching the boundary number of cycles of $2 \cdot 10^6$.

Figure. 1. Reinforced-concrete steel 8, 12 and 20 mm diameter bars.

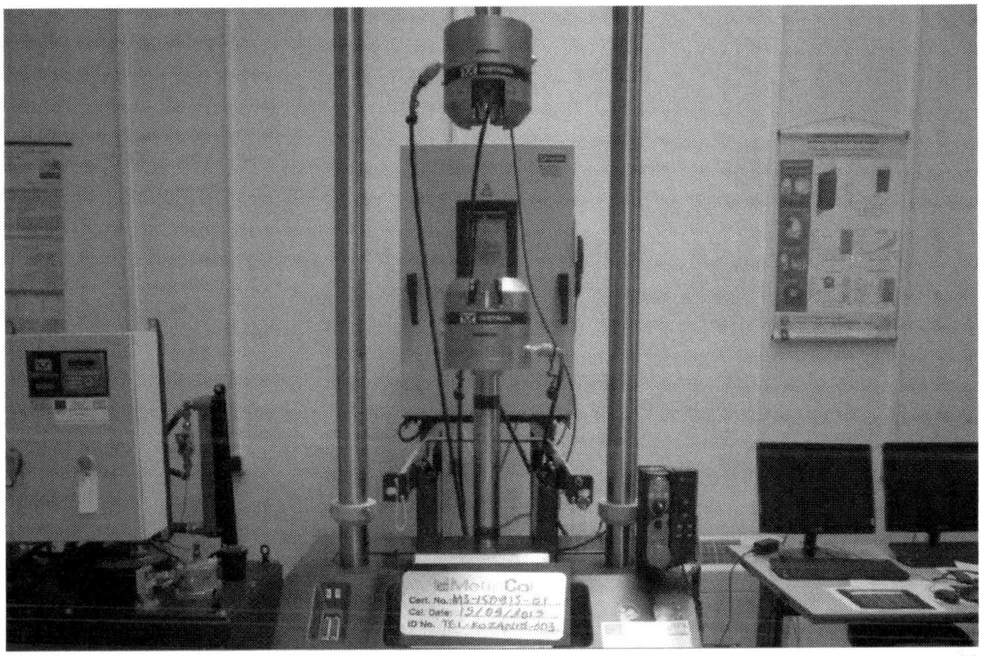

Figure. 2. 100kN capacity Instron fatigue testing machine

The maximum and minimum loads were calculated so as to agree with the EN specifications of maximum stresses of σ_{max}=300 MPa and minimum stresses of σ_{min}=150 MPa. The load cycle frequency was set at 25 Hz.

Hardness and tensile tests were also carried out so as to determine whether the material was in agreement with the EN standards concerning concrete steel bars. The chemical composition of the material was determined using optical emission spectroscopy and the steel microstructures were studied using optical and scanning electron microscopy (SEM) to investigate the presence of inclusions that could lead to premature failure during testing. The fracture surfaces of the mechanical tests were further studied using scanning electron microscopy to obtain a better picture of the failure process.

3. Results and Discussion

The chemical composition of the steel bars tested is presented in Table 1 and the mechanical test results are given in Table 2. Both are in agreement with the EN specifications so any failure during testing could not be attributed to poor quality of the material.

Table 1. Chemical composition of steel bars

Bar diameter (mm)	%C	%Mn	%Si	%S	%P
8	0.22	0.57	0.16	0.045	0.024
12	0.21	0.56	0.15	0.043	0.023
20	0.21	0.57	0.15	0.043	0.022

Table 2. Mechanical tests results

Bar diameter (mm)	Yield Strength (MPa)	Tensile Strength (MPa)	% Elongation	VPN
8	531	641	29	213
12	534	634	30	211
20	497	625	28	208

The fatigue tests of bars in the as-delivered state, that is without any preparation of the specimen grip parts, varied depending on the bar diameter Tables 3-5. The testing parameters were calculated according to EN 1421-3 and the tests were carried out at a load cycle frequency of 25 Hz for a maximum of $2*10^6$ cycles.

Table 3. Fatigue Test of 8mm bars
(Fmax =15.072 N, Fmin=7.536 N, Fmean=11.304 N, Famplitude=3.768 N)

Specimen	Nominal diameter (mm)	Length (mm) 140,14d	Failure (yes/no)
Φ 8-1	8	140	no
Φ 8-2	8	140	no
Φ 8-3	8	140	no
Φ 8-4	8	140	no
Φ 8-5	8	140	no

Table 4. Fatigue Test of 12mm bars
(Fmax =33.912N, Fmin=16.956N, Fmean=25.434N, Famplitude=8.478N)

Specimen	Nominal diameter (mm)	Length (mm) 140,14d	Failure (yes/no)
Φ 12-1	12	168	yes 934331 cycles
Φ 12-2	12	168	no
Φ 12-3	12	168	no
Φ 12-4	12	168	no
Φ 12-5	12	168	no
Φ 12-6	12	168	yes 934331 cycles
Φ 12-7	12	168	yes 964063 cycles
Φ 12-8	12	168	no

Table 5. Fatigue Test of 20mm bars
(Fmax =94.200, Fmin=47.100N, Fmean=70.650, Famplitude=23.550N)

Specimen	Nominal diameter (mm)	Length (mm) 140,14d	Failure (yes/no)
Φ 20-1	20	280	yes 1088570 cycles
Φ 20-2	20	280	yes 1854310 cycles
Φ 20-3	20	280	yes 1115141 cycles
Φ 20-4	20	280	yes 675589 cycles
Φ 20-5	20	280	yes759420 cycles

As can be seen all five bars of eight millimetre diameter withstood the required number of 2·106 cycles when tested at a frequency of 25Hz. However, when the 12 mm bars were tested a number of specimens failed usually within or very near the grip jaws (Figure 3). Moreover, all (10/10) specimens of 20 mm diameter failed and this caused suspicion about the quality of the bar material.

Figure 3. Premature failure within or near grip jaws

To exclude any such possibility the steel microstructures were studied using optical microscopy (Figure 4). As can be seen all diameter bars exhibited a ferrite - pearlite microstructure and no martensite or other phases were present that could account for the premature failure observed.

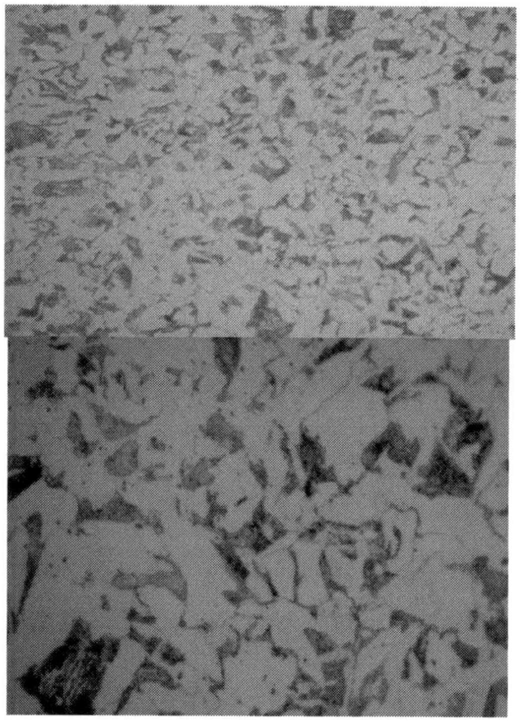

Figure 4. Steel microstructure, top x400, bottom x1000

As expected, the fracture surface observed using scanning microscopy revealed the presence of manganese sulphide inclusions (Figure 5) but this cannot by itself explain why premature fracture occurred only in the 12 and 20 mm bars. Studying the literature showed that fatigue strength of concrete steel is diameter dependent [1, 7-9]. Steel bars of smaller diameter exhibit higher fatigue strength than steel bars of the same quality but bigger diameter. In German standards for concrete steel a gradual reduction in the requirements for fatigue strength is reported and in BS 4449 [9] the specifications differ depending on group diameter as shown in Table 6.

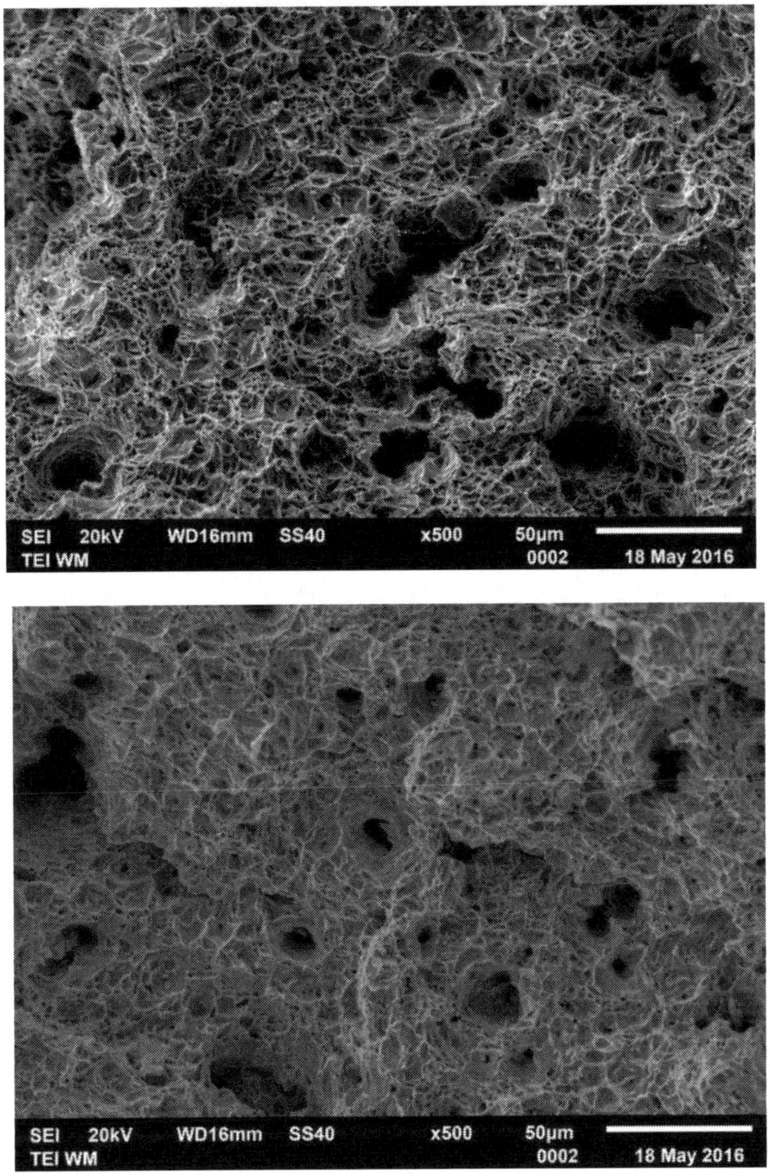

Figure 5. Fracture surfaces observed using SEM showing manganese sulphide inclusions

Table 6. Fatigue test specifications according to BS 4449:2005

Bar diameter (mm)	$2\sigma_A$ (MPa)	σ_{max} (MPa)	$2\sigma_m$ (MPa)
≤16	200	250	150
>16, ≤20	185	230	137,5
>20, ≤25	170	212,5	127,5
>25, ≤32	160	200	120
32	150	187,5	112,5

In the present study the results appear to agree with the above and originally the differences in fatigue strength of the 12 and 20 mm bars in relation to that of the 8 mm ones was attributed to their larger diameters.

However, a closer look at the bibliography [2-4, 10-12] reveals that premature failure inside or close to the fatigue machine grips is a common problem with fatigue testing of reinforcing bars. This is due to the difficulties related to the fact that fatigue cracks are usually generated in the area where test specimens are fixed in the jaws of a testing machine, i.e. where significant notches are present. Thus, the results of such tests cannot be regarded as reliable and should or may be rejected according to the EN 1421-3.

Tilly [2] reports information about numerous attempts to properly prepare the grip parts of specimens so as to eliminate the problem of crack initiation in the grip part of the specimens; such attempts were always connected with using an additional layer between a specimen and a testing machine jaw. He [2] mentions the use of leather belts wrapped around the specimen grip part, low melting point metallic fillings, resins and aluminium pads. However, Tilly [2] does not provide any information concerning the effectiveness of the measures mentioned.

Krzysztof Krasnowski [3] mention a successful method that was used to eliminate the problem but do not give any information of the steps taken. In addition, in the NordTest method for fatigue testing of concrete reinforcement steel bars [4] four different methods for preparing the grip parts of specimens are described in detail. However, they all seem rather complicated and/or timeconsuming.

In order to locally change stresses in the specimen grip area and to make the notches caused by the jaws too small to lead to the premature failure the following technique was used with surprising success in the present study. For the 20 mm bars, firstly, a 50 mm length of the specimen ends was turned to 19.5 mm diameter, until the relief on the surface was almost completely removed. There was also a 4 mm long curved part machined past the grip part (Figure 6).

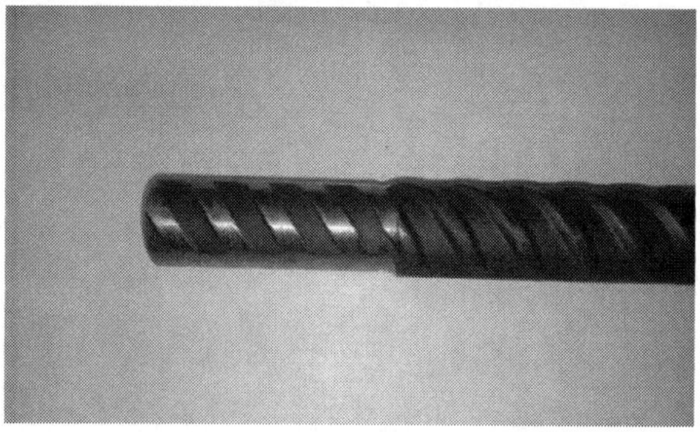

Figure 6. Machined specimen ends.

This resulted in a larger surface area being gripped by the machine jaws and eliminated the applied stress concentrating on only four or five points that may cause failure. Then, approximately 0.5 mm of aluminium, 5 cm wide, tape was wrapped around the specimen ends to reduce the depth of the notches caused on the specimen surface by the machine jaws part (Figure 7).

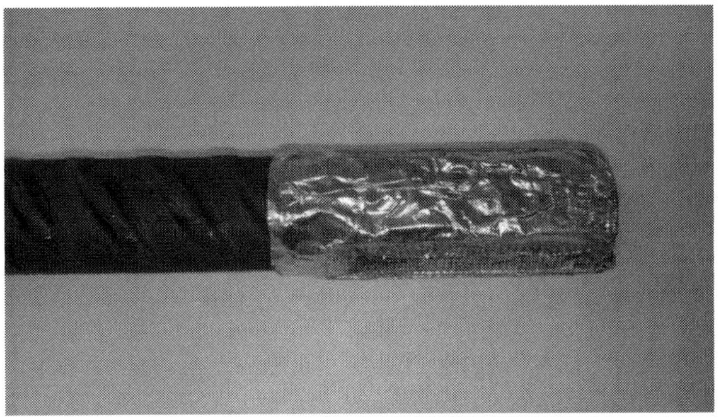

Figure 7. Specimen ends wrapped with aluminium tape.

The fatigue test results of such specimens (Figure 8) showed that none of the required five, as per EN 1421-3, failed. Therefore, the problem with specimen failure was not due to material quality but specimen surface morphology that results in stress concentration and notch creation caused by the machine jaws.

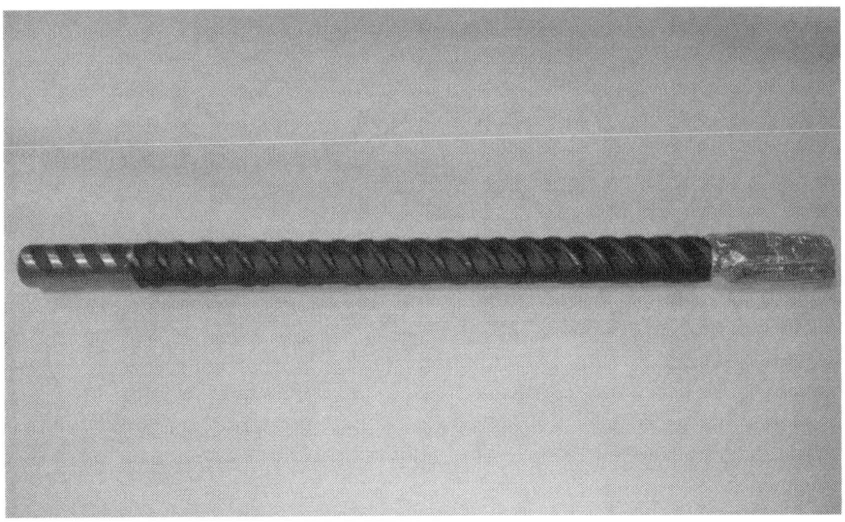

Figure 8. Specimen after preparation

The etiology of the bars' premature failure can be investigated through Finite Element Modelling techniques which is a widely accepted approach for engineering components and materials [13] if the model is properly verified/validated [14]. The use of non-linear properties [15] allows in this case not only the determination of critical stress concentrations but also the propagation of the initiated failure.

4. Conclusions

High cycle fatigue tests conducted on reinforced-concrete steel ribbed rods in accordance with the specifications in EN 10080 are difficult to perform due to premature failure appearing within or near grip jaws. Specimen pretreatment [16] has been effective in prolonging the life cycle of several engineering components. In the same way, the presented preparation method described above will aid researchers on fatigue testing to obtain accurate test results and save on material and time.

References

[1] Sismanis P. and Mastorakis A., 5-7 Nov 2008, *Fatigue strength of concrete-reinforcing steel bars*, Panhellenic Conference of Anti-Sesmic Mechanics, **article 1785**

[2] Tilly G. P., 1979, *Fatigue on steel reinforcement bars in concrete: a review.* Fatigue of Engineering Materials and Structures, **no. 2**, pp. 251-68

[3] Krasnowski K., 2015, *Concrete-reinforcing Steel Bars – Applications and Fatigue Tests*, BIULETYN INSTYTUTU SPAWALNICTWA, **No. 1**, pp.13-20

[4] NordTest method, 1989, *STEELS FOR CONCRETE REINFORCEMENT:AXIAL LOAD FATIGUE TESTING OF BARS*, **NT mech 014**

[5] EN 10080, 2005, *Steel for the reinforcement of concrete - Weldable reinforcing steel*

[6] EN 1421-3, 2007, *Steel for the reinforcement of concrete – Weldable reinforcing steel* – **Part 3**: Technical class B500C

[7] DIN 1045-1:2001-07, *Concrete, reinforced and pre-stressed concrete structures* - **Part 1**: Design and construction

[8] DIN 488-1: 2009-08 *Betonstahl* - **Teil 1**: Stahlsorten, Eigenschaften, Kennzeichnung

[9] BS 4449:2005+A2:2009, *Steel for the reinforcement of concrete. Weldable reinforcing steel. Bar, coil and de-coiled product*

[10] Bannister J. L. 1975, *Fatigue resistance of reinforcement for concrete.* Proc. Conf. Underwater Construction Technology, Department of Civil and Structural Engineering Report, University College, Cardiff

[11] Burton K. T. and Hognestad E., 1967, *Fatigue tests on reinforcing bars - tack welding of stirrups.* **ACIJ. 64**, pp. 244-52

[12] Walker E. F., Austen I. M., Harrison T. C. and Morley J., 1975, *Fatigue and corrosion fatigue of reinforcement bars,* Proc. Conf. Underwater Construction Technology, Department of Civil and Structural Engineering Report, University College, Cardiff

[13] Bouzakis K. D., Efstathiou K., Paradisiadis G., Tsouknidas, A: Experimental and FEM-supported investigation of wet ceramic clay extrusion for the determination of stress distributions on the applied tools' surfaces. Journal of the European Ceramic Society, 2008, vol. 28, issue 11, pp. 2117-2127.

[14] Tsouknidas A., Savvakis S., Asaniotis Y., Anagnostidis K., Lontos A., Michailidis N.: The effect of kyphoplasty parameters on the dynamic load transfer within the lumbar spine considering the response of a bio-realistic spine segment. Clinical Biomechanics, 2013, vol. 28, issue 9-10, pp. 949-55.

[15] Bouzakis K.D., Maliaris G., Tsouknidas A.: A FEM supported semi-solid high pressure die casting process optimization based on rheological properties by isothermal compression tests at thixo temperatures extracted. Computational Materials Science, 2012, vol. 59, pp. 133-139.

[16] Bouzakis K. D., Bouzakis E., Skordaris G., Makrimallakis S., Tsouknidas A., Katirtzoglou G., Gerardis S.: Effect of PVD films wet micro-blasting by various Al2O3 grain sizes on the wear behaviour of coated tools. Surface and Coatings Technology, 2011, vol. 205, issue suppl. 2, pp. S128-S132.

IOP Conf. Series: Materials Science and Engineering **161** (2016) 012068 doi:10.1088/1757-899X/161/1/012068

Microwave sintering of ceramic materials

V G Karayannis

Department of Environmental Engineering, Technological Education Institute of Western Macedonia, 50100, Kozani, Greece

E-mail: vkarayan@teiwm.gr

Abstract. In the present study, the potential of microwave irradiation as an innovative energy-efficient alternative to conventional heating technologies in ceramic manufacturing is reviewed, addressing the advantages/disadvantages, while also commenting on future applications of possible commercial interest. Ceramic materials have been extensively studied and used due to several advantages they exhibit. Sintering ceramics using microwave radiation, a novel technology widely employed in various fields, can be an efficient, economic and environmentally-friendlier approach, to improve the consolidation efficiency and reduce the processing cycle-time, in order to attain substantial energy and cost savings. Microwave sintering provides efficient internal heating, as energy is supplied directly and penetrates the material. Since energy transfer occurs at a molecular level, heat is generated throughout the material, thus avoiding significant temperature gradients between the surface and the interior, which are frequently encountered at high heating rates upon conventional sintering. Thus, rapid, volumetric and uniform heating of various raw materials and secondary resources for ceramic production is possible, with limited grain coarsening, leading to accelerated densification, and uniform and fine-grained microstructures, with enhanced mechanical performance. This is particularly important for manufacturing large-size ceramic products of quality, and also for specialty ceramic materials such as bioceramics and electroceramics. Critical parameters for the process optimization, including the electromagnetic field distribution, microwave-material interaction, heat transfer mechanisms and material transformations, should be taken into consideration.

1. Introduction

Due to their advantages, ceramics have been extensively studied, marketed and used so far, starting from various raw materials and possessing specific characteristics for being serviceable in multiple environments. In ceramic industry as well as in powder metallurgy, sintering is one of the most important technological processes and has been developed for over 70 years. It takes places when packed particles are heated to a temperature where there is sufficient atomic motion to grow bonds between the particles. Driving force for the process is the particle surface tension leading to minimizing the surface area. Sintering mainly depends on the type of material under consideration, its melting temperature, particle size (volume, surface) and several other processing parameters. Fundamental mechanisms of the process include grain boundary diffusion, viscous flow, evaporation and recondensation. Sintering conditions determine the development of a microstructure to provide physico-mechanical and chemical properties required for the ceramic body. Solid-state sintering of clayey raw materials through conventional thermal treatment has long been considered for the development of various ceramic materials and is traditionally a preferred manufacturing technique for industrial ceramics [1-9].

On the other side, microwave radiation is emerging as an innovative environmentally benign technology, widely employed in various fields, such as in green chemistry, biosciences, biomass processing, waste, sewage sludge and wastewater treatment (especially for oxidation/degradation of pollutants of poor biodegradability), synthesis of nanoparticles, processing of nanostructured composites (nanoconsolidation) and nanoporous materials, as well as in metallurgical processes including drying of agglomerations and reduction of ore concentrates. The goals are the reduction of processing time and equipment size to significant extent and also the increase of selectivity and product yield and purity in case of synthesis or purification reactions [10-20].

Particularly, microwave-assisted sintering can be an efficient, economic and valuable approach for eco-friendlier processing of various powdery oxide and non-oxide materials [21-24].

In the present study, an overview of the potential of microwave technology as viable and energy-efficient alternative to conventional heating procedures in ceramic manufacturing is attempted, by reviewing advances published in the scientific research literature in the field, and commenting on the advantages/disadvantages reported and also on future applications of possible commercial interest. It should be noted here that fast heating is required for sintering advanced ceramics, in order to avoid excessive grain growth that is detrimental to mechanical performance. The role of microwave processing conditions on the development of ceramic microstructures and properties is discussed and analyzed.

2. Microwave irradiation

Microwave irradiation can produce efficient internal heating, as energy is supplied directly and penetrates the material through molecular interaction with the electromagnetic field. Since energy transfer occurs at a molecular level, its interaction with a dielectric material results in translational motions of free or bound charges and rotation of the dipoles. The resistance of the induced motions causes heat losses, and thereby heat is generated throughout the material resulting in volumetric heating, thus reducing the processing time.

The power absorbed per unit volume, P ($W\ m^{-3}$), is expressed as follows (see Equation 1) [15]:

$$P = \sigma\,|E|^2 = 2\,\pi\,f\,\varepsilon_0\,\varepsilon'_r\,\tan\delta\,|E|^2 \qquad (1)$$

where E ($V\ m^{-1}$) is the magnitude of the internal field,
σ the total effective conductivity ($S\ m^{-1}$),
f the frequency (usually 2.45 GHz: industrial scientific and medical frequency)
ε_0 the permittivity of free space ($\varepsilon_0 = 8.86\times10^{-12}\ F\ m^{-1}$),
ε'_r the relative dielectric constant and
$\tan\delta$ is the loss tangent.

The microwave field distribution within a ceramic body, and the effect on the material, is determined by the dielectric parameters (ε'_r and $\tan\delta$). Specifically, the dielectric constant measures the ability of a material to store microwave energy. The loss tangent provides an indication of how well a material can be penetrated by electric field and how it dissipates energy into heat [25].

3. Microwave sintering of ceramic materials

3.1. Bulk ceramics

Microwave processes can improve the densification efficiency and considerably reduce the processing cycle-time for rapid and uniform heating, especially of thick ceramic materials, thus resulting in substantial energy and cost savings. Hence, there is sufficient motivation to promote the use of microwaves in the sintering of various ceramics, and also glass-ceramics, nano-ceramics and bioceramics, and to investigate the effect of microwave sintering conditions on the microstructure, phase composition and properties of the materials produced compared to conventionally sintered ones.

Significant developments have already been made in the use of microwaves in the sintering of ceramics. Particularly for the development of advanced ceramics, very fast sintering is required to avoid grain growth. By microwave heating, densification of ceramics can be attained in a very short time. Rapid and enhanced consolidation achieved upon microwave processing of ceramics should mainly be attributed to a microwave effect, assumed to be a non-thermal phenomenon and possibly explained by the concentration of electrical energy into the closed pores of green (non-sintered) ceramics [26]. However, other study on a microscopic scale of microwave sintering of compacted ceramic powders shows that local electric fields disproportionately intense close to grain boundaries can be generated and rough surfaces due to strong focusing can be produced, which can lead to a highly non-uniform energy deposition [27]. Therefore, in order to increase the understanding and control the sintering progress, by determining the optimal parameters of microwave irradiation for obtaining an expected microstructure, an attractive method was proposed for in-situ shrinkage measurement of a microwave irradiated sample during sintering in a specific microwave cavity appropriately designed. The sintering kinetics are found to be extraordinarily fast for achieving satisfying density, when microwave sintering of CuO was chosen as a test material [27]. Furthermore, in order to design suitable cavity and cells for both hybrid and direct microwave sintering of fine-grained pure α-Al_2O_3 powder and to allow reliable comparison with conventional sintering, finite element simulation was employed. A significant enhancement of densification under microwave irradiation during the first and intermediate stages is evidenced upon sintering the alpha alumina powder compacts along an identical thermal cycle, without use of any doping element to initiate heating [29]. On the other hand, no difference between the two heating modes during the late stages of sintering is indicated, when comparing densification and grain growth of microwave sintered Al_2O_3 ceramic materials to conventionally sintered ones, by analyzing microwave power absorption of alumina ceramics and heating profiles for both sintering modes [30]. In nanostructured Al_2O_3 ceramics in particular, with regard to the annealing of nanoporous alumina-based membranes under microwave heating, the influence of microwave processing on mass transport phenomena and phase transformations is highlighted. Actually, faster mass transport, significantly depending on the microwave field intensity, is observed upon microwave sintering of alumina powder compacts. In addition, phase transformations can be characterized quantitatively and also a preferred orientation of pores in ceramics can be predicted [31].

The sintering process of glass powder (CaO-ZrO_2-SiO_2), microwave heated to produce a glass-ceramic material, is also accelerated in comparison to conventional sintering, although the densification and microstructural evolution studied during crystallization of the glass-ceramics obtained is not affected differently when compared to conventionally-treated ones, and exhibits a negligible porosity as a result of micropores [32].

It should further be noted that, for ceramics development, several secondary resources have also been microwave processed. In particular, after rapid microwave sintering of compacts composed of highly calcareous (Class C) lignite combustion fly ash and bottom ash for 30 min (at 1000°C), dense ceramic microstructures, mainly composed of gehlenite, quartz and kyanite, are achieved [33]. Also, microwave sintered (800-1000°C) siliceous (Class F) coal fly ash samples appear to be denser, and thus stronger, than conventionally sintered ones at the same temperature and time [34]. In order to reduce the volume of municipal solid waste incineration fly ash, microwave irradiation has better sintering efficiency than traditional sintering with electro-furnace (800-1100°C) for the stabilization and transformation of washed fly ash with calcium carbonate into ceramic blocks [35,36]. Moreover, rice husk ash, SiO_2-rich material rapidly sintered (800-1200°C) by microwave energy using a multi-mode system, is shown to contain SiO_2-cristolbalite and α-SiO_2 as major ceramic phases, with amount of crystalline silica phase increased after microwave sintering, leading to enhanced mechanical performance [37].

3.2. Bioceramics

Submicrometer tricalcium phosphate (β-TCP) powder produced in-situ by wet chemical precipitation, and sintered to obtain dense β-TCP ceramics by microwave technology for 15 min at 1100°C, shows a better densified microstructure with average grain size of 3μm, and higher hardness, than samples conventionally sintered for 2h at the same temperature, leading to the conclusion of a superior mechanical performance for the microwave sintered β-TCP powder [38]. Also, porous biphasic HAP/TCP bioceramics were obtained by microwave sintering of two microspherical agglomerated HAP powders. The porous microstructures of the obtained bioceramics consist in biphasic mixtures of HAP and a-TCP crystalline phases developed during the sintering process, while spherical intra-agglomerate pores and shapeless inter-agglomerate pores are detected. The microstructure obtained appears to have only a minor effect on the indentation fracture toughness [39]. The potential of a microwave heating was further demonstrated in the sintering of bioceramics from stabilized zirconia (3Y-TZP), another excellent susceptor of microwave energy, due to its large concentration of point defects. In fact, the mechanical properties including hardness and fracture toughness of high density and ultrafine microstructures with monomodal grain size distribution resulting from short time processing in a domestic microwave oven, compare very well with 3Y-TZP ceramics conventionally sintered for a long duration [40].

3.3. Electroceramics

Microwave sintering has also widely been considered for processing electroceramics, a specific category of ceramic materials primary used for their electrical properties, including ferroelectric materials, ferrites, solid electrolytes and piezoelectrics:

In particular, microwave sintering of strontium bismuth titanate powder prepared by solid state route leads to higher densification (97% of the theoretical density), fine microstructure, and good mechanical and ferroelectric properties in much shorter duration of time, in comparison to that heated in conventional furnaces [41]. Calcium-doped barium titanate ($Ba(1\text{-}x)Ca_xTiO_3$) ferroelectric ceramics synthesized by microwave processing technique, optimized at 1100°C for 1h to form a single perovskite phase, show dense and homogeneous packing of sub-micrometer size grains [42]. Also, preparation of SiO_2 added $BaTiO_3$ ceramics was successfully attained by microwave sintering at low temperatures. A homogeneous microstructure with small grains appears for 0.5wt% SiO_2 addition, while an enhanced densification and appearance of $Ba_2TiSi_2O_8$ second phase with columnar grains for further increase of SiO_2 addition to 1and 2wt% are stated, leading to the improvement of the electrical break-down strengths [43].

Lead zirconate titanate are ceramic perovskite materials showing marked piezoelectric effect, being used in a number of practical applications in the area of electroceramics. Higher density and more uniform grain distribution are found for ceramic pellets from lanthanum-doped PZT fine powders (PLZT) processed by high energy mechanical ball milling, and then microwave sintered at 1150◦C for much shorter sintering time than for conventionally sintered ones. Comparable dielectric and piezoelectric properties are obtained for microwave sintered PLZT ceramics [44].

For lanthanum gallate electrolytes, which generally present a superior ionic conductivity at intermediate temperatures than YSZ ones, especially when lanthanum is substituted by alkaline rare earths such as strontium and magnesium (LSGM), activated microwave sintering at lower temperatures makes possible to overcome problems related to grain coarsening and secondary phases formation occurring at elevated temperatures upon conventional heating. Hence, dense stable electrolyte layer for applications in intermediate temperature-solid oxide fuel cells are developed, due to heat generation in situ, leading to faster sintering kinetics [45].

Furthermore, microwave sintering can be a promising technology even for the development of materials such as MnZn-ferrites to be used in high magnetic permeability applications, as it results to higher magnetic permeability values, higher densities and coarser microstructures, in comparison to conventional firing [46-48].

4. Concluding remarks

- Microwave sintering emerges as a new field of processing and synthesis of ceramic materials, and appears to be a powerful method for sintering advanced ceramics in a short time.
- Significant progress in application and commercialization of microwave technology, particularly in specialty ceramics processing, has already been made.
- Comparison with conventional sintering reveals a number of benefits of microwave processing, in terms of microstructural design and physico-mechanical properties.
- Rapid heating with accelerated densification is enabled, while excessive grain coarsening is limited, leading to higher densities and uniform and fine-grained microstructures, with enhanced mechanical performance.
- By microwave sintering of ceramic materials, significant temperature gradients between the surface and the interior that are frequently encountered upon conventional sintering of a ceramic body at high heating rates can be avoided, which is of importance especially for manufacturing large-size ceramic products of quality.
- The distribution of the electromagnetic field, the interaction between microwaves and the material, heat transfer mechanisms and material transformations appear to be critical for the process optimization.

References

[1] Kingery W D, Bowen H K and Uhlmann D R 1976 *Introduction to Ceramics* 2nd ed. (New York: Wiley)
[2] Gotoh K, Masuda H and Higashitani K 1997 *Powder Technology Handbook* 2nd ed., rev. and expanded (New York: Marcel Dekker, Inc.)
[3] Matrenin S V, Ovechkin B B, Zenin B S and Tayukin R V 2015 *IOP Conference Series: Materials Science and Engineering* **93** (1) 012040
[4] Domopoulou A, Spiliotis X, Koukouzas N, Gudulas K, Papapolymerou G and Karayannis V 2014 Proc. IOC – 46th International October Conference, Bor Lake, Serbia
[5] Kirchhof M J, Forster H, Schmid H-J and Peukert W 2012 *J. Aerosol Sci.* **45** 26–39
[6] German R M 2010 *Crit. Rev. Solid State* **35** 263–305
[7] Olevsky E A, Tikare V and Garino T 2006 *J. Am. Ceram. Soc.* **89** (6) 1914–1922
[8] Ch'ng, H N and Pan J 2007 *Acta Mater.* **55** 813–824
[9] Olevsky E A 1998 *Mat. Sci. Eng. R.* **23** 41–100
[10] Motuzas J, Drobek M, Diniz da Costa J C and Julbe A 2014 *Chem. Eng. J.* **240** 155–160
[11] Bassyouni F A, Abu-Bakr S M and Rehim M A 2012 *Res. Chem. Intermed.* **38** 283–322
[12] Remya N and Lin J-G 2011*Chemical Eng. J.* **166** 797–813
[13] Budarin V L, Clark J H, Lanigan B A, Shuttleworth P and Macquarrie D J 2010 *Bioresource Technol.* **101** 3776–3779
[14] Leonelli C and Mason T J 2010 *Chem. Eng. Process.* **49** (9) 885–900.
[15] Das S, Mukhopadhyay A K, Datta S and Basu D 2009 *Bull. Mater. Sci.* **32** (1) 1–13
[16] Yuen F-K and Hameed B H 2009 *Adv. Colloid Interfac.* **149** 19–27
[17] Bai C, Ren W, Qiu G, Hu M, Chen D and Wen L 2007 *ISIJ Int.* **47** (4) 528–532
[18] Viswanathan V, Laha T, Balani K, Agarwal A and Seal S 2006 *Mat. Sci. Eng. R.* **54** 121–285
[19] Tompsett G A, Conner W C and Sigfrid Yngvesson K 2006 *Chem. Phys. Chem.* **7** 296 – 319
[20] Appleton T J, Colder R I, Kingman S W, Lowndes L S and Read A G 2005 *Appl. Energ.* **81** 85–113
[21] Zhao X, Yan L and Huang K 2011 in "Advances in Induction & Microwave Heating of Mineral & Organic Materials" 28–48
[22] Oghbaei M and Mirzaee O 2010 *J. Alloy. Compd.* **494** (1-2) 175–189
[23] Mascarenhas J, Marcelo T, Inverno A, Castanho J and Vieira T 2008 *Metal Powder Report* **63** (11) 12–15
[24] Riedel H and Svoboda J 2006 in "Simulation of Microwave Sintering with Advanced Sintering

Models. Advances in Microwave and Radio Frequency Processing", edited by M. Willert-Porada (Springer, New York) 210–216
[25] Al-Harahsheh M and Kingman S W 2004 *Hydrometallurgy* **73** 189–203
[26] Agrawal D K 1998 *Curr. Opin. Solid St. M.* **3** (5) 480–486
[27] Birnboim A, Calame J P and Carmela Y 1999 *J. Appl. Phys.* **85** (1) 478–482
[28] Marinel S and Savary E 2009 *J. Mater. Process. Tech.* **209** 4784–4788
[29] Croquesel J, Bouvard D, Chaix J-M, Carry C P and Saunier S 2015 *Mater. Des.* **88** 98–105
[30] Chatterjee A, Basak T and Ayappa K G 1998 *AIChE J.* **44** (10) 2302–2311
[31] Bykov Y V, Egorov S V, Eremeev A G, Kholoptsev V V, Plotnikov I V, Rybakov K I, Semenov V E and Sorokin A A 2010 *Powder Metall. Met. C+*, **49** (1-2) 31–41
[32] D'Arrigo M C, Siligardi C, Leonelli C, So J Y and Kim H S 2002 *J. Porous Mat.* **9** 299–305
[33] Karayannis V G, Moutsatsou A K and Katsika E L 2013 *J. Ceram. Process. Res.* **14** (1) 45–50
[34] Fang Y, Chen Y, Silsbee M R and Roy D M 1996 *Mater. Lett.* **27** 155–159
[35] Chen C-L, Lo S-L, Kuan W-H and Hsieh C-H 2012 *Sustain. Environ. Res.* **20** (6) 381–385
[36] Chou S-Y, Lo S-L, Hsieh C-H and Chen C-L 2009 *J. Hazard. Mater.* **163** 357–362.
[37] Makul N and Agrawal D K 2010 *Mater. Lett.* **64** 367–370
[38] Mirhadi B 2014 *Sci. Sinter.* **46** (2) 185–193
[39] Veljovic Dj., Palcevskis E, Dindune A, Putic S, Balac I, Petrovic R and Janackovic Dj. 2009 *J. Mater. Sci.* **45** 3175–3183
[40] Upadhyaya D D, Ghosh A, Dey G K, Prasad R and Suri A K 2001 *J. Mater. Sci.* **36** 4707–4710
[41] Angalakurthi R and Raju K C J 2011 *IOP Conference Series: Materials Science and Engineering* **18** (Symposium 6) 092028
[42] Sonia, Patel R K, Prakash C and Kumar P 2011 *J. Ceram. Process. Res.* **12** (6) 634–639
[43] Yan Y, Liu L, Ning C, Yang Y, Xia C, Zou Y, Liu S, Wang X, Liu K, Liu X and Liu G 2016 *Mater. Lett.* **165** 135–138
[44] Kumar A, Reddy Emani S, Bhanu Prasad V V, James Raju KC and James A R 2016 *J. Eur. Ceram. Soc.* **36** (10) 2505–2511
[45] Kesapragada S V, Bhaduri S B, Bhaduri S and Singh P 2003 J. Power Sources **124** (2) 499–504
[46] Tsay C-Y, Liu K-S, Lin I-N 2001 *J. Eur. Ceram. Soc.* **21** (10-11) 1937–1940
[47] Thota S, Kashyap S C, Gupta H C, Nath T K 2014 *Journal of Superconductivity and Novel Magnetism* **28** (1) 131–136
[48] Tsakaloudi V, Papazoglou E and Zaspalis V T 2004 *Mat. Sci. Eng. B* **106** 289–294

ENVIRONMENTAL DEFECTS AND ECONOMIC IMPACT ON GLOBAL MARKET OF RARE EARTH METALS

G Charalampides, K Vatalis, V Karayannis and A Baklavaridis

Department of Environmental Engineering, Technological Education Institute of Western Macedonia, 50100, Kozani, Greece

E-mail: chara@teiwm.gr

Abstract. Rare earth elements include the 14 lanthanides as well as lanthanium and often yttrium. Actually, most of them are not very rare and occur widely dispersed in a variety of rocks. Rare earth metals are vital to some of the world's faster growing industries: catalysts, Nd-magnets, ceramics, glass, metallurgy, battery alloys, electronics and phosphors. Worldwide, the main countries for distribution of rare earths deposits include China, USA, Russia, Brasil, India, Australia, Greenland and Malaysia. The mining and processing of rare earth metals usually result in significant environmental defects. Many deposits are associated with high concentrations of radioactive elements such as uranium and thorium, which requires separate treatment and disposal. The accumulation of rare earth elements in soils has occurred due to pollution caused by the exploitation of rare earth resources and the wide use of rare earths as fertilizers in agriculture. This accumulation has a toxic effect on the soil microfauna community. However, there are large differences in market prices due to the degree of purity determined by the specifications in the applications. The main focus of this article is to overview Rare Earth Metals' overall impact on global economy and their environmental defects on soils during processing techniques and as they are used as fertilizers.

1. Introduction

The term 'rare earths' is misleading, as it does not refer to their abundance in the earth's crust, but to the inconspicuous appearance of the minerals from which they were originally isolated. Almost all rare earth elements (REE) in the earth's crust are more abundant than gold, silver or platinum, while Cerium (Ce) is the most abundant of all REEs and is more common in the Earth's crust than copper or lead. Admittedly, rare earths are not present in equal amounts in REE ores. They are generally divided into the light rare earth elements (LREEs) and the heavy rare earth elements (HREEs), with the HREEs being much less abundant and thus much more valuable. The attributions to these groups are not distinct, but in general Lanthanum to Gadolinium are called LREEs while Terbium to Lutetium are called HREEs.

Despite their similar basic chemical properties, each REE displays unique characteristics for specific applications and usually cannot be substituted one for another. This has resulted in a 'criticality' classification which is based on a REE importance for specific applications (e.g. renewable energy), lack of comparable and reliable substitutes, and the monopolization of supply sources. According to the U.S. Department of Energy, the group of critical REEs comprises five elements: neodymium, europium, terbium, dysprosium, and yttrium [1].

REEs have gained visibility to the general public through the crisis of 2010 and the price spike of 2011 [2]. Numerous headlines about REEs have appeared in the media, and suddenly the world became alarmed that China was about to crush high-tech industries of Western economies due to imposition of export restrictions. China at that time was holding a share of 95 per cent of global production of REEs. There are fewer than twenty large companies that trade REEs, and these are distributed to a handful of countries: China, Canada, Australia, USA, Russia, India and Japan. China is the biggest supplier of REEs in the world, while United States, Japan and Germany are the biggest importers [3].

The REEs are found, usually several together, in a variety of accessory minerals, such as phosphates, carbonates, fluorides and silicates. They rarely form more continuous ore bodies.

REEs do not occur as native elemental metals in nature, only as part of the host mineral's chemistry. For this reason, the recovery of REMs must be accomplished through complex processing methods to chemically break down the minerals containing the REEs.

Despite the abundance of more than 200 known REE-bearing minerals, only three of them are considered to be the principal REE mineral ores most feasible for the extraction of REMs: bastnäsite [(Ce,La)(CO_3)F], xenotime (YPO_4), and monazite [(Ce,La)PO_4)] [4].

Rare earth elements (REEs) have previously been used as geochemical tracers in groundwater flow systems owing to the commonly reported similarities between their input-normalized REE patterns and those of the aquifer rocks/sediments through which they flow and react [5,6].

REEs have been used in fertilizers in the agriculture of China for about 20 years. They have been shown to be beneficial elements for plants. For example, they have improved the yield and quality for several kinds of crops. Although they are widely distributed geographically, they are chiefly mined, concentrated, and separated in China, which holds the largest reserves in the world (80%) and is a major producer of REEs for the world market. Since 1990, REE fertilizer has been widely used in more than 20 Chinese provinces [7].

This paper reviews the current literature on environmental impact of REMs after mining, processing and disposal, mainly due to high concentrations of radioactive elements such as uranium and thorium, as well as their use as fertilizers in agriculture. It also studies in brief their economic impact on global market.

2. Environmental defects deriving from processing and soil applications of REMs

REEs may possibly also be extracted from alternative sources, such as industrial wastes, and by recycling end-of-life REM-bearing products such as consumer products or lighting. Substitution especially of the less common REEs is another important option. An entirely different but equally relevant issue concerns the environmental impact of REE mining and beneficiation. Of particular concern is the common association of REE with thorium and uranium and thus the need to strictly control the dispersion of radioactive elements into the environment [8]. This pertains in particular to the treatment and disposal of tailings. Rather lax legislation in China has resulted in significant pollution, especially related to illegal mining [9].

2.1. Processing techniques

The REE processing is characterized by high levels of water consumption, energy inputs, and chemicals use [10]. The land allocation can be also significant for both mining and processing operations, as well as for the tailings dams, and long-term storages of the radioactive waste materials.

The method of physical beneficiation of rare earth bearing minerals depends on the mineralogy of the deposit. In most cases, the deposits are presented as hard rock, requiring the ore to be initially comminuted to liberate the valuable mineral grains [11].

Conventional physical separation methods such as gravity separation, magnetic separation, electrostatic separation and froth flotation are employed to concentrate rare earth bearing minerals [12]. In placer, mineral sands deposits, gravity separation (spirals) is typically used to remove the silicate gangue and multiple magnetic and electrostatic separation stages progressively concentrate the monazite from the other heavy minerals-ilmenite, rutile, zircon and leucoxene. Due to their nature, ion-adsorption clays require no physical beneficiation and can be processed by direct hydro-metallurgical techniques [13].

The concentration of pure REEs from mined rock is complex, involves many stages, and impacts the economic decisions of the industry. First, the rare earth containing minerals are recovered from the host rock via comminution and physical separation. The concentrated minerals are subsequently chemically leached into a solution in a process commonly referred to as cracking. The individual elements are selectively removed from the mixed REE solution via hydrometallurgical techniques such as solvent extraction and ion exchange. The precipitated products can either be sold as pure metal oxides or reduced to pure metal products depending on the required end purpose (Figure 1).

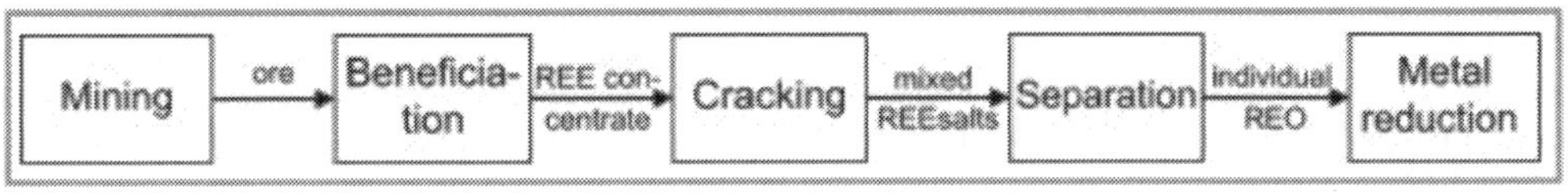

Figure 1. Schematic of REE production technology [11].

The development of recycling and reuse of REEs from waste materials or mine tailings is generally recognized as a more environmentally friendly activity than establishing a new REE mine [14]. However, recycling techniques for REEs have a number of environmental consequences including high energy use, consumption of large amounts of chemicals and the generation of waste chemicals and water. Hydrometallurgical methods to recycle REE magnets require large amounts of strong mineral acids and non-recyclable reagents such as H_2SO_4, NaOH and HF, and generate large quantities of waste water [14].

3.2 Environmental Degradation due to radioactivity of REMs

Most of the rare earth deposits have the presence of radioactive thorium and in some cases uranium. The concentration of radioactive elements, usually being relatively benign for human health in the ore body, rises significantly during beneficiation. This could be of serious concern for the waste by-product, emissions or tailings after the cracking stage in the processing of rare earths bearing minerals [15].

The concentration of thorium and uranium varies significantly for different ore bodies. Intermediate concentrates and final waste products contain radioactive elements depending on the processing route and potential co-extraction of Th/U. A higher concentration of thorium/uranium in the ore body makes the co-production of radioactive elements more feasible. However, the current market for thorium is very limited [16] and most of it has to end up either at the long-term storage or permanent disposal. The separation, treatment and disposal of radioactive materials in the latter case would result in significant additional costs, as well as high human health and environmental risks. The low radioactive ore bodies are highly preferable for the new REE projects, but even in this case there are environmental issues that have to be carefully addressed.

It is likely that the radioactivity aspect is overlooked in many new REE project proposals, both from the environmental and human health risk perspectives, and from the economic point of view (i.e. costs associated with treatment, disposal, and future land rehabilitation). Nevertheless, China has banned mining of pure monazite due to high-level radioactive elements and recently revised emission guidelines for the rare earths industry, setting standards that are in some cases more stringent than industrial nations [16]. Global awareness of these developments suggests that new REE projects in other jurisdictions may require revision of radioactive waste management before they can successfully go into production.

3.3. Soil defection by using REEs as fertilizers

Recently, the environmental pollution of rare earth elements (REE) has increased dramatically as a result of wide application of REEs in agricultural production as microelement fertilizers as well as in industry. Therefore, the biogeochemical cycling of REEs and their ecological and environmental effects are greatly concerned by many researchers. Rock weathering is an important REE source for an ecological system and a key process of global biogeochemical cycling of REE [17]. REE enrichment characteristics and mechanism analysis reveal that the released REE from weathering of carbonate soil pollution, and rocks (mainly limestone), sillicalite and shale are mainly stored in profiles.

In addition to their use in industry, REEs are more widely applied to cropland as microelement fertilizers due to their abilities to increase yields and improve qualities of crops. As a result, more and more REEs are moving into the ecosystems. They accumulate in soils, bioaccumulate in crops, and enter the food chain, causing a problem of REE environmental pollution special to China. In recent years, REE environmental effects have become of great concern [18]. Some work should be done to and a method should be developed that can reliably estimate bioavailability of REEs to plants and thereby evaluate the potential health risk of REEs in soils and predict their impact on the ecosystem.

Mixtures of REEs in fertilisers are nowadays widely used in Chinese agriculture to improve crop nutrition, and have been in common use for about 20 years [19]. In most cases, however, La, several lanthanides, and probably Y are contained in the mixtures, making it difficult to conclude whether any single element is of particular importance or whether the effects are less element-specific, which seems most likely. The capacity of rare earth uptake by the roots might be limited, mainly due to the self-protection mechanisms of the plants against adverse effects of rare earths [20]. The results of several studies [20, 21] showed that the accumulation of REEs in soil was significant in the study area and its concentration was strongly correlated with the distance from the pollution source.

Gillet and Ponge [22] have indicated that soil fauna is very sensitive to the changes in the soil environment. The soil fauna community responds to soil degradation and soil heavy metals pollution, such as Cd, Cu, Pb and Hg, in many ways. Thus, soil fauna can be an indicator of soil degradation and is often used to evaluate soil quality.

Heavy metal pollution of soil can lead the soil fauna community to change, but the response of different soil fauna to contamination is different. Almost all the deposits of REE contain the radioactive element thorium, and therefore environmental cost should be related to market competitive force.

3. Economic Impact On Global Market

The balance of demand and supply in the world market of REMEs was always rather unstable. After a rapid increase in demand in the 1980s by developed countries, it somewhat decreased in 1991–1993. In that moment, the REMs market was affected, first, by a tremendous increase in their production in China and, second, by the presence of large reserves in countries of the former Soviet Union [23].

The most significant increase of prices took place during the years 2009-2011 as it is shown by the Table 1. More recently (2012), according to the Metal-Pages quotations [24], there was a sudden and substantial fall in prices. These changes of trend are likely due to different reasons: the actual, persistent heavy economic crisis of the industrialized countries; the contraction of hi-tech consumption; the placing on the market of large stocks of REOs by some Chinese private operators fearful of possible inspections and confiscations by the Chinese Government [25].

REEs	2009	2010	2011	2012	Price fall from 2011 to 2012 (in %)
Ce	4,5	61	158	42,5	73
La	6,25	60	151,5	36	76
Pr	14	86,5	248,5	175	30
Nd	14	87	318	154	52
Dy	100	295	2510	1500	40
Eu	450	630	5870	4010	32
Tb	350	605	4410	3400	23

Table 1: Prices of REEs (in US$/kg) for the four-years period 2009 – 2012, showing a significant increase between 2009-2011, and thereafter a sudden and substantial fall for 2012, with the corresponding percentage lost (data based on Metalpages Inc., [24]).

Germany has already made diplomatic efforts to secure a diversified supply of REEs, including with other resource rich countries such as Australia [26, 27]. Countries such as China and India "have now given a strategic orientation to their raw materials policies and have taken measures to meet their needs for raw materials", and that "in the medium term, this can impact on German and European companies' access to sources of raw materials" [28].

In Greece significant reserves of rare earth elements are found in alluvial deposits in the coastal and groundwater environment of Strymona gulf between Strymona river and Kavala region, Northern Greece. Specific ore deposits research estimates ore reserves of 485 million t with an average content of 1.7% rare earths. Detailed and systematic investigations also show that bauxites and lateritic bauxites in

Central Greece contain representative concentrations, ranging overall from 3,275-6,378 g/t REE. It must be stressed that there is still an emerging interest for the red mud from the aluminum metallurgy [29]. Generally, in Greece they occur in a number of environments and are found in igneous (primary types), sedimentary, or metamorphic rocks (secondary types) of different ages. In secondary rocks the REE have been further concentrated from a primary enrichment through sedimentary processes or weathering. Secondary types are economically the most important types in Greece [30].

4. Conclusions

Due to high demand and limited availability of REEs, Europe is unable to meet today its industrial needs for the manufacturing sector. Therefore the EU has included them in the group of 14 critical minerals. Namely, China currently controls completely the mining activity (almost 95% of world production) and thereafter the production, the enrichment technologies and metallurgy and end-metal products of rare earths, resulting for both Europe and the U.S.A. in full industrial dependency, reaching almost to 100 %. This is happening at a time when the demand and needs tend globally constantly growing at an annual rate of 8-11%. However, Canada and Australia, for example are currently running significant mining projects to enter production in 2019, while in Sweden there are plans to begin production in 2018. Also, Greenland holds some interesting exploration projects which are in early exploration stage. Almost 50,000 tons of rare earths are needed per year out of China at present time. With a rate of 15% demand increase, then in 2018, the total demand out of China should be at least 100,000 tons on the basis of the world economy continuous increase.

There are large differences in market prices of rare earth oxides. Also, the prices depend on the degree of purity determined by the specifications in the applications. The price ranges from 11 \$/kg La_2O_3 to 1,600 \$/kg Eu_2O_3 [31]. It has to be underlined however, that China controls 95% of world production.

The distribution of metals in the aerial parts of plants is the result of both the fixation and the transfer mechanisms. The complexation of organic ligands in the xylem vessels plays an important role, and elements with strong complexation capacities (strong motilities) are preferentially transported to the top. The REEs concentrations, as well as total nitrogen, total potassium, and pH, had some effect on the soil macrofauna community. A small amount of REEs can promote soil macrofauna diversity, but a large amount of REEs can reduce the soil macrofauna diversity.

In order to fully understand the effects of REEs in agricultural application, environment and human health, it is suggested that future research in several areas is needed, which would shed new light on the effects of REEs on agriculture, environment and human health.

Finally, it has to be emphasized that environmental concerns must be paid attention; almost all the deposits contain the radioactive element thorium. Environmental cost is related to the market competitive force. It is very important that current rare earth market with profitable prices is greatly stimulated by the strict quota control of China and curb of smuggling to limit the supply. Therefore, all producers should pay great attention on the adjustment of future quota amount [32].

References

[1] US DoE 2011 *Critical Materials Strategy* U.S. Department of Energy
[2] Massari S and Ruberti M 2013 *Resour.* Policy 38, 36–43
[3] Roskill 2015 *Rare earths: Market outlook to 2020*, 15th ed. (London: Roskill)
[4] Tyler G 2004 *Sci. Total Environ.* 329 (1-3) 231–239
[5] Banner J L, Wasserburg G J, Dobson P F, Carpenter A B and Moore C H 1989 *Geochim. Cosmochim. Ac.* 53, 383–389
[6] Smedley PL 1991 *Geochem. Cosmochim. Acta* 55, 2767–2779
[7] Yu Z and Chen M (Eds.) 1995 *Rare Earth Elements and Their Applications* (Beijing: Metallurgical Industry Press), 286–294
[8] Barakos G, Mischo H and Gutzmer J 2015 'Rare earth underground mining approaches with respect to radioactivity control and monitoring strategies', in I B De Lima & W Leal (eds), *Rare Earths Industry: Technological, Economic, and Environmental Implications*, (Elsevier: Amsterdam, pp. 121-1138. (In An outlook on the rare earth elements mining industry).
[9] Kiggings R D, 2015 *The political economy of rare earth elements* (New York: Palgrave MacMillan)

[10] EPA 2012 *Rare earth elements: a review of production, processing, Recycling, and Associated Environmental Issues* (Cincinnati, OH: US Environmental Protection Agency).
[11] GolevA, Scott M, Erskine P D, Ali S H and Ballantyne G R 2014 *Resour. Policy* 41 52-59 [12] Jordens A, Cheng Y P and Waters K E 2013 *Miner. Eng.* 41, 97–114
[13] Chi R, Xu S, Zhu G, Xu J, Qju X 2001 *Beneficiation of Rare Earth ore in China*. Light Metals 2001as Held at the 130th TMS Annual Meeting 1159-1165
[14] Binnemans K, Jones P T, Blanpain B, Van Gerven T, Yang Y, Walton A, Buchert M, 2013 *J. Cleaner Prod.* 51 1–22
[15] Gambogi, J., 2013. Thorium [Advance Release], 2011: Minerals Yearbook U.S. Geological Survey
[16] Schüler D, Buchert M,, Liu D-I R, Dittrich D-G S and Merz D.-I.C. 2011 *Study on Rare Earths and Their Recycling*, Öko-Institut eV Darmstadt
[17] Song Zhaoliang, Liu Congqiang, Han Guilin and Wang Zhongliang 2006 *Journal of Rare Earths* 24, 491–496
[18] Cao Xindea, Wang Xiaorong and Zhao Guiwen 2000 *Chemosphere* 40, 23–28 [19] Pang X, Li D C and Peng A 2002 *Environ. Sci. Pollut. R.* 9, 143–148
[20] Diatloff E, Smith F W and Asher C J 1995 *J. Plant Nutr.* 18, 1963–1976
[21] Hu X, Ding Z H, Chen Y J, Wang X R and Dai L M 2002 *Chemosphere* 48, 621–629
[22] Gillet S and Ponge J-F 2006 *Appl. Soil Ecol.* 26, 219
[23] Charalampides G., Vatalis I.K., Baklavaridis A., Benetis P-N, 2015, Rare Earth Elements: Industrial Applications and Economic Dependency of Europe, International Conference on Applied Economics (ICOAE) 2015, Procedia, Economics and Finance, Elsevier.
[24] Metalpages,2011–2012./http://www.metal-pages.com/metalprices/rare earths (accessed 5.10. 11 and 11.06.12)
[25] Lian, R.,Stanway,D.,2011,China's Baotou Rare-Earth suspends facilities for one month, By Reuters,October18./http://www.reuters.com/article/2011/10/18/ baotou-rareearth-idUSL3E7LI05120111018S (accessed 20.10.11).
[26] Fiott D., 2011, Europe and Rare Earths: Dependable Diplomacy or Strategic Scarcity? Madariaga Paper – Vol. 4, No. 9
[27] GFMET, German Federal Ministry of Economics and Technology (BMWi), 2010a, *State Secretary Dr. Bernd Pfaffenbach Visits Australia*. Available at: www.bmwi.de/English/Navigation/ Press/press-releases,did=372338.html> [Accessed 25 February 2011].
[28] GFMET, German Federal Ministry of Economics and Technology (BMWi),2010b, *The German Government's Raw Materials Strategy:Safeguarding a Sustainable Supply of Non-Energy Mineral Resources for Germany* (Berlin).
[29] Arvanitidis N. & Papavasileiou K., 2011, Mineral Wealth, *Institute of Geology and Mineral Exploration*. (In Greek)
[30] Eliopoulos D., Economou G., Tzifas I., Papatrachas, C., 2014, The Potential of Rare Earth Elements in Greece, ERES2014, 1st European Rare Earth Resources Conference, Milos, 04-07-2014.
[31] www.lynascorp.com
[32] CHEN Zhanheng, 2011, Global rare earth resources and scenarios of future rare earth industry, *Journal of Rare Earths, Vol. 29, No. 1, p. 1*

Production and mechanical properties of Al-SiC metal matrix composites

K Karvanis[1], D Fasnakis[2], A Maropoulos[3] and S Papanikolaou[2]

[1]Department of Electrical Engineering, Technological Education Institute of West Macedonia, Greece
[2]Department of Mechanical Engineering, Technological Education Institute of West Macedonia, Greece
[3]Department of Mechanical Engineering, Aristotle University of Thessaloniki, Greece

E-mail:karvanisk@gmail.com

Abstract. The usage of Al-SiC Metal Matrix Composites is constantly increasing in the last years due to their unique properties such as light weight, high strength, high specific modulus, high fatigue strength, high hardness and low density. Al-SiC composites of various carbide compositions were produced using a centrifugal casting machine. The mechanical properties, tensile and compression strength, hardness and drop-weight impact strength were studied in order to determine the optimum carbide % in the metal matrix composites. Scanning electron microscopy was used to study the microstructure-property correlation. It was observed that the tensile and the compressive strength of the composites increased as the proportion of silicon carbide became higher in the composites. Also with increasing proportion of silicon carbide in the composite, the material became harder and appeared to have smaller values for total displacement and total energy during impact testing.

1. Introduction

Metal Matrix Composites consist of a metal matrix reinforced with particles or fibers of a second phase. The mechanical properties of the Metal Matrix composites depend on many factors such as matrix material, weight fraction, reinforcement phase material and the production method of the composite.

Al-SiC is a metal matrix composite consisting of an aluminium matrix and silicon carbide as the reinforcement phase. Silicon carbide is a low cost material with a high elastic modulus and it is very often used as a reinforcement phase. It can be found in particle or whisker form. [1] The properties of Al-SiC are notable such as light weight, high strength, high specific modulus, high fatigue strength, high hardness and low density [2]. One of the most important points in the production of these composites is the achievement of a good distribution of the reinforcement phase in the matrix.

For better distribution of the reinforcement phase a novel variation method of stir-casting is used. The matrix is in semi-solid condition when the reinforcement particles are added and then it is completely melted followed by mechanical stirring. A better distribution of particles in the microstructure is achieved with this method compared to conventional stirring. In addition, the thin gas layer of particles is broken which improves wetting between matrix and reinforcement. [3] Wetting of the reinforcement can also be improved by adding alloy elements to the molten metal, such as Cu, Mg and Fe.

In the present study material production methods, mechanical behavior and microstructure of the composite were studied in order to determine the optimum carbide % in the metal matrix composites and to investigate the distribution of the reinforcement material.

2. Experimental

2.1. Metal Matrix

Aluminium alloy is used as the matrix material. The chemical analysis of the aluminium matrix is shown in Table 1 and Figure 1 shows the pure aluminium alloy used as the matrix material.

Table 1. Chemical analysis of the aluminium alloy matrix.

Title	Si	Fe	Cu	Mn	Mg	Cr	Zn	Ti
% percentance	0,84	0,72	0,04	0,03	0,02	0,12	0,05	0,03

Figure 1. Part of the aluminium alloy used as matrix.

2.2. Reinforcement phase

The material used as the reinforcement phase was α-SiC particles of 325 mesh size. 0.8% Copper (150 mesh) was also used as a wetting agent. Both the silicon carbide and the copper particles were produced by Alfa Aesar Company.

2.3. Experimental details

The tensile – compressive strength tests were carried out at room temperature on Instron 5989 Floor Model Testing System, using the Bluehill 3 software, Figure 2. The impact tests were carried out on Instron CEAST 9350 Drop Tower Impact System and the CEAST VisualIMPACT software, Figure 3. The impact tests were performed at a speed of 5m/sec. In the experiments for the compressive strength, the specimens were compressed to 30% of their original length.

Figure 2. Tensile - Compressive test machine.

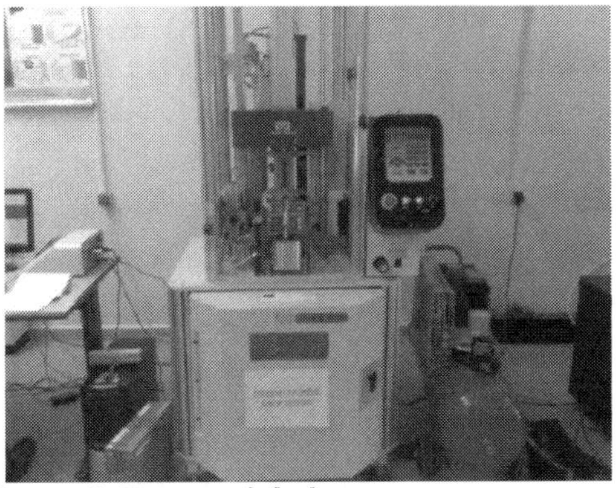

Figure 3. Drop-weight impact tester.

3. Production method

The method used for the production of the composites was a combination of melt-stirring and compo-stirring [3]. This combination method was used in order to obtain a better dispersion of the particles and to avoid particle agglomeration and sedimentation in the melt. Casting took place in a centrifugal machine. Figure 4 shows the molds which were produced by the lost wax casting method. An electric furnace equipped with an electric drill was used for the stirring process.

Figure 4. Molds produced by the lost wax casting method.

The silicon carbide particles were heated to 1076 °C for one hour and 30 minutes in order to remove the absorbed hydroxide and other gases leading to improved wetness properties. In addition, the molds and the stainless steel stirrer were preheated and the aluminum matrix was melted in an electric furnace at 720 °C. 0.8% copper was added as a wetting agent. For the slurry being in a semi-solid condition, the furnace temperature was adjusted below the melting point of the aluminum and it was stirred with an electric drill at 50 rpm. When the reinforcement particles were added the matrix was in a semi-solid condition. After the addition of the SiC particles, the mixture was completely melted (above 720 °C). The stirring process was carried out at 150 rev/min by the stainless steel stirrer for 20 min. Then the slurry was poured in the molds, which were on the centrifugal machine and centrifugal casting took place. The rpm of the centrifugal machine was adjusted to 280 rpm, since segregation of the particles either at the inner or the outer periphery of the casting was noticed at higher rpm [4].

The dimensions of the specimens were 135mm x 20mm for the tensile strength experiments and 30mm x 20mm for the compressive strength, impact and hardness tests. All the specimens were of cylindrical shape (Figures 5,6). The composite contents were adjusted so as to contain 3%, 6%, 9%, 12%, and 15% of silicon carbide. For reliability in the measurements, three samples for every experiment were made and the average values of were used.

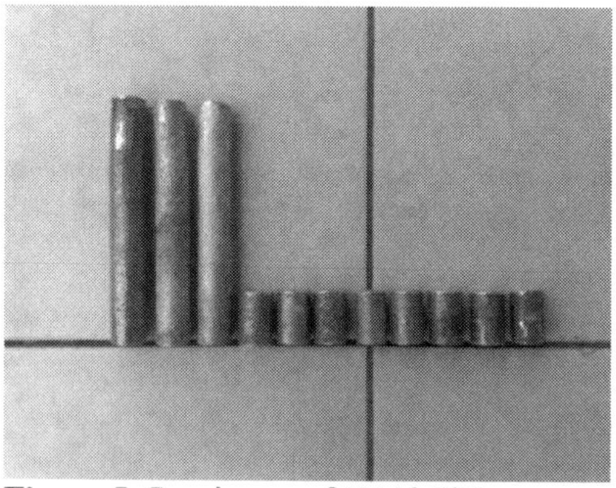

Figure 5. Specimens of 3%Al-SiC.

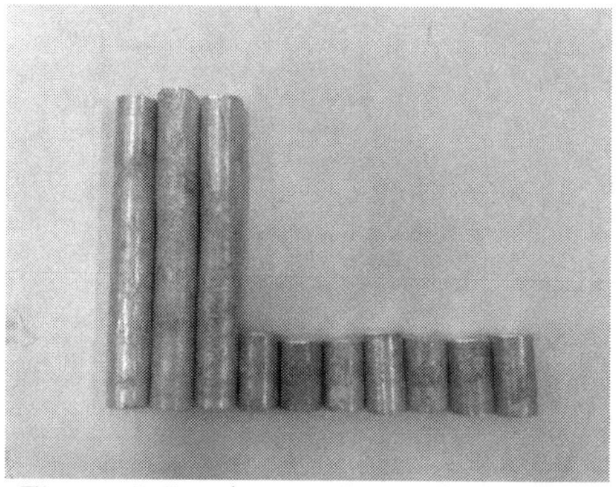

Figure 6. Specimens of 6% Al-SiC.

4. Results and Discussion

The mechanical test results are given in Tables 2-4 and Figure 7 shows the specimen surfaces as observed in the scanning electron microscope.

Table 2. Tensile strength results.

	σ_y MPa	σ_{ts} MPa
aluminium	12.42	62.97
3% average value	15.37	72.36
6% average value	18.03	81.32
9% average value	20.79	92.23
12% average value	23.38	105.44
15% average value	25.21	119.38

Table 3. Compressive strength results.

	σ_y MPa	σ_{cs} MPa	R %
3% average value	32.60	140.86	30
6% average value	35.63	146.98	30
9% average value	40.81	158.63	30
12% average value	47.13	181.35	30
15% average value	51.90	196.32	30

Table 4. Impact test results.

	Peak force (kN)	Total Energy (j)	Total Displacement (mm)
3% average value	36.16	67.068	3.040
6% average value	34.73	66.595	3.029
9% average value	36.45	65.930	2.823
12% average value	36.80	64.106	2.578
15% average value	37.20	63.257	2.471

IOP Conf. Series: Materials Science and Engineering **161** (2016) 012070 doi:10.1088/1757-899X/161/1/012070

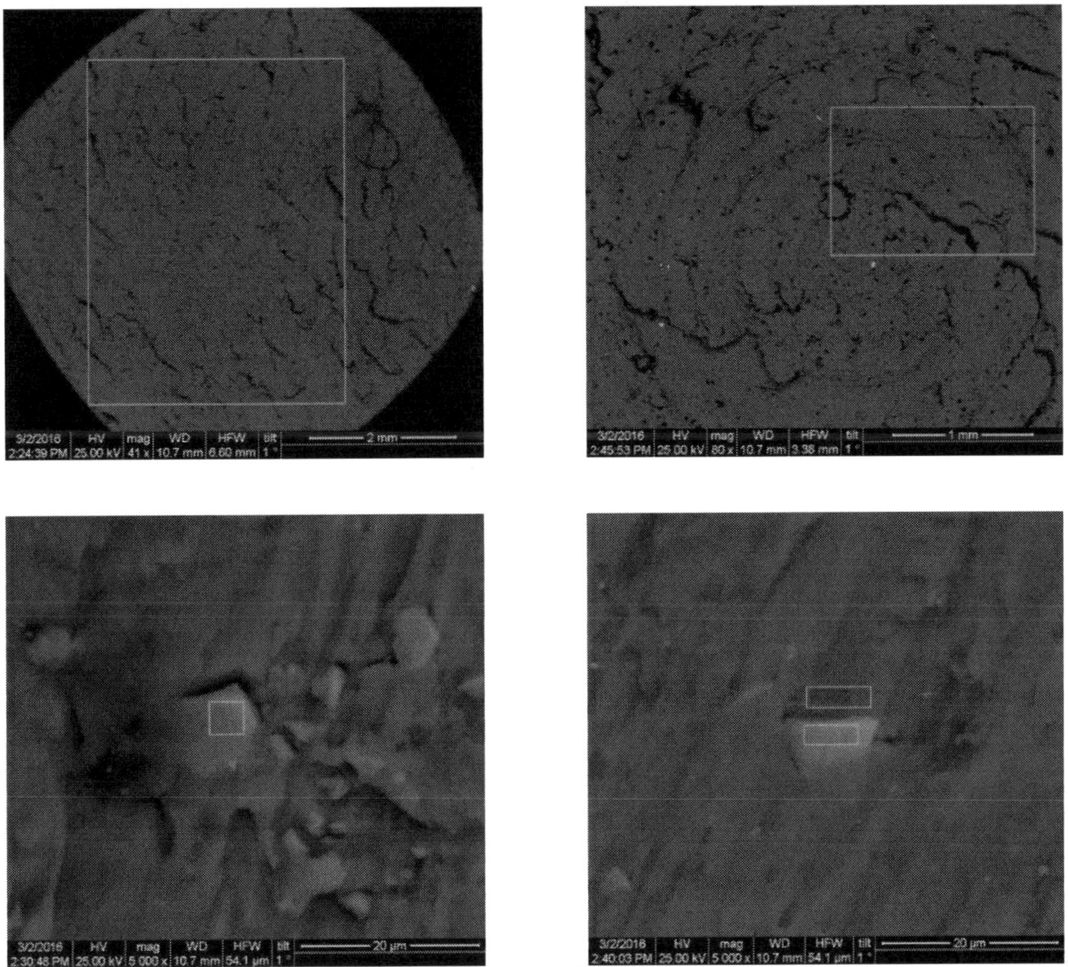

Figure 7. SEM Microscopy on Al-SiC_3% tensile test specimens.

Figures 8 and 9 show the yield and tensile strength curves where it is observed that as the silicon carbide particles have higher concentrations in the composite, the specimens improve their resistance to tensile stress. Notably the tensile strength value for the Al-SiC_15% specimen was almost doubled in comparison to the value of pure aluminium. The highest value for the tensile strength was exhibited by the specimen with 15% volume fraction of SiC particles. Other researchers [2,5] also noticed an increase in the values of tensile strength as the silicon carbide concentrations became higher in the composites.

Figure 10 shows that with increasing proportion of silicon carbide in the composite, the material improves its compressive strength. Maximum compressive strength was achieved with 15% of silicon carbide particles. The same result, improvement in the values of compressive strength as the % of SiC particles increased was obtained in other work where Al-SiC composites were fabricated by stir casting" [6].

Also, the composite yield strength increased as the proportion of the silicon carbide particles became higher in the composites, Figure 11.

Figures 12 to 14 show that as the weight fraction of the SiC particles in the composites increased, the values for total displacement and for total energy are reduced in drop weight impact test whereas the peak force remained more or less the same. This can be attributed to the rise in hardness observed with increasing weight fraction of the SiC particles in the composite.

From Figure 15 it can be concluded that as the SiC particles have a higher weight ratio in the composite, the specimens become harder. Furthermore, the same was observed in other work [2,6] where an increase in the hardness values of the composites as the weight fraction of the SiC particles in the composite increased is reported.

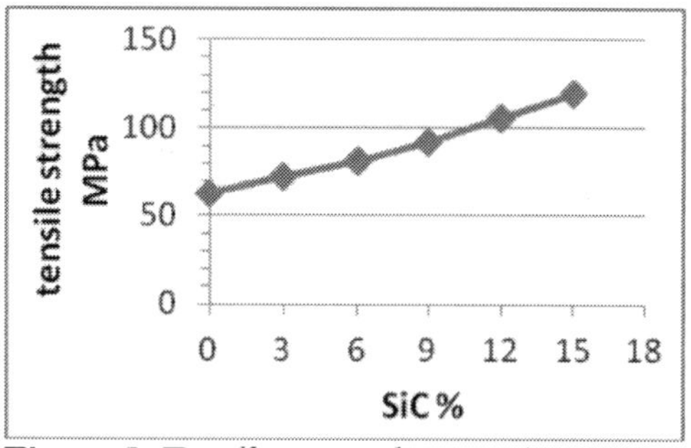

Figure 8. Tensile strength curve (MPa).

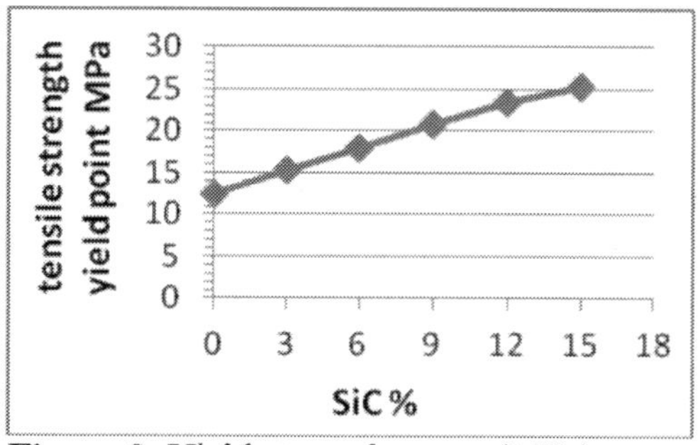

Figure 9. Yield strength curve (MPa).

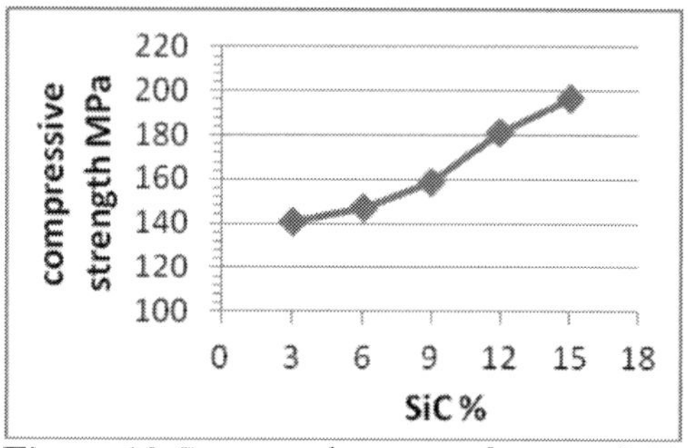

Figure 10. Compressive strength curve (MPa).

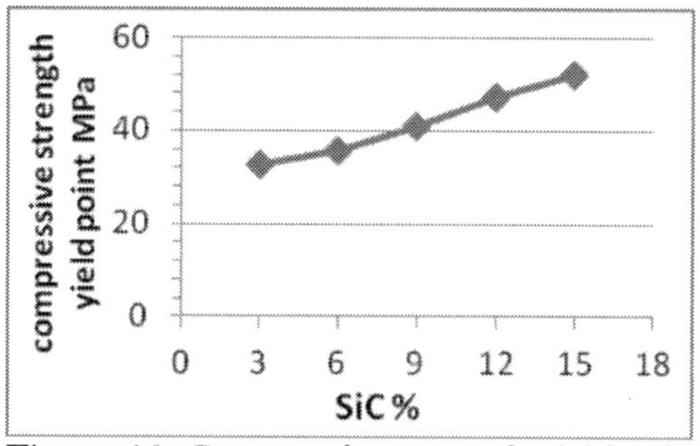

Figure 11. Compressive strength yield point curve (MPa).

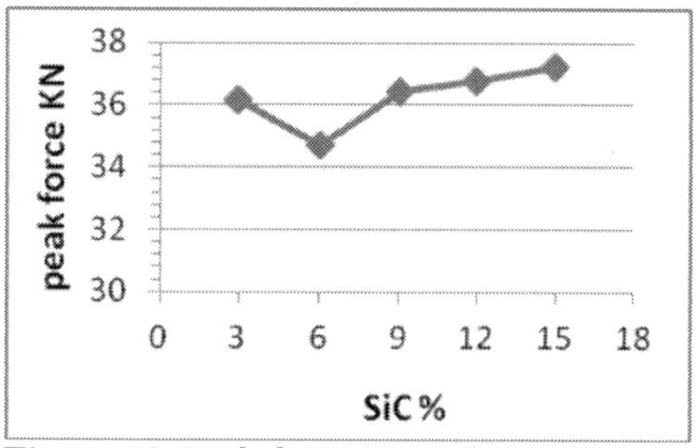

Figure 12. Peak force curve (kN).

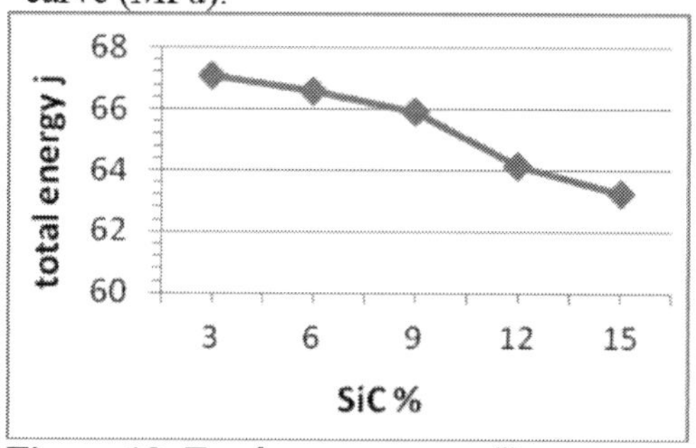

Figure 13. Total energy curve (j).

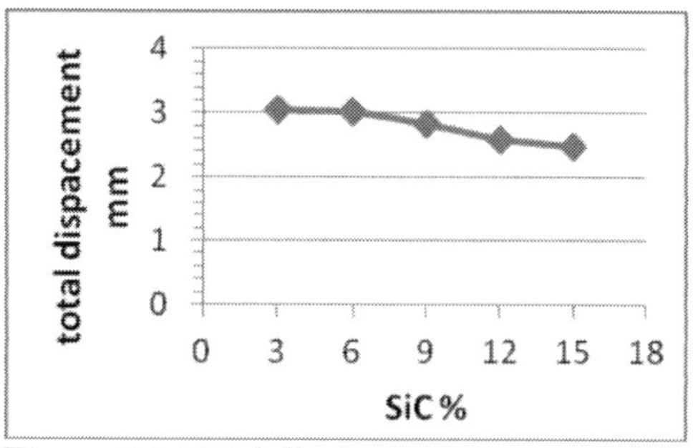

Figure 14. Total displacement curve (mm).

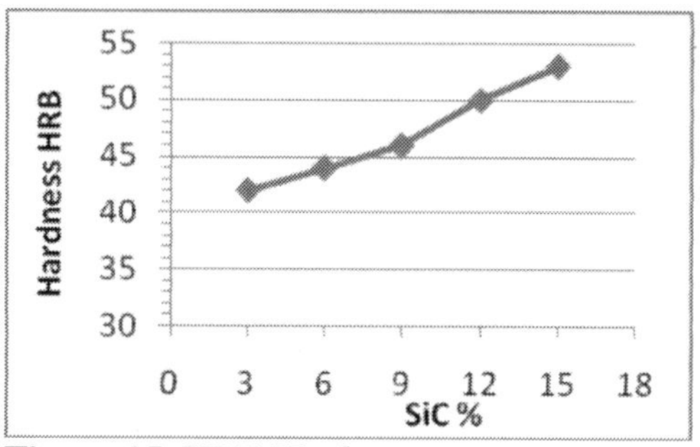

Figure 15. HRB Hardness.

5. Conclusions

1. The experimental values for the three specimens of every experiment were similar in every proportion and this suggests that a very good distribution of the silicon carbides in the composites was achieved. The variation of the stir-casting method used is a low cost but very efficient method for the production of Al-SiC composites.
2. The mechanical properties of the Al-SiC composites are improved as the silicon carbide particles obtain higher volume fractions in the composites.
3. From the photos of the SEM microscopy, it is clear that many of the SiC particles are broken in smaller pieces. This has been caused by the stirring method which was used.

Acknowledgments
We would like to thank the company "ALUMINIUM OF GREECE" and especially Mr. Karava Ioanni for the supply of the aluminium alloy.

References

[1] Deborah D L Chung, 2010, *Composite Materials Second Edition,* Springer

[2] K L Meena Dr. A Manna, Dr. S S Banwait and Dr. Jaswanti, 2013, An analysis of mechanical properties of the developed Al/SiC-MMC's, *American Journal of Mechanical Engineering*, Vol. 1, No. 1, 14-19

[3] Bala G Narasimha, Vamsi M Krishna and Dr. Anthony M Xavior, 2013, A review on processing of particulate metal matrix composites and its properties, *International Journal of Applied Engineering Research*. Volume 8, Number 6, pp. 647-666

[4] Emila Panda, Dipak Mazumdar and S P Mehrotra, MAY2016, Mathematical modeling of particle segregation during centrifugal casting of metal matrix composites, *Metallurgical and Materials Transactions,* Volume 37A, 1675-1687

[5] Neelima Devi. C, Mahesh.V and Selvaraj. N, 2011, Mechanical characterization of Aluminium silicon carbide composite, *International Journal of Applied Engineering Research, Dindigul,* Volume 1, No 4, 793-799

[6] Farouk Shehata, Nahed ElMahallawy, Mohamed Arab and Mohamed Agwa, 2013, Equal Channel Angular Pressing of Al-SiC Composites Fabricated by Stir Casting, *Open Journal of Metal*, 3, 26-33

A proposed procedure for expressing the behavior of a full engine cycle by identifying its critical load timings

Mihalache Marius Andrei, Nagit Gheorghe, Musca Gavril, Merticaru Vasile Jr., Ripanu Marius Ionut

"Gheorghe Asachi" Technical University of Iasi, Department of Machine Manufacturing Technology, Blvd. D. Mangeron no. 59A, Iasi, Romania

andrei.mihalache@yahoo.com, nagit@tcm.tuiasi.ro, musca@tcm.tuiasi.ro,

merticaru@tcm.tuiasi.ro, ripanumariusionut@yahoo.com

Abstract. In the present study the authors propose a new algorithm for identifying the right loads that act upon a functional connecting rod during a full engine cycle. The loads are then divided into three categories depending on the results they produce, as static, semi-dynamic and dynamic ones Because an engine cycle extends up to 720°, the authors aim to identify a method of substitution of values that produce the same effect as a previous value of a considered angle did. In other words, the proposed method aims to pin point the critical values that produce an effect different as the one seen before during a full engine cycle. Only those values will then be considered as valid loads that act upon the connecting rod inside FEA analyses. This technique has been applied to each of the three categories mentioned above and did produced different critical values for each one of them. The whole study relies on a theoretical mechanical project which was developed in order to identify the right values that correspond to each degree of the entire engine cycle of a Daewoo Tico automobile.

1. Introduction

Studies that highlight the behavior of connecting rods are being published periodically. They are fitted to a wide range of parts as it is considered that there are a few minor particularities that do not interfere with the generally accepted conclusions. The present study is structured as a case study as it is developed solely to a single type of an engine, the Daewoo Tico's one. In order to highlight the full behavior of a part under certain loads, one should consider the entire working cycle of that part under all known loads that act upon it. This is a tedious work in case of a connecting rod because of the sheer number of data that would have to be taken into account directly dependent on all 720° degrees of an engine cycle. Considering that previous studies showed that the behavior of the connecting rod are different depending on the type of loads considered to act upon them, this leads to a significant amount of work and data's to be processed [1].

The proposed method presents an algorithm which will identify only the values corresponding to engine cycle angles that produce a significant change when assessing the overall behavior characteristic to the functioning connecting rod inside the working engine; values that will override the ones that do not. Otherwise, all 720 values corresponding to an entire engine cycle would have to be taken into account each one as an individual load even though most of them will produce the same results in terms of the connecting rod's behavior [2]. The values of the loads are only theoretical as they are extracted from a technical project of an engine design especially developed for this study. The resulted values were compared to similar ones considered by other researches in their studies and were found to be appropriate. The loads on the other hand were divided into three categories which have better highlighted the overall behavior of the considered connecting rod [3].

The method applies to all and has been adjusted accordingly. This shows that the principle of it may be used to a wide range of similar studies easing up consistently the logistics involved in such types of studies.

2. Method description

The loads are divided into three categories depending on the type of regimes that they are used in: *static, semi-dynamic* and *dynamic*. The division takes into account the analytical vector approach as loads are depicted as vectors which are applied to both ends of the connecting rod.

The distribution of loads produces different results as it acts upon the connecting rod's head and body. The two main loads which were considered in the present study are the tensile and the compressive ones. In case of the axial load their values are not at almost any point identical when applied either to the crank end or to the piston end of the connecting rod.

The tensile load extends to the entire upper section of the connecting rod's head spreading over 180° in an obvious opposition with the compressive load that only acts upon a 120° wide section (*Fig. 1*) [5].

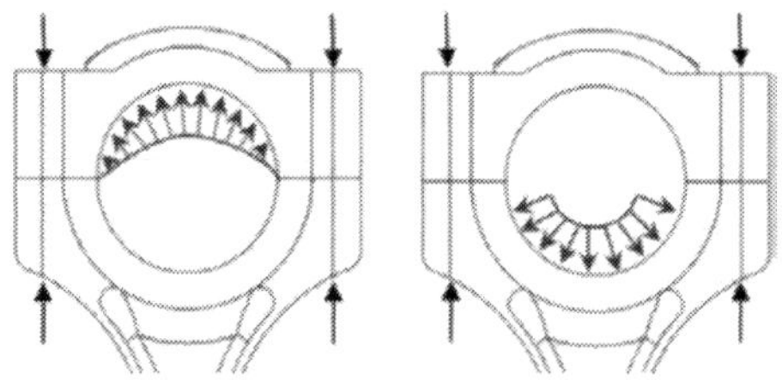

Fig. 1. Axial loads distribution on connecting rod's head:

a) Tensile load at 180°,

b) Compressive load at 120°.

2.1. Static load regime

In this case the load that acts upon the piston end of the connecting rod is being produced by the piston itself and the O-rings which are subjected to a full inertia load which drives the contact surfaces between the connecting rod and the bolts to be constrained. The loads that act upon the crank end of the connecting rod are divided into the tensile component of the axial load which tends to extend the part and the compressive one. These loads represent an effect of the tightening mounting method with the anti-friction bush and the bolts [2].

As stated above, this study uses the analytical vector approach (*Fig. 2*). Upon the body of the connecting rod (*the piston end of it*) we have both the tensile and the compressive components of the axial load that act each one at a time in a full engine cycle. The middle section of the connecting rod is being subjected to these loads as well as to bending in the swinging plane of the part and the plane perpendicular to it. These values were not considered as they are not making a significant change in the functional connecting rod's overall behavior compared to those that were taken into account for the part's crank end. This end named the connecting rod's head, is being also subjected to an axial load. Because the head of the part is widely filleted with the body, the compressive component is very much diminished in this case and thus only the tensile component is being taken into account.

The optimization procedure from the mass reduction point of view which has been carried out on the original 3D scan of the connecting rod showed that exactly this section may be improved with a larger fillet that will not only respect the original's behavior under the same loads but will even enhance the stress distribution and thus its diminishment in that particular area. It also pointed out that although the optimization produces a loss of a few grams the stress and compression peaks and distribution do change for better when assessing the entire connecting rod's behavior. That is considered a breakthrough by the authors of the research that makes subject of a different scientific paper [4].

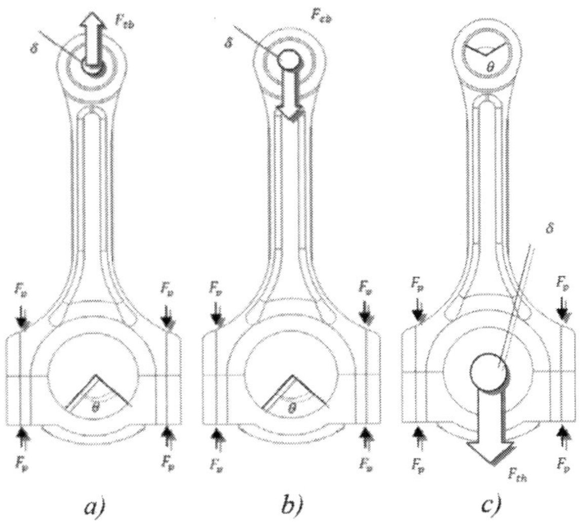

Fig. 2. Components of axial loads that act upon the connecting rod's ends:

a) Body's corresponding tensile load,

b) Body's corresponding compressive load,

c) Head's corresponding tensile load.

The meanings of the symbols (*Fig. 2*) are as follows: F_{tb} is the tensile component of the axial load that acts upon the connecting rod's body, F_{cb} is the compressive component of the axial load that acts upon the connecting rod's body, F_{th} is the tensile component of the axial load that acts upon the connecting rod's head, F_p represents the pressure load produced by the bolts tightening, Θ is the constrain angle and δ represents the movement under the imposed load.

2.2. Semi-dynamic load regime

In this case we will take into account that the load that acts upon the connecting rod is an inertia type of force which breaks down into two components: axial and normal. The axial component is also divided into the tensile and compressive type of forces. This type of load regime is different from the static one because it is considering the alternation of forces during an entire engine cycle and not at a given point in time as static regime did. Considering each individual value corresponding to every angle out of the 720 ones would be time consuming and very unproductive as we will demonstrate that certain critical values produce the same results. It is well known that the loads that act upon a certain point from within the connecting rod whole geometric surface have two major components: a bending and an axial form (*Fig. 3*) [2].

The bending load depends on the bending momentum and it is expressed as a function of the load normal to the part's weight center axis related to both angular and linear acceleration corresponding values. The variation of these loads during a half engine cycle from 0° to 360° is almost the same as the one from 360° to 720°. The F_i symbol from figure 3 represents the inertia type of force distributed on the entire surface of the connecting rod.

As presented above, the values of the axial loads on each end of the connecting rod are not identical at any time due to different inertia loads. This makes it equal in terms of deciding which end of the connecting rod to choose since axial components distribution is the same at both ends. In the present study, the authors applied the load to the connecting rod's body or the piston end of it (*Fig. 3*). Results from the first half of an engine cycle had to be compared with the ones obtained from the other half. We have therefore overlapped the distributions under axial load characteristic to the first half of engine cycle with the one from the second half. Then we have divided these distributions into three different regions: "*i*", "*ii*" and "*iii*". Each segment of curve from each region received a letter from "*a*" to "*f*". By comparing those segments we have been able to choose the ones which showed different variations and eliminate the ones which were overlapping (*Fig. 4*).

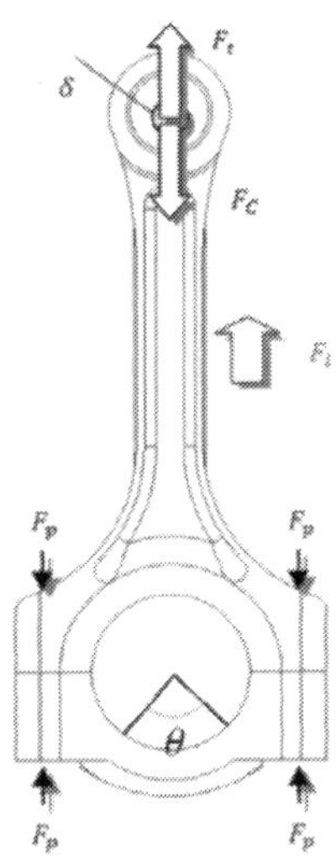

Fig. 3. Axial loads distribution on connecting rod's body.

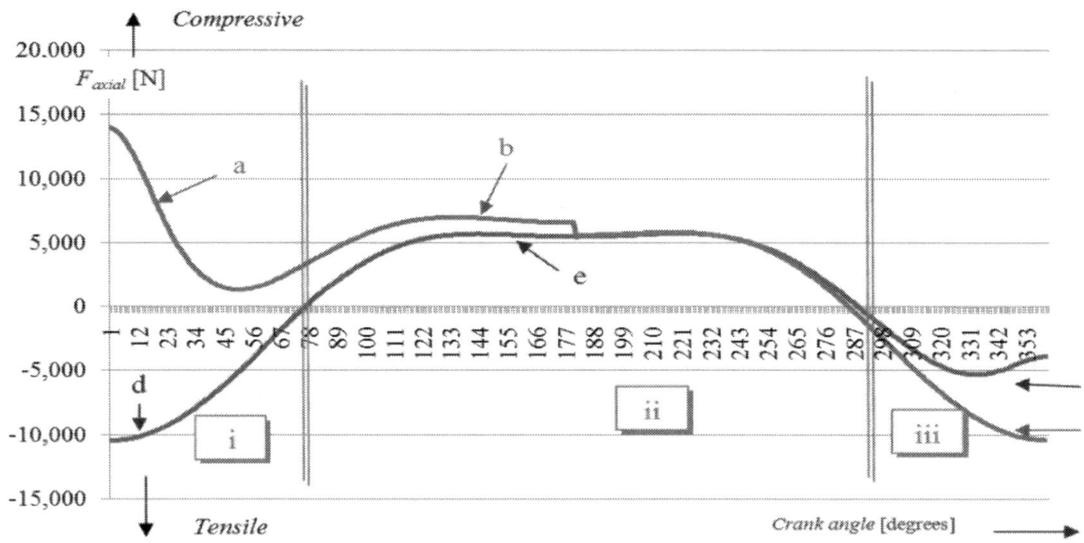

Fig. 4. Axial load distribution from 0° to 360° overlapped with the one from 360° to 720°.

The "*i*" region has two segments: "*a*" and "*d*". Both display different variations and do not overlap at any time.

The "*ii*" region has two segments: "*b*" and "*e*". The "*b*" segment shows a bigger variation in the compressive domain than "*e*" does up to 180° then both being almost identical which led to the conclusion that segment "*e*" may be eliminated from the process. The lowest value from the variation of segment "*e*" was considered and then compared with the lowest one from segment "*b*" in order to validate the decision.

The "*iii*" region has also two segments: "*c*" and "*f*". Since segment "*c*" displays a bigger variation in the tensile domain than segment "*f*" does the last one was eliminated. The lowest value from the

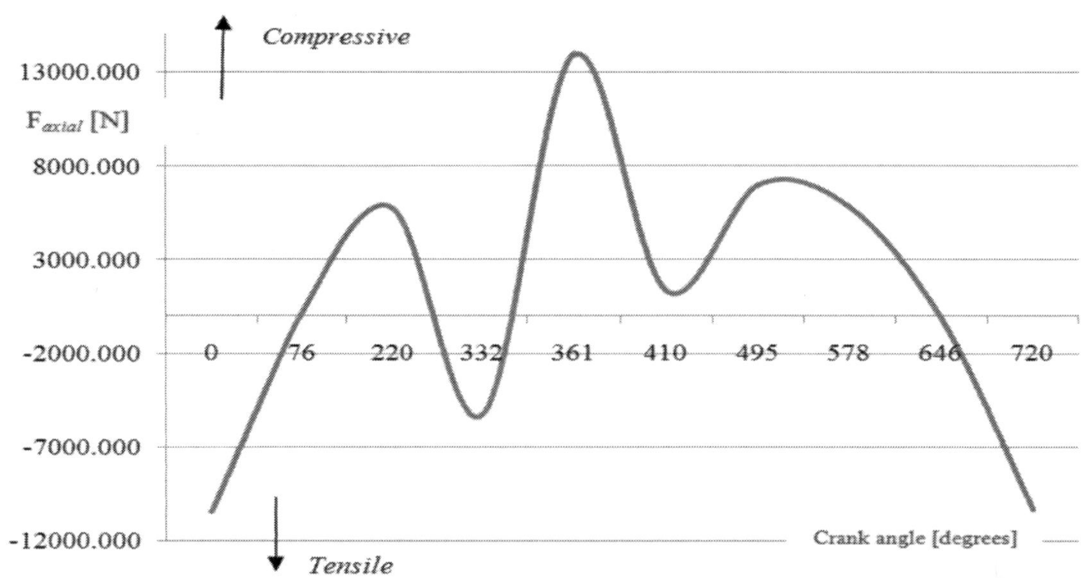

Fig. 5. Axial load distribution at proposed angles.

variation of segment "*f*" was considered and then compared with the lowest one from segment "*c*" in order to validate the decision.

In order to highlight the behavior of the connecting rod on a full engine cycle our method states that it is enough to consider only the segments "*a*", "*b*", "*c*" and "*d*". Since the first change of sign is being recorded for the 76° angle the "*d*" segment will extend from 0° to 75°. In the same manner "*e*" will extend from 76° to 287° being divided around the 180° angle. The "*f*" segment will extend from 288° to 360°. Region "*a*" will extend from 361° to 435°. Region "*b*" segment will extend from 436° to 647° being divided around the 540° angle. In the end, region "*c*" will extend from 648° to 720°.

The critical values which will be considered in the evaluation process of the connecting rod on a full engine cycle are corresponding to the following angles: 0°, 76°, 361°, 410°, 495°, 578°, 646° and 720°. In addition to those, values that are corresponding to the 220° and 332° angles from segments "*e*" and "*f*" were also considered in order to validate the method. The resulted distribution is presented in figure 5.

2.3. Dynamic load regime

In this case we will consider the assembly which consists of the connecting rod, the piston and the crank (*Fig. 6*). The setup was developed in order to simulate as much as possible the real environment in which the process takes place. Thus it performs as it should under an inertia type of load. The difference from the semi-dynamic regime consists in the type of load that has been applied to the piston's head in order to generate the motion of the considered assembly. This load is considering also the normal component of the inertia type of force thus making it suitable for a full dynamic FEA simulation [3]. The inertia force or the total force as it is also known results from summing up the forces that appear due to the accumulated pressure of gases inside the cylinder, the translation inertia forces of moving masses, weight related forces and friction related forces. We will follow the same logic and apply the same working principle. The distribution under inertia load over the first half of the engine cycle has been overlapped with the one from the second half. Then we have divided these distributions into three different regions: "*i*", "*ii*" and "*iii*". Each segment of curve from each region received a letter from "*a*" to "*f*" naming particular areas (*Fig. 7*).

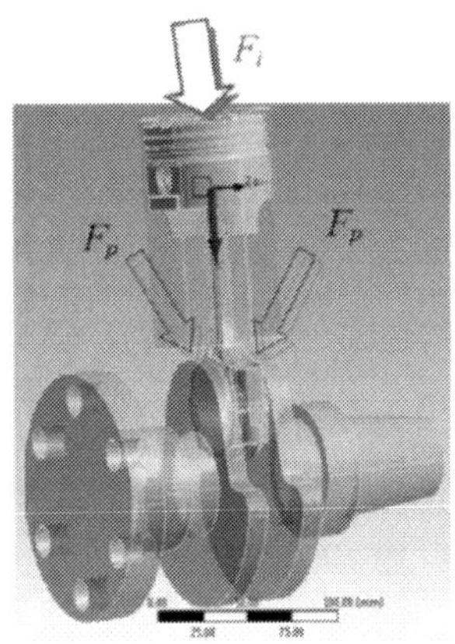

Fig. 6. The assembly: the connecting rod, the piston and the crank.

The "*i*" region has two segments: "*a*" and "*d*". Both display different variations and do not overlap at any time.

The "*ii*" region has two segments: "*b*" and "*e*". The "*b*" segment shows a bigger variation in the compressive domain than "*e*" does up to 180° then both being almost identical which led to the conclusion that segment "*e*" may be eliminated from the process. The lowest value from the variation of segment "*e*" was considered and then compared with the lowest one from segment "*b*" in order to validate the decision.

The "*iii*" region has also two segments: "*c*" and "*f*". Since segment "*c*" displays a bigger variation in the tensile domain than segment "*f*" does the last one was eliminated. The lowest value from the variation of segment "*f*" was considered and then compared with the lowest one from segment "*c*" in order to validate the decision.

In order to highlight the behavior of the assembly on a full engine cycle it is enough to consider only the segments "*a*", "*b*", "*c*" and "*d*". Since the first change of sign is being recorded for the 76° angle the "*d*" segment will extend from 0° to 75°. In the same manner "*e*" will extend from 76° to 287° being divided around the 180° angle. The "*f*" segment will extend from 288° to 360°. Region "*a*" will extend from 361° to 435°. The "*b*" segment will extend from 436° to 647° being divided around the 540° angle. In the end, "*c*" will extend from 648° to 720°.

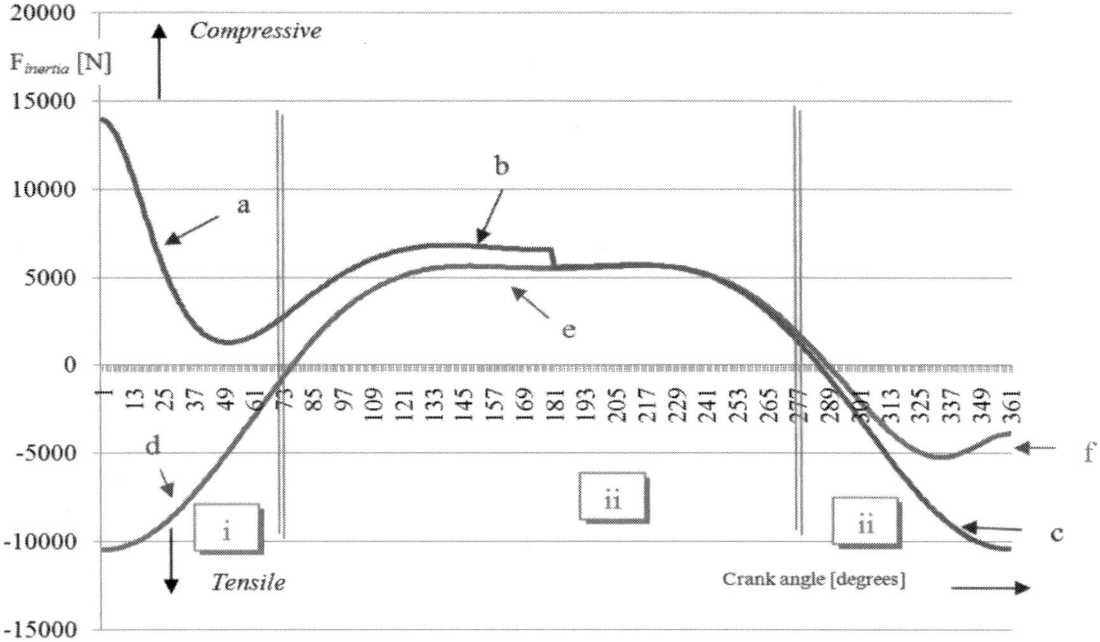

Fig. 7. Inertia load distribution from 0° to 360° overlapped with the one from 360° to 720°.

The critical values which will be considered in the evaluation process of the connecting rod on a full engine cycle are corresponding to the following angles: 0°, 76°, 361°, 410°, 499°, 574°, 646° and 720°. In addition to those, values that are corresponding to the 215° and 333° angles from segments "*e*" and "*f*" were also considered in order to validate the method. The resulted distribution is presented in figure 8.

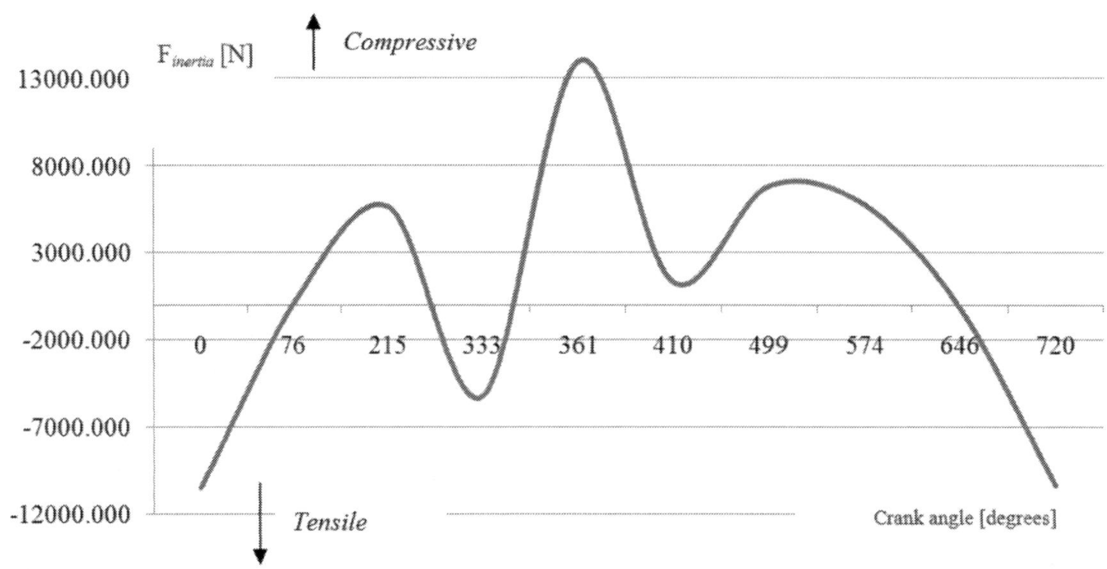

Fig. 8. Inertia load distribution at given critical angles.

3. Conclusions

In the case of the static regime the loads are represented by the two sub-components of the axial load, namely: the compressive load and the tensile load. In case of the semi-dynamic regime the loads are represented by one of the two components of the inertia load, namely the axial load along a full engine cycle (*this gives it the semi-dynamic character*). In case of the dynamic regime the loads are represented by the inertia load. Those three regimes give us a complete perspective over the overall behavior of the connecting rod in a full engine cycle [6].

The proposed method applies to the regimes that stretch over an entire engine cycle. It aims to identify the values considered as critical that would simulate the behavior of the connecting rod just as if we would had to consider the all 720 of them that are corresponding to the angles from a full engine cycle.

In order to perform FEA analyses for all three regimes, constraints and loads should be set from the beginning. In the case of the two last regimes it would be a tedious work to perform FEA analyses for each value corresponding to the angles from a full engine cycle. Thus, the need for a method that would identify several values that would respect the overall distribution of the connecting rod under a load, arisen.

In the case of the semi-dynamic regime we have identified eight critical values that do respect the trend line imposed by all 720. In addition we have considered two more values from the dismissed curve segments in order to show that they do not interfere with the final result and the method's working principle. We can observe that values corresponding to the two additional angles are not affecting the trend line imposed by the other values as they are smaller.

In the case of the dynamic regime we have also identified eight critical values that do respect the trend line imposed by all 720. We have also considered two more values from the dismissed curve segments in order to show that they do not interfere with the final result. We can observe that values corresponding to the two additional angles are not affecting the trend line imposed by the other values as they are smaller.

The method proves that in the case of FEA analyses considering only 8 critical values out of 720 would provide the user with the same perspective over the connecting rod's behavior under given loads. The method may be extended to other types of working cycles as well not to mention all forms of connecting rods.

References

[1] Pouplana I, Oñate E, *Combination of a non-local damage model for quasi-brittle materials with a mesh-adaptive technique*, Finite Elements in Analysis and Design vol. 112, 2016, pp. 26–39, available at http://www.sciencedirect.com/science/article/pii/S0168874X15001997

[2] Webster W.D, Coffell R, Alfaro D, A three dimensional finite element analysis of a high speed diesel engine connecting rod, SAE Technical Paper 831322, 2007, pp. 83-96, available at papers.sae.org/831322/

[3] Mustapha L, Benyebka B.S, *Couple Stress Effects on the Dynamic Behavior of Connecting Rod Bearings in Both Gasoline and Diesel Engines*, 2008, pp. 44-56, available at http://www.tandfonline.com/doi/abs/10.1080/10402000701739305

[4] Mihalache M.A, Nagit G. and Ripanu M.I, *Optimization procedure of a scanned solid body through FEA by means of reverse engineering*, 2013, Proceedings of the 6th International Conference of Manufacturing Science and Education, pp. 175-178, available at http://conferences.ulbsibiu.ro/mse/MSE2013/index.htm

[5] Dyniewicz B, *Space–time finite element approach to general description of a moving inertial load*, Finite Elements in Analysis and Design, vol. 62, 2012, pp. 8–17, available at http://www.sciencedirect.com/science/article/pii/S0168874X12001461

[6] Masching H, Bletzinger K.U, *Parameter free structural optimization applied to the shape optimization of smart structures*, Finite Elements in Analysis and Design, vol. 111, 2016, pp. 33–45, available at http://www.sciencedirect.com/science/article/pii/S0168874X15001900

Aspects Concerning the Use of Recycled Concrete Aggregates

I Robu[1], C Mazilu[1] and R Deju[2]

[1]Technical University of Civil Engineering Bucharest, Romania

[2]National Institute for Physics and Nuclear Engineering "Horia Hulubei", Magurele, Romania

E-mail: claudiumazilu@yahoo.com

Abstract

Natural aggregates (gravel and crushed) are essential non-renewable resources which are used for infrastructure works and civil engineering. Using recycled concrete aggregates (RCA) is a matter of high priority in the construction industry worldwide. This paper presents a study on the use of recycled aggregates, from a concrete of specified class, to acquire new cement concrete with different percentages of recycled aggregates.

1. Introduction

In European Union member states from Southeast Europe, it is estimated that the construction industry will grow by 4.2% thereafter complicating aggregate supply management. In addition, a significant additional problem that can be associated to the aggregates industry is wasting potential resources through waste dumping of inert waste, especially waste from construction and demolition activities, [1].

In 2012, in Romania, less than 10% of construction and demolition waste (including concrete) are valorized, while the European Union requires that by 2020 this proportion should be at least 70% (Directive 2008/98 / EC on waste, transposed into Romanian legislation by Law 211/2011).

Depending on the efficiency of waste processing and the quality of recycled aggregate concrete (RCA) obtained, poor quality aggregate can be used as foundation material for roads and at the high quality for new concrete on construction, [2].

To obtain good quality concrete using recycled aggregate is necessary to meet the minimum requirements defined by the rules for the manufacture of concrete with natural aggregate. Properties of recycled aggregate (density, granulosity, granule shape, water absorption, weight loss to Los Angeles test, attached mortar content [3] etc.) are the basis for concrete quality; also establishing appropriate proportions between components and the concrete production methods are extremely important for its quality, [4,5].

In the literature [6-8], it was shown that the recycled aggregate concrete (RAC) are less resistant, with 15 to 40% versus concrete with natural aggregates. The factors influencing reducing mechanical resistance include:

- mechanical strength of the original concrete and respectively for recycled aggregates obtained;
- the origin of the original concrete from multiple sources versus once source;
- the amount of attached mortar to recycled aggregates;
- the percentage of replacing natural aggregates with recycled aggregates;
- use of fine recycled aggregates;

The present study aimed the obtaining of some recycled concrete aggregate from established class concrete and their use in different proportions to achieve new concrete. The new concretes are characterized by the point of view of density, mechanical strength, modulus of elasticity etc.) and the factors affecting their properties analyzed.

Should be noted that the use of recycled aggregates in concrete is made on the basis of National/European specific prescriptions (SR EN 206: 2014), with limits for type A or B recycled aggregates. Aggregates obtained by crushing of hardened concretes unused in structural elements are considered reclaimed crushed aggregates, [9].

2. Experimental

2.1 Obtaining of the recycled concrete aggregate

To achieve recycled aggregates several batches of concrete class C16/20, C25/30 and C35/45 were made, the compositions calculation being made according NE012/2007 and CP012/2007. Tests for producing recycled aggregate was carried out using concrete samples of the established three classes after 28 days of storage under the above conditions, [10]. Cubes with 150mm side were crushed in a first stage with a jaw crusher Liebherr type set at 50 mm nominally. The resulting material was separated by sieving on granulometric sorts and 10-50 sort was used for preliminary tests of crushing in the second stage with a jaw crusher BB 200 Retsch model, respectively a hammer crusher Buffalo Shuttle WA-12-H model.

The fine aggregate, sort 0-2, was analyzed by granulometric point of view getting elementary sorts 0 - 0.063; 0.063 - 0.125; 0.125 - 0.25; 0.25 - 0.5; 0.5 - 1; 1 - 2; > 2 for all three concrete classes for the two crushing variants in the second stage. Comparing granulometric compositions for recycled aggregates (see granulometric curves of the figure 1) obtained with an impact crushing to those obtained with a jaw crusher, generally, a higher proportion of coarse parts(sorts 1-2; > 2) in the aggregate obtained with a jaw crusher is observed, particularly for concrete aggregates from larger class. This is in accord with the literature data [10] according to which the hammer crusher produce more fine part than the jaw crusher and recycled aggregate derived from a higher class concrete is coarser than that coming from a lower concrete.

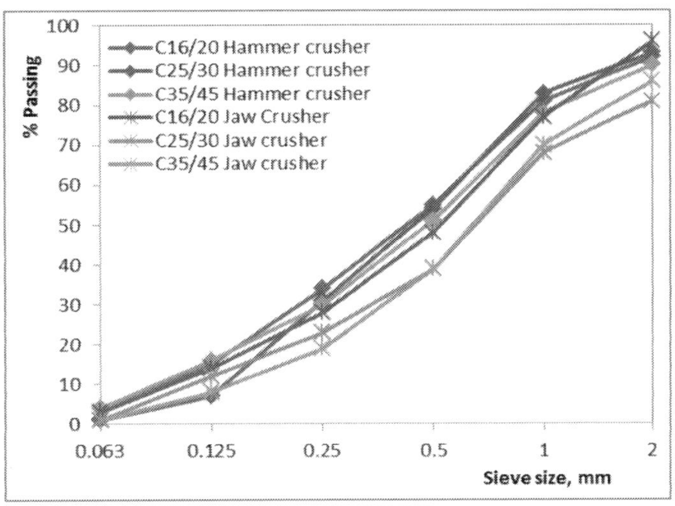

Figure 1 Granulometric curve of the fine aggregate

Finally, for the production of recycled aggregate concrete and to compare their properties with those of the original concrete, through the successive crushing and screening, there were obtained three granulometric classes of recycled aggregate 0-4; 4-8 and 8-16. In this study, for new concrete, only recycled aggregate derived from C25/30 concrete class, obtained with a hammer crusher, was used.

2.2 Obtaining the recycled aggregate concrete

The amounts of mixing water, cement, aggregates and additives were established according to NE012/2007 and CP012/2007, following the same steps as in the case of a natural aggregate concrete [11].

The proportions in the mixture were calculated so as to satisfy the following conditions:

- the recycled aggregate to replace the natural aggregate in a proportion of 0% (reference), 20%, 40%, 60%, 80% and 100%, for all sorts used;
- the recycled aggregate was used dry, the necessary water for saturation of the recycled aggregate was added during the components' mixing;
- river aggregate and recycled from crushed concrete, 0/16 mm on 3 granulometric sorts: 0/4; 4/8 and 8/16 mm, respectively;
- slump class of the fresh concrete S_2 (50 ÷ 90 mm);
- superplasticizer additive / high range water reducer.

As with concrete from which the recycled aggregate was obtained, CEM V A S-V 42,5 N cement was used and as an additive: superplasticizer Glenium 27 (1% weight of cement). Obtaining, preserving and preparation of samples for testing was done according to the rules adopted for concrete with natural aggregate.

In table 1 are shown the compositions for 1m^3 concrete with different proportions of recycled aggregate (R1-R6) and some data of interest such as: fresh concrete slump, A/C global ratio, density of fresh concrete, experimental observations.

Table 1 Recycled aggregate concrete compositions

Code samples	Cement, Water, Effective A/C ratio, Glenium 27 Additive	Aggregates on sorts, for 1m^3 concrete, kg	Fresh concrete slump, mm	A/C global (real)	Density of fresh concrete, kg/m^3	Observations
R1 0% recycled	Cement: 345 kg for 1 m^3 concrete; Water: 190 liters for 1 m^3 concrete; Glenium 27 Additive: 3.3 liters for 1m^3 concrete; A/C=0.55	0/4 - 815 4/8 - 454 8/16- 545	65	0.47	2380	Workable
R2 20%recycled		0/4 - 790 4/8 - 440 8/16- 530	80	0.51	2350	Workable
R3 40%recycled		0/4 - 770 4/8 - 430 8/16- 510	70	0.54	2340	Workable
R4 60%recycled		0/4 - 740 4/8 - 410 8/16- 500	90	0.57	2290	Slight water separation
R5 80%recycled		0/4 - 720 4/8 - 400 8/16- 480	60	0.58	2270	Low cohesion, Nonworkable
R6 100%recycled		0/4 - 695 4/8 - 385 8/16- 465	50	0.62	2240	Low cohesion, Nonworkable

According to experimental observations, recycled aggregate concrete loses its cohesion and becomes more nonworkable as the percentage of recycled aggregate increases.

If to the establishment of theoretical composition the A/C ratio was 0.55, the real (experimental), for carrying out the slump class the A/C ratio was initially used small, in the case of compositions with 0, 20 and 40% recycled aggregate and over the theoretical value in the case of recipes with a lot of recycled aggregate (80 and 100%). This is due, of course, to the increase of need for absorption water when increasing the amount of recycled aggregates.

2.3 Characterization of recycled aggregate concrete

All concrete samples, cubes (d=15cm) and cylinders (dxh=15x30cm), were demoulding after a day and kept in water at 20°C±2°C, until the date of 28 days necessary for determinations on hardened concrete. Preparing and maintaining test specimens was done according to standards.

Compressive strength was determined following a request by friction using a hydraulic press 3000kN with a loading speed of 0.6 ± 0.2 N/mm^2·s, according to SR EN 12390-3. Previously, concrete samples (cubes) were weighed and geometrically measured to calculate the apparent density.

On polyhedral samples with fresh surface (carved from cubes with sides of 15cm), with surface area approximately equal, it was determined water absorption (the open porosity) after 24 hours in which the samples of concrete were kept in water at 22°C.

It has been calculated the volumic absorption, A_v (%), with the relation (1),

$$A_v = [(m_1 - m)/(m_1 - m_2)]\cdot 100, \qquad (1)$$

where m is the mass of dry material to constant weight, m_1– the mass of saturated material with water, after 24hours and m_2 – the mass of saturated material hydrostatic weighed.

As a nondestructive method for analyzing hardened concrete was used ultrasonic method using an betonoscop UTD 1004 NAMICON type which measure the ultrasonic propagation time (t in seconds) through concrete. The dynamic modulus of elasticity was calculated using the equation (2),

$$E_d = V_L^2 \cdot \rho_a \cdot \frac{(1+\nu_d)\cdot(1-2\cdot\nu_d)}{(1-\nu_d)} \qquad (2)$$

in which:

- V_L is the ultrasonic pulse velocity, V_L = 0,15/t (m/s);
- ν_d - Poisson coefficient (ν_d =0,22);
- ρ_a - apparent density of the hardened concrete, Kg/m^3.

Static modulus of elasticity was determined by means of electronic press by 3000kN, equipped with load and displacement sensors, for cylindrical concrete samples with different proportions of recycled aggregate.

3. Results and discussions

In figure 2 are represented on the same graph the variations of apparent density and compressive strength depending on the percentage of recycled aggregate in hardened concrete. A decline in concrete density was observed with the increase in content of the recycled aggregate, situation also evidenced in the case of densities obtained for fresh concrete. Obviously, this behavior is due to the smaller density of the recycled aggregate compared to the natural aggregate.

The compressive strength of concrete has a surprising variation which was expected to decline by increasing the percentage of the recycled aggregate. This was not confirmed by the destructive tests carried out, much more so, in some compositions seeing a slight increase in strength to concrete with natural aggregate, figure 2.

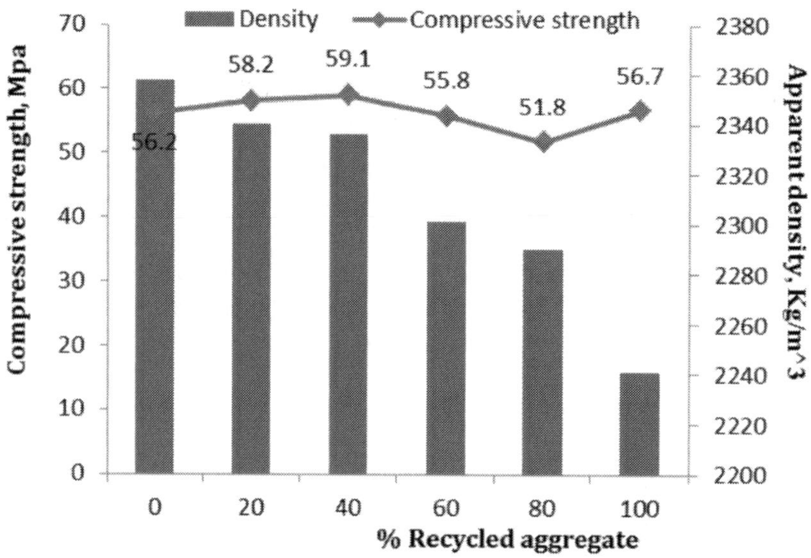

Figure 2 The variation of apparent density and compressive strength of concrete, depending on the proportion of recycled aggregate content.

The explanations for such behavior arise from the following:

- concrete from which the recycled aggregate comes has not been used, its quality is very good, not containing impurities as it usually happens to the demolition aggregat;
- recycled aggregate was obtained by crushing, it has "fresh" surfaces, roughness, which provides a better adhesion to the cement paste compared to the river aggregate;
- unhydrated cement granules of recycled aggregate allow the formation of additional amounts of hydrocomponents with binding properties, during setting and curing of the new concrete, which leads to a strengthening of the structure.

The water absorption of recycled aggregate concrete is greater with the higher proportion of recycled aggregate in concrete (which has a high porosity compared to natural aggregate), correlated with increasing real A/C ratio, used. This is evidenced in figure 3.

Nondestructive tests indicates the decreased of the longitudinal ultrasonic propagation velocity by increasing the proportion of recycled aggregate in concrete (table 2), respectively with porosity increasing and density decreasing of the concrete (V_L=340m/s in air and approx. 5000m/s in metallic bar for calibration).

With the help of stress-strain diagram, drawn using an loading and unloading program of the specific charges, it was determined the static modulus of elasticity by the slope of the graphic (figure 4), after stabilization of the deformations.

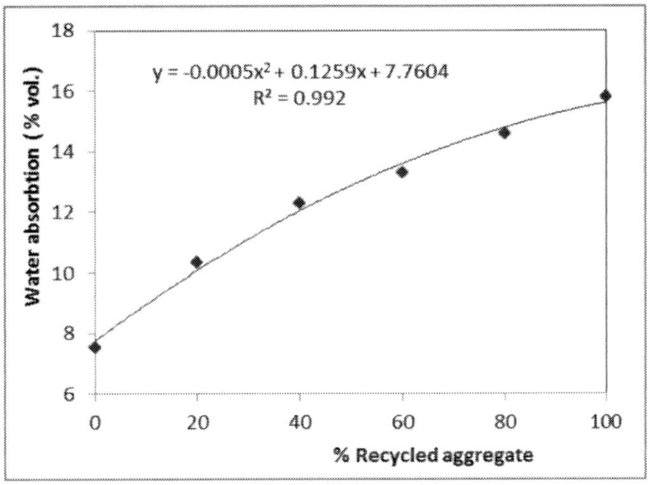

Figure 3 The variation of water absorbtion (% vol.), depending on the proportion of recycled aggregate content.

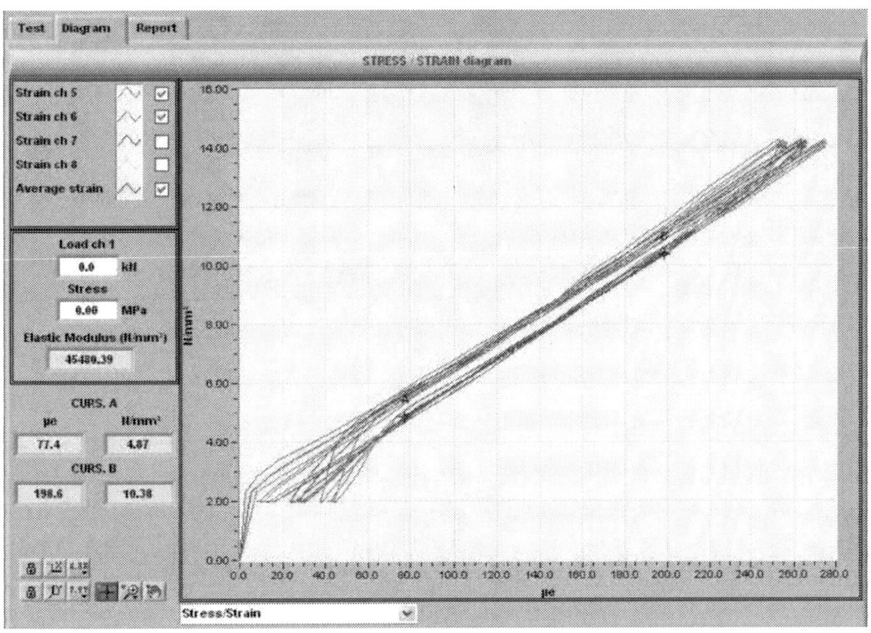

Figure 4 Software interface used for determination of the static modulus of elasticity for the R3 concrete.

Table 2 Recycled aggregate concrete characteristics

Characteristics	R1, 0% recycled	R2, 20% recycled	R3, 40% recycled	R4, 60% recycled	R5, 80% recycled	R6, 100% recycled
Ultrasonic pulse velocity, V_L (m/s)	4688	4505	4491	4386	4333	4261
Dynamic modulus of elasticity, E_d (MPa)	45345	41554	41226	38731	37620	35602
Static modulus of elasticity, E_s (MPa)	48254	47059	45480	43952	38333	32983

The static modulus of elasticity and the dynamic modulus of elasticity, obtained by nondestructive ultrasonic method, shows a decrease values by increasing the proportion of recycled aggregate in concrete, respectivelly by increasing the A/C ratio (table 2 and figure 5). This is attributed to a lower modulus of elasticity of the recycled concrete aggregates compared to natural aggregate and modulus of elasticity of the new cement stones from concrete [12-13].

Generally, dynamic modulus of elasticity is greater than the static modulus [14], different behavior in this case are due to the fact that for E_s determinations were carried out at 365 days and 28 days for E_d.

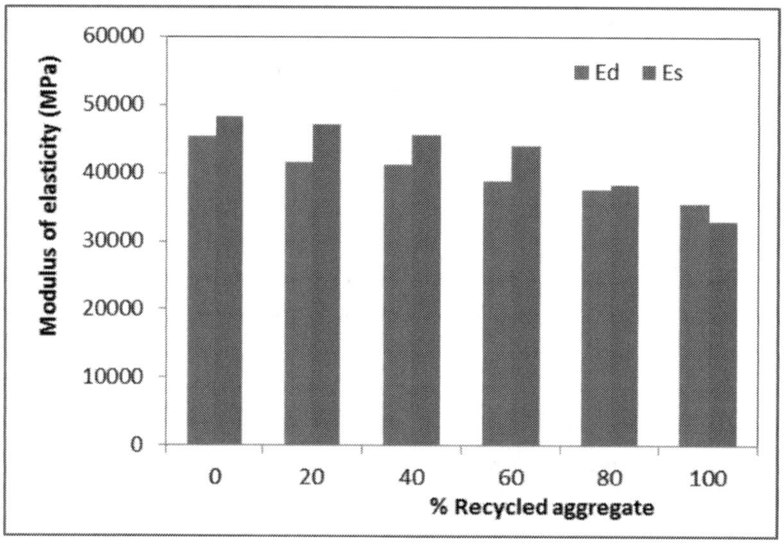

Figure 5 The variation of modulus of elasticity (% vol.), depending on the proportion of recycled aggregate content.

4. Conclusions

Increasing the use of recycled aggregates in new concrete will help eliminate demands on natural resources, reduce landfill disposal, and reduce energy costs associated with transport. The aim to characterize the recycled aggregates is their valorification in new concrete used in construction.

In this regard have been made a series of concrete in which the recycled aggregate content was varied from 0 to 100%. The new concrete were characterized by point of view of the change in the density, compressive strength, water absorbtion, ultrasonic pulse velocity and modulus of elasticity with the proportion of recycled aggregates.

The majority of these properties shows lower values by increasing the share of recycled aggregate in concrete. This is attributed, particularly, to the presence of adhered mortar in recycled aggregate which causing increased porosity and leads to lighter concretes.

Also, it has been shown that an increase in recycled aggregate content not necessarily mean a reduction in compressive strength, quality of the aggregate having a decisive role.

Acknowledgement

This work was supported by the Romanian National Authority for Scientific Research, CNDI-UEFISCDI, in the framework of 145/02.07.2012 project.

References

[1] Cement & Concrete Association of New Zealand 2011 *Best Practice Guide for the use of Recycled Aggregates in New Concrete*

[2] Robu I 2009 Aspecte privind utilizarea de agregate reciclate si subproduse industriale in beton *Simpozionul National "Noi reglementari pentru beton (Produse-Proiectare-Executie)"* Bucuresti 191-195

[3] Sanchez de Juan N and Gutierrez P A 2009 *Study on the influence of attached mortar content on the properties of recycled concrete aggregate* Construction and Building Materials, **23(2)** 872-877

[4] UEPG, 2009-2010, *Sustainable Development in the European Aggregates Industry* - Annual Review

[5] Marinkovic S *et al* 10-12 october 2010 Recycled aggregate concrete for structural use – an overview of technologies, properties and applications, *Aces workshop – Innovative materials and techniques in concrete construction* Corfu

[6] Tabsh S W and Abdelfatah A S 2009 Influence of recycled concrete aggregates on strength properties of concrete, *Construction and Building Materials* **23(2)** 1163–1167.

[7] Safiuddin M *et al* 2011 Properties of high-workability concrete with recycled concrete aggregate, *Materials Research* **14** 248-255.

[8] Shehata M H *et al* 2000 Reactivity of reclaimed concrete aggregate produced from concrete affected by alkali–silica reaction, *Cement and Concrete Research*, **40(4)**, 575-582

[9] Organismul Național de Standardizare din România 2014 *Romanian Standard: Beton. Specificație, performanță, producție și conformitate* SR EN 206:2014

[10] Deju R *et al* 2015 Selection tests for recycled radioactive sand obtaining method *Romanian Reports in Physics* **67(2)** 673-692

[11] Robu I and Mazilu C 2015 *Materiale de Construcție* **1** Ed. Conspress

[12] Xiao J et al 2005 Mechanical Properties of Recycled Aggregate Concrete Under Uniaxial Loading *Cement and Concrete Research* **35(6)** 1187-1194

[13] Rao M C *et al* 2011 Influence of Field Recycled Coarse Aggregate on Properties of Concrete *Materials and Structures*, **44(1)** 205-220

[14] Mindess S 2003 *Concrete* 2nd Edition, Pearson Education Inc. Upper Saddle River NJ 644

IOP Conf. Series: Materials Science and Engineering **161** (2016) 012073 doi:10.1088/1757-899X/161/1/012073

Thermal design of a natural gas - diesel dual fuel turbocharged V18 engine for ship propulsion and power plant applications

S Douvartzides and I Karmalis

Laboratory of Internal Combustion Engines, Department of Mechanical and Industrial Engineering, Technical Education Institute (TEI) of Western Macedonia, Koila-Kozani, Greece.

E-mail: sdouvar@teiwm.gr; karmalis@teiwm.gr

Abstract. A detailed method is presented on the thermal design of a natural gas - diesel dual fuel internal combustion engine. An 18 cylinder four stroke turbocharged engine is considered to operate at a maximum speed of 500 rpm for marine and power plant applications. Thermodynamic, heat transfer and fluid flow phenomena are mathematically analyzed to provide a real cycle analysis together with a complete set of calculated operation conditions, power characteristics and engine efficiencies. The method is found to provide results in close agreement to published data for the actual performance of similar engines such as V18 MAN 51/60DF.

1. Introduction

For more than a century, internal combustion engines (ICEs) have revolutionized both human transportation and habits. Spark (Otto) and compression (Diesel) ignition engines have spanned everyday life applications from cars, trucks and small airplanes to agriculture and manufacturing equipment, trains, ships and power plants. In view of the future increase of both global population and energy demand, internal combustion engines are currently attracting a renewed research interest aiming at greater thermal efficiencies and lower fuel consumptions. Diesel engines, especially, are found to play a critical role in the high power range from 0.5 to 100MW. Such engines typically operate below 1000 rpm, have up to 18 cylinders in line or V configurations, and find applications in ship propulsion, driving of machinery and power generation plants [1-3].

Thermal evaluation of ICEs has always been one of the fundamental criteria for engine design and performance. Theoretical and experimental methods on this field have been documented through the years in numerous books and reports [3-7]. In theory, the accuracy of a thermal design depends on the number and strength of the applied assumptions. Ideal air cycle analysis is commonly employed as fast and easy to provide the maximum thermal efficiency of an engine at a given pressure ratio. More accurate but complex methods need to account for real cylinder compositions, real valve timing, heat transfer, fluid flow dynamics, real geometry considerations and real combustion phenomena. Such real cycle methods are able to closely simulate actual performance and are valuable for preliminary modeling, cost effective engine development and energy management improvements.

In the present work, a detailed method is presented for the thermal design of a large turbocharged compression ignition engine. Real cycle analysis is accomplished based on fundamental science and empirical relations and is found able to accurately predict engine performance in comparison to actual data documented for the MAN 51/60DF (dual fuel) engine [2, 8]. The study was undertaken to formulate a basis for future work on internal combustion engine simulations and exhaust heat recovery. The method employed provide a set of engine operation conditions exploitable for energy management and efficiency improvements in diesel engine power plants.

2. Engine and Fuel Characteristics

The MAN 51/60DF engines are offered in 6-to-9 in-line and V12 to 18-cylinder models covering a power range up to 18000 kW at 500 rpm with a brake mean effective pressure rating of 1.9 Mpa [2]. As an actual example the four stroke V18 cylinder MAN 51/60DF engine has been considered. A cross section of the engine is shown in Fig. 1. This is a natural gas-diesel dual fuel engine equipped with a MAN TCA77 (four stroke version) turbocharger. Engine and turbocharger characteristics are summarized in Table 1 and Table 2, respectively [8, 9]. Engine speed is set at the maximum value of 500rpm and operation is considered steady at full load.

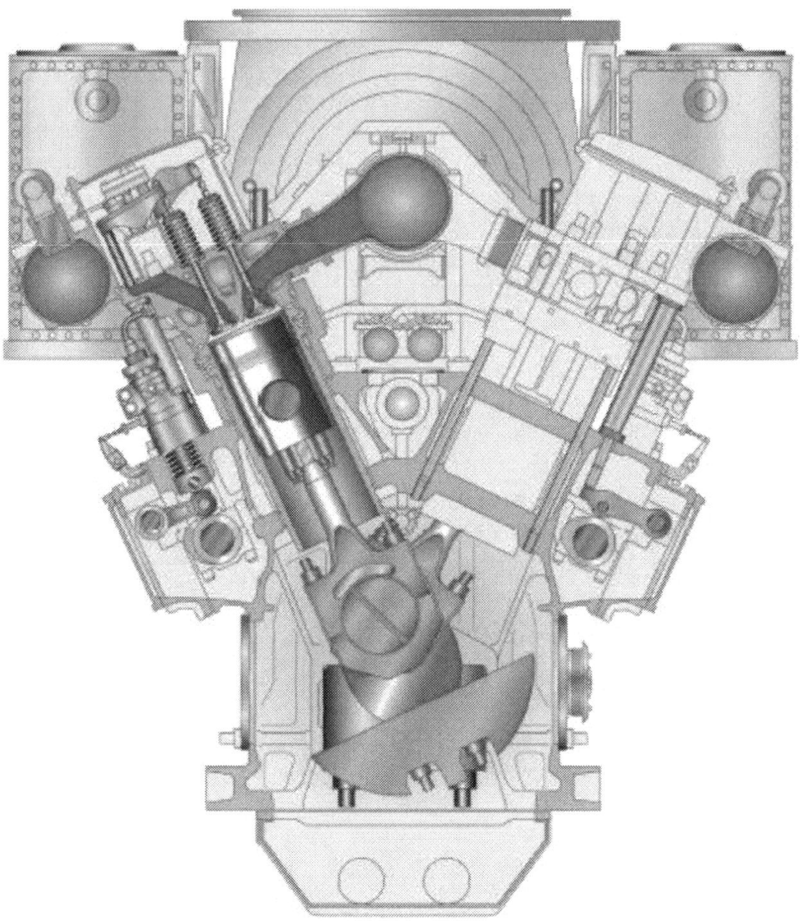

Figure 1. Cross section of the V18 MAN 51/60DF engine [8].

Table 1. 18 cylinder MAN 51/60DF engine characteristics [8]

Parameter	**Value**

Number of cylinders	18
Cylinder bore	510 mm
Piston stroke	600 mm
Displacement per cylinder	122.5 lt
Compression ratio	13.3
Vee angle	50°

Table 2. MAN TCA77 (four stroke version) turbocharger characteristics [9]

Parameter	**Value**
Turbine type	Axial flow turbine
Pressure ratio	up to 5.5
Maximum permissible temperature	650°C
Maximum permissible engine power	20900 kW
Air flow rate	about 15-35 m^3/s

The engine may operate with either diesel oil or natural gas. In the course of the analysis two modes of operation have been considered, one with heavy oil with an approximate composition of 87 wt. %. carbon and 13 wt. %. hydrogen, and one with natural gas with an approximate composition of 95% mol CH_4, 4% mol C_2H_6 and 1% mol C_3H_8. Lower heating values (LHV) were calculated to be 40 MJ/kg and 49.8MJ/kg, respectively. Combustion typically takes place with lean fuel mixtures and the analysis was conducted assuming a lambda λ ratio equal to 2.

3. Method

Atmospheric conditions were assumed at $T_0 = 293$ K and $p_0 = 0.1$ MPa. Turbocharger pressure ratio was set at the maximum value of $\Pi = 5.5$, and compressor outlet temperature and pressure were calculated as

$$p_C = \Pi p_0$$

$$T_C = T_0\left(1 + \frac{\Pi^{(\gamma-1)/\gamma} - 1}{\eta_C}\right)$$

where $\gamma = c_p/c_v$ is the average specific heat capacity ratio of air during compression (equal to 1.39) and η_C is the isentropic efficiency of the compressor taken 0.8. Air density may also be taken as

$$\rho_\alpha = \frac{10^6 p_C}{RT_C}$$

where $R = 287$ J/kgK is air gas constant.

For a turbocharged engine at the nominal speed of maximum brake power, the pressure of the residual gases inside the cylinder is typically 75-98% of the p_C. Assuming a mean percentage of 85% this was calculated as

$$p_r = 0.85 p_C$$

Then the temperature of the residual gases was assumed equal to $T_r = 825$ K. The assumed value will later be calculated and corrected if necessary.

Air exiting from the turbocharger typically undergoes a temperature change inside the intake manifold before entering the cylinder. This is of the order of -5 to +10 K and an average value of

$\Delta T = 5$ K was considered. Moreover, pressure loss in the intake manifold can be estimated according to Bernoulli equation as

$$\Delta p = 10^{-6}(\beta^2 + \xi_{\text{in}})\left(\frac{V_{in}^2}{2}\right)\rho_\alpha$$

Here β is a coefficient of velocity decrease inside the manifold, ξ_{in} is a coefficient of local pressure loss due to intake valve port restriction and V_{in} is the mean velocity of air at the valve port restriction. Empirical values of $(\beta^2 + \xi_{\text{in}})$ vary between 2.5 and 5, while V_{in} typically vary between 50 and 130 m/s. In the present study, assumed values were taken equal to 3 and 70 m/s, respectively. Then, air pressure at the end of intake stroke is

$$p_1 = p_C - \Delta p$$

The mass ratio of residual gases to fresh air inside the cylinder is given as

$$x = \left(\frac{T_C + \Delta T}{T_r}\right)\left(\frac{p_r}{rp_1 - p_r}\right)$$

where r is compression ratio, and the fraction of the residual gases in the total mass trapped is

$$x_r = \frac{x}{1 + x}$$

Air temperature at the end of intake stroke is approximately given by

$$T_1 = \frac{T_C + \Delta T + xT_r}{1 + x}$$

Cylinder gas composition during compression is calculated according to the relations in Table 3 [5, 6]. Based on this composition, the specific heat of the reactants may be estimated as a function of temperature through correlations given in literature [5, 6, 10]. During compression, a mean polytropic exponent n_c is considered. This is calculated through an iterative converging scheme following the relations

$$n_c = \gamma + \Delta\gamma$$

$$T_2 = T_1 r^{n_c - 1}$$

$$T_{av} = \frac{T_1 + T_2}{2}$$

$$\gamma = 1 + \frac{8.314}{c_{v,R@T_{av}}}$$

where $c_{v,R@T_{av}}$ is the reactant mixture specific heat in the average temperature T_{av} of the compression process, and $\Delta\gamma$ is assumed equal to -0.02. For Diesel engines $\Delta\gamma$ is typically in the range of -0.02 to +0.02 [7]. This take into consideration the overall heat transfer between the reactants and the cylinder walls during compression. Negative $\Delta\gamma$ values correspond to heat transfer from the gases to the cylinder wall while positive values correspond to heat transfer in the opposite direction. After convergence, temperature at the end of the compression stroke is calculated and pressure at the same state is given by

$$p_2 = p_1 r^{n_c}$$

Compression is followed by fuel injection, combustion, and rapid increase in temperature and pressure. Cylinder gas composition after combustion is calculated according to the relations in Table 4 [4, 5]. Given the compositions of the reactants and products one can calculate the coefficient of molar multiplication μ as

$$\mu = \frac{N_P}{N_R}$$

where N_R, N_P are the total number of moles of the reactants and products, respectively.

Table 3. Composition of the reactant mixture in moles per mole of O_2 reactant for $\lambda \geq 1$. For a hydrocarbon fuel C_aH_b it is $\kappa = 4/[4 + (b/a)]$. M_f is the fuel molecular weight.

Reactant	**moles/moles of O_2**
Fuel	$\frac{4(1-x_r)(1+2\kappa)/\lambda}{M_f}$
CO_2	$x_r\kappa/\lambda$
$H_2O_{(g)}$	$2x_r(1-\kappa)/\lambda$
CO	0
H_2	0
O_2	$1 - x_r/\lambda$
N_2	3.773
Moles of reactant mixture: N_R	$(1-x_r)\left[\frac{4(1+2\kappa)/\lambda}{M_f} + 4.773\right] + x_r N_P$

Table 4. Composition of the burned mixture in moles per mole of O_2 reactant for $\lambda \geq 1$. For a hydrocarbon fuel C_aH_b it is $\kappa = 4/[4 + (b/a)]$.

Product	**moles/moles of O_2**
CO_2	κ/λ
$H_2O_{(g)}$	$2(1-\kappa)/\lambda$
CO	0
H_2	0
O_2	$(\lambda-1)/\lambda$
N_2	3.773
Moles of product mixture: N_P	$[(1-\kappa)/\lambda] + 4.773$

3.1. Diesel fuel mode analysis

In the Diesel fuel mode of operation, the temperature of the combustion products T_3 at the end of the combustion process may be calculated by the first law of thermodynamics, written as

$$\zeta\frac{\text{LHV}}{N_R} + \left(c_{v,R@T_2} + 8.314\alpha\right)T_2 + 2270(\alpha - \mu) = \mu c_{p,P@T_3}T_3$$

Here, T_2 and T_3 are set in Celsius degrees, ζ is a factor of heat utilization, LHV is in kJ/kg, N_R is the number of moles of the reactants in kmol/kg of fuel, and α is the combustion pressure ratio. For a turbocharged Diesel engine at full load operation ζ is about 0.86 and α is about 1.5 [7]. The energy balance is solved iteratively to provide T_3 and the maximum pressure of the cycle is given as

$$p_3 = \alpha p_2$$

The ratio of pre-expansion is given by

$$\varepsilon = \frac{\mu T_3}{a T_2}$$

and the ratio of main expansion is

$$\delta = \frac{r}{\varepsilon}$$

During the expansion stroke, a mean polytropic exponent n_e is considered. This is calculated through an iterative converging scheme according to the relations

$$n_e = \gamma + \Delta\gamma$$

$$T_4 = \frac{T_3}{\delta^{n_e - 1}}$$

$$T_{av} = \frac{T_3 + T_4}{2}$$

$$\gamma = 1 + \frac{8.314}{c_{v,P@T_{av}}}$$

where $c_{v,P@T_{av}}$ is the product mixture specific heat in the average temperature T_{av} of the expansion process, and $\Delta\gamma$ is assumed equal to -0.02. After convergence, temperature at the end of the expansion stroke is calculated and pressure at the same state is given by

$$p_4 = \frac{p_3}{\delta^{n_e}}$$

Now the temperature of the residual gases inside the cylinder may be calculated and compared to the value assumed before during the analysis. The new T_r value is taken by

$$T_r = \frac{T_4}{\sqrt[3]{\frac{p_4}{p_r}}}$$

and an iterative calculation can be applied until the convergence of T_r value.

3.2. Natural gas fuel mode analysis

In the natural gas mode, the engine operates according to a lean-burn Otto combustion cycle. The pre-mixed lean gas–air mixture is ignited by the compression ignition of a small quantity of marine diesel pilot fuel injected into the main combustion space. A pilot injection of less than 1% of the normal liquid fuel quantity ensures very low NOx emissions in gas mode [2].

In an Otto cycle the temperature of the combustion products T_3 at the end of the combustion process may be calculated by the first law of thermodynamics, written as

$$\zeta \frac{\mathrm{LHV}}{N_R} + c_{v,R@T_2} T_2 = \mu c_{v,P@T_3} T_3$$

where ζ is again a factor of heat utilization set at 0.86 and T_2, T_3 are set in Celsius degrees. The energy balance is solved iteratively and give T_3. For a gas engine this typically varies between 2200 and 2500K. The maximum pressure of the cycle is given as

$$p_3 = \frac{0.85 p_2 \mu T_3}{T_2}$$

During the expansion stroke, a mean polytropic exponent n_e is considered. This is calculated through an iterative converging scheme according to the relations

$$n_e = \gamma + \Delta\gamma$$

$$T_4 = \frac{T_3}{r^{n_e - 1}}$$

$$T_{av} = \frac{T_3 + T_4}{2}$$

$$\gamma = 1 + \frac{8.314}{c_{v,P@T_{av}}}$$

where $c_{v,P@T_{av}}$ is the product mixture specific heat in the average temperature T_{av} of the expansion process, and $\Delta\gamma$ is assumed equal to -0.02. After convergence, temperature at the end of the expansion stroke is calculated and pressure at the same state is given by

$$p_4 = \frac{p_3}{r^{n_e}}$$

Then, the temperature of the residual gases inside the cylinder may be calculated again and iteration may be followed until the convergence of the T_r value.

3.3. Cycle evaluation

Volumetric efficiency is given for both fuel modes as

$$\eta_v = \frac{T_C}{T_C + \Delta T} \frac{1}{r - 1} \frac{1}{p_C} (r p_1 - p_r)$$

Indicative mean effective pressure is given for diesel cycle by

$$imep = \frac{0.95 p_C}{r - 1} \left[\frac{a\varepsilon}{n_e - 1} \left(1 - \frac{1}{\delta^{n_e - 1}}\right) - \frac{1}{n_c - 1} \left(1 - \frac{1}{r^{n_c - 1}}\right) + a(\varepsilon - 1) \right]$$

and for natural gas (Otto) cycle by

IOP Conf. Series: Materials Science and Engineering **161** (2016) 012073 doi:10.1088/1757-899X/161/1/012073

$$imep = \frac{0.96p_C}{r-1}\left[\frac{p_3}{p_2(n_e-1)}\left(1-\frac{1}{r^{n_e-1}}\right)-\frac{1}{n_c-1}\left(1-\frac{1}{r^{n_c-1}}\right)\right]$$

Friction mean effective pressure was calculated for both fuel cases by the correlation given by Kolchin and Demidov [7] for direct injection four stroke diesel engines as

$$fmep = 0.089 + 0.0118\bar{u}_p$$

where $\bar{u}_p = 2Ns$ is piston mean velocity at engine speed N. Hence, brake mean effective pressure is given as,

$$bmep = imep - fmep$$

Indicative power is given for both fuel modes in kW as

$$P_i = \frac{imepV_D N}{n_R}$$

where $imep$ is indicative mean effective pressure in MPa, V_D is overall engine displacement volume in liters, N is crankshaft rotational speed in 1/s and n_R is 2 for four stroke engines.

Indicative cycle efficiency is given for diesel cycle as

$$\eta_i = \frac{\lambda\left(\frac{A}{F}\right)_{st} imep}{\rho_f \eta_v \mathrm{LHV}}$$

where λ is lamda ratio, $(A/F)_{st}$ is stoichiometric air-fuel mass ratio, ρ_f is diesel fuel density, and LHV is diesel fuel lower heating value in MJ/kg. For the natural gas cycle η_i is given by

$$\eta_i = \frac{\lambda\left(\frac{A}{F}\right)_{v,st} T_C imep}{\rho_f \eta_v \mathrm{LHV}}$$

where $(A/F)_{v,st}$ is stoichiometric air-fuel molar or volume ratio and LHV is natural gas lower heating value in MJ/m^3.

Mechanical efficiency is given by

$$\eta_m = \frac{bmep}{imep}$$

and brake mean effective pressure is given by

$$P_b = \frac{bmepV_D N}{n_R}$$

The thermal efficiency of each cycle is calculated as

$$\eta_{th} = \eta_i \eta_m$$

Finally, brake specific fuel consumption is calculated by the expression

$$bsfc = \frac{3600}{\eta_{th}\text{LHV}}$$

4. Results and Discussion

The cycle analysis of both fuel modes has been conducted according to the method described. Operation pressure and temperature conditions have been estimated in all salient cycle points. These are summarized in Table 5. As can be observed, both modes provide similar operation conditions and therefore are expected to give about the same power and efficiency characteristics. Since both fuels provide similar cycle operation the selection criteria of engine materials are about the same in both cases. However, one can observe that maximum engine temperature is 2430.93 K when operation takes place with natural gas. This temperature is about 150 K higher than the maximum temperature of the Diesel cycle.

Table 5. Calculated operated conditions for the Diesel and Natural Gas operation modes.

Parameter	Diesel mode	NG mode
Ambient temperature, T_0 (K)	293	
Ambient pressure, p_0 (MPa)	0.1	
Turbocharger outlet temperature, T_C (K)	517.63	
Turbocharger outlet pressure, p_C (MPa)	0.55	
Fraction of residual gases, x_r	0.0436	0.0425
Temperature of residual gases, T_r (K)	825	848
Temperature at the end of intake stroke, T_1 (K)	535.83	536.47
Pressure at the end of intake stroke, p_1 (MPa)	0.522	0.522
Temperature at the beginning of combustion, T_2 (K)	1332.66	1334.09
Pressure at the beginning of combustion, p_2 (MPa)	17.29	17.29
Temperature at the end of combustion, T_3 (K)	2284.35	2430.93
Pressure at the end of combustion, p_3 (MPa)	25.93	26.82
Temperature at the end of expansion, T_3 (K)	1095.54	1128.05
Pressure at the end of expansion, p_3 (MPa)	1.09	1.10

Results for power characteristics and engine efficiencies are given in Table 6 in comparison with data published for the actual V18 MAN 51/60DF engine. As shown, results for brake mean effective pressure, torque and brake power are in satisfying agreement with the values reported for the actual engine. Mean effective pressure is found around 1.9 MPa and brake power is estimated close to the reported value of 17550 kW. Overall, the method employed is found to be highly accurate and promising for implementation in future works.

Table 6. Calculated performance characteristics for the Diesel and Natural Gas operation modes in comparison to real values published in [2, 8].

Parameter	Diesel mode	NG mode	Real value
Engine speed	500 rpm		
Fuel mass flow rate, $\dot{m}_f$ (kg/h)	4030.66	3412.82	
Air mass flow rate, $\dot{m}_a$ (kg/h)	116332.9	116336.2	
Volumetric efficiency, η_v (%)	94.95	94.95	
Brake specific fuel consumption, $bsfc$ (g/kWh)	230.66	190.40	183.5[8]†
Indicated mean effective pressure, $imep$ (MPa)	2.107	2.156	

Friction mean effective pressure, $fmep$ (MPa)	0.207	0.207	
Brake mean effective pressure, $bmep$ (MPa)	1.900	1.949	1.900[2]
Indicated thermal efficiency, η_i (%)	43.26	41.99	
Mechanical efficiency, η_m (%)	90.17	90.40	
Brake (real) thermal efficiency, η_{th} (%)	39.01	37.96	
Torque, τ (kNm)	333.7	342.3	335.2[8]
Indicated power, P_i (kW)	19376.81	19826.78	
Friction power, P_f (kW)	1902.88	1902.88	
Brake (real) power, P_b (kW)	17473.91	17923.89	17550[8]

† for Diesel mode without attached pumps

5. Conclusions

This paper focuses on the thermal design and evaluation of an 18 cylinder four stroke turbocharged dual fuel reciprocating engine with the characteristics of the commercial V18 MAN 51/60DF model. Assuming constant engine speed at 500 rpm and both Diesel and natural gas modes of operation, an analytical method was presented in detail to calculate real cycle operation conditions, power characteristics and engine efficiencies. Results have shown that both fuel modes provide similar cycle operation with a maximum pressure of 26.82MPa and a maximum temperature of 2430.93K.

Brake mean effective pressure was estimated 1.900 MPa and 1.949 MPa at Diesel and natural gas operation, respectively. Engine power was estimated 17473.91 kW and 17923.89 kW at Diesel and natural gas operation, respectively. These results are found in close agreement with the actual values of $bmep$ =1900 MPa and P_b = 17550 kW which are reported by the MAN manufacturer. Overall, the applied method was found to be highly accurate and promising for future application in combustion engine research.

References

[1] Kuiken K 2008 *Diesel Engines for Ship Propulsion and Power Plants, Vol. 1* (Target Global Energy Training)
[2] Woodyard D 2009 *Pounder's Marine Diesel Engines and Gas Turbines* 9th edition (Elsevier Ltd.)
[3] Taylor C F 1985 *The Internal Combustion Engine in Theory and Practice* Vol. 1 (The MIT Press)
[4] Ferguson C R and Kirkpatrick A T 2001 *Internal Combustion Engines: Applied Thermosciences* 2nd edition (New York: John Wiley & Sons Ltd.)
[5] Heywood J B 1988 *Internal Combustion Engine Fundamentals* (McGraw Hill)
[6] Caton J A 2016 *An Introduction to Thermodynamic Cycle Simulations for Internal Combustion Engines* (UK: John Wiley & Sons Ltd.)
[7] Kolchin A and Demidov V 1984 *Design of Automotive Engines* (Mir Publishers)
[8] *51/60DF Project Guide - Marine: Four-stroke dual-fuel engines compliant with IMO Tier II* 2014 (Augsburg: MAN Diesel & Turbo SE)
[9] *Exhaust Gas Turbochargers* 2011 (Augsburg: MAN Diesel & Turbo SE)
[10] McBride J B, Gordon S and Reno M A 1993 *Coefficients for calculating thermodynamic and transport properties of individual species* (USA: National Aeronautics and Space Agency (NASA))

Waste reduction possibilities for manufacturing systems in the industry 4.0

P Tamás, B Illés and P Dobos

Institute of Logistics, University of Miskolc, Hungary

E-mail: alttpeti@gmail.com

Abstract. The industry 4.0 creates some new possibilities for the manufacturing companies' waste reduction for example by appearance of the cyber physical systems and the big data concept and spreading the „Internet of things (IoT)". This paper presents in details the fourth industrial revolutions' more important achievements and tools. In addition there will be also numerous new research directions in connection with the waste reduction possibilities of the manufacturing systems outlined.

1. Introduction

The customer needs can be only satisfied with creation of the appropriate manufacturing technologies and logistics processes [1, 2]. The realization of the process improvement becomes necessary for the both areas in the interest of the competitiveness' observing and/or increase. Nowadays the improvement of the up to date manufacturing technologies [3, 4] (turning, injection moulding, …, etc.) has less improvement potential than the improvement of the logistic systems. This is the result of the industry 4.0 which was created some new possibilities (Internet of Things (IoT), cyber physics systems, big data, …, etc.) in the formation, improvement of the manufacturing logistics systems [5, 6].

In the past the continuous production was widely applied but recently the intermitted manufacturing has come to the front because of the continually changing customer needs. In our opinion those companies will be competitive in the future who are able to satisfy the unique customer needs before the order deadline and lower cost than theirs competitors.

The process developmental methods' reconsideration has become necessary because of increase the processes' complexity, such as joint application of the value stream mapping's method and the simulation modeling techniques [7]. The new efficiency improvement possibilities have become available with using the devices of the industry 4.0 and lean philosophy. In this essay are introduced the essence of the lean philosophy, history of the industry 4.0 and its more important devices and finally there is outlined the research and waste reduction possibilities.

2. Lean philosophy in the manufacturing

There are some definitions in connection with the lean philosophy's explanation. In our opinion the essence of this philosophy was the best expressed by Taichi Ohno. He said that the essence is the lead time's reduction between the ordering and the money receipt with the elimination of the wastes [8].

Basically the philosophy has two principles namely the reduction of the wastes and respect for human [9]. In the last period the lean philosophy starting from the TPS has developed into a well-defined device and rule system, which is usable for both production and services fields' reduction of logistics wastes. The lean basically distinguishes the 3 MUs, namely the fluctuation (Mura), the overload (Muri) and the 8 essential wastes (Muda) [9]. We can state that the muri and the muda can result in all cases muda, therefore a lot of literature speaks only from the muda's elimination. The 8 wastes are the followings: overproduction, unnecessary inventory, -transport, -motion, -waiting, -defects, - over-processing, not used skills [9].

There are appeared numerous very important new solutions in connection with the lean philosophy at the manufacturing areas, which are the followings:

- kanban system,
- heijunka (production leveling),
- jidoka,
- andon,
- Poka-Yoke,
- SMED,
- Kaizen,
- 5S.

Formation the workers' problem solving ability is very important to create the continuous improvement inside the examined area. The Toyota has been realizing this with application the improvement- and the coaching Kata [10].

The coaching kata's aim is teaching the improvement kata. The improvement kata is a routine for realization of the efficient process improvement. We can reach easier the aims of the process improvement with practice of the mentioned katas.

Table 1. Industrial revolutions

Industrial revolution 1 Beginning: from the 18th century's sixties More important things: • steam-engine, • mechanization of the textile plants, • steamships, steam railway, …, etc.	**Industrial revolution 2** Beginning: from the 19th century's seventies More important things: • electricity, • oil industry, • steel industry, • technological equipments' manufacturing (lathe, drill, miller, …, etv.), • invention of the internal combustion engines, • formation of the service industry, • mass production.
Industrial revolution 3 Beginning: from the 20th century's thirties More important things: • nuclear power, • new technologies, • CAD/CAM systems, • CIM systems, • processes, networks, • closed material flow systems.	**Industrial revolution 4** Beginning: from today More important things: • Internet of Things (IoT), • cyber physics systems, • industry 4.0, • logistics 4.0, • manufacturing 4.0, • lean 4.0., • hospital logistics 4.0., …, etc.

3. Formation of the industrial revolution

Basically the industrial resolutions are related to changing the social, economic and technological. It can be told that the appropriate economic and social environment is necessaries in the interest of the technologies' invention and spread. We can define the beginning of the first industrial revolution from invention of the steam-engine (1769, James Watt). The industrial revolutions' characteristics are introduced with table 1.

4. More important devices of the industry 4.0

Nowadays the industry 4.0 has numerous devices from which the more important are the Internet of Things (IoT), cyber physics systems and the big data. These devices will transform the whole world according to a lot of experts [6, 15]. These will create new possibilities for the improvement of the production and services processes. This chapter's aim the presentation of the more important terms which will be used in the next chapters.

Internet of things (IoT): Firstly this term was used in 1999 by Kevin Ashton [5]. The IoT enables the access of the different equipment through internet/some networks, as well as in certain cases the communication between these equipment. In the last decades the peoples were recorded the majority of the data what we can find on the internet. In essence this was significantly determined the available data's quantity. In the interest of the realized logistics systems' more efficient improvement we need to collect more information about the systems' things (e. g. products, machines, material handling equipment, humans, etc.) with use of the IoT. On the basis of these data we can analyze more information about our system and we can optimize more efficient as before. For example if we put some sensors on a technological equipment's more important parts which will send signals from the status of the parts, then we can get information before the failure of the technological equipment. The eNet company made an online research through the internet about that what kinds of IoT solutions do the population know. These solutions are the followings:

- smart alarm at home (we can get a message in the case of abnormality),
- smart car (the car can sense its environment and give a message or push a brake in the case of emergency),
- smart reading device (we can get real time information about the status of the gas- or the electric meter through a network, consequently the provider can work efficiently),
- smart home (we can control the equipment in our house from our mobile phone e. g. heating, lighting, …, etc.),
- online cash register (communication between the National Tax and Customs Administrations and cash registers).

Cyber physics systems: Development of the informatics and automation, as well as the increase theirs cohesion enabled the application of the cyber physics systems (if an electronic device contains a control and network connection then we can call this system as cyber physics system). The cyber physics systems are able to collect data from their environment, as well as after analyzing of these data they are able to modify their positions. The cyber physics systems are connected through network, their significant part are also connected with each other because of this we can apply the swarm intelligence, which can result more efficient work. In the practical life there are numerous areas where application of the cyber physics systems can result the increase of the efficiency. These areas are the followings:

- we can reduce the air resistance, consumption and the number of the accidents in the traffic with the coupling of the cars,
- we can reduce the manufacturing wastes with coupling the manufacturing system's elements,
- at the formation and actuation of the supply chains.

Big Data concept: The data amount in the world is approximately doubled in every two years [6], which result a huge amounts of data in the different areas of the life (astronomy, logistics, trade, stock exchange, …, etc.). We can create new services and useful conclusions with the elaboration of the data's correlation. Such service can be e. g. forecast for the flight prices' with a software, which is able to determine the estimated flight price on the basis of the previous period's data (in this cases we don't need to know the process of the price determination) [6, 11]. The big data's essence is determination of the probabilities with mathematical methods and procedures. In according to a lot of experts the big data will change significantly the future, because of that we can make decisions on the basis of the huge amount of the data without know of the causes-effects.

The expressions which were explained in this chapter are related to each other. We can't speak from cyber physics systems and big data without IoT.

5. More important research directions in the manufacturing systems' process improvement

Basically the aim of the industry 4.0 is the realization of the intermitted manufacturing with mass production's productivity and specific cost. If we would reach this aim then we can satisfy the unique customer needs quickly and efficiently. This looks like unreachable but the significant part of the technologies are available nowadays. In the interest of this aim's reach there is needed development of two logistics areas. These are:

- improvement of the value stream mapping's method,
- elaboration of new intelligent logistics solutions.

5.1. Improvement of the Value Stream Mapping's Method

The value stream mapping's method was created with use of the Toyota's material and information flow diagram. Firstly this was published in the "Learning to see" in 1999 by Mike Rother & John Shook [8, 9]. This method's relevant aim is the reduction of the wastes with improvement of the logistics processes. The value stream mapping's method has been simultaneously used for improvement of one product line's logistics processes so far.

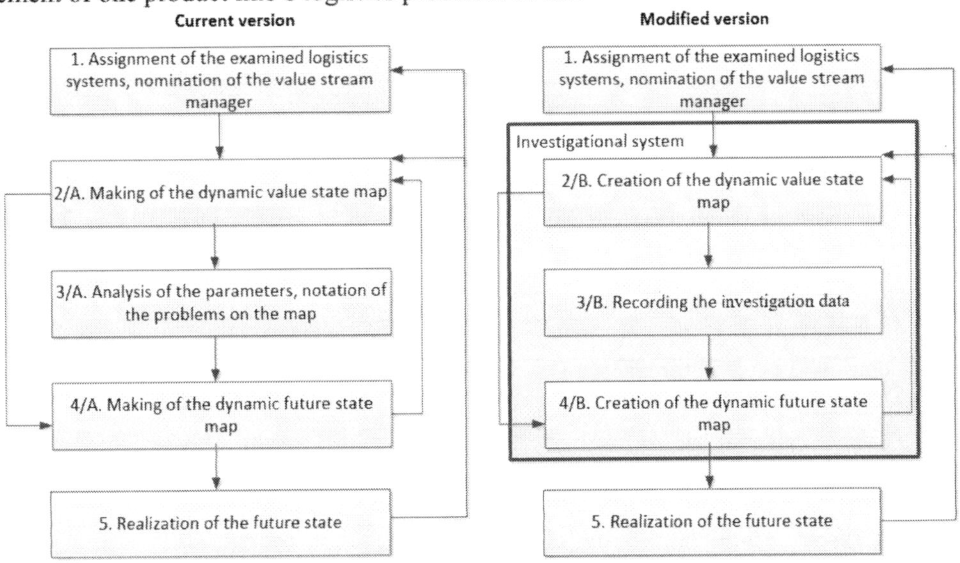

Figure 1. Steps of the dynamic value stream mapping

The basic method can't be used with appropriate efficiency in the case of complex logistics systems [2]. We can distinguish the value stream mapping's two types that are the static and the dynamic value stream mapping's method. The static value stream mapping can be used effectively at examination of one product line's processes. The other method can be used effectively at examination of complex logistics systems [12, 13]. The dynamic value stream mapping's method come to the front because of the increasing complexity of the processes. This method's steps was clearly determined but the process improvement's mode and measure is only realized on the basis of trial and error, anticipation, experience (we examine the possibilities through an simulation model). In our opinion there is a significant potential in elaboration of an examination system, which is able to the creation of the best system version on the basis of the predefined data structures, goal functions and optimization algorithms. The dynamic value stream mapping's current and modified versions are introduced on the figure 1. Application of the dynamic value stream mapping is realized according to the next steps:

1. Assignment of the examined logistics systems, nomination of the values stream manager: Firstly we have to determine the product lines. After that we have to assign the value stream managers to the product lines. The value stream manager is responsible for the improvement of the product lines' value stream [9].

2/A. Making of the dynamic value stream map: The value stream managers have to know their product lines' material- and information flow and have to collect the data for the current state map [9]. After this activity the experts have to make a dynamic value stream map with a simulation framework.

2/B. Creation of the dynamic value stream map: This step differ from the step 2/A therein that the collected data we have to record in a predefined data structure and on the basis of this data are created the simulation investigational model by a simulation framework. We have to elaborate the different data structures, algorithms, and the necessary adaptation for a selected simulation framework regarding the manufacturing systems.

3/A. Analysis of the parameters, nomination of the problems on the map: We can analyze the effects of the logistics parameters' modification regarding the examined logistics system by the dynamic value stream map (for example: effects of the changeover time's reduction for the work in process and the lead time, effects of the equipments' failure for the production plan, ..., etc.). We can determine the main problems on the dynamic value stream map regarding the assigned logistics system's material- and information flow by this parameter analysis [9].

3/B. Recording the investigation data: This steps differ from the step 3/A. therein that we have to give the process improvement's data, conditions and goal functions in a predefined data structure. We have to elaborate this the data model's structure to be applied which is adequate for recording of the analysis possibilities, conditions and the goal functions' data. Of course we have to realize the adaptation of this data structure.

4/A. Making of the dynamic future state map: We have to create the future state map without the determined problem.

4/B. Creation of the dynamic future state map: The dynamic future state map is automatically created on the basis of the investigational data with take into consideration the investigational aims. We have to elaborate to this an algorithm which is able to generation of the system variations and the selection of best variation.

5. Realization of the future state: Firstly we have to mark the value stream map loops on the future state map (pacemaker loop, additional loop). After that we have to create a task list regarding every loop.

After we have made the task list then we have to make a yearly value stream plan with the following contents:

- scheduling the assigned tasks' realization,
- responsible(s) for the implementation,
- to be reached target indicators,
- monitoring of the implementation.

The next step is the realization and its control [9]. The value stream mapping contains basically five sequential steps where feedback can occur between the steps in the next cases:

- if the current state map doesn't contain such information that will be necessary to the future state map,
- if the future state was realized then we can execute a new examination from the second step.

The current and the modified versions differ from each other in the next ways:

- the modified version' use is easier because in the case we can use elaborated data structure, algorithms (the current version determine only the investigational frames),
- the modified version's use require less lead time, because the examination' steps were determined exactly,
- determination of the future state map is realized with more efficiency because in this case a simulation optimization algorithm will chose the best future state map instead of our trial and error and/or experience.

We can state if we can realize the above concept then we can get a faster and more cost efficient process improvement method than in earlier. This method can increase significantly the user companies' competitiveness.

5.2. Elaboration of New Intelligent Logistics Solutions

In our explanation the intelligent logistics solutions those devices/(part)systems which are able to respond to changes of the external environment (with transmission of automatic information, control of the own operation) [14, 15]. It can be possible the more significant waste reduction with application and improvement of these solutions. The wastes' reduction will be realized because of the reduction of the lead time of the tasks to be realized, respectively the increase of the optimized collaboration between the systems objects (source and drain objects, material handling equipment, staff, …, etc.). The optimized collaboration will result more efficient human and machine resource utilization. The more important logistics solutions in the production area are the followings:

- Intelligent identification devices (this device will send a message from the external environment's disadvantageous changes e. g. if the temperature or humidity will reduce under a defined value).
- Intelligent technological equipment (e. g. automated manufacturing on the basis of the identify devices' data, failure forecast with use of the sensors, …, etc.).
- Intelligent quality control (e. g. the machine will realize the quality control on the basis of the RFID chip's data which can be found on the product).
- Intelligent material handling equipment (e. g. automated working on the basis of the environmental data).
- Intelligent warehouse (automated ordering on the basis of the stock reduction with use of the sensors).
- Intelligent logistics systems (e. g. Total or partial automated logistics systems with application of device system of the industry 4.0 which is based on the human-machine and/or machine-machine communication).

The industry 4.0 will have a relevant effect on the formation and actuation of the manufacturing systems. According to the forecast the next changes are possibilities:

- There can be realized the communication between the different devices (technological equipment, material handling equipment, parts, unit loads, …, etc.) and the central unit, respectively between the devices because of these the central control will be changed to decentralized central control in the future.
- The complex decision making will be changed to real time decision making with use of the simulation.
- The narrowly planned production systems will be changed to such production systems which are based on modularization. This can be possible because of the increase the technological equipment's flexibility (e. g. spread of the 3D printing).
- The passive parts will be changed to intelligent parts which will be able to influence their environmental with use of the preprogrammed information.

- There will be possible larger systems' optimization if we will use such devices' data which have network connection. Actually there will be possible the optimized creation and actuation of a company's total supply chain.
- New business models will be created with the processing of the huge amounts of collected data from the logistics systems (e. g. we will forecast the equipment's failure or the human's disease on the based on analysis of the previous period's data in the future consequently the production downtime will be reduced).

6. Summary

The industry 4.0 has been making a relevant changing at the manufacturing systems' formation and actuation. The appearance of the IoT and the cyber physics systems, as well as the big data have created a significant research potential regarding the more efficient actuation and continuous improvement of the logistics systems. The communication between the devices, the information which are derive from the product's tracking, as well as the possibilities in the network collaboration will provide a more widespread optimization possibilities for the manufacturing companies. This paper introduced in details the process of formation of the industry 4.0, as well as its current devices, possible improvement directions. In addition in this paper were also outlined the more relevant research possibilities in the case of the unit loads' manufacturing process' improvement/waste reduction.

Acknowledgment

"This project has received funding from the European Union's Horizon 2020 research and innovation programme under grant agreement No 691942". "This research was (partially) carried out in the framework of the Center of Excellence of Mechatronics and Logistics at the University of Miskolc."

References

[1] Bohács G, Frikker I, Kovács G 2013 Intermodal logistics processes supported by electronic freight and warehouse exchanges *Transport and telecommunication* **14**:(3) pp 206-213

[2] Tamás P 2016 Application of value stream mapping at flexible manufacturing systems *Key Engineering Materials* **686** pp 168-173 doi:10.4028/www.scientific.net/KEM.686.168

[3] Shukla P, Lawrence J 2010 Mathematical modelling of the fibre laser surface processing of a zirconia engineering ceramic by means of three-dimensional finite-element analysis, *Proc. IMechE* **225** doi: 10.13140/RG.2.1.2769.9923

[4] Felhő Cs, Kundrák J 2014 Investigation of the Topography of Machined Surfaces *Applied mechanics and materials* **693** pp. 412-417

[5] Ashton K 2009 That 'Internet of Things' Thing, in the real world things matter more than ideas *RFID Journal* (http://www.rfidjournal.com/articles/pdf?4986)

[6] Mayer V, Kenneth Cukier S 2013 Big Data, A revolution that will transform how we live, work, and think, ISBN 978-0-544-00269-2

[7] Mahfouz A, Arisha A 2013 Lean distribution assessment using an integrated framework of value stream mapping and simulation, *Proceedings of the 43rd Winter Simulation Conference, IEEE Computer Society* pp 3440–3449

[8] Womack J P, Jones D T 2008 Lean thinking *Simon & Schuster Inc.* ISBN 978-963-9686-83-0

[9] Rother M, Shook J 2003 Learning to See: Value Stream Mapping to Add Value and Eliminate Muda, *Lean Enterprise Institute* ISBN 0966784308

[10] Rother M 2009 Toyota Kata - Managing People for Improvement, Adaptiveness and Superior Results, ISBN 978-0-0-7163985-9

[11] Golosova M V, Grigorieva M A, Klimentov A A, Ryabinkin E A, Dimitrov G, Potekhin M 2015 Studies of Big Data metadata segmentation between relational and non-relational databases *Journal of Physics: IOP Conference Series* Volume **664** doi:10.1088/1742-6596/664/4/042023

[12] Solding P, Gullander P 2009 Concepts for simulation based value stream mapping, *Proceedings of the 2009 Winter Simulation Conference* pp 2231-2237
[13] Košťál P, Velíšek K 2011 Flexible manufacturing system *In: World Academy of Science, Engineering and Technology* **77** ISSN 2010-376X. pp 825-829
[14] Illés B, Trohák A, Bíró Z 2013 Can message filter algorithms for remote diagnostics of vehicles *Applied mechanics and materials* **309** pp 213-220
[15] Gubán A, Kása R 2014 Conceptualization of fluid flows of logistificated processes, *Advanced logistic systems* **7** pp 27-24

Effect of Hydrogen Use on Diesel Engine Performance

A Cernat[1], C Pana[1], N Negurescu[1], C Nutu[2], I Mirica[1], D Fuiorescu[1]

[1]Department of Thermotechnics, Engines, Thermical and Frigorific Equipment, University Politehnica of Bucharest, Blvd. Splaiul Independentei no. 313, Romania
[2]Department of Automotive Engineering, University Politehnica of Bucharest, Blvd. Splaiul Independentei no. 313, Romania

Email: cernatalex@yahoo.com

Abstract. Necessity of pollutant emissions decreasing, a great interest aspect discussed at **2015 Paris Climate Conference**, highlights the necessity of alternative fuels use at diesel engines. Hydrogen is considered a future fuel for the automotive industry due to its properties which define it as the cleanest fuel and due to the production unlimited sources. The use of hydrogen as fuel for diesel engines has a higher degree of complexity because of some hydrogen particularities which lead to specific issues of the hydrogen use at diesel engine: tendency of uncontrolled ignition with inlet backfire, in-cylinder combustion with higher heat release rates and with high NO_X level, storage difficulties. Because hydrogen storing on vehicle board implies important difficulties in terms of safety and automotive range, the partial substitution of diesel fuel by hydrogen injected into the inlet manifold represents the most efficient method. The paper presents the results of the experimental researches carried on a truck diesel engine fuelled with diesel fuel and hydrogen, in-cylinder phenomena's study showing the influence of some parameters on combustion, engine performance and pollutant emissions. The paper novelty is defined by the hydrogen fuelling method applied to diesel engine and the efficient control of the engine running.

1. Introduction

Necessity of pollutant emissions decreasing, a great interest aspect discussed at the **21st Session of the Conference of the Parties (COP21)** at **2015 Paris Climate Conference**, and the gradual diminution of the worldwide oil reserves highlights the necessity of alternative fuels use at diesel engines, in context of searching of alternative energy from durable and renewable resources.

Classic fuel replacement with alternative fuels was started many decades ago and nowadays also represents one of the most actual preoccupations connected by necessity of fossil fuel saving and the possibility of diesel engine pollutant emissions decrease and efficiency increase. These aspects impulse the research's to develop and use new fuel sources that allow the classic fuels substitution, even in a partial way. Many researchers' consider hydrogen a future fuel for the automotive industry due to its properties which define it as the cleanest fuel and due to its the unlimited production sources [1], [2], [3], [4], [5]. Hydrogen use as fuel for diesel engines is a high complexity aspect also because of some hydrogen particularities, such as: very wide flammability limits, very high combustion velocity, lower minimum ignition energy, lower caloric power per volume unit, higher diffusivity coefficient versus diesel fuel. These properties lead to a higher degree of complexity of hydrogen use as fuel at diesel engine which is reflected in some hydrogen particularities which lead to specific issues of its use at diesel engine: tendency of uncontrolled ignition with inlet backfire, in-cylinder combustion with higher heat release rates and even knock (higher maximum pressure rise rate, brutal engine running) and with high NO_X level, storage difficulties[1], [2], [3], [4], [5], [6], [7], [8], [9], [10]. Because of higher autoignition of hydrogen, defined by a lower cetane number, an important issue is related to the difficulty of the autoignition without require major modification of the standard engine design. This is the most critical issue. Due to improvement of the combustion at hydrogen use as addition in the inlet air to diesel engine, the level of CO, HC and smoke from the exhaust gases will decrease [2], [3], [4], [5]. At low and medium partial loads (often used in exploitation) and at use of low hydrogen energetic ratios, the NO_X emission level can decreases comparative to classic diesel engine. Due to combustion improvement and reduction of carbon content, the hydrogen use as fuel for diesel engine leads to the improving of the energetic and emissions engine performance [3], [4], [5], [10]. At dual fuelling, with diesel fuel and hydrogen, the CO, HC and smoke emissions has a much lower level comparative to diesel engine fuelled only with diesel fuel due to improvement of the combustion, a lower carbon content in air-fuel mixture and a higher homogeneity of air-hydrogen mixture [1], [2], [3]. Slightly increase of the CO_2 emission level can be registered [3]. At different regimes, the level of the NO_X emission increases at the rise of the hydrogen addition, especially at large engine load because the in-cylinder gases temperature is much higher comparative to the NO_X formation temperature [3], [4]. Different researcher's, N. Saravanan [5], Younkins M. [3] and Tomita [6] show that the NO_X emissions forming is avoided due to a shorter duration in which the high temperature reached inside the engine cylinder exists; thus, the NO_X emissions level decreases with almost 14% at hydrogen fuelling in relative low additions (up till 15% energetically substitution of diesel fuel) at small engine loads. At big hydrogen additions, the level of NO_X emissions increases, comparative to the emissions level of a standard engine, because increases the duration of maintaining high temperatures inside the cylinder [6], [4]. Talibi [7] shows the CO, HC, smoke emissions level decrease due to improvement of the combustion and a CO_2 emission slight increase, at hydrogen use in addition with diesel fuel [7]. Santoso shows the increase of NO_X emissions with the rise of hydrogen addition percent only when the gas temperature is over the temperature of NO_X formation [11], for higher engine loads. Researchers highlight the decrease of brake specific fuel consumption (BSFC) at the increase of hydrogen amount which replaces the diesel fuel due to the increase of brake thermal efficiency [3], [4], [12], [10]. Pollutant emissions are dependent on engine operating conditions (engine load, engine speed etc), fuel composition, air/fuel equivalence ratio and oxygen content. At use of the hydrogen as addition in inlet air, hydrocarbons and smoke emissions level decreases due to improvement of the combustion and of lower carbon content in air-fuel mixture [7]. For small hydrogen quantities, CO emission level decreases due to lower oxygen content in air-fuel mixture, but if the substitute percent of diesel fuel by hydrogen increases, the CO emission level increases, especially at higher engine loads [4]. At high engine loads, the NO_X emissions level increases with hydrogen addition percent increase, only when the gas temperature is over the temperature of NO_X formation [10], [13], [14]. Authors like Lambe, Talibi or Pechlivanoglou, show that the NO_X emissions level decreases at partial loads of engine operation and when small hydrogen quantities in addition in air are used [7], [13], [14]. The paper presents, in a comparative way, results of the experimental researches carried on a truck diesel engine fuelled with diesel fuel and in dual mode. The results obtained during experimental investigations show improvement of the energetic and pollution performance of the engine fuelled with hydrogen as addition in air comparative to classic engine. The influences of hydrogen addition are analyzed and shown.

2. Methodology

The research work has been carried out on automotive truck diesel engine fuelled with diesel fuel and dual fuelled with diesel fuel and hydrogen at different hydrogen flows, up till 40.3 L/min, in order to develop efficient methods of control of the in-cylinder complex processes from hydrogen fuelled diesel engine, which crucial influence the engine performance.

The study of in-cylinder phenomena's was developed and the influence of some parameters on combustion, engine performance and pollutant emissions are shown. The paper novelty is defined by the hydrogen fuelling method applied to diesel engine and the efficient control of the engine running. A novelty aspect is defined also by establishment of optimal correlation between *engine load-hydrogen addition-diesel fuel injection timing-EGR rate-exhaust gases temperature*, in order to control the combustion process and to obtain the best energetic and pollution engine performance. Thus, an important contribution in the knowledge field of hydrogen use at diesel engines is achieved, the professional literature offering extreme few information's at the present time. Hydrogen storing on vehicle board implies important difficulties in terms of safety and automotive range and that why the most efficient method of hydrogen use is the partial substitution of diesel fuel by hydrogen, which is injected into the inlet manifold. The used fuelling method was diesel gas.

The D2156 MTN8 diesel engine was mounted on a test bed equipped with adequate instrumentations: Hoffmann eddy-current dynamometer, AVL data acquisition system, thermometers, thermocouples, thermo resistances and manometers for monitoring the engine functional parameters, air flow meter, hydrogen flow meter, diesel fuel consumption device and AVL gas analyzer. The figure 1 presents the scheme of the experimental test bed.

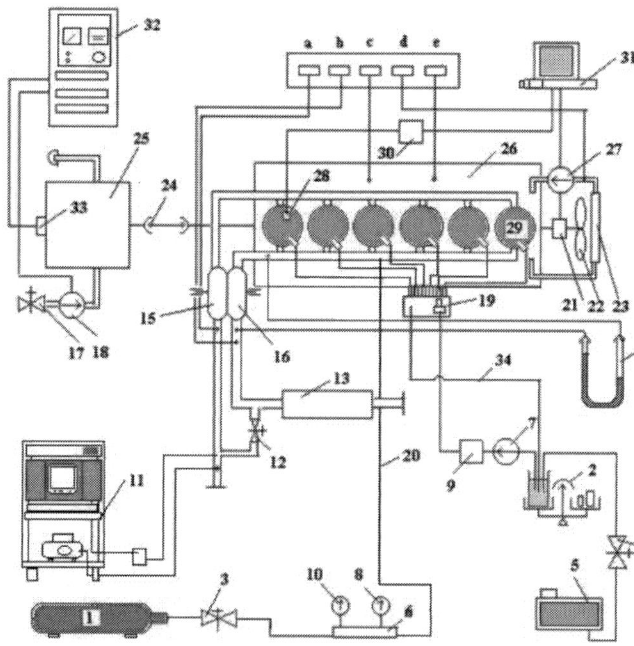

Figure 1. The diesel-hydrogen experimental engine test bed scheme.

The test bed has the following components: 1. hydrogen bottle; 2.weight balance; 3. hydrogen valve; 4. diesel fuel valve; 5. diesel fuel tank; 6. pressure regulator; 7. diesel fuel supplying pump; 8. manometer high pressure; 9. diesel fuel filter; 10. low pressure manometer; 11. AVL DiCom 4000 gas analyzer; 12. gas recirculation valve; 13. Meriam air flowmeter; 14. differential manometer for turbocharging pressure measuring; 15-16. turbo-compressor group; 17. dynamometer water cooling valve; 18. water circulation pump for dynamometer cooling; 19. diesel fuel injection pump; 20. hydrogen pipe; 21. incremental speed transducer; 22. ventilator 23. cooling liquid radiator; 24. coupler; 25. Hoffman eddy-current dynamometer; 26. D2156 MTN8 diesel engine; 27. water circulation pump for the engine cooling; 28. Kistler piezoelectric pressure transducer; 29. injector; 30. AVL charge amplifier; 31. PC with AVL 617 Indimeter acquisition board; 32. dynamometer controller; 33. dynamometer power cell; 34. diesel fuel pipe; a) inlet air temperature indicator; b) exhaust gas temperature indicator; c) oil temperature indicator; d) oil pressure indicator; e) water cooling temperature indicator.

The experimental research has been followed for different hydrogen flows at the operating regimen of 70% engine load, 1450 rev/min engine speed and at normal thermal regimen (80^{0}C cooling agent temperature). The engine was fuelled firstly only with diesel fuel then with diesel fuel and hydrogen in addition, at different hydrogen rates between 10.2 L/min and 40.3 L/min, hydrogen being injected into the engine inlet manifold at the pressure of 2.5 bar. For measurement of the volumetric flow rate of hydrogen a special hydrogen flow meter was used. The quantity of the hydrogen in inlet air that substitutes the diesel fuel is established by an energetic substitute ratio value which takes into consideration the energetically value of the both fuels, x_c [%]:

$$x_c = \frac{Ch_H \cdot Hi_H}{Ch_{DF} \cdot Hi_{DF} + Ch_H \cdot Hi_H} \cdot 100 \tag{1}$$

where: Hi_{DF}, Hi_H represents the lower heating value of diesel fuel and hydrogen in [kJ/kg] and Ch_{DF}, Ch_H are diesel fuel and hydrogen consumptions in [kg/h].

For only diesel fuelling the x_c value is zero.

In order to explain the tendency of variation for the smoke emission from the exhaust gases at dual fuelling, the exhaust smoke quantity was calculated with the thermodynamic model, zero-dimensional, two-zone, developed by authors. To establish a correlation with the measured values of smoke emission, the thermodynamic model uses the difference between speeds of soot formation and soot oxidation, the model being calibrated by in-cylinder pressure diagrams acquisitioned for 200 consecutive cycles. In order to evaluate the forming mass velocity of soot, at the same engine running regime, the Kuo correlation was used [15], [16], [17], [18]:

$$m_f = C_f \cdot m_c \cdot p^{0.5} \cdot \exp\left(-\frac{6290}{T}\right) \tag{2}$$

where: coefficient $C_f = (2.5....5.0)\cdot 10^{-3}$, m_c mass of fuel in gaseous state, p, T in-cylinder pressure and temperature.

The combustion of the soot particles is controlled by chemical reactions that take into consideration the phenomena's of adsorption of oxygen at the particle surface and desorption of the reaction products from the particle surface. Velocity's of adsorption and desorption of carbon monoxide at the particle surface depend in an exponential way by temperature [16], [17], [18]. The particle surface contains a very reactive zone, partial full of oxides in chemical equilibrium, and a less reactive zone. Practically, the global mass speed on surface unit for soot oxidation is established by summarizing of desorption velocities of the considered reactive zones, [16], [17], [18]:

$$v = Mc \cdot \left[x_A \cdot \frac{K_{Aa} \cdot p_O}{1 + K_Z \cdot p_O} + K_{Ba} \cdot p_O \cdot (1 - x_{Ao}) \right] \tag{3}$$

where coefficient Mc=12 is the molecular mass of the carbon.

The x_A is a surface fraction related with a very reactive zone and the 1-x_A fraction represents the less reactive zone. The surface fraction filled by oxides x_{Ao} is defined by the equilibrium between oxygen adsorption velocity from gaseous state (for oxides forming) and carbon monoxide desorption velocity from the droplet surface. With x_{Ao} is determinates the velocity of carbon monoxide release after the oxidation into the very reactive zone takes place [13], [14], [15]. Into the less reactive zone, the oxygen participate to a slow oxidation reaction that leads to the generation of a new active zone, followed by desorption of carbon monoxide with the velocity $K_{Ba} \cdot p_O \cdot (1-x_A)$. Also, during a phenomenon of chemical arrangement by thermal activation, the very reactive zone creates a new low reactive zone, with velocity of $K_T \cdot x_A$ [16], [17], [18]. The expressions semi-empirical relations of the velocity's are:

$$K_{Aa} = 20 \cdot \exp\left(-\frac{15098}{T}\right) \tag{4}$$

$$K_{Ba} = 4.46 \cdot 10^{-3} \cdot \exp\left(-\frac{7650}{T}\right) \tag{5}$$

$$K_T = 1.51 \cdot 10^{5} \cdot \exp\left(-\frac{48817}{T}\right) \tag{6}$$

$$K_Z = 21.3 \cdot \exp\left(-\frac{2063}{T}\right) \tag{7}$$

$$K_{Ad} = \frac{K_{Aa}}{K_Z} \tag{8}$$

evaluated in units of K_{Aa}, K_{Ba} [$g \cdot cm^{-2} \cdot s^{-1} \cdot bar^{-1}$], K_T [$g \cdot cm^{-2} \cdot s^{-1}$], K_Z [bar^{-1}], [16], [17], [18]

At relative low temperatures and partial pressures of oxygen, the entire surface is defined as a very reactive zone, x_A=1, the term K_T being negligible. The fraction x_A decreases once with the temperature increasing and becomes negligible comparative to the coefficient of desorption velocity, K_{Ba}, at high temperatures [16], [17], [18]. With these relations the velocity of soot formation and oxidation are determined and calculated by integration for determination of soot formation and oxidation laws.

From the soot balance, defined by the difference between formation and oxidation of soot quantities, for diesel fuelling and dual fuelling, is established the smoke relative final quantity (SFQ [%]) in the exhaust gases, for each substitute ratio x_c.

3. Results

From in-cylinder pressure diagrams, the heat release diagrams were calculated for the regime of 1450 rev/min and 70% engine load. The rate of heat release characteristics for all substitute ratios are presented in the figure 2. At hydrogen fuelling, the peak values of heat release rate raised for all substitute ratios comparative to diesel fuelling, with an increase of 11.4% in maximum heat release values. The increased maximum values of cycle heat release rates are in correlation with a sooner achieve per cycle of maximum heat release rate values. The increase of inlet hydrogen quantity bring closer to TDC (Top Dead Centre) the moment of peak value. The heat release rise rate increases at the hydrogen use comparative to diesel fuelling due to a better combustion, figure 2, fact that is in correlation with a sooner achievement per cycle of moment when maximum values of heat release rate occurs and influence the aspect of heat release laws, figure 3, with an significant increase of cycle heat quantity for x_c=3.11%.

Figure 4 shows the continuous decreasing in variation of brake specific energetic consumption (BSEC) with substitute ratios. The BSEC continuously decreases at the increase of hydrogen addition in inlet air.

The lowest BSEC of 9500 kJ/kWh is achieved for x_c values up till 2.42, corresponding to a hydrogen rate of 10.3...30.7 L/min, versus the value registered only for diesel fuelling of 9800 kJ/kWh, due to achievement of a much more homogeneous fuel-air mixture and the combustion improvement. The percentage of BSEC reduction is around 3% versus diesel fuelling. Comparative to diesel fuelling at hydrogen-diesel fuelling, due to combustion improvement and lower carbon content in the air-fuel mixture, the smoke emission level decreases for all substitute ratios, figure 5.

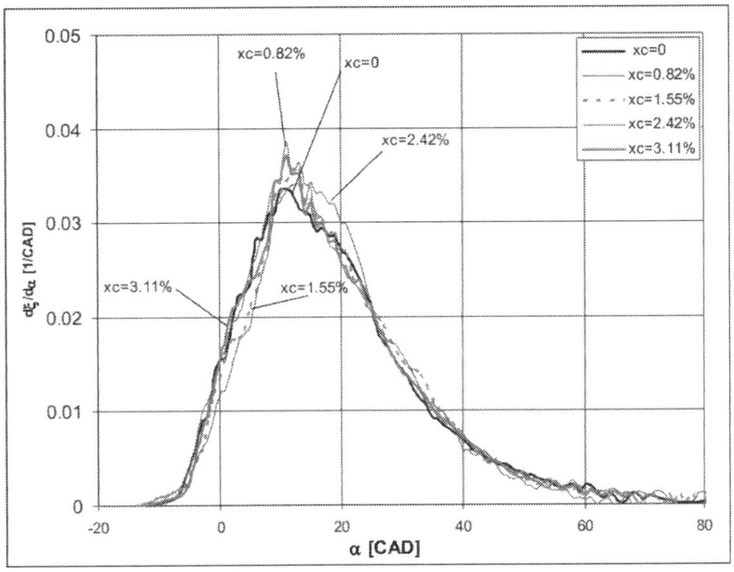

Figure 2. Heat release rate for different x_c at 70% load and 1450 rev/min regime.

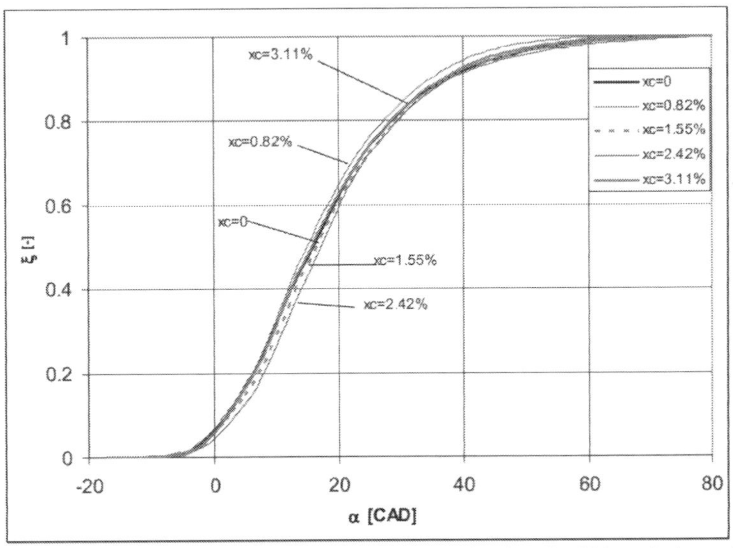

Figure 3. Heat release law for different x_c at 70% load and 1450 rev/min regime.

During combustion process hydrogen forms a homogeneous mixture versus heterogeneous mixture which is specific to diesel engine, fact that contributes to the reduction of the smoke emission, evaluated by K smoke number. Forward increase of hydrogen inlet quantity, which affect the combustion temperature, leads to a slightly increase tendency for smoke number with out exceeding the standard engine values, figure 5. At hydrogen cycle quantity increase, combustion improvement and the lower carbon content of the air-fuel mixture lead to the decrease of the level for CO_2 and CO emissions, as figure 6 and figure 7 shows. The CO emission decreases also due to higher burning velocity of hydrogen which accelerates the diesel fuel combustion at engine regimes of high loads.

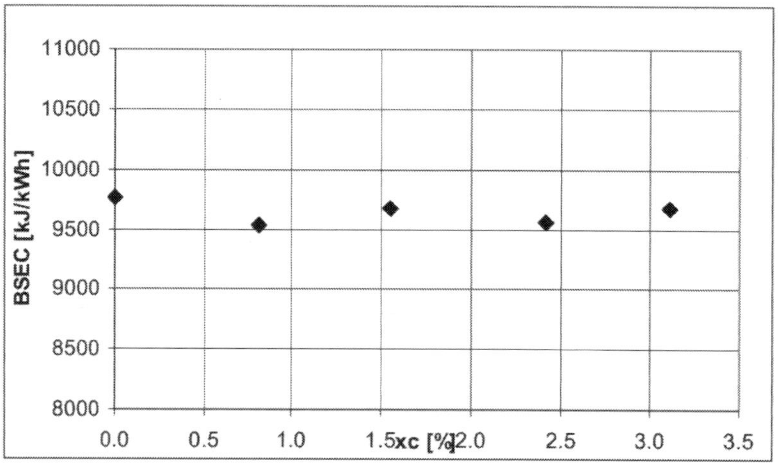

Figure 4. BSEC versus x_c at 70% load and 1450 rev/min regime.

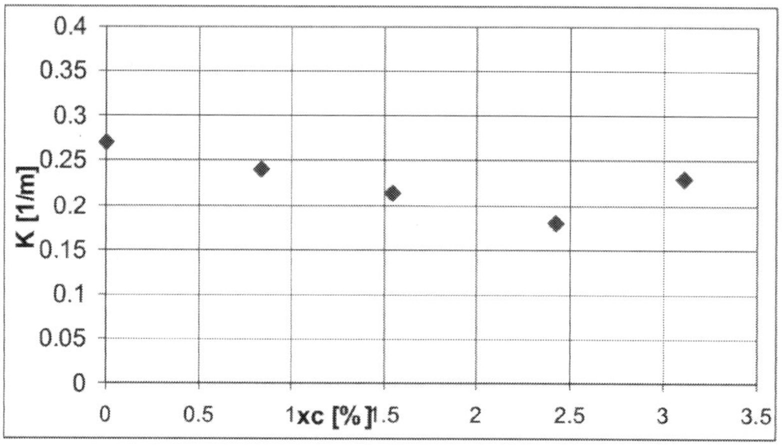

Figure 5. Smoke number versus x_c at 70% load and 1450 rev/min regime.

Figure 8 presents the variation of NO_X emissions level with the substitute ratios x_c. Comparative to classic engine, in hydrogen-diesel dual fuelling operation mode, the NO_X emissions level decreases with 5.6% at x_c=2.42 or inside the interval x_c= 2…3%.

At engine small loads and hydrogen use the diesel-fuel cycle dose is reduced and the high effect of combustion temperature on the NO_X formation is reduced. At the hydrogen adding, at the same engine load regime, the cycle dose of diesel fuel and the combustion duration decreases and the heat release rate slightly decreases at the beginning of the combustion, in the interval of 0…10 CAD, leads to global temperature and NO_X decreasing. In consecutive cycles, the combustion duration with higher heat release rate comparative to diesel fuelling is shorter. The values of in-cylinder global temperature, calculated with the thermodynamic model, figure 10, are below the value registered at x_c=0, at list in 0…20 CAD interval.

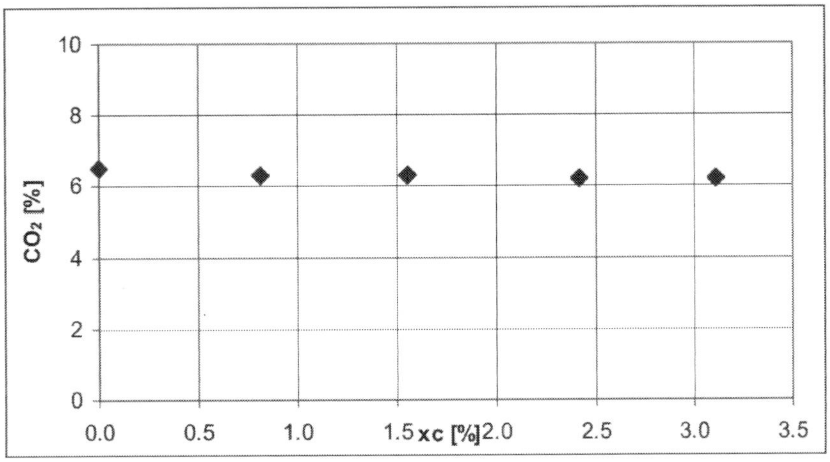

Figure 6. CO_2 versus x_c at 70% load and 1450 rev/min regime.

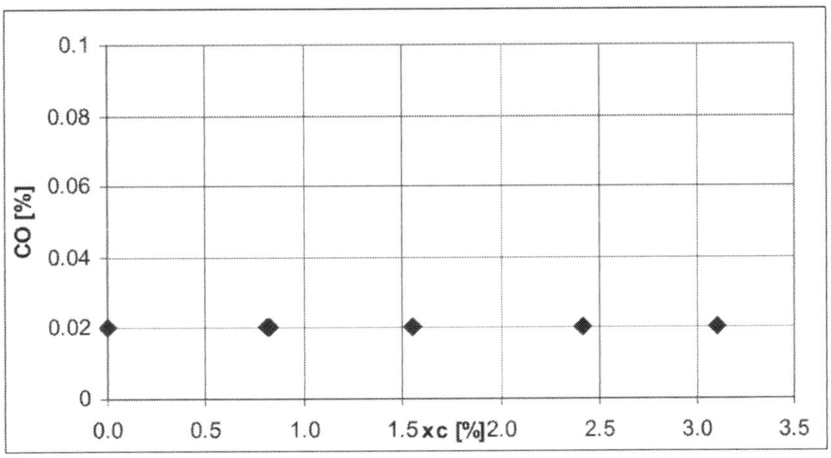

Figure 7. CO versus x_c at 70% load and 1450 rev/min regime.

Also the maximum values decreases with ~2.5% comparative to reference engine, for example from 1800 K at x_c =0 down till 1755 K at x_c = 3.11, below the values of NO_X thermal formation, fact that can explain the NO_X emission reduction [6], [7], [19]. In case of fuelling with small hydrogen quantities in addition in inlet air, the explication for NO_X emissions level decrease is the fact that hydrogen burns fast, leading to the avoiding of NO_X formation due to a shorter duration of the combustion and high temperatures (registered in some consecutive cycles or for different x_c=2.42) are registered only for short time ~1.8 ms – 2 ms, ("the combustion is so rapid that the high temperatures exist only for approx. 2ms." after Georgios Pechlivanoglou [13]). Whatson [14] and Thalibi et al. [7] explains NO_X decrease at the hydrogen adding in air through the increase of the mole fraction of water vapours in the combustion products produced at the hydrogen combustion which absorb energy released from combustion and thus the peak combustion temperatures decrease [7], which is in correlation with the anterior mentioned reduction for the maximum values of gases temperature.

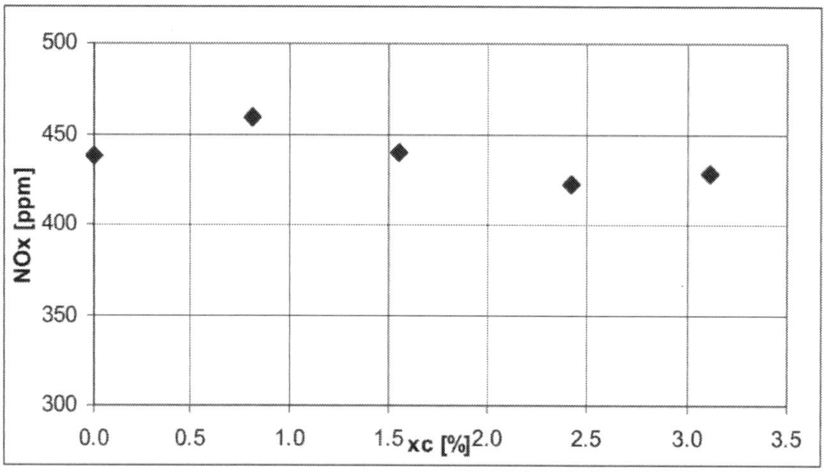

Figure 8. NO_X versus x_c at 70% load and 1450 rev/min regime.

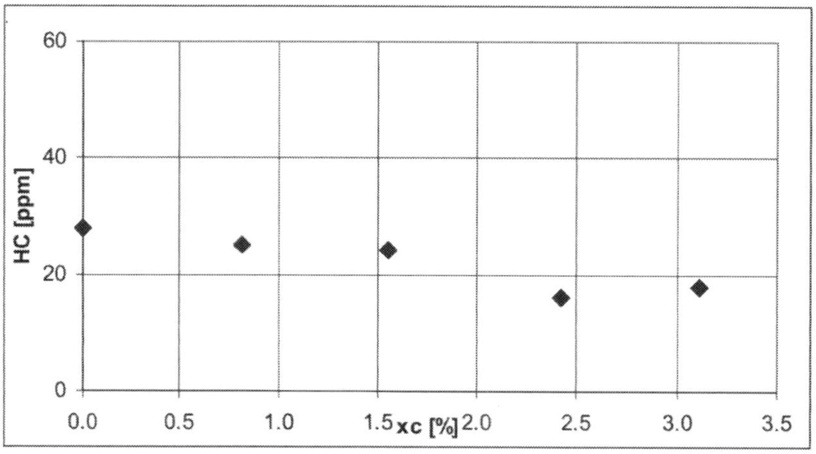

Figure 9. HC versus x_c at 70% load and 1450 rev/min regime.

Due to the general combustion improvement and of the fact that hydrogen don't contain carbon and has higher burning velocity, comparative to diesel fuel, the HC emissions level decreases at hydrogen use, figure 9, especially for raised values of x_c. For x_c=0.82…1.55 the HC emission level remains in the same area as classic fuelling with a slightly decrease tendency for low x_c values. The HC emissions level decreases comparative to diesel fuelling not only due to combustion improvement, but also due to lower carbon content in the final fuel-air mixture. The higher burning velocity of hydrogen accelerates the diesel fuel combustion at small engine loads, the HC emission being reduced with almost 50% for x_c=2…3.11.

The smoke variation tendency registered during experimental investigation, for a limited domain of substitute ratio variation, can be explained thru the combined effect of in-cylinder pressure and temperature. Soot forming speed is evaluated with the previous formula and represented in the figure 11 for different values of substitute ratio. The peak value for maximum soot forming speed is achieved for classic fuelling solution, at x_c=0. At dual fuelling, for x_c =2.42, the maximum speed of soot forming increases with 10%; even for small hydrogen quantities, for substitute ratio of x_c=1.55, the maximum speed of soot forming decreases with 10% and with 20% for x_c =3.11, comparative to diesel fuelling. For other x_c values the soot forming speed is comparable with values registered of x_c =0. Soot oxidation speeds are presented in the figure 12.

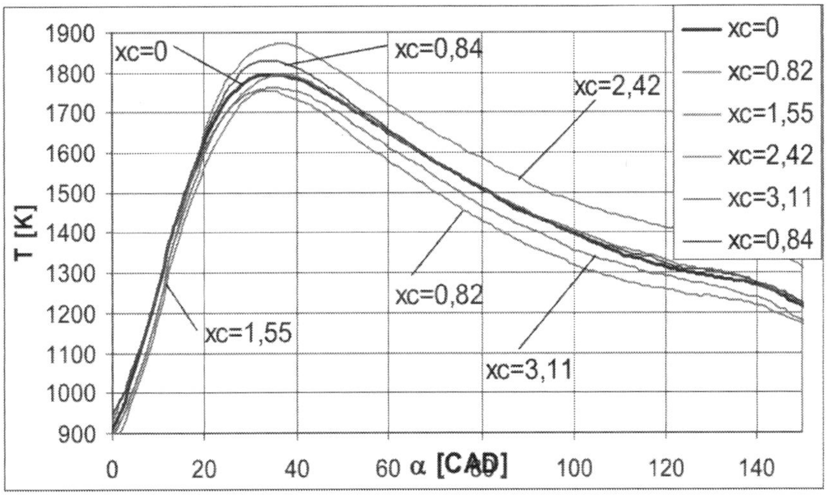

Figure 10. Combustion global temperature for different substitute ratios x_c.

At hydrogen fuelling the soot oxidation speed starts to increase even for low hydrogen quantities. The highest value of soot oxidation speed is reached for dual fuelling at x_c=2.42. For x_c = 2.42 the soot oxidation is with 70% more efficient comparative to diesel fuelling. In the presence of hydrogen, the soot oxidation speed increases with 18% for x_c=0.82 and with 30% for x_c= 0.84, the soot oxidation being improved. The oxidation speed decreases with 20% for x_c=3.11, but in condition of a much lower initial carbon forming speeds, figure 11 and figure 12.

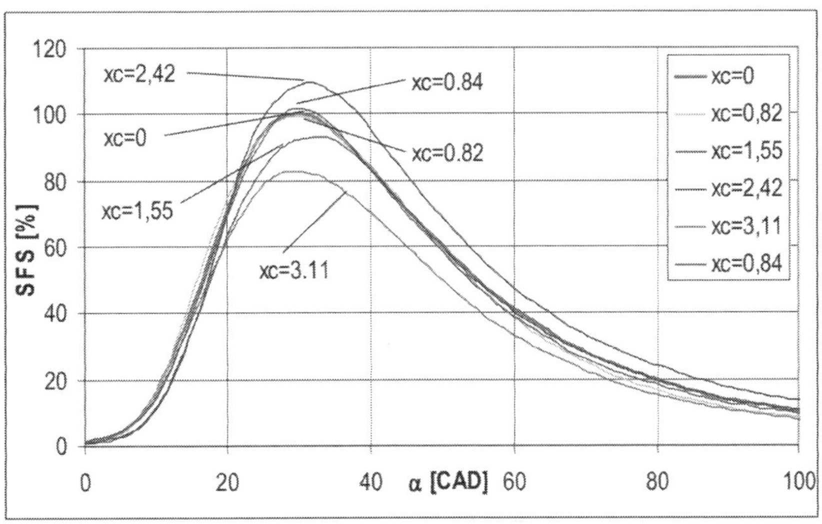

Figure 11. Soot forming speed at different substitute ratio x_c.

Even for small x_c values, x_c=0.82,..., 2.42, the soot oxidation is with 18%...70% more efficient and is achieved in initial condition of a 20% less lower speed of soot forming, comparative to classic fuelling, figure 11. Forward integration of soot forming and oxidation speeds leads to achievement of soot forming and oxidation laws presented in the figure 13 and figure 14 for different x_c. For only diesel fuelling the soot formation is maximum (100%) and is superior for x_c= 0.82, 0.84, 1.55 and 3.11, as figure 13 shows. At dual fuelling, for x_c=0.82…1.55 the quantity of formatted soot is lower with ~10%, and decreases with 20% for x_c=3.11. For x_c =2.42 the soot quantity formed inside the cylinder increase with 10%, but the improvement of the oxidation process, figure 11 and 14, will lead to the reduction of the in-cylinder final soot quantity.

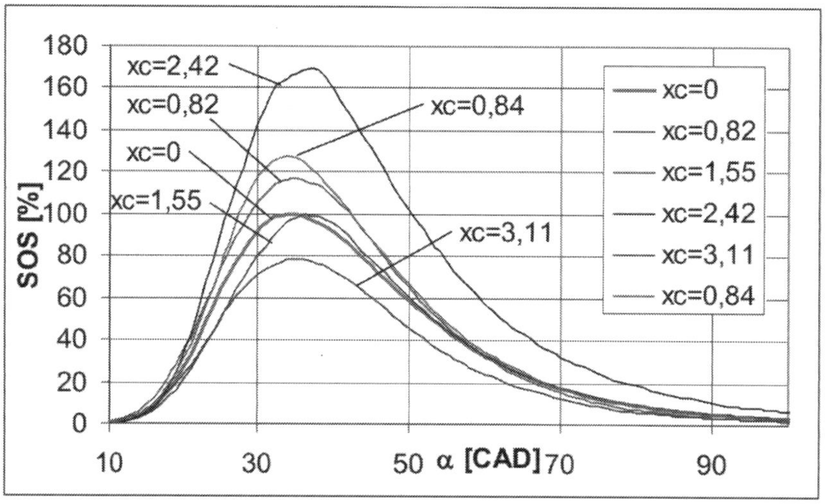

Figure 12. Soot oxidation speed at different substitute ratio x_c.

At x_c=1.1 the soot formation represents ~60%, initially being obtained with 34% less soot inside the cylinder, but the oxidation of formed soot being with 49% less efficient comparative to diesel fuelling.

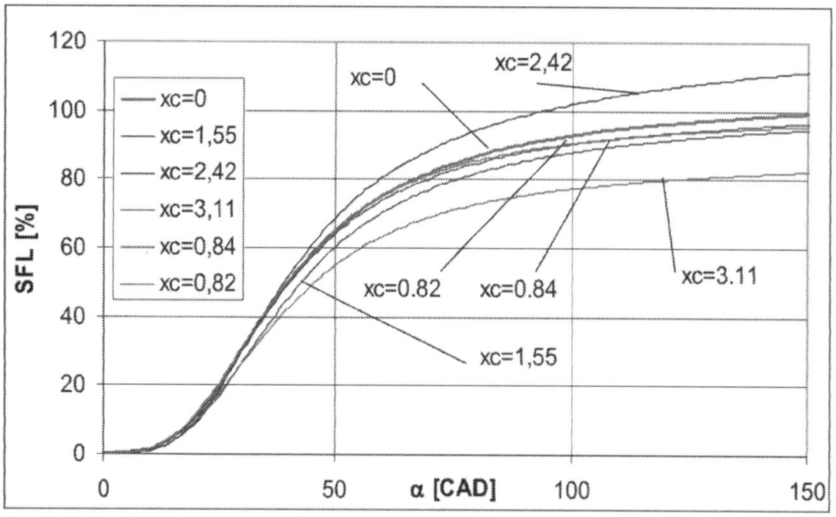

Figure 13. Soot forming law for different substitute ratios x_c.

For x_c =2.2 soot formation represent 70% and the soot quantity formed inside the engine cylinder being with 30% lower comparative to classic engine. The soot oxidation is assured in the same terms of the previous lower value of x_c. The improved soot oxidation process at dual fuelling, which is developed with a much higher speeds comparative to diesel fuelling, figure 12, leads to the increase of the soot quantity which is burned during combustion process. Thus, the soot oxidation efficiency increases with 10% at x_c=0.82, with 15% for x_c=0.84 and with 62% for x_c=2.42. For substitute ratios up till x_c =1.55 the soot oxidation is comparable with situation registered for diesel fuelling; for x_c=3.11 the quantity of soot oxidation is decreased with 22%.

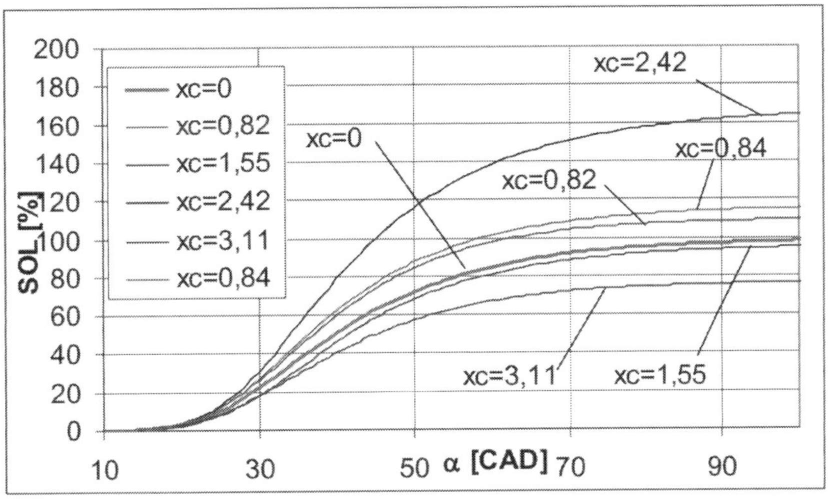

Figure 14. Soot oxidation law for different substitute ratios x_c.

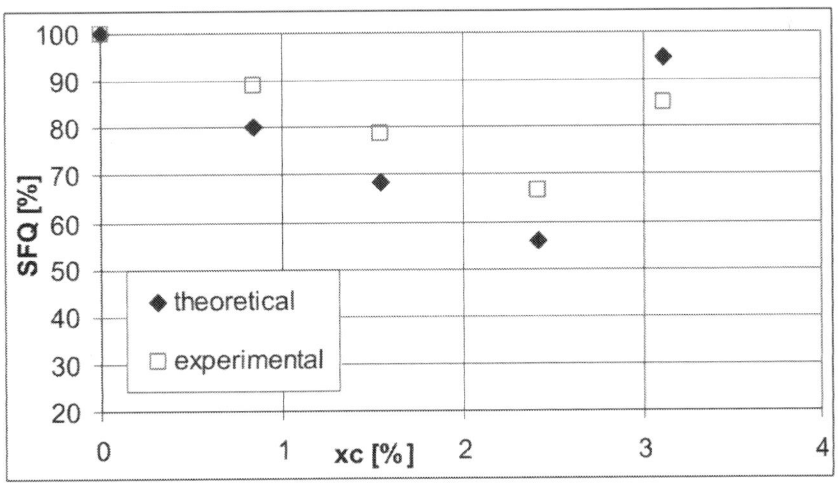

Figure 15. Smoke relative emission for different substitute ratios x_c.

Thus, at dual fuelling with diesel fuel and hydrogen, the soot oxidation process is accelerated, the soot oxidation efficiency increases up till 63%, in conditions of a much diminished process of soot formation since the hydrogen cycle quantity increases, the initial quantity of formatted soot being reduced with 20% comparative to diesel fuelling. Dual fuelling with diesel fuel and hydrogen reduces the soot forming process and influences in a much significant way the soot oxidation process, which is significantly improved comparative to diesel fuelling. At the increase of the hydrogen dose, acceleration of the soot oxidation process, in conditions of a much lower soot formed quantities, leads to a continuously decrease of the calculated smoke emission. A similar tendency is registered also in the variation of the experimental measurements. Thus, figure 15 shows the variation of the smoke relative emission, theoretically evaluated and compared with the experimental values, presented in relative values of smoke number reported to the condition x_c=0. The soot emission variation tendency is explained thru speeds and laws of formation and oxidation of the soot, according to the variation of the pressure and temperature regimes established inside the cylinder at dual fuelling.

4. Conclusions

The experimental investigations results, obtained for a truck diesel engine fuelled with diesel fuel and hydrogen at the regime of 70% load and 1450 rev/min, lead the following main conclusions:

- due to higher hydrogen burning velocity the heat release rate increases with the rise of the hydrogen cycle dose and the registered values are reached sooner per cycle comparative to diesel fuelling
- hydrogen use in dual fuel operation leads to a decrease of almost 3% in terms of brake specific energetic consumption (BSEC) comparative to standard diesel engine
- at the hydrogen-diesel fuel dual fuelling operation mode, the NO_X emissions level decreases with 5.6% comparative to diesel engine for a 2.42% percent of substitute ratio of diesel fuel by hydrogen
- comparative to classic engine, the HC and smoke emissions level decreases at the hydrogen addition increase
- the CO and CO_2 emissions are in the same domain registered for diesel fuelling or slightly reduced
- the tendency of variation registered at the reduction of the smoke emission from the exhaust gases is explained by the combined effect of pressure and temperature of the in-cylinder gases
- a 20% decreases in speed of soot forming at dual fuelling comparative to classic fuelling
- soot quantity formed inside the engine cylinder decreases at the increase of hydrogen quantity, for all substitute ratios, the reduction being up till 20% comparative to diesel fuelling
- at dual fuelling, hydrogen accelerate the soot oxidation process and even for small x_c values, x_c=0.82…2.42, the soot oxidation is with 18%...70% more efficient
- increase of the hydrogen cycle quantity reduces the soot forming process and influences in a much significant way the soot oxidation process, which is significantly improved
- hydrogen, used in addition to diesel fuel, is a promising alternative fuel for diesel engines and the application of diesel-gas method don't require major modifications of the engine design.

References

[1] Younkins M, Boyer B, Wooldridge M 2013 Hydrogen DI Dual Zone Combustion System, *SAE Int. J. Engines* **6** 2013:10.4271/2013-01-0230 1 pp 45-53

[2] Naber J 1998 Hydrogen combustion under diesel engine conditions *Int J Hydrogen Energy* **23** 3 (1998) 63-71

[3] Mirica I, Cernat A, Pana C, Negurescu N, Nutu C 2015 Performance comparison between hydrogen and diesel fuel fuelled compression ignition engine *U.P.B. Scientific Bulletin Series D* Iss. 4 Vol. **77** ISSN 1454-2358 Bucharest pp 217-228

[4] Senthil R, Kumar M, Loganathana E, Gunasekaran J 2015 Performance, Emission and Combustion Characteristics of CI Engine Fuelled With Hydrogen *International Journal of*

Innovative Research in Science, Engineering and Technology **4** Special Issue 2 ISSN (Online): 2319 – 8753

[5] Saravanan N, Nagarajan G, Dhanasekaran C, Kalaiselvan K M 2007 Experimental Investigation of Hydrogen Fuel Injection in DI Dual Fuel Diesel Engine *SAE World Congress and Exhibition SAE Paper 2007-01-1465* **Automotive** 10

[6] Eiji Tomita, Nobuyuki Kawahara, Zhenyu Piao, Shogo Fujita 2001 Hydrogen Combustion and Exhaust Emissions Ignited with Diesel Oil in a Dual Fuel Engine, *SAE Paper 2001-01-3503 SAE International Fall Fuels and Lubricants Meeting and Exhibition* **Automotive** 10

[7] Talibi M, Hellier P, Balachandran R, Nicos Ladommatos N 2014 Effect of hydrogen-diesel fuel co-combustion on exhaust emissions with verification using an in-cylinder gas sampling technique *International Journal of Hydrogen Energy* **39** pp 15088-15102

[8] Probir Kumar Bose, Rahul Banerjee, Madhujit Deb 2012 Hydrogen and Diesel Combustion on a Single Cylinder Four Stroke Diesel Engine in Dual Fuel mode with Varying Injection Strategies *International Journal of Mechanical, Aerospace, Industrial, Mechatronic and Manufacturing Engineering* **6** No:10

[9] Rottengruber H, Berckmüller M, Elsässer S, Brehm N, Schwarz C 2004 A High-efficient Combustion Concept for Direct Injection Hydrogen Internal Combustion Engine, *15th World Hydrogen Energy Conference* Paper nr. 28 J-01 (Yokohama Japan)

[10] Bose P K, Rahul Banerjee, Madhujit Deb 2013 Effect of hydrogen-diesel combustion on the performance and combustion parameters of a dual fuelled diesel engine *International journal of energy and environment Issue 3* **4** pp 497-510

[11] Santoso W B, Nur A, Ariyono S, Bakar R A, COMBUSTION CHARACTERISTICS OF A DIESEL-HYDROGEN DUAL FUEL ENGINE 2010 *Editors: M.M. Rahman, M.Y. Taib, A.R. Ismail, A.R. Yusoff and M.A.M. Romlay©Universiti Malaysia Pahang, National Conference in Mechanical Engineering Research and Postgraduate Studies (2nd NCMER 2010), Faculty of Mechanical Engineering, UMP Pekan* ISBN: 978-967-0120-04-1 (Kuantan, Pahang, Malaysia) pp 23-32

[12] Cernat A, Pana C, Negurescu N, Nutu C, Mirica I 2015 Experimental investigations of the hydrogen use at the automotive diesel engine *Processing ESFA 2015*

[13] Georgios Pechlivanoglou 2007 Hydrogen Enhanced Combustion History, Applications and Hydrogen Supply by Plasma Reforming *PPRE 2005-2007-University of Oldenburg* 35

[14] Lambe S, Watson H 1993 Optimizing the design of a hydrogen engine with pilot diesel fuel ignition *International Journal Vehicles Design* 14:370e89

[15] Elkotb M M 1982 Fuel Atomisation for Spray Modeling *Program Energy Combustion Science* **8** pp 61-91

[16] Kuo T W 1978 Evaluation of a phenomenological spray-combustion model for two open-chamber diesel engines *SAE Paper 872057*

[17] Appleton J P 1973 Soot oxidation kinetics at combustion temperatures *AGARD Propulsion and Energetics Panel 41 st Meeting "Atmospheric Pollution by Aircraft Engines*

[18] Merker G, Schwarz C, Teichmann R 2011 *Internal Combustion Engines Fundamentals: Running, Simulation, Measuring* (Germany: Springer) p 800

[19] Fanos Christodoulou, Athanasios Megaritis 2014 Experimental Investigation of the Effects of Simultaneous Hydrogen and Nitrogen Addition on the Emissions and Combustion of the Diesel Engine *International Journal of Hydrogen Energy* **39** pp 2692-2702

Acknowledgements

The authors address thanks to the AVL GMBH Graz Austria for equipments. Partial, the work has been funded by the Sectoral Operational Programme Human Resources Development 2007-2013 of the Ministry of European Funds through the Financial Agreement POSDRU/159/1.5/S/134398. This work has been funded by University Politehnica of Bucharest, through the "Excellence Research Grants" Program, UPB-GEX. Identifier: UPB-EXCELENTA-2016 Experimental researches of hydrogen use at diesel engine, Contract number 38/26.09.2016.

The Depression Produced by Impellers and the Similarity Problem

Lucian MÂNDREA[1], Corina Alice BĂBUȚANU[2], Gabriela OPRINA[3], Gheorghe MILITARU[4]

[1]Politehnica University of Bucharest, Str. Splaiul Independenței nr.313, sector 6, Bucharest, Romania, [2,3]National Research Institute for Electrical Engineering, Str. Splaiul Unirii nr.313, sector 3, Bucharest, Romania, [4]Politehnica University of Bucharest, Str. Splaiul Independenței nr.313, sector 6, Bucharest, Romania

e-mail mandrea_lucian@hotmail.com, lucian.mandrea@upb.ro

Abstract. The authors intend to apply original analytical formulas to estimate the depression produced by different types of impellers. The problem has been approached in different studies, especially by numerical simulations and not from the analytical point of view. The study is useful to estimate not only the magnitude of the depression, but also the dimensions of the depression zone. Unpleasant effects can be produced if cavitations appear, leading to propelling unit damage and consequently increasing the operation costs of different applications, such as power industry, chemical industry equipment or naval transport. The authors test impellers of different sizes in two experimental setups in order to characterize the phenomenon.

1. Introduction

Analytical formulas are always preferred by engineers because they are easier to use. The previous original contributions of the authors [1; 2] are used in this paper for other kind of impellers and practical conclusions are obtained. A short numerical example is also presented and a similarity study is added. Experimental results are used to complete the theoretical original formulas to estimate the behavior of impellers in different practical cases.

2. The theoretical characteristics of the depression zone

For a depression zone as in figure 1, one can see the equilibrium of the forces involved in defining the depression zone, filled usually by a two phase fluid. The parameters which can characterize the depression zone were described in detail in [1, 2]. These parameters are the radius R and the height H of the parabolic depression zone and also the pressure inside the depression zone.

As previously computed [1; 2], the radius of the depression zone is:

$$R = \sqrt{\frac{2}{\rho \varepsilon^2 t^2}(p_e - p_v)} \quad (1)$$

In the previous formula, at the radius R the pressure becomes equal with the vaporisation pressure p_v, the exterior liquid (water) has the pressure p_e, ε is the tangential acceleration, ρ the liquid density and t the time variable.

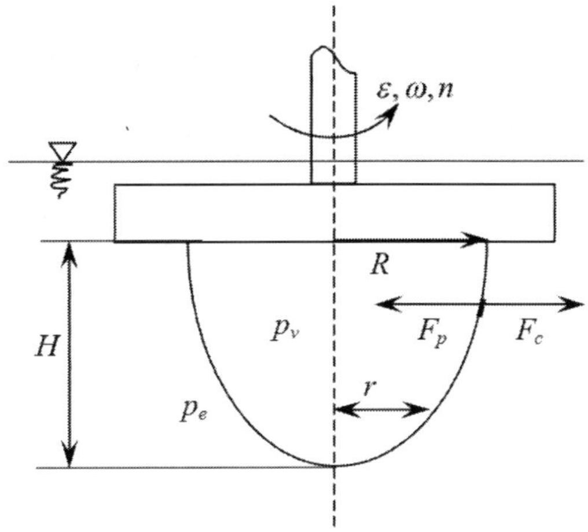

Figure 1.The equilibrium of the forces and the characteristics of the depression zone

The pressure inside the depression zone can be expressed depending on the pressure in the exterior with:

$$p = p_e - \frac{\rho \varepsilon^2 t^2}{2} r^2 \tag{2}$$

The height of the parabolic depression zone is given by:

$$H = \frac{\varepsilon^2 t^2}{2g} R^2 \tag{3}$$

3. Case study

The following amounts for a hydropower plant are considered: $p_e = 10^5$ Pa, $p_v = 2 \cdot 10^3$ Pa, $\rho = 1000$ kg/m^3, the total necessary time to bring the turbine to synchronism $t_{total} = 5$ min = 300 s and the final rotational speed as the rated speed of the runner n = 750 rot/min.

One can determine the rotational speed corresponding to synchronism, respectively the acceleration of the circular movement. So, the final radius R can be obtained by applying Eq. (1).

Retaining two decimal fractions, R and D, respectively (the diameter of the depression zone) become: R = 0.18 m, D = 0.36 m.

$$\omega_{final} = \frac{\pi n}{30} = 25\pi \text{ rad/s}, \quad \varepsilon = \frac{\omega_{final}}{t_{total}} = \frac{\pi}{12} \text{ rad/s}^2.$$

For the numerical values previously specified one obtains:

$$p = 10^5 \left(1 - \frac{25\pi^2}{8} r^2\right),$$

so the pressure is a two degree function of the radius r.

Using the previous numerical values and substituting in (3), from figure 1 it results the depression zone's height, H = 10 m.

4. Experimental results

The first experimental measurement was made on an experimental setup (figure 2) operating at a rated speed of the impeller n = 2520 rot/min. The angle of the blades is 45°, so the rotating impeller has two effects.

The first is a centrifugal force which produces the depression zone and the second is an increased height of the parabola caused by an increased vertical flow.

Figure 2. The first experimental setup and the parabolic depression zone

After reaching the rated rotational speed, the movement of the impeller is permanent, so

$$\varepsilon t = \omega, \quad \omega = \frac{\pi n}{30} \cong 264 \text{rad/s}.$$

Substituting in (3), for example, a correspondence between the radius and the depression zone height is obtained:

$$H = 3552 \cdot R^2 \tag{4}$$

By comparing the results of the carried out experiments, namely the values $R = 5\,\text{mm}$ and $H = 9\,\text{cm}$, with the theoretical value $H = 0.089\,\text{m}$ obtained by using the relation (4), a good agreement of the results can be noticed.

The second experimental setup (figure 3) consists of a transparent cylindrical vessel equipped with a variable speed impeller, electronic speed control and speed measurement system.

The aim of the second set of experiments was to asses the influence of the width of impeller blades used to circulate the fluid (figure 4), on the depression magnitude.

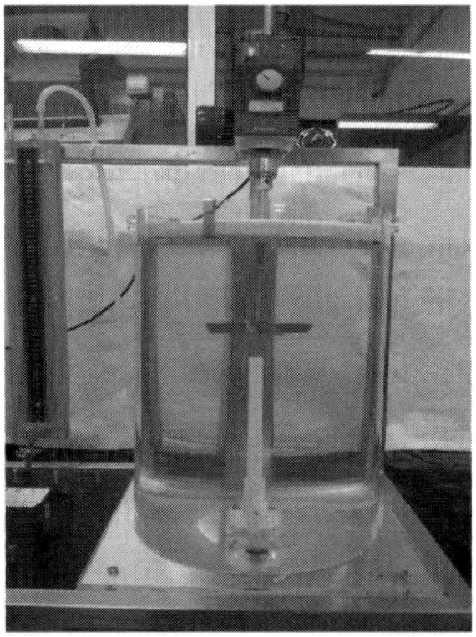

Figure 3. The second experimental setup

Figure 4. The four types of impellers

The measured depression values and friction moments at different speeds for the four types of impellers are presented in Table 1.

Table 1.

	Impeller with 2 thin blades (10 mm width)		Impeller with 2 medium blades (20 mm width)		Impeller with 2 wide blades (40 mm width)		Impeller with 2 shorter blades (20 mm width)	
n [rot/min]	ΔH [mmH_2O]	M [N·cm]	ΔH [mmH_2O]	M [N·cm]	ΔH [mmH_2O]	M [N·cm]	ΔH [mmH_2O]	M [N·cm]
60	0	2.3	0	2.5	1	2.4	0	2.1
100	1	2.4	1	3.0	5	2.9	0	2.3
150	2	2.8	5	3.5	11	3.7	0	2.5
200	4	3.2	11	4.3	21	4.7	0	2.8
250	8	3.8	18	5.3	35	6.0	2	3.0
300	11	4.3	27	6.3	49	7.4	7	3.2
350	16	5.0	38	7.6	67	9.3	10	3.5
400	20	5.7	52	9.2	88	11.3	15	3.8
450	28	6.6	64	10.8			18	4.1
500	36	7.6	80	12.6			23	4.4
550	48	8.5	96	14.4			28	4.7
600	58	9.7					33	5.0
650	69	11.0					39	5.4
700	81	12.3					46	5.9
750	93	13.8					55	6.3

The impellers have a blade length of 100 mm at a hub diameter of 20 mm. The impeller with shorter blades has the length of 60 mm at a width of 20 mm. All the blades are vertical, the same as the hub. The value of the depression ΔH, expressed in mmH_2O, is shown in figure 5.

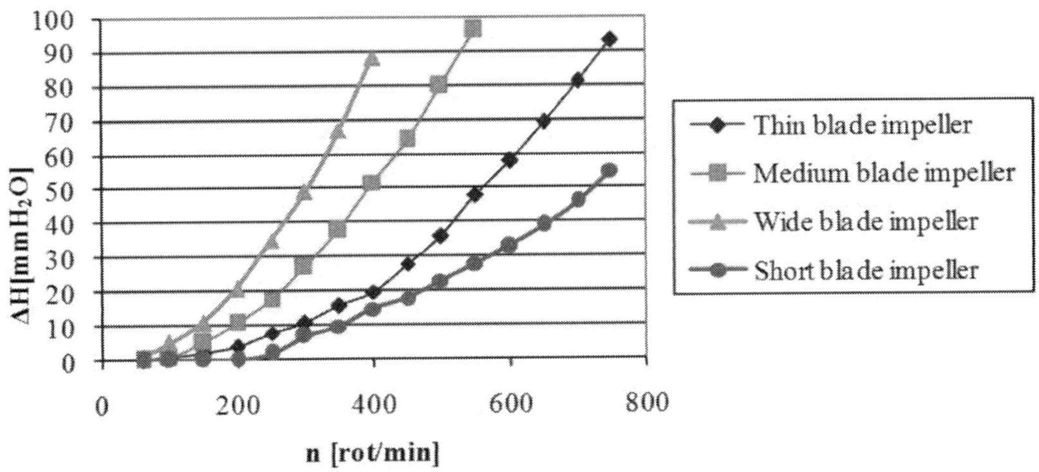

Figure 5.The depression produced by the four types of impellers

By analyzing figure 5, it results that the depressions are parabola functions and correspond with the theoretical difference of pressure as a function of second degree of the variable radius r, as in (2).

The centrifugal force is increasing with the width of the blade and, as a consequence, the depression follows the same trend. Also, it can be noticed that the impeller with shorter blades realizes the smallest depression, as in figure 5.

The momentum has a similar variation, as in figure 6.

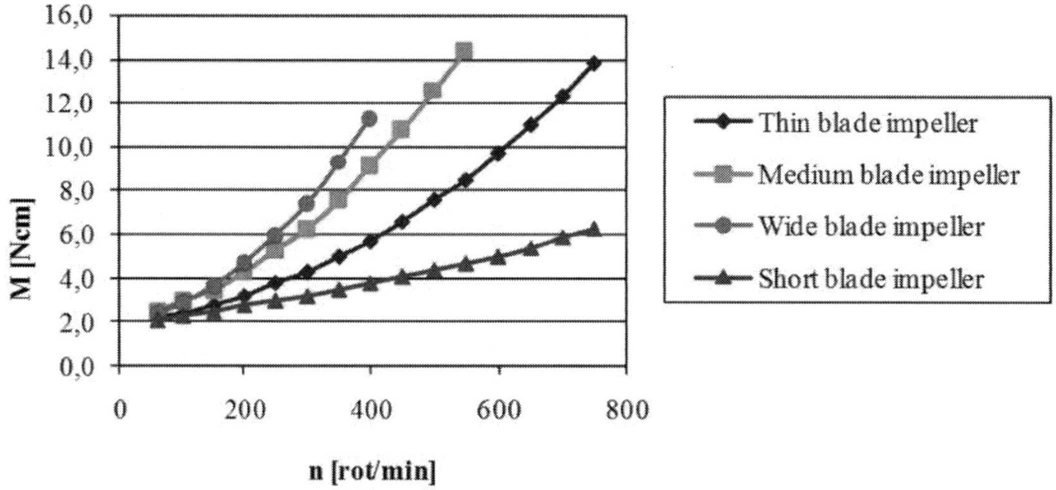

Figure 6.The momentum produced by the four types of impellers

These types of graphs can be used at the computation of the power needed by impellers to perform certain types of tasks. As previously seen, the momentum grows with the increase of the blades width and decreases in the case of a shorter blade impeller.

5. Similitude criteria

The centrifugal forces are preponderant in the studied phenomenon and they depend on the tangential acceleration. They act as inertia forces. For this reason, it is appropriate to chose as a similitude criterion the Newton criterion:

$$\mathrm{Ne} = \frac{F}{\rho V^2 l^2} \tag{5}$$

Then, one can obtain:

$$\mathrm{Ne} = \mathrm{Ne}', \text{ respectively } \frac{F}{\rho V^2 l^2} = \frac{F'}{\rho' V'^2 l'^2}. \tag{6}$$

Because water is used in both experimental setups, we consider that $\rho = \rho'$; substituting $V = \omega r$ and replacing the characteristic length with the variable radius r, it results:

$$\frac{F}{\omega^2 r^4} = \frac{F'}{\omega'^2 r'^4} \tag{7}$$

By considering ω and r for the turbine, the centrifugal force can be obtained as:

$$F = m\omega^2 r \tag{8}$$

Taking into account formula (7), for a certain ω' used in the model installation, it results r'; thus, by using the similitude criterion, one can obtain the force F':

$$F' = F\left(\frac{\omega'}{\omega}\right)^2 \left(\frac{r'}{r}\right)^4 \tag{9}$$

Now, all the amounts for the model installation are known. Obviously, if the amounts used in the model installation are known, the amounts corresponding to the prototype installation can be calculated.

6. Conclusions

Using different types of analyses one can conclude that the original theoretical simplified formulas of the authors, (1), (2) and (3) correspond with the phenomenon and, as a consequence, they can be used in practical applications.

The magnitude of the depression under the impellers and the magnitude of the resistant moment can be increased by varying the length and width of the blades, as shown in figure 5 and figure 6.

The results of the research are useful in practical applications which require a preliminary estimate of the phenomena that characterize the real case operation. Moreover, the means for performing the necessary assessment of the real applications are provided by the current research, through the original similitude calculus synthesized by formula (9).

Acknowledgement The research was conducted within NUCLEUS Program, contract no. 14N/2016, *Innovative solutions regarding the design and implementation of hydrokinetic turbines* financed by the Romanian Executive Agency for Higher Education, Research, Development and Innovation Funding.

References

[1] L.Mândrea, G.Oprina, C.Băbuţanu, D.Marin, G.Negreanu 2015 *Applied Mechanics and Materials* ed A Olaru vol. 841 (Pfäffikon Switzerland: Trans Tech Publication) pp 116-121

[2] L.Mândrea, C.Cipu, S.Cănănău 2015 World Renewable Energy Congress Bucharest Romania

[3] J. Bin, L. Xian-wu, W. Yu-lin, L. Shu-hong, X. Hong-yuan, O. Akira, *J. of Hydrodynamics* 2010 **22**(5) supplement: 747-752, DOI: 10.1016/S1001-6058(10)60025-X

[4] H. Fang-wen, D. Shi-tang *J. of Hydrodynamics* 2010 **22**(5) supplement: 457-461 DOI: 10.1016/S1001-6058(09)60236-5

[5] Hsiao, C., Jain, A., Chahine, G.L. 2006 *Proc. of 26th Symposium on Naval Hydrodynamics (Rome)* (Dynaflow Inc. USA) available at http://www.dynaflow-inc.com/publications/pdf_documents/2006/26Naval_Hydrodynamic_BubbleEntrainment_Chahine_Hsiao.pdf .

The Analysis of Renewable Energy Researches in Turkey

S O Tan[1], T Toku[2] and İ Türker[3]

[1]Department of Electric and Energy, TOBB Technical Sciences Vocational School, Karabuk University, Karabuk, Turkey
[2]Deparment of Information Processing Management, Karabuk University, Karabuk, Turkey
[3]Department of Computer Engineering, Faculty of Engineering, Karabuk University, Karabuk, Turkey

E-mail: serhatorkuntan@karabuk.edu.tr

Abstract. The rapid consumption of limited conventional energy resources mobilizes many countries in the world against global energy crisis. As well as the energy crisis, the environmental pollution caused by existing energy sources also encourages the researchers to study in new energy technologies and also renewable energy resources. From this point of view, it is important for each country to identify its wind, solar, geothermal, biomass, hydro and other renewable energy potentials. Considering this urgent energy requirement, the researches and especially the academic studies have been increased on renewable energy resources to meet the energy demand by means of indigenous resources in each country. Consequently, the main purpose of this study is to analyze the academic studies in Turkey to find out the increment rate of researches, their publication years and the more focusing branch on renewable energy by illustrating the statistical distribution of these data. Automated Data Retrieval Methods have been employed to achieve data from Web of Science database and statistical analyses have been made by SQL server management studio program. The academic studies in all variety of renewable energy areas have a tendency to increase which indicates the importance ratio of renewable energy in Turkey.

1. Introduction

The energy requirement in the world is increasing day by day with the rapid consumption of limited conventional energy resources [1]. Since the reserves of gas, oil and coal are sepulchered deep in the sea or ground, it is difficult, expensive and also dangerous to identify and unearth these reserves [2,3]. Furthermore, frequentative crises in energy and also global warming influence the quality of life adversely and this situation promotes the researchers to find out new energy sources or benefit from renewable energy resources [4]. People have been used renewable energy resources for different purposes from past to present. For instance, while the energy of wind and water were used simply for mechanical drive many centuries ago, they are still being used with high power turbines today [5]. Although technological developments are necessary to produce energy to compensate the energy requirements, both technological developments and energy requirements have been increased in the world for years. This situation is also valid for Turkey and the production of energy increases with the consumption of energy as can be seen from Table 1. Turkey is highly dependent on imported energy resources which influence its economy unfavorably. In other aspects, air pollution also constitutes an environmental concern in Turkey. Considering these disadvantages, renewable energy resources seems to be the most productive and efficient for clean sustainable energy development. Turkey has quite convenient geographical location for using distinct kinds of renewable energy sources. The major resources are hydropower, biomass, biofuels, wind power, solar energy, and geothermal energy which will have provided most of Turkey's renewable energy as for many of the world countries in the future [6,7].

Table 1. Actual and target values of energy production and consumption of Turkey in ktoe units.

Energy sources	Energy production and consumption (in parenthesis) of Turkey by years					
	2000	2005	2010	2015	2020	2025
Hard coal and lignite	17 202 (20 256)	21 259 (30 474)	28 522 (50 311)	31 820 (83 258)	39 385 (129 106)	45 944 (296 997)
Oil and natural gas	3408 (59 250)	2127 (73 256)	1735 (92 637)	1516 (112 993)	1604 (136 365)	1455 (179 765)
Central heating	253 (253)	495 (495)	884 (884)	1336 (1336)	2018 (2018)	2748 (2748)
Hydropower	3763 (3763)	5845 (5845)	7520 (7520)	8873 (8873)	9454 (9454)	10 445 (10 445)
Wood and waste	6963 (6963)	6760 (6760)	6446 (6446)	6029 (6029)	5681 (5681)	5393 (5393)
Geothermal	432 (432)	1380 (1380)	3760 (3760)	4860 (4860)	4860 (4860)	5400 (5400)
Nuclear	0.0 (0.0)	0.0 (0.0)	3657 (3657)	9143 (9143)	18 286 (18 286)	29 200 (29 200)
Solar	204 (204)	459 (459)	907 (907)	1508 (1508)	2294 (2294)	3248 (3248)
Wind	55 (55)	250 (250)	620 (620)	980 (980)	1440 (1440)	2134 (2134)

Hydropower is the major energy supply for many of the countries as it provides one-fifth of the world's power generation. Although it has a crucial role in electricity generation, the remaining potential in Turkey is almost vast at present [8]. Taking advantage of direct combustion process, biomass production, especially as wild plant growth, is quite common in many of the world countries [9]. Biomass energy is also one of the major resources in Turkey for about 52% of Turkey's domestic energy consumption [10]. Biomass is a promising energy type for Turkey with its great potential providing from rural energy services. Several academic studies have been presented on biomass energy resources and its potentials in Turkey [11-13].

Although there are many challenges when environmental factors are taken into consideration in the site selection of the wind turbines, wind energy production has a significant place among other energy resources [14]. Among European countries, the highest share in wind energy potential belongs to Turkey but this share is relatively low when compared with the average wind energy potential of world [15]. Turkey has also a high potential in solar-energy when considering the yearly average solar-radiation and total yearly radiation period with the values of 3.6 kWh/m^2 day and ~2610 h respectively [5]. The prediction of solar-radiation in different cities of Turkey has been presented by many academic studies [16–19]. Finally, geothermal energy which contained as heat in the Earth's interior, [20] is a widely used renewable energy source with a projected life of 30–50 years. [21]. It has a considerable potential in Turkey and is used generally in electricity generation which will be estimated as the most important field for using geothermal energy in the future [22].

Eventually, with the worry of finding alternative energy sources, the academic studies converge to renewable energy which has a vital importance to solve the energy crisis and environmental pollution in the world. [23]. The goal of this study is to analyze the academic researches to find out the increment rate of researches, their publication years and the more focusing branch on renewable energy by illustrating the statistical distribution of these data.

2. Methodology

A data retrieval method was used to provide automatic retrieval of information based on data placed into Web of Science database. This data retriever program reads the clipboard periodically to see whether new data has been placed therein. The program detects whether the data satisfies a predetermined criterion during new data search in the clipboard. The data retriever program also starts a data-retrieval operation as a web-based search based on data if the criterion is satisfied. But if it is not satisfied, the program does not perform any data-retrieval action [24]. This operating logic of Automated Data Retrieval Method provides us to achieve data from Web of Science database and statistical analysis has been made by utilizing SQL server management studio. Analysis values were transferred to Excel Power Pivot to extract the statistical distributions of the increment rate of the researches, their publication years and the more focusing branch on renewable energy.

3. Results and Discussion

The statistical distribution of the increment rate of academic researches on renewable energy, their publication years and the more focusing branch of the renewable energy in Turkey has been analyzed. The distribution of the total number of researches versus research years is shown in Figure 1.

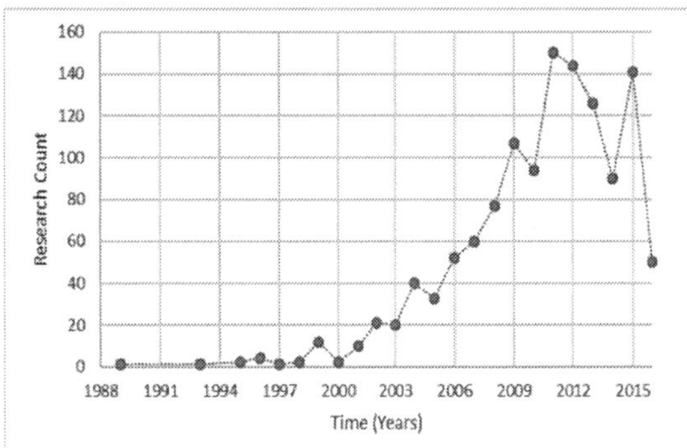

(a)

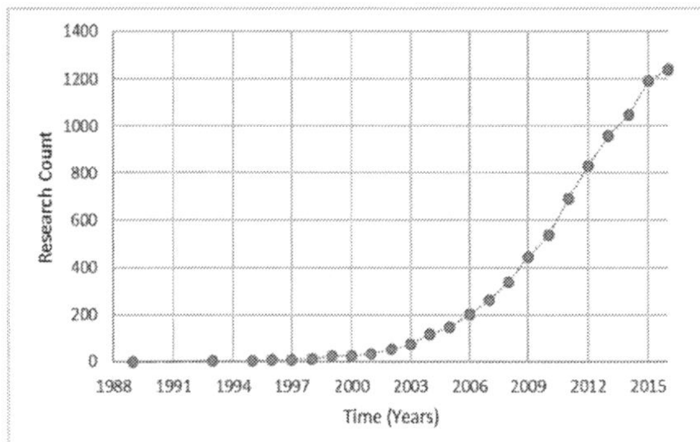

(b)

Figure 1. The distribution of the academic researches versus years on renewable energy is presented in a. The same data is presented cumulatively in b.

As seen from Figure 1, the academic studies in all types of renewable energy areas increase with the passing years. The increment rate of the academic researches by years can be attributed to the energy requirement increase in Turkey. Identifying the most focusing type of renewable energy among the total number of academic researches is also important when the increment of the academic researches over the years is considered. Figure 2 represents the total number of academic researches on different types of renewable energy in Turkey. The most studying research area up to now is bioenergy as seen from Table 2, but it must be noted that bioenergy also includes the energy types such as biofuel, biomass, biogas and etc. For this reason, the number of academic studies on wind energy and hydropower appear to be remarkable, and even impressive.

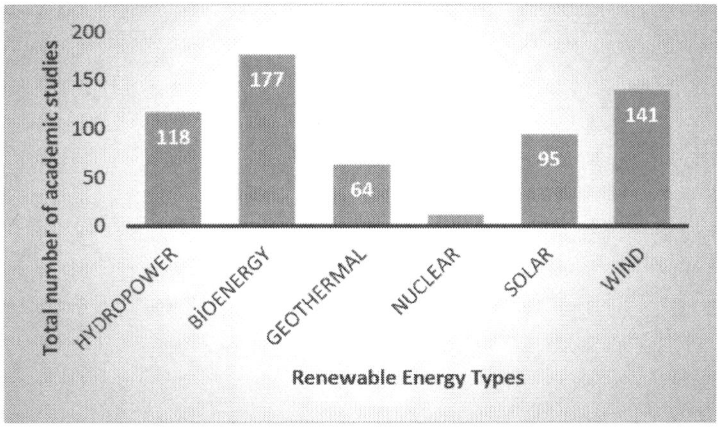

Figure 2. The total number of academic researches on different types of renewable energy in Turkey.

The distribution academic studies over years on distinct types of renewable energy forms are given separately in the following figures given with Figure 3 label.

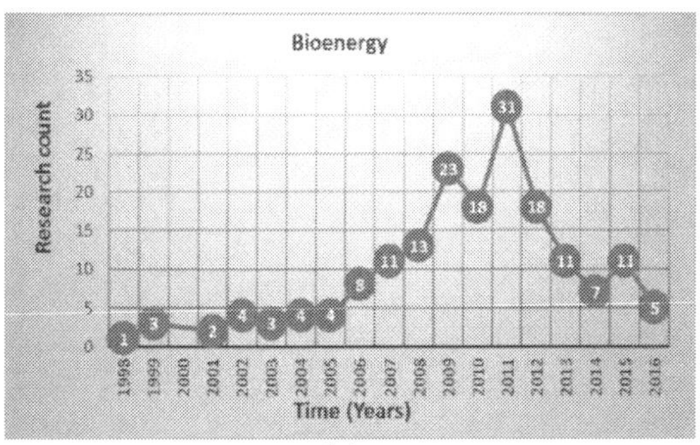

(a)

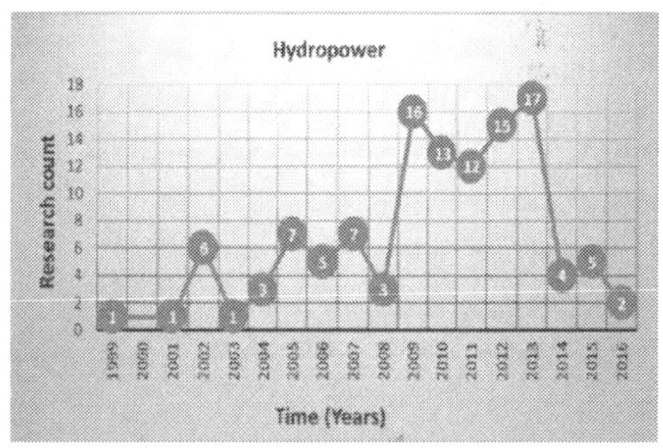

(b)

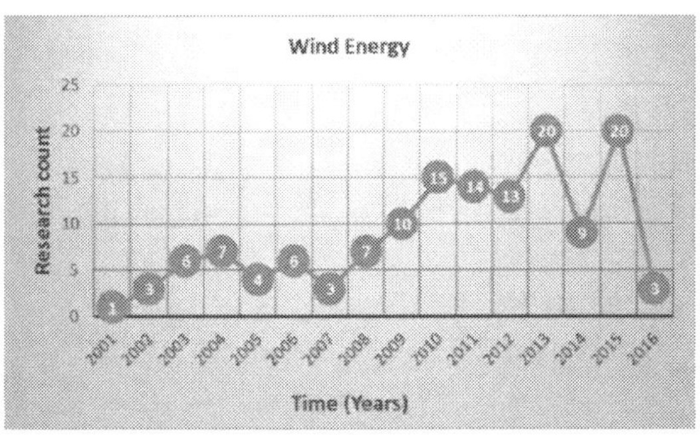

(c)

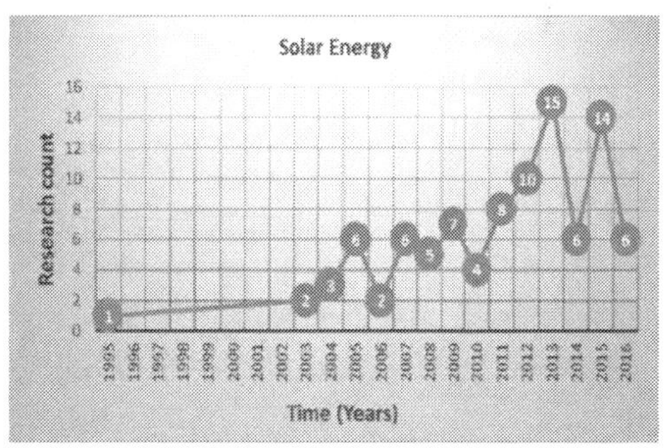

(d)

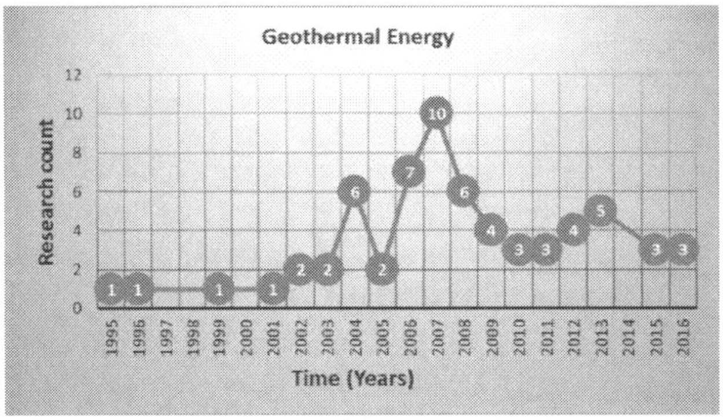

(e)

Figure 3. The distribution of academic studies over years on distinct types of renewable energy forms a) Bioenergy b) Hydropower c) Wind Energy d) Solar Energy e) Geothermal Energy.

The number of academic researches on all types of renewable energy forms increases by years as presented in Figure 3. This situation mostly valid for especially solar energy, wind energy and also hydropower and this can be interpreted as the energy requirement in Turkey emerges to these areas in recent years.

The environmental pollution caused by existing energy sources is also critical issue which should be handled in academic manner. The studies on environmental pollution were also statistically analyzed and the distribution of these studies by years is given in Figure 4. The distribution in Figure 4 can be defined as the sensitivity to environmental pollution increases in direct proportion to the interest in renewable energy sources by years.

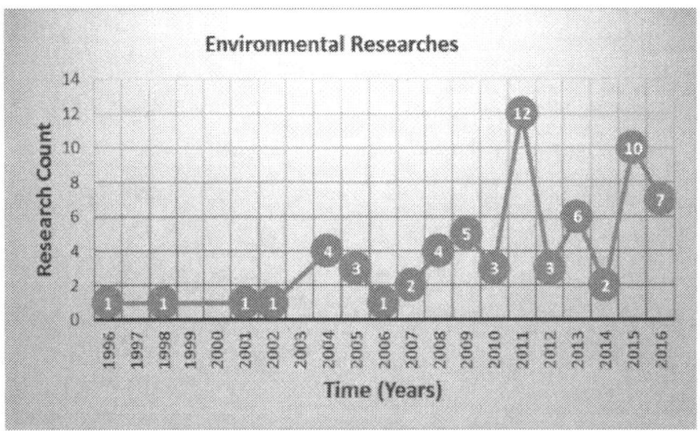

Figure 4. The distribution of environmental researches over years.

Conclusion

The analysis of renewable energy resources in Turkey is strategically important when considering its variety of utilizable energy sources. The sustainable economic development and also the standard of living quality are directly related with energy resources. Turkey has not efficiently evaluated many of its renewable energy sources for some reasons such as technical, economic and also lack of researches. The increment of energy requirement with each passing day in the world and also in Turkey will also require an increase in researches on renewable energy resources. From this point of view, the academic studies on the subject of renewable energy in Turkey have gained a great importance to meet and support this energy requirement. This study concluded the statistical data analyses of the researches and academic studies in Turkey which were extracted from Web of Science database to reveal the importance academically given to the renewable energy resources. The academic studies in all types of renewable energy areas have a tendency to increase which indicates the importance ratio of renewable energy in Turkey. The scientific correlation of the data uncovered the importance order of these areas by analyzing the statistical distribution of data. Consequently, the academic studies in Turkey on renewable energy forms and also environmental pollution caused by existing energy sources, increase with the increment of energy requirement in recent years.

References

[1] Huesemann M H 2003 *Clean Technologies and Environmental Policy* **5:(1)** 21–34
[2] Pacesila M, Burcea S G and Colesca S E 2016 *Renewable and Sustainable Energy Reviews* **56** 156–170
[3] Zhang K, Nieto A and Kleit A N 2014 *Mineral Economics* 1–12
[4] Abedin M J, Masjuki H H, Kalam M A, Sanjid A, Ashrafur Rahman S M and Masum B M 2013 *Renewable and Sustainable Energy Reviews* **26** 20–33
[5] Kaygusuz K and Sarı A 2003 *Energy Conversion and Management* **44** 459–478
[6] EREC (European Renewable Energy Council) 2006 *Renewable energy scenario by 2040 Brussels: EREC Statistics*
[7] Evrendilek F and Ertekin C 2003 *Renewable Energy* **28** 2303–2315
[8] Yuksel I 2008 *Renewable and Sustainable Energy Reviews* **12** 1622–1640
[9] Cuff D J and Young W J 1980 *US Energy Atlas. Free Press/McMillan Publishing Co*, New York.
[10] Demirbas A 2000, *Energy Education Science and Technology* **5** 21–45
[11] Demirbas A 2008 *Energy Policy* **36** 834–842
[12] Gokcol C, Dursun B, Alboyaci B and Sunan E 2009 *Energy Policy* **37** 424–431
[13] Kaygusuz K and Turker M F 2002 *Renewable Energy* **26** 661–678
[14] Keller J 1990 *Town Plann Rev* **91** 139–54
[15] Kenisarin M, Karsli V M and Caglar M. 2006 *Renewable and Sustainable Energy Reviews* **10** 341–69
[16] Şaylan L, Şen O, Toros H and Arısoy A 2003 *Energy Conversion and Management* **43**:**18** 29–37.
[17] Tırıs M, Tırıs Ç and Türe IE 1996 *Energy* **21:1** 67–70
[18] Dinçer İ, Dilmaç Ş, Türe I E and Edin M 1996 *Energy Conversion and Management* **37:2** 183–98
[19] Sözen A, Arcaklıoğlu E, Özalp M and Kanıt E G 2004 *Applied Energy* **77** 273–86
[20] Bilgen S, Kaygusuz K and Sarı A 2004 *Energy Sources* **26:11** 19–29
[21] Hepbaslı A 2003 *Energy Sources* **25** 667–77
[22] Köse R 2005 *Renewable Energy* **30** 67–79
[23] Yongxiu H, Yang X, Yuexia P, Huiying T and Rui W 2016 *Renewable Energy* **89** 695-705
[24] Waggoner S E 2010 *US8762401 B2*

Analysis of critical operating conditions for LV distribution networks with microgrids

M A Zehir[1,4], A Batman[1], M A Sonmez[1], A Font[1], D Tsiamitros[2], D Stimoniaris[2], T Kollatou[2], M Bagriyanik[1], A Ozdemir[1] and E Dialynas[3]

[1]Istanbul Technical University, Ayazaga Campus, 34469, Maslak, Istanbul, Turkey
[2]Western Macedonia University of Applied Sciences, Koila Kozanis 1, Kozani, 50100, Greece
[3]National Technical University of Athens, Iroon Polytechneiou 9, Zografou, 15780, Athens, Greece
[4]E-mail: zehirm@itu.edu.tr

Abstract. Increase in the penetration of Distributed Generation (DG) in distribution networks, raises the risk of voltage limit violations while contributing to line losses. Especially in low voltage (LV) distribution networks (secondary distribution networks), impacts of active power flows on the bus voltages and on the network losses are more dominant. As network operators must meet regulatory limitations, they have to take into account the most critical operating conditions in their systems. In this study, it is aimed to present the impact of the worst operation cases of LV distribution networks comprising microgrids. Simulation studies are performed on a field data-based virtual test-bed. The simulations are repeated for several cases consisting different microgrid points of connection with different network loading and microgrid supply/demand conditions.
Keywords. microgrids, smart grids, co-simulation, distributed generation, distribution networks.

1. Introduction

Penetration of renewable energy resources into the power system has a rising trend in many countries. DG from renewables provides technical, environmental and economic benefits both for the utility and for the customers [1]. In spite of its numerous advantages, it may have some negative impacts on the system. DG facilities are mostly connected to the grid through current source inverters, which follow network voltage and supply current in relation with the generated power. The prospective impacts of an inverter-based generation facility on a network are voltage limit violations and increase in line losses. The problems may be more significant in LV distribution networks, due to bidirectional power flow both in a radial topology and over the lines with high R/X ratios.

In order to supply power from a local generation unit to the utility in a radial network, the voltage of its connection node should be higher than the substation. The amount of voltage difference between the utility grid and the microgrid depends on the supplied power, substation voltage and line parameters.

There are several studies in the literature focused on analysing DG impact on LV networks. The dependency of voltage levels to active power flow in LV networks is stated with the related equations in [2]. Scenarios to investigate the local generation impact are proposed in [3]. Voltage variation and its mitigation by generation reduction are studied in [4]. In addition to solely DG centred analysis, there is a need to investigate the renewable-based microgrid impacts over LV networks.

Microgrids are one of the leading areas of Smart Grid technologies. Comprising storage and manageable load units close to distributed generators and advanced local control infrastructures, they may enable penetration of more renewable generations into the power systems. According to a comprehensive survey study done with 460 smart grid experts from different countries, microgrids will primarily be deployed in rural areas, military facilities, hospitals and public buildings that have mostly MV and LV distribution networks [5].

This study focuses on the critical operating conditions for LV networks with microgrids. A virtual test bed that represents the steady state operational behaviour of a small scale microgrid and a rural LV network model is used in the analysis. Both the bus voltages and the line losses are investigated for different points of connection, local generation output and network loading cases.

2. The Virtual Test Bed

The virtual test bed used in the analysis is previously developed as a part of a bilateral research project (Smart Electric Energy Management and Buildings Energy Efficiency Technologies-SEEMBEET) led by authors [6]. It represents the steady state operational behaviour of a small scale microgrid laboratory. The real system has two groups of PV panels with 2.2 kW power output in total, one 1 kWp wind turbine, a lead-acid battery group with 600 Ah storage capacity and 5 groups of loads with approximately 2.3 kW total demand.

The reason for developing a virtual test bed is to conduct research beyond the pilot applications. A laboratory has some operational constraints, such as single point of connection to the distribution grid, limited number and capacity of devices in the system and dependency on local conditions (like solar irradiation, wind speed, temperature and etc.). On the other hand, a virtual test system that has identical behaviour with the real microgrid can allow analysis with different points of connection to a larger system, including different scenarios with several operating conditions. Moreover, long term operation can easily be investigated. With combined simulation solutions, it is possible to implement energy management actions and a virtual test bed can provide in depth analysis of different algorithms.

The virtual test bed used in this study is developed using PSCAD software [7]. At the first stage, individual models of local generations (PV panel groups and wind turbine), storage (lead-acid batteries) and manageable loads are developed. Steady state power output of PV panel groups, P_{pv}, for their whole operational range is modeled by (1). In (1), I_r denotes the solar irradiation in W/m^2 and T is the temperature in °C. During modeling, a data set that is generated with the existing PV block in the software library together with the developed MPPT and inverter models, is used and a first order equation is derived (1).

$$P_{pv} = 8.542173377 - 0.1719887T + 0.170425778I_r \qquad (1)$$

A similar approach is preferred for wind turbine model. The test data including all power outputs at all operational wind speeds (taking into account tip speed ratio), is used to derive a fourth order equation (2) [8].

$$P_w = 0.0768v_w^4 - 3.7371v_w^3 + 55.024v_w^2 + 189.02v_w + 186.76 \qquad (2)$$

R-squared fitting performance for PV and wind turbine models are found to be 0.9968 and 0.9969, respectively. Comparison of real data with the model output for the whole operational range can be found in Figure 1 and Figure 2.

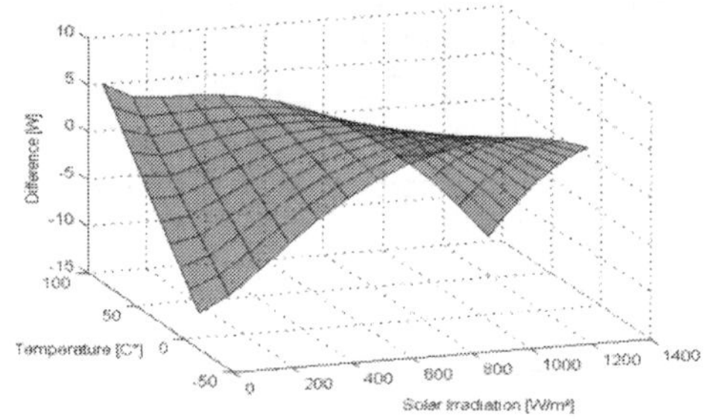

Figure 1. The difference between the real data and the fitted curve for PV model

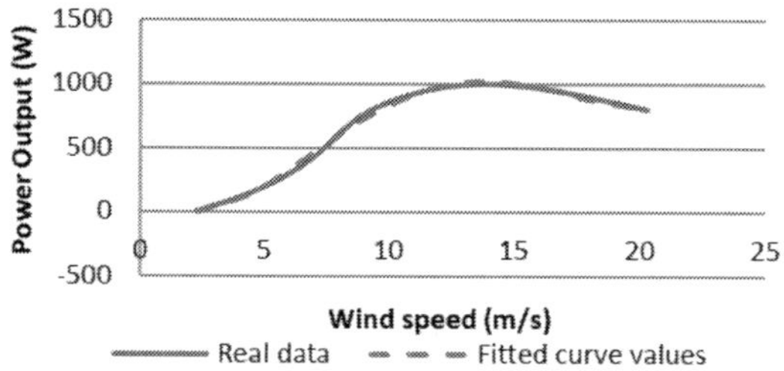

Figure 2. The difference between the real data and fitted curve for the wind turbine model

The authors have another publication regarding the development of the virtual test bed. Interested audiences can find the details of the modelling process in [7].

3. Case Study

In this study the test bed is connected to a LV distribution network to analyse the critical operating conditions. The chosen network is a 50 kVA, 400/230 V, 50 Hz system previously proposed in [2] (Figure 3). It is a small scale distribution network with long lines and high R/X ratios (Table 1). Its two feeders with 10 consumer nodes are modelled in PSCAD.

Table 1. LV distribution network line types with R/X ratios

Line Types	R/X Ratio (per km)
AMKA 3x35+50	8.35
AMKA 3x25+30	11.32
AMKA 3x16+25	17.69
AMKA 4x50S	7.28
AMKA 3x10+10	20.8
AMKA 3x6+6	34.22

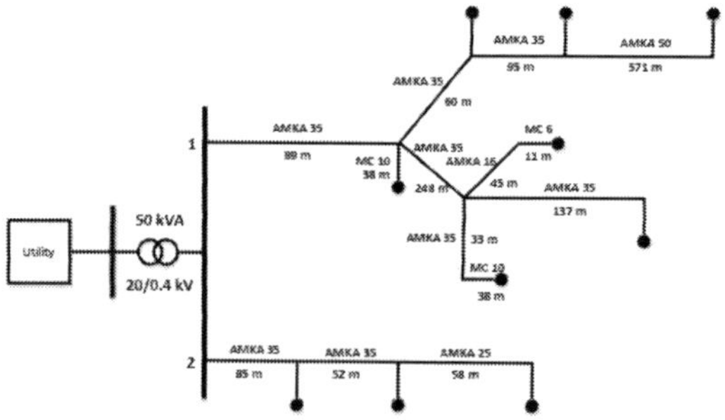

Figure 3. Sample LV distribution network with two feeders and 10 consumer nodes [2]

3.1. Scenarios

In order to investigate the impact of a microgrid on a distribution network, the test bed is connected to the three different nodes; namely, close point of connection, medium point of connection and far point of connection. Four different microgrid supply/demand cases are simulated for each connection point. The first case refers to maximum microgrid local generation and minimum microgrid demand profile. The second case is just the complement of the first one, where the local generation is minimum and microgrid demand is maximum. The last two cases are maximum local generation with maximum demand and minimum local generation with minimum demand. Note that the first two cases and the last two cases correspond to high and low power exchanges between the microgrid and the network, respectively. All those four cases, are simulated together with maximum and minimum total demand in the LV network. Table 2 illustrates the simulation cases for each point of connection to the grid. Combining with three different points of connection, 24 different scenarios are investigated. Moreover, each scenario includes switching from grid-connected mode operation to islanded mode operation to compare the bus voltage magnitudes and the line losses.

Table 2. Scenarios for each microgrid point of connection

Scenario No	Network Load	Microgrid Load	Microgrid Supply
1 (max. demand)	High	High	Low
2	High	High	High
3	High	Low	High
4	High	Low	Low
5	Low	Low	Low
6 (max. supply)	Low	Low	High
7	Low	High	High
8	Low	High	Low

3.2. Simulation Results

The most critical cases for each point of connection are illustrated in Figure 4, Figure 5 and Figure 6. In each figure, the bus with the worst voltage level is shown in a red circle, while the most loaded line is shown in a blue circle. Bus voltages, line loadings and total % line losses are given for the critical scenarios. The network loading during scenario 1 is 2.7 kW per connection point, the average of daily peaks faced with in the field. For scenario 6, the amount of loading per consumer is 0.5 kW, which represents continuously working loads like refrigerators and air conditioners, together with stand-by consumption of electronic devices.

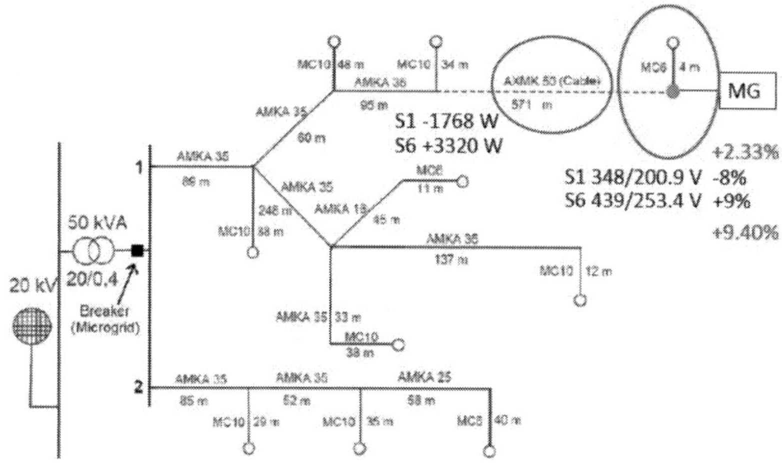

Figure 4. The most critical impacts for far point of connection

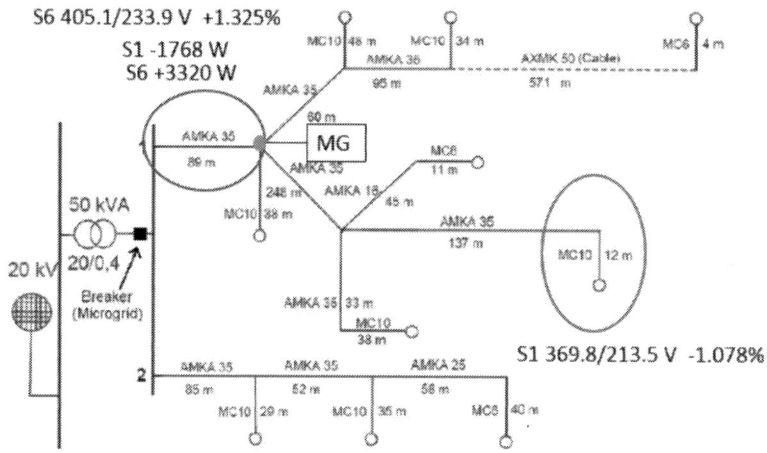

Figure 5. The most critical impacts for near point of connection

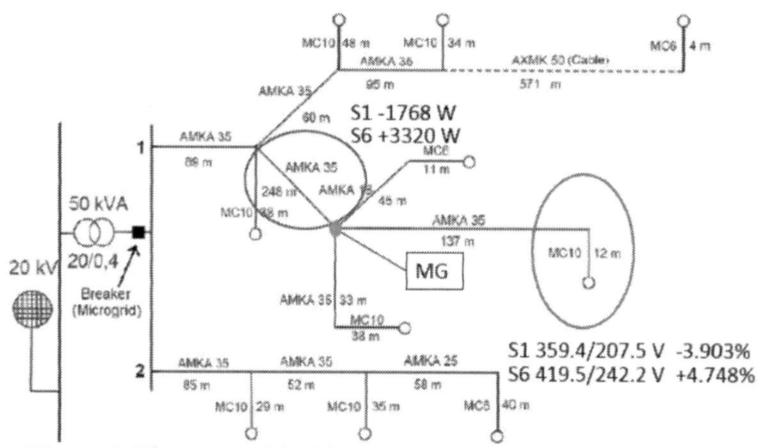

Figure 6. The most critical impact for middle point of connection

Among 24 representative scenarios for three point of connections, the most critical voltage magnitude and line loadings are faced for the maximum demand (scenario 1) and for the maximum supply (scenario 6) cases. According to the numerical results, far point of connection caused highest negative impacts, while near point of connection resulted in the lowest impacts. One of the main reasons of this result is the value of the impedance between the microgrid and the distribution substation. Higher impedance increases the voltage drop over that path and consequently, the microgrid needs to increase its connection point voltage to higher levels to supply the utility grid. Another reason is the availability of a single line between the microgrid and the neighboring busses. However, there are several other alternative lines to other costumers which reduce the line loadings for near point of connections. In other scenarios, power exchanges between the microgrid and the network are not too high and therefore, the impact of the microgrid is low.

4. Discussions and Conclusion

In this study, the most critical operating conditions for LV distribution networks with microgrids are investigated. Using previously developed virtual test bed and a small scale LV network model, several scenarios are simulated for different points of connection.

The study provides several valuable outcomes that can be used both for network planning and for network operation. During the planning stage, the main aim should be the minimization of the impedance between the microgrid and the network. In order to achieve that, it is more preferable to connect the microgrid closer to the substation. In addition, the number of connections between the microgrid and the neighbouring consumers can be increased.

After deciding the location of connection, the microgrid equipment capacities can be scaled considering the two most extreme cases, which are maximum supply/minimum demand and maximum demand/minimum supply in both the microgrid and the network.

It should be noted that the microgrid used in this study is a three phase balanced system. Additional analysis is required for unbalanced loadings and for single phase microgrid connection cases. Transient analysis can also provide more valuable insight about new advanced operational approaches. Further studies are planned for long-term operation of the microgrid with the network by integrating daily supply/demand profiles. It can be done to see the occurrence rates of critical cases in daily operation.

Acknowledgments

This study is funded as a part of "Smart Electric Energy Management and Buildings Energy Efficiency Technologies-SEEMBEET" project under the framework of the "Bilateral Research and Technology Cooperation Turkey-Greece 2013-2014" organized by "The Scientific and Technological Research Council of Turkey (TUBITAK)" and "General Secretariat for Research and Technology (GSRT)".

References

[1] Colmenar-Santos A, Reino-Rio C, Borge-Diez D and Collado-Fernandez E 2016 A review of factors that can contribute most to achieve a scenario of DG units embedded in the new distribution networks. *Renewable and Sustainable Energy Reviews* RSER(2016)59

[2] Laaksonen H, Saari P and Komulainen R 2005 Voltage and frequency control of inverter based weak LV network microgrid. *International Conference on Future Power Systems* Amsterdam Netherlands

[3] Mahmud MA, Hossain MJ and Pota HR 2014 Voltage variation on distribution networks with distributed generation: worst case scenario. *IEEE Sysems Journal* (8) 4

[4] Doumbia ML, and Agbossou K 2007 Voltage variation analysis in interconnected electrical network-distributed generation. *IEEE Canada Electrical Power Conference* Montreal Canada

[5] Zpryme and IEEE 2012 Power Systems of the Future: The Case for Energy Storage, Distributed Generation and Microgrids, *https://building-microgrid.lbl.gov/sites/all/files/the_case_for_energy_storage_distributed.pdf*

[6] Istanbul Technical University (ITU) and Western Macedonia University of Applied Sciences (TEIWM) 2015 Smart Electric Energy Management and Buildings Energy Efficiency Technologies (SEEMBEET), *http://www.edalab.itu.edu.tr/seembeet*

[7] Zehir MA, Batman A, Sonmez MA, Font A, Tsiamitros D, Stimoniaris D, Kollatou T, Bagriyanik M, Ozdemir A and Dialynas E 2015 Development of a field data-based virtual test bed for microgrid integration of building automation technologies. *ELECO Conference*, Bursa, Turkey

[8] Wholesalesolar 2016 Whisper 200 Technical Specifications, *http://pdf.wholesalesolar.com/wind%20pdf%20folder/Whisper_100_200_Spec_Sheet.pdf*

IOP Conf. Series: Materials Science and Engineering **161** (2016) 012079 doi:10.1088/1757-899X/161/1/012079

Wind turbines: current status, obstacles, trends and technologies

E I Konstantinidis and P N Botsaris

Democritus University of Thrace, School of Engineering, Department of Production Engineering and Management, Section of Materials, Processes and Engineering Vas. Sofias str. Building I, Office 103, Central University Campus, 67100 Xanthi, Greece

E-mail: ekonstan@pme.duth.gr

Abstract. The last decade the installation of wind farms around the world is spreading rapidly and wind energy has become a significant factor for promoting sustainable development. The scope of the present study is to indicate the present status of global wind power expansion as well as the current state of the art in the field of wind turbine technology. The RAM (reliability/availability/maintenance) section is also examined and the Levelized Cost of Energy for onshore/ offshore electricity production is presented. Negative consequences that go with the rapid expansion of wind power like accidents, environmental effects, etc. are highlighted. Especially visual impact to the landscape and noise pollution are some factors that provoke social reactions. Moreover, the complicated and long permitted process of a wind power plant, the high capital cost of the investment and the grid instability due to the intermittent nature of wind, are also significant obstacles in the development of the wind energy production. The current trends in the field of research and development of onshore and offshore wind power production are analyzed. Finally the present study is trying to achieve an estimation of where the wind industry targets for the years to come.

1. Introduction

The increasing energy demand globally together with the emission of harmful gases (NOx, SOx etc) caused by the conventional fossil fuel resources, have made the need of increasing the use of renewable energy sources in the energy balance of each country imperative. Surely, green growth should not be perceived as a turn to an irrational and at any cost use of renewable energy sources, but as an investment in policies and technologies which would turn renewable energy sources into competitive and sustainable energy sources. According to the European Energy Roadmap 2050 [1] the share of renewable energy rises substantially in all scenarios, achieving at least 55% in gross final energy consumption in 2050, an increase of 45 percentage points from today's level of around 10%.

Wind power, i.e. the kinetic energy of the wind, is a renewable energy source that is used among others mainly for the production of electrical power. The global wind resources (land and near-shore) are estimated to be 72 TW which is seven times the world's electricity demand and five times the world's energy demand. In contrast to the global pollution issues and the significant cost of fossil fuels, wind seems to be a clean, affordable and inexhaustible source of energy and its exploitation can meet directly both the global demand for renewable and clean energy and the need to secure new energy sources. The wind turbines are installed either onshore (on land) or offshore (beyond the coast). Offshore wind farms seem to be advantageous due to the enormous energy potential associated with the large continuous areas and the stronger winds that imply greater power generation, even though they have higher initial investment, operation and maintenance costs.

2. Wind power in numbers

By the end of 2015 the global cumulative installed wind capacity has reached 432 GW (Figure 1) with offshore installations to have a total power of 12 GW [2]. In the same time there is 142 GW of installed wind power capacity in the EU: 131 GW onshore and 11 GW offshore [3].

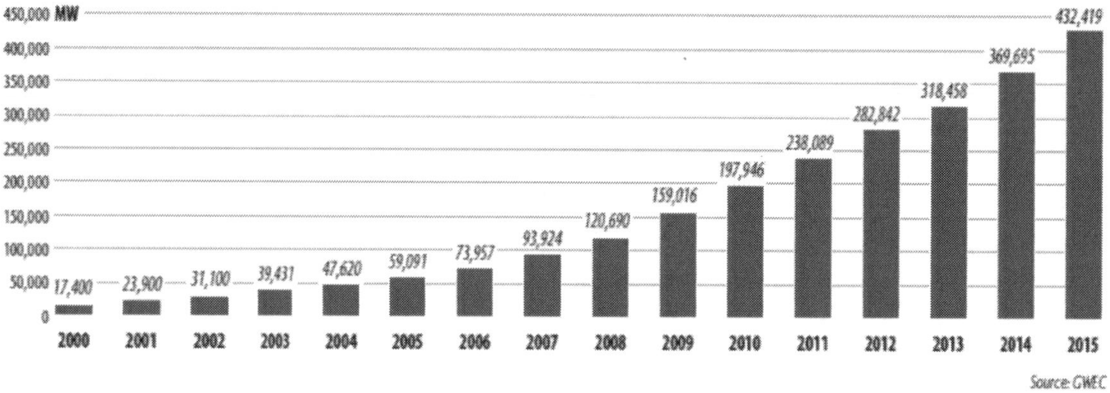

Figure 1. Global cumulative installed wind capacity 2000-2015
(Source: Global Wind Energy Council)

According to a Caduff et al. study [4] the bigger the wind turbine is, the greener the produced electricity is. Of course a bigger wind turbine means mainly producing more power more efficiently. The last decade the wind turbine size has more than doubled. In 2005 the maximum operational wind turbine was 3,6 MW and now is 8 MW. The largest operational wind turbine models based on capacity and their manufacturers, are the following: Vestas V164 8MW (with 164m rotor diameter), Enercon E126 7.5MW, Samsung S7.0 171 7MW, MHI SeaAngel 7MW, Repower 6M 6.2MW, Siemens SWT-6.0 150, Alstom Haliade 6.0MW, Sinovel SL6000 6.0MW, Areva M5000 5.0MW, Gamesa G5MW [5]. According to BTM Navigant survey [6] regarding wind turbine installation market share, in 2014 Vestas (Denmark) had 12.3% of the wind energy market, Siemens (Germany) 9.9%, GE Wind (USA) 9.1%, Goldwind (China) 9%, Enercon (Germany) 7,8%, Suzlon (India) 5,8%, United Power (China) 5,1%, Gamesa (Spain) 4,7%.

The largest operational onshore wind farm in the world is Gansu Wind Farm in China. It has a current capacity of power of 6,800 MW and a target capacity of 20,000 MW by 2020. The first world's offshore wind power plant was installed in Vindeby (Denmark) in 1991 and had a total capacity of 4,95 MW (11 X 0,45 MW – BONUS wind turbines). At this moment the largest offshore wind farm in the world (and the largest wind farm in Europe by megawatt capacity) is the London Array. It is a 630MW wind farm consisted of 175 Siemens Wind Power SWT-3.6 turbines and located 20 km off the Kent coast in the United Kingdom. In February 2016 Dong Energy has announced the Hornsea project, a 1,2 GW offshore plant which will be made up of 7 MW turbines and will occupy more than 400 square kilometers, situated about 120km off the Yorkshire coast also in the United Kingdom.

Among the EU members, Germany remains the country with the largest installed capacity (45 GW), followed by Spain (23 GW), the UK (14 GW) and France (10 GW). The total wind power capacity installed at the end of 2015 could produce 315 TWh and cover 11.4% of the EU electricity consumption in a normal wind year [3].

3. Wind turbine technologies

There are two main types that modern wind turbines can be categorized in: the horizontal axis wind turbines and the vertical axis wind turbines. The majority of the large scale wind turbines are horizontal axis due to their greater efficiency and power output. The main parts that a large scale horizontal wind turbine consists of are: rotor (blades, hub and pitch system), nacelle (rotor shaft, bearings, gearbox/ generator, mechanical brake and yaw system), tower, foundation and the electrical (power feed cables, lightning protection, power converter, transformer) and control system (sensors, actuators, system consisting of hardware and software).

There are two different drive systems for converting rotational power into electrical: indirect and direct drive. Indirect drive system uses a gearbox to increase the rotational speed of the shaft that drives the generator. The indirect drive configuration often uses asynchronous squirrel-cage induction generator (SCIG), wound rotor induction generator (WRIG) or doubly fed induction generator (DFIG). The direct drive does not use gearbox because there is a full-scale power converter with multi-pole generators: excited synchronous generator (EESG) or permanent magnet synchronous generator (PMSG). There are also configurations that use gearboxes with a full-scale power converter and thus, bigger and more expensive generators are unnecessary. These indirect drive types use smaller generators: wound rotor synchronous generator (WRSG), permanent magnet synchronous generator (PMSG) and asynchronous squirrel-cage induction generator (SCIG). The configuration uses an excited synchronous generator (EESG) is the most reliable, powerful and expensive configuration. Wind turbines with DFIG configuration is the most commonly offered type by the major manufacturers. The trend is towards full scale power converter configurations with multi – pole permanent magnet synchronous generator (PMSG), because they reduce losses and they are lighter than types with excited synchronous generator [7].

The condition monitoring systems, which comprise combinations of sensors and signal processing equipment, are used to monitor the status of operating major components. Monitoring may be: on-line (and thus provide instantaneous feedback of condition) and off-line (the data are collected at regular time intervals using measurement systems that are not integrated into the wind turbine). A detailed view of a wind turbine condition monitoring system will typically present the following data: wind turbine data (wind speed, active and reactive power, yaw angle, etc. and command, operational and fault status), electrical and mechanical data, meteorological data, grid data, statistical data. Different monitoring techniques in wind turbines are used in order to prevent damage or failure of components. The most common condition monitoring techniques that are used are the following: process parameters, performance monitoring, vibration analysis, acoustic emission, ultrasonic testing (UT), oil analysis, strain measurement, electrical effects, shock pulse method, radiographic inspection and thermography [8].

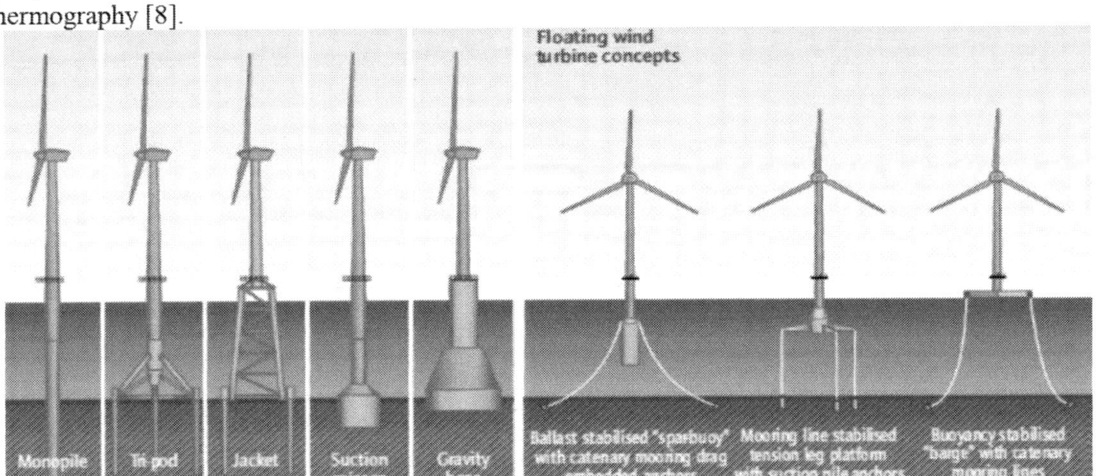

Figure 2. Fixed bottom foundations and floating offshore concepts [9]

There are different types of foundations that are used for offshore wind turbines (Figure 2). The two main categories are the fixed bottom (for depths up to 50 m) and the floating concepts (for deeper waters). The monopole is the simplest and cheapest type of foundation that is used commonly in the North Sea into the sea bed to a depth of 0–25 m. A gravity-based foundation is normally fabricated in reinforced concrete with ballast and it uses its own weight to anchor to the seabed at water depth between 0 and 25m. An alternative to these two is the suction caisson, in which by removing the water from a caisson onto which the turbine is situated the suction force thus created is used to promote easy installation. A tripod structure is a relatively lightweight three-legged steel structure, in which the frame is submerged in the water, providing good stability and stiffness and it is used for depths between 20 and 50m. Jacket is often used in water depths of about 20–50 m and usually consists of a three or four-legged lattice structure made from tubular steel.

The need for higher quality wind resources unconstrained by water depth has turned the wind industry developers in the direction of floating support options (which were already experienced in the offshore Oil and Gas industry). The main floating concepts are: the spar buoy, the tension-legged platform, the buoyancy stabilized and the semi - submersible. Statoil together with Siemens Wind Energy installed in 2009 the world's first full-scale floating spar type wind turbine (Hywind) of 2,3 MW power output at the coast off Norway and at a depth of 200m [10]. This wind turbine survived 11 meter waves with seemingly no wear. In summer of 2015 a 7MW floating offshore wind turbine (semi-submersible concept) developed by Mitsubishi Heavy Industries (MHI) has been moored at the Fukushima demonstration site, 20 kilometres offshore. [11]

4. Maintenance issues and wind power related costs

Wind turbines, just like every other technical asset, must be serviced regularly and repaired in case of defects. The main reasons that make the maintenance of wind turbines of special significance are the ambient conditions that are unusually tough and the extremely high dynamic loading of the components. These two factors are considered carefully in the design and selection of materials, but conventional components still require maintenance. Common causes of wind turbines failure can be: components failure, control system failure, high wind, waves, lightning, grid failure, icing, etc. The commonest reason which is responsible for wind turbines failure is components breakdown. Figure 3 presents the average failure rates for wind turbine components from different sources [7].

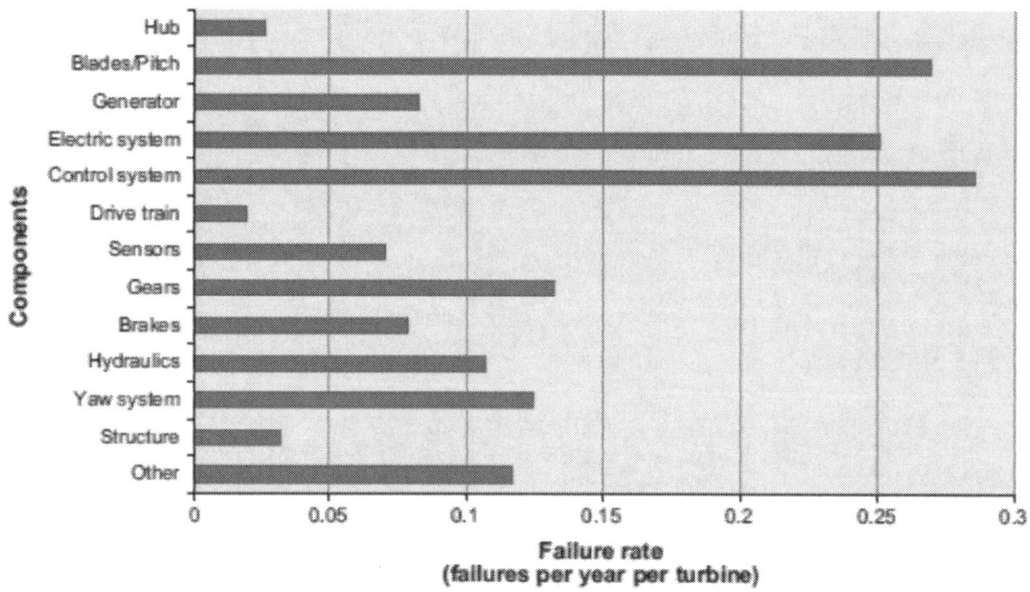

Figure 3. Average failure rates for WT components from different studies [7]

To achieve the best possible power production efficiency of the installed wind turbine, a high reliability level should be reached. The objective is to increase the turbine availability, by improving the wind turbine reliability or their maintenability or both, if this is possible. The wind turbine reliability is a pivotal factor in the successfully function of a wind power plant. High reliability can be achieved by understanding and minimizing the failures of the system. The reliability of a wind turbine does not depend exclusively on the reliability of its subsystems, but also on external, indirect factors like the maintenance strategy used, the existence of spare parts, the time needed to repair, the existence of control systems of the subsystems, the type of wind turbine related to the operational environment (onshore, nearshore, offshore) and the training of the personnel. Maintenance planning could make maintenance more efficient and could lead to a reduction of failure events. In order to decide the most effective maintenance strategy, the most critical components of the wind turbines, which are prone to failures, should be identified. The maintenance strategy should take into account both the reliability improvement and the reduction of maintenance cost. The operation and maintenance strategy of wind turbines is directed towards enhanced condition monitoring systems.

The levelized cost of energy (LCOE) is the price of electricity required for a project where revenues would equal costs, including making a return on the capital invested equal to the discount rate. An electricity price above this would yield a greater return on capital, while a price below it would yield a lower return on capital, or even a loss. According to IRENA (International Renewable Energy Agency) [12] for 2014 the LCOE range was for onshore projects 0,06-0,12 USD/kWh and for offshore projects 0,10-0,21 USD/kWh. The regional weighted average installed costs in 2014 for onshore wind range from around 1.280 USD to 2.290 USD/kW (with China and India having weighted average installed costs 35% to 44% lower than in other regions). The average installed cost ranges in 2014 for offshore wind projects was 2.700-5.070 USD /kW. The variable operation and maintenance costs for onshore wind projects in 2014 for Austria was 0,04 USD/kWh, for the Netherlands 0,013 - 0,018 USD/kWh, for Norway 0,021 - 0,039 USD/kWh, for Spain 0,028 USD/kWh, for Sweden 0,010 - 0,035 USD/kWh, for Switzerland 0,045 USD/kWh, for the United States 0,01 – 0,024 USD/kWh and for Denmark 0,015 - 0,019 USD/kW.

5. The less popular side of wind power and the obstacles

Visual impact, noise pollution, ecosystem-related issues (environmental hazards for flora and fauna), health hazards (electromagnetic fields, possible cause of accidents), impact on tourism and agriculture, the reduction in the value of land and the end of life (wind turbine waste treatment), are the main drawbacks of wind farms. These effects are also sources of rejection of wind turbine installation from the inhabitants of the planned wind farms. Noise pollution and the degree of visual disturbance could be avoided by selecting wind turbine sites far from residential areas. Offshore wind farms seem to be advantageous as they have reduced noise impact and visual impact. From a visual disamenity point of view, the socially optimal location for installing offshore wind farms is unlikely to be greater than ca 18 km from the shore [13].

Caithness Windfarm Information Forum published a detailed table that includes all documented cases of wind turbine related accidents and incidents which could be found and confirmed through press reports or official information releases up to 31 March 2016 [14]. The data include human accidents (fatal, injuries, human health related), fire accidents, wind turbine components failures that result in severe damages (structural components failures, blade failures etc), construction and transportation damages, ice throw, environmental damages (wildlife damages, pollution from oil leakages etc). This data clearly shows that blade failure is the most common accident with wind turbines, closely followed by fire. Figure 4 shows a chronological summary of the previous mentioned accident table. The trend is as expected - as more turbines are built, more accidents occur.

In a survey of Zimmerling et al [15] based on 2.955 installed turbines (the number installed in Canada by December 2011), an estimated 23.300 birds would be killed from collisions with turbines each year in Canada; (c.a.270 million birds died each year in Canada, ~200 million were killed by cats and ~25 million by building collision [16]). The problem of bird mortalities due to wind turbine collision could be reduced by careful site selection, the use of brightly colored sections of rotors or the use of some night time lighting (for the bats).

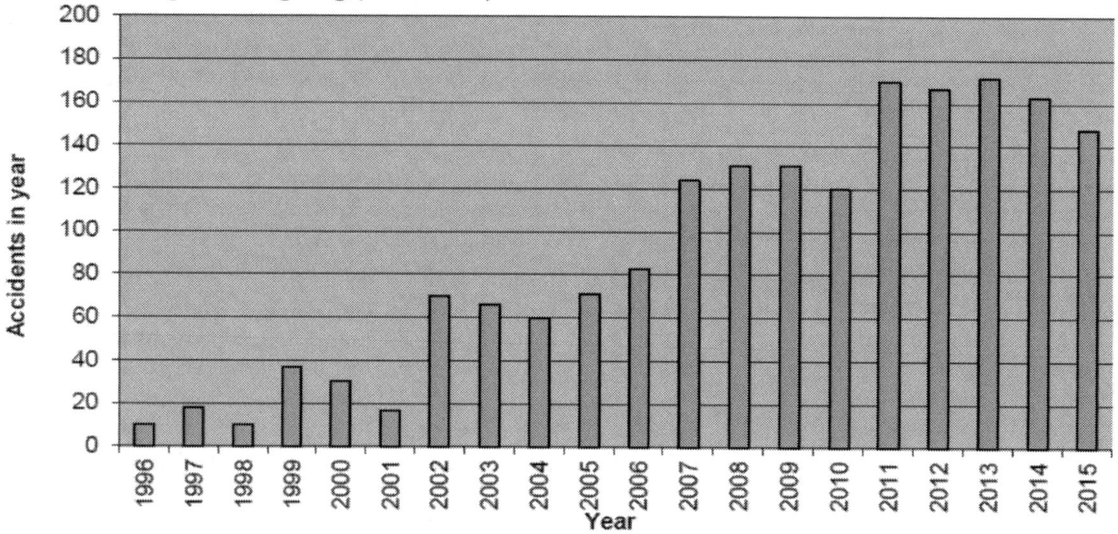

Figure 4. Wind turbine accidents and incidents [14]

Nevertheless, the average general attitude towards any kind of renewable energy installations (including wind power) is positive. The majority of people support any action which yields to cleaner environment and to sustainable development. Despite the general approval of wider society with respect to wind power, when it comes to local communities, the negative reactions have often been especially intense. A major cause for these reactions has been a fear of property values falling due to proximity to a wind farm because of the visual impact, noise and other issues mentioned before. Public information, consulting and allowance of the local people (local community organisations, environmental societies, wildlife trusts, local politicians) for participation and involvement in the planning and decision making procedures, could considerably contribute in the prevention of local resistance towards wind power investments.

The wind turbine end of life is an issue that is going to be of high importance in the near future. A large amount of already installed turbines will reach the end of their service life shortly. A modern wind turbine is designed to work for an estimated time of 20 years. After this point it has to be either renewed or recycled. The main problem regarding wind turbines decommissioning process is that except steel, copper and aluminum which are fully recyclable, the glass reinforced plastics (based on polyester or epoxy) used in the rotor blades have proven challenging to recycling. Options of blade material waste treatment are: mechanical recycling (which is a labor intensive process and uses the produced material as a filler in artificial wood, cement or asphalt production), incineration, pyrolysis and landfill, which is the worst option [17].

The complicated and long permitted procedures of a wind power plant in many countries, together with the long planning phase (including among others environmental, engineering, feasibility and site-specific studies) that are required, are challenges that an investor has to face in the early beginning of a wind power project. The high investment cost and the high levelized cost of wind energy compared to energy market prices are also characteristics that have to be taken into consideration for estimating the viability of the investment. Of course the high feed-in tariffs and the policies of subsidies that most of the governments follow in order to promote wind energy, make wind a competitor with the existing conventional energy sources. The grid instability due to the intermittent nature of wind is another disadvantage as this practically means power losses in transmission lines. Additionally the variable or the intermittent nature of wind supply is probably the biggest obstacle wind power has to overcome, as the power grids require a reliable energy supply and energy companies are often forced to apply a cost of the extra gas turbine plant required to make up the missing guaranteed supply to intermittent of suppliers, in order to ensure integration of variable renewables sources like wind into a grid [18]. Alternatives to the instability of wind power production would be implementing energy storage configurations and also the use of wind forecasts for preparing grid operators for extreme events which power generation is unusually high or low.

6. Research & Development trends and the future of wind industry

The large offshore models constructed for extreme weather and sea conditions and designed to reduce installation time and maintenance frequency, is the field of research and development (R&D) competition for the wind turbine developers. These huge constructions will logistically force in the next decade the supply chain towards the establishment of coastal manufacturing and assembly facilities to avoid constraints associated with road/rail transport. The floating offshore models are also something that the wind turbine manufacturers focus. The floating wind turbines have the advantage that all construction (and also maintenance) could be done on land and then transported and hooked up to the seabed. Advanced drivetrain designs, enhanced condition monitoring systems, larger rotors made by light-weight advanced materials and taller towers with new design architectures and materials, suggest a whole new field for innovations in the near future. Totaro [19] highlights that the new technologies and their patents will be likely directed towards six core areas: reducing component weight and manufacturing costs, transport and assembly, monitoring and control, turbine reliability, grid integration and optimizing performance. The most ambitious R&D is trying to create a rotor blade longer than 200m for a 50-MW offshore wind turbine. This is 2,5 times longer and 6 times more powerful than the largest blades and turbines currently in operation [20].

7. Conclusions

The last decade the installation of wind turbine farms around the world is spreading rapidly onshore as well as offshore. The present study has examined the current status of wind power and presented the related technologies, the maintenance issues, the costs, as well as the negative effects and obstacles of the wind industry. Despite some local negative reactions, the majority of people embraces and supports renewable energy installations. In the future we will see taller towers with larger rotors from advanced materials that could capture greater amounts of wind energy. The wind turbine technology will be focused on higher reliability with the implementation of advanced control and monitoring systems. The operation and maintenance strategy will be directed towards enhanced condition monitoring systems and advanced maintenance strategies. The annual amount of waste from wind turbines due to their end of life will grow substantially and should be managed sustainably.

As an epilogue to this study it should be emphasized that the domination of environment friendly energy sources, which are economically competitive against fossil fuel energy sources, is something the whole mankind could benefit from.

Acknowledgments

The Caithness Windfarm Information Forum (www.caithnesswindfarms.co.uk) is kindly acknowledged for providing accident data used in this study.

References

[1] European Commission, *Energy Roadmap 2050*, December 2011
[2] Global Wind Energy Council, *Global Wind Statistics 2015*
[3] The European Wind Energy Association, *Wind in power: 2015 Statistics*
[4] Caduff M., Huijbregts M.A.J, Althaus H.J., Koehler A. and Hellweg S., 2012 *Environmental Science & Technology* (ACS) **46**, 4725-4733
[5] http://www.windpowermonthly.com/10-biggest-turbines, accessed 04 Mai 2016
[6] https://www.navigantresearch.com/research/world-wind-energy-market-update-2015, accessed 04 Mai 2016

[7] Perez J.M.P., Marquez F.P.G., Tobias A. and Papaelias M., 2013 Wind turbine reliability analysis *Renewable and Sustainable Energy Reviews* (Elsevier) **23**, 463-472

[8] Márquez F.P.G., Tobias A., Pérez J.M.P. and Papaelias M., 2012 Condition monitoring of wind turbines: Techniques and methods *Renewable Energy* (Elsevier) **46**, 169-178

[9] Edenhofer o. et al, *IPCC Special Report on Renewable Energy Sources and Climate Change Mitigation*, 2012 (Cambridge University Press)

[10] http://www.statoil.com/en/technologyinnovation/newenergy/renewablepowerproduction/pages/ default.aspx, accessed 05 Mai 2016

[11] http://www.windpoweroffshore.com/article/1358221/worlds-largest-floating-turbine-installed-fukushima, accessed 05 Mai 2016

[12] The International Renewable Energy Agency (IRENA), 2015 *Renewable power generation costs in 2014*

[13] Ladenburg J. and Dubgaard A., 2007 Willingness to pay for reduced visual disamenities from offshore wind farms in Denmark *Energy Policy* (Elsevier) **35**, 4059-4071

[14] http://www.caithnesswindfarms.co.uk/AccidentStatistics.htm, accessed 07 Mai 2016

[15] Zimmerling J.R., Pomeroy A.C., d' Entremont M.V. and Francis C.M., 2013 Canadian estimate of bird mortality due to collision and direct habitat loss associated with wind turbine developments *Avian Conservation and Ecology* **8** (2), Art.10

[16] Calvert A.M.C., Bishop C.A., Elliot R.D., Krebs E.A., Kydd T.M., Machtans C.S., and Robertson G.J., 2013 A synthesis of human-related avian mortality in Canada Avian Conservation and Ecology **8** (2), Art.11

[17] Andersen N., *Wind turbine end of life: characterization of waste material*, 2015 Master Thesis, (Sweden: University of Gavle) p 14-15

[18] IEA *World energy outlook 2012 Renewable energy outlook,* OECD/IEA 2012 (http://www.worldenergyoutlook.org/media/weowebsite/2012/WEO2012_Renewables.pdf)

[19] Totaro P. by Totaro & Associates, Six trends shaping future turbine designs *Wind Power Engineering*, 06 May 2011

[20] Pelosi R., The next generation in wind power technology, *Renewable Energy World*, 30 March 2016

Optimal route scheduling-driven study of a photovoltaic charging-station (parking lot) for electric mini-buses

V Zacharaki[1], S Papanikolaou[1], Ch Voulgaraki[1], A Karantinos[1], D Sioumpouras[1], D Tsiamitros[1], D Stimoniaris[1], S Maropoulos[1] and Y Stephanedes[2]

[1]Western Macedonia University of Applied Sciences, Koila Kozanis 1, Kozani, 50100, Greece
[2]University of Patras, Ypatias 1, 26500, Rio, Patras, Greece

E-mail: dtsiamit@teiwm.gr

Abstract: The objective of the present study is threefold: To highlight how electro-mobility can: (a) Contribute to the promotion of the environmental conservation of the rural areas (through an integrated solution for reducing the carbon footprint of road facilities and transport), (b) Enhance tourism-based economical development, (c) facilitate students in their daily transport and residents (elderly, disabled, distant-residents) in their daily on-demand transport. The overall goal is to design an energy-efficient, regional intelligent transportation system with innovative solar-energy charging-stations for e-vehicles in municipalities with many geographically scattered small villages and small cities. The innovative character of the study is that it tries to tackle all three specific objectives simultaneously and with the same means, since it utilizes Intelligent Transportation Systems (ITS). The study is adapted and applied to an area with the above characteristics, in order to demonstrate the proof of concept.

1. Problem description

The Western Macedonia Region offers a great number of tourist attractions; therefore, an intelligent network of electric mini-buses will bring added-value to these attractions. Additionally, a cross-border interconnection of these touristic sites can be enhanced by a regular cross-border route of a "touristic electric mini-bus". Such an electric mini-bus interconnection can facilitate tourists and the local population to identify many cultural common characteristics between the Western Macedonia region and the neighboring country (FYROM).

Moreover, especially in Prespa Municipality, which consists of small and district residential areas (villages), students have to use "traditional/conventional" public transport or taxis in their daily trip to school. Replacing the current transport modes with more energy-efficient ones within a regional intelligent transportation system will support the efficient realisation of both the tourist promotion of the area and the students' daily transportation.

Furthermore, due to the heavy winter periods, residents in this region, and especially elderly, disabled, distant-residents, face great difficulties in their daily transport. This concept faces this challenge and facilitates residents in these Municipalities with the on-demand use of utility electric vehicles (e-vehicles).

The concept is compatible with the broad EU energy policy context such as Climate-Energy packages,

Energy Union and decarbonisation of transport, since it creates synergies between Renewable Energy Sources, electricity grid and transport users (e.g. services to the grid with smart charging). Further, the solar availability in the region, which is indicated for high performance of energy supply through photovoltaic panels, could be exploited for replacing the conventional energy used by current transport, with renewable solar energy.

Therefore, the main objective of the paper is to present a study of an optimal route scheduling in coordination with the design of a photovoltaic charging-station (external parking lot) for the charging of electric mini-buses.

2. Method

The methodology is adapted to the regional needs and is based on the European Union's requirements on the promotion of energy efficiency towards "green" transport and mobility, and the EU Directive for Intelligent Transportation Systems (ITS). The first step of the method includes the study and design of the optimal route scheduling in the region under the newly imposed constraints. This dictates the configuration of the electric mini-buses, so that their electrical specifications serve as the basis for estimating the power needed to fully charge them. The next step is the study regarding the specifications of all the ancillary equipment needed to support the operation of a solar charging-station, i.e. PV panels, inverter, batteries and charging controller specifications. Finally, a cost analysis for the solar charging-station will be performed. The methodology is intended to be applied to the Prespa Municipality and the Florina Municipality in Greece as well as to the Bitola Municipality in the Former Yugoslav Republic of Macedonia.

2.1. Optimal route scheduling principles

The Intelligent Transportation System consists of the supply of real-time information, regarding the e-vehicles' availability through the day, booking, payment, etc., to the e-vehicles' users (tourists, disabled persons, students) through a sensors-data collector-server-client application as well as information on user needs and network state. Smart phone apps tailored to the specific types of users will be developed.

Depending on the trip objective, the e-vehicles will be subject to dynamic and adaptive routing, thus reducing travel time and enhancing reliability. Especially in cases of incidents, this will be supported by optimal adaptive routing [1] in the stochastic time-dependent network of the new service. The network-level impact from adaptive routing will be estimated from a user-equilibrium traffic assignment model [2], in which users make adaptive routing decisions based on online information. In particular, users' choice sets of routing policies will include link travel times as random variables with time dependent distributions, and the equilibrium will be in terms of distribution of link travel times, flow and other quantities related to the trip [3].

2.2. Determination of the specifications of the e-vehicles

The indicative technical specifications regarding the battery of the electric mini-bus are presented in Table 1:

Table 1. Electric bus battery technical specifications.

Electric bus Battery	
Type	**Lithium**
Total Nominal Voltage	**120V**
Maximum Battery Charging Current	**50A**

Battery Capacity	**540Ah**
Charging time	**9 hours**

The charging station will have a capacity of up to two electric buses, while being capable of charging one bus at a time. The consumption per bus is calculated as follows:

According to the specified technical characteristics of the electric bus (Table 1), the battery voltage of the electric bus is 120V DC and the battery capacity is 540Ah. Moreover, the battery charging rate is 9 hours for full charging, and the per-hour charging rate of the electric car is 50A. Therefore, according to the above specifications, in order to fully charge the battery of the electric bus, the installed power of the photovoltaic panels should be at least:

$\mathbf{P_{bus}\ (W_p) = V\ (V)\ x\ I\ (A) = 120\ (V)\ x\ 50\ (A) = 6000\ W_p}$

Thus, the total installed power of the charging station so as to be capable of providing a full charge for the electric bus should be: $P_{bus}\ (W_p) = 6000\ W_p$

2.3. Specifications of the photovoltaic charging-stations (externalparking-lots)

The Inverter is selected on the basis of the power system requirements. Therefore, the Inverter's power output should be at least 6000 $\mathbf{W_p}$, i.e. 8000 $\mathbf{W_p}$. Moreover, the Inverter's AC current output should be greater than or equal to the AC current input of the electric bus. The latter is approximately equal to:

$$I_{AC\ input-bus}\ (A) = \frac{P_{bus}(W_p)}{V_{AC}\ (V)} = \frac{6000\ (W_p)}{230\ (V)} \cong 26\ (A)$$

Therefore, the Inverter's AC current output should be greater than or equal to 26 A.

Finally, the Inverter's rated DC input voltage should be 48V, and, thus the Inverter's maximum DC input current should be:

$$I_{DC\ input-Inverter}\ (A) = \frac{P_{Inverter}(W_p)}{V_{DC}\ (V)} = \frac{8000\ (W_p)}{48\ (V)} \cong 160\ (A)$$

Summarizing, the Inverter's required specifications are presented in Table 2:

Table 2. Inverter's technical specifications.

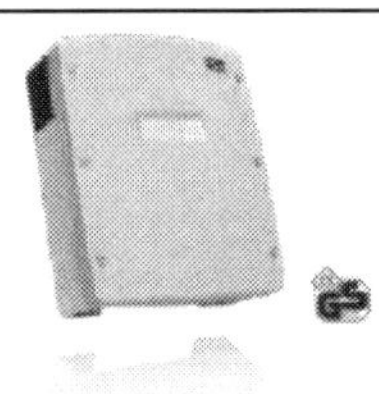

Rated Power Output	**8000W**
Output Voltage / Frequency	**230VAC / 50Hz**
Minimum Current AC	**26A**

Input Voltage	**48VDC**
Maximum DC input current	**160A**
Protection	
From short-circuit	√
From overloads	√
From high temperatures	√

The charging regulator is selected based on the power system requirements. Therefore, the charge controller's power should be greater than 6000 Wp. Moreover, the charging controller's output current should be less than the Inverter's DC input current, i.e. less than 160A.

Regarding the battery side, the charging controller's nominal battery voltage should be 48V.
Finally, regarding the photovoltaic array side, the charging controller's current input should accept low currents so that the cable dimensions are small as well. Therefore, indicative current values should be in the range 20A – 30A. In turn, the voltage input should be high enough so as to have a power of 6000W, i.e. the voltage input should be in the range 200V – 300V.

Summarizing, the charging controller's required specifications are presented in Table 3.

Table 3. Charge Controller's technical specifications.

Minimum Power	**6000W**
Output Current	**< 160A**
Nominal Battery Voltage	**48VDC**
Minimum Charging Power	**6000W**
Input Current	**20A – 30A**
Input Voltage	**200V – 300V**
Protection	
From short-circuit	√
From overloads	√
From high temperatures	√
From hypertension & hypotension	√

From reverse polarity	√

The required batteries' capacity is determined by the capacity requirements of the electric bus, which, according to Table 1, is 540Ah. Therefore, the batteries' total capacity should be greater than 540Ah. Iin order to achieve the required voltage and system capacity, the batteries should be connected as presented below:

Two (2) parallel arrays of twenty four (24) batteries of 2V, connected in series, with a total output voltage:
V_{batout} (V) = 24 x 2 = 48V
and capacity:
I_{batout} (Ah) = 323Ah x 2 arrays = 646 Ah
The batteries will be placed next to the charging station in an open-type concrete building (natural ventilation system through louvers).

The system's total required power for the charging of the electric bus is 6000 Wp. Therefore, the following circuitry is performed:

- 27 photovoltaic panels with total power 6210W (27 panels x 230W = 6210W), which will be divided into 3 arrays of 9 panels each.
- Total output voltage per array is V_{out} (V) = V_{max} X panel number in the array = 30V X 9 = 270V, and total output voltage is V_{out} (V) = 270V as well.
- Total output current per array is I_{out} (A) = I_{max} = 8A, and total output current is I_{out} (A) = I_{max} X arrays' number = 8 x 3 =24A.

An indicative view of the charging station is shown in Figure 1:

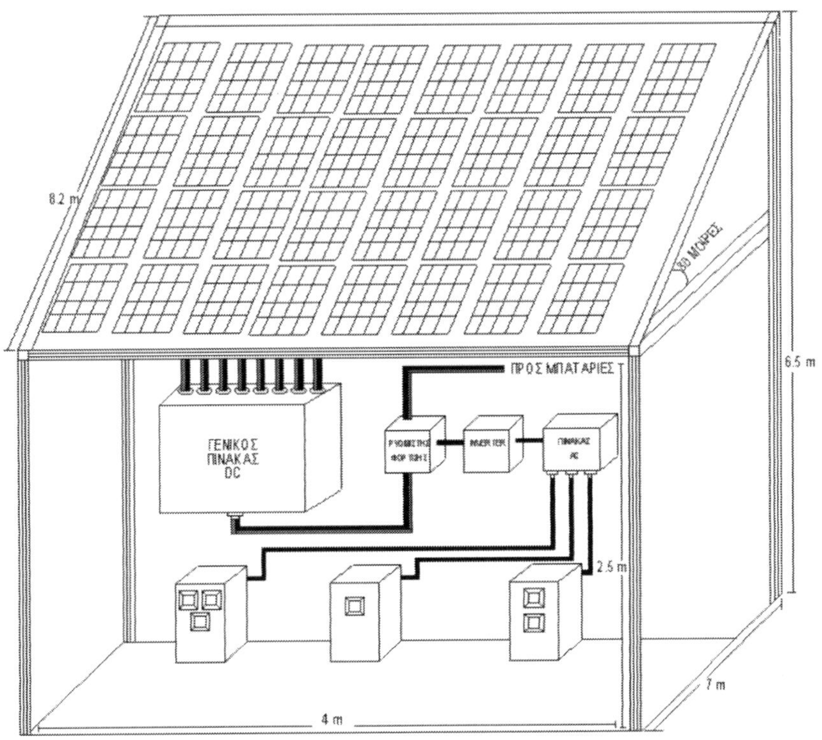

Figure 1: View of the charging station.

IOP Conf. Series: Materials Science and Engineering **161** (2016) 012080 doi:10.1088/1757-899X/161/1/012080

2.4. Cost analysis

In this section, an indicative cost of the charging station (parking) is presented, according to the previously specified/determined system requirements. More specifically, the indicative cost per system component, as well as the whole charging station's (external parking-lot's) indicative cost are presented in Table 4.

Table 4. Indicative component cost and whole station's cost.

Photovoltaic Panels	4.500 €
Batteries	4.500 €
Charging Regulator	800 €
Inverter	4.200 €
Cables and ancillary parts	1000 €
Infrastructure costs	50.000€
Total Indicative cost	**65.000 €**

3. Results

This study will significantly contribute to regional tourism development through the optimization of the road network for tourist attractions. Moreover, the study will provide efficient data regarding the design and cost of a solar charging-station (parking lot) for the charging of electric mini-buses. The mini-buses are intended to provide a student-transport and tourist-transport services within an intelligent transportation system in the regional route network.

The realization of the Intelligent Transportation System along with the development of a smart phone application, and the utility vehicles can facilitate people with special needs (elderly, disabled, distant-residents) in effectively realizing their daily on-demand transport and accessibility needs, particularly during the heavy winter periods. The scheduling and realization of the student transport by electric minibuses that are charged by RES will substantially reduce the cost of their transportation while increasing trip reliability for the new service. Service expandability and transferability will assure reduced costs in future applications, and will support capitalisation of the results across the region and in other locations.

All the outputs involve electric vehicles charged by RES, therefore contributing directly to the indicator CO34: Annual decrease of GHG. More specifically, the estimated annual decrease of GHG is 164,839 tons of CO2 equivalents/year. Moreover, the surface of improved cross-border road and infrastructure is estimated to be 1.000,00 square meters.

4. Conclusions

The realization of the above concept will result in the following:

(1) The provision of four electric minibuses and four utility cars that will transfer tourists and students at the routes that will be defined, as well as residents, especially elderly, disabled and distant-residents, on-demand.

(2) Four Renewable energy sources (RES)-supported external parking-lots will be designed and implemented.

(3) Through the utilization of smart phones, an intelligent Transportation, Client-Server application, will be deployed to the parking lots-e-minibuses.

(4) An optimal route scheduling for the transportation of tourists and students through a 20-year horizon, will facilitate solution of a yearly faced problem of the Municipalities. The reduced travel time is estimated to be 10 minutes, while the average border crossing travel time is estimated to be 30 minutes. Finally the energy efficiency awareness barometer is expected to reach the 80%.

References

[1] Gao, S. and Chabini, I. 2006 *Transportation Research Part B*, **40(2)**:93–122.

[2] Gao, S. 2008, *TRISTAN.*

[3] Stephanedes, Y. J. and Yoldasis, C. 2016. *Schedule Update. PHAROS Integrated Eco-routing and Fleet Management System Project*, ENPI/2014/343-446.

Bucket wheel rehabilitation of ERC 1400-30/7 high-capacity excavators from lignite quarries

Fl Vîlceanu[1], C Iancu[2]

[1]University of Petroşani, Romania
[2]Department IMST, University "Constantin Brâncuşi" of Târgu-Jiu, Romania

E-mail: ccggnew@gmail.com

Abstract. The existence of bucket wheel equipment type ERC 1400-30/7 in lignite quarries with lifetime expired, or in the ultimate life period, together with high cost investments for their replacement, makes rational the efforts made to rehabilitation in order to extend their life. Rehabilitation involves checking operational safety based on relevant expertise of metal structures supporting effective resistance but also the replacement (or modernization) of subassemblies that can increase excavation process productivity, lowering energy consumption, reducing mechanical stresses.
This paper proposes an analysis of constructive solution of using a part of the classical bucket wheel, on which are located 9 cutting cups and 9 chargers cups and adding a new part so that the new redesigned bucket-wheel will contain 18 cutting-chargers cups, compared to the classical model. On the CAD model of bucket wheel was performed a static and a dynamic FEA, the results being compared with the yield strength of the material of the entire structure, were checked mechanical stresses in the overall distribution map, and were verified the first 4 vibrating modes the structure compared to real loads. Thus was verified that the redesigned bucket-wheel can accomplish the proposed goals respectively increase excavation process productivity, lowering energy consumption and reducing mechanical stresses.

1. Introduction

The existence of bucket wheel equipment type ERC 1400-30/7 (figure 1) in lignite quarries with lifetime expired, or in the ultimate life period, together with high cost investments for their replacement, makes rational the efforts made to rehabilitation in order to extend their life.

Rehabilitation involves checking operational safety based on relevant expertise of metal structures supporting effective resistance but also the replacement (or modernization) of subassemblies that can increase excavation process productivity, lowering energy consumption, reducing mechanical stresses [1].

An essential subassembly of these machines is the drum shedder, component of driveline composed of bucket-wheel, shaft and gearbox. So this subassembly is subject to rehabilitation and modernization by redesign, since is the main part of the equipment that can increase excavation productivity. Drum shedder type excavator machines with high capacity for lignite quarries, model Erc1400 were installed first time in Romania in 1970. So they are widely used and proved that are reliable equipment [2].

Consequently the drum shedder and its driveline will be subject to CAD modeling and FEM analysis as steps to modernization and replacement.

Figure 1. Drum shedder excavator type ERC 1400-30 / 7

2. Modelling of the drum shedder subassembly

2.1. Subassembly CAD modelling

For correct modeling of studied subassembly must be analyzed the main cinematic chain [3]. Acting drum shedder powertrain consists of an electric motor – cardan joint shaft - gearbox, as shown in figure 2.

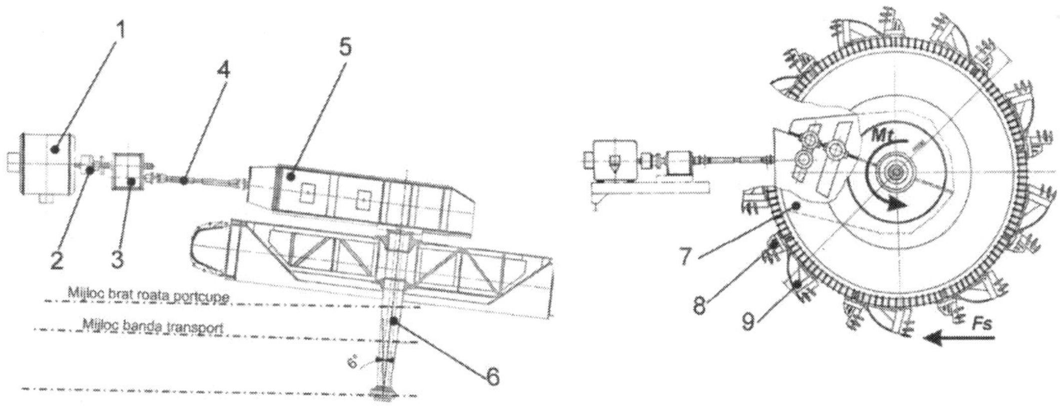

Figure. 2 wheel shedder driveline

1-electric motor, 2-flexible coupling, 3- gearbox, 4- shaft, 5- wheel shedder gearbox, 6-wheel shedder axle, 7- drum shedder, 8- cutting cups, 9- chargers cups.

The main element that gives machine productivity is drum shedder and its main component elements, respective cutting cups (pos. 8), and chargers cups (pos. 9), cf. figure 2 [4]. Drum shedder is a subassembly with a mass of about 30 tons (figure 3) and that in the early stages of the modernization process is to totally removed, meaning drum shedder, its cups and wheel shaft.

Figure 3. Classic drum shedder

1- drum shedder ring, 2 – wheel cone, 3- wheel hub, 4- wheel shaft

This paper proposes the analysis of a new constructive solution based on the use of part of the classic drum shedder (figure 3), with its components and bearings supporting the shaft and placing a new ring (figure 4), containing combined cutting / chargers cups, attached to the classic rotor.

Figure. 4 drum shedder ring with cutting / chargers cups

Before machining, after removing the subassembly of drum shedder and placing of ring with cutting / chargers cups, will be performed an expertise on all drum shedder subassembly, by non-destructive testing of welds, but also analysis and tests necessary to determine the capable load of material.

Thus was CAD modeled the classic drum shedder, and wheel shaft also. Modelling shaft is also necessary because the classic conical-cylindrical gearbox will be replaced with a new one conical-planetary, whose main advantage is that is lighter than the classic one.

CAD modeling has led to the geometric model (figure 5), where 1-shedder wheel shaft assembly, modeled with new gearbox as well as classic model of assembling rotor bearings; 2- wheel ring with cutters/chargers cups, that will be fitted on the wheel cone (pos. 3), using elements 4 - 4` that will finally permits attachment to the classic part of the rotor (pos. 5). Fasteners 4 and 4` ensure both the connection between the old and the new part of the wheel but also the protection against deposits of excavated material inside the rotor.

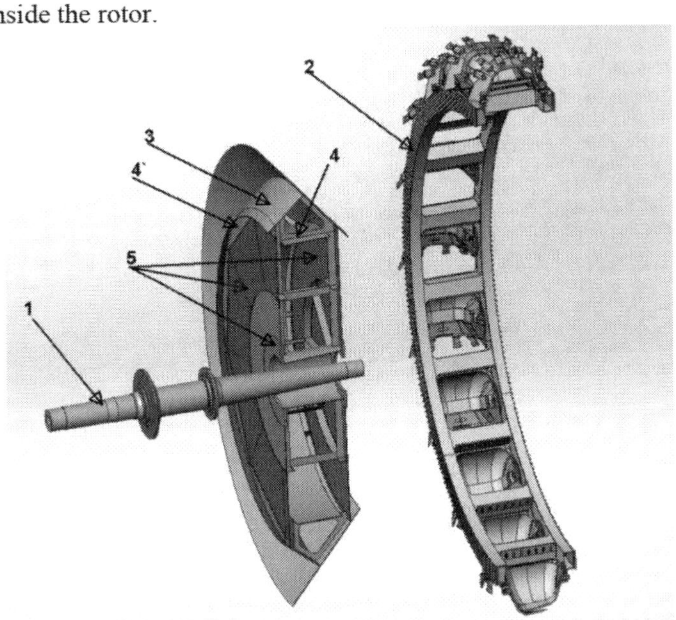

Figure. 5 Modernized rotor

After completing shedder wheel subassembly (item 2, figure 6), this is mounted on the arm shedder (1), through two bearings: one fixed on classic part of the arm and one mobile (item 3) on the auxiliary metallic construction (pos. 4) on which is arranged the cinematic rotor drive. This auxiliary metallic construction may have a form optimized for fulfilling the criterion of stability and structural strength but also for optimum gripping of cinematic assembly.

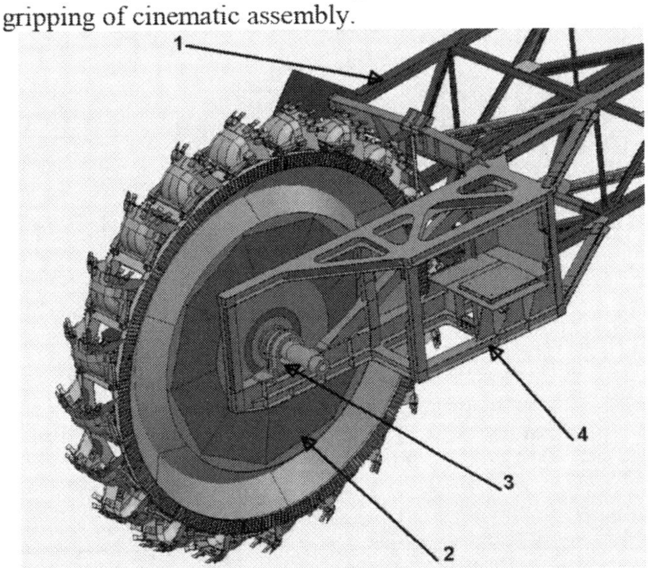

Figure 6. Attaching wheel on the arm shedder

2.2. MEF analysis of drum shedder subassembly

The second stage contains static and dynamic analysis, highlighting their modes of vibration for modernized drum shedder [5].

In the process of excavation, on the assembly tooth - bucket – rotor - shaft acts cutting force components, which are shown in figure 7, considering the reference system of the rotor.

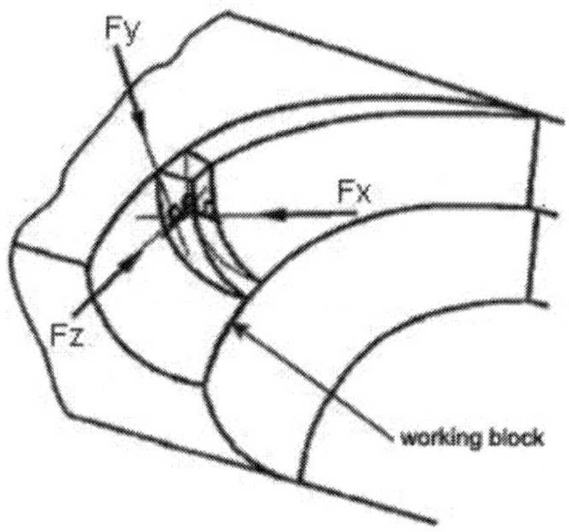

Figure 7. Cutting force components

Where:

- Fy - tangential force (cutting) - excavation force;
- Fx - radial force, imposed by the moving mechanism and maintaining consistently the equipment in excavation;
- Fz - lateral force to advance.

The main source of vibration to an excavator is the time fluctuations of the three components of excavation force, being not synchronous to the fundamental frequency of coming into excavation of the cups.

Time variation of excavation component forces is caused by variations in mechanical properties of the layer of coal and uneven wear of cups tooth, which makes the fundamental frequency of perturbation to approach the bucket wheel rotation.

It's been calculated the excitation frequency (excavation) for two types of rotor, classic and upgraded (modernized), depending on the number of cups placed, but also shedder wheel speed n_{rc} with the relationship [5]:

$$f_c = \frac{n_{rc}}{60} \cdot z_c \cdot (\pm 20\%)\ [Hz] \tag{1}$$

It can be established the following frequencies, table 1

Table 1. Excitation frequencies for classic and modernized rotor.

	Drum shedder turation [rot/min]	Number of cups	Excitation frequency [Hz] (Excavation frequency)
Classic Rotor	4.32	9 cutting cups, 9 chargers cups	$f_c = \frac{n_{rc}}{60} \cdot z_c = \frac{4.32}{60} \cdot 18 = 1,296\,Hz \cdot (\pm 20\%) = (1.037...1.555)\,Hz$
Modernized Rotor –var I	4.68	18 cutting / chargers cups	$f_c = \frac{n_{rc}}{60} \cdot z_c = \frac{4.68}{60} \cdot 18 = 1,56\,Hz \cdot (\pm 20\%) = (1.123...1.684)\,Hz$
Modernized Rotor–var II	4.68	20 cutting / chargers cups	$f_c = \frac{n_{rc}}{60} \cdot z_c = \frac{4.68}{60} \cdot 20 = 1,56\,Hz \cdot (\pm 20\%) = (1.248...1.872)\,Hz$

This calculation was performed to see the range where to integrate FEM model of the modernized wheel shedder, i.e. the number of cups to be arranged.

For static structural calculus are considered the following hypotheses:

- To the hub bucket, namely the bushing used for mounting driveshaft, is applied the output torque of the gearbox, considered as acting torque for drum shedder;

For this modernized gearbox the torque is approx. 1337 KNm, which according to DIN 22261/2 must be multiplied by a factor of 1.5, so the calculus torque is:

$$M = 1.5 \times 1337 = 2005.5 \text{ KNm} \quad (2)$$

- Disposal of fixed supports on the cutting side of drum shedder for 1/4 of the total number of cups (for 18 cups and 20 cups). We apply this condition on the same number of cups, respectively 5 cups.

Structural-modal analysis by applying FEM involves classical three steps: pre-processing, structural-modal calculus and post-processing, by highlighting results [6].

1. Preprocessing consists of:

1.1 - Establishing the type of material for the structure - S355 J0 [7];

1.2 - Verifying CAD model and importing in FEM software, figure 8.a, b;

1.3 - Meshing the model with tetrahedral higher order elements since are discretized curved surfaces, figure 8. c;

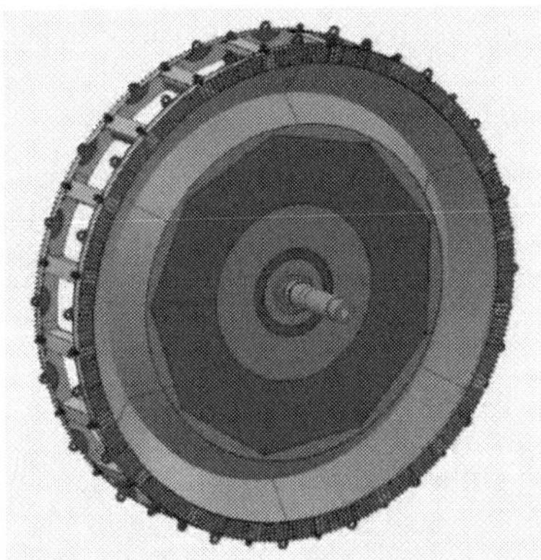

Figure 8.a CAD model (phase)

Figure 8.b CAD model - detail

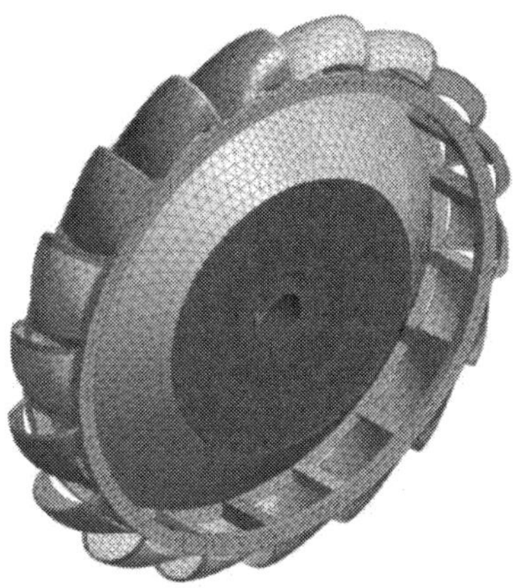

Figure 8.c Meshing the model with tetrahedral higher order elements

1.4 - Specifying supports layout:
- It's considered fixed seating on the driveshaft bearings for modal analysis, figure9.a;
- It's considered fixed seating on excavating part for 5 cutters cups for static analysis, figure9.b

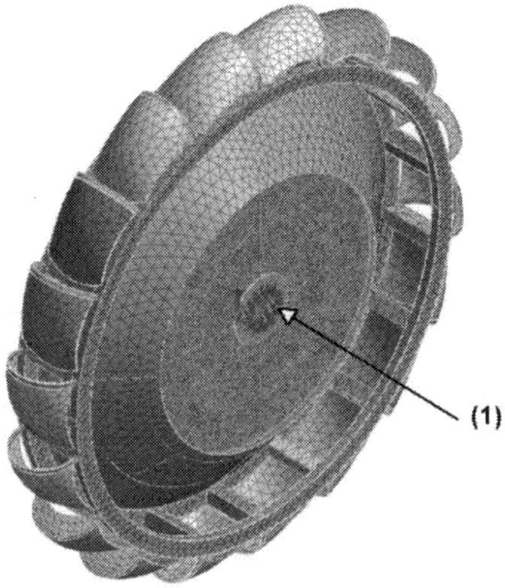

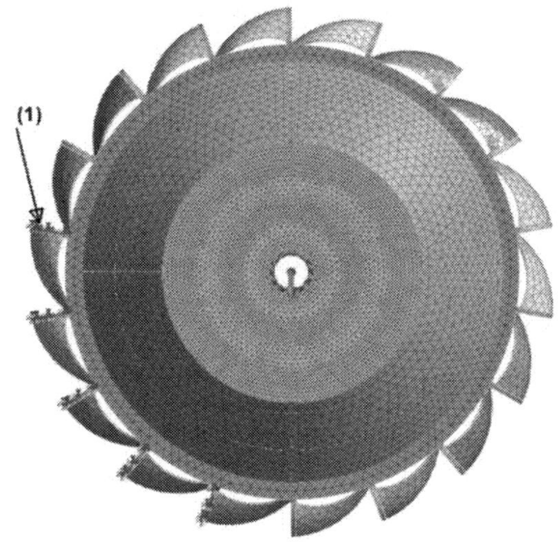

Figure 9.a Supports layout - modal analysis Figure 9.b Supports layout - static analysis

1.5 - Specifying loads:
- For modal analysis is specified gravitational loading G = 9.81 m / s^2, figure 10.a (pos.1);
- For static analysis is specified torque M = 2005.5 KNm (pos.1) and gravitational acceleration (pos. 2), figure 10 b.

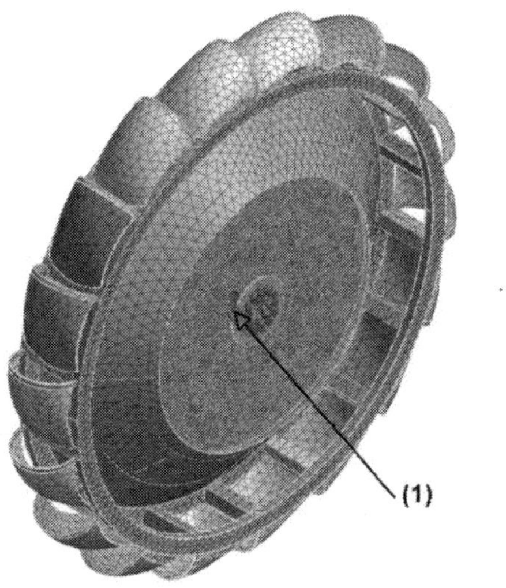

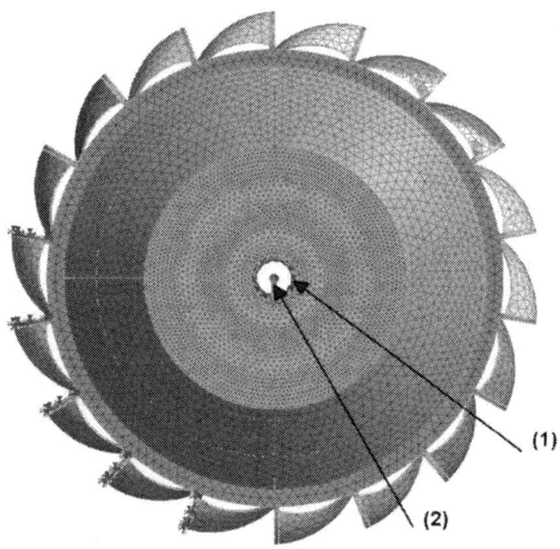

Figure 10.a Loads - modal analysis Figure 10.b Loads - static analysis

2. Structural-modal calculus (static and dynamic analysis) is achieved through the specific FEM computer software, with a specified degree of convergence.

3. Postprocessing - highlighting and interpretation of results for the two models considered respectively with 18 and 20 cups cutters / loading and interpretation of distributions of stresses and deformations.

Further it was adopted following criteria:

- for static analysis - Establishing equivalent Von Misses stress and comparing the maximum value with minimum yield for S 355 J0, the material used for making analyzed subassembly.
- for modal analysis - frequency of first vibration mode to be higher than the frequency calculated, according to table 1.

The Von Misses stress distributions are presented in figure 11 a, b for drum shedder with 18 cutters / loading cups respectively for drum shedder with 20 cutters / loading cups, and in figure 12 a, b the displacement distribution for the 2 variants of drum shedder (18 cutters / loading cups and 20 cutters / loading cups).

Figure 11.a Stress distribution – 18 cups

Figure 11.b Stress distribution - 20 cups

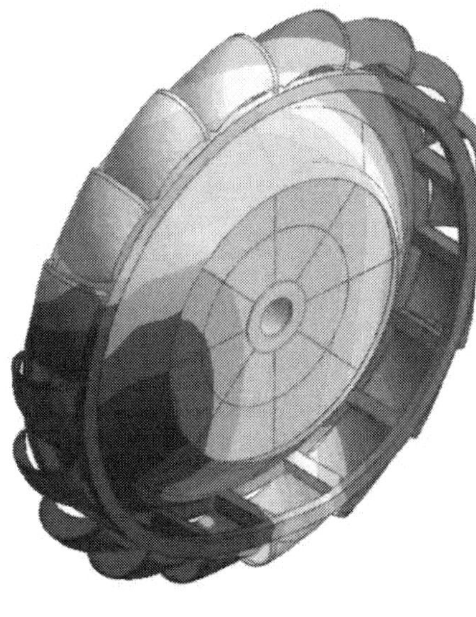

Figure 12.a Displacement distribution – 18 cups

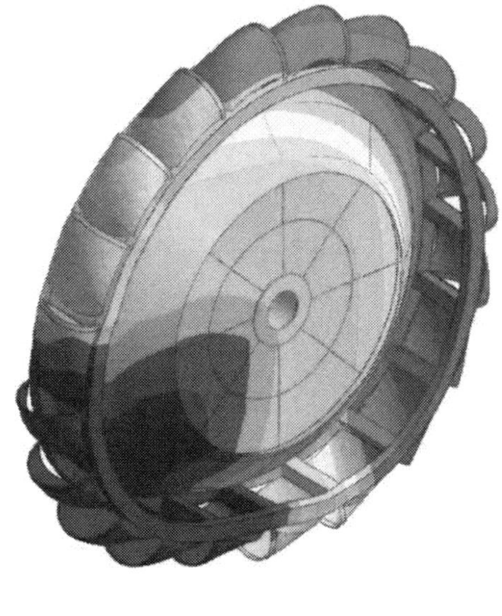

Figure 12.b Displacement distribution - 20 cups

IOP Conf. Series: Materials Science and Engineering **161** (2016) 012081 doi:10.1088/1757-899X/161/1/012081

Values of equivalent stress and displacement for the 2 variants are:

$\sigma_{VM\ 18} = 60.85\ N/mm^2$ $\sigma_{VM\ 20} = 144\ N/mm^2$ (3)

$\delta st_{\ 18} = 1.246\ mm$ $\delta st_{\ 20} = 3.08\ mm$ (4)

Results for modal analysis, meaning deformed shape and frequency for the first three natural vibration modes are presented in figures 13 a, b, c for 18 cutters / loading cups and figure 14 a, b, c for the second variant of drum shedder with 20 cutters / loading cups).

Figure 13.a Mode 1=5.357 Hz

Figure 13.b Mode 2=1.909 Hz

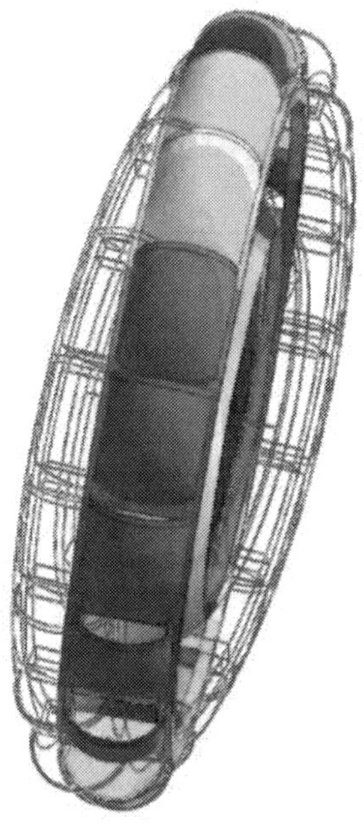
Figure 13.c Mode 3=1.191 Hz

Figure 14.a Mode 1=5.487 Hz

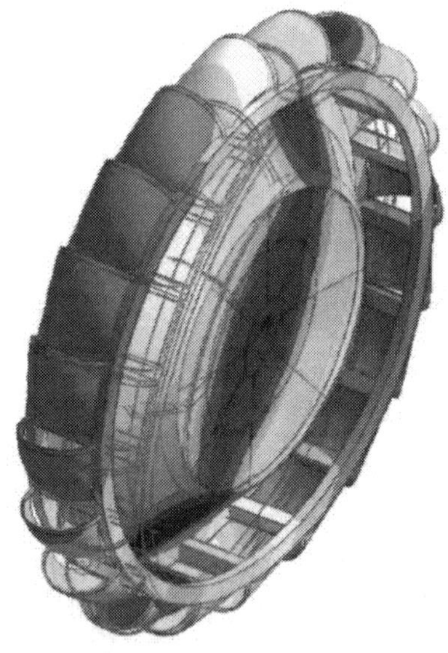
Figure 14.b Mode 2=1.940 Hz

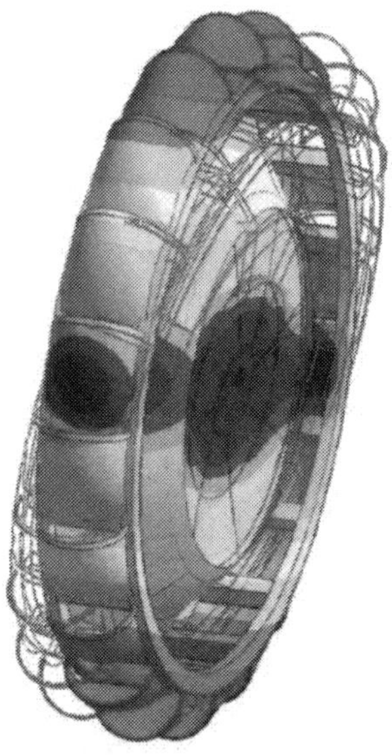
Figure 14.c Mode 3=1.196 Hz

Conclusions

a. Static analysis – it can be seen that for both variants the equivalent Von Misses stresses do not exceed the yield strength of the material of drum shedder.

Instead, for the 20 cups rotor, the equivalent stress is significantly larger:

$$\sigma_{VM.20} = 144\ \text{N/mm}^2 > \sigma_{VM.18} = 60.85\ \text{N/mm}^2 \quad (5)$$

This mechanical tension, for a large number of operating cycles, can lead in the end to cracking in places with high stress [8].

In practice, checking the rotors that are made with caisson type rings, like the one considered for our case, it was observed that cracks appears exactly in the clamping area of the cups on this ring, as shown in figure (photo) 15.

Figure 15. Cracks in areas heavily strained ring rotor:

1- grip ear cup, 2- rotor ring crack

b. Modal analysis - considering the way of functioning during excavation and deformed shape of the rotor, it will be considered frequencies of vibration modes 2 and 3. From this point of view there aren't large differences between the two constructive types of rotors, and the displacements for respective modes are small.

Comparing the results obtained for the two types of rotors we can conclude that the rotor with 18 cups is the constructive version that will be adopted. Even if tensions in the ring zone are small compared with the 20 cups version must be considered the clamping mode and thickness required for ring caisson.

Some other advantages of this constructive version are:

- Recovery of classic rotor material about 50-60%;
- Reduction of rotor vibration that occurs for membrane constructive version.
- Decreased deposition of material inside the rotor by disposing inside a protective ring.

References

[1] Iancu C and Vîlceanu Fl 2015 A Method of Calculus of Residual Lifetime for Lifting Installation, *American Journal of Mechanical Engineering*, ISSN (Print): 2328-4102, ISSN (Online): 2328-4110, **3**, no.4, (www.sciepub.com), pp. 115-121

[2] Rasper L 1975 *The bucket wheel excavator: development-design-application*, Clausthal-Zellerfeld, (Bay Village, Ohio: Trans Tech Publications)

[3] Durst W and Vogt W revision 2008 *Bucket Wheel Excavator*, Series on Mining Engineering, **7**, ISBN 978-0878490752, (Switzerland: Trans Tech Publications)

[4] Kunze G Göhring H Jacob K 2002 *Baumaschinen Erdbau und Tagebaumaschinen Herausgegeben* (Construction digging/earth moving and mining equipment), ISBN 978-3-528-06628-4, (Wiesbaden: Springer Fachmedien)

[5] Schlecht B 2014 Schlussbericht zu dem IGF - *Vorhaben Untersuchung und Optimierung des dynamischen Verhaltens von Schaufelradantrieben der Forschungsstelle(n)*, (Investigation and optimization of the dynamic behavior of bucket wheel), (Dresden: Technische Universität IMM)

[6] Nitin S G *et al* 2008 *Practical Finite Element Analysis*, ISBN 978-81-906195-0-9 (New Delhi: Ed. Finite to Infinite)

[7] http://www.mencinger.biz/SteelData/Equivalent%20Carbon-%20Stainless-%20Offshore %20Steel.htm

[8] Scheffler M 1994 *Grundlagen der Fördertechnik - Elemente und Triebwerke* (Basics of handling equipment - elements and engines), ISBN 978-3-322-96882-1, (Wiesbaden: Springer Fachmedien)

One of possible variants of the organization for recycling lubricate cooling of technological means for small businesses

I Rusica[1], A Toca[2], I Stingaci[3], S. Scaticailov[4], I. Scaticailov[5], O. Marinescu[6] and P. Kosenko[7]

[1] Associate professor, Technical University of Moldova, Chisinau, RM
[2] Associate professor, Technical University of Moldova, Chisinau, RM
[3] Senior lecturer, Technical University of Moldova, Chisinau, RM
[4] Senior lecturer, Technical University of Moldova, Chisinau, RM
[5] Senior lecturer, Technical University of Moldova, Chisinau, RM
[6] Senior lecturer, Technical University of Moldova, Chisinau, RM
[7] Associate professor, Technical University of Moldova, Chisinau, RM

E-mail: ion.stingaci@gmail.com

Abstract. In the paper we analyze the application lubricate cooling technological environment in the processing of various materials in the past century greatly have increased cutting speed and respectively, has increased productivity [1]. Today, none of production in which anyway is used metal cutting machines of all types (milling, turning, grinding, drilling, etc.) is not without lubricant cooling technological liquid which in turn are designed to reduce cutting force and the load on metal cutting machine tools and machined parts in order to increase durability machine tools and reduce errors of processing details and also in resource energy saving. When using lubricate cooling technological environment reduces the temperature in the cutting zone resulting in higher tool life and the preservation of the surface structure being treated reducing wear of metal parts of the machine. Typically, lubricant cooling process fluids is used without replacing as long as possible not yet beginning to negatively affect the quality of process. However life expectancy lubricate cooling technological environment is limited. According to existing normative acts every kind of lubricate cooling technological environment through certain time must be deleted by from the system and subjected to a recycling. Lubricate cooling technological environment must be disposed of for the following reasons: occurs the microbial and the mechanical pollution cutting fluid, free oil impairs operational characteristics cutting fluid and increases consumption.

1. Problem

Through the use of cutting fluids in the processing of different materials over the past century have significantly increased cutting speed and therefore increased productivity [1]. Nowadays no one manufacture in which, one way or another, used cutting machines of all kinds (milling, turning, grinding, drilling, etc.) can't do without, the so-called COTS - lubricating and cooling process fluids, which in its turn are intended to reduce the cutting force and the load on the metal-cutting machine tools and the work pieces as to increase durability and reduce the error machine processing parts, as well as energy savings. When using cutting fluids - decreases the temperature in the cutting zone, which leads to increased tool life and preservation of structure of the treatment surface, reduced wear of metal parts of the machine. Usually cutting fluids used without replacement as long as possible until the quality of cutting fluids does not begin to adversely affect the quality of the process. However, life expectancy is limited to cutting fluids. According to existing normative acts every kind cutting fluids should be removed after a certain time of the cutting fluids feed system and subjected to recycling. Workers who are in regular contact with contaminated cutting fluids can get serious diseases of internal organs and skin, cause allergies. Spent cutting fluids causes corrosion processes machined parts and manufacturing equipment. Cutting fluids role in the processing of materials is not reduced at the beginning of the 21st century. Data appeared about new cutting speeds fluctuating within the 1000-6000 m/min. The recommendations on the use of new cutting materials, there is a requirement that the use of these cutting materials is not permitted without cutting fluids. Given the use of high speeds and capabilities of modern machine tools (machining centers with a large number of tools), you can create small and medium-sized businesses with a small number of machinery equipment and production of a large range of products. But this is business of the future.

In the past 25 years after the Soviet collapse, most large industrial enterprises in the CIS countries went bankrupt or were transformed into small enterprises of small or medium-sized businesses. The number of machinery equipment in enterprises decreased significantly. For example, in Moldova, the biggest company in the field of engineering is no more than 100 machines. For disposal of cutting fluids [1, 4] proposed the following methods: 1) chemical, physico-chemical; 2) Thermal, evaporation; 3) The membrane purification, filtration; 4) centrifugation; 5) The sorption; 6) combinations thereof. If we analyze these methods we can see that they are based on different principles and implement the necessary recycling processes on different hardware. So a number of companies [2] offer several methods of disposal of cutting fluids and accordingly the supply of various equipment for the disposal of various types of cutting fluids. And from the enterprise requires a clear answer, what components must be disposed of. We believe that a small business with 20 - 40 does not have a working professional, thoroughly versed in matters of recycling cutting fluids.

By studying the problem of disposal of cutting fluids, we found [3] an example of a decision problem of recycling cutting fluids in a big city like Moscow. The volume of annual deliveries cutting fluids in Moscow is over 25,000 tons. In the Moscow Government is concerned that small businesses are not able to deal with the disposal of cutting fluids and accepted the proposal to organize a single center for the development of scientific research and experimental - design works and the development of new technologies for emulsion production, as well as equipment for disposal, recovery and recycling of used lubricating -ohlazhdayuschih liquids. In the Moscow Government is concerned that small businesses are not able to deal with the disposal of cutting fluids and accepted the proposal to organize a single center for the development of scientific research and experimental - design works and the development of new technologies for emulsion production, as well as equipment for disposal, recovery and recycling of used cutting fluids. In this article, we do not set the task to assess the effectiveness of a particular method. We just emphasize that none of these methods does not provide a complete recycling of waste cutting fluids [1, 4]. It follows that for the full utilization of cutting fluids in the enterprise need to use several methods for disposal. To use several methods, the company should have trained professionals, both in art and in the field of chemistry and different technologies. It is necessary to set up a laboratory to determine the maximum permissible concentration of the various components. It is known that there is a disposal problem worldwide. And usually means for waste management are invested in large cities and major regional centers. But a large part of the population lives in small towns and the problem of waste disposal for small towns is relevant, as well as for large cities. Due to the small enterprises waste recycling capacity for small towns is not economically profitable to build. On this basis, the Republic of Moldova initiated several projects on waste disposal for several areas (covering about 200,000 people.). Within the framework of these projects is planned to build a unified ambulance drives in one place for a number of settlements. With all the localities vehicles using waste will be dumped in the common sanitary storage. The idea is to save material resources at the disposal of household waste.

2. Research methods

To study the problem of disposing of cutting fluids in the Republic of Moldova, we used a statistical method, which is one of the known methods of research. The statistical authorities of the Republic of Moldova given that in the Republic there are about 5000 companies in the manufacturing industry. More than 4200 companies are privately owned, including 200 enterprises with foreign capital. These enterprises employ about 100,000 workers.

On the basis of these data, using different calculation methods aggregated amount of equipment, we came to the conclusion that the Republic of Moldova can work about 25 000 - 40 000 machine tools. Naturally, all of these machine tools use different types of cutting fluids. This can be judged on the basis of that is in the Republic, in addition to pre-existing companies were more than 200 companies with foreign capital. For each type of cutting fluids need to develop its recycling technology. And we put the question-whether a small company is able to organize the disposal of cutting fluids collected from the work of 20 - 30 machines?

3. Research results

As a result, the statistical data were obtained on the number of investigations of the imported quantity concentrates for preparing cutting fluids to the Republic for 7 years. These data are shown in Fig. 1.

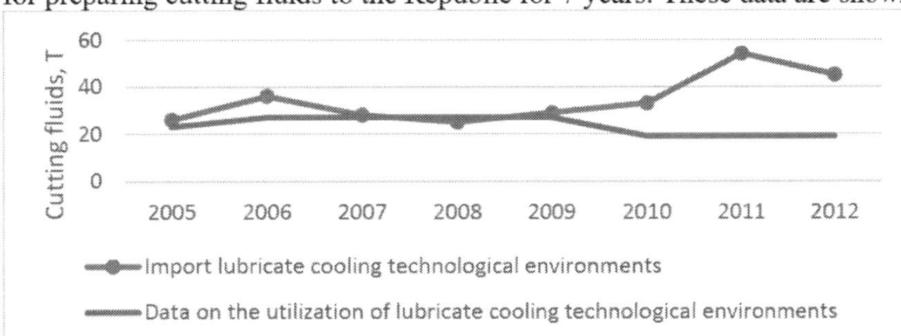

Figure 1. Imports into the Republic of Moldova concentrates for preparing cutting fluids data and reports on their utilization.

As you can see, in the republic for a long time imported an average of 35 tons concentrates for preparing cutting fluids (curve points). Consequently, each year the same number of concentrates for preparing cutting fluids must be disposed of and must not fall into the water basins of the rivers Prut and Nistru. Discharge of waste into sewer cutting fluids harms the environment and ultimately to all residents of Moldova, Ukraine and Romania using the water of the two rivers for drinking. The rivers Dniester and Prut provide more than 80% of the needs of the population and economic complex of the Republic in the water. These are the only river water that is safe to drink. Although there is some evidence that drinking water is contaminated in Moldova by 90%, without further explanation what the components. On the other hand, we used statistical data on utilization of cutting fluids (curve without points. Fig. 1), which shows that not all volumes imported cutting fluids recycled. Analysis of the data plots reveals that part of the Republic received cutting fluids not disposed (curve without points) concentrates. There is in our view a few options: 1) option - part of the enterprises do not know what it is necessary to account for the disposal of industrial waste. 2) option - part of the cutting fluids concentrates has not been used. On the other hand it is possible to account for the disposal, but do nothing for recycling. Since we know the situation in most industries, we can say that no one entity is not engaged in recycling of used cutting fluids. All imported cutting fluids in volume, according to the graph in Fig. 1 merges into storm sewers or fecal.

This we can confirm the research conducted by ecologists [5, 6] in the implementation of the alloy expeditions group of environmentalists on the rivers Prut and Nistru. We've simplified graphics part of the studies authors [5, 6] without prejudice to the science. From these studies it follows that in the analysis of water, after all the cities 1, 2, 3, 4 (in our charts here cities replaced the figures) along the rivers Prut and Nistru (Figure 2 and Figure 3) increase the weight content of the following components: heavy metals, oil products and phenols.

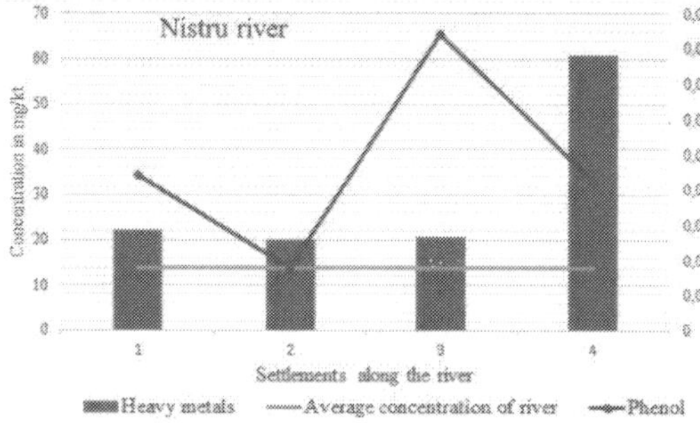

Figure 2. The content of heavy metals and phenols.

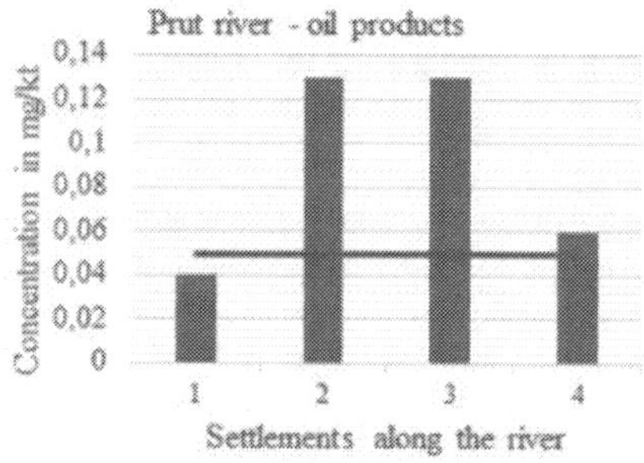

Figure 3. The content of oil products

These data demonstrate that in the cities at the expense of the productive activities in the rivers merge water production waste without recycling. On the other hand, based on analysis of water in the Dniester and Prut can not give an answer to the question of what the company throws in waste water basins cutting fluids. It is necessary to take into account the fact that the expedition carried out tests for harmful substances in the water for a limited range of substances. The data on the components included in the cutting fluids omitted. According to [4] in cutting fluids include the following components: mineral oil, emulsifiers, corrosion inhibitors, biocides, antifoaming agents and phenol. These studies showed that the water in the river increases the concentration of a number of harmful substances such as phenols and oil products during the course of the rivers, which are part of the cutting fluids.

4. Conclusion

We propose to establish in the Republic of Moldova within a number of regional centers for waste cutting fluids waste. In the initial conditions, the job should head the Ministry of Environment, which is due to the budget-could be built to start one such facility. In the future, studying the work of all industrial companies could collect certain batch of spent cutting fluids and bring them to process in this area. It is understood that this processing will continue to be made for a fee. But we believe that this fee will be less than if every small business would be organized recycling cutting fluids at home. On the other side of the joint efforts to contribute to the purification of the water basin from the harmful chemical elements and compounds. This would lead to the purification of drinking water for residents of Moldova, Romania and Ukraine.

References

[1] Smazochno-ohlajdajushhie tehnologichesckie sredstva dlja obrabotki metallov rezaniem: Spravochnik / Pod. Red. S. G. Ehntelisa, E. M. Berlinera. -M .: Mashinostroenie, 1986, -352 p.

[2] ZAO Ionoobmennye tehnologii, Moskva, www.ioteh.ru

[3] Rasporiajenie Pravitel'stva Moskvy ot 13.01.99 nr. 20-Pn "O merah po uporjadocheniju rabot v oblasti Smazochno-ohlajdajushhih jidkostejj (SOJ), primenjaemyh v promyshlennosti g,Moskvy

[4] Smazochno-ohlajdajushhie tehnologichesckie sredstva i ih primenenie pri obrabotke rezaniem: Spravochnik L.V.Hudobin, A,P,Babichev, E,M,Bulyjov i dr./Pod. Obshh. red. L.V.Hudobina.-M.: Mashinostroenie, 2006,-544 p.

[5] http://www.meteo.md/mold/01072011.htm

[6] http://www.meteo.md/mold/15072011.htm

IOP Conf. Series: Materials Science and Engineering **161** (2016) 012083 doi:10.1088/1757-899X/161/1/012083

Biomass district heating methodology and pilot installations for public buildings groups

N Chatzistougianni[1], E Giagozoglou[1], K Sentzas[2], E Karastergios[2], D Tsiamitros[1], D Stimoniaris[1], A Stomoniaris[1] and S Maropoulos[1]

[1]Western Macedonia University of Applied Sciences, Koila Kozanis 1, Kozani, 50100, Greece
[2]Municipality of Deskati, Deskati, 51100, Grevena, Greece

E-mail: dtsiamit@teiwm.gr

Abstract: The objective of the paper is to show how locally available biomass can support a small-scale district heating system of public buildings, especially when taking into account energy audit in-situ measurements and energy efficiency improvement measures. The step-by-step methodology is presented, including the research for local biomass availability, the thermal needs study and the study for the biomass district heating system, with and without energy efficiency improvement measures.

1. Problem

In Western Macedonia Region, a large number of mainly rural households use lignite for heating purposes, and due to the financial crisis lignite and fire wood are displacing heating oil also in urban areas. Moreover, in the same region, three operational district heating networks based on the nearby lignite-fired plants are active. The intense agricultural and forestry activities in the area provide an abundance of different types of biomass according to national statistical authorities [1]. The main idea behind the paper is to present a methodology for setting up a biomass district heating network for small public building groups that can later be evolved in the core of an extensive district heating network of a city. Such a network is considered a renewable energy resource system that can bring the following benefits to the rural and urban areas of the region:

1) The supply of firewood in bigger quantities and at central collection points is expected to reduce the supply cost, both for the local biomass, as well as for the biomass that is purchased by the market.

2) The efficiency of the burner-boiler system is controlled much more effectively, leading to even more savings in economical and energy resourcesterms.

3) The use of properly maintained filters contributes to the environmental conservation of the areas.

4) Increase of comfort indicators for the inhabitants of theareas.

2. Method

The methodology is adapted to the regional needs and is also based on the Greek National legislation on Energy Efficiency in Buildings. Therefore, the first step is the estimation of the local availability and supply of biomass. Then, the public buildings' thermal needs drive the study for the design and estimation of capacity and cost for the biomass unit and the district heating network. The next step is the energy efficiency study of the public buildings and the proposal for targeted energy efficiency improvements that will decrease the buildings' thermal needs. At this phase, the necessity to conduct in-site measurements during the energy audits of buildings, rather than rely on tables and statistics, is emerged. Then, the reduced thermal needs drive the study for the re-design and re-estimation of capacity and cost for the biomass unit and the district heating network. The methodology is applied to four public buildings at the Karpero village of the municipality of Deskati.

2.1. Local availability and supply ofbiomass

Since almost all of the inhabitants already use fire wood for heating purposes, due to the local rights on forest woods, the availability of biomass and the cost of supply are secured. Moreover, more extensive, quantified studies for the whole Grevena Prefecture [2]-[5] reveal that the available biomass of the prefecture is enough to cover also the entire city of Grevena. More specifically, if the residues of the local wood industries are taken into account and the availability of the dry lignite of the Western Macedonia region, it is deduced by Table I that the attempt of biomass district heating of the public buildings is possible, in terms ofresources.

Table 1. Available biomass for the Grevena prefecture.

Biomass type	Availability in tons
Wheat cultivation residues	30.000
Corn cultivation residues	10.000
Energy plants cultivation	500-1.000
Wood biomass	21.000
Forestry residues	2.000-4.000

At the following tables, the supply cost of biomass is also shown.

Table 2. Supply cost for the prefecture of Grevena (per ton).

Biomass type	Cost (€/tn) (including transport, not VAT)
Pellet	200,00
Wheat cultivation residue	60,00
Corn cultivation residue	40,00
Forestry residue	75,00 – 81,00

Table 3. Supply cost per energy resources (The boiler efficiency is assumed equal to 85 %) [5].

Biomass type	Cost (€/MWh) (including transport, not VAT)

Pellet	56,14
Wheat cultivation residue (**max cost**)	24,00
Corn cultivation residue	17,00
Forestry residue – (**max cost**)	23,00 – 24,50
Oil	76,00
Natural Gas	50,00
Lignite (Public Power Corp.)	16,00
Dry lignite	23,50

2.2. Design and estimation of the biomass unit and the district heating network for public buildings

The methodology is adapted to the needs of a Greek rural municipality, the Municipality of Deskati. More specifically, the village of Karpero is selected, since it already involves utilisation of wood for heating by the inhabitants. Five municipal buildings are indicated by the local authorities.
Their total surface is 3.100,00 m^2. The installed total capacity of the boilers is almost 700 kWs. According to the relative data sheets, the annual oil consumption for heating is 25,7 tns (306 MWh).
At the following table, the necessary collected data are broken down to each building.

Table 4. Collected data per building.

	A	B	C	D	E
Usage	Gymnasium	Municipality	Primary school	Kindergarten	Medical center
Construction date	1980	2003	1972	1996	1997
Thermal Isolation	NO	YES	NO	YES	YES
Type of transparent parts	Aluminum (double glass) 2000	Aluminum (double glass)	Aluminum (double glass) 2010	Aluminum (double glass)	Aluminum (double glass)
Number of	2	2	2	1	1
Heating surface	1605,08	748,09	603	100,8	50
Installed capacity	350	150	70	30	0
Oil consumption 2010-2014 [tn]					
2010					
2011	14-15 t	2,5	-	2,5 - 3,0 t	
2012	14-15 t	2-2,5 t	6,7 t	2,5 - 3,0 t	
2013	14-15 t	2-2,5 t	5,0 t	2,5 - 3,0 t	
2014	14-15 t	2-2,5 t	5,2 t	2,5 - 3,0 t	
Normalized installed capacity [kcal/h,m2]	*218*	*201*	*116*	*298*	-

Due to the occasional use of building E, this building is excluded from the study.
As shown from Table 4, the Normalized installed capacity [kcal/h,m2] oft he four buildings varies between 116 and 298 [kcal/h,m2].
Assuming an efficiency rate of 85 % of the existing boilers, the following table is derived:

Table 5. Normalized characteristics of the four public buildings.

Building	A	B	C	D
efficiency	85%	85%	85%	85%
Boiler capacity [kcal/h,m^2]	185	170	99	253
Boiler capacity [kcal/h,m^3]	62	57	33	84
Normalized energy consumption [KWh/m^2,a]	95	34	87	**302**

Comparing these buildings with similar from the same climate zone (Zone D, according to the National Regulation for buildings energy efficiency – KENAK), the normalized energy consumption is similar. E.g. a building with similar building envelope and usage has normalized energy consumption equal to 107[KWh/m^2,a]. The only exception is buildingD.
For the calculation of the thermal needs of the buildings and then for the calculation of the nominal proposed installed capacity, the local thermal conductance coefficients are considered. Then the thermal conductance coefficients of the National Regulation (KENAK) for old buildings without thermal isolation are considered. Consequently, two scenarios were developed: the optimistic / realistic evaluation and the pessimistic evaluation scenarios.
For both cases, the real thermal needs of all the buildings are expected to be lower than 700 kWs, as shown in the following table:

Table 6. Total thermal needs.

Building	A	B	C	D	**Total**	
	Gymnasium	Municipality Hall	Primary school	Kindergarten	Mcal/h	
Thermal needs Mcal/h	154	58	72	9	**293**	Optimistic scenario
Thermal needs Mcal/h	**232**	**62**	**88**	**10**	**392**	Pessimistic scenario

For the calculation of the losses of the district heating network, the relative manuals have been used.
The efficiency of the biomass boiler is again assumed equal to 85%.
Considering the district heating network losses and the thermal needs, the following table is derived:

Table 7. Proposed installed capacity of the biomass unit.

	Optimistic		Pessimistic	
	MCAL/H	KW	MCAL/H	KW
Buildings thermal needs	**293**	**340**	**392**	**456**
Losses	15	18	16	18

Total	308	358	408	474
Boiler efficiency	85%	85%	85%	85%
Installed boiler capacity	362	**421**	480	**558**

Of course, it is proposed that the 420-558 kW capacity should be covered by at least two boilers with 200 kW and 300 kW, in order to use only one boiler during the periods with low heating demand.
The configuration of the system is shown in Figure 1.

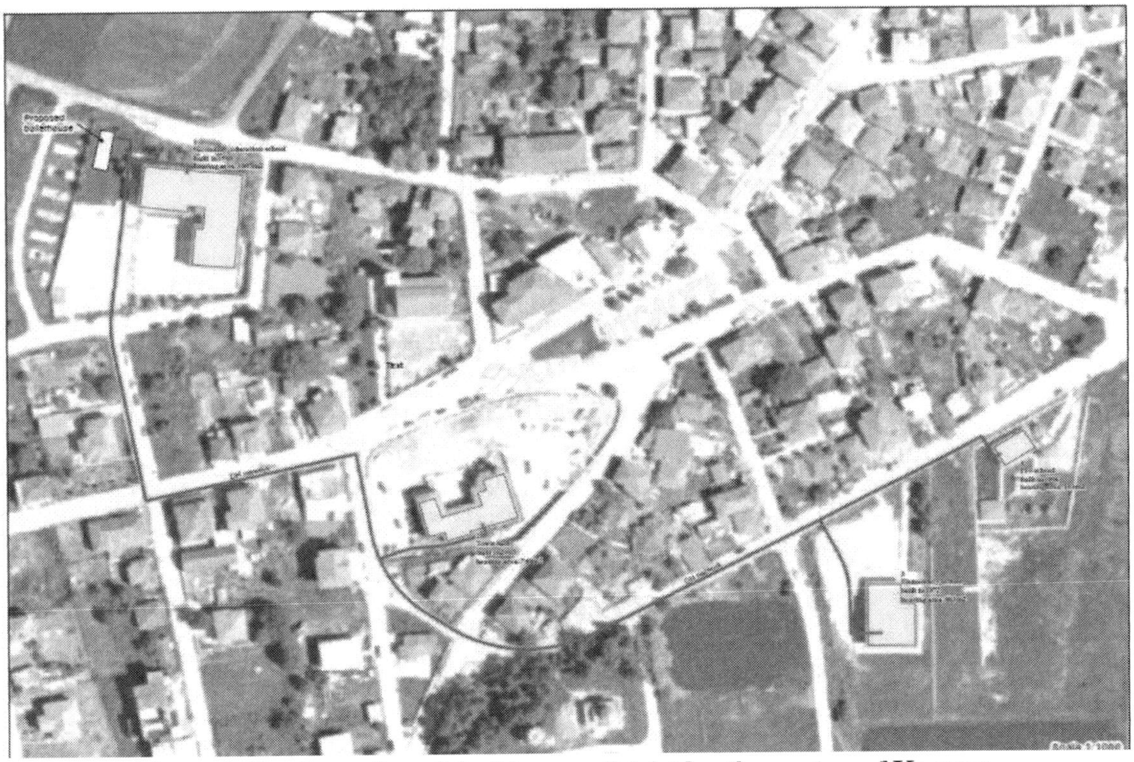

Figure 1. Configuration of the biomass district heating system of Karpero.

For the network, the installation of underground preinsulated EN253 pipes is proposed, as in the cases of the Kozani, Ptolemais and Amyntaio. The network involves 670 m of excavation works, if the boiler unit is installed close to the Gymnasium, as shown in Figure 1. For the primary heating network, the following temperatures of operation are proposed: Inlet: 90° C and Temperature difference: DT=30° C (primary circuit).
For the easier management and automated supply of the unit with biomass, the biomass material should be supplied as wood chips. Consequently, enough space should be available next to the boiler for storing of the biomass.
The 26 tns that are consumed for space heating needs, will be now replaced with wood based biomass, leading also to the following environmental benefits:

Table 8. Comparison of emissions between the oil and biomass for the specific buildings.

Emissions [$\lambda_{\kappa,\nu}$] (g/kg) for 26 tns of OIL					
CO2	**SO2**	**CO**	**NOx**	**HC**	**Particles**
3142	0.7	0.572	2.384	0.191	0.286
Emissions (tn) for wood-based biomass					
CO2	**SO2**	**CO**	**NOx**	**HC**	**Particles**
81.69	0.02	0.01	0.06	0.00	0.01

For the cost estimation of the building, the following main parts are considered:

- Installation of the biomass unit in a container 15X5m (including 500KW biomass boiler, pumps, biomass storage, ancillaryequipment).
- Underground network of EN253 pipes of approximately 670m of length.
- Thermal substations for the consumers (totally4).

Total Cost: ~160.000 - 190.000 € (no VAT included), ~210.000-240.000€ (VAT included).

2.3. Energy efficiency improvements of public buildings and recalculation of the thermal needs.
In order to further improve the efficiency of the whole system (biomass district heating system and buildings), detailed energy audit for the four buildings is conducted. According to each building's weakness, targeted improvements are applied, like external thermal isolation, building energy management systems, etc and the new classification is produced. The new classification leads to revised thermal needs of the buildings and a revised study for the biomass district heating network and the biomass boiler.
The results of the energy audit of each building and the targeted proposed improvements are shown in the following tables:

Table 9. Results of the energy audits (according to the KENAK software)

Building	A	B	C	D
	Gymnasium	Municipality Hall	Primary school	Kindergarten
Thermal needs kWh/m2	**194,1**	**65,6**	**221,1**	**243,2**
Total needs kWh/m2	**255,2**	**256,6**	**281,1**	**367,4**
Energy class	**Z**	**E**	**Z**	**E**

If the results of Table 9 are calculated in terms of consumed heating oil, the results show great differences than the actual consumed fuel. For example, for the case of building A (Gymnasium), a simple calculation gives:
194,1 KWh/m2*1605m2=311.530 kWh=311,53MWh
which equals to almost 30 tns of heating oil per annum. However, the consumed oil, as shown in Table 4, is 14-15 tns/a. This deviation is justified only partly due to the assumptions made at the KENAK software. However, these deviations also highlight the need of conducting in-site energy audit measurements and not rely exclusively on theoretical values from tables, since the construction uncertainties may lead to different results in energy classification [8].

Table 10. Targeted interventions at the four buildings and gains.

Building	A	B	C	D
	Gymnasium	Municipality Hall	Primary school	Kindergarten
energy interventions	Placement of heat insulation in the nutshell and the roof of the building	Upgrade cooling system	Placement of heat insulation in the nutshell and the roof of the building	No interventions
Total cost of interventions (euros)	80.000	4.000	36.000	The problem in this building is in its heating system.
% saving of primary energy	**51,6**	**10,1**	**52,8**	
damping of cost (years)	**4,6**	**5,2**	**4,8**	

Based on the percentage savings in primary energy from Table 10, the thermal needs of the four buildings are revised, leading to respective revision of the biomass district heating system. The comparison between the previous biomass district heating system and the revised one is outlined in the following results section.

3. Results

3.1 Comparative analysis after the re-design and re-estimation of the biomass unit and the district heating network for public buildings.

The application of the energy interventions described above will decisively contribute to energy savings for the proposed buildings. Therefore it is expected that the building's installed thermal load and the biomass boiler capacity will be reduced, compared to the initial estimations presented above.
Taking into consideration the thermal insulation to be applied on the shell and roof of the buildings, it is estimated that the actual buildings thermal demand will be reduced to 280KW, therefore the biomass boiler nominal capacity is estimated to 350KW.
The district heating system installation cost is estimated to 150,000€ (no VAT incl.), consisting of the boiler and boiler house cost, the preinsulated network cost and the district heating substations cost.

4. Conclusions

A proposed methodology of energy improvements in public buildings and their support with biomass district heating systems is described in this paper. The benefits for rural municipalities are impressive, in terms of operation cost and the environment, whereas the installation costs are not very high. Moreover, the results of the energy efficiency studies point out the need for detailed energy audits with in-situ measurements at the buildings.

References

[1] Reference to webpage: http://www.statistics.gr/ .
[2] Giagozoglou Eythymios,*Utilisation of biomass for district heating networks: The case of Grevena –MSc Theses*.
[3] Technical report of the project PROARES – Larisa University of Applied Sciences.

[4] Laspopouloy Tarina, *District heating study for the Grevena city with biomass,Graduate Theses*, National Technical University ofAthens.
[5] Final report of ALTENER project 2002: *Concerted Actions to accelerate the district heating of Grevena city by Biomass Investment*, ANKOSA.
[6] Reference to webpage: BIOFOSS project:http://biofoss-project.eu/ .
[7] TEE-KENAK software, available atwww.tee.gr.
[8] Dimitrios Stimoniaris, Dimitrios Tsiamitros, V. Zacharaki, F. Dialynas, T. Kottas, S. Maropoulos, M. Stefanovski, Z. Stepanovski, I. Milosovska, K. Micalevski, V. Karagiannis, T. Papotis, E. Dialynas 2014 *MedPower 2014 Conference*, Athens, Greece.

IOP Conf. Series: Materials Science and Engineering **161** (2016) 012084 doi:10.1088/1757-899X/161/1/012084

Interstitial Condensation Risk at Thermal Rehabilitated Buildings

I Baran[1], I Bliuc[1], A Iacob[1], L Dumitrescu[1], R A Pescaru[1] and C Helepciuc[1]

[1] Technical University "Gheorghe Asachi" from Iaşi, Faculty of Civil Engineering and Building Services, B-dul Dimitrie Mangeron 1, Iaşi 700050, Romania

E-mail: irina.baran9@gmail.com

Abstract. The increasing thermal insulation degree of existing residential buildings, aiming to reduce the energy requirements for ensuring the indoor comfort, has as expected effect the elimination of condensation risk. However, in some cases this phenomenon occurs, both on the inner surface of the closing element and also in its structure. The surface condensation causes can be identified and can be easily removed. Instead, the causes and even the presence of interstitial condensation are more difficult to be observed. But the moistening of the insulation materials and the reduction of thermal insulation capacity or even its total degradation, contravene into a large extent or totally to the main purpose of the additional thermal protection. To avoid such situations, it is necessary to respect some principles concerning the structure, resulted from the knowledge of the water vapour diffusion behaviour of various materials.

It is known that condensation vulnerability is higher for the additional thermal protection solutions by disposing the insulating material on the inside surface of the closing element. But practice has shown that the condensation phenomenon is not totally excluded neither in the case of outside thermal insulation – which is the current solution applied to the rehabilitation works - if the principles mentioned above are not known and respected.

In this paper two models are compared on which the risk of interstitial condensation can be checked. The analysis made on two structures of exterior walls with thermal insulation demonstrates the need for additional verifications before proposing a solution for thermal rehabilitation of the envelope elements.

1. The interstitial condensation mechanism and the verification of the condensation developing risk

The condensation of the water vapour can take place in the mass of building elements that separate the environments with different temperatures and vapour concentrations, as a result of diffusion of water in the gaseous form that is produced from the interior heated space to the outside - in winter conditions - and from the outside in summer conditions, in the case of basements or other rooms with low temperatures [1].

Basically, in order that interstitial condensation phenomena to not affect the normal operation of a building, there must be a balance between the amount of vapour that accumulates in the structure of the element in the cold season, and that which could be eliminated by drying during the hot weather.

In case when this balance is not achieved, there is a gradual accumulation of water that usually is stationing in the pores of the insulating material, which, by wetting, considerably diminishes the thermal insulation capacity.

The favouring factors which determine the interstitial condensation phenomenon and influence its intensity are:

- characteristics of indoor climate, respectively vapour flow sources, ventilation rate and temperature;
- characteristics of the outdoor climate, respectively air temperature and humidity;
- characteristics of the materials that make up the element, respectively porosity, hygroscopicity and the vapour permeability;
- degree of exposure to aggressive climatic factors, rain, wind, solar radiation;
- the existence of other sources of moisture with synergistic effect on the phenomenon of diffusion (external water seepage caused, for example, by malfunctioning of devices for rainwater drainage).

In a graphical interpretation of the phenomenon of vapour diffusion, the condition that the condensation should not meet is that the partial pressures diagram does not have a common point with the saturation pressure diagram. There are three possible distinct situations:

- the partial pressures diagram has no common point with that of pressures of saturation, respectively, in any section of the element, the following condition is fulfilled: $p_v<p_{vs}$. **The condensation does not appear;**
- the partial pressures diagram is tangential to the saturation pressure diagram – **a surface condensation appears;**
- the partial pressures chart diagram intersects the saturation pressure - **a condensation zone appears** (Figure 1).

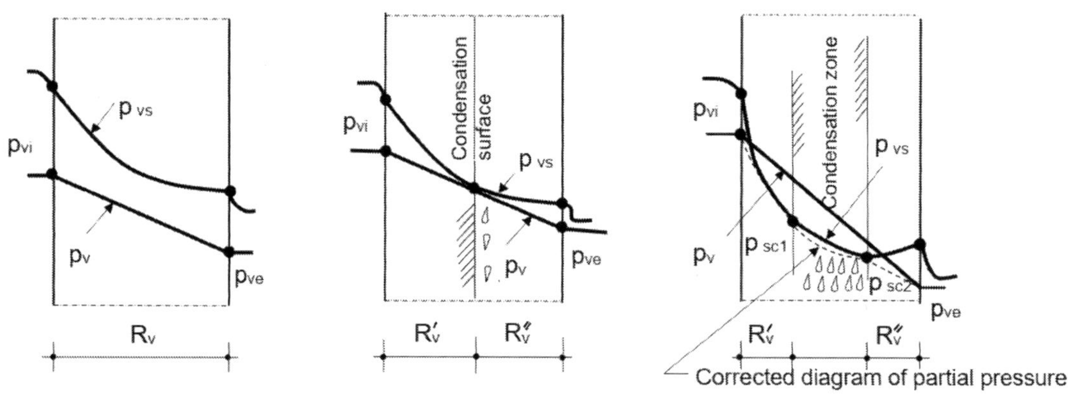

Figure 1. Verification of condensation risk of in the mass of closing elements; graphical representation.

In reality, the phenomenon is more complex, in most cases, to the diffusion transport of water is added the water transport in liquid form by capillary action - that modifies the properties of materials - and the surface diffusion that occurs as an intermediate stage between the liquid and gas ones. These aspects can be considered if we use complex computer programs capable of simultaneous simulation of moisture transfer by diffusion and capillary action and the action of external climatic factors such as rainfall and solar action [2].

2. Methods for assessing the risk of interstitial condensation

An accurate assessment of the interstitial condensation risk in the stage of projects development for new buildings or for rehabilitation works is very important not only to avoid the adverse effects of moisture on building materials, but also for the assurance of indoor environmental quality.

The verification way of the risk of interstitial condensation is regulated by SR EN ISO 13788:2013 - *Hygrothermal performance of building components and building elements - Internal surface temperature to avoid critical surface humidity and interstitial condensation - Calculation methods* [3].

The verification is based on analysis of moisture annual balance sheet by comparing the total amount of water accumulated through condensation and the one that can be removed by drying, on the base of monthly vapour flow calculation. For this is considered the mean monthly air temperature and humidity specific for the building site. It is considered exclusively the moisture transfer by diffusion and the analysis is based on Glaser method that simplifies the phenomenon of heat and mass transfer by accepting the following assumptions:

- the condensation occurs only at the interface between layers of different materials and remains at that interface;
- the thermal conductivity is independent of the moisture content of the material;
- the transfer of moisture in a liquid state and the migration by capillary action do not occur in the closing element;
- there is no transfer by convection inside the structure;
- the boundary conditions are the values of mean monthly temperature and humidity;
- the transfer of heat and humidity is unidirectional;
- the effect of rain and solar radiation are neglected.

The complexity of the phenomenon is reflected with much more accuracy in the method that is based on the calculation program WUFI 2D, which takes into account the transitional regime concerning the boundary conditions and more data on their characteristics, such as:

- dependence on moisture of the vapour resistance factor;
- dependence on moisture of the thermal conductivity coefficient of materials;
- coefficient of transport and redistribution of liquid water;
- moisture storage function defined for hygroscopic porous materials.

The boundary conditions variable in time are defined by the following parameters:

- indoor and outdoor air temperature;
- humidity inside and outside;
- impact of rain on the exterior wall;
- incidence of solar radiation on the exterior wall surface.

3. Interstitial condensation risk analysis on the basis of two models; results, discussion.

The risk was analysed for two solutions of thermal rehabilitation of the exterior walls, which are estimated to have a greater vulnerability to interstitial condensation. There were considered the climatic conditions of outside air (temperature and humidity), specific for the area of Iasi and the characteristics of the indoor climate, specific for a normal use of dwellings, respectively the indoor air temperature T_i = 20 °C and a relative humidity φ_i = 60%.

a. Walls of brick masonry of 37.5 cm thickness, with thermal insulation of 10 cm expanded polystyrene disposed to the inner surface.

The geometric and physical characteristics are presented in Table 1.

Table 1. Geometric and physical characteristics of the external wall

Layer	Material	Layer Thickness d [m]	Resistance factor to water vapour permeability μ [-]	Thermal conductivity λ [W/(m·K)]	Resistance to thermal transfer R [m²·K/W]	Equivalent thickness s_d=μ d [m]
1	Reinforced lime cement plaster	0.005	100	0.93	0.005	0.500
2	Expanded polystyrene	0.100	30	0.04	2.500	3.000
3	Brick masonry	0.375	6.1	0.80	0.469	2.287
4	Cement plaster	0.025	7	0.87	0.029	0.175
		TOTAL			3.003	

The verification of the interstitial condensation risk according to SR EN ISO 13788:2013 consists in assessing the moisture balance by evaluating the monthly flow of water accumulation from condensation (Table 2).

Table 2. The flow of condensation and accumulation in the wall

Month	Interface 2	
	Condensation flowrate g_c [kg/m²]	Accumulated moisture content M_a [kg/m²]
November	0	0
December	0.05900500	0.5900500
January	0.06596900	0.1249740
February	0.03650500	0.1614790
March	-0.00791460	0.1535644
April	-0.13040352	0.0231608
May	-0.40223550	0

It is found that the accumulation of water condensation begins in January, and the drying process begins in March. In May the eliminated water flow is higher than the accumulated one.

The section in which the condensation occurs is that of contact between the interior thermal insulation and the existing wall (warm face of the insulation).

The element behaviour in transient regime was analysed using minimum, maximum and average values of the relative humidity in characteristic sections of the element, figure 2, obtained by using WUFI 2D software.

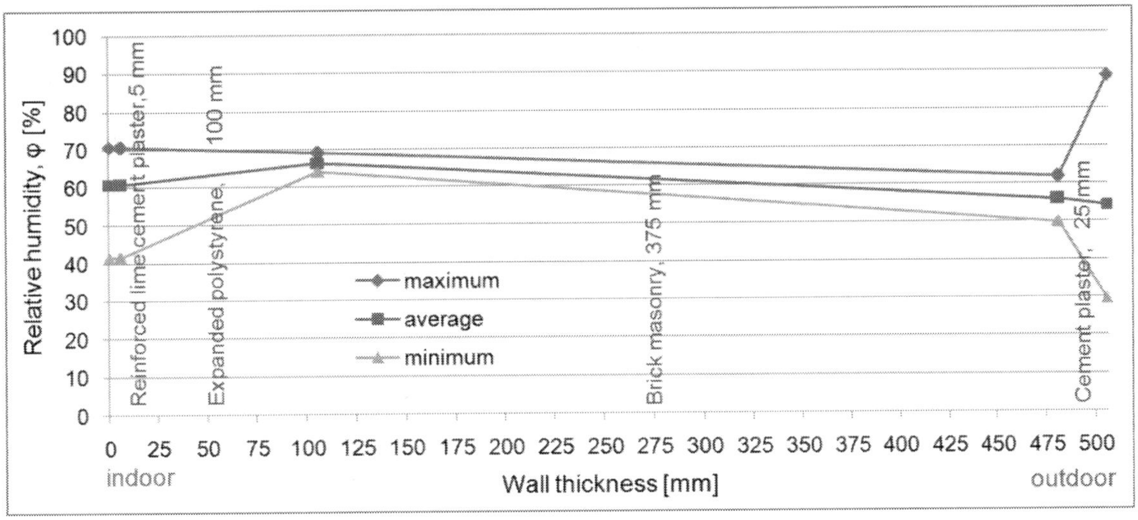

Figure 2. Mean values of relative humidity in characteristic sections of masonry wall, additionaly insulated with polystyrene at the interior

The results of the analysis in transient regime show that:

- the relative humidity values in characteristic sections reach a peak of 70% in the section separating the polystyrene insulation and the brickwork, confirming the results of the calculation according to SR EN ISO 13788:2013 regarding the absence of the risk of progressive accumulation of condensation water and the section where condensation appears;
- higher values of relative humidity, close to 90% occur on the external surface, as a result of the action of rain, taken into account by the program.

b. External wall made of masonry of autoclaved aerated concrete blocks with 30 cm thickness and a thermal insulation of 10 cm expanded polystyrene thermal insulation disposed at the outer surface.

The geometric and physical characteristics are presented in Table 3.

Table 3. Geometric and physical characteristics of the external wall

Layer	Material	Layer Thickness d [m]	Resistance factor to water vapour permeability μ [-]	Thermal conductivity λ [W/(m·K)]	Resistance to thermal transfer R [m²·K/W]	Equivalent thickness s_d=μ d [m]
1	Reinforced lime cement plaster	0.020	7	0.87	0.023	0.14
2	Autoclaved aerated concrete masonry	0.300	3.9	0.30	1.00	1.17
3	Expanded polystyrene	0.100	30	0.04	2.50	3.00
4	Cement plaster	0.005	100	0.93	0.005	0.50
		TOTAL			3.528	

The results of the risk assessing of interstitial condensation according to SR EN ISO 13788:2013 by evaluating the monthly flow of water accumulation through condensation are shown in Table 4.

Table 4. The flow of condensation and accumulation in the wall

Month	Interface 3	
	Condensation flowrate g_c [kg/m^2]	Accumulated moisture content M_a [kg/m^2]
November	0	0
December	0.000179	0.000179
January	0.009588	0.009767
February	-0.055109	-0.045342
March	-0.209878	0

It can be noticed that the accumulation of condensation water begins in December and the drying process begins in February, the month in which the eliminated water flow is higher than that of the accumulated one.

The section in which condensation occurs is at the contact between the exterior thermal insulation and the existing wall.

The element behaviour in transient regime was analysed using minimum, maximum and average values of the relative humidity in characteristic sections of the element, figure 3, obtained by using WUFI 2D software.

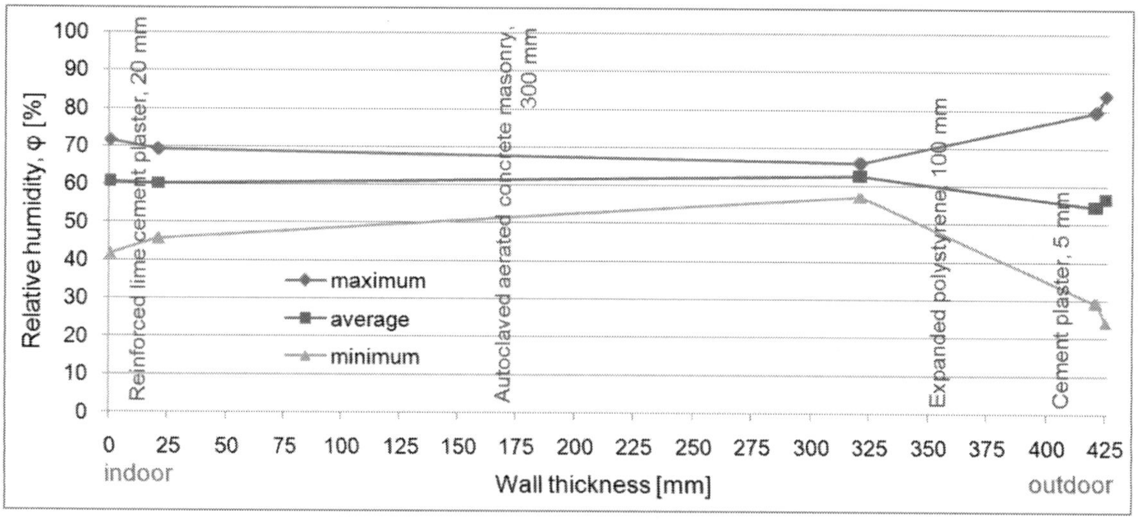

Figure 3. Mean values of relative humidity in characteristic sections of masonry wall made of autoclaved aerated concrete blocks additionally insulated with polystyrene at the exterior side

The results of the analysis in transient regime show that:

- the relative humidity values in characteristic sections reach a peak of 70% in the section that separates the inner plaster from the masonry autoclaved concrete blocks, unconfirming the calculation results made according to SR EN ISO 13788:2013 concerning the section where is present the most pronounced risk of condensation ;
- higher values of relative humidity, about 80%, occur on the external surface, as a result of the action of rain, which is taken into account by the program.

Discussions on the results of the analysis can be done on two distinct aspects. One concerning the accuracy and concordance of results obtained for the two models and the other on the composition principles of layered elements from the point of view of behaviour to the interstitial condensation.

In general, the two models - based on steady state, respectively transient regime, give similar results in assessing the risk of interstitial condensation. It should be noted, however, that taking into account the effects of rain and sunlight is very important, highlighting the need for a waterproofing layer (but permeable for vapours) on the outside surface of the wall. In addition, in practice are met situations when the swinging rain effect, especially on North façade, is manifested by the intensification of the interstitial condensation, which is becoming visible on the inside surface of the wall.

Studies reported in the research area, reflected in recommendations concerning the avoiding of condensation risk in layered structure, emphasize the main factors involved in the appearance and development of this phenomenon:

- the arrangement of layers of different materials from the inside to the outside in order to increase the vapour permeability;
- the use of vapour barriers on the warm face of the thermal insulating material;
- the use of outside protection systems which include a ventilated air layer.

The second structure, that was the subject of the analysis, does not comply with the first recommendation regarding the composition of the layered structures, the outside layer of thermal insulation, the expanded polystyrene, being much less permeable than the existing structure made of autoclaved concrete. However, there is no risk of gradual water accumulation through condensation, the behaviour being even better than the first structure, correctly made. The explanation can be found only in the different thermal resistance of the two structures, the second having a thermal resistance R = 3.695 m^{2}K / W> R = 3.17 m^{2}K / W.

Obviously, firm conclusions may not be drawn only on the basis of two situations. But a thorough study concerning the weight of different factors involved in the production of interstitial condensation seems to be necessary. This study would determine to what level of thermal protection the order of arrangement of layers or the presence of vapour barrier is no longer so important.

In all circumstances, a correct and economic solution of additional thermal protection involves a risk analysis of interstitial condensation, preferably using a model for conducting of the transfer phenomena in transient regime, taking into account the action of rain and solar radiation.

3. Conclusions

The condensation verification according to SR EN ISO 13788:2013, having as model the stationary transfer phenomena, provides realistic information concerning the risk of interstitial condensation. The significant amount of calculation made by hand and the use of the graphical method can easily generate errors that may distort the results of an analysis.

Computer programs based on the transient regime, able to take into account the variability with moisture of the materials characteristics, as well as the complex action of climatic factors, provide more information even more truthful, useful in developing some optimal solutions of additional thermal protection. The results of a systematic research program and using numerical modelling could be summarized in a catalogue of building closing solutions with interstitial condensation risk.

References

[1] Australian Building Codes Board 2014 Condensation in Buildings. Handbook.

[2] Ramos N M, Delgado J Q, Barreira E and de Freitas V P 2009 Hygrothermal properties applied in numerical simulation: Interstitial condensation analysis, *Journal of Building Appraisal* 5 161–170. doi:10.1057/jba.2009.27.

[3] SR EN ISO 13788:2013 - Hygrothermal performance of building components and building elements - Internal surface temperature to avoid critical surface humidity and interstitial condensation - Calculation methods (ISO 13788:2012).

IOP Conf. Series: Materials Science and Engineering **161** (2016) 012085 doi:10.1088/1757-899X/161/1/012085

An optimal sizing method for energy storage system in wind farms based on the analysis of wind power forecast error

R L Ye[1], Z Z Guo[1], R Y Liu[1], and J N Liu[2]

[1]Department of Electrical Engineering, Harbin Institute of Technology, Harbin, 150001, China
[2]State Grid AC Engineering Construction Company, Beijing, 100052, China
[1]E-mail: 18911267942@163.com

Abstract: Energy storage system (ESS) in a wind farm can effectively compensate the fluctuations of wind power. How to determine the size of ESS in wind farms is an urgent problem to be solved. A novel method is proposed for designing the optimal size of ESS considering wind power uncertainty. This approach uses non-parametric estimation method to analysis the wind power forecast error (WPFE) and the cumulative wind power deviation (CWPD) within the scheduling period. Then a cost-benefit analysis model is established to obtain the optimal size of ESS based on the analysis of WPFE and CWPD. A series of wind farm data in California are used as numerical cases, which presents that the algorithm presented in this paper has good feasibility and performance in optimal ESS sizing in wind farms.

1. Introduction

Given the high environment sensitivity and low predictability, large-scale integration of wind power accompanies great influence on power system and big challenges to grid dispatching operation[1-3]. Therefore, it is of great significance to increase wind power forecast accuracy so as to ensure the safe operation of power system. As of right now, prediction results in large-scale wind farms are not always satisfying, for example, the average value of mean absolute error of wind power prediction around 12 hours varies between 10% (flat terrain) to 22% (complex terrain)[4], which cannot meet the requirements for generation scheduling. Hence, how to cope with the influence caused by wind power forecast error on generation scheduling and guarantee safe and stable operation of the system deserve further research.

Due to the obvious advantages of energy storage system (ESS) over traditional thermal power generating unit in response speed and regulation accuracy, combined operation of energy storage unit and wind power generation system has already become an effective way of solving the problems of large-scale wind power integration[5-8]. Specifically, to satisfy the requirements of wind power grid-integration, for a same capacity it requires a thermal power generating unit with a capacity that is 1.6-2 times of rated capacity of the wind farm, while a system equipped with energy storage device requires a power capacity that is merely 10-20% of rated capacity of the wind farm[9].

Recent years have seen lots of scholars and engineers constantly develop large-scale energy storage device and technology. Huge breakthrough has been made in energy storage technology, particularly in compressed-air energy storage[10-13] , which prompts the mass application of energy storage device in wind farms. Equipping ESS in wind farms has become the most direct and efficient way for large-scale wind power integration areas to restrain the fluctuation of output power and make it satisfy the requirements specified in relevant standards[14-18].

The size of an ESS, i.e. the rated capacity and peak output of an ESS, directly affects the project cost and its application in wind farms. Many papers have proposed methods to optimize the size of ESS for different ESS working conditions[15-28]. HAN et al.[18] pointed out that the application of ESS could help wind farms keep constant and stable output, but require wind farms to equip a bigger energy storage reserve capacity, which will lead to higher project cost of ESS. In [23], the risk theory was harnessed to evaluate the size of ESS in different scenes of electricity market environment and in accordance with the evaluation results, wind power manufacturers needed to buy appropriate energy storage source from the market, which was a dynamic process and of poor maneuverability. LIANG et al.[24] took into account the fluctuations of wind power, and conducted some simulation analysis based on wind farms with different capacities and ESS sizes, and drew the conclusion that the optimal ratio of ESS capacity and wind farm capacity was 1:4. HE et al.[26] focused on an independently operated wind-solar hybrid generation system and set the optimal sizing of ESS in the system as their research target. The maximum load power shortage rate and the maximum instantaneous load power shortage were also considered in determining the capacity of ESS. BREKKEN et al.[27] proposed a power flow control strategy by installing zinc-bromine cell in wind farms and found out the influences of different energy storage control strategies on ESS sizing. The calculation results indicated that the strategy of artificial neural network was slightly better than the simple energy storage response strategy. Drawing lessons from [27] and trying to simplify the process, this paper employs the simple strategy for power control.

In essence, ESS in a wind farm is applied to cope with the random fluctuations of wind power and there is strong correlation between ESS sizing and wind power forecast error (WPFE), while in [15-27], when calculating the optimal size of ESS, the economic benefits was usually taken as the basic consideration and few researches have taken WPFE into consider. Based on the analysis of WPFE, this paper optimizes the size of ESS in a wind farm with cost-benefit analysis method. ESS with larger capacity will have a better performance in smoothing wind power fluctuations,but more cost is required to be paid for energy storage devices.Conversely, ESS with limited capacity or lower peak power can help reduce the cost,but the fluctuation-smoothing effect will be significantly weakened and will in addition lead to more wind curtailment loss or higher reserve capacity compensation.

Given the analyses above, an optimal ESS sizing method based on the analysis of wind power prediction error is presented in this paper. By methods of statistics and probability, the distribution functions of WPFE and the cumulative wind power deviation (CWPD) with respect to dispatching temporal dimension are established, based on which the relation function of the size of ESS and wind energy losses caused by insufficient storage capacity is built. Then the cost-benefit curves of the ESS in a wind farm are finally achieved. Based on the curves,wind farms are able to measure the cost and benefit of an ESS and determine its optimal size.

2. Wind power forecast error analysis

2.1. Calculation of wind power forecast error

Expression (1) gives the calculation of wind power forecast error:

$$\varepsilon_t = \frac{P_t - \hat{P}_t}{P_{cap}} \tag{1}$$

where P_t is the actual value of wind farm output power at time t, $\hat{P}_t$ is the predicted value of wind farm output power at time t, and P_{cap} is rated installed capacity of the wind farm.

In order to describe the relation between wind power fluctuations and the capacity of ESS, the cumulative wind power deviation (hereinafter referred to as "CWPD"), namely the wind power deviation caused by cumulative forecast errors within one dispatching period, is used to measure the capacity of an ESS, as is shown in expression (2).

$$e_{c,t} = \sum_{\tau=1}^{t} \varepsilon_i \cdot \Delta t, \quad \varepsilon_i = \begin{cases} \varepsilon_i, & \varepsilon_i \le P_{ESS} \\ P_{ESS}, & \varepsilon_i > P_{ESS} \end{cases} \tag{2}$$

where $e_{c,t}$ is the CWPD caused by cumulative wind power forecast error from period 1 to t; Δt is the time variation and P_{ESS} is the peak output of ESS.

Normalization is conducted to facilitate the comparison of CWPDs in different forecast periods, as shown in expression (3).

$$e_{c,t}^{*} = \frac{e_{c,t}}{T} \tag{3}$$

where T is the dispatching period.

2.2. Principles of Non-parametric Estimation

As its estimating functions usually have free forms and few constrains and usually require little for data distribution, non-parametric estimation method has stronger applicability and higher robustness in solving nonlinear and non-homogeneous cases that can't be described by certainty model, and the whole regression model is totally data driven. Frequently-used non-parametric estimation methods include the kernel density estimation (KDE) , the nearest neighbor density estimation(NNDE), etc. KDE method is adopted in this paper.

The fundamental of non-parametric estimation on probability density function is: assuming x_1, x_2, ..., x_n are n discrete random samples, with the probability density function $f(x)$ unknown, and the density function estimation is derived from empirical distribution function as shown in expression (4):

$$f_h(x) = \frac{1}{n}\sum_{i=1}^{n} K_h(x - x_i) \tag{4}$$

where $K(*)$ is the kernel function, h is the bandwidth coefficient. Common kernel functions include uniform kernel function, Gaussian kernel function, cosine kernel function, triangle kernel function, etc. According to the statistical and experimental findings of Epanechikov and Scott, different kernel functions have equivalent effects when the bandwidth coefficient is optimal[29]. Therefore the standard Gaussian kernel function and the cross-validation methodology (CV) are employed in this paper to obtain the optimal bandwidth h, and expression (4) is used for density estimation.

The data of a wind farm in Texas, U.S. during the years of 2004-2006 are taken as example for analysis. Wind power forecasts with 1 day in advance are conducted following non-parametric estimation and normal distribution estimation, and the results are shown in figure 1. The kernel function adopted in non-parametric estimation is standard Gaussian kernel function, $K(x) = exp(-x^2/2)/\sqrt{2\pi}$.

From figure 1, it can be found that the probability density estimating function obtained by non-parametric estimation greatly fits the distribution of wind power forecast error, and the best bandwidth is 0.02434. The maximum value of wind power forecast error is 0.8959, the minimum value is -0.7592, and the distribution curve is not symmetrical about 0. The probability of negative error is higher than that of positive. When adopting the normal distribution method to fit the wind power forecast error, the variance is 0.03854 and the fitting effect is not ideal.

Setting the dispatching period T=24h and normalizing the CWPD by expression (3), the maximum value of cumulative wind power forecast deviation -11.2348 and the minimum value 8.7886 are obtained. Non-parametric estimation and normal distribution estimation methods are respectively conducted to estimate CWPD, as is shown in figure 2. In non-parametric estimation, the optimal bandwidth is 0.1718; while in normal distribution estimation, the variance is 2.928.

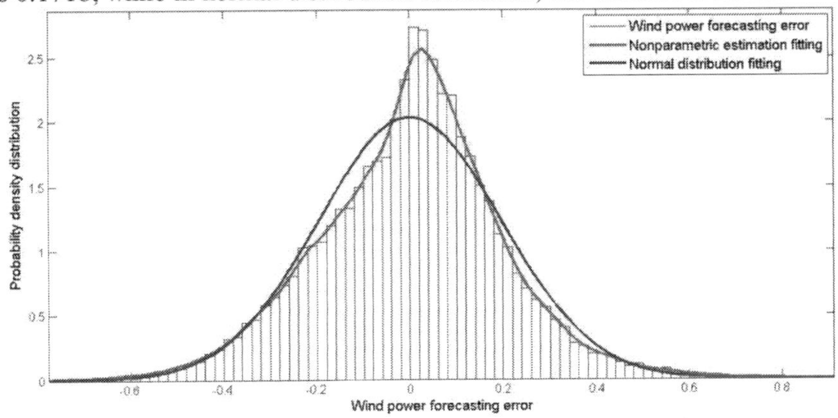

Figure 1. Probability density distribution of wind power forecast error

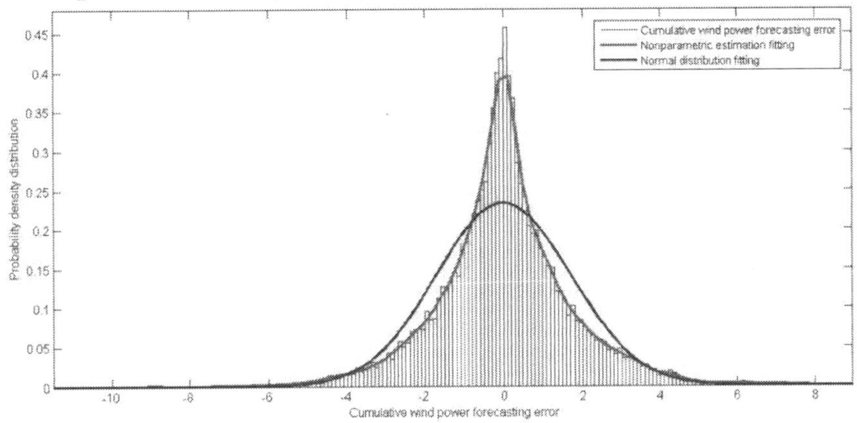

Figure 2. Probability density of cumulative wind power forecast deviation

3. Determination of Energy Storage System Scale

Energy storage devices installed in wind farms can not only compensate the deviation between forecast and actual value of wind power, but also effectively restrain the wind power fluctuations. In this paper, the scale of ESS is determined from the perspective of compensating the wind power forecast error. The probability method is harnessed to describe WPFE and CWPD; based on the distribution function of forecast error, a method of determining the scale of ESS in wind farms is proposed; the model of peak output and rated capacity of energy storage device is established; furthermore, in accordance with the cost and benefit curves, the optimal scale of ESS is worked out.

In order to measure the benefit of ESS, variable W_{loss} is introduced to denote the wind energy loss caused by shortage of ESS capacity and the relations of peak output P_{ESS} and rated capacity C_{ESS} with W_{loss} are established respectively. When analyzing the relationship between P_{ESS} and W_{loss}, C_{ESS} is presumed as infinite; while when analyzing that between C_{ESS} and W_{loss}, P_{ESS} is presumed as infinite. Eventually, a mathematical model, in which the wind power loss is the objective function and the peak output and rated capacity of ESS are variables, is established and the relationships among the three are shown in a three-dimensional graph.

3.1. Determination of ESS Rated Power

When ESS is short in maximum charge-discharge power, wind energy loss or reserve compensation would occur even if ESS capacity is enough. The loss of wind energy caused by the lack of power of ESS can be calculated by expression (5)

$$W_{loss,P} = W_{loss1,P} + W_{loss2,P} \tag{5}$$

where $W_{loss1,P}$ is the loss of wind energy caused by the lack of charging power, $W_{loss2,P}$ is the loss of wind energy caused by the lack of discharging power. $W_{loss1,P}$ and $W_{loss2,P}$ can be calculated as follows:

$$W_{loss1,P} = \int_{P_{ESS}}^{1} f(\varepsilon_t)(\varepsilon_t - P_{ESS})d\varepsilon_t \tag{6}$$

$$W_{loss2,P} = \int_{-1}^{-P_{ESS}} f(\varepsilon_t)(\varepsilon_t - P_{ESS})d\varepsilon_t \tag{7}$$

3.2. Model of ESS Rated Capacity

Generally speaking, determining the rated capacity of ESS is much more complex than determining the peak output of it. The usage of stored energy is normally processed in certain intervals to ensure the continuous operation of ESS. In order to ensure ESS capacity meet the requirement of compensating the cumulative energy errors, the most conservative method is to set forecast error ε as 100%, and in the meantime set ESS capacity as 2 times of the capacity of the wind farm in one dispatching period, i.e. $C_{ESS} = 2*T*P$. We can find from the distribution of forecast errors that the probability of forecast error being extremity (ε=100%) is relatively low, hence the conservative method of determining the capacity of ESS will waste many resources. For this reason, this paper proposes a new method of determining the relation between ESS capacity and wind energy loss through analyzing the CWPD $e^*_{c,t}$ caused by wind power forecast error ε_t in one dispatching period.

The method harnesses the basic energy storage response strategy to evaluate the capacity of ESS. In other words, presuming that the grid dispatches wind farms in accordance with the forecast wind power values, and putting the forecast errors as input of ESS, and using the cumulative forecast energy deviation in one dispatching period as the the capacity of ESS. Firstly, according to the requirements of a specific energy storage dispatching period, calculate CWPD in each period, and find out the maximum CWPD $e_{c,max}$. Secondly, analyze the cumulative wind power forecast errors based on non-parametric estimation and empirical distribution model so as to obtain the probability density function $g(e^*_{c,t})$ and cumulative distribution function $G(e^*_{c,t})$ with respect to the cumulative forecast errors in the dispatching period.

If ESS is required to compensate all wind power deviations, the rated capacity of it should be set as the difference of maximum and minimum value of CWPD, as shown in expression (8):

$$C_{ESS,0} = e_{c,max} - e_{c,min} \tag{8}$$

As a matter of fact, the distribution of CWPD mostly concentrates around value 0, and the probability of big deviation is tiny. Hence installing an ESS with big capacity to compensate the small probability of wind power deviation would inevitably lead to huge waste of resources.

As the value of wind power forecast error could be either positive or negative, and the charging process or discharging process of ESS goes reversely with the polarity of the forecast error, the cases that CWPD is positive or negative need be considered separately.

a) When the CWPD is positive, the actual wind power is higher than forecast. If the charging capacity of ESS is insufficient, wind curtailment or system reserve compensation would occur; since the loss of wind curtailment equals to unit cost of reserve compensation, the loss caused by insufficient chargeable capacity of ESS can be described by expression (9):

$$W_{loss1,C} = \int_{\bar{C}_{ESS}}^{1} g\left(e_{c,t}^{*}\right)\left(e_{c,t}^{*} - \bar{C}_{ESS}\right)de_{c,t}^{*} \tag{9}$$

b) On the contrary, when CWPD is negative, the actual wind power is lower than forecast. If the discharging capacity of ESS is insufficient, the output power deviation needs to be balanced by system reserve. Although no direct wind energy loss occurs in the case, yet the safe and stable operation of grid would be severely harmed, as shown in expression (10):

$$W_{loss2,C} = \int_{-1}^{\underline{C}_{ESS}} \alpha \cdot g\left(e_{c,t}^{*}\right)\left(\underline{C}_{ESS} - e_{c,t}^{*}\right)de_{c,t}^{*} \tag{10}$$

Wind energy loss caused by insufficient capacity of ESS can be calculated by expression (11):

$$W_{loss,C} = W_{loss1,C} + W_{loss2,C} = \int_{(1-\beta)C_{ESS}}^{1} g\left(e_{c,t}^{*}\right)\left(e_{c,t}^{*} - (1-\beta)C_{ESS}\right)de_{c,t}^{*} + \int_{-1}^{\beta C_{ESS}} \alpha \cdot g\left(e_{c,t}^{*}\right)\left(\beta C_{ESS} - e_{c,t}^{*}\right)de_{c,t}^{*} \tag{11}$$

$$C_{ESS} = \bar{C}_{ESS} + \underline{C}_{ESS} = (1-\beta)C_{ESS} + \beta C_{ESS} \tag{12}$$

where α is the compensation coefficient, denoting the compensation for the loss generated by insufficient power output of the wind farm; $\bar{C}_{ESS}$ is the charging capacity of ESS; $\underline{C}_{ESS}$ is the discharging capacity of ESS; $\beta(0 \le \beta \le 1)$ is the sliding coefficient, denoting the charging or discharging status of ESS.

For each dispatching period, the initial status of ESS (charging/discharging) is determined by the distribution of cumulative forecast error. For a same ESS capacity, different initial statuses correspond to different effective operation times of stored energy. Presuming that an ESS is at its extreme charging status, if at this moment a negative forecast error occurs, the ESS can't release the power, which means that the ESS can't work effectively.

When optimizing the initial operation status of ESS, presuming the total energy storage capacity is fixed and we hope that the effectively-operating time of ESS lasts as long as possible, the model can be described by expression (13):

$$\max\ t_{ESS} = G\left((1-\beta)C_{ESS}\right) - G\left(\beta C_{ESS}\right) \tag{13}$$

Wind energy losses in wind farms correlate simultaneously with the peak output and rated capacity of ESS, based on which, the relations between W_{loss} and P_{ESS}, W_{loss} and C_{ESS} are established and can be described as shown in expression (14):

$$W_{loss}\left(C_{ESS}, P_{ESS}\right) = \int_{\beta C_{ESS}}^{1} \Phi(e_{c,t}^{*}, P_{ESS})de_{c,t}^{*} + \int_{(1-\beta)C_{ESS}}^{1} \Phi(e_{c,t}^{*}, P_{ESS})de_{c,t}^{*} \tag{14}$$

where $\Phi(e_{c,t}^{*}, P_{ESS})$ is the probability density function of $e_{c,t}^{*}$ when P_{ESS} is given. In application, the peak output and capacity of ESS are normally integral multiples of unit power and unit capacity of it; hence they could be derived through interpolation method. Calculation processes are as follows:

1) Determine the peak output value of the ESS and calculate the loss of wind energy caused by insufficient peak output in accordance with expression (5);

2) Calculate the CWPD $e_{c,t}^{*}$ based on the peak output determined in 1);

3) Calculate the loss of wind energy caused by insufficient capacity of ESS in accordance with expression (12);

4) The total wind energy loss under the specific capacity of ESS is the sum of the two losses calculated in 1st and 3rd step.

3.3. Cost-Benefit Analysis of Energy Storage System

The benefit of installing an ESS in a wind farm is that the safe and stable operation of the grid would be influenced without the ESS. When the wind power forecast deviation is negative, reserve usage fee must be paid to the grid; when the wind power forecast deviation is positive, wind energy loss occurs. The benefit B_{ESS} that an ESS brings to a wind farm is equivalent to the wind energy loss saved by installing the ESS minus the wind energy loss caused by not installing the ESS minus the installation cost of ESS, as is shown in expression (15):

$$B_{ESS} = \left(W_{loss,inst} - W_{loss,uninst}\right) LC\ P_{wind} - C_{inst} \tag{15}$$

Where $W_{loss,inst}$ and $W_{loss,uninst}$ are respectively the wind energy losses before and after the installation,and both can be derived by expression (14); LC is the lifespan of the ESS; P_{wind} is the rated capacity of the wind farm; C_{inst} is the cost of ESS installation, which is normally a linear function of peak output and rated capacity of ESS.

4. Analysis of Examples

To verify the effectiveness and feasibility of the optimal ESS sizing method proposed in this paper, a wind farm in Texas, U.S. as well as the data it released during 2004-2006 is taken as example for analysis. When the CWPD is positive, ESS should be charged; while when it is negative, ESS should change into the discharge status. In order to ensure the ESS can operate effectively in a long time, the initial status of ESS in each dispatching period are presumed to be fixed. The CWPDs in each dispatching period within the three years are collected and analyzed and the distribution function is obtained, as shown in figure 3, where the x-axis is CWPD and the y-axis is the distribution function of it.

It can be seen from figure 3 that different sliding coefficients β lead to different probabilities of effective operation of ESS. For a same ESS capacity C_{ESS}, when β is B1 and B2 respectively, the effective action times of ESS are correspondingly $t_{ESS,1}$ and $t_{ESS,2}$, $t_{ESS,1}$ is obviously bigger than $t_{ESS,2}$. Hence it can be derived from above that the determination of the initial status of ESS does influence the adjustment of CWPD, and the initial status of ESS needs to be optimized.

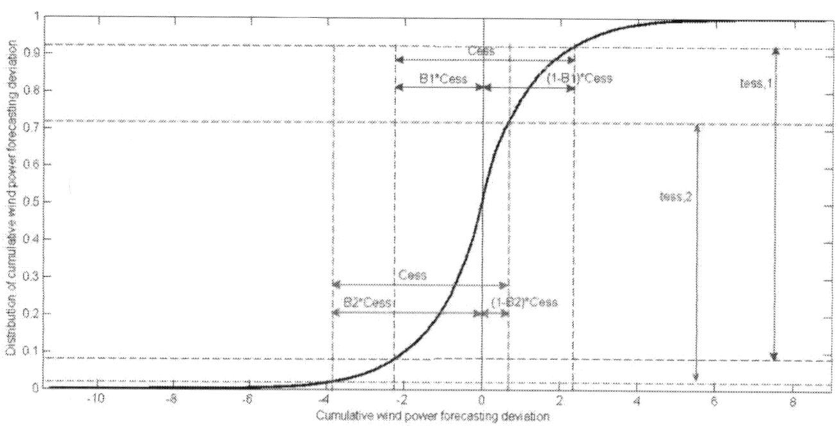

Figure 3. Distribution function of cumulative wind power forecast deviation

According to expression (13), when the capacity of ESS is given, we can establish the model targeting at the effective working time of ESS to be the longest, and work out optimal sliding coefficients and the corresponding possibilities λ_{ESS} of ESS acting effectively under different conditions by interpolation, as shown in figure 4, and finally gain the optimum charging/discharging sliding coefficient values of ESS at the beginning of each dispatching period.

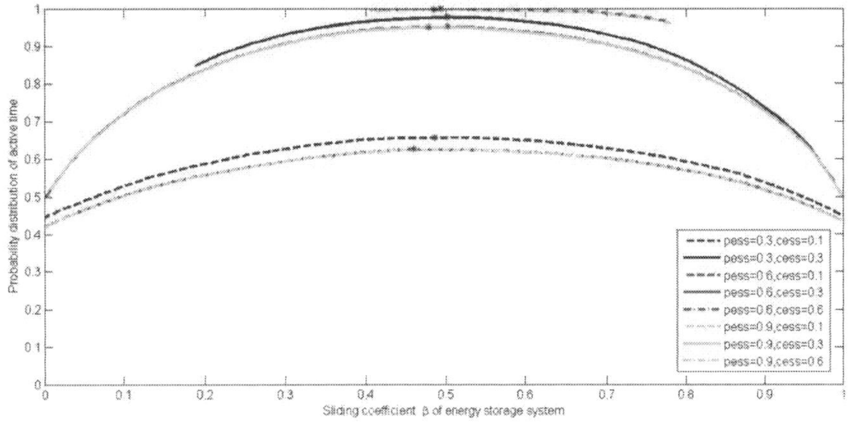

Figure 4. Probability distribution of storage system's effectiveness times

From figure 4 it can be seen that there isn't a linear relation between ESS's initial sliding coefficient and its effective action time. Table 1 gives the optimal sliding coefficients when P_{ESS} is 0.9p.u.(per unit), 0.6p.u., 0.3p.u., and C_{ESS} is 0.6p.u., 0.3p.u., and 0.1p.u.

Table 1. Optimal sliding coefficients and λ_{ESS}

P_{ESS}	C_{ESS}	Sliding coefficient β	λ_{ESS}
0.3	0.1	0.486023	0.656715
0.3	0.3	0.502373	0.977021
0.6	0.1	0.459388	0.625912
0.6	0.3	0.502769	0.951459
0.6	0.6	0.494132	0.998177
0.9	0.1	0.459388	0.52577
0.9	0.3	0.47965	0.950471
0.9	0.6	0.484902	0.997797

It can be observed from Table 1 that the optimal sliding coefficients vary corresponding to the values of P_{ESS} and C_{ESS} .When P_{ESS} and C_{ESS} are respectively in different values, the corresponding optimal sliding coefficients are not identical.

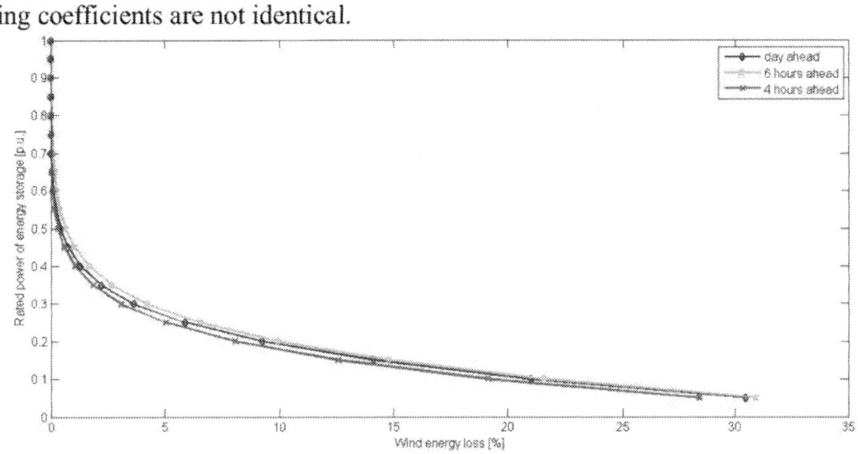

Figure 5. Relation between peak output of energy storage system and wind energy loss

Based on the expression (5), presuming the capacity of energy storage system is infinite, the relation between the peak output of ESS and wind energy loss is calculated, as shown in figure 5.It can be derived from the figure 5 that the shorter the forecast period, the more precise the forecast, and the lower the wind energy loss. Analyzing each of the forecast curves, it can be seen that wind energy loss saved by the increase of peak output of unit storage energy goes down with increase of the output.

Likewise, presuming the peak output of ESS is infinite, and the relation between rated capacity of storage system and wind energy loss could be calculated out by the expression (11), as is shown in figure 6.

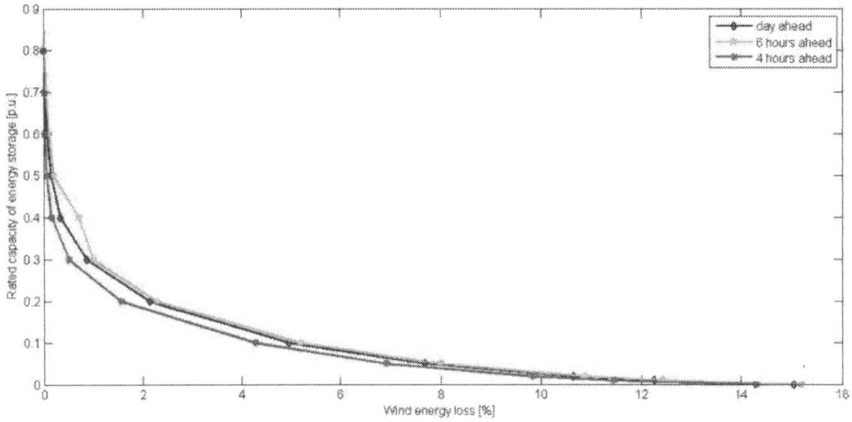

Figure 6. Relation between rated capacity of energy storage system and wind energy loss

It can be observed in figure 6 a relation similar to that in figure 5: when the rated capacity of ESS rises, wind energy loss correspondingly reduces, and the wind energy loss saved by the increase of unit storage system capacity decreases gradually.

Based on the forecast (with 1 day in advance) data and the expression (14), the relation between ESS scale and wind energy loss is worked out as shown in figure 7.

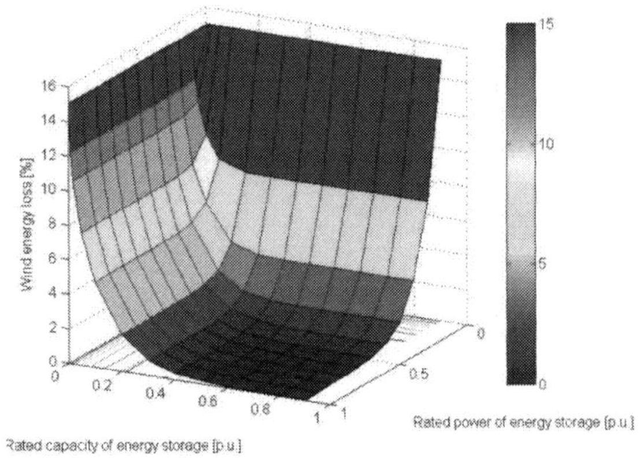

Figure 7. 3D diagram of relation between scale of energy storage system and wind energy loss

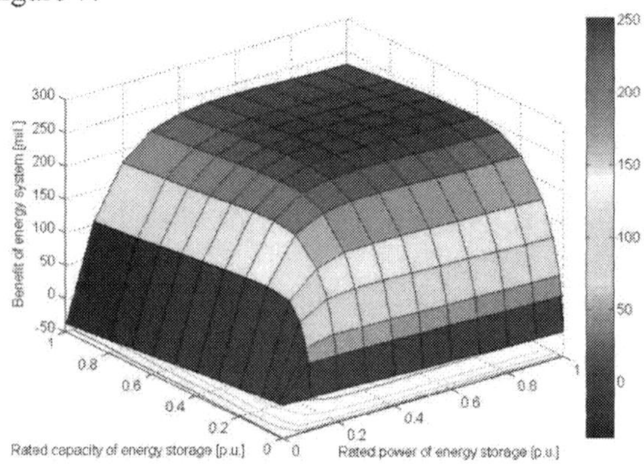

Figure 8. Relation between scale and benefit of energy storage system

We can find in figure 7 that when the peak output of ESS is 0.8959p.u. and the rated capacity is 0.8343p.u., the wind energy loss is 0, which can ensure a full integration of wind power into grid, but the cost is extremely high.

Compressed-air energy storage device can constantly generate electricity for over 24 hours with low energy consumption, long energy-storing time and working hours. Specifically, a compressed-air energy storage station could work for 30-40 years after installation. Therefore, a compressed-air ESS is adopted in this paper to calculate the cost of ESS.

Based on the cost-benefit optimization principle, the capacity of ESS is optimized according to expression (15). Presuming the electricity price is 0.15$/kWh, and the reserve cost and wind curtailment cost are 1.5 times of normal electricity price, namely 0.225$/kWh. According to the studies worked out by the Institute of Engineering Thermal Physics, Chinese Academy of Sciences, the cost of compressed-air energy storage is 400$/kW, 50$/kWh, with 30 years of lifetime, and the results of this example are shown in figure 8. When the rated power of ESS is 0.5p.u. and the capacity is 0.4p.u., ESS' s installation cost is $214,880,000 and the wind energy loss saved is $476,101,187, which means a maximum net benefit (income) of $261,221,187 for wind farm. The results indicate that the existence of ESS in a wind farm can greatly reduce wind energy loss and the installation cost of ESS can easily be recovered.

5. Conclusions

With combination of non-parametric estimation and empirical distribution estimation, the forecast error of wind farms in different forecast periods is analyzed; based on which models of peak output, rated capacity and wind energy loss of installing an ESS are established; furthermore, in accordance with cost-benefit optimization principle, the scale of ESS is optimized, and the optimum ESS matching scheme is worked out. In the end, through the analysis on an example of a wind farm in Texas, the optimal ESS sizing method proposed in this paper is verified to be of feasibility and effectiveness.

References

[1] Wind Power Grid Integration.Available online:https://en.wikipedia.org/wiki/Wind_power_grid_integration (accessed on 8 Oct 2015).

[2] Kabouris J and Kanellos F D 2009 Impacts of large scale wind penetration on energy supply industry *Energies* 2 1031-1041

[3] Chang Y J and Yu J L 2009 Analysis on long-term dynamic behavior of power grid containing wind farms *Power technology* 33(18) 144-147.

[4] Kariniotakis G N, Martí I, Casas D, Pinson P and Nielsen T 2004 What performance can be expected by short-term wind power prediction models depending on site characteristics *Proceedings of European Wind Energy Conference* London UK 22-25

[5] Marwali M K, Ma H, Shahidehpour S M and Abdul-Rahman K H 1998 Short-term generation scheduling in photovoltaic-utility grid with battery storage *IEEE Trans. Power System* 13 1057–1062

[6] Barton J and Infield D 2004 Energy storage and its use with intermittent renewable energy *IEEE Trans. on Energy Conversion* 19(2) 441-448

[7] Lu B and Shahidehpour M 2005 Short-term scheduling of battery in a grid-connected PV/battery system *IEEE Trans. Power System* 20 1053–1061

[8] Pascual J, Sanchis P and Marroyo L 2014 Implementation and control of a residential electrothermal microgrid based on renewable energies, a hybrid storage system and demand side management *Energies* 7 210-237

[9] Xiao C Y, Wang N B and Ding K 2010 System power regulation scheme for Jiuquan wind power base *Proceedings of the CSEE* 30(10) 1-7

[10] Kim Y M, Lee J H, Kim S J and Favrat D 2012 Potential and evolution of compressed air energy storage: energy and exergy analyses *Entropy* 14 1501-1521

[11] Gu Y, McCalley J, Ni M and Bo R 2013 Economic modeling of compressed air energy storage. *Energies* 6 2221-2241.

[12] SedighNejad H, Iqbal T and Quaicoe J 2014 Compressed air energy storage system control and performance assessment using energy harvested index *Electronics* 3 1-21

[13] Liu W, Liu L, Zhou L, Huang J, Zhang Y, Xu G and Yang Y 2014 Analysis and optimization of a compressed air energy storage-combined cycle system *Entropy* 16 3103-3120

[14] Alt J T, Anderson M D and Jungst R G 1997 Assessment of utility side cost savings from battery energy storage *IEEE Trans. Power System* 12 1112–1117
[15] Lee T Y and Chen N 1993 Optimal capacity of the battery energy storage system in a power system *IEEE Trans. Energy Convers* 8 667–673
[16] Lo C H and Anderson M D 1999 Economic dispatch and optimal sizing of battery energy storage systems in utility load-leveling operations *IEEE Trans. Power System* 14 824–829
[17] Shu Z and Jirutitijaroen P 2014 Optimal operation strategy of energy storage system for grid-connected wind power plants *IEEE Transactions on Sustainable Energy* 5(1) 190-199
[18] Han T, Lu J P and Qiao L 2010 Optimized scheme of energy-storage capacity for grid-connected large-scale wind farm *Power System Technology* 34(1) 169-173
[19] Korpaas M, Holen A T and Hildrum R 2003 Operation and sizing of energy storage for wind power plants in a market system *Int. J. Elect. Power Energy Syst.* 25 599–606
[20] Nelson D B, Nehrir M H and Wang C 2006 Unit sizing and cost analysis of stand-alone hybrid wind/PV/fuel cell power generation systems *Renew Energy* 31 1641–1656
[21] Brunetto C and Tina G 2007 Optimal hydrogen storage sizing for wind power plants in day ahead electricity market *Renew Power Gener* 1 220–226
[22] Ekren O and Ekren B Y 2008 Size optimization of a PV/wind hybrid energy conversion system with battery storage using response surface methodology *Appl. Energy* 85 1086–1101
[23] Pinson P, Papaefthymiou G, Klöckl B and Verboomen J 2009 Dynamic sizing of energy storage for hedging wind power forecast uncertainty. *Proceedings of IEEE Power & Energy Society General Meeting* Canada July 26-30
[24] Liang L, Li J L and Hui D 2011 Optimization configuration for capacity of energy storage system in large-scale wind farm *High Voltage Engineering* 37(4) 930-936
[25] Li J, Wei W and Xiang J 2012 A simple sizing algorithm for stand-alone PV/Wind/Battery hybrid microgrids *Energies* 5 5307-5323
[26] He Y Q and Zhang J C 2012 Research on optimizing approach of storage capacity in standalone wind/PV hybrid system *Electric Power Science and Engineering* 4 9-13
[27] Brekken T K, Yokochi A and Jouanne A 2011 Optimal energy storage sizing and control for wind power applications *IEEE Transactions on Sustainable Energy* 2(1) 69-77
[28] Scott D W 1992 Multivariate density estimation: theory practice and visualization Wiley series in probability and mathematical statistics *John Wiley and sons* New York

IOP Conf. Series: Materials Science and Engineering **161** (2016) 012086 doi:10.1088/1757-899X/161/1/012086

Waste Management in the Circular Economy. The Case of Romania.

Anca N IUGA [1]

[1] Department of Engineering and Industrial Management, Transilvania University, Mihai Viteazul street no.5 , 500036, Braşov , România

Email: [1]anca.iuga@unitbv.ro, ancadezv@yahoo.com

Abstract. Applying the principles of sustainable development in Romania involves a new approach to ecological waste using basic concepts of circular economy to weigh accurately the proposed projects in this area taking into account existing environmental resources and zero waste objectives. The paper is focused on: quantitative and qualitative measures of waste prevention in Romania, the changing status of the waste by selling it as product, the mechanisms for paying for treatment and / or disposal which discourage waste generation and the use of financial resources obtained from secondary raw materials for the efficiency of waste management.

1. Introduction

According to Executive Director of the European Environment Agency, Hans Bruyninck"circular economy concept has advanced in European policy decision as a positive perspectives, based on solutions for achieving economic development while respecting environmental limits". Romanian European Environment Agency is ready to support the transition towards a circular economy through analyses and evaluations.

Unlike traditional perspective of linear economy, circular economy is trying to respect environmental limits by increasing distribution of renewable or recycled resources and reducing the consumption of raw materials and energy. Thus, both the emissions and resources wasting will be reduced. Also, concepts such as eco design, distribution, reuse, repair, recycling of products and materials will play an important role in maintaining the use and value of products, components and materials. It is essential to move to a circular economy in order to meet the resource efficiency agenda established under the Europe 2020 strategy for smart, sustainable and inclusive growth [1]. It can get higher performance and sustainable efficient use of resources, which can bring major economic benefits. In the industrial sector, is already recognizes that improving resource productivity is a strong business case. It is estimated that improving resource efficiency along the value chain could reduce material needs with 17% -24% by 2030 [2] and a better use of resources should be a general potential saving of 630 billion per year for European industry [3] .

2. Waste generation

Waste generation is highly dependent on local consumer habits, the type of buildings, general living conditions and the type of industry and commerce. For example, in areas with individual fireplaces with wood, coal or coke, waste is considerably more and denser in winter compared to buildings with central heating. In rural areas, the amount of waste collected per capita is lower than in urban areas due to in-situ composting. In some societies waste are very wet in the autumn time and has a high content of organic matter due to conservation of vegetables. This can be problematic for waste incineration installations which may need additional fuel. Homes, with a relatively low income, produce less waste but more organic than homes with a relatively high income. The single measurement of waste generation there is made by weighing platforms from the landfill, recycling stations and treatment facilities with quality recording of waste type and transport origin. Based on these statistics, there are recorded quarterly/ or seasonal waste variations and can do prognosis.

2.1. Waste classification

In terms of nature and places of production, waste is classified as follows:

1. Household waste - waste from the household sector or other similar sectors (including hazardous waste they contain) and can be taken by our current pre-collection systems or villages collection systems;

2. Street waste - specific waste of public traffic routes, from the daily activity of the population, city green spaces, animals, deposition of solids in the atmosphere;

3. Similar waste with municipal waste - waste from small or large industry, commerce, public and administration sectors, which presents similar composition and properties to household waste and can be collected, transported, processed and stored with them;

4. Bulky waste - solid waste from different sources, which due to its size, cannot be taken with the usual pre-collection or collection systems, but requires a differentiated treatment.

5. Construction waste - waste from demolition of industrial and civil constructions;

6. Hazardous waste - toxic, flammable, explosive, infectious waste or otherwise, which after was introduced into the environment, can harm plants, animals or humans;

7. Agricultural waste - waste from agricultural and livestock units (manure, animal waste from slaughterhouses and meat industry);

8. Industrial waste - waste from technological processes;

9. Hospital waste - waste from the hospitals activities and health units which are incinerated in crematoria;

Depending on specific local conditions, could be other types of waste that requires special attention, for example, marine ballast (polluted with oil / chemicals) and mining waste.

2.2. The main objectives of solid waste management

The main objectives of solid waste management are: protect public health; protect the environment; maintaining cleanliness in public for these places to be aesthetically acceptable; conservation of natural resources through waste reduction policies and recycling. All these objectives are achieved through a good collection and waste safe treatment, a proper waste storage and disposal. By my opinion integrated waste management is vital for the community, for the following reasons:

1. The deposits capacity decreases continuously -Location and construction of new deposits is a difficult and expensive process.

2.Many waste materials are rare natural sources, imposing their recovery, decreasing the environmental impact and increasing humans' life quality.

3.The materials found in waste volume may be an opportunity to start a business.

4.A system that does not rely on one alternative is more flexible to economical, technological and legislative changes.

5.The investors or creditors favour capital projects that are part of a carefully eco -designed strategy.

Local authorities are in an advantageous position in assessing the proposals for a new facility, when they have the chance to thoroughly examine the entire system. Pollutant emissions from industry are very rich and very diverse. In very vast area, some principles are advanced only to reduce emissions of pollutants that need to be adapted for each industrial branch separately.

3. Industrial Product Life Cycle Stages

Life cycle approach means considering the environmental impacts and resources used throughout the lifecycle of products (goods or services).

Approaching life cycle helps us to identify sensitive points and aspects of a product which can be improved as lower environmental impact, a reduced use of resources along the life cycle stages or compromises between different options on products. Sensitive issues can arise at any stage of the life cycle, from raw material extraction and conversion, and manufacturing and distribution, to customer use and / or consumption.

It ends with reuse, recycling, energy recovery and final disposal. The key objective of lifecycle based approach is to contribute to decisions or express any managerial transfer of responsibility. This means minimizing impacts on the environment in a life cycle stage or in a geographic region or a certain category of impacts, while helping to avoid environmental impacts increase elsewhere. For example, saving energy during the use phase of a product without simultaneously increase the amount of material required for its delivery and impacts associated with this supply.

At each stage of the life cycle there are raw materials consumed and chemicals are released as emissions. They contribute to different environmental impacts and issues like resource scarcity.

4. Measures to reduce waste, in order to achieve a circular economy in Romania

Circular economy approaches, exclude concept of waste and usually involve innovation along the value chain, rather than rely on solely solutions for the life cycle end of a product. For example, these approaches may include:

1. Reducing the amount of material required to provide a given service;
2. Extending the useful product life cycle (durability);
3. Reducing the use of energy and materials within production of use phases (efficiency);
4. Reducing the use of hazardous materials or difficult to recycle into new products and production processes (substitution);
5. The creation of markets for secondary raw materials, based on standards and public acquisitions (recycling);
6. Designing products that are easily maintained, repaired, refurbished, remanufactured or recycled (eco-design);
7. Development of services that consumers need to reduce waste (service maintenance / repair, etc.);
*8.*Stimulating and supporting the activities of waste reduction and quality separation, made by consumers;
9. Stimulating the separation and collection systems that minimize the costs of recycling and reuse;
*10.*Facilitating group activities to prevent the transformation in waste of industrial products (industrial symbiosis) and encouraging the expansion and consumers choice improvement. In this case the customer could use leasing and exchange of services as an alternative to holding of products. We need to protect consumers' interests (in terms of cost, protection, information, contract terms, issues related to insurance, etc.).

In order to show the relationship between industrial product lifecycle and industrial emissions, I have realized a matrix named "*Eco-friendly methods for reduction of industrial pollutant emissions–Industrial Product Life Cycle Stages Matrix*"table 1.

According with Luminiţa I. POPA and VasileN.POPA [4],the Industrial Product Life Cycle Stages are the follows: market needs; research and development; idea generation; opportunity identification and concept definition; research design and development; prototype &production; distribution and manufacturing; marketing; sales; maintenance &service; products feed-back; removal& disposal. As we can see in the matrix" research and development" and "research design and development "lifecycle stages have the highest score: 7.

Another two lifecycle stages ("prototype &production" and "distribution and manufacturing"), have a high score: 6. In this case, the industrial company managers have to take in consideration that these three metrics are very important, because they could affect the entire industrial product life cycle. Also another three metrics,"Eco-friendly production output" (score 10),'Eco-friendly operational life cycle" (score 10) and"Implementing TQM method" (score 8) are the most important ways to reduce the cost of green acquisitions.

Table 1. Eco-friendly methods for reduction of industrial pollutant emissions– Industrial Product Life Cycle Stages Matrix (Realized by the author) Source: [4].

No.	Industrial Product Life Cycle Stages	Eco-friendly methods for reduction of industrial pollutant emissions								
		Raw materials replacement	Industrial processes liquids recycling	Eco-friendly techno-logical process	Use closed techno-logical processes	Reduction of time required for techno-logical processes	Eco-friendly production output	Eco-friendly operational life cycle	**Total scores**	**Weighted score (%)**
(1)	(2)	(3)	(4)	(5)	(6)	(7)	(8)	(9)	(10)	(11)
1	Market needs	X	-	-	-	-	X	X	3	**6.66**
2	Research and development	X	X	X	X	X	X	X	7	**15.55**
3	Idea generation	-	-	X	X	X	X	X	5	**11.11**
4	Opportunity identification and concept definition	-	-	-	-	X	X	X	3	**6.66**
5	Research design and development	X	X	X	X	X	X	X	7	**15.55**
6	Prototype, production	X	X	X	X	X	X	-	6	**13.33**
7	Distribution and manufacturing	X	X	X	X	X	X	-	6	**13.33**
8	Marketing	-	-	-	-	-	X	X	2	**4.44**
9	Sales	-	-	-	-	-	X	X	2	**4.44**
10	Maintenance /service	-	-	-	-	-	X	X	2	**4.44**
11	Products feed-back	-	-	-	-	-	-	X	1	**2.22**
12	Removal& disposal and recycling	-	-	-	-	-	-	X	1	**2.22**
	Total scores	5	4	5	5	6	10	10	45	**100**
	Weighted score (%)	**11.11**	**8.88**	**11.11**	**11.11**	**13.33**	**22.22**	**22.22**	**100**	**100**

5. European Union Resource efficient Scoreboard

The Resource Efficiency Scoreboard presents indicators covering themes and subthemes of the *Roadmap to a Resource Efficient Europe*. The scoreboard aims to monitor the implementation of the roadmap, to communicate the link between resources and economy and to engage stakeholders. Indicators are arranged in three groups – lead, dashboard and theme-specific indicators. For several indicators it is more meaningful to view the data at country level rather than at EU-28 level. The scoreboard by default presents the data for Belgium, the first country in the Member States list (sorted in protocol order). Data for the EU and other Member States were selected from the country list.

6. EU indicators: "Turning waste into a resource"

6.1. Waste excluding major mineral wastes

The indicator shows trends in waste generation, both EU-wide and for individual EU countries. It covers both non-hazardous and hazardous waste from: all sectors of the economy (production); households (consumption). It does not cover mineral wastes or soil. Over 90% of these come from the mining and construction sectors, which are subject to considerable fluctuation over time. Waste generation from which major mineral wastes are excluded *reflects general trends more accurately* than statistics on total waste generated. The indicator shows the amount of waste generated annually in the EU as a whole and in individual countries, expressed in kilos per inhabitant. It is based on data collected in the way stimulated by the Waste Statistics Regulation and is available for every second year as of reference year 2004 [5] (source: *http://ec.europa.eu/eurostat/web/waste/generation-of-waste-excluding-major-mineral-wastes*).

6.2. Landfill rate of waste excluding major mineral wastes

The indicator is defined as the volume of waste landfilled (directly or indirectly) in a country per year divided by the volume of the waste treated in the same year. Waste taken into account excludes major mineral wastes, dredging spoils and contaminated soils. This exclusion enhances comparability across countries, as mineral waste accounts for high quantities in some countries due to economic activities such as mining and construction. One exception, however, is that the indicator explicitly includes combustion wastes and solidified, stabilised and vitrified wastes, despite them being completely or partly mineral. The indicator is derived from the two-yearly reporting of the countries according to the Waste Statistics Regulation. It covers landfilling of hazardous (hz) and non-hazardous (nh) waste from all economic sectors and from households, including waste from waste treatment (secondary waste).[5](source: *http://ec.europa.eu/eurostat/cache/metadata/FR/t2020_rt110_esmsip.htm*).

6.3. Recycling rate of municipal waste

The recycling rate is the tonnage recycled from municipal waste divided by the total municipal waste arising. Recycling includes material recycling, composting and anaerobic digestion. Municipal waste consists to a large extent of waste generated by households, but may also include similar wastes generated by small businesses and public institutions and collected by the municipality; this latter part of municipal waste may vary from municipality to municipality and from country to country, depending on the local waste management system. For areas not covered by a municipal waste collection scheme the amount of waste generated is estimated.[5] (*http://ec.europa.eu/eurostat/cache/metadata/FR/t2020_rt120_esmsip.htm*).

6.4. Electrical and electronic equipment waste (WEEE)

WEEE poses on the one hand a risk to the environment (hazardous components), on the other hand it has a high potential for recycling to replace raw materials by secondary raw materials, such as precious metals and other highly valuable special materials. For the calculation of recycling rates it is crucial to know the volume of end-of-life electrical and electronic equipment. As this is for many devices and many countries difficult to deduct, the volume of put on the market during the previous 3 years (considered as easier to deduct) is considered as proxy for the volume of WEEE in the reference year (see statements in Article 7 of the WEEE-Directive 2012/19/EU). The collection rate is calculated as the collected volume of WEEE in the reference year, divided by the average sum of EEE put on the market in the three previous years. The 'recycling rate of e-waste' (this indicator) equals the 'total collection' in the present year divided by the average of the 'put on the market' of the three preceding years multiplied with the 'reuse and recycling rate' (of treatment facilities), considering that the total amount of collected e-waste is sent to treatment / recycling facilities.[5](*http://ec.europa.eu/eurostat/cache/metadata/FR/t2020_rt130_esmsip.htm*).

7. Comparison of "Turning waste into a resource" indicator. Romania vs. European Union.
Using the indicator"*Recycling rate of e-waste*" (table 2) for Romania and EU, it is made an average of three (2010-2012) as following: Romania: 13,125; European Union: 31,433; it is chosen this period of time because there is after 2007, the year in which Romania has become full member of European Union.This relevant indicator shows that Romania is 2,4 times less efficient in the field of „Recycling rate of e-waste" than EU countries. Using the indicator *"Recycling rate of municipal waste"* (table 2) for Romania and EU, it is made an average of three years (2010-2013) as following: Romania: 13,125; European Union: 56,425; it is chosen this period of time because there is after 2007, the year in which Romania has become full member of European Union.This relevant indicator shows that Romania is 4.3 times less efficient in the field of „Recycling rate of municipal waste" than EU countries [5].

Table 2 *"Turning waste into a resource" indicator. Romania vs. European Union Comparison [5]. Romania/ European Union*

LEAD INDICATOR

THEMATIC INDICATORS

Transforming the economy

Turning waste into a resource

Indicators	*Romania/ European Union*	Measure-ment units	2000	2005	2010	2011	2012	2013	2014	2015	**Source**
Generation of waste excluding major mineral wastes	Romania	Kg/ capita	(:)	(:)	2050	(:)	2041	(:)	(:)	(:)	Eurostat
	EU	Kg/ capita	*(:)*	*(:)*	*4293*	*(:)*	*4999*	*(:)*	*(:)*	*(:)*	Eurostat
Landfill rate of waste excluding major mineral wastes	Romania	%	(:)	(:)	53	(:)	49	(:)	(:)	(:)	Eurostat
	EU	%	*(:)*	*(:)*	*9*	*(:)*	*8*	*(:)*	*(:)*	*(:)*	Eurostat
Recycling rate of municipal waste	Romania	%	0.00	01.80	12.80	11.70	14.80	13.20	(:)	(:)	Eurostat
	EU	%	48.40	53.80	57.70	57.40	55.70	54.90	55.10	(:)	Eurostat
Recycling rate of e-waste	Romania	%	(:)	(:)	12.0	10.3	14.5	(:)	(:)	(:)	Eurostat
	EU	%	(:)	(:)	30.4	31.9	32.0	31.7	(:)	(:)	Eurostat

Last update: 10/05/2016 11:07:01

Source: Eurostat [5] (http://ec.europa.eu/eurostat/web/europe-2020-indicators/resource-efficient-europe)

8. Comparison of domestic material consumption indicator. Romania vs. European Union
Domestic Material Consumption (DMC) is based on the Economy-wide Material Flow Accounts (EW-MFA) [5]. Comparison of domestic material consumption indicator,"Romania vs. European Union"is shown in the table 3. The classification of materials used in EW-MFA and for which DMC is calculated is a Eurostat based system. EW-MFA includes the material categories:
*1.*Biomass and biomass products;
*2.*Metal ores and concentrates, raw and processed;
*3.*Non-metallic minerals, raw and processed;

*4.*Petroleum resources, raw and processed;
*5.*Other products;
*6.*Waste imported for final treatment and disposal.
It is important to note that the term "consumption" as used in DMC denotes apparent consumption and not final consumption. DMC does not include upstream hidden flows related to imports and exports of raw materials and products. The indicator is a Resource Efficiency Indicator. It has been chosen as a lead indicator presented in the Resource Efficiency Scoreboard for the assessment of progress towards the objectives of the Europe 2020 flagship initiative on Resource Efficiency. The DMC is defined as the total amount of material directly used in an economy and equals direct material input (DMI) minus exports Domestic material consumption –is represented in "tonnes per capita".
(Source: http://ec.europa.eu/eurostat/cache/metadata/DE/t2020_rl110_esmsip.htm)

Table 3 Comparison of domestic material consumption indicator. Romania vs. European Union [5]

Time geo	1990	1991	1992	1993	1994	1995	1996	1997	1998	1999	2000	2001
Romania	:	:	:	:	13.685	14.463	11.251	9.940	10.820	8.700	7.684	12.372
EU (28 countries)	..	..	..	..	..	..	..	..	..	..	15.485 (s)	15.51 (s)

Table 3 Comparison of domestic material consumption indicator. Romania vs. European Union [5] (cont.)

Time geo	2002	2003	2004	2005	2006	2007	2008	2009	2010	2011	2012	2013	2014
Romania	12.069	13.255	14.265	15.705	17.020	20.557	26.843	21.262 (s)	19.748 (s)	22.325 (s)	21.845 (s)	22.056 (s)	21.327(ps)
EU (28 countries)	15.287	15.138 (s)	15.835 (s)	15.857 (s)	16.132 (s)	16.575 (s)	16.362 (s)	14.39 (s)	13.939 (s)	14.456 (s)	13.427 (s)	13.040 (s)	13.079 (ps)

Flags are codes added to the data and defining a specific characteristic:
p = provisional; s = Eurostat estimate (phased out)

Domestic material consumption indicator for Romania before has joined EU (table 4)
Time considered: 13 years before Romania has joined EU (from 1994 to 2006); Average: 161229/13 years = 12.402 tonnes per capita.

Table 4 Domestic material consumption indicator for Romania, before has joined EU (13 years time period)

Year	1994	1995	1996	1997	1998	1999	2000	2001
DMC (tonnes per capita t/c)	13.685	14.463	11.251	9.940	10.820	8.700	7.684	12.372
Year DMC (tonnes per capita t/c)	2002	2003	2004	2005	2006	-	-	-
	12.069	13.255	14.265	15.705	17.020	-	-	-

Domestic material consumption indicator for Romania after has joined EU (table 5)
Time considered: Eight years after Romania joined the EU (2007-2014)
Average: 175963/ 8 years = 21.995 tonnes per capita

Table 5 Domestic material consumption indicator for Romania after has joined EU

Year	2007	2008	2009	2010	2011	2012	2013	2014
DMC (tonnes per capita)	20.557	26.843	21.262 (s)	19.748 (s)	22.325 (s)	21.845 (s)	22.056 (s)	21.327 (ps)

After joining the EU, the domestic material consumption in Romania, has increased 1.77 times, compared with the previous period 1994-2006. That means that Romania has benefited from EU membership (reduced fees, transfer of know-how).

9. Domestic material consumption indicator for EU (28 countries) after Romania has joined EU (table 6)
Time considered: Eight years after Romania joined the EU (2007-2014)

EU Average: 102.189/ 8 years = 12.773 tonnes per capita at EU level (28 countries) ;
After joining the EU (2007-2014), the domestic material consumption in Romania, has increased 1.72 times, compared with EU average in the same period of time (8 years). That means that Romanian people has consumed as a consuming society, as is shown in the figure 1.

Table 6 Domestic material consumption indicator for EU (28 countries) after Romania has joined EU

Year	2007	2008	2009	2010	2011	2012	2013	2014
DMC (tonnes per capita)	16.575 (s)	16.362 (s)	14.390 (s)	13.939 (s)	14.456 (s)	13.427 (s)	13.040 (s)	13.079 (ps)

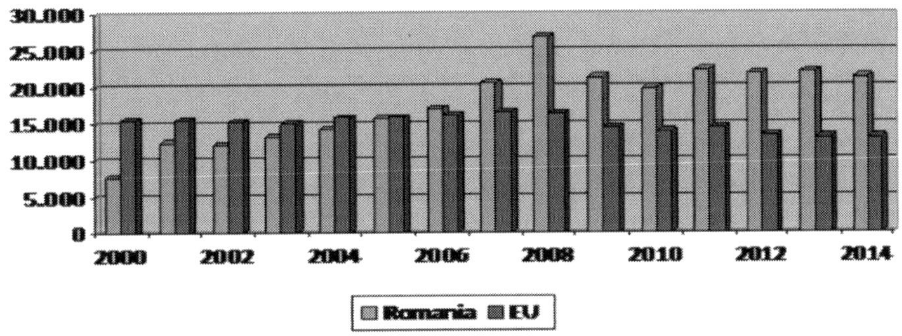

Figure 1. Domestic material consumption indicator. Romania vs. European Union Comparison

Source: **[5]** .
http://ec.europa.eu/eurostat/tgm/printTable.do?tab=table&plugin=1&language=en&pcode=t2020_rl110&printPreview=true.

10. Conclusions

The transition benefits towards a circular economy in Europe could be considerable by reducing environmental pressures reducing drastically Romanian economy dependence on imports which, if gradual increases, can become a source of national vulnerability.

Increasing global competition for natural resources has contributed to increased prices and volatility. Also circular economy strategies applied could have as effect of reducing costs and increased competitiveness of Romanian industry by net benefits which consist of job opportunities.
Creating a circular economy in Romania requires fundamental changes in the value chain, from product design and production processes to new models of circularity business and consumption patterns.

In this manner, recycling will turn waste into a resource, and product life extension facility will contribute to reducing natural resources consumption.

Some Romanian companies are already are experimenting new circularity business models such as those based on functions and services of collaborative consumption model specific for circular economy. For the future Romanian managers could take some measures for waste reduction: the changing status of the waste by selling it as product; the mechanisms for paying for treatment and / or disposal which discourage waste generation; the use of financial resources obtained from secondary raw materials for the efficiency of waste management.

References

[1] A resource-efficient Europe – Flagship initiative under the Europe 2020 Strategy COM (2011) *Communication from the commission to the European parliament, the council, the European economic and social committee and the committee of the regions* (Brussels, 26.1.2011)

[2] Meyer, B. et al 2011, *Macroeconomic modelling of sustainable development and the links between the economy and the environment.*(Study for European Commission) available at address: http://ec.europa.eu/environment/enveco/studies_modelling/pdf/report_macroeconomic.pdf

[3] Towards a circular economy: A zero waste programme for Europe.*Communication from the Commission to the European parliament, the council, the European economic and social committee and the committee of the regions(*Bruxelles, 2.7.2014). http://cor.europa.eu/en/activities/stakeholders/Documents/COM%282014%29%20398%20final.pdf

[4] Luminiţa I. POPA and VasileN.POPA, 2013 *PLM Innovation Matrix for a Complex Product Development Process* (*Applied Mechanics and Materials* Vol. 371) ed Trans Tech Publications (Switzerland) p 862-866

[5] http://ec.europa.eu/eurostat/web/waste/generation-of-waste-excluding-major-mineral-wastes Eurostat. http://ec.europa.eu/eurostat/cache/metadata

IOP Conf. Series: Materials Science and Engineering **161** (2016) 012087 doi:10.1088/1757-899X/161/1/012087

Working fluid selection for the Organic Rankine Cycle (ORC) exhaust heat recovery of an internal combustion engine power plant

S Douvartzides and I Karmalis

Laboratory of Internal Combustion Engines, Department of Mechanical and Industrial Engineering, Technical Education Institute (TEI) of Western Macedonia, Koila-Kozani, Greece

E-mail: sdouvar@teiwm.gr; karmalis@teiwm.gr

Abstract. Organic Rankine cycle technology is capable to efficiently convert low-grade heat into useful mechanical power. In the present investigation such a cycle is used for the recovery of heat from the exhaust gases of a four stroke V18 MAN 51/60DF internal combustion engine power plant operating with natural gas. Design is focused on the selection of the appropriate working fluid of the Rankine cycle in terms of thermodynamic, environmental and safety criteria. 37 candidate fluids have been considered and all Rankine cycles examined were subcritical. The thermodynamic analysis of all fluids has been comparatively undertaken and the effect of key operation conditions such as the evaporation pressure and the superheating temperature was taken into account. By appropriately selecting the working fluid and the Rankine cycle operation conditions the overall plant efficiency was improved by 5.52% and fuel consumption was reduced by 12.69%.

1. Introduction

Increased fuel prices and environmental pollution concerns force research interest on the enhancement of the efficiency of existing power systems and the discovery of solutions leading to rational energy utilization. For a typical internal combustion engine (ICE), approximately 30% of the chemical energy of the fuel is brought into the environment by the exhaust gases. This energy can be usefully exploited by an appropriate waste heat recovery system leading in an overall efficiency improvement.

There are three typical approaches of ICE waste heat recovery including thermoelectric generation [1, 2], electrical or mechanical turbo compounding [3] and steam or organic Rankine cycles. The organic Rankine cycle (ORC) is a technology for low-grade heat to power conversion. It is similar to the conventional steam Rankine cycle used in power generation plants, but steam is replaced by an appropriate organic fluid. Hence, ORC is capable to realize more efficient expansion at low power and show better thermodynamic performance at low temperatures. Such cycles have been successfully applied in waste heat recovery, biomass energy plants and geothermal or solar energy applications [4-6].

Research on organic Rankine cycles for ICE waste heat recovery has started after the oil crisis of the '70s. BMW, Honda and Cummins have installed ORC waste heat recovery systems in specific car or truck models and reported overall efficiency improvements in the range of 4-10% [6-8]. In these cases water, ethanol and R245fa have been selected as working fluids and heat was recovered from both exhaust gas and engine cooling systems. Similar results have been reported in numerous research studies [9-27] using a large variety of other organic fluids. Hung et al. [24] demonstrated that the major properties affecting the system performance included latent heat and specific heat of liquid and vapor. Chen et al. [25] and Maizza and Maizza [26] suggested that fluids with high latent heat and low specific heat are preferable while Yamamoto et al. [27] suggested that low latent heat fluids are preferable.

In the present investigation ORC exhaust heat recovery is applied in a large Diesel-natural gas dual fuel engine typically used for ship propulsion and power plant applications. This is the four stroke turbocharged V18 MAN 51/60 DF engine with maximum brake power of 17550 kW at 500 rpm [28, 29]. Exhaust combustion gases have been considered as the only heat source of the Rankine cycle and numerous candidate working fluids were compared upon thermodynamic, environmental and safety criteria.

2. Engine and fuel characteristics

Exhaust gas ORC heat recovery is considered in a V18 MAN 51/60DF engine, shown in Fig. 1. This is a natural gas-diesel dual fuel engine equipped with a MAN TCA77 (four stroke version) turbocharger. Engine and turbocharger characteristics are summarized in Table 1.

Figure 1. The V18 MAN 51/60DF (dual fuel) engine.

Table 1. 18 cylinder MAN 51/60DF engine characteristics [28, 29]

Parameter	Value
Number of cylinders	18
Cylinder bore	510 mm
Piston stroke	600 mm
Displacement per cylinder	122.5 lt
Compression ratio	13.3
Vee angle	50°
Turbocharger type	MAN TCA 77 (four stroke version)[23]
Turbocharger pressure ratio	5.5
Maximum engine speed	500 rpm
Maximum engine power	17550 kW at 500 rpm

The real cycle analysis of the engine has been conducted elsewhere [30] and operation conditions are summarized in Table 2. The engine is considered to steadily operate with either Diesel or natural gas fuel at the maximum speed of 500 rpm and at full load. Maximum temperature of the exhaust gases was estimated about 1128 K for natural gas mode. This value is about 30K higher than the maximum temperature of the Diesel cycle and for consistency natural gas operation will be considered. Exhaust gas temperature after the turbocharger is 633K as given by the MAN manufacturer [28]. Exhaust gas composition is given in Table 3.

Table 2. Calculated operated conditions for the Diesel and Natural Gas operation modes. Engine speed is at the maximum value of 500 rpm [30].

Parameter	Diesel mode	NG mode
Ambient temperature, T_0 (K)	293	
Ambient pressure, p_0 (MPa)	0.1	
Turbocharger outlet temperature, T_C (K)	517.63	
Turbocharger outlet pressure, p_C (MPa)	0.55	
Fuel mass flow rate, $\dot{m}_f$ (kg/h)	4030.66	3412.82
Air mass flow rate, $\dot{m}_a$ (kg/h)	116332.9	116336.2
Fraction of residual gases, x_r	0.0436	0.0425
Temperature of residual gases, T_r (K)	825	848
Temperature at the end of intake stroke, T_1 (K)	535.83	536.47
Pressure at the end of intake stroke, p_1 (MPa)	0.522	0.522
Temperature at the beginning of combustion, T_2 (K)	1332.66	1334.09
Pressure at the beginning of combustion, p_2 (MPa)	17.29	17.29
Temperature at the end of combustion, T_3 (K)	2284.35	2430.93
Pressure at the end of combustion, p_3 (MPa)	25.93	26.82
Temperature at the end of expansion, T_3 (K)	1095.54	1128.05
Pressure at the end of expansion, p_3 (MPa)	1.09	1.10
Exhaust gas temperature at the exit of the turbocharger, T_l (K)	631[28]	633[28]
Brake (real) power, P_b (kW)	17473.91	17923.89
Brake (real) thermal efficiency, η_{th} (%)	39.01	37.96

Table 3. Composition of the engine burned mixture for natural gas mode and λ=2 [30].

Product	Molar fraction
CO_2	0.050522
$H_2O_{(g)}$	0.098183
CO	0
H_2	0
O_2	0.099613
N_2	0.751682
Moles of product mixture: N_P	1

In the natural gas mode, the engine operates according to a lean-burn Otto combustion cycle. Natural gas is considered with an approximate composition of 95% mol CH_4, 4% mol C_2H_6, 1% mol C_3H_8 and with a lower heating value (LHV) of 49.8MJ/kg. Combustion typically takes place with lean fuel mixtures and the analysis was conducted assuming a lambda λ ratio equal to 2 [30]. The pre-mixed lean gas–air mixture is ignited by the compression ignition of a small quantity of marine diesel pilot fuel injected into the main combustion space. A pilot injection of less than 1% of the normal liquid fuel quantity ensures very low NOx emissions in gas mode [31].

3. Rankine cycle analysis

In the present study, exhaust gas energy is exploited for power generation in a bottoming Rankine cycle. An ORC system consists of a working fluid pump, an evaporator, an organic turbine or expander and a condenser, as shown in Figure 2. Such a cycle consists of the following four steady state processes:

- Process 1-2: compression of liquid fluid in a pump.
- Process 2-3: isobaric heat addition to the fluid in an exhaust gas heat recovery exchanger.
- Process 3-4: expansion in a turbine or expander.
- Process 4-1: isobaric heat rejection in a condenser.

These four processes may form a simple Rankine cycle or a superheated Rankine sycle as shown in the *T*-*s* diagrams of the Figure 3. In the former case heat addition process 2-3 evaporates the fluid up to the condition of saturated vapor while in the latter case heat addition proceeds further to provide high temperature superheated vapor at 3'.

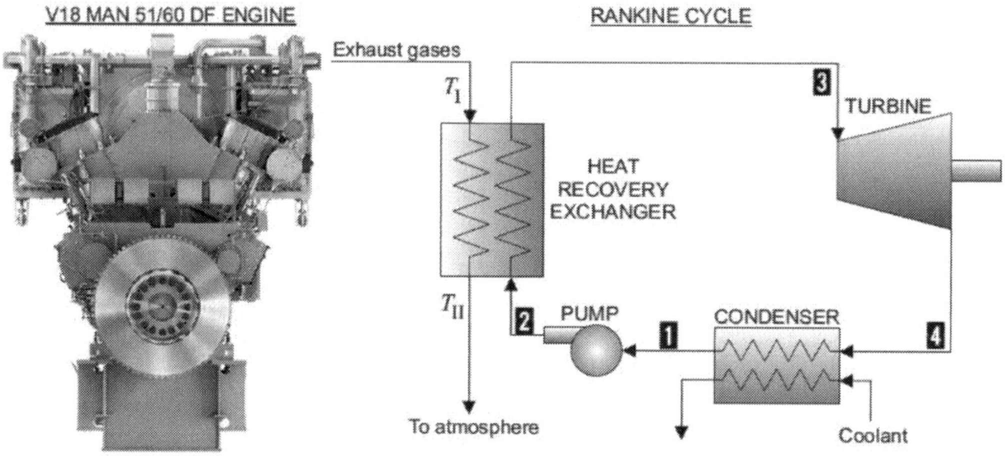

Figure 2. Schematic diagram of the ORC exhaust heat recovery system.

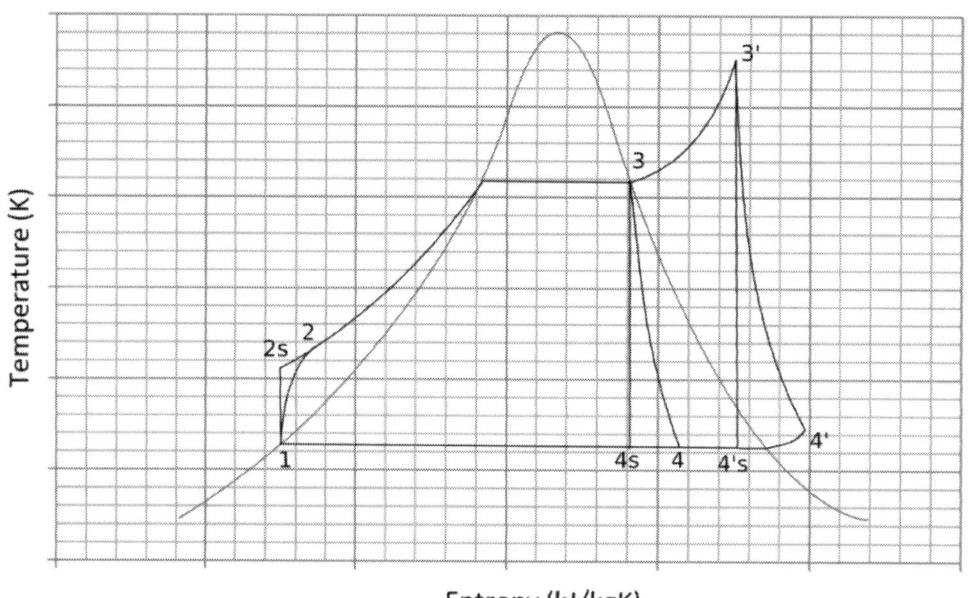

Figure 3. T-s diagrams of a simple (1-2-3-4-1) and a superheated (1-2-3'-4'-1) Rankine cycle.

Process 1-2 is carried out by a pump which receives the working fluid in the state of saturated liquid at the pressure of the condenser and compress it to the pressure of the isobaric heat recovery exchanger. Deviation from ideal isentropic compression is taken into account by the pump isentropic efficiency

$$\eta_p = \frac{w_{ps}}{w_{pa}} = \frac{h_{2s} - h_1}{h_2 - h_1}$$

where w_{ps} is the ideal isentropic specific pump work, w_{pa} is the actual non-isentropic specific pump work, h_1 is the specific enthalpy of the saturated liquid at the inlet of the pump, h_2 is the specific enthalpy of the compressed liquid at the exit of the actual pump and h_{2s} is the specific enthalpy of the compressed liquid at the exit of the ideal isentropic pump. The pressure of the condenser has been considered 0.1MPa and the pressure of the heat exchanger has been considered in all base cases 3MPa. The isentropic pump efficiency was taken 80%.

Heat is recovered from the engine exhaust gases inside an insulated isobaric heat recovery exchanger. This results in the evaporation of the organic fluid which is taken at the saturation or superheated state 3. The rate of heat recovery by the exhaust gases is given by

$$\dot{Q}_r = \dot{m}_G(h_\text{I} - h_\text{II}) \approx \dot{m}_G c_{v,T_\text{I}} T_\text{I} - \dot{m}_G c_{v,T_\text{II}} T_\text{II}$$

where $\dot{m}_G$ is the mass flow rate of the combustion gases. Subscripts "I" and "II" correspond to the states of the exhaust gases before and after the heat recovery process, respectively. If sulfur is present in the fuel it transforms into sulfur dioxide during combustion and later into sulfur trioxide through further oxidation. Suflur trioxide can then combine with steam vapor to form corrosive sulfuric acid if the temperature of the gas effluents is lower than the sulfuric acid dew point. Thus, T_II at the outlet of the heat recovery exchanger must be kept above the limiting value of 373.15K. In the present study, temperature T_I is considered 633K as suggested by the MAN manufacturer [28] and temperature T_II is assumed 378K. These values are kept invariable to set a constant temperature difference for all comparative fluid cases. Since the heat exchanger is insulated, heat losses are negligible and all heat recovered is absorbed by the organic fluid. That is,

$$\dot{Q}_r = \dot{m}_F(h_3 - h_2)$$

where $\dot{m}_F$ is the mass flow rate of the organic fluid.

The superheated organic fluid expands into a turbine to produce work. Expansion takes place in a single stage from the pressure of the recovery exchanger to the pressure of the condenser. Deviation from ideal isentropic expansion is taken into account by the turbine isentropic efficiency

$$\eta_t = \frac{w_{ta}}{w_{ts}} = \frac{h_3 - h_4}{h_3 - h_{4s}}$$

where w_{ta} is the actual non-isentropic specific turbine work, w_{ts} is the ideal isentropic specific turbine work, h_3 is the specific enthalpy of the superheated fluid at the inlet of the turbine, h_4 is the specific enthalpy of the expanded fluid at the exit of the actual turbine and h_{4s} is the specific enthalpy of the expanded fluid at the exit of the ideal isentropic turbine. The turbine isentropic efficiency was considered 80%. The fluid leaving the turbine may be either superheated or a saturated gas-liquid mixture. In the latter case, the percentage of gas in the mixture is given by the mixture quality x as

$$x = \frac{h_4 - h_1}{h_{fg}}$$

where h_{fg} is the latent heat of evaporation at the pressure of the condenser. Quality x varies between 0 and 1. Finally, the expanded fluid returns at the initial state 1 by passing through the condenser. Heat rejected during condensation is given by

$$\dot{Q}_{out} = \dot{m}_F(h_4 - h_1)$$

The net power produced by the Rankine cycle is calculated as

$$P_{\mathrm{ORC}} = \dot{m}_F\left(w_{ta} - w_{pa}\right) = \dot{Q}_r - \dot{Q}_{out}$$

and the thermal efficiency of the Rankine cycle is given as

$$\eta_{\mathrm{ORC}} = \frac{P_{\mathrm{ORC}}}{\dot{Q}_r} = 1 - \frac{\dot{Q}_{out}}{\dot{Q}_r}$$

The overall efficiency of the ICE-ORC power plant is then calculated as

$$\eta = \frac{P_b + P_{\mathrm{ORC}}}{\dot{m}_f \mathrm{LHV}}$$

where P_b is the brake power of the engine taken as 17923.89 kW, $\dot{m}_f$ is the mass flow rate of natural gas calculated as 3412.82 kg/h and LHV = 49.8 MJ/kg is the lower heating value of natural gas. Efficiency improvement is finally given by

$$\Delta\eta = \eta - \eta_e$$

where η_e is the thermal efficiency of the engine which has been calculated 37.96% [30].

4. Working fluids

The working fluid selection is of key importance on the thermal performance of the ORC recovery system. Generally, the working fluids can be classified into the following three categories:

- wet fluids,
- dry fluids, and
- isentropic fluids.

This classification is based on the slope dT/ds of the saturation vapor curve of each fluid in a T-s diagram, as shown in Figure 3. The type of the working fuel can be determined through the following simplified equation [22]

$$E = \frac{C_p}{T_H} - \frac{\frac{nT_{rH}}{1-T_{rH}} + 1}{T_H^2}\Delta\mathrm{H}_{vap}$$

where E is the reciprocal of the saturation vapor curve slope (ds/dT), C_p is the specific heat of the fluid at constant pressure, T_H is the evaporation temperature of the fluid, n is suggested to be 0.375 or 0.385 by Poling et. al. [32], T_{rH} is the reduced evaporation temperature ($T_{rH} = T_H/T_{cr}$, where T_{cr} is the critical temperature), and $\Delta\mathrm{H}_{vap}$ is the latent heat of vaporization. Fluids having $E > 0$ have a negative saturation vapor curve ($dT/ds < 0$) and are classified as wet fluids. Instead, fluids having $E < 0$ have a positive saturation vapor curve ($dT/ds > 0$) and are classified as dry fluids. Finally, fluids having $E \approx 0$ have an almost vertical saturation vapor curve and are classified as isentropic.

Conventional Rankine power plants usually employ water as working fluid. Water has a small molecular weight (18 kg/kmol), a high boiling point of 100°C at atmospheric pressure and a critical temperature of 347°C. It is chemically very stable, cheap, abundant, non-toxic and environmentally friendly but also has some disadvantages. Water is a typical wet fluid and when it is used in Rankine cycles requires high superheating to avoid condensation and erosion of the turbine blades due to the presence of liquid at the final stages of expansion. Steam turbines are also complex and expensive as they require to handle large steam volumes at the turbine exit. Due to these drawbacks water and other wet fluids are suitable only in high temperature applications like in plants for central power generation. In ORC recovery systems capturing heat from low-grade sources dry or isentropic organic fluids are suggested.

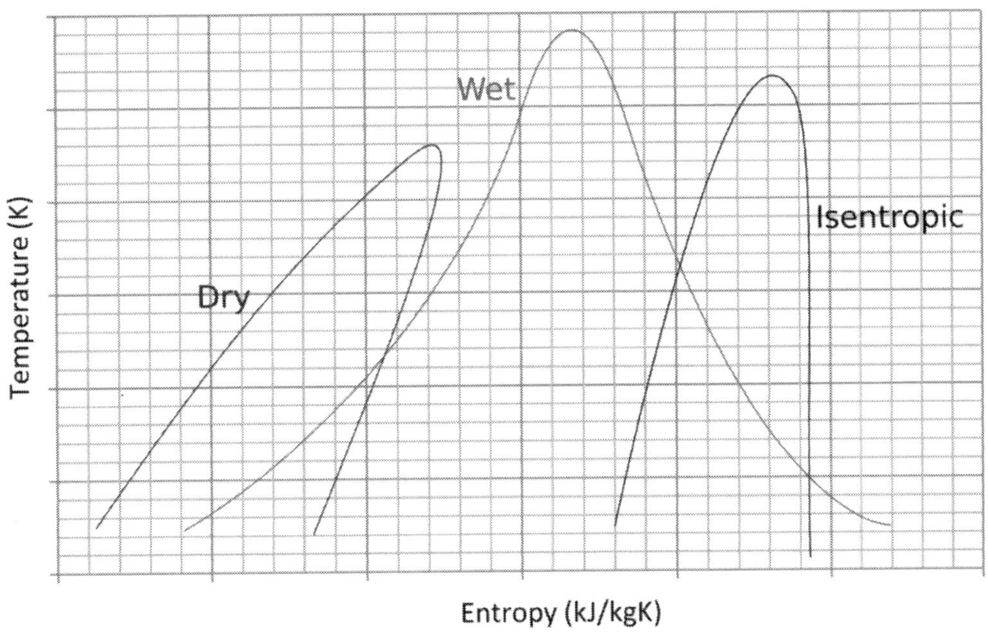

Figure 4.Typical T-s diagrams of dry, wet and isentropic fluids. Classifications is based on the slope of the saturation vapor curve.

4.1. Thermodynamic selection criteria

In terms of thermodynamics, appropriate fuels are usually those having high latent heats, high density and low specific heats [25, 26]. Higher latent heats are found to provide higher unit work output during expansion and higher densities are found to provide the same power outputs with smaller sized equipment. In the present study the thermodynamic comparison of 37 candidate fluids was conducted considering the same operation conditions of the V18 MAN 51/60DF exhaust heat recovery Rankine cycle described earlier. The most important thermodynamic properties of these fluids are summarized in Table 4. All selected fluids have critical pressure above 3MPa and therefore all Rankine cycles considered are subcritical.

Table 4. Thermodynamic properties of various fluids [33-37].

ASHRAE Code	Substance Name	MW (kg/kmol)	Latent heat of evaporation	Critical temperature (K)	Critical pressure (MPa)	Type of fluid

			at 0.1MPa (kJ/kg)			
R11	Trichloromonofluoromethane	137.37	181.49	471.11	4.4076	Isentropic
R12	Dichlorodifluoromethane	120.91	166.3	385.12	4.1361	Isentropic
R13	Chlorotrifluoromethane	104.46	149.47	302	3.879	Wet
R14	Tetrafluoromethane	88.01	134.54	227.51	3.75	Wet
R21	Dichlorofluoromethane	102.92	217.76	451.48	5.1812	Wet
R22	Chlorodifluoromethane	86.468	233.92	369.295	4.99	Wet
R23	Trifluoromethane	70.01	239.552	299.293	4.832	Wet
R32	Difluoromethane	52.02	382.14	351.255	5.782	Wet
R41	Fluoromethane	34.03	489.164	317.28	5.897	Wet
R113	1,1,2-Trichloro-1,2,2-Trifluoroethane	187.38	144.45	487.21	3.3922	Dry
R114	1,2-Dichloro-1,1,2,2-Tetrafluoroethane	170.92	136.06	418.83	3.257	Dry
R115	Chloropentafluoroethane	154.47	116.49	353.1	3.12	Isentropic
R116	Hexafluoroethane	138.02	117.09	293.03	3.048	Wet
R123	2,2-Dichloro-1,1,1-Trifluoroethane	152.93	170.35	456.831	3.6618	Dry
R124	1-Chloro-1,2,2,2-Tetrafluoroethane	136.5	165.99	395.425	3.6242	Wet
R125	Pentafluoroethane	120	164.25	339.173	3.6177	Isentropic
R134a	1,1,1,2-Tetrafluoroethane	102.03	217.16	374.21	4.0592	Isentropic
R141b	1,1-Dichloro-1-Fluoroethane	116.9	222.88	477.5	4.212	Dry
R142b	1-Chloro-1,1-Difluoroethane	100.5	223.43	410.26	4.055	Isentropic
R143a	1,1,1-Trifluoroethane	84.04	226.82	345.857	3.761	Isentropic
R152a	1,1-Difluoroethane	66.05	330.18	386.411	4.5167	Wet
R236ea	1,1,1,2,3,3-Hexafluoropropane	152	165.32	412.44	3.5019	Dry
R236fa	1,1,1,3,3,3-Hexafluoropropane	152	160.48	398.07	3.2	Dry
R245ca	1,1,2,2,3-Pentafluoropropane	134	201.15	447.57	3.925	Dry
R245fa	1,1,1,3,3-Pentafluoropropane	134	196.88	427.2	3.64	Dry
R417a	46.6% R125, 50% R134a, 3.4% R600 (by weight)	106.75		360.19	4.036	Isentropic
R422a	85.1% R125, 11.5 R134a, 3.4% R600a (by weight)	113.6		344.9	3.747	Isentropic
R422d	65.1% R125, 31.5% R134a, 3.4% R600a (by weight)	109.94		352.71	3.903	Isentropic
R423a	52.5% R134a, 47.5% R227ea (by weight)	125.96		372.64	3.587	Isentropic
R170	Ethane	30.07	489.79	305.33	4.8718	Wet
R290	Propane	44.09	426.232	369.825	4.2476	Isentropic
R600	Butane	58.12	386.01	425.125	3.796	Dry
R601	Pentane	72.1	357.89	469.7	3.37	Dry
	Cyclohexane	84.16	356.3	553.64	4.075	Dry
	Benzene	78.11	394.96	562.05	4.894	Dry
	Toluene	92.14	361.007	591.75	4.1263	Dry
R717	Ammonia	17.02	1370.25	405.4	11.333	Wet

4.2. Environmental and safety selection criteria

Organic fluids may also be classified according to their chemical structure into the following groups:

- CFCs: Chlorofluorocarbons (CFCs) are organic compound that contain only carbon, chlorine and fluorine. They are also known by the DuPont brand name "Freon". Such organic fluids are R11, R12, R13, R113, R114 and R115.
- HCFCs: Hydrochlorofluorocarbons (HCFCs) contain carbon, chlorine, fluorine and hydrogen atoms. Such organic fluids are R21, R22, R123, R124.
- HFCs: Hydrofluorocarbons (HFCs) contain carbon, fluorine and hydrogen atoms. Such organic fluids are R23, R32, R41, R125, R134a, R143a, R152a, R236ea, R236fa, R245ca, R245fa, R417a, R422a, R422d, R423a.
- PFCs: Perfluorocarbons (PFCs) or just fluorocarbons contain only atoms of carbon and fluorine. Such organic fluids are R14 and R116.
- HCs: Hydrocarbons (HCs) are known substances containing only carbon and hydrogen atoms. Such fluids are R170 (ethane), R290 (propane), R600 (butane), R601 (pentane), cyclohexane, benzene and toluene.

In addition to the criteria for thermodynamic performance, selection criteria must also be considered on the effect of the fluids to the natural environment and on the safety during handling and operation. Environmental criteria are of major importance and focus on the effect of each fluid on ozone layer depletion and global warming. Quantification of these effects takes place using the indices of Ozone Depletion Potential (ODP) and Global Warming Potential (GWP). The ODP is a relative measure of the degradation effect of each fluid on the ozone layer on the assumption that the ODP value for R11 is equal to 1. The GWP is a relative measure of how much heat the fluid gases trap in the atmosphere compared to the heat absorbed by an equal amount of carbon dioxide. It is usually calculated over the time interval of 100 years. Based on these definitions the higher the value of ODP or GWP the more detrimental is the effect of the fluid on the ozone layer and global warming respectively. ODP values are considered medium below about 0.1 and high above 0.1. GWP values are considered low below 150, medium between 150-2500 and high above 2500. ODP and GWP values are summarized for the majority of the candidate fluids in Table 5.

An indication of the dangerousness of a fluid during handling and operation is provided by the ASHRAE 34 safety classification standard. Accordingly, fluids are classified into two health effect groups A and B as well as into three flammability groups 1, 2 and 3. Class A rating indicates no evidence of toxicity below 400ppm while Class B indicates evidence of toxicity below this concentration limit. On the other hand, class 1 indicates a fluid which does not propagate a flame in open air under normal conditions, class 2 indicates a fluid that may propagate a flame in open air under certain conditions and class 3 indicates a fluid of high flammability. Toxicity and flammability classifications are jointly providing a fluid index which can be either A1, A2, A3, B1, B2 and B3. Recently two new safety classes A2L and B2L have been added for fluids that are mildly flammable. For the majority of the candidate fluids of the present investigation ASHRAE 34 classification is given also in Table 5.

Table 5. Environmental parameters and safety classification of various fluids.

ASHRAE Code	Substance Name	Ozone depletion potential	Global warming potential - 100 yr	ASHRAE 34 Safety Group
R11	Trichloromonofluoromethane	1	4750	A1
R12	Dichlorodifluoromethane	1	10900	A1
R13	Chlorotrifluoromethane	1.7	14400	A1
R14	Tetrafluoromethane	0	7390	A1
R21	Dichlorofluorometane	0.04	151	B1

R22	Chlorodifluoromethane	0.05	1810	A1
R23	Trifluoromethane	0	14800	A1
R32	Difluoromethane	0	675	A2L
R41	Fluoromethane	0		
R113	1,1,2-Trichloro-1,2,2-Trifluoroethane	1	6130	A1
R114	1,2-Dichloro-1,1,2,2-Tetrafluoroethane	1	10000	A1
R115	Chloropentafluoroethane	0.44	7370	A1
R116	Hexafluoroethane	0	12200	A1
R123	2,2-Dichloro-1,1,1-Trifluoroethane	0.02	77	B1
R124	1-Chloro-1,2,2,2-Tetrafluoroethane	0.022	609	A1
R125	Pentafluoroethane	0	3500	A1
R134a	1,1,1,2-Tetrafluoroethane	0.055	1430	A1
R141b	1,1-Dichloro-1-Fluoroethane	0.12	725	A2
R142b	1-Chloro-1,1-Difluoroethane	0.12	2310	A2
R143a	1,1,1-Trifluoroethane	0	4470	A2L
R152a	1,1-Difluoroethane	0	124	A2
R236ea	1,1,1,2,3,3-Hexafluoropropane	0	1370	
R236fa	1,1,1,3,3,3-Hexafluoropropane	0	9810	A1
R245ca	1,1,2,2,3-Pentafluoropropane	0	693	
R245fa	1,1,1,3,3-Pentafluoropropane	0	1030	B1
R417a	46.6% R125, 50% R134a, 3.4% R600 (by weight)	0	2346	A1
R422a	85.1% R125, 11.5 R134a, 3.4% R600a (by weight)	0	3143	A1
R422d	65.1% R125, 31.5% R134a, 3.4% R600a (by weight)	0	2729	A1
R423a	52.5% R134a, 47.5% R227ea (by weight)	0	2280	A1
R170	Ethane	0	5.5	A3
R290	Propane	0	3.3	A3
R600	Butane	0	4	A3
R601	Pentane	0	4	A3
	Cyclohexane			A3
	Benzene			B2
	Toluene			A3
R717	Ammonia	0	0	B2L

CFCs and later HCFCs have been used extensively as refrigerants during the first part of the last century. These compounds were found to be relatively safe for handling and operation but their environmental effect was not addressed until the mid '70s. At the time, the severe consequences of the continuous utilization of the CFCs became obvious and this led to the international phase out treaty of 1987 which is known as the Montreal Protocol. Under this treaty CFCs are currently banned. Replacement has taken place mainly by HCFCs which were equally safe but their widespread utilization eventually became a major global warming source. In view of this threat HCFCs fall also into phase out plans aiming at a gradual step down of their utilization after 2016. In this category, R22 was the most applicable fluid in residential and commercial air conditioning and refrigeration. It has been already phased out in most developed countries but it is still used as a refrigerant in many other developing. DuPont suggests R22 substitution by HFC refrigerant blends of similar physical properties such as R417a, R422a and R422d [34-36]. HFCs are currently adopted in preference to HCFCs since they don't contain chlorine and they don't harm the ozone layer. However, as can be observed in Table 5, they form greenhouse gases with a high GWP. Due to this effect international discussions are open on the future reduction of HFCs also. PFCs, finally, were also used as replacement fluids of CFCs. They are generally very inert and non-toxic but have high GWP values between 6000 and 10000.

To replace the substances mentioned above there exist some interesting hydrocarbon choices. Propane (R290), butane (R600) and pentane (R601) are commonly used as refrigerants or in ORCs. They are also mixed in hydrocarbon blends containing also ethane (R170). They have excellent thermodynamic properties, and in this respect they are as good as or better than HCFs or HCFCs in most applications. They are environmentally friendly with zero ODP and extremely low GWP values. However, they are highly flammable and must be handled with care. In order to ensure safety, hydrocarbon applications are governed by various international, regional and national standards and regulations. They may pose an explosion risk if the concentration is between the lower and upper flammability limits but following the safety standards the concentration of any leaked hydrocarbon will not increase enough even in extreme situations.

The list of the selected fuels concludes with ammonia (R717), the only non-organic fluid candidate. Ammonia is renowned for its favorable thermodynamic properties, it has zero ODP and zero GWP. However, it has a number of disadvantages that have so far prevented its use in commercial applications, e.g. material compatibility, toxicity, and flammability. Even under strict safety regulations there is pressure on ammonia installations to reduce charge levels, especially when situated in populated areas.

5. Results and discussion

All 37 selected fluids have been comparatively examined for the ORC exhaust heat recovery of the V18 MAN 51/60 natural gas engine. Initially, ORC operation conditions were kept similar in all cases with the pressure of the condenser constant at 0.1MPa and the pressure of the evaporator constant at 3MPa. Since all fluids have critical pressures above 3MPa every Rankine cycle examined was subcritical. Considering simple non-superheated operation the comparative characteristics of all fluids are summarized in Table 6. As shown, under these conditions exhaust heat recovery was found capable to increase power plant efficiency by 3.61-4.93%. These terms are translated into additional power outputs of 1700-2325kW just by exploiting the otherwise wasted engine gases, and also mean fuel consumption improvements by 8.68-11.45%. Higher efficiency improvements are taken for R22 (4.93%), ethane (4.82%), R143a (4.71%), R41 (4.64%), R14 (4.53%), propane (4.5%), R13 (4.48%), R32 (4.42%), ammonia (4.41%), R23 (4.40%), R12 (4.40%), R11 (4.32%), R21 (4.28%), R152a (4.25%).

Table 6. Comparative characteristics of the simple (non-superheated) exhaust heat recovery Rankine cycle and overall plant efficiency improvement for all candidate fluids.

ASHRAE Code	Fluid mass flow (kg/s)	Pump power (kW)	Turbine power (kJ/kg)	Maximum cycle temperature (K)	Expanded fluid quality	ORC Power (kW)	ORC efficiency (%)	Power plant efficiency (%)	$\Delta\eta$ (%)
R11	44.18	108.19	2146.99	445.07	Superheated	2038.80	19.65	42.28	4.32
R12	51.21	124.77	2199.50	376.25	0.9748	2074.73	19.99	42.36	4.40
R13	61.64	146.76	2257.97	290.67	0.8972	2111.21	20.35	42.44	4.48
R14	75.31	170.18	2306.90	219.98	0.8133	2136.72	20.59	42.49	4.53
R21	36.72	94.66	2158.96	417.06	0.9446	2064.30	19.89	42.34	4.38
R22	39.94	102.68	2428.86	343.19	0.8962	2326.18	22.42	42.89	4.93

R23	41.95	105.13	2179.91	280.11	0.8262	2074.78	19.99	42.36	4.40
R32	26.52	79.21	2163.85	321.17	0.8182	2084.64	20.09	42.38	4.42
R41	21.11	86.82	2272.56	287.94	0.7931	2185.74	21.06	42.60	4.64
R113	47.57	114.26	1921.25	478.88	Superheated	1806.99	17.41	41.79	3.83
R114	53.70	128.14	1963.65	414.01	Superheated	1835.51	17.69	41.85	3.89
R115	65.26	151.71	2003.06	350.69	Superheated	1851.35	17.84	41.89	3.93
R116	76.28	172.18	2018.98	292.4	0.955	1846.79	17.80	41.88	3.92
R123	44.64	111.01	1981.46	444.45	Superheated	1870.45	18.03	41.93	3.97
R124	48.39	118.97	2003.16	385.26	Superheated	1884.19	18.16	41.96	4.00
R125	52.21	124.97	2015.99	330.64	0.9895	1891.02	18.22	41.97	4.01
R134a	40.02	105.32	2007.00	359.35	0.9751	1901.68	18.33	41.99	4.03
R141b	34.14	101.38	2083.23	454.73	Superheated	1981.85	19.10	42.16	4.20
R142b	36.58	111.13	2100.45	393.07	Superheated	1989.32	19.17	42.18	4.22
R143a	35.67	110.79	2334.01	359.35	Superheated	2223.22	21.42	42.67	4.71
R152a	27.44	98.31	2102.53	365.22	0.9241	2004.22	19.31	42.21	4.25
R236ea	43.90	107.26	1856.37	405.98	Superheated	1749.10	16.86	41.67	3.71
R236fa	49.16	123.26	1826.30	394.75	Superheated	1703.04	16.41	41.57	3.61
R245ca	36.50	95.41	1862.00	432.09	Superheated	1766.58	17.02	41.71	3.75
R245fa	38.37	101.78	1876.49	416.57	Superheated	1774.70	17.10	41.72	3.76
R417a	44.29	112.39	2040.08	346.87	0.9801	1927.69	18.58	42.05	4.09
R422a	49.27	125.03	2036.64	334.97	0.9854	1911.60	18.42	42.01	4.05
R422d	46.74	118.59	2047.83	341.07	0.9804	1929.24	18.59	42.05	4.09
R423a	47.13	119.59	1961.77	364.09	Superheated	1842.18	17.75	41.87	3.91
Ethane	19.26	128.30	2401.82	282.89	0.8589	2273.52	21.91	42.78	4.82
Propane	19.78	123.35	2246.60	350.84	0.979	2123.25	20.46	42.46	4.50
Butane	19.08	114.98	2092.89	410.85	Superheated	1977.92	19.06	42.16	4.20
Pentane	18.61	110.60	1902.99	462.13	Superheated	1792.39	17.27	41.76	3.80
Cyclohexane	17.10	86.11	1837.94	529.43	Superheated	1751.82	16.88	41.68	3.72
Benzene	17.66	78.63	2014.62	523.55	0.1294	1935.99	18.66	42.07	4.11
Toluene	17.43	81.07	1826.40	566.22	Superheated	1745.33	16.82	41.66	3.70
Ammonia	7.24	38.45	2118.90	338.87	0.8366	2080.44	20.05	42.37	4.41

The list of thermodynamically favorable fluids which was selected in the previous discussion may be greatly narrowed on the basis of their environmental footprint. Among these choices R22, R14, R13, R12 and R11 have already been prohibited as CFCs. R143a is an HFC used mainly as a blend component in refrigerants. It has zero ODP, high GWP at 4470 and is mildly flammable at the A2L class. R41 is also an HFC used mainly in semiconductor manufacturing but with unknown GWP and safety behavior. It may be considered for further evaluation but it can't be selected until further research. R14 is a PFC with a high GWP value of 7390. Instead, R32 presents a good performance with medium to low GWP at 675. R23 is an HFC with a very high GWP at 14800. R21 is an HCFC with medium ODP and is under phase out. R152a is a promising and environmentally friendly choice with good performance. Finally the list above includes ethane and propane, two well known HCs, and ammonia, the only non organic candidate. These fluids have a low environmental impact but are highly flammable and need special attention and safety measures. Ammonia is also dangerously toxic.

Based on this discussion the list of selected fluids narrows only to ethane, R41 (under lack of data), propane, R32, ammonia and R152a.

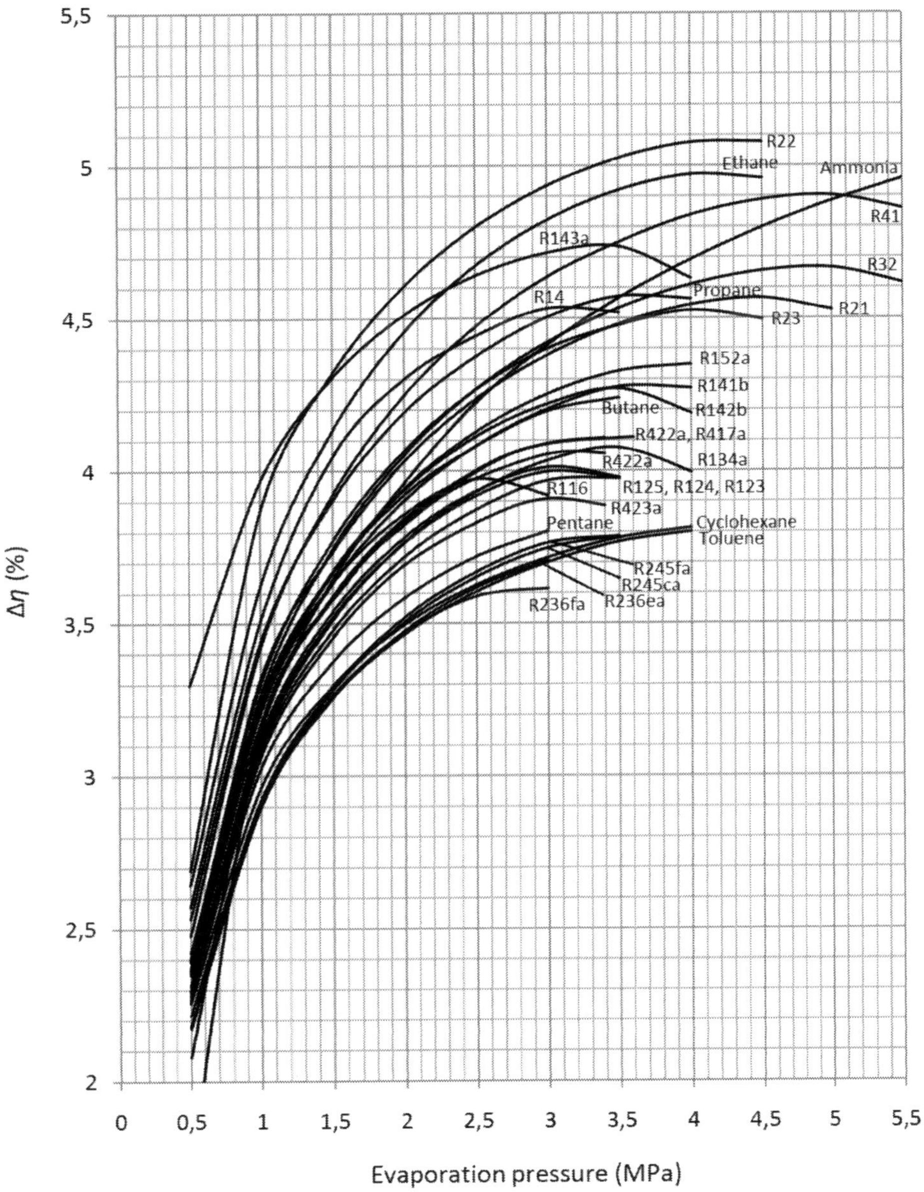

Figure 5. Effect of the ORC evaporation pressure on the overall efficiency improvement of the V18 MAN 51/60 DF natural gas power plant for selected working fluids. Condenser pressure is 0.1MPa in all cases and Rankine cycles are simple without superheating.

The effect of the ORC evaporation pressure on the efficiency improvement of the power plant may be shown for selected fuels in Figure 5. Results are given for subcritical pressures and simple non-superheated cycles in all cases. Given the Montreal Protocol prohibition CFCs are not considered any further and Figure 5 contains only R22 for comparison. As shown, an increase of the evaporation pressure is always beneficial leading to better plant efficiencies for all fluids. It can be seen that the effect of the evaporation pressure is more dramatic at low pressures. As the evaporation pressure increases the slope of the curve of the efficiency improvement $\Delta\eta$ reduces and even negative improvement values can be taken when attaining the critical pressure of a fluid. In this respect one can set a limiting upper evaporation pressure at about 85-95% of the critical pressure value of each fluid. The optimum evaporation pressure for a fluid is therefore dictated by its critical pressure and fluids with higher critical pressures have greater potential for efficiency enhancement. Accordingly, ethane, R41, R32 and ammonia may further improve their performance at elevated pressures. Maximum pressure is set for ethane at 4.5MPa, for R41 at 5MPa and for R32 at 5MPa. Ammonia has an extremely high critical pressure of 11.333 MPa but results show that above 5MPa further improvement of about 1% is only possible. To avoid risk of leakages and increased costs for special safety measures the maximum pressure for ammonia was set also at 5MPa. Finally, propane is found to maximize performance at 3.5MPa and R152a at 4MPa.

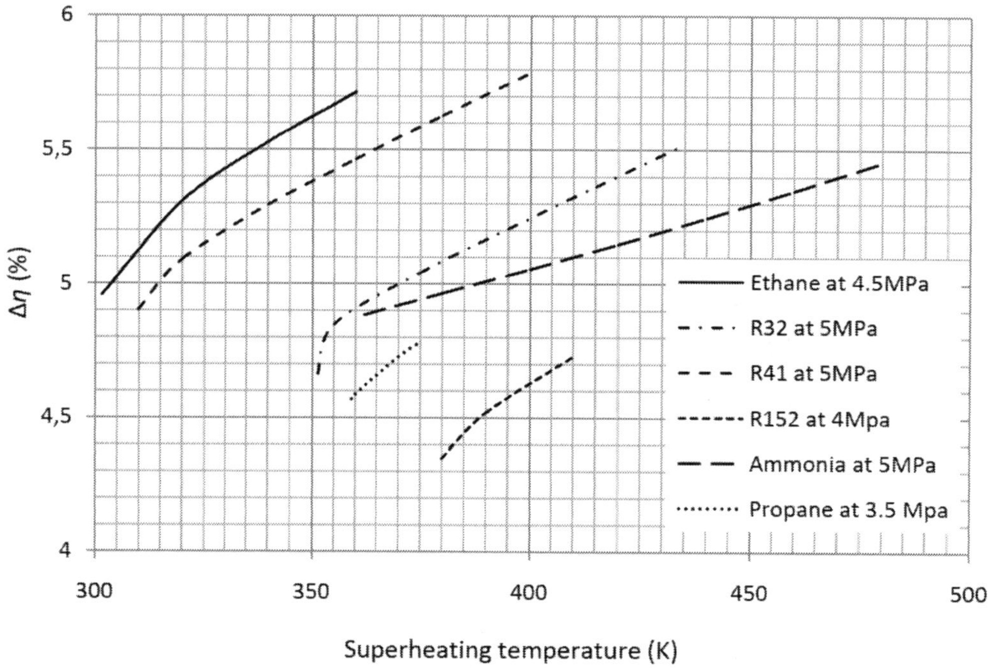

Figure 6. Effect of the ORC superheating temperature on the overall efficiency improvement of the V18 MAN 51/60 DF natural gas power plant for selected working fluids.

Except propane which is an isentropic fluid, all other fluids selected so far are wet. For a wet fluid and a simple non-superheating Rankine cycle an increase of the evaporation pressure results in a reduction of the quality of the saturated gas-liquid mixture taken after expansion. This means an increase of the percentage of liquid state at the final stages of turbine expansion and a danger for turbine blade erosion and wear. In all these cases simple Rankine cycle is inappropriate and superheating of the fluid vapor above the saturation temperature must be followed. The effect of the superheating temperature on the efficiency improvement of the power plant is illustrated for the cases of ethane, R41, propane, R32, ammonia and R152a at Figure 6. Each fluid is assumed at its optimum pressure mentioned before and superheating temperature is increased so that after expansion the fluid to be slightly superheated. As can be seen, an increase of the superheating temperature always leads to an overall efficiency improvement. Maximum efficiencies are attained at about 360K for ethane, 400K for R41, 375K for propane, 435K for R32, 480K for ammonia and 410K for R152a. The characteristics of these optimum ORC cycles are summarized at Table 7. Among the selected fluids the ORC cycle of R32 is selected as the most appropriate due to high thermodymamic performance, low environmental impact and low flammability-toxicity. Such a cycle may provide an overall efficiency improvement of the order of 5.52%, as shown in Table 7. This improvement is also translated into a reduction of the fuel consumption by 12.69%.

Table 7. Comparative characteristics of superheated exhaust heat recovery Rankine cycles and overall plant efficiency improvements for the selected working fluids.

ASHRAE Code	Fluid mass flow (kg/s)	Pump power (kW)	Turbine power (kJ/kg)	Maximum cycle temperature (K)	Evaporation pressure (MPa)	ORC Power (kW)	ORC efficiency (%)	Power plant efficiency (%)	$\Delta\eta$ (%)
Ethane	14.72	148.72	2844.99	360	4.5	2696.27	25.98	43.68	5.71
R41	15.63	108.58	2838.96	400	5.0	2730.38	26.31	43.75	5.79
Propane	17.85	130.51	2386.34	375	3.5	2255.83	21.74	42.74	4.78
R32	19.68	99.32	2703.64	435	5.0	2604.32	25.10	43.48	5.52
Ammonia	5.69	51.04	2623.21	480	5.0	2572.14	24.79	43.41	5.45
R152a	23.40	112.73	2343.77	410	4.0	2231.04	21.50	42.69	4.73

Conclusions

In the present work the ORC exhaust heat recovery of a V18 MAN 51/60DF internal combustion engine power plant was examined. The plant was assumed to operate with natural gas at 500rpm and design focused on the selection of an appropriate ORC working fluid on the basis of thermodynamic, environmental and safety criteria. The study was undertaken for 37 working fluids under subcritical Rankine cycle conditions and examined the effect of key operation conditions such as evaporation pressure and superheating temperature. After an appropriate selection of these operation conditions for each candidate substance, R32 was selected as the most appropriate working fluid given its high thermodynamic performance, its low environmental impact and its low flammability-toxicity characteristics. Analysis has shown that R32 may provide an overall efficiency improvement of the power plant by 5.52% and a reduction of fuel consumption by 12.69%. Given the high power rating of the V18 MAN 51/60DF engine at about 18000kW, these improvements are translated into enormous cost savings during operation. The methodology employed in the present study may also serve as a guideline for future optimization of other energy systems.

References

[1] Hussain Q E, Brigham D R, Maranville C W 2009 *SAE Int. J. Engines* **2**(1) 1132
[2] Yu C and Chau K T 2009 *Energy Conversion and Management* **50**(6) 1506
[3] Hountalas D T, Katsanos C O and Lamaris V T 2007 *SAE technical paper* 2007-01-1563.
[4] Li J 2015 Structural optimization and experimental investigation of the organic Rankine cycle for solar thermal power generation (Berlin Heidelberg: Springer - Verlag)
[5] Quoilin S and Lemort V 2009 *Proc. 5th European Conference on Economics and Management of Energy in Industry* (Algarve-Portugal)
[6] Schuster A, Karellas S, Karakas E and Spliethoff H 2009 *Applied Thermal Engineering* **29**(8) 1809
[7] Ringler J, Seifert M, Guyotot V and Hübner W 2009 *SAE technical paper* 2009-01-0174
[8] Endo T *et al* 2007 *SAE technical paper* 2007-01-0257.
[9] Nelson C 2009 *Proc. Directions in energy efficiency and emissions research (DEER) conference* (Dearborn-Michigan, USA)
[10] Arias D A, Shedd T A and Jester R K 2006 *SAE technical paper* 2006-01-1605
[11] Ringler J, Seifert M, Guyotot V and Hübner W 2009 *SAE technical paper* 2009-01-0174.

[12] Srinivasan K K, Mago P J, Krishnan S R 2010 *Energy* **35**(6) 2387
[13] Bombarda P, Invernizzi C M and Pietra C 2010 *Applied Thermal Engineering* **30** 212
[14] Vaja I and Gambarotta A 2010 *Energy* **35**(2) 1084
[15] Briggs T E, Wagner R, Edwards K D, Curran S and Nafziger E 2010 *SAE technical paper* 2010-01-2205.
[16] Hountalas D T, Mavropoulos G C, Katsanos C, Knechtet W 2012 *Energy Conversion and Management* **53** 19
[17] Boretti A 2012 Applied Thermal Engineering **36** 73
[18] Larsen U, Pierobon L, Haglind F and Gabrielii C 2013 *Energy* **55** 803
[19] Yu G, Shu G, Tian H, Wei H and Liu L 2013 *Energy* **51** 281
[20] Zhang H, Wang E, Fan B 2013 *Applied Energy* **102** 1504
[21] Zhang X, He M, Zeng K and Zhang Y 2010 *Journal of Engineering Thermophysics* **1** 15
[22] Amicabile S, Lee J I and Kum D 2015 *Applied Thermal Engineering* **87** 574
[23] Liu B, Chien K and Wang C 2004 *Energy* **29** 1207
[24] Hung T, Shai T and Wang S 1997 *Energy* **22** 661
[25] Chen H, Goswami D Y and Stefanakos E K 2010 *Renewable and Sustainable Energy Reviews* **14** 3059
[26] Maizza V and Maizza A 1996 *Applied Thermal Engineering* **16** 579
[27] Yamamoto T, Furuhata T, Arai N and Mori K 2001 *Energy* **26** 239
[28] *51/60DF Project Guide - Marine: Four-stroke dual-fuel engines compliant with IMO Tier II* 2014 (Augsburg: MAN Diesel & Turbo SE)
[29] *Exhaust Gas Turbochargers* 2011 (Augsburg: MAN Diesel & Turbo SE)
[30] Douvartzides S and Karmalis I 2016 Thermal design of a natural gas - diesel dual fuel turbocharged V18 engine for ship propulsion and power plant applications, *Proc. of the Imanee International Conference* (Chalkidiki-Greece) submitted.
[31] Woodyard D 2009 *Pounder's Marine Diesel Engines and Gas Turbines* 9th edition (Elsevier Ltd.)
[32] Poling B, Prausnitz J and Connell J O 2000 *The properties of gases and liquids* (McGraw- Hill Professional)
[33] NIST National Institute of Standards and Technology, Thermophysical properties of fluid systems, USA, 2016 (http://webbook.nist.gov/chemistry/fluid/).
[34] Thermodynamics properties of DuPont™ ISCEON® MO29, DuPont™ ISCEON® 9 Series: Refrigerants, 2005 (https://www.chemours.com).
[35] Thermodynamics properties of DuPont™ ISCEON® MO59 (R417a), DuPont™ ISCEON® 9 Series: Refrigerants, 2005 (https://www.chemours.com).
[36] Thermodynamics properties of DuPont™ ISCEON® MO79 (R422a), DuPont™ ISCEON® 9 Series: Refrigerants, 2005 (https://www.chemours.com).
[37] Thermodynamics properties of DuPont™ ISCEON® 39TC (R423a), DuPont™ ISCEON® 9 Series: Refrigerants, 2005 (https://www.chemours.com).

Studies concerning the decontamination of hydrocarbons-polluted soil areas using bioremediation techniques

C Deac[1], A Barbulescu[2], A Gligor[1], M Bibu[1], V Petrescu[1]

[1] Lucian Blaga University of Sibiu, Bd. Victoriei 10, 550024 Sibiu, Romania
[2] CONPET S.A., Str. Anul 1848 1-3, 100559 Ploiesti, Romania

E-mail cristian.deac@ulbsibiu.ro

Abstract. The accidental or historic contamination of soils with hydrocarbons, in areas crossed by oil pipelines or where oil- or gas-extraction installations are located, is a major concern and has significant financial and ecological consequences, both for the owners of those areas and for the oil transportation or exploitation companies. Therefore it is very important to find the optimal method for removing the pollution. The current paper presents measures, mainly involving bioremediation, recommended and applied for the depollution of a contaminated area in Romania. While the topic of dealing with polluted soils is well-established in the Romanian speciality literature, bioremediation is a relatively novel approach and this paper presents important considerations in this regard. Contaminated soil samples were taken from 10 different locations within the targeted area and subjected to a thorough physical and chemical analysis, which led to determining a specific scoring table for assessing the bioremediation potential of the various samples. This has allowed the authors to establish for each of the sampled areas the best mix of factors such as nutrients (nitrogen, phosphorus, potassium), gypsum, microelements etc., that would lead to obtaining the best results in terms of the contaminants' biodegradation.

1. Introduction

The contamination of soils with oil-derived products is a very important problem nowadays, because the presence of hydrocarbons in the soil tends to affect the normal biological circuits, blocking the development of plants [1, 2].

This type of pollution can have several sources, among which can be mentioned:

- Accidental failures of hydrocarbons pipelines;
- Exploitation of research and exploitation wells;
- Leakage from hydrocarbons storages or losses during their transportation;
- Road and railway accidents which involve oil tanks;
- Marine pollution due to the activity of oil rigs;
- Pollution of water courses and their shores by intentional or accidental spilling. [1]

Pipeline failures alone have accounted for the spilling of around 6.4 million liters of oil on land, in 2001. [3]

While there have been proposed and developed several types of solutions for the depollution of soils contaminated with hydrocarbons – among others chemical methods, thermal methods, pedological methods etc. - bioremediation is a relatively new approach that is gaining more and more attention and appreciation due to its benefits.

Bioremediation is a technique that uses live microorganisms to generate enzymes that would attack and degrade environmental contaminants into less toxic forms. The microorganisms can originate from the polluted environment itself, or they can be brought there from other areas. The process often involves manipulating the environmental parameters so that the microbial growth is encouraged and the degradation is accelerated. [1, 4]

Bioremediation has several advantages over the „conventional" depollution technologies. Firstly, its application is relatively cheap. Also, it involves reducing oils to mineral compounds, while the chemical methods, for example, merely transfer the contaminant from one environmental factor to another one. Also, it can be considered as a „green" technology, being based on a natural depollution process.

Currently there can be identified two main directions for the bioremediation of soils:

- bioaugumentation, that involves adding bacteria known to efficiently contribute to the degradation of oil are added to the already existing microbial population;
- biostimulation, that involves stimulating the growth of indigenous oil-degrading microorganisms by adding nutrients and other materials. Recent field studies have proven that biostimulation is more efficient that bioaugmentation because adding more microorganisms does not seem to significantly improve the degradation of hydrocarbons more than the simple addition of nutrients [5, 6]

The main steps in selecting the bioremedation treatment and the response plan include:

- Pre-treatment assessment; this step analyses whether bioremediation is a viable option, based on the type and concentration of contaminating hydrocarbons, on the presence of biodegrading microorganisms, on the type of affected soil and not least on environmental factors such as the pH and the temperature;
- Setting up a plan for treatment and monitoring; this step involves selecting agents that would limit the treatment's speed (for example nutrients) and determining the strategies for applying these agents;
- Assessing the treatment's evolution and ending it; after implementing the treatment, analyses have to be carried out periodically to assess its efficiency and to determine the optimal moment for ending the treatment. [7]

Bioremediation can be applied in-situ, to the soil directly on the pollution site, or ex-situ, on soil that has been removed from the polluted area by excavation.

While the in-situ bioremediation techniques are generally more desirable because they imply less costs and less disturbance to the environment, they are limited by the depth of the soil that can be treated efficiently. Often, an optimal diffusion of oxygen in soil, a very important factor in the successful development of biodegrading microorganisms is possible only for depths up to 0.3 m. Also, on-site treatments require a maximal content of 8% TPH in the soil.

Ex-situ techniques involve excavating the contaminated soil and placing it on an impermeable barrier. They require airing equipment, equipment for providing nutrients and generally a large volume of work. The main applicable ex-situ treatment types are landfarming, composting and bio-stacking.

Starting from analyzing the various factors influencing the biodegradation rate of contaminants and thus the success of bioremediation, in this paper the authors present an attempt at a systematic approach to applying a bioremediation treatment, based on an actual pollution case.

2. Assessment of the importance of the main factors influencing the selection of bioremediation treatments. Development of an assessment matrix

Based on earlier studies and on the speciality literature [3, 4, 6, 8, 9], the authors have determined that the main factors that need to be taken into account for selecting the bioremediation approach include:

- The type of contaminating oil, the total petroleum hydrocarbons (TPH) in the soil and the "age" of pollution (expressed using the ratio between hydrocarbons and phytane as reference);
- The type of soil, its salinity, its pH, temperature and humidity;

- The presence of nutrients (N, P, K), of oxygen and of microelements in the soil;
- The existence of a microbial population capable of degrading the pollutants.

The microbial development and activity are affected by the environment's pH, temperature and humidity. The bacteria usually associated with quick biodegradation of organic matter require a pH-neutral environment, so it is important to create optimal conditions in this regard. For example, if the soil is too acid, the pH value can be increased by adding lime.[8]

The hydrocarbons are more easily degraded in aerobic conditions, so providing enough oxygen in the soil is another important factor. This is why bioremediation is usually applied together with soil loosening methods.

The soil's structure can decide the efficiency of supplying the air, water and nutrients that are needed by the microorganisms. Therefore, among other materials, straw or sawdust can be added to the soil to improve its structure.

An important factor is also the so-called pollution aging. This refers to the fact that oils tend to change their composition over time due to chemical reactions, biotransformation, volatilization etc. However, some hydrocarbon molecules such as phytane or pristine are more resistant to the microbial degradation than for example C18-hydrocarbons, so over time the ratio between C18 hydrocarbons and phytane will decrease as the oil is biodegraded by the microorganisms.

The soil's salinity can be expressed by means of the electrical conductivity. The balance between the content of Na and Ca+Mg in the soil can be corrected by adding gypsum. Gypsum, however, cannot be added in amounts larger than 10 t/ha every 6 months, so if the soil's salinity would require a larger amount of gypsum, it needs to be added in several batches over this amount of time.

Starting from the characteristic values for each of the factors described above and on previous researches in this regard [4], the authors have developed an assessment matrix, presented in table 1. This assessment matrix also allows to define global biodegradability ranges based on the total number of points for each of the cases mentioned above and, in correlation with the initial TPH values, there can be determined the optimal duration of bioremediation treatments.

Table 1. Pollutant biodegradability assessment matrix

Parameter	**Very easily biodegradable (4 points)**	**Easily biodegradable (3 points)**	**Average biodegradability (2 points)**	**Difficultly biodegradable (1 point)**	**Very difficultly biodegradable (0 points)**
TPH (%)	1-3	3-5	5-10	10-15	>15
EC, µS/cm	<2000	2000-4000	4000-8000	8000-14000	>14000
Microorganisms (MPN/g soil)	≥ 107	≥ 105	≥ 104	≥ 103	< 103
Clays (%)	< 10	10 - 20	20 - 40	40 - 60	> 60
API density	> 30	25 - 30	20 - 25	15 - 20	< 15
Pollution age	(C18/phytane) soil ≈ (C18/phytane) oil	(C18/phytane) soil ≈ 75% (C18/phytane) oil	(C18/phytane) soil ≈ 50% (C18/phytane) oil	(C18/phytane) soil ≈ 25% (C18/phytane) oil	(C18/phytane) soil almost insignificant
pH	6.81-7.20	5.81-6.80 7.21-7.80	5.41-5.80 7.81-8.40	5.01-5.40 8.41-9.00	<5.00 >9.00
Global biodegradability	28 points	21-27 points	14-20 points	7-13 points	0-6 points

3. Target area. Analyses carried out on the contaminated soil

In order to check the applicability of the assessment matrix presented above, the authors have focused their researches on an area that has been subjected to an accidental pollution with hydrocarbons near Poarta Alba, Constanta county, Romania.

From this area, in order to better assess the pollution and to be able to provide specific bioremediation solutions for different degrees of contamination, there have been analysed soil samples from 10 different locations within the visibly polluted area.

These soil samples were analysed from a physical, chemical and microbiological point of view.

The physical analyses targeted elements such as the texture, porosity, hydraulic conductivity and apparent density of the soil.

The chemical analyses in turn referred to determining the pH, the content of nutrients (N, P, K), the content of clays, the total petroleum hydrocarbons (TPH), the API gravity of the contaminating oil etc.

Microbiological analyses were carried out in order to determine the amount of hydrocarbons-oxidating bacteria and to isolate efficient strains that can be grown and multiplied ex situ and then reintroduced in the contaminated soil in order to carry out the actual biodegradation processes.

4. Results and discussion

The main results of the chemical and microbiological analyses for the 10 locations are presented in table 2.

Table 2. Results of the chemical and microbiological analyses

Soil sample no.	TPH-IR, mg/kg	Bacteria, MPN/g soil	pH	Electrical conductivity µS/cm	Fe, mg/kg	Mn, mg/kg	Clays, %	API gravity of the oil	C18/ phytane oil	C18/ phytane soil
A1	13900	$\geq 10^7$	8.2	2060	50	280	1.9	45	2.5	2.5
A2	21000	$\geq 10^7$	8.31	2910	1000	360	7.8	34	2.31	2.3
A3	119600	$\geq 10^7$	8.45	590	2900	480	0.2	34	2.31	1.68
A4	100900	$\geq 10^7$	8.06	11200	3200	270	0.4	34	2.31	2.26
A5	55300	$\geq 10^7$	8.49	5960	120	120	3.6	31	2.07	1.14
A6	203900	$\geq 10^7$	8.37	1300	550	160	1.6	31	2.07	1.16
A7	142000	$\geq 10^7$	8.28	7980	910	210	5.7	28.5	2.25	1.64
A8	65000	$\geq 10^7$	8.36	10360	18	230	6.9	34	-	-
A9	9700	$\geq 10^7$	7.62	17100	2420	440	10.5	28.8	5.3	3.05
A10	19300	$\geq 10^7$	7.81	11200	1220	490	9.6	28.8	5.3	1.88

These values were then compared to those comprised in the global assessment matrix described above and the resulting matrix is presented in table 3.

Based on this matrix, there have been then designed specific bioremediation treatments for each of the 10 locations, as shown in table 4.

Table 3. Assessment matrix applied to the 10 analysed locations

Soil sample no.	**Score for TPH**	**Score for EC**	**Score for bacteria**	**Score for clays**	**Score for API density**	**Score for pollution age (C18/Phytane)**	**Score for pH**	**Total score**	**Assessment of global biodegradability**
A1	4	3	4	4	4	4	2	25	Easy
A2	4	3	4	4	4	4	2	25	Easy
A3	1	4	4	4	4	3	1	21	Easy
A4	1	1	4	4	4	4	2	20	Average
A5	2	2	4	4	4	2	1	19	Average
A6	0	4	4	4	4	2	2	20	Average
A7	1	2	4	4	3	3	2	19	Average
A8	2	1	4	4	4	0	2	17	Average
A9	4	0	4	3	3	2	3	19	Average
A10	4	1	4	3	3	1	2	18	Average

Table 4. The bioremediation treatments prescribed for each of the 10 locations in the targeted area

Sample no.	Total treatment duration [months]	NPK treatment duration [months]	N [mg/kg soil]	P [mg/kg soil]	K [mg/kg soil]	Gypsum, stage I [g/m^2]	Gypsum, stage II [g/m^2]	Gypsum, stage III [g/m^2]	Straw [kg/m^2]	Surfactant [l/m^2]
A1	24	6	596	119	119	0	0	0	10	3
A2	24	9	900	180	180	0	0	0	10	3
A3	84	51	5126	1025	1025	0	0	0	10	3
A4	96	43	4324	865	865	1000	170	0	10	3
A5	72	24	2370	474	474	590	0	0	10	3
A6	108	87	8739	1748	1748	0	0	0	10	3
A7	96	60	6086	1217	1217	470	0	0	10	3
A8	72	28	2786	557	557	1000	170	0	10	3
A9	24	4	416	83	83	1000	1000	900	10	3
A10	24	8	827	165	165	1000	560	0	10	3

As can be seen from table 4, the total treatment duration depends both on the global biodegradability and on the initial TPH value of the contaminated soil. Even if the global biodegradability index points to an easy biodegradation, a large initial TPH value (e.g. 119600 mg/kg for A3) can impose a long total treatment duration (up to 7 years), whereas if the initial TPH value is very large and the global biodegradability is only average, the treatment duration may surpass 8 years.

The addition of a surfactant is necessary in order to increase the oil's mobility [4] and thus to facilitate a better contact between the microorganisms and the contaminant.

The duration of the treatment with nutrients depends on the contents in nitrogen, phosphorus and potassium already present in the soil, on the requirement to achieve an optimal recommended C:N:P:K ratio that is comprised between 100:10:1:1 and 100:5:1:1 [10] and on following recommended monthly doses:

N - 30 g/m^2;

P_2O_5 - 12 g/m^2;

K - 7 g/m^2.

In all cases, when the TPH value drops below 3% (30000 mg/kg), there can be applied a phytoremediation treatment to complement bioremediation, by planting specific plant species for the respective terrain type (willow and poplar for wooded areas or alfalfa and oat for agricultural lands. [1]

5. Conclusions

The current paper has presented a scientific, methodical approach to applying a bioremediation treatment in order to eliminate or at least reduce the contamination of soils with hydrocarbons.

The assessment matrix developed by the authors can, in correlation with other considerations and factors, taken as a fundament for unfolding a successful bioremediation treatment.

Also, an area accidentally polluted with hydrocarbons has been assessed from a physical, chemical and biological point of view and the assessment matrix was applied to it for determining the optimal response.

In future, the authors intend to continue their researches and if possible refine the assessment matrix.

References

[1] Avram L, Dussap C G, Troquet M and Troquet J 2006 *Applied Ecology* (in Romanian), Publishing House of the Petroleum-Gas University of Ploiesti, Romania;

[2] Onutu I and Tanasescu C 2004 *Pollutants in oil and petrochemistry* (in Romanian), Publishing House of the Petroleum-Gas University of Ploiesti, Romania;

[3] Salanitro J P 2001 Bioremediation of petroleum hydrocarbons in soil. *Advances in agronomy*, 72, pp.53-105.

[4] Barbulescu A 2014 *Contributions regarding the optimisation of the decontamination of soils and subsoils polluted with hydrocarbons* (in Romanian),, Petroleum-Gas University Ploiesti, Romania

[5] Ionescu C et al 1999 Pollution and environmental protection, Briliant Publishing House, Bucharest;

[6] Kingston P F 2002 Long-term environmental impact of oil spills, *Spill Sci. Technol. Bulletin*, 7;

[7] Shaheen E I 1992 *Technology of Environmental Pollution Control*, 2nd ed., Calgary, 1992;

[8] Vidali M 2001 Bioremediation. An overview, *Pure Appl. Chem.*, 73, pp. 1163-1172.

[9] Riser-Roberts E 1998 *Remediation of Petroleum Contaminated Soils. Biological, Physical and Chemical Processes*, Lewis Publishers, London, 1998.

[10] U.S. Environmental Protection Agency 2003 *Aerobic Biodegradation of Oily Wastes, A Field Guidance Book for Federal On-scene Coordinators*;

IOP Conf. Series: Materials Science and Engineering **161** (2016) 012089 doi:10.1088/1757-899X/161/1/012089

Determining the explosion risk level and the explosion hazard area for a group of natural gas wells

A Gligor, V Petrescu, C Deac, M Bibu

Lucian Blaga University of Sibiu, Bd. Victoriei 10, 550024 Sibiu, Romania

E-mail cristian.deac@ulbsibiu.ro

Abstract. Starting from the fact that the natural gas engineering profession is generally associated with a high occupational risk, the current paper aims to help increase the safety of natural gas wells and reduce the risk of work-related accidents, as well as the occurrence of professional illnesses, by applying an assessment method that has proven its efficiency in other industrial areas in combination with a computer-aided design software. More specifically, the paper focuses on two main research directions: assessing the explosion risk for employees working at natural gas wells and indicating areas with a higher explosion hazard by using a modern software that allows their presentation in 3D. The appropriate zoning of industrial areas allows to group the various functional areas function of the probability of the occurrence of a dangerous element, such as an explosive atmosphere and subsequently it allows also to correctly select the electrical and mechanical equipment that will be used in that area, since electrical apparatuses that are otherwise found in normal work environments cannot generally be used in areas with explosion hazard, because of the risk that an electric spark, an electrostatic discharge etc. ignites the explosive atmosphere.

1. Introduction

The natural gas producing and distributing industry is characterised by a high level of occupational risk. Therefore, the assessment of the explosion and fire hazard has to play an important role when selecting the type of equipments that can be used in specific areas of natural gas engineering facilities. [1]

For example, the considerations regarding the usage of electrical equipments in areas with explosion hazard are significantly different from those employed in the case of normal-risk areas.

In order to ensure the explosion safety of the operating personnel and of the installations that function in such industrial environments, it is necessary, even before implementing any technical solutions, to analyse thoroughly whether it would be possible for flammable substances to accumulate and generate an explosive atmosphere.

Also, it is necessary to determine the probability of an explosive atmosphere to occur within the considered industrial space and to identify all types of employed flammable substances so as to facilitate the grouping of that area within one of the standardised types of explosion hazard areas.

Whereas the possibility to assess the explosion risks in various other parts of a natural gas extraction, storage and transportation system has been discussed by various authors ([2, 3, 4]), in the current paper, the authors use a consecrated method to determine the explosion risk for the specific case of a group of natural gas wells; furthermore, they also used a specialised 3D computer-aided design software to indicate the main problem areas within this facility.

2. The analysed facility

For an optimal exploitation and monitoring of the surface installations corresponding to natural gas extraction wells, these are usually grouped spatially, forming so-called groups of wells.

The spatial grouping of the wells also allows a better possibility for the operating personnel to intervene in the extraction process.

The technological installations of the considered group of natural gas wells has been modelled by the authors using the software package AutoCAD Plant 3D and consist of following component elements: measurement polygons for each well (isolation faucets for instruments, measurement diaphragm, pressure regulator, pressure valves, horizontal subterraneous separators), wastewater storage tank, pumping station, natural gas metering and regulating station etc.

Figure 1 presents a general overview of the analysed group of natural gas wells, while figure 2 presents some individual equipments from the group of natural gas wells.

Figure 1. Overview of the analysed group of natural gas wells

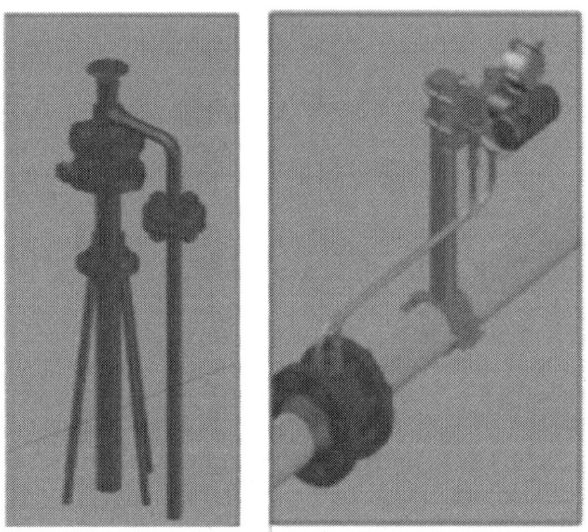

Figure 2. Individual equipments from the analysed group of wells

3. The general risk assessment methodology

In the following, the authors used for determining the explosion risk level within the analysed group of natural gas wells, a method based on the one elaborated by the National Institute for Research and Development for Environmental Protection Bucharest, Romania (I.N.C.D.P.M.). This method targets the quantitative determination of the risk/security level for a workplace, sector, section or plant, based on the systemic analysis and the assessment of the risks for accidents and professional illnesses.

The method implies identifying all explosion risk factors within the analysed system (or workplace) based on predetermined checklists and then quantifying the magnitude of the risk based on a combination of the gravity and frequency of the maximal predictable consequence. It is then possible to assess the security level for a workplace as being inversely proportional to the risk level.

In order to apply the method, several mandatory steps have to be followed:

- defining the analysed system (workplace);
- identifying the risk factors in the system;
- assessing the explosion risks;
- ranking the risks and determining the prevention priorities;
- proposing prevention measures.

In turn, these steps require the preparation of several important work instruments:

- a list for the identification of risk factors;
- a list of possible consequences of the action of risk factors on the human body;
- a scale for ranking the gravity and probability of consequences;
- a risk assessment grid;
- a risk level scale and a safety level scale, respectively;
- a workplace assessment form (centralising document);
- a list of proposed measures.

Based on the individual risk factors identified during these steps, there can be defined a global risk level for a workplace (N_r), as a weighted average of the risk levels determined for the risk factors. The rank of each risk factor has been used as weighing element (it being equal to the risk level), so as to make sure that the obtained result reflects reality as closely as possible.

The formula for calculating the global risk level is given in equation 1:

$$N_r = \frac{\sum_{i=1}^{n} r_i \cdot R_i}{\sum_{i=1}^{n} r_i} \qquad (1)$$

where:
N_r is the global risk level for a workplace;
r_i - the rank of the risk factor „i";
R_i - the risk level for the risk factor „i";
n - the number of risk factors identified for the considered workplace.

4. Assessing the explosion risk in the analysed for the group of natural gas wells

In order to identify the explosion risk factors for the natural gas wells, the authors analysed the tasks indicated in the job description files for the personnel unfolding its activity within this system, as well as the corresponding technological processes. The risk factors were identified and grouped by the generating element within the work system (operator, work task, production means and work environment - figure 4). In all, there were identified 58 risk factors. Based on this, the authors then determined the partial risk level for each risk factor. Taking these into account, there was then determined the global risk level, using the formula:

$$N_r = \frac{\sum_{i=1}^{58} r_i R_i}{\sum_{i=1}^{58} r_i}, \tag{2}$$

that, when using the corresponding numerical values, leads to:

$$= \frac{0\cdot(7\cdot 7)+0\cdot(6\cdot 6)+3\cdot(5\cdot 5)+6\cdot(4\cdot 4)+25\cdot(3\cdot 3)+16\cdot(2\cdot 2)+8\cdot(1\cdot 1)}{(0\cdot 7)+(0\cdot 6)+(3\cdot 5)+(6\cdot 4)+(25\cdot 3)+(16\cdot 2)+(8\cdot 1)} = \frac{468}{154} = 3.03$$

Thus, the global risk level calculated for the workplace "Group of natural gas wells operator" is 3.03, a value that places it among the workplaces with small to medium risk levels.

From figure 3, showing the partial risk levels for the risk factors at the considered group of natural gas wells, it can be noticed that of the total of 58 identified risk factors, 9 surpass, as partial risk level, the value of 3, thus being considered medium to high risk factors.

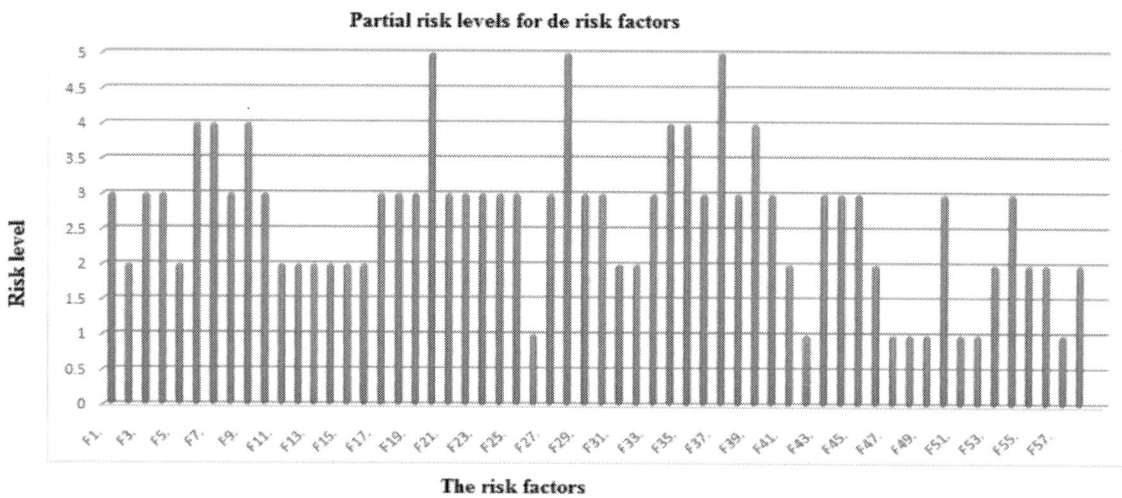

Figure 3. Partial risk levels for the risk factors at the group of natural gas wells

Figure 4 presents a distribution of the identified risk factors for the group of natural gas wells, by generating sources. Of the 58 risk factors:

- 25 factors are related to the operator (43%);
- 17 factors are related to the production means (29%);
- 8 factors are related to the work task (14%);
- 8 factors are related to the work environment (14%).

By determining the explosion risk level, it is possible to determine the priorities in prevention measures for the workplace, but also to optimise the usage of resources allocated for this purpose.

An advantage of the method elaborated at the I.N.C.D.P.M. Bucharest, Romania is the fact that its application is not limited by the condition of the physical existence of the system that needs to be assessed. It can be used in all stages related to the lifecycle of a work system or of an element of this work system: conceiving and design, physical creation, putting into production, unfolding of the work processes.

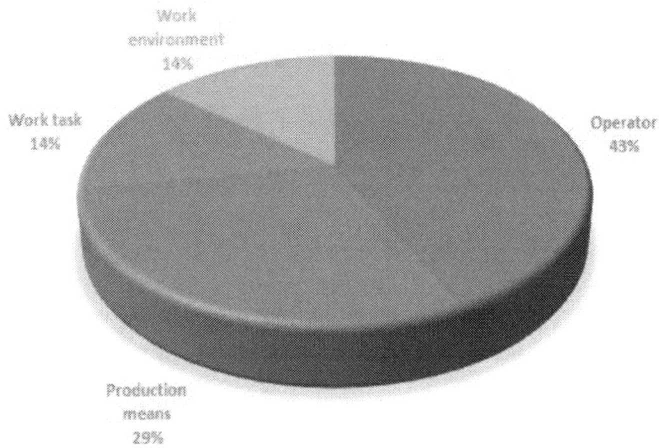

Figure 4. Distribution of the risk factors by generating sources

5. Indication of the explosion risk areas

An explosion risk area represents an industrial space in which, under normal functioning conditions, there can accumulate, permanently or accidentally, gases, vapours of flammable liquids, dusts or powders in sufficient amounts to give birth to an explosive atmosphere. Explosion hazard areas can be classified into several types, function of the length of the periods during which the explosive atmosphere is present. [5]

- area 0 - is the area in which the explosive atmosphere is present during the normal functioning of the technological installations, permanently or for a period of more than 1000 hours per year.
- area 1 - is the area in which the explosive atmosphere can occur during the normal functioning of the technological installations for intermittent periods with durations comprised between 10 and 1000 hours per year.
- area 2 - comprises the industrial areas within which an explosive atmosphere cannot occur during the normal functioning, but only in cases of predictable defects or for periods between 0.1 and 10 hours per year.

For the case of the analysed system (a group of natural gas wells), no area was identified that would be consistent with the requirements for inclusion in area 0.

The zoning of industrial spaces represents the process of classification of spaces within an industrial installation function of the probability of occurrence of an explosive atmosphere and is important for the subsequent correct selection of the electrical and mechanical equipment that will be used in that area. The electrical apparatuses for normal environments cannot be used, in general, in areas with explosion hazard due to the risk of the explosive atmosphere igniting due to electric sparks or arcs, hot surfaces of equipment, electrostatic discharges etc. [3, 4]

Therefore, at the execution of the electrical apparatuses and equipment for potentially explosive areas, there have to be applied certain supplemental constructive measures that would offer them safety during functioning and in the presence of the explosive atmosphere.

For the zoning of the group of natural gas wells, the authors have used the software package AutoCAD Plant 3D, that allowed to clearly emphasise the areas with explosion hazard, as shown in figure 5 and figure 6.

6. Conclusions

The current paper has presented the results of applying a tested, general assessment method to the specific work area of a group of natural gas wells.

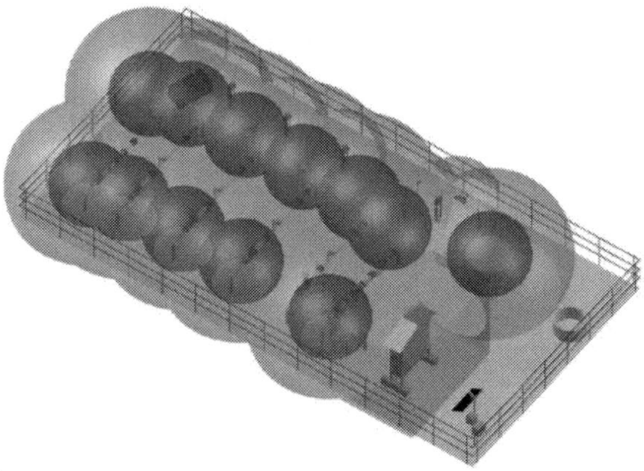

Figure 5. Zoning plan showing the physical location of areas with explosion hazard

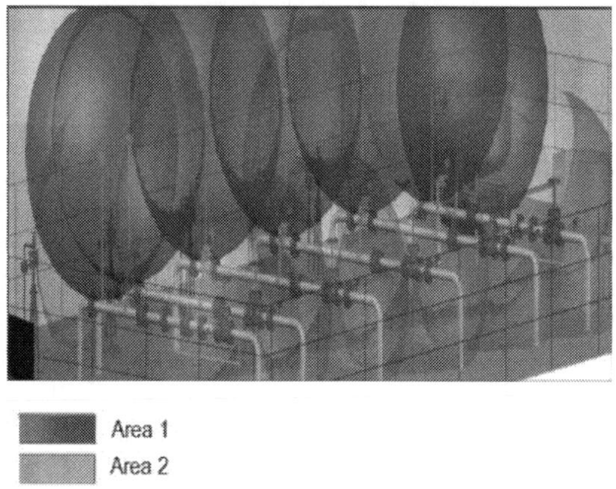

Figure 6. Detail of the zoning plan for the analysed group of natural gas wells

One of the main advantages of the method presented in this paper is that its application is not limited by the condition of a physical existence of the system to be assessed. It can be used in all stages related to the lifecycle of a work system or of an element of such a system: conceiving and design, physical creation, putting into production, unfolding of the work processes.

The application of the gravity/probability method for the analysed workplace has allowed the identification of explosion risk factors and to determine the dimension of risks for the occurrence of an explosion at this workplace. For this, for each identified risk factor there will be indicated technical and organisational measures that need to be taken in order to keep the risk level in the "area of acceptable risks".

After the division of the space into specific hazard areas, the electrical installations can be conceived either through the usage of electrical equipment certified for such areas, or, if no such equipment is available, through conceiving new equipment and then certifying it.

In future, the authors intend on the one hand to try to refine the method so as to better take into account all intervening factors and on the other hand to apply it to further similar systems.

References

[1] Nolan D P 2014 *Handbook of fire and explosion protection engineering principles for oil, gas, chemical and related facilities, 3rd edition*, Elsevier

[2] Cioca L I, Popescu Stelea M and Moraru R 2014 *Occupational safety and health risk assessment in Romanian surface gas extraction facilities*, Proceedings of the 14th International Multidisciplinary Scientific GeoConference SGEM 2014

[3] Moraru R I et al 2014 *Safety and health at work risk assessment in a compression station from the major Romanian natural gas company*, Proceedings of the 18th Conference on Environment and Mineral Processing, p. 37-42

[4] Jo Y-D and Crowl D A 2008 Individual risk analysis of high-pressure natural gas pipelines, *Journal of Loss Prevention in the Process Industries*, 21(8), p. 589-595

[5] Huidan A-S 2008 *Electrical equipment for automation and control in environments with explosion hazard* (in Romanian), Technical Publishing House, Bucharest, Romania

IOP Conf. Series: Materials Science and Engineering **161** (2016) 012090 doi:10.1088/1757-899X/161/1/012090

Size Differentiation Of A Continuous Stream Of Particles Using Acoustic Emissions

E Nsugbe,[1] A Starr,[2] P Foote,[3] C Ruiz-Carcel[4] and I Jennions[5]

School Of Aerospace Transport And Manufacturing Building 50, College Rd, Cranfield University ,Cranfield, MK43 0AL, United Kingdom.[1,2,3,4,5]

Email:e.nsugbe@cranfield.ac.uk

Abstract. Procter and Gamble (P&G) require an online system that can monitor the particle size distribution of their washing powder mixing process. This would enable the process to take a closed loop form which would enable process optimisation to take place in real time. Acoustic Emission (AE) was selected as the sensing method due to its non-invasive nature and primary sensitivity to frequencies which particle events emanate. This work details the results of the first experiment carried out in this research project. This experiment involved the use of AE to distinguish between the sizes of sieved polyethylene particle (53-250microns) and glass beads (150-600microns) which were dispensed on a target plate using a funnel. By conducting a threshold analysis of the impact peaks in the signal, the sizes of the particles could be distinguished and a signal feature was found which could be directly linked to the sizes of the particles.

1. Introduction and Problem Statement

Procter and Gamble (P&G) have certain quality standards that need to be upheld before their washing powder product can be sold. These checks comprise of a wet chemistry test to determine solubility properties, inspect the chemical properties of the powders and physical inspection to determine the particle size distribution. These methods are slow and reliant on human error and as a result this has given rise to the need for in-process monitoring sensors which would help apply a control loop to the process. This research will focus on the use Of Acoustic Emission (AE) sensors to estimate the size distribution of the powders online. Due to the size ranges of the particles being monitored (63-1000microns) and noisy production workspace, acoustic sensors in the ultrasonic region are used in this research as they are primarily sensitive to frequencies which particle events occur. [1]

2. Related Works Summary and Knowledge Gap

AE has been used previously to monitor the sizes of particles and in some case was used to estimate the overall flow rate of the particles.[1][2] Leach et al conducted pioneering studies using an ultrasonic microphone and a rotating drum, and he was able to establish a linear correlation between particle size and AE.[3] Buttle et al used a quantitative sizing technique that comprised of a deconvolution technique to estimate the sizes of particles falling on a target plate due to the deconvolution required the technique would struggle to work in real-time.[4] Ivantsiv et al designed a model capable of estimating particle size and mass flow rate, although his system was designed to work offline. [1] Papp et al investigated offline the relationship between the changes in size of particles with the change in

AE occurring during a granulation process. Her results showed that an inverse relationship existed between the AE frequency and particle size. [5] Neural networks were used by Bastari et al and Chen et al to classify different sized particles online, despite being able to perform a classification on the particles no indication of the physical behaviour of the process can be deduced with this system. [6][7] Hu et al designed an online model that linked the AE output voltage to particle size, his model was unable to detect particles smaller than 90microns and this was attributed to the overall signal noise in the process. [8] Ren et al used the wavelet transform to estimate the respective energies of differently sized particles and used this approach to estimate the mass fraction of the various particles in a mixture. [9] Although using Ren's model to estimate the size distribution of a mixture with different particle sizes, would lead to a very large and complex algorithm thereby slowing online computation time.

Comparing this signal processing architecture to the existing systems in literature, Bastari et al and Chen et al designed online particle classification systems using a data driven approach, this method would not suffice in this research problem as further information about the process behaviour cannot be inferred from a data driven system.[6][7] In terms of a more hybrid signal processing approach, Ren used the wavelet transform to identify particle size distribution online, but due to the algorithm complexity, a high model computation time would be required which would not be acceptable for a process which is aimed at being optimised online.[9]

In this research we would be looking to design for the first time, a hybrid signal processing model capable of identifying particle size distribution online. The work carried out in this report details the first experiment taken to design an online particle size distribution estimation model.

3. Research Methodology

The design space approach is being employed as the research methodology as it helps in the understanding of how a product behaves under various process conditions.[10] A designs space approach is a useful tool used to help comprehend the interaction between input variables and process parameters, and how they influence final product quality. [10] It is also useful in developing a versatile process which can comfortably deal with variability in input materials and process parameters. [10] A supporting statistical method used in conjunction with the design space approach is the Design Of Experiment (DOE).[12] Supporting the experiments with statistics helps in tracing root causes of variability and also distinctions between causal and correlative relationships. [11]This research problem would be investigated with the Design Space Approach and supported with the DOE technique.

4. Experimental Details

The washing powder compound comprises of an array of different particles types, so in order to simplify the problem a single experimental particle type was chosen and sieved into different seizes. This experiment detailed here involved the differentiation of particles of various sizes using their AE signal. This experiment would serve as the first stage in helping to solve our research problem of identifying the size distribution of a compound that comprises different sized particles. For this experiment, Medium Density Polyethylene and acid washed glass beads were used as the experimental particles.

4.1. AE Sensors

The sensor used was the PCI-2 Physical Acoustic sensors by Mistras. The sensor bandwidth spans from 100K-1MHz and a sampling rate of 1Ms was used during the acquisition. Figure 1 shows the block diagram of the signal processing chain. The acquired signal goes through a preamplifier before being sent through a band pass filter. This filtered signal is sent to the Analogue to Digital Converter where this signal is digitised and passed to a Field Programmable Gate Array(FPGA) where sampling and signal averaging takes place. The final digitised waveform is continuously streamed to the hard-disk of the computer. [13]

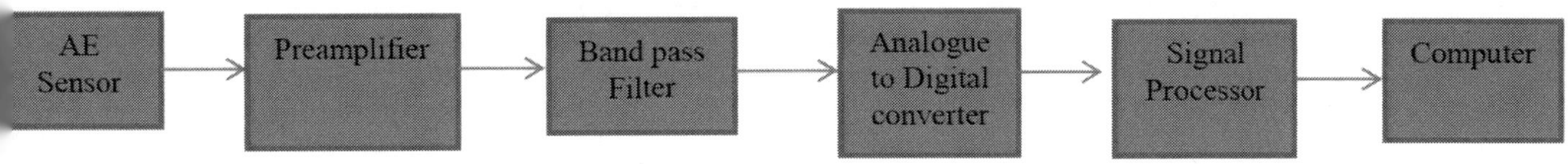

Figure 1: Block Diagram Of The Signal Processing Chain

4.2. Experimental Setup

Figure 2 shows an image of the experimental setup used here. It comprises a funnel of dimensions 8x11x18mm^3 and an Aluminium sheet of 0.7mm thickness with the sensor attached at the back of the plate using beeswax adhesive as this ensures a secure but gentle coupling between sensor and surface. A funnel was selected as part of the experimental setup in to ensure repeatability in the dispensation rate of the powders. The funnel exit was blocked using a finger while a measured mass of powder was poured into it. Then the finger was removed and the full powder mass was dispensed on the target plate through the funnel and the acoustic emissions were recorded. Assuming there are no sources of interferences,(through humidity, temperature and electromagnetic sources) by having a repeatable source, maintaining a constant height, and keeping the plate thickness constant throughout the experimentation, the variable being measured becomes the mass/size of the particles of interest. [2]

Figure 2: Final Experimental Setup

4.3. Powder Sieving

To further reduce experimental variability, the polyethylene powders were sieved into narrow size distribution bands by means of dry sieving as shown in Table 1 using an Endecotts layer sieve with an aperture size range from 53-1000microns. The glass beads came in pre-sieved classes so further sieving was not required for the particles.

Table 1: Particle Class Information.

Particle Type	Particle Class	Size Distribution	Average Mass In Grams(g)	Bulk Density In g/cm^3
Polyethylene	Class1	151-250microns	0.0047	0.3435
	Class2	126-150microns	0.00053	0.2522
	Class3	64-125microns	Particles could not be separated	0.2158

	Class4	53-63microns	Particles could not be separated	0.2113
Glass Beads	Class1	150-212microns	0.0049	1.5850
	Class2	425-600microns	0.1578	1.5938
	Class3	212-300microns	0.0091	1.5910

4.4. Environmental Scanning Electron Microscope Images
To produce the size distribution of each particle class, 10mg was taken from each class and viewed under the electron microscope .This was repeated six times and each time five particles were isolated and their dimensions(length and width) were measured.

4.4.1. Polyethylene Particles
Figures 3 show Environmental Scanning Electron Microscope (ESEM) image of the class 4 polyethylene particles. It can be observed that the polyethylene particles are highly irregular shaped.

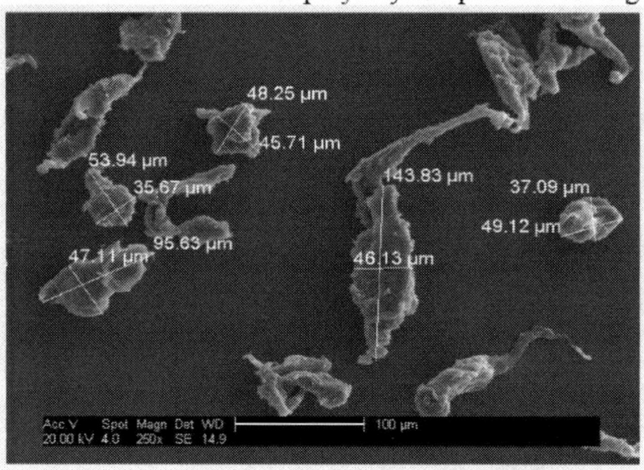

Figure 3: Class 4 Particles (53-63microns)

Due to the apertures of the sieves being square and the particles being irregular, a generalised mean was calculated by way of the Root Mean Squared (RMS) to the acquired dimensions to produce the histogram in figure 4.

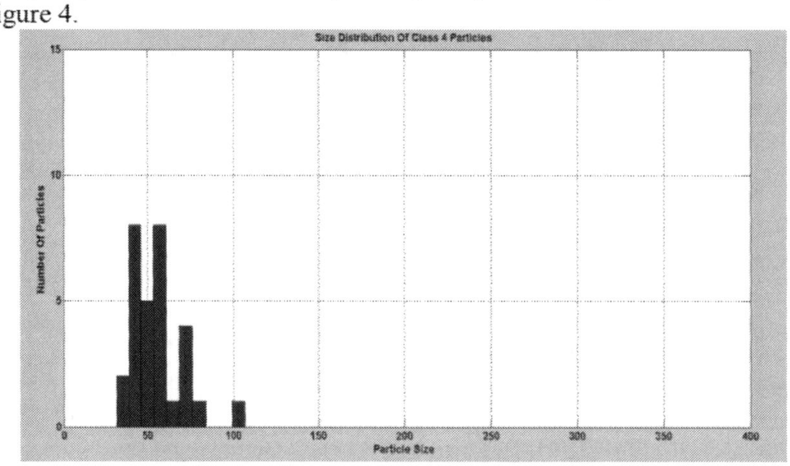

Figure 4: Size Distribution Of Class 4 Particles

4.4.2. Glass Beads

The ESEM image of the class1 glass beads can be seen in figures 5. And unlike the PE particles the glass beads have a tighter distribution and a spherical geometry.

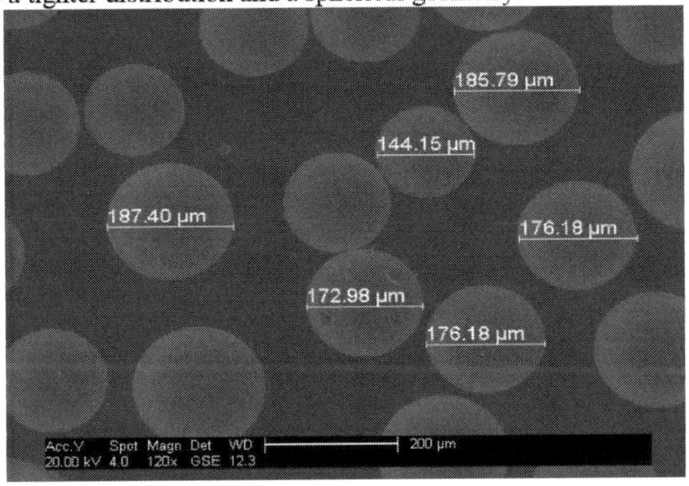

Figure 5: Class1 Glass Beads (150-212 microns)

A respective histograms of the glass bead particles can be seen in figures 6, and in comparison to the PE particles the distribution of the glass beads are within their sieved bands thereby limiting the number of outliers present.

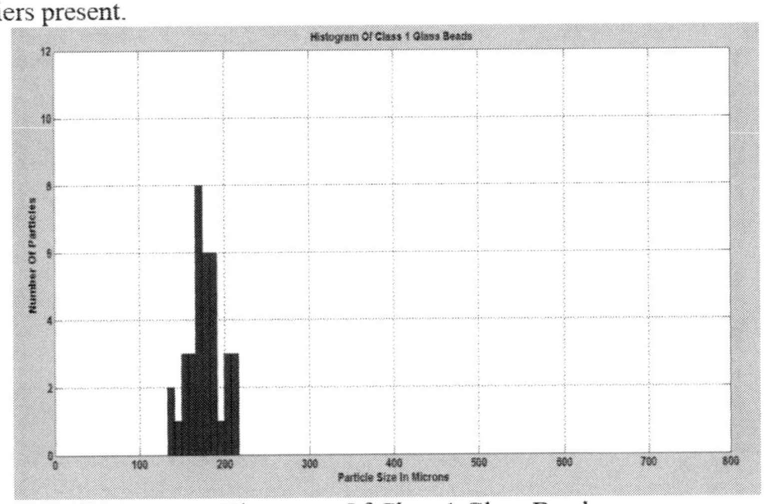

Figure 6: Histogram Of Class 1 Glass Beads

5. Experimental and Data Analysis Procedure

For each particle group, 2g was weighed and dispensed into the funnel while the exit was blocked using a finger as the measured mass of powder was poured into it. This procedure was repeated four times for each particle class. The acquired data was analysed using a thresholding method. This method was used by Hu et al to differentiate particle sizes and also by Ivanstiv et al to not only differentiate particle sizes but also measure flow rate, the thresholding method was chosen due to its simplistic and effective nature [1] [8]

The thresholding method works with the signal shaping chain which can be seen in figure 7, it states that a linear relationship exists between particle size and output sensor voltage. The signal shaping chain was from the pioneering work carried out by Leach et al and validated by Buttle et al. [3] [4] It shows that a convolution between the source function, wave propagation function and instrument response function yields the AE Voltage Signal.

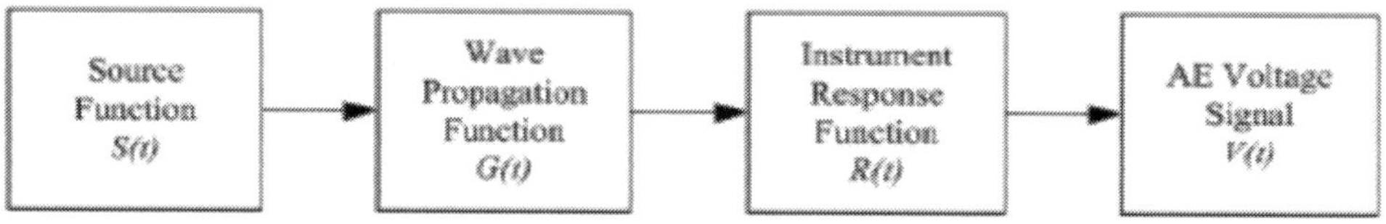

Figure 7: Signal Shaping Chain [8]

When a particle of a certain size drops on a target medium, it gives off an impulse which is related to the size of the impinging particle. This impulse is represented by a peak in the time domain and with a suitable threshold capable of detecting these peaks; particles can be identified and differentiated using their impact peaks.

5.1.1. Designed Threshold Method and Particle Sizing Approach

A varying threshold method was implemented and details of each step can be seen below;

-Step1: Take absolute values of signal to eliminate non-negative signal values.

-Step2: Identify maximum peak from experimental repetitions of the same size particles.

-Step3: Reduce threshold by a factor of 10% each time and take arithmetic mean of peaks above the threshold level.

-Step4: Calculate linear correlation co-efficient for each threshold level (e.g 90%,80%)

-Step5: Select threshold which provides best linear correlation for data

With the dispensation of the particles being through a funnel, this caused the amplitude of the acquired AE signal to vary, this can be seen by the highly varying amplitude of the AE signal in figure 8. So for the data analysis procedure, 20,000 data points were extracted from a section of the data where the flow rate was maximum and constant.

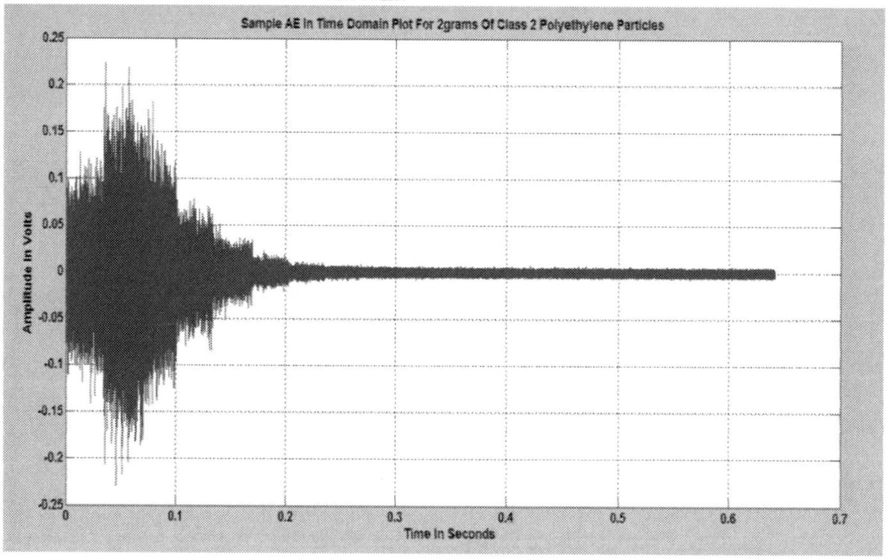

Figure 8: Acoustic Emission Plot Of Class 2 Polyethylene Particles

The plot in figure 9 shows a simple example of an isolated constant segment of 20,000 data points along with an example of how a 60% threshold would work.

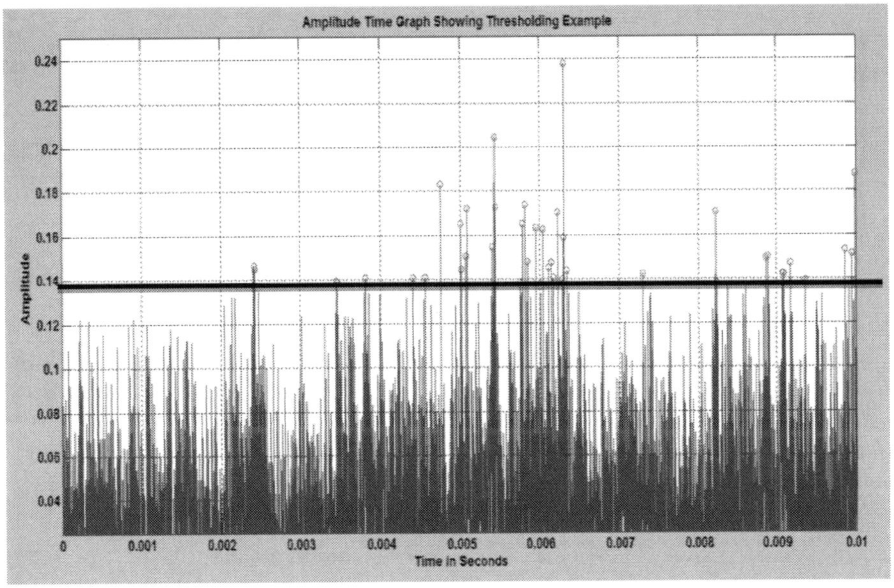

Figure 9: Amplitude-time Graph with a threshold level of 60% (peaks above 0.138)

6. Results

6.1. Polyethylene results

Once the data sections were isolated, the threshold method was applied and the linear correlation coefficient was calculated for each threshold level. The thresholding exercise began with a 90% threshold this reduced by a factor of 10 each time and stopped at the 10% threshold. The threshold varying stopped at the 10% level because below this threshold we would enter the sensor noise region and in order to ensure the best data quality, the selected threshold should be selected to be greater than the noise level. Although for this simple experimental setup, setting the threshold to include the noise would still give a good correlation (greater than 0.90) due to the negligible noise contribution and high signal to noise ratio. But in a more practical rig setup where there may be greater noise contribution, the selected threshold would need to be above the noise level in order to ensure maximum signal to noise ratio.

Figures 10 and 11 show correlation figures of particle size against mean of threshold peaks for 90% and 10% threshold levels, with their respective correlation co-efficient being 0.87 and 0.97. From the correlation co-efficient score, it can be seen that the deepest threshold level(10%) provides the best linear correlation. A possible reason for this could be that the linear correlation increased with the higher number of data points considered.

An explanation for the outliers at the zero mark on the x-axis in figure 10 is due to the maximum peak identification method. For each particle class, the experiment was repeated four times. The maximum possible peak value was identified by inspecting all four repetitions of the experiment, this meant for the high threshold there was the possibility of the particle sizing algorithm to identify zero peaks for some experimental data set.

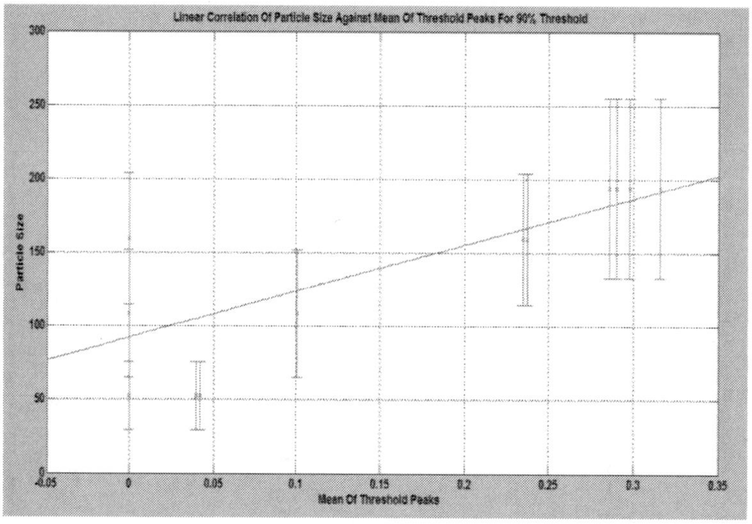

Figure 10: Linear Correlation of Particle Size against Mean of Threshold Peaks For 90% Threshold (correlation factor 0.87)

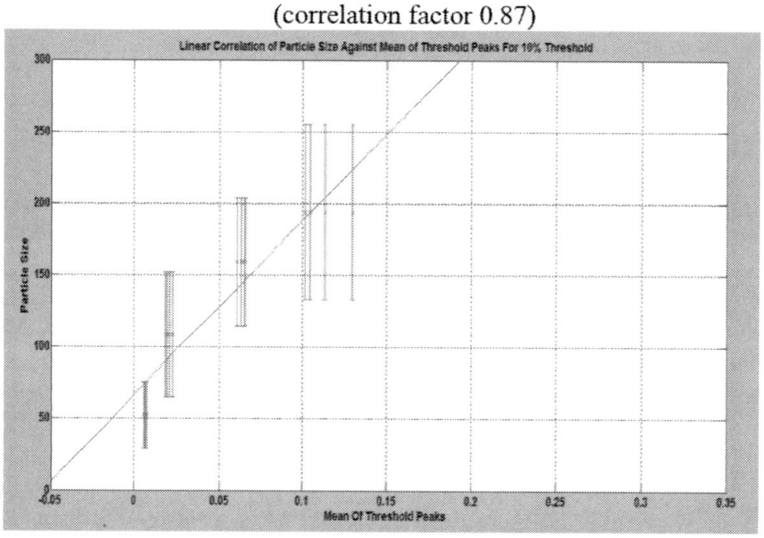

Figure 11: Linear Correlation of Particle Size against Mean of Threshold Peaks For 10% Threshold (correlation factor 0.97)

It can be noted that the mean of threshold peaks showed some variability for the bigger particles, this was due to the wider distribution in the particle class due to sieve sizes. For example, the particles in class 4 ranged from 54-63microns accounting for a variation of 9microns. Whereas the class 1 particles ranged from 151-250microns,which equals a variation of 99microns.

The plot in figure 12 was obtained using the 16 collected data points and it illustrates how the linear correlation between the particle size and mean of threshold peaks increases with the lowering of the threshold level.

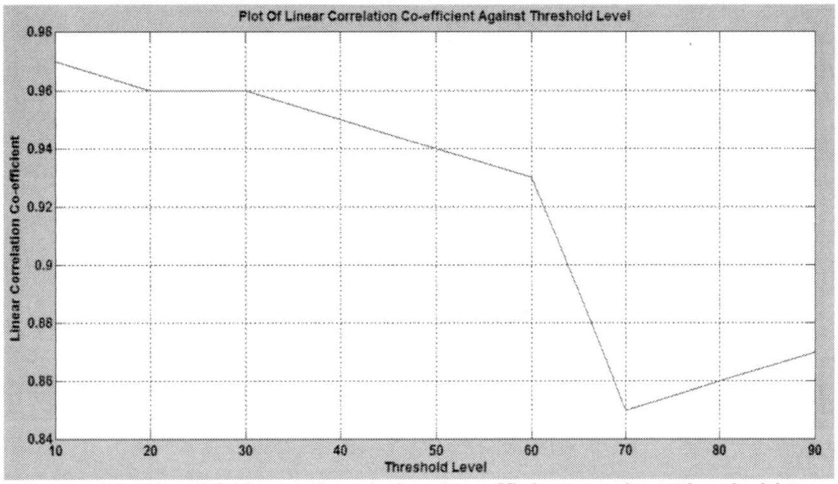

Figure 12: Plot Of Linear Correlation Co-efficient Against Threshold Level

6.2. Glass Beads Results

The threshold analysis was tested with glass beads as validation that the technique can work for more than just one particle type, as the washing powder compound comprises of different types of particles, so it was important to make sure that this technique was not unique to just one particle type. Figure 13 shows a sample AE plot for class 2 glass beads and unlike the polyethylene, the glass beads produced more constant amplitude when the flow was constant.

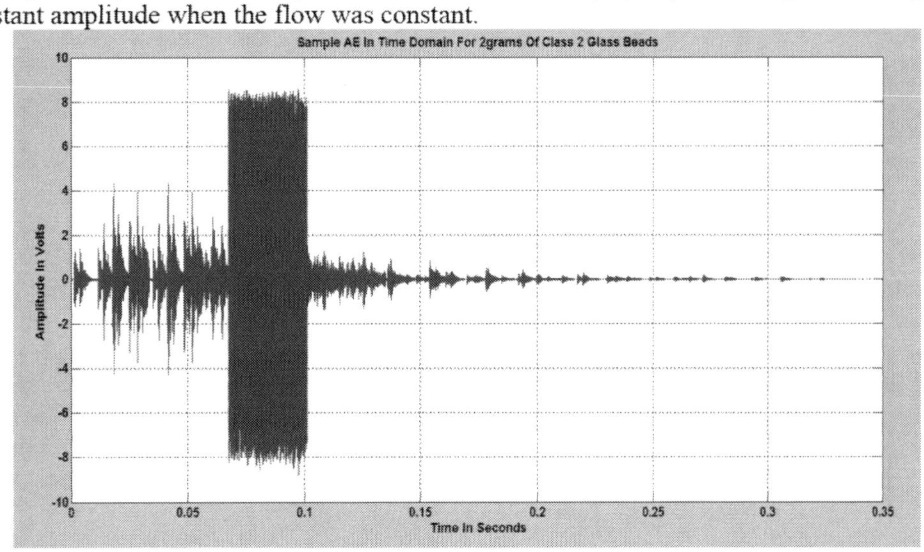

Figure 13: Sample AE In Time Domain Plot For 2g Of Class 2 Glass Beads

The glass beads sizes were classified using the same signal processing method used for the polyethylene and with a 10% of maximum threshold. For each class of glass bead, the dispensation exercise was repeated 5 times each giving a total of 15 data points. For all 5 experimental runs from each particle class and with a 10%, the AE amplitudes were averaged to obtain a representative value for each particle group. A correlation between these values and the median particle size of each class of glass bead was calculated and can be seen in figure 14.

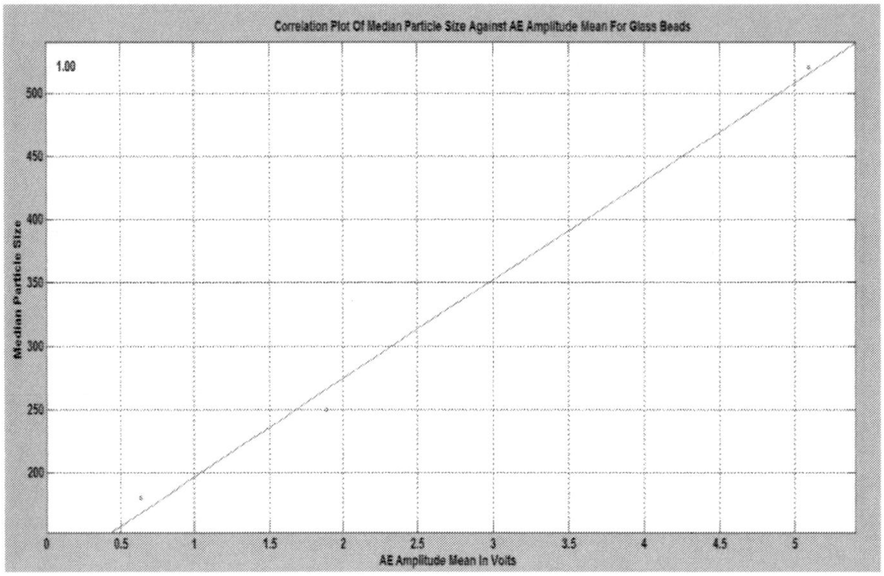

Figure 14: Correlation Plot Of Median Particle Size Against AE Amplitude Mean For Glass Beads(R=1)

From figure 14, it can be seen that a good correlation exist between the median particle size and AE amplitude mean. This would suggest that the signal processing technique can be used to effectively differentiate between particles of various sizes under the same condition used in this experimental setup.

7. Conclusion

From the literature, enough evidence exists to support the assumption that particles sizes can be monitored online. The experiment carried out in this work was aimed at the determination of the particle sizes from their acoustic emissions using sieved sets of polyethylene powder and glass beads. The objective was to successfully differentiate particles of different sizes using their AE and to find a signal feature which was could be directly correlated with particle size. In this experiment, a known mass of particles were dispensed into a funnel which released a continuous stream of particle unto a target plate which an acoustic emission sensor was placed on. By implementing a threshold and evaluating the amplitude of the peaks on each threshold, the particle sizes were distinguishable and the amplitude mean of the threshold were able to be correlated to the various particle sizes. A linear correlation was seen to exist between the AE and particle size, thereby validating the results obtained in previous literature. The next experiment would now involve the mixing of particles of various sizes to see how they interact when in a mixture.

Acknowledgement

The authors gratefully acknowledge the support of partners Procter & Gamble, IIT, Ajax, Centre for Process Innovation, the University of Birmingham, the University of Leeds, and the University of Durham. This research is part of Chariot and The Advance Manufacturing Supply Chain Initiative (AMSCII) which is a government supply chain fund which is helping to rebuild British manufacturing prowess.

References

[1] Ivantsiv V, Spelt J and Papini M 2009 Mass flow rate measurement in abrasive jets using acoustic emission, *Meas. Sci. Technol.* **20** 095402.

[2] Pecorari C 2013 Characterizing particle flow by acoustic emission, *Journal of Non-destructive Evaluation*, vol. **32**, no. 1, pp. 104-11, .

[3] Leach M, Rubin G, and Williams J 1977 Particle size determination from acoustic emissions, *Powder Technology* **16** 153–58

[4] Buttle D, Martin S and Scruby C 1991 Particle sizing by quantitative acoustic emission, *Research in Non-destructive Evaluation*, vol. **3**, no. 1, pp. 1-26.

[5] Papp M, Pujara C and Pinal R, 2008 Monitoring of high-shear granulation using acoustic emission: predicting granule properties. *J Pharm Innov*. **3**:113–22

[6] Bastari A, Cristalli C, Morlacchi R and Pomponi E 2011 Acoustic emissions for particle sizing of powders through signal processing techniques *Mechanical Systems and Signal Processing*, vol. **25**, no. 3, pp. 901-16.

[7] Chen X and Chen D, 2008. Measuring average particle size for fluidized bed reactors by employing acoustic emission signals and neural networks. *Chem. Eng. Technol.* **31**, 95–102

[8] Hu Y, Huang X, Qian X, Gao L, and Yan Y, 2014 Online particle size measurement through acoustic emission detection and signal analysis, in *Proc. IEEE Int. Instrum. Meas. Technol. Conf.*, pp. 949–53

[9] Ren C, Wang J, Song D, Jiang B, Liao Z, Yang W 2011 Determination of particle size distribution by multi-scale analysis of acoustic emission signals in gas-solid fluidized bed. *J. Zhejiang Univ., Sci.,* **12** (4), 260–67

[10] Box G, Hunter W and Hunter J 1978. Statistics for experimenters: an introduction to design, data analysis and model building. New York: John Wiley & Sons Inc.

[11] Hansuld E and Briens L 2014 *International Journal of Pharmaceutics* **472** 192–201 193

[12] Soravia S and Orth A 2009. *Design of experiments*. Ulmann's encyclopedia of

[13] industrial chemistry. Retrieved from onlinelibrary.wiley.com.

[14] *PCI-2 Based AE System User's Manual.*

An overview of bird related issues in electrical power systems

Ö Polat[1], K Yumak[1], N E Atilla[2] and M Bağrıyanık[3]

[1] ÅF Consult, ODTU Teknokent Met Alani, Mustafa Kemal Mah. Dumlupinar Blv. No: 208, D Blok No: 3, 06510, Cankaya, Ankara, Turkey
[2] Trakya EDCo., 100. Yıl Mahallesi, Barboros Caddesi No: 24 Tekirdag, Turkey
[3] Istanbul Technical University, Electrical Engineering Dept., Sariyer, Istanbul, Turkey

E-mail: onder.polat@afconsult.com

Abstract. Power system equipments, especially overhead lines and utility poles are posing great risk to avian species. Main purpose of this paper is to draw attention of distribution companies and community on conserving bird species and their habitats. The paper also provides regulatory framework about protecting bird species, impacts of birds on utility equipments and remedial actions that can be taken for problem specific cases. Furthermore, a sector review of industrial applications and effectiveness of mitigation measures are given. Distribution companies may apply a systematic approach for reducing electrocutions and collisions by developing and implementing Avian Protection Plans (APPs).

1. Introduction

One of the most crucial problems in distribution networks is the bird related outages although they are generally underestimated. Main factors that cause power outages are investigated in Kansas, Manhattan for a five year period (1998-2002) and results show that %19 of the outages is directly related with the animals [1]. Also IEEE has conducted a survey among 114 utility companies and %86 of respondents stated "birds" as the primary source of animal-intrusion outages [2]. Replacement, repair and maintenance cost of damaged equipments, unserved energy and production loss of consumers forces utilities to take proper actions on bird related outages. According to the calculations on economic impacts of wildlife – power line interactions in California, annual cost of animal-related outages is estimated between \$32 million - \$317 million [3].

Main causes of bird mortalities in power lines can be classified into two categories as electrocutions and collisions [4]. Electrocutions occur when bird touches two different phases of power line or one phase and grounded part, whereas birds not noticing wires and flying directly to them is defined as collision. A quantitative review about bird mortalities indicated that each year 12-64 million birds die at US power lines. Electrocution mortalities is roughly estimated between 0.9-11.6 million and collision victims is approximately in the interval of 8-57 millions [5].

Bird mortalities at power lines has been identified as a problematic issue since 1970s. In each following year this issue drew more attention of utilities, wildlife protection organisations and public. Edison Electric, US Fish and Wildlife Services (USFWS), National Audubon Society and 10 utility companies formed Avian Power Line Interaction Committee (APLIC) to define bird related outages and design mitigation techniques in 1989 [6]. APLIC documents [4,6,7] helped many utility companies create their own Avian Protection Plans (APPs) to minimize bird mortalities thus enhancing system reliability. In addition, IEEE has published two standards; the one is a guideline about reducing bird-related outages [8] and the other one describes animal related problems in electric power supply substations and explains preventive methods [2].

Moreover, utility equipments are posing risks to endangered species. This issue raises an environmental concern on public and governments. Laws and regulations are enacted to provide protection of migratory and endangered species. For this reason, utility companies are obliged to take corrective actions to avoid punishments and fines.

Having considered environmental and economic benefits, utility companies are encouraged to improve a systematic approach to mitigate bird related outages. Problem solving process starts with fully understanding nature of the problem and determining root causes. Morphological and behavioural inspections on birds must be carried out to choose optimum solution methodology. Energized equipments will be less attractive and dangerous for birds undertaking a series of activities:

- Implementing avian outages reporting system
- Analysis of bird species in researched area
- Looking up for bird related outage mitigation products
- Avian safe designs of utility constructions and equipments
- Scheduled inspection programs focusing on avian safe and retrofitted equipments

In a realistic point of view, there exists no practical solution to keep the birds out from utility constructions permanently because birds are incredibly productive to meet their needs in an exceptional range of area. Intelligent, observant and adaptive capabilities of birds separates them from all other outage cause phenomenas like lightings or switching surges. Therefore, abilities of birds should be taken into account when developing avian safe solution strategies.

In this study, preliminary studies of the R&D project "Bird-related outages in power systems" in Trakya EDCo. are given in a comprehensive manner. The project is financially supported by Turkey Energy Market Regulatory Authority (EMRA). The interaction between birds and power lines are evaluated through regulatory framework, bird-related issues, mitigation techniques and effectiveness of utility applications.

2. Regulatory Framework

Industrialization, global warming, rapidly growing consumption of natural resources and over-hunting diminish population of birds and other species each passing day. Governments regulate their legal system to limit man-kind activities which have negative impacts on the diversity of bird species. However, domestic laws seem inadequate to protect migratory birds because they fly over long distances passing through many countries. Thus, regional and international scaled conventions have signed with the cooperation of governments.

2.1. Migratory Bird Treaty Act (MBTA - 1918)

MBTA is an essential legal basis for conservation and protection of North American migratory birds. It is forbidden to take, hunt, possess, kill, sell, buy, import or export any kind protected birds, their parts, nests or eggs [9]. MBTA has a strict liability statue and has authority to fine individuals (person, corporation, organisation etc.) up to $15.000 and/or up to six months imprisonment for misdemeanour conviction. If the law is knowingly violated then fine raises up to $250.000 and/or up to two year imprisonment can be possible. Moon Lake Electric Association was the first utility company sentenced guilty in the context of MBTA for causing death of 38 raptors and fined $100.000 [10]. After the punishment Moon Lake has agreed to sign MoU and taking necessary steps to prevent bird deaths.

2.2. Bald and Golden Eagle Protection Act (BGEPA – 1940)

BPEGA provides an additional protection to North American eagle species. Utility companies are obliged to meet requirements of BGEPA in the roosting and feeding areas of eagles prior to facilitate any construction activities. Individuals harming eagle species, nest or eggs could be fined up to $100.000 and/or up to one year imprisonment. If the crime is intentional penalties reach $250.000 and/or up to two years imprisonment [11].

2.3. The Endangered Species Act (ESA – 1973)
Primary objective of ESA is to conserve native flora and fauna species that are in danger of extinction or threatened [12]. ESA's conservation framework also covers habitats of animals and plants.

2.4. International Regulatory Framework on Bird Protection
Apart from the US regulations, birds living in various countries are preserved under legislative work. Regulations includes both protection of birds and their habitat needs. Following agreements are signed in this scope:

- The Birds Directive (1979): European Union (EU) countries published this directive to protect all endangered or threatened species [13]. Directive draws a general outline about how to manage, protect and control the bird population. Any intentional killing, hurting, removal of nests, taking eggs and give disturbance to birds are prohibited. Contracting parties should allocate Special Protection Areas (SPAs) for birds according to this directive.
- The Habitats Directive (1992): EU enacts this directive as a complementary legal support to Birds Directive. It requires Special Area of Conservation (SCAs) sites from member states for species apart from birds. Designated SPAs and SCAs comprise Natura 2000 Network [14].
- Berne Convention (1979): It was the first product generated by international cooperation to ensure conservation of flora and fauna species. Articles of agreement are quite similar with the Birds Directive. Convention principles are applied over a large geographical area because contracting parties are all over the world.
- Convention on International Trade in Endangered Species (CITES - 1975): Specimen of wild animals and plant are separately listed according to vulnerability status. International trade rules (import, export of dead and live animals etc.) of species are regulated with this agreement [15].
- Bonn Convention (1983): It provides a conservation framework in which contracting parties may act to protect migratory birds and their habitats. Birds are categorised in two appendices either as endangered or threatened. Endangered birds are fully protected with strict rules.
- Convention on Biological Diversity (1992): Contracting parties prepared a legal instrument to sustain biological diversity and key components creating diversity and share of genetic resources [16]. It encourages parties to develop projects that will help conserve biological resources. Processes and activities that has possible negative impacts on diversity will be examined via conducting scientific researches.
- Ramsar Convention (1971): Ramsar is an international treaty promising preservation of wetlands with the aid of decisive domestic sanctions and international cooperation. Contracting parties designate at least one ecological sensitive site specifically includes waterbirds.

Turkey has signed following international agreements to conserve bird species and habitats located inside national borders:

- Berne Convention
- CITES
- Convention on Biological Diversity
- Ramsar Convention

3. Bird Related Issues and Mitigation Techniques
Various bird species cause various types of problems in distribution networks. While birds having large wingspans contact directly to phases resulting a short circuit, large flocks of small birds gather on poles and may create both contamination issues and bridging gaps along energized wires. Common bird related problems are as follows:

- Electrocution Outages
- Bird Collision
- Streamer Outages

- Bird Contamination Issues
- Nesting Problems
- Woodpecker Damage

3.1. Electrocution Outages

Electrocutions occur when a bird simultaneously touches two energized conductors with phase differences or one phase and a grounded equipment. There are several environmental factors contributing to increase electrocution susceptibility:

- Bird species: Large birds such as eagles, hawks and storks tend more to be electrocuted due to their high wrist-wrist distance.
- Age: Juvenile birds lacking of flight control are susceptible to electrocutions
- Weather Conditions: Dry feather provides insulation up to 70 kV whereas wet feather can only withstand 5-7 kV.
- Seasonal Fluctuation: Utility poles are valuable sit and wait places to look for prey in winter season. In spring time, birds use utility poles to nest and birds rest under the shade of crossarms in summer.

Utility pole configuration extremely impacts electrocution risk beside from environmental factors. Corner poles, double dead end poles and the poles containing energized equipments such as transformer, capacitor or arrester are relatively more dangerous. Bird mortalities due to electrocution can be identified if some of indicators like burn spots, open wounds, burnt feathers and damaged claws are found on carcasses.

General mitigation techniques for electrocution phenomena are summarized below:

- Reconfiguration: Utility pole are redesigned in a manner that it will be avian safe. APLIC recommends 150 cm phase separation for eagle protection. Steel poles with steel crossarms should be avoided because touching only phase will possibly cause birds to be electrocuted. Instead, fiberglass crossarm or wooden pole could be preferred if possible. Also suspension insulators provides a safer platform for birds to roost and decreases electrocution risk.
- Retrofitting with Insulation: Hazardous contact points of utility equipments are coated with insulation materials when adequate separation or other alternative reconfiguration techniques are not achieved. Bushing covers for transformers, arresters or capacitors and insulation of wires are the different kinds of this method.
- Perch Management: Triangle anti perch devices, plastic and metallic spikes and derivatives of these equipments are located on crossarms to prevent birds from roosting. However, if equipments are not placed properly, bird electrocution risk will increase dramatically. Besides, this method only shifts existing problems to other poles of utility so it should be the last choice.

3.2. Bird Collisions

Collisions generally emerge in areas where roosting, nesting and feeding activities of birds are frequent. If overhead lines lie across between feeding and roosting sites then numerous routine flights enhance possibility of collision risk. The thin overhead ground wires are hardly noticed by birds and, mostly birds could not find enough time to change flight direction resulting a collision. Following methods are used to minimize collision events:

- Line orientation: Parallel lines can be clustered so that birds can flyover risky area with just one ascend. Also utility poles may be situated nearby ridges (treelines, hills etc.) which has more visibility for attracting attentions of birds.
- Habitat modification: Habitats of bird species are reorganized to be safer against utility pole threats. Some of the possible measures are; trees having same or higher altitude with the utility poles may be planted, power lines are situated so as not to divide roosting and feeding areas of birds and factors that causes stress in birds are eliminated (shooting, speeding etc.).

- Removing shield wire: Occurrence of collision possibility is the highest on overhead ground or shield wire since small radius wires are hardly noticeable for birds. Hence, removing shield wire will dramatically reduce collision events. However, shield wires protects the system against lightning strokes so this application may not be feasible where lighting strokes are intensive.
- Burying power lines: Although burying lines is the most precise solution, technical and economic constraints may hamper implantation of this method.
- Power line marking devices: Birds can react earlier if they notice power line before coming too close. For this purpose there exists some devices to make lines thicker, colourful and somehow more distinguishable. Main devices are in use are; aerial marker spheres, spiral vibration dampers, flight diverters and bird flappers.

3.3. Nesting Issues
Birds prefer utility poles to mate and raise their youngs in the absence of natural nesting sites. Nest size and position on the poles differ with respect to specimen of the birds. White stork (Ciconia ciconia) is one of the most important stick nest builder on transmission and distribution poles predominantly in Europe. A white stork nested on low voltage tangent pole in Trakya region is shown in figure 1.

Figure 1. Ciconia ciconia nest.

Nesting create outages in various types of ways:

- Birds drop materials and preys which may contact energized equipments when they take off and land to their nests,
- Nests attract predators and animals climbing over top of utility poles,
- Bird feces accumulating on insulators cause flashovers

It has been recorded that monk parakeet nests caused 198 outages and as a result 10.000 consumers are left unsupplied in South Florida in 2001 [17].

Removing bird nests generally do not provide a permanent solution because most species have instincts to conserve its habitat and they will reconstruct in place of removed area. Moreover, companies can be confronted with large fines unless nest removal permits are taken. Applicable solution methods to overcome nesting issues are given below:

- Installing alternative nest poles,
- Artificial nesting platforms on top of utility poles ensuring necessary electrical safety clearances
- Perching discourages such as triangle anti perch devices, plastic spikes or micro electrical shock systems,
- Placing stick deflectors to bounces off nest materials.

3.4. Streamer Outages

One of the experienced problems with raptors is the streamer outages. Raptors perched on utility poles commonly drops feces before taking off. Feces are mixed with highly conductive uric acid which could partially or completely bridges the distance between conductors. This leads flashovers ending up serious bird injuries or mortalities and outages. Classifying an outage as "streamer" do not always reflect the reality since detection of this kind of problem is not easy. Some of the following clues can be tracked down for investigations of streamer outages:

- Existence of large birds in the region such as eagles, herons and ospreys.
- Outages recorded in 23:00 – 06:00 times of interval [18].
- Food-rich areas like agricultural fields or wetlands.
- Successful reclosing actions after overcurrent relay trip.
- Insulators having burn spots on a few disks of strings which are close to live ends.

134 streamer outages were recorded in China's transmission system in 2004 and each year estimated financial loss is about millions of dollars [19]. Perching discouragers are widely preferred to diminish streamer outages whereas insulator shields or alternative tower configurations are other applicable choices.

3.5. Bird Contamination

Birds perching on utility equipments leave their droppings onto insulators and accumulation of excrements lower surface resistivity of insulators which can cause outages. Highly conductive humid feces create arc-flash hazards even in nominal voltage ratings. On the contrary of streamer outage that can only be induced by large bodied birds, bird contamination issue could be generated by both large bodied birds and large flocks of small birds.

Perching discourages are not always the best option to avoid from contamination issue since discourages made up for large bodied birds may not be effective against small birds and vica versa. Therefore, insulator shields presents a more generic solution compared to discouragers. Additional shield weight attached to structure must be calculated and periodic cleaning of shield needs to be performed. Sudden flashovers related to bird contamination typically appear in short or long term. Hence, periodic cleaning of insulators are beneficial to maintain insulation resistance. IEEE Std. 957-2005 [20] gives methodology regarding to maintenance of ceramic and non-ceramic insulators.

If an outage occurs when thousands of birds perch together over a night, there is little to do to prevent such a unique case.

3.6. Woodpecker Damage

Woodpeckers drill holes in utility poles aiming to find insect, produce nest cavities to feed their youngs and also drums to attract females for breeding. These rituals speed up degradation of wood strength and decreases pole's lifespan substantially. Besides holes around crossarms may cause instant outages. Thus, cost of woodpecker damage to utilities reaches millions of dollars in each year. Especially in North America where wooden poles and woodpeckers are sharing the same habitat, distribution companies encounter severe financial and labour force losses. An American utility company spent 385.000$ to repair and replace the 100 poles damaged by woodpeckers [21]. Damage size depends on several factors such as; habitat availability, pole configuration, woodpecker species and insect population.

Applying wire mesh to poles is the common and effective solution in order to reduce woodpecker damage. Artificial cavities are not a promising approach because cavity creating is an important component of breeding activities. Ultrasonic devices are generally inefficient to deter birds from utility poles as the most birds do not hear high frequencies and even if they hear after a period of time they habituate the voice. However, a state of art sonic mitigation device called "Sonic Dissuader" [22] is designed specifically to scare woodpecker out from poles using alarm signals of woodpecker species. Device is only activated by the detection of vibration frequency created with pecking activity so that habituation may be prevented. Test result show that alarm signals did not change time birds spending on pole though they froze right after hearing the alarm call.

4. Sector Review

Utility companies can be separated into two groups considering their perspectives to deal with bird related outages. First group cope with the issue professionally by employing avian reporting systems in which electrocuted birds are recorded with pole numbers, locations and pole configurations. These companies give adequate training to field personnel about what to do when faced with bird mortalities or injuries and how to report these events. Data gathered from the field are transferred to a database and evaluated for determining the poles to be retrofitted. Retrofitting techniques are then checked according to their effectiveness. Second group generally does not have a systematic corporate policy and avian reporting systems, they only retrofit hazardous poles based on field experience also follow up controls of retrofitted poles are not executed.

APLIC recommends utility companies to develop and implement APP. Many companies mostly located in USA and Canada have established their APP [23-28] aiming to; reduce utility related bird mortalities, decrease unserved energy cost, diminish repair and replacement cost of utility equipments and ensure legal compliances.

APLIC gives a general outline scheme of reducing bird related outages risk in figure 2 that can be used by the utilities.

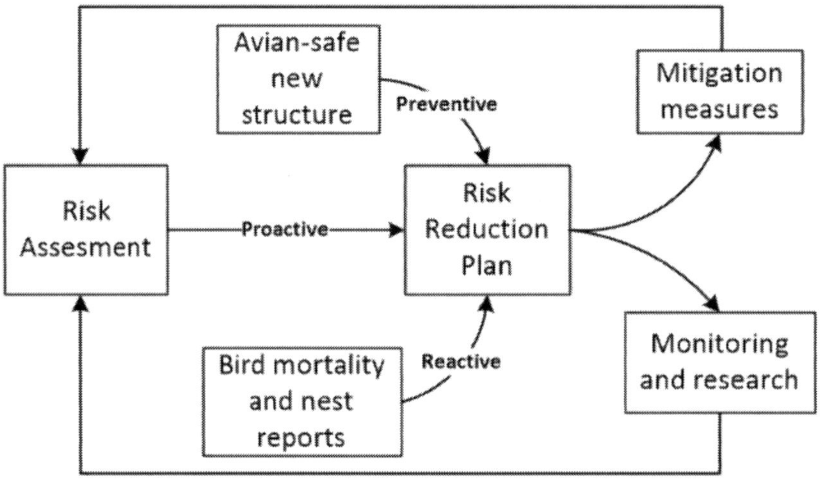

Figure 2. Risk reduction plan [7].

Roles of different type of measures in the risk reduction plan are briefly explained below:

- Preventive measures include designing new and reconfigured constructions so as to be avian safe. Providing 150 cm phase-phase clearance between conductors, using suspended insulators, and utilizing crossarms made of non-conductive materials are the examples of this kind.
- Reactive measures are defined as reporting bird mortalities and risky nests to retrofit hazardous poles and lines. Insulation covers, perching discouragers and line marking devices can be applied.
- Proactive measures cover risk assessment and providing sources to train and increases awareness of operation and maintenance staff of the company.

Transmission and distribution companies which have taken mitigation measures obtained effective results in the aspect of reducing bird mortalities and outages. The results of industrial applications are summarized in Table 1.

Table 1. Industrial applications and effectiveness of mitigation measures

Company	Mitigation Measure	Result
National Grid (USA) [29]	500 steel poles are retrofitted using insulation covers and plastic perching discouragers	Annual average outages decreased by %80.
Seattle City Light (USA) [28]	Nesting platforms were situated on utility structures, new and reconfigured structures are designed with avian-safe clearances.	Bird deaths has diminished by %53 over a 20 year period.
Eskom (South Africa) [30]	360 transmission poles were retrofitted by 7800 bird guards	Voltage dip cost of $3,26 million is saved.
RWE (Germany) [31]	13.000 bird flappers installed on earth wires.	%90 reduction in bird collision rate.

5. Conclusions

Since many bird species are threatened and faced with extinction, bird related outages have a distinguishing environmental importance compared to other types of faults in power systems. Universities, experts, governmental and non-governmental organizations have been working on reducing bird related outages since 1970s. However, annual electrocution and collision rates still seem to be high because most of the distribution companies do not have a systematic approach to deal with the issue. It is critically important to implement an avian outage reporting system and examine bird fauna in the area prior to implementations of mitigation products and designing avian safe equipments. Hence, distribution companies need to develop and implement their own site-specific avian protection plans to reduce bird mortalities as enhancing service reliability. APLIC guidelines are the key resources for a holistic approach in the development of APPs. However, the studies on each APP should be evolved to meet the particular needs of the power system service area. Taking preventive, reactive and proactive measures will not only provide cost savings also raise a public awareness.

Acknowledgements

This research was financially supported by Turkey Energy Market Regulatory Authority (EMRA) as a part of TREDAS EDCo. “Bird-Related Outages in Power Systems” R&D Project. Special thanks to Mehmet Dönmez and TREDAS EDCo. project team for their contributions. Also, Dr. Mustafa Kaya from Trakya University Biology Department deserves an appreciation for providing his experiences on avian species. Finally, we thank our ÅF Consult colleagues, Gökhan Tosun and Tanju Kahyaoglu for their support during the project.

References

[1] Sahai S and Pahwa A 2006 A probabilistic approach for animal-caused outages in overhead distribution systems *Probabilistic Methods Applied to Power Systems Int. Conf. on, Stockholm, Sweden*
[2] IEEE Power and Energy Society 1993 *IEEE Std. 1264 - Guide for Animal Deterrents for Electric Power Supply Substations*
[3] Energy and Environmental Economics Inc 2005 *The Cost of Wildlife-Caused Power Outages to California's Economy*
[4] APLIC, Edison Electric Institute and California Energy Commission 2006 *Suggested Practices for Avian Protection On Power Lines: The State of the Art in 2006* Washington DC USA
[5] Loss S R, Will T, Marra P P 2014 Refining estimates of bird collision and electrocution mortality at power lines in the United States *PLoS ONE*
[6] APLIC and Edison Electric Institute 2012 *Reducing Avian Collisions with Power Lines: The State of Art in 2012 Washington DC USA*
[7] APLIC and USFWS 2005 *Avian Protection Plan Guidelines*
[8] IEEE Power and Energy Society 2011 *IEEE Std 1651-2010 - IEEE Guide for Reducing Bird-Related Outages*
[9] 50 FR 13710 (Code of Federal Regulations) 1985 *List of Migratory Birds*
[10] United States v. Moon Lake Electric Ass'n Inc 1999 45 F. Supp. 2d 1070
[11] 16 U.S. Code § 668 2011 *Bald and golden eagles*
[12] USFWS 2005 *Habitat Conservation Plans – Section 10 of Endangered Species Act*
[13] Stroud D A 2003 *The Status and Legislative Protection of Birds Of Prey and their Habitats in Europe* Edinburgh pp 54-56
[14] European Commission 2013 *Guidelines on Wilderness in Natura 2000*
[15] Convention on International Trade in Endangered Species of Wild Fauna and Flora 1973 *Appendices I, II and III* Washington DC USA
[16] United Nations 1992 *Convention on Biological Diversity*
[17] Goodrich-Mahoney J, Abrahamson L and Ballard J 2008 *Environment Concerns in Rights-of-Way Management* Burlington: Elsevier
[18] Sundararajan R, Burnham R, Carlton R et al. 2004 Preventive measures to reduce bird related power outages – Part II– Streamer and contamination outages *IEEE Transactions on Power Delivery* Vol 19 No 4 pp 1848-1853
[19] Zhou Y, Xu X, Nie Q, Liang X and Guan Z 2009 Bird streamer caused flashover in EHV transmission line *IEEE Trans. On Dielectrics and Electrical Insulation* Vol 16 No 1
[20] IEEE Power Engineering Society 2005 *IEEE Std 957- IEEE Guide for Cleaning Insulators*
[21] Tupper S K, Andelt W F, Cummings J L, Weisner C and Harness R 2010 Polyurea Elastomer protects utility pole crossarms from damage by pileated woodpeckers *The Journal of Wildlife Management* Vol 74 No 3 pp 605-608
[22] Tupper S K, Cummings J L, Andelt W F, Werner S, Harness R 2011 Evaluation of sonic dissuader to reduce damage by pileated woodpeckers *Wildlife Damage Management*
[23] Clean Line Energy Partners 2014 *Avian Program*
[24] Turley N J and Dumas B 2015 *Idaho Power Avian Protection Plan*
[25] HDR Engineering 2010 *Marine Corps Base Camp Pendleton Avian Power Line Protection Plan*
[26] EDM International 2010 *Powder River Energy Corporation Avian Protection Plan*
[27] Insignia Environmental 2013 *San Diego Gas and Electric Company East County Substation Project Avian Protection Plan*
[28] Seattle City Light 2014 *Avian Protection Plan*
[29] Reynolds B, Dupre R and Eccleston D 2014 To protect and deter *T&D World Magazine*
[30] Vosloo H, Van Rooyen C 2001 Guarding Against Bird Outages *T&D World Magazine*
[31] AEWA Technical Committee and CMS Scientific Council 2011 *Guidelines on How to Avoid or Mitigate Impact of Electricity Power Grids on Migratory Birds in the African - Eurasian Region*

IOP Conf. Series: Materials Science and Engineering **161** (2016) 012092 doi:10.1088/1757-899X/161/1/012092

Energy Payback Time of a Rooftop Photovoltaic System in Greece

E.Rachoutis[1]**, D. Koubogiannis**[1*]

[1]Department of Energy Technology Engineering, Technological Educational Institute of Athens, Agiou Spyridonos Street, 122 10 Egaleo, Athens, Greece

[*]e-mail: dkoubog@teiath.gr

Abstract. Life Cycle Analysis (LCA) is an important tool to quantitatively assess energy consumption and environmental impact of any product. Current research related to energy consumption in buildings moves towards Nearly Zero Energy Building (NZEB). In such a building, an important issue concerns the energy production by renewable sources, including on-site production. The most feasible way to achieve renewable energy utilization in a building level in Greece is by using rooftop Photovoltaic (PV) systems, also promoted in the last decade by the national legislation concerning energy conservation measures. Apart from cost-related issues and payback times, Embodied Energy (EE) and Embodied CO_2 (ECO_2) emissions have also to be considered against the anticipated corresponding savings. Using a particular PV system as a case study, its basic constitutive materials are determined and their masses are calculated. Embodied energy values are estimated by using embodied energy coefficients available in the international literature. Considering a specific geographic location in Greece for the building on which the PV is installed, the annual energy generated by the system is estimated based on its performance data and curves. The Energy and CO_2 Payback Times (EPBT and CO_2PBT) are estimated and assessed, as well as future work is suggested.

1. Introduction

The rapid increase in the energy needs worldwide and the growing concern about global warming has led scientists during the last decades to search for new environmentally conscious electricity generation technologies. Given the unlimited supply of sunlight and the old idea of the conversion of sun power into electricity, one of the technologies that have met a rapid development is arguably the photovoltaic energy generation devices. Although such devices cause no greenhouse gas emissions and do not consume energy during their operation, they actually affect the environment and consume an amount of energy during their life. This energy and the related environmental impact concerns the export, transportation and processing of their constitutive raw materials, the manufacturing and construction of the devices, their transportation, distribution and installation to site, maintenance and replacement actions during their operational stage of life and finally their removal and disposal and/or recycling of materials.

In order to completely assess the sustainability of such systems fairly and suggest possible improvements, the amounts of energy and emissions involved with them, which are known as Embodied Energy (EE) and Embodied CO_2 (ECO_2) emissions, have to be taken into account. Such an

IOP Conf. Series: Materials Science and Engineering **161** (2016) 012092 doi:10.1088/1757-899X/161/1/012092

analysis is usually realized in the context of a cradle-to-grave Life Cycle Assessment (LCA). In addition such data are valuable for the LCA of the whole building, since current research concerning energy consumption in buildings moves towards the Nearly Zero Energy Building (NZEB), in which an important issue is the energy production by renewable sources, including on-site production. The most feasible way to achieve renewable energy utilization in a building level in Greece is by using rooftop Photovoltaic (PV) systems. Such PV systems and motivation for their use have also been included in national legislation concerning energy conservation measures.

A sufficient number of comprehensive studies which are concerned with the LCA of different PV systems and technologies can be found in the literature. Some representative works are presented below. In [1], the LCA of a poly-crystalline silicon PV system was presented and it was found that the gross energy required through the life cycle of the system was 1494 MJ/panel and the global warming potential 80 $kgCO_2$/panel. The Energy Pay Back Time (EPBT) of the system was found to be from 3.3 to 6.5 years depending on the geographic conditions. In [2], the greenhouse gas emissions of a rooftop multi-crystalline silicon PV under average Southern-Europe insolation (1700 $kWh/m^2/year$) were calculated to be about 25-35 $grCO_2/kWh_e$. Moreover, in [3], the LCA of the components of a 3.5 MW multi-crystalline PV system was presented an embodied energy value of 542 MJ/m^2, an amount of greenhouse gas emissions of 29 $kgCO_2/m^2$ and an energy payback time of 0.37 years for average USA insolation and temperature conditions were estimated.

In the present work, a typical commercial rooftop PV system is selected and installed to be considered in a particular location in Greece. Simulating its performance by means of freely available software, annual energy and corresponding emissions savings due to its operation are calculated. In addition, the amount of the initial embodied energy and emissions of this PV system are quantified and the Energy Pay Back Time (EPBT) and the CO_2 emissions Pay Back Time (CO_2PBT) of the system are estimated and assessed. Comparisons with corresponding results from the literature are provided and future research is suggested.

2. Case study - Methodology

In the present work, a typical commercial PV system was selected as a case study. This system was considered to be installed at a particular location in Greece. In order to calculate and assess the EPBT and CO_2PBT quantities related to it, the following tasks had to be accomplished:

- Quantification of the performance of the PV system, i.e. calculation of the energy savings (E_{SAV}) and emissions ($CO_{2,SAV}$) annual savings due to the utilization of the system. E_{SAV} is calculated in the form of electrical energy (kWh_e).
- Estimation of the Embodied Energy (EE) and the Embodied CO_2 savings (ECO_2) of the PV system. EE is calculated in the form of primary energy (MJ_p).
- Calculation of the Energy Pay Back Time (EPBT), according to the formula

$$EPBT[yr] = EE[MJ_p] / E_{SAV}[MJ_p/yr] \quad (1)$$

- Calculation of the CO_2 emissions Pay Back Time (CO_2PBT), according to the formula

$$CO_2PBT[yr] = ECO_2[kgCO_2] / CO_{2,SAV}[kgCO_2/yr] \quad (2)$$

It should be remarked that in order to apply equation (1), the energy savings E_{SAV} have to be calculated in the form of primary energy (in MJ_p).

The PV system that was selected as a case study herein consists of the following parts:

- Ten poly-Si panels, each of them having a nominal power of 0.3 kW. Thus the nominal power of the whole system is 3 kW. These panels are accompanied by a 10-year guarantee and assurance that their efficiency will not be reduced below 90% during the first 10 years and 80% during the first 25 years of operation.
- An inverter with a nominal output power of 3 kW and a five-year guarantee.

- Two aluminum bases, one for four and one for six PV panels, respectively.
- Two types of cables, namely 10m of solar cable for the DC part of the system and 20m of power supply cable for the AC part.

The efficiency of such a PV system is generally influenced by the following parameters:

- The latitude of the place where it is installed (the lower the latitude, the higher the intensity of the solar radiation and the higher the efficiency of the panel).
- The sun height (higher production of electricity during summer than winter period).
- The orientation of the PV panel (south for the north hemisphere).
- The slope of the panel (ranging from 20° to 40°, depending on the location).
- The panel temperature (the lower the temperature, the higher the efficiency).

The above PV system was considered to be installed at a particular location in Greece, namely on the roof of an apartment building at Piraeus, where latitude is 37°57'35" north, longitude is 23°37'38" east and elevation is about 34m above sea level.

3. Performance of the PV system

In order to estimate the performance of the PV system, the free Software PVGIS [4] was utilized. Some data were required by the software for the simulation of the present case study. The data values used in the present study are provided for the sake of completeness below:

- The geographical data mentioned in the previous section.
- Built-in PVGIS database for the solar radiation, according based on the particular geographical data.
- Nominal power of the PV system: 3 kW.
- Losses due to temperature and low irradiance: 10.2% (using local ambient temperature).
- Loss due to angular reflectance effects: 2.7%.
- Other losses (concerning cables, inverter etc): 14.0%.
- Combined PV system losses: 24.8%.

Figure 1 shows a screen printout of the PVGIS software during simulation. In particular, PV geographical information and performance data for the case under consideration are depicted.

Figure 2 presents data for the PV system concerning the average daily (E_d) and the average monthly (E_m) electricity production (in kWh), as well as the average daily (H_d) and the average monthly (H_m) specific global irradiation (in kWh/m^2) received by the modules of the given system. In the same figure, E_m and H_m have also been plot per month of the year.

According to the simulation results shown in Figure 2 (left), it was estimated that the annual energy production by the selected PV system in the particular location is 4010 kWh$_e$/year, i.e. E_{SAV}=4010 kWh$_e$/yr. This has to be transformed into primary energy in MJ$_p$/yr. According to the national regulation in Greece, the conversion factor that is used (in energy performance of buildings) to calculate primary energy from final energy use is 2.9 for electricity. Thus, E_{SAV} = 4010 kWh$_e$/yr * 3.6 MJ$_e$/ kWh$_e$ * 2.9 MJ$_p$/MJ$_e$ = 41864.4 MJ$_p$/yr, i.e. E_{SAV}=41864.4 MJ$_p$/yr.

Similarly, due to the fact that the conversion factor to calculate CO_2 emissions from final energy use is 0.989 kgCO_2/kWh$_e$ for electricity, $CO_{2,SAV}$ = 0.989 kgCO_2/kWh$_e$ * 4010 kWh$_e$/yr = 3965.9 kgCO_2/yr, i.e. $CO_{2,SAV}$ = 3965.9 kgCO_2/yr.

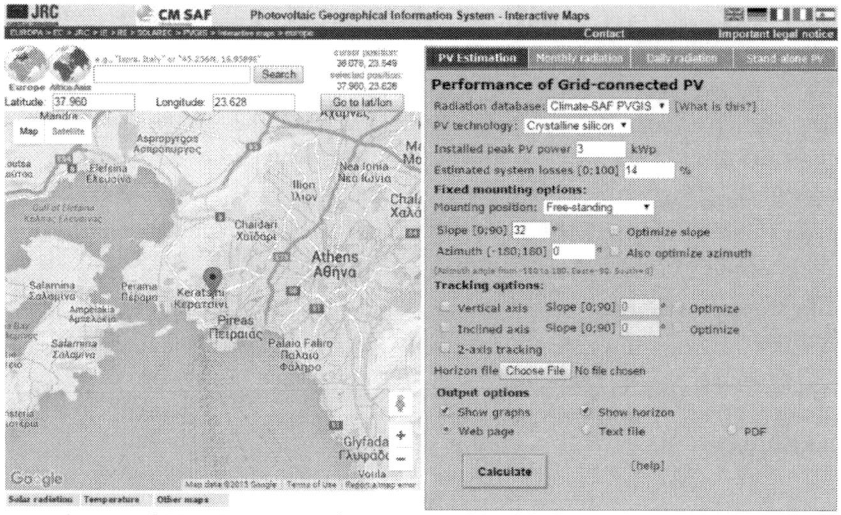

Figure 1. Screen printout of PVGIS concerning geographical information and performance data for the case under consideration.

Fixed system: inclination=31°, orientation=0° (Optimum at given orientation)				
Month	E_d	E_m	H_d	H_m
Jan	7.39	229	3.09	95.8
Feb	8.22	230	3.49	97.6
Mar	10.60	329	4.57	142
Apr	12.70	380	5.57	167
May	13.60	421	6.07	188
Jun	14.40	433	6.59	198
Jul	14.30	442	6.61	205
Aug	13.90	430	6.43	199
Sep	13.20	395	5.98	179
Oct	10.10	313	4.46	138
Nov	7.14	214	3.08	92.3
Dec	6.33	196	2.67	82.6
Annual average	**11.0**	**334**	**4.89**	**149**
Annual total	**4010**		**1780**	

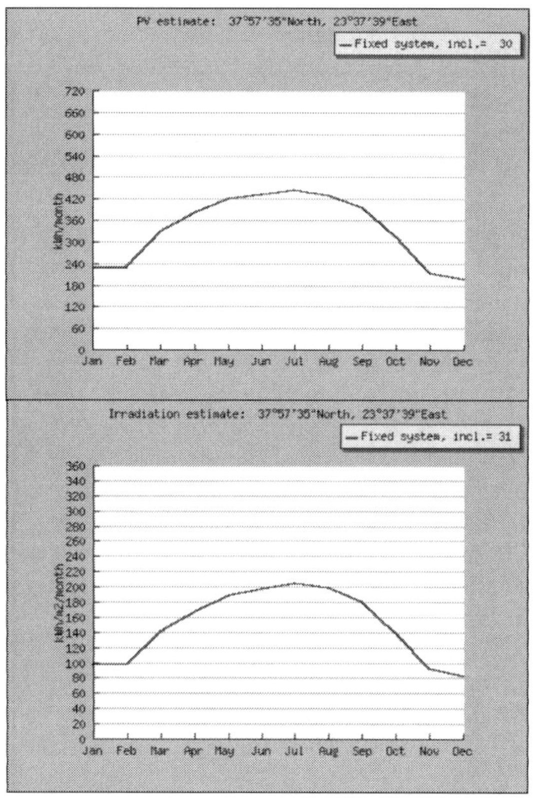

Figure 2. Left: Average daily (E_d) and average monthly electricity production (E_m) (in kWh), average daily sum (H_d) and average monthly (H_m) specific global irradiation received by the modules of the given system (in kWh/m^2). Right: Monthly evolution of E_m and H_m.

4. Estimates for the embodied energy and CO_2 of the PV system

In this section, the concepts of Embodied Energy (EE) and Embodied CO_2 (ECO_2) emissions are briefly presented. Then a reference is made to the methods of embodied energy calculation. Finally estimates for the EE and ECO_2 of the PV system under consideration are provided.

4.1 The concepts of embodied energy and CO_2

Three main stages can be distinguished during the life time of a product; the initial stage concerning its production, i.e. the period of time from the beginning of its construction till the moment it becomes operational, the period of time that the product is operational and the final stage of its life concerning its withdrawal and removal of its materials. The amounts of energy involved in the above stages sum up to the so called Embodied Energy (EE) of the product. In particular, the energy consumed during the first stage is the Initial EE (IEE). The energy consumed during the operational stage is the Recurrent EE (REE). Similarly let the energy concerning its withdrawal named after Withdrawal EE (WEE).

Referring to a PV system, the IEE consists of two parts. The first one is the energy consumed directly (IEE_D) for actions related directly to site like transportation and installation. The second one is the energy consumed indirectly (IEE_I), upstream these activities for the manufacturing of the PV system, as well as for the production of all the materials required for it. REE concerns the energy to be consumed for the construction/materials/processes/equipment related to maintenance/replacement actions during the operational period of the PV system (e.g. for the replacement of the inverter). Finally WEE concerns energy related to withdrawal and disassembly actions. Like IEE, REE and WEE consist of two parts, one concerning direct and one concerning indirect energy consumption. Again the direct parts refer to energy consumed on site by equipment, tools, etc or to transport materials and products to the site, while the indirect parts refer to energy consumed to acquire process and manufacture the materials involved at the corresponding stage.

In the light of the above, the total EE of the product can be written as $EE = IEE_D + IEE_I + REE + WEE$, where the first term concerns on site actions for the PV system, the second term concerns energy consumed upstream this activity, while the rest terms concern energy to be consumed downstream (with respect to time).

A companion concept concerning the environmental impact of Embodied Energy is that of the embodied carbon emissions or Embodied CO_2 (ECO_2). This is accordingly defined for each of the EE terms mentioned above, to be the corresponding amount of CO_2 emissions to the atmosphere. In this context, ECO_2 of a product is defined to be the CO_2 quantity corresponding to its EE.

4.2 Methods of embodied energy calculation

According to the literature [5], the two basic categories of methods for calculating EE rely either on the Process Analysis or the Input-Output (IO) Analysis. Each of them has its own pros and cons [5], so Hybrid Analysis methods have been proposed in order to combine the advantages of both approaches and facilitate more comprehensive and accurate analysis. Furthermore, hybrid analysis has been proposed in the forms of process-based and IO-based hybrid analysis [5]. Unfortunately, independent the method of analysis used, EE databases suffer from problems of variation and incompatibility [6].

In previous works by the second author, the focus was placed on the Initial indirect EE (IEE_I) of the materials and equipment comprising typical Hellenic dwellings. For example in [7] the EE of the Electro-Mechanical Installations items were estimated for typical Hellenic dwellings, while in [8] similar work has been conducted for the corresponding Civil Engineering (building construction) materials. In these works, a three-step methodology was followed. According to this, the first step is a material analysis, i.e. the breakdown of the product under consideration to its constitutive single materials. The calculation of the corresponding masses was the second step, while at a third step, EE and ECO_2 values were obtained by multiplying mass values by the corresponding EE and ECO_2 coefficients of the materials. Although, these coefficients are nationally dependent parameters, available values from the open literature were used [9], since there is lack of a comprehensive Hellenic database. Since the coefficients used by [9] rely on process analysis, such an approach suffers from truncation errors [10]; the latter are related to the definition of the method boundary. So, although not being a rigorous approach, it provides the capability to obtain important practical results for initial guidance.

4.3 Calculation of the PV system EE and ECO_2

A typical PV panel consists of the following materials:

- Aluminum Frame, which is responsible for the stability of the whole construction.
- Solar Glass, for the protection of the solar cells.
- Polymer Material EVA (Ethylane Vinyl Acetate), placed in the front and back part of the panel, for the encapsulation of the solar cells.
- Solar Cells, constructed by a semiconductor like silicon, for the absorption of the photons.
- Back sheet.
- Junction Box.

Apart from the panel, the other important components of a PV system are:

- The base, which is responsible for the proper support of the PV panel with the proper slope.
- The inverter to convert the DC output of the PV panels into AC, so that the connection with the electrical grid can be possible.
- The cables for the connection between the solar panels and the DC part of the inverter, as well as the AC part of the inverter with the electrical grid.

In [11], a detailed Life Cycle Inventory of the PV system under consideration has been conducted. In what follows, the main results are presented. The EE and ECO_2 coefficients [9] that were utilized are presented in Table 1. Due to the fact that the ECO_2 of silicon was not available, the value 98.08 $kgCO_2/kg$ from the literature [12] was used.

Table 1 EE and ECO_2 coefficients from [9] utilized in the present work.

Material	EE [MJ/kg]	ECO_2 [$kgCO_2/kg$]
Aluminium	155	8.24
Cermics	10	0.66
Copper	42	2.6
Plastic	80.5	2.73
Iron	25	1.91
PVC	77.2	2.61
Rubber	91	2.66
Stainless Steel	56.7	6.15
Silicon	2355	-
Glass	15	0.86
EVA	128	2.3

Table 2 summarizes results for the whole set of materials used in the PV system. The results concern the mass, EE and ECO_2 contribution of each material to the system, both in terms of absolute value and percentage. The latter, namely the percentage contributions of each material in terms of mass, EE and ECO_2 are schematically shown in Figure 3 (left). It is obvious that glass and aluminum are the dominant materials in terms of mass, while silicon is ranked third. However, in terms of EE and ECO_2, silicon has the greater impact due to the high corresponding coefficients, while glass is ranked third due to low such values. Aluminum is at the second position both in terms of mass and EE / ECO_2, contributing about one third of the total PV system values. The contribution of these three materials (glass, aluminum, silicon) is summed up to about 90% for mass, 95% for EE and 96% for ECO_2.

In Figure 3 (right) the percentage contribution of the PV system basic items to the total mass, EE and ECO_2 values are depicted. According to this, among the basic items of the PV system, the PV panels are dominant in mass and exhibit the greater energy and environmental impact.

Table 2 Mass, EE and ECO_2 contribution of each material to the whole PV system, both in terms of absolute value and percentage.

Material	**Mass [kg]**	**EE [MJ]**	**ECO_2 [kgCO_2]**	**Mass [%]**	**EE [MJ]**	**ECO_2 [%]**
Solar Glass	130.8	1962.0	112.5	45.7	3.4	4.3
Aluminum	110.4	17118.2	910.0	38.6	29.8	34.9
Silicon	15.3	36055.1	1501.6	5.3	62.8	57.5
Copper	10.0	421.3	26.1	3.5	0.7	1.0
EVA	7.7	985.6	17.7	2.7	1.7	0.7
PVC	7.0	542.7	18.4	2.5	0.9	0.7
Galvanized Steel	3.2	181.4	19.7	1.1	0.3	0.8
Plastic	1.0	80.5	2.7	0.3	0.1	0.1
Rubber	0.3	19.1	0.2	0.1	0.0	0.0
Iron	0.3	6.3	0.5	0.1	0.0	0.0
Ceramics	0.2	3.0	0.6	0.1	0.0	0.0
Total	**286.3**	**57375.1**	**2610.0**	100.0	100.0	100.0

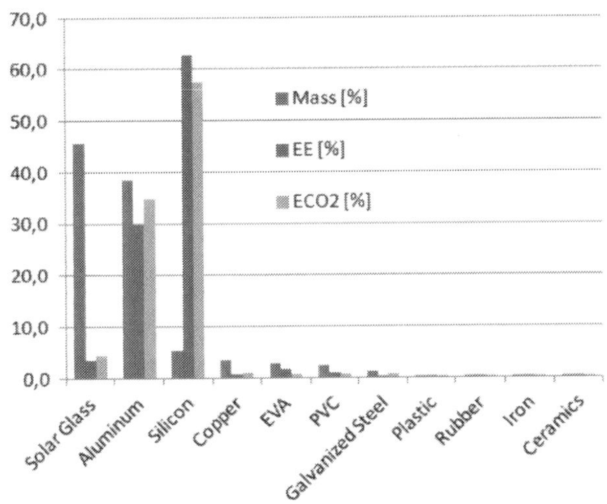

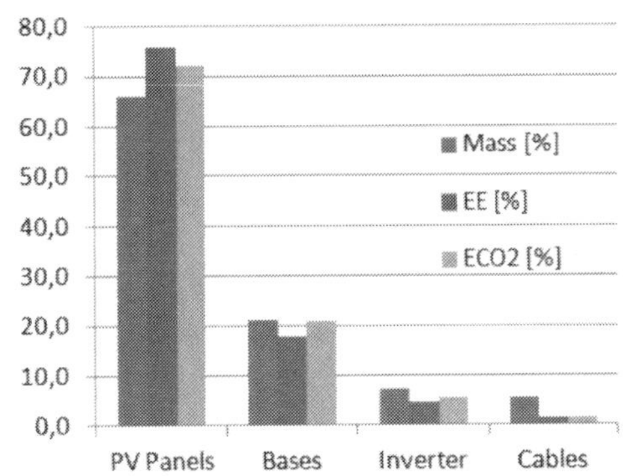

Figure 3. Left: Percentage contribution of each material in the PV system in terms of mass, EE and ECO_2. Right: Percentage contribution of the PV system basic items to the total mass, EE and ECO_2.

4. Results and discussion

Energy production by renewable energy technologies is an environmentally friendly procedure, as CO_2 and other GHGs are not emitted and energy is not used during the stage of operation. However, such technologies like PV panels do emit CO_2 and other gasses and consume energy during the stage of their construction, transportation, collection of the required raw materials, etc. This is exactly what the concepts of the Energy Pay-back Time and Carbon Dioxide Pay-back Time represent. Based on the results obtained in sections 3 and 4 and according to equations (1) and (2), these two quantities for the case under consideration give the following values:

$$EPBT = (57375.1\ MJ_p) / (41864.4\ MJ_p/yr) = 1.4\ yr = 16.8\ months$$

$$CO_2PBT = (2610\ kgCO_2) / (3965.9\ kgCO_2/yr) = 0.7\ yr = 8.4\ months$$

The payback times obtained above are judged to be rather low. In order to assess them, they were compared with corresponding values from similar studies in the literature [13], [14]. Table 3 presents this comparison, as well as the location of the place where each study has been conducted and the corresponding annual received energy intensity. According to it, EPBT is lower in Southern Europe compared to that of Switzerland (Central Europe). This is logical, since at lower latitudes (Southern Europe) the solar radiation is greater, so the denominator at equation (1) becomes greater and EPBT is reduced. Moreover, the present result is lower than that of [14], although the locations exhibit rather similar received energy values (actually, the denominator is slightly greater due to the higher received energy of 1780 instead of 1700 $kWh/m^2/yr$). What is important is that in the present study an underestimated EE value at the numerator of equation (1) has been used. This is due to the fact that only material production and not manufacturing EE has been taken into account (due to lack of information), as well as energy consumed for transportation and installation of the PV system to site (i.e. direct energy) has not been considered.

Table 3 Comparison of EPBT with corresponding values from similar studies.

Reference	Location	Energy received [$kWh/m^2/yr$]	EPBT [yrs]
[13]	Switzerland	1117	3.5
[14]	South Europe	1700	2.1
Present	Greece	1780	1.4

Assuming that for the PV system under consideration producing 4010 kWh_e/yr, the efficiency of the PV panels remains unaffected, the following environmental impact can be estimated: 32.5 gCO_2/kWh_e for 20-years operation, 26 for 25-years operation and 21.7 for 30-years operation. The latter value is compared in Table 4 with corresponding values from similar studies (also concerning 30-years operation). As it can be seen, the emissions per unit of annual energy production are underestimated. This is attributed to the same reasons explained before.

Table 4 CO_2 emissions per unit of annual energy production.

Reference	[2]	[15]	[16]	Present
[gCO_2/kWh_e]	25-35	104	40	~22

5. Conclusions – Future research

A rooftop photovoltaic system was studied. Assuming the location of installation and using open simulation software, the annual energy production was calculated. In addition, the energy embodied in the system was estimated and the energy payback time was evaluated. Similarly, the corresponding CO_2 payback time was quantified. Both values were underestimated compared to the literature. This was attributed to the greater energy production in Greece with respect to locations used in the literature, but mainly to the underestimation of the embodied energy. Among the basic issues that are missing in the present procedure is to take into account the energy required for the manufacturing of the PV system, not only the energy related to the production of its constitutive materials.

As a continuation to this research, it is planned to estimate the manufacturing energy consumed in the production line of a typical photovoltaic system, make a thorough literature review on the topic to perform comparisons and study how energy payback time is affected by different installation sites in Greece. The aim in the long-term is to obtain in a more reliable way representative values of the embodied energy and emissions concerning PV systems operating in Greece (for example in order to use them in LCA studies of Near Zero Energy Buildings).

Acknowledgement

The authors would like to acknowledge company "Aleo Solar Greece" and especially Mr Panagiotis

Fragkos for providing practical information and valuable data.

References

[1] Stoppato, A., Life cycle assessment of photovoltaic electricity generation, Energy, 33(2):224-232, 2008.
[2] Fthenakis, V. and Alsema, E., Photovoltaics Energy Payback Times, Greenhouse Gas Emissions and External Costs: 2004-early 2005 Status, Progress in Photovoltaics: Research and Applications, 14(3):275-280, 2006.
[3] Mason, J.E., Fthenakis, V.M., Hansen, T. and Kim H.C., Energy Pay-Back and Life Cycle CO2 Emissions of the BOS in an Optimized 3.5 MW PV Installation, Progress in Photovoltaics: Research and Applications, 14:179-190, 2006.
[4] www.photovoltaic-software.com/pvgis.php
[5] Dixit, M.K. A Framework for an Improved Input-output-based Hybrid Method for Embodied Energy Calculation. Proc. 51st ASC Annual International Conference hosted by Texas A&M University in College Station, Texas April 22-25, 2015; http://ascpro0.ascweb.org/archives/cd/2015/paper/CPRT314002015.pdf.
[6] Dixit, M.K., Fernandez-Solis, J.L., Lavy, S., Culp, C.H., Need for an embodied energy measurement protocol for buildings: A review Paper. Renewable and Sustainable Energy Reviews, 16(6):3730-3743, 2012.
[7] Koubogiannis, D.G. and Balaras, C.A. Embodied Energy in Electro-Mechanical Installations of Hellenic Dwellings. Proc. International Conference Energy in Buildings, Athens: ASHRAE Hellenic Chapter and Technical Chamber of Greece; 2014. http://ashrae.gr/EinB2014/EinB2014_Proceedings.pdf.
[8] Syngros G. Embodied energy and embodied CO_2 estimation of building constructive materials in typical Hellenic dwellings, MSc Diploma Thesis, MSc in Energy, TEI Athens and Heriot-Watt University, 2016.
[9] Hammond, G.P. and Jones, C.I., Inventory of Carbon and Energy (ICE) Version 1.6a. Sustainable Energy Research Team, Department of Mechanical Engineering, University of Bath; 2008. http://perigordvacance.typepad.com/files/inventoryofcarbonandenergy.pdf.
[10] Lenzen, M., Errors in conventional and input-output–based life-cycle inventories. Journal of Industrial Ecology, 4:127-148, 2000.
[11] Rachoutis, E., Life Cycle Assessment of a Rooftop Photovoltaic System, Diploma Thesis (in Greek), Energy Technology Engineering Department, Technological Educational Institute of Athens, Greece, 2015.
[12] Kemmoku, Y., Ishikawa, K., Nakagawa, S., Kawamoto, T. and Sakakibara, T., Life Cycle CO_2 emissions of Photovoltaic/Wind/Diesel Generating System, Electrical Engineering in Japan, 138(2):14-23, 2002.
[13] Alsema, E.A. and de Wild-Scholten, M.J. Environmental impacts of crystalline silicon photovoltaic module production, LCE2006, 13th CIRP International Conference on Life Cycle Engineering, Leuven, Belgium, 2006.
[14] Jungbluth, N. Tuchschmid, M. And de Wild-Scholten, M., Life Cycle Assessment of Photovoltaics: Update of ecoinvent data v2.0, ESU-services Ltd. Working Paper, download from: www.esu-services.ch.
[15] Tripanagnostopoulos, Y., Souliotis, M., Battisti, R. and Corrado, A. Energy, Cost and LCA Results of PV and Hybrid PV/T Solar Systems, Prog. Photovolt: Res. Appl., 13:235–250, 2005.
[16] National Renewable Energy Laboratory (NREL), USA, Life Cycle Greenhouse Gas Emissions from Solar Photovoltaics http://www.nrel.gov/docs/fy13osti/56487.pdf.

Sustainable operation of a biological wastewater treatment plant

E Trikoilidou[1], G Samiotis, D Bellos and E Amanatidou

[1]Western Macedonia University of Applied Sciences, Environmental Engineering and Pollution Control Department, Environmental Chemistry and Wastewater Treatment Laboratory, Koila, Kozani, 50100, Greece

E-mail: waterlab3@teiwm.gr

Abstract. The sustainable operation of a biological wastewater treatment plant is significantly linked to its removal efficiency, cost of sludge management, energy consumption and monitoring cost. The biological treatment offers high organic removal efficiency, it also entails significant sludge production, which contains active (live) and inactive (dead) microorganisms and must be treated prior to final disposal, in order to prevent adverse impact on public health and environment. The efficiency of the activated sludge treatment process is correlated to an efficient solid-liquid separation, which is strongly depended on the biomass settling properties. The most commonly encountered settling problems in a wastewater treatment plant, which are usually associated with operating conditions and specific microorganisms growth, are sludge bulking, floating sludge, pin point flocs and straggler flocs. Sustainable management of sludge and less energy consumption are the two principal aspects that determine the operational cost of wastewater treatment plants. Sludge treatment and management accumulate more than 50% of the operating cost. Aerobic wastewater treatment plants have high energy requirements for covering the needs of aeration and recirculations. In order to ensure wastewater treatment plants' effective operation, a large number of physicochemical parameters have to be monitored, thus further increasing the operational cost. As the operational parameters are linked to microbial population, a practical way of wastewater treatment plants' controlling is the microscopic examination of sludge, which is proved to be an important tool for evaluating plants' performance and assessing possible problems and symptoms. This study presents a biological wastewater treatment plant with almost zero biomass production, less energy consumption and a practical way for operation control through microbial manipulation and microscopic examination.

1. Introduction

Problems caused by wastewaters: Human activities, such as agricultural including livestock, as well as urban and industrial development, create large quantities of wastewater that have to be treated prior their discharge in water bodies or land. If they are not treated properly, their pollutant load can cause serious environmental deterioration, with direct impact on human health. Agricultural and urban wastewater accumulate large quantities of pollutants, such as organic, nitrogen and phosphorus compounds, in surface waters, underground waters and soils. These pollutants are responsible for eutrophication and are nutrients for a variety of microbial species.

Activated sludge principle: Since decades, a wide range of technologies, processes and techniques has been developed and implemented for the treatment of municipal and/or industrial wastewaters. The most commonly applied wastewater treatment application is the biological processes, i.e. the use and exploitation of bacteria species for removing pollutants. The principle idea is to cultivate a large number of bacteria colonies (activated sludge, AS), which utilize pollutants for their own needs (growth and energy production), by creating the necessary conditions for their development. Different conditions, such as aerobic, anaerobic or anoxic, enhance the growth of specific bacteria that can be manipulated for the treatment of different substrates. Biological process is followed by solids/liquid separation achieved commonly by gravitational sedimentation. A fraction of the separated solids is recirculated, while the rest is wasted.

The problem of waste sludge: While biological treatment offers significant performance with relatively low cost, it also entails significant sludge production due to the accumulation of biomass and inactive sludge in the wastewater treatment plant (WWTP). The excessive accumulation of sludge hinders sedimentation efficiency and can also create problems concerning mixing and aeration of the bioreactors. Consequently, large quantities of accumulated sludge have to be disposed (waste sludge) frequently from a biological treatment plant, in order to maintain acceptable effluent quality. The quantity of waste sludge depends on influent composition and volumetric load, as well as on WWTP's operating conditions. Higher loads result in higher biomass yields, while critical operating conditions, such as hydraulic retention time (HRT), solids retention time (SRT), biomass age, food to microorganisms ratio (F/M), dissolved oxygen concentration (DO) and alternating conditions (aerobic, anoxic, anaerobic etc.), significantly affect biomass production.

Landfill disposal of sludge can create negative environmental, economical and social impacts, such as soil and water pollution, threats on human health, land devaluation. Sludge treatment increases wastewater treatment cost more than 50%.[1-3] Consequently, sludge minimization is a key towards sustainable operation and design of WWTPs.

Sludge minimization approaches: Recently, approaches for minimizing sludge production have intrigued scientists and engineers, as it can resolve the problem of sludge management at its source. It is well known that WWTP operating conditions are linked to the biological and morphological characteristics of activated sludge and to microbial manipulation that influence sedimentation process, treatment efficiency and net sludge production.[3-5] *High SRT:* One approach for minimizing sludge production is the amplification of microbial cell lysis and the growth of biomass from the utilization of cell lysis products, defined as cryptic growth.[6-7] Starvation conditions that occur in high SRT processes enhance cryptic growth, due to the high biomass concentration and the low F/M.[5,8] *Microbial manipulation*: Successful microbial manipulation can minimize sludge production up to where almost no sludge is disposed.[5,9] Microbial manipulation is the guided growth of desired microbial species by imposing specific operating condition in the bioreactors. The successful microbial manipulation towards the growth of predator microbial species can minimize sludge production, as they feed on bacteria and particulate organics. Growth of predator microbial species can be achieved at high SRT and different operating regimes (extended aeration, cyclic alteration between anaerobic anoxic and oxic conditions etc.) that provoke metabolic changes.[8,10] Consequently, in high SRT AS processes, the new cells production rate becomes equal to the decay rate. The phenomenon of solids' accumulation reduction is assisted by the activity of predator species that consume various complex substrates containing influent particulate organics, bacteria, cell debris, etc. In such cases, the net biomass production rate is minimal and tending to zero values.[11]

Complete solids retention process' principal idea: One of the most promising approaches towards sustainable WWTP operation and sludge management is the complete solids retention AS process, which incorporates both high SRT and microbial manipulation for predator microbial species growth. The principal idea of this technology is the almost complete recirculation of sludge in the WWTP, resulting: (a) SRT almost equal to the days of WWTP operation (b) increased sludge age, (c) high biomass concentration in bioreactors, (d) extended starvation conditions (low F/M ratio), (e) low substrate utilization rates (SUR) and (f) growth of predator microbial species. All the above trigger biochemical and biological processes that minimize sludge production, such as bacteria decay/lysis, predation phenomena and the degradation of some of the considered unbiodegradable particulate organics (influent particulate organics and cell debris).[9,12] Other studies conducted on membrane bioreactors (MBR) operating at high SRT up to complete retention of solids also indicated that biomass production was equilibrated by biomass loss (endogenous metabolism, death, lysis and predation) or that influent substrate was completely oxidized for cell maintenance (Henze et al., 2000; Spérandio Sperandio et al., 2013).[13-14]

Economical footprint: The sustainable design and operation of a WWTP is inextricably linked to its environmental and economical footprint. High treatment efficiency and low energy consumption are the keys for sustainability. Energy demands for operation of wastewater pumps, air blowers, mixers and separators etc, are analogous to influent load and significantly differ between treatment processes. Energy consumption ranges between 1 to 25 Kwh/kg $COD_{removed}$ and 0,25 to 13 Kwh/kg $COD_{removed}$ for aerobic and anaerobic processes respectively.[15-19] Successful microbial manipulation, thus efficient treatment and less sludge production, as well as on low energy consumption should be the goal for engineers towards sustainable wastewater treatment plants' design and operation.

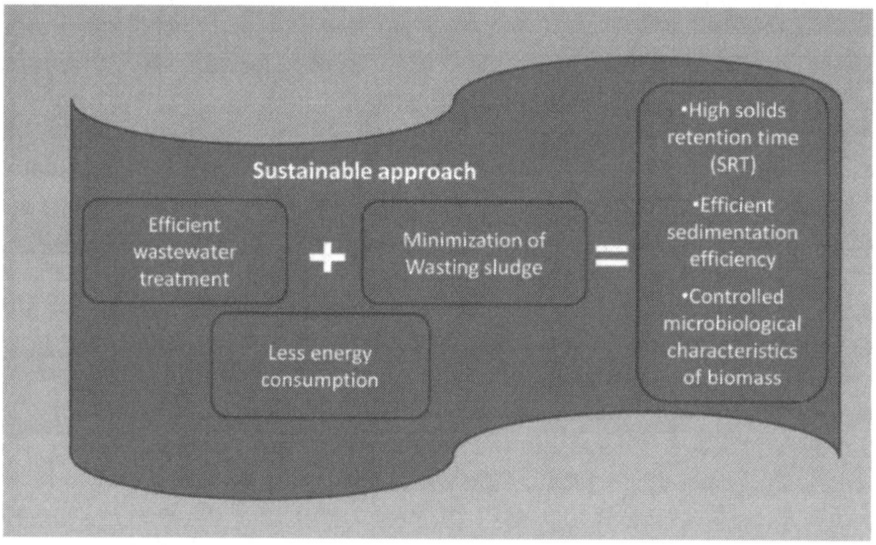

Figure 1. Key parameters towards sustainability of wastewater treatment technologies

This work presents a successful application of a novel activated sludge WWTP, designed for treating high strength wastewaters with minimal waste sludge production. The studied slaughterhouse WWTP operated beyond the typical literature limits concerning biomass concentration, recirculation rates and tank design, while achieving excellent treatment efficiency with low energy cost and negligible waste sludge quantities. The work provides engineers with the information needed for the biological process control through microbial manipulation and offers an opportunity to re-evaluate some critical parameters in WWTP design and operation, towards more efficient and sustainable wastewater treatment.

2. Materials and Methods

The studied activated sludge WWTP is installed in Almopia Prefecture, Pella, Greece. This patented WWTP is designed for treating high strength wastewaters by applying the complete solids retention activated sludge process, with low economical and environmental footprint.[20] The WWTP consists of two subsystems (Figure 1): (a) a preliminary simultaneous nitrification/denitrification system (SNdN) and (b) a pre-anoxic, complete mix, complete SRT, extended aeration system (PCMAS). The preliminary system has a sedimentation tank with internal recycle of solids, ie it operates as a biological selector, where younger biomass that utilizes readily biodegradable organics thrives. The effluent and the waste sludge is transferred (inflow) in PCMAS system. Before the SNdN system, primary treatment, (automated bar screen, decanter, aerated flow and load equalization tank) is performed.

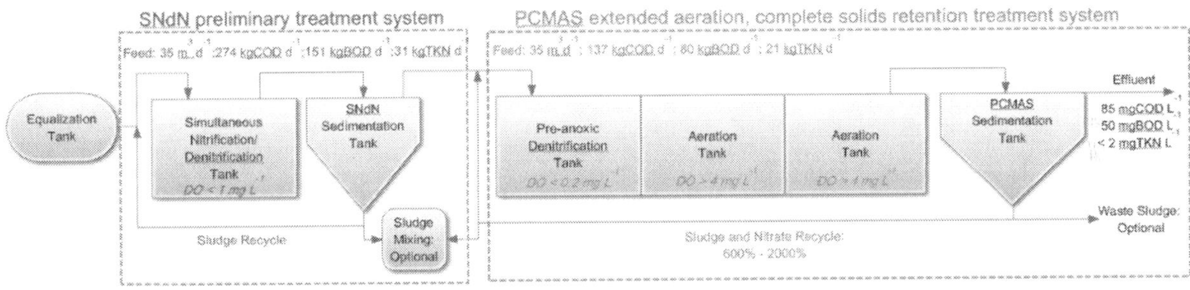

Figure 1. Flowchart of the complete solids retention WWTP and relating operation conditions

This complete solids retention AS treatment system appears significant differences in comparison to conventional AS WWTPs, such as bioreactors' configuration, settling tank geometry and dimensions, solids and mixed liquor recycle points and rates. The wastewater was treated in alternating anoxic, semi-anoxic and aerobic bioreactors, maintaining low biomass growth yields and increased presence of specific microbial species in activated sludge. In order to control sedimentation process, the WWTP operates with high RAS rate (over 600%), thus causing "forced sedimentation" (Grady et al. 1999; Maine, 2003; Mercalf and Eddy, 2003; Amanatidou et al., 2015a).[1,3,21-22] Wastewaters of this slaughterhouse have relatively low volumetric load (approximately 35 m^3 d^{-1}), high organic load (8250 mgCOD L^{-1} and 4538 $mgBOD_5$ L^{-1}) and high nitrogen load (1250 mgTKN L^{-1}). Due to its origin, the influent organics are highly biodegradable, with an average COD / $BOD_{ultimate}$ ratio of 0,969. Particulate matter in influent is approximately 1100 mgTSS L^{-1}, consisting mainly of organics (1008 mgVSS L^{-1}) and a small fraction of inert solids (0,027). Not all particulate organic matter is considered biodegradable, but previous studies has demonstrated that under the specific WWTP operating conditions and design, this considered unbiodegradable fraction of particulate organics is eventually utilized by the microorganisms.[9]

3. Results and discussion

In the studied slaughterhouse WWTP, the removal of organic compounds and nitrogen is completed in two stages. An initial removal of COD and TN was achieved in the SNdN system, but the main treatment took place in PCMAS system. Effluent quality was in accordance to the demands of the enforced legislation. The overall WWTP removal efficiency was approximately 99% for organics removal and 90% for nitrogen removal.

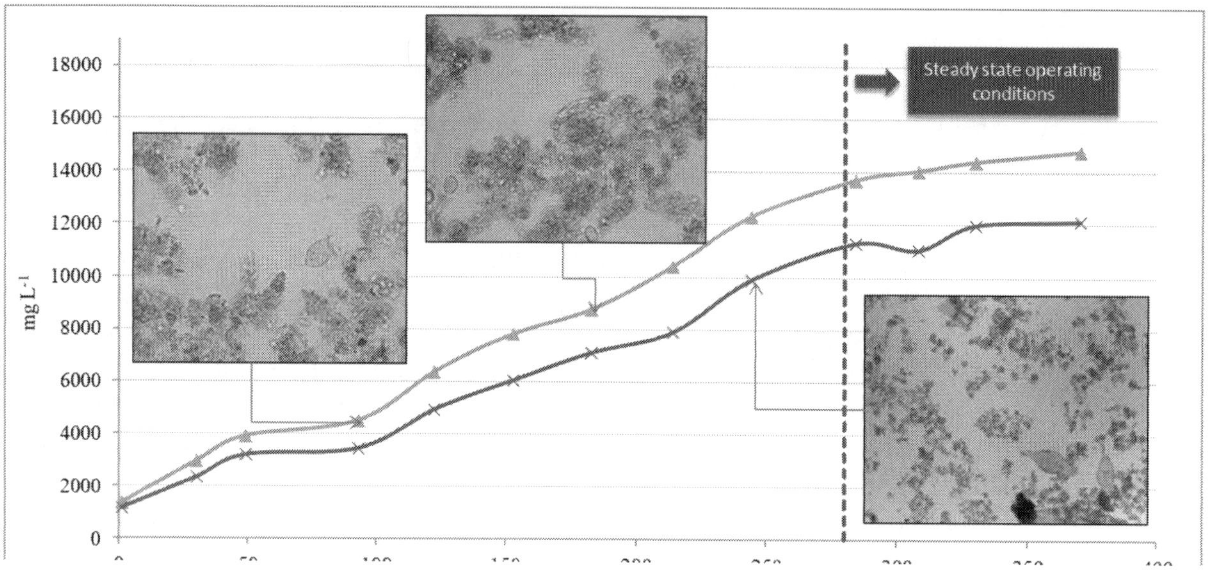

Figure 2. Solids accumulation in the studied WWTP and representative microscopic examination images; MLSS concentration; MLVSS concentration

The mixed liquor suspended solids (MLSS) concentration, as well as their volatile fraction (MLVSS), were increasing until the 280th day of WWTP operation, a time when plateau phase was reached, considered as steady state operating conditions (Figure 2). At steady state operating conditions, almost zero net biomass production and constant MLVSS to MLSS ratios were observed, attributed to WWTP operating conditions and biomass characteristics. These observations lead to the conclusion that by complete solids retention and specific operating conditions, most of the considered unbiodegradable particulate COD, of influent and of endogenous decay, are consumed by microorganisms. In this study, the microorganisms that are considered responsible for consuming these considered unbiodegradable fractions are rotifer and ciliate species (Figure 2).

Satisfactory solids/liquid separation in sedimentation tank was achieved despite the high MLSS concentrations. This was achieved due to the sedimentation tank's design and the fact that it operated with high sludge recycling rates (RAS 1000-2000%), thus creating forced sedimentation in the clarifiers and enchasing process efficiency.[3] Worth mentioning that air-pumps were used for sludge and nitrate recirculations instead of electrical pumps, reducing that way the energy requirements of the WWTP. Efficient control of sedimentation efficiency was performed, based on the periodically conducted settling velocity tests and state point analysis.[3] The critical RAS rates were estimated in order to control sludge blanket level, as well as the overall sedimentation process efficiency.

Energy and labour cost conservation in the studied WWTP is achieved mainly because of the minimized quantities of waste sludge and therefore fewer needs for equipment (sludge dehydration units etc.) and personnel. In conventional activated sludge (CAS) aerobic systems, the observed biomass production rate (Y_{obs}) usually ranges between 0,12 and 0,45 kgMLVSS $kgCOD_{removed}^{-1}$.[1] Consequently, in our case one will expect a waste sludge production of 35 to 130 kgMLSS day^{-1} or 12,77 to 47,45 tonnes of sludge annually, corresponding to approximately 546 to 2028 m^3 year^{-1}. However, the studied system gradually obtained almost negligible Y_{obs} values. At steady state operating conditions, a Y_{obs} mean value of approximately 0,025 kgVSS kgCOD^{-1} was recorded, resulting in an annual sludge disposal of 2,6 tones or 111 m^3 year^{-1}, corresponding to waste sludge minimization rates between 79,5 to 94,5 %.

Under the imposed WWTP operating conditions, successful microbial manipulation and efficient sedimentation were the keys for achieving efficient treatment and negligible quantities of waste sludge, while the use of air pumps for recirculations and the absence of chemical additives (coagulants, flocculants etc.) minimized energy consumption and operating cost. In our case, more than 90% of WWTP's operating cost stems from the energy demands of the aeration system, which provides the necessary air oxygen for the highly aerobic bioreactors (DO > 4 mg L^{-1}) and for the recirculation air pumps. The rest energy demands of the WWTP are because of the submerged anoxic tank mixer, the 2 electrical pumps and of the SCADA system.

Due to the nature of the slaughterhouse wastewater (high organic and nitrogen load) and the high biomass concentrations, the oxygen consumption in the WWTP is relatively high. Oxygen consumption in the WWTP occurs due to organics and nitrogen oxidation, as well as due to biomass respiration. Based on literature[1] and by taking into account the relatively small amount of oxygen that is released in anoxic bioreactors due to denitrification, the calculated oxygen consumption values were approximately 130 kgO_2 d^{-1}, 90 kgO_2 d^{-1} and 190 kgO_2 d^{-1} for organics oxidation, ammonia oxidation and biomass respiration respectively, leading to a total oxygen demand of 410 kgO_2 d^{-1}. The actual amount of oxygen required is approximately 940 kgO_2/d, calculated by taking into account the effects of salinity, surface tension, temperature, elevation, diffused depth, the desired oxygen operating level (4 mg L^{-1}), the effects of mixing intensity and the basin configuration.

A 7,5 kW electrical air blower is used in order to feed both the aeration system and the recirculation air pumps, consuming approximately 168 kWh d^{-1}. Nitrate and sludge recycle is performed using air pumps connected to the aeration system, thus no additional energy is demanded. The only additional energy consumption is due to the use of two 250 watt submersible electrical feed pumps (maximum consumption of 12 kWh d^{-1}). Consequently, the total energy consumption is approximately 180 kWh d^{-1}.

It must be noted that energy consumption in a WWTP can be expressed in relation to volumetric feed, to COD or BOD_5 removal as well as to TKN removal (when separate electromechanical equipment is used). In our case, the energy cost is referred to both carbon and nitrogen elimination because carbon oxidation and nitrification processes are performed in the same bioreactor by common electromechanical equipment. Consequently, by expressing kWh $kgBOD^{-1}$ or kWh $kgCOD^{-1}$, nitrogen removal is included. In this study, the specific energy consumption is 1,13 kWh $kgBODremoved^{-1}$ and 0,62 kWh $kgCOD^{-1}$.

These values are lower than most of the reported specific energy consumptions of conventional activated sludge (CAS) processes, which range between 0,95 to 3,3 kWh $kgBODremoved^{-1}$ or 0,66 to 3,65 kWh $kgCOD^{-1}$.[1,23,25] Furthermore, the studied WWTP consumes significantly lower energy than membrane bioreactors (MBR), which are considered to demand 150% to 300% more energy (Williams, 2007).[1,24-25] Only membrane aerated biofilm reactors and anaerobic processes were found to have lower specific energy consumption, with values approximate 0,25 kWh $kgCOD^{-1}$ and 0,1 kWh $kgCOD^{-1}$ respectively (Pitas et al., 2010; WEF, 2009; Mizuta and Shimada, 2010; Yand et al., 2010; Bodík and Kubaská, 2013).[15-19] Anaerobic treatment can also produce energy by biogas exploitation, thus any comparison with this anaerobic processes is somehow meaningless. Worth mentioning that the SNdN, which is a simultaneous nitrification/denitrification bioreactor, can alternatively be used as an anaerobic bioreactor and thus can utilize a portion of influent organics for biogas valorization, if the necessary electromechanical equipment is installed (membrane cover, biogas filters etc).

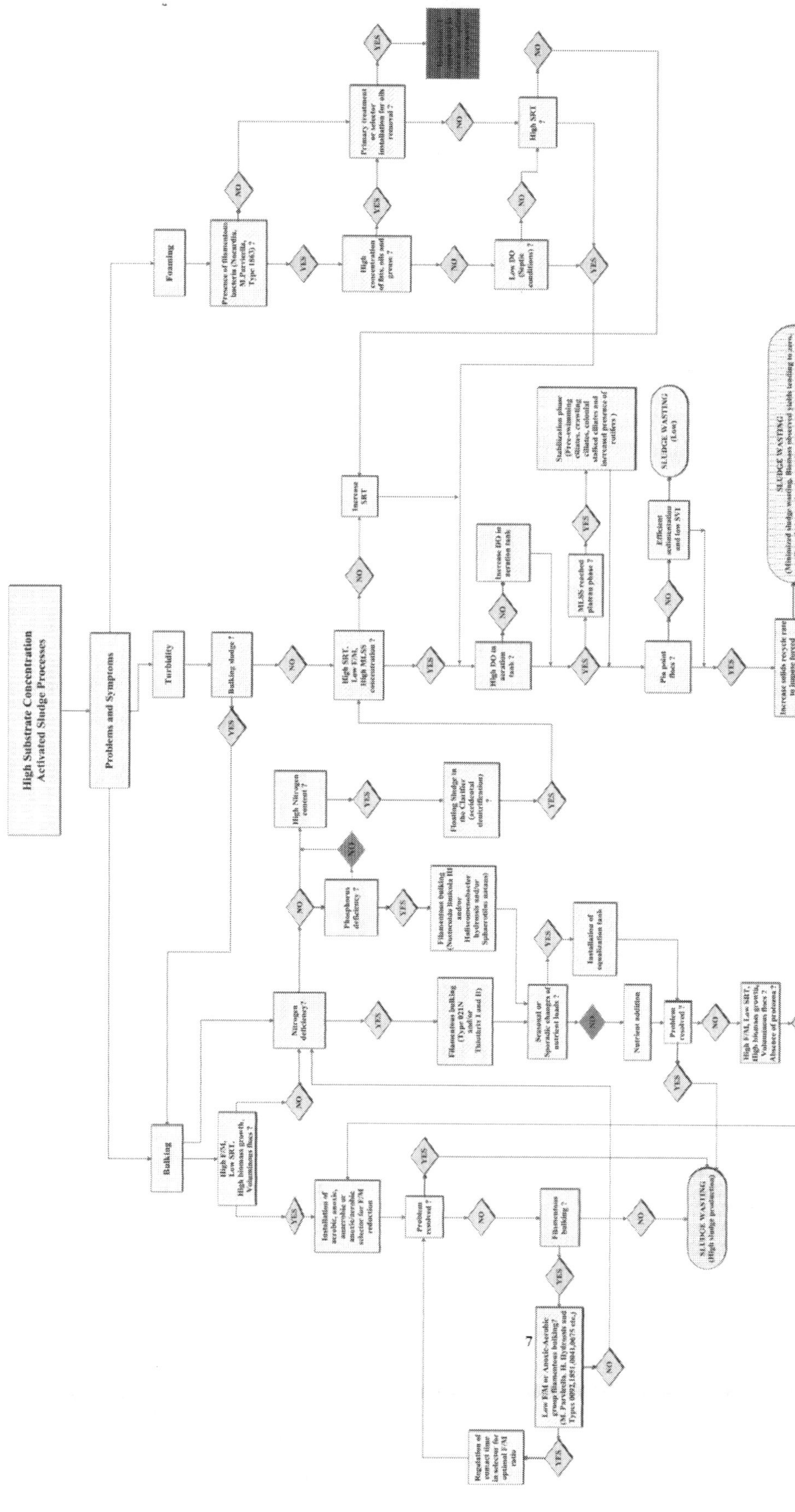

Figure 3. Flowchart of WWTP processes control – Logical steps towards the best solution of WWTP operating problems

During the operation of an AS WWTP and depending on imposed operating conditions, biomass biological and morphological characteristics may alternate significantly. The growth of undesired microbial species, such as filamentous bacteria, fungi etc., could create foaming problems, reduce sludge flocs' cohesiveness (high sludge volume index, SVI) and hinder settling ability of sludge in the clarifiers. The main operating conditions that affect biomass characteristics are activated sludge age (t_c), hydraulic retention time (HRT), DO concentration, food to microorganisms ratio (F/M) and availability of nutrients (C,N,P).

In order to avoid problems related to undesired microbial species growth, such as sludge bulking due to filamentous bacteria growth, successful microbial manipulation and regular monitoring of biomass characteristics had to be performed. It is well known that high concentrations of soluble readily biodegradable organics enhance filamentous bacteria growth.[26] The excessive growth of filamentous bacteria hinders sedimentation efficiency and deteriorates effluent quality. Therefore, the utilization of influent soluble, readily biodegradable COD in SNdN system prevented filamentous growth in the WWTP, thus SNdN system operated as a microbial selector. The monitoring of biomass by microscopic analysis proved to be a rapid and accurate tool for controlling microbiological characteristics of biomass. Alternations of biomass characteristics can be easily observed with microscopic examination, which in turn can help identify their cause and rapidly suggest a WWTP operating conditions related solution. In Figure 3, a flowchart of the WWTP processes control, with the logical steps towards the best solution, is presented.

4. Conclusions

By successfully imposing complete solids retention AS process, through microbial manipulation, extended starvation conditions, highly aerobic bioreactors, high recirculation rates and controlled sedimentation efficiency, efficient treatment of high strength wastewaters can be performed with negligible waste sludge production. Sludge minimization can reach up to 95%, leading to an over 50% reduction of WWTP construction and operational cost. Furthermore, the limited use of electrical pumps reduces energy consumption resulting in a highly energy efficient WWTP, having lowest specific energy consumption than most of the treatment systems. While energy consumption of the studied WWTP is comparable even with the considered more energy efficient anaerobic processes, the environmental and economic benefit of minimizing waste sludge production renders this technology sustainable.

References

[1] Metcalf and Eddy, 2003. Wastewater Engineering: Treatment and Reuse. Metcalf and Eddy Inc., McGraw Hill, Boston, USA (revised by: Tchobanoglous G., Burton L.F., Stensel H.D.).

[2] Guo, W.Q., Yang, S.S., Xiang, W.S., Wang, X.J., Ren, N.-Q., 2013. Minimization of excess sludge production by in-situ activated sludge treatment processes - a comprehensive review. Biotechnol. Adv. 31, pp. 1386-1396.

[3] Amanatidou Elisavet, Samiotis Georgios, Trikoilidou Eleni, Pekridis George, Taousanidis Nikolaos, 2015. Evaluating sedimentation problems in activated sludge treatment plants operating at complete sludge retention time. Water Research J, Volume 69, pp. 20–29.

[4] Y. Wei, R.T. van Houten, A.R. Borger, D.H. Eikelboom, and Y. Fan, 2003. Minimization of excess sludge production for biological wastewater treatment, Water Research 37, pp. 4453–4467.

[5] Amanatidou Elisavet, Samiotis Georgios, Bellos Dimitrios, Pekridis George, Trikoilidou Eleni, 2015. Net biomass production under complete solids retention in high organic load activated

sludge process. Bioresource Technology J., 182, pp. 193–199.
[6] Rocher, M., Goma, G., Begue, A.P., Louvel, L., Rols, J.L., 1999. Towards a reduction in excess sludge production in activated sludge processes: biomass physicochemical treatment and biodegradation. Appl. Microbiol. Biotechnol. 51, pp. 883-890.
[7] Liu, Y., Tay, J.H., 2001. Strategy for minimization of excess sludge production from the activated sludge process. Biotechnology Advances 19, pp. 97-107.
[8] Loosdrecht, M. and Henze, M. 1999. Maintenance, endogenous respiration, lysis, decay and predation, Water Sci. Technol. 39, pp. 107–117.
[9] Amanatidou Elisavet, Samiotis Georgios, Trikoilidou Eleni, Tsikritzis Lazaros, 2016. Particulate organics degradation and sludge minimization in aerobic, complete SRT bioreactors. Water Research (94), pp. 288–295.
[10] Foladori, P., Andreottola, G., Ziglio, G., 2010. Sludge Reduction Technologies in Wastewater Treatment Plants. IWA Publishing.
[11] Amanatidou Elisavet, Samiotis Georgios, Trikoilidou Eleni, Pekridis George, Tsikritzis Lazaros, 2016. Complete Solids Retention Activated Sludge Process. Water Science and Technology, 73.6, pp. 1364 –1369.
[12] M. Henze, M.C.M. van Loosdrecht, G.A. Ekama, and D. Brdjanovic, 2008. Biological Wastewater Treatment: Principles, Modelling and Design. IWA Publishing.
[13] Henze, M., Gujer, W., Mino, T., van Loosdrecht, M.C.M., 2000. Activated Sludge Models ASM1, ASM2, ASM2d and ASM3. Scientific and Technical Report No.9. IWA Publishing.
[14] Sperandio, M., Labelle, M.-A., Ramdani, A., Gadbois, A., Paul, E., Comeau, Y., Dold, P.L., 2013. Modelling the degradation of endogenous residue and unbiodegradable influent organic suspended solids to predict sludge production. Water Science & Technology 67, pp. 789-796.
[15] Pitas, V.; Fazekas, B.; Banyai, Z.S.; Karpati, A., 2010. Energy efficiency of the municipal wastewater treatment. Journal of Biotechnology 150, pp. 163-164
[16] Water Environment Federation (WEF), 2009. Energy conservation in water and waste water facilities, 1st Ed., WEF Press, McGraw Hill, New York.
[17] Mizuta, K.; Shimada, M., 2010. Benchmarking energy consumption in municipal wastewater treatment plants in Japan, Water Science and Technology 62, pp. 2256-2262.
[18] Yang, L.; Zeng. S.; Chen, J.; He, M.; Yang W., 2010. Operational energy performance assessment system of municipal waste water treatment plants. Water Science and Technology 62, pp. 1361-1370.
[19] Bodík, I.; Kubaská, M., 2013. Energy and sustainability of operation of a wastewater treatment plant. Environment Protection Engineering 39, pp. 15-24.
[20] Bellos, D., 2012. Special compact activated sludge wastewater treatment plant for large and medium size applications. Patent number: 1007711, International Classification (INT.CL8):C02F 9/00, C02F 11/00.
[21] Grady, C.P.L., Daigger Jr., G.T., Lim, H.C., 1999. Biological Wastewater Treatment, second ed. Marcel Dekker, Inc. Revised and expanded.
[22] Maine department of environmental protection, (USA), 2003. Document Describing the State-point Analysis Process. Maine DMR O&M News.
[23] Williams, R., Schuler, P., Comstock, K., Pope R., 2008. Large Membrane Bioreactors of Georgia: A Guide and Comparison. Water Environment Federation Membrane Technology Conference. Atlanta Georgia. pp 548-561.
[24] Chudoba P., Rosenbergová R., Beneš O., 2010. Benchmarking of large wastewater treatment plants, Proc. from Conference Wastewater, STU Bratislava, I. Bodík (Ed.), pp. 385.
[25] Malcolm B., Middlenton R., Wheale G., Schulting F., 2011. Energy efficiency in the water industry, a global research project, Wat. Proc. Techn. 6, pp. 221.
[26] Amanatidou Elisavet, Samiotis Georgios, Trikoilidou Eleni, & Michailidis Avraam, 2016. Influence of Wastewater Treatment Plants' Operational Conditions on Activated Sludge Microbiological and Morphological Characteristics. Environmental Technology J. 37, pp. 265-278.

IOP Conf. Series: Materials Science and Engineering **161** (2016) 012094 doi:10.1088/1757-899X/161/1/012094

How much Energy is Embodied in your Central Heating Boiler?

D. Koubogiannis[1*], C. Nouhou[1]

[1]Department of Energy Technology Engineering, Technological Educational Institute of Athens, Agiou Spyridonos Street, 122 10 Egaleo, Athens, Greece

[*]e-mail: dkoubog@teiath.gr

Abstract. Life Cycle Analysis (LCA) is an important tool in current research to quantitatively assess energy consumption and environmental impact of a building. In the context of LCA, the Embodied Energy (EE) related to the building and the corresponding Embodied CO_2 emissions are valuable data. In such a case, these data concern the constitutive materials of the building and any subsystem, component or equipment in it. Usually, after calculating the mass of these materials, embodied energy values are estimated multiplying them by the corresponding EE coefficients concerning the production of these materials (EE_{MP}). However, apart from transportation energy costs, another part of EE is that consumed for the manufacturing of any item as a finished product. The present work focuses on the manufacturing EE (EE_{MFG}) of central heating boilers in Hellenic dwellings. Six typical boilers of different types are studied. Each of them is analyzed to its constitutive materials and its EE_{MP} is estimated. For four of them, the boiler house where it was constructed in Greece was visited and data were collected. Based on them the corresponding boiler EE_{MFG} values are estimated. The results concerning the EE for material production and manufacturing, as well as the results concerning the corresponding ECO_2 values are discussed and assessed. Benchmark values correlating EE and ECO_2 with the mass or the heat rate of the boiler are extracted.

1. Introduction

Buildings are responsible for about 40% of the total energy consumption and about one third of total CO_2 emissions in Europe [1]. Current practice and legislation focuses on the Near Zero Energy Building (NZEB). Furthermore, current research points to the concept of Life Cycle Zero Energy Building (LCZEB) [2]. In the context of performing Life Cycle Analysis of a building, its Life Cycle Energy (LCE) has to be estimated. LCE comprises of two major parts; its Embodied Energy (EE) and its Operational Energy (OE) [3]. Operational energy refers to the amount of energy consumed by the building to fulfill the requirements concerning water supply, heating, cooling, ventilation, lighting, hot water, sanitation and operation of various appliances and equipment. On the other hand, Embodied Energy (EE) refers to the energy consumption related either directly or indirectly to the building facility, like material production, facility construction, maintenance, replacement, demolition, recovery of each material involved in the building.

1.1 Embodied energy and embodied CO_2 emissions

Three main stages can be distinguished during the life time of a building; the initial stage concerning the period of time from the beginning of its construction till the moment it becomes operational, the period of time that the building is operational (normally this is the more lasting period) and the final stage concerning demolition of the building and removal of its materials. EE of a building is accumulated along with time during the above three stages of its life. The energy consumed during the first stage is the Initial EE (IEE), the energy consumed during the operational stage of the building is the Recurrent EE (REE) and the energy consumed during the third stage of its life is the Demolition EE (DEE). This fact is expressed by the following equation:

$$EE = IEE + REE + DEE \quad (1)$$

IEE consists of two parts, namely the energy consumed directly (IEE_D) for the construction of building facility on site for construction, fabrication, transportation and administration and the energy consumed indirectly (IEE_I) upstream these activities for the production of all the required materials. REE concerns the energy that will have been consumed for the construction, materials, processes and equipment related to maintenance/replacement and retrofit/refurbishment actions during the operational period of the building facility. Demolition EE concerns the energy that will have been consumed for the demolition of the facility, transportation of materials off site, their recycling/reuse and the relevant administration after the end of its life. Like IEE, each of the other two terms, namely REE and DEE consists of two parts, one concerning direct EE and one concerning indirect EE [4]. The direct parts refer to energy consumed on site by equipment, tools, etc or to transport materials and products to the site, while the indirect parts refer to energy consumed to acquire process and manufacture the materials involved at the corresponding stage. Taking the above into account, equation (1) now becomes

$$EE = (IEE_I + IEE_D) + (REE_I + REE_D) + (DEE_I + DEE_D) \quad (2)$$

According to equation (2), EE can be separated in direct (IEE_D+REE_D+DEE_D) and indirect (IEE_I+REE_I+DEE_I) amounts of energy. From the six terms appearing in equation (2), the second one IEE_D is the one concerning on site construction of the building facility. The term left of it concerns energy consumed upstream this activity, while the terms to the right concern energy consumed (or to be consumed) downstream (with respect to time).

A companion concept concerning the environmental impact of Embodied Energy is that of the embodied carbon emissions or Embodied CO_2 (ECO_2), defined for each of the EE terms defined above, to be the corresponding amount of CO_2 emissions to the atmosphere. In this context, the life cycle CO_2 of the building is defined to be the CO_2 quantity corresponding to the life cycle EE of the building.

According to the literature [5], [6], the two basic categories of methods for calculating EE rely either on the Process Analysis or the Input-Output (IO) Analysis. Each of them has its own pros and cons [4], so Hybrid Analysis methods have been proposed in order to combine the advantages of both approaches and facilitate more comprehensive and accurate analysis. Furthermore, hybrid analysis has been proposed in the forms of process-based and IO-based hybrid analysis [4].

Unfortunately, independent the method of analysis used, EE databases suffer from problems of variation and incompatibility. According to [3], most of the previous studies either followed the International LCA standards, or they did not follow any standards, so the authors refer to the necessity of the development of a global database and compiled a list of parameters, mostly local, that are responsible for the lack of a global database. Such parameters are the methods of EE estimation, the building design, the construction methods, the kind and quantity of the constructive materials, the system boundaries or the geographical locations.

1.2 Previous and present work
In previous works by the first author, the focus was placed on the Initial indirect EE (IEE_I) of the materials and equipment comprising typical Hellenic dwellings. In these works, the constitutive materials met in such a building were classified in two main categories; the Electro-Mechanical Installations (EMI) and the Civil Engineering Materials (CEM). In [7], [8], [9] the EE of the EMI items were estimated for typical Hellenic dwellings, while in [10] similar work has been conducted for the CEM of typical Hellenic buildings.

In order to estimate the total EE and the related ECO_2 emissions of a whole building or even a component of it, a good material analysis is necessary. In all the above mentioned works, the material analysis, i.e. the breakdown of the set of items under consideration (mechanical or civil engineering) in their constitutive materials was the first step. The calculation of their corresponding masses was the second step, while at a third step, their EE / ECO_2 were obtained by multiplying mass values by the corresponding EE / ECO_2 coefficients of the materials. Although, these coefficients are nationally dependent parameters, available values from the open literature were used [11], since there is lack of a comprehensive Hellenic database. Although not being a rigorous approach, it provides the capability to obtain important practical results for initial guidance.

Since the coefficients used by [11] rely on process analysis, such an approach suffers from truncation errors [6]. The latter are related to the definition of the method boundary. Among the issues that are missing in such a procedure is to take into account the energy required for the manufacturing of the main compound items. Such an item is for example, the boiler in the EMI set. In the previous works, the contribution of the boiler to the IEE concerned the value corresponding to the energy contained per kg for each of its constitutive materials. However, in order to have a finished product in the form of a complete boiler, energy was also consumed in its production line and this contribution was missing. The assessment of the latter part is the objective of the present work.

In the present work, the focus is on the indirect EE of central heating boilers as a standard appliance of building services. This EE is considered as a sum of two terms, namely the EE concerning the production of materials constituting the boiler (EE_{MP}) and the EE concerning the manufacturing process of the boiler in its production line at the boiler construction house (EE_{MFG}). There are some works in the literature referring for example to the manufacturing energy of various products [12] or the derivation of metrics for product assembly and equipment [13]. The three-step methodology mentioned in the previous section is used herein to calculate EE_{MP}. Regarding the second term, this is calculated by estimating the energy consumption at each process of the production line. A way to accomplish this is by multiplying the nominal electrical power of the machine used at each stage by the corresponding average processing time.

In the light of the above, the present work focuses on the energy consumed for the manufacturing of typical domestic central heating boilers. The aim is to assess the contribution of this kind of energy to the overall EE by comparing it with the energy concerning the production of the constitutive boiler materials. Some standard oil/gas boilers, either made by steel or cast iron, as well as an autonomous compact boiler unit are studied. Their EE_{MP} and EE_{MFG} values are calculated and compared each other. Benchmark energy values concerning the manufacturing process of typical central heating boilers in Greece are extracted and their significance in terms of contribution in EE is assessed. Conclusions are made and future research directions are described.

2. Methodology - Test cases
In order to calculate the energy consumed for the production of materials used to construct a boiler (EE_{MP}), a three-step methodology [9] is implemented. According to this, the first step is a material analysis, i.e. the breakdown of the product under consideration to its constitutive single materials. The calculation of the corresponding masses was the second step, while at a third step, EE and ECO_2 values were obtained by multiplying mass values by the corresponding EE and ECO_2 coefficients of the materials. Although, these coefficients are nationally dependent parameters, available values from the open literature were used [11], since there is lack of a comprehensive Hellenic database. Since the coefficients used by [11] rely on process analysis, such an approach suffers from truncation errors [6];

the latter are related to the definition of the method boundary. So, although not being a rigorous approach, it provides the capability to obtain important practical results for initial guidance.

On the other hand, in order to calculate the energy consumed for the manufacturing of a boiler (EE_{MFG}), its production line is analyzed to stages. At each stage, this energy is estimated by means of multiplying the nominal power of the machine used by the average processing time. The required information can be either estimated by examining the production line at the boiler house or directly provided by the manufacturers. Electrical energy so obtained, is then transformed to primary energy for the sake of compatibility with EE concerning material production. According to the national regulation in Greece, the conversion factor that is used (in energy performance of buildings) to calculate primary energy from final energy use is 2.9 for electricity.

Similarly, the CO_2 emissions due to EE_{MP} will be denoted by $ECO_{2,MP}$, while that due to EE_{MFG} will be denoted by $ECO_{2,MFG}$. $ECO_{2,MP}$ values are computed by the same three-step methodology mentioned above for EE_{MP}, where the corresponding ECO_2 coefficients [11] are used. $ECO_{2,MFG}$ values are computed by the EE_{MFG} values (in kWh_e) and the fact that, according to the national regulation (in energy performance of buildings), the conversion factor to calculate CO_2 emissions from final energy use is 0.989 $kgCO_2/kWh_e$ for electricity.

Central heating boilers are generally classified to three categories according to their heating power, namely small boilers (<50 kW_h), medium boilers (50-400 kW_h) and large boilers (>400 kW_h). Furthermore, according to their main material of construction, they are classified to steel boilers and cast iron boilers. Three steel boilers and three cast iron boilers are studied herein. All of them are gas/oil central heating boilers. Independent their material of construction, the three of them are small boilers, while the other three are medium ones.

Table 1 presents the six boilers selected herein as test cases and the names to be used for them.

Table 1 Typical small and medium size central heating boilers.

Case	Boiler Type	Heat Rate [kW_{th}]	Size
B1	Steel oil-gas boiler	67.7	Medium
B2	Steel oil-gas boiler	46.7	Small
B3	Steel oil-gas unitary unit	35.0	Small
B4	Cast iron oil-gas boiler	67.7	Medium
B5*	Cast iron oil-gas boiler [14]	60.0	Medium
B6*	Cast iron oil-gas boiler [15]	35.0	Small

*Only for the calculation of energy for material production (EE_{MP}) (no data for manufacturing, EE_{MFG}).

To obtain the required information, in order to estimate the energy consumption at each stage of the production line, as well as to analyze boilers B1-B4 to their constitutive materials, two boiler construction houses in Athens (Greece) were visited and the relevant information was recorded. Details on the description of the production line and the processes taking place, as well as the calculation of the manufacturing energy are described in [16].

3. Results and discussion

In what follows, results are presented concerning the two parts of embodied energy under consideration herein, namely EE_{MP} and EE_{MFG}, as well as the corresponding results for ECO_2. It is reminded that, according to Table 1, among the six cases studied, B1, B2, B3 are steel boilers, while B4, B5, B6 are cast iron boilers.

3.1 Material production EE and ECO_2

Based on the information by the manufacturers, each of the four boilers B1-B4 was analyzed to its constitutive materials. The corresponding masses were calculated and then transformed to embodied energy and emissions. In addition, corresponding results for two more boilers presented in [14], [15] are also presented.

Table 2 presents the average contribution of the main materials used for steel and cast iron boilers (right), in terms of mass, EE and ECO_2. These data have been extracted from the corresponding data concerning each of the six boilers B1-B6. Due to the fact that similar results concerning steel boilers were obtained and the same happened for the cast iron ones, it was decided to present herein averaged values for each of the two kinds of boilers.

Table 2 Average contribution in terms of mass, EE and ECO_2 of the main materials used for steel boilers (left) and cast iron boilers (right).

Steel boilers (B1, B2, B3)				Cast iron boilers (B4, B5, B6)			
Material	**Mass [%]**	**EE [%]**	**ECO_2 [%]**	**Material**	**Mass [%]**	**EE [%]**	**ECO_2 [%]**
Steel	94.1	94.9	95.6	Cast iron	92.1	85.4	89.2
Cement	4.8	1.0	2.3	Steel	1.1	1.1	1.0
Plastic	0.9	2.9	1.3	Cement	0.2	0.0	0.1
Aluminum	0.2	1.2	0.9	Plastic	1.0	2.9	1.3
				Aluminum	0.0	0.1	0.1
				Copper	5.0	8.8	7.4
				PEF	0.6	1.6	0.9

According to Table 2, it is obvious that steel and cast iron are the dominant materials in terms of mass, EE and ECO_2, since they are the main construction material in each boiler kind. Steel contributes with about the 95% of the total in the three quantities in case of steel boilers. Cement is the second material in terms of mass contribution in steel boilers. Copper is ranked second in cast iron ones. The EE and ECO_2 contribution of the various materials are generally different from their mass contribution. This is due to the fact that materials participating in low quantities may have great EE coefficient, like for example in the cases of aluminum, copper and plastic. It should be marked that the EE coefficients are expressed in terms of primary energy [MJ_p].

Table 3 presents EE results concerning material production for each of the typical boilers studied herein. Apart from absolute values, energy values normalized by the boiler mass and the boiler heat rate are also provided. According to this table, it is interesting to notice that EE_{MP} is about 24 times the boiler mass in case of steel boilers or 27 times in case of cast iron boilers (thus an average of 25 independent the kind of boiler). This is rather logical, since in both kinds of boilers, there is one dominant material. Concerning EE_{MP} as a function of the boiler heat rate, it can be noticed that as the heat rate becomes lower the ratio EE_{MP}/Heat Rate becomes greater, meaning that EE_{MP} does not scale with heat rate. The average values are about 60 MJ_p/kW_{th} for steel boilers and about double 122 MJ_p/kW_{th} for cast iron boilers.

Table 3 EE_{MP} for typical steel and cast iron boilers.

Case	**Heat Rate [kW_{th}]**	**Mass [kg]**	**EE_{MP} [MJ_p]**	**EE_{MP}/Heat Rate [MJ_p/kW_{th}]**	**EE_{MP}/Mass [MJ_p/kg]**
B1	67.7	202	4826.6	71.3	23.9
B2	46.7	88.3	2124.3	45.5	24.1
B3	35.0	92.9	2235.6	63.9	24.1
Average				**60.2**	**24.0**
B4	67.7	313	7847.8	115.9	25.1
B5	60.0	218	6118.3	102.0	28.1
B6	35.0	185.7	5214.5	149.0	28.1
Average				**122.3**	**27.1**

Table 4 presents ECO_2 results concerning material production for each of the typical boilers studied herein. Absolute, as well as normalized CO_2 values by the boiler mass and heat rate are also provided.

According to this table, the ratio of $ECO_{2,MP}$ to the boiler mass is 1.7 for steel boilers and 2 in the case of cast iron boilers. Average values of the $ECO_{2,MP}$ to heat rate ratio are about 4.4 $kgCO_2/kW_{th}$ for steel boilers and about double 8.9 $kgCO_2/kW_{th}$ for cast iron boilers.

Table 4 $ECO_{2,MP}$ for typical steel and cast iron boilers.

Case	Heat Rate [kW_{th}]	Mass [kg]	$ECO_{2,MP}$ [$kgCO_2$]	$ECO_{2,MP}$/Heat Rate [$kgCO_2/kW_{th}$]	$ECO_{2,MP}$/Mass [$kgCO_2$/kg]
B1	67.7	202	352.9	5.2	1.7
B2	46.7	88.3	153.6	3.3	1.7
B3	35.0	92.9	161.8	4.6	1.7
Average				**4.4**	**1.7**
B4	67.7	313	596.0	8.8	1.9
B5	60.0	218	438.1	7.3	2.0
B6	35.0	185.7	373.2	10.7	2.0
Average				**8.9**	**2.0**

3.2 Manufacturing EE and ECO_2

The information concerning boilers B1-B4 and required to calculate the energy consumption at each stage of their production line was obtained by two different boiler construction houses in Athens. An example of such information and the relevant calculation of the electrical energy required for the manufacturing of boiler B3 is provided in Table 5. As mentioned in a previous section, the total electrical energy has to be transformed to primary energy [MJ_p] for the sake of compatibility with the EE related to the production of materials.

Table 5 Electrical energy required for the manufacturing of boiler B3.

Process machine	Time [min]	Power [kW]	Electrical Energy [kWh_e]
Metal cutter	7.4	12	1.5
Cord positioning	3.7	5	0.3
Cylinder	8.3	10	1.4
Press	1.5	6	0.2
Rotating shears	5.2	5	0.4
Welding	10	150	25.0
Bending	6	21	2.1
Total			**30.9**

Table 6 presents EE results concerning manufacturing for each of the four boilers B1-B4, for which data were made available. According to this table, EE_{MP} is 2.1 times the boiler mass in case of steel boilers or 0.7 times in case of cast iron boilers. In addition, In addition, average values of the ration EE_{MP}/Heat Rate are about 5.6 MJ_p/kW_{th} for steel boilers and about 3.2 MJ_p/kW_{th} for cast iron boilers.

Table 6 EE_{MFG} for typical steel and cast iron boilers.

Case	Heat Rate [kW_{th}]	Mass [kg]	EE_{MFG} [MJ_p]	EE_{MFG}/Heat Rate [MJ_p/kW_{th}]	EE_{MFG}/Mass [MJ_p/kg]
B1	67.7	202	340.6	5.0	1.7
B2	46.7	88.3	114.1	2.4	1.3
B3	35.0	92.9	322.1	9.2	3.5
Average				**5.6**	**2.1**
B4	67.7	313	217.1	**3.2**	**0.7**

Table 7 presents ECO_2 results concerning manufacturing for each of the four boilers B1-B4. The ratio of $ECO_{2,MP}$ to the boiler mass is 0.2 for steel boilers and 0.1 in the case of cast iron boilers. Average values of the $ECO_{2,MP}$ to heat rate ratio are about 0.5 $kgCO_2/kW_{th}$ for steel boilers and about double 0.3 $kgCO_2/kW_{th}$ for cast iron boilers. Opposite to what happens with the EE_{MP} values per unit heat rate or mass, where cast iron boilers had double value compared to steel ones, in the case of EE_{MFG} per unit heat rate or mass, cast iron boilers have about the half value compared to steel ones.

Table 7 $ECO_{2,MFG}$ for typical steel and cast iron boilers.

Case	Heat Rate [kW_{th}]	Mass [kg]	$ECO_{2,MFG}$ [$kgCO_2$]	$ECO_{2,MFG}$/Heat Rate [$kgCO_2/kW_{th}$]	$ECO_{2,MFG}$/Mass [$kgCO_2/kg$]
B1	67.7	202	32.3	0.5	0.2
B2	46.7	88.3	10.8	0.2	0.1
B3	35.0	92.9	30.5	0.9	0.3
Average				**0.5**	**0.2**
B4	67.7	313	20.6	**0.3**	**0.1**

3.3 Overall EE and ECO_2 results

Table 8 summarizes EE_{MP} and EE_{MFG} for the four typical steel and cast iron boilers studied above. Two quantities are computed for each boiler, namely the ratio of the boiler mass to heat rate and the ratio of EE_{MFG} to EE_{MP}. Average mass to heat rate is 2.5 kg/ kW_{th} in average for steel boilers, while in case of cast iron ones this is greater, with a value 4.62.5 kg/ kW_{th}. The EE related to manufacturing is about the 9% of the EE related to production of the boiler materials for steel boilers, while this is lower, about 3%, in case of cast iron boilers.

Table 8 EE_{MP} and EE_{MFG} for typical steel and cast iron boilers.

Case	Heat Rate [kW_{th}]	Mass [kg]	Mass/ Heat Rate [kg/ kW_{th}]	EE_{MP} [MJ_p]	EE_{MFG} [MJ_p]	EE_{MFG}/ EE_{MP}
B1	67.7	202	3.0	4826.6	340.6	0.07
B2	46.7	88.3	1.9	2124.3	114.1	0.05
B3	35.0	92.9	2.7	2235.6	322.1	0.14
Average			**2.5**			**0.09**
B4	67.7	313	**4.6**	7847.8	217.1	**0.03**

Table 9 summarizes absolute and normalized values of $EE=EE_{MP}+EE_{MFG}$ for typical steel and cast iron boilers. According to it, the EE of a steel boiler (in MJ_p) can be estimated multiplying its heat rate (in kW_{th}) by 65.8. In case of a cast iron boiler, the corresponding multiplier is about the half (119 MJ_p/kW_{th}). Alternatively, and perhaps more reliably, the EE of a boiler (in MJ_p), either steel or cast iron, can be estimated multiplying its mass (in kg) by an average value of 26.

Table 9 Values of $EE=EE_{MP}+EE_{MFG}$ for typical steel and cast iron boilers.

Case	EE [MJ_p]	EE/Heat Rate [MJ_p/kW_{th}]	EE/Mass [MJ_p/kg]
B1	5167.2	76.3	25.6
B2	2238.4	47.9	25.3
B3	2557.7	73.1	27.5
Average		**65.8**	**26.2**
B4	8064.9	**119.1**	**25.8**

Table 10 summarizes absolute and normalized values of $ECO_2= ECO_{2,MP}+ ECO_{2,MFG}$ for typical steel and cast iron boilers. According to it, the ECO_2 of a steel boiler (in $kgCO_2$) can be estimated multiplying its heat rate (in kW_{th}) by 4.9. In case of a cast iron boiler, the corresponding multiplier is about double (9.1 $kgCO_2/kW_{th}$). Alternatively, and perhaps more reliably, the ECO_2 of a boiler (in $kgCO_2$), either steel or cast iron, can be estimated multiplying its mass (in kg) by a value of about 2.

Table 10 Values of $ECO_2 = ECO_{2,MP} + ECO_{2,MFG}$ for typical steel and cast iron boilers.

Case	ECO_2 [kg]	ECO_2/Heat Rate [kg/kW_{th}]	ECO_2/Mass [kg/kg]
B1	385.2	5.7	1.9
B2	164.4	3.5	1.9
B3	192.3	5.5	2.1
Average		**4.9**	**1.9**
B4	616.6	**9.1**	**2.0**

4. Conclusions – Future work

Several central heating boilers for Hellenic dwellings were studied. By utilizing data that were kindly made available from two boiler construction houses, the initial indirect Embodied Energy (EE) and Embodied ECO_2 (ECO_2) emissions were calculated. These data refer to the main constitutive materials of the boilers, as well as to the manufacturing stages along their production line. The mass quantities of the materials were used to estimate the part of EE concerning the production of the materials used for the construction of the boiler (EE_{MP}). The data related to the stages in the production line were used to estimate the part of EE concerning the manufacturing of the boiler as a finished product (EE_{MFG}). Correlations of these EE values, as well as of the corresponding ECO_2 values, with the heat rate and mass of the boiler were searched and relevant benchmark values were extracted for both steel and cast iron boilers. The practical utility of such values could be the estimation of the EE and ECO_2 quantities of a boiler by means of knowing its heat rate (e.g. from the heating technical study of the building) in order to be utilized in the context of Life Cycle Assessment of the building. As a continuation to this research, it is planned to apply similar studies in a greater range of boilers (heat rates, boiler houses, etc) in order to assure or modify the benchmark values found herein.

Acknowledgement

The authors would like to acknowledge boiler construction houses "THERMA-E.G. & P. Mandilaras" and "SAMARAS, Central heating natural gas applications", for providing practical information and valuable data.

References

[1] Energy in Figures – Statistical Pocketbook 2015 edition, Brussels: European Commission. https://ec.europa.eu/energy/en/statistics/energy-statistical-pocketbook.
[2] Hernandez P and Kenny P. From net energy to zero energy buildings: Defining life cycle zero energy buildings (LC-ZEB). Energy and Buildings 42 (2010) 815-821.
[3] Dixit MK, Fernandez-Solis JL, Lavy S, Culp CH. Need for an embodied energy measurement protocol for buildings: A review Paper. Renewable and Sustainable Energy Reviews 2012;16 (6):3730-3743.
[4] Dixit MK. A Framework for an Improved Input-output-based Hybrid Method for Embodied Energy Calculation. Proc. 51st ASC Annual International Conference hosted by Texas A&M University in College Station, Texas April 22-25, 2015; http://ascpro0.ascweb.org/archives/cd/2015/paper/CPRT314002015.pdf.
[5] Treloar GJ. Extracting embodied energy paths from Input-Output tables: Towards an Input-Output-based Hybrid Energy Analysis Method, Economic Systems Research, 9(4), 375-391, 1997.
[6] Lenzen M. Errors in conventional and Input-Output-based Life-Cycle Inventories, Journal of Industrial Ecology, 4(4), 127-148, 2001.

[7] Koubogiannis DG, Daskalaki A and Balaras CA. A contribution to Building Lifecycle Analysis: Embodied energy analysis of mechanical installations for a typical urban Greek dwelling. 3rd International Exergy, Life Cycle Assessment, and Sustainability Workshop & Symposium (ELCAS3), 7-9 July, Nisyros–Greece, 2013.
[8] Koubogiannis DG, Lavoutas A, Lekkas A and Balaras CA. Estimation of Embodied CO2 in Electro-Mechanical Installations for an Urban Hellenic Dwelling. International Conference on Buildings Energy Efficiency and Renewable Energy Sources (BEE RES 2014), 1-3 June, Kozani–Greece, 2014.
[9] Koubogiannis, DG and Balaras CA. Embodied Energy in Electro-Mechanical Installations of Hellenic Dwellings. Proc. International Conference Energy in Buildings, Athens: ASHRAE Hellenic Chapter and Technical Chamber of Greece; 2014. http://ashrae.gr/EinB2014/EinB2014_Proceedings.pdf.
[10] Syngros G, Balaras CA and Koubogiannis DG. Embodied CO2 Emissions of Building Construction Materials in Hellenic Dwellings, submitted to the SBE16 International Conference "Sustainable Synergies from Buildings to the Urban Scale", to be held in Thessaloniki, October 17-19 (2016).
[11] Hammond GP and Jones CI. Inventory of Carbon and Energy (ICE) Version 1.6a. Sustainable Energy Research Team, Department of Mechanical Engineering, University of Bath; 2008. http://perigordvacance.typepad.com/files/inventoryofcarbonandenergy.pdf .
[12] Ciceri ND, Gutowski TG and Garetti M. A tool to estimate materials and manufacturing energy for a product, IEEE International Symposium on Sustainable Systems and Technology (ISSST), June 2010.
[13] Feng SC, Senthilkumaran and Brown CU. Energy metrics for product assembly equipment and processes, , Jounal of Cleaner Production; 2014 65: 142:151.
[14] Daskalaki A. Towards the Embodied Energy Analysis of typical urban Greek dwellings. The case of space heating & hot water electromechanical installations, MSc Diploma Thesis, MSc in Energy, Heriot-Watt University; 2012.
[15] Lavoutas A and Lekkas A. Recording and percentage estimation of materials used in the basic electromechanical installations in urban Hellenic dwellings, Diploma Thesis (in Greek), Department of Energy Technology Engineering, Educational Technological Institute of Athens, Greece, 2014.
[16] Nouhou K. Embodied energy in central heating boilers, Diploma Thesis (in Greek), Department of Energy Technology Engineering, Educational Technological Institute of Athens, Greece, 2016.

IOP Conf. Series: Materials Science and Engineering **161** (2016) 012095 doi:10.1088/1757-899X/161/1/012095

An analysis of photovoltaic irrigation system for olive orchards in Greece

N Taousanidis[1] and K Gavros[2]

[1]Mechanical Engineering & Industrial Design Dpt, Western Macedonia University of Applied Sciences, Koila Kozani, Kozani, 50100, Greece

[2]Electrical Engineering Dpt, Western Macedonia University of Applied Sciences, Koila Kozani, Kozani, 50100, Greece

Abstract. Olive tree cultivation is of major importance in Greece. It has been proved that irrigation of olive orchards increases their production. The classic method followed is diesel pump irrigation. Since Greece favours high insolation the alternative of photovoltaic pumping is proposed. A case study for an olive orchard in Crete is studied with the two alternatives. The photovoltaic pumping system is a direct pumping system as olive trees tolerate even deficit irrigation and storage tank increases the cost. A comparison using the Life Cycle Costing method is proposed. Considerations about climate and economic conditions are taken into account and the study concludes with the profound advantage of the renewable system over the conventional one in strict economic terms.

1. Introduction

Olive trees were, traditionally, cultivated without artificial irrigation, exploiting rainfall conditions, even in the Mediterranean climate, characterized by low rainfall. However researchers [1] have shown that drip irrigation systems with a total discharge of 4000 to 9000 m^3 per ha, during the irrigation period, increased the yield by 30 to 52%. Irrigated olive trees apart from the higher production use less field area and demand increased fertilization. Consequently demand more energy than the traditional ones.

Greece is the third country in olive oil production and first in black olive production cultivating around 150 million of olive trees, comprising 87.21% of its tree cultivation. Their contribution to the local and national economy is of major importance.

The above mentioned characteristics of this cultivation underline the necessity of abundant energy for irrigation purposes with no grid topology limitation, thus call for solar energy which through photovoltaics is a well established practice since 1977 [2]. Greece is situated between 41°45′ and 34°48′north latitude with an average insolation of 4.56 $kWh/m^2/day$ but ranging from 4.21 (lowest) to 6.61 (highest) during the irrigation period (April to September for olive trees), offering ideal conditions for solar energy utilisation.

The vast majority of off-grid irrigation systems in Greece use diesel pumps. However as the photovoltaic industry develops, photovoltaic module costs present a 22% annual cost reduction in the last few decades [3]. Meanwhile the unstable future of oil prices (significant rise between 1999-2008, ups and downs between 2009-2016) [4] and the possibility of bankruptcy in one third of gas and oil companies [5] does not support a stable future for diesel as a fuel. Already in 2013 ISES Solar World

Congress, R.Foster and Costa [6] presented case studies with great success in use of PV in water pumping applications with new technological innovations. Proving the economic viability of such a system tailored to the Greek climate, type of olive orchards and market environment will push the productivity and reduce the production cost in this sector. The aim of this paper is to prove that exists a viable solar solution for the irrigation of olive orchards as an alternative to a Diesel one.

2. System analysis

The system chosen is a direct pumping system that pumps directly the water to the distribution system avoiding the use of an elevated storage tank, which distributes the water exploiting the gravity.

The main components of a PV irrigation system are the power station (PV array, controller and possibly inverter) the pumping station (DC/AC motor, pump) and the distribution system (network, emitters) (figure 1). These three systems have to cooperate in the most efficient way to tackle the variable needs of olive trees in water. This is not as simple as it sounds, as the power source is variable and intermittent and the irrigation needs are also varying but not in the same mode. The storage tank could also act as a buffer, smoothing the time differences, however increasing the system cost and eliminating the flexibility (even the portability for small scale) of the system. For this reason a thorough analysis of the factors affecting the sizing of the system should follow.

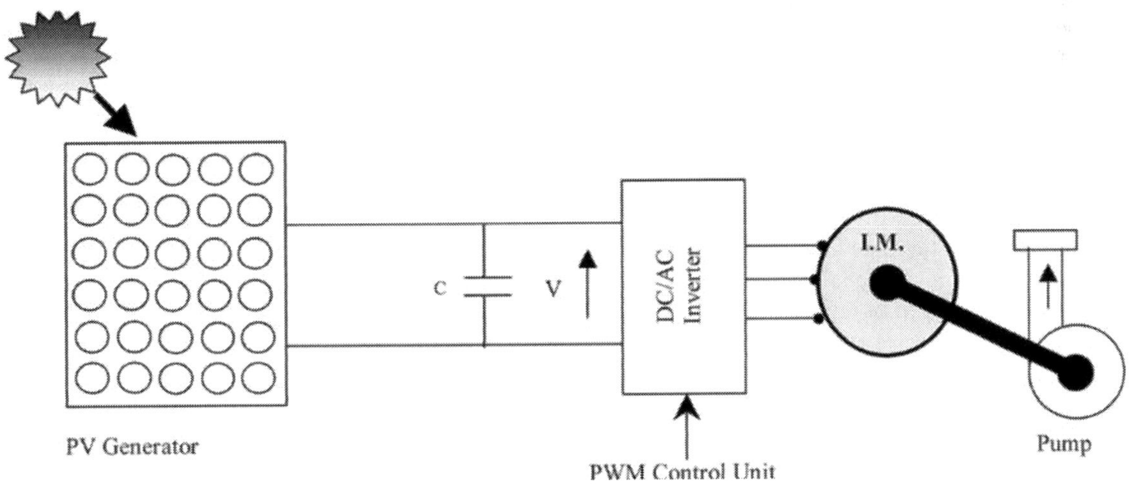

Figure 1 PV direct pumping system [7]

2.1 The distribution network

The most appropriate irrigation systems are the drip irrigation and the small water sprinklers system [1]. Drip irrigation is the most widely applied system and our research focuses on this system. In such a system two are the crucial parameters for network's proper operation: the total pumping head H and the volume of water pumped V. The total pumping head equals to the sum of static head which is the suction head (height from suction point till pump) and the discharge head (height from pump to storage inlet) and of dynamic head which is the frictional losses and the discharge pressure at the irrigating outputs (equation (1)). On the other hand the necessary volume of the water is calculated through a complicated process based on the crop evapotranspiration as proposed by FAO [8].

$$H = h_{stat} + h_{dyn} \quad (1)$$

where

h_{stat} : the static head

h_{dyn}: the dynamic head which can be expressed as a function of the water volume V [9]

2.2 The pumping station

Assuming that the amount of water for a certain day is known [8] the pump and consequently the motor, calls for the optimum operation (speed) which obviously is variable and thus an electronically variable speed pump is necessary. This kind of operation can best be met by variable speed AC motor as 3-phase asynchronous motor with a generic frequency inverter [7]. The produced electrical energy E should cover the hydraulic energy E_h taking into account the pumping station efficiency η_{ps} (equation (2)),

$$E = E_h \ \eta_{ps} \tag{2}$$

the hydraulic energy for a certain day is given by the following equation

$$E_h = H\, V\, \rho\, g \tag{3}$$

where

ρ: water density

g: standard gravity

2.3 The power station

Obviously the sizing of the power station has to satisfy the amount of energy E, thus the energy produced by the photovoltaic system has to match that amount. However the PV array energy production depends on parameters related to the array and on climate conditions, mainly insolation. The variability of the weather conditions and size optimization demands the use of a maximum power point tracker (MPPT). In order to minimize the cost of PV system, we have chosen a non sun-tracking system with a tilt equal to the local latitude

3. Economic analysis

The methodology chosen to evaluate the feasibility of the direct pumping PV system over a Diesel one takes into account all parameters affecting on these systems such as; capital cost of the components, installation, structures, fuels, operating and maintenance (O&M) costs.

The method of the Life Cycle Costing (LCC) is widely used to evaluate the financial viability of a system [11, 12, 13]. To evaluate the LCC of a photovoltaic system of water pumping the period of analysis is supposed to be 25 years.

The life-cycle cost of a solar PV system consists of the initial capital investment C_0 and the present value of operation and maintenance costs OM_{pv}.

$$LCC_{PV} = C_{0PV} + OM_{pv} \tag{4}$$

The three major elements of the initial capital investment C_{0PV} are the PV array, motor-pump-control unit, and costs of installation. These investments depend on the peak power rating of the PV array.

Operation and maintenance costs include taxes, insurance, maintenance, recurring costs, etc. It is generally specified as a percentage t% of the initial capital cost. All operating costs are escalated at a rate e_0 and discounted at rate d. The life-cycle maintenance for a lifetime of N (25) years is:

$$OM_{PV} = OM_{0PV}\,[(1+e_0)/(d-e_0)]\,(1 - [(1+e_0)/(1+d)^N]) \tag{5}$$

where

$$OM_{0PV} = t\, C_{0PV}/100 \tag{6}$$

For the relevant diesel system the total life-cycle cost LCC_d is the sum of capital cost C_{0d}, LCC_f of fuel cost, LCC_m of maintenance cost and LCC_r of replacement cost.

$$LCC_d = C_{0d} + LCC_f + LCC_m + LCC_r \tag{7}$$

where

$$LCC_f = \text{annual fuel cost}\,[(1+FE)/(d - FE)]\,(1 - [(1+ FE)/(1+d)^N]) \tag{8}$$

with FE the mean annual fuel escalation

$$LCC_m = \text{annual maintenance cost}\,[(1+e_0)/(d-e_0)]\,(1 - [(1+e_0)/(1+d)^N]) \tag{9}$$

$$LCC_r = \Sigma_i\,(\text{item cost}\,[(1+e_i)/(1+d)^{N*i/v+1}]) \tag{10}$$

where

e_i the escalation rate of item prices

N lifetime of the project

i, the number of replacement (1 to v)
v the number of replacements over lifetime of items

We have to underline here that the lifetime of diesel-generator set is around 6 years. Thus equations (8) and (9) present recurring costs over project's lifetime, whereas equation (10) are the non-recurring life-cycle costs which are presented in the 6th, 12th and 18th year of the project (3 replacements).

As we can see there are some initial values depending on the size of the system as C_{OPV}, C_{Od}, annual fuel cost, annual maintenance cost and item cost. Thus is necessary to size the system for the case chosen and later calculate the LCC and compare the systems.

4. Case study

The olive orchard chosen is located in the island of Crete which favours high insolation values (figure 3). The island has a Mediterranean semiarid climate, with an average monthly rainfall from 1 to 11 mm during the irrigation period. Average monthly temperatures range from 11°C in January to 26°C in July, which means that winters are mild and the risk of frost is very low as the absolute minimum temperature is above 0°C. The analysis has been performed per hectare

The PV system is located at 35°12' North latitude and 24°54' Western longitude. The tilt angle of the PV cells is 30°, its azimuth is 0° and. It is assumed that the olive orchard (Olea Europea, var. Koroneiki) is already established and aged approximately 25 years. The trees are spaced 5X5 m and the density is 300 trees per hectare. The water for each tree is estimated to 7 l/hr or 2.1 m^3/hr. The pumping head is taken 40m.

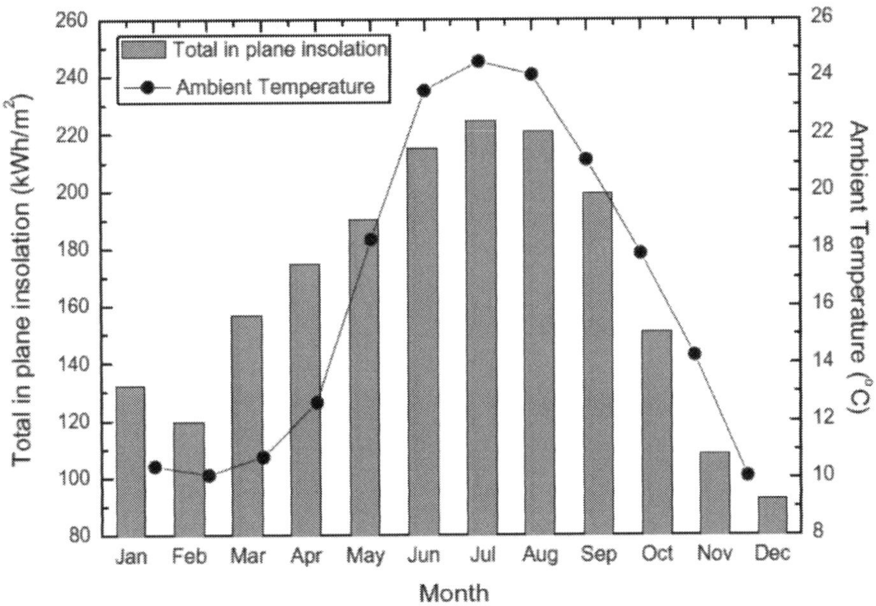

Figure 2 Monthly averaged total in-plane insolation and ambient temperature averaged during the daytime hours [10]

As far as the economic data are concerned according to Oil Institute (http://www.oliva.net/), the price of olive oil varies throughout the year and from one year to another. In this work, an average price of 3.1 €/kg (for extra virgin oil), has been considered. Cultivation costs for an intensive irrigated olive orchard (300 trees per ha) were taken from the report issued by the International Olive Council [14] resulted equal to 2.13 €/kg or for a production of 1000kg/ha, 2130 €/ha. Indirect cost (managerial, insurances, olive mill, etc.) were assumed equal to 10% of total cost. The drip irrigation system investment cost was considered equal to 2500 € /ha. The above data are useful to understand the necessity of irrigation, no matter the system chosen.

According equations (2) and (3), with a pumping station efficiency η_{ps}=0.85 and a daily irrigation time of 8hr the necessary daily energy is 215kWh/ha.

PV arrays can operate in standard conditions (insolation 1000W/m^2, temperature 25°C) with an efficiency of 15%. However due to different losses (temperature, soiling, degradation, internal network, inverter etc) the whole system's efficiency amounts to 10%. With a mean daily insolation on the 30° tilted surface around 6 kWh/m^2, we need a 300m^2 of a 4.5kW PV system, we can satisfy the hydraulic needs of the irrigation system and losses of the pumping station. The initial capital investment of the PV stand alone system was assumed (based on market prices) equal to 1.5 € per peak Watt, or C_{OPV} = 6750€ and with t=1% OM_{OPV} = 675€. Thus for PV maintenance escalation rate e_0=2% (due to gradual ageing) and a discount rate value d equal to 3%, as it seems reasonable in the current economic situation in the European context, equation (6) gives OM_{PV} = 14900€ and the LCC for PV reaches 22325€ throughout the 25 years system's lifetime.

The diesel generator system similarly has to have a rated power of 4,5 kVA. However the initial capital investment of such a system is only C_{0d}=1550€ but the annual maintenance cost arises to 10% of this or 155€. The fuel consumption is around 1.1 l/h or about 1600 l for the irrigation period. With a current price for agricultural diesel around 0.65€, the annual fuel cost, equals to 1040€. Even for the most advantageous case with a zero escalation in fuel price (FE=0), from equation (8) LCC_f =18110€. With the annual maintenance cost of 155€ equation (9) results in LCC_m= 3422€. Finally considering 4 replacements for all the system equation (10) produces a result of LCC_r = 3922€. All these values result to LCC_d of 27000€ throughout the 25 years project's lifetime

5. Results and Discussion

The economic viability of a stand alone direct pumping system against a diesel one for an olive orchard in Crete has been proven. Although the parameters chosen are in favour of diesel (zero annual escalation in price, subsidised agricultural diesel) the PV system is the lower cost option. We have to underline also that environmental parameters, which probably are in favour of PV systems haven't been considered. Also the adoption of a subsidy system for agricultural purposes (similar to that of agricultural diesel) is proposed to be taken seriously into consideration. Finally, the future of PV is much more promising than the "grey" one of diesel and the earliest extended adoption of an innovative and promising technology will lower its cost even more and lead to more efficient systems.

6. References

[1]http://www.moa.gov.cy/moa/da/da.nsf/All/0D925864C995E21BC2257A23004AA26D/$file/ArdevomenesKaliergies.pdf? OpenElement accessed 25 August 2016

[2] Buresch M 1983 *Photovoltaic energy systems* (New York: Mc Graw-Hill)

[3] Cengiz M S, Mamis M S, Akdag M and Cengiz C 2015 A review of prices for photovoltaic systems, *Int. J. on "Technical and Physical Problems of Engineering (IJTPE)*, **7** issue 24 no 3 pp 8-13

[4] https://en.wikipedia.org/wiki/Price_of_oil, accessed 30 August 2016

[5] Deloitte Center for Energy Solutions 2015 *The crude downturn for exploration & production companies, One situation, diverse responses*, http://www2.deloitte.com/gr/en/pages/energy-and-resources/articles/the-crude-downturn-for-exploration-and-production-companies-deloitte-greece.html

[6] Foster R and Cota A 2014 *Solar water pumping advances and comparative economics* Energy Procedia, **57**, pp. 1431-36

[7] Betka A and Moussi A 2004 *Performance optimization of a photovoltaic induction motor pumping system*, Renewable Energy **29** pp2167–81

[8] Allen RG Pereira LS Raes D and Smith M 1998 *Crop evapotranspiration. Guidelines for computing crop water requirements* FAO Irrigation and Drainage Paper 56. FAO. Roma

[9] Reca J Torrente C Lopez-Luque R and Martínez J 2016 *Feasibility analysis of a standalone direct pumping photovoltaic system for irrigation in Mediterranean greenhouses*, Renewable Energy **85** pp 1143-54

[10] Kymakis E Kalykakis S and Papazoglou T 2009 *Performance analysis of a grid connected photovoltaic park on the island of Crete*, Energy Conversion and Management **50** pp433–8
[11] Brandemuehl, MJ and Beckman, WA 1979 *Economic evaluation and optimization of solar heating systems* Solar Energy **23**, pp1-10
[12] Purohit, P 2007 *Financial evaluation of renewable energy technologies for irrigation water pumping in India* Energy Policy **35** (6), pp3134-44
[13] Bouzidi, B Haddadi, M and Belmokhtar, O 2009 *Assessment of a photovoltaic pumping system in the areas of the Algerian Sahara*, Renewable and Sustainable Energy Reviews **13**, pp879-86
[14] http://www.internationaloliveoil.org/documents/index/388-economia/1815-international-olive-oil-production-costs-study/ accessed 30 August 2016

Measurements of electric and magnetic fields, in heavy vehicles parking space, in the vicinity of a power station with 150kv to 20kv transformers

Stavros Gkanatsios[1], S Grigorescu[1], A Pliatsios[2], M Gkanatsiou[2], E Panagiotou[2], E Boukouvala[2], K Gavros[2], D Mitropoulos[2]

[1] Polytechnic University of Bucharest, Romania,
[2] Technological Educational Institute of Western Macedonia, Greece

E-mail:stavros.gkan@gmail.com

Abstract. The present paper investigates the electric and magnetic fields of extremely low frequencies in substations which step down voltage in Kozani, which is a city of about 65.000 inhabitants. In the substation, apart from the presence of voltage distribution transformers there are also power cables of 150 kV generated in pillars as well as power cables of 20 kV and 380V for the power supply of the city. Pillars with high, medium and low voltage power cables cross the parking space of heavy vehicles.

1. Introduction

The problem of power consumption and energy saving solutions has been one of the most significant global issues in contemporary society. As conditions seem to be worsening in the last decades, owing to high percentages of urbanization and industrialization, the use of energy reduction methods and efficient power management systems is essential. However, apart from the increase in demand for energy, there is also created an energy hub of concentration points, such as the points in electrical nodes. Obviously, the larger the urban area, the greater or more energy nodes should be built.

Needless to say, the energy nodes, which are built, are closely associated with the emitted pollutants (i.e. air, soil, water pollutants, etc.). In this case, as energy nodes are electric power nodes, the resulting pollution will be electric and magnetic field pollution and the potential soil pollution will result from fluid leakage from voltage distribution transformers.

The Electric fields occur either when there is charge concentration under certain conditions (static case), or whenever they exist in a conductor. In the case of a rectilinear current-carrying conductor of infinite length, the electric field, E, is constant, depending on the distance of the measuring point from the conductor, in which there is an electron flow.

E is given either by $$\vec{E} = \frac{1}{4\pi\varepsilon_0} \sum_{i=1}^{N} \frac{q_i}{r_i^2} \vec{r_i} \quad \text{(Newton/Coulomb)} \quad (1)$$

$$\text{or} \quad E = \frac{V}{r} \quad (\text{Volt / m}) \quad (2)$$

The magnetic flux, which is created around the conductor of infinite length, is given by the Biot - Savart law

$$dB = \kappa_\mu \frac{I \cdot dL \cdot \sin\alpha}{r^2} \quad (3)$$

Where I is the current intensity in A, dL is the elementary length of the straight conductor in meters, r is the distance of point A from the primary conductor, α is the angle between r and dL, because if the distance between two supporting pillars are large, then the wires are curved due to the weight (Figure 1.a)

$$\text{Where } \kappa_\mu = \frac{\mu_0}{4\pi} = 10^{-7} \frac{N}{A^2} \quad (4)$$

K_μ is the magnetic constant, and μ_0 - is the magnetic permeability of vacuum

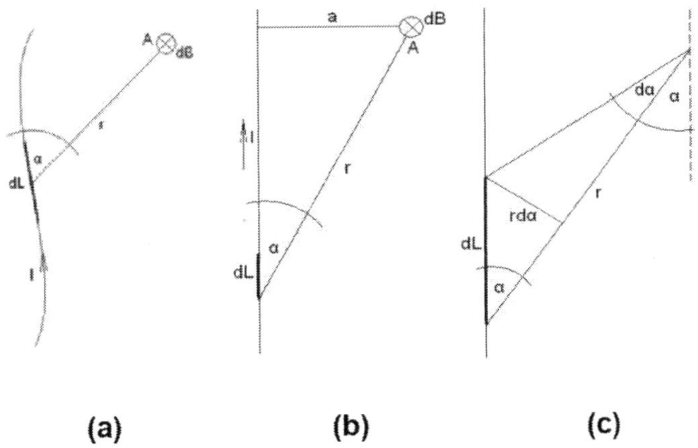

Figure 1: *a*. Infinite length case, *b, c* Infinite length, straight conductor case

In the first approach, if we consider a conductor of infinite length, the equation is simplified, and taking into account the cases of Figure 1.b and Figure 1.c [1]

$$r = \frac{a}{\sin\alpha} \quad \text{and} \quad dL = \frac{rd\alpha}{\sin\alpha} = \frac{ad\alpha}{\sin^2\alpha} \quad (5)$$

$$B = \int_0^\pi \kappa_\mu \frac{I}{a} \sin\alpha d\alpha = \kappa_\mu \frac{I}{a} \int_0^\pi \sin\alpha d\alpha = -\kappa_\mu \frac{I}{a} \cos\alpha \Big|_0^\pi = -\kappa_\mu \frac{I}{a}(-1-1) = \kappa_\mu \frac{2I}{a} \quad (6)$$

$$B = \kappa_\mu \frac{2I}{a} = \frac{\mu_0}{2\pi} \frac{I}{a} \quad \text{with measurement unit for the } [B] = \left[\frac{V \cdot s}{m^2}\right] \quad \text{Tesla (T)} \quad (7)$$

In this particular case, we were asked to study a place that belongs to the municipality of Kozani. To check if there were elevated intensities of electric and / or magnetic fields due to the direct proximity of such an energy hub that will probably affect the people who will be active there.

According to World Health Organization [2]:

The bodies of humans significantly perturb the spatial distribution of ELF electric field. At low frequencies the human body is a good conductor. The key features of dosimetry for humans to expose to ELF electric fields are as follow.

• The electric field inside the body is normally five to six orders of magnitude smaller than the external electric field.

• When exposure is mostly to the vertical field, the predominant direction of the induced fields is also vertical.

• For a given external electric field, the strongest induced fields are for the human body in perfect contact through the feet with ground (electrically grounded) and the weakest induced fields are for the body insulated from the ground (in "free space").

• The total current flowing in a body in perfect contact with ground is determined by the body size and shape (including posture), ratherthan tissue conductivity.

• The distribution of induced currents across the various organs and tissues is determined by the conductivity of those tissues.

• The distribution of an induced electric field is also affected by the conductivities, but less so than the induced current.

• There is also a separate phenomenon in which the current in the body is produced by means of contact with a conductive object located in an electric field.

For magnetic fields the permeability of the tissue is the same as that of the air, so the field in tissue is the same as the external field. The main interaction of magnetic fields is the Faraday induction of electric fields and is associated with current densities in the conductive tissue. The key features of dosimetry for the exposure of humans to ELF electric fields are as follow.

• The induced electric field and current depend on the orientation of the external field. Induced fields in the body as a whole are greater when the field is aligned from the front to the back of the body, but for some individual organs the highest values are for the field aligned from side to side.

• The weakest electric fields are induced by a magnetic field oriented along the vertical body axis.

• For a given magnetic field strength and orientation, higher electric fields are induced in larger bodies.

• The distribution of the induced electric field is affected by the conductivity of the various organs and tissues. These have a limited effect on the distribution of induced current density.

Proximity to high-low voltage transformers, and especially the high-low-voltage cables of ELF waves, for which, we must notice that can penetrate seawater to a depth of hundreds of meters, always creates phobias in the general population since sometimes they are responsible for perturbation in:

Biophysical mechanisms, Neurobehaviour, Neuroendocrine system, Neurodegenerative disorders, Cardiovascular disorders, Immunology and Haematology, Reproduction and Development and Cancer.

The theme cannot be exhausted in any case and for this reason there are continuous research and recommendations for further research as in the example that is indexed in references [2],[3] and [4].

In diverse studies for frequencies above 0 to 100Hz we notice that by far the majority of studies have been conducted on power frequency (50 on 60 Hz) magnetic fields, with a few studies using power frequency electric fields.

Furthermore, fear and mistrust is increased because in terms of dosimetry there is a non-linear dose-effect relationship (cause / effect) according to the hitherto experimental data that are related to biological / health effects particularly in ELF frequencies. In addition, while we can accurately estimate the energy absorbed, either in whole or in part, this is not possible in the microscopic (cellular) level. Because of the above it is possible for an EM field of a much lower frequency than that of microwaves, not to create a problem due to thermal energy, kT, caused by oscillations, but to create other serious biological effects without the effect of heating [6].

2. Technical details

Technical data of the area interest and the requested area.
The object of the study is the measurement of electrical and magnetic fields that belong to the SLF (super low frequency), including EM waves in the frequency between 30 and 300 Hertz in which also belong the frequencies of AC power grids (50Hz and 60 Hz). There is also a conflict designation which includes this frequency range in the extremely low frequency (ELF) in some contexts.

For the particular case, we refer to the electric energy supply of the capital city of West Macedonia, Kozani which is powered by a substation in the immediate vicinity, with the parking space having the following characteristics.

Power station and parking space characteristics:

- 2 great transformers from 150kV to 20 kV of 20 MW each or 25 each under cooling
- More transformers to 380 V.
- Pillars with cables of high voltage of 150kV, 20 kV and 380 which transverse the parking station.

The area of the parking is 41000 m^2 and only a wall separates the power station and the parking area as shown in the following maps. Both the power station and the vehicle parking are placed at a distance of 3 km out of the city of Kozani.

Figure 2. Google map of the station and the parking

IOP Conf. Series: Materials Science and Engineering **161** (2016) 012096 doi:10.1088/1757-899X/161/1/012096

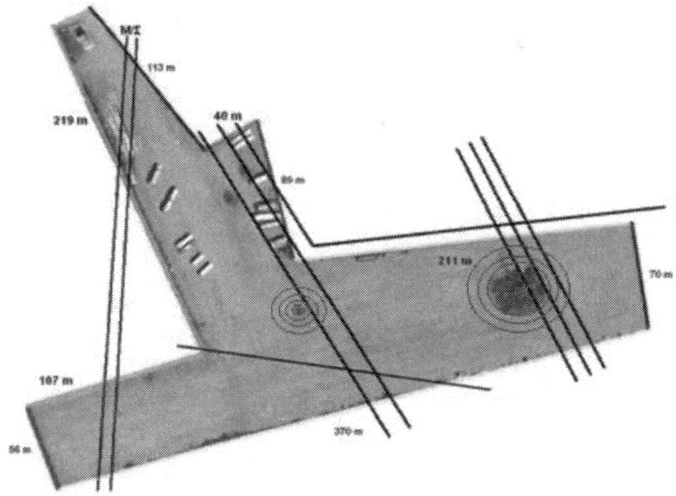

Figure 3. The Parking dimensions and the pillars with the cables of high and medium voltage.

In the municipality of Kozani a problem has occurred due to the existence of this power station, which is connected with the parking, in which there are daily people working. For this reason, we studied the intensities of EM fields in the respective area and compared with those provided by the World Health Organization, as well as those provided by the Greek legislation through the Greek Atomic Energy Commission and listed in the following tables so as to ascertain the conditions for its operation.

Data related with permissible values of electric and magnetic fields, which led various organizations and states.

Table 1. E.U. proposal for the E.M. Fields/Greek law nr.512 /25/4/2002 about limitations for Electric, Magnetic and EM fields.

Frequency Range	**Electric Field E Intensity (V/m)**	**Magnetic Field H Intensity (A/m)**	**Magnetic Inductance B (μT)**
0-1 Hz	---	$3{,}2x10^4$	$4x10^4$
1-8 Hz	10000	$3{,}2x10^4/f^2$	$4x10^4/f^2$
8-25 Hz	10000	4000/f	5000/f
0.025-0.8 kHz	250/f	4/f	5/f
0.8-3 kHz	250/f	5	6,25
3-150 kHz	87	5	6,25

Table II. Values in the frame of the Greek Law (the one with the 50Hz).

General population		**Professional employees**	
Electric Field	**Magnetic Field**	**Electric Field**	**Magnetic Field**
5 kV/m (50 Hz)	100 μT (50 Hz)	10 kV/m (50 Hz)	500 μT(50Hz)
4,2 kV/m (60 Hz)	84 μT (60 Hz)	8.3 kV/m (60 Hz)	420μT(60Hz)

Instrumentation and Methodology.

The apparatus that was used for the measurements was the EFA-300 model equipped with an external B-field Probe for the measurement of the magnetic field values, and with an electric field unit in order to measure electric fields (Figure 4) for which more precautions must be taken because in addition to the magnetic field measurements, the measurements of the electric fields are easily influenced by people or objects near the sensor.

EFA-300 Basic Unit **B-Field Probe 100 cm^2** **EFA-300 E-Field Unit**

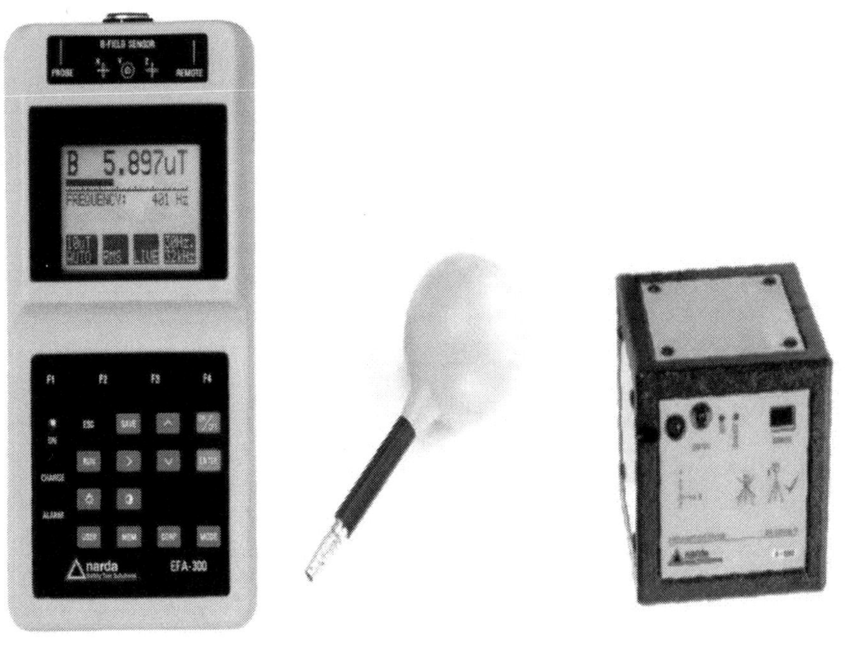

Figure 4. EFA apparatus with its electric and magnetic components

The measurements were carried out in 768 places in the area of interest. A lot of them, particularly the electrical ones were repeated because the area was already used by several cars, which influence the spatial distribution of the electric field.

Measurements were taken in the morning, because the period was in the mid-July and the temperatures during midday in the open parking space were very high.

The measurements were made at 1 meter from the ground which corresponds approximately to the height of the main vital organs in the average human

- Measurements at 0; 1; 5; 10 and 15 meters distance from the separation wall.
- Measurements at 0; 1; 5; 10 and 15 meters distance from the high and medium voltage pillars circularly, within the space of interest.
- Measurements from 0 to 1; 5; 10 and 15 meters away from the high voltage cables and pillars of media interest in the interior space.

3. Results

Sample measurements and graphs

Measurement results for magnetic fields B in μT and for electric fields E in Volt/m are displayed in the following graphs:

Magnetic Fields

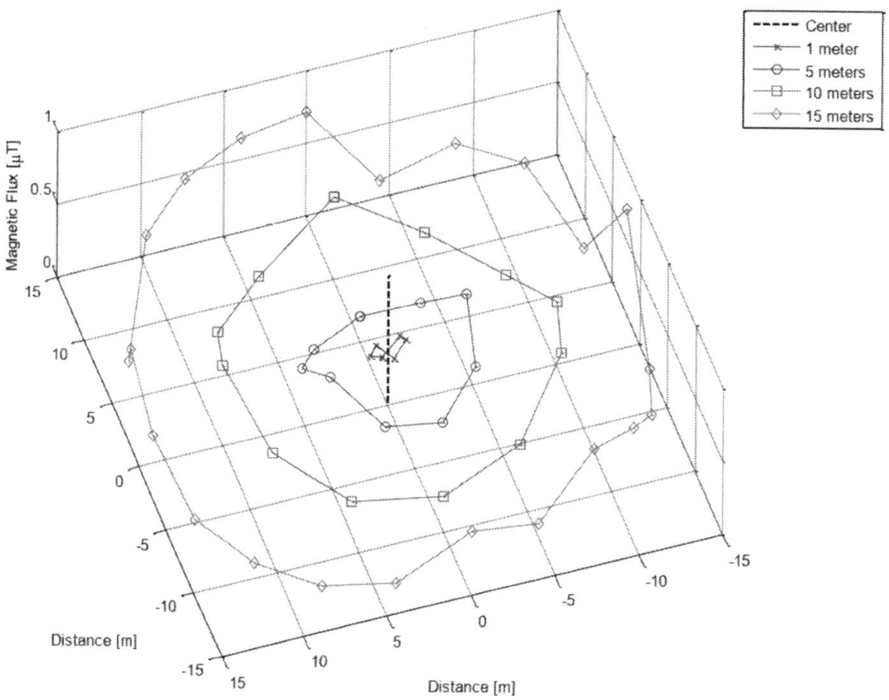

Figure 5. Magnetic field around medium voltage pillar (G point)

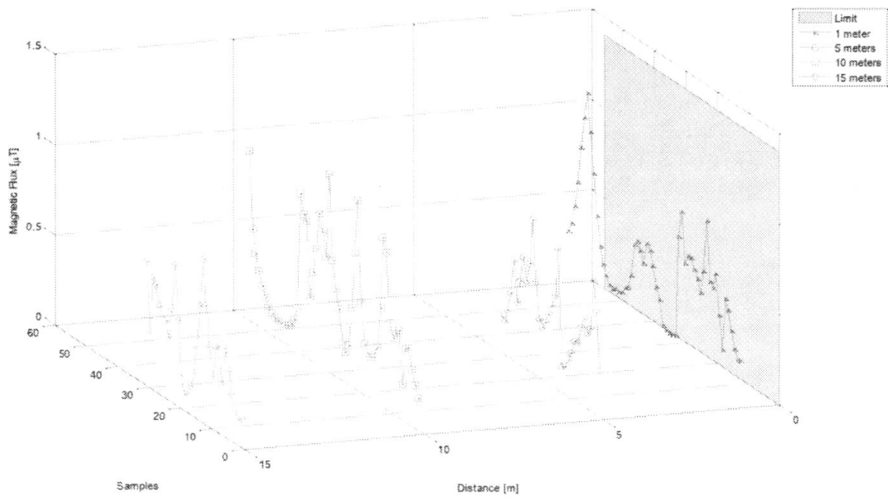

Figure 6. Magnetic Field around the walls (Points A-E)

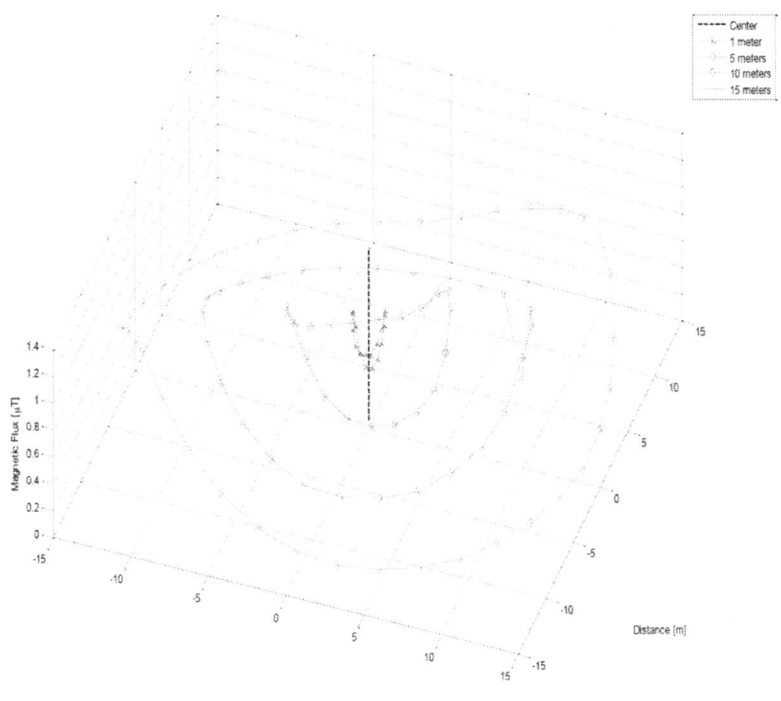

Figure 7. Magnetic field around a pillar of high voltage. (Point F)

Electric fields

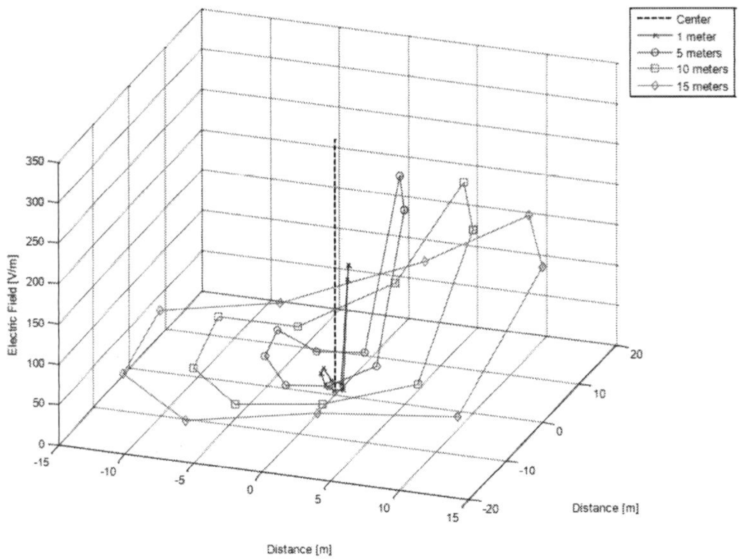

Figure 8. Electric field around the medium voltage pillar (G point)

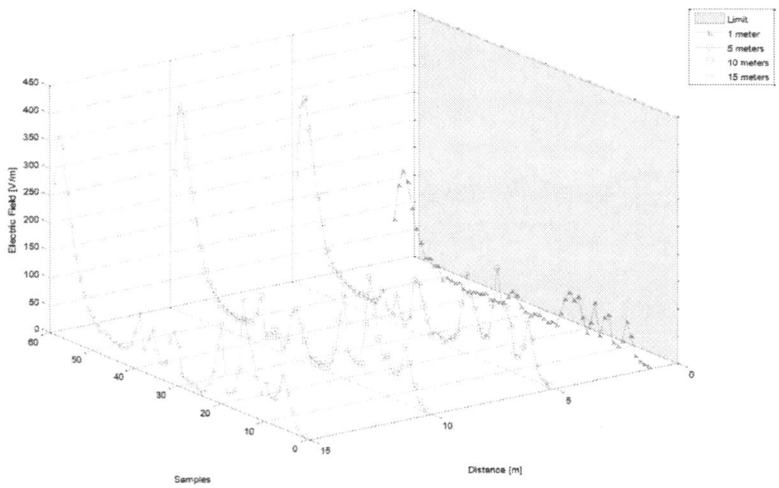

Figure 9. Electric Field around the walls (Points A-E)

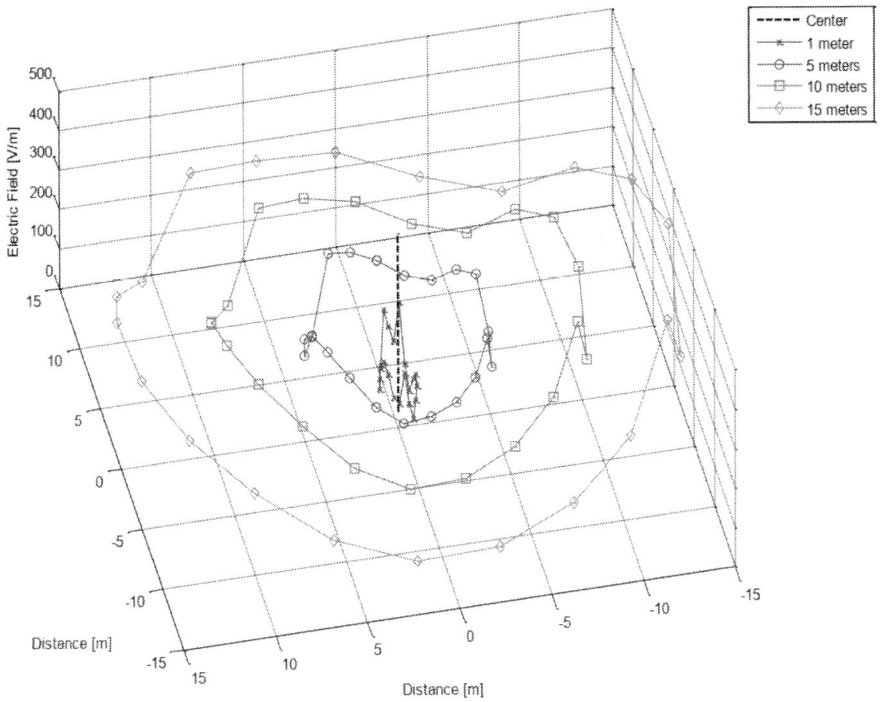

Figure 10. Electric field around the pillar of high voltage (Point).

4. Discussion and Conclusions.

Based on the measurements made on the magnetic fields elevated values were observed near the high voltage line and near the low and / or medium voltage cables where their values do not exceed the permissible limits (Table 2). Also we noticed that the point on the transformer from Hellenic Public Power Corporation (HPPC) (the entrance to the site) generated locally elevated magnetic field, but fell in a short distance (<5 meters) without even exceeding the permissible limits (Table 2). However, the low or medium voltage cables starting from the transformers that are on HPPC side substation, created elevated magnetic fields due to the leakage of a high amount of current which is modulated according to the load of the feeding line.

As for the electric field it is mainly due to the high voltage power lines. The field is stronger under the transmission lines whereas the electric field generated only by the transformer and it is outside the field of the power lines, is lower.

This is true for the measurements shown in the figures above and the same applies for measured values. Increased electric field values were recorded outdoors near the high or medium voltage lines. We observed that in the case of the electric fields the highest values obtained are lower than the statutory lower value than the Greek Atomic Energy Commission (GAEC) set. We report that during the measurements there were parked vehicles which hindered measurements.

An attempt was made to compare the measured values measured with those arising from the theory, using where necessary formulas and fields consisting of two lines, but in many cases there were divergences. The reason is due to several intersecting lines. This shows how to explore in the near future the possibility of the possible creation of a model that responds to the values that were measured.

We have attempted to attribute S.A.R. values (Specific Absorption Rate) since the operation of heavy vehicles parking is constant. The first performance tests have already been conducted using the values measured and based on the following relationship:

$$SAR = \frac{1}{V}\int_{sample} \frac{\sigma(r)\left|E(r)\right|^2}{\rho(r)} d(r) \qquad \text{and finally} \qquad SAR = \frac{\sigma E^2}{\rho} \tag{8}$$

Where σ–is the sample conductivity, E - is the RMS generated internal electric field ρ – is the sample density and V – is the sample volume.

We considered the height at which measurements were approximately equivalent to the height of the main vital organs of man. Also we considered an average human tissue density and an average conductance value. Since the values of the density and conductivity differ too much, e.g. density of fat is 0,9 g/cm^3 while the bones' 1,8 g/cm^3 and the conductivity of muscle from ~ 1.000 S/m reaches ~ 0.005 S/m for the fat, we decided to conclude it in a future study.

References

[1] https://eclass.teiath.gr/modules/document/file.php/.../3%20Μαγνητικό%20πεδίο.doc
[2] ICNIPR 2003, ICNIPR; 2004, Litvak, Foster and Repacholi 2002
[3] Geneva World Health Organization, 2007 WHO *Extremely Low Frequency fields. Environmental Health Criteria*, Vol 238.
[4] Health Council of the Netherlands 2006 *Proposals for research info health effects of electromagnetic fields (0-300Hz) from the Hague* Publications no2006/11E.
[5] EMF from Scientific Comities from 9th plenary meeting on 27 January 2015 *Opinion in Potential health effects of exposure to electromagnetic fields* (SCHENHR).
[6] Dimitris Panagopoulos et.al. PLoS 2003 *Evaluation of Specific Absorbtion Rate as a dosimetric Quantity for EMF Bioeffects* **8(6)** e62663.

The use of the bicycle compatibility index in identifying gaps and deficiencies in bicycle networks

A Ilie[1], C Oprea[2], D Costescu[3], E Roşca[4], O Dinu[5] and F Ghionea[6]

[1,2,3,4,5,6] Department Transport, Traffic and Logistics, Faculty of Transport, University Politehnica of Bucharest, Spl. Independenţei, No.313, Bucharest, 060042, România

E-mail: anamariailie@yahoo.com

Abstract. Currently, no methodology is widely accepted by engineers, planners, or bicycle coordinators that allow them to determine how compatible a roadway is in providing efficient operation of both bicycles and motor vehicles. Previous studies reported a number of approaches to obtain an appropriate level of service; some authors developed the bicycle level of service (BLOS) and other authors developed the bicycle compatibility indexes (BCI). The level of service (BLOS) for a bicycle route represents an evaluation of safety and commodity perceived by a bicyclist reported to the motorized traffic, while running on the road surface. The bicycle compatibility index (BCI) is used by bicycle coordinators, transportation planners, traffic engineers to evaluate the capability of specific roadways to accommodate both motorists and bicyclists and to plan for and design roadways that are bicycle compatible. After applying BCI and BLOS models for the designed bicycle infrastructure network in the city of Dej, one can see that only few streets are Moderately Low compatible compared to the others with a high degree of compatibility that recommends to include them in the bicycle infrastructure network.

1. Introduction

A sustainable transport system must provide mobility and access to all urban inhabitants, offering a safe, cozy and comfortable transport. Achieving this goal is difficult because the transport demand is generated by social groups that are not only different but often have conflicting interests. The same road space is claimed by motorists, public transport and non-motorized users [1].

Despite the advantages offered by non-motorized travels by bicycle in local perception, they are often undervalued, considered to be unsophisticated compared to motorized travels, being viewed as a symbol of impoverishment, as a ludic element or aimed at those with adequate physical condition. Non-motorized travels contribute to social harmonization and integration, mitigate inequalities and disparities (between drivers and pedestrians, expensive and cheap vehicle keepers, driver and passenger etc.), facilitate the mobility of the socially disadvantaged and promote people contact, interaction and social inclusion, dialogue and human relationships in a greater way than the travels by car [2]. If one analyses the transport modes from ecological point of view (pollution, risk of accidents, energy and space consumption), without requiring a calculation of the economic impact of the different transport modes' advantages or disadvantages, it is reasonable to pay attention to bicycle travels and to provide them financing as they deserve [3],[4].

The evolution of the last years' studies in relation to awareness of the role the sustainable mobility plays in increasing the residents' quality of life in urban areas led to the need for implementing

Sustainable Urban Mobility Plans. Also, their development is necessary because, on European Union's demand, the existing of a Sustainable Urban Mobility Plan becomes mandatory for local governments wishing to access European funds in Transport. A Sustainable Urban Mobility Plan is consonant with the investigation of European experience and with the research in mobility integrated field that fundaments a new model of practice and institutional, legislative and regulatory reformulations, technological innovations, creating new regulatory and planning instruments in all areas involved in managing mobility: urbanism and urban planning, transport, traffic, economy, sociology, ecology etc., according to European standards that were not implemented until now.

The Bicycle Level of Service (BLOS) for a route is an evaluation of comfort and safety perceived by a bicyclist regarding the motorized traffic while he runs on a roadway corridor. He identifies the quality of service on the roadway. Previous studies showed a series of approaches to obtain this level of service; some developed the Bicycle Level of Service (BLOS) and some the Bicycle Compatibility Index (BCI).

In the frame of the Sustainable Urban Mobility Plan for the city of Dej elaboration we analyzed the properties of the bicycle network. Severe constrains were given in the planning of the bicycle network due the geometric characteristics of the road infrastructure and the existing built environment. Consequently, the length, the density and the connectivity of the bicycle network are very low. In this paper we present a model based on Bicycle Compatibility Index (BCI) and Bicycle Level of Service (BLOS) developed to evaluate the gaps and deficiencies in bicycle network designed for city of Dej.

Previous studies showed a series of approaches to obtain the quality of service on the roadway for bicycle facilities; some developed the Bicycle Level of Service (BLOS) and some the Bicycle Compatibility Index (BCI) [5], [6]. Both measures offer an evaluation of comfort and safety perceived by a bicyclist regarding the motorized traffic while he runs on a roadway corridor.

The Bicycle Compatibility Index (BCI) and The Bicycle Level of Service (BLOS) can be applied to all categories of roads and contain those variables that the bicyclists regularly use.

BCI and BLOS have been used in various applications:

- Evaluating the "friendship" between the roadway and the bicycle;
- Assisting the bicyclists in selecting a direct and safe travel route;
- Identifying the gaps and deficiencies in a designed bicycle network.

2. The Bicycle Compatibility Index and The Level of Service of a bicycle route

The BCI model was developed to evaluate safety and comfort in bicycle travels, perceived by the user, taking into consideration the traffic on the road. It has been applied to plan the modernization of road infrastructure, ensuring the un-motorized trips in many American cities from States as Florida, New York, Maryland, Virginia and Delaware etc.

There are six BLOS, from A to F, that are determined taking into consideration the following indicators: daily traffic volume, number of lanes, vehicles' speed limit, number of freight vehicles, type of the running surface and average width of cross section. The last levels indicate the need of constructing new infrastructures dedicated to bicycles [7].

The Bicycle Compatibility Index (BCI) was first developed by Road Safety Research Center from North Carolina University for Federal Road Research Administration. BCI is promoted as a procedure to evaluate the "bicycle compatibility" with the road. The main variables like bandwidth, traffic volume and travel speed in addition to the secondary variables are introduced in an equation to obtain a number that is evaluated on a linear scale to determine the compatibility of the bicycle with the road from A (the best compatibility level) to F (the worst compatibility level).

The disadvantage of this approach is that the intersections and discontinuous routes are not taken into consideration. The level of compatibility for a bicyclist varies from Very High to Very Low. In table 1 there are presented the values of BCI associated to LOS [6].

The BCI reflects the comfort level of a bicyclist on the basis of observed road geometrics, surrounding land use and operational characteristics of roadways. From the table 1, it is clear that the lower the BCI value the greater will be the level of comfort a bicyclist experiences. LOS A (represented by an index ≤1, 50) indicates that a roadway is extremely compatible (or comfortable) for the average adult bicyclist while LOS F (represented by an index > 5, 30) is an indicator that the roadway is extremely incompatible (or uncomfortable) for the average adult bicyclist.

Table 1. Bicycle Compatibility Index (BCI) ranges associated with level of service (LOS) designations and compatibility level qualifiers [6]

LOS	BCI range	Compatibility Level
A	≤1,50	Extremely High
B	1,51-2,30	Very High
C	2,31 – 3,40	Moderately High
D	3,41 – 4,40	Moderately Low
E	4,41 – 5,30	Very Low
F	>5,30	Extremely Low

This model predicts the overall comfort level rating of a bicyclist using eight significant variables and an adjustment factor (AF) to account for three additional operational characteristics. The variable with the largest effect on the index is the presence or absence of a bicycle lane or paved shoulder (BL); the presence of a bicycle lane (paved shoulder) that is at least 0.9 m wide reduces the index by almost a full point, indicating an increased level of comfort for the bicyclist. Increasing the width of the bicycle lane or paved shoulder (BLW) or the curb lane (CLW) also reduces the index as does the presence of residential development along the roadside (AREA). On the other hand, an increase in traffic volume (CLV and OLV) or motor vehicle speeds (SPD) increases the index, indicating a lower level of comfort for the bicyclist. The presence of on-street parking (PKG) also increases the index [6].

The other three variables with a potential impact on the comfort level of bicyclists are influenced by the percentage of large trucks or buses, vehicles turning right into driveways, and vehicles pulling into or out of on-street parking spaces.

To determine these variables one must know the following entry data [8],[9]: number of lanes, lane widths, curb lane width or width of the motor vehicle travel lane closest to the curb, bicycle lane (paved shoulder) width, land use adjacent to the roadway, motor vehicle speed, hourly traffic volume by lane in one direction of travel, large truck volume, presence and density of on-street parking, parking lane and percentage of parking occupancy, parking time limits and percentage of right turning traffic.

Thus, the equation used to determine BCI, presented in Eq. 1, is based on Ravadá model [10]. To develop BCI there were selected eight independent variables linked to bicyclists' levels of comfort and three variables considered to be adjustment factors.

$$BCI = 3{,}67 - 0{,}966BL - 0{,}410BLW - 0{,}498CLW + 0{,}002CLV + 0{,}0004OLV + 0{,}22SPD + 0{,}506PKG - 0{,}264AREA + AF \quad (1)$$

Where:

BL represents the presence of a bicycle lane or paved shoulder > 0.9 metres (No = 0, Yes = 1)
BLW - bicycle lane width [m]
CLW - curb lane width [m]
CLV - curb lane volume – vehicles per hour in one direction [veh/h]
OLV - other lane(s) volume same direction [veh/h]
SPD - 85th percentile speed [km/h]
PKG - presence of parking lane with more than 30% occupancy (No = 0, Yes = 1)
AREA - type of roadside development (Non-residential = 0, Residential = 1)
AF - adjustment factor, AF = Ft + Fp + Fr
Ft - adjustment factor for truck volume (ranges from 0 to 0.5)

Fp - adjustment factor for parking turnover (ranges from 0 to 0.6)
Fr - adjustment factor for right turning (ranges from 0 to 0.1)

3. The use of the bicycle compatibility index in identifying gaps and deficiencies in bicycle network designed for Dej city

Dej City is situated on a varied landscape, with altitudes between 220 and 360 meters. The city's streets with the largest traffic flows are generally narrow with tight curves with poor visibility, with slopes between 0.00% and 15.19% [11].

To identify the mobility degree, in October 2015, Transport, traffic and logistics Department from Transport Faculty, University Politehnica of Bucharest together with Dej City Municipality conducted a survey with 660 questionnaires at respondents' residence. The purpose of the survey was to identify the main needs and problems that Dej population that uses cycling as a daily means of transport is facing.

Taking into account the outcome of the survey and given that most streets from Dej city have a declivity greater than the maximum allowable slope for designing cycle lanes (7% on short distances), one proposed to place the bicycle lanes only on the streets that connect the city center to its suburban areas, to high schools and to recreational areas.

Thus, in Sustainable Urban Mobility Plan for the city of Dej, it was proposed a street network for the bicycle lanes (see Figure 1) taking into consideration: the number of users, the type of the streets, the number of lanes, the possibility of placing facilities specific to bicycles on the sidewalk and on the road, the existence of parking spaces at the boundary between the roadway and the sidewalk, the vehicles' speed (from bicyclists safety reasons) and the vehicles' flow in pick hours [12].

Finally, one proposed six bicycle routes that connect the city's main streets (with slopes under 4%), thus uniting the four high schools in the city with the recreational areas, with the suburban areas and with the city center.

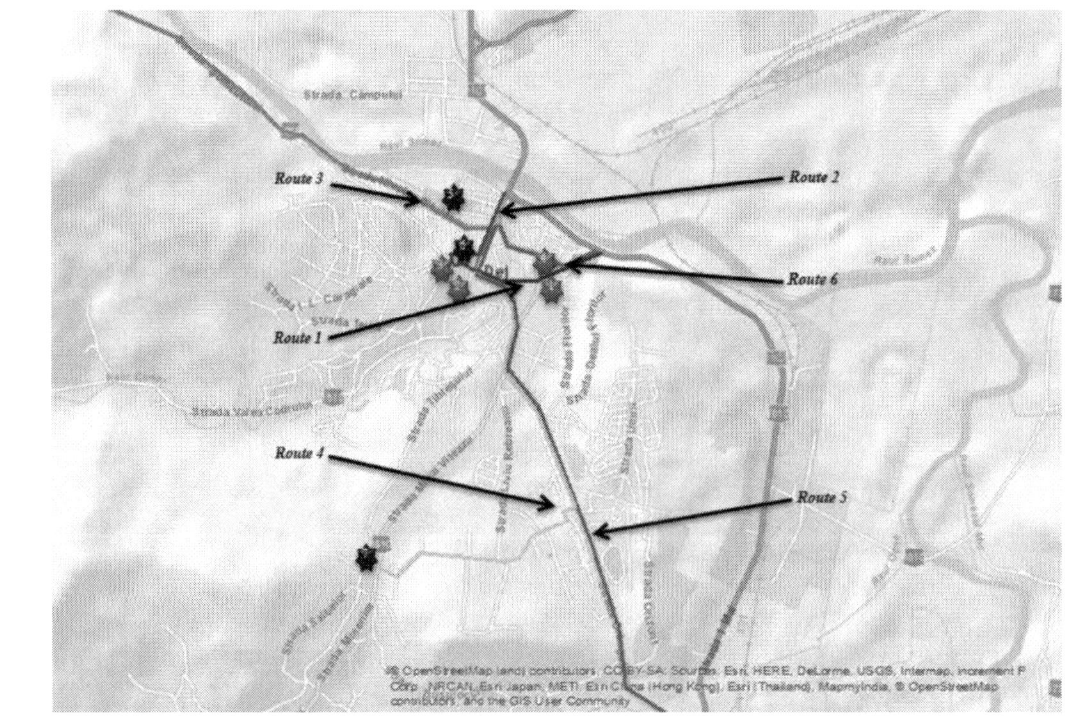

Figure 1. Bicycle lanes routes proposed for the city of Dej (1 - Andrei Mureşanu highscool, 2 - Alexandru Papiu Ilarian highscool, 3 - Constantin Brancuşi highscool, 4 - Someş Dej highscool, 5 – The Great Park, 6 –Bobâlna Square, 7 –Toroc Balnear Park) ([11] [12][13])

IOP Conf. Series: Materials Science and Engineering **161** (2016) 012097 doi:10.1088/1757-899X/161/1/012097

The suggested routes cover about 18 km of bicycle lanes located on the following streets: 1 Mai Street, Mărăşeşti Street, 22 Decembrie 1989 Street, Bobâlna Street, Nicolae Titulescu Street, Şomcutului Street, Avram Iancu Street, Ecaterina Teodoroiu Street, Crângului Street, Nichita Stănescu Street (Figure 2).

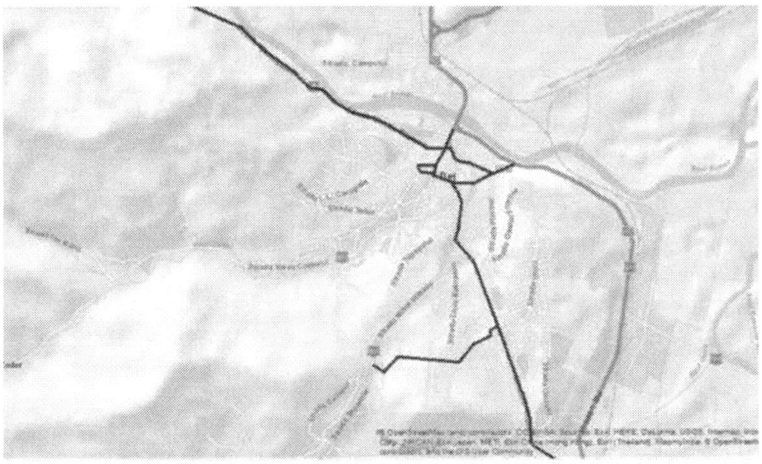

Figure 2. Road network dedicated to bicycle use for the city of Dej

For the ten streets one applies the BCI model to identify the deficiencies and the gaps in bicycle infrastructure network for Dej city. In order to obtain the necessary attributes for BCI and BLOS computation [6],[8],[9], [10], we used GIS procedures (see Figure 3).

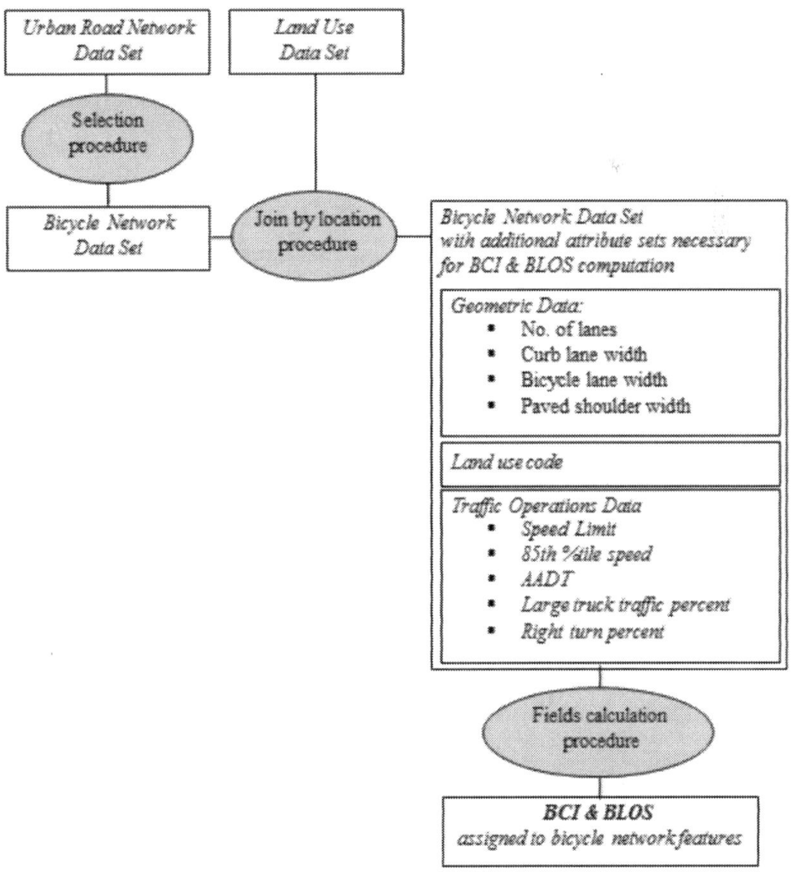

Figure 3. Flow diagram of BCI and BLOS data preparation process

The resulted attributes for the selected streets witch are proposed for bicycle network are summarized in table 2 and table 3.

Table 2. Geometric and roadside attributes for streets witch are proposed for bicycle network in city of Dej ([11], [12])

Feature		Geometric and Roadside Data				
FID	Name	No. of Lanes (one direction)	Curb lane Width [m]	Bicycle Lane Width [m]	Paved Shoulder Width [m]	Residential Area [y/n]
1	Nicolae Titulescu Street	1	3,5	1	0	y
2	Şomcutului Street	1	3,5	1	0	y
3	Nichita Stănescu Street	1	3	0	0	y
4	Crângului Street	1	4	0	0	y
5	Mărăşeşti Street	1	5	0	0	y
6	Bobâlna Street	1	3	0	0	y
7	1 Mai Street	2	3,5	0	0	y
8	Ecaterina Teodoroiu Street	1	5	0	0	y
9	Avram Iancu Street	2	3,5	0	0	y
10	22 Decembrie 1989 Street	2	3,5	0	0	y

Table 3. Traffic operations and parking attributes for streets witch are proposed for bicycle network in city of Dej (source: [11, 12])

Feature		Traffic Operations Data					Parking Data		
FID	Name	Speed Limit [km/h]	85th %tile Speed [km/h]	AADT [veh/day]	Large Truck (HV) [%]	Right Turn (R) [%]	Parking Lane [y/n]	Occup. [%]	Time Limit [min]
1	Nicolae Titulescu Street	50	60	1993	0,015	0,1	n	0	0
2	Şomcutului Street	50	60	2700	0,015	0,1	n	0	0
3	Nichita Stănescu Street	40	50	2560	0	0,1	n	0	0
4	Crângului Street	50	60	10019	0.020	0,1	n	0	0
5	Mărăşeşti Street	40	50	2736	0	0,1	n	0	0
6	Bobâlna Street	30	40	1853	0	0	n	0	0
7	1Mai Street	40	50	5770	0,020	0,1	n	0	0

8	Ecaterina Teodoroiu Street	40	50	11591	0.020	0,1	y	0,30	0
9	Avram Iancu Street	40	50	13705	0,020	0,1	n	0	0
10	22 Decembrie 1989 Street	50	60	7783	0,020	0,1	y	0,30	0

After applying the BCI model for every street from the designed bicycle infrastructure network in the city of Dej using data from table 2 and 3 and Eq.1, the BCI values, the level of service and the bicycle compatibility level can be seen in table 4 and figure 4 which shows the bicycle compatible level for each link based on BCI and BLOS scores.

Table 4. The bicycle compatibility level for the designed bicycle infrastructure network ([6], [8], [9], [10], [11], [12])

Street	BCI	Level of service	Bicycle Compatibility Level
Nicolae Titulescu Street	1,83	B	Very High
Şomcutului Street	1,90	B	Very High
Nichita Stănescu Street	3,29	C	Moderately High
Crângului Street	2,46	C	Moderately High
Mărăşeşti Street	2,32	C	Moderately High
Bobâlna Street	3,00	C	Moderately High
1 Mai Street	3,14	C	Moderately High
Ecaterina Teodoroiu Street	3,90	D	Moderately Low
Avram Iancu Street	3,77	D	Moderately Low
22 Decembrie 1989 Street	4,00	D	Moderately Low

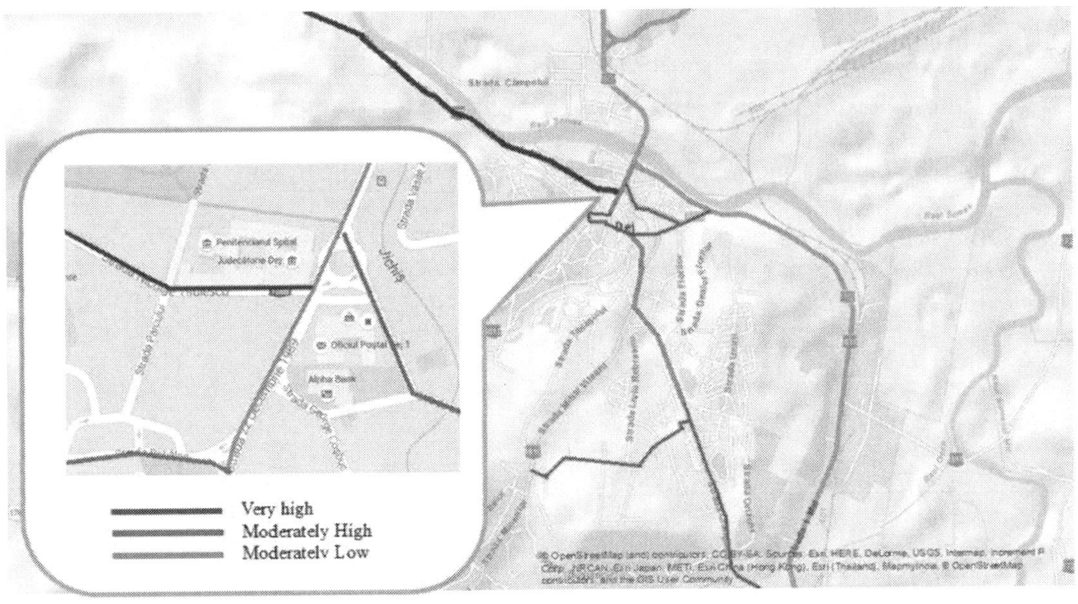

Figure 4. Bicycle Compatibility Level with the streets from the designed bicycle infrastructure network in the city of Dej

4. Conclusions

BCI and BLOS are useful indicators in measuring the appropriateness of roadways for bicycle travel. After applying BCI and BLOS models for the designed bicycle infrastructure network in the city of Dej, one can see that only three streets are Moderately Low compatible compared to the other seven that have a high degree of compatibility that recommends including them in the bicycle infrastructure network.

Ecaterina Teodoroiu Street, 22 Decembrie 1989 Street şi Avram Iancu Street have a level of service D which means that the bicyclist will not enjoy a high comfort on these, thus being considered as deficiencies of the bicycle infrastructure network. Therefore, if the bicycle lanes can't be placed on these streets, they will be placed on the asphalted sidewalks that have a width of approximately 3.50 meters and can support both the pedestrians and the bicyclists.

References

[1] Raicu Ş and Costescu D 2013 Mobility, transport and traffic – Major themes of the sustainable development *Proceedings of Days of the Technical Sciences Academy of Romania* pp 408 - 414

[2] Roşca E, Ruscă A, Ilie A, Ruscă F 2010 Non-motorized transportation. An educational challenge for urban communities *TERUM* **8 (17)** pp 5-13

[3] Ilie A, Ştefănică (Oprea) C, Dinu O, Ruscă F, Burciu Ş 2015 Development of non-motorized infrastructure network in Oradea city using Dijkstra algorithm in ArcGIS, *Applied Mechanics and Materials* **809-810** pp 1067-72

[4] Ilie A and Ştefănică(Oprea) C 2012 On non-motorized movements in large urban areas *AGIR Newsletter* **1** pp 69-74

[5] Shawn M, Stewart W P, Turner S 1997 *Bicycle Suitability Criteria: Literature Review and State-of-the-Practice Survey* (Texas, United States) 3988-1 p 53

[6] Federal Highways Administration 1998 *The Bicycle Compatibility Index: A Level of Service Concept Implementation Manual* (Washington, DC, United States) FHWA-RD-98-095 p 22

[7] Landis B W et al 1997 Real-time human perceptions: toward a bicycle level of service *Transportation Research Record: Journal of the Transportation Research Board* **1578**

[8] Harkey D et al 1998 *The Bicycle Compatibility Index: Level of Service* (Washington, DC, United States) FHWA-98-072 p 116

[9] Somenahalli S 2008 Finding a bicycle route that offers a high level of service in Adelaide *31st Australasian Transport Research Forum* **31** pp 153-164

[10] Ravadá C 2004 Bicycle compatibility indices *East Central Intergovernmental Association*

[11] *General Urbanistic Plan of Dej Municipality* 2015 available at http://www.primariadej.ro/ , accessed: 1.10.2015

[12] Roşca E et al 2015 *Sustainable Urban Mobility Plan for Dej City* (SUMP) Transport, traffic and logistics Department (Bucuresti, Romania)

[13] ArcGIS platform available at http://www.esri.com , accessed: 10.10.2015

The quality of service in passenger transport terminals

C Oprea[1], E Roşca[2], M Popa[3], A Ilie[4], O Dinu[5] and M Roşca[6]

[1,2,3,4,5,6] Department Transport, Traffic and Logistics, Faculty of Transport, University Politehnica of Bucharest, Spl. Independenţei, No.313, Bucharest, 060042, România

E-mail: cristina_stefanica@yahoo.co.uk

Abstract. The quality of service in transport terminals is differently perceived by engineers, economists, transport operators and sociologists. The traveler's perception is nevertheless decisive. The quality of service is well connected with the inside design of terminals, with the facilities in terminals and with the provided service standards. In order to provide a high level of service, the activities taking place in the public transport terminal and the maximum travelers flow size must be carefully analyzed and dimensioned. The purpose of modelling is to find the best route for each traveler from origin (entrance) to destination (exit) through all the intermediate service points, taking into consideration the instant network conditions. In developing the model we consider the walking, the waiting and the serving time. Using a simulation program written in ARENA we determine the waiting time. For validation, the model is used to evaluate the performance level in Bucharest Basarab station. By comparing the total walking distance for the possible routes and the utility function that describes the utility of all activities from entrance to exit we can find the optimal route.

1. Introduction

Transport terminals represent junction points between local, regional and interregional transport [1]. The main objective of a terminal is to serve transport demand by maintaining the travelers' safety, convenience and comfort [2], [3]. When a terminal is designed, all the actors and activities that contribute to the size of the flows into the terminal must be taken into consideration. Despite the modern means of purchasing tickets prior to travel, there are still traditional travelers that buy tickets from tickets offices in terminal. Furthermore, plenty of other activities in the terminal, other than purchasing tickets, take place prior to travel and at the end of it, the terminal becoming a multifunctional center. The study from the present paper is limited to the transport-related activities and not to other functions that modern passenger terminals have today. Future researches will expand the model presented in this early stage.

A pedestrian flow in a terminal is formed by: travelers (for whom the terminal represents their origin; for whom the terminal represents their destination; for whom the terminal represents their transfer station); attendants; employees (employees in the commercial and technical area, cleaning staff etc.).The different areas of the terminal are linked through corridors (walkways), stairs, escalators, lifts or moving walkways. In table 1 the activities from a transport terminal, the actors that execute them (the users) and the needed facilities are centralized.

Table 1. Activities in a passenger transport terminal [4]

Activity	Actors	Facilities
Access, entrance	travelers, attendants, employees	doors, gates, tourniquets, train/bus/plain/etc.
Information	travelers, attendants, employees	boards, electronic displays, signs, arrows
Movement	travelers, attendants, employees	corridors, stairs, escalators, lifts
Waiting	travelers, attendants	waiting areas, platforms, queues
Buying tickets	travelers	tickets offices, tickets machines
Shopping	travelers, attendants	commercial areas
Exit	travelers, attendants, employees	doors, gates, train/bus/plain/etc.

2. Improving service quality in transport terminals – psychological and physiological aspects

The terminal operator must maintain a high level of service inside the terminals reflected through the fluidity of travelers flow and by inducing a sense of concern for their safety, convenience and comfort.

A pleasant waiting environment must be created focusing on the users' psychological (waiting/in service time perception, information and equity) and physiological aspects (noise, lighting, climate and ventilation, available space) [5].

Psychological aspects

The most important cause of stress in our days is induced by the lack of control [6]. Because the waiting time in a queue can't be controlled by users, it represents a very unpleasant experience for them. By giving them the control and by offering them activities that they can do while waiting (something to read, to eat or to watch), information regarding the estimated waiting time, providing FIFO serving discipline avoiding queue re-ordering, their stress can be diminished. A proper social behavior is the key to a waiting tolerated by users [7], [8].

Physiological aspects

The waiting space can contribute to the traveler's mood. The waiting area should have a pleasant architecture and should be nicely decorated. Furthermore, this should be designed so that noise, illumination, climate and ventilation are adequate.

Noise: A noisy environment, especially one with intermittent, high frequencies or low frequencies sounds, is not only annoying but it also can influence health. Studies have shown that noise favors aggressive behavior of passengers [9], [10], [11]. In a waiting area the accepted noise level is 40-48dB. The ambient noise resulted from the conversations can be covered by background music.

Climate and ventilation: The travelers comfort also depends on air ventilation and on temperature and humidity. For the travelers that wait the indicated temperature in winter is between 200 and 240C and in summer between 220 and 270C (taking into account the clothes corresponding to each season). The humidity in summer should be between 20% and 30% and in winter between 50% and 80%.

Illumination: Can influence the travelers' state of mind and their ability to use constructively the waiting time. To create a pleasant environment for the travelers, the illumination in a big room should be irregular for relaxation, while in a small room it should be uniform to create the space impression..

Space: Every person that stands needs space for not feeling embarrassed by the others. The level of service for queuing areas is presented in table 2 [12].

Table 2. Level-of-service criteria for queuing areas (P/m^2)[12]

LOS	Space (P/m^2)
A	≤ 0.83
B	$0.83 - 1.11$
C	$1.11 - 1.67$
D	$1.67 - 3.33$
E	$3.33 - 5.0$
F	≥ 5.0

3. Travelers' behavior models

3.1. Choosing the servers

To complete activities in terminals (table 1), the passengers have two options in choosing the servers [13]: pre-trip choices and en-route choices. For a pre-trip choice, the total route is established by the traveler at the moment he enters into the system. For en-route choice, the optimal route may change during the walking trip, according to the step by step experience of traveler.

A choice model is characterized by:

- *Choice's characteristics* - the entrance and the exit of the travelers, a list of activities that they are going to perform and the alternative locations of each of these activities;
- *The set of alternatives* - all the O-D routes;
- *The choice variables* - walking time between the activities, waiting time, service time, total walking distance, and attractiveness and familiarity of the activity

The utility function UR(t) that describes the utility of all activities from entrance to exit, for a traveler that enters the system at time t is (considering the simplified version in which we take into account only the walking time to a certain service „point"/server, the waiting time in queue and the serving time in a certain service „point"/server):

$$U_R(t) = \alpha \sum_{\substack{a\in A\cup M\cup L \\ a\in R}} T_a^{walk}(t) + \beta \sum_{a\in(C^i\cup P\cup B)} T_a^{wait}(t) + \gamma \sum_{\substack{a\in C^i\cup P\cup B \\ a\in R}} T_a^{serve}(t) \quad (1)$$

Where a represents a link of R route;

T_a^{walk} - walking time on a link;

T_a^{wait} - waiting time;

T_a^{serve} - serving time;

α, β, γ - the weights given to each of the corresponding attributes;

3.2. En-route choice model

The model identifies the optimal route from entrance to exit taking in consideration the instantaneous system conditions. The route choice is based on a variety of factors. It aims the modeling of the movements in a public transport terminal at peak hours and to estimate the travelers' density and comfort in limited capacity conditions [14].

The en-route choice model is characterized by:

- *Choice's characteristics* - consists in identifying the actual position of a traveler, his exiting point and all locations of intermediate stops in terminal at a given moment;
- *The set of alternatives* - consists of all feasible routes from his actual position to exit.;
- *The choice variables* (time-dependent travel time) - the walking time, the waiting time and the service time. Other potential variables are traveled distance (L_r), pleasantness (p), habit (h), pollution and noise levels (L_{noise}), safety (F), trip motive (M_i^{trip}) of pedestrian etc.

The en-route choice model consists of three separate modules:

•Shortest path module - this module calculates all shortest paths at moment t, taking in consideration the travel times at that moment - it may be called "instantaneous shortest path".

• Individual route module - this module determines the next link to be followed by a traveler arriving at a route decision point (a node in the network).

• Walking module - this module calculates individual travel times for every traveler, depending on the current conditions in the link.

The logical diagram for en-route choice model, in which the interaction of these modules is indicated, is shown in figure 1.

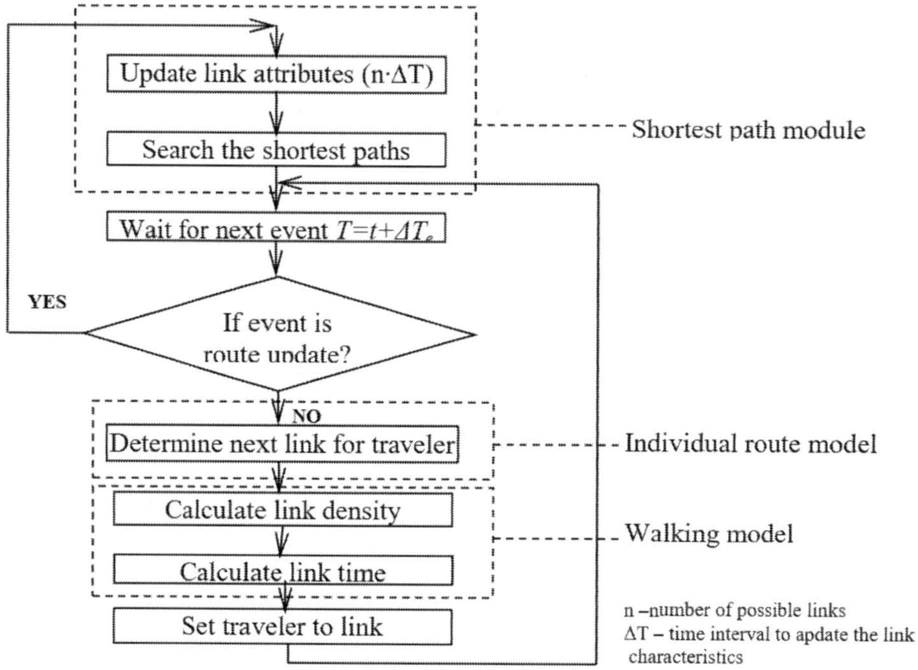

Figure 1. The logical diagram for en-route choice model (Source: [12])

4. Case study – Bucharest Basarab railway terminal

We consider Bucharest Basarab railway terminal with its facilities (figure. 2). To distinguish the areas inside the terminal, different hatchings are used. For modeling the flows inside the station one considers the station hall, the location of the facilities (figure 2) and the travelers flow formalization (figure 3).

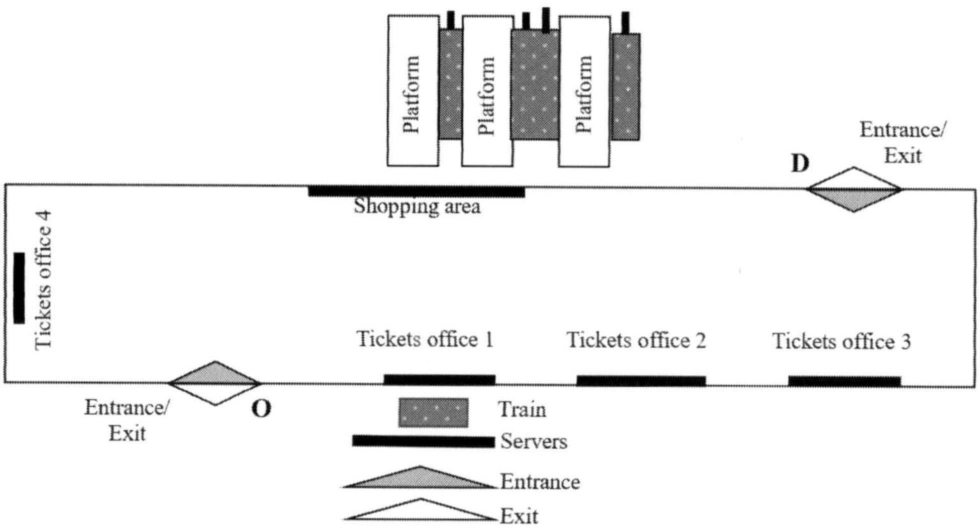

Figure 2. Bucharest Basarab railway terminal

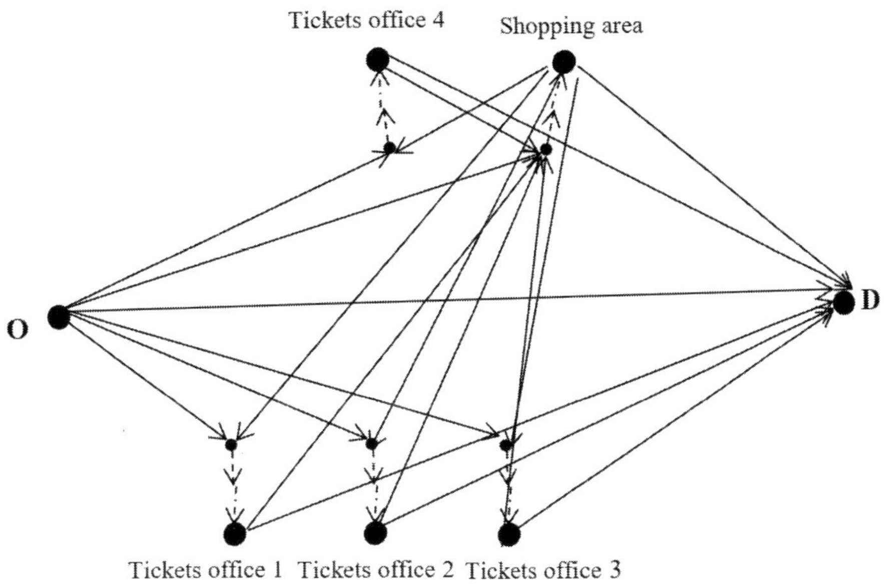

Figure 3. Travelers flow paths

One considers a traveler that enters into the system in node O and exits through node D. He has to fulfill two activities: to buy a ticket from the tickets offices and to buy a cup of coffee from the shopping area. He has four alternatives for the first activity and one alternative for the second activity. The purpose of the model is to choose the servers so that the total time from entrance (origin) to exit (destination) to be minimal. The first step is to identify the possible paths between origin and destination so that the traveler will buy a ticket and a coffee.

The network resulting from these activities (executed in mentioned order) is presented in figure 4.

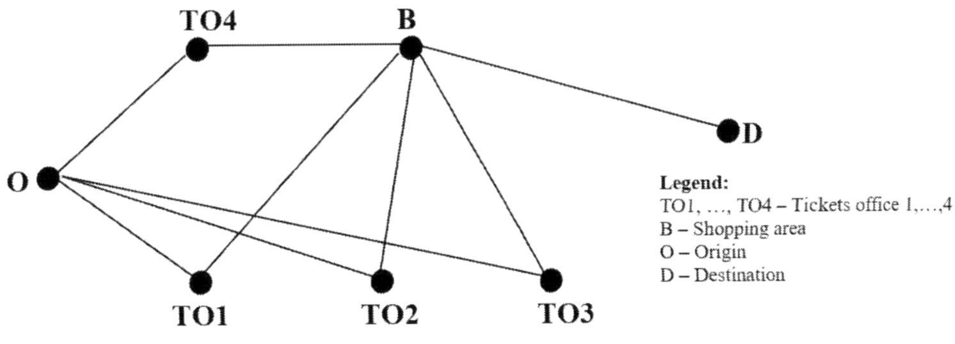

Figure 4. Travelers flow network

Further one identifies the possible paths between origin and destination and determines the total time for these paths.

A traveler needs time to walk from a server to another and because of the changes that appears in the system, the initial minimal route might change. That is the reason why the model should be repeated after every activity.

An important factor is the weight given to each of the corresponding attributes (walk, wait, serve). These weights are considered to be equals.

4.1. The walking time

The walking time at moment *t*, for the traveler *i* that walks with speed v_i (dependent on density *k*) on a link *a* characterized of distance L_a is:

$$T_i^{walk}(t) = \frac{L_a}{v_i(k_a(t))} \tag{2}$$

If the travelers' density is less than 0.5 travelers/m^2, the walking speed is considered 1.6m/s [15]. The distance and time matrixes are shown in tables 3 and table 4.

Table 3. The distance matrix

d[m]	TO1	TO2	TO3	TO4	B	D
O	6	8	10	5		
TO1					9	
TO2					10	
TO3					12	
TO4					6	
B						8

Table 4. The time matrix

t[s]	TO1	TO2	TO3	TO4	B	D
O	4	5	7	4		
TO1					6	
TO2					7	
TO3					8	
TO4					4	
B						5

For determining the waiting times and the serving time at servers a computer simulation model using ARENA software was realized (figure 5). The travelers arrive random (exponential with intensity of 2 travelers/minute) and 30% of the travelers are supposed to buy a coffee in the shopping area. Two situations were considered:

a) the tickets office is randomly chosen (figure 6.a));
b) the tickets office with minimum queue is chosen (figure 6.b)).

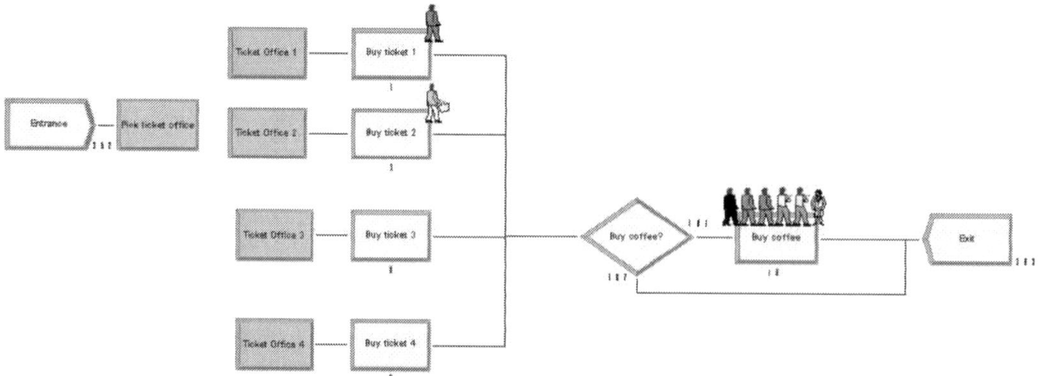

Figure 5. ARENA simulation

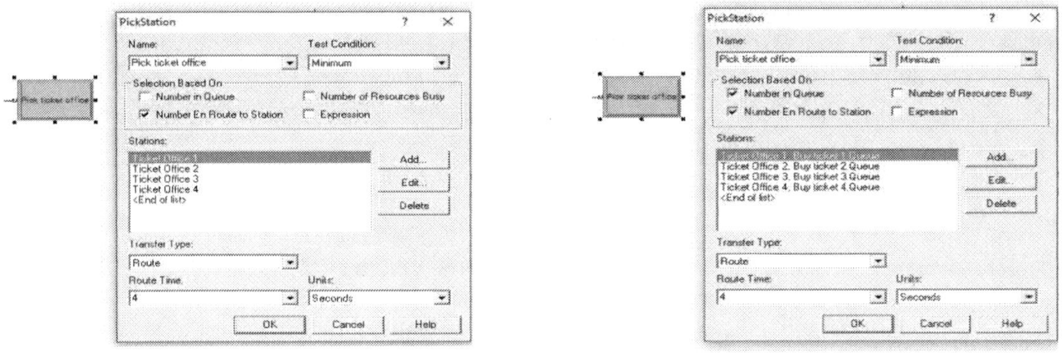

a) b)

Figure 6. ARENA simulation
a) – choosing the tickets office randomly
b) – choosing the tickets office with minimum queue

4.2. The serving time

The servings follow a uniform repartition with minimum and maximum intervals of 0.5 minutes and 1.5 minutes at tickets offices and 1 minute and 2 minutes at shopping area. In the model the average values $T_{serve}^{TO1...TO4} = 60\ s$, $T_{serve}^{B} = 90\ s$ are considered.

4.3. The waiting time

The waiting times after simulation (25 replication in peak hours: 7 a.m. – 10 a.m.) are:

(i) the tickets office is randomly chosen:

$T_{wait}^{TO1} = 324\ s;$
$T_{wait}^{TO2} = 360\ s;$
$T_{wait}^{TO3} = 360\ s;$
$T_{wait}^{TO4} = 396\ s;$
$T_{wait}^{B} = 792\ s$

(ii) the tickets office with minimum queue is chosen:

$T_{wait}^{TO1} = 288\ s;$
$T_{wait}^{TO2} = 324\ s;$
$T_{wait}^{TO3} = 360\ s;$
$T_{wait}^{TO4} = 396\ s;$
$T_{wait}^{B} = 792\ s$

As it can be seen, the waiting time values are smaller in the second situation. Those values will be considered further on. The utilities (Eq.1) for the possible paths between origin and destination are:

$U_{O-TO1-B-D}(t) = 415\ s;$
$U_{O-TO2-B-D}(t) = 427\ s;$
$U_{O-TO3-B-D}(t) = 440\ s;$
$U_{O-TO4-B-D}(t) = 450\ s$

The total walking distance measured in travelers × meters in the morning peak hours (3 h) is:

O-TO1-B-D: 7291; O-TO2-B-D: 8242;
O-TO3-B-D: 9510; O-TO4-B-D: 6023.

IOP Conf. Series: Materials Science and Engineering **161** (2016) 012098 doi:10.1088/1757-899X/161/1/012098

The best O-D route from utility point of view is O-TO1-B-D and from total walking distance point of view is O-TO4-B-D.To improve these parameters one proposes the shopping area to be adapted so that two sellers can work simultaneously. Going through the same steps as above we obtain:

$$U_{O-TO1-B-D}(t) = 263\ s;$$
$$U_{O-TO2-B-D}(t) = 276\ s;$$
$$U_{O-TO3-B-D}(t) = 301\ s;$$
$$U_{O-TO4-B-D}(t) = 310\ s$$

The total walking distance in the morning peak hours will be the same because the distance between the facilities and the number the travelers are the same.

5. Conclusions

As it was described above, the quality of service is connected with the terminals' inside design, illumination, ventilation and provided service standards (promptly information, personnel attitude, waiting time). Moreover, the quality of the service results from the time spent into the system by the traveler. Two quality indicators were determined: total walking distance and the utility.

A solution to improve these quality indicators (the shopping area should be adapted so that two sellers can work simultaneously) was proposed. The utility indicator improved for all four possible routes. The best O-D route from utility point of view remained O-TO1-B-D.The weights given to each of the corresponding attributes (walk, wait, being served) in the utility function were considered to be equals. Taking into account some papers in literature [9],[16], where the waiting time is indicated to be the most unpleasant one, a study that considers this should be developed.

It is obvious that by increasing the number of servers, the waiting times decrease. It is interesting to know if in the time intervals between trains' departures and arrivals, the operating regime of the servers is perhaps one with inactivity. In this case the simulation should be adapted to suit an operating regime with interruptions.

Operating on flexible intervals may be a step in future research. Furthermore, different types of intervention (investing in new facilities, using the existing spaces for different activities etc.) analyzed separately or in combinations can lead to reorganization decisions to increase service quality in the passenger terminal.

6. References

[1] Raicu S, Dragu V, Popa M, Burciu Ş. 2009 About the high capacity public transport networks territory functions *15th International Conference on Urban Transport and Environment* (Bologna) Ed. C.A. Brebia (Wessex Institute of Technology Press, Southampton, U.K.) pp 41-50

[2] Anas A, Arnott R, Small K A 1997 Urban spatial structure *UCTC* **357**, http://www.socsci.uci.edu/~ksmall/JEL%20Paper.pdf , accessed: 20.02.2016.

[3] Ştefănică (Oprea) C, Dragu V, Burciu Ş, Ilie A, Dinu A 2015 Planned traffic stability on a rail network *Applied Mechanics and Materials* **809-810** pp 1073-1078

[4] Ştefanică (Oprea) C 2012 *Integration Of Transport Modes And Technologies In Large Urban Areas* PhD theses (Bucharest)

[5] Dragu V, Roşca E, Ruscă F 2004 Service quality in the terminals joining magistral and urban transport *CODATU XI* pp 123-129

[6] Bateson J E G 1985 Perceived control and the service encounter *The service encounter, managing employee/customer interaction in service businesses* J.A. Czepiel, M.R, Solomon and C.F. Surprenant (Eds.) (Mass.: Lexington Books, Lexington) pp 67-82

[7] Larson R C 1987 Perspective son queues: social justice and psychology of queuing *Operations Research* **35** pp 895-905

[8] Maister D H 1985 The psychology of waiting lines available at: http://davidmaister.com/wp-

content/themes/davidmaister/pdf/PsycholgyofWaitingLines751.pdf accessed: 14.11.2015
[9] Hall W R 1991 *Queuing Methods* (New York: Prentice-Hall Inc) p 476
[10] Jones D M and Broadbent D E 1987 Noise *Handbook of human factors* ed G Salvendy (New York) pp 623-649
[11] TRB 1999 *Transit Capacity and Quality of Service Manual* (Washington, D.C.) p 35
[12] Daamen W 2004 *Modelling Passenger Flows in Public Transport Facilities* (Delft: DUP Science) p 404
[13] Cascetta E 2001 *Transportation Systems Engineering: Theory and Methods* (United States: Springer) p 710
[14] Banos A and Charpentier A 2007 *Simulating Pedestrian Behavior In Subway Stations With Agents* Proceedings of the 4th European Social Simulation Association Toulouse (France) pp 611-621
[15] Mahmassani H S et al 2009 *Incorporating Weather Impacts in Traffic Estimation and Prediction Systems* (University of Virginia) DTFH61-06-D-00005 p110
[16] Guo Z and Wilson N H M 2004 Assessment of the transfer penalty for transit trips: A GIS-based disaggregate modeling approach *Transportation Research Record* **1872** pp 10-18

IOP Conf. Series: Materials Science and Engineering **161** (2016) 012099 doi:10.1088/1757-899X/161/1/012099

A new model of Ishikawa diagram for quality assessment

Luca Liliana

University "Constantin Brâncuşi" of Targu-Jiu, Str. Eroilor, No.30, Romania

E-mail: lylyanaluca@yahoo.com

Abstract. The paper presents the results of a study concerning the use of the Ishikawa diagram in analyzing the causes that determine errors in the evaluation of theparts precision in the machine construction field. The studied problem was"errors in the evaluation of partsprecision" and this constitutes the head of the Ishikawa diagram skeleton.All the possible, main and secondary causes that could generate the studied problem were identified. The most known Ishikawa models are 4M, 5M, 6M, the initials being in order: materials, methods, man, machines, mother nature, measurement. The paper shows the potential causes of the studied problem, which were firstly grouped in three categories, as follows: causes that lead to errors in assessing the dimensional accuracy, causes that determine errors in the evaluation of shape and position abnormalities and causes for errors in roughness evaluation. We took into account the main components of parts precision in the machine construction field. For each of the three categories of causes there were distributed potential secondary causes on groups of M (man, methods, machines, materials, environment/ medio ambiente-sp.). We opted for a new model of Ishikawa diagram, resulting from the composition of three fish skeletons corresponding to the main categories of parts accuracy.

1. Introduction

Most organizations use quality tools for various purposes related to controlling and assuring quality. Although a good number of quality tools specific are available for certain domains, fields and practices, some of the quality tools can be used across such domains. These quality tools are quite generic and can be applied to any condition. There are seven basic quality tools used in organizations. These tools can provide much information about problems in the organization assisting to derive solutions for the same.

The seven tools are: histogram, cause-effect diagram, Pareto diagram, correlation diagram, control chart, data stratification, Brainstorming.

Ishikawa diagrams were popularized in the 1960s by Kaoru Ishikawa, who pioneered quality management processes in the Kawasaki shipyards, and in the process became one of the founding fathers of modern management. It is known as a fishbone diagram because of its shape, similar to the side view of a fish skeleton.

Dr. Kaoru Ishikawa (1915 – 1989) was a Japanese professor, advisor and motivator with respect to the innovative developments within the field of quality management. Kaoru Ishikawa is best known for the development of the concept of the fishbone diagram, which is also known as the 'Ishikawa diagram'. This diagram is still used in many organizations for making diagnoses or taking concrete actions in which the root cause of the problem is identified.

With his cause and effect diagram (also called the "Ishikawa" or "fishbone" diagram), management leader made significant and specific advancements in quality improvement.

The design of the diagram looks much like a skeleton of a fish. Fishbone diagrams are typically worked right to left, with each large "bone" of the fish branching out to include smaller bones containing more detail.

The technique uses a diagram-based approach for thinking through all of the possible causes of a problem. This helps you to carry out a thorough analysis of the situation. There are four steps to using the tool:

1. Identify the problem.
2. Work out the major factors involved.
3. Identify possible causes.
4. Analyze your diagram.

Causes are usually grouped into major categories to identify these sources of variation. The categories typically include:

- People: Anyone involved with the process;
- Methods: How the process is performed and the specific requirements for doing it, such as policies, procedures, rules, regulations and laws;
- Machines: Any equipment, computers, tools, etc. required to accomplish the job;
- Materials: Raw materials, parts, pens, paper, etc. used to produce the final product;
- Measurements: Data generated from the process that are used to evaluate its quality;
- Environment: The conditions, such as location, time, temperature, and culture in which the process operates.

Ishikawa diagram is being defined as a graphic representation that schematically illustrates the relations between a specific result and its causes, [1,2]. The studied effect or negative problem is "the fish head" and the potential causes and sub-causes define the "fish bone structure".

Therefore, the diagram clearly reveals the relations between a problem identified in a product and its potential causes.

Ishikawa Diagram is a simple graphical instrument to understand the causes that produce quality defects and is used to analyze the relation between a problem and all possible causes. All categories of causes start with the letter M (machines, methods, man, materials, maintenance, mother nature - environment, management) for the productive domains. 4M, 5M, 6M, 7M Ishikawa diagram were performed like this.

In [3] it is shown that obtaining a correct diagram is possible only through working in a team with experience. An interesting model of Ishikawa diagram was developed in the case of some automotive defects [4,5].

In [6] it is presented a method for assessing the quality of welding by applying one of the classic instruments of quality management.

Ishikawa diagram application areas are continuously expanding. For example, nowadays the method is also being applied in the medical field [7].

In [8] it is presented a study regarding tracing the cause-effect diagram concerning the tolerance dimensions by using software instruments.

Many specialized work in the fields of Quality Management and Quality Engineering show different patterns of Ishikawa diagrams. We can illustrate with a few categories of main cases which were the basis of some existing Ishikawa diagrams, [3,9,10,11,12,13,14]:

Man, Method, Measurement, Machine, Material;

-Material, Personnel, Process, External factors, Management;

- Software, Users, Hardware, Environment;
- Equipment, Process, People, Materials, Environment, Management;
- People, Material, Method, Environment, Machine;
- Working conditions, Raw materials, Management, Technology, Machine, Workers;
- Measurements, Materials, Personnel, Environment, Methods, Machines;

- Transport, Premises, People, Clients, Finance, IT;
- Physical evidence, People, Place, Service, Process, Promotion, Price, Productivity & quality
-Equipment, Policies, Procedures, People, Environment, Measurement;
-Measurement, Material, Man, Mother nature, Machine, Method;
-Environment, Material, People, Equipment, Process;
-Technology, Procedures, Policies, People.

The paper presents the results of a study concerning the use of the Ishikawa diagram in analyzing the causes that determine a non-quality problem in the evaluation of the parts precision.

The studied problem was "errors in the evaluation of the parts precision in the machine construction field".

The development of the Ishikawa diagram in a detailed form for determining the possible causes of a problem has the advantage of giving the possibility of identifying and analyzing all the factors connected to the problem.

2. Study concerning the use of the Ishikawa diagram in analysing the causes that determine errors in the evaluation of the parts precision

The non-quality problemstudiedin this paperis"errors in the evaluation of the parts precision in the machine construction field". This paperproposesIshikawadiagramby covering the stepssetforthbyDalein[15], namely the following:
- It is defined very clearly the effect of the problem considered,
- It is written the effect in the right and it is drawn a line from right to left,
- It is checked if each team member has understood well the problem,
- They are determined the main categories of causes which are the main branches of the diagram,
- It is organized a brainstorming session to determine possible secondary causes,
- It is organized another brainstorming session in order to discuss in detail the causes and to determine those who have the major degree of probability for producing the studied effect,
- They are traced and recorded the appropriate sub-branches.

Following the brainstorming session conducted with specialists from the technical measurements domain, potential causes were identified coming from 3 directions.

The study identified three directions from which derive the causes:

A) Causes that lead to errors in evaluating the dimensional accuracy, fish skeleton -(5MA):

Man: tired and nervous operator; untrained operator, inexperienced operator; carelessness within measuring;

Methods: inadequate measuring method; inaccurate measuring scheme; error of the position of the measured object; error of the position of the device; errors of the regularization procedure; inaccurate regularization to the nominal size; number of the realized measurements; suppression of gross errors; blocks with inaccurately chosen amount of scales; not applying the corrections generated by systematical errors;

Machines: A. devices with an accuracy inadequate to the tolerance; devices with inadequate measuring limits; attrition of the devices; not observing the periodical metrical checks; errors of the device limiting the measurement force; theoretical errors of the devices; abnormalities of the measuring surfaces; inaccurate choice of the sensitive contacts; inaccurate choice of the changeable elements (tips, calibrated wires);

Materials: patterns realized with flaws; scales not clinging; opening of the position prisms; caliber attrition; spatial variations of the piece;

Medio ambiente (sp.) / Environment: temperature, pressure, humidity, vibrations, noise, light, air composition.

B) Causes that determine errors in the evaluation of shape and position abnormalities, fish skeleton -(5MB):

Man: tired and nervous operator; untrained operator, inexperienced operator; carelessness within measuring.

Methods: inaccurate choice of the measuring base; the position of the verified surfaces; errors of the regularization procedure; inadequate measuring method; inaccurate measuring scheme; error of the position of the measured object; error of the position of the device.

Machines: Errors within the movement system of the measured object; errors within the movement system of the device; errors of the design of the measuring device; errors within the fabrication of the measuring device; devices with inadequate measuring limits; not observing the periodical metrical checks;

Materials: Spatial variation of the surfaces; scales not clinging;

Medio ambiente (sp.) / Environment: temperature, pressure, humidity, vibrations, noise, light, air composition.

C) Causes for errors in roughness evaluation, fish skeleton - (5MC):

Man: visual acuity; eye sensitivity and adaptation, tired and nervous operator; untrained operator, inexperienced operator; carelessness within measuring.

Methods: A. inadequate measuring method; inaccurate measuring scheme; error of the position of the measured object; error of the position of the device; number of the realized measurements.

Machines: Flaws of the printing system of the roughness graph; altered tip of the sensitive contact; errors of the movement device of the sensitive contact; inaccurate settings of the working parameters of the devices; flaws of the optic systems; not observing the periodical metrical checks;

Materials: Bending of the piece surfaces; inadequate roughness samples;

Medio ambiente (sp.) / Environment: temperature, pressure, humidity, vibrations, noise, light, air composition.

3. A new model for Ishikawa diagram

We determined many possible causes and potential sub-causes and we grouped them into three main causes of the defect:

A. Causes that lead to errors in evaluating the dimensional accuracy,

B. Causes that determine errors in the evaluation of shape and position abnormalities,

C. Causes for errors in the evaluation of roughness.

For each of the three categories of causes there were distributed potential secondary causes on groups of M (man, methods, machines, materials, Medio ambiente (sp.)/ Environment). We opted for a new model of Ishikawa diagram, resulting from the composition of three fish skeletons corresponding to the main categories of parts accuracy. This new model with the formula (5MA + 5MB + 5MC) adds itself to the list of multiple choice Ishikawa diagrams that have been created so far. We opted for a new model of Ishikawa diagram, resulting from the composition of three fish skeletons corresponding to the main categories of parts accuracy. This diagram is presented in the paper, figure 1.

4. Conclusions

Above seven basic quality tools help you to address different concerns in an organization. Therefore, use of such tools should be a basic practice in the organization in order to enhance the efficiency.

Performing the Ishikawa diagram in a more detailed form in order to determine the potential causes of a found defect has the advantage that it offers the possibility to identify and analyze all factors, which relate to the problem studied.This tool is excellent for capturing team brainstorming output and for filling in from the 'wide picture'. Helps organize and relate factors, providing a sequential view. This diagram deals with time direction but not quantity. It can become very complex and can be difficult to identify or demonstrate interrelationships.

IOP Conf. Series: Materials Science and Engineering **161** (2016) 012099 doi:10.1088/1757-899X/161/1/012099

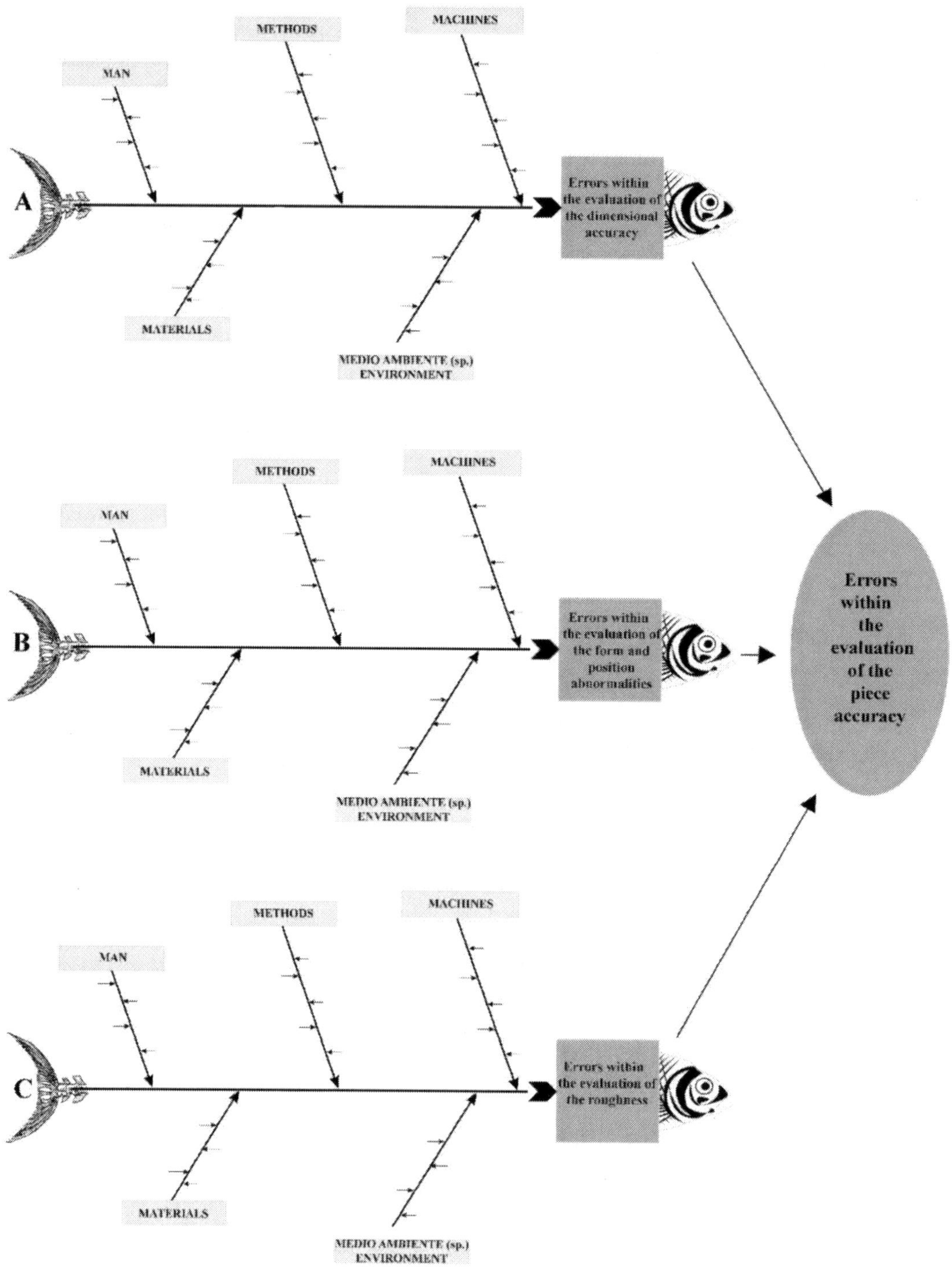

Figure 1. A new model of Ishikawa diagram, ($5M_A$+$5M_B$+$5M_C$)

Benefits of Using a Cause-and-Effect Diagram: helps determine root causes; encourages group participation; uses an orderly, easy-to-read format; indicates possible causes of variation; increases process knowledge; identifies areas for collecting data. The paper presented a new formula for the Ishikawa diagram was determined, (5MA + 5MB+5 MC). The determined Ishikawa diagram provides a complete picture of all potential causes that produce the studied failure.

The application of control and quality assessment techniques proves the important role that the customer with his requirements has. The traditional tools of quality management are the base in many organizations where improving quality is desired and they must be known and applied.

References

[1] American Society for Quality, Fishbone diagram. http://www.org/learn-about-quality/cause-analysis-tools/overview/fishbone.html

[2] Ishikawa K, Loftus J H (Eds)1990 *Introduction to quality control* , Tokyo Japan: 3A Corporation.

[3] Paraschivescu A O 2006 *Managementul calitatii,* Tehnopress Publishing (Iasi) pp.155-156.

[4] Luca L 2011 Study on the determination and classification of the causes that determine the faulty operation of a vehicle fuel pump. Recent Researches in Manufacturing Engineering. 3-rd WSEAS International Conference On Manufacturing Engineering, Quality and Production System (MEQAPS'11). Published by WSEAS Press ISBN 978-960-474-294-3 pp. 21-24.

[5] Luca L 2015 *The Study of Applying a Quality Management Tool for Solving Non-Conformities in an Automotive. Applied Mechanics and Materials* ISSN: 1662-7482 Vols. **809-810**Trans Tech Publications (Switzerland) pp 1257-1262.

[6] Luca L 2014 *Study on Identification and Classification of Causes which Generate Welds Defects. Applied Mechanics and Materials* ISSN: 1662-7482 Vol. **657** Trans Tech Publications (Switzerland) pp 256-260.

[7] Kam Cheong Wong 2011Using an Ishikawa diagram as a tool to assist memory and retrieval of relevant medical cases from the medical literature. *Journal of Medical Case Reports* 5:120. http://www.jmedicalcasereports.com/content/5/1/120.

[8] Cirtina L M 2012 *Improving the quality of processes by drawing cause-effect diagram using software instruments* 12h International Multidisciplinary Scientific Geoconference (SGEM), (Albena) Vol. **IV**, pp. 1017-1021.

[9] Cananau N 1998 *Sisteme de asigurarea a calitatii* Junimea Publishing (Iasi) pp121-126.

[10] Ionita I 2008 *Managementul calitatii si ingineria valorii* ASE Bucuresti Publishing pp186-187.

[11] Pruteanu O 1998 *Managementul calitatii totale,* Junimea Publishing (Iasi) pp.166-167.

[12] Ilie G, Ciocoiu C N 2010 *Application of Fishbone diagram to determine the risk of an event with multiple causes. Management Research and Practice, Vol. 2 Issue 1*pp: 1-20.

[13] http://www.cityprocessmanagement.com/Downloads/CPM_5Ys.pdf

[14] http://www.nursingtimes.net/Journals/2013/04/12/k/x/z/Using-fishbone-analysis--to-investigate-problems-160413.pdf

[15] Dale, B G 1994 *Managing quality* (2-nd Ed.), Prentice Hall, (Horls).

IOP Conf. Series: Materials Science and Engineering **161** (2016) 012100 doi:10.1088/1757-899X/161/1/012100

Decision method for optimal selection of warehouse material handling strategies by production companies

P Dobos, P Tamás and B Illés

Institute of Logistics, University of Miskolc, Hungary

E-mail: peterdobos30@gmail.com

Abstract: Adequate establishment and operation of warehouse logistics determines the companies' competitiveness significantly because it effects greatly the quality and the selling price of the goods that the production companies produce. In order to implement and manage an adequate warehouse system, adequate warehouse position, stock management model, warehouse technology, motivated work force committed to process improvement and material handling strategy are necessary. In practical life, companies have paid small attantion to select the warehouse strategy properly. Although it has a major influence on the production in the case of material warehouse and on smooth costumer service in the case of finished goods warehouse because this can happen with a huge loss in material handling. Due to the dynamically changing production structure, frequent reorganization of warehouse activities is needed, on what the majority of the companies react basically with no reactions. This work presents a simulation test system frames for eligible warehouse material handling strategy selection and also the decision method for selection.

1. Introduction

Due to the extending customer orientation and the fourth industrial revolution promptness, flexibility and increased self-tailoring have initiated a launch in product diversity. This process gives intense challenges for production and supply chain management operating in mass production [1]. Increase in number of product variety required by customers, decrease of product lifecycles, seasonally changing customer requirements have leaded to an increase in material stock level and/or to an increasing uncertainty in production planning. The costumers would like to receive the ordered products as soon as possible. For that reason decrease in production lead time means serious market advantage for the companies [2]. Production lead time can be interpreted as a sum of the following components:

- lead time of material purchasing,
- lead time of material stock-in into the warehouse,
- lead time of material stock-out of the warehouse, commissioning of materials and delivery to the technological containers,
- lead time of the technological operations of products and lead time of storage between two operations,
- lead time of packaging, composing unit loads and shipment to the costumers.

Decrease of production lead time come to material storage-in, -out and relocation is important by the materials with increased product spectrum, means short life cycle and small amount of order. This can be ensured by adequate selection of material handling strategies and optimized conducting of material handling tasks [2, 7-8]. Presently applied storage-in, storage-out and relocation startegies and their optimal selection concept will be presented in the thesis.

2. Presently applied warehouse storage material handling strategies

This part contains a certain overview in professional literature and material handling strategies presently applied in practice and the possible relations can be defined between them (Figure 1).

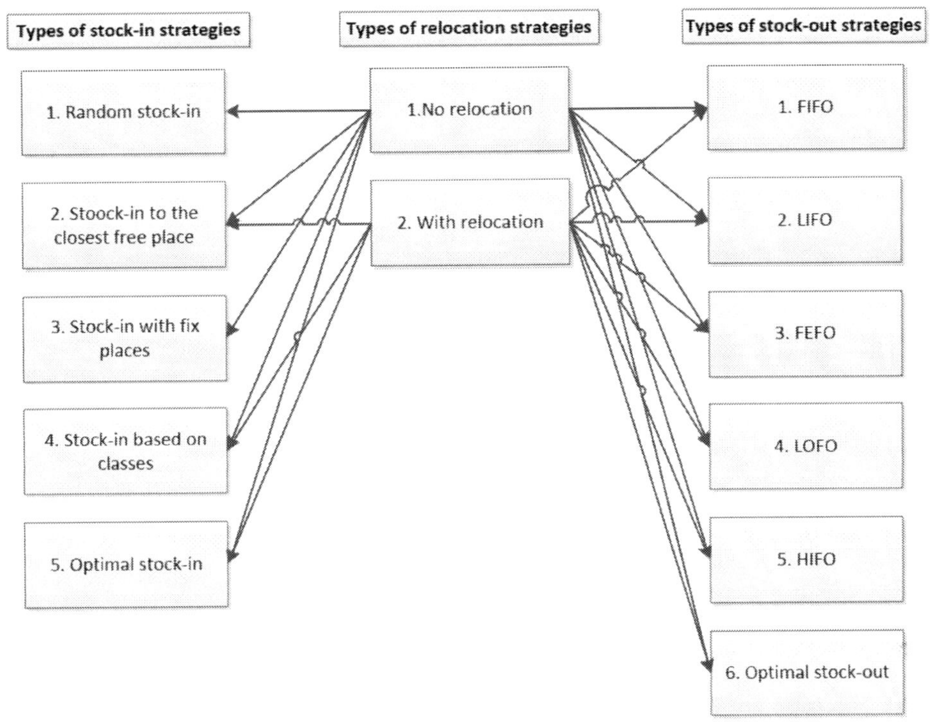

Figure 1. Relation possibilities of material handling strategies

Introduction of stock-in strategies:

- Random stock-in: materials to be stocked-in are placed to the free store places on a randdom way [3].
- Stock-in to the closest free store place: Placement of materials happens to the closest free store place [3].
- Stock-in with fix places: Materials to be stocked in are placed to a fixed store places according to product types. In case of constantly changing environment, this model can be applied only with heavy losses [4].
- Stock-in based on classes: Based on a chosen aspect, the stored materials are entitled to groups, ABC categories and the ppossible store zones are determined for them according to their groups. Forming of groups can happen based on: popularity, turn-over, volume, Pick Density, Cube per Order Index [5].
- Optimal stock-in: The goal of optimal stock-in is to have a mixture of stock-in strategies. It means a costumized application of different stock-in strategies within certain zones of the warehouse. This allows to stock-in the surplus arose within a day.

Introduction of stock-out strategies:

- FIFO (First In First Out): The firstly stocked-in product will be stocked-out firstly according to product type.
- LIFO (Last In First Out): The latest stocked-in product will be firstly stocked out according to product type.
- FEFO (First Expired First Out): Product with the shortest expiry will be stocked-out firstly according to product type.
- LOFO (Low First Out): the product purchased on the lowest price will be stocked-out firstly according to product type.
- HIFO (High First Out): the product purchased on the highest price will be stocked-out firstly according to product type.
- Optimal stock-out: The goal of the optimal stock-out is to have a mixture of stock-out strategies. It means a costumized application of different stock-out strategies within certain zones of the warehouse. This allows to stock-out the surplus arose within a day.

Introduction of relocation stock strategies:

- Without relocation: materials are not moved from one place to another place while they are in the warehouse waiting.
- With relocation: In a dynamic stock environment, it is possible to move materials among warehouse places while they are in the warehouse waiting. The goal of the relocation is to have the materials closer to the storage-out point, which results to a decrease in the distance of material moving routes. (ensures opportunity to take away material moving tasks from the peak times) [6-7]. Relocation can be applied in case of all stock-in strategies except by the stock-in with fix places. The advantage; simple systematization on fixed shelf system would be elliminated with relocation in case of the stock-in with fix places strategy.

The thesis examines the cases, where the material handling within the warehouse happens in a form of whole unit loads (there is no commissioning).

3. Selection concept of optimal warehouse material handling strategy

For a production company, the planning of mid-term and long-term strategy with an eye to the future is a crucial process besides the daily oprational management. The base for determining strategy is an environment with such conditions, that is able to provide comperable proposals, refered to same basis and dimensions, depending on the changing parameters. These data are used for the return, time planning of possible investments determined by mid-term and lon-term strategy. In a production company, row material management , stock strangulation are essential parts of strategy forming, for founding which is needed the selection concept for optimal material handling strategy introdduced in this chapter. Selection of optimal warehouse material handdling stratetgy version is planned to achieve by using a simulation model, which „simulates" the opration of materialflow-system according to the diffferent straetgy versions by using up data of specific time-period in the past, than it selects the version with the best objective function value based on the logistics inddicator values gained per stratetgy versions. The process steps are shown in figure 2. The introduced conception outlines the frames of a research work topic.

Selection concept of adequate material handling strategy consists of the following steps:

1.) Recording the examination data: Hereinafter, data basis and their content needed for the examination is introduced.

- Store system data base:
  - storing system size data (overall dimension, size of storing boxes, size of delivering corridors, etc.),
  - occupied store places, empty store places,
  - type of materials on store places,

- distance between warehouse store places and out- and in-stock points.
- Material handling system data base:
  - number of material handling machines,
  - position of material handing machines at the start of simulation,
  - fuel level of material handling machines at the start of simulation,
  - acceleration, working spead of material handling machines,
  - fuel usage, filling of material handling machines,
  - maintenance of material handling machines,
  - delivering capacity of material handling equipment,
  - time for picking up and handing down materials in case of material handling machines,
  - material handling routs,
  - delimination of applied roads of material handling equipment.
- Product data base:
  - store position of starting stock per product types,
  - placing possibilities of product types,
  - assembly matrix of parts,
  - overall dimension, filling amount of a unit load per product types,
- Data base for material handling tasks:
  - formation time and execution time-frame of stock-in tasks per product types,
  - formation time and execution time-frame of stock-out tasks per product types,
- Data base for evaluation:
  - weigh of objective function components,
  - specific cost functions based on data gained from the past.

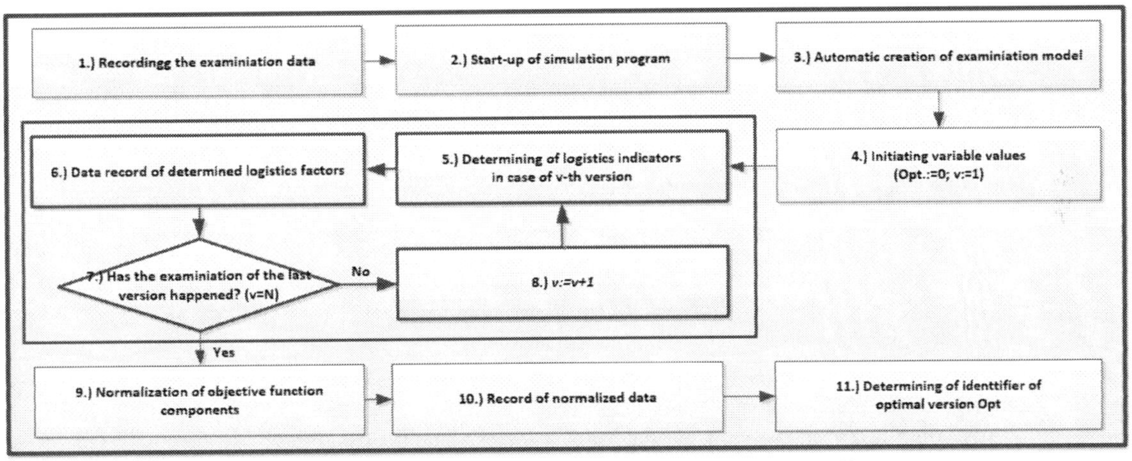

Figure 2. Selection concept of optimal material handling

2.) Start up of simulation program: After the record of data of sheets, the simulation program is started, which ensures the possibility to evaluate all together 60 kind of warehouse material handling strategy versions (5 stock-in, 2 relocation stock and 6 stock-out strategy, that means 60 versions in total) based on given data gained from the past and also to select the best version. Basic requirements towards the simulation proram based on the data gained from the past are automatic creation of examined system, run of allmaterial handling strategy versions and also selection of adequate version.

3.) Automatic creation of examination model: Based on the examination data, the examined logistics system will be established automattically (material handling system, store system, start-up stocks).

4.) Initiating of variable values: In the simulation examination model, initialization of two basic variables are done, namely the v, which contains the examined material handling strategy versions and the Opt, which contains the optimal verrsion identifier. Certainly, there are other technical kind of variables applied in programming, that are needed to be initialized, but now they arre out of consideration.

5) *Determining of logistics indicators in case of v-th version:*

- Warehouse material handling cost [EURO]: The warehouse material handling cost is defined as a product (multiplication) of specific cost function determined by the data gained from the past time-period and all the material handling rout distance determined by the simulation program. The specific cost function will be defined by linear regression calculation by applying the least-square method. Dependent variables of the function are daily-executed route distance in average and its standard deviation. The dependent variable values are defined also by the simulation proggram (essentially, the specific cost is gained from substitution of dependent varible values into the specific cost function).
- Total stock-out delay time [sec]: Total deviation from stock-out deadlines within the examined time-period.
- Stock-in perforamnce [ER/Hour]: Maximum amount of stocked in products by hours within the examined time-period in the past.
- Stock-out performance [ER/Hour]: Maximum amount of stocked out products by hours within the examined time-period in the past.
- Subjective factor: total impression on the material handling sttrategy version (e.g. low chances for mistakes, stable, easy to apply, etc.). The value of the factor needs to be determined between 1 and 10 (10 is the best value).

6.) Data record of determined logistics factors: The logistics factor values defined according to each material handling staretgy version will be recorded in a predefined date sheet. Based on these data, the normalization of objecttive function components (Step 9.) and also the establishment of objective function can happen at a later point of time.

7.) Examination whether each of the possible version has been examined?: In case, it happened, than selection process of the optimal version takes place. In case, it did not happen, the next version will be chosen for examination.

8.) Incrementation of identifier for material handling staretgy version.

9.) Normalization of objective function components: definition of objective function value of v-th strategy version of z, that happens based on the logistics indicattor values determined by the simulation model. In order to ensuer that indicators with different dimensions could be applied as objective function components, they need to be normalized. After normalization, all the values will range between 1-10 in our case. Besiides that we will aim to select a material handdling strategy version, where the goal is to minimize. Steps for normalization:

- Normalization of objective function components to be minimzed (warehouse material handlingg cost, total stock-out deelay time):
  - determination of mean value of logistics features to be minimalized,
  - spliting parts between the minimum value and the mean value, and also between the maximum value and the mean value for 5-5 intervals,
  - scoring the values of examined versions by help of gained scale (1-10 scores).
  - the smaller the examined component value is, the better the objective function value is.
- Normalization of objective function components to be maximized (in- and out-stock performance, subjective factor):
  - consideration of mean value of logistics features to be maximized,
  - spliting parts between the minimum value and the mean value, and also between the maximum value and the mean value for 5-5 intervals,
  - scoring the values of examined versions by help of gained scale (1-10 scores),
  - for the aim of standard handling, execution of objective functions' transformation to be maximized

- egységes kezelhetőség érdekében a maximalizálandó célfüggvények transzformációját el kell végezni (subtruct from 11, the objective function to be maximized will be formed to a component),
- after the ttransformattion it can be achieved that the goal will become to reach the smallest value possible also by the objective function compoonents to be maximized.

10.) Record of normalized data: The normalizedd versions of logistics indicattor values determined by diffferent material handlingg staretgy versions will be recorded in a predifined data sheet.

11.) Determining the identifier of optimal version: The only parameter of the objective function determined in the (1) relation is the identifier of the material handling strategy version. Actually, the objective function selects the best version (*Opt*) based on the weighted sums of normalized objective function components determined in accordance to the material handling strategy versions, and defines its objective function value (*C*).

$$C = \min_{v} \left\{ \sum_{l=1}^{5} w_l \cdot Y_{vl} \right\}, \quad 0 \le w_l \le 1 \quad and \quad \sum_{l=1}^{5} w_l = 1, \tag{1}$$

where:

- C : objective function value belonging to optimal material handdling strategy version,
- v : identifier of examined material handling strategy version,
- l : identifier of objective function component,
- W_l : weigh of objective l function component,
- Y_{vl} : value of objective l function component value of version v.

4. Summary

Selection and operation of adequate row material warehouse material handdling strategy have a major influance on the companies' competitiveness by executing internal warehouse material handling tasks more effectively and increasing efficiency in production service. The curently applied in-, out- and relocation stocking material handling straetegies are reviewed in this work. In professional literature, sophisticated examination system cannot be found based on which adequtae material handling strategies could be chosen. Due to the dynamically changing costumers' needs, review of warehouse material handling strategies applied in the present is needed in order to increase effectiviness. In the publication, an examination system conception for selecting the adequate material handling startegies is introduced. This will be worked out in details in our future reasearch work.

Acknowledgment

"This project has received funding from the European Union's Horizon 2020 research and innovation programme under grant agreement No 691942". "This research was (partially) carried out in the framework of the Center of Excellence of Mechatronics and Logistics at the University of Miskolc."

References

[1] Industry 4.0 – Challenges and solutions for the digital transfomation and use of technologies, Deloitte, http://www2.deloitte.com/ (2014)

[2] Womack J P, Jones D T 2008 Lean thinking *Simon & Schuster Inc.* ISBN 978-963-9686-83-0

[3] Manzini R 2012 Warehouse in the Global Supply Chain *Springer* pp 31-51

[4] Chen L, Langevin A, Riopel D 2011 A tabu search algorithm for the relocation problem in a warehouse system, *Int. J. Production Economics* pp 147-156

[5] Guerriero F, Musmanno R, Pisacane O, Rende F 2013 A mathematical model for Multi-Levels Products Allocation Problem in a warehouse with compatibility constraints. *Applied Mathematical Modelling* **37** pp 4395-4398

[6] Chen L, Langevin A, Riopel D 2010 The storage location assigment and interleaving problem in an automated storage/retrieval system with shared storage *International Journal of Production Research* **45** (4) pp 991-1011
[7] Chen L, Langevin A, Riopel D 2009 Minimizing the paek load in a shared storage system based on the duration-of-stay of unit loads *International Journal of Shipping and Transport Logistics* **1** pp 20-36
[8] Costa J, Sameiro Carvalho M, Nobre A 2015 Implementation of Advanced Warehouses in a Hospital Environment - Case study P*ublished under licence by IOP Publishing Ltd* doi: 10.1088/1742-6596/616/1/012005

IOP Conf. Series: Materials Science and Engineering **161** (2016) 012101 doi:10.1088/1757-899X/161/1/012101

A Study on Improving Logistics in a Production Enterprise in the Automotive Domain

A Muntean[1], M Inţă [1] and I A Stroilă [1]

[1]Lucian Blaga University of Sibiu, Victoriei B-dul, no 10, Sibiu, Romania

Email: marinela.inta@ulbsibiu.ro

Abstract: In order to remain on the market and to overcome competition, companies need to develop and implement a strategy of continuous improvement. This is based on customer satisfaction needs, management based on realities and respect for people. In order to meet market requirements global road Faurecia Company uses a system called "Faurecia Excellence System", a system that promotes continuous improvement. FES system controls all activities of Faurecia, from research and development to sales including production and company functions and implements all 114 main procedures.

This paper examines the current activity and introduce the new methods to improve manufacturing processes. These new methods are based on some fundamental concepts, namely: a high level of stocks blocks resources and can mask other problems decreasing visibility and communication; product quality is essential and the goods must be produced only when needed using Kanban principle. Additional expenditure with purchased raw materials regarding storage until they are needed, identification and packaging for storage, inventory costs were reduced using a JIT based principle in purchasing.Thus, the paper proposes reorganizing production by reducing the level of stocks using JIT principle and production time. The application made in software Arena refers to simulating the fabrication process to reduce throughput time.

Introduction

Currently both customers and manufacturing companies put more emphasis on quality to remain on the market and to overcome the competition, they need to develop and implement a strategy on continuous improvement of processes, starting from JIT system [1].

Among most widespread continuous improvement models there are qualitative model PDCA (plan, do, check and act) figure 1, where:

- Plan (planning) - in the planning there is the identification of opportunity and plan for change;
- Do - implementing change (continuous improvement) on a small scale;
- Check - for checking data is used to analyze the results of changes and to see if it was made with some benefit;
- Act: if the change was successful it must be implemented on a large scale and continuous evaluate the results. If the change fails, the cycle must start again.

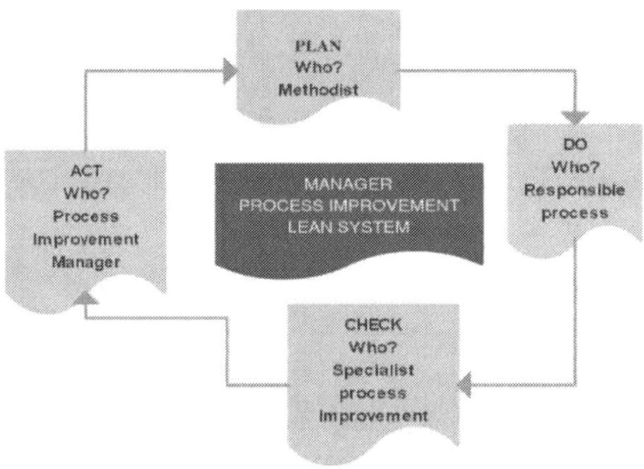

Figure 1 Pan, do, check, alct

Even if the difficulty of applying the concept of "Just In Time" represents a risk, instead it can be successful and bring competitive advantages. One can say that "Just In Time- JIT" is an organizational philosophy, which assumes that production activities should be done with maximum precision so the inventories could be minimized, that being regarded as the greatest harm in the company because they lead to cost increase.

A high stock in a company hide production and communication problems. Thus, by applying the concept of JIT results in continuous improvement. Faurecia is a multinational conglomerate, operating worldwide and a European leader in manufacturing auto parts. The Group has 160 plants, approximately 60,000 employees worldwide and covers 35% of the European market in the field therefore the means of production optimization is an ongoing process, directly supporting the competitiveness of the automotive industry.

Faurecia Company specializes in the design, development and production of six main modules representing about 15% of the value of a car: seats, dashboard, doors, soundproofing, front bumper and exhaust systems, [2] table 1.

Table 1. Auto parts produced by Farecia Group

Dashboard	Door cover	Exhaust system
	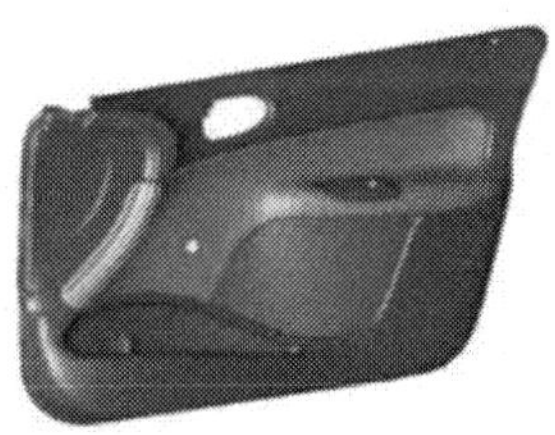	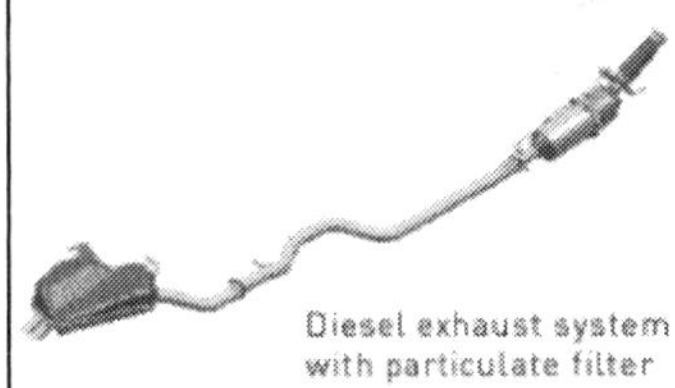
Seats	Front bumpers	Modules and systems

Recognized as an architect of the car seat, Faurecia excels in designing and manufacturing components: frames, parts, foam covers, which then assembles them into complete seats, delivered in the shortest time customers (33 of factories producing chairs deliverer in Just in Time system). At Headquarters in the town Tălmaciu hires more than 800 employees producing for the most popular models of cars: Audi, BMW, Toyota and Peugeot [3].

2. Quality system in the company

To meet automotive market requirements Faurecia Company uses a system called Faurecia Excellence System, a system that promotes continuous improvement [4]. Faurecia Excellence System presented in figure 2 is based on key attributes such as: leadership, development, production and customers and seeks to achieve excellence in terms of quality, cost and execution. It can be said that the system FES lead all activities of the company from R & D to sales including production.

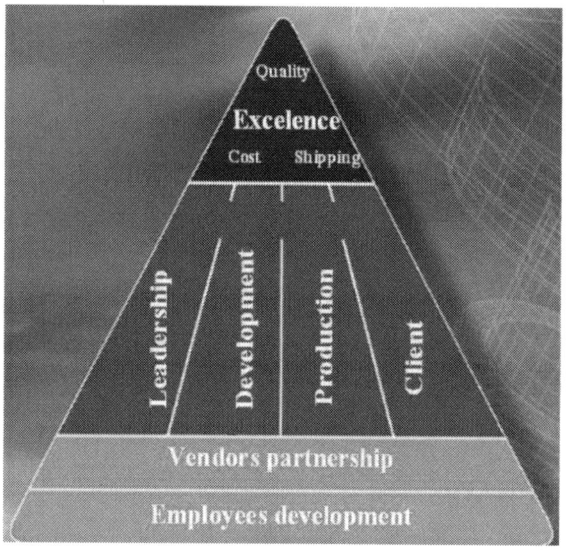

Figure 2 Faurecia Excellence System [4]

Using this work system, it is organized into autonomous teams making partnerships with suppliers, defining customer requirements and shipping at time quality products. By applying FES system each of the 238 production units strive to achieve the highest level.

The excellence is the result of integrating the latest Faurecia Group's expertise in the 5 Business Groups by applying best practices implemented by them. Faurecia elements of staff commitment to achieve business excellence consist of objectives for meeting the expectations of all stakeholders.

These objectives are:

- Key performance indicators system;
- Quality System Efficiency;
- Efficiency in Production;
- System Management and Development Program;
- Empower employees to achieve excellence.

In other words, Faurecia Excellence System is so constituted as to ensure the Group's development in all areas, focusing on customers, employees and shareholders, and job security is one of the key elements of the Faurecia Excellence System.

3. Application of the concept "Just in Time" at S.C. FAURECIA SEATING TĂLMACIU S.R.L

To adopt a JIT operational environment, the company must revise its current operating activities and introduce new ways of manufacturing. At the base of these methods are a few fundamental concepts, namely:

- product quality is essential;
- high level of stocks is an intangible resource;
- eliminate activities that do not add quality product;
- making products only when needed.
- well trained workers so that they participate in product quality and efficiency.

JIT concept applied to Faurecia Seating Talmaciu company has created an operational system containing the following elements:

- Maintaining a minimum level of stocks;
- Creation of a planning and production scheduling pull-through;
- The purchase of raw materials and manufacture as required in pools of small size;
- Fast and inexpensive adjustment of equipment;
- Creating flexible work cells;
- Maintaining high levels of product quality;
- The introduction of an effective system of preventive current repairs.

So:

✓ Maintaining a minimum level of stocks.

One of the objectives of the JIT operating environment is to maintain a minimum level of stocks. Faurecia Excellence System implemented at Faurecia Seating Talmaciu does not focus heavily on reducing and maintaining a minimum number of stocks. Raw materials are purchased and stored until they are needed, thus achieving additional expenditure to identify products with their packaging for storage, inventory and inspection personnel responsible for these tasks.

To reduce inventories, were eliminated factors leading to the increase in stocks such as quality defects, malfunctions of equipment, time control, etc. Improving performance at each of these levels is not a goal in itself, but rather a necessary condition for reducing stocks

- ✓ The creation of a planning and production scheduling pull-through.

At Faurecia Seating Talmaciu raw materials and subassemblies are not always purchased as needed, long-term products are manufactured and stored before receiving orders, thereby creating stocks. Production Application pull-through will help eliminate these stocks bringing us one step closer to JIT.

- ✓ Purchase of raw materials and manufacture as required in pools of small size.

In the pull-through method, the volume of orders received from customers determines production volume. Low inventory levels are maintained, but it requires more frequent adjustment of the equipment, resulting in more activity interruptions.

Long production processes, previously considered more efficient in terms of costs, are no longer appropriate in the JIT environment where goods are not produced to be stored.

- ✓ Fast and inexpensive adjustment of equipment

Equipment must be located in places most effective and similar products must be scheduled on common groups of equipment, and installation time required for adjustment can be minimized. Time control machinery can be reduced by creating work cells that will produce similar products while minimizing the adjustment of machinery.

- ✓ Creating flexible work cells

The biggest problem in the implementation of work cells to Faurecia Seating Talmaciu is to divide the whole production system. This problem can be solved by design and investment. The whole company has rearranged and equipment (machinery) are repositioned, modified or altered to allow production in cells.

The cost due to setup time can be reduced considerably by implementation of cells with similar products. An alternative to minimize this cost of cells implementation is to make a gradual transition in the layout. The cost must be justified by the economy that we do after the modified layout of cells.

These costs and risks can be evaluated, which is more difficult is to calculate and control the costs with the employees. Implementation of these cells involves employees training at Faurecia Seating Talmaciu.

Each of the members of a working cell should be able to work on every machine in the cell, this often means that employees should be "more functional". Therefore, cell implementation is better to be done in steps, both cellular and systems and the traditional operating simultaneously for a period of time.

There are many benefits of cellular production Faurecia Seating Talmaciu if the cells have been implemented correctly. Production processes become more balanced and productivity increases due to reorganization and cellular ordering.

During the crossing parts from one process to another, setup time and time between operations are reduced, resulting in a release of inventory. Production work cells combined with JIT also helps eliminate overproduction, producing goods only when necessary.

The result is reduced costs and better control of operations.

Figure 3 Working group

✓ At Faurecia Seating Talmaciu there is a quality control department pretty well structured, which brought to the company its leadership in the market, so the raw materials and finished products are of the highest quality.

Faurecia launched the "Breakthrough Quick Response Quality Control Quality Hovering" (QRQC), that advanced planning quality control and rapid control of quality, figure 3.

But quality control is improved by adopting JIT environment thus providing a continuous improvement of products which are checked throughout the production process.

✓ Introduction of an effective preventive current repairs.

When a company reorganizes its equipment in flexible cells, each unit of the cell becomes a part of it. If a machine breaks down, the entire cell stops.

Because products cannot be transferred to another machine while the machine is repaired, JIT operations require an efficient application of preventive current maintenance. Machine operators should take advantage of the work stoppage to perform routine repairs.

4 Case Study. Analysis of final assembling process agility by applying Just In Time in S.C. Faurecia Seating Talmaciu S.R.L

By applying a JIT system resulting not only reduce the level of inventory or production time, but also get high quality products.

Seat assembly is done on 2 different lines, one for the front seats and the other for the rear seats. Thus, the front seats assembly line includes 10 workstations for slides and control lever assembly on the production pallet to electrical harness, visual inspection and removing wrinkles and scratches in IR cabin to checking the electrical systems of seat. All this activity is served by 11 workers.

In the case of assembly line of rear seats we have 7 workstations served by 3 workers, where it performs the following operations: handling reinforcement backrest, "lach" assembly, pivot pin, seatbelt retractor and "feld-pad" by screwing, assembles backrest cover on the reinforcement and clasps it, removing creases by ironing, visual inspection for delivery to the customer.

The two assembly lines were designed for an output target of 1000 seats sets / day, i.e. 21,000 sets seats/ month, which means that every 104 seconds will be performed the covers for a set of seats.

Supply of the line lasts 60 minutes and move the trolley from one post to another lasts 10 seconds.

For the analysis agility of weaving (assembly of the seat covers) and final assembly line of the car seats value stream map was realized [5].

Analyzing the initial value stream map, it was found that:

- In case of weaving: $Lead\ time = 3{,}2\ days$ (1)
- In the case of the assembly line: $Lead\ time = 7{,}5\ days$ (2)

and the line takt is: $$Takt\_time = \frac{5days \times 4week \times 2exchanges \times 450\,min}{21000} = 0{,}857\,min \qquad (3)$$

There are still points of bottleneck even if the assembly does not last too long. Given these identified nonconformities, the organization decided the creation of working group to solve these issues in a Kaizen site, proposing to improve solutions by implementation of Kanban and production cells.

As a solution for improvement is proposed in parallel manufacture of subcomponents for 8 hours, storing them in a supermarket and then transferred to the final assembly line and achieve production cell, as shown in the figure below.

For the analyze of the solution a model of the assembly line was realized in Arena software, figure 4 to simulate the process and compare it with the target of the company [6].

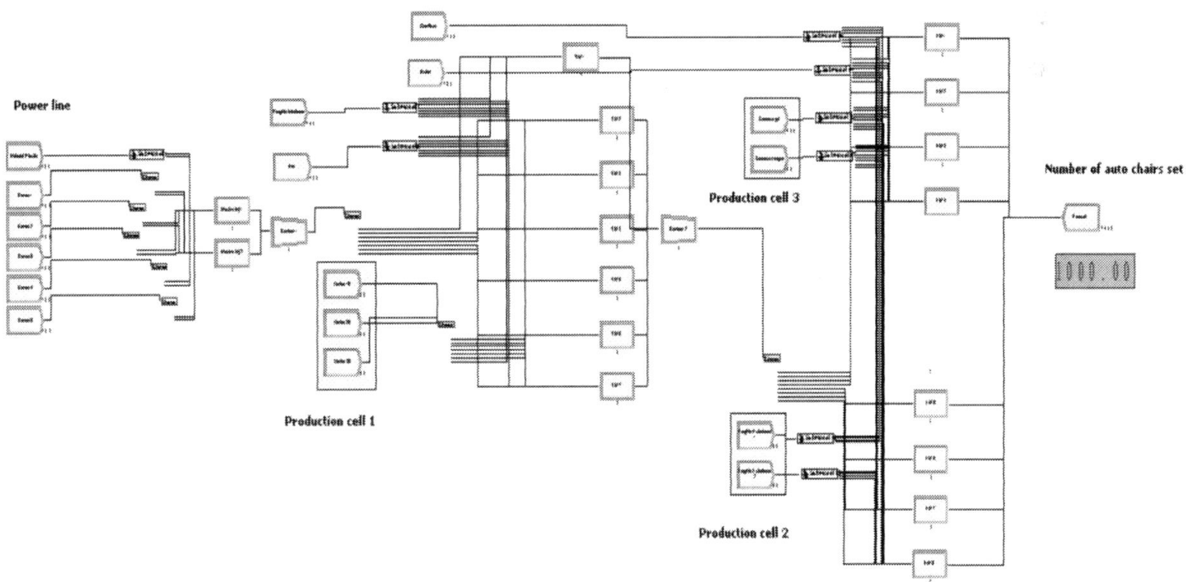

Figure 4 Simulation model of improved assembly line using Arena software

Restoring value stream map for the 2 lines resulted in a Cycle time = 79 sec line weaving, resulting in a 25% improvement with a respective Cycle time = 419 sec for the assembly line and an improvement of 12.5%, values are shows in figure 5.

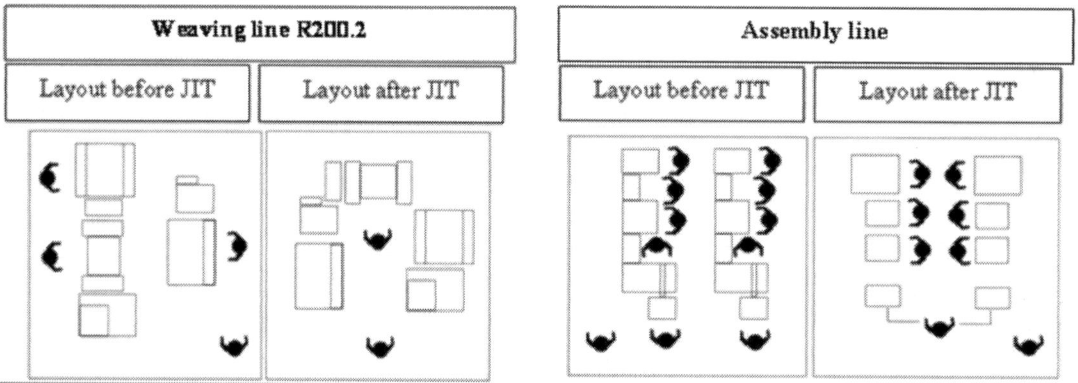

Setup time	Unit	Before	After	Improvement
Tranzit time	min	60	2	95%
Cycle time	sec	104	79	25%
Employees	-	3	1	65%
Parts per employee	Per hour	22	48	+115%
Surface	m²	25	15	-40%
Productivity	%			+115%

Setup time	Unit	Before	After	Improvement
Tranzit time	min	50	4	-90%
Cycle time	sec	478	419	-12.5%
Employees	-	11	8	-30%
Parts per employee	Per hour	90	103	+14%
Intermediate stocks	Pcs	80	10	-90%
Surface	m²	90	77	-15%
Productivity	%			+14%

(a) Seat covers *(b) Final assembly car seats*

Figure 5 Changes of location and improvements resulting from implementation of JIT

The simulation was performed for the final assembly line, length of the run for a week (five days, twoo shifts, 450 minutes/shift) resulting a number of five replicas with the results shown in table 2.

Table 2 Results of the simulation model

Replication length, (min)	Replication number	Number out Number of auto chairs set	Half width
450	1	376	26.89
900	1	874	114.35
900	2	908	165.18
2250	1	4326	936.31
2250	2	4619	1269.65
2250	3	5008	1317.29

Queuing time for the replications are 0.14 – 1.69 minutes. Regarding the confidence level in Arena software this has the value of 95%.

The conclusion for the simulation is that the target could be achieved in a work program in twoo shifts, five days / week and overtime added.

5. Conclusions

This study is part of the completion and analysis of the current car seat assembly by simulating the production process using software Arena and making improvements to the manufacturing process.

By applying JIT under the two processes improved resource management at Faurecia Seating Talmaciu by reducing and maintaining a minimum level of inventories which led to lower cost storage, deterioration of product quality during storage, so the risk to produce inventories that can not be sold they have been avoided.

It was noted that after the integration of all assembly processes in flexible manufacturing cells decreased time during both assembly and auxiliary transport semi-assembly line at the final assembly line. This measure also led to lower surplus production and the elimination of intermediate stocks.

The new reorganization of assembly-line production allows operators to move from the main station where they operate any of the other positions when needed.

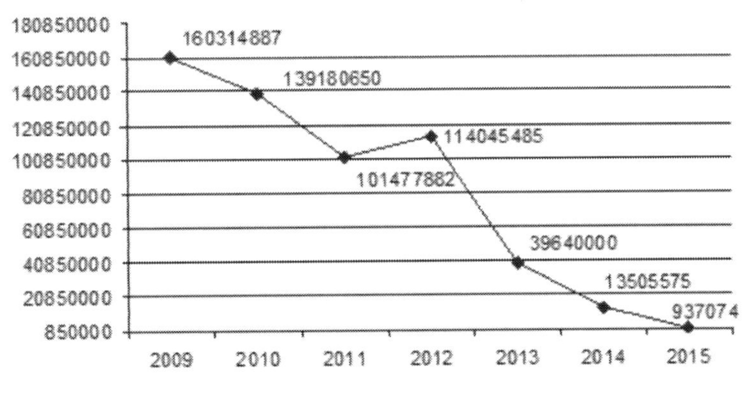

Figure 6 Inventory costs reduction

Can be seen from the graph shown in figure 6, that, expenses of payable inventories were reduced by at least 70% by applying JIT, in that they are received only when needed.

References

[1] Cheng T C, Podolsky S, 2007 *Just in Time Manufacturing. An introduction*, Ed. Springer-Verlag, New York, LLC, pp 44-62.

[2] *Internal documentation from S.C.Faurecia Seating Tălmaciu S.R.L*

[3] *Manual SMI Faurecia TAL-I-LSS-2016/RO*

[4] *Book of quality - S.C.Faurecia Seating Tălmaciu S.R.L.*

[5] Beju L D, 2013 *Principiile fabricatiei agile*, Editura Univerității "Lucian Blaga" din Sibiu, Sibiu, pp.29-61.

[6] Altiok T and Melamed M, 2007 *Simulation Modeling and Analysis with Arena*, Elsevier Inc., ISBN 10: 0-12-370523-1, pp. 141- 163, 223-249

Acknowledgement

Article financed of research grants LBUS, no 1506/17.04.2015 with S.C. LINE SOFT S.R.L Sibiu company.

The use KPI's to determine the waste in production process

G Borsos[1,2], **C C Iacob**[1,3], **G Calefariu**[1,4]

[1]Department of Engineering and Industrial Management, Transylvania University of Brasov, BV 500174, Romania

[2]E-mail: gabriel.borsos@unitbv.ro

Abstract. In theory and practice of management is well-known Lean approach about forms of waste from production processes (Muda) and the method VSM (Value Stream Map), one of the most effective methods for determining the activities generating value within industrial companies. It is also obvious concern of the specialists for performance measurement regardless of purview of the organizations. The literature review has shown that the link between performance indicators and the objectives of the companies is researched in detail. However, the correlation between indicators and the forms of waste that generate deviations from the setpoints is rather nature practical and it depends on the talent and managerial skills of those directing production processes. The paper presents the results of a applied study, performed by the authors, through which it was has sought to will create a system of performance indicators specific to manufacturing activity that to be a useful tool to quantify the losses and to determining ways to improve default losses.

1. Introduction

The current economic environment obliges the managers of the industrial company to seek permanent optimization of activity of the organizations that they lead.

In addition, an ever more difficult access to resources whether financial or material, requires special attention with regard to minimize wastage (in all its forms - as defined in the Lean philosophy).

The correlation between organizational performance and reducing waste is unequivocal and there are many books and papers by renowned theorists and practitioners of management, which support this claim [1, 2, 3].

Concepts such as Lean Manufacturing, Value Stream Map (VSM), Kaizen, well known by specialists, emerged in response to the issue of eliminating waste from production processes and became, in time, the essential management tools for obtaining performance by industrial companies and not only.

The industrial performance is another very important topic in production management. In this case, literature abounds with works on the subject of the performance of organizations and systems to measure them.

Some researchers have studied the link between performance management and performance measurement system (SMP), showing that SMP contribute to increasing managerial performance through relevant information that it gives managers in decision-making [4].

A standard SMP is represented by the system of performance indicators (Key Performance Indicators - KPIs). KPI provides information on what to do to increase performance substantially [5].

The researches made by the authors of this article have revealed that PMSs have only the role to highlight how the results meet initial expectations of the organization, yet without to determine the link between indicators and forms of waste.

Starting from these premises, we sought to develop a performance measurement system of the production process that highlights not only deviations of measured values of indicators from the permissible limits but also the amount of the main forms of waste that causes these deviations.

2. Theoretical considerations about wasting and performance in industrial processes

2.1. The waste in industrial processes

Wasting (loss) in industrial processes is represented by those activities that consume resources but do not add value to the final product [6, 7, 8]. In addition, the corrections, reprocessing pending are classified as forms of waste [8].

According to the Lean philosophy, within an enterprise are three types of activities [8]: activities that add value, activities that do not add value and can be eliminated and activities that add no value but are necessary.

The most non-value activities are the losses generators in one or more forms.

Lean defines seven forms of waste (Muda) [6, 7, 8, 9]: overproduction, waiting, movement, transport, reprocessing, areas, stocks.

To the seven forms can be added: personal inefficiency and overloading.

2.2. Performance processes within the enterprise

Performance is the main concern of managers being an element of enterprise competitiveness.

Peter Drucker [10] shows the connection between objectives and performance and states that the objectives of organizations should established in certain areas of performance such as productivity, profitability, innovation, attitude and work, performance, management performance, resources, market position.

Other researchers [4] states that characteristics such as the specific objectives, the difficulty, comprehensibility, tangibility affects the organizations performance.

Increase process performance and hence the company is the main task of managers. For this there is a whole panoply of tools and methods: Lean, Kaizen, TQM (Total Quality Management), Lean Six Sigma, etc. [11].

The performance measurement is essential in the management process. Numerous studies have shown that, to adopt the best decisions regarding development of the organization, managers must have accurate and actual data about the performance of processes taking place within the company. Thus, Rylková and Bernatik [12] shows that for maximum efficiency, performance measurement must made in five areas: finance, marketing, human resources, processes, standards for the future.

It is recognized that a production process not only aims to produce quantities of goods but also the production of these goods at a quality level as high as possible, with the least cost and in the shortest time (high productivity). In conclusion, we can say that the output performance is a multivariables function:

$$Pp = f\{N, Q, C, T\} \tag{1}$$

Where:

Pp – the performance of the process;
N – The quality of the products;
Q – Amount produced;
C – Cost of production;
T – Duration of production (productivity).

Performance measurement systems therefore must include tools and methods capable of assessing evolutionary trend at all four variables from the equation (1).

2.3. Performance indicators

A fundamental tool for measuring performance is the performance indicators system (KPIs) [13].

A classification of KPIs is presented by Fitzgerald et al. [14]. They classifies the indicators in the result / driver indicators and lead / lag indicators. Parmenter [5] complements this classification by dividing indicators of lead / lag into three categories: indicators based on the past, present and future.

To be used effectively, performance indicators must submit three main features [5, 13, 15]: responsibility, to be easily assimilated, to be timely.

The required number of KPIs for a more precise measurements of the organization's performance is, according to experts comprised between 10 and 20 [5, 13, 16, 17]. In our opinion, the number of indicators should be determined according to the size of the organization, its peculiarity, the complexity of the processes taking place in the company and of course, depending on the number and characteristics of the goals to be achieved.

3. Use of performance indicators for determining the losses in production processes

Research by Kloviené and Speziale [18] concludes that, although the issue of performance measurement systems is analyzed in detail, performance measurement in companies (especially in SMEs) is not done in a systematic way.

Starting from this, we set the first objective for our research:

O1 - developing a system of performance indicators that reflect as closely developments of the four variables that define the performance of the production process and which to have an emphasized universality.

It is obvious that the existence of waste in production processes generate counter-performance.

The counter-performance is not directly measured by indicators, being considered as the difference between the expected and the realized outcome.

This has a major negative impact on the processes lead by managers with little experience.

To eliminate this drawback, it is necessary both to establish the causality between the measured indicators and forms of waste but also quantification of losses.

This goal is the second objective of the research:

O2 - establishing the correlations between performance indicators and forms of waste.

To achieve the objectives, we conducted a applied study about production process from two industrial companies.

Based on the observation, we creat a system of non-financial performance indicators, that to be able to reflect, as accurately as, the activity of the production's departments from the companies studied.

The study was conducted in several stages:

- E1 Analysis of the existing PMS;
- E2 Develop a system of indicators that meet the requirements of universality according with the first objective;
- E3 Establishing the correlations between KPIs and existing forms of waste in the production process.

In the study, we used the symbols shown in table 1:

Table 1. The symbols used in the study.

Symbol	Signification	Symbol	Signification	Symbol	Signification
I_s	stocks index	N_{def}	Number of defects	Sp	Overproduction
I_C	costs index	St	Stocks	I_T	Time utilization index
I_{ts}	Overtime index	Rp_r	Rework / reprocessing	T_{oa}	Duration of the unplanned stops
P_r	Production achieved	M	Movement	C_r	Unit cost achieved

Symbol	Signification	Symbol	Signification	Symbol	Signification
T_D	Available time	A	Waiting	C_t	Theoretical unit cost
S_s	Safety stock	Tr	Transport	ε	Scrap index
I_{Nm}	Index of executing the work rate	T_{ls}	Overtime	N_{mr}	Achieved work rate
I_{pr}	Production achieved index	Q_c	Ordered quantity	N_{mt}	Theoretically work rate
T_{ef}	Time actually worked	Q_{nec}	Necessary quantity	S_e	Existing stock
Ov	Overload index	r	Scrap rate	T_{ol}	Duration of the stops due logistics

3.1. E1. Analysis of the existing PMS

The production processes and systems for measuring the performance of the two companies under study, are similar.

Performance indicators used in the production departments of the two companies were classified in terms of the four variables of the organizational performance and are presented in Table 2:

Table 2. The existing performance indicators system.

Category	Symbol	Description	Target value (ideal)	Formula
N	FTR	First time right	100%	$\frac{P_r - N_{def}}{P_r} \times 100$
Q	Q_{nec}	Quantity to be produced (planned)	-	$Q_c - (S_e - S_s)$
	P_r	Production achieved	-	
	G_{rp}	Production's realizing rate	100%	$\frac{P_r}{Q_{nec}}$
T	T_{ef}	Time actually worked	100%	$T_D - T_{oa}$
	N_{mr}	Achieved work rate	-	$\frac{P_r}{T_{ef}}$
C	C_{prr}	Manufacturing costs achieved	-	-
	C_r	Unit cost achieved		$\frac{C_{prr}}{P_r}$

The above indicators are additional to the indicators used at company level (that were not surveyed).

Analyze of the PMS from Table 2, it is obviously that it does not provide information about the types of waste present in the production process.

This leads to difficulties in adopting optimization measures activity, because the production managers are forced to make decisions based solely on intuition and personal experience.

3.2. E2. Development a new system of indicators

Starting from equation (1) and based on direct observations, we have developed a system of indicators that can be used in any company regardless of activity area.

The indicators proposed by us are non-financial and are additional to the performance measuring system of the business.

KPI system developed by us is presented in table 3:

Table 3. The system of performance indicators developed in the study

Category	KPI	Formula	Target value (ideal)	The form of waste measured
N	r	$\frac{N_{def}}{P_r} \times 100$	*0*	$Rp_r = \lvert 1-\varepsilon \rvert \times 100$
	ε	$\frac{1}{1-r}$	*1*	
Q	I_s	$\frac{S_e}{S_s+Q_c}$	*1*	$St = \lvert 1-I_s \rvert \times 100$
	Q_{nec}	$Q_c-(S_e-S_s)$	-	$Sp = \lvert 1-I_{pr} \rvert \times 100$
	P_r	$(S_s+Q_c)\times\varepsilon-S_e$	-	
	I_{pr}	$\frac{P_r}{Q_{nec}}=\frac{\varepsilon-I_s}{1-I_s}$	*1*	
T	T_{ef}	T_D-T_{oa}	*1*	$A = \lvert 1-I_T \rvert \times 100$
	I_T	$\frac{T_{ef}}{T_D}$	*1*	
	N_{mr}	$\frac{P_r}{T_{ef}}$	-	$M = \lvert 1-I_{Nm} \rvert \times 100$
	I_{Nm}	$\frac{N_{mr}}{N_{mt}}$	*1*	
	T_{ol}	-	-	$Tr = \frac{T_{ol}}{T_D} \times 100$
	I_{ls}	$\frac{T_{ls}}{T_D}$	*0*	$Ov = \lvert 0-I_{ls} \rvert \times 100$
C	I_C	$\frac{C_r}{C_t}$	*1*	$Sp+Rp_r+Tr+M+St+Tr = \lvert 1-I_C \rvert \times 100$

3.3. E3. Establishing the correlation between the indicators and forms of waste

In the third stage of the study, we associated the indicators with one or more forms of waste possible developments highlighted by the KPI.

Of course, each KPI measures the performance from a certain perspective, but also the proposed system indicates the size of the counter-performance (the difference between the target value and the value obtained).

It is noted that at the level of productivity (time) are manifested most forms of waste.

Starting from the idea that the wasting increases the costs, we considered cost index variation as the result of the sum of all losses in the production process.

In the study we considered as homogeneous production, but if production is heterogeneous, then the indicators proposed by us become partial indicators (for an assortment from the products range), and the total is calculated as sum of partial indicators.

For example, the index of stocks I_S, for a product, has the formula shown in table 3, and when there are multiple varieties, the formula is:

$$I_s = \sum_{i=1}^{n} I_{s_i} = \sum_{i=1}^{n} \frac{S_{e_i}}{S_{s_i} + Q_{c_i}} \tag{2}$$

Where:

$S_{e_i}, S_{s_i}, Q_{c_i}$ means existing stock, safety stock and ordered quantity for *i* product

4. Conclusion

The study presented in the article led us to the conclusion that measuring the performance of production processes within industrial enterprises is a common practice.

However, the PMSs used not marked the counter-performance, which leads to difficulties in optimizing the activity.

Measuring waste improves the management process easier, in the sense that gives managers accurate indications of the direction in which should be acted to achieve operational excellence.

Therefore, we sought to develop a system of indicators that can be used to quantify the losses (in the Lean manufacturing acceptation) of industrial production.

The system developed by us has a high degree of universality and can be adopted, with minor adaptations, by companies in all fields, not just those with industrial specific.

Performance indicators shown cannot be used to characterize the work of the whole organization, so it is necessary for them to complete performance measurement systems business, existing within organizations.

References

[1] Imai M 1986 *Kaizen. The Key to Japan's Competitive Succes* (New York: McGrew-Hill Education)
[2] Womack J P, Jones D T, and Roos D 1990 *The Machine That Changed The World* (New York: Macmillan Publishing Company)
[3] Ohno T 1988 *Toyota Production System. Beyond Large-Scale Production* (Portland: Productivity Inc.) pp 45-75
[4] Aunuraffiq R, Sari R N and Sari Y M. Basri 2015 *The Moderating Effect of Goal Setting on Performance Measurement System-Managerial Performance Relationship* Procedia Economics and Finance **31** (2015) 876 – 84
[5] Parmenter D 2010 *Key performance indicators (KPI): developing, implementing, and using winning KPIs*, (New Jersey: John Wiley & Sons)
[6] Lonnie W 2010 *How to Implement Lean Manufacturing* (New York: McGrew-Hill).
[7] Filip F C 2012 *Contributions on implementing the efficient management methods specific for production systems* PhD thesis (Brasov: Transylvania University of Brasov)
[8] https://leanromania.wordpress.com/lean-in-romania /accessed: 21.01.2016
[9] Stoianovici G V 2011 *Waste Elimination Utilising Lean Manufacturing Techniques* (Sebes:11th National Multidisciplinary Conf. "Prof. Dorin Pavel - founder of Romanian Hydropower")
[10] Drucker P 2007 *The Practice of Management* (Burlinghton: Elseviere)
[11] Pawliczek A and Piszczur R 2012 *Relevance of management systems ISO 9000 and ISO 14000 for stabilization of business in the sustainability context*, Proceedings of 19th IBIMA Conference (Barcelona Norristown: IBIMA) 1338–54
[12] Rylková Ž and Bernatík W 2014 *Performance measurement and management in Czech enterprises* Procedia - Social and Behavioral Sciences **110** 961 – 68
[13] Graham I, Goodall P, Peng Y, Palmer C, West A, Conway P, Etienne J, Mascolo and F.U. Dettmer

2015 *Performance measurement and KPIs for remanufacturing*, Journal of Remanufacturing, **5:10**

[14] Fitzgerald L, Brignall S, Silvestro R, Voss C and Robert J 1991 *Performance measurement in service businesses*, (London: Chartered Institute of Management Accountants)

[15] Sorovou D, Politou A, Kalligeris S and Topouzidou 2001 *D3-KPI Manual Part I -General Methodology* **IST-2000-28760**

[16] Kaplan R S and Norton D P 1996 *The balanced scorecard: translating strategy into action*. (Boston: Harvard Business Press)

[17] Hope J and Fraser R 2013 *Beyond budgeting: how managers can break free from the annual performance trap*. (Boston: Harvard Business Press)

[18] Kloviené L and Speziale M T 2015 *Is Performance Measurement System Going Towards Sustainability in SMEs?* Procedia - Social and Behavioral Sciences **213** (2015) 328 – 33

Education For Sustainability - Experiences From Greece

Athena Baronos

Professor of Applications
Department of Mechanical Engineering and Industrial Design
Technological Education Institute of West Macedonia
Koila Kozani 50100 Greece

Email abarono@teiwm,gr

Abstract. One of the main issues involved is changing people's attitudes, values, behavior and consumption patterns. University education in Greece aims to educate engineers so that besides acquiring theoretical knowledge, they also learn to show competences and are motivated to act accordingly. Educating for sustainable development also entails the development of critical capacities and the necessary skills to be able to identify and formulate problems. This paper outlines the way in which an approach to teaching sustainability has been embodied in the Industrial Design, West Macedonia Greece. More specifically, it describes a course to develop comprehensive case studies and support material in order to aid, Industrial Design students in understanding the sustainability concepts and how solutions can be developed

1. Introduction

Sustainable Development, as defined by the Brundtland Commission in 1987, is "development that meets the needs of the present without compromising the ability of future generations to meet their own needs according to the World Commission on Environment and Development [1]. In a sustainable society, environmental protection and economic objectives belong to a common framework. The President's Council on Sustainable Development's definition of sustainable development has been broadened to include social equity (United Nations Conference on Environment and Development [2] In a sustainable world, environmental protection, economic objectives, and social justice should be linked in harmony.

In Greece nowadays sustainable development is a big challenge to universities, although incentives and barriers for implementation of the curriculum are addressed. In this respect 'sustainability' had moved from being a discrete coverage of 'environmental issues' to embedding social, economic and environmental aspects of sustainability alongside other courses. Students should learn about and view sustainability as an integrated part of civil engineering education, not as an added extra; sustainability had become part of the mainstream curriculum. This paper aims at developing in the way in which an approach to teaching approach using role playing that has been embodied in the Industrial Design, West Macedonia Greece.

1.1. The Department of Industrial Design of TEI West Macedonia

The Department of Industrial Design is one of the two Divisions, Mechanical Engineering and Industrial Design, belongs to the School of Engineering of TEIWM. It is a large department with a significant record of exceptional teaching and research.

The structure of the curriculum of the Industrial Design degree program is such that graduates of the department develop integrated scientific and technical knowledge across the spectrum associated with the Industrial product design.

The philosophy of the whole program is based on the principle: Graduates are qualified and capable personnel who will staff the productive units, combining informatics, operations research and integrated design - development systems, would contribute effectively to economic growth.
Industrial product design is one of the main characteristics of the graduate that maintains after completion of the course, preparing a thesis and six-month internship in the profession.

The course Product Design II and is taught in the 5th semester of the curriculum
The course learning outcomes, specific knowledge, skills and competences of an appropriate level, which the students will acquire with the successful completion of the course are described.
• Description of the level of learning outcomes for each qualification cycle, according to the Qualifications Framework of the European Higher Education Area
• Descriptors for Levels 6, 7 & 8 of the European Qualifications Framework for Lifelong Learning and Appendix B
• Guidelines for writing Learning Outcomes
Upon successful completion of the program, students will be able to:
1. Design products.
2. Use CAD systems in the process of designing products.
3. Solve problems arising during the process of designing a product.
4. Apply knowledge gained in courses in ergonomics, marketing, materials science and technology, manufacturing processes and cost analysis.
5. Analyze the relationship between cost and technical solutions.
6. Be creative in solving design problems.
7. Work as part of a designing team.
8. Write and present reports.

1.1.1 General Competences

Taking into consideration the general competencies that the degree-holder must acquire (as these appear in the Diploma Supplement and appear below), at which of the following does the course aim are:
Search for, analysis and synthesis of data and information, with the use of the necessary technology
- Adapting to new situations
- Decision-making
- Working independently
- Team work
- Working in an international environment
- Working in an interdisciplinary environment
- Production of new research ideas

Upon successful completion of the program, students have the basic theoretical and practical knowledge in the fields of the subject area of Industrial Design Engineering and the related profession and are able to properly apply the theoretical and practical knowledge acquired during the study period - be able to creatively implement the scientific knowledge and modern technology in understanding and solving problems in industrial Praxis, and to promote innovation in design and development of new products and services in Greece and internationally. Have gained the necessary competencies to proceed to their second cycle study

1.1.2 The Framework for product design is

• Process design of new products.
• New Product Marketing Plan.
• Product life cycle.
• Product design and technology

1.2. Methodology - Innovation in teaching Industrial Design The Department of Industrial Design in TEI West Macedonia

Teaching communication skills to industrial design students is not always an easy task, however, it is one of the skill sets employers are increasingly looking for in graduates. When an industrial design course this year I wanted to use fun and innovative teaching methods to create a stand-out example course in keeping with the connected curriculum vision. [3]. New coursework components in which students were asked to make a role playing in small teams to explain a point of view themselves (seller buyer) was selected by each team. The report forwarded to the relevant lecturer to be included in the teaching material, to benefit future cohorts. As well as helping students to learn from participation rather than traditional teaching, they got the opportunity to practice teamwork, and unusually, to showcase their creativity.

1.3. Using Teamwork in a project based case study

Team work is a way to train the demands of work life of professionals, especially to learn how to communicate, to learn self-organization and to cooperate From the other hand Project based case studies, simulation are useful to stimulate effective competences for solving problems, to provide a basis for lifelong learning and to increase the students' motivation to learn.

2. Roles in the coursework

Characterisation of the role playing-an active role in designing the course where the students usually learn better than passively listening students, especially social competences, conflicting roles while lectures can be useful for developing complex technical knowledge – and it should be mixed with physical attendance

2.1The role of the teacher

Teachers' role is changing from a broadcaster of knowledge to a coach that supports the students' process of learning and transfers knowledge from practice to lecture halls

2.2 Role playing Description

The role playing is a synthetic view of the three dimensions of sustainability namely, economic, social and environmental.

Before playing, Students are required to work as a team for about a month and to retrieve information concerning a role in the project organization that selected and assigned to them... At the end of this preliminary phase, they have to hand a report assignment, which contains:
(i) Sustainable development issues,
(ii) A short review of the role that is related to sustainable development and the organization that has been assigned to them
(iii)(iii) A critical assessment of the information collected.

Studying a specific concept associated with sustainable development deepens the students' knowledge around sustainable development issues. The document does not reflect the workload associated with the information gathering and analysis tasks. Instead, it only acts as a memo that has to be clear and concise. Such an exercise is designed both to develop the students' synthesis skills, and to help them in the perspective of the role-play.

Before the role-play starts, the teacher prepares the classroom to materialize the three dimensions of sustainable development with chairs and stickers that are posted on the participants, acting as decision makers, have to prepare a five-minute speech emphasizing how their organization is addressing sustainable development issues and including the related concept that they have previously selected.

The teacher does not intervene at all except to make sure that the time limits are respected. The last sequence of the session is devoted to debriefing. The teacher asks the students to wrap up what they have learned about the concept of sustainable development as well as its related concepts. (S) he makes sure that what the students' state is correct, and stresses important issues

3. An example –Production of a Sustainable –Green Product -Case Studies

The Case Study is the production of several industrial design green products The agents stakeholders involved are

- Government
- Supervision departments
- Design companies, suppliers,
- Users
- Partners
- Contractors

The students are separated into interesting groups and are asked to describe their role as well the Project environment (organization) Objectives (points of view) in a report after a Relative team brainstorming. The objectives of the role playing where

- To accomplish the requirements of the project role descriptions;
- To establish a project team with high-level performance;
- To accumulate experience for following real life projects;

Most importantly In the Project Team To increase job skills To get experience and broader career opportunity

Reports were prepared in terms of Government to promote local economic development and to assure Quality meets the requirements of technical codes; through Government supervision departments to keep up with Related national regulations regarding fire prevention, safety, environmental protection, labor and health and other aspects should be avoided.

In terms of Local Government was To create new job opportunities; and to lift up city competitiveness.

In terms of the User Product quality should be eligible at a Reasonable price.

In terms of Designing Companies To get economic benefit; To obtain successful cases in terms of technology and design; To train and foster designing teams; and To establish strategic cooperative partners

In terms of Partners the aim was to implement and complete supervisor's job within the agreed project term, to ascertain that Quality meets the requirements on design and Project investment should be controlled within the budget scope, To get economic benefit and Safety should be under control;

The Contractors point of view is To get economic benefit to train technical teams, and introduce and master new technologies via project applications to get a successful case in order to upgrade corporate brand reputation and two to make the owner satisfied and keep a long term cooperative relationship with the owner.

As for the Suppliers the main issue was to promote corporate reputation; to get economic benefit; To get a successful case regarding related technologies to lift up products quality, R&D level and efficiency and get more cooperation opportunities.

An interesting part was the Implementation coordination meeting system where the Coaching professor adopted an implementation, coordination meeting system to enable positive interaction and efficient communication [4]

Conflicts were faced After a consultation was held to reach a consensus, whether to make changes or how to change was decided. The deriving project objectives were imparted, realized, checked and adapted The aim of the project was Reaching strategic not reacting objectives [5]

4. Assessment- The Students' Opinion

A questionnaire survey similar to Katz [6] is conducted to show students' opinion. The questionnaire consisted of the following questions:

5. Results

There were 57 questionnaires answered at the end of the course.
The results were as follows:

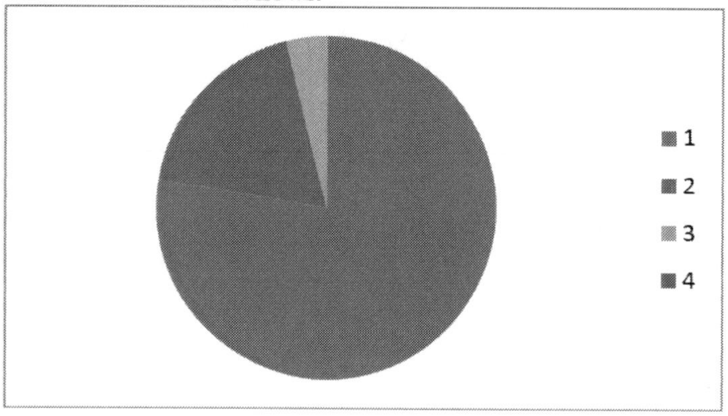

Figure 1 Students opinion about the course
1. Good, 2. Improvement, 3. I do not answer, 4. Comments

- 76% of the students think that the course was good

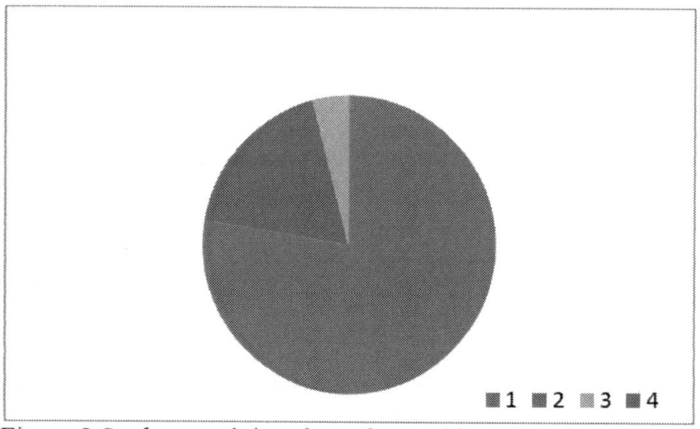

Figure 2 Students opinion about the teaching with the proposed method
1. Good, 2. Improvement, 3. I do not answer, 4. Comments

- 56% of the students answered that the teaching of the course with the method presented in this topic was good while 36% think that it was significantly improved.

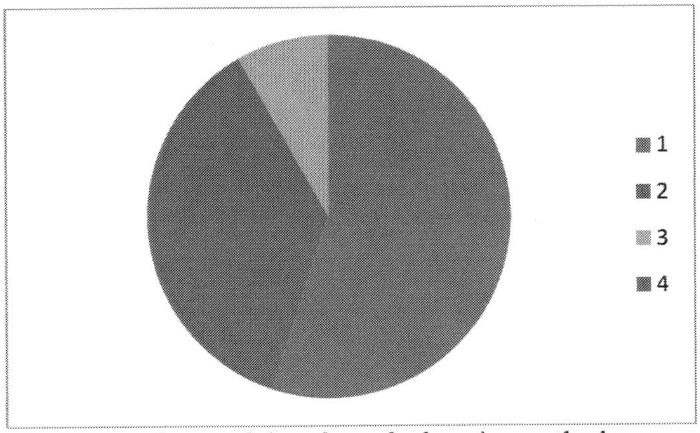

Figure 3 Students opinion about the learning method
1. Good, 2. Improvement, 3. I do not answer, 4. Comments

- 56% of the students answered that the learning method presented in this topic was good and 34% that was significantly improved

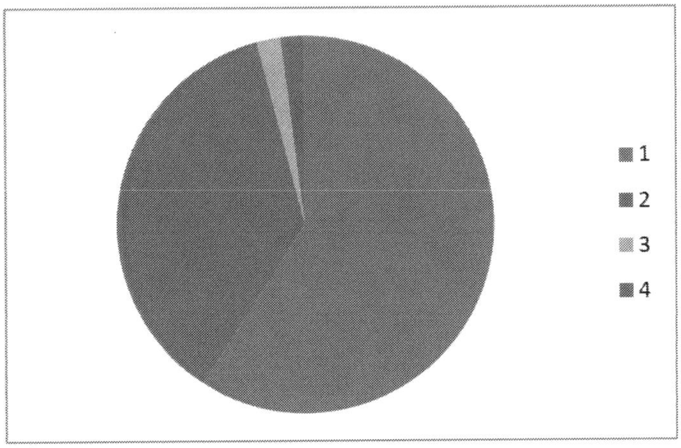

Figure 4 Students opinion comparison the proposed teaching method vs the traditional method
1. Good, 2. Improvement, 3. I do not answer, 4. Comments

- 60% of the students prefer this method than the traditional while 22% think that it was a significant improvement.

5.1 Comparison with similar studies

A study using similar methodology –role playing- for sustainable education have been selected for comparison. Perdan et al [6] is the description of a project on current research and teaching material in engineering at the University of Surrey, UK. It also incorporates a multimedia course already developed there and taught at the postgraduate level. The module is also incorporated into modules in the Engineering Doctorate Program. Another is that of [7] at Carnegie Mellon The main difference between the above mentioned programs and ours at the University of Thessaly is that the subjects are taught by a multidisciplinary team in Surrey and it was IT based.

6. Conclusions

A team-based design project with a problem-based learning approach is outlined. There is an added value in implementing of competence based case studies on project management courses. Didactical skills are strengthened through role playing *Role playing* increases students' motivation and enthusiasm to learn. Conclusively the methodology can be used after modifications for students in other engineering departments of other universities. The interactive models developed may be used to enhance their quantitative understanding.

References

[1] World Commission on Environment and Development (Brundtland Commission), (1987), Our Common Future: Oxford University Press, p.43.
[2] United Nations Conference on Environment and Development (UNCED), (1992) Agenda 21: Programme of Action for Sustainable Development (New York: United Nations,)
[3] Gunn, I. W. (1996). "Education for Sustainability through Engineering", Environmental Engineering Education and Training, . T. V. Duggan and C. A Brebbia, Eds Computational Mechanics Publications, 45-56.
[4] Jahan K. and Y. Mahta (2007). "Sustainability Across the Curriculum" International Journal of Engineering Education, 23 (2) 209-217.
[6] Perdan S., Azapagic A., and. Clift R., (2000). "Teaching sustainable development to engineering students", International Journal of Sustainability in Higher Education 1(3)267-279 Davidson C. I., [7]
[7] Hendrickson C. T., and Matthews H. S. (2007), "Sustainable Engineering: A Sequence of Courses at Carnegie Mellon University". International Journal of Engineering Education, 23(2) 287-293
[7] Katz, L., Sutherland, J. and. McLaughlan R. G (2007). "Instructional Strategies to Educate for Sustainability in Technology Assessment" International Journal of Engineering Education, 23 (2), 200-208.

EAF Management Optimization

M COSTOIU[1], A IOANA[1], A SEMENESCU[1] and D MARCU[1]

[1]University Politehnica of Bucharest, Spl. Independentei 313, Bucharest, Romania

Email: adyioana@gmail.com

Abstract. The article presents the main advantages of electric arc furnace (EAF): it has a great contribution to reintroduce significant quantities of reusable metallic materials in the economic circuit, it constitutes itself as an important part in the Primary Materials and Energy Recovery (PMER), good productivity, good quality / price ratio, the possibility of developing a wide variety of classes and types of steels, including special steels and high alloy. In this paper it is presented some important developments of electric arc furnace: vacuum electric arc furnace, artificial intelligence expert systems for pollution control Steelworks. Another important aspect presented in the article is an original block diagram for optimization the EAF management system. This scheme is based on the original objective function (criterion function) represented by the price / quality ratio. The article presents an original block diagram for optimization the control system of the EAF. For designing this concept of EAF management system, many principles were used.

1. Introduction

Electric Arc Furnace (EAF) is a complex and effective aggregate used in steels elaboration. The importance of this aggregate is the fact that it uses waste materials as base metal load. For this reason, EAF has a great contribution to the reintroduction of significant quantities of reusable metallic materials in the economic circuit.

EAF is considered an important part in the Primary Materials and Energy Recovery (PMER).

Other advantages of steel in EAF development are: good productivity, good quality / price ratio, the possibility of developing a variety of classes and types of steels, including special steels and high alloy.

As [1] „Steelmaking processes generate fume that must be captured and controlled to prevent emissions to atmosphere. This is done with large extraction fans and bag filter cleaning systems. The operation of these systems is critical to ensure good environmental performance of the steelmaking process and optimization is necessary to ensure that the fume extraction systems are run economically."According to the National Strategy for Sustainable Development of Romania (SNDDR) the development of Romanian steel industry, with its primary segment, sets the product development rolled steels , as a main priority [3]. An important development of the electric arc furnace is represented by the EAF vacuum solution of this complex operation unit. In figure 1 such a variant is presented – Vacuum Arc Remelting (VAR), [2].

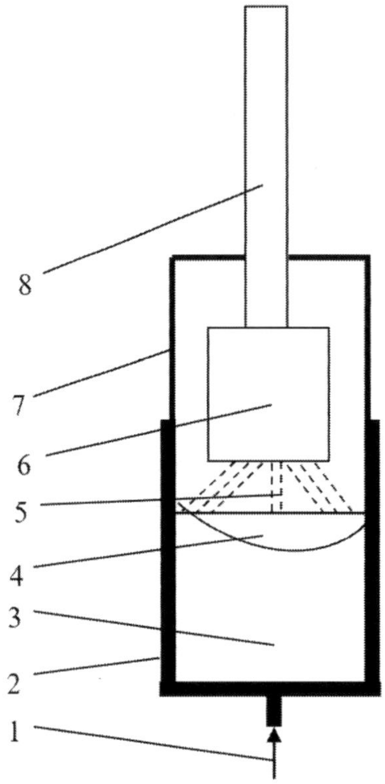

Figure 1. Vacuum Arc Remelting (VAR) Scheme
1 – Cooling water; 2 – Mold ingot; 3 – Ingot; 4 – Liquid metal; 5 – Electric arc; 6 – Electrod; 7 – Vacuum chamber; 8 – Driving rod electrod

Another development of the electric arc furnace is the artificial intelligence expert systems for steelworks pollution control (Brimacombe, Jackson and Schofield, 2001). Figure 2 presents the scheme of expert system.

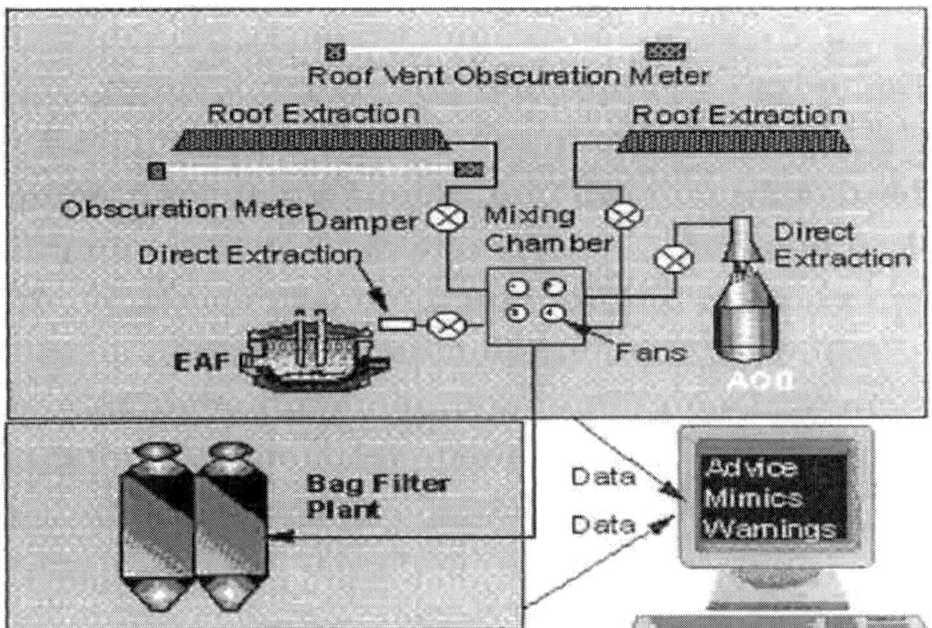

Figure 2. Scheme of expert system of model for a EAF fume extraction (after [1])

2. Elements of EAF management optimization

In figure 3 it is presented an original block diagram for optimization the control system of the EAF [3, 4, 5].

For designing this concept of EAF management system, the following principles were used [6, 7]:

A. The principle of analogy - requires competent observation and analysis of the modelled reality using both analogy with other areas of research and logical homology.

According to this principle, for developing mathematical models, the following stages were taken:

• defining the modelled objective – this is the primary stage of the modelling analysis; This stage must satisfy both the aim and objectives of the system while ensuring their compatibility;

• defining criteria for efficiency - it is a step conditioned by the correct setting of the objectives of the system and it allows the optimization of modelling solutions;

• developing options - based on accessing realistic, effective and original solutions;

• evaluating alternatives – according to the criteria of efficiency previously set;

• determining the final solution - based on comparative analysis of different solutions to which modelling has led.

B. The Concepts Principle - is based on systems theory concepts including the concept of feed - back.

C. The Principle of Hierarchy - implies the need for designing a hierarchical models system, in order to structure decision and coordinate interactive subsystems.

D. The Uncertainty Principle - is generated mainly by the high complexity of the processes related to CAE. In addition, the existence of interaction between subsystems components - interaction that cannot always be exactly predetermined - and nonlinearities in the system, the subjectivity of choice and prescribing the objective function contributes to increasing uncertainty factors.

It is useful to note that, based on this modelling principle, as system complexity increases, requiring a hierarchical structure, the precision of developed models decreases but the degree of relevance increases.

E. The Internal Model Principle - establishes that a dynamic system is structured in a stable way only if:

• it uses negative reaction of controlled sizes;

• it incorporates into a feedback loop a reduplicate model of the dynamic structure of exogenous signals; this internal model provides the signals meant to compensate asymptotically the disturbances of the considered CAE system.

The Modeling System designed for EAF processes consists of six subsystems represented by the following mathematical models:

- the Prescribing Mathematical Model of the Objective Function (PMMOF);
- the Mathematical Model for Load Calculation (MMLC.);
- the Mathematical Model of Effective Control of Melting (MMECM);
- the Mathematical Model Load of Preheating Process (MMLPP.);
- the Mathematical Model of Reactive Powder Injection (MMRPI);
- the Mathematical Model for Calculating the Design of Recuperative Burners corresponding to the Preheating Process (MMCDRBPP).

Below it is presented the stages and computational algorithms related to the main models, as well as the modeling results including the validation experiments of the designed model.

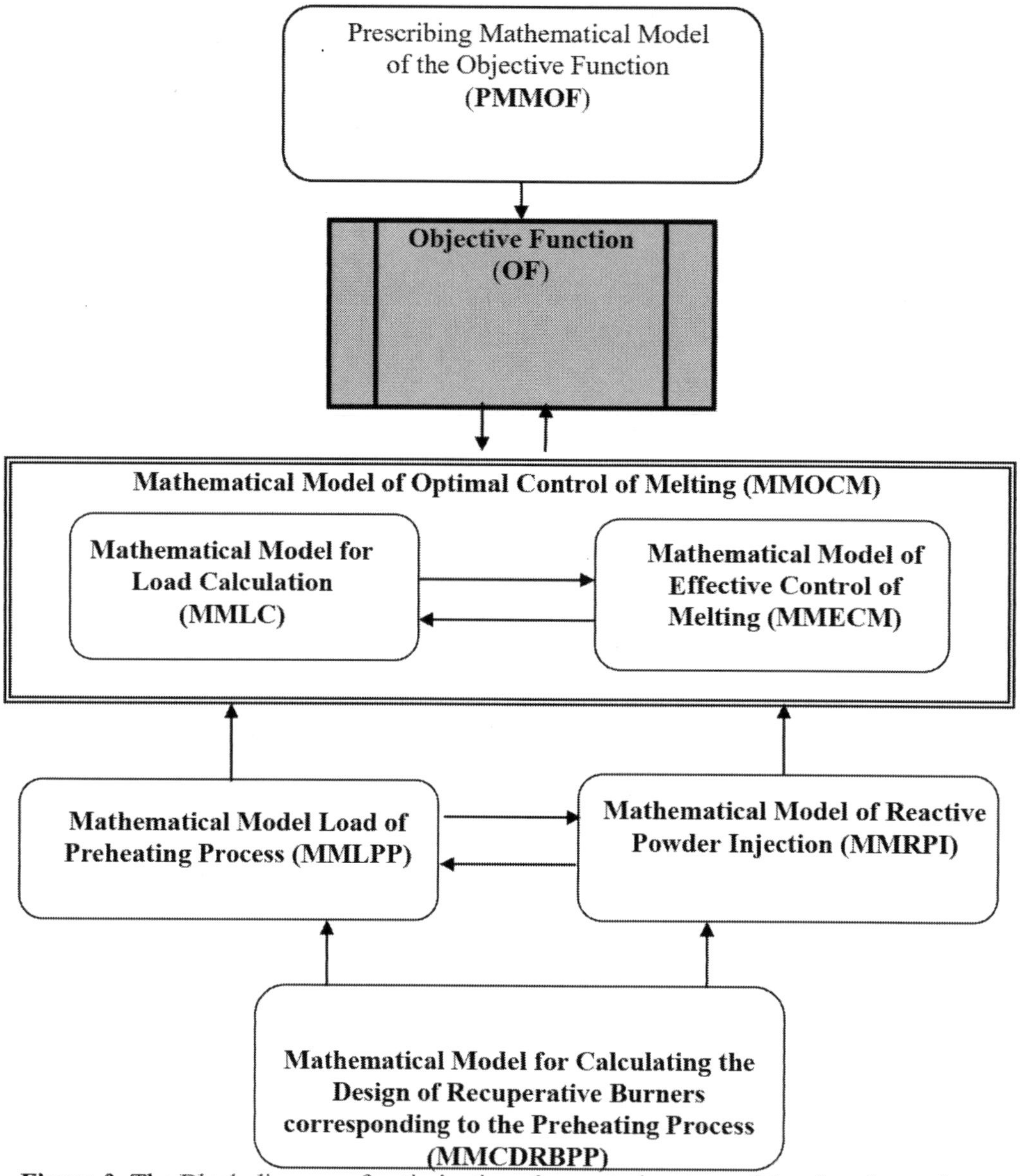

Figure 3. The Block diagram of optimization the control system of the electric arc furnace

3. Mathematical model of establishment (prescription) of objective function (MMOFE)

The Prescribing (establishing) of the objective function (OF) of modelling system related to EAF processes is based on a qualitative - economic analysis of these processes. In this respect, the developing of a new product (in this case a mark of quality steel) must equally provide both profitability and the ability to ensure quality conditions.

The Prescribing Mathematical Model design of objective function is based on quantication of the Objective Function (OF) as a qualitative - economic matrix M_{QE}, as in the diagram shown in Fig. 4.

The levels of prescribing the Objective Function are obtained by applying an algorithm relating three vectors:

- vector $\overline{T}$ – the technical parameters vector (t_i);

• vector $\overline{E}$ – the economic parameters vector (e_j);

• vector $\overline{P}$ – the shares vector (p_k).

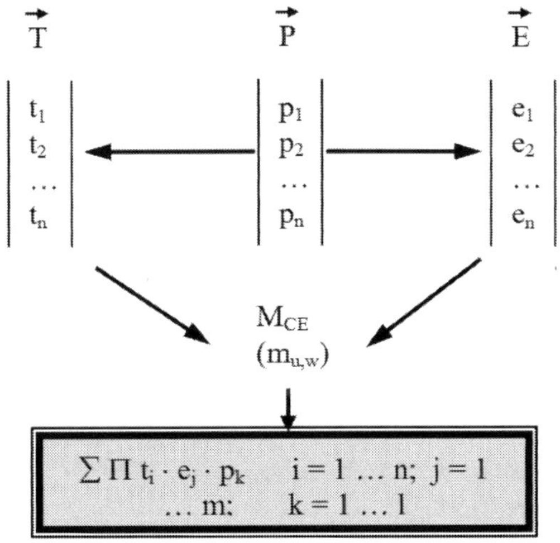

Figure 4. The quantification of the objective function of the modeling system

The components of the two vectors $\overline{T}$ and $\overline{E}$ considered with significant shares in the quantification (prescription) of the objective function OF are:

• t1 - chemical composition of steel (precision fit trademark);
• t2 - steel purity (in gas);
• t3 - steel purity (the inclusions);
• e1 - the specific consumption of raw materials;
• e2 - the specific energy consumption;
• e3 - melting productivity of EAF.

The assignment of the values to the two vectors was made on a relative scale having as reference an Optimal Level (OL), as follows:

• OL = value 10
• OL ± 10 % = value 9
• OL ± 20% = value 8
• OL ± 30% = value 7
• OL ± 100% = value 0

The Optimal Level (OL) for each component of the two vectors is:
- for t1 – the arithmetic mean of the prescribed limits of variation of the steel composition mark;
- for t2 - the prescribed minimum gas content;
- for t3 - the prescribed minimum content of inclusions;
- for e1 - specific minimum standards consumption of raw materials;
- for e2 - the specific prescribed minimum consumption of energy;
- for e3 - the normal maximum productivity of the steel elaboration process.

4. Conclusions

The electric arc furnace for steel elaboration is a complex and special utility aggregate. Therefore to optimize its control is of great importance.

The constructive and functional upgrades "suffered" by the electric arc furnace over time, objectively argues the importance of this complex aggregate. Among the upgrades, one can mention: the vacuum electric arc furnace, the DC electric arc furnace.

The original optimization scheme of EAF control is based on more specific correlations and mathematical models. We mention the main ones: the Prescribing Mathematical Model of the Objective Function (PMMOF); The Mathematical Model for Load Calculating (MMLC.); The Mathematical Model of Effective Control of Melting (MMECM);The Mathematical Model Load Preheating Process (MMLPP.), linked by dual correlation between them.

The objective function of the Modelling System (Control) is the price / quality ratio. The purpose of the component models of the control system is to maximize this ratio.

References

[1] Brimacombe L, Jackson C and Schofield N 2001 Artificial intelligence expert systems for steelworks pollution control *La Revue de Métallurgie-CIT*, **1** 111-116

[2] Chapelle P, Bellot J.P, Jardy A, Czerwiec T, Robbe X, Champin B and Ablitzer D 2001 Behaviour of the electric arc in the VAR process *La Revue de Métallurgie-CIT*, **12** 1129-1138

[3] Ioana A, Nicolae A, Predescu C, Sandu I.F, Sohaciu M, and Calea G.G 2002, *Optimal Managing of Electric Arc Furnaces* (Fair Partners Publishing, Bucharest) 37-172

[4] Ioana A 2007 *Production Management of Metallic Materials Industry. Theory and Applications*, (PRINTECH Publishing Bucharest) 18-143

[5] Ioana A and Marinică A 2014 *Electric Arc Furnace – Application In Efficiency of the Energy, Advances in Energy Research and Development* (ORIC Publications Little Rock, Arkansas, USA) 417- 433

[6] Ioana A and Marinică A 2014 *Advances in Energy Research and Development*, *Electric Arc Furnace – Application In Efficiency of the Energy* (ORIC Publications, , Little Rock, Arkansas, USA), 417- 433

[7] Ioana A, Semenescu A and Preda C.F. 2014 Economic Analysis and Management Models for Metallic Materials Industry *The Journal of American Business Review* **2** 159-165

[8] Ioana A and Bălescu C 2009 Environmental Study of the Formation of Evacuated Burnt Gases from a Steels Making Plant *Revista de Chimie* **5** 468-471

IOP Conf. Series: Materials Science and Engineering **161** (2016) 012105 doi:10.1088/1757-899X/161/1/012105

Marketing Management Elements

M COSTOIU[1], A IOANA[1], A SEMENESCU[1] and D MARCU[1]

[1]University Politehnica of Bucharest, Spl. Independentei 313, Bucharest, Romania

adyioana@gmail.com

Abstract. The article takes as a starting point several complementary definitions of marketing. Also present several key points related to marketing policy: defining the position that the company will occupy on the market with respect to competition, fixing the image with what will approach the market, achieving a balance between turnover rates profit and expenses as a condition for the survival of the company. In below article it is proposed an original concept of marketing management. This concept takes into consideration on the one hand correlations between company position in the market definition and on the other hand the image received by customers. The ultimate goal of this original concept of marketing management is to ensure an effective balance between: the turnover rate of profit and expenses of the company.

1. Introduction

As Stanton and Miller, [1] „Marketing is a whole system of economic activities related to pricing, promotion, distribution of products and services designed to meet the requirements of actual and potential consumers".

As Marcu, Semenescu and Ioana, [2] "The complete definition of marketing should take into account that marketing is related more to the concept. Is a matter of attitude and action. Marketing concept or vision is addressing to marketing as a system of ideas in order to satisfy the customers and consumers. "

On the most general plan, conceptual approach to marketing is done through marketing policy. Marketing policy is a concept that integrates the functions of the enterprise with which they can reach their market through the mobilization and allocation of all resources for this purpose [3]

The concept of marketing policy and his role in market economy can be understood only if it takes into consideration the following key elements [2-5]:

- Defining the position on the market that the company will have in balance with the competition;
- Fixing the perception of the company on the market.
- Making a balance between turnover and profit rates and expenses as a condition for surviving on the market.
- Determine the optimal reports between offer and what is the company actual doing on the market with internal and external sources.
- Relationships that are established between all the participants to the company life and marketing policy.

Marketing policy can be:

- General – When takes into account entire marketing strategy at corporate level;
- Specific – when has specific points related to objectives and specialized points of view from the expertise.

In the actual economical market conditions, a company cannot survive and evolve only if have efficient relations with all the economic factors from the areas it is performing, first of all with their consumers

2. About Marketing Management

In figure 1 it is presented one of the first marketing concept.

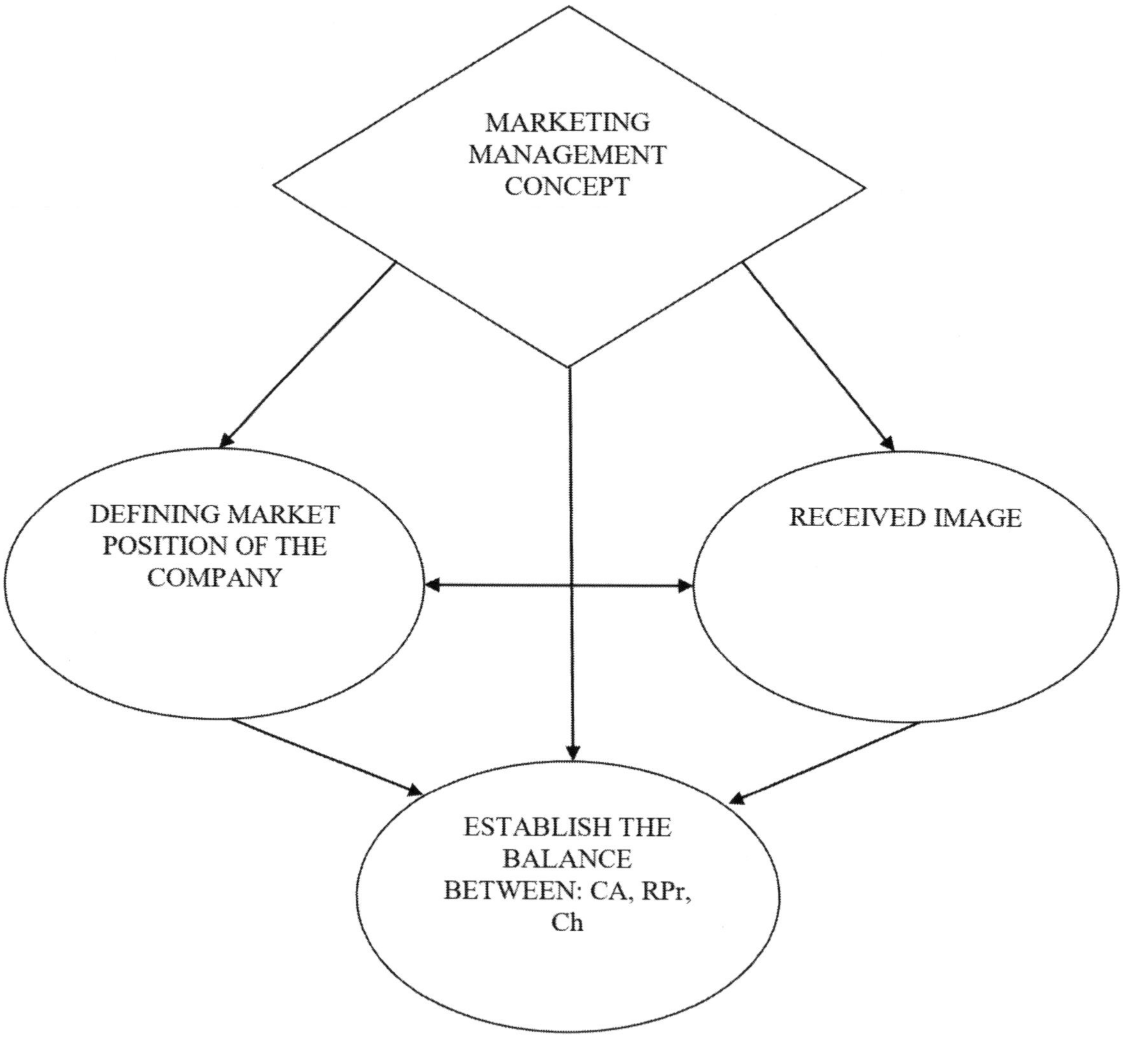

Figure1. Marketing management concept
CA – Turnover; RPr – Profit rate; Ch - Expenses

Marketing strategy is the general line, the orientation which the enterprise wants to follow in marketing environment in order to set the long term objectives. In order to be accurate, marketing strategy needs to consider the following concepts: existence, characteristics, structure of things that the company has in her disposition to achieve their goals.

Below, in figure 2, it is presented the perception of a marketing strategy.

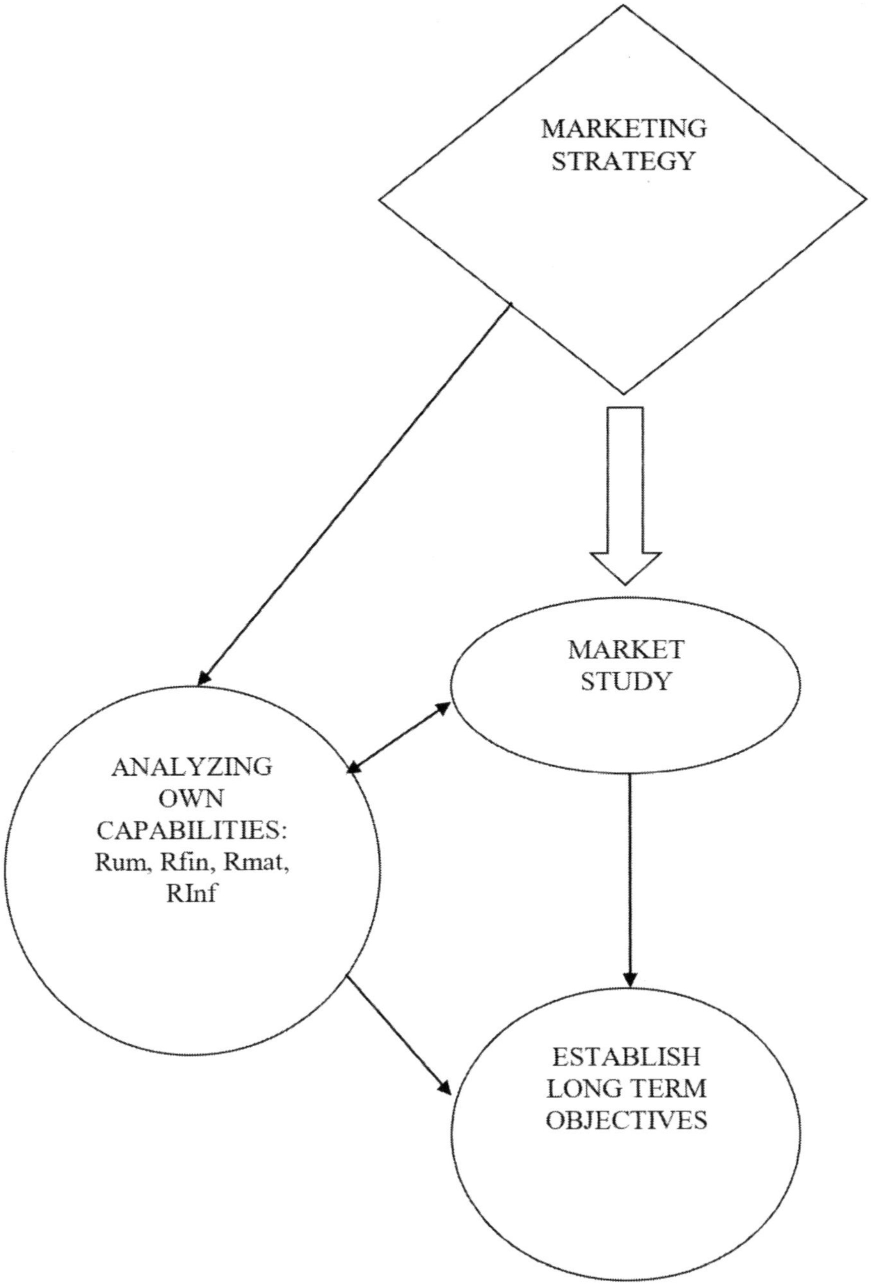

Figure 2. Marketing strategy concept
Rum – Human resource; Rfin – Financial resource;
Rmat – Material resource; Rinf – Informatic resource

In figure 3 it is presented the correlations between the 4th types of resources, very important correlations with every marketing manager should have on his hands. It is necessary to draw attention upon human resource and informational resource which governess the other 3 resources types.

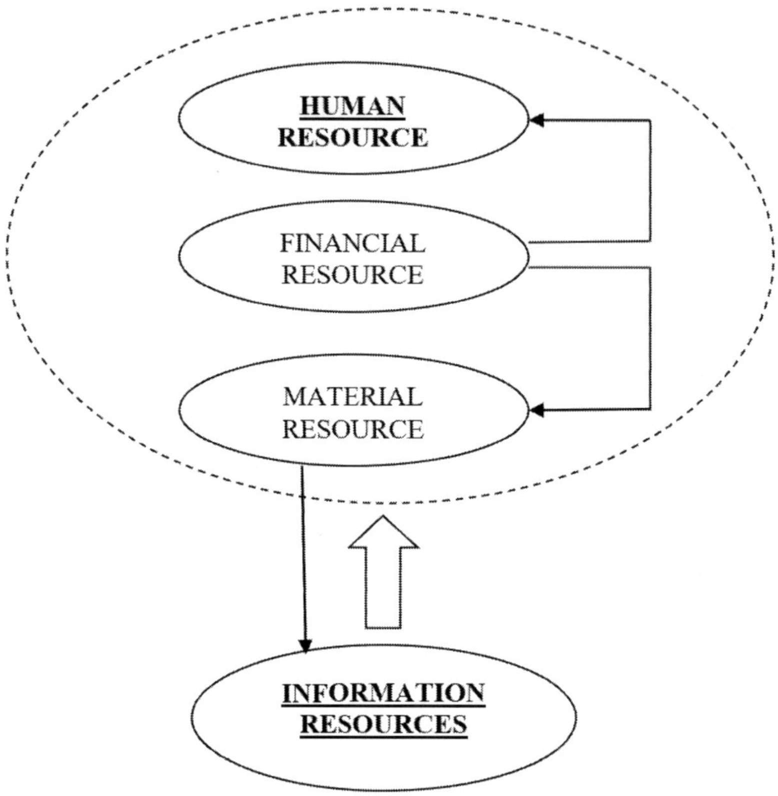

Figure 3. Correlations between necessary resources of marketing

An important role is played by the distribution (in terms of quantity, but especially in terms of quality) of financial resources on top of the other three resources, human, material and informational.

In case of a scientific marketing strategy market segmentation process has an important role [7,8]. Segmentation is the base of marketing program elaboration focus on a target audience. Segmentation process is described in below scheme (figure 4):

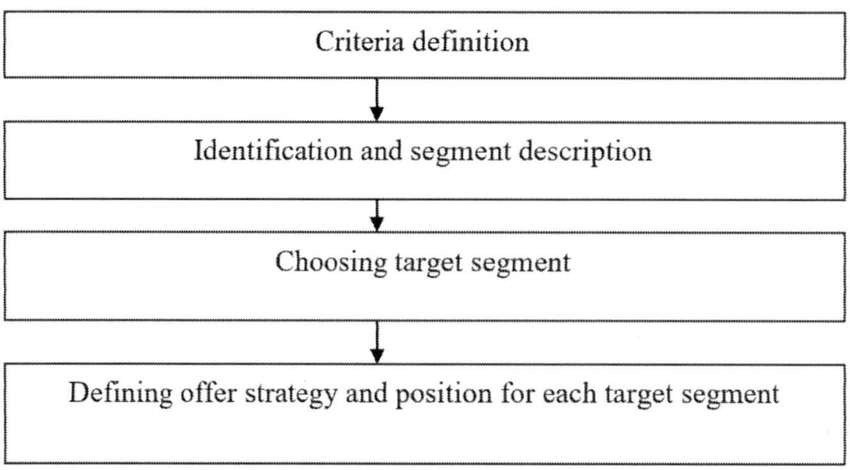

Figure 4. Market segmentation process

Below, it is necessary to make an explanation of the main steps of this process of market segmentation.

3. Criteria definition

At this stage it is necessary to answer the following questions:

- Who are the base consumers of the product?
- Which are the differences and similarities between them?

Information about the evaluation of consumer preferences can be achieved by various scale methods such as:

- semantic differential
- Stapel 's scale
- Likert 's scale
- method of paired comparisons (comparing similar products, products of different companies)
- method based on Fishbein -Rosenberg 's model.
- other direct methods : observation, testing, investigation

4. Identification and description of market segment

Identification of market segments goals is a necessary step in gathering information that will serve in deciding on how to sell the product.

People belonging to different social and economic groups often are inclined to buy products in various types of stores, read different publications, watch various television programs and respond to arguments in favor of the sale.

5. Choosing market segment

At this stage it is estimated the attractiveness and is decided the number of segments on which the company need to focus. According to the chosen segments, the company must determine strengths and weaknesses of competitors, market size, relationships with distribution channels, profitability and company mission

Any production company that manufactures and sells a product has three alternatives in choosing target segment:

- Develop one marketing program specifically for each market segment (differentiated marketing).
- Produce a single size and attract buyers through a unique program of promotion that is based on the tastes of the largest number of customers (undifferentiated marketing).
- To focus efforts (limiting the number of sizes and programs) in several market segments, giving up some buying groups considered less interesting (concentrated marketing).

Concentrated marketing enjoys the benefits of the three alternatives. Companies that practice it gain a strong position in segments over which they focused and they know very well, achieving significant benefits and savings in their specialization

On the other hand, narrow specialization, on a small number of products has huge risks in terms of modification of dimensions or segment characteristics being followed by market lose.

In choosing the most suitable alternative should be taken into account available resources, homogeneity of product and product market, the product stage in its life cycle, and competitors strategy related to the product

Undifferentiated marketing treated as a homogenous market, without any segments, relying on what is common to use and by the advertising. There are advantages in terms of production cost, logistic of goods and advertising.

Because the needs of consumers are different this takes to partial satisfaction of a big part of these ones.

Undifferentiated marketing has been widely used in consumer goods, food and beverages, but is currently in decline. The main purpose of this is to maximize sales marketing

Differentiated marketing , used in machinery and equipment has the advantage of adapting better to the needs of customers (which usually leads to increased sensitivity of sales) and the disadvantage of increasing production costs , supply, and inventory .

6. Conclusions

Marketing involves activities in three areas: design, attitude and action. Marketing priority should be meeting customers and consumers satisfaction. Unfortunately this adage is often a mere assertion, considering that the main purpose is making profit.

A correct (scientifically) and efficient marketing, on top of specific market studies, should have from the beginning till the end of each production cycle and sales the capacity of undertaking. This capability is quantified by quantitative and qualitative level of the four types of resources: human resources (most important), financial resource, material resource and informational resource.

Also effective marketing must take into consideration the correlations between the four types of his resources (in terms of quality and quantity): the financial resources on top of the human and material resources, and the subordination of the 3 types of resources (human, financial, material) over the informational resource.

References

[1] Stanton W.J., Miller K 1991 *Fundamentals of Marketing* (New York, McGraw Hill, USA).

[2] Marcu D, Semenescu A, Ioana A 2014 *Marketing-Teorie şi Aplicaţii* (Editura Martix Rom, Bucureşti).

[3] Bruce A, Longton K 2003 *Cum să gândim strategic? (*Ed. Teora, Bucureşti).

[4] Demetrescu M.C. 2004 *Metode de analiză în marketing* (Ed. Teora, Bucureşti).

[5] Fanlkener C, Browman K 2004 *Elemente de strategie concurenţială* (Ed. Teora, Bucureşti).

[6] Kotler Ph, Armstrong G 2004 *Principiile marketingului, ediţia a treia* (Ed. Teora, Bucureşti).

[7] Niculescu E 2000, *Marketing modern* (Ed. Polirom, Bucureşti).

[8] Porter M.E. 2003 *Strategie concurenţială* (Ed. Teora, Bucureşti).

IOP Conf. Series: Materials Science and Engineering **161** (2016) 012106 doi:10.1088/1757-899X/161/1/012106

Investigations regarding the evaluation of specific intellectual property production risks within Quality Management System

R Pakocs[1], N B Lupulescu[2]

[1,2]Transilvania University of Braşov, Faculty of Technological Engineering and Industrial Management, Department of Manufacturing Engineering, 500036 B-dul Eroilor 29, Romania

E-mail: ramona.pakocs@unitbv.ro

Abstract. This paper is a theoretical research concerning methods for risk assessment of specific intellectual property production risks that are identified in the product achievement stage within the Quality Management System. In order to realize this, we will start by identifying the specific intellectual property production risks and by proposing some new calculating formulas for minimalizing their negative effects. The theoretical model proposed assessment of specific intellectual property production risks, will be realized based on 3 hypothetical situations. This study intends to reduce the intellectual property risks identified in the production process of commercial societies that have an industrial profile.

1. Introduction

Within every commercial society that has an industrial profile, the application of a Quality Management System is sought together with some continuous improvements for the product's quality without taking into consideration the intellectual property risks faced by the society after the release of the product.

Before the product is launched on the market, it has been thought, created, projected and executed based on some plans, sketches/drawings, projects that are protected against internal and external intellectual theft through different intellectual property forms of protection (such as Law number 8/1996) or by different documents, certificates, patents. After the product has been launched on the market it becomes vulnerable against its competitors, becoming a source of imitation, denigration etc. [2]

Previous national and international investigations from the management and intellectual property domains [2, 3, 4, 5, 6, 7], have approached the risk problem only from the perspective of abiding some norms, principles and an organizational structure in the industrial process. Furthermore, these investigations are handling the risk problem only from within the system and in a punctual matter.[2]

By studying the risk sources and analyzing some expertise in intellectual property right breaches [2, 3, 4, 5, 6], results that a risk category is often treated with superficiality and sometimes even neglected, namely: the specific intellectual property risk category. For this aspect we propose the identification of specific intellectual property risks in the product's realization process within QMS.[4]

Based on this study we will suggest some calculating formulas that can be applied to the intellectual property risks with the purpose to reduce its negative effects that can spread on all industrial companies.

2. Detecting the specific intellectual property production risks within the product`s realization process

The realization processes are referring to the product's Production Processes and are executed in the following order, according to Figure 1:

Figure 1. Product realization processes. [4]

In order to obtain optimum results, we suggest a working scheme in which we will identify the specific intellectual property production risks from the product's realization process point of view in the exact order of their application in every industrial profile company. [4]

Identifying the specific IP production risks within the product's realization process represents a first step in obtaining the intended results, namely reducing its negative effects.

The working scheme in which the specific intellectual property production risks will be identified from the product's realization processes point of view is represented in Figure 2.

As it can be observed from Figure 2, the main possible intellectual property risks from the production process were identified. These risks will be noted as follows:

RUCP - Risk of unfair competition in production;
TRUI - Technological risk of unprotected inventions;
TRUD/M - Technological risk of unprotected drawings / models;
RPN-S - Risk of products non-standardization;
RPP - Risk of piracy of products.

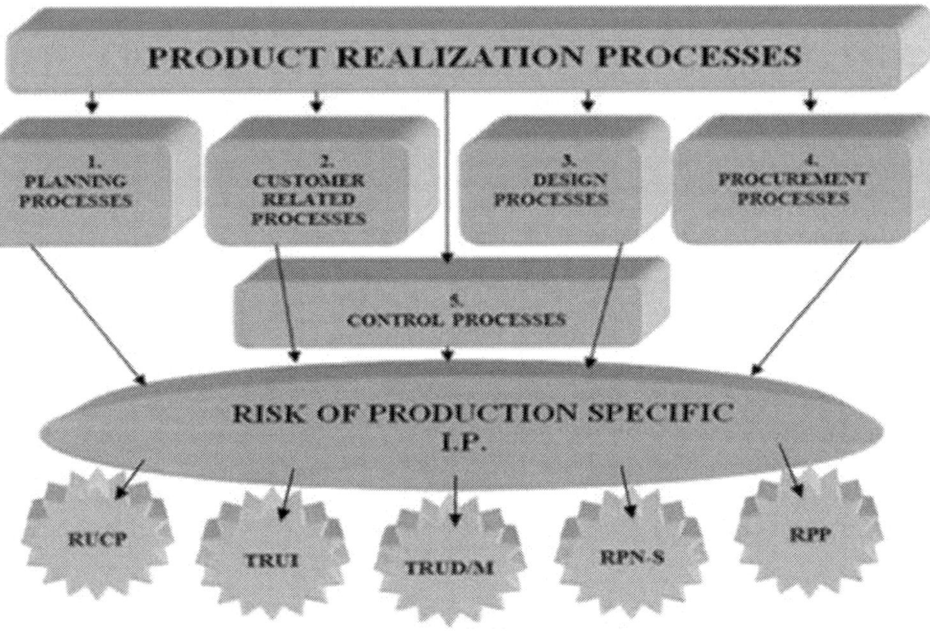

Figure 2. Production risks specific to IP, identified in the realization of the product process.

3. Theoretical model proposed assessment of specific intellectual property production risks

The intellectual property risks are influenced by the following criteria:
AP- IP risks apparition probability;
IRpi- IP risk impact;
TRpi- time (moment) of detecting the apparition of IP risks.

From a mathematical point of view [2, 5], the IP risks apparition probability is of 50%. When we are talking about the impact of IP risks on commercial societies with an industrial profile, they can have values ranging between: $IR_{PI} = \{1,\ldots,100\}$. When the impact value is getting closer to 100, the company's bankruptcy risk is growing.

The impact of the IP risks depends directly on their detection time - *TRpi*. The sooner the IP risks are detected, the lower their impact will be.

It has been demonstrated from previous research [2, 8] that the *IRpi* differs based on *TRpi*.

As such, a first general formula for estimating the specific IP production risks within the production process is [4]:

$$R_{PI} = PA * \frac{TR_{PI}}{IR_{PI}} \tag{1}$$

Detecting the IP risks apparition time (moment) is directly proportionate with their impact. The IP risks also depend on their apparition frequency.

For this aspect we use the following annotation:
FRpi- IP risks apparition frequency

In mathematics, the IP risks frequency can be of 3 types:

a) Descending cumulated IP risk frequency, which represents the sum of all IP risks that have appeared at *TRpi min* moment.

$$FR_{PI}D = \sum (R_{PI}1 + R_{PI}2 + \ldots + R_{PI}N) TR_{PI}\min \tag{2}$$

b) Ascending cumulated IP risk frequency, which represents the sum of all IP risks that have appeared until the *TRpi max* moment and also including it.

$$FR_{PI}C = \sum_{X1,X\max}(R_{PI}1 + R_{PI}2 + \ldots + R_{PI}N)TR_{PI}\max \quad (3)$$

c) Average cumulated IP risk frequency, which represents the arithmetic sum between the ascending cumulated frequency and the descending cumulated frequency and appears in the *TRpi medium* moment.

$$FR_{PI}M = \frac{FR_{PI}D + FR_{PI}C}{2} \quad (4)$$

Where:
FRpiD- descending cumulated IP risk frequency;
FRpiC- ascending cumulated IP risk frequency;
FRpiM- average IP risk frequency;
TRpi min- time (moment) of detecting the IP risks in the anticipation stage;
TRpi max- time (moment) of detecting the IP risks in the whole production process;
TRpi medium- time (moment) of detecting the IP risks in the in the merchandising production process.

4. Possible hypotheses for estimating the specific intellectual property production risks based on their detection (apparition) moment

There are 3 situations for detecting the specific intellectual property risks identified in the production process (Figure 3) namely:
1. In the anticipation stage of the production process.
2. In the production process.
3. In the production merchandising process.

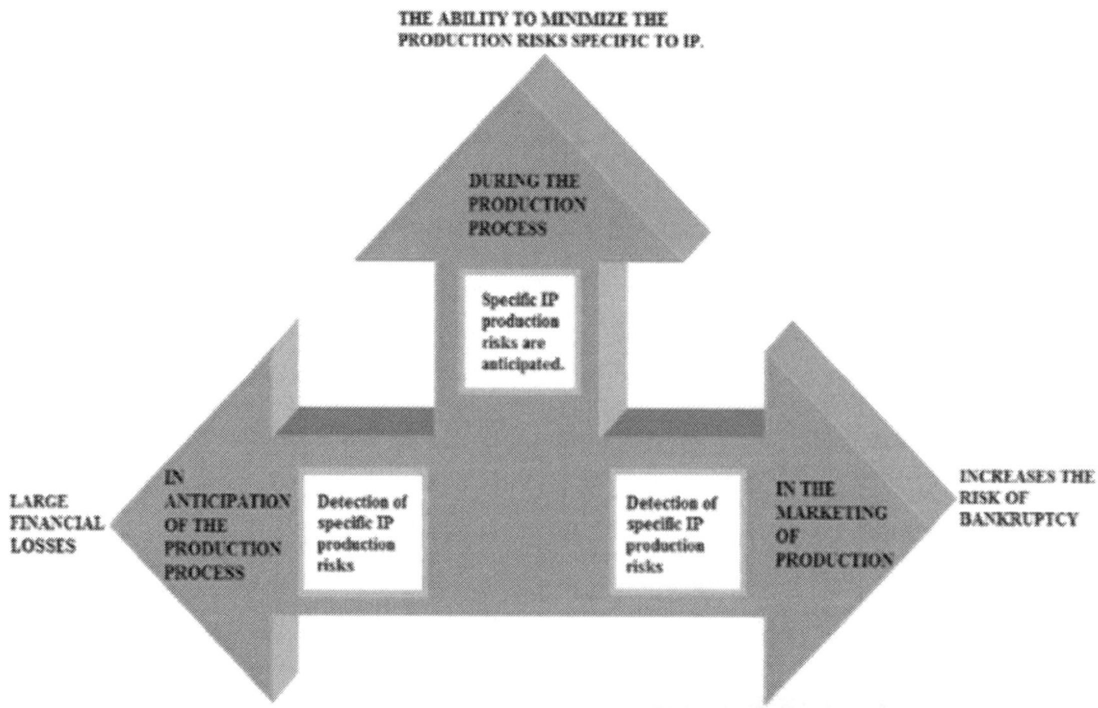

Figure 3. Hypothesis tests for IP risks depending on their discovery time. [4]

Hypothesis 1: Detecting the specific IP production risks in the anticipation stage of the production process.

This is the stage in which:

TRpi ≤ 0 => IRpi » 0 => FRpiD=0

The sooner the detection of the IP risk apparition is (TRpi min), the smaller their impact is on the company.

This situation offers the possibility of reducing to a minimum the IP risk.

Hypothesis 2: Detecting the specific IP production risks during the production process.

This is the stage in which:

TRpi mediu ≥ 50 % => IRpi > 50% => FRpiM ≥ 50

In this stage, detecting the IP risks leads to larger financial losses, with the possibility of reducing them up to 50%.

Hypothesis 3: Detecting the specific IP production risks in the merchandising stage.

This is the stage in which:

TRpi max aims towards 100 => IRpi aims towards the value of 100 => FRpiC apparition aims towards 100 repetitions => BANKRUPTCY risks grows.

In this stage of detecting the specific IP production risks, the commercial society with an industrial profile records significant loses so that the impact of IP risks is up to 100%, a fact that can increase the bankruptcy risk.

In order to evaluate the specific IP production risks identified in the product's realization process, we will consider the 3 hypothesis presented in Figure 3, as 3 possible hypothesis, each one depending on the IP risks detection time (moment).

As such, we will analyze the specific IP risks that we have identified in the production process through the 3 time hypothesis and we will see what happens with each IP risk identified in different production time moments and the way in which they are affecting the well-functioning of industrial enterprises.

Taking into consideration the fact that the specific IP production risks (noted with *RPpi*) depend on:

PA- IP risks apparition probability;

IRpi- IP risk intensity;

FRpi- IP risk frequency;

TRpi- time (moment) of detecting the apparition of IP risks.

We are suggesting the following calculating formula:

$$RP_{PI} = PA * \frac{FR_{PI} * IR_{PI}}{TR_{PI}} \tag{5}$$

Based on this formula and on the 3 risk detection hypothesis, the specific IP production risks identified in the product's production stage will be evaluated.

For example:

In case of *Risk of unfair competition in production (noted RUCP)* we propose the following assumptions:

$$RUCP = PA * \frac{FR_{PI} * IR_{PI}}{TR_{PI}} \tag{6}$$

Hypothesis 1: Detecting RUCP in the anticipation stage of the production process.

In this stage:

PA = 50%

$$FR_{PI}D = \sum (R_{PI}1 + R_{PI}2 + ... + R_{PI}N) TR_{PI}\min \tag{7}$$

TRpi min = 0

IRpi min = 0%
Results that:

$$RUCP = PA * \frac{FR_{PI}D * IR_{PI}\min}{TR_{PI}\min} \tag{8}$$

In this situation *RUCP* = 0 => detecting *RUCP* in the anticipation stage leads to the possibility of reducing this risk to 0.
Hypothesis 2: Detecting RUCP in the production process time.
In this stage:
PA = 50%

$$FR_{PI}M = \frac{FR_{PI}D + FR_{PI}C}{2} \tag{9}$$

FRpiM = 50
TRpi medium = 50
IRpi medium = 50%
Results that:

$$RUCP = PA * \frac{FR_{PI}M * IR_{PI}medium}{TR_{PI}medium} \tag{10}$$

In this situation *RUCP* = 50 => detecting *RUCP* within the production process time can lead to a great loss of the enterprises capital ≥ 50%.
Hypothesis 3: Detecting RUCP in the production's merchandising process.
In this stage:
PA = 50%

$$FR_{PI}C = \sum_{X1,X\max}(R_{PI}1 + R_{PI}2 + \ldots + R_{PI}N)TR_{PI}\max \tag{11}$$

FRpiC = 100
TRpi max = 100
IRpi max = 100%
Results that:

$$RUCP = PA * \frac{FR_{PI}C * IR_{PI}\max}{TR_{PI}\max} \tag{12}$$

In this situation the *RUCP* impact on the company can be of 100% => detecting *RUCP* in the merchandising stage can lead to losses of up to 100% => the company's bankruptcy risk grows considerably.

5. Conclusions

Evaluating the specific intellectual property production risks, identified in the product's creation process from the Quality Management System can be a long and costly process as it implies the hiring of intellectual property experts that will offer personalized expertise for each detection hypothesis and that will analyze their impact on the company's financial situation.

The present paper represents a theoretical proposal that can be improved or modified based on other factors and indicators for specific intellectual property production risks.

6. References

[1] Reilly R F and Schweihs R P 2004 *The Handbook of Business Valuation and Intellectual Property Analysis* (USA: The McGraw – Hill Companies)
[2] Fântână R S 2008 *Research on integrated management of intellectual property risk technical and economic* (Romania: The Publishing House of "Transilvania" University of Brasov)
[3] Pakocs R and Lupulescu N B 2014 *Risk management and risk type analysis specific to intellectual property in industrial profile companies* (Romania: AFASES)
[4] Pakocs R 2015 *Quality and intellectual property risk management in the industrial companies* (Romania: The Publishing House of "Transilvania" University of Brasov)
[5] Secară G and Fântână R S 2010 *General Management* (Bucharest: ProUniversitaria Publishing House)
[6] Smith G V and Parr R L 2005 *Intellectual Property. Evaluation, exploitation and damages for infringement* (Hoboken, New Jersey: IRECSON Publishing House)
[7] Ursachi I 2001 *Management* (Bucharest, Romania: ASE Publishing House)
[8] Bârsan P N and Popescu I 2003 *Risk management. Concepts. Methods. Applications* (Romania:The Publishing House of "Transilvania" University of Brasov)

A concept of technology for freight wagons modernization

M Płaczek[1], A Wróbel[1] and A Buchacz[1]

[1] Silesian University of Technology, Faculty of Mechanical Engineering,
Institute of Engineering Processes Automation and Integrated Manufacturing Systems,
Konarskiego 18A, 44-100 Gliwice, Poland

E-mail: marek.placzek@polsl.pl

Abstract. Rail transport is a very important part of the modern economy, one of the components determining its dynamic development. It is therefore important to conduct research and taking action aimed at the development and refinement of this branch of industry. Such actions directly translate into an increase in its effectiveness, safety, reduction of burden on the environment and society. Nowadays numerous studies are conducted, aimed at introducing new technologies and solutions, both in terms of railway infrastructure and logistics management systems, as well as in traction vehicles themselves. Introduction of modern technology helps eliminate or reduce nuisance problems associated with the implementation of any kind of transport or the operation of the used technical means. Presented paper concerns an issue of freight wagon modernization using composite materials. It presents a part of a research project that aim is to develop a technology of freight wagons modernization during their periodic repair. The main problem during exploitation of concerned types of freight wagons designed for coal transport is corrosion of the wagon's body shell. The goal of the project is to elongate the period between periodic repairs (by better corrosion protection) and improve conditions of exploitation of modernized freight wagons (for example easier unloading during winter conditions – no freezes of the charge to the freight wagon body shell). One of elements of the project is also to develop a system for diagnosing the technical condition of the modernized shell of wagon body during operation. For this purpose the use of non-destructive testing methods of technical state of constructions will be used, including methods that use the analysis of dynamic response of the object. Application of the composite panels to the freight wagon's body shell was proposed as the solution that can solve mentioned problems during exploitation of freight wagons. The composite panels composed of fiberglass and epoxy resin were proposed. They will be mounted on the body shell using blind rivet nuts. What is more the body shell of the modernized freight wagon will be painted using an anti-corrosion agent. The research project is realized by a consortium composed by research units and industry and supported by the Polish National Centre for Research and Development.

1. Introduction

Railway transport is very important for the development of the modern economy this is why many research works concerning with problems occurring in this way of transport and its development are being carried out all the time [1-3]. The goal of those works are usually to develop the infrastructure that is used for transport of goods and people, make it more cost-effective, safe and less burdensome for the environment [4-6]. Very important are problems of freight wagons dynamics during exploitation in variety driving conditions [7-11]. In all of those works the problem of modelling of real object has a strong influence on obtained results and has to be solved using precise methods [8, 12-18]. This is why computer-aided methods, include Finite Element Method, are used very often [19-24]. On the other hand, experimental analysis and tests are also very important in order to verify results of calculations and simulations [25]. Nowadays, computer aided methods, new materials and technology are widely used in development of all technical devices [26-33].

Presented paper concerns the possibilities of freight wagons modernizations using composite materials [24-37]. Work presents a concept of technology for modernization of freight wagons using composite panels and a report of the process of modelling and computer aided analysis of freight wagon type 1415 A3 designed for coal and aggregate transport. It is a part of the research project that is carried out by the consortium that consists of research unite – Institute of Engineering Processes Automation and Integrated Manufacturing Systems of the Silesian University of Technology and enterprises – DB Schenker Rail Poland SA and Germaz. The aim of this research project is to modernize the analysed freight wagon during its renovation using new materials. Effects that are to be achieved by the modernization are:

- better corrosion protection of the wagon elements,
- easier unloading of the wagon in winter conditions (no freezing of the cargo to the floor and sides),
- reduction of the weight of the wagon while its payload increases,
- easier management of the freight wagons during exploitation.

In this work the process of modelling of the analysed freight wagon using Siemens NX software is presented. The 3D model was created on the basis of very incomplete documentation and measurements of the real object. Precision of the created model was verified by checking its mass after defining material properties of wagon components and juxtaposing it with the mass of the real wagon. The created 3D model is very detailed and obtained discrepancy is about 5% of the real freight wagon mass. After its verification the model was used in strength analysis using the Finite Element Method in NX software. Obtained results – stress and displacements of elements were calculated.

2. The analysed types of freight wagons – problems during exploitation

The presented work is continuation of the previous research project concerned with modernization of the freight wagons using new, non-classical materials. The proposed solution for modernization of the freight wagons shells was to use the composite materials connected to the old elements made of steel. In this research project the scope of works is wider. Not only tasks concerned with use of new materials in the freight wagons body in order to obtain better corrosion protection and easier unloading are considered. The possibility of reduction of the weight of the wagon while its payload increases is also considered as well as introduction of systems for easier management of the freight wagons during exploitation (monitoring, management of the freight wagons life circle).

In the present research project four types of freight wagons were considered:

- type 1415 A3 of open freight wagon series EAOS, produced by BREC Belgium,
- hopper freight wagon type 418V,
- hopper freight wagon type FALS Slovak production,
- hopper freight wagon type FALS Bulgarian production.

In figure 1 created CAD models of all considered freight wagons are presented. The freight wagon type 1415 A3 was designed for coal and aggregate transport and was designed for tipping unloading. Examinations of real objects were performed and photographic documentation of damages of freight wagons prepared for repair was created. Also employees of repair facility were interviewed about typical defects and operation problems. Recurring damage and their causes were indicated. In case of all considered types of freight wagons the main problems occurring during their exploitation are: corrosion of the freight wagons shells of body and floors and freezing of the cargo to the sides and floor of the wagons. In case of the type 1415 A3 there is also a problem with damages of the vehicle body shells that occurs as the result of inappropriate methods of unloading (using scoop, excavator, etc.). For the wagon type 1415 A3 it can be also noticed that it has a very durable wagon supporting

IOP Conf. Series: Materials Science and Engineering **161** (2016) 012107 doi:10.1088/1757-899X/161/1/012107

frame and frame of the body – there is almost no damages in repaired wagons. It means that there is a possibility to reduce the weight by the use of smaller cross-sections profiles.

Figure 1. CAD models of considered freight wagons.

In the presented research project an application of composite panels to the freight wagon's body shell was proposed as the solution that can solve mentioned problems that occur during exploitation of freight wagons [34-37]. The applied composite panels will be composed of fiberglass and epoxy resin and will be mounted on the body shell using rivet nuts. The body shell of the modernized freight wagon will be painted using an anti-corrosion agent. The composite panels mounted on the modernized freight wagon's doors and walls are presented in figure 2. The composite material that is to be applied is glass-epoxy composites based on epoxy resin belong to the group of chemically hardenable materials. They are produced as the result of fiberglass saturation and compressed into a layer. This type of composite was chosen because it is distinguished by the following features: good mechanical properties and insulation, high parameters of bending and compressive strength with low weight, the possible wide range of applications and also low water absorption.

Figure 2. The modernized freight wagon 1415A3 and composite panels mounted on the modernized freight wagon's door and wall.

On the other hand the system for structural health monitoring should be developed as a system that includes elements from different science areas, such as mechanics, electronics and informatics. Such connection brings new possibilities and new effects, so these systems can be called a mechatronic system [38-41]. Piezoelectric transducers were proposed to be used in the developed system. Taking into account their possibilities, piezoelectric transducers can be used in various modern devices as sensors or actuators [42-49]. A process of freight wagons qualification to repair can be supported by the system designed for inferred from the dynamic response of the wagon about the changes in its technical condition. At this moment all freight wagons are verified about their technical condition at a certain time interval and during this qualification process a standard qualifying protocol is used. The protocol that is used by the DB Schenker Rail Poland SA Company during qualification process includes such data as:

- the analysed freight wagon's type;
- the drawings including views of all side walls and the floor of the wagon;
- the identification number of the wagon;
- the date and type of the previous and actual repair;
- the list of wagon's elements with the possibility to mark whether the item is suitable for repair, replacement, or it is in good technical condition;
- the responsible people signatures and the place for other notes.

During carried out research project a statistic analysis of data from qualifying protocols were analysed in order to verify if some wagon's elements are the most exposed to damage during exploitation of freight wagons. Qualifying protocols from years 2012 up to 2014 were analysed and the number of analysed protocols in case of BDŻE, CFRE, CSDE, 401W, 408W and 412W types of freight wagons was 221. The percentage of repairs of individual wagons elements was specified on the basis of protocols analysis. Obtained results for mentioned types of wagons are presented on chart in figure 4. In order to analyse the data from protocols, all elements of the freight wagon's body shell were named and marked on the scheme of the freight wagon. Designations of elements of the BDŻE, CFRE, CSDE, 401W, 408W and 412W wagons types are presented in figure 3.

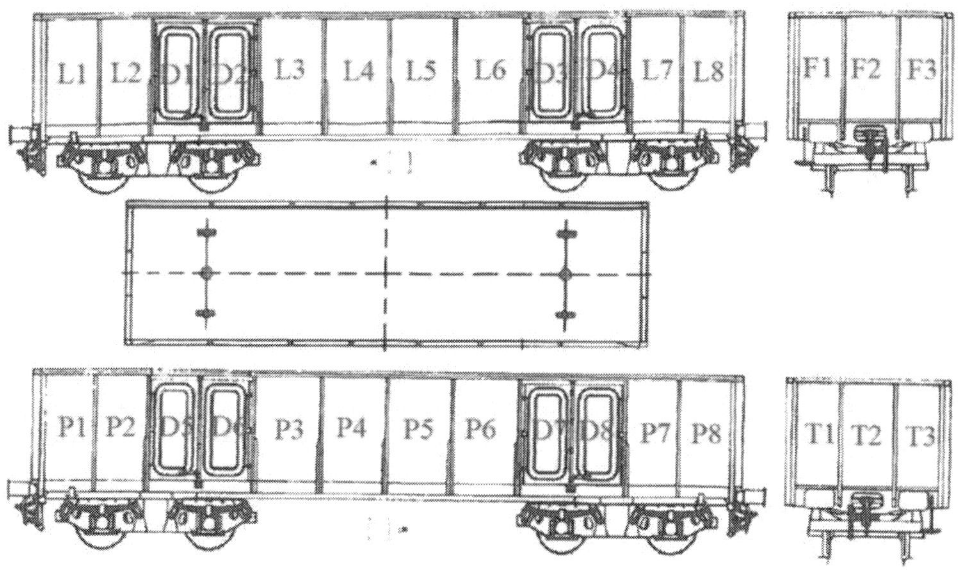

Figure 3. Designation of BDŻE, CFRE, CSDE, 401W, 408W and 412W wagon's elements.

The most important, taking into account the aim of the research work – the modernization of the freight wagons using composite panels mounted to their body shell, is the information about the damage of its elements. Percentage of repairs of individual elements of the shell of the body in case of BDŻE, CFRE, CSDE, 401W, 408W and 412W types of freight wagons is presented in figure 4.

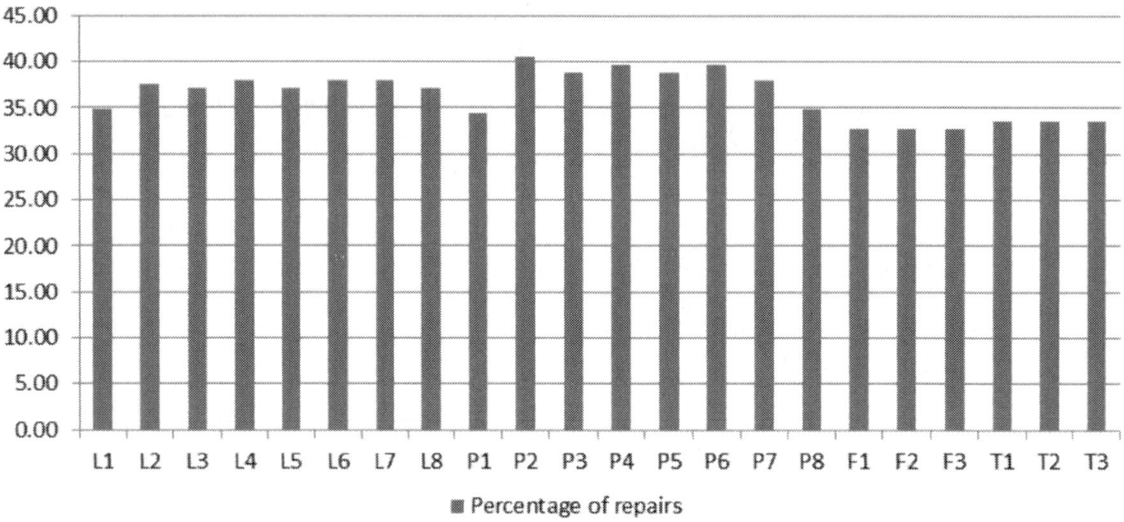

Figure 4. Percentage of individual elements repairs of the body shell of freight wagons types BDŻE, CFRE, CSDE, 401W, 408W and 412W.

Analysis of qualifying protocols shows that there is no part of the freight wagon's body shell that is more exposed to damage during exploitation of the wagons then the other elements. It is necessary to carry out a more accurate operational documentation in order to obtain more precise information about the process of damage of individual elements of the freight wagons body shell during their exploitation. This documentation should be created starting from the production of wagons and it should take into account operating conditions of carriage, such as the type of transported cargo, as well as all carried out repairs. The mechatronic system for structural health monitoring of freight wagons that is to be developed will also provide data about individual wagon's operation conditions and it should be storage in wagon's documentation. The system should inform about destruction of the wagon's body shell during operation of the wagon. By using the system it will be possible to detect major defects of wagon's body shell without transporting it to the service station. It is necessary to verify the possibility of inference from the dynamic response of the wagon about the changes in its technical condition during its standard operation in order to develop such system [34].

3. The CAD model of the considered freight wagon type 14151 A3

The presented in figure 1 CAD models of all considered in the research project types of freight wagons were created using the NX 8.5 software. In case of type 1415 A3 all documentation supplied by consortium member – DB Schenker Rail Poland SA, consists of:

- operation and maintenance manual,
- technical conditions of repair and receipt after repair of 4-axle freight wagon series Eaos type 1415 A3 production BREC Belgium,
- calculations of air brake,
- program of performance tests,
- report on the performance tests,
- 11 construction drawings including assembly drawing, drawings of support, wagon body, floor, cargo doors with overall dimensions, without giving details of the used profiles, sheet thickness, etc.

The incomplete documentation of the considered freight wagon supplied by consortium member forced the visual inspection of real objects and measurements. Selected elements of created CAD model of the considered freight wagon are presented in figures 5 to 7.

Figure 5. CAD models of the considered freight wagon's elements – the support frame and floor.

Figure 6. CAD model of the considered freight wagon's elements – the wagon's body.

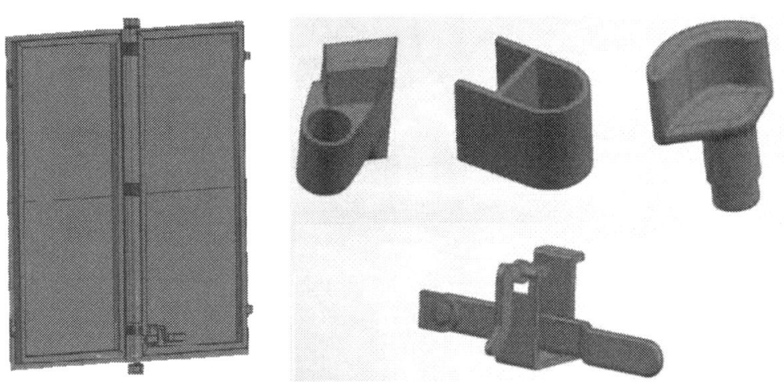

Figure 7. CAD models of the considered freight wagon's elements – the wagon's doors.

By juxtaposing the weight of the real object and the CAD model after the material parameters were applied, the created CAD model of the 1415 A3 type of freight wagon was verified. The weight of the considered freight wagon is 20600 kg, while the weight of the CAD model is 19500 kg. The difference in weight is 5,34 % it is caused by omission of the braking system (air pipes, brake pads, etc.). It can be assumed that the created CAD model of the considered freight wagon is very precise and can be successfully used in next analysis.

4. The Strength analysis

As the next step of the presented work the strength analysis of the body of the considered freight wagon type 1415 A3 was carried out. It was realized using Siemens NX 8.5 software and Finite Element Method. In accordance with norm EN 12663-2:2010: Railway applications - Structural requirements of railway vehicle bodies - Part 2: Freight wagons [Required by Directive 2008/57/EC] the load of the model was assumed [50]. The structural design and assessment of freight wagon bodies depend on the loads they are subject to and the characteristics of the materials they are manufactured from in agreement with this norm. It is intended to provide a uniform basis for the structural design and assessment of the vehicle body within the scope of this European Standard. Based on proven experience supported by the evaluation of experimental data and published information the loading requirements for the vehicle body structural design and assessment are. The aim of this European Standard is to allow the supplier freedom to optimize his design whilst maintaining requisite levels of safety considered for the assessment [50].

The way of verification described by the norm is that the following tests are to be carried out:

a) Forcing outwards in the horizontal direction at a level of 1,5 m above the floor:

1) force of 100 kN applied at four centre posts of each side wall;

2) force of 40 kN applied at the corner posts of wagons equipped with drop ends.

The significant permanent deformation at the point where the force is applied shall not exceed 1 mm. In addition, the elastic deformation observed during the test shall not result in any encroachment of the loading gauge. In figure 8 the way of the freight wagon forcing is presented [50].

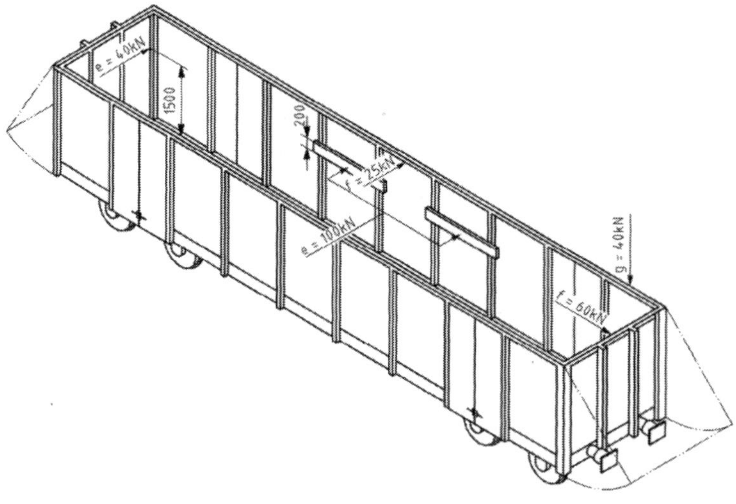

Figure 8. The way of the freight wagon forcing [50].

The CAD model of the considered freight wagon was forced by a system of forces presented in figure 9. Those forces are transferred to the wagon's body by additional rigid beams.

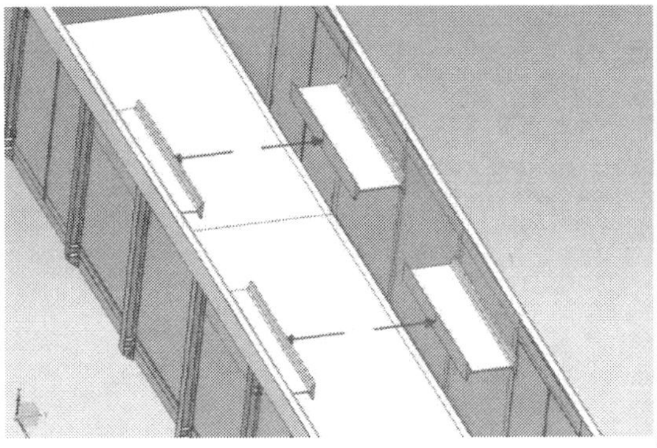

Figure 9. The way of the CAD model forcing.

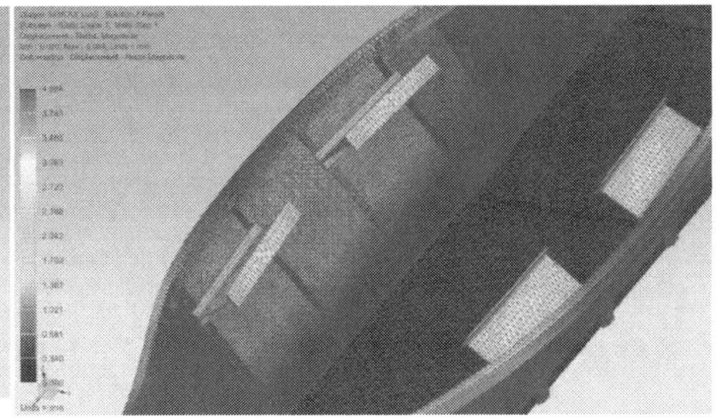

Figure 10. Displacement of the considered freight wagon body.

Results obtained using the Finite Element Method in NX 8.5 software are presented in figures 10 and 11 as the freight car displacements and stress respectively.

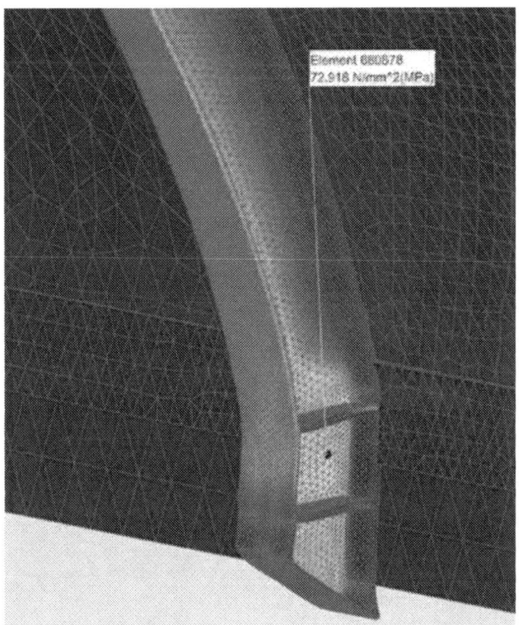

Figure 11. Stress of the considered freight wagon's body.

Maximum value of the displacement was equal 4,084 mm and it occurred on the upper side beam. Average stress obtained in strength analysis is 50-65 MPa while the yield strength is 235 MPa. The maximum value is 72 MPa.

The support frame analysis was carried out twice. In the first case only the uniformly distributed load of 58 tons (the maximum weight of cargo) was being taken into consideration, while in the second case also compressive force of 2 MN in the axis of bumpers was applied. In the first case the maximum value of displacement was 4,32 mm, and maximum stress 170 MPa, while its average values were about 40 to 60 MPa. In the second case the maximum displacement was 4,49 mm in the vertical axis and 0,74 mm in the axis of the bumpers. Maximum value of stress was 180 MPa and is average value was also about 60 MPa. Selected results are presented in figures 12 to 14.

Figure 12. Displacement of the considered freight wagon's support frame.

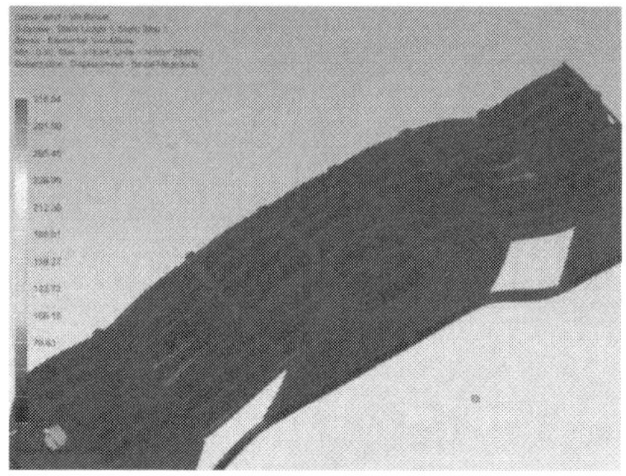

Figure 13. Stress of the considered freight wagon's support frame.

Figure 14. Displacement of the considered freight wagon's support frame in the axis of the bumpers.

Also strength analysis of the cargo door was carried out. Obtained results are presented in figures 15 and 16. The maximum value of displacement is 4,49 mm and maximum stress is 160 MPa.

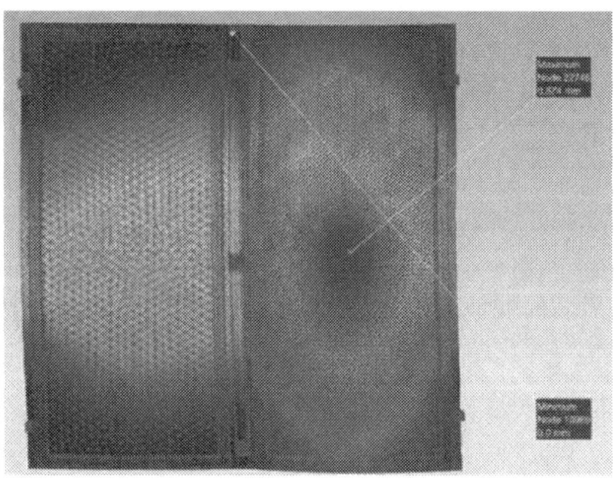

Figure 15. Displacement of the considered freight wagon's door.

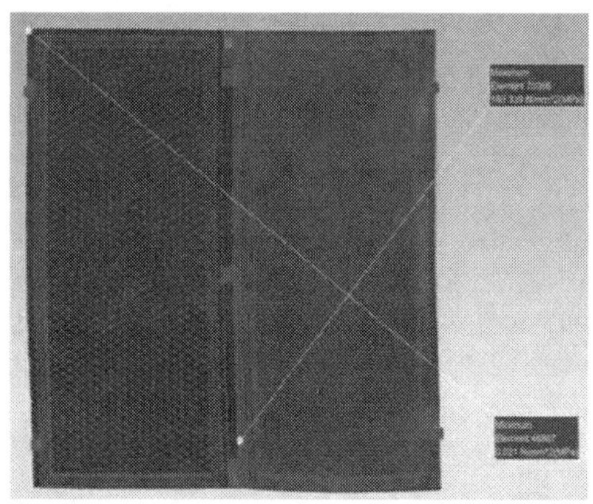

Figure 16. Stress of the considered freight wagon's door.

Presented results of the strength analysis will be verified during the future works, while tests on the real objects and prepared laboratory stands will be carried out.

5. Conclusions

Railway transport is very important for the development of the modern economy. This is why nowadays there are a lot of research works concerned with this mean of transport that main aim is its modernization and development. It should be noted that there are a lot of problems with old infrastructure in railway transport that cases a lot of loss and not necessary costs. The goal of those works are usually to develop the infrastructure that is used for transport of goods and people, make it more cost-effective, safe and less burdensome for the environment.

Presented paper concerns the possibilities of freight wagons modernization during their periodic repairs. The possibility of composite materials application in the construction of the freight wagon body is considered as well as reduction of the weight of the wagon while its payload increases. In order to do it a series of analysis of the real object and its model in its present form had to be carried out, especially the strength analysis. The incomplete documentation of the considered freight wagon supplied by consortium member forced the visual inspection of real object and measurements. However, after the verification of the created CAD model it can be assumed that it is very precise and can be successfully used in such kind of analysis. The carried out analysis proved that there is a possibility to use smaller cross-sections profiles and thickness of the wagon body sheets could be reduced in the modernized wagons. What is more, the additional composite panels mounted to the wagon's body shell will also bring an additional stiffness to the structure. It means that it will not be necessary to replace the corroded panels during repairs of the wagons while composite panels will be applied. The presented concept of technology for freight wagons modernization should solve mentioned problems that occur during exploitation of freight wagons. The proposed technology of mounting composite panels to the wagon's body shell during its renovation is easy to apply and will extend the period of time between necessary repairs of freight wagons what will reduce costs of their exploitation.

Acknowledgments. The work was carried out under the project number PBS2/A6/17/2013 agreement implemented under the Applied Research Program, funded by the National Centre for Research and Development.

References

[1] Berghuvud A and Stensson A 2001 Consequences of Nonlinear Characteristics of a Secondary Suspension in a Three-Piece Freight Car Bogie, *Vehicle System Dynamics: International Journal of Vehicle Mechanics and Mobility* **36 (1)** 37-55

[2] Jönsson P A, Andersson E and Stichel S 2006 Influence of link suspension characteristics variation on two-axle freight wagon dynamics, *Vehicle System Dynamics* **44** 415-423

[3] Johnson M, Welch R and Yeung K 1977 Analysis of Thermal Stresses and Residual Stress Changes in Railroad Wheels Caused by Severe Drag Braking, *J. Manuf. Sci. Eng.* **99 (1)** 18-23

[4] Andersson E, Häggström J, Sima M and Stichel S 2004 Assessment of train-overturning risk due to strong cross-winds, *Proceedings of the Institution of Mechanical Engineers Part F: Journal of Rail and Rapid Transit* **218 (3)** 213-223

[5] Hecht M 2009 Wear and energy-saving freight bogie designs with rubber primary springs: principles and experiences, *Proceedings of the Institution of Mechanical Engineers Part F: Journal of Rail and Rapid Transit* **223 (2)** 105-110

[6] Song Z, Fang S, Zhang Y and Xie J 2013 Cracking analysis of bolster cover plate in C70 freight wagons, *Engineering Failure Analysis* **30** 43-60

[7] Berghuvud A. 2002 Freight car curving performance in braked conditions, *Proceedings of the Institution of Mechanical Engineers, Part F: Journal of Rail and Rapid Transit* 23-29

[8] Bruni S, et al. 2011 Modelling of suspension components in a rail vehicle dynamics context, *Vehicle System Dynamics* **49 (7)** 1021-1072

[9] Connolly D. P, et al. 2014 Assessment of railway vibrations using an efficient scoping model,

Soil Dynamics and Earthquake Engineering **58** 37-47.

[10] Dukkipati R V and Dong R 1999 The Dynamic Effects of Conventional Freight Car Running over a Dipped-joint, *Vehicle System Dynamics: International Journal of Vehicle Mechanics and Mobility* **31 (2)** 95-111

[11] Zakharov S and Goryacheva I 2005 Rolling contact fatigue defects in freight car wheels, *International Journal on the Science and Technology of Friction, Lubrication and Wear* **258 (7-8)** 1142-1147

[12] Jönsson P A, Stichel S and Persson I 2008 New simulation model for freight wagons with UIC link suspension, *Vehicle System Dynamics* **46** 695-704

[13] Rusiński E, Koziołek S and Jamroziak K 2009 Quality assurance metod for desing and manufacturing process of armoured vehicles, *Eksploatacja i Niezawodność – Maintenance and Reliability* **3** 70-77

[14] Zhai W M 2004 Modelling and experiment of railway ballast vibrations, *Journal of Sound and Vibration* **270 (4-5)** 673–683

[15] Jamroziak K, Bocian M and Kulisiewicz M 2013 Energy consumption in mechanical systems using a certain nonlinear degenerate model, *Journal of Theoretical and Applied Mechanics* **51 (4)** 827-835

[16] Jamroziak K and Kosobudzki M 2012 Determining the torsional natural frequency of underframe of off-road vehicle with use of the procedure of operational modal analysis, *Journal of Vibroengineering*, **14 (2)**, 472-476

[17] Wrobel A 2012 Kelvin Voigt's model of single piezoelectric plate, *Journal of Vibroengineering*, **14 (2)**, 534-537

[18] Monica Z 2015 Virtual modelling of components of a production system as the tool of lean engineering IOP Conf. Series: Materials Science and Engineering **95** 012109

[19] Iacob-Mare C and Manescu T S 2013 Study of the freight wagon body through the method of finite elements, *Metalurgia* **65 (7)** 13

[20] Kovalev R, et al. 2009 Freight car models and their computer-aided dynamic analysis, *Multibody System Dynamics* **22 (4)** 399-423

[21] Lee S H 2005 A CAD–CAE integration approach using feature-based multi-resolution and multi-abstraction modelling techniques, *Computer-Aided Design* **37 (9)** 941-955

[22] Herbus K and Ociepka P 2015 Analysis of the Hexapod Work Space using integration of a CAD/CAE system and the LabVIEW software, *IOP Conf. Series: Materials Science and Engineering* **95** 012096

[23] Bojović N.J 2002 A general system theory approach to rail freight car fleet sizing, *European Journal of Operational Research* **136 (1)** 136-172

[24] Connolly D. P, et al. 2014 Assessment of railway vibrations using an efficient scoping model, *Soil Dynamics and Earthquake Engineering* **58** 37-47.

[25] Jönsson P A, Andersson E and Stichel S 2006 Experimental and theoretical analysis of freight wagon link suspension, *Proceedings of the Institution of Mechanical Engineers Part F: Journal of Rail and Rapid Transit* **220 (4)** 361-372

[26] Paprocka I, Kempa W. M, Grabowik C and Kalinowski K 2014 Predictive and reactive scheduling for a critical machine of a production system, *Advanced Materials Research* **1036,** 909-914

[27] Gwiazda A, Herbus K, Kost G and Ociepka P 2015 Motion analysis of mechatronic equipment considering the example of the Stewart platform, *Solid State Phenomena* **220/221**, 479-484

[28] Jamroziak, K, Kosobudzki M and Ptak J 2013 Assessment of the comfort of passenger transport in special purpose vehicles, *Ekspl. i Niezawodnosc - Maintenance and Reliability* **15 (1)**, 25-30

[29] Rusinski E, Dragan S, Moczko P and Pietrusiak D 2012 Implementation of experimental method of determining modal characteristics of surface mining machinery in the modernization of the excavating unit, *Archives of Civil and Mechanical Engineering* **12**,

471-476
[30] Klarecki, K, Rabsztyn D and Hetmanczyk M 2015 Analysis of pulsation of the sliding-vane pump for selected settings of hydrostatic system, *Maintenance and Reliability* **17 (3)** 338-344
[31] Grabowik C, Cwikla G and Jnik W 2014 The New Approach to Design Features Identification, *Applied Mechanics and Materials* **657** 750-754
[32] Kalinowski K, Grabowik C and Kempa W 2014 The Graph Representation of Multivariant and Complex Processes for Production Scheduling, *Advanced Materials Research* **837** 422-427
[33] Klarecki K, Rabsztyn D and Hetmanczyk M 2015 Influence of setting the selected parameters of hydraulic systems on pressure pulsation of gear pumps, *Diagnostyka* **16** 49-54
[34] Placzek M, Buchacz A and Wrobel A 2015 Use of piezoelectric foils as tools for structural health monitoring of freight cars during exploitation, *Eksploatacja i Niezawodnosc – Maintenance and Reliability* **17 (3)** 443-449
[35] Placzek M, Wrobel A and Baier A 2015 Computer-aided strength analysis of the modernized freight wagon, *IOP Conference Series-Materials Science and Engineering* **95** Art. No 012042
[36] Wrobel A, Placzek M, Buchacz A and Majzner M 2015 Study of mechanical properties and computer simulation of composite materials reinforced by metal, *International Journal of Materials & Product Technology* **50 (3-4)** 259-275
[37] Zolkiewski S 2011 Testing composite materials connected in bolt joints, *Journal of Vibroengineering* **13 (4)**, 817-822
[38] Buchacz A and Galeziowski D 2012 Synthesis as a designing of mechatronic vibrating mixed systems, *Journal of Vibroengineering*, **14 (2)**, 553 -559
[39] Banaś W, et al. 2015 Determination of the robot location in a workcell of a flexible production line IOP Conf. Series: Materials Science and Engineering **95** 012105
[40] Bialas K, Buchacz A and Galeziowski D 2015 Comparison of active and semi-active damping in synthesis of various mechatronic discrete systems, *Int. J. of Materials and Product Technology* **50 (3/4)** 340-355
[41] Buchacz A and Galeziowski S 2015 Designing of discrete mechatronic vibration systems with negative value parameters, *Mechanical Systems and Signal Processing* (in press) DOI: dx.doi.org/10.1016/j.ymssp.2015.02.003
[42] Tuma J, et al. 2011 Active Vibration Control of Hydrodynamic Journal Bearings, Springer Proceedings in Physics **139** 619-624
[43] Tuma J 2009 Gearbox Noise and Vibration Prediction and Control, *Int. J. of Acoustic and Vibration* **14 (2)** 99-108
[44] Buchacz A, Placzek M and Wrobel A 2014 Modelling of passive vibration damping using piezoelectric transducers – the mathematical model, *Eksploatacja i Niezawodnosc – Maintenance and Reliability* **16 (2)** 301-306
[45] Placzek M 2013 Dynamic Characteristics of a Piezoelectric Transducer with Structural Damping, *Solid State Phenomena* **198** 633-638
[46] Placzek M 2015 Modelling and investigation of a piezo composite actuator application, *International Journal of Materials & Product Technology* **50 (3-4)** 244-258
[47] Buchacz A, Banas W and Placzek M 2015 Influence of the excitation parameters of the mechanical subsystem on effectiveness of energy harvesting system, *IOP Conference Series-Materials Science and Engineering* **95** Article number 012052
[48] Tuma J, Suranek P and Mahdal M 2014 Simulation of the parametric excitation of the cantilever beam vibrations, 15Th International Carpatian Control Conference 588-591
[49] Suranek P, Mahdal M and Tuma J 2014 Modelling and Simulation of an Active Damped Structure, 15Th International Carpatian Control Conference 588-591
[50] EN 12663-2:2010: Railway applications - Structural requirements of railway vehicle bodies – Part 2: Freight wagons.

IOP Conf. Series: Materials Science and Engineering **160** (2016) 012108 doi:10.1088/1757-899X/160/1/012108

Automated Decisional Model for Optimum Economic Order Quantity Determination Using Price Regressive Rates

M M ROŞU[1], C I TARBĂ[2] and C NEAGU[3],

[1,2,3] POLITEHNICA University of Bucharest, Splaiul Independenţei, no. 313, building CB, sector 6, Bucharest, Romania

e-mail: magdalenarosu@yahoo.com

Abstract. The current models for inventory management are complementary, but together they offer a large pallet of elements for solving complex problems of companies when wanting to establish the optimum economic order quantity for unfinished products, row of materials, goods etc. The main objective of this paper is to elaborate an automated decisional model for the calculus of the economic order quantity taking into account the price regressive rates for the total order quantity. This model has two main objectives: first, to determine the periodicity when to be done the order n or the quantity order q; second, to determine the levels of stock: lighting control, security stock etc. In this way we can provide the answer to two fundamental questions: How much must be ordered? When to Order? In the current practice, the business relationships with its suppliers are based on regressive rates for price. This means that suppliers may grant discounts, from a certain level of quantities ordered. Thus, the unit price of the products is a variable which depends on the order size. So, the most important element for choosing the optimum for the economic order quantity is the total cost for ordering and this cost depends on the following elements: the medium price per units, the stock cost, the ordering cost etc.

1. Introduction

For a company who wants to reduce production costs and the stock generated costs, the inventory management is a necessary condition but it is not enough. The inventory management must be harmonised with the business strategy, with the financial management, with the information management and with the other traditional disciplines for management but it depends significantly on human resource factors that are responsible for the forecast on medium and large term.

The development of the decisional model for determining the optimal amount of supply involves the analysis of the entire supply system [1], Figure 1, within a company with all the three main subsystems which make up the forecasting, the tracking and the launch.

The forecast subsystem focuses on specifying the conditions under which orders should be launched (times and quantities). It is very important to know at what point the orders should be launched and in the same time it is necessary to provide for the use of each item. For detecting any changes in rates of use, the forecasts are drawn up relatively frequently, usually once a month or twice a month.

---

[1] magdalenarosu@yahoo.com or magdalena.rosu@upb.ro

The tracking subsystem focuses on the inventory control so as to verify its actual situation and to justify the ordering. If the satisfactory stocks still cover the needs, the order is not launched. Tracking the stocks should be rigorously given the permanent consumption of stock. It is not uncommon that in addition to tracking the operations conducted weekly or even twice a week, some items control the situation after each stock release from the warehouse.

The launching orders subsystem focuses on establishing the order quantities, taking into account the changing consumption and costs. The strategy for launching the orders must ensure a minimum level of expenditure for the supply and storage.

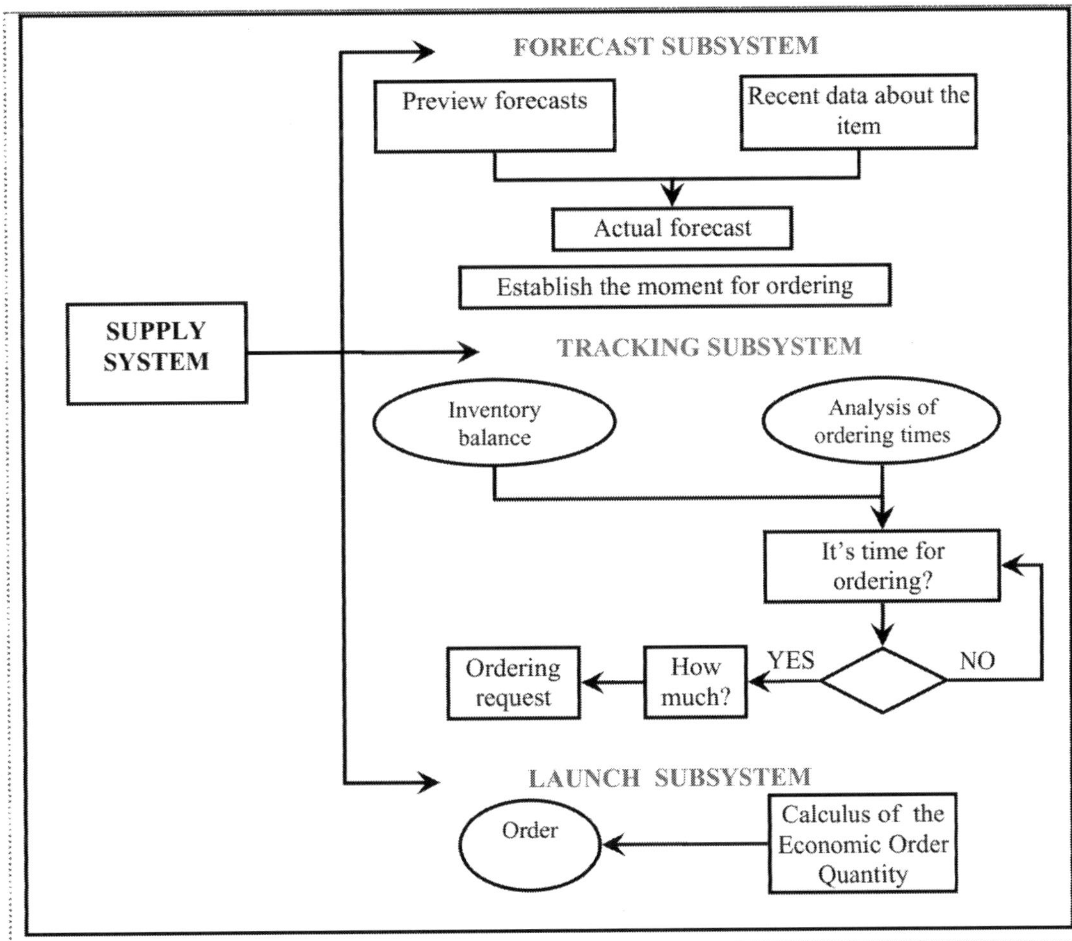

Figure 1. Supply system [1]

2. Model for the Inventory Optimisation

This model follows two main objectives: the first one is to calculate an optimum relative for time (n) and the second one is to calculate the economic order quantity (Qo). This is the way to answer the following questions: „how much to order?" and „when is it necessary to order?". This model is based on the Wilson model which is a determinist model relying on the following elements: constant demand, linear variation of use, equal time periods for demand, the order launching at start of each time period, quick supply and continuum process. By taking into account all these elements, the model follows the minimum costs for the inventory.

So, the optimum quantity, q_o, will be calculated with the formulas below taking into account the price for different order quantities [1], [3]:

For the first price interval, q_{o1}, it is calculated by formula (1):

$$q_{o1} = \sqrt{\frac{2 \cdot D \cdot c}{p_1 \cdot h}} \tag{1}$$

where:
D is total quantity for demand;
c – the inventory cost;
p – the price per item;
h – the rate for hold cost.

For the second price, the interval q_{o2}, it is calculated by formula (2). In this case, it will be taken into account the medium price between the first and second price interval and the limit point for price, L:

$$q_{o2} = \sqrt{\frac{2 \cdot D[L_1(p_1 - p_2) + c]}{p_2 \cdot h}} \tag{2}$$

For the third price, the interval q_{o3} is calculated by formula (3):

$$q_{o3} = \sqrt{\frac{2 \cdot D[L_1(p_1 - p_2) + L_2(p_2 - p_3) + c]}{p_3 \cdot h}} \tag{3}$$

If we have an n price interval for price, the optimum quantity will be calculated by formula (4):

$$q_{on} = \sqrt{\frac{2 \cdot D[L_1(p_1 - p_2) + L_2(p_2 - p_3) + ... + L_{n-1}(p_{n-1} - p_n) + c]}{p_n \cdot h}} \tag{4}$$

The total cost for demand, TC, will be calculated taking into account the medium price per unit, MPU, between the adjacent price interval and the quantity. For the calculus of the total cost for the demand in the first price interval, there will be used the procurement price. For the second price interval, the medium price per unit is calculated by means of the following formula (5):

$$MPU_1 = \frac{L_1(p_1 - p_2)}{q} + p_2 \tag{5}$$

For the third price interval, the MPU is calculated with formula (6) shown below:

$$MPU_2 = \frac{L_1(p_1 - p_2) + L_2(p_2 - p_3)}{q} + p_3 \tag{6}$$

For the n price interval, the MPU is calculated with formula (7) shown below:

$$MPU_{n-1} = \frac{L_1(p_1 - p_2) + L_2(p_2 - p_3) + ... + L_n(p_{n-1} - p_n)}{q} + p_3 \tag{7}$$

The total cost for the demand will be determined by means of the formula below, but taking into account the medium price per unit [1], [3], [4].

$$TC(q) = D * MPU + \frac{D}{q}c + \frac{1}{2}MPU * q * h \tag{8}$$

For this application the input data is shown in Table 1. These are total quantities for the demand (D), the price (p) for procurement and the limits for price (L), the ordering cost (c) and the rate for the hold cost (h). For running the automated model there will be established the following input data:

Table 1. Input data for the automated decisional model of the economic order quantities determination

Input data	Notation	Values
Total quantity for demand [items]	D	
Ordering cost [units]	c	
Rate for hold cost [%]	h	
First interval for price [units/item]	p_1	
minimum quantity [items]	L_0	
maximum quantity [items]	L_1	
Second interval for price [units/item]	p_2	
minimum quantity [items]	L_1+1	
maximum quantity [items]	L_2	
Third interval for price [units/item]	p_3	
minimum quantity [items]	L_2+1	
maximum quantity [items]	L_3	

The application model was developed with Microsoft Excel and for the generated random values of the quantity and for the calculus of the economic order quantity, Qo, there was written a code in Microsoft Visual Basic for the Application using specific functions and methods for doing this calculus.

The data from Figure 2 is obtained by running the application model for the automated calculus of the economic order quantities. For each price interval there will be calculated the minimum point of the costs curve by means of the formulas 1, 2, 3 and it is easy to show if these values are in the established price interval. For our example, only the first value of the optimum quantity is the price interval.

Input data			Output data		
Input dates	Notation	Value	Optimum quantity [items]	Medium unitary price [units]	Total cost for demand [units]
Total quantity for demand [items]	D	980400			
Ordering cost [units]	c	1000			
Rate for hold cost [%]	h	0.25			
First interval for price	p1	0.98			
minimum quantity	L0	1			
maximum quantity	L1	300000	89460.97	0.98	982709.94
Second interval for price	p2	0.95			
minimum quantity	L1+1	300001			
maximum quantity	L2	600000	287332.56	0.9813	1000746.48
Third interval for price	p3	0.92			
minimum quantity	L2+1	600001			
maximum quantity	L3	980400	479771.6894	0.9763	1017733.96

Figure 2. Automated Model for the Calculus of the Economic Order Quantity, Qo

For calculating the total cost for some different quantities, as shown in Figure 3, the model will analyse different aleatory values of quantity between imposed limits of price. For the random quantities generated by the model there will be calculated the medium price for the interval of the supply price and then the total cost (TC) will be generated.

All these values will be obtained while running the model by pushing the grey button of the application. The economic order quantity, Qo, will be displayed below the grey button, as shown in Figure 3. The value of the economic order quantity must be the divisor of the total demand and must have the minimum total cost of the ordering. The optimum relative of time, n, is the result of dividing the total demand to the economic order quantity. This value will be displayed in the same cell with the economic order quantity.

Output dates for first price interval			Output dates for the second price interval			Output dates for the third price interval		
Quantities	Medium unitary price [units]	Total cost for demand [units]	Quantities	Medium unitary price [units]	Total cost for demand [units]	Quantities	Medium unitary price [units]	Total cost for demand [units]
6812	0.98	1105548.96	316056	0.9785	1001056.47	605244	0.9646	1020301.65
21576	0.98	1008874.44	316342	0.9785	1001062.39	608393	0.9644	1020429.03
29062	0.98	998086.87	317008	0.9784	1001076.36	613866	0.9640	1020656.14
54666	0.98	985422.95	317485	0.9783	1001086.54	664383	0.9606	1023065.38
68127	0.98	983528.33	340711	0.9764	1001739.55	670593	0.9603	1023396.90
97308	0.98	982787.46	391638	0.9730	1004045.33	679623	0.9597	1023891.45
123765	0.98	983874.68	398170	0.9726	1004410.34	715083	0.9578	1025966.38
125051	0.98	983950.75	404007	0.9723	1004747.74	730035	0.9570	1026899.61
135967	0.98	984658.53	419665	0.9714	1005701.71	771997	0.9550	1029681.34
136225	0.98	984676.48	425413	0.9712	1006068.63	800992	0.9537	1031728.58
143574	0.98	985208.35	472546	0.9690	1009367.02	812344	0.9532	1032555.14
150580	0.98	985748.87	481520	0.9687	1010046.03	816642	0.9531	1032871.56
174312	0.98	987769.62	497786	0.9681	1011312.30	832643	0.9524	1034065.70
176627	0.98	987979.49	501680	0.9679	1011621.84	842048	0.9521	1034779.04
190656	0.98	989289.61	519097	0.9673	1013034.41	842077	0.9521	1034781.25
199862	0.98	990180.48	529937	0.9670	1013935.33	869181	0.9511	1036881.65
205624	0.98	990748.87	542151	0.9666	1014968.95	883791	0.9506	1038039.70
208781	0.98	991063.50	551534	0.9663	1015775.54	927360	0.9491	1041590.85
212834	0.98	991470.57	554163	0.9662	1016003.40	933009	0.9489	1042061.26
215243	0.98	991714.12	579204	0.9655	1018212.15	943125	0.9486	1042909.02
259945	0.98	996406.83	579493	0.9655	1018238.03	944312	0.9486	1043008.93
263199	0.98	996758.82	597131	0.9651	1019832.81	972387	0.9478	1045398.24

Generate random quantities and find the quantity for the minimal cost

For the range L0 - L3, the minimum cost is 982800.25, that corresponds to an economic order quantity (Qo) of 81700 items and the optimum relative for time (n) of 12

Figure 3. Automated Model for the Calculus of the Economic Order Quantity Qo

3. Results and Discussions

The model generated a report with the economic order quantity (Qo) and with the optimum relative time (n) for the supply, as shown in Figure 3. All the data is automatically stored in a database, at each model iteration.

The graph of the total cost for the supply depending on the quantity ordered, is shown in Figure 4. This graph is generated automatically by the model after each iteration of the model.

It is easy to show on the graph where the minimum cost for the supply is but the model will generate the exact amount for the supply and the number of the time relative for the supply.

For this model there can be added other features to generate reports using the average unit price for calculating the optimal economic order quantity compared to other methods for determining the optimal amount of the supply such as using the uniform price method reduction. These features could provide better tracking of the supply process and of the manufacturing process.

In the future there will be developed a model which will make an analysis of both known methods for determining the economic order quantity, firstly by uniformly reducing the price of the supply and secondly by using the regressive price of the supply orders.

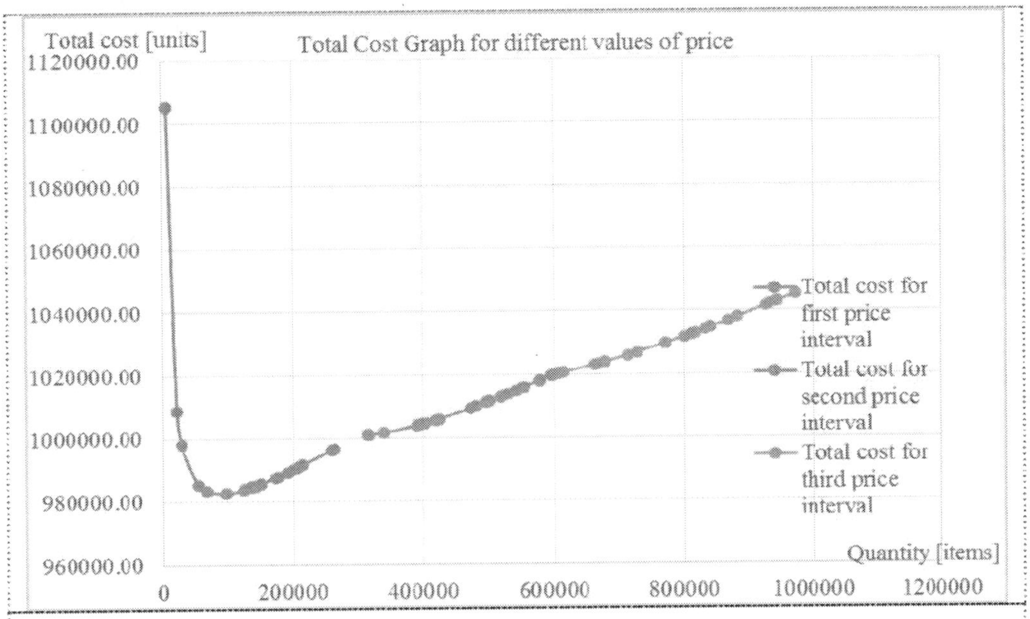

Figure 4. Total Cost Graph

4. Conclusions

The automated model tool presented enables the optimisation of the economic order quantity, Qo for the supply system management by meeting the critical criterion such as the order in time and the minimum of the stock level. By using this model tool, the supply process is simplified and made accessible to all levels of personnel from the supply department in the case of using the regressive rate for the procurement price.

The model above can be used for different total quantities and for many price intervals and it also is very useful for production process planning.

Future developments of this model will include more features from the production process planning, such as the calculus of the economic batch size which many times depends on the total quantity of the unfinished products demands. Other supplementary data, instructions or even messages, while running the model, can be added if there is a requirement for it.

5. References

[1] C. Neagu, *Ingineria şi Managementul Producţiei*, Editura Didactică şi Pedagogică, 2005, Bucuresti.

[2] C. Neagu, M. G. Catană, E. Niţu, M. M. Roşu, *Ingineria si managementul productiei. Aplicatii.* BREN, Bucharest.

[3] S. Nahmias, *Production and Operation Analysis*, Sixth Edition, McGraw Hill Education, New York, 2013.

[4] R. D. Reid., N. Sanders, *Operation Management an integrated approach*, 5th ed., Willey, New York, 2013, Inventory Management, pp 451-504.

[5] S. N. Chapman, *The Fundamentals of Production Planning and Control*, Prentice Hall, 2005

IOP Conf. Series: Materials Science and Engineering **161** (2016) 012109 doi:10.1088/1757-899X/161/1/012109

The optimal design support system for shell components of vehicles using the methods of artificial intelligence

M Szczepanik and A Poteralski

Institute of Computational and Mechanical Engineering, Faculty of Mechanical Engineering, Silesian University of Technology, Poland

Email: miroslaw.szczepanik@polsl.pl

Abstract. The paper is devoted to an application of the evolutionary methods and the finite element method to the optimization of shell structures. Optimization of thickness of a car wheel (shell) by minimization of stress functional is considered. A car wheel geometry is built from three surfaces of revolution: the central surface with the holes destined for the fastening bolts, the surface of the ring of the wheel and the surface connecting the two mentioned earlier. The last one is subjected to the optimization process. The structures are discretized by triangular finite elements and subjected to the volume constraints. Using proposed method, material properties or thickness of finite elements are changing evolutionally and some of them are eliminated. As a result the optimal shape, topology and material or thickness of the structures are obtained. The numerical examples demonstrate that the method based on evolutionary computation is an effective technique for solving computer aided optimal design.

1. Introduction

Optimal properties of shell components of vehicles can be provided by using the computer aided optimization techniques. For example, an appropriate strength of structures can be established by changing their shape, topology and material properties. The choice of optimal shape and topology are the key elements defining the effectiveness of structures, and thus finding them is the problem of great practical interest. Shape and topology structural optimization is a very active research area. Several competing approaches for topology optimization exist [1][2][3]. More recently, intelligent optimal design techniques based on bio-inspired methods like the evolutionary algorithm (EA), the particle swarm optimizer (PSO) and artificial immune system (AIS) have found applications to structural optimization problems. The main feature of evolutionary method is to simulate of biological processes based on heredity principles (genetics) and the natural selection (the theory of evolution) to creating of optimal individuals (solutions) presented by single chromosomes. Evolutionary algorithms are usually applied in the situations when the optimization problems are too complicated for the traditional gradient optimization methods. The task considered in the present work is related to such a problem. This task consists in creating an effective optimization algorithm for shell-structures in respect of topology [4], shape and material or thickness arrangement. The main advantage of the evolutionary algorithm is the fact that this approach does not need any information about the gradient of the fitness function and gives a strong probability of finding the global optimum. The fitness function is calculated for each chromosome in each generation by solving the boundary-value problem by means of the finite element method (FEM). In order to solve the optimization problem the fitness function, design variables and constraints should be formulated. Coupling the finite element method and the evolutionary algorithm gives an effective and efficient alternative optimization tool, which enables solving a large class of the optimization problems of mechanical structures for example structure of car wheel. The main feature of the first proposed optimization method is the evolutionary distribution of the material, changing its material properties or by changing its thickness in the construction of car wheel. This process leads to the elimination of the part of material from the wheel construction and as a result the new shape and the topology of the car wheel emerges. Using interpolation surfaces (hyper surface) reduces the number of the design variables and shortens the time of the computation. The application of the professional program of the finite element method MSC NASTRAN in this method enables the optimization of the complex mechanical systems. The numerical examples confirm the efficiency of the proposed optimization method and demonstrate that the method based on evolutionary computation is an effective technique for solving computer aided optimal design problems.

2. Formulation of evolutionary optimization problem

Evolutionary algorithms are algorithms searching the space of solutions and they operate on populations of individuals. Each individual which consists of chromosomes in population represents a single solution. All chromosomes consist of genes which are design variables in optimization problems. Next the fitness function is computed in order to adaptation each individual in population. All genes of an individual decide about the fitness function value. The algorithm of evolutionary optimization has a several stages. First an initial population of individuals is created. The design variables (genes) in randomly way are generated. In the next step, the fitness function value for each chromosome is computed. In the next step genes are changed using evolutionary operators and working on the parent population individuals. The offspring population is created on the bases of modified individuals with previous step. The algorithm is continuing iteratively till the end of the computation. The stop condition of the computation can be formulated as the maximum number of iterations. Two types of operators are used: mutation and crossover. Mutation is represented by uniform mutation, mutation with Gaussian distribution and boundary mutation. The last operator is simple and arithmetical crossover.

Consider a 2D structure which, at the beginning of an optimization process, occupies a domain $\Omega_0\left(in\ E^3\right)$, bounded by a boundary Γ_0. The domain of this structure Ω_0 is filled by a homogeneous and isotropic material (parameters: Young's modulus E_0 and a Poisson ratio ν). The thickness of the structure g_0 is also constant at the beginning of the evolutionary process. The shells are considered in the framework of theory of elasticity. During the optimization process the domain Ω_t, its boundary Γ_t and the field of Young's modulus $E(\boldsymbol{x})=E_t, \boldsymbol{x}\in\Omega_t$ or the thickness $g(\boldsymbol{x})=g_t$ can change for each itaration t (for t=0, E_0=const, g_0=const). The optimization process proceeds in an environment in which the structure fitness is describing by the objective is to minimize the stress functional

$$J=\int_{\Omega}\psi(\sigma)d\Omega \tag{1}$$

where ψ is an arbitrary function of stress tensor σ, with a constraint imposed on the volume of the structure

$$V\equiv\left|\Omega\right|\leq V_{\max} \tag{2}$$

In order to solve the formulated problem Finite Element models of the structures are considered [5]. The structure is divided into finite elements $\Omega_e, e=1,2,...,R$, and node displacements are calculated by solving a system of linear algebraic equations

$$\mathbf{KU=F} \tag{3}$$

where **U** is a column matrix of unknown displacements, **F** is a known column matrix of acting forces and **K** is a known global stiffness matrix of the structure whose elements are given as follows:

$$\mathbf{K^e=B^TDBV^e} \tag{4}$$

where **D** and **B** are the known elasticity and geometrical matrices, respectively, $\mathbf{V}^e$ represents the volume of the finite element.

3. The idea of evolutionary optimization

The distribution of the Young's modulus or thickness $E(x), x\in\Omega_t$ in the shell is described by a hipersurface $W(x), x\in H^3$. The hipersurface $W(x)$ is stretched under $H^3\subset E^3$ and the domain Ω_t is included in H^3, i.e. $(\Omega_t\subseteq H^3)$ [6].

The shape of the hipersurface $W(x)$ is controlled by genes h_j, j=1,…,N, which create a chromosome

$$ch=\langle h_1,h_2,...,h_j,...,h_N\rangle, \qquad h^{\min}\le h_j\le h^{\max} \tag{5}$$

where $h^{\min}$ - the minimum value of the gene and $h^{\max}$ - the maximum value of the gene.

Genes are values of the function $W(x)$ in interpolation nodes x_j, i.e. $h_j=W(x_j)$, j=1,2,…,N.

The assignation of Young's moduli or thickness to each finite element $\Omega_e, e=1,2,...,R$ is adequately performed by the mapping:

$$E_e=W(x_e),\ x_e\in\Omega_e,\ e=1,2,...,R \tag{6}$$

$$g_e=W(x_e),\ x_e\in\Omega_e,\ e=1,2,...,R \tag{7}$$

It means that each finite element can have different material. When the value of the Young's modulus or thickness for the e-th finite element is included in:

- the interval $0\le E_e<E_{\min}$ (or $0\le g_e<g_{\min}$), the finite element is eliminated and the void is created,
- the interval $E_{\min}\le E_e<E_{\max}$ (or $g_{\min}\le g_e<g_{\max}$), the finite element remains having the value of the Young's modulus from this material (figure 1).

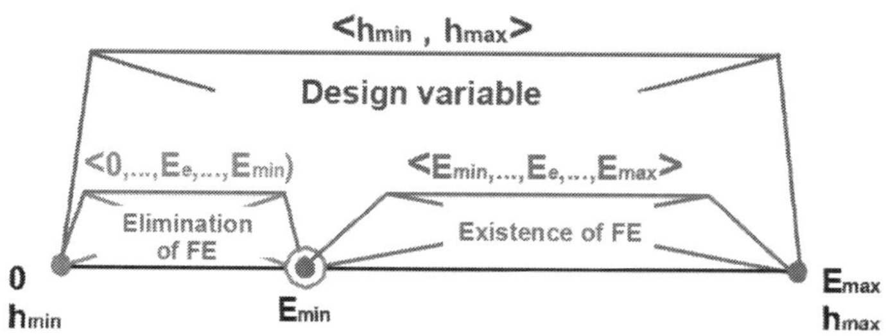

Figure 1. Requirements for elimination and existence of the finite elements

Two types of interpolations which aim at the appropriate selection of mass densities are used. The first of them is the multinomial interpolation, the other the interpolation bases on the neighborhood of elements [13].On the basis of obtained results the interpolation bases on the neighbourhood of elements is better than multinomial interpolation. In addition to second interpolation bases on the neighbourhood of elements the optional number of control points can be loaded and all working space always is used. It is following advantage of this method. In addition to increase number of control points permit to obtain more accurate distribution of hyper surface but for bigger number of control points, number of individuals in each generation must be increased.

In order to improve optimization results two different additional have been introduced:

- the additional procedure aiding the topology optimization [9][10],
- the assignation of materials [13].

Using the first method (the additional procedure aiding the topology optimization), one can change material properties of finite elements during the immune optimization process and some elements are eliminated. As a result, the optimal shape, topology and material of structure are obtained [9][10].

Using second method the Young's modulus value should be devided into the enforced number of subintervals (equal to the number of materials). Each subinterval represents another material whose the Young's modulus belongs to this subinterval. The subintervals should have the same length and their centres correspond with Young's modulus values of suitable materials [13].

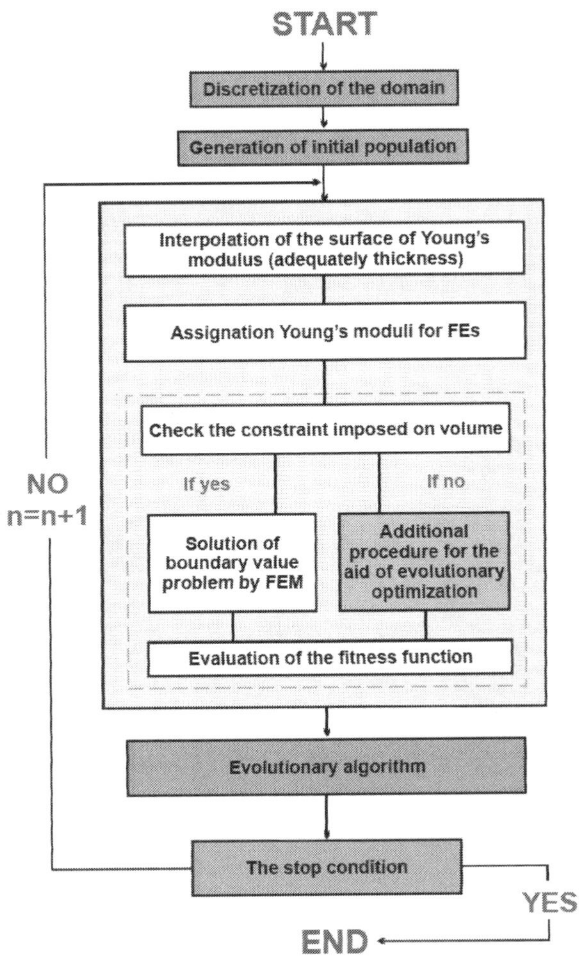

Figure 2. Evolutionary optimization of shells

4. Evolutionary optimization algorithm of topology, shape and material or thickness

The minimization of the fitness functions (1) or (2) with respect to the chromosome (5) is performed by means of an evolutionary algorithm [7][8] with the floating point representation (figure 2). After the FEM discretisation the starting population of chromosomes is randomly generated. At the next step the main loop of the optimization algorithm is performed. Operations included in the main loop lead to the calculation of the fitness function. It requires that the boundary-value problem should be solved by the FEM. After the calculation of the fitness function for all of the chromosomes in the population the evolutionary algorithm is applied. The evolutionary algorithm contains the following operators: the ranking selection, the simple and arithmetical crossovers, the uniform and boundary mutations and the cloning. As the result a new offspring population is created. The end of the algorithm's work, i.e. a break in the main loop activity, occurs after the declared generation number. The algorithm can be also stopped, when after the specified iteration number the change of the fitness function is very small [9].

5. Numerical example – car wheel optimization

The problem is to find optimal shape, topology and thickness of a car wheel by the minimization of the stress functional and by the volume constraint. The geometry of a car wheel with characteristic dimensions (Table 1) is presented. Three surfaces of revolution built the structure (figure 3): the central surface with the holes destined for the clamping screws, the surface of the ring of the wheel and the surface connecting the two mentioned earlier. The connecting surface is optimized. The shell-structure is loaded with a pressure c^0 and the tangent force s^0 (figure 4b). The support is applied on the central surface in the direction of the rotation axis of the wheel. The nodes in the areas near the holes destined for the clamping screws are fixed (figure 4b). The symmetry of the car wheel is imposed (revolution of the 1/5 part of the structure). Figure 4a presents distribution of the control points of the interpolation hipersurface. In this way the number of design variables (genes) are decreased and the symmetrical results are obtained. The input data and the parameters of evolutionary algorithm are presented in the tables 2 and 3, respectively. The maps of thickness presents the results of the optimization after 100th generations (figure 5). All the presented solutions are included in the search space and are optimal – the second optimization criterion is the esthetic aspect dependent on the car user. The results correspond very well with the car wheels produced by different automotive companies (figure 5).

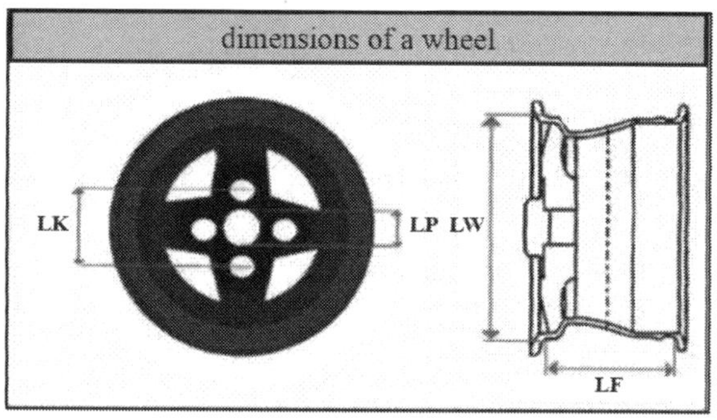

Figure 3. Geometry and characteristic dimensions of a car wheel

Table 1. Characteristic dimensions of a car wheel

diameter of the wheel LW	355.6 mm
width of a tyre LF	175 mm
diameter of the wheels spacing LK	110 mm
diameter of a wheel hub LP	60 mm
thickness of the wheel hub	30 mm
thickness of a tyre	8 mm

Table 2. Input data to the optimization task

tangent force s^0 [N]	pressure c^0 [Mpa]	number of design variables	number of control points	number of the chromosomes
500	0.22	23	86	100
material	range of the change of the genes [mm]	existing of an element	elimination of an element	V_{max} [cm^3]
aluminium	4.0÷20.0	$4 \leq g_e < 10$	$10 \leq g_e \leq 20$	5 500

Table 3. Parameters of evolutionary algorithm

cloning	uniform mutation	boundary mutation	simple crossover	arithmetical crossover
2%	5%	5%	10%	10%

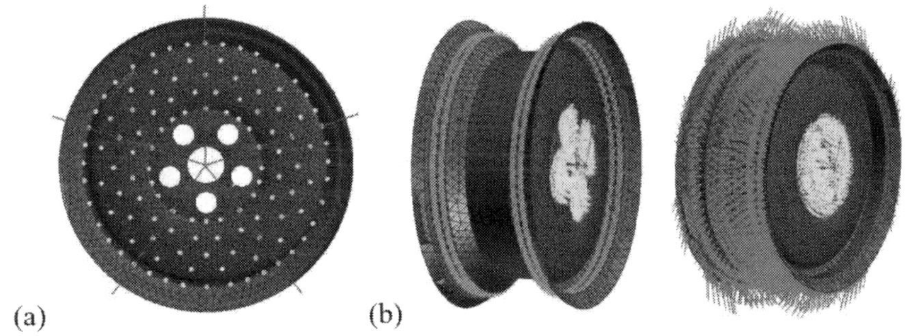

Figure 4. Car wheel: (a) distributing of the control points of the interpolation hipersurface; (b) boundary conditions

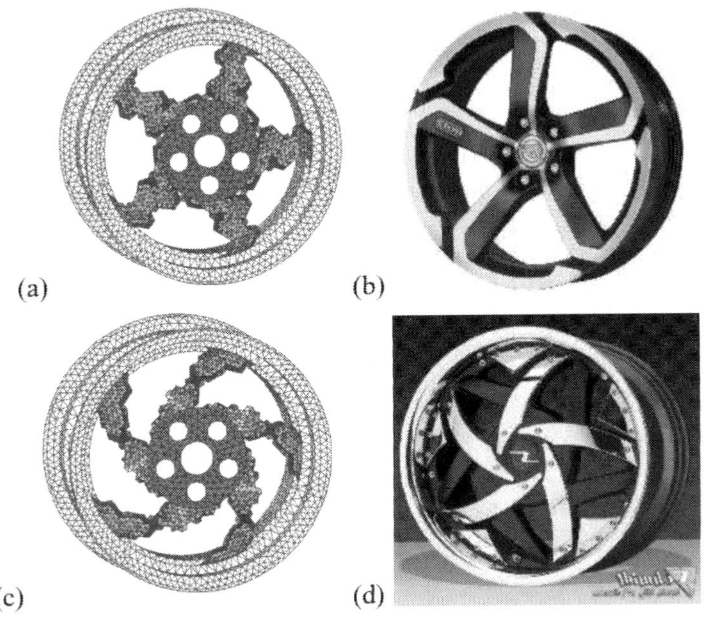

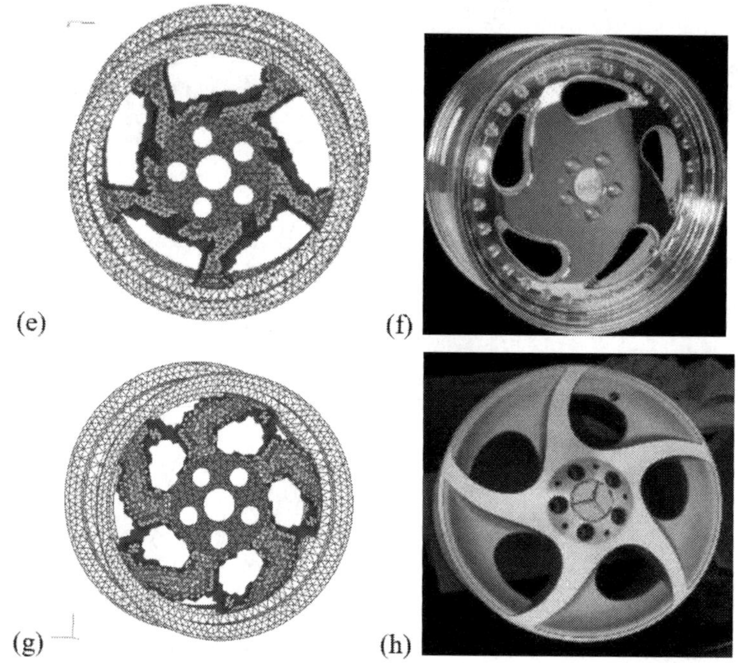

Figure 5. The results of the car wheel optimization: (a), (c), (e), (g), the best obtained solution - maps of thickness, (b), (d), (f), (h) real car wheels which are in good coincidence with obtain solutions

6. Conclusions

An effective tool of evolutionary optimization of shells has been presented. Using this approach shape, topology and material or thickness optimization are performed simultaneously. The important feature of this approach is a strong probability of finding the global optimal solutions received by implementation of the evolutionary algorithm and its universality consisting in application of this approach to the optimization of the 2-D structures in plane stress/strain [10]. Evolutionary method belong to methods based on population of solutions and they have some interesting features which can be considered as alternative to artificial immune system [11][12][13][14] or particle swarm optimizers [10][15][16]. Comparison between AIS, PSO and EA in the paper [6][17][18][19] is presented.

References

[1] Bendsøe M P 1989 Optimal shape design as a material distribution problem Struct. Optim.
[2] Burczyński T and Osyczka A 2003 *Evolutionary Methods in Mechanics* Proc. IUTAM Symposium Kluwer Dordrecht
[3] Xie Y M and Steven G P 1997 Evolutionary Structural Optimization London Springer
[4] Yamada T et al. 2010 A topology optimization method based on the level set method incorporating a fictitious interface energy Comput. Methods Appl. Mech. Engrg.
[5] Zienkiewicz O C and Taylor R L 2000 *The Finite Element Method* Butterworth Heinemann Oxford
[6] Burczyński T Poteralski A and Szczepanik M 2015 Immune and Swarm Optimization of Structures Advances in Evolutionary and Deterministic Methods for Design Optimization and Control in Engineering and Sciences Chapter no 19
[7] Arabas J 2001 Lectures on Evolutionary Algorithms WNT Warszawa (in polish)
[8] Michalewicz Z 1996 *Genetic algorithms + data structures = evolutionary algorithms* Springer-Verlag Berlin
[9] Burczyński T and Szczepanik M 2013 Intelligent optimal design of spatial structures Computer

& Structures Elsevier
[10] Szczepanik M and Burczyński T 2012 Swarm optimization of stiffeners locations in 2-D structures Bulletin of the Polish Academy of Sciences Technical Sciences vol 60 no 2 pp 241-246
[11] Poteralski A Szczepanik M Ptaszny J Kuś W and Burczyński T 2013 Hybrid artificial immune system in identification of room acoustic properties Inverse Problems in Science and Engineering Taylor & Francis
[12] Poteralski A 2014 Optimization of mechanical structures using artificial immune algorithm Beyond Databases Architectures and Structures Communications in Computer and Information Science vol 424 pp 280-289
[13] Poteralski A Szczepanik M and Burczyński T 2015 Immune Optimal Design of 2-D and 3-D Structures International Conference on Artificial Intelligence and Soft Computing (ICAISC) Lecture Notes on Artificial Intelligence vol 9120 Springer pp 471-482
[14] Poteralski A 2015 Data processing in immune optimization of the structure, Beyond Databases Architectures and Structures (BDAS) Communications in Computer and Information Science vol 521 Springer pp 309-319
[15] Poteralski A Szczepanik M and Burczyński T 2015 Swarm and immune computing of dynamically loaded reinforced structures International Conference on Artificial Intelligence and Soft Computing (ICAISC) Lecture Notes on Artificial Intelligence vol 9120, Springer pp. 483-494
[16] Szczepanik M Poteralski A Ptaszny J and Burczyński T 2012 Hybrid Particle Swarm Optimizer and Its Application in Identification of Room Acoustic Properties Swarm and Evolutionary Computation - International Symposia SIDE 2012 and EC 2012 Held in Conjunction with ICAISC 2012 Zakopane Poland Lecture Notes in Computer Science vol 7269 Springer
[17] Poteralski A Szczepanik M Dziatkiewicz G and Kuś W 2013 Comparison between PSO and AIS on the basis of identification of material constants in piezoelectrics Springer-Verlag Berlin Heidelberg ICAISC 2013 Part II LNAI vol 7895 pp 569–581
[18] Poteralski A Szczepanik M Beluch W and Burczyński T 2014 Optimization of composite structures using bio-inspired methods Artificial intelligence and soft computing ICAISC 2014 vol 8468 pp 385-395
[19] Szczepanik M Poteralski A Długosz A Kuś W and Burczyński T 2013 Bio-inspired optimization of thermomechanical structures Springer-Verlag Berlin Heidelberg ICAISC 2013 Part II LNAI vol 7895 pp.79-90

E-learning implementation in superior technical educational system

MUSCA Gavril[1a], MIHALACHE Andrei[1b], MUSCA Elena[2c]

[1] "Gheorghe Asachi" Technical University of Iasi, Department of Machine Manufacturing Technology, Blvd. D. Mangeron, No.59A, Iasi-700050, Romania

[2] Institute of Computer Science Iasi

[a]musca@tcm.tuiasi.ro, [b]andrei.mihalache@yahoo.com, [c]coca.musca@gmail.com

Abstract. E-learning methods apply to most modern and various domains but also represent a great tool for the mechanical educational system where there are a lot of sustained efforts for its implementation. Using, administrating and maintaining an e-learning system for a certain field of study requires knowledge related to computation system's utilization but also the understanding the working mechanisms behind it that allows the system to be fully customized in order to be perfect fitted to the user's needs and requirements. A Moodle based test is evaluated from several points of views such as coherence clarity, concise content, information synthesis capacity and the presentation mode which makes the difference between clear or fuzzy graphical representations or terms. The authors appreciate that the ability of managing information in real time by the professor is a decisive decision in order to successfully implement an e-learning web platform. Updating information and structuring trainee's activities from thoroughgoing study up to their individual proposals for conceived applications leads to a better understanding and practical knowledge of theory.

Keyword: e-learning, engineering education, Moodle, analysis of activities.

1. Introduction.

Training students in mechanical engineering and a continuous formation for specialists from the industrial environment requires multiple sustained efforts from both the trainers and the trainees. The actual performances of CAD, CAM, CAE, PDM and PLM software impose a continuous adaptation to new requirements needed by the engineered product's quality and performance. That is why we appreciate that e-learning systems represent a modern solution that answers to few of the current issues.

Keeping in mind that project management represents a tedious and very intricate activity that requires people with good knowledge and skills we consider that training them has to be done in the most complete and comprehensive way possible. In order to assess the benefits of collaborative teamwork performed for project development and engineering design we need to first train our students and get them familiar with collaborative learning procedures. We think about the learning curve as a sum of various instruments used in the process. That is why traditional learning such as paper based courses and basic lab experiments or seminaries are considered somehow obsolete but may be better acknowledged as part of the whole system. Nowadays students interact online far more quickly that anyone could have ever imagine a few decades ago and look for their information from various sources some more reliable than others. That is why we do refer to above mentioned traditional forms of education as to passive education whereas the online environment is considered the true active part of the system. That being said, not all the online content is adequate and appropriate to be used in educational purposes. Today efforts are more focused whether to better understand the factors which affect the quality of engineering education. Different educational systems struggle to understand changes brought by the web and try to fit them in their programs as complementary. The authors feel that there lies the problem because the online content and the means to access it should represent the first priority that everything else revolves on. That would lead inevitably to excellence in education and efficiency because of the much more object related structure of courses and training materials available online. The goal of every educational system is to prepare the student for a career. That being said we cannot ignore the true benefits of web content being delivered anywhere at any time. Real life examples, debates, quizzes and forum topics are very powerful instruments that students may attend in the learning process which researches have shown to bring more effectiveness, interest and competitiveness among students. They feel that being constantly pushed to solve problems, having granted access to meaningful and reliable up to date information helps them to quickly understand the mechanisms and be aware of the issues that may appear in the process. Thus, students may gain knowledge in engineering design, processing and manufacturing. That is why the authors consider that in addition to traditional courses, CAD, CAM, PLM training is a must.

2. E-learning system application in engineering education.

In Romania and abroad there are many e-learning systems known approaches. One of them, nationally implemented is the one from the Polytechnic University of Bucharest [12] whom functions among others for skill improvement of engineer professors from pre university technical educational system and as a benchmark of the project's success its participant's results [2] were considered as reference in the management of resources related to mechanical computer assisted design. Targeting the development of several tools used in sheet metal processing the authors developed a collaborative system [7] which involved teams from different universities. The paper [7] presents the team efforts of the CAD/CAM/PLM laboratory of Machine Manufacturing Technology department for collaborative development in case of several industrial projects. The project development has been achieved due to collaboration between mechanical student's team, researchers and engineers from manufacturing based enterprises. Because the level of competences was different for every team involved that problem meant that training sessions were necessary in order to get team members to be familiar with CAD/CAM software and PLM related technical data administration. The teams have developed and sustained a series of working modules based mainly on the documentation they had at their disposal as well as the one found online. Software specific tools and tutorials were frequently accessed and it allowed the project to grow. A step forwards was made by designing the system's database in MySQL as well as its interface in PHP language [7]. The decision was in favor of multiple working modules because past experiences shown that module based testing systems are more efficient because they provide convergence of training processes allowing main issues to be clearly highlighted within tests that being the very foundation of testing to begin with. The project expanded on several branches and directions and it also included quizzes, clearness and efficient content management for ease of training, problem identification and solving, on place verifications and evaluation for the evolution and the degree of understanding during training sessions, adaptability and dynamicity. Of course the authors appreciate that there is room for improvement constantly trying to fine tune and make all the necessary adjustments in such a way that quizzes may be more objective orientated inside trainings and in such a way that they could be easily adapted to every team's specific needs and degree of knowledge and professional experience. The need for adaption arises when members of even the same team started to respond differently to the same training content thus making it clearly that the questionnaires have to be addressed to certain aspects individually. Experience showed that question formulation had to be different for every team due to their degree of understanding, knowledge, and experience and implication level. The team of researchers was the most involved ones being very capable and responsive. It also had been the team which learned the quickest. The student team was very receptive but it lacked experience and professional knowledge thus making it the one which had less questions and for that matter, debates regarding discussed matters within the project. That being said it is very much alike with what Moodle proposes and succeeds by its implementation in technical superior education systems. Moodle widely implemented as a management system knows new utilization domains but also new means of connection such as PC's networks, cloud systems, mobile OS's as well as other. Its adaptability and scalability to those environments as well as its property as open sourced software make it so appealing [1], [11]. It can provide maximum output when correctly configured and may lead to rapid development and improved knowledge of the trainees those being key ingredients for their competiveness on an ever changing and dynamic professional market [9], [12].

E-learning methods apply to most modern and various domains but also represent a great tool for the mechanical educational system where there are a lot of sustained efforts for its implementation. A very interesting direction is that of the m-learning platforms which are complementary to e-learning ones in terms of future development of engineering teaching methods or measurements dissemination [3]. As an a immediate application Moodle may be used as a comparison form between results that were obtained traditionally in the field of science of materials with the ones obtained by e-learning methods for graduate students starting with their first year of study [4].

Another interesting application is that of a distance based educational system through e-learning dedicated to technical universities students or to economical profiled universities, or to the employees of a company such as production focused companies that are interested and have the capability to implement such a system on their intranet or it may be addressed to technical and economic data management systems MRP/ERP [8].

3. Considerations over a Moodle based system for training in design and manufacturing in mechanical engineering.

Using, administrating and maintaining an e-learning system for a certain field of study requires knowledge related to computation system's utilization but also the understanding the working mechanisms behind it that allows the system to be fully customized in order to be perfect fitted to the user's needs and requirements.

The transformation process involved in turning a traditionally presented course into one that uses an e-learning system requires total implication, dedication and a considerable amount of work for it to provide maximum output in optimal conditions [5]. Professors tend to divide the evaluation process of a certain learning system into technological, pedagogical and organizational components. This division on components does not exclude an internal analysis as well as the functioning product's quality.

An evaluation process of an online based learning system performed professionally will take into account as an expert analysis the one that considers multi criteria of evaluation for the analyzed software components which grade the highest in their utility domains for which they have been designed. For example a Moodle test will be evaluated from several points of views such as coherence clarity, concise content, information synthesis capacity and the presentation mode which makes the difference between clear or fuzzy graphical representations or terms. However the test will not be assessed from the student grades point of view because this is not a defining feature in the evaluation process of the software's capability but is more about the professor's capability to order the information and present it properly.

4. Moodle system assessment

Figure 1. General view over the e-learning system CAD/CAM/PLM orientated with details about course studies that it contains.

Keeping in mind the above presented reasons and benefits, the Machine Manufacturing Technology Department from the Technical University of Iasi has implemented inside the CAD/CAM/PLM laboratory an e-learning educational system based on Moodle (v.2.6) as shown in figure 1. The first variant has been developed and used back in 2013 and at the present time the platform contains 7 courses studies and has been used by more than 650 students from both graduate and master studies. Even though subjects are related each course study is original because the authors have adapted them to trainee's specific activities and knowledge. The available resources for students contain manuals elaborated by the authors, tutorials and product documentation for each software solution used in the laboratory (Solid Edge, NX CAM, ESPRIT, and CIMCO), scientific articles and links to relevant content website.

In order to appreciate the quality of student knowledge a special attention was given to the dynamic development of trainee's activity area. Continuous evaluation of the trainee for the given course time of 14 weeks is carried out by:

- The completion of practical sessions highlighted by the student's personal reports,
- The completion of a number of 3 or 4 simple tests, multiple answers tests or essay like tests that contain questions from the main topics of the presented course,

- Solving two home works (mini projects) having topics about designing technologies for the machining on CNC driven lathes and milling machines, proposed machining processes simulation and development of the specific G code,

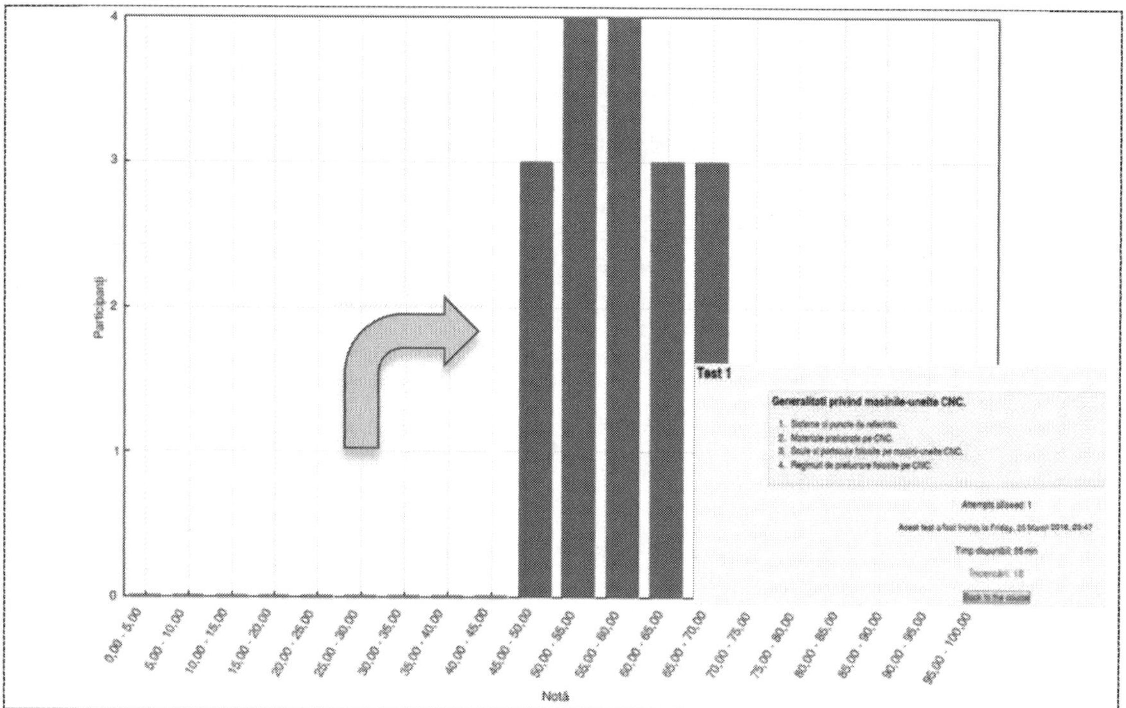

Figure 2. Measuring the quality of trainee's instruction sessions and the normal curve of distribution for their test results.

- Participation in forums through interventions on specially proposed themes by the course leading professor.

For assessing the correctness of testing activities the following Moodle features have been used:

- The elaboration of questions having the same subject but in a differently presented form (one or two occurrences at almost 50 questions per test),
- Random occurrences of question order for each student,
- Random occurrences of answer order for the same question,
- Verification of grades distribution normal curve (figure 2 illustrates grade repartition for one class of registered users).

5. Conclusions.

Analyzing statements from the same profile literature it results that the management of learning systems knows new application domains and different approaching directions. Using cloud systems, specialized networks, smart phones or tablets is a very actual and daily habit of the common user and thus it represents the future perspective of every e-learning system.

Considering the performances of new intercommunication technologies e-learning systems will become more attractive as they will encapsulate up to date information addressed to every type of student in

particular dependent on its profile and its interface and presentation form will be more intuitive and adjustable to any screen or OS environment.
The authors appreciate that the ability of managing information in real time by the professor is a decisive decision in order to successfully implement an e-learning web platform. Updating information and structuring trainee's activities from thoroughgoing study up to their individual proposals for conceived applications leads to a better understanding and practical knowledge of theory.
In the case of computer assisted design classes as well as assisted manufacturing there are requirements for applications which will ensure a strong knowledge of CNC machines, tooling and their tool holders, manufacturing materials characteristics and those of machining regimes. There are is also a need for CNC machine programming languages knowledge as well as for used CAM software in order to be able to simulate the manufacturing processes and thus to generate specific NC machining operations in G code accepted language.

References
[1] M. J. Casany, M. Alier, E. Mayol, et.al. Moodbile: A Framework to Integrate m-Learning. Applications with the LMS, Journal of Research and Practice in Information Technology, Vol. 44, No. 2, May 2012
[2] G. Constantin, R. Constantin, G. Taşcă, G. Muscă, M. Vasilescu, *Proiectare asistata 2D şi 3D a maşinilor şi sistemelor de producţie*, Editura Academiei Oamenilor de Ştiinţă din România, 138 pag., Bucureşti, 2011, ISBN 978-606-8371-46-7
[3] Drago Cmuk, Optimization of a m-learning and e-learning Methods in Metrology Education, Faculty of Electrical Engineering and Computing, Zagreb, Croatia, Faculty of Engineering, Benevnto, Italy
[4] L.A. Dobrzański, F. Brom, E-learning on the example of materials science, Journal of Achievements in Materials and Manufacturing Engineering, Volume **29 Issue 1** July 2008
[5] O. Dodun, E. Panaite, N. Seghedin, et.all., Analysis of an e-learning platform use by means of the axiomatic design, in 9th International Conference on Axiomatic Design – ICAD 2015, Procedia **CIRP 34** (2015) 244 – 249
[6] E. Kurilovas, V. Dagiene, Multiple Criteria Evaluation of Quality and Optimisation of e-Learning System Components, Electronic Journal of e-Learning Volume 8 **Issue 2** 2010 (141 - 150)
[7] G. Muscă, E. Muscă, V. Merticaru, Contributions to Computer aided Training in CAD, CAM, PLM, Academic Journal of Manufacturing Engineering, p.72-77, VOL. 9, ISSUE 4/2011
[8] W. Torbacki, E-learning for manufacturing enterprises and universities based on ISOF Academy, in Journal of Achievements in Materials and Manufacturing Engineering, Volume **22 Issue 1** May 2007
[9] C. Silvestru, I. Burcezan (Precup), R. Bere, M. Lupescu, Concept: Using E-Learning Platforms in Economical and Entrepreneurial Development p.90-99, DOI: 10.12948/issn14531305 /19.2.2015 .09, *Informatica Economică* vol. 19, no. 2/2015
[12] M. Saracevic, M. Kozic, F. Selimovic, B. Fetaji, Improving University Teaching Using Application for Mobile Devices Based on Netbiscuit Platform and Integration with the LMS, p.87-92, TEM Journal – **Volume 2 / Number 1**/ 2013.
[11] M. Wang, Y. Chen, M. J. Khan, Mobile Cloud Learning for Higher Education: A Case Study of Moodle in the Cloud, In The International Review of Research in Open and Distance Learning, p. 255-267, Vol 15 | No 2 April/14Vol 15 | No 2 April/14
[12] ***, Proiect CONCORD, POSDRU/87/1 .3/5/61397, "Reţea naţională de formare continua a cadrelor didactice din învăţământul preuniversitar, profesional şi tehnic", Universitatea Politehnica din Bucureşti, 2011

IOP Conf. Series: Materials Science and Engineering **161** (2016) 012111 doi:10.1088/1757-899X/161/1/012111

Load curve classification for the evaluation of demand side management programs

K Yumak[1], G Tosun[1], B Varlık[1] and M Bağrıyanık[2]

[1] AF Consult, ODTÜ Teknokent Met Alani, Mustafa Kemal Mah. Dumlupinar Blv. No: 208, D Blok No: 3, 06510, Cankaya, Ankara, Turkey

[2] Istanbul Technical University, Electrical Engineering Dept., Sariyer, Istanbul, Turkey

E-mail: kahraman.yumak@afconsult.com

Abstract. The demand side management of electrical power systems is an issue of growing concern. In this study, annual load curves those have been obtained by field measurements on MV (31.5kV) feeders in an electricity distribution company in Turkey have been studied. Feeders supplying different type of customers (Agricultural irrigation, Commercial & Industrial and Residential) are classified through Peak and Night Ratio for the assessment of convenient incentive or price based methods. Statistical parameters, such as peak-to-average ratio, load factor and standard deviation are calculated. Active power deviations on annual and daily load curves are discussed.

1. Introduction

Historically, electrical power systems consist of limited number of large power plants feeding and balancing whole demand in a centralized concept. The system is top-down designed and the power is supplied through transmission and distribution systems in unidirectional way [1]. In 1973, the oil embargo led to increase the costs in the generation of electricity. Subsequently, instead of increasing generation capacity to supply the growing demand, the idea of reducing and/or managing the demand started to be considered in power systems [2, 3].

The electrical power demand varies over the course of the day, week and year due to several factors, such as seasonal and environmental changes. Load management refers to a wide category of methods and instruments to control power demand in electrical systems to be in balance with the electricity supply. It is defined as any activity done by customer and/or electricity supplier to modulate the demand fluctuation with the objectives of reducing total peak load, increasing load factor, improving capacity and utilization of power system [4].

In the literature, the terms of load management (LM), demand side management (DSM) and demand response (DR) can be found as overlapping definitions for similar concepts [5, 6]. Load management actions can be installing energy-efficient equipment, load curtailment, load shifting, peak clipping, valley filling, and using distributed generation and energy storage technologies for the benefit of the whole utility system.

Load management methods can be divided into two subgroups, such as direct and indirect methods. In direct programs, utilities have the option to control customers' equipment based on an agreement. In indirect programs, it is customers' decision to involve or not to the program. If customers participate in the program, then they get the benefits which is offered by the utility. The LM methods are schematically illustrated in figure 1.

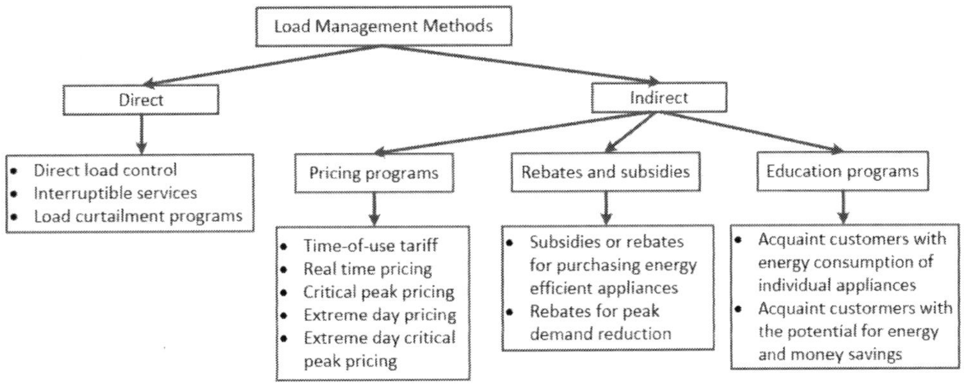

Figure 1. Load management methods [6]

Direct load management is based on remote control of the end-users' specific equipment to turn on and off during peak demand periods and critical events [7-9]. Water heaters, pool pumps, air conditioners, washing machines and dishwashers are examples of the appliances commonly used in residential customers.

Interruptible tariffs [10, 11] is a pricing option that customer permits the utility to reduce the demand in peak loading or emergency conditions. An agreement between the customer and the utility is required for assessment of the interruptible load (IL). Demand limiters or remote signaling can be used for load reduction.

Load curtailment programs offer payment or discounted rate in return for reduction on the peak demand to a given amount when there is an urgent unbalance in the power system. This program commonly used for large industrial customers.

The pricing program is a subsection under indirect load management methods. There are tariff related programs containing price discounts for demand reduction [12]:

- Time-of-use (TOU) pricing: Different prices for different time durations over a day, week or season. The prices are constant for the time periods.
- Real time pricing (RTP): Prices vary over short time periods (commonly hourly) enabling a better reflection of the supply costs.
- Critical peak pricing (CPP): Dynamic pricing based on TOU and RTP, pre-set higher rates during for a limited number of critical days or hours.
- Extreme day pricing (EDP): Similar to CPP, higher price is for the whole 24 hours of the extreme day (unknown until a day-ahead).
- Extreme day critical peak pricing (ED-CPP): CPP rates are used during peak and off-peak periods during extreme days. A flat rate is used for the other days.

Rebates and subsidies are offered as a motivation to the customers to increase the usage of energy efficient equipment or participation in demand reduction. Education programs are for increasing customers' awareness on load management programs.

The direct programs (direct load control, interruptible and curtailable services) given in figure 1 can also be categorized as incentive based programs. These services depends on the offered incentives to the customers and based on contracts designed by policymakers, grid operators, and utilities. Customer participation is voluntary and there could be penalties in case of failures on the agreed commitments. The price based programs can be implemented indirectly and designed to reflect the costs of electricity production. It fully depends on customer choice. Customers are expected to change their electricity usage to low priced time periods by an internal economic decision.

It is needed to classify load curves in order to analyze the demand variations. Classification would help to decide if incentive (direct methods) or price based methods is suitable for the specific loading conditions. Furthermore, the investigation of load characteristics with load management perspective helps the utility for the assessment of load management feasibility. In this study, field measurements of MV feeders in a distribution company service area in Turkey with different load characteristics are classified using Peak and Night Ratios. And also customer characteristics are evaluated through statistical parameters.

2. Load curve classification and evaluation

Load curve classification can be performed using Peak Ratio, Night Ratio and statistical parameters, such as peak-to-average-ratio (PAR), load factor and standard deviation. The definitions of the related parameters are reviewed in this section.

2.1. Peak ratio and night ratio

Peak Ratio (PR) is defined as the ratio of average power absorbed during peak hours in a certain time duration, to average power absorbed during off-peak day hours in the same period. Similarly, Night Ratio (NR) is defined as the ratio of average power absorbed during night hours in a certain time duration, to average power absorbed during off-peak day hours in the same period. [13, 14] The defined certain time duration can be a day, a month or a year.

In Turkey, there are two options in the national retail electricity tariff programs: flat and time variant (time-of-use, TOU) pricing [15]. The TOU tariff program defines three time durations, such as day period ($06{:}00 \le t < 17{:}00$), peak period ($17{:}00 \le t < 22{:}00$) and night period ($22{:}00 \le t < 06{:}00$). As expected, the price is highest in the peak period and the price is lowest in the night period. The definitions of PR and NR values based on Turkish TOU tariff program are given in equation (1) and (2).

$$Peak\ Ratio = \frac{Average\ power\ absorbed\ during\ peak\ period}{Average\ power\ absorbed\ during\ day\ period} \tag{1}$$

$$Night\ Ratio = \frac{Average\ power\ absorbed\ during\ night\ period}{Average\ power\ absorbed\ during\ day\ period} \tag{2}$$

If PR value is greater than 1, it means the average power consumption during peak period is higher than the average power consumption during day hours. Likewise, if NR value is greater than 1, it means the average power consumption during night period is higher than the average power consumption during day hours.

According to the calculated PR and NR values, load curves can be classified and analyzed for the determination of convenient demand side management methods. The expected output from the application of selected method is to decrease the PR value and increase the NR value. Figure 2 represents an approach for the classification according to PR and NR values. The explanation for the classes are as follows:

- **Class 1:** Permanent use. This customer type represents a flat consumption. There is no high deviation in the load curve.
- **Class 2:** Off peak and small peak period usage. This customer type have relatively high PR and NR values. The peak period usage could be reduced using demand side management methods. Mandatory or voluntary tariff programs is suitable for this kind of customers.
- **Class 3:** Off peak disconnection usage. In this area, NR is very low and PR is lightly modulated. This means the power consumption in night period is very low, the power consumption in day and peak periods is modulated. NR could be increased and PR could be decreased. Incentive programs are suitable for this type of load curves. Energy storage, ice

storage or standby generators could be utilized for decreasing consumption in day and peak periods.

- **Class 4:** Peak hour disconnection usage. In this type of customers, PR is very low and NR is very high. Mandatory or tariffs programs are not convenient methods for this kind of load curves. Very strong incentive based programs are recommended.
- **Class 5:** Average off-peak usage. Incentive program is recommended in order to increase NR and decrease PR values.

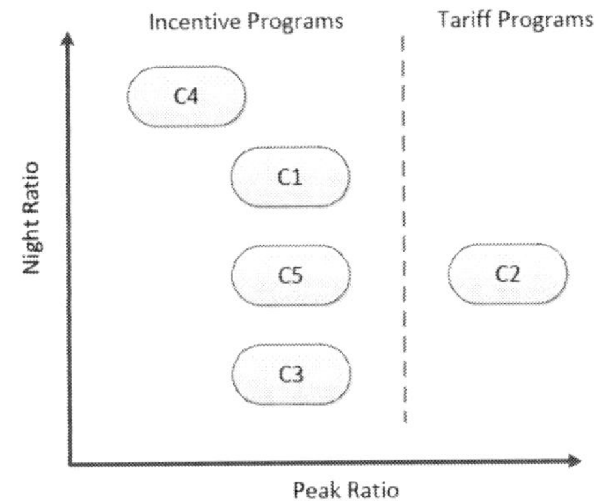

Figure 2. Classification regarding to PR and NR values [13].

2.2. Statistical parameters [16]

In this subsection, the definitions of statistical parameters which can be used to evaluate load curves of different customer types are given.

- **Peak-to-average-ratio (PAR):** It is defined as the ratio of the peak load to the average load which is given in equation (3). In load management applications, it is expected to reduce PAR value [17].

$$PAR = \frac{P_{peak}}{P_{average}} \tag{3}$$

- **Load factor (LF):** It is defined as the ratio of average load to peak load during a time period. If the load curve is in smoother characteristic, LF value becomes higher. The highest possible LF value, which presents a flat load profile, is 1. The higher LF means power consumption is comparably constant and the utilization rate of infrastructure is also higher. LF is the inverse of PAR.
- **Standard deviation (σ):** It is given in equation (4) and defined as the square root of the variance. Variance is the average of the squared difference of the values from their average value. It is a measure of the deviation from mean value. The higher standard deviation indicates the more spread apart from mean value. The unit of this parameter is same with the data. In the scope of this study, it is kW.

$$\sigma = \sqrt{\frac{1}{N}\sum_{i=1}^{N}(P_i - P_{average})^2} \tag{4}$$

3. Field measurements

The active power demands are measured at the beginning of selected MV (31.5 kV) feeders using a multifunction power analyser "UMG 96RM-E" and the measurements are taken through grid visualisation software "GridVis" which are both provided by Janitza. The measurements are in 10 minute periods and covers one year starting from 15 January 2015 and ending on 15 January 2016. There are three kind of load types: agricultural irrigation, commercial&industrial and residential.

3.1. Agricultural irrigation feeders

These feeders are supplying mostly agricultural irrigation pumps (above %80 percent of the total loads). Rest of the loads are residential and commercial customers. The annual load curves can be seen in figure 3. It can been that there are high seasonal deviations in the load demand. Demand of irrigation pumps becomes dominant between "April last week – June first week" and "June last week – September last week". Maximum demand for both feeders is higher than 15 MW, while base load for the feeder 1 and feeder 2 is 1.3 MW and 0.5 MW, respectively. During heavy agricultural irrigation, base load is also increases up to 5-8 MW. Daily load curves are given in figure A1 in Appendix. On daily basis, the demand of irrigation pumps starts from 7 a.m. and increases till noon. After noon the demand decreases to base load continuously until 5 p.m. A sudden increase occurs after 10 p.m. and continues till midnight. Then, demand gradually decreases to base load.

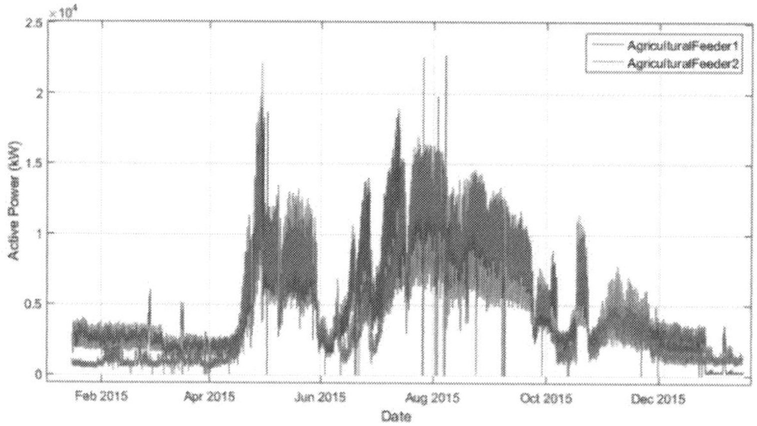

Figure 3. Annual load curves of agricultural irrigation feeders.

3.2. Commercial and industrial feeders

Feeders are fully supplying (100%) commercial and industrial loads. The annual load curves can be seen in figure 4. Daily load curves are given in figure A2 in Appendix. Different from the agricultural irrigation feeders, there is no fluctuation depending on seasons. But during national holidays in Turkey between 16-19 July 2015 and 23-27 September 2015, the demand is in the lowest level for a few days. Maximum demand for both feeders is higher than 25 MW and base load is about 3 MW. According to daily load curves, the demand of utilities start in the morning 8 a.m. and the demand continuously increases till noon. There is a temporal decrease during lunch break for 1 hour and demand gradually decreases to base load until 7 p.m. It is holiday on Sundays, thus the demand is on base load level.

3.3. Residential feeders

Figure 5 shows the annual load curves of the feeders. These feeders are supplying 100% residential customers at the city center. Maximum demands of the feeders are higher than 6.5 MW and 3.5 MW. Base loads are 1.8 MW and 0.8 MW. Figure A3 in Appendix shows the daily load curves for residential feeders. The demand of the customers starts increasing from the base level at 6 a.m. to the first peak point at noon. Until 7 p.m., there is relatively flat consumption. Around 9 p.m., the second (also higher) peak occurs. Afterwards, demand gradually decreases until 6 a.m.

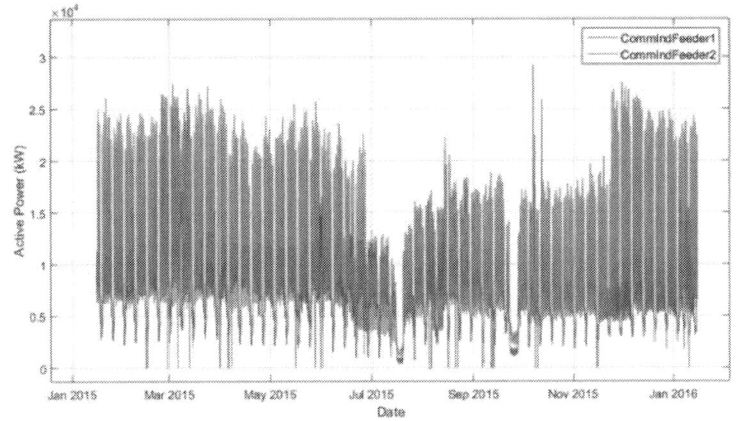

Figure 4. Annual load curves of commercial and industrial feeders.

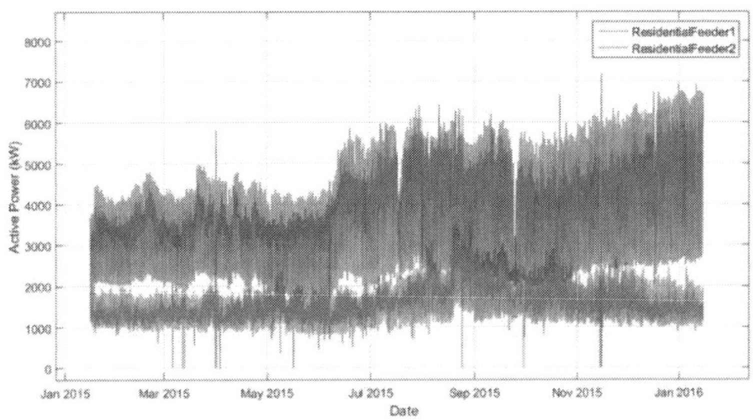

Figure 5. Annual load curves of residential feeders.

4. Analysis results

The results of PR and NR ratios of the feeders according to the equations (1) and (2) can be seen in figure 6. PR ratios of residential feeders are greater than 1, which means average power absorbed during peak hours is higher than the day hours. For the rest of the feeders, PR ratios are less than 1. NR values of the agricultural irrigation feeders are the highest.

According to figure 2, residential customers fall into Class 2 and tariff programs are suitable for them. The peak period usage could be reduced by mandatory or voluntary tariff programs. Agricultural irrigation feeders fall in Class 4. In this type of customers, mandatory or tariff programs are not convenient methods. Very strong incentive based programs are recommended to shift the peak loads. Commercial and industrial feeders are fall between Class 3 and 5. The power consumption in night period is very low and the power consumption in day and peak periods is modulated. Load shifting or peak clipping can be implemented using energy storage technologies or standby generators. Incentive programs are suitable for these customers.

Table 1 lists the results of statistical parameters. The smallest maximum and average demand occurs in residential feeders. Average demand is the highest in commercial and industrial feeders. PAR and LF are inverse each other. The highest LF is in residential feeders, meaning utilization rate of infrastructure is higher than the other types. As it can be seen, the lowest LF is in agricultural feeders. Standard deviation of the residential customers is quite low. The highest deviation occurs in commercial and industrial feeders.

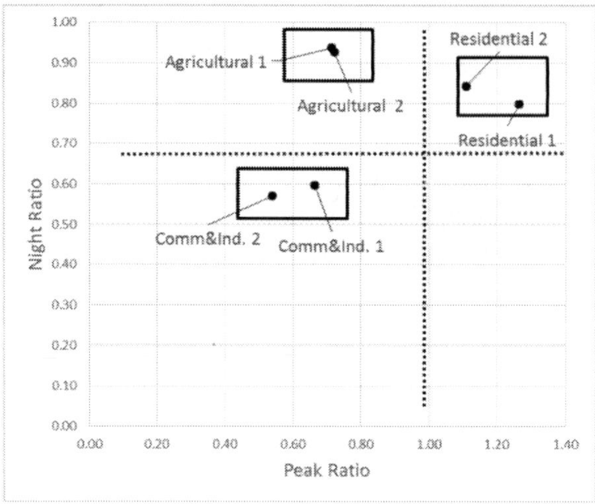

Figure 6. Peak and night ratios of the feeders.

Table 1. Results of statistical parameters

	Agricultural		Residential		Comm&Ind.	
	Feeder 1	Feeder 2	Feeder 1	Feeder 2	Feeder 1	Feeder 2
Min. (MW)	1.3	0.5	1.8	0.8	2.5	3.0
Max. (MW)	22.78	18.29	7.17	3.66	27.44	27.55
Average (MW)	5.13	4.61	3.79	1.55	8.58	10.12
PAR	4.44	3.97	1.89	2.36	3.20	2.72
Load Factor	0.23	0.25	0.53	0.42	0.31	0.37
Std. Dev. (MW)	3.77	4.16	1.26	0.41	4.57	5.96

5. Conclusions

Active power measurements are performed on several MV feeders in a distribution company in Turkey. The annual and daily load curves are given and demand deviations are discussed. Feeders with different type of customers (Agricultural irrigation, Commercial& Industrial and Residential) are classified and reviewed using PR and NR ratios. Since the PR values of residential feeders are higher than 1, tariff programs are found to be suitable for demand side management purposes. For agricultural feeders, instead of price based programs, strong incentive based methods are recommended. For commercial and industrial feeders, incentive programs with energy storage or local generation technologies are appeared to be suitable. According to numerical analysis, load factor is found to be the lowest in agricultural feeders. In residential feeders, LF is the highest and standard deviation is the lowest. The maximum and average power demand of commercial and industrial feeders are found to be higher than the other feeders.

Appendix

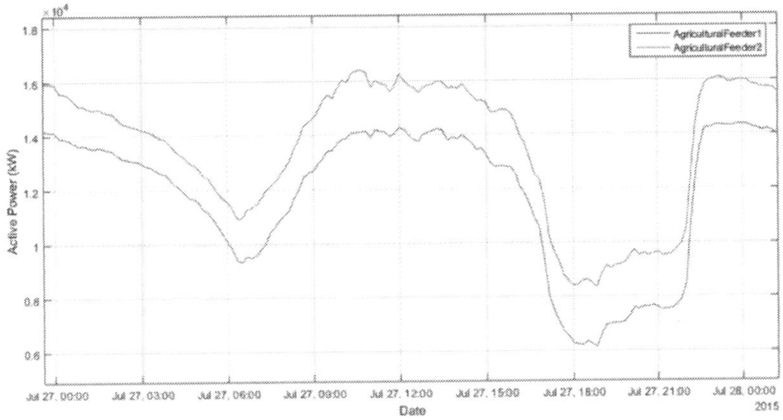

Figure A1. Daily load curves of agricultural irrigation feeders.

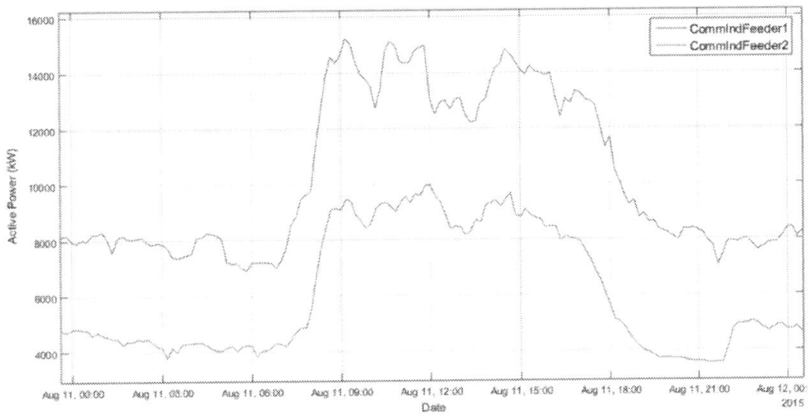

Figure A2. Daily load curves of commercial and industrial feeders.

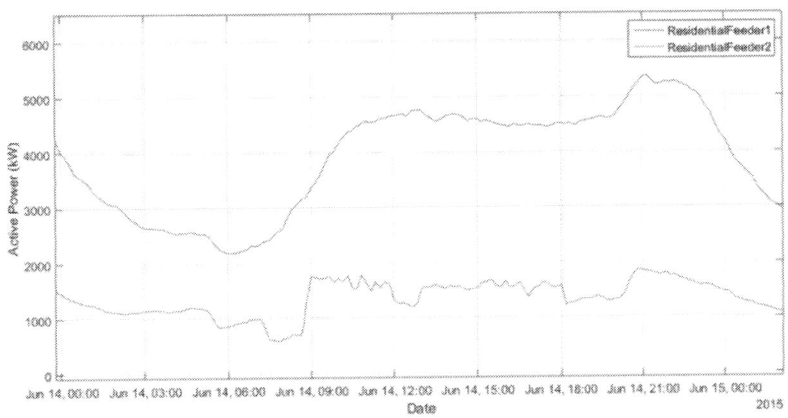

Figure A3. Daily load curves of residential feeders.

References

[1] Palensky P 2011 Demand side management: demand response, intelligent energy systems, and smart loads *IEEE Trans. on Indust. Informatics*, Vol. 7, No. 3

[2] Lampropoulos I, Kling W L, Ribeiro P F and van den Berg J 2013 History of demand side management and classification of demand response control schemes *IEEE Power & Energy Society General Meeting, Vancouver BC*

[3] Hu Z, Han X and Wen Q et al. 2013 *Integrated resource strategic planning and power demand-side management* (China Electric Power Press and Springer-Verlag Berlin Heidelberg) ISBN 978-3-642-37083-0

[4] AlOwaifeer M, AlMuhaini M and AlSaggaf A 2015 The impact of load management and renewable dg on the service restoration of future power distribution systems *IEEE 15th Int. Conf. on Environment and Electrical Engineering (EEEIC)*

[5] Mahmood A, Javaid N, Khan M A and Razzaq S 2015 An overview of load management techniques in smart grid *Int. Journal of Energy Research*

[6] Kostková K, Omelina L, Kycina P and Jamrich P 2012 An introduction to load management *Electric Power Systems Research, Elsevier*

[7] Finn P, O'Connell M and Fitzpatrick C 2013 Demand side management of a domestic dishwasher: Wind energy gains, financial savings and peak-time load reduction *Sustainable Development of Energy, Water and Environment Systems, Elsevier*

[8] Puranik S 2014 Demand side management potential in Swedish households *MSc. Thesis Chalmers University of Technology, Göteborg, Sweden*

[9] Teive R C G and Vilvert S H 2002 Demand side management for residential consumers by using direct control on the loads *5th Int. Conf. on Power System Management and Control, IET*

[10] Bhattacharya K, Bollen M H J and Daalder J E 2000 Real time optimal interruptible tariff mechanism incorporating utility-customer interactions *IEEE Trans. on Power Systems* Vol 15 No 2

[11] Caves D W, Herriges J A, Hanser P and Windl R J 1988 Load impact of interruptible and curtailable rate programs: evidence from ten utilities *IEEE Trans. on Power Syst.* Vol 3 No 4

[12] Albadi M H, El-Saadany E F 2008 A summary of demand response in electricity markets *Electric Power Systems Research, Elsevier*

[13] Electric load management industry 2009 *Leonardo Energy European Copper Institute*

[14] Eissa M M 2010 Demand side management program evaluation based on industrial and commercial field data *Proc. of the 14th Int. Middle East Power Systems Conf. (MEPCON'10) Cairo University Egypt*

[15] Electricity market tariffs regulation *Turkey Energy Market Regulatory Authority (EMRA)*

[16] Felea I, Dan F and Dzıţac S 2012 Consumers load profile classification corelated to the electric energy forecast *Proc. of the Romanıan Academy* Series A Vol 13, N 1

[17] Manasseh E, Ohno S, Kalegele K and Tariq R 2014 Demand side management to minimize peak-to-average ratio in smart grid *Proc. of the 46th ISCIE Int. Symp. on Stochastic Systems Theory and Its Applications*

IOP Conf. Series: Materials Science and Engineering **161** (2016) 012112 doi:10.1088/1757-899X/161/1/012112

Green Acquisitions And Lifecycle Management Of Industrial Products In The Circular Economy

Vasile N. POPA [1] **, Luminita I. POPA** [2]

[1]Regional Department for Defense Resources Management Studies, Mihai Viteazul street no.130, Brasov, Romania, Associate Professor

[2]Transilvania University of Brasov, Department of Automation and Information Technologies, Mihai Viteazul street no.5 , Braşov , România

Email:[1] vasilepopa2001@yahoo.com

Abstract. The article addresses the issue of green acquisitions which occur within an industrial company focused on reducing inputs while maintaining output (substitution and efficiency). These processes characterize a circular economy oriented on resource efficiency (costs saved by reducing purchasing inputs). In our article, we focus on the industrial procurement practice which can help businesses save money and materials. Besides the possibility to negotiate prices, buyers can influence suppliers to offer products and services in an efficient manner in terms of green resources. The life cycle of industrial products is used to demonstrate the environmental advantages and disadvantages of various options for acquisitions and initiatives to totally reuse them.

1. Introduction

Green acquisition refers to purchasing sustainable products with specific environmental or energy attributes. The connection between public acquisition and the environment is presented in figure 1.

Figure 1. The connection between public acquisition and the environment.

Public authorities are major consumers in Europe, spending some 16% of EU gross domestic product (a sum equivalent with a half of the Germany GDP). Using their purchasing power to choose goods and services that respect the environment, they can make an important contribution to sustainable development. Green public acquisitions cover areas such as: purchase of computers and efficient buildings in terms of energy, wood production office equipment in compliance with environmental rules, recyclable paper, electric cars, public transport more environmentally friendly, organic food in canteens, electricity from renewable energy sources, air conditioning systems complying with state of advanced environmental solutions. Green acquisition must be regarded as a concept or a notion that does not concern only the award (including activities in developing specifications and before signing the contract), but a concept that covers a set of activities that begins with identifying the need to eliminate the subject contract concluded to meet the actual need (this approach is specified otherwise in the new generation of green public acquisition legislation). It is necessary to make an appropriate distinction between green acquisition process and green acquisition procedures. Green public acquisition procedures are formal, specific and structured processes, described as "practice" for the award of a contract or the conclusion of a framework agreement therefore represents a sequence of tasks set by reference, to conduct a green acquisition process in a effective and efficient manner. Romanian public acquisition law treats a green acquisition as a process and refers to the following acquisition procedures: open tender, restricted tender, negotiated with or without publication of a contract notice, competitive dialogue and competitive solutions.

2. Methods

Addressing the issue of green acquisitions which occur within an industrial company , I have gained my results from knowledge of the acquisitions specific literature. Based on my experimental observation I have obtained an empirical model named 'Eco – Industrial Product Life Cycle Stages and ways to reduce the cost of green acquisitions matrix". I have discovered that there are there metrics that the industrial company managers have to take in consideration: distribution and manufacturing; sales; maintenance &service.

Based on my experimental observation I have obtained an empirical model named "Product Life Circular Economy Key Principles and Eco and Industrial Product Life Cycle Stages Matrix".

I have discovered that there are two metrics that the industrial company managers have to take in consideration: research and development and research eco-design and development". Also the metrics named "Optimize resource yields by circulating products, components, and materials at the highest utility at all times in both technical and biological cycles" and "More efficient industrial systems reducing pollution and negative environmental impacts "are the most important Circular Economy Key Principles in this matrix.

3. Results

3.1. Energy Efficient Green Acquisitions (EEGA)

Romanian energy efficiency resource acquisition programs seek to purchase energy savings in the public interest, often through financial or technical assistance. Although the decision process for creating these programs varies and can include government, utilities, consumer groups, and other stakeholders, it is usually a Romanian government entity that gives final approval of the volume of energy savings to be acquired and how it will be funded. Energy Efficient Green Acquisitions process has some activities as follows:

- Start event: to be equated with "*the identified and documented need for products, services or energy efficient works* ".
- end event: to be equated with " *implementation assessment of the achieved contract and extracted lessons learned*" in order to be included in the next cquisition process;
- involvement of all factory parties in the acquisitions process (not just the acquisitions department);
- Achieving the expected outcomes must help to proceedings starting with the EEA planning stage.

The activities included in the Energy Efficient Green Acquisitions EEGA are described in the table 1, below:

3.2. Green acquisitions strategies

Basically, the political decision to buy green should not pose problems for Romanian public authorities.In fact, a green acquisition process does not normally require any structural changes by the contracting authority. However, putting the policy into practice will first require some strategic

planning: organizing appropriate training for acquisition departments, ensuring the acquisition departments with access to environmental information, and setting priorities when choosing the contracts most suitable for "greening". Once this is in place, Romanian contracting authorities will be able to proceed with the proper organization of green public acquisitions procedure . The Romanian managers must assess the staff training needs and ensuring access to environmental information. Acquisitions department should be given the legal, financial and environmental knowledge they need to decide to what extent and where they can be introduced effectively, environmental conditions in the acquisitions procedures if they are set at level to get the best value for money and whether they match the environmental priorities of the contracting authority.
It is important that green purchasing policy to be communicated to a large number of stakeholders, including suppliers, contractors or service providers, for them to consider new requirements. Cooperation between acquisitions authorities is another way to improve access to knowledge and know-how in the field of environment and communicate this policy to the stakeholders.

3.3. Setting general priorities for greening your acquisitions
The acquisitions department staff needs to adopt a phased approach. This staff will start with a small range of products and services whose environmental impact is clear or where greener alternatives are easily available and not more expensive (eg recycled paper, energy- efficient office equipment).

Table 1 The activities included in the Energy Efficient Green Acquisitions (EEGA)

EEGA steps	No.	Activities
EEGA Planning	1	Identify and analyze the need for green acquisition - Gathering information
	2	**Assignment documentation** Defining the subject of contract; selection, team formation; preparation strategy.
	3	Defining and developing technical specifications and selection criteria
	4	Definition of the criteria
	5	Defining and contract development
EEGA Ongoing	6	Publication of announcement procedure
	7	Examination of tenders - Receiving and opening tenders - Checking qualification criteria
	8	Evaluation of tenders - Evaluation of technical capacity - Assignment criteria application (choosing the contracts most suitable tenders for "greening") - Life cycle cost calculation
EEGA End	9	Contract award
	10	Acquisition outcome communication
	11	Signing the contract
EEGA Post ongoing	12	Contract Management
	13	Monitoring implementation of the contract in terms of quality and quantity

3.4. Considering environmental impact
The acquisitions department staff will select those products (such as vehicle fleet) or services (such as cleaning) that have a major environmental impact. They will consider availability and cost of

IOP Conf. Series: Materials Science and Engineering **161** (2016) 012112 doi:10.1088/1757-899X/161/1/012112

environmentally superior alternatives and availability of data. In this case they must be able to get scientific and environmental data necessary to establish criteria for this product. After this, acquisitions department staff will decide what they want technically, and will express it in a call for tender.

3.5. The complex business environment in which company acquisition department is breaking into a new market

A company breaking into a new market with its current product range might need to use innovation in order to manage EEGA contracts (table 1, activity 12) and acquire new manufacturing or sales capabilities (figure 2).

3.6. Green acquisitions goals and ways to reduce costs

Green acquisitions goals are as following: increased profitability or energy efficiency products / services purchased; reducing the cost of green acquisition. Green acquisitions goals and ways to reduce costs are as following:
Reducing prices: merging orders; negotiating more favourable agreements; buying larger quantities; providing long-term contracts; sharing information; adopt reciprocity method;
Reducing costs: reducing the number of suppliers; improving energy efficiency systems; increase employee efficiency; implementing TQM method; implementing JIT method; Implementing E-Acquisitions.

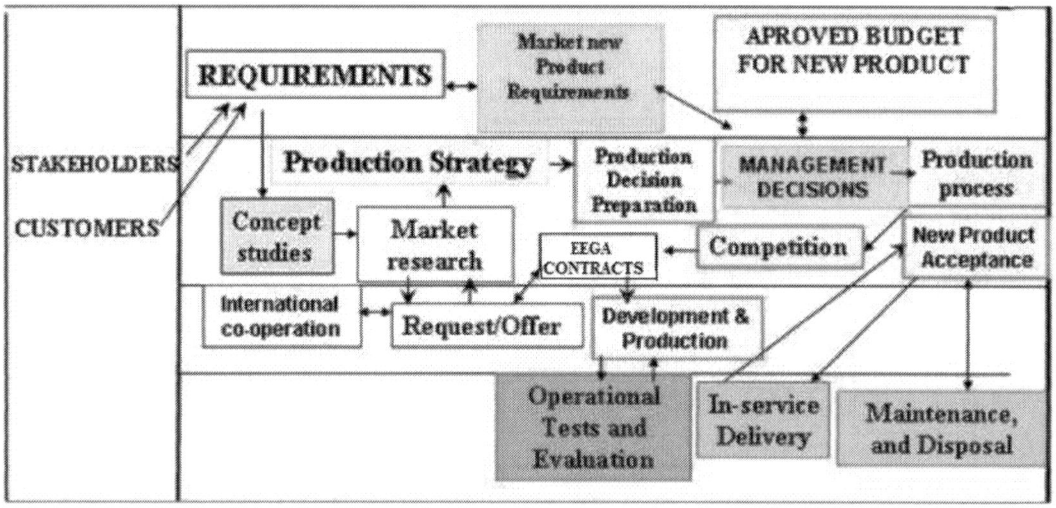

Figure 2. The complex environment in which a company acquisition department is breaking into a new market (source [1] figure 2, page 3)

3.7. Eco – Industrial Product Life Cycle Stages

Recent decades have brought with intense technological development and about the impact of material production and services on the environment, on human health of all living beings and all environmental components. For this reason, all entities producing goods and services, must pay particular attention to how the product works, provide services, mode of operation / use, disposal or recovery / recycling after disposal. This involves analyzing each stage of manufacture of an industrial product and to find out ways to reduce the cost of green acquisitions. The analysis must yield to the environmental impact assessment and the community throughout the product chain of existence. It is improperly called life cycle, concerned not only a material thing or service, but suggestively, was adopted the same formula as in the case of living beings. Right it should be formulated as "Eco – Industrial Product Life Cycle" of the product, from manufacturing to disposal, after use. If outline the

IOP Conf. Series: Materials Science and Engineering **161** (2016) 012112 doi:10.1088/1757-899X/161/1/012112

life cycle of a material good, we can better understand and assess the impact on the environment and community at stages of production, operation and decommissioning then.
We have realized " Eco – Industrial Product Life Cycle Stages and ways to reduce the cost of green acquisitions Matrix" (table 2) in order to identify the relationship between the ways to reduce the cost of green acquisitions (reducing prices and reducing costs) and Eco – Industrial Product Life Cycle Stages. The Eco – Industrial Product Life Cycle Stages are the follows: market needs for green industrial products; research and development; idea generation; opportunity identification and concept definition; research eco-design and development; prototype &production; distribution and manufacturing; eco-marketing; sales; maintenance &service; products feed-back; removal, disposal and recycling.
As we can see in the matrix" distribution and manufacturing" lifecycle stage" has the highest score: 11. Another two lifecycle stages ("Sales; maintenance &service") have a high score: 9. In this case, the industrial company managers have to take in consideration that these three metrics are very important, because they could affect the entire eco –industrial product life cycle. Also the metrics named" Sharing information" (score 9) and" Implementing TQM method" (score 8) are the most important ways to reduce the cost of green acquisitions.

Table 2 Eco – Industrial Product Life Cycle Stages and ways to reduce the cost of green acquisitions Matrix (source [1] figure 2, page 4)

		Ways to reduce the cost of green acquisitions													
		Reducing prices						**Reducing costs**							
	Eco – Industrial Product Life Cycle Stages	Merging orders	Negotiating more favourable agreements	Buying larger quantities	Providing long-term contracts	Sharing information	Adopt reciprocity method	Reducing the number of suppliers	Improving energy efficiency systems	Increase employee efficiency	Implementing TQM method	Implementing JIT method	Implementing E-Acquisitions	**Total scores**	**Weighted score (%)**
(1)	(2)	3	4	5	6	7	8	9	10	11	12	13	14	15	16
1	Market needs for green industrial products	-	**X**	-	**X**	**X**	**X**	-	-	**X**	**X**	**X**	**X**	**8**	**12.13**
2	Research and development	-	-	-	-	**X**	**X**	-	**X**	-	**X**	**X**	**X**	**6**	**09.09**
3	Idea generation	-	-	-	-	-	-	-	**X**	-	**X**	**X**	-	**3**	**04.54**
4	Opportunity identification and concept definition	-	-	-	-	**X**	**X**	-	**X**	-	-	-	**X**	**4**	**06.06**

5	Research eco-design and development	-	-	-	-	-	-	-	X	-	X	X	X	4	06.06
6	Prototype& production	-	-	X	-	-	-	-	X	X	X	X	X	6	09.09
7	Distribution and manufacturing	X	X	X	X	X	X	X	X	X	X	X	-	11	16.66
8	Eco-Marketing	-	X	-	-	X	-	-	-	-	-	-	-	2	03.03
9	Sales	X	X	-	X	X	X	-	-	X	X	X	X	9	13.64
10	Maintenance &service	X	X	X	X	X	-	X	-	X	X	-	X	9	13.64
11	Products feed-back	-	-	-	-	X	-	-	-	X	-	-	-	2	03.03
12	Removal& disposal and recycling	-	-	-	-	X	-	-	-	X	-	-	-	2	03.03
	Total scores	**3**	**5**	**3**	**4**	**9**	**5**	**2**	**6**	**7**	**8**	**7**	**7**	**66**	**100**
	Weighted score (%)	**04.54**	**7.57**	**04.54**	**06.06**	**13.64**	**7.57**	**03.03**	**09.09**	**10.61**	**12.13**	**10.61**	**10.61**	**100**	**100**

3.8. Circular economy key principles (table 3, source [3])

In the case of industrial products, circular economy is based on three key principles:

Principle 1: Preserve and enhance natural capital by controlling finite stocks and balancing renewable resource flows. This starts by dematerializing utility - delivering utility virtually, whenever optimal. When resources are needed, the circular system selects them wisely and chooses technologies and processes that use renewable or better-performing resources, where possible. A circular economy also enhances natural capital by encouraging flows of nutrients within the system and creating the conditions for the regeneration of, for example, soil.

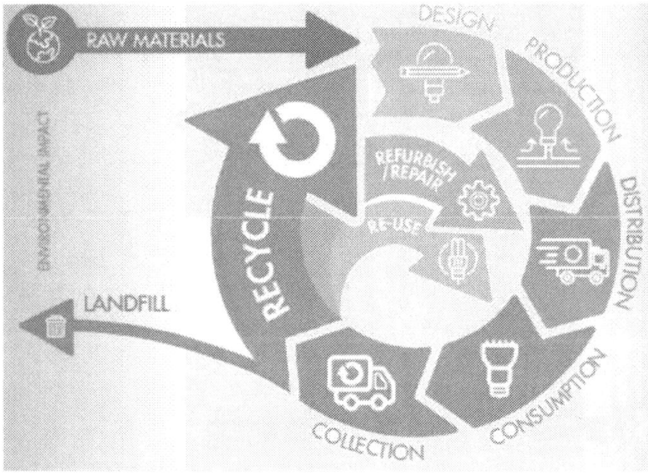

Figure 3. Circular economy concept. Source [2]

Principle 2: Optimize resource yields by circulating products, components, and materials at the highest utility at all times in both technical and biological cycles.
This means designing for remanufacturing, refurbishing, and recycling to keep technical components and materials circulating in and contributing to the economy. Circular systems use tighter, inner loops (e.g. maintenance, rather than recycling) whenever possible, thereby preserving more embedded energy and other value. These systems also maximize the number of consecutive cycles and/or the time spent in each cycle, by extending product life and optimizing reuse. Sharing in turn increases product utilization. Circular systems also encourage biological nutrients to re-enter the biosphere safely for decomposition to become valuable feedstock for a new cycle. In the biological cycle, products are designed by intention to be consumed or metabolized by the economy and regenerate new resource value. For biological materials, the essence of value creation lies in the opportunity to extract additional value from products and materials by cascading them through other applications. As in any linear system, pursuing yield gains across all these levers is useful and requires continued system improvements. But unlike a linear system, a circular one would not compromise effectiveness.
Principle 3: Foster system effectiveness by revealing and designing out negative externalities. This includes reducing damage to systems and areas such as food, mobility, shelter, education, health, and entertainment, and managing externalities, such as land use, air, water and noise pollution, and the release of toxic substances. In our industrial case that means more efficient industrial systems by reducing pollution and negative environmental impacts.
We have realized "Product Life Circular Economy Key Principles and Eco and Industrial Product Life Cycle Stages Matrix" (table 3) in order to identify the relationship between Circular Economy Key Principles and Eco – Industrial Product Life Cycle Stages.
As we can see in the matrix " Research and development" and "Research eco-design and development" lifecycle stages" have the highest score: 3. In this case, the industrial company managers have to take in consideration that these two metrics are very important, because they could affect the entire eco –industrial product life cycle. Also the metrics named "Optimize resource yields by circulating products, components, and materials at the highest utility at all times in both technical and biological cycles" (score 9) and "More efficient industrial systems reducing pollution and

negative environmental impacts" (score 10) are the most important Circular Economy Key Principles in this matrix.

Table 3. Product Life Circular Economy Key Principles and Eco and Industrial Product Life Cycle Stages Matrix (Adapted by the authors from source: [1]; [3])

No.	Eco – Industrial Product Life Cycle Stages	Circular Economy Key Principles (industrial products): **Preserving and enhancing natural capital stocks** by controlling finite and renewable resources balancing flows	**Optimise resource yields** by circulating products, components, and materials at the highest utility at all times in both technical and	**More efficient industrial systems** reducing pollution and negative environmental impacts	**Total scores**	**Weighted score (%)**
(1)	(2)	(3)	(4)	(5)	(6)	(7)
1	Market needs for green industrial products	X	-	X	2	**8.69**
2	Research and development	X	X	X	3	**13.05**
3	Idea generation	-	X	X	2	**8.69**
4	Opportunity identification and concept definition	-	X	X	2	**8.69**
5	Research eco-design and development	X	X	X	3	**13.05**
6	Prototype, production	-	X	X	2	**8.69**
7	Distribution and manufacturing	X	X	X	3	**13.05**
8	Eco-Marketing	-	-	X	1	**4.35**
9	Sales	-	-	X	1	**4.35**
10	Maintenance /service	-	X	-	1	**4.35**
11	Products feed-back	-	X	-	1	**4.35**
12	Removal& disposal and recycling	-	X	X	2	**8.69**
Total scores		4	9	10	23	**100**
Weighted score (%)		**17.40**	**39.13**	**43.47**	**100**	**100**

4. Conclusions

Green acquisitions is also about influencing the green market. By promoting green acquisitions, public authorities can provide industry with real incentives for developing green technologies. In some sectors of industrial productions, the impact can be significant, because public purchasers command a large share of the market (computers industry, energy efficient buildings, public transportation industry etc.). If we take into account the costs of an entire Eco – Industrial Product Life Cycle of a contract, green public acquisitions give the opportunities to save public money and to protect the environment at the same time. By acquiring wisely, we can save materials and energy, we can reduce waste and pollution and encourage sustainable patterns of consumer behavior.

References

[1] Luminiţa I. POPA and Vasile N. POPA, 2013 *PLM Innovation Matrix for a Complex Product Development Process*, *Applied Mechanics and Materials* (Vol. 371) ed Trans Tech Publications (Switzerland) pp 862-866

[2] Managementul deşeurilor de echipamente de iluminat. Bucureşti, 7 aprilie 2016; Forumul industriei de iluminat 2016, Govnet. Downloaded from http://govnet.ro/uploads/files/54_Govnet_Recolamp.pdf

[3] Towards a circular economy: business rationale for an accelerated transition , November 2015 Published by the Ellen MacArthur Foundation. Charity Registration No. 1130306; OSCR registration no. SC043120; EU transparency register N°389996116741-55. Downloaded from https://www.ellenmacarthurfoundation.org/assets/downloads/TCE_Ellen-MacArthur-Foundation-9-Dec-2015.pdf

[4] European Commission Buying green 2016 *A handbook on green public procurement*, 3rd Edition February Luxembourg: Publications Office of the European Union

[5] United Nations 2014 *Sustainable procurement in the UN system* (Annual Statistical Report on United Nations Procurement)

Public funding - a solution for technological SMEs and entrepreneurs

Calefariu Emilia[1, a], **Buda Traian Alexandru**[2, b]

[1] Braşov, Romania. Morii Str., Nr. 8, Jud. Brasov, Loc. Cristian Cod: 507055.
[2] Braşov, Romania. Aleea Lăcrămioarelor, Nr. 3, Bl. 10, Sc. C, Ap. 14, Brasov, Cod 500218

E-mail: [a] emiliacalefariu@yahoo.com, [b] traian.buda@unitbv.ro

Abstract. Technological SMEs are in a constant struggle for growth or sometimes maintaining the production capacity, increase market share, supporting tax burden, ensuring employees' salaries, profit growth. They constantly consider short-term survival of the company, with trying to maintain a long-term uptrend for the business. Entrepreneurs are again in a position to access public financing under quite favorable conditions. The paper aims to analyze the opportunity of accessing these external financing options, which have both advantages and a series of long-term constraints which should not be excluded before the final decision to access this funding. New research is required, thus sustainable development can be maintained for the business environment in order to increase efficiency, competitiveness, sustainable development strategies that generate job security, regional growth and rewarding the risk-taking of the entrepreneur.

1. Review of SME Financing Possibilities on Projects / Grants

Companies have the possibility to access different possible sources of funding, which are coming from institutions or governments, such as: the European Union (Structural Funds and Cohesion, research and development , European Defense Agency); North Atlantic Treaty Organization (initiatives for cooperation and development, science for world peace and security); Romania / East-European countries (public-private partnerships and other national programs) and other different possible regional sources (grants programs from Nordic countries, bilateral cooperation).

European Union cohesion policy involves implications for SMEs through research and innovation, economic competitiveness, information and communication technology or environmental protection [1]. These implications were drawn for 2014-2020 year interval, by the entrepreneurial operational programs, involving increased competitiveness through innovation [2,3].

1.1. Considerations on public funding

European funds represent a complex and very appealing financial instrument. Identifying funding solutions implies fallowing strict rules by the laws and European practices. By applying the Europe 2020 strategy, it is aimed to stimulate competitiveness for SMEs / companies, states, and even small businesses, sustainable development and research with emphasis on rewarding the performance

[4,5,6]. Expanding financing solutions represents, at the macroeconomic level, an element of flexibility and efficiency.

New lines of research should be directed towards the most captivating segments of interest, for example: highlighting the specific moments that are most likely for strategic moves, including a detailed analysis of the pros and cons for investing in the specifier area; the potential for obtaining financing; identifying funding source for obtaining higher yields; identifying the role of utility functions for value creation; underlining how business models affect strategic success and sustainability by designing an inappropriate manner the number of hierarchical levels.

Management structures of SMEs are characterized by low bureaucracy compared with the large organizations and a team of stakeholders to support innovation within the company. Spin-offs help SMEs go beyond research and development encountered. The knowledge generated in large organizations is used and developed in the spin-off, which is marketed. Entrepreneurial nature of the organization and its ability to innovate include: the type of innovation (products, services) industry, generated by innovation and technological change management process that supports innovation. Measuring technological change can be effected by various indicators such as R & D. SMEs can more innovation in areas such as IT, while large firms, which have funds, are leading innovators in the pharmaceutical industry or aviation. For both, finding the best funding solution is mandatory in order to succeed on the market.

1.2. The purpose of public funding

Europe 2020 priorities relate to: intelligent development (digital agenda, innovation); sustainable growth (resource efficiency); inclusive growth (new skills). EU cohesion policy have implications on companies through research and innovation, economic competitiveness, information and communication technology, environmental protection, in terms of regional development and social robust implications.

Table.1 Important European objectives for 2020

Nr. crt.	Objectives	Implications for SMEs
1	Raising the employment rate	- raising the employment rate for above 75% of the active population - reducing unemployment after graduation - assisting people over 50 in finding a job
2	Raising R & D	- allocation 3% of EU GDP for R & D - stimulating R & D by creating special programs for researchers and fiscal facilities
3	A more efficient use of energy	- a minimum of 20% reduction in emissions of greenhouse gases - a minimum of 20% increase in energy efficiency - increasing the share of renewable energy sources to 20% both for companies, but also for private houses
4	Education	- reducing early school leaving rate to under 10% - increasing the share of graduates to 40% for people aged 30-35 - easier access to professional reconversion - ease of access to education for the rural area population
5	Poverty reduction and social exclusion limitation	- reducing by 20 million the number of people suffering from poverty or at risk of suffering - cultural awareness and limitation of discriminatory treatment

Proposals aimed at improving the entrepreneurial business environment, entrepreneurial culture, education, access to finance and contracts awarded by state regulation and taxation, coordinated support. Possible solutions are considering substantial improvement in the absorption of European funds, developing public-private partnerships PPP, reducing the share of social spending in the budget to increase fiscal space for public investment dedicated, multi-annual budgeting and prioritization of investments [7]. Entrepreneurs' investments can be founded in various manners. The ones that are often used are obtaining financing from a credit institution / bank, obtaining a loan from a business angel, from the government, from the European programs. There is a connection between bank lending and financing through European funds in order to provide co-financing; eligible costs can be funded by grants (up to cashing them from the managing authority), expenses representing customer contribution; there is the possibility of co-financing. For start-ups, bank financing conditions are more drastic given that this history cannot be presented. In those cases, other solutions need to be looked for.

2. Time frames regarding public funding

In order to create a more quantitative view over the process of public funding accessibility, we propose a model that analyses the time frames that the process deals with. These time intervals are very valuable to the SME / company that is aiming at obtaining funds in this manner, because of the speed that information, technology and culture face.

Figure 1. Time intervals that occur when aiming at financing through public funds

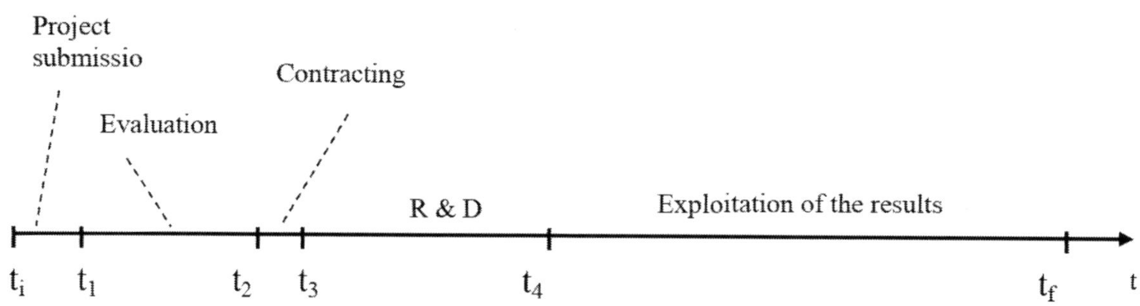

The moments that occur during the process of accessing public funds are:

- t_i – the initial time, the one when the company takes the decision to research and develop a new project and needs financing it from public funds. This moment is called "the call" and is the same as the launch of the competition for projects; this moment must be preceded by market evaluation and opportunity studies of the project;
- t_1 – represents the deadline, the moment when the project must be submitted (announced from the beginning of the programme);
- t_2 – is the moment when the results are communicated on the online platform;
- t_3 – represents the contracting and the actual start of research and development;
- t_4 – is the time at which the planned R & D is completed;
- t_f – is a constants moment in time; it is the moment when the results of the project can be effectively commercialized on the market.

2.1. Funding efficiency

Choosing to obtain public funding and not looking for other financing solutions is a matter of intense reflection and close look on the efficiency of the project. Different other criteria need to be considered, such as:

- Difficulties in accessing bank loans;
- Access requirements to credits: guarantees (sometimes including personal possessions), unfair terms / constrictions imposed by banks, high penalties etc.
- The chances of success in obtaining public funding for the project, considering the funds allocated and the number of projects that will be proposed for funding on the same field of interest;
- Analysis on the difficulties involved with project management;
- Capacity of handling multiple controls and dealing with a number of bureaucratic papers;
- Consideration of the public authority's decision to reduce funds, following the signing of the contract;
- A realistic planning on the SMEs expansion (capital growth, increasing market share, turnover, profit).

Public funding is not in all cases, contrary to appearances, the most advantageous solution for a company. An entrepreneur aims a fundamental condition for the project to be a success and for an efficient spending of public money, regardless of their source. He's target is to obtain profit. Therefore, including verification as the one proposed, for the data in the assessment criteria for publicly funded projects would lead to a more accurate assessment of them.

2.2. The need to implement this solution for SMEs

Technological entrepreneurship represents the sequence of processes through which organizational resources are assembled, and technological systems with the strategies of targeting the opportunities. Technological entrepreneurship involves the identification of high potential, of commercial opportunities associated with intensive technologies, superior resource fructification and prudent management of fast dynamics through elite decision-making skills. Technological Entrepreneurship fructifies unique abilities (the agility in a constantly changing environment, team spirit, openness to innovation, passion for top areas) (Roberts, 1989). Motivation is essential and is based on the desire for independence (liberty and control, overcoming bureaucratic barriers, pursuit of target achievement), wealth and exploitation (Oakey, 2003).

Modern technological SMEs are characterized by: innovation capabilities and creative thinking encouragement; balanced organizational structure that encourages flexibility and experimentation of new architectures; quality of human resources; a creative environment based on innovative structures at all levels; a good orientation on customers and investors. A volatile external environment implies special attention on adaptability and flexibility. Entrepreneurship and promoting its growth is an important goal for many governments. A more unified perspective for how studies and samples fit together needs to be studied in order to satisfy the need to define for policy makers [11].

The development of technological SMEs is sustained by innovation (at the level of products and services, as well as organizational management) and value creation takes place by leveraging differential performance and focus on opportunities - efficient use of resources and the formulation of strategic options, the creation of new markets or business.

Technological innovation of SMEs takes into account the type of innovation (products, services), the industrial sector, and the technological change generated by innovation and the process of management of innovation and is expressed by the total number of innovations made relative to the number of employees. MacPherson (1997) indicates a directly proportional relationship between the performance of a SME and the external support, of scientific nature, technical and vocational. In contrast, Didd (2009), criticizes the existence of a directly proportional relationship between CDI and the development or the profitability, identifying other relevant factors, such as: the level of managerial experience and the technological know-how of the entrepreneur, the propensity to stimulate innovation or the proactive activities.

3. ***Problems encountered on public financing and solutions* to solving them**

Grant beneficiaries have complained about problems related to delays in the granting of advances and repayments and there were cases where beneficiaries have resorted to bank loans. Procurement procedures have created some problems, such as: the discriminatory criteria used to procurement, misuse of selection criteria and qualification, communicating the proceedings. Also attainment of the objectives attracted reduction or even cessation of funding.

Implementation issues were considered: the verification process of public procurement, shifting deadlines of the stages, the difficulty of obtaining loans necessary to pay for work that did not receive pre-financing, inconsistencies between the information included in projects execution and the actual situation on the ground, multitude of riders, not updating documentation etc.

For a SME, understanding the concrete funding opportunities is vital in strategic decision context at the time, namely selection and blending of the most effective variants adapted to the peculiarities of the company.

Another problem that could occur is the use as a rating factor in the duration of execution of works in the public procurement procedure works. This problem leads to financing the projects with small execution times that are unrealistic, leading to addenda extending the duration of execution of the works contract. This situation may involve a violation of principles of equal treatment and nondiscrimination, which may lead to ineligibility of expenses reimbursed under the contract.

The development of tender documentation with incomplete outdated technical information that is not consistent with the overall situation on the ground since launching the procurement procedure is another aspect that needs to be taken into consideration. This situation leads to an issuance by the manufacturer, right from the start, of additional orders notes or waiver notes. Depending on the outcome of the review required in settlement amounts, if there is additional material / maneuvers that were not tendered or not traded in an addendum become ineligible. The ineligible costs when accessing public funding need to be extremely careful supervised, because these costs are entirely supported by the company.

4. Conclusions

Productive enterprises financing problem involves an objective analysis of the parameters involved in accessing such a program. The analysis method proposed in this paper implies a comparative analysis of the advantages and disadvantages of accessing public funds and an analysis of the temporal opportunity to obtain financing. Adding value involves the integration of diverse management skills, taking into account the available resources portfolio, the extraordinary dynamics of the market, but also analyzing rationally the possibility of future development of the company. These opportunities involve both an analysis of the field of production, technological equipment required and the possible development on a highly competitive market. Development strategies and market penetration are permanently looking to improve its approach in terms of financial and achieving optimal positioning for obtaining value, given the continuing diversification opportunities.

The proposed method is a decision support for technological SMEs, especially in the South-eastern Europe, an area that benefits from financial support at European level. The approach shows an easy way to understand the problems faced by SMEs and proposes new ways of solving the technological needs and the financial ones, both long term and short term.

References

[1] Buda T, Calefariu E, Gheorghe C and Calefariu G 2014 *Noi aspecte privind decizia de investiție în sistemele de fabricație*, RECENT, Vol. 15, no. 2(42), pp 64-72.
[2] Nicolescu O. 2000 *Management methods and techniques of systems organization (Sisteme metode si tehnici manageriale ale organizatiei)*, Ed. Economica.
[3] Graevenitz G. von, D. Harhoff R. Weber. 2010. *The effects of entrepreneurship education*, Journal of Economic Behaviour & Organization, 76, 90–112.
[4] Calefariu G 2000 *Production systems - theory and applications* (in Romanian), Lux Libris, Brasov, 2011.
[5] Nicolescu O. *Management methods and techniques of systems organization* (Sisteme metode si tehnici manageriale ale organizatiei), Ed. Economica.

[6] Singh M, Shankar R, Narain R and Agarwal A 2003. *An interpretive structural modeling of knowledge management in engineering industry*. Journal of Advances in Management Research, 1(1), 28-40.

[7] Prelipean G, Boscoianu M, Calefariu E and Lupan M 2014 *Innovative instruments for SME financing in Romania - a new proposal with interesting implications on markets and institutions*, The 3rd International Conference "Emerging Market Queries in Finance and Business", Universitatea Petru Maior Tg. Mureş, Bucureşti, 24th-25th, Octombrie.

[8] Fritsch M, Noseleit F and Schindele Y 2011, *Success or failure? Business- , industry- and region- specifi c determinants of survival – A multidimensional analysis for German manufacturing*, mimeo, Friedrich Schiller University, Jena.

[9] Lofsten H and Lindelof P 2005, *R&D networks and product innovation patterns: academic and nonacademic new technology-based firms on Science Parks*, Technovation, 25 (9), 1025–37.

[10] Van Stel A, Storey D, Thurik AR. 2007, *The effect of business regulations on nascent and young business entrepreneurship*, Small Business Economics, 28 (2), 171–86.

[11] Kerr WR and Ramana Nanda 2011 *"Financing constraints and entrepreneurship."* in Audretsch, D. B., Falck, O., & Heblich, S. (Eds.). Handbook of research on innovation and entrepreneurship. Edward Elgar Publishing.

IOP Conf. Series: Materials Science and Engineering **161** (2016) 012114 doi:10.1088/1757-899X/161/1/012114

Support system for generating statistics of existing railway rolling stock

A Wróbel[1], M Płaczek[1] and A Buchacz[1]
[1]Silesian University of Technology, Faculty of Mechanical Engineering,
Institute of Engineering Processes Automation and Integrated Manufacturing Systems,
Konarskiego 18A, 44-100 Gliwice, Poland

E-mail: andrzej.wrobel@polsl.pl@polsl.pl

Abstract. Rail transport is a very important element of the transport system in European countries. Assumptions of rail transport has become more and more attractive in both passenger and freight transport. An example of this may be just noisy action "lorries on track", the promotion of intermodal transport. The principal factors affecting the condition of railway traffic safety are: the technical condition of the railway infrastructure, the technical condition of the rolling stock, the operation of crossings. The main purpose of this article is to present a tool to assist generate statistics operation railway wagons. As part of the project telemetric module allows the location of the freight wagon and the state of its load. Computer program supports GPS module will enable the location of a wagon with an accuracy of 1 meter, generate graphs of speed, road grade and load freight wagon.sheet".

1. Introduction

Rail transport is one of the most important modes of transport in Poland. In 2011, rail carriers transported more than 250 million passengers and 250 million tons of goods.

They are carried mainly coal, coke, oil and metal ores. The advantage of rail transport compared with passenger automotive are small costs on the assumption longer routes, and the speed of execution of transport. An additional advantage of rail transport compared with the electric motor transport is lower emissions of air pollutants. Rail transport also has drawbacks for example not hand over the goods to any place. An additional advantage of rail transport compared with the electric motor transport is lower emissions of air pollutants.

Very popular in recent years has become the carriage of goods using more modes of transport, the so-called intermodal transport. A transport unit, for example a lorry uses the road on the start and end part of the route, while the rest part of the route is performed by, for example rail transport.

Utilitarian objective of combined transport is primarily to protect the environment. At the Faculty of Mechanical Engineering Technical University of Silesia are conducted research on the analysis and synthesis of mechanical systems and mechatronics [2-4, 6, 9, 10], concern both theoretical [3,4,12,16] considerations how practical industrial tasks [7, 17]. Recent works undertaken relate to research the properties of composite materials [9,10] as components and assemblies railway wagons. Some of the work is conducted in collaboration with other research centers [5,7]. In this article, authors use also the knowledge and experience of other research centers [8, 13-15].

Recently, the authors of the article are working on a project carried out by the consortium composed of: Institute of Engineering Processes Automation and Integrated Manufacturing Systems from the Silesian University of Technology, and DB Schenker and Germaz companies. The aim of this research project is to modernize the analyzed wagon during its renovation using new materials. The main objective of the work is modernize of the wagon during its renovation using new innovative composites in order to protect sheathing of the wagon against corrosion.

Modern cars are manufactured from standard materials such as steel plates and profiles. During meetings with representatives of the leading transport companies the following issues concerning the application of the existing structure were reported. The material being carried in the winter term freezes to the surface plates, and makes it impossible to rapid unloading. In order to unload the wagon it is necessary to warm up the walls or use mechanical tools for separation of the material. However, the biggest problem is the corrosion of metal parts. This concerns mainly plating the walls and floor of the wagon. As part of its work use of innovative composite materials as components or assemblies freight wagon in order to eliminate the presented problems.

This work is a continuation of the previous research works concerned with analysis of new, composite materials application if freight wagons.

One of the tasks was undertaken to establish a system for monitoring and collection of data on freight wagons. Of course, such systems exist, but the data to the database are entered manually by the staff. This article proposes a method of automatic data collection and provide statistics crossings railway wagons. The most important data from the point of view of the representatives of the railway lines were long trips, travel time, the degree of loading wagon.

2. The device for automatic data archiving

Proposed a system called telemetry system is capable for locating freight wagon with an accuracy of one meter, also it has the ability to collect from piezoelectric sensors. Piezoelectric sensor which is glued to the frame of the wagon can indicates if the car is moving or stopped on a siding. It also has another task, on the basis of the signal of this sensor, we can conclude whether the car is moving loaded or empty. During the tests voltage was simulated by potentiometer P1 at the measuring input analog AN3 (figure 1). System after 5 minutes of inactivity goes into sleep mode, which allows for significant extension of battery life. Analog input marked on the diagram AN3 has a measuring range of 0-30V and input resistance 230kΩ. Optocoupler input with series resistor 2,7kΩ a minimum value of switching voltage of 3.5V and a minimum current of 0.5mA optocoupler switching. For higher voltages, to limit the current, must-core select the appropriate additional series resistor.

IOP Conf. Series: Materials Science and Engineering **161** (2016) 012114 doi:10.1088/1757-899X/161/1/012114

A) B)

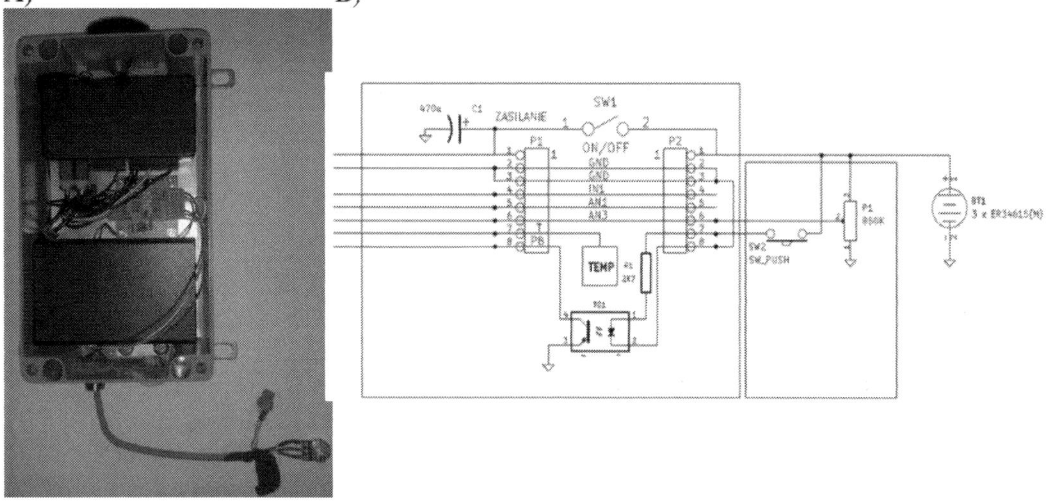

Figure 1. Telemetric system (A), wiring diagram analog sensor (B)

As part of the first attempts it was assumed that the data collected by the system will be as presented on figure 2.

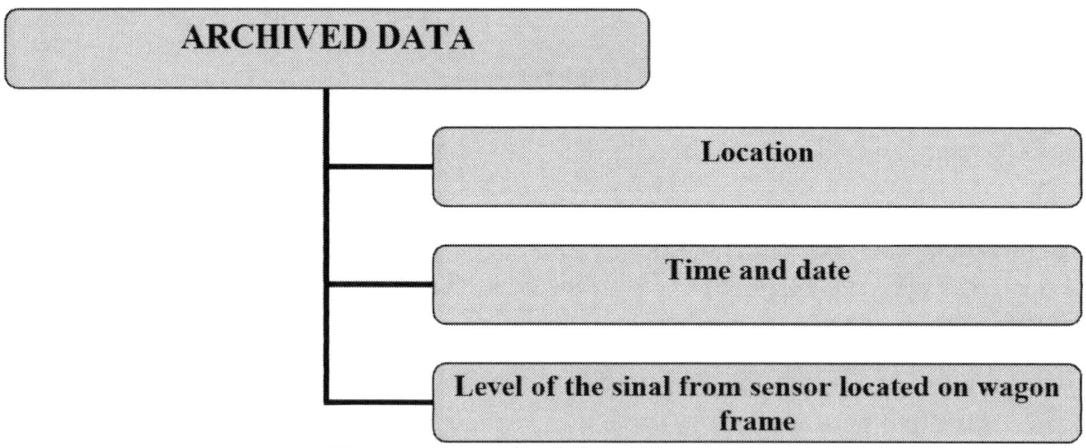

Figure 2. Archived data by system

Entering the the server address in any web browser to obtain the view of the map. Then user can choose which car will be monitored and in what period of time. After indicating the dates the system automatically generates a route as shown in figure 3. The positioning accuracy is dependent on the factory settings, but it is worth noting that increasing the sampling frequency significantly reduces battery life.

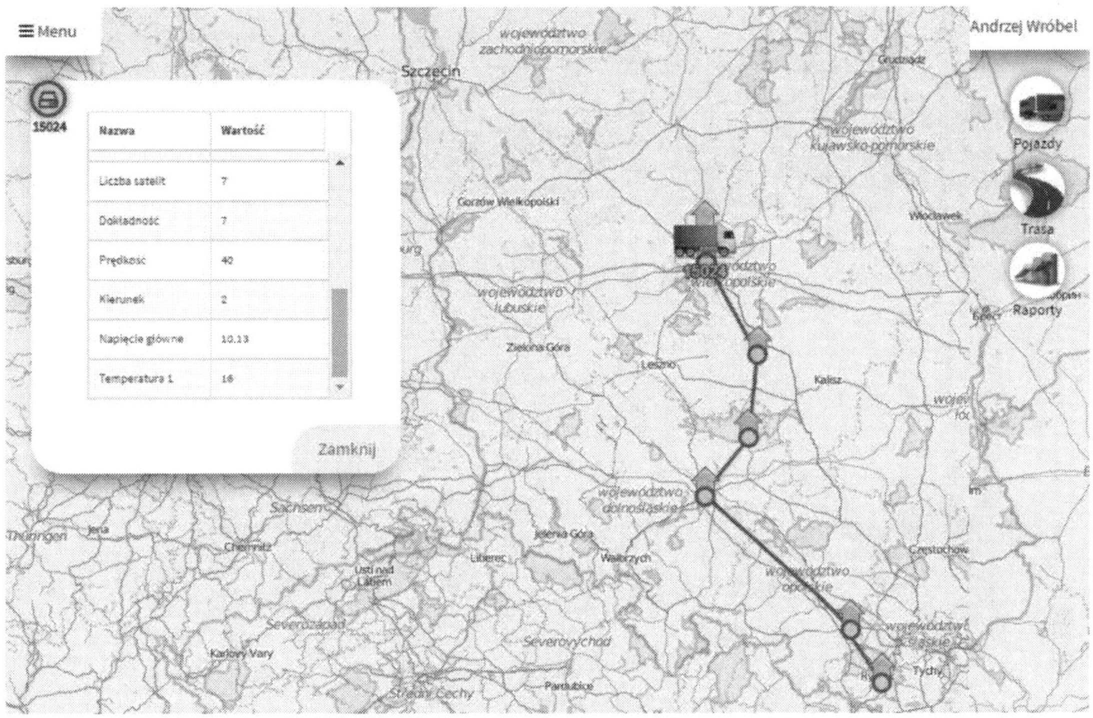

Figure 3. Sample route monitored wagon

On figure 3 characteristic points of the route was marked. After clicking on the selected point left window (figure 3) with main parameters of the route. In this window is indicated value of the signal from sensor for example 10,13V, which means that the car is about 33% loaded.

3. Laboratory stand for first tests

The first tests the device to collect telemetry data were carried out on a special stand made to scale. This the stand was prepare for study the possibility of changing the shell of a freight wagon was presented. In order to confirm or deny the possibility of using these materials model car and drive unit was created. The wagon was designed to allow easy change his shell. To change a travel parameters as: acceleration, delay, variable speed travel, the electrical control box with inverter were established. In figure 4 model for the analysis of observed driving was shown.

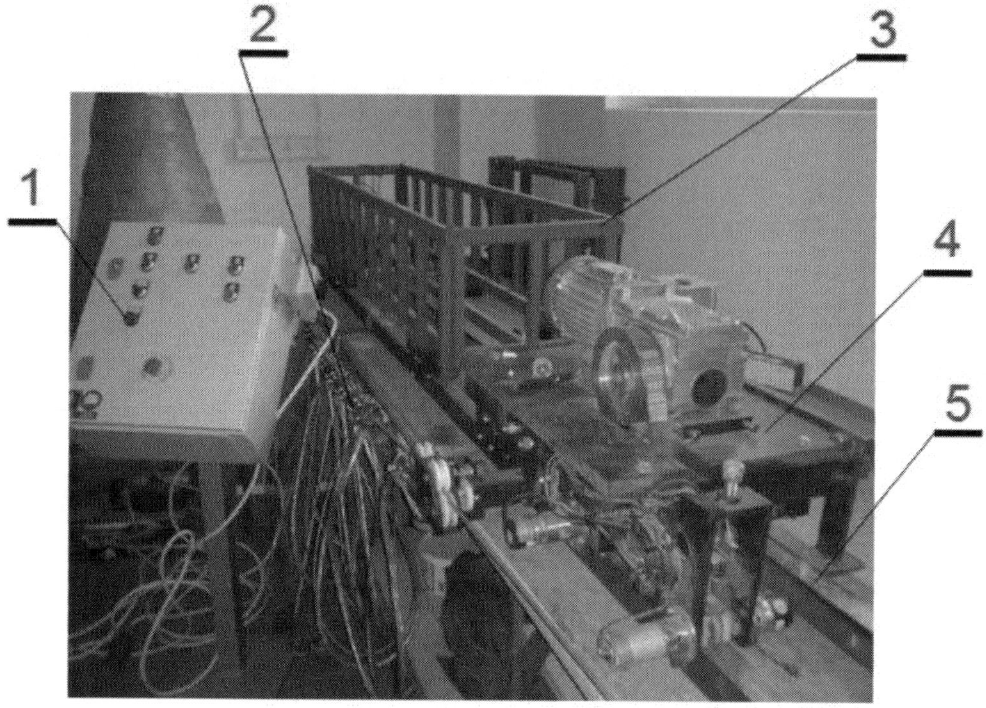

Figure 4. Sample route monitored wagon

At presented figure 4 it was indicated the main components of the laboratory stand by the numbers. The control unit (1) contains residual current devices and overcurrent protection necessary for safe operation of the station. In order to determine the running speed, acceleration and deceleration times and the number of cycle travel LG inverter was used. Guides for power cables are marked (2). The drive unit with an electric motor (3) and a model railway carriage in the scale (4) which are moving on tracks (5)

4. Summary

This article shows how to monitoring by GPS system location and load condition of the freight wagon. The obtained statistics can be automatically transferred to the database at any time and collected by the service. Presented location system is not a new concept, but only indicates the possibility of a new use of previously used solutions in the industry.

Acknowledgments. The work was carried out under the project number PBS2/A6/17/2013 agreement implemented under the Applied Research Program, funded by the National Centre for Research and Development.

REFERENCES

[1] Baier A. et al 2012 Experimental synthesis and analysis of geometric and structural properties of chosen elements of railway wagons, *Silesian University of Technology Publishing House*, Gliwice.

[2] Białas K 2012 Mechanical and electrical elements in reduction of vibrations, *Journal of Vibroengineering*, **14/1** 123-128.

[3] Bruni S et al 2011 Modelling of suspension components in a rail vehicle dynamics context, *Vehicle System Dynamics*, **49/7** 1021-1072.

[4] Buchacz A Płaczek M and Wróbel A 2014 Modelling of passive vibration damping using piezoelectric transducers – the mathematical model, *Eksploatacja i Niezawodnosc – Maintenance and reliability*, **16/2** 301–306.

[5] Connolly D P et al 2014 Assessment of railway vibrations using an efficient scoping model, *Soil Dynamics and Earthquake Engineering*, **58** 37-47.

[6] Dymarek A and Dzitkowski T 2012 Method of active synthesis of discrete fixed mechanical systems, *Journal of Vibroengineering*, **14/2** 458-463.

[7] Klarecki K Hetmańczyk M and Rabsztyn D 2015 Influence of the selected settings of the controller on the behavior of the hydraulic servo drive, *Mechatronics - Ideas for Industrial Application, Advances in Intelligent Systems and Computing* **317** (2015) 91-100.

[8] Kovalev Ret al 2009, Freight car models and their computer-aided dynamic analysis, *Multibody System Dynamics*, **22/4** 399-423.

[9] Płaczek M 2015 Modelling and investigation of a piezo composite actuator application, *Int. J. Materials and Product Technology*, **50/3-4** 244-258.

[10] Płaczek M Wróbel A and Baier A 2015 Computer-aided strength analysis of the modernized freight wagon, *IOP Conf. Series: Materials Science and Engineering*, **95** 012042.

[11] Rusiński E Koziołek S and Jamroziak K 2009 Quality assurance method for design and manufacturing process of armoured vehicles, *Eksploatacja i Niezawodność – Maintenance and Reliability*, **3** 70-77.

[12] Wróbel A. et al 2015, Study of mechanical properties and computer simulation of composite materials reinforced by metal, *Int. J. Materials and Product Technology*, **50/3-4** 259-275.

[13] Zakharov S and Goryacheva I 2005 Rolling contact fatigue defects in freight car wheels, *International Journal on the Science and Technology of Friction, Lubrication and Wear*, **258/7–8** 1142-1147.

[14] Zhai W M 2004 Modelling and experiment of railway ballast vibrations, *Journal of Sound and Vibration*, **270/4–5** 673–683.

[15] http://www.smart-material.com/

[16] Buchacz A and Wróbel, A 2008. Homogenization of mechatronical systems in modeling piezoelectric layer. *Machine Dynamics Problems*. 25, **Vol.32, No 3**, pp.7-11

[17] Wróbel A 2015 Analysis of possibility of applying the PVDF foil in industrial vibration sensors. W: Modern technologies in industrial engineering *Bristol: Institute of Physics Publishing,* s. 1-5, bibliogr. 11 poz. (*IOP Conference Series ; Materials Science and Engineering* **vol. 95 1757-8981**)

Institute of Physics
Dirac House, Temple Back
Bristol BS1 6BE UK

ISSN: 1757-8981
ISBN 978-1-5108-3383-8

CURRAN ASSOCIATES INC.
proceedings.com